OCEAN SEISMO-ACOUSTICS

Low-Frequency Underwater Acoustics

NATO CONFERENCE SERIES

I Ecology
II Systems Science
III Human Factors
IV Marine Sciences
V Air–Sea Interactions
VI Materials Science

IV MARINE SCIENCES

Recent volumes in this series

Volume 8 Structure and Development of the Greenland–Scotland Ridge
edited by Martin H. P. Bott, Svend Saxov, Manik Talwani, and Jörn Thiede

Volume 9 Trace Metals in Sea Water
edited by C. S. Wong, Edward Boyle, Kenneth W. Bruland, J. D. Burton, and Edward D. Goldberg

Volume 10A Coastal Upwelling: Its Sediment Record
Responses of the Sedimentary Regime to Present Coastal Upwelling
edited by Erwin Suess and Jörn Thiede

Volume 10B Coastal Upwelling: Its Sediment Record
Sedimentary Records of Ancient Coastal Upwelling
edited by Jörn Thiede and Erwin Suess

Volume 11 Coastal Oceanography
edited by Herman G. Gade, Anton Edwards, and Harald Svendsen

Volume 12 Hydrothermal Processes at Seafloor Spreading Centers
edited by Peter A. Rona, Kurt Boström, Lucien Laubier, and Kenneth L. Smith, Jr.

Volume 13 Flows of Energy and Materials in Marine Ecosystems: Theory and Practice
edited by M. J. R. Fasham

Volume 14 Mechanisms of Migration in Fishes
edited by James D. McCleave, Geoffrey P. Arnold, Julian J. Dodson, and William H. Neill

Volume 15 Heterotrophic Activity in the Sea
edited by John E. Hobbie and Peter J. leB. Williams

Volume 16 Ocean Seismo-Acoustics: Low-Frequency Underwater Acoustics
edited by Tuncay Akal and Jonathan M. Berkson

OCEAN SEISMO-ACOUSTICS

Low-Frequency Underwater Acoustics

Edited by

Tuncay Akal

and

Jonathan M. Berkson

SACLANT ASW Research Center
La Spezia, Italy

Published in cooperation with NATO Scientific Affairs Division

PLENUM PRESS · NEW YORK AND LONDON

Library of Congress Cataloging in Publication Data

Ocean seismo-acoustics.

(NATO conference series. IV, Marine sciences; v. 16)
Proceedings of the Symposium on Ocean Seismo-Acoustics, organized by SACLANT ASW Research Centre, held June 10–14, 1985, at La Spezia, Italy.
Includes bibliographies and indexes.
1. Underwater acoustics—Congresses. 2. Ocean bottom—Congresses. I. Akal, Tuncay. II. Berkson, Jonathan M. III. Symposium on Ocean Seismo-Acoustics (1985: La Spezia, Italy) IV. SACLANT ASW Research Center. V. Series.
QC242.024 1986 551.46'01 86-4940
ISBN 0-306-42266-2

Proceedings of a SACLANT ASW Research Center symposium,
held June 10–14, 1985, at San Terenzo di Lerici, La Spezia, Italy

A Division of Plenum Publishing Corporation
233 Spring Street, New York, N.Y. 10013

Printed in the United States of America

FOREWORD

Seafloor investigation has long been a feature of not only seismology but also of acoustics. Indeed it was acoustics that produced depth sounders, giving us the first capability of producing both global and local maps of the seafloor. Subsequently, better instrumentation and techniques led to a clearer, more quantitative picture of the seabed itself, which stimulated new hypotheses such as seafloor spreading through the availability of more reliable data on sediment thickness over ocean basins and other bottom features. Geologists and geophysicists have used both acoustic and seismic methods to study the seabed by considering the propagation of signals arising from both natural seismic events and man-made impulsive sources. Although significant advances have been made in instrumentation, such as long towed geophysical arrays, air guns and ocean bottom seismometers, the picture of the seafloor is still far from complete.

Underwater acoustics concerns itself today with the phenomena of propagation and noise at frequencies and ranges that require an understanding of acoustic interaction at both of its boundaries, the sea surface and seafloor, over depths ranging from tens to thousands of meters. Much of the earlier higher frequency (>1 kHz) work included the characterization of the seafloor in regimes of reflection coefficients which were empirically derived from surveys. The results of these studies met with only limited success, confined as they were to those areas where survey data existed and lacking a physical understanding of the processes of reflection and scattering. The development in recent years of systematic methods of measuring the physical and acoustical characteristics of the seabed have led to a much better understanding of propagation. However these techniques are exceedingly slow and expensive and do not fulfill the needs of present day acoustics. The relatively recent recognition of the importance of the role that shear and evanescent waves in the seabed have on water borne sound has been a major factor in the recognition of the need for acousticians and seismologists to exchange and discuss their common goals and developments.

The realization of the benefits that could be achieved from such an exchange led the Centre to organize the Symposium on Ocean Seismo-Acoustics. In their entirety these papers give both the acoustician and the geophysicist a good picture of the other's interests, approaches and techniques. The authors represent a wide span of interests and institutions and, therefore, many different points of view.

That the Symposium on Ocean Seismo-Acoustics was organized and hosted by the SACLANT ASW Research Centre is fitting for two reasons. Firstly, one of the Centre's key missions is to promote the exchange of scientific and engineering information for the common benefit and understanding of the nations of the NATO community. Secondly, the Centre, throughout its twenty-six years as a research institution, has actively and significantly contributed to seafloor acoustics.

I have little doubt therefore that this volume will serve experienced scientists as well as those who are contemplating future research in the difficult environment of the sea.

Ralph R. Goodman
Director
SACLANT ASW Research Centre

PREFACE

Because of the ease with which sound can be transmitted in seawater, acoustic techniques have provided the widest range of tools for accumulating knowledge of the environment below the ocean surface. Consequently, the fields of underwater acoustics and marine seismology have both used sound waves for research purposes. Traditionally, the marine seismology community has used earthborne propagation of elastic waves to study the sub-oceanic solid earth; the underwater acoustics community on the other hand has concentrated on studying waterborne, compressional-wave propagation phenomena in oceanic waters and in shallow sub-bottoms.

In recent years, the research interests of these two communities have come closer together; in some areas they have overlapped. Research in underwater acoustics has been extended to the lower part of the frequency spectrum (one to a few hundred hertz), and thus overlaps the spectral domain of the seismologists. Additionally, studies of propagation of earthborne energy into the water column, and of shear-waves, interface waves, and Love waves, have increased. Coincidentally, research in marine seismology has extended to higher frequencies and to the velocity-depth structure of the uppermost part of the seafloor, overlapping the domain of underwater acousticians. These trends have been driven by the need to understand both water column and sub-bottom phenomena and by advances in technology, particularly the development of ocean bottom seismometers and multi-sensor hydrophone arrays. The overlapping interests have therefore stimulated increased activity that has benefitted both disciplines.

The Symposium on Ocean Seismo-Acoustics, sponsored by SACLANT ASW Research Centre, with partial support from the Office of Naval Research, London, was held at La Spezia, Italy, June 10-14, 1985. Its purpose was to enhance cross-disciplinary dialogue and generate ideas for new research directions. This volume contains papers originating from the Symposium and includes several additional features such as subject and author indexes.

The papers are organized into eight topics:

- Propagation: Theoretical Developments
- Propagation: Modelling and Experiments
- Propagation: Fluctuations
- Boundary Scattering
- Sediments: Seismo-Acoustic Waves and Geoacoustic Characteristics
- Seismo-Acoustic Waves: Propagation, Attenuation, and Inverse Techniques
- Seismo-Acoustic Noise
- Technology and New Approaches in Seismo-Acoustic Measurements

Categorization of papers covering a diversity of topics is at best inexact and ofttimes subjective. Many of the papers were broad enough in scope to fit easily into categories other than the one chosen, and in this respect alone the subject and author indexes should prove valuable.

This volume contains 75 research papers and 8 review papers by authors from many nations. The papers cover diverse areas of ocean seismo-acoustics including theory, instrumentation, signal processing, experimental techniques, and experimental results. The number of quoted references is over 1000, an indication of the diversity of topics and a reflection of the fact that the papers might normally be scattered among many acoustical, seismic, geological, and geophysical journals, published in several languages. Thus, this volume presents the current status of both fields, combining papers that report recent work as well as review papers that provide perspective. As such it consolidates the research activities of these two fields. These characteristics should make this volume not only useful to seismo-acoustic specialists but also to other researchers whose work may be influenced by the interaction of marine seismology and underwater acoustics.

In the preparation of this volume, we received assistance from a large number of individuals at SACLANTCEN; regrettably all of their names cannot be listed. However, we would particularly like to thank O. Toso for secretarial assistance and H. Ali for suggesting improvements to several papers. We are especially grateful to R. Nekritz for assistance in preparing the volume, and to Dr. R. Goodman for his guidance and support.

Tuncay Akal
Jonathan M. Berkson

La Spezia, Italy
October, 1985

CONTENTS

PROPAGATION: THEORETICAL DEVELOPMENTS

PROPAGATION: MODELLING AND EXPERIMENTS

* Indicates Review Paper

PROPAGATION: FLUCTUATIONS

BOUNDARY SCATTERING

SEISMO-ACOUSTIC WAVES: PROPAGATION, ATTENUATION, AND INVERSE TECHNIQUES

SEISMO-ACOUSTIC NOISE

TECHNOLOGY AND NEW APPROACHES IN SEISMO-ACOUSTIC MEASUREMENTS

SOUND PULSE PROPAGATION IN A WEAKLY RANGE DEPENDENT SHALLOW OCEAN

E. Topuz* and L. B Felsen+

*Technical University of Istanbul
Electrical Engineering Faculty
Ayazaga, Istanbul, Turkey

+Polytechnic Institute of New York
Dept. of Electrical Engineering and Computer Science
Route 110, Farmingdale, New York 11735 U.S.A.

ABSTRACT

In a weakly range dependent shallow ocean environment, sound propagation can be described effectively in terms of local intrinsic modes [J. M. Arnold and L. B. Felsen, to be published in Wave Motion], which follow lateral trajectories, behave like adiabatic modes where these can be defined, but also accommodate uniformly the transition through cutoff from the adiabatic to the radiative regime. Intrinsic modes yield exact solutions away from the coastline in an isovelocity wedge shaped ocean above an isovelocity fluid bottom [J. M. Arnold and L. B. Felsen, J. Acoust. Soc. Am. 76, 850-860], and therefore render that configuration important for establishing the validity of various analytical and numerical models. Line source excited pressure field calculations by intrinsic modes in the frequency domain [E. Topuz, L. B. Felsen, and J. Yaniv, J. Acoust. Soc. Am. Suppl. 1, 76, S11, 1984] are now generalized to accommodate pulse signals with Gaussian envelope. As in the time harmonic case, the transient intrinsic modes behave adiabatically downslope from cutoff but pass smoothly through the upslope cutoff transition. Their behavior is compared with that of pulsed normal modes in a constant-depth ocean and provides insight into the effect of weak range dependence on transient modal propagation. [Work supported by ONR Ocean Acoustics].

I. INTRODUCTION

Realistic models for sound propagation in weakly range dependent shallow oceans, wherein the bottom structure substantially influences the transmission properties, present non-separable boundary value problems which in general are not amenable to exact analytic solutions. Conventionally, these problems are attacked either by purely numerical methods or by approximate techniques which take advantage of the weak range dependence. Due to lack of dependable experimental data, the accuracy of these results is usually judged by comparing them with independently generated solutions. In most cases the nature of the different approximations involved in each approach obscures the physics underlying the propagation phenomena and makes it difficult to account with some confidence for discrepancies between alternative solutions. Availability of reliable canonical solutions of proven accuracy thus becomes important since those may serve as test problems for assessing relative accuracies of various techniques.

The idealized two dimensional ocean wedge presents such a canonical problem. For weak range dependence, adiabatic modes which adapt locally to the changing environment provide an effective description of the propagation in the wedge for ranges wherein the adiabatic mode energy remains well trapped within the water column [1]. This description fails, however, beyond the ranges where adiabatic modes pass through cutoff into the radiating regime, and it must be augmented there with appropriate transition functions [2,3]. A recently developed rigorous self-contained analytical solution eliminates the need of approximate transition formulations [4,5]. This solution has as its building blocks uniformized local modes-the intrinsic modes-which are not hampered by the phenomenological approximations that accompany the adiabatic mode theory. The full spectral integral synthesizing an intrinsic mode satisfies the wave equation and boundary conditions (with the exception of the edge condition at the coastline (apex)) exactly. Away from the apex, the intrinsic mode field is uniformly valid and results in a field distribution that, for small bottom slopes, agrees with the adiabatic mode where the latter is well guided within the water column, and that is displaced into the bottom region upslope far beyond the cutoff range of the corresponding adiabatic mode, with a smooth transition in between. In an earlier contribution we have demonstrated that the intrinsic mode in a wedge-shaped ocean can be computed efficiently anywhere in the water and the bottom [6]. Here, utilizing intrinsic modes and their asymptotically approximated forms, we undertake the task of evaluating pulsed sound pressure fields in the water waveguide. Again, intrinsic modes prove to be efficiently computable. We present examples of typical pulse data and compare these with results for a range independent channel to identify some effects of weak range dependence on the propagation of pulsed sound signals.

II. FORMULATION

We consider the idealized ocean wedge depicted in Fig. 1. Both the water column of slowly varying depth and the semi-infinite bottom fluid are assumed to have constant sound speeds $v_1, v_2 > v_1$ and constant densities ρ_1, ρ_2, respectively. The wedge is excited by a y-directed unit amplitude line source located in the water waveguide and emitting a modulated gaussian pressure pulse.

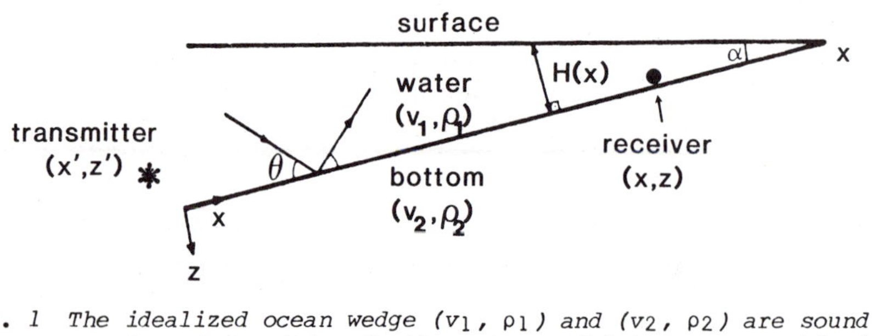

Fig. 1 The idealized ocean wedge (v_1, ρ_1) and (v_2, ρ_2) are sound speeds and densities in the water and bottom, respectively.

A. Time Harmonic Green's Function

For small values of the non-separability parameter α (the wedge angle), the m-th intrinsic mode which satisfies the homogeneous wave equation and the boundary conditions except at the coastline can be expressed in the water as [6]:

$$u_m(x,z) = (-1)^m \sqrt{\frac{\rho_1}{i\alpha\pi}} \int_C \cos\left[\frac{\phi}{2} - kz\sin\theta\right] e^{iq_m(\theta,x)} d\theta \tag{1}$$

where

$$q_m(\theta,x) = \frac{1}{2\alpha}\left\{\int_{\theta_c}^{\theta} \phi(\theta')d\theta' - 2kH(x)\cos\theta - (2m-1)\pi\theta\right\} \tag{1a}$$

$$\phi(\theta) = -2\tan^{-1}\left[\frac{b\sqrt{\sin^2\theta_c - \sin^2\theta}}{\sin\theta}\right] \tag{1b}$$

with

$$b = \rho_1/\rho_2, \quad k = \omega/v_1, \quad \theta_c = \cos^{-1}(v_1/v_2) \tag{1c}$$

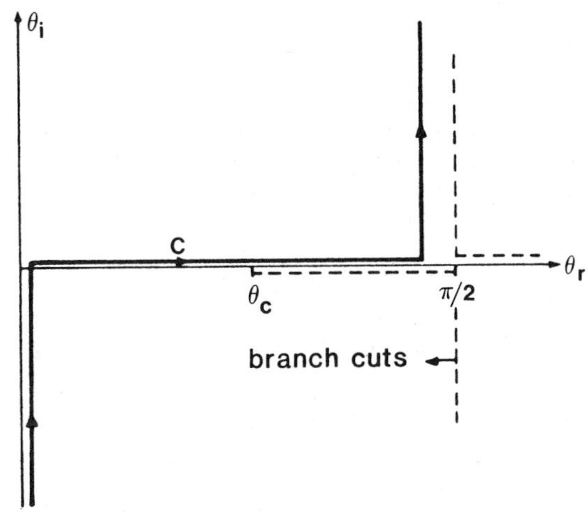

Fig. 2 Integration contour C in the complex θ plane. Spectral branch cuts are also shown for small losses in the bottom medium.

The integration contour C is depicted in Fig. 2. The Rayleigh phase $\phi(\theta)$ describing reflection, at the bottom interface, of plane wave spectra characterized by the angle of incidence θ (see Fig. 1) contains a branch point at the critical angle of total reflection, $\theta = \theta_c$.

The integral (1) can be evaluated asymptotically in inverse powers of the large parameter $1/\alpha$. The saddle point equation

$$\left.\frac{\partial q}{\partial\theta}\right|_{\theta_m} = 0 \rightarrow \phi(\theta_m) + 2kH(x)\sin\theta_m = (2m-1)\pi; \quad \theta_m \equiv \theta_m(x) \tag{2}$$

yields precisely the transverse resonance condition for a plane parallel layer of depth H(x). The corresponding adiabatic mode cutoff range is at $x = x_m$ where $\theta_m(x_m) = \theta_c$ and the water depth decreases to

$$H(x_m) = (2m-1)\pi\cdot(2k\sin\theta_c)^{-1} \tag{3}$$

Downslope sufficiently away from x_m, θ_m is well removed from the branch point θ_c. Evaluating (1) asymptotically at the isolated saddle point θ_m yields a leading term contribution that is found to agree completely with the normalized field of the m-th adiabatic mode

$$\sqrt{\cos\theta}\cdot u_m(x,z) \sim \bar{u}_m(x,z) = \sqrt{2\rho_1} N_m^{1/2} \sin\{k[z+H(x)]\sin\theta_m\} e^{iq_m(\theta_m)} \quad (4)$$

where

$$N_m = Q_m[b\sin^2\theta_c + kH(x)Q_m]^{-1} \quad (4a)$$

$$Q_m = \delta\sqrt{\sin^2\theta_c - \sin^2\theta_m}\left[\sin^2\theta_m + b^2(\sin^2\theta_c - \sin^2\theta_m)\right] \quad (4b)$$

$$\delta = 1 \quad (4c)$$

As one proceeds upslope towards x_m, θ_m approaches the branch point θ_c in the phase, and the simple saddle point evaluation becomes inapplicable. Here a uniform asymptotic evaluation of (1) is required. This can be achieved by introducing the transformation

$$\eta = \sqrt{\theta_c - \theta} \quad (5)$$

which removes the branch point and introduces a second stationary point at $\theta = \theta_c$ in addition to the one at $\theta = \theta_m$ [5]. Uniform asymptotic evaluation of (1) then yields [7]

$$u_m(x,z) \sim \hat{u}_m(x,z) = N_m^{1/2} \sin\{k[z+H(x)]\sin\theta_m\}$$
$$\cdot\ \delta[\psi^{-1/4}\mathrm{Ai}'(\psi) - \delta\psi^{1/4}\mathrm{Ai}(\psi)]\exp\{i\tfrac{1}{2}[q(\theta_m,x)+q(\theta_c,x)]\} \quad (6)$$

where

$$\delta = \begin{cases} +1 & \text{for } x < x_m \\ -1 & \text{for } x > x_m \end{cases} \quad (6a)$$

$$\psi = \left\{\tfrac{3}{4} i[q(\theta_c,x) - q(\theta_m,x)]\right\}^{2/3} \quad (6b)$$

q and N_m are given by (1a) and (4a, 4b), respectively, and the prime on the Airy function Ai denotes differentiation with respect to the argument.

The time harmonic Green's function G_ω of the wedge (at least, sufficiently far upslope from the source) can be constructed as a superposition of intrinsic modes. The important contribution then arises from those modes that, at the source range, are trapped in the water column for all relevant components of the source frequency spectra, whence

$$G_\omega(x,x',z,z') \cong \sum_{m=1}^{M} A_m \bar{u}_m^*(x',z') u_m(x,z) \quad (7)$$

where $*$ denotes complex conjugation

$$A_m = \frac{1}{2\rho_1} \frac{1}{\sqrt{\cos\theta_m(x')}} \tag{7a}$$

(x', z') and (x, z) refer to source and receiver locations, respectively, and the coefficients A_m are determined by requiring (7) to reduce to the adiabatically approximated Green's function at the source range [1]. In (7), the excitation coefficients of the intrinsic modes are taken as the adiabatic mode fields at the source range. This approximation introduces some simplification in the numerical computations and is well justified when $x_m \gg x'$ and $\alpha \ll 1$. These conditions are satisfied in all test cases considered in this paper. The intrinsic mode fields $u_m(x, z)$ can be calculated from the spectral integral formulation in (1). The numerical evaluation of (1) can be performed fairly efficiently since both of the integrands in (1) and (1a) are smoothly varying functions of θ, and the dominant contribution in (1) arises from the segment of the integration path which runs along the real θ axis (see Fig. 2). Intrinsic mode fields can also be evaluated asymptotically via (1). These two approaches are in excellent agreement for the small values of the wedge angle α considered here, and the latter approach, which results in considerable improvement in computational efficiency, will be preferred.

The corresponding approximate Green's function for the constant depth ($H = H_o$) layer can be expressed as

$$\bar{G}_\omega(x, x', z, z') = \sum_{m=1}^{M} E_m \psi_m^*(x', z')\psi_m(x, z) \tag{8}$$

where

$$E_m = \left[\sqrt{i}\cos\theta_m\right]^{-1} \tag{8a}$$

$$\psi_m(x, z) = \sin[k(z + H_o)\sin\theta_m] N_m^{1/2} e^{ikx\cos\theta_m} \tag{8b}$$

In (8) the asterisk on ψ denotes complex conjugation, the summation runs over the M guided normal modes, and θ_m, N_m are to be obtained from (2) and (4a-4c) by replacing $H(x)$ with H_o.

B. Time Dependent Green's Function

We now assume that the y-directed line source at $(0, z')$ in the water column emits a pulse with truncated gaussian envelope and carrier frequency ω_o

$$p(t) = \begin{cases} \exp\{-i\omega_o(t-t_o) - (t-t_o)^2/(4a^2)\} & \text{for} \quad 0 \le t \le 2t_o \\ 0 \quad \text{elsewhere} & \end{cases} \tag{9}$$

At receiver locations (x, z) sufficiently removed from the source location, the time dependent pressure field in the water region can be recovered from (7) by Fourier inversion

$$g(x, z, z', t) = \frac{1}{2\pi} \int_{-\infty+i\varepsilon}^{\infty+i\varepsilon} P(\omega) G_\omega(x, z, z') e^{-i\omega t} d\omega \tag{10}$$

where $\varepsilon > 0$ shifts the integration path so that it lies above the singularities of the integrand to ensure causal solutions [8]. $P(\omega)$ is the Fourier transform of the source function

$$P(\omega) = 2\alpha\sqrt{\pi}\exp\{i\omega t_o - a^2(\omega-\omega_o)^2\}\cdot \mathrm{Re}\{\mathrm{erf}[t_o/2a - ia(\omega-\omega_o)]\} \tag{11a}$$

where Re indicates "real part" and erf is the error function.

Numerical evaluation of (10) is by no means straightforward since it represents essentially a triple integral and involves the computation of (a) the phase integral in (1a), (b) the angular spectrum integral (1), and (c) the Fourier integral (10). To obtain an efficient algorithm, one may exploit the weak range dependence of the ocean wedge and the non-violent behavior of the phase integral, as well as the smooth frequency dependence of the eigenvalues θ_m in (2). Thus, because of the assumed small values of the wedge angle α, we may utilize the uniform asymptotic expressions $\hat{u}_m(x,z)$ in (6) for the intrinsic modes $u_m(x,z)$ in (1). These asymptotic approximations are found to be in excellent agreement with the intrinsic mode field everywhere in the water waveguide for wedge angles $\alpha \lesssim 3^o$, as considered in this paper. Their utilization improves numerical efficiency by a factor of approximately 10 without noticeably degrading the overall accuracy.

The numerical evaluation of the spectral integral (1) is now reduced to the much simpler problem of evaluating the eigenvalues of (2). Here, one may take advantage of the smooth dependence of θ_m on frequency. By introducing a new spectral variable β via

$$\cos\beta_m(x) = \sin\theta_c \sin\theta_m(x) \quad , \tag{12}$$

the branch point $\theta = \theta_c$ in (1b) is removed, and a fast Newton-Raphson algorithm can be employed in (2) to evaluate the eigenvalues, with the value of θ_m at one frequency serving as the initial guess for the next frequency. For input pulses which have a fairly narrow frequency spectrum, one can approximate the phase integrals in (1a) with the first few terms of an expansion around the pulse modulating frequency ω_o. Retaining the first three terms, one obtains

$$\int_{\theta_c}^{\theta_m} \phi(\theta')\,d\theta' \cong \int_{\theta_c}^{\bar{\theta}_m} \phi(\theta')\,d\theta' + \Delta C_1 + \frac{\Delta^2}{2} C_2 \tag{13}$$

where

$$\Delta = (\omega_o - \omega)/\omega_o \tag{13a}$$

$$C_1 = H(x)\, N_m(\bar{\theta}_m) \tan\bar{\theta}_m\, \phi(\bar{\theta}_m) \tag{13b}$$

$$C_2 = H^2(x)\, N_m^2(\bar{\theta}_m)\, \phi(\bar{\theta}_m) \tan\bar{\theta}_m \Big\{ 2 + \tan^2\bar{\theta}_m + b\sin^2\theta_c \sin^2\bar{\theta}_m \cdot \left[(b^2-1)(\sin^2\theta_c - \sin^2\bar{\theta}_m) - \sin^2\theta_c\right] \frac{N_m(\bar{\theta}_m)}{\sqrt{\sin^2\theta_c - \sin^2\bar{\theta}_m Q_m^2(\bar{\theta}_m)}} \Big\} \tag{13c}$$

with $\theta_m \equiv \theta_m(\omega)$, $\overline{\theta}_m \equiv \theta_m(\omega_o)$, and N_m, Q_m given by (4a), (4b), respectively. In the test cases considered in this paper, $\Delta < 0.5$ is satisfied for all relevant components of the source frequency spectra, and (13) has been found to be very accurate except when $\overline{\theta}_m$ is close to θ_c; in this event, the integration interval $(\theta_m(\omega) - \theta_c)$ is small for all ω, and direct numerical integration of the phase function is preferred and efficient. Use of (13), whenever applicable, results in an additional overall improvement of the computational efficiency by a factor of approximately 10. In the numerical evaluation of the time harmonic Green's function G_ω in (10) the intrinsic mode fields have been computed from the uniform asymptotic formulas in (7) by use of the approximate expressions for the phase integrals; the frequency inversion was done via FFT.

III. DISCUSSION OF THE NUMERICAL DATA

Typical examples of the pulse response in the ocean wedge are depicted in Figs. 3 and 4. Referring to Fig. 1, we have used water and bottom speeds $v_1 = 1500$m/sec, $v_2 = 1704.5$m/sec and densities $\rho_1 = \rho_2 = 1$, respectively. Both media are taken to be nondispersive and lossless. The source is located in the water column 7.39 km downslope from the coastline, and emits a pressure pulse at frequency $f_o = \omega_o/2\pi$ with a Gaussian envelope of (1/e)-width equal to (4a).

Figure 3 depicts the pulse broadening due to intramodal dispersion. The parameters for this figure are $\alpha = 1.55^o$, $f_o = 25$ Hz and $a = 50$ msec. The height of the water column at the source range ($x = 0$) is 200m, and the source is located at a depth of 20m. For these parameters, three intrinsic modes yield pressure field distributions that are trapped inside the water at the source range. To demonstrate the effects of the intramodal dispersion,only the lowest numbered of these modes is considered in the computations of Figs. 3a and 3b. At the center frequency the corresponding adiabatic mode undergoes cutoff at the range $x_1 = 6.22$ km, where the water depth decreases to 31.5m. The receiver range is varied from 1 km to 7 km at a constant depth of $H = 20$m. For most of this path, the receiver moves essentially along a constant pressure contour (see Fig. 3b). As seen from Figs. 3a and 3b, the intrinsic mode intramodal dispersion remains small and is not noticeable within the time resolution of the figure drawings until the cutoff range of the corresponding adiabatic mode is approached. Thereafter it increases rapidly as one continues upslope towards the coastline. However, in this region, the trapped intrinsic mode energy is transferred into beamed radiation in the bottom region [6], with a resulting rapid decay of the received energy as depicted in Fig. 3b.

Figure 3a also has the arrival times of the lowest mode in constant depth layers with $H = H(x') = \text{const.}$ and $H = H(x) = \text{const}$, whenever $x < x_1$. $H(x')$ and $H(x)$ refer to water depths at the source and receiver ranges, and the corresponding arrival times in these layers are designated by t and $\overline{t}$, respectively. In regions where the intrinsic mode is well trapped, the pulse response of the $H = H(x') = \text{const.}$layer as obtained from (8) and (10) is found to represent the intrinsic mode response rather closely. As may be expected, the actual arrival time of the pulse in the wedge falls between t and $\overline{t}$. However, it is always much closer to t, which arrival time may therefore be used for a good estimate of mode arrivals everywhere in the wedge.

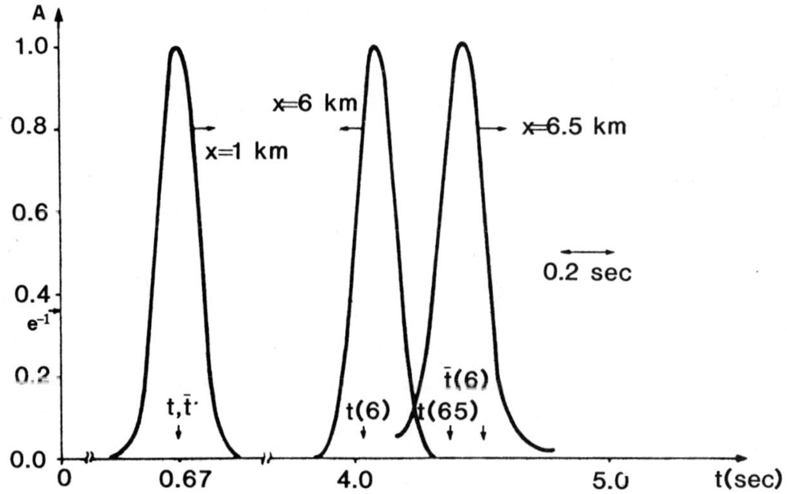

Fig. 3a Normalized pulse response (A) for lowest intrinsic mode at ranges k = 1 km, 6 km and 6.5 km. The cut off range of the corresponding adiabatic mode is at 6.22 km. t(x) and t(x) denote arrival times in corresponding constant depth layers with H = H(x') = const and H = H(x) = const, respectively. The time origin is set at the pulse maxima at the source range (x = 0), where (1/e) pulse width is 0.2 sec. Other parameters are given in the text.

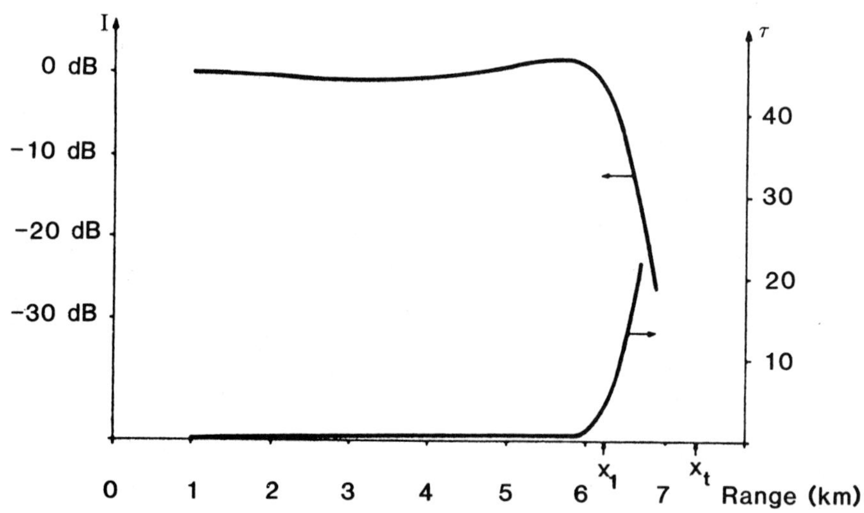

Fig. 3b Variation of the relative pulse maxima and pulse widths with range for the lowest intrinsic mode, for the same parameters as in Fig. 3a. x_1 and x_t are the cutoff range of the corresponding adiabatic mode and the location of the coastline, respectively. I measures the relative value of the pulse maxima in dB, normalized with respect to the value at x = 1 km. $\tau = 100.[\Delta t(x) - \Delta t(0)]/\Delta t(0)$ measures relative pulse broadening in percent, where $\Delta t(0)$ and $\Delta t(x)$ are the (1/e) pulse widths at the transmitter and receiver locations, respectively.

Pulse broadening due to intermodal dispersion is demonstrated in Fig. 4. The parameters here are $\alpha = 3.1^o$, $f_o = 50$ Hz and a = 30 msec. At the source range, the water depth is 400m, and the source is located at the depth of 75m. Twelve intrinsic modes are now trapped at the source range x = 0, for all relevant constituents of the source frequency spectra. At the center frequency the corresponding adiabatic modes undergo cutoff at ranges which vary from $x_1 = 7.1$ km for the first mode to $x_{12} = 0.7$ km for the last mode. The pulse response is depicted in Fig. 4 at two ranges x = 3 km and x = 4 km, and for a receiver depth of 65m. Since the intrinsic mode energy trapped in the water waveguide is transferred predominantly into the bottom region as one proceeds upslope beyond the cutoff range of the corresponding adiabatic mode, the contributing number of intrinsic modes decreases with increasing receiver range. Thus it is sufficient to consider 9 and 7 modes at the ranges x = 3 and 4 km, respectively. The successive disappearance of higher order mode fields from the water column as one proceeds towards the coastline makes the wedge an inherently low dispersion waveguide as evidenced by the pulse data in Fig. 4. The surprisingly smooth behavior of this response for a multimode environment results from the fact that at any range, the delay differences are determined essentially by the wave velocities of the contributing modes at the source range. To demonstrate this point we have indicated the arrival times t_i of the most strongly excited intrinsic modes together with the corresponding arrival times $\tilde{t}_i$ in the constant depth layer H = H(x') = const. Clearly, the latter furnish a rather good approximation to the intrinsix mode arrivals. Since at large ranges even the last contributing intrinsic mode has a field distribution which is well guided at the source range, the intermodal dispersion remains small everywhere in the wedge.

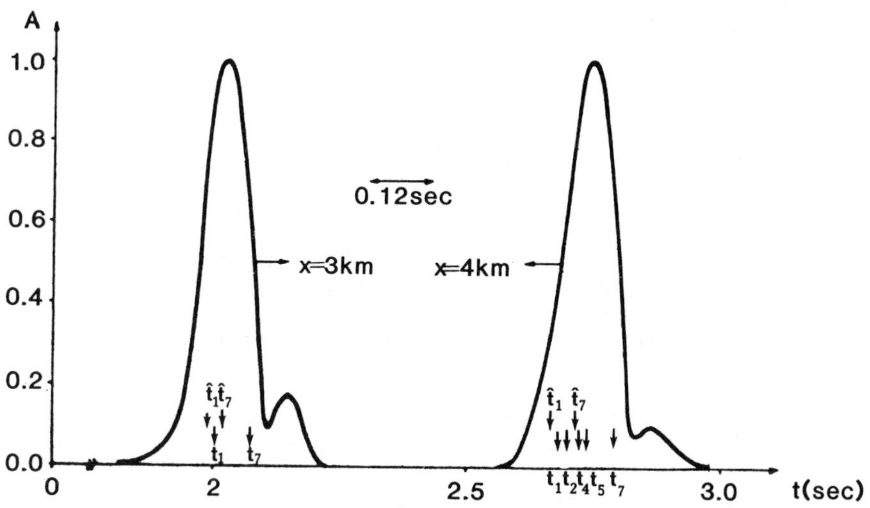

Fig. 4 Normalized pulse response (A) in a multimode environment at x = 3 km and x = 4 km, where the number of contributing intrinsic modes is 9 and 7, respectively. Arrival times t_i and $\hat{t}_i$ of the strongly excited intrinsic modes and the corresponding modes in the constant depth layer with H = H(x') = 400 m, respectively, are indicated. The time origin coincides with the pulse maxima at the transmitter location (x = 0), where the (1/e) pulse width is 0.12 sec. Other parameters are listed in the text.

As noted earlier, the uniform asymptotic forms in (6) were used to compute the intrinsic mode fields, together with the approximate expressions (13) for the phase integrals. These approximations were found to improve the computational efficiency substantially without appreciably deteriorating the overall accuracy. In the present nonoptimized form, our code requires approximately 2 sec of CPU time on an IBM 4341 mainframe for a sampling interval of 0.25 Hz for each mode at any receiver location.

IV. CONCLUSION

We have presented a method of calculating time dependent pressure fields in a two-dimensional, weakly range dependent, idealized ocean wedge. The problem has been formulated in terms of intrinsic mode fields which satisfy the wave equation and the boundary conditions, with the exception of the edge condition on the coastline, exactly. We have demonstrated that this representation can be reduced to accurate and computationally efficient forms by taking advantage of the assumed small bottom slope. For the parameter ranges investigated here, it has been found that the ocean wedge constitutes an inherently low dispersion waveguide wherein, even for large transmitter-receiver separations, the delay differences between contributing intrinsic modes are determined essentially by their wave velocities at the transmitter location where they represent pressure field distributions which are well trapped within the water column. More detailed studies are in progress to shed further light on the pulsed wave behavior in this canonical range dependent propagation environment.

ACKNOWLEDGMENT

This work was sponsored by the Office of Naval Research under Contract No. N-0014-79-C-0013.

REFERENCES

1. A. D. Pierce, "Extension of the method of normal modes to sound propagation in an almost-stratified medium," J. Acoust. Soc. Am., Vol. 37, pp.19-27 (1965).

2. A. D. Pierce, "Guided mode disappearance during upslope propagation in variable depth shallow water overlying a fluid bottom," J. Acoust. Soc. Am., Vol. 72, pp. 523-531 (1982).

3. A. D. Pierce, "Augmented adiabatic mode theory for upslope propagation from a point source in variable depth shallow water overlying a fluid bottom," J. Acoust. Soc. Am., Vol. 79, pp. 1837-1847 (1983).

4. J. M. Arnold and L. B. Felsen, "Rays and local modes in a wedge-shaped ocean," J. Acoust. Soc. Am., Vol. 73, pp. 1105-1119 (1983).

5. J. M. Arnold and L. B. Felsen, "Intrinsic modes in a nonseparable ocean waveguide," J. Acoust. Soc. Am., Vol. 76, pp. 850-860 (1984).

6. E. Topuz, L. B. Felsen and J. Yaniv, "Numerical implementation of intrinsic mode Green's functions for oceans with weakly sloping penetrable bottom," paper presented at the 108th Acous. Soc. Meeting (1984).

7. L. B. Felsen and N. Marcuvitz, Radiation and Scattering of Waves, Englewood Cliffs, N. J., Prentice Hall, pp. 410-416 (1973).

8. L. B. Felsen, ed., Transient Electromagnetic Fields, Springer, pp. 19 (1976).

INTRINSIC MODES IN A WEDGE-SHAPED OCEAN WITH STRATIFIED ELASTIC BOTTOM

I.T. Lu and L.B. Felsen

Polytechnic Institute of New York
Dept. of Electrical Engineering and Computer Science
Route 110, Farmingdale, N.Y. 11735 U.S.A.

ABSTRACT

Propagation of low frequency acoustic signals in an ocean environment with weak lateral variation in the sound speed and (or) bottom profile can be modeled effectively in terms of local intrinsic modes (Arnold and Felsen, to be published in Wave Motion). At any lateral location (x,y), these local modes have a depth (z) variation like a normal mode in a laterally homogeneous ocean defined by conditions at (x,y), and they propagate in the (x,y)-plane along lateral ray trajectories. Local intrinsic modes differ from adiabatic modes, first, by incorporating some adiabatic mode coupling effects and, second, by uniformly accommodating the transition through cutoff from trapped to radiating during upslope propagation. It has previously been shown that local intrinsic modes become exact for the prototype problem of a homogeneous water column separated from a homogeneous fluid bottom by a plane slanted interface. The model is now generalized to allow for stratification of the slanted bottom along planes parallel to the water-bottom interface, and for elastic properties in these layers. Examined in detail is the special case where the bottom is an elastic half space. Upslope propagation in the water, perpendicular to the coast line, now involves transformation of a locally trapped mode downslope into a partly leaky and then into a totally leaky mode due to the separately occurring compressional and shear wave transitions. The numerically computable generalized intrinsic mode spectral integral for the acoustic field in the water and the seismic field in the bottom describes these transitions, as well as the reduction to adiabatic modes in the totally trapped regime.

INTRODUCTION

Analytical modeling of sound propagation in a range-dependent ocean and bottom environment has been beset with difficulties engendered by the non-separability of the resulting wave equations and (or) boundary conditions. Even numerical approaches based on finite element or finite difference methods, on numerical treatment of an integral equation formulation, or on a rigorously applied mode coupling scheme, are unsuitable when the range-dependent propagation interval extends over long distances compared to the local wavelength. Thus, one has had to resort to various approximate procedures built around "weak" variations of certain parameters that describe the propagation channel. While the restrictions imposed can usually be stated in terms of asymptotic order estimates, the actual range of parameters validated thereby can be assessed only by numerical comparison with another reliable reference solution. The dilemma has been that such reference solutions have involved either range-independent models or simple coordinate separable range-dependent models (such as a wedge with free or rigid boundaries). This has hampered the testing of range-dependent algorithms allowing access to the bottom.

The situation has improved with the recent construction of source-free solutions for a wedge-shaped ocean with homogeneous water and (fluid) bottom media. These solutions, which satisfy the water and bottom wave equations and boundary conditions exactly, have been termed intrinsic modes[1] because they exhibit mode-like behavior. The observation region must exclude the range domain near the wedge apex (coast line) but there is no restriction (in principle) on the bottom slope. However, the tractable algorithm constructed so far is based on a weak slope approximation. The resulting intrinsic mode representation involves a spectral integral of global validity, which reduces in its lowest order asymptotic approximation to an adiabatic mode when that can be supported by the local ocean height but, unlike the adiabatic mode, remains uniformly valid in the water and the bottom as that mode propagates upslope through cutoff, thereby transferring its energy from the water into the bottom. Intrinsic mode shapes for various mode numbers and various ocean and bottom parameters, together with a description of the algorithm have been presented elsewhere[2].

The intrinsic modes are synthesized by global self-consistency of plane wave spectra reflected between the slanted sides of the water wedge, due account being taken of the boundary conditions on the wedge faces. In the analysis so far, the ocean surface was treated as a perfectly reflecting free boundary, and the bottom as an infinite homogeneous half space with incidence angle dependent plane wave reflection coefficient $R(\theta)$ and transmission coefficient $T(\theta)$. The analysis can be formally extended without change if the perfect reflection conditions for the upgoing spectra at the surface are replaced by interface conditions $R_u(\theta)$ and $T_u(\theta)$, with the conditions for the downgoing spectra at the bottom now labeled $R_d(\theta)$ and $T_d(\theta)$. Moreover, $R_{u,d}$ and $T_{u,d}$ need not describe exterior homogeneous half spaces but they can express the effect of any discrete or continuous stratification in the direction perpendicular to the respective boundaries. Furthermore, the exterior layering can involve multiwave media, with the consequent coupling between various wave species at layer interfaces and through continuous stratification. These generalizations do not disturb the principle of self-consistent superposition of plane wave spectra in the ocean fluid which remains a one-species medium even when the exterior is not.

Thus, intrinsic modes can be defined more generally in a dipping fluid layer surrounded by an exterior that is plane stratified with respect to its wedge boundary but otherwise arbitrary. This model can serve as a useful test for geological structures satisfying these requirements. Here, we apply it to the ocean environment, keeping the top surface free but allowing the bottom to assume fluid and (or) elastic stratification, with the latter giving rise to p-SV coupled seismic events. Because the fluid theory has been presented elsewhere[1,2] and applies without change (except for redefinition of $R_d(\theta)$ and $T_d(\theta)$), we merely list the result. We then specialize to an elastic homogeneous half space for the first numerical implementation of the generalized algorithm, assuming intrinsic mode propagation perpendicular to the coast line. Due to the excitation of p and SV motion in the bottom, an originally trapped upslope propagating intrinsic mode passes through two cutoff regions, the first being located at the critical range for SV-radiation, and the second being related to the leakage of p-radiation, into the bottom. The computed intrinsic mode shapes in the water and the bottom are examined in this context.

FORMULATION OF INTRINSIC MODE PROBLEM

Referring to Fig. 1, intrinsic modes are source-free solutions of the scalar wave equation for the acoustic pressure field u(x,z) in the water wedge

$$\left(\frac{\partial^2}{\partial x^2} + \frac{\partial^2}{\partial z^2} + k^2\right) u(x,z) = 0, \quad -H(x) < z < 0 \tag{1}$$

We restrict ourselves to the two-dimensional (x,z) space perpendicular to the wedge apex, and exclude the region near the apex from consideration. The wavenumber k equals (ω/v), the ocean fluid is described by the sound speed v and density ρ, and a time dependence $\exp(-i\omega t)$ is suppressed.

At the ocean surface,

$$u = 0 \text{ at } z = -H(x), \tag{2}$$

while at the lower interface z = 0, the boundary conditions depend on the nature of stratification of the bottom. Across a fluid-fluid interface, the boundary conditions require continuity of pressure and of its weighted normal derivative, whereas across a fluid-solid boundary, one requires continuity of the normal components of stress and displacement, and also of tangential stress. Since the tangential stress vanishes in the liquid, the last condition reduces simply to the requirement that the tangential components of the stress tensor be zero at the boundary of the solid medium. In each fluid or solid layer, the pressure or pressure and shear wave fields, respectively, satisfy their appropriate wave equations[3,4].

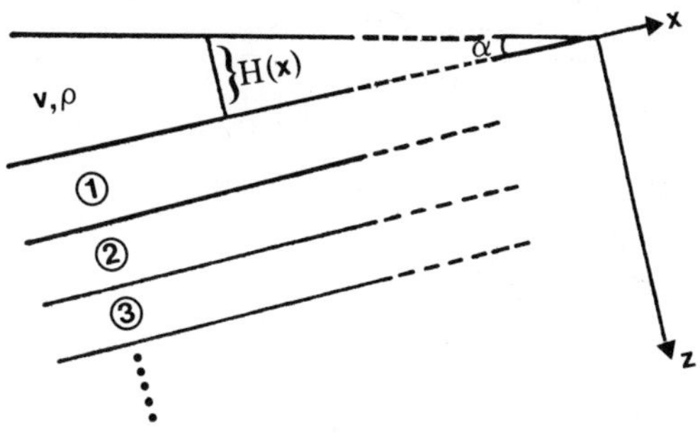

Fig. 1 - Idealized fluid wedge with density ρ, sound speed v and bottom slope α, on top of a z-stratified half-space comprising arbitrary fluid or multiwave (for example, elastic) media.

Since the solution of (1) subject to the above-stated boundary conditions is to be constructed by superposition of plane wave spectra, the boundary conditions at the stratified bottom can be satisfied in terms of the plane wave reflection coefficient $R_{pp}(\theta)$, where the subscripts denote incident and reflected species, respectively. This reflection coefficient can be synthesized systematically by passing through the stack of layers from the last (deepest) interface to the one bounding the water channel. Details may be found in reference 5. For the numerical treatment here, we restrict ourselves to a homogeneous solid bottom which constitutes the simplest generalization to seismic phenomena of the previous fluid bottom intrinsic mode calculations[2]. Therefore, we only show the reflection coefficient for this special case, and also the corresponding transmission coefficient, which is required for evaluation of the p-SV field in the bottom.

Again referring to reference 2, retaining the same small bottom slope assumptions which justify omission of the remainder $E(\theta)$ in (8) of reference 2, we express the m-th intrinsic mode pressure field in the water as

$$U_m(x,z) = B_m \int_C f(\theta,z) e^{iQ_m(\theta,x)} d\theta \tag{3}$$

$$Q_m(\theta,x) = \frac{1}{2\alpha}\left\{-2kH(x)\cos\theta - 2m\pi\theta + \int_{\theta_o}^{\theta} [\phi_u(\theta'+\alpha) + \phi_d(\theta')]d\theta'\right\} \tag{3a}$$

$$f(\theta,z) = e^{-i\frac{1}{2}\phi_u(\theta+\alpha)} \cos(\tfrac{1}{2}\phi_d(\theta) - k\, z\sin\theta)\ , \quad k = \frac{\omega}{v} \tag{3b}$$

where α is the wedge angle and $\phi_{u,d}$ is the phase of the reflection coefficient $R_{u,d}$,

$$\phi_{u,d} = \frac{1}{i}\, \ell n(R_{u,d}) \tag{4}$$

The excitation strength B_m of the m-th intrinsic mode, left unspecified here, depends on the source configuration. The spectral variable θ represents the propagation angle of a spectral plane wave with respect to the x-axis, θ_o is an arbitrary reference value, and the contour of integration C is shown in Fig. 2.

The fields in the bottom can be derived from two potentials ϕ,ψ corresponding to p, SV motion respectively[3-5]. One has the components of particle displacement

$$u_x = \frac{\partial\phi}{\partial x} - \frac{\partial\psi}{\partial z}, \quad u_z = \frac{\partial\phi}{\partial z} + \frac{\partial\psi}{\partial x} \tag{5}$$

and the components of the stress tensor

$$\tau_{zz} = \lambda\left(\frac{\partial u_x}{\partial x} + \frac{\partial u_z}{\partial z}\right) + 2\mu\frac{\partial u_z}{\partial z} \tag{6}$$

$$\tau_{zx} = \mu\left(\frac{\partial u_x}{\partial z} + \frac{\partial u_z}{\partial x}\right)$$

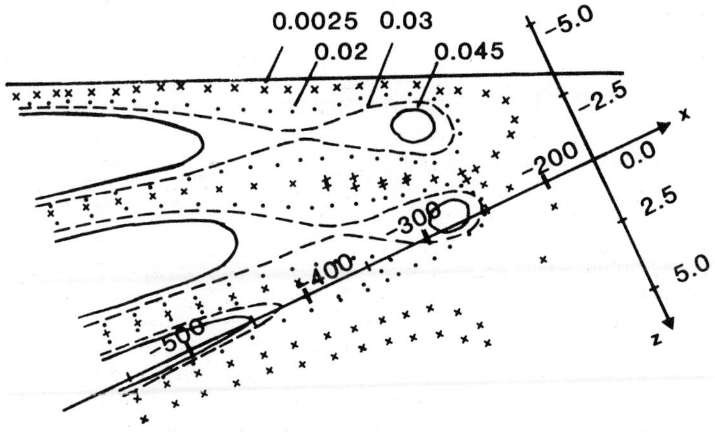

Fig. 6 - Equal amplitude contours for the x-displacement field of Fig. 3. The constant in (3) and (7) is chosen as B_m=1.

The x-displacement field pattern of the third mode, m=3, is shown in Fig. 3. Downslope from the first cutoff at x_s=-466.9, where SV waves begin to leak into the bottom, the mode is well trapped; the stationary point $\hat{\theta}$ of the integrand in (3) is real in that region and is located with respect to the bottom shear wave branch point $\bar{\theta}_s$ at $\hat{\theta} < \bar{\theta}_s$.

After cutoff, for $x > x_s$, the m=3 mode is no longer totally trapped, as evidenced analytically by motion of the saddle point across the branch cut due to $\bar{\theta}_s$ into the bottom sheet. Here, $\hat{\theta}$ becomes complex and essentially represents a leaky mode (see Fig. 4). Because of the strong contrast between the liquid and the solid for the parameters in (12), most of the energy is reflected back into the wedge, with consequent small leakage as indicated by the small imaginary part of the complex saddle point. Note that the p-wave in the bottom is essentially evanescent for $x < x_p$, x_p=-410.2 being the p-wave cutoff range that corresponds to the p-wave branch point spectral angle $\bar{\theta}_p$. Since the stationary point $\hat{\theta}$ is complex, it does not cross the branch cut associated with $\bar{\theta}_p$ at this range but moves toward it, with Re $\hat{\theta} > \bar{\theta}_p$, until x=-324.6 (see Fig. 4). For $x > -324.6$, $\hat{\theta}$ moves through the $\bar{\theta}_p$ cut into the bottom sheet, becoming strongly complex, with consequent strong leakage of both p and SV wave energy into the bottom. The steepest descent path through the saddle point, as a function of range, remains to be charted in detail. If interception of branch points is involved, this should furnish an interpretation of the field behavior in the water by lateral wave shedding.

Two enlarged versions of Fig. 3 are shown in Figs. 5(a) and 5(b), and a contour plot in Fig. 6. One may note that the peak amplitude in the water, after decreasing initially due to SV leakage, increases near the upslope range $x \approx -325$. This may be due to the fact that the SV leakage is small, thereby confining essentially the same mode energy within a successively smaller cross section.

The numerical results presented here, as well as their sparse interpretation, should be regarded as preliminary. More detailed studies are in progress.

CONCLUSIONS

We have shown that the previously reported intrinsic mode construction for a homogeneous fluid wedge overlying a homogeneous fluid bottom can be extended to fluid or solid bottom stratification along plane layers parallel to the fluid bottom boundary. As the simplest example, we have presented in detail the results for a homogeneous solid bottom; for small water wedge angles α, they are as before[1] in the form of a spectral integral whose analytic and numerical properties resemble those for the fluid bottom[2], but with some additional complication engendered by two cutoff and leakage phenomena associated with SV and p waves, respectively. Preliminary numerical results for one of the intrinsic modes, m=3, have revealed an adiabatic behavior downslope where the mode is trapped inside the water and is evanescent in the bottom. Further upslope, there occur transitions ascribable to SV and p wave cutoff in the water. These transitions are fully accommodated by the numerical evaluation of the spectral integral but various aspects of the physical mechanism of energy transfer into the bottom, and of the remaining energy in the water, remain to be clarified. This is the subject of future investigation.

ACKNOWLEDGEMENT

This work was supported by the Office of Naval Research, Ocean Acoustics Branch, under Contract No. N-00014-79-C-0013.

REFERENCES

1. J.M. Arnold and L.B. Felsen, Intrinsic modes in a wedge-shaped ocean, J. Acoust. Soc. Am., 76, 850-860 (1984).

2. E. Topuz and L.B. Felsen, Intrinsic modes: numerical implementation in a wedge-shaped ocean, to be published in J. Acoust. Soc. Am.

3. W.M. Ewing, W.S. Jardetsky and F. Press, "Elastic Waves in Layered Media", McGraw-Hill, New York (1957).

4. L.M. Brekhovskikh, "Waves in Layered Media", Academic Press, New York (1960).

5. B.L.N. Kennett, "Seismic Wave Propagation in Stratified Media", Cambridge University Press, New York (1983).

6. J.M. Arnold and L.B. Felsen, Local intrinsic modes--layer with non-planar interface, to be published in Wave Motion.

A GREEN'S FUNCTION METHOD FOR ONE-WAY WAVE PROPAGATION IN A RANGE-DEPENDENT OCEAN ENVIRONMENT

K. E. Gilbert and R. B. Evans

Naval Ocean Research and Development Activity
Ocean Acoustics and Technology Directorate
NSTL, Mississippi 39529-5004

ABSTRACT

A Green's function method (e.g., the Fast Field Program) is an accurate and flexible way to compute wave propagation in a horizontally stratified environment. However, there has been no exact theory for applying the Green's function approach to one-way wave propagation in a range-dependent environment. This paper derives a generalized Green's function method for exactly solving the one-way wave equation in an environment that varies discretely in range. The derivation leads to a depth-separated inhomogeneous elliptic wave equation, which has a vertical source distribution for the inhomogeneous term. On a given range step, the source distribution is the acoustic field at the end of the previous step. It is shown numerically that the method gives an accurate solution for one-way propagation.

INTRODUCTION

Historically, one-way wave propagation in range-dependent environments has been treated by various parabolic equation methods.[1] The application of Green's function methods to such environments has not been possible because of the lack of a theory that generalizes the Green's function method to include a "marching" solution to the one-way wave equation.

In this paper we obtain an operator solution by formally integrating the one-way equation. An explicit marching solution is obtained by using a spectral representation of the operator solution. The final formalism is similar to the Green's function approach in the "Fast Field Program" (FFP) and could, in fact, be thought of as a "Range-Dependent" FFP (RDFFP). In the FFP, the source function is a delta function, while in the RDFFP the source function is the field at the end of the previous range step.

We show the numerical accuracy of the Green's function approach by comparing the RDFFP solution with a coupled mode solution[2] and with a wide-angle parabolic equation solution.[3,4]

THEORETICAL ANALYSIS

Operator Solution for the One-Way Wave Equation

Consider the two-way (outgoing and incoming) wave equation for the acoustic pressure P in a cylindrically symmetric waveguide:

$$\frac{\partial^2 P}{\partial r^2} + \frac{1}{r}\frac{\partial P}{\partial r} + \frac{\partial^2 P}{\partial z^2} + k^2 P = 0 \tag{1}$$

Here, the wavenumber k is a function of both range, r, and depth, z, and is given by $\omega/c(r,z)$, where ω is the angular frequency and c is the sound speed. The associated far-field equation for the variable $\psi = P/r^{1/2}$ is

$$\frac{\partial^2\psi}{\partial r^2} + \frac{\partial^2\psi}{\partial z^2} + k^2\psi = 0 . \tag{2}$$

By formally solving Eq. (2) for $\partial\psi/\partial r$, we obtain a one-way wave equation for ψ:

$$\frac{\partial\psi}{\partial r} = i\left(\frac{\partial^2}{\partial z^2} + k^2\right)^{1/2} \psi , \tag{3}$$

where we have chosen the positive square root to propagate an outgoing wave. If k is independent of range, Eq. (3) is exact; that is, a solution of Eq. (3) is also a solution of Eq. (2). For a range-varying medium, Eq. (3) accurately propagates an outgoing wave provided the backward-going energy is small compared to the outgoing energy.

For convenience we define the operator $(\partial^2/\partial z^2 + k^2)$ as Q. Then Eq. (3) becomes

$$\frac{\partial\psi}{\partial r} = iQ^{1/2}\,\psi . \tag{4}$$

The formal operator solution to Eq. (4) is

$$\psi(r) = exp(i\Delta r Q^{1/2})\,\psi(r_o) , \tag{5}$$

where $\Delta r = r - r_o$. In Eq. (5) we write the r dependence explicitly but for now suppress the z dependence.

Explicit Solution for ψ

To obtain an explicit solution from Eq. (5), we use the spectral representation for a function of an operator Q.[5] Given a function of the operator Q, we can write

$$F(Q) = \frac{1}{2\pi i} \oint \frac{F(x)}{xI - Q}\, dx , \tag{6}$$

where I is the unit operator. In Eq. (6) the operator $1/(xI - Q)$ denotes the inverse of $xI - Q$, and the integral over x must enclose the entire spectrum of Q. (The spectrum of Q is the set of complex numbers, x, where the inverse of $xI - Q$ is not defined.) Applying Eq. (6) to Eq. (5) we obtain

$$\psi(r) = \frac{1}{2\pi i} \oint \frac{exp(i\Delta r x^{1/2})}{x - Q}\, \psi(r_o)\, dx , \tag{7}$$

where the unit operator I has been suppressed. Defining $x = \varkappa^2$, where $\varkappa$ is the horizontal wavenumber, we have

$$\psi(r) = \frac{1}{i\pi} \oint \frac{exp(i\varkappa\Delta r)}{\varkappa^2 - Q}\, \psi(r_o)\, \varkappa d\varkappa . \tag{8}$$

We now define

$$\frac{1}{\varkappa^2 - Q}\,\psi(r_o) = \Phi(\varkappa,r_o) . \tag{9}$$

Hence, the equation for $\Phi(\varkappa, r_o)$ is

$$(Q - \varkappa^2)\, \Phi(\varkappa, r_o) = -\psi(r_o)\,. \tag{10}$$

Explicitly writing the z dependence we have

$$\left[\frac{\partial^2}{\partial z^2} + k^2(r,z) - \varkappa^2\right] \Phi(\varkappa, r_o, z) = -\psi(r_o, z)\,. \tag{11}$$

The solution for $\psi(r,z)$ is then

$$\psi(r,z) = \frac{1}{i\pi} \oint exp(i\varkappa\Delta r)\, \Phi(\varkappa, r_o, z)\, \varkappa d\varkappa\,. \tag{12}$$

Given the field $\psi(r_o,z)$ we use Eq. (11) to generate the spectral decomposition $\Phi(\varkappa,r_o,z)$ relative to the "new" environment in the interval r_o to r. (We assume the environment is independent of range between r_o and r.) In Eq. (12) the spectral components are carried forward by multiplying by $exp(i\varkappa\Delta r)$; the full field, $\psi(r,z)$, is recovered by integrating over $\varkappa$.

Vertical Density Dependence

To account for vertical variations in density, we replace $\partial^2/\partial z^2$ with $\varrho(\partial/\partial z)\varrho^{-1}(\partial/\partial z)$. Equation (11) then becomes

$$\left(\frac{\partial}{\partial z}\varrho^{-1}\frac{\partial}{\partial z} + \omega^2/\beta - \varkappa^2/\varrho\right)\Phi(\varkappa, r_o, z) = -\psi(r_o, z)/\varrho\,, \tag{13}$$

where β is the bulk modulus: $\beta = \varrho c^2$. On a range step from r_o to r we take β and ϱ to be functions of z only.

NUMERICAL IMPLEMENTATION

Finite Element Discretization

Solving the inhomogeneous wave equation for $\Phi(k,r_o,z)$ is the main difficulty in obtaining $\psi(r,z)$ in Eq. (12). The approach taken here is to discretize the z dependence of the inhomogeneous wave equation in Eq. (13). This approach converts Eq. (13) into a matrix-vector equation.

In Eq. (13) we expand ψ and Φ in terms of linear one-dimensional finite element functions:

$$\psi(r_o,z) = \sum_j \psi_j\, h_j(z) \tag{14}$$

and

$$\Phi(\varkappa,r_o,z) = \sum_j \Phi_j\, h_j(z)\,, \tag{15}$$

where the basis functions $h_j(z)$ are the "hat" functions shown in Figure 1. It is easy to show that for hat functions, the coefficients ψ_j and Φ_j are

$$\psi_j = \psi(r_o, z_j) \tag{16}$$

and

$$\Phi_j = \Phi(\varkappa, r_o, z_j)\,. \tag{17}$$

It is important to note that expanding ψ and Φ in terms of hat functions is equivalent to linear interpolation between the grid points z_j. The advantage of the basis function formalism is that we can obtain a convenient matrix representation for the operators in Eq. (13).

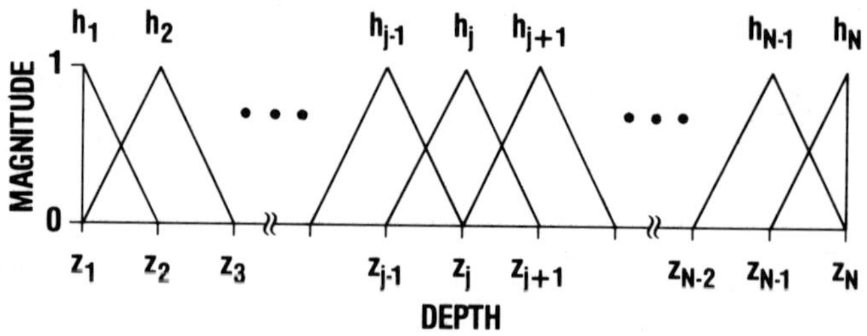

Figure 1. Basis functions (''hat'' functions) used to discretize the vertical dependence of the functions ψ and Φ in Eqs. (14) and (15), respectively.

To obtain the matrix representation, we substitute the hat function expansions for ψ and Φ into Eq. (13), multiply by $h_i(z)$, and integrate over z:

$$\sum_j \left[\int\left(h_i \frac{\partial}{\partial z} \varrho^{-1} \frac{\partial}{\partial z} h_j\right) dz + \omega^2 \int (h_i\ \beta^{-1} h_j)\ dz - \varkappa^2 \int (h_i\ \varrho^{-1}\ h_j)\ dz\right] \Phi_j$$

$$= -\sum_j \left[\int (h_i \varrho^{-1} h_j)\ dz\right] \psi_j\ . \tag{18}$$

In Eq. (18) the z dependence is not written explicitly. Note, however, that because a hat function $h_i(z)$ is nonzero only over z_{i-1} to z_{i+1}, the sum over j goes from $i-1$ to $i+1$, and the integration is from z_{i-1} to z_{i+1}. The integrals in Eq. (18) define the following matrix elements:

$$M_{1ij} = -\int \dot{h}_i(z)\ \varrho^{-1}\ \dot{h}_j(z)\ dz\ .\ z) \tag{19}$$

In Eq. (19) integration by parts is used to obtain the form of the matrix elements with $\dot{h}$, where $\dot{h}$ denotes $\partial h/\partial z$. The other matrix elements are

$$M_{2ij} = \int h_i(z)\ \beta^{-1}(z)\ h_j(z)\ dz \tag{20}$$

and

$$M_{3ij} = \int h_i(z)\ \varrho^{-1}(z)\ h_j(z)\ dz\ . \tag{21}$$

If $\varrho^{-1}(z)$ and $\beta^{-1}(z)$ are represented as piecewise linear functions, simple analytic expressions can be obtained for the matrices M_1, M_2, and M_3. Using the above definitions for M_1, M_2, and M_3, we can write Eq. (18) in matrix-vector form

$$(M_1 + \omega^2 M_2 - \varkappa^2 M_3) \bullet \vec{\Phi} = -M_2 \bullet \vec{\psi}\ , \tag{22}$$

where the column vectors $\vec{\Phi}$ and $\vec{\psi}$ are $(\Phi_1, \Phi_2, \ldots, \Phi_n)^T$ and $(\psi_1, \psi_2, \ldots, \psi_n)^T$, respectively ($T$ denotes transpose).

The matrices M_1, M_2, and M_3 are tridiagonal, so we have an easily solved system of linear equations[6] of the form

$$\begin{pmatrix} \ddots & \ddots & \ddots & & O \\ & A_j & B_j & C_j & \\ O & & \ddots & \ddots & \ddots \end{pmatrix} \begin{pmatrix} \vdots \\ \Phi_j \\ \vdots \end{pmatrix} = \begin{pmatrix} \vdots \\ D_j \\ \vdots \end{pmatrix} \tag{23}$$

Boundary Conditions

The boundary conditions on Φ at the top and bottom of the computational grid were imposed by prescribing the ratios $\Phi_1/\Phi_2 = r_1$ and $\Phi_n/\Phi_{n-1} = r_2$. Setting $r_1 = 0$ gives a pressure-release surface. An impedance condition was used for the boundary condition at the bottom of the computational grid.

The impedance condition was set by prescribing the ratio $\Phi/(\partial\Phi/\partial z) = Z$ at the bottom of the grid $(z + L)$. Numerically this condition gives

$$r_2 = \Phi_n/\Phi_{n-1} = Z/(Z - \Delta z). \tag{24}$$

Two different values of Z were used: $Z = 0$ and $Z = (ik_v)^{-1}$, where k_v is the vertical wavenumber at $z = L$. The value $Z = 0$ gives a pressure release boundary at $z = L$; the value $Z = (ik_v)^{-1}$ matches Φ and $\partial\Phi/\partial z$ to an outgoing plane wave of the form $exp(ik_v z)$.

For both bottom boundary conditions, an artificial absorbing layer was added to the physical bottom so that $\psi \cong 0$ at $z = L$. The pressure release boundary condition $(Z = 0)$ perfectly reflects all incident waves so the absorbing layer is needed to attenuate spurious reflections from the end of the computational grid. For the second boundary condition, $\psi(L) = 0$ is required for a plane-wave condition to be valid at $z = L$. The wavenumber spectrum is completely discrete for the first boundary condition. For the second boundary condition the spectrum consists of a discrete part plus a continuous part.

Boundary conditions at interfaces are handled automatically in the finite element matrices. The linear finite element scheme used here is second-order accurate, even with discontinuities in ϱ and c.

Starting Field

An exact starting field can be generated by slightly modifying Eqs. (12) and (13). We replace $-\psi(r_o,z)$ with $-\delta(z - z_o)$ and we replace $(1/i\pi)exp(i\varkappa\Delta r)$ with $(2/i\pi\varkappa)^{1/2}\, exp(i\varkappa\Delta r)$, which is the asymptotic form of $i\pi(r)^{1/2}H_0^{(1)}(\varkappa\Delta r)$. With these replacements we can generate an exact starting field and simultaneously take the first range step.

Integration Contours

Equation (12) indicates that the integration should completely enclose the spectrum of Q. For the case with $Z = (ik_v)^{-1}$ this requirement is equivalent to enclosing the poles of Φ and the branch cut associated with k_v (see Fig. 2). For $Z = 0$ the branch cut is replaced by many closely spaced poles. In the actual numerical calculation, the contours C_1 and C_2 in Figure 2 were used. The integrals were of the form

$$\psi(r,z) = (1/i\pi)exp(\alpha_1\Delta r)\int_{\varkappa_{min}}^{\varkappa_{max}} exp(i\varkappa\Delta r)\ \Phi(\varkappa - i\alpha_1,r_o,z)\varkappa d\varkappa$$

$$- (1/i\pi)exp(\alpha_2\Delta r)\int_{\varkappa_{min}}^{\varkappa_{max}} exp(i\varkappa\Delta r)\ \Phi(\varkappa + i\alpha_2,r_o,z)\varkappa d\varkappa \tag{25}$$

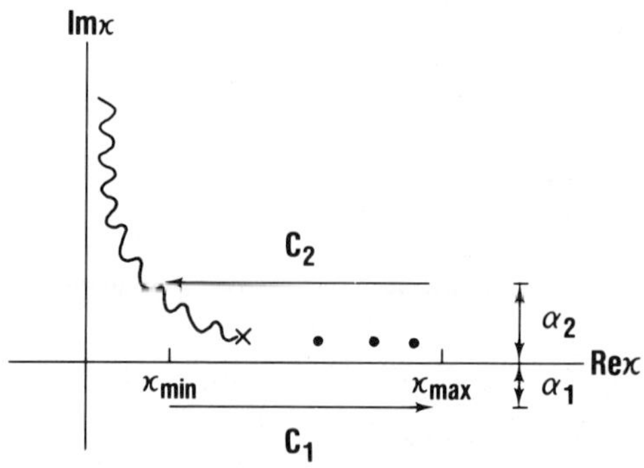

Figure 2. Integration contours used to evaluate the integral in Eq. (12).

where in Eq. (25) $\varkappa$ is the real part of the wavenumber, and the imaginary part has been written explicitly. The integrals in Eq. (25) were computed with a fast Fourier algorithm. For large values of Δr the contribution from the integral over C_2 is negligible. For small values, however, it cannot be neglected.

Note that because attenuation displaces the poles and the branch point upward, the value of α_2 must be large enough that the contour C_2 is above the poles and the branch cut. Otherwise, contributions from poles and the branch cut will be lost.

NUMERICAL EXAMPLE

Environmental Inputs

Figure 3a shows the ocean environment used in the numerical example. The water column is homogeneous ($c_w = 1500$ m/s, $\varrho_w = 1.0$ gm/cm^3, $\alpha_w = 0$), and the bottom is a homogeneous half-space ($c_b = 1700$ m/s, $\varrho_b = 1.2$ gm/cm^3, $\alpha_b = 7.35 \times 10^{-3}$ dB/m). The environment is divided into regions I, II, and III. In regions I and III the depth, D_1, is 200 m; in region II the depth, D_2, is 65 m.

Transmission Loss

In Figure 3b the transmission loss is shown for a 25-Hz source at a depth, z_s, of 100 m with the receiver at a depth of 50 m. The solid line was calculated with RDFFP. (Both impedance conditions gave the same transmission loss.) The dashed line was calculated in two different ways. First, the coupled-mode model CUPYL[2] was used (with backscatter turned off) to solve the one-way wave equation. Next, the wide-angle version of the IFD parabolic equation model[3,4] was used. Since the CUPYL and IFD results were indistinguishable on a transmission loss plot, we only show the single dashed line. The excellent agreement between RDFFP and CUPYL/IFD shows that RDFFP can give an accurate solution to the one-way wave equation.

Horizontal Wavenumber Spectrum

Figure 3c shows the magnitude of $\Phi(k,r,z)$ at the start of each of the three environmental sections. (We show only the result for the half-space impedance condition, $Z = (ik_v)^{-1}$. With the pressure release boundary condition, $Z = 0$, the branch cut is replaced by a series of closely spaced poles.) The path in the complex $\varkappa$ plane is along C_1 (see Fig. 2). Figure 3d is the same as Figure 3c, except the path is along C_2. Along C_1 the imaginary component of wavenumber is $\alpha_1 = 1 \times 10^{-4}$, while along C_2 it is $\alpha_2 = 1.1 \times 10^{-3}$.

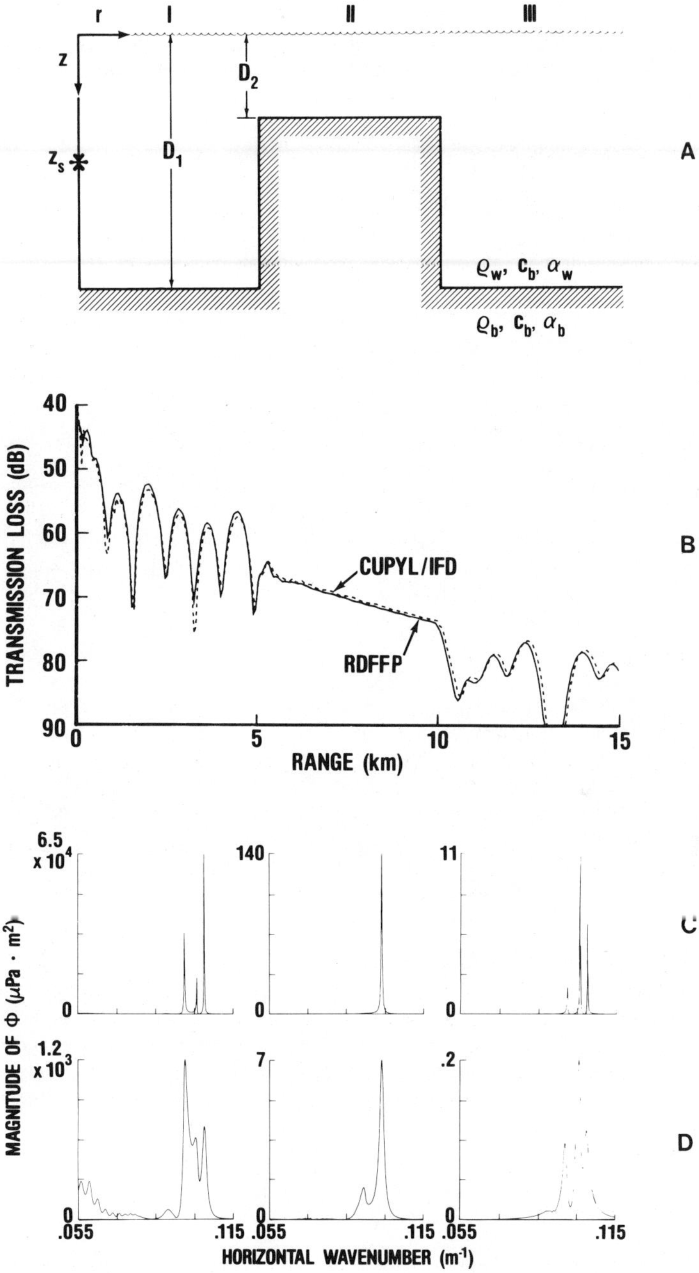

Figure 3. Test case results. (a) Three-region range-dependent environment. (b) Comparison of transmission loss computed with RDFFP and CUPYL/IFD. The source and receiver are at depths of 100 m and 50 m, respectively; the frequency is 25 Hz. (c) Horizontal wavenumber spectrum at the start of each of the three environmental regions, evaluated along contour C_1 in Figure 2. (d) Same as (c), except the contour is C_2.

It is clear from Figure 3c that three modes propagate in region I. In region II only one mode propagates, while in region III there are again three propagating modes. The single propagating mode in region II couples most strongly to the second mode in region I and in region III. As a result of the coupling, the second mode goes from the weakest mode in region I to the strongest mode in region III.

The contribution from the continuum is apparent from Figure 3d. Although the continuum contribution is not a dominant effect, it must be included to obtain an accurate result at the ''break points'' between the environmental regions.

SUMMARY AND CONCLUSIONS

A Green's function method (RDFFP) for solving the one-way wave equation was derived and applied to propagation in a range-dependent ocean environment. RDFFP calculations were in excellent agreement with coupled mode and wide-angle parabolic equation calculations.

The method presented here has a number of attractive features. For example, since the RDFFP deals directly with the horizontal wavenumber spectrum, it can be easily extended to range-dependent elastic wave propagation. The RDFFP computes both the total field and its spectral components on each range step, so with the extension to elastic waves, RDFFP could provide valuable information on phenomena such as range-dependent compressional-shear conversion and elastic mode coupling.

The main drawback of the method is that with its present numerical implementation, it is computationally intensive. For compressional wave calculations, the existing numerical model is considerably slower than conventional models, such as coupled modes and the parabolic equation. Computational techniques for efficiently implementing the method are currently being investigated. If successful, these techniques will make it practical to compute elastic wave propagation in a range-dependent medium.

REFERENCES

1. J. A. Davis, DeWayne White, and R. C. Cavanagh (eds.), ''NORDA Parabolic Equation Workshop,'' Naval Ocean Research and Development Activity, NSTL, Miss., TN 143 (1982).
2. R. B. Evans, ''A coupled mode solution for acoustic propagation in a waveguide with stepwise depth variations of a penetrable bottom,'' *J. Acoust. Soc. Am.* 74, 188-195 (1983).
3. D. Lee and K. E. Gilbert, ''Recent progress in modeling bottom-interacting sound propagation with parabolic equations,'' *OCEANS 82 Conference Record*, Washington D.C., 172-177 (1982).
4. G. Botseas, D. Lee, and K. E. Gilbert, ''IFD: Wide Angle Capability,'' Naval Underwater Systems Center, Technical Report No. 6905 (1983).
5. E. R. Lorch, *Spectral Theory*, Oxford University Press, New York, pp. 102-103 (1962).
6. J. F. Claerbout, *Fundamentals of geophysical data processing*, McGraw-Hill Co., New York, pp. 188-189 (1976).

FACTORIZATION AND PATH INTEGRATION OF THE

HELMHOLTZ EQUATION: NUMERICAL ALGORITHMS

Louis Fishman*, and Stephen C. Wales[+]

*Department of Civil Engineering
The Catholic University of America
Washington, D.C. 20064 USA

[+]Acoustics Division
Naval Research Laboratory
Washington, D.C. 20375 USA

ABSTRACT

The reduced scalar Helmholtz equation is addressed. The previously introduced factorization and path integration analysis is outlined. The resulting computational algorithm is briefly discussed and illustrated with several numerical examples appropriate for ocean seismo-acoustic modeling. Several extensions and further thematic developments are highlighted.

INTRODUCTION

The reduced scalar Helmholtz equation plays a significant role in studies of electromagnetic, seismic, and acoustic direct wave propagation processes. For homogeneous media, the classical methods of Fourier analysis are quite appropriate for analyzing the n-dimensional Helmholtz equation. The extension of these Fourier methods to the study of wave propagation in inhomogeneous media provides the principal thrust of this work and the specific focus on microscopic, or phase space, methods. It is in phase space, rather than in configuration space, that, from a mathematical perspective, the interesting geometry takes place.

For transversely inhomogeneous environments, implying medium homogeneity with respect to a single distinguished direction, the n-dimensional Helmholtz equation can be exactly factored into separate forward and backward one-way wave equations. Phase space path integral constructions provide for a representation of the wave field and, subsequently, for a computational basis of the pseudo-differential wave equation. In providing the principal means to extend the homogeneous Fourier constructions, the path (functional) integral formulation is distinct from a normal-mode generalization. The resulting marching algorithm generalizes the Tappert/Hardin split-step FFT algorithm to the full one-way (factorized Helmholtz) wave equation. Moreover, the manner of marching the radiation field is independent of the medium and any approximation to the square root Helmholtz operator, resulting in a modular code architecture and highly versatile propagation program.

The factorization-/path integration-based code can readily accommodate density profiles and range updating, and further provides the basis for a weak-backscatter perturbation theory incorporating the "one additional dimension" associated with backscatter. In particular, the interplay between phase space- and configuration space-based path integrals provides a complementary stochastic framework for the arbitrary inhomogeneity Helmholtz wave propagation process. The factorization/weak-backscatter approach is conceptually distinct from the more familiar weak-scattering perturbation theories and is, in a sense, the most natural approach for physical problems, such as sound propagation in the ocean, where there is an obvious distinguished direction and weak range dependence.

FACTORIZATION AND PATH INTEGRATION

The n-dimensional Helmholtz operator can be approximately factored in the form

$$L = \partial_x^2 + (\bar{k}^{-2}K^2(\underset{\sim}{x}) + \nabla_\perp^2) \simeq (\partial_x - \lambda(\underset{\sim}{x},-i\nabla_\perp))(\partial_x + \lambda(\underset{\sim}{x},-i\nabla_\perp)) + \text{(smooth error)} \qquad (1)$$

where the operator $\lambda(\underset{\sim}{x},-i\nabla_\perp)$ is defined in an appropriate asymptotic fashion and $K(\underset{\sim}{x})$ is the refractive index field and $\bar{k}$ is a reference wave number.[1] For a transversely inhomogeneous environment, $K^2(\underset{\sim}{x}) = K^2(\underset{\sim}{x}_\perp)$, a normal-mode (spectral) analysis readily illustrates that this factorization is, in principle, exact, resulting in separate, physical forward and backward, one-way wave equations. The forward evolution (one-way) equation

$$(i/\bar{k})\,\partial_x\,\phi^+(x,\underset{\sim}{x}_\perp) + (K^2(\underset{\sim}{x}_\perp) + (1/\bar{k}^2)\nabla_\perp^2)^{1/2}\,\phi^+(x,\underset{\sim}{x}_\perp) = 0 \qquad (2)$$

is the formally exact wave equation for propagation in a transversely inhomogeneous half-space supplemented with appropriate outgoing wave radiation and initial-value conditions. One-way wave equations have recently been reviewed in greater detail.[2,3]

The Fourier, or phase space, construction of the formal square root operator in Eq.(2) is motivated by a generalization of the well-known homogeneous medium (Fourier optics) form. The mathematical framework is provided by pseudo-differential operator theory and the construction procedure is enhanced by the analogy between the forward wave propagation problem and quantum mechanics.[1,2,4] The relevant result[1,2] is that the formal wave equation (2) now takes the explicit form

$$(i/\bar{k})\,\partial_x\,\phi^+(x,\underset{\sim}{x}_\perp) + (\bar{k}/2\pi)^{n-1}\int_{R^{2n-2}} d\underset{\sim}{x}'_\perp\, d\underset{\sim}{p}_\perp \cdot \Omega_{\mathbf{B}}(\underset{\sim}{p}_\perp,(\underset{\sim}{x}_\perp+\underset{\sim}{x}'_\perp)/2)\,\exp(i\bar{k}\underset{\sim}{p}_\perp\cdot(\underset{\sim}{x}_\perp-\underset{\sim}{x}'_\perp))\,\phi^+(x,\underset{\sim}{x}'_\perp) = 0. \qquad (3)$$

In Eq.(3), the symbol $\Omega_{\mathbf{B}}(\underset{\sim}{p},\underset{\sim}{q})$ associated with the square root Helmholtz operator $\mathbf{B} = (K^2(\underset{\sim}{q}) + (1/\bar{k}^2)\nabla_{\underset{\sim}{q}}^2)^{1/2}$ satisfies the Weyl composition equation

$$\Omega_{\mathbf{B}}2(\underset{\sim}{p},\underset{\sim}{q}) = K^2(\underset{\sim}{q})-\underset{\sim}{p}^2 = (\bar{k}/\pi)^{2n-2}\int_{R^{4n-4}} d\underset{\sim}{t}d\underset{\sim}{x}d\underset{\sim}{y}d\underset{\sim}{z}\,\Omega_{\mathbf{B}}(\underset{\sim}{t}+\underset{\sim}{p},\underset{\sim}{x}+\underset{\sim}{q}) \cdot \Omega_{\mathbf{B}}(\underset{\sim}{y}+\underset{\sim}{p},\underset{\sim}{z}+\underset{\sim}{q})\,\exp(2i\bar{k}(\underset{\sim}{x}\cdot\underset{\sim}{y}-\underset{\sim}{t}\cdot\underset{\sim}{z})) \qquad (4)$$

with $\Omega_{\mathbf{B}^2}(p,q)$ the symbol associated with the square of $\mathbf{B}, \mathbf{B}^2 = (K^2(q) \mp (1/k^2)\nabla_q^2)$. The construction procedure for the square root Helmholtz operator can be summarized pictorially by the following correspondence diagram

$$\begin{array}{ccc} \mathbf{B}^2 & \longleftrightarrow & \Omega_{\mathbf{B}^2} \\ \uparrow & & \updownarrow \\ \mathbf{B} & \longleftrightarrow & \Omega_{\mathbf{B}} \end{array}$$

where the arrows symbolize the one- and two-way mappings between the appropriate quantities.

Exact solutions of the Weyl composition equation (4) can be constructed in several cases.[5] For example, in Figure 1, $\Omega_{\mathbf{B}}(p,q)$ for the two-dimensional (n = 2) linear medium, $K^2(q) = K_0^2 + 2\delta q$, is plotted. Nonuniform and uniform perturbation solutions corresponding to definite physical limits (frequency, propagation angle, field strength, field gradient) recover several well-known approximate wave theories (ordinary parabolic, range refraction parabolic, rational linear) and systematically lead to several new full-wave, wide-angle approximations.[1,2,5]

The form of Eq.(2) and the resultant Markov property of the propagator suggest the basis for a path integral construction. In an operator notation, then,

$$\exp(i\bar{k}\mathbf{B}x) = \lim_{N \to \infty} \prod_{j=1}^{N} \exp(i\bar{k}\mathbf{B}\Delta x_j) \tag{5}$$

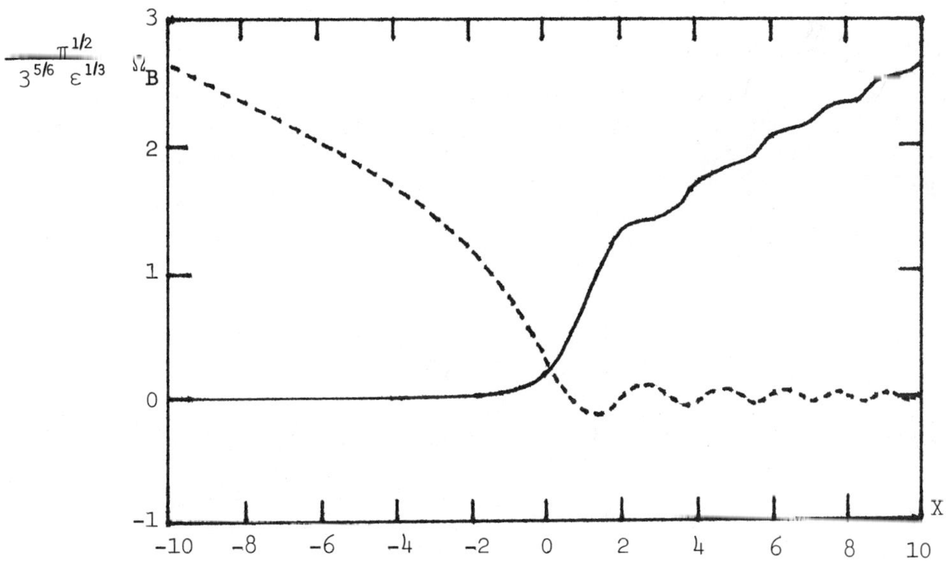

Fig. 1. The real (——) and imaginary (---) parts of $\Omega_{\mathbf{B}}(p,q)$ for the two-dimensional (n = 2) linear medium, $K^2(q) \equiv K_0^2 + 2\delta q$. $X = (3^{1/3}/2\varepsilon^{2/3})(K_0^2 + 2\delta q - p^2)$ with $\varepsilon = |\delta|/k$.

where $\Delta x_j = x/N$, symbolically representing the propagator in terms of the infinitesimal propagator. The detailed Fourier construction results in a parabolic-based Hamiltonian phase space path integral representation of the propagator in the form[1,4]

$$G^{+}(x,\underset{\sim}{x}_{\perp}|0,\underset{\sim}{x}'_{\perp}) = \lim_{N\to\infty} \int_{R^{(n-1)(2N-1)}} \prod_{j=1}^{N-1} d\underset{\sim}{x}_{j\perp} \prod_{j=1}^{N} (\bar{k}/2\pi)^{n-1} d\underset{\sim}{p}_{j\perp}$$

$$\cdot\exp(i\bar{k}\sum_{j=1}^{N}(\underset{\sim}{p}_{j\perp}\cdot(\underset{\sim}{x}_{j\perp}-\underset{\sim}{x}_{j-1\perp}) + (x/N)\, H(\underset{\sim}{p}_{j\perp},\underset{\sim}{x}_{j\perp},\underset{\sim}{x}_{j-1\perp}))) \tag{6}$$

where

$$H(\underset{\sim}{p},\underset{\sim}{q}'',\underset{\sim}{q}') = (\bar{k}/2\pi)^{n-1} \int_{R^{2n-2}} d\underset{\sim}{s}d\underset{\sim}{t}\; F(\underset{\sim}{q}'-\underset{\sim}{q}'',\underset{\sim}{s})$$

$$\cdot h_{\mathbf{B}}(\underset{\sim}{p},((\underset{\sim}{q}''+\underset{\sim}{q}')/2)-\underset{\sim}{t})\;\exp(i\bar{k}\underset{\sim}{s}\cdot\underset{\sim}{t}). \tag{7}$$

In Eq.(7), $F(\underset{\sim}{u},\underset{\sim}{v})$ and $h_{\mathbf{B}}(\underset{\sim}{p},\underset{\sim}{q})$ are related to the operator symbol $\Omega_{\mathbf{B}}(\underset{\sim}{p},\underset{\sim}{q})$ through

$$\hat{\Omega}_{\mathbf{B}}(\underset{\sim}{u},\underset{\sim}{v}) = F(\underset{\sim}{u},\underset{\sim}{v})\;\hat{h}_{\mathbf{B}}(\underset{\sim}{u},\underset{\sim}{v}) \tag{8}$$

where $\hat{\Omega}_{\mathbf{B}}$ and $\hat{h}_{\mathbf{B}}$ are the corresponding Fourier transforms.[1,2,4]

While the parabolic-based path integral construction proceeds from the n-dimensional extended parabolic (Weyl pseudo-differential) equation, elliptic-based path integral constructions associated with Feynman and Fradkin, Feynman and Garrod, and Feynman and DeWitt-Morette result from an imbedding in an (n+1)-dimensional parabolic (Schrödinger) equation. The path integral analysis allows for a global perspective of the wave theory in addition to providing a unifying framework for dynamical approximations, resolution of the square root operator, and the concept of an underlying stochastic process.[1,4]

COMPUTATIONAL ALGORITHM

The phase space path integral provides for a computational basis for the pseudo-differential wave equation (3). Choosing the standard ordering, $F(\underset{\sim}{u},\underset{\sim}{v}) = \exp(-i\bar{k}\underset{\sim}{u}\cdot\underset{\sim}{v}/2)$, in Eqs.(6) and (7) results in a post-point marching algorithm in the form

$$\phi^{+}(x+\Delta x,\underset{\sim}{x}_{\perp}) \simeq \int_{R^{n-1}} d\underset{\sim}{p}_{\perp}\;\exp(i\bar{k}\underset{\sim}{p}_{\perp}\cdot\underset{\sim}{x}_{\perp})(\exp(i\bar{k}\Delta x h_{\mathbf{B}}(\underset{\sim}{p}_{\perp},\underset{\sim}{x}_{\perp}))\;\hat{\phi}^{+}(x,\underset{\sim}{p}_{\perp})) \tag{9}$$

where

$$h_{\mathbf{B}}(\underset{\sim}{p}_{\perp},\underset{\sim}{x}_{\perp}) = (\bar{k}/\pi)^{n-1} \int_{R^{2n-2}} d\underset{\sim}{s}d\underset{\sim}{t}\;\Omega_{\mathbf{B}}(\underset{\sim}{s},\underset{\sim}{t})\;\exp(-2i\bar{k}(\underset{\sim}{x}_{\perp}-\underset{\sim}{t})\cdot(\underset{\sim}{p}_{\perp}-\underset{\sim}{s})) \tag{10}$$

and $\hat{\phi}^{+}$ is the Fourier transformed wave field. Equation (9) provides an increasingly better approximation in the limit of infinitesimal range step size Δx. Further, in the parabolic limit, the algorithm reduces to the Tappert/Hardin split-step FFT algorithm.[6]

The path integral formulation interprets the wave theory in terms of an infinitesimal propagator summed over all phase space paths. For the Helmholtz theory, the exact infinitesimal propagator is not, in general, given by the locally homogeneous medium propagator, as in the ordinary parabolic (Schrödinger) propagator construction. The approximate extended parabolic wave theories then correspond to approximate infinitesimal propagators summed over the complete phase space. In retaining the "sum over all paths," diffraction, or full-wave, effects are incorporated. Equations (9) and (10) further illustrate that the manner of marching the radiation field is independent of the medium and any approximation to the square root Helmholtz operator. With this data contained in the matrix input function $I_D = h_{\mathbf{B}}(\underset{\sim}{p}_\perp, \underset{\approx}{x}_\perp)$, the resulting modular code architecture allows for a highly versatile direct propagation program. Moreover, the propagation models constructed and computed correspond to singular integro-differential equation as well as partial differential equation approximations to the one-way wave equation.

For a two-dimensional model ocean/bottom propagation environment with a perfectly reflecting ocean surface, the construction of the computational algorithm is straightforward. Two perfectly reflecting surfaces, corresponding to the ocean surface and the bottom of the Fourier transform regime, are introduced. A "false bottom" with attenuation is provided to absorb the wave field as it approaches the bottom boundary. With this approximation to the radiation condition, the Fourier transform of the wave field in Eq.(9) is replaced by a discrete fast sine transform. The careful construction of the attenuating "false bottom" is essential to minimize spurious reflection, or wrap-around, effects. The bottom layer must be sufficiently deep with an attenuation profile which is sufficiently gradual to avoid "ringing" effects and, ultimately, sufficiently strong to completely damp out the wave field. Employing a Simpson's rule integration for the inverse transform in Eq.(9) enables the propagated wave field to be expressed in the matrix form

$$\phi^+(x+\Delta x, z_n) = \sum_m A_{nm} \hat{\phi}^+(x, p_m) \tag{11}$$

for each depth point z_n. In Eq.(11), ϕ^+ and $\hat{\phi}^+$ are column vectors and the matrix $\underset{\approx}{A}$ is defined by its matrix elements

$$A_{nm} = \eta \sin(p_m z_n + \bar{k}\Delta x h^o_{\mathbf{B}}(p_m, z_n)) \exp(i\bar{k}\Delta x h^e_{\mathbf{B}}(p_m, z_n)) \tag{12}$$

where $h^e_{\mathbf{B}}$ and $h^o_{\mathbf{B}}$ are the even and odd parts with respect to p of $h_{\mathbf{B}}(p,z)$ in Eq.(10) and η is an appropriate transform normalization constant. The initial start up field is provided by a normal-mode start-up program or a Gaussian field when applicable. A detailed discussion of these and other numerical considerations can be found elsewhere.[7-9]

The high-frequency approximation to the operator symbol results in a particularly fruitful computational model. This specific algorithm corresponds to a standard pseudo-differential operator with the symbol $(K^2(\underset{\sim}{q})-\underset{\sim}{p}^2)^{1/2}$ with $h_{\mathbf{B}}(\underset{\sim}{p}_\perp, \underset{\approx}{x}_\perp)$ in Eq.(10) being just the square root function. It represents the most naive generalization of the homogeneous medium result,[1,2] corresponding to the locally homogeneous medium approximation to the infinitesimal propagator, and is quite distinct from geometric, or semiclassical, approximations on the wave field. The high-frequency algorithm does not, in general, strictly conserve energy. However, the slightly modified algorithm corresponding to a Weyl pseudo-differential operator with the symbol Re $(K^2(\underset{\sim}{q})-\underset{\sim}{p}^2)^{1/2}$ is energy conserving. For this approximation, in the two-dimensional case, for example, $h_{\mathbf{B}}(p,z)$ in Eq.(10) takes the specific form

$$h_{\mathbf{B}}(p,z) = (1/2) \int_{R^1} dt\, t^{-1} \exp(2i\bar{k}pt)\, K(t+z)\, J_1(2\bar{k}tK(t+z)) \tag{13}$$

where $J_1(y)$ is the appropriate Bessel function. Equation (13) is efficiently computed by an FFT algorithm. Further, the high-frequency, or "classical" limit,[2,4] approximation to the operator symbol provides the starting point for a uniformized high-frequency development[5] leading to a more widely applicable computational model.

NUMERICAL RESULTS

The extended parabolic wave theories derived from the factorization analysis and computed by the path integral-based marching algorithm have been applied to ocean acoustic environments, seismological environments, and extreme model environments designed to establish their range of validity and manner of breakdown. The propagation model environments include strongly refracting sound speed profiles, multiple-well profiles, and rapidly changing environments, some, extreme enough to simulate an ice/water interface. The propagation frequencies ranged over two orders of magnitude from 2.5 to 250 Hz. By design, the computation of transmission loss (dB re 1 m) as a function of range (km) curves requires wide-angle codes, often, with capabilities in excess of 40 degrees. Reference transmission loss curves are provided by a Fast Field Program algorithm (FFP) and additional comparisons are made with several rational-linear codes.

Several representative model ocean/bottom environments and propagation experiments are summarized in Figure 2. The corresponding computed transmission loss curves are displayed in Figures 3-5. For the extreme double well ocean model of case A at 25 Hz, the high-frequency approximation is in excellent agreement with the FFP result as shown in Figure 3. This is also found to be the case for all of the range-independent models in the NORDA Parabolic Equation Workshop.[7-10] Figure 1 illustrates that

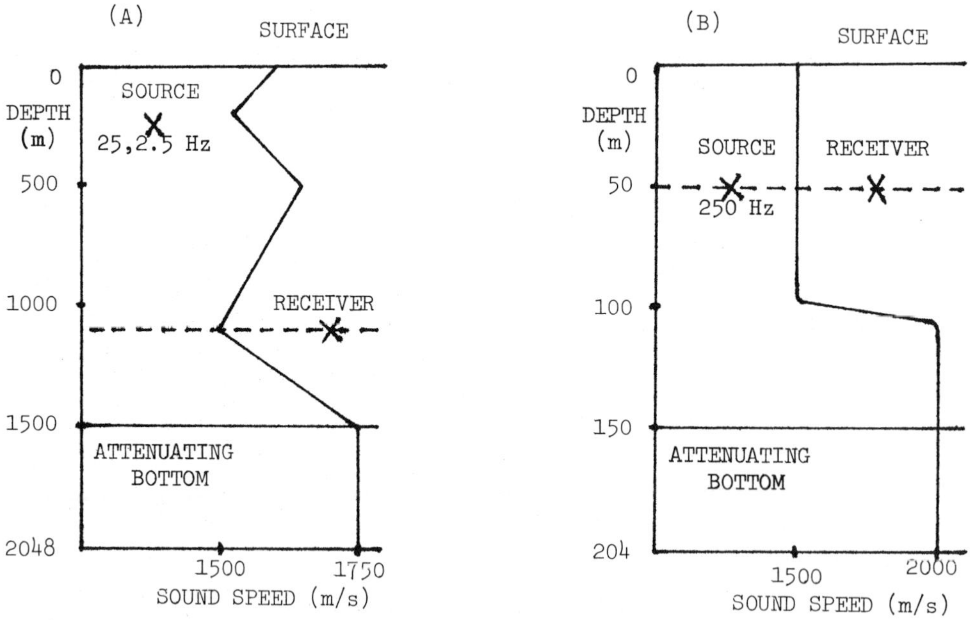

Fig. 2. Model ocean/bottom environments and propagation experiments. Case A is an extreme double well ocean model. Case B is a rapidly changing shallow water model environment.

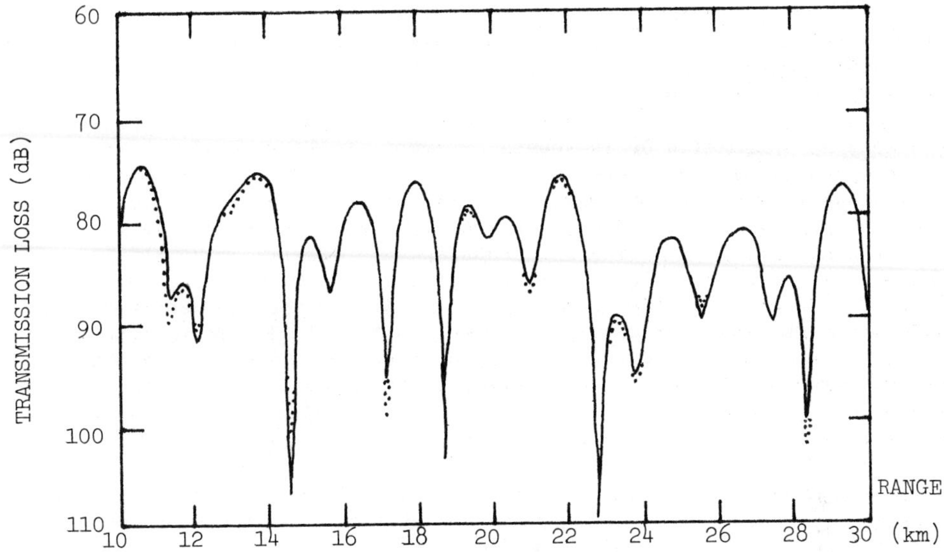

Fig. 3. Transmission loss versus range for case A at 25 Hz.
FFP (——) High-Frequency (····)

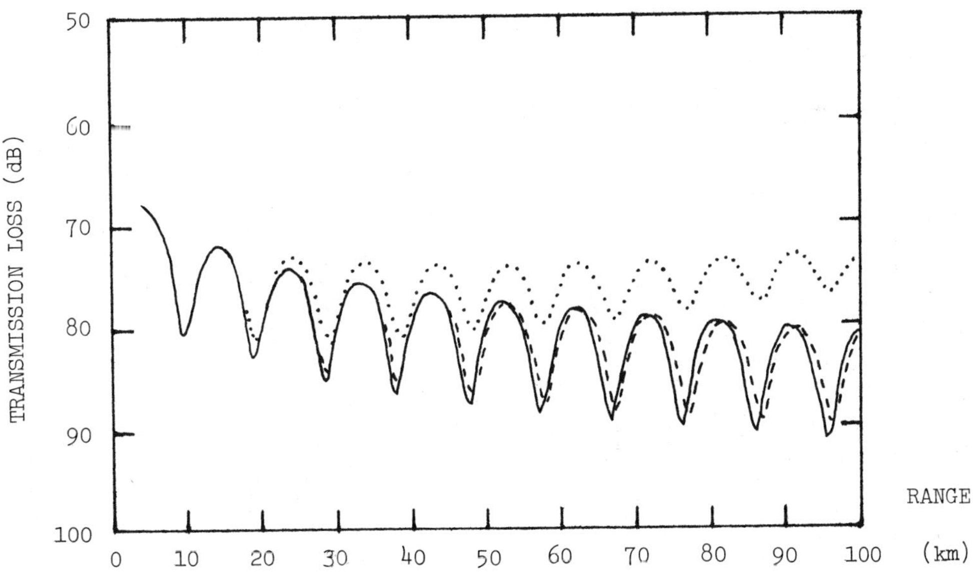

Fig. 4. Transmission loss versus range for case A at 2.5 Hz.
FFP (——) High-Frequency (····) Weyl Algorithm (----)

$(K^2(q)-p^2)^{1/2}$ is the leading algebraic term in an outer-scale asymptotic expansion of the operator symbol. As such, the high-frequency propagation algorithm can be generally expected to break down at sufficiently large ranges, in addition to becoming increasingly inadequate with decreasing frequency and increasing sound speed gradient. This is illustrated in Figure 4 for case A at 2.5 Hz where the high-frequency propagation algorithm now rapidly breaks down, violating conservation of energy. The Weyl algorithm of Eq.(13), however, restores energy conservation and is in very good agreement with the FFP result. In general, the uniformized high-frequency approximation of the operator symbol is expected to provide a more accurate and widely applicable propagation algorithm.[5] For the rapidly changing shallow water environment of case B at 250 Hz, the high-frequency, Weyl, and FFP results are virtually identical as shown in Figure 5. At greater ranges or with increasing sound speed gradient, the high-frequency approximation begins to fail. Again, the Weyl algorithm, and, in particular, the uniformized high-frequency approximation are more appropriate propagation models. The detailed, extensive numerical computations and subsequent evaluations of the propagation models, in addition to a more complete discussion of phase space symbol approximations, are presented elsewhere.[7,8]

DISCUSSION

The principal idea underlying the practical implementation of the phase space marching algorithm is the construction of a small number of approximate operator symbols, which when taken together, allow for wave field computations over a very wide range of model environments and propagation parameters. In addition to the analytical perturbation approaches, an essentially numerical solution to the composition equation can be addressed. It is believed that such a numerical approach will be

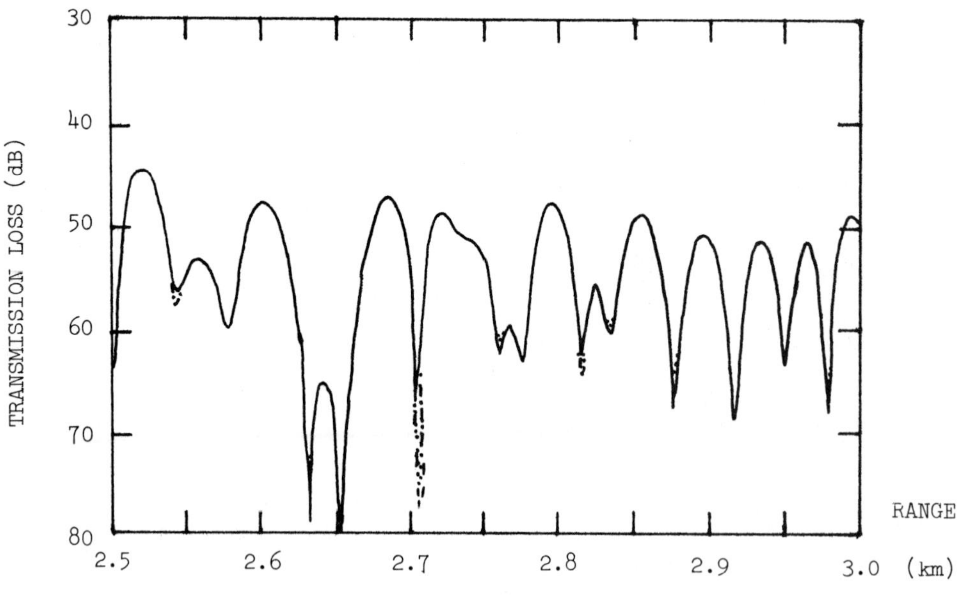

Fig. 5. Transmission loss versus range for case B at 250 Hz. FFP (——) High-Frequency (····) Weyl Algorithm (----)

particularly appropriate in regimes primarily complementary to those addressed by the high-frequency-based perturbation analysis. Again, the manner of marching the radiation field is independent of the particular approximate operator symbol and the detailed analytical or numerical computation of the matrix input function.

A detailed error and stability analysis of the path integral algorithm remains to be done. Moreover, while the phase space path integral construction summarized in Eqs.(6)-(8) is formally correct in the $N \to \infty$, or functional, limit, its use as the basis of the computational algorithm for a fixed, finite N does not necessarily insure, in any sense, a best approximation. In this regard, it would be instructive to compare the path integral computation with an appropriate finite difference calculation of the singular integro-differential wave equation.

The formulation readily accommodates variable density $(\rho(\underset{\sim}{x}_\perp))$, resulting, for example, in the appropriate composition equation[11]

$$K^2(\underset{\sim}{q})-\underset{\sim}{p}^2 + (1/2\bar{k}^2)\nabla_{\underset{\sim}{q}}\cdot\underset{\sim}{\alpha}(\underset{\sim}{q}) - (i/\bar{k})\underset{\sim}{\alpha}(\underset{\sim}{q})\cdot\underset{\sim}{p}$$

$$= (\bar{k}/\pi)^{2n-2} \int_{R^{4n-4}} d\underset{\sim}{t}d\underset{\sim}{x}d\underset{\sim}{y}d\underset{\sim}{z} \quad \Omega_{\mathbf{B}}(\underset{\sim}{t}+\underset{\sim}{p},\underset{\sim}{x}+\underset{\sim}{q})$$

$$\cdot\Omega_{\mathbf{B}}(\underset{\sim}{y}+\underset{\sim}{p},\underset{\sim}{z}+\underset{\sim}{q}) \exp(2i\bar{k}(\underset{\sim}{x}\cdot\underset{\sim}{y}-\underset{\sim}{t}\cdot\underset{\sim}{z})) \tag{14}$$

with the identification

$$\underset{\sim}{\alpha}(\underset{\sim}{x}_\perp) = \nabla_\perp(\ln(\rho(\underset{\sim}{x}_\perp))). \tag{15}$$

Range variability can be immediately accommodated at the level of range updating as in general parabolic equation codes.[3] More fundamentally, the treatment of wave propagation in generally inhomogeneous environments can be addressed by factorization in conjunction with invariant imbedding methods.[2,11,12] The detailed construction of the stochastic-based path integral would include backscatter effects and thus would provide a complementary framework for the study of general environments. It is further hoped that such a path integral would have computational applications. In particular, the construction of a true functional of the path should allow for a well-defined "sum over paths" computation, thus taking explicit advantage of the multi-parallel processing computer architectures presently being designed and built. The interplay between parabolic- and elliptic-based path integrals leading to such a construction has been illustrated for the homogeneous medium limit.[4]

The phase space-based construction of the square root Helmholtz operator provides the basis for the exact inversion of the inherently nonlinear relationship between wave field data and the refractive index field in the multidimensional formulation. While the direct algorithm proceeds around the correspondence diagram in a clockwise fashion, the inverse algorithm is based on a counterclockwise traversal. For transversely inhomogeneous environments, this leads to oceanic sound speed profile inversion. For range-dependent environments, the inclusion of backscatter effects, even in an approximate manner, would provide the basis for a generalized acoustic tomography, extending the diffraction algorithms based on the Born, Rytov, and distorted-wave Born approximations.[1,2]

Extensions to time domain wave equations, vector formulations, coherence theory, and stochastic modeling have been previously discussed.[1,2,4]

ACKNOWLEDGMENTS

This work was supported under grants from the National Science Foundation (EAR 83-12745), the Office of Naval Research (N00014-83-K-0094), and the U.S. Army Research Office (DAAG 29-85-K-002).

REFERENCES

1. L. Fishman and J.J. McCoy, Factorization, path integral representations, and the construction of direct and inverse wave propagation theories, IEEE Trans. Geosc. Rem. Sens., GE-22 (6): 682 (1984).
2. L. Fishman and J.J. McCoy, Derivation and application of extended parabolic wave theories. Part I. The factorized Helmholtz equation, J. Math. Phys., 25 (2): 285 (1984).
3. D. Lee, "The State-Of-The-Art Parabolic Equation Approximation As Applied To Underwater Acoustic Propagation With Discussions On Intensive Computations," rep. no. TD 7247, Naval Underwater Systems Center, New London Laboratory, New London (1984).
4. L. Fishman and J.J. McCoy, Derivation and application of extended parabolic wave theories. Part II. Path integral representations, J. Math. Phys., 25 (2): 297 (1984).
5. L. Fishman and A. Whitman, Exact and uniform perturbation solutions of the Helmholtz composition equation, J. Math. Phys., submitted for publication (1985).
6. F.D. Tappert, The parabolic approximation method, in: "Wave Propagation in Underwater Acoustics," J.B. Keller and J.S. Papadakis, eds., Springer, New York (1977).
7. L. Fishman and J.J. McCoy, A new class of propagation models based on a factorization of the Helmholtz equation, Geophys. J. R. Astr. Soc., 80: 439 (1985).
8. L. Fishman, J.J. McCoy, and S.C. Wales, Factorization and path integration of the Helmholtz equation: numerical algorithms, J. Acoust. Soc. Am., submitted for publication (1985).
9. J.A. Davis, D. White, and R.C. Cavanagh, "NORDA Parabolic Equation Workshop," NORDA tech. note 143, Naval Ocean Research and Development Activity, NSTL Station (1982).
10. L. Fishman, J.J. McCoy, and S.C. Wales, Factorization and path integration of the Helmholtz equation: numerical algorithms, in: "Congress Proceedings: 11th IMACS World Congress," R. Henriksen, ed., NFA, Oslo (1985).
11. J.J. McCoy, L. Fishman, and L.N. Frazer, Reflection and transmission at an interface separating transversely inhomogeneous acoustic half-spaces, Geophys. J. R. Astr. Soc., submitted for publication (1985).
12. J.J. McCoy and L.N. Frazer, Propagation modelling based on wave field factorization and invariant imbedding, Geophys. J. R. Astr. Soc., submitted for publication (1985).

RANGE DEPENDENT PROPAGATION CODES BASED ON WAVE FIELD FACTORIZATION AND INVARIANT IMBEDDING

John J. McCoy*, Louis Fishman*,
and L. Neil Frazer+

*Department of Civil Engineering
The Catholic University of America
Washington, D.C. 20064 USA

+Hawaii Institute of Geophysics
2525 Correa Road
Honolulu, Hawaii 96822 USA

ABSTRACT

In the absence of range variation the Helmholtz equation can be exactly factored into two one-way equations. These factored equations provide the bases for a number of computationally efficient codes, the most familiar of which is the ordinary parabolic wave theory. In this paper we discuss the reintroduction of range dependence to enable analysis of media with arbitrary variation. Two calculations will be presented. One is the reflection and transmission of an arbitrary field at an interface separating two transversely inhomogeneous media. The second is the reflection and transmission of an arbitrary field by a transition region separating two transversely inhomogeneous media. The second problem is addressed by an invariant imbedding procedure and taken together the solutions to the two problems represent an exact and complete generalization of Kennett's method, in the context of scalar waves.

INTRODUCTION

We are actively engaged in the development of propagation models that seek to exploit the effectiveness of automatic computation for iterating, or updating, a data set. In the context of propagation through a range-independent environment, it is intuitively clear that there is no coupling between radiation fields propagating in opposite range directions, i.e., there is no coupling between a right-going wave field and a left-going wave field. This suggests the development of propagation models that are described by one-way wave equations. These one-way wave equations admit of a solution algorithm based on marching the radiation field, starting from an initial range plane that is near to the source of acoustic energy.

The above described program has most frequently been accomplished with the "parabolic wave equation"[1] accepted to be governing. Recently, a number of extended parabolic wave theories,[2-4] which are intended to better approximate the Helmholtz equation accepted to be governing, have been proposed and implemented. In a number of papers,[5-10] we have presented both a framework for constructing an exact one-way wave equation, based on a governing Helmholtz equation, and an algorithm for marching the radiation field. In another contribution to these proceedings,[9] we discuss this work and present results from an extensive numerical study that we are presently completing.

In this paper we outline two studies that reintroduce range dependence into the environment. The first study considers an environment described by two (different) range-independent acoustic half-spaces, joined at a planar interface. The second considers the half-spaces to be separated by a planar transition layer of arbitrary thickness. Within the transition layer the wave number field can vary with position change in any directon. The theoretical development required by the two problems are presented, in detail, elsewhere.[11,12] Computer implementation of numerical algorithms based on this work is in its beginning stages. Some of this is touched on in another contribution to these proceedings.[13]

Conceptually, the reduction of the planar interface problem to a solution updating algorithm is straightforward since the coupling of the right- and left-going wave fields occurs at the planar interface. Thus, assuming an acoustic source located in the left range-independent half-space, an available marching algorithm appropriate for a right-going one-way wave equation can be used to march the acoustic field from an initial plane up to the planar interface. The interaction of this incident field and the interface results in appropriate "initial plane" data for marching the acoustic field away from the interface. This marching is based on an appropriate right-going one-way wave equation for the half-space to the right of the interface and an appropriate left-going one-way wave equation for the half-space to the left of the interface. The solution of the interaction problem is provided by reflection and transmission operators which, when applied to the incident acoustic field, result in the appropriate initial plane data for marching away from the interface to the left and to the right, respectively. We outline the theory for determining these two operators and present three approximate solutions for them, which are appropriate for the parabolic wave theory, an extended parabolic wave theory, and for our high frequency one-way wave theory.

The solution of the finite transition layer problem is also to be based on the construction of reflection and transmission operators. For a calculated acoustic field incident on the transition layer from the left, the reflection operator determines the initial data for marching back to the left and the transmission operator determines the initial data for marching to the right, away from the other end of the transition region. To make this program computational requires construction of the transition layer reflection and transmission operators. A solution algorithm based on invariant imbedding[14] can be interpreted as iterating, or updating, the solution field. We describe this algorithm.

REFLECTION AND TRANSMISSION AT A PLANE INTERFACE

Consider two range-independent half-spaces positioned to the left and to the right of a plane interface at x = 0. Consider further a source of acoustic energy located in the left half-space at a point removed from the interface. The source is to introduce a prescribed wave field that propagates to the right, i.e., toward the interface. This right-going wave field is denoted by $\phi_L^+(x,\underset{\sim}{q})$, where the subscript L locates the half-space, the superscript + determines the direction, the range direction coordinate is x, and the cross-range direction coordinates are $\underset{\sim}{q}$. The ϕ_L^+ field satifies an appropriate right-going, one-way wave equation,

$$\left(\frac{i}{k_o}\right)\partial_x\phi_L^+ + \mathbf{H}_L\phi_L^+ = 0 \tag{1}$$

where k_o is a reference wavenumber and $\mathbf{H}_L$ is an operator described over the $\underset{\sim}{q}$ coordinate.

On interacting with the planar interface, the incident ϕ_L^+ field results in a reflected wave field, $\phi_L^-(x,\underset{\sim}{q})$, and a transmitted field, $\phi_R^+(x,\underset{\sim}{q})$. These fields likewise satisfy appropriate left-going, and right-going, one-way wave equations, respectively.

$$\left(\frac{i}{k_o}\right)\partial_x\phi_L^- - \mathbf{H}_L\phi_L^- = 0 \tag{2a}$$

and

$$\left(\frac{i}{k_o}\right)\partial_x\phi_R^+ + \mathbf{H}_R\phi_R^+ = 0 \tag{2b}$$

Thus, using Eqs.(2) we need only determine "initial plane" data to calculate the ϕ_L^- and ϕ_R^+ fields at all points in their respective half-spaces. The conditions that determine this initial plane data are continuity conditions at the interface. We write

$$\phi_L^+(o,\underset{\sim}{q}) + \phi_L^-(o,\underset{\sim}{q}) = \phi_R^+(o,\underset{\sim}{q}), \tag{3a}$$

and

$$\rho_L^{-1}(\underset{\sim}{q})\partial_x\phi_L^+(o,\underset{\sim}{q}) + \rho_L^{-1}(\underset{\sim}{q})\partial_x\phi_L^-(o,\underset{\sim}{q}) = \rho_R^{-1}(\underset{\sim}{q})\partial_x\phi_R^+(o,\underset{\sim}{q}), \tag{3b}$$

where the $\rho(\underset{\sim}{q})$ are mass density fields. Using Eqs.(1) and (2), Eq.(3b) is written

$$\boldsymbol{\rho}_L^{-1}\mathbf{H}_L\phi_L^+(o,\underset{\sim}{q}) - \boldsymbol{\rho}_L^{-1}\mathbf{H}_L\phi_L^-(o,\underset{\sim}{q}) = \boldsymbol{\rho}_R^{-1}\mathbf{H}_R\phi_R^+(o,\underset{\sim}{q}), \tag{4}$$

where $\boldsymbol{\rho}^{-1}$ is now interpreted as an operator, that which multiplies by a function $\rho^{-1}(\underset{\sim}{q})$.

Equations (3a) and (4) can be solved by introducing reflection and transmission operators, defined by

$$\phi_L^-(o,\underset{\sim}{q}) = \mathbf{r}\phi_L^+(o,\underset{\sim}{q}), \tag{5a}$$

and

$$\phi_R^+(o,\underset{\sim}{q}) = \mathbf{t}\phi_L^+(o,\underset{\sim}{q}), \tag{5b}$$

These operators are to satisfy operator equations, obtained from Eqs.(3a) and (4),

$$\mathbf{1} + \mathbf{r} = \mathbf{t} \tag{6a}$$

and

$$\mathbf{a}_L - \mathbf{a}_L\mathbf{r} = \mathbf{a}_R\mathbf{t}, \tag{6b}$$

where $\mathbf{1}$ is the identity operator, and

$$\mathbf{a} = \rho^{-1}\mathbf{H} \tag{7a}$$

From Eqs.(6) it immediately follows that

$$(\mathbf{a}_L + \mathbf{a}_R)\mathbf{r} = (\mathbf{a}_L - \mathbf{a}_R) \tag{8}$$

with **r** determined, Eq.(6a) determines **t**.

The above calculations are formal. To make them computational requires a representation of the various operators, which allows their specification in terms of ordinary functions. The cases in which the two half-spaces are homogeneous are special, with **H** operators, convolutions, and ρ values, independent of $\underset{\sim}{q}$. This suggests representing the operators in a Fourier space in which they are described by multiplications by ordinary functions. Thus, we have the **H** operator described as a multiplication in $\underset{\sim}{p}$ space, by $(K_0^2 - \underset{\sim}{p}^2)^{\frac{1}{2}}$, a plane wave "slowness". Further, the **a** operator is described by a multiplication in $\underset{\sim}{p}$, by $(K_0^2 - \underset{\sim}{p}^2)^{\frac{1}{2}}/\rho$, a plane wave "admittance". Finally, the **r** and **t** operators are described by multiplications in $\underset{\sim}{p}$, by reflection and transmission coefficients, respectively.

With the introduction of cross-range variability the several operators are no longer multiplicative in Fourier space. This significantly complicates the problem, but a formalism does exist that enables us both to specify the operators in terms of ordinary functions, termed operator symbols, and to combine the symbols of two successively applied operators to obtain the symbol of the combination. The most studied of these operator symbol representations is referred to as a Weyl symbol; in terms of it the operator H acting on a generic f(q) is given by

$$\mathbf{H}f(\underset{\sim}{q}) \leftrightarrow \left(\frac{k_o}{2\pi}\right)\int\left(\int \Omega_H\left(\underset{\sim}{p}, \frac{\underset{\sim}{q}+\underset{\sim}{q}'}{2}\right)\exp(ik_o\underset{\sim}{p}\cdot(\underset{\sim}{q}-\underset{\sim}{q}'))d\underset{\sim}{p}\right) f(\underset{\sim}{q}')d\underset{\sim}{q}', \tag{9}$$

where $\Omega_H(\underset{\sim}{p},\underset{\sim}{q})$ is the Weyl symbol of the operator **H**. Notice that the symbol is defined on a combined position/Fourier space that may be suggestively termed a phase space in recognition of a principal application of this representation, providing quantum descriptions of dynamical systems. We also write without further discussion the "composition equation" for combining the Weyl symbols of two operators, **A** and **B** to form **AB**

$$\Omega_{AB}(\underset{\sim}{p},\underset{\sim}{q}) = \left(\frac{k_o}{\pi}\right)^4 \iiiint \Omega_A(\underset{\sim}{p}+\underset{\sim}{t}, \underset{\sim}{q}+\underset{\sim}{u})\Omega_B(\underset{\sim}{p}+\underset{\sim}{r}, \underset{\sim}{q}+\underset{\sim}{v})$$

$$\exp(2ik_o(\underset{\sim}{u}\cdot\underset{\sim}{r} - \underset{\sim}{t}\cdot\underset{\sim}{v}))d\underset{\sim}{r}d\underset{\sim}{t}d\underset{\sim}{u}d\underset{\sim}{v} \tag{10}$$

The interested reader is referred to Refs. [5-10] for a discussion of the representation of a class of operators termed pseudodifferential, in the context of factoring the Helmholtz equation into one-way wave equations and to the references cited therein for discussions in a more general context.

Applying the composition equation to Eq.(7a), we write for the symbol of the admittance operator, a,

$$\Omega_a(\underset{\sim}{p},\underset{\sim}{q}) = \left(\frac{k_o}{\pi}\right)^2 \iint (\Omega_H(\underset{\sim}{p}+\underset{\sim}{r},\underset{\sim}{q})/\rho(\underset{\sim}{q}+\underset{\sim}{u}))\exp(2ik_o\underset{\sim}{u}\cdot\underset{\sim}{r})d\underset{\sim}{r}d\underset{\sim}{u}, \tag{11}$$

where $\Omega_H(\underset{\sim}{p},\underset{\sim}{q})$ is the symbol of the **H** operator that describe the one-way wave equation. Accepting for the governing equation,

$$((\partial_x^2 + \nabla_q^2) - \underset{\sim}{\alpha}(\underset{\sim}{q})\cdot\nabla_q + k_o^2K^2(\underset{\sim}{q}))\phi = 0, \tag{12}$$

where $\underset{\sim}{\alpha}(\underset{\sim}{q}) = \nabla_q \ln\rho(\underset{\sim}{q})$, then, $\Omega_H(\underset{\sim}{p},\underset{\sim}{q})$ is determined from the equation,

$$K^2(\underset{\sim}{q}) - \underset{\sim}{p}^2 + \frac{1}{2k_0^2} \nabla_q \cdot \underset{\sim}{\alpha}(\underset{\sim}{q}) - \frac{i}{k_o} \underset{\sim}{\alpha}(\underset{\sim}{q}) \cdot \underset{\sim}{p}$$

$$= \left(\frac{k_o}{\pi}\right)^4 \iiiint \Omega_H(\underset{\sim}{p}+\underset{\sim}{r}, \underset{\sim}{q}+\underset{\sim}{v}) \Omega_H(\underset{\sim}{p}+\underset{\sim}{T}, \underset{\sim}{q}+\underset{\sim}{u})$$

$$\exp(2ik_o(\underset{\sim}{u} \cdot \underset{\sim}{r} - \underset{\sim}{t} \cdot \underset{\sim}{v}))d\underset{\sim}{r}d\underset{\sim}{t}d\underset{\sim}{u}d\underset{\sim}{v}. \tag{13}$$

Equation (13) is obtained from the requirement that applied twice the operator **H** gives

$$\mathbf{H}^2 = \nabla_q^2 - \underset{\sim}{\alpha}(\underset{\sim}{q}) \cdot \nabla_q + k_o^2 K^2(\underset{\sim}{q}). \tag{14}$$

Applying the composition equation to Eq.(8) gives the equation that determines the symbol for the reflection operator, i.e.,

$$\Omega_{a_L}(\underset{\sim}{p}, \underset{\sim}{q}) - \Omega_{a_R}(\underset{\sim}{p}, \underset{\sim}{q}) = \left(\frac{k_o}{\pi}\right)^4 \iiiint (\Omega_{a_L}(\underset{\sim}{p}+\underset{\sim}{t}, \underset{\sim}{q}+\underset{\sim}{u}) + \Omega_{a_R}(\underset{\sim}{p}+\underset{\sim}{t}, \underset{\sim}{q}+\underset{\sim}{u}))$$

$$\Omega_r(\underset{\sim}{p}+\underset{\sim}{r}, \underset{\sim}{q}+\underset{\sim}{v})d\underset{\sim}{r}d\underset{\sim}{u}d\underset{\sim}{t}d\underset{\sim}{v} \tag{15}$$

An analogous equation can be written to determine the symbol of the transmission operator. A number of the intermediate steps required in writing Eqs.(11)-(15) are given in Ref. [11].

The complexity of the task of determining the various operator symbols using Eqs.(11)-(15) is a consequence of the very general problem for which the determined symbols would constitute an exact solution. Practical computational algorithms would be based on approximate symbols that can be justified in certain asymptotic regimes. We reproduce here three such limits since these have all received some attention in the literature. All of these are written for two dimensions and refer to propagation environments in which ρ is a constant; this is the case most frequently considered in ocean acoustics. Reference [11] considers a varying mass density field.

The first limit is the "high frequency" limit, which can be justified provided variations in the wavenumber field $K^2(q)$ is slow on the scale of a wavelength. In this case

$$\Omega_H(p,q) = (K^2(q) - p^2)^{1/2} \tag{16}$$

and

$$\Omega_r(p,q) = \frac{\Delta((K^2(q) - p^2)^{1/2})}{\Sigma((K^2(q) - p^2)^{1/2})} \tag{17}$$

where $\Sigma(\ )$ and $\Delta(\ )$ indicate the sum and difference of the indicated argument, as evaluated on the two sides of the interface. Propagation models based on this approximation are discussed in detail in Ref. [9]. The second limit is the ordinary "parabolic equation" limit which can be justified for weak inhomogeneity, slow variations, and narrow angles. In this case

$$\Omega_H(p,q) = 1 + \tfrac{1}{2}((K^2(q) - 1) - p^2) \tag{18}$$

and

$$\Omega_r(p,q) = \tfrac{1}{2}(K_L^2(q) - K_R^2(q)). \tag{19}$$

Finally, we have an extended parabolic theory which can be justified for arbitrary inhomogeneity, but for slow variations and narrow angles. In this case

$$\Omega_H(p,q) = K(q) - \left(\frac{p^2}{2K(q)} + \frac{1}{k_o^2}\,\frac{\partial q^2 K(q)}{K^2(q)}\right), \tag{20}$$

and

$$\Omega_r(p,q) = (\Omega_r^{(0)}(q) + \Omega_r^{(1,0)}(q)) + \Omega_r^{(1,1)}(q)p + \Omega_r^{(1,2)}(q)p^2, \tag{21a}$$

where

$$\Omega_r^{(0)}(q) = \frac{\Delta(K(q))}{\Sigma(K(q))}, \tag{21b}$$

$$(\Sigma(K(q)))\Omega_r^{(1,2)}(q) = \tfrac{1}{2}\Big((\Sigma(K^{-1}(q)))\Omega_r^{(0)}(q) - \Delta(K^{-1}(q))\Big), \tag{21c}$$

$$(\Sigma(K(q)))\Omega_r^{(1,1)}(q) = \frac{i}{2k_o}\Big(\Sigma(K^{-1}(q)))\partial_q\Omega_r^{(0)}(q)$$
$$+ 2(\Sigma(\partial_q K(q)))\Omega_r^{(1,2)}(q)\Big), \tag{21d}$$

and

$$(\Sigma(K(q)))\ {}_r^{(1,0)}(q) = \frac{1}{8k_0^2}\Big((\Sigma(\partial_q^2 K(q)/K^2(q))\ (\Omega_r^{(0)}(q) - \partial_q^2\Omega_r^{(0)}(q))$$
$$- \Delta(\partial_q^2 K(q)/K^2(q)) + 2\ (\Sigma(\partial_q^2 K(q)))\Omega_r^{(1,2)}(q)\Big) \tag{21e}$$
$$+ \frac{i}{2k_o}(\Sigma(\partial_q K(q)))\Omega_r^{(1,1)}(q)$$

A propagation model for a range independent environment, corresponding to an $\Omega_H(p,q)$ given by Eq.(20) has recently been suggested as a range refraction parabolic equation.[15] The designation range refraction is apparently suggested by a numerical comparison of the predictive capability of this model with the ordinary parabolic wave theory, in a limited number of test cases in which the sound field is propagated through a front.

REFLECTION AND TRANSMISSION AT A PLANAR TRANSITION LAYER

The solution to the finite transition layer problem can also be based on constructing reflection and transmission operators. The advantage of this approach is the idea of invariant imbedding,[14] in which the reflection and transmission operators are to be interpreted as functions of the thickness of the transition region. Thus, the thickness of the transition region is now viewed as a variable and we construct the reflection and transmission as functions of this variable. The power of this interpretation is that it enables our constructing a transition layer of any thickness, in increments, starting with a transition region of zero extent, i.e., a planar interface. The construction process is, therefore, by iterating or updating a data set.

Consider that the transition layer is approximated by N component layers, all of which are range-independent. Further introduce $R_R^{(n)}$, $R_L^{(n)}$, $T_R^{(n)}$, $T_L^{(n)}$ to denote reflection and transmission operators for a transition layer comprised of the first n-1 component layers. The subscript locates the direction of the incident signal to the transition layer. Adding another layer results in increments to these operators. We write, without derivation, the following expressions for the incremented reflection and transmission operators,

$$\mathbf{R}_R^{(n+1)} = \mathbf{R}_R^{(n)} + \mathbf{T}_L^{(n)}\mathbf{P}_L^{(n)}\mathbf{r}_R^{(n+1)}\mathbf{P}_R^{(n)}(\mathbf{1} - \mathbf{R}_L^{(n)}\mathbf{P}_L^{(n)}\mathbf{r}_R^{(n+1)}\mathbf{P}_R^{(n)})^{-1}\mathbf{T}_R^{(n)},$$

$$\mathbf{T}_L^{(n+1)} = \mathbf{T}_L^{(n)}\mathbf{P}_L^{(n)}(\mathbf{r}_R^{(n+1)}\mathbf{P}_K^{(n)}(\mathbf{1} - \mathbf{R}_L^{(n)}\mathbf{P}_L^{(n)}\mathbf{r}_R^{(n+1)}\mathbf{P}_R^{(n)})^{-1}\mathbf{R}_L^{(n)}\mathbf{P}_L^{(n)}+\mathbf{1})\mathbf{t}_L^{(n+1)},$$

$$\mathbf{T}_R^{(n+1)} = \mathbf{t}_R^{(n+1)}\mathbf{P}_R^{(n)}(\mathbf{1}+\mathbf{R}_L^{(n)}\mathbf{P}_L^{(n)}(\mathbf{1}- \mathbf{r}_R^{(n+1)}\mathbf{P}_R^{(n)}\mathbf{R}_L^{(n)}\mathbf{P}_L^{(n)})^{-1}\mathbf{r}_R^{(n+1)}\mathbf{P}_R^{(n)})\mathbf{T}_R^{(n)},$$

and (22)

$$\mathbf{R}_L^{(n+1)} = \mathbf{r}_L^{(n+1)}+ \mathbf{t}_R^{(n+1)}\mathbf{P}_R^{(n)}\mathbf{R}_L^{(n)}\mathbf{P}_L^{(n)}(\mathbf{1}-\mathbf{r}_R^{(n+1)}\mathbf{P}_R^{(n)}\mathbf{R}_L^{(n)}\mathbf{P}_L^{(n)})^{-1}\mathbf{t}_L^{(n+1)}.$$

In these expressions, the **P**'s denote propagation operators over the added layer and the **r**'s and **t**'s are reflection and transmission operators for the interface that locates the right-side of the added layer. A derivation of Eq.(11) is contained in Ref. [12].

Equations (22) will be computational only after a representation that expresses the various operators in terms of ordinary functions, and a formalism for combining operators is given. Again, this is simply accomplished in the absence of cross-range variability since all of the operators are multiplicative in Fourier space in this special case. In the seismology literature, this reduced algorithm is referred to as Kennett's method.[16,17] Kennett's method is routinely applied to the much more complicated vector wave problem. In the presence of cross-range variability the Fourier space representation of the operators can be extended with the introduction of a phase space, and operator symbols. Further, the composition equation of the Weyl calculus, for example, does provide a formalism for combining operators. However, the computational task required is significantly complicated, since the formalism does not reduce to simply multiplying symbols. Further, Eq.(21) requires the construction of the formal inverse operators contained therein. In Ref. [12], we briefly discuss a local "single scatter", actually a "double scatter", approximation, obtained on writing

$$(\mathbf{1} - \mathbf{R}_L^{(n)}\mathbf{P}_L^{(n)}\mathbf{r}_R^{(n+1)}\mathbf{P}_R^{(n)})^{-1} \approx 1 + \mathbf{R}_L^{(n)}\mathbf{P}_L^{(n)}\mathbf{r}_R^{(n+1)}\mathbf{P}_R^{(n)}, \quad (23)$$

and contrast it with a global single scatter approximation.

As noted in the Introduction computer implementation of the transition layer models based on Eqs.(22) is in its beginning stages. Some further ideas are presented in Ref. [13] of these proceedings.

ACKNOWLEDGEMENT

This research was supported by the National Science Foundation, contracts EAR-8312745 and EAR-8318926, and by the Office of Naval Research, N00014-83-K-0094.

REFERENCES

1. F.D. Tappert, The Parabolic Approximation Method, in: "Wave Propagation in Underwater Acoustics," J.B. Keller and J.S. Papadakis, eds., Springer-Verlag, New York (1977).

2. R.R. Greene, The Rational Approximation of the Acoustic Wave Equation with Bottom Interaction, J. Acoust. Soc. Am., 76(6):1764-73 (1984).
3. D.J. Thomson, and N.R. Chapman, A Wide-Angle Split-Step Algorithm for the Parabolic Equation, J. Acoust. Soc. Am., 74:1848-1854 (1983).
4. D. Lee, G. Botseas, and J.S. Papadakis, Finite-Difference Solution to the Parabolic Wave Equation, J. Acoust. Soc. Am., 70:795-800 (1981).
5. L. Fishman, and J.J. McCoy, Derivation and Application of Extended Parabolic Wave Theories I. The Factorized Helmholtz Equation, J. Math. Phys., 25:285-296 (1984).
6. L. Fishman, and J.J. McCoy, Derivation and Application of Extended Parabolic Wave Theories II. Path Integral Representations, J. Math. Phys. 25:297-308, (1984).
7. L. Fishman, and J.J. McCoy, Factorization, Path-Integral Representations, and the Construction of Direct and Inverse Wave Propagation Theories, IEEE Trans. Geosc. Remote Sensing GE-22, 6:682-692 (1984).
8. L. Fishman, and J.J. McCoy, A New Class of Propagation Models Based on a Factorization of the Helmholtz Equation, Geophys. J. R. Astr. Soc., 80:439-461 (1985).
9. L. Fishman, and S.C. Wales, Factorization and Path Integration of the Helmholtz Equation, in: "Proc. Ocean Seismo-Acoustics Conf.", Plenum Publishing Corporation, New York (1985).
10. L. Fishman, J.J. McCoy, and S.C. Wales, Factorization and Path Integration of the Helmholtz Equation, J. Acous. Soc. Am., submitted for publication.
11. J.J. McCoy, L. Fishman, and L.N. Frazer, Reflection and Transmission at an Interface Separately Transversely Inhomogeneous Acoustic Half-spaces, Geophys. J. R. Astr. Soc., submitted for publication.
12. J.J. McCoy, and L.N. Frazer, Propagation Modelling Based on Wave Field Factorization and Invariant Imbedding, Geophys. J. R. Astr. Soc., submitted for publication.
13. L.N. Frazer, and J.J. McCoy, An Acoustic Reflectivity Method for Laterally-Varying Layered Media, in: "Proc. Ocean Seismo-Acoustics Conf.", Plenum Publishing Corporation, New York (1985).
14. R. Bellman, and G.M. Wing, "An Introduction to Invariant Embedding," Wiley, New York (1975).
15. F.D. Tappert, and D. Lee, A Range Refraction Parabolic Equation, J. Acoust. Soc. Am., 76(6):1797-1803 (1984).
16. B.L.N. Kennett, Reflections, Rays and Reverberations, Bull. Seism. Soc. Am., 64:1685-96 (1974).
17. B.L.N. Kennett, "Seismic Wave Propagation in Stratified Media," Cambridge University Press, Cambridge (1983).

AN ACOUSTIC REFLECTIVITY METHOD FOR LATERALLY VARYING LAYERED MEDIA

L. Neil Frazer

The University of Hawaii
Hawaii Institute of Geophysics
Honolulu, Hawaii 96822

John J. McCoy

The Catholic University of America
Department of Civil Engineering
Washington, D.C. 20064

INTRODUCTION

To compute the response of a stratified earth or sea-bed model it is often convenient to approximate the model by a stack of homogeneous plane parallel layers. The equations of motion are Fourier transformed with respect to time and Fourier-Bessel transformed with respect to the horizontal coordinates. Then one computes the exact response of the layer stack using Haskell[1] matrices, or Kennett[2] iteration, or the global matrix method of Schmidt[3] and Jensen. Such approaches are referred to in seismology as "reflectivity methods" after the work of Fuchs and Muller[4] in the early 1970's. The Kennett[2,5] form of the reflectivity method is the one that we shall generalize here. As a form of invariant imbedding it has the advantage that each layer of the model is needed only once; thus the effective dimensionality of the model is reduced by one. In our new method the earth or sea-bed is still described as a stack of plane parallel layers; however, within each layer the velocity, density, and quality factor, Q, are now functions of horizontal position x. These functions are all independent from layer to layer so any earth or sea-bed model can be parameterized in this way. The two foundations of the new method are factorization[7,8] and imbedding[6]. Both of these topics are explored in detail elsewhere in this volume so the mathematics in this paper will be kept at a very low level. To simplify notation we treat propagation in two spatial dimensions.

FACTORIZATION

The Helmholtz equation for pressure $\phi(x,z)$ in a medium whose sound speed $c(x)$ and density $\rho(x)$ are functions of horizontal distance, x, is

$$(\partial_z^2 + \partial_x^2 - \alpha(x)\partial_x + k^2(x))\phi = 0 \tag{1}$$

in which $\alpha(x) = \partial \ln \rho(x)/\partial x$ and $k^2(x) = \omega^2/c^2(x)$ where ω is temporal frequency. Let us regard the pressure field ϕ as an N-component column vector of pressure samples, one sample at each value of x. Since each component is a function of depth z so is the vector ϕ. We regard the operator $H^2 = \partial_x^2 - \alpha(x)\partial_x + k^2(x)$ as a matrix. For example: the matrix form of ∂_x^2 has -1 for each element on the diagonal and 1/2 for each element next to the diagonal; the matrix form of $k^2(x)$ is $\mathrm{diag}\{k^2(x_1), k^2(x_2), \ldots, k^2(x_N)\}$, the matrix form of $-\alpha(x)\partial_x$ has elements $-\alpha(x_i)/2$ on the diagonal above the main diagonal and elements $\alpha(x_i)/2$ on the diagonal below the main diagonal; the matrix for H^2 is the sum of those three matrices. We can (almost) always compute the square root of a matrix and very often we can obtain it in the form

$$H = U^{-1}DU \tag{2}$$

where the matrix D is diagonal. Now, knowing H, we may factor equation (1) as

$$(i\partial_z + H)(-i\partial_z + H)\phi = 0 \tag{3}$$

In this equation it is clear that the first factor annihilates that part of ϕ travelling in the +z (down) direction while the second factor annihilates that part of .z travelling in the -z (up) direction. Since their product annihilates the total field ϕ it follows that we can write $\phi = \phi_U + \phi_D$, where the subscripts indicate direction of propagation, and that

$$(i\partial_z + H)\phi_D = 0 \quad \text{and} \quad (-i\partial_z + H)\phi_U = 0 \tag{4a,b}$$

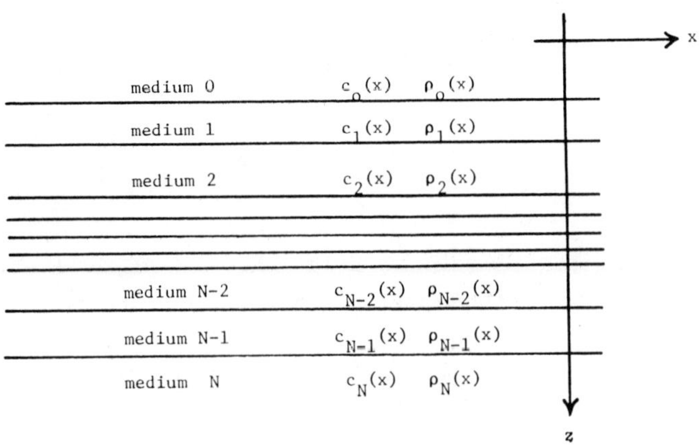

Fig. 1 A layered model. Velocity and density vary with x in each layer.

PROPAGATION

For any matrix A, or operator A, we now define $\exp(A)$ by its series.

$$\exp(A) = \sum_{n=0}^{\infty} \frac{1}{n!} A^n \tag{5}$$

With this definition it follows immediately that for any given $\phi_D(z_o)$ equation (4a) is solved by $\phi_D(z) = P_D(z,z_o)\phi_D(z_o)$ where the propagator matrix P_D is given by

$$P_D = \exp(i(z-z_o)H). \tag{6}$$

If equation (2) holds we get the computationally more viable form

$$P_D = U^{-1} \exp(i(z-z_o)D)U \quad . \tag{7}$$

Similar results obtain for equation (4b).

Note that since (6) is exact it allows us to jump from z_o to z in one step no matter how large $z-z_o$ may be; however, when H is not diagonalizable equation (7) is not available and so we must calculate (6) using the series definition (5). In that case it may be much more efficient to calculate P_D using a path integral[8,9].

REFLECTION AND TRANSMISSION

As shown in Fig. 2 an incident wave ϕ_i encounters the boundary between an upper half-space (subscript -) and a lower half-space (subscript +) resulting in a reflected wave ϕ_r and a transmitted wave ϕ_t. These waves

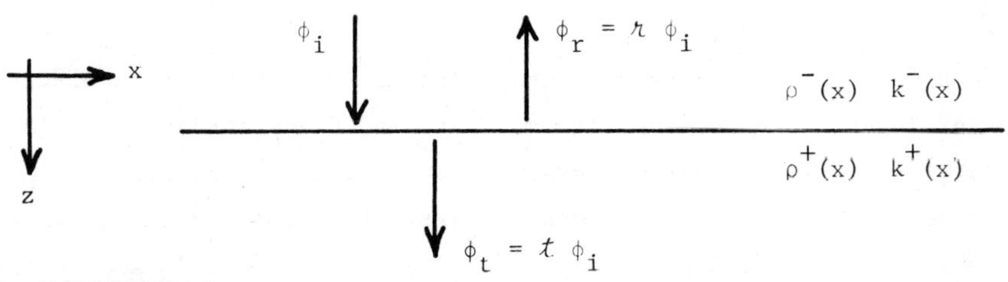

Fig. 2 Incident and reflected waves at the boundary between two laterally variable half-spaces.

satisfy versions of equations (4):

$$(i\partial_z + H^-)\phi_i = 0, \quad (-i\partial_z + H^-)\phi_r = 0, \quad (i\partial_z + H^+)\phi_t = 0; \tag{8a,b,c}$$

in which H^- and H^+ are, respectively, the square roots of the matrices for:

$$(H^-)^2 = \partial_x^2 - \alpha^-(x)\partial_x + k^-(x)^2 \quad \text{and} \quad (H^+)^2 = \partial_x^2 - \alpha^+(x)\partial_x + k^+(x)^2. \tag{9a,b}$$

We assume that the reflected and transmitted waves are related to the incident wave by matrices r and t, respectively. Thus

$$\phi_r = r\,\phi_i \quad \text{and} \quad \phi_t = t\,\phi_i \tag{10a,b}$$

To determine r and t we recall that pressure ϕ and vertical displacement u must be continuous across the boundary. Thus

$$\phi_i + \phi_r = \phi_t \quad \text{and} \quad u_i + u_r = u_t. \tag{11a,b}$$

The first of these conditions gives immediately $1 + r = t$; to use the second condition recall the momentum equation $\rho\ddot{u} = -\lambda\partial_z\phi$ and substitute for ∂_z using equations (8) to get $(1/\rho^-)H^-(1-r) = (1/\rho^+)H^+t$. In this last relation $1/\rho^-$ is the matrix with diagonal elements $1/\rho^-(x)$ and $1/\rho^+$ is the matrix with diagonal elements $1/\rho^+(x)$. For convenience we introduce the admittance matrices

$$A^- = (1/\rho^-)\,H^- \quad \text{and} \quad A^+ = (1/\rho^+)\,H^+ \tag{12a,b}$$

and then solve for the r and t matrices in the form

$$r = (A^- + A^+)^{-1}(A^- - A^+) \quad \text{and} \quad t = (A^- + A^+)^{-1}\,2A^-. \tag{13a,b}$$

IMBEDDING

We now show how to construct the solution of a stack of laterally inhomogeneous layers from the matrices r, t, P_U and P_D derived above. As shown in figure 3 we have a stack of layers whose reflection operator is R_D; and we want to add another layer and another interface to the top of the stack. The new layer is characterized by propagation matrices P_U and P_D and the new interface by reflection and transmission operators r_U, r_D, t_U, and t_D. Intuitively it is clear that R'_D, the reflection operator of the incremented stack is given by

$$\begin{aligned} R'_D &= r_D + t_U P_U R_D P_D t_D \\ &\quad + t_U P_U R_D P_D r_U P_U R_D P_D t_D + \ldots \\ &= r_D + t_U P_U (1 - R_D P_D r_U P_U)^{-1} R_D P_D t_D \end{aligned} \tag{14}$$

The first two terms in this series are primary waves and the later terms are multiples. A wonderful feature of the imbedding approach is the naturalness of the final result. For example, to verify the correctness of (14) simply read each term in the series from right to left.

Note that the problem posed above has now been formally solved with no use of mathematics more sophisticated than matrix algebra. Also the connection between the laterally varying problem solved here and the traditional laterally homogeneous problem is immediate: if ρ and c are independent of x then all the matrices in equation (14) are simultaneously diagonalizable by Fourier transformation with respect to x; equation (14) becomes a scalar equation to be evaluated once for each horizontal wavenumber k_x.

PSEUDODIFFERENTIAL OPERATORS

The problem with the numerical procedure outlined above is that it is not very fast. Fortunately, the calculus of pseudodifferential operators[11] provides us with a way to generate and compose all the matrices needed above that is very fast provided we do not ask for perfect accuracy. The key to

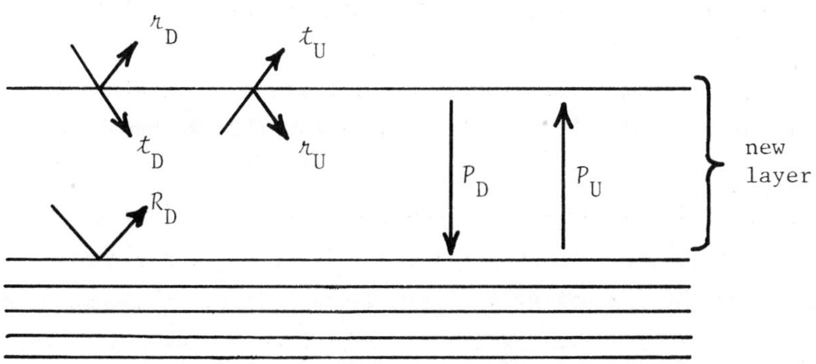

Fig. 3 The imbedding problem. We wish to add another layer and interface to a stack of layers with a known response.

using pseudodifferential operators is that the "symbol" of the pseudodifferential operator is always the quantity that is familiar from calculations in laterally homogeneous media. For example, at high frequency the symbol of the interface reflection operator $\mathcal{R}$ is just a local plane wave reflection coefficient allowed to vary with x.

If $\mathcal{B}$ is a pseudodifferential operator on wavefields $\phi(x)$ then we denote the kernel of $\mathcal{B}$ by $B(x,x')$ and the symbol of $\mathcal{B}$ by $b(p,x)$. Thus applying $\mathcal{B}$ to ϕ we have

$$\mathcal{B}\phi(x) = \int_{-\infty}^{\infty} dx' \; B(x,x') \; \phi(x') \tag{15}$$

where the kernel B is related to the Weyl symbol b by

$$B(x,x') = \frac{1}{2\pi}\int_{-\infty}^{\infty} dp \; e^{ip(x-x')} \; b(p,\frac{x+x'}{2}) \tag{16}$$

Now consider the operator $H^2 = \partial_x^2 - \alpha(x)\partial_x + k^2(x)$. By ordinary Fourier analysis it can be shown[12] that the symbol of H^2 is $k^2(x) - p^2 + \alpha'(x)/2 - i\alpha(x)p$. To get the symbol of H from the symbol of H^2 we use the first high frequency rule: "At high frequency the symbol of a function of an operator is the function of the symbol." According to this rule: The symbol of H is simply

$$h(p,x) = \sqrt{k^2(x) - p^2 + \alpha'(x)/2 - i\alpha(x)p}. \tag{17}$$

Similarly, the symbols of the operators $P_U(z,z_o)$ and $P_D(z,z_o)$ are, by (6),

$$p_U(p,x) = \exp(i(z-z_o)h(p,x)) \quad \& \quad p_D(p,x) = \exp(i(z_o-z)h(p,x)). \tag{18a,b}$$

When we extend this rule to include functions of several operators we infer that "At high frequency the symbol of a function of operators is the function of the operator symbols." Thus from equations (12) we have

$$a^-(p,x) = (1/\rho^-(x))h^-(p,x) \quad \text{and} \quad a^+(p,x) = (1/\rho^+(x))h^+(p,x) \tag{19a,b}$$

and from equations (13) we have

$$r(p,x) = \frac{a^-(p,x) - a^+(p,x)}{a^-(p,x) + a^+(p,x)} \quad \text{and} \quad t(p,x) = \frac{2a^-(p,x)}{a^-(p,x) + a^+(p,x)}. \tag{20a,b}$$

Now that we have the symbols of all our operators we can construct their kernels by Fourier transformation according to equation (16). For example: to construct the kernel of $\mathcal{r}$ we first generate a matrix of r(p,x) values; then we Fourier transform the columns of that matrix to get $\hat{r}(y,x)$; then we generate the kernel R(x,x′) by interpolation in the transformed matrix according to

$$R(x,x') = \hat{r}(x-x',(x+x')/2) \tag{21}$$

We have just seen how the use of the operator symbol gives a rapid and direct method for constructing the kernel. Now we will see how the use of symbols gives us a rapid method for composing operators which are local in scope. In figure 3 the operators $\mathcal{r}_U$, $\mathcal{r}_D$, $\mathcal{t}_U$, $\mathcal{t}_D$, $\mathcal{P}_U$ and $\mathcal{P}_D$ are "local in scope" if the scale length of the variation of $\rho(x)$ and c(x) within the layer is greater than the thickness of the layer. On the other hand the operator $\mathcal{R}_D$ is "global in scope" because it contains the effects of the whole stack of layers. To rapidly evaluate the imbedding equation (14) we utilize the second high frequency rule: "At high frequency the symbol of the product of local operators is the Hadamard product of their symbols." Thus in equation (14) the symbol of the (product) operator $\mathcal{t}_U\mathcal{P}_U$ is simply $t_U(p,x)p_U(p,x)$, the symbol of $\mathcal{P}_D\mathcal{r}_U\mathcal{P}_U$ is $p_D(p,x)r_U(p,x)p_U(p,x)$, and the symbol of $\mathcal{P}_D\mathcal{t}_D$ is $p_D(p,x)t_D(p,x)$. Note that we cannot use the second high frequency rule to compose a local operator with a global operator. However it can be seen that the number of operator compositions (matrix multiplications) needed to evaluate (14) has been reduced to four or five depending on how many multiples one wishes to include in the series. The use of the pseudodifferential operator calculus together with the high frequency rules gives a significant saving in computation time over the direct linear algebra approach outlined earlier in the paper.

EXAMPLES

The velocity model of figure 4(a) consists of a homogeneous high velocity block embedded in a low velocity homogeneous whole space. The block begins 2 km below the source/receiver surface and extends to infinite depth. Figure 4(b) shows a single-channel record section for this model, using a smooth 1-sided pulse as a source function; and it can be seen that the diffractions from the edge of the block behave as expected from physical modeling experiments[13]. Figure 4(c) shows a refraction record section for the same model with a source at x = 20 km. There the head wave is seen to emerge from the reflection at about 5 km from the source and then dissipate past the end of the block.

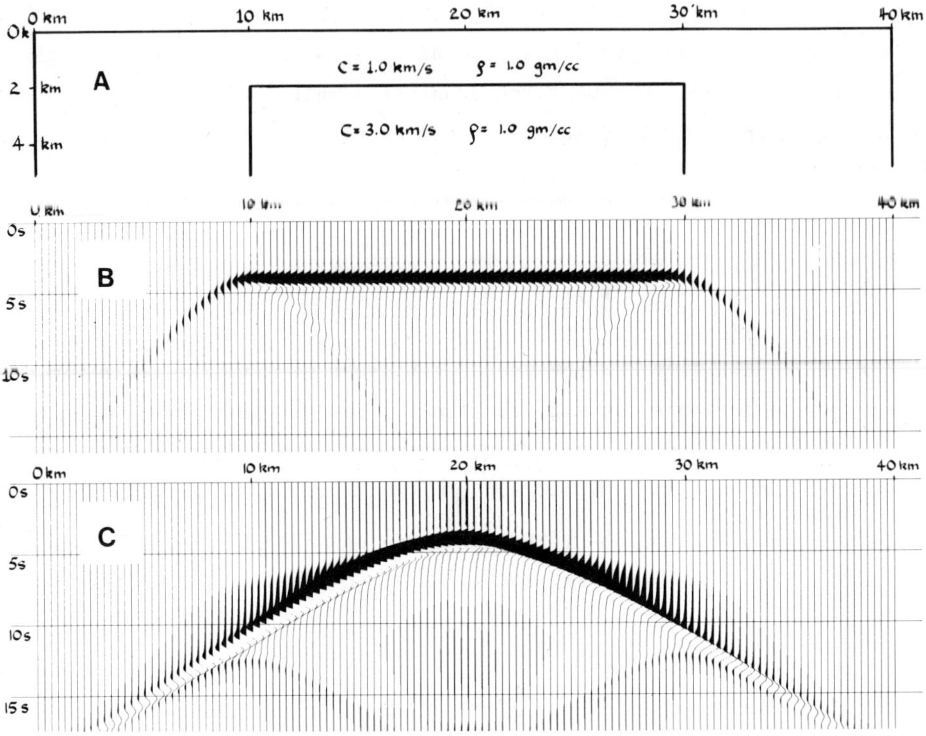

Fig. 4 (a) Block model (b) single-channel record section
(c) Refraction record section

The velocity model of figure 5(a) shows an irregular body about 20 km long and 3 km thick with a velocity of 1.2 km/s, embedded in a homogeneous whole space of velocity 1.0 km/s. The actual velocity model used in calculations consisted of twenty-three layers with velocity and density specified at sixty-four points in each layer. Figure 5(b) shows a single-channel record section for this model with the source/receiver surface 2.5 km above the highest point of the body and a smooth one-sided pulse as a source function. Notice the amplitude behavior of the reflections from the upper and lower surfaces of the body and the diffractions from the edges of the body.

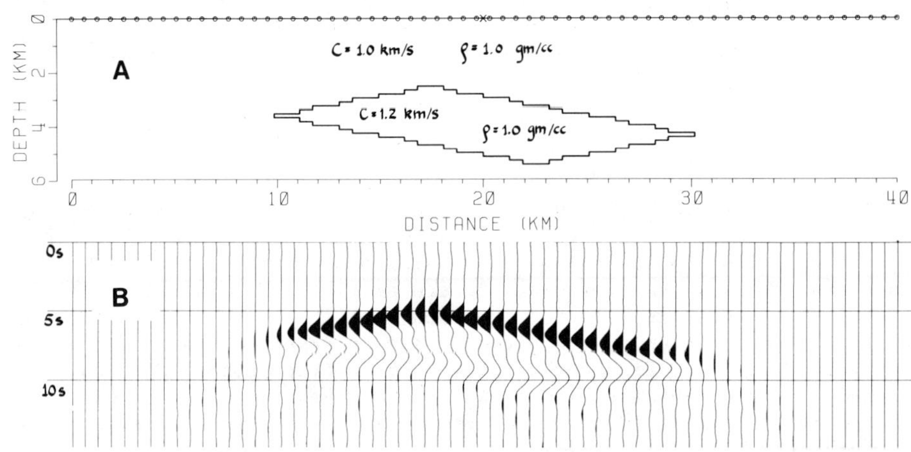

Fig. 5 (a) Embedded body model (b) Single-channel record section

ACKNOWLEDGMENTS

D. A. Myhre wrote substantial portions of the computer code. This work was supported by the U.S. Office of Naval Research and by the National Science Foundation under grant EAR-8318926. Hawaii Institute of Geophysics contribution No. 1610.

REFERENCES

1. N. A. Haskell, The dispersion of surface waves on multi-layered media, Bull. seism. Soc. Am., 43:17 (1953).
2. B. L. N. Kennett, Reflections, rays and reverberations, Bull. seism. Soc. Am., 64:1685 (1974).
3. H. Schmidt and F. B. Jensen, An efficient numerical solution technique for wave propagation in horizontally stratified ocean environments, SM-173, SACLANT ASW Research Centre (1984).
4. K. Fuchs and G. Muller, Computation of synthetic seismograms with the reflectivity method and comparison with observations, Geophys. J. R. astr. Soc., 23:417 (1971).
5. B. L. N. Kennett, "Seismic wave Propagation in stratified media," Cambridge University Press, Cambridge (1983).
6. R. E. Bellman and G. M. Wing, "An introduction to invariant imbedding," Wiley, New York (1975).
7. L. Fishman and J. J. McCoy, Derivation and application of extended parabolic wave theories. I The factorized Helmholtz equation, J. Math. Phys., 25:286 (1984).
8. L. Fishman and J. J. McCoy, Derivation and application of extended parabolic wave theories. II Path integral representations, J. Math. Phys., 25:297 (1984).
9. L. N. Frazer and J. J. McCoy, Phase-space Feynman path integrals, operator orderings and the propagation of scalar wavefields, Geophys. J. R. astr. Soc., (submitted).
10. J. J. McCoy and L. N. Frazer, Propagation modelling based on wave field factorization and invariant embedding, Geophys. J. R. astr. Soc., (submitted).
11. L. Hormander, The Weyl calculus of pseudodifferential operators, Comm. Pur. Appl. Math., 32:359 (1979).
12. J. J. McCoy, L. Fishman and L. N. Frazer, Reflection and transmission at an interface separating transversely inhomogeneous acoustic half spaces, Geophys. J. R. astr. Soc., (submitted).
13. F. J. Hilterman, Three-dimensional seismic modelling, GEOPHYSICS, 35:1020 (1970).

A VECTOR PARABOLIC EQUATION MODEL FOR ELASTIC PROPAGATION

Stephen C. Wales

Acoustics Division
Naval Research Laboratory
Washington, DC, 20375 USA

ABSTRACT

Some aspects of vector parabolic equations are examined as a means of propagating waves in linear elastic solids. A revised derivation of a previously published model, based on a Born approximation for scattering from an inhomogeneous layer, enables parabolic equation (PE) computations in elastic media with different longitudinal and shear speeds. This method of derivation is shown to reduce to the normal PE for the scalar case. The elastic PE model has been implemented using finite-difference techniques. Computed results for a channeled environment are consistent with physical considerations. Analysis reveals that hard and soft boundaries do not lend themselves to a straightforward implementations in a potential formulation.

INTRODUCTION

The PE model in underwater acoustics has attracted considerable attention over the last decade for several reasons, among which are its accuracy, capability of computing wave-theoretic effects in a range dependent environment and ease of propagation using the split-step method.[1,2] Considerably less interest has been paid to vector PE methods primarily because of the added complexity of two coupled propagation modes with different characteristic speeds. Four theoretically distinct models[3-8] have been developed for PE computations in elastic media, assuming potential or displacement representations, and one or two wave speeds. Only one of these (a potential representation, single reference wave speed) has been computationally implemented prior to this paper.[3,4]

In this paper, a vector PE model for elastic solids, with different longitudinal and shear speeds, is rederived[8] by first developing a set of equations describing wide-angle propagation in a weakly inhomogeneous elastic medium, and then reducing the equations to a narrow-angle vector PE model. The wide-angle equations are examined in the context of the scalar limit and exact factored Helmholtz equation results.[9] This model has been implemented on a VAX computer; some limited results are provided and discussed. Problems, due to the coupled nature of the equations at boundaries, were encountered and are briefly described. The propagation computation results presented appear to be consistent with basic physical considerations.

BORN APPROXIMATION FOR ELASTIC WAVES

The propagation medium is assumed to be infinite in extent and to be homogenous at all points exterior to the slab described by $x_o < x < x_o + L$. A time harmonic radiation field is incident upon the slab at $x = x_o$. Interior to the slab, the property fields of the medium are assumed to be weakly inhomogeneous, a condition necessary to justify the use of a perturbation calculation for a range of the order of L.

The incident radiation field is represented as displacement potential fields across the plane $x = x_o$; these fields are denoted by $\Phi_o(\underset{\sim}{x})$ and $\underset{\sim}{\Psi}_o(\underset{\sim}{x})$, and are related to the displacement, $\underset{\sim}{U}_o$, by

$$\underset{\sim}{U}_o(\underset{\sim}{x}) = \nabla \Phi_o(\underset{\sim}{x}) + \nabla \times \underset{\sim}{\Psi}_o(\underset{\sim}{x}) \tag{1}$$

Thus, in the absence of the region of inhomogeneity, the displacement field at all points in $\underset{\sim}{x}$ is given by the plane wave expansion

$$\underset{\sim}{U}_o(\underset{\sim}{x}) = \frac{1}{4\pi^2}\int \Big\{ i\underset{\sim}{k}'_L \hat{\Phi}_o(x_o,\underset{\sim}{k}'_\perp) \exp[i\underset{\sim}{k}'_L\cdot(\underset{\sim}{x}-x_o\underset{\sim}{e}_x)] + i\underset{\sim}{k}'_s\times\hat{\underset{\sim}{\Psi}}_o(x_o,\underset{\sim}{k}'_\perp) \exp[i\underset{\sim}{k}'_s\cdot(\underset{\sim}{x}-x_o\underset{\sim}{e}_x)]\Big\}\, d\underset{\sim}{k}'_\perp \; . \tag{2}$$

Here, $\hat{\Phi}_o(x_o,\underset{\sim}{k}'_\perp)$ and $\hat{\underset{\sim}{\Psi}}_o(x_o,\underset{\sim}{k}'_\perp)$ are the two dimensional spatial Fourier transforms of $\Phi_o(\underset{\sim}{x})$ and $\Psi_o(\underset{\sim}{x})$ at x_o respectively and

$$\underset{\sim}{k}'_L = [k_L^2 - k'^2_\perp]^{\frac{1}{2}}\underset{\sim}{e}_x + \underset{\sim}{k}'_\perp \; , \qquad \underset{\sim}{k}'_s = [k_s^2 - k'^2_\perp]^{\frac{1}{2}}\underset{\sim}{e}_x + \underset{\sim}{k}'_\perp \; , \tag{3}$$

where k'_L and k'_s are the longitudinal and transverse wavenumbers for the homogeneous background medium.

The inhomogenous property fields within the slab are described by the scalar perturbation fields $\epsilon_\lambda(\underset{\sim}{x})$, $\epsilon_\mu(\underset{\sim}{x})$, and $\epsilon_\rho(\underset{\sim}{x})$, corresponding to the Lamé constants and the mass density. Thus, $\lambda[1+\epsilon_\lambda(\underset{\sim}{x})]$, $\mu[1+\epsilon_\mu(\underset{\sim}{x})]$, and $\rho[1+\epsilon_\rho(\underset{\sim}{x})]$ represent the spatially varying material properties. It is necessary that the inhomogeneities be everywhere small relative to unity for perturbation theory to be valid. Using a single-scatter, or Born, approximation, we write the following expressions for the perturbed potential fields (See, for example Morse and Feshbach[10] or deHoop[11] for details.),

$$\Phi_1(\underset{\sim}{x}) = -\int \Bigg[\epsilon_\rho(\underset{\sim}{x}')\, \underset{\sim}{U}_o(\underset{\sim}{x}')\cdot\nabla\left(\frac{\exp(ik_L r)}{4\pi r}\right) + \frac{\lambda}{\rho\omega^2}\epsilon_\lambda(\underset{\sim}{x}')\,[\nabla\cdot\underset{\sim}{U}_o(\underset{\sim}{x}')]\;\nabla^2\left(\frac{\exp(ik_L r)}{4\pi r}\right) + \frac{\mu}{\rho\omega^2}\epsilon_\mu(\underset{\sim}{x}')\,[\nabla\underset{\sim}{U}_o(\underset{\sim}{x}') + \underset{\sim}{U}_o(\underset{\sim}{x}')\nabla] : \nabla\nabla\left(\frac{\exp(ik_L r)}{4\pi r}\right)\Bigg]\, d\underset{\sim}{x}'$$

and (4)

$$\underset{\sim}{\Psi}_1(\underset{\sim}{x}) = -\int \Bigg[\epsilon_\rho(\underset{\sim}{x}')\, \underset{\sim}{U}_o(\underset{\sim}{x}')\times\nabla\left(\frac{\exp(ik_s r)}{4\pi r}\right)$$

$$+ \frac{\mu}{\rho\omega^2} \epsilon_\mu(\underset{\sim}{x}') \, [\nabla \underset{\sim}{U}_o(\underset{\sim}{x}') + \underset{\sim}{U}_o(\underset{\sim}{x}')\nabla \,] \dot{\times} \nabla\nabla \left(\frac{\exp(ik_s r)}{4\pi r} \right) \Bigg] \, d\underset{\sim}{x}' \, .$$

The integrations are over the domain of the slab, the gradients are with respect to $\underset{\sim}{x}'$, and r is the magnitude of the vector from the scattering point, $\underset{\sim}{x}'$, to the observation point, $\underset{\sim}{x}$. As written, (4) is valid for any value of $\underset{\sim}{x}$. It represents the effects due to a thin scatterring layer in an otherwise homogeneous medium.

Next substitute into Eqs. (4) the plane wave expansion representations for both the unscattered field, Eq. (2), and the spherical wave expansion,[12]

$$\frac{\exp(ik_o r)}{4\pi r} = \frac{i}{8\pi^2} \int \exp[iK_x(x-x')+i\underset{\sim}{K}_\perp\cdot(\underset{\sim}{x}_\perp-\underset{\sim}{x}'_\perp)] \, \frac{d\underset{\sim}{K}_\perp}{K_x} \, , \tag{5}$$

(valid only for x > x'). The resultant expressions for perturbed fields, in terms of seven dimensional integrals, over $\underset{\sim}{K}_\perp$, $\underset{\sim}{k}'_\perp$ and $\underset{\sim}{x}'$, are;

$$\Phi_1(\underset{\sim}{x}) = \int \Bigg\{ \Bigg[(\underset{\sim}{k}'_L \cdot \underset{\sim}{K}_L) \, \epsilon_\rho(\underset{\sim}{x}') - \frac{k_L'^2 \, K_L{}^2}{k_F{}^2} \epsilon_\lambda(\underset{\sim}{x}') - \frac{2\,(\underset{\sim}{k}'_L\cdot\underset{\sim}{K}_L)^2}{k_s{}^2} \epsilon_\mu(\underset{\sim}{x}') \Bigg]$$

$$\hat{\Phi}_o(x_o,\underset{\sim}{k}'_\perp) \exp[ik'_{xL}(x'-x_o)]$$

$$+ \Bigg[\epsilon_\rho(\underset{\sim}{x}') - \frac{(\underset{\sim}{k}'_s\cdot\underset{\sim}{K}_L)}{k_s{}^2} \epsilon_\mu(\underset{\sim}{x}') \Bigg] \underset{\sim}{K}_L \cdot [\underset{\sim}{k}'_s \times \hat{\underset{\sim}{\Psi}}_o(x_o,\underset{\sim}{k}'_\perp)]$$

$$\exp[ik'_{xs}(x'-x_o)] \Bigg\} \frac{i\exp[i\underset{\sim}{K}_L\cdot(\underset{\sim}{x}-\underset{\sim}{x}')+i\underset{\sim}{k}'_\perp\cdot\underset{\sim}{x}'_\perp]}{32\pi^4 \, K_{xL}} \, d\underset{\sim}{K}_\perp \, d\underset{\sim}{k}'_\perp \, d\underset{\sim}{x}'$$

and (6)

$$\underset{\sim}{\Psi}_1(\underset{\sim}{x}) = \int \Bigg\{ \Bigg[\epsilon_\rho(\underset{\sim}{x}') - \frac{2\,(\underset{\sim}{k}'_L\cdot\underset{\sim}{K}_s)}{k_s{}^2} \epsilon_\mu(\underset{\sim}{x}') \Bigg] (\underset{\sim}{k}'_L \times \underset{\sim}{K}_s)$$

$$\hat{\Phi}_o(x_o,\underset{\sim}{k}'_\perp) \exp[ik'_{xL}(x'-x_o)]$$

$$+ \Bigg[\epsilon_\rho(\underset{\sim}{x}') - \frac{(\underset{\sim}{k}'_s\cdot\underset{\sim}{K}_s)}{k_s{}^2} \epsilon_\mu(\underset{\sim}{x}') \Bigg] [\underset{\sim}{k}'_s \times \hat{\underset{\sim}{\Psi}}_o(x_o,\underset{\sim}{k}'_\perp)] \times \underset{\sim}{K}_s$$

$$- \frac{(\underset{\sim}{k}'_s\times\underset{\sim}{K}_s)}{k_s{}^2} \epsilon_\mu(\underset{\sim}{x}') \, [\underset{\sim}{k}'_s \times \hat{\underset{\sim}{\Psi}}_o(x_o,\underset{\sim}{k}'_\perp)] \cdot \underset{\sim}{K}_s$$

$$\exp[ik'_{xs}(x'-x_o)] \Bigg\} \frac{i\exp[i\underset{\sim}{K}_s\cdot(\underset{\sim}{x}-\underset{\sim}{x}')+i\underset{\sim}{k}'_\perp\cdot\underset{\sim}{x}'_\perp]}{32\pi^4 \, K_{xs}} \, d\underset{\sim}{K}_\perp \, d\underset{\sim}{k}'_\perp \, d\underset{\sim}{x}' \, .$$

The quantity k_F is defined as,

$$k_F = \omega / C_F \qquad , \quad C_F^2 = \lambda / \rho \tag{7}$$

where the subscript F stands for fluid since this represents a wavenumber in an elastic fluid.

The integrals over $d\underset{\sim}{x}'_\perp$ may be viewed as Fourier transforms of the inhomogeneity strengths. Further, by Fourier transforming with respect to $\underset{\sim}{x}_\perp$, and making use of a delta function relation of exponentials, the integrals over over $\underset{\sim}{K}_\perp$ yield,

$$\hat{\Phi}_1(x,\underset{\sim}{k}_\perp) = \frac{i \exp[ik_{xL}(x-x_0)]}{8\pi^2 k_{xL}} \int \Bigg\{ \Bigg[(\underset{\sim}{k}'_L \cdot \underset{\sim}{k}_L)\, \hat{\epsilon}_\rho(x',\underset{\sim}{k}_\perp - \underset{\sim}{k}'_\perp)$$

$$- \frac{k_L^4}{k_F^2} \hat{\epsilon}_\lambda(x',\underset{\sim}{k}_\perp - \underset{\sim}{k}'_\perp) - \frac{2(\underset{\sim}{k}'_L \cdot \underset{\sim}{k}_L)^2}{k_s^2} \hat{\epsilon}_\mu(x',\underset{\sim}{k}_\perp - \underset{\sim}{k}'_\perp) \Bigg] \hat{\Phi}_0(x_0,\underset{\sim}{k}'_\perp)$$

$$+ \Bigg[\hat{\epsilon}_\rho(x',\underset{\sim}{k}_\perp - \underset{\sim}{k}'_\perp) - \frac{2(\underset{\sim}{k}'_s \cdot \underset{\sim}{k}_L)}{k_s^2} \hat{\epsilon}_\mu(x',\underset{\sim}{k}_\perp - \underset{\sim}{k}'_\perp) \Bigg]$$

$$[\underset{\sim}{k}'_s \times \hat{\underset{\sim}{\Psi}}_0(x_0,\underset{\sim}{k}'_\perp)] \cdot \underset{\sim}{k}_L \Bigg\} \, d\underset{\sim}{k}'_\perp \, dx'$$

and (8)

$$\hat{\underset{\sim}{\Psi}}_1(x,\underset{\sim}{k}_\perp) = \frac{i \exp[ik_{xs}(x-x_0)]}{8\pi^2 k_{xs}} \int \Bigg\{ \Bigg[\hat{\epsilon}_\rho(x',\underset{\sim}{k}_\perp - \underset{\sim}{k}'_\perp)$$

$$- \frac{2(\underset{\sim}{k}'_L \cdot \underset{\sim}{k}_s)}{k_s^2} \hat{\epsilon}_\mu(x',\underset{\sim}{k}_\perp - \underset{\sim}{k}'_\perp) \Bigg] (\underset{\sim}{k}'_L \times \underset{\sim}{k}_s) \; \hat{\Phi}_0(x_0,\underset{\sim}{k}'_\perp)$$

$$+ \Bigg[\hat{\epsilon}_\rho(x',\underset{\sim}{k}_\perp - \underset{\sim}{k}'_\perp) - \frac{(\underset{\sim}{k}'_s \cdot \underset{\sim}{k}_s)}{k_s^2} \hat{\epsilon}_\mu(x',\underset{\sim}{k}_\perp - \underset{\sim}{k}'_\perp) \Bigg]$$

$$[\underset{\sim}{k}'_s \times \hat{\underset{\sim}{\Psi}}_0(x_0,\underset{\sim}{k}'_\perp)] \times \underset{\sim}{k}_s$$

$$- \frac{(\underset{\sim}{k}'_s \times \underset{\sim}{k}_s)}{k_s^2} \hat{\epsilon}_\mu(x',\underset{\sim}{k}_\perp - \underset{\sim}{k}'_\perp) \; [\underset{\sim}{k}'_s \times \hat{\underset{\sim}{\Psi}}_0(x_0,\underset{\sim}{k}'_\perp)] \cdot \underset{\sim}{k}_s \Bigg\} \, d\underset{\sim}{k}'_\perp \, dx'$$

Referring to Eqs. (8), the transformation from input to output potential fields can be characterized as a nonlocal operation defined on the inverse space of $\underset{\sim}{x}_\perp$, accompanied by a phase shift. Since the inverse space representations directly give the resolution of the respective fields into plane waves, the nonlocal nature of the input-output transformation is intuitive. A plane wave incident in one direction results, upon scattering, in a spectrum of plane waves spanning a range of directions. Further, Eqs. (8) demonstrate that the spectrum of plane waves generated by a given incident wave is a function of the absolute direction of that wave. That is, the nonlocal transformation is not a convolution. Thus, a Fourier inversion back to physical space transforms the input fields into output fields, which are also described by nonlocal operations.

The differential limit is accomplished in a straightforward fashion by considering the layer to reach to the observation point, x, that is $x = x_0 + L$. Taking the range derivative with respect to the upper limit of integration, and letting the limit approach zero yields the following results,

$$
\begin{aligned}
\partial_x \hat{\Phi}(x,\underset{\sim}{k}_\perp) = {} & i\, k_{xL} \hat{\Phi}(x,\underset{\sim}{k}_\perp) + \frac{i}{8\pi^2 k_{xL}} \iint \Bigg\{ \Bigg[(\underset{\sim}{k}'_L \cdot \underset{\sim}{k}_L)\, \hat{\epsilon}_\rho(x,\underset{\sim}{k}_\perp - \underset{\sim}{k}'_\perp) \\
& - \frac{k_L^4}{k_F^2} \hat{\epsilon}_\lambda(x,\underset{\sim}{k}_\perp - \underset{\sim}{k}'_\perp) - \frac{2(\underset{\sim}{k}'_L \cdot \underset{\sim}{k}_L)^2}{k_s^2} \hat{\epsilon}_\mu(x,\underset{\sim}{k}_\perp - \underset{\sim}{k}'_\perp) \Bigg] \hat{\Phi}(x,\underset{\sim}{k}'_\perp) \\
& + \Bigg[\hat{\epsilon}_\rho(x,\underset{\sim}{k}_\perp - \underset{\sim}{k}'_\perp) - \frac{2(\underset{\sim}{k}'_s \cdot \underset{\sim}{k}_L)}{k_s^2} \hat{\epsilon}_\mu(x,\underset{\sim}{k}_\perp - \underset{\sim}{k}'_\perp) \Bigg] \\
& [\underset{\sim}{k}'_s \times \hat{\underset{\sim}{\Psi}}(x,\underset{\sim}{k}'_\perp)] \cdot \underset{\sim}{k}_L \Bigg\} \, d\underset{\sim}{k}'_\perp ,
\end{aligned}
$$

and (9)

$$
\begin{aligned}
\partial_x \hat{\underset{\sim}{\Psi}}(x,\underset{\sim}{k}_\perp) = {} & i\, k_{xs} \hat{\underset{\sim}{\Psi}}(x,\underset{\sim}{k}_\perp) + \frac{i}{8\pi^2 k_{xs}} \iint \Bigg\{ \Bigg[\hat{\epsilon}_\rho(x,\underset{\sim}{k}_\perp - \underset{\sim}{k}'_\perp) \\
& - \frac{2(\underset{\sim}{k}'_L \cdot \underset{\sim}{k}_s)}{k_s^2} \hat{\epsilon}_\mu(x,\underset{\sim}{k}_\perp - \underset{\sim}{k}'_\perp) \Bigg] (\underset{\sim}{k}'_L \times \underset{\sim}{k}_s)\, \hat{\Phi}(x,\underset{\sim}{k}'_\perp) \\
& + \Bigg[\hat{\epsilon}_\rho(x,\underset{\sim}{k}_\perp - \underset{\sim}{k}'_\perp) - \frac{(\underset{\sim}{k}'_s \cdot \underset{\sim}{k}_s)}{k_s^2} \hat{\epsilon}_\mu(x,\underset{\sim}{k}_\perp - \underset{\sim}{k}'_\perp) \Bigg] [\underset{\sim}{k}'_s \times \hat{\underset{\sim}{\Psi}}(x,\underset{\sim}{k}'_\perp)] \times \underset{\sim}{k}_s \\
& - \frac{(\underset{\sim}{k}'_s \times \underset{\sim}{k}_s)}{k_s^2} \hat{\epsilon}_\mu(x,\underset{\sim}{k}_\perp - \underset{\sim}{k}'_\perp)\, [\underset{\sim}{k}'_s \times \hat{\underset{\sim}{\Psi}}(x,\underset{\sim}{k}'_\perp)] \cdot \underset{\sim}{k}_s \Bigg\} \, d\underset{\sim}{k}'_\perp .
\end{aligned}
$$

Here we have added the homogeneous field to the first term on the righthand side of each equation to obtain the total fields

These two equations describe the propagation of the displacement potentials in an inhomogenous elastic medium. The following assumptions were made in deriving these equations: (1) The fields were assumed to be defined on an initial surface at x_0 and the effects of the inhomogeneities down range from x_0 at x were assumed to be entirely due to the region between the two planes. (2) The inhomogeneities and their gradients are assumed to be sufficiently weak that a Born approximation can be made. (3) The material properties must be continuous in range.

THE SCALAR LIMIT

Before continuing with Eqs. (9) to obtain an elastic parabolic theory, it is instructive to reduce these equations to the scalar limit and examine their relationship to a formal reduction of the Helmholtz equation. The equation,

$$\partial_x \hat{\Phi}(x,\underset{\sim}{k}_\perp) = i\, k_{xF} \hat{\Phi}(x,\underset{\sim}{k}_\perp) + \frac{i\, k_F^2}{4\pi^2} \int \frac{\hat{\epsilon}(x,\underset{\sim}{k}_\perp - \underset{\sim}{k}'_\perp)}{2\, k_{xF}} \hat{\Phi}(x,\underset{\sim}{k}'_\perp)\, d\underset{\sim}{k}'_\perp \tag{10}$$

represents the solution to the scalar wave equation,

$$\nabla^2 \Phi(\underset{\sim}{x}) + k_F^2 \,[\, 1 + \hat{\epsilon}(\underset{\sim}{x})]\, \hat{\Phi}(\underset{\sim}{x}) = 0 \tag{11}$$

Equation (10) may be compared with the results of Fishman and McCoy[9] to obtain some sense of the restrictions on the current theory. These authors give as a wide-angle, weak-inhomogeneity, and weak-gradient theory, the equation,

$$\partial_x \hat{\Phi}(x,\underset{\sim}{k}_\perp) = i\, k_{xF} \hat{\Phi}(x,\underset{\sim}{k}_\perp) + \frac{i\, k_F^2}{4\pi^2} \int \frac{\hat{\epsilon}(x,\underset{\sim}{k}_\perp - \underset{\sim}{k}'_\perp)}{(k_{xF} + k'_{xF})} \hat{\Phi}(x,\underset{\sim}{k}'_\perp)\, d\underset{\sim}{k}'_\perp \tag{12}$$

This equation differs from the one derived in this paper only in the presence of the $(k_{xF} + k'_{xF})$ term in the denominator vis-a-vis the $2k_{xF}$ term obtained in this paper. This difference is atributable to the use by Fishman and McCoy of the half-space Green's function, $G^+ = -2\,\partial_x G$, which is the proper propagator to use in a forward propagation problem. Obviously, the differences only manifest themselves in the coupling between the spectral components of the waves. Further, the differences are readily seen to be proportional to $(k_\perp^2 - k_\perp'^2)/4k_F^2$, and the theory may be termed narrow-angle about each spectral component, which propagates correctly at all angles.

For further comparison the ordinary parabolic equation can be written as

$$\begin{aligned}\partial_x \hat{\Phi}(x,\underset{\sim}{k}_\perp) &= i \left[k_F - \frac{k_\perp^2}{2k_F} \right] \hat{\Phi}(x,\underset{\sim}{k}_\perp) \\ &\quad + \frac{i\, k_F^2}{4\pi^2} \int \frac{\hat{\epsilon}(x,\underset{\sim}{k}_\perp - \underset{\sim}{k}'_\perp)}{2\, k_F} \hat{\Phi}(x,\underset{\sim}{k}'_\perp)\, d\underset{\sim}{k}'_\perp ,\end{aligned} \tag{13}$$

in the current notation. It is readily seen to be both narrow-angle in the sense of coupling between modes and in the propagation of a spectral component.

While the more formal analysis can not yet be carried through for the elastic theory, Eqs. (9) are the elastic equivalent to Eq. (10) and therefore some confidence may be attached to the results.

AN ELASTIC PARABOLIC APPROXIMATION

The elastic parabolic equations are determined from the first-order elastic differential equations by expanding k_{xL} and k_{xS} in power series of $k_\perp^2$ and then dropping all terms on the order $k_\perp^4$ and $\epsilon\, k_\perp^2$ yielding the following pair of equations,

$$\begin{aligned}\partial_x \hat{\Phi}(x,\underset{\sim}{k}_\perp) &= i \left[k_L - \frac{k_\perp^2}{2k_L} \right] \hat{\Phi}(x,\underset{\sim}{k}_\perp) \\ &\quad + \frac{i}{8\pi^2\, k_L} \int \left\{ \left[k_L^2 \hat{\epsilon}_\rho(x,\underset{\sim}{k}_\perp - \underset{\sim}{k}'_\perp) - \frac{k_L^4}{k_F^2} \hat{\epsilon}_\lambda(x,\underset{\sim}{k}_\perp - \underset{\sim}{k}'_\perp) \right.\right.\end{aligned}$$

$$- \frac{2k_L^4}{k_s^2} \hat{\epsilon}_\mu(x, \underset{\sim}{k}_\perp - \underset{\sim}{k}'_\perp) \Big] \hat{\Phi}(x, \underset{\sim}{k}'_\perp)$$

$$+ \left[\hat{\epsilon}_\rho(x, \underset{\sim}{k}_\perp - \underset{\sim}{k}'_\perp) - \frac{2k_L}{k_s} \hat{\epsilon}_\mu(x, \underset{\sim}{k}_\perp - \underset{\sim}{k}'_\perp) \right]$$

$$[(\underset{\sim}{k}_\perp - \underset{\sim}{k}'_\perp) k_s + \underset{\sim}{k}'_\perp (k_s - k_L)] \cdot \underset{\sim}{e}_x \times \hat{\underset{\sim}{\Psi}}(x, \underset{\sim}{k}'_\perp) \Big\} \; d\underset{\sim}{k}'_\perp$$

and (14)

$$\partial_x \hat{\Psi}(x, \underset{\sim}{k}_\perp) = i \left[k_s - \frac{k_\perp^2}{2k_s} \right] \hat{\Psi}(x, \underset{\sim}{k}_\perp)$$

$$+ \frac{i}{8\pi^2 k_s} \int \Big\{ \left[\hat{\epsilon}_\rho(x, \underset{\sim}{k}_\perp - \underset{\sim}{k}'_\perp) - \frac{2k_L}{k_s} \hat{\epsilon}_\mu(x, \underset{\sim}{k}_\perp - \underset{\sim}{k}'_\perp) \right]$$

$$[(k_s - k_L) \underset{\sim}{k}'_\perp - k_L (\underset{\sim}{k}_\perp - \underset{\sim}{k}'_\perp)] \times \underset{\sim}{e}_x \hat{\Phi}(x, \underset{\sim}{k}'_\perp)$$

$$+ [\hat{\epsilon}_\rho(x, \underset{\sim}{k}_\perp - \underset{\sim}{k}'_\perp) - \hat{\epsilon}_\mu(x, \underset{\sim}{k}_\perp - \underset{\sim}{k}'_\perp)] \; [k_s^2 \hat{\underset{\sim}{\Psi}}(x, \underset{\sim}{k}'_\perp)$$

$$- k_s (\underset{\sim}{e}_x \underset{\sim}{k}'_\perp + \underset{\sim}{k}'_\perp \underset{\sim}{e}_x) \cdot \hat{\underset{\sim}{\Psi}}(x, \underset{\sim}{k}'_\perp) - k_s (\underset{\sim}{k}_\perp - \underset{\sim}{k}'_\perp) \cdot \hat{\underset{\sim}{\Psi}}(x, \underset{\sim}{k}'_\perp) \underset{\sim}{e}_x] \Big\} \, d\underset{\sim}{k}'_\perp .$$

As in the fluid case the following expansions are introduced for the longitudinal and transverse fields

$$\Phi(x, \underset{\sim}{k}_\perp) = \phi(x, \underset{\sim}{k}_\perp) \exp(ik_L x)$$

and (15)

$$\underset{\sim}{\Psi}(x, \underset{\sim}{k}_\perp) = \underset{\sim}{\psi}(x, \underset{\sim}{k}_\perp) \exp(ik_s x)$$

and their inverse transforms in the cross-range coordinates. Substitution of these expansions into Eqs. (14) results in the cancellation of $ik_L \hat{\Phi}(x, \underset{\sim}{k}_\perp)$ in the first equation and of $ik_s \hat{\underset{\sim}{\Psi}}(x, \underset{\sim}{k}_\perp)$ in the second equation. The cross-terms in each equation retain an exponential phase factor after the principal phase has been removed. The ψ term in the first equation retains an $\exp[i(k_s - k_L)x]$ and the ϕ term in the second equation retains an $\exp[i(k_L - k_s)x]$, thus the equations are range dependent. Otherwise Eqs. (14) remain unchanged.

Equations (14) are to be recognized as convolutions, where the $\underset{\sim}{k}'_\perp$ and $(\underset{\sim}{k}_\perp - \underset{\sim}{k}'_\perp)$ terms represent in inverse, k, space differential operators on the field and the inhomogeneities respectively in real, x, space. Performing the convolution results in the parabolic equations for an elastic medium,

$$2ik_L \partial_x \phi(\underset{\sim}{x}) + \nabla^2 \phi(\underset{\sim}{x}) + k_L^2 \epsilon_L(\underset{\sim}{x}) \phi(\underset{\sim}{x})$$

$$= i [\underset{\sim}{k}_s \nabla_\perp \epsilon_{Ls}(\underset{\sim}{x}) + (\underset{\sim}{k}_s - \underset{\sim}{k}_L) \epsilon_{Ls}(\underset{\sim}{x}) \nabla_\perp] \cdot \underset{\sim}{e}_x \times \underset{\sim}{\psi}(\underset{\sim}{x}) \exp[i(k_s - k_L)x]$$

and (16)

$$2ik_s\,\partial_x \underset{\sim}{\psi}(\underset{\sim}{x}) + \nabla_\perp^2 \underset{\sim}{\psi}(\underset{\sim}{x}) + k_s^2 \epsilon_s(\underset{\sim}{x})\,\underset{\sim}{\psi}(\underset{\sim}{x})$$

$$= -ik_s\epsilon_s(\underset{\sim}{x})\,(\underset{\sim}{e}_x\nabla_\perp + \nabla_\perp\underset{\sim}{e}_x)\,\underset{\sim}{\psi}(\underset{\sim}{x}) - ik_s\nabla_\perp\epsilon_s(\underset{\sim}{x})\cdot\underset{\sim}{\psi}(\underset{\sim}{x})\,\underset{\sim}{e}_x$$

$$+\, i\,[(k_s - k_L)\epsilon_{Ls}(\underset{\sim}{x})\,\nabla_\perp\times\underset{\sim}{e}_x - k_L\nabla\,\epsilon_{Ls}(\underset{\sim}{x})\times\underset{\sim}{e}_x]\,\phi(\underset{\sim}{x})\,\exp[i(k_L - k_s)x] \quad ,$$

where the following definitions have been introduced for convenience:

$$\epsilon_L(\underset{\sim}{x}) = \epsilon_\rho(\underset{\sim}{x}) - \frac{k_L^2}{k_F^2}\,\epsilon_\lambda(\underset{\sim}{x}) - \frac{2k_L^2}{k_s^2}\,\epsilon_\mu(\underset{\sim}{x}) \quad ,$$

$$\epsilon_{Ls}(\underset{\sim}{x}) = \epsilon_\rho(\underset{\sim}{x}) - \frac{2k_L}{k_s}\,\epsilon_\mu(\underset{\sim}{x}) \quad ,$$

and (17)

$$\epsilon_s(\underset{\sim}{x}) = \epsilon_\rho(\underset{\sim}{x}) - \epsilon_\mu(\underset{\sim}{x}) \quad .$$

This model has been implemented on a Digital Equipment Corp. VAX-11/750 minicomputer. The equations were solved using standard finite-difference techniques. The central-difference, Crank-Nicholson, technique was modified along the lines of the Douglas method for improved stability at larger step-sizes.[13] The resulting set of equations was written as a band matrix of width 7 with the boundary conditions at either end. The equations were solved using an LU decomposition technique for band matrices.[13] The method was not fast, principally because the equations must be resolved at each range due to the range-dependent coupling terms. A typical run was 400 depth points, 5 m spacing, to 100 km in 12.5 m steps taking about 10 hrs. A fluid version of the same model was considerably faster.

Boundary conditions, currently available, are either radiative or Dirichlet, where either the potential or its derivative may be specified at a boundary for each potential. Free and hard surfaces are not Dirichlet in an elastic potential theory and represent a point of current effort. Further, Eqs. (16) are analytically unstable at a free boundary due to coupling between the two potentials. Specifically, if a surface wave is inserted into Eqs. (16) and the boundary conditions, the solution will have two parts representing exponentially increasing and decreasing solutions. This is to be expected since the boundary terms were ignored in the initial integral representations. Landers[4] reported a similar problem in the treatment of his theory at free boundaries which he removed by assuming an additional solution in the depth direction and dropping the remaining ∂_{zz} terms.

Figure 1 presents the results of a run using the model presented in this paper. On the left-hand side of the figure is the profile used in terms of the longitudinal and shear speeds. The frequency used is 10 Hz. A Gaussian longitudinal potential source was specified at zero range. As we proceed in range, the following effects may be observed: (1) the longitudinal potential generates a shear wave potential field in this inhomogeneous medium; (2) the shear wave potential is antisymetric, as it must be from the equations and physical considerations; (3) both the longitudinal and shear waves are repeatedly focused by the profile as expected; and (4) although it is not readily apparent in this figure, there is coupling between the two potentials produced by the inhomogeneous environment. This produces rapid fluctuations of the potential at the wavelength level and in some regions the longitudinal potential may be said to be forcing the shear potential and vice-versa.

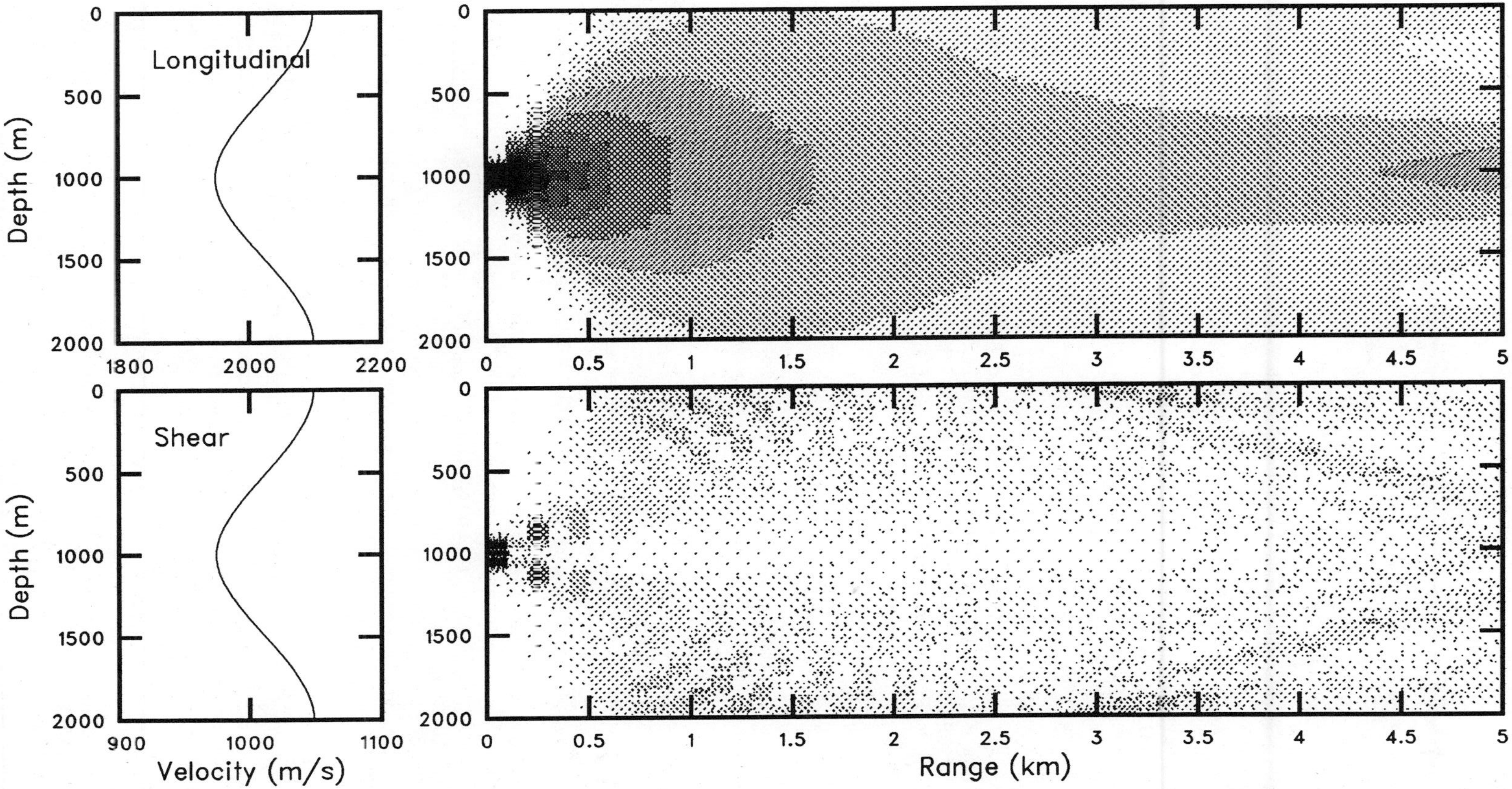

Fig. 1. Vector PE propagtion run. Longitudinal and shear sound speed profiles (left) and the amplitude of the longitudinal and shear displacement potentials (right). Potential amplitudes are shown on a grey scale in arbitrary units from zero level as white to high levels as black. The longitudinal grey scale levels are 20 times the shear levels. The frequency is 10 Hz and a Gaussian longitudinal source was used at 1000 m.

CONCLUSIONS

This paper has examined a vector PE model for displacement potential propagation in elastic solids. The method, based on a Born approximation, resulted in a set of equations representing wide-angle propagation in an elastic medium. These wide-angle equations were examined for the scalar case for which exact results to the factored Helmholtz equation are available. The wide-angle equations propagated each spectral component correctly in a wide-angle sense and coupled the spectral components correctly in a narrow-angle sense, providing some confidence to the derivation method. From the elastic wide-angle equations a narrow-angle vector PE model was derived. This model was observed to propagate the potentials in accordance with physical expectations. Boundaries were observed to represent a difficulty for potential formulations since they are not Dirichlet-like and result in coupling between the two potentials with attendent stability problems. It was further observed that the use of the soft boundary conditions with the vector PE model, results in unstable surface wave conditions, viz. a pair of solutions, one of which increases exponentially with range.

Current work on this model is focused on resolving computational issues at the boundaries, coupling elastic to fluid PE models, and testing the model against rigorous calculations in idealized scenarios. This work was supported by ONR funds and will be submitted by the author to The Catholic University of America in partial fulfillment of the requirements of the Ph.D. degree.

REFERENCES

1. F. D. Tappert, The Parabolic Approximation Method, in: "Lecture Notes in Physics 70: Wave Propagation and Underwater Acoustics," J. B. Keller and J. S. Papadakis, eds., Springer Verlag, Berlin (1977).
2. F. D. Tappert and R. H. Hardin, in: "A synopsis of the AESD workshop on acoustic modeling by non-ray tracing techniques," C. W. Spofford, AESD TN-73-05, Arlington, Va. (1973).
3. T. Landers and J. F. Claerbout, Numerical calculations of elastic waves in laterally inhomogeneous media, J. Geophys. Res. 77:1476 (1972).
4. T. Landers, "Elastic waves in laterally inhomogeneous media," Ph.D. thesis, Stanford Univ. (1971).
5. J. J. McCoy, A Parabolic Theory of Stress Wave Propagation Through Inhomogeneous Linearly Elastic Solids, J. Appl. Mech. 44:462 (1977).
6. J. A. Hudson, A parabolic approximation for elastic waves, Wave Motion 2:207 (1980).
7. J. P. Corones, B. DeFacio, and R. J. Krueger, Parabolic approximations to the time-independent elastic wave equation, J. Math. Phys. 23:557 (1982).
8. S. C. Wales and J. J. McCoy, A comparison of parabolic wave theories for linearly elastic solids Wave Motion 5:99 (1983).
9. L. Fishman and J. J. McCoy, Derivation and application of extended parabolic wave theories. I. The factorized Helmholtz equation, J. Math. Phys. 25:285 (1984).
10. A. T. deHoop, "Representation theorems for the displacement in an elastic solid and their application to elastodynamic diffraction theory," Eng.Ing. thesis, Technische Hogeschod, Delft (1958).
11. P. Morse and H. Feshbach, "Methods of Theoretical Physics," McGraw-Hill, New York (1953).
12. L. Brekhovskikh, "Waves in Layered Media," p. 239, Academic Press, New York (1960).
13. A. R. Mitchell and D. F. Griffiths, "The Finite Difference Method in Partial Differential Equations," John Wiley & Sons, Chichester (1980).

INFLUENCE OF BOTTOM REFRACTION ON THE PROPAGATION OF UNDERWATER SOUND

J. Miller, A. Nagl, and H. Überall

Department of Physics
Catholic University
Washington DC 20064, USA

ABSTRACT

Experimental measurements [J. S. Hanna, J. Acoust. Soc. Am. 53: 1686 (1973); R. E. Christensen et al., J. Acoust. Soc. Am. 57: 1421 (1975)] of the transmission loss of underwater sound propagating over an ocean bottom containing unconsolidated sediments have presented evidence on the presence of bottom-refracted energy being returned into the water column. The bottom refractions manifest themselves either via their pulse arrival times (indicating bottom-refracted paths), or via their interference with steady-state transmission loss data attributable to bottom reflections alone. The theoretical interpretation of these experiments has so far been carried out in the form of transmission-loss estimates based on ray theory and ray spreading. In a previous study [A. Nagl et al., J. Acoust. Soc. Am. 63: 739 (1978)], we have devised a normal-mode program of underwater sound propagation containing water layers of linear velocity gradients. This program has been extended to a bottom containing unconsolidated sediment layers of different densities and velocity gradients, and shown to reproduce bottom-refracted fields interfering with reflected sound. The interference minima can be used to determine the sound speeds and their gradients in the bottom layers.

EVIDENCE FOR BOTTOM REFRACTION

Sound traveling from a transducer source (T) to a receiver (R), both submerged in the ocean, can normally take several different paths: a direct path TR, a path including an extra surface reflection (TSR), and three paths involving a reflection from the ocean bottom B (including no, one, and two extra reflections from the sea surface S, i.e., TBR, TBSR, or TSBR, and TSBSR). If, however, the bottom consists of unconsolidated sediments (silt-clay surficial sediments[1]) which show a positive velocity gradient (and additionally, a surface sound velocity less than that in the water), refraction of sound occurs which can direct bottom-penetrating sound energy back into the water, hence establishing a fourth bottom-interacting path which involves bottom refraction. More recent experimental evidence for such bottom-penetrating signals has been discussed by Hanna[2] (on the basis of its interference with bottom-reflected sound), and by Christensen et al[3] (on the basis of the travel times of short sound pulses); earlier evidence for such bottom refractions is quoted in the

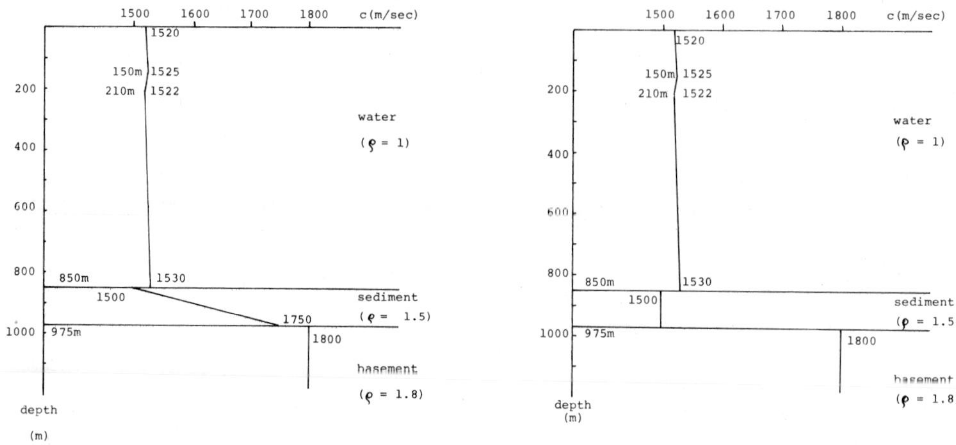

(a) (b)

Fig. 1. Sound speed profiles in water column, sediment layer, and basement: (a) sediment with velocity gradient, (b) isovelocity sediment.

mentioned publications[2,3]. The bottom refractions are stronger at lower frequencies and at shorter penetration depths, since less bottom absorption takes place in these cases. Hamilton[1] states that conditions for non-reflective bottom returns should exist over most deep-sea ocean floors, and in some shallow-water areas.

The mentioned study by Hanna[2] analyzes experimental transmission-loss data on the basis of a (plane-wave) bottom-reflection coefficient $r(\theta)$ which includes the bottom-refraction effects, and which is used to obtain a theoretical estimate of the transmission loss as a function of range up to about seven water depths (18 NM) that contains the effect of interferences between bottom-reflected and bottom-refracted paths. He showed that the location in range of these interference features, by comparison with experimental data, could provide information on the velocity gradient in the sediment and on the sound velocity ratio at the water-sediment interface, but not on the density ratio at the interface.

NORMAL MODE CALCULATION

We here present a mode-theoretical study of the transmission loss through an ocean of intermediate depth with variable sound velocity, overlaying a sediment layer with a velocity gradient that lies on top of an isovelocity basement (Fig. 1a). The sound velocity variation in the water is chosen comparable to that of Hanna[2], as well as the 2% sound speed ratio at the surface of the sediment, and the sound speed gradient of $g = 2\ \text{sec}^{-1}$ ($1.5\ \text{sec}^{-1}$ for Hanna). Our water depth is chosen as 850m, and the frequency as $f = 15$ Hz (in order to keep the modes down to a manageable number of ten), the sediment depth as 125m (density 1.5 gm/cm^3), and the sound velocity in the basement as 1800 m/sec (density 1.8 gm/cm^3). The transmission loss calculated from this model is then compared with that corresponding to an isovelocity ($c = 1500$ m/sec) sediment which does not give rise to any backward-refracted energy (Fig. 1b). We find indeed a clear set of additional interference features in the transmission loss corresponding to a sediment with velocity gradient (Fig. 1a), which show a similar character to those in the transmission loss data that were analyzed by Hanna[2], and that can be interpreted as being caused by the interference between bottom-reflected and bottom-refracted sound.

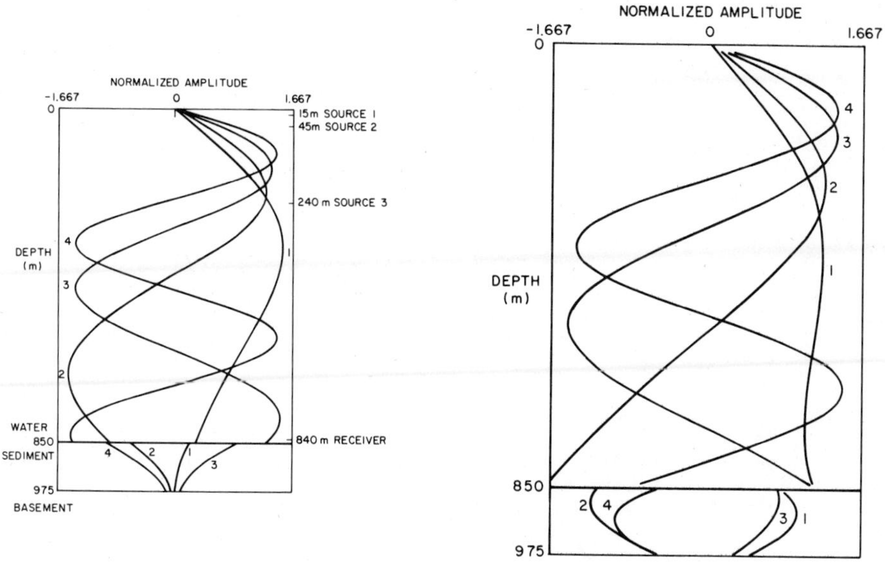

(a) (b)

Fig. 2. First four normal modes (depth functions) for the velocity profile of Fig. 1, for (a) sediment with gradient, (b) isovelocity sediment.

Our normal-mode is that developed by us earlier[4], but applied to a stratified geometry, and extended so as to include (fluid) bottom layers of densities constant within each layer, but differing from that of other layers. The depth-function modal solutions are given by Airy functions in each layer of variable sound speed.

Figure 2a shows the normalized modal depth functions Nos. 1-4 for the $g \neq 0$ case of Fig. 1a, and Fig. 2b the depth functions for the $g = 0$ case of Fig. 1b. The diminution of sound energy in the lower part of the sediment, due to back-refraction into the water by the gradient, is evident in Fig. 2a. In the following figures, we present our mode-theoretical calculation of transmission loss up to a range of 10.5 km. Fig. 3 corresponds to a source depth of 15m, and Fig. 4 to a source depth of 45m. In both cases, the receiver is placed on the bottom, as in Hanna's example[2]. Figures 3a and 4a correspond to the $g = 2$/sec sound speed gradient in the sediment, and Figs. 3b and 4b correspond to $g = 0$.

Considering first Fig. 3b which only contains bottom reflections, we see a set of simple interference features which in the calculation stem from the modal interferences, but which may be physically interpreted by the interferences of rays along the different water-borne paths mentioned at the beginning (including the "Lloyd's Mirror" interference between the TR and TSR paths). Comparing to this the results of Fig. 3b for the case $g \neq 0$, we see that the bottom refraction which is present in this case has introduced an additional set of interference features which did not exist before, viz. those at ranges of 2 km, 6 km, and probably also at 10 km, and others. Comparing similarly the two corresponding figures, Fig. 4a and Fig. 4b for the 45-m source depth, similar features are visible including extra bottom-refraction interference minima at the same ranges. Another calculation carried out at a source depth of 240 m showed less clear-cut results.

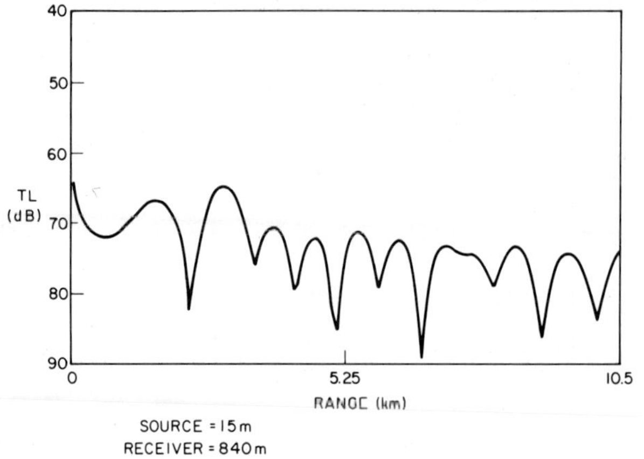

Fig. 3a. Transmission loss vs. range, source depth 15m. Gradient case.

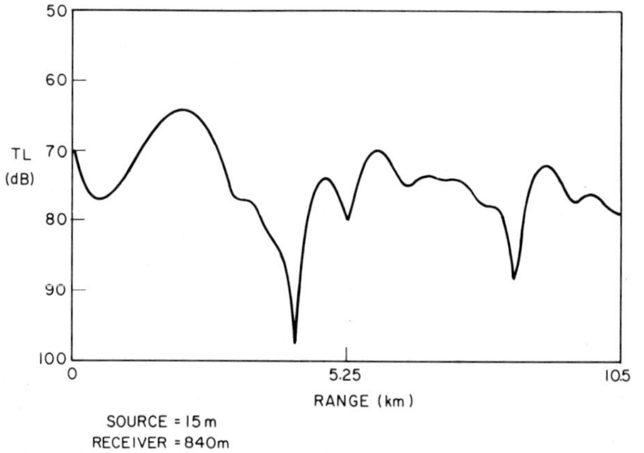

Fig. 3b. Transmission loss vs. range, source depth 15m. Isovelocity case.

HANNA'S MODEL

Hanna[2] has interpreted the extra interferences introduced by the bottom refractions in terms of the zeros of the bottom reflection coefficient $r(\theta)$ where θ is the grazing angle of an incident plane wave. He has shown that interference nulls should be located at grazing angles θ determined by the vanishing of the argument of the Airy function, i.e.,

$$\cos \theta_i = (c/c_1) \left[1 + a_i (g/\pi f)^{2/3}\right]^{1/2} , \tag{1}$$

where c and c_1 are the water and sediment sound velocities, respectively, at the water-sediment interface, and a_i the zeros of the Airy function. At the first zero, we find a grazing angle $\theta_1 = 30.55°$ for the sound speed profile in the water column of Fig. 1; the further values are $\theta_2 = 43.70°$, $\theta_2 = 54.17°$ etc.

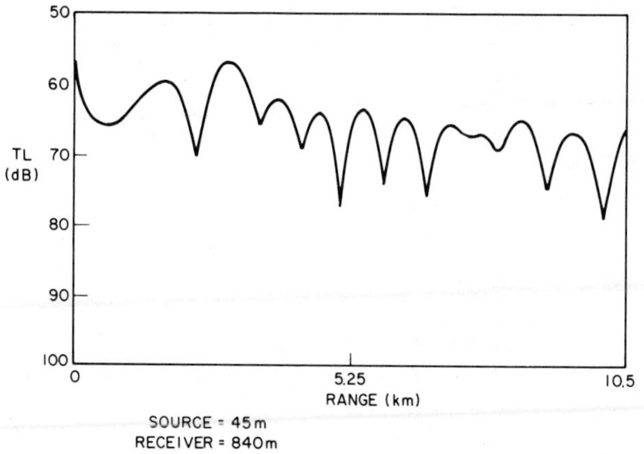

Fig. 4a. Transmission loss vs. range, source depth 45 m. Gradient case.

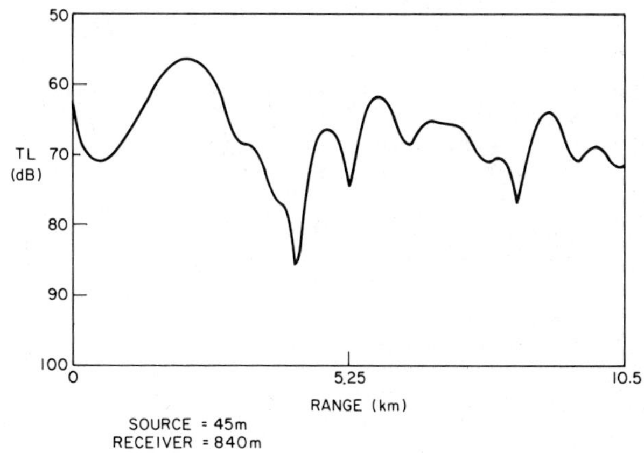

Fig. 4b. Transmission loss vs. range, source depth 45 m. Isovelocity case.

In order to identify interference zeros due to bottom refraction in the transmission loss curves of Figs. 3 and 4, a relation between grazing angle and range must be established, so that Eq. (1) can be applied. This can be obtained by using the equivalent ray angles θ_n of our normal modes (labeled by the mode number n):

$$\theta_n(z) = \cos^{-1}(k_n/k(z)) \; , \tag{2}$$

where $k(z) = 2\pi f/c(z)$, with c(z) taken from Fig. 1. The mode eigenvalues k_n for our velocity profile are listed in Table 1 for the ten propagating modes that this profile admits. The corresponding equivalent ray paths can be traced starting from the source S, and they consist of circular arcs in the three fluid layers as shown schematically in Fig. 5. In this way, the bottom grazing angles and the corresponding range values shown in Table 1 are obtained. It is seen that only the last propagating mode in our profile (n = 10) has a bottom grazing angle large enough (30.9°) to fall into the region which correspond to Hanna's interference zeros, and

Table 1. Eigenvalues k_n of nth mode, equivalent ray grazing angle θ_n at water-sediment interface, and corresponding range R.

n	k_n	θ_n(bott.)	R(m)
1	0.061717		
2	0.061426	4.31°	8025
3	0.060980	8.11°	5234
4	0.060346	11.58°	3865
5	0.059502	14.99°	3036
6	0.058448	18.41°	2478
7	0.057197	21.8°	2079
8	0.055814	25.0°	1789
9	0.054400	28.0°	1578
10	0.052879	30.9°	1406

in fact it coincides quite well with θ_1 = 30.55°. The corresponding range where thus Hanna's theory predicts an interference minimum due to bottom returns is R = 1406 m, which agrees only qualitatively with the lowest new interference minimum of our figures 3a or 4a, located at R ≅ 2000 m. The higher-order minima predicted by Eq. (1) correspond to larger bottom angles $\theta_2, \theta_3 \ldots$, and would thus be located at ranges less than 1400 m. These minima cannot exist, however, since they would correspond to modes n 10 which our profile does not support. Accordingly, no further minima appear to the left of that at R 2000 m in Figs. 3a or 4a. What appears as extra minima at larger ranges in these figures, as compared to the isovelocity-sediment case of Figs. 3b and 4b, respectively, can thus not be explained by Hanna's theory. These extra minima may still be caused by more complicated types of interfering bottom returns, for example by a redistribution of basement reflections due to the velocity gradient in the sediment. They may, however, also be caused by boundary waves[5] on the sediment and/or basement interface. This point remains yet to be investigated.

ACKNOWLEDGEMENTS

This work was supported in part by the Office of Naval Research. One of us (J.M.) wishes to thank Singer Co. (Link Simulation Systems Division) for their aid. Further support from the Computer Center of Catholic University is acknowledged.

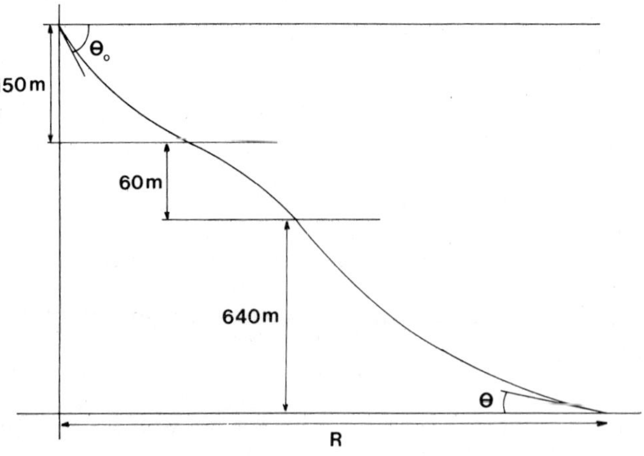

Fig. 5. Equivalent ray path, with bottom grazing angle determining range.

REFERENCES

1. E. L. Hamilton, Sound channels in surficial marine sediments, J. Acoust. Soc. Am. 48: 1296 (1970).

2. J. S. Hanna, Short-range transmission loss and the evidence for bottom-refracted energy, J. Acoust. Soc. Am. 53: 1686 (1973).

3. R. E. Christensen, J. A. Frank, and W. H. Geddes, Low-frequency propagation via shallow refracted paths through deep ocean unconsolidated sediments, J. Acoust. Soc. Am. 57: 1421 (1975).

4. A. Nagl, H. Überall, A. J. Haug and G. L. Zarur, Adiabatic mode theory of underwater sound propagation in a range-dependent environment, J. Acoust. Soc. Am. 63: 739 (1978).

5. D. M. F. Chapman, Acoustic interface waves at the seabed, in "Proceedings of the 22nd. Meeting of the Society of Engineering Science," Pennsylvania State University, Oct. 7-11, 1985.

BOTTOM INTERACTION EFFECTS ON NORMAL MODES:

AN ALGEBRAIC APPROACH

David H. Wood*+, Mark D. Duston*+, and Ghasi R. Verma*+

*Code 3332, New London Laboratory
US Naval Underwater Systems Center
New London, Connecticut 06320, USA

+Department of Mathematics
University of Rhode Island
Kingston, RI 02881, USA

ABSTRACT

We assume that the speed of sound in the water and in the bottom of the ocean is a function of only the depth, and not the range. We also assume that the ocean and its bottom is eventually underlaid with a rigid interface. This problem can be solved by the method of normal modes, involving the eigenvalues and eigenfunctions of a depth dependent ordinary differential equation. We investigate the changes in these eigenvalues and eigenfunctions that result from changes in the depth dependent sound speed within the ocean and its bottom, using a perturbation approach. We formulate the perturbation in terms of an algebraic eigenvalue problem, and we show that it is equivalent to the usual power series expansion in a small parameter.

INTRODUCTION

The speed of sound in the ocean is nearly uniform, but the small variations in speed give many very important effects. We want to exploit the fact that if the sound speed were uniform, the wave propagation could be modeled using an equation for which we know (Ref. 1, pg. 34) the exact solution in a convenient form. In this paper, small changes in the equations governing an ocean with uniform speed are thought of as "perturbations" and we seek the equations that govern changes in the resulting solutions. We expand the perturbed solutions as infinite "Fourier" series in the normal modes of the unperturbed problem. This gives us an eigenproblem for the matrix of Fourier coefficients. We then go on to show that this formulation is consistent with another commonly used formultation (Ref. 2, Chapt XIX). To do this, we expand our matrix of Fourier coefficients in a power series in a small parameter that expresses the size of the perturbation. We obtain expressions for the leading terms of the perturbed solution, and note that they agree with known results of this type (Ref. 3).

We believe that the algebraic formulation given in this paper has some important advantages over the one involving power series in a small parameter. The first advantage is that in the algebraic formulation there is no error due to truncating the power series. The power series method requires that the perturbation is uniformly small. It can be useful that our approach permits the perturbation to be larger in size. These larger perturbations, however, would be small in the sense that they would have very little effect on the high-order normal modes. The second advantage that is important to us is that the set of algebraic equations for the eigenvalues should be solvable more quickly than usual because it has a special structure that can be exploited (Refs. 4 - 7). Of course, our Fourier series must also be truncated after a finite number of terms, and this effect will need to be estimated (Ref. 8). We require a higher order of accuracy in the perturbations of the eigenvalues than in the perturbations of the eigenfunctions. This is required to maintain a uniform accuracy in graphs of propagation loss as a function of range.

THE UNPERTURBED PROBLEM

Since we assume that the sound speed in the ocean depends only on the depth z, the excess pressure p(r,z) in an inviscid, non-heat conducting fluid is cylindrically symmetric and satisfies the Helmholtz equation [Ref. 1, pg. 19],

$$p_{rr} + \frac{1}{r} p_r + p_{zz} + k^2 p = 0 , \tag{1}$$

where literal subscripts denote partial differentation. We choose the boundary conditions

$$p = 0 \text{ at } z = 0 , \tag{2a}$$

and

$$p_z = 0 \text{ at } z = h . \tag{2b}$$

This last boundary condition assumes that the ocean and its bottom are eventually underlaid with a rigid interface at depth $z = h$. To solve this equation by method of separation of variables we take $p(r,z) = \phi(z) R(r)$ and we obtain two ordinary differential equations

$$\phi_{zz} + k^2\phi = \lambda \phi , \tag{3}$$

and

$$R_{rr} + \frac{1}{r} R_r = -\lambda R , \tag{4}$$

where λ is a constant, independent of r and z. For these equations the boundary conditions (2a) and (2b) take the form

$$\phi(0) = 0 \tag{5a}$$

and

$$\phi_z(h) = 0 . \tag{5b}$$

It is known [Ref. 1, pg. 23] that there are countably many solutions of (3) satisfying (5a) and (5b) namely,

$$\phi_n(z) = \sqrt{\frac{2}{h}} \sin\left(\frac{(2n-1)\pi z}{2h}\right) , \quad n = 1, 2, 3, \ldots . \tag{6}$$

Each of these requires its corresponding value of λ to be given by

$$l_n = k^2 - \left[\frac{(2n-1)\pi}{2h}\right]^2 , \quad n = 1, 2, 3, \quad \ldots . \tag{7}$$

The lower case letter used to denote these values differs slightly from the numeral 1, and will always be subscripted. We want to represent this set of solutions in vector notation. Let Φ be a column vector defined by

$$\Phi = [\phi_1(z), \phi_2(z), \ldots\ldots \quad]^t , \tag{8}$$

where the superscript t denotes the matrix transpose. Let L be the diagonal matrix

$$L = \mathrm{diag}(l_1, l_2, l_3, \ldots\ldots \quad) . \tag{9}$$

Then Φ satisfies the vector equation

$$\Phi'' + k^2\Phi = L\Phi \tag{10}$$

and the orthogonality condition

$$\int_0^h \Phi\Phi^t dz = I , \tag{11}$$

where I is the identity matrix.

ALGEBRAIC FORMULATION OF THE PERTURBED PROBLEM

In place of (10) we will consider the perturbed vector equation

$$\Psi'' + k^2 \, (q(z) + \varepsilon \, s(z)) \, \Psi = \Lambda\Psi \tag{12}$$

with boundary conditions

$$\Psi(0) = 0 \tag{13}$$

and

$$\Psi_z(h) = 0 . \tag{14}$$

In (12), the quanities within parentheses are scalars, but Λ is a diagonal matrix. Although we will be only treating the case $q(z) = 1$, the argument given below remains true for more general $q(z)$.

Now, let the matrix D be defined by the equation

$$\Psi = D\,\Phi \quad , \tag{15}$$

with the condition

$$I = \int_0^h \Psi\Psi^t \, dz = \int_0^h D\,\Phi\Phi^t\, D^t \; dz \;, \tag{16}$$

which implies that

$$D\,D^t = I \;. \tag{17}$$

Substituting (15) into (12) we obtain

$$D\Phi'' + k^2 (1 + \varepsilon\, s(z))\, D\Phi = D\Phi \quad . \tag{18}$$

Inserting the value of Φ'' from (10) into (18) we obtain

$$D\,L\,\Phi + k^2 \varepsilon\; s(z)\; D\Phi = D\Phi \quad . \tag{19}$$

Multiplying by Φ^t on the right and integrating over the interval (0,h) we obtain

$$D\,L + \varepsilon\, D\,A = \Lambda\, D \;, \tag{20}$$

Where the matrix A is given by

$$A = \int_0^h k^2\; s(z) \quad \Phi\Phi^t \, dz \;. \tag{21}$$

Taking the transpose of (20), we have

$$(L + \varepsilon\, A)\; D^t = D^t \Lambda \quad . \tag{22}$$

This completes our derivation of the algebraic eigenproblem for D, the matrix of Fourier coefficients from the Fourier series expansions of the perturbed eigenfunctions, ψ_n. Notice that this actually is an algebraic eigenproblem. All of the indicated matricies are arrays of constants, nothing depends on the variable z. In the usual expression for an eigenvalue problem, D^t would be a vector and lambda would be a scalar. The above expression is a standard extention of this usual notation. Here D^t is a matrix with columns which are the eigenvectors and Λ is a diagonal matrix made up of the eigenvalues. Notice that it is important that the diagonal matrix Λ multiplies from the right.

COMPARISON TO EXPANSIONS IN POWER SERIES IN EPSILON

Recall that the parameter epsilon multiples the entire perturbation that we have taken. One of the standard methods for treating perturbation problems of this type is to expand everything in power series in the (presumably small) parameter epsilon. As explained in the introduction, we think that there are advantages in attacking the algebraic problem directly, but we want to now show that the algebraic formulation derived above agrees with the power series method if we simply expand all of our matricies as power series in epsilon.

Now, let the power series expansions in epsilon be given by

$$D = I + \varepsilon\, D^{(1)} + \varepsilon^2\, D^{(2)} + \;\; \ldots . \tag{23}$$

and

$$\Lambda = L + \varepsilon\Lambda^{(1)} + \varepsilon^2\Lambda^{(2)} + \ldots\ldots \tag{24}$$

The condition $D\,D^t = I$ implies

$$D^{(1)} + D^{(1)t} = 0\ , \tag{25}$$

$$D^{(1)}\,D^{(1)t} + D^{(2)} + D^{(2)t} = 0\ , \tag{26}$$

and

$$D^{(n)} + D^{(n)t} + \sum_{i=1}^{n-1} D^{(i)}\,D^{(n-i)t} = 0\ . \tag{27}$$

Substituting the expansions (23) and (24) into (20) and equating coefficients of like powers of ε to zero yields

$$D^{(1)}\,L + A = \Lambda^{(1)} + L\,D^{(1)}\ , \tag{28}$$

$$D^{(1)}\,A + D^{(2)}\,L = \Lambda^{(1)}\,D^{(1)} + L\,D^{(2)} + \Lambda^{(2)}\ , \tag{29}$$

and

$$D^{(n)}\,L + D^{(n-1)}\,A = L\,D^{(n)} + \Lambda^{(n)} + \sum_{i=1}^{n-1} \Lambda^{(i)}\,D^{(n-i)}\ . \tag{30}$$

Rearranging the equations (28)-(30) we get

$$D^{(1)}\,L - L\,D^{(1)} = \Lambda^{(1)} - A, \tag{31}$$

$$D^{(2)}\,L - L\,D^{(2)} = \Lambda^{(1)}\,D^{(1)} - D^{(1)}\,A + \Lambda^{(2)}\ , \tag{32}$$

and

$$D^{(n)}\,L - L\,D^{(n)} = \sum_{i=1}^{n-1} \Lambda^{(i)}\,D^{(n-i)} - D^{(n-1)}\,A + \Lambda^{(n)}\ . \tag{33}$$

Equation (25) implies that $D^{(1)}$ is antisymmetric, therefore $D^{(1)}\ L - L\ D^{(1)}$ has all zeros on the diagonal, and using (31), we obtain

$$\Lambda^{(1)} = \text{diagonal part of } (A)\ . \tag{34}$$

For the non-diagonal terms of $D^{(1)}$, we obtain

$$d_{ij}^{(1)} = \frac{-a_{ij}}{l_j - l_i} \text{ for } i \neq j\ , \tag{35}$$

and for the diagonal terms of $D^{(1)}$ we obtain

$$d_{nn}^{(1)} = 0 \;, \qquad n = 1, 2, 3, \ldots \;.$$

To solve for the second order perturbation of the eigenvalue matrix, (32) is rearranged and the diagonal part is taken:

$$\text{diagonal part of } (D^{(1)} A + D^{(2)} L - L D^{(2)} - \Lambda^{(1)} D^{(1)}) = \Lambda^{(2)} \;. \tag{36}$$

Now $D^{(2)} L - L D^{(2)}$ is antisymmetric and has zeros on the diagonal. The matrix $\Lambda^{(1)} D^{(1)}$ also has zeros on the diagonal, resulting in

$$\text{diagonal part of } (D^{(1)} A) = \Lambda^{(2)} \;. \tag{37}$$

This gives

$$\lambda_n^{(2)} = \sum_{p \neq n} d_{np}^{(1)} a_{pn} = \sum_{p \neq n} \frac{a_{pn}^2}{l_n - l_p} \;, \tag{38}$$

which is identical to (19.3.13) of Titchmarsh.

The second order perturbation of the eigenfunctions is found by first expressing $D^{(2)}$ as the sum of a symmetric matrix and an antisymmetric matrix:

$$D^{(2)} = H^{(2)} + S^{(2)} \;, \tag{39}$$

where $H^{(2)}$ is symmetric (Hermitian) and $S^{(2)}$ is antisymmetric (skew-symmetric). Equation (26) implies that

$$H^{(2)} = -\frac{1}{2} D^{(1)} D^{(1)t} \;. \tag{40}$$

This yields

$$h_{ij}^{(2)} = -\frac{1}{2} \sum_{p} d_{ip}^{(1)} d_{pj}^{(1)t} = -\frac{1}{2} \sum_{\substack{p \neq i \\ p \neq j}} \frac{a_{ip} a_{pj}}{(l_i - l_p)(l_j - l_p)} \;. \tag{41}$$

Substituting equation (39) into equation (29) and isolating the terms containing $S^{(2)}$ gives us

$$S^{(2)} L - L S^{(2)} = \Lambda^{(1)} D^{(1)} - D^{(1)} A + \Lambda H^{(2)} - H^{(2)} + \Lambda^{(2)} \;. \tag{42}$$

The left hand side of this equation is symmetric and may be obtained by finding the symmetric part of right hand side. The matrix $\Lambda H^{(2)} - H^{(2)} \Lambda$ drops out because it is antisymmetric, therefore,

$$S^{(2)} L - L S^{(2)} = \text{symmetric part of } (\Lambda^{(1)} D^{(1)} - D^{(1)} A) + \Lambda^{(2)}$$

$$= \frac{(\Lambda^{(1)} D^{(1)} - D^{(1)} A) + (\Lambda^{(1)} D^{(1)} - D^{(1)} A)^t}{2} + \Lambda^{(2)}$$

$$= \frac{(\Lambda^{(1)} + A) D^{(1)} - D^{(1)} (\Lambda^{(1)} + A)}{2} + \Lambda^{(2)} . \tag{43}$$

Using c_{ij} to denote an element of the right hand side of (43), we obtain

$$\sum_p s_{ip}^{(2)} l_{pj} - \sum_p l_{ip} s_{pj}^{(2)} = c_{ij} . \tag{44}$$

The first summation simplifies to $s_{ij}^{(2)} l_j$ and the second simplifies in a similar manner, giving us

$$s_{ij}^{(2)} (l_j - l_i) = c_{ij} . \tag{45}$$

From (39) it can be observed that

$$d_{ij}^{(2)} = h_{ij}^{(2)} + s_{ij}^{(2)} . \tag{46}$$

When i = j = n, there is no contribution from the antisymmetric part and this equation simplifies to

$$d_{nn}^{(2)} = h_{nn}^{(2)} = - \frac{1}{2} \sum_p \frac{a_{np}^2}{(l_n - l_p)^2} . \tag{47}$$

This is equivalent to (19.3.16) of Titchmarsh.

For the off-diagonal terms, $i \neq j$,

$$d_{ij}^{(2)} = h_{ij}^{(2)} + s_{ij}^{(2)} , \tag{48}$$

where it can be shown that

$$s_{ij}^{(2)} = \frac{- \sum_{p \neq j} \frac{\lambda_{ip}^{(1)} a_{pj}}{l_j - l_p} - \sum_{p \neq j} \frac{a_{ip} a_{pj}}{l_j - l_p} + \sum_{p \neq i} \frac{a_{ip} \lambda_{pj}^{(1)}}{l_p - l_i} + \sum_{p \neq i} \frac{a_{ip} a_{pj}}{l_p - l_i}}{2 (l_j - l_i)} . \tag{49}$$

This gives an expression for $d_{ij}^{(2)}$ of the form

$$d_{ij}^{(2)} = \frac{1}{2} \sum_{\substack{p \neq i \\ p \neq j}} \frac{a_{ip}\, a_{pj}[(1_j - 1_i) - (1_p - 1_i) + (1_j - 1_p)]}{(1_p - 1_i)(1_j - 1_i)(1_j - 1_p)}$$

$$+ \frac{\lambda_j^{(1)}\, a_{ij}}{(1_j - 1_i)^2} - \frac{\lambda_i^{(1)}\, a_{ij}}{(1_j - 1_i)^2}$$

$$= - \frac{1}{(1_j - 1_i)} \sum_{\substack{p \neq i \\ p \neq j}} \frac{a_{ip}\, a_{pj}}{1_p - 1_i} + \frac{1}{1_j - 1_i} \frac{\lambda_j^{(1)}\, a_{ij}}{(1_j - 1_i)} - \frac{\lambda_i^{(1)}\, a_{ij}}{(1_j - 1_i)^2} \; . \quad (50)$$

Noting that $\lambda_{nn}{}^{(1)} = a_{nn}$, the above expression simplifies to

$$d_{ij}^{(2)} = \frac{1}{(1_j - 1_i)} \sum_{p \neq i} \frac{a_{ip}\, a_{pj}}{(1_p - 1_i)} - \frac{a_{ii}\, a_{ij}}{(1_j - 1_i)^2} \; . \quad (51)$$

Interchanging the indices i and j gives us

$$d_{ji}^{(2)} = \frac{1}{(1_i - 1_j)} \sum_{p \neq j} \frac{a_{ip}\, a_{jp}}{1_p - 1_j} - \frac{a_{jj}\, a_{ij}}{(1_i - 1_j)^2} \; .$$

$$= \frac{1}{(1_i - 1_j)} \sum_{p \neq j} - \frac{a_{ip}\, a_{jp}}{1_j - 1_p} - \frac{a_{jj}\, a_{ij}}{(1_j - 1_i)^2}$$

$$= \frac{1}{1_j - 1_i} \sum_{p \neq j} \frac{a_{ip}\, a_{jp}}{1_j - 1_p} - \frac{a_{jj}\, a_{ij}}{(1_j - 1_i)^2} \; , \quad (52)$$

which is equivalent to (19.3.15) of Titchmarsh.

AN OCEAN SOUND CHANNEL EXAMPLE

This example is an idealized model of ocean sound channel with its axis at one-fifth of the depth of the ocean. Further examples can be found in Ref. 3. For the example we will treat here,

$$s(z) = 2 \left(\sin\left(\frac{5\pi}{4h} z + \frac{\pi}{4}\right) - \frac{2\sqrt{2}}{5\pi}\right) \; . \quad (53)$$

This perturbation determines the matrix A which has entries given by

$$a_{mn}= \int_0^h k^2(2\ (\sin(\frac{5\pi}{4h}\ z + \frac{\pi}{4}) - \frac{2\sqrt{2}}{5\pi}) \sqrt{\frac{2}{h}}\ \sin(\frac{(2m-1)\pi}{2h}z)\sqrt{\frac{2}{h}}\ \sin(\frac{(2n-1)\pi}{2h}z)\ dz$$

$$= \frac{2k^2}{h} \int_0^h (\sin(\frac{5\pi}{4h}\ z + \frac{\pi}{4}) - \frac{2\sqrt{2}}{5\pi})\ [\cos(\frac{(m-n)\pi}{h}\ z)\ \ -\cos(\frac{(m+n-1)\pi}{h}z)]\ dz$$

$$= \frac{2k^2}{h}\left[\frac{h\ \frac{5\pi}{4}\cos(\frac{\pi}{4})}{(\frac{5\pi}{4})^2 - [(m-n)\pi]^2} - \frac{h\ \frac{5\pi}{4}\cos(\frac{\pi}{4})}{(\frac{5\pi}{4})^2 - [(m+n-1)\pi]^2}\right]$$

$$= \frac{40k^2}{\pi}\cos(\frac{\pi}{4})\left(\frac{1}{25\ -[4(m-n)]^2} - \frac{1}{25\ -[4(m+n-1)]^2}\right). \tag{54}$$

The perturbations in the eigenvalues and eigenfunctions to first order in epsilon are the matrices $\Lambda^{(1)}$ and $D^{(1)}$ respectively. The diagonal matrix $\Lambda^{(1)}$ has entries of the form

$$\lambda_n^{(1)} = a_{nn} = \frac{40k^2}{\pi}\cos(\frac{\pi}{4})\left(\frac{1}{25} - \frac{1}{25\ -[4(2n-1)]^2}\right). \tag{55}$$

The matrix $D^{(1)}$ has zeros on the diagonal,

$$d_{nn}^{(1)} = 0\ , \tag{56}$$

and nondiagonal entries

$$d_{mn}^{(1)}= -\ \frac{160\ h^2\ k^2\ \cos(\pi/4)}{\pi^3\ [(2m-1)^2 - (2n-1)^2]}\left(\frac{1}{25\ -[4(m-n)]^2} - \frac{1}{25\ -[4(m+n-1)]^2}\right). \tag{57}$$

The perturbations of the eigenvalues to second order in epsilon are given by the diagonal matrix $\Lambda^{(2)}$ which has entries of the form:

$$\lambda_n^{(2)} = \sum_{p\neq n} d_{np}^{(1)}\ a_{pn}$$

$$= -\ \frac{40k^2}{\pi}\cos(\frac{\pi}{4}) \sum_{p\neq n} d_{np}^{(1)}\left(\frac{1}{25\ -[4(n-p)]^2} - \frac{1}{25\ -[4(n+p-1)]^2}\right)$$

$$= \frac{4h^2k^2}{\pi^2}\left(\frac{40\ \cos(\pi/4)}{\pi}\right)^2 \times$$

$$\sum_{p\neq n} \frac{1}{(2n-1)^2 - (2p-1)^2}\left(\frac{1}{25\ -[4(n-p)]^2} - \frac{1}{25\ -[4(n+p-1)]^2}\right)^2. \tag{58}$$

The second order perturbation of the eigenfunctions are determined by the entries of $D^{(2)}$ which has diagonal elements of the form

$$d_{nn}^{(2)} = -\frac{1}{2}\sum_{p \neq n} \frac{a_{np}^2}{(l_n - l_p)^2}$$

$$= -\frac{1}{2}\left(\frac{4h^2k^2}{\pi^2}\right)^2\left(\frac{40\cos(\pi/4)}{\pi}\right)^2 \times$$

$$\sum_{p \neq n}\left(\frac{1}{(2p-1)^2 - (2n-1)^2}\right)^2\left(\frac{1}{25 - [4(p-n)]^2} - \frac{1}{25 - [4(p+n-1)]^2}\right)^2, \quad (59)$$

while the non diagonal elements have the form

$$d_{mn}^{(2)} = \left(\frac{4h^2k^2}{\pi}\right)^2\left(\frac{40\cos(\pi/4)}{\pi}\right)^2 \frac{1}{(2m-1)^2 - (2n-1)^2}\left\{\sum_p \frac{1}{(2m-1)^2 - (2p-1)^2} \times\right.$$

$$\left[\left(\frac{1}{25 - [4(m-p)]^2} - \frac{1}{25 - [4(m+p-1)]^2}\right)\left(\frac{1}{25 - [4(n-p)]^2} - \frac{1}{25 - [4(n+p-1)]^2}\right)\right]$$

$$\left. - \frac{1}{(2m-1)^2 - (2n-1)^2}\left(\frac{1}{25} - \frac{1}{25 - [4(2m-1)]^2}\right)\left(\frac{1}{25 - [4(m-n)]^2} - \frac{1}{25 - [4(m+n-1]^2}\right)\right\}.$$

GRAPHS OF THE NORMAL MODES

Shown below are graphs of the perturbed and unperturbed normal modes for the example.

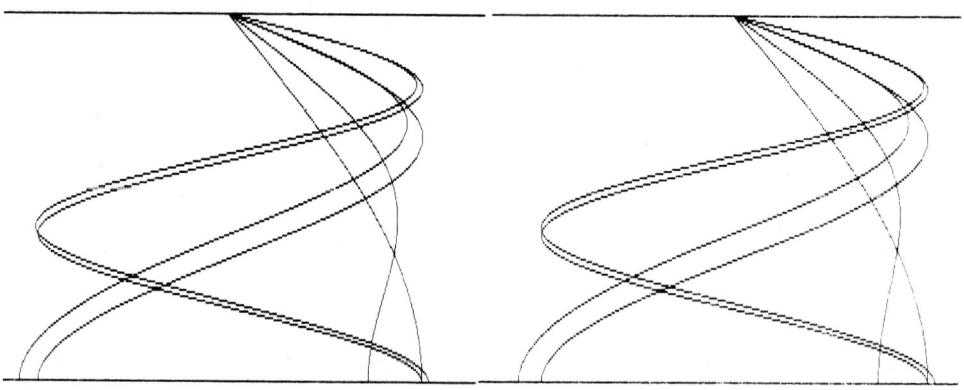

Fig. 1. Graphs of both perturbed and unperturbed normal modes comparing results to first order in ε, on the left, and second order in ε, on the right. For these graphs $\varepsilon = .001$, which produces a weak effect. The perturbed normal modes are here approximated by ten terms of a Fourier Series.

SUMMARY

We assumed that the bottom of the ocean is eventually underlaid with a rigid subbottom. We exploited the fact that the sound speed within the ocean and its bottom varies by only a few percent. We thought of this variation, which we assumed is a function of only the depth, as a small perturbation of an otherwise uniform sound speed.

Since the normal modes in an ocean with a uniform sound speed and a rigid subbottom are already known in terms of sine functions, we sought an expansion of each of the normal modes of the perturbed ocean as an infinite series of sine functions, that is, as a Fourier series. We obtained an algebraic eigenproblem for the infinite matrix of the Fourier coefficients that we need: infinitely many Fourier coefficients for each of infinitely many normal modes.

This algebraic formulation is not standard. The usual approach is to expand the normal modes as power series in the parameter epsilon that measures the smallness of the perturbation, then to further expand each of these corrections in a Fourier series. The main effort of this paper was expanding everything in the algebraic system that determines the Fourier coefficients as power series in epsilon and confirming that the results of the usual approach were reobtained.

REFERENCES

1. D. J. Ahluwalia and J. B. Keller, Exact and asympotic representations of the sound field in a stratified ocean, in Wave Propagation and Underwater Acoustics, J. B. Keller and J. S. Papadakis, eds, Lecture Notes in Physics, vol. 70, Springer Verlag, New York, 1977.

2. E. C. Titchmarsh, Eigenfunction Expansions Associated with Second-Order Differential Equations, Part II, Oxford University Press, Oxford, England, 1958.

3. D. H. Wood, M. Duston, and G. Verma, Changes in eigenvalues due to bottom interaction using perturbation theory, Proceedings of the 11th IMACS World Congress, Oslo, Norway, August 5 - 9, 1985 (to appear).

4. G. A. Merchant and T. W. Parks, Efficient solution of a Toeplitz-plus-Hankel coefficient matrix system of equations, IEEE Trans. Acoustics, Speech, and Signal Processing, Vol. ASSP-30, No. 1, 1982, pp 40-44.

5. H. Lev-Ari, Cholesky factorization of structured matrices, Second SIAM Conference on Applied Linear Algebra, Raliegh, North Carolina, April 29 - May 2, 1985.

6. P. J. Davis, Circulant Matrices, John Wiley and Sons, New York, 1979.

7. U. Grenander and G. Szego, Toeplitz Forms and Their Applications, Chelsea Publishing Company, New York, 1984.

8. D. Gottlieb and S. A. Orszag, Numerical Analysis of Spectral Methods: Theory and Applications, CBMS-NSF Regional Conference Series in Applied Mathematics, vol. 26, Society for Industrial and Applied Mathematics, Philadelphia, 1977.

GAUSSIAN BEAMS AND 3-D BOTTOM INTERACTING ACOUSTIC SYSTEMS

Homer P. Bucker and Michael B. Porter*

U.S. Naval Ocean Systems Center, Code 541

San Diego, California 92152

ABSTRACT

The analysis of the performance of current array systems requires accurate propagation modeling. In addition, future signal processing algorithms may incorporate a propagation model in order to obtain improved target tracking. The acoustic field received at an array of sensors located on or near the ocean bottom is strongly affected by the local bathymetry and by the physical properties of the ocean subbottom. At very low frequencies, the acoustic field may even be affected by the physical properties of the basement underlying the top sediment layers. This paper describes two techniques based on Gaussian beam tracing for computing the acoustic field in such an environment. The first method employs empirically derived formulas governing the spread of the beams and has the advantage of great simplicity. In the second method the beam curvature and width are obtained formally from an ordinary differential equation along the central ray. This latter method has recently received a lot of attention in the seismological community. Both methods are free of the difficulties at caustics and in shadow zones which afflict standard ray tracing algorithms. Comparisons are presented between the standard ray tracing, simplified beam tracing, formal beam tracing and the exact solution for a difficult negative sound speed gradient problem previously examined by Pedersen and Gordon. Finally, a detection system is proposed that employs Gaussian beam tracing convolved with target tracking.

---

* Now at the Naval Research Laboratory, Washington, D.C. 20375.

INTRODUCTION AND REVIEW

Scientific investigations in bottom interacting acoustic systems dates back to the 1930s. However, a researcher today can begin with the shallow water experiments of Worzel and Ewing [1] and the theoretical analysis of a shallow water sound field in terms of normal modes and a branch line integral by Pekeris [2].

Interest in bottom effects was stimulated in the late 1950s with the introduction of bottom bounce sonar systems. Large experimental programs (Marine Geophysical Survey in U.S. and NAVADO in U.K.) were undertaken to define bottom loss vs. grazing angles at different frequencies and areas of the ocean. Theoretical analysis [3] of the problem showed strong interference effects due to reflections from various sediment layers. Also, it was very difficult to determine bottom losses at low grazing angles because of sound paths that refract back into water because of strong gradients in the sediment. It eventually became clear that it is better to describe bottom interaction by a geoacoustic model that details the true thickness and properties of the sediment and rock layers in the real sea floor. The general subject of geoacoustic modeling (with compilation of many measured values) was recently summarized by Hamilton [4]. Additional measured properties of sea floor sediments are listed in Hamilton and Bachman [5] and Hamilton et al. [6].

In the late 1960s and 1970s, many of the loose ends of range-independent bottom interacting propagation had been cleaned up and range-dependent propagation models had been introduced. Special note might be given to shallow water work of Hastrap [7] and Akal [8] at SACLANTCEN. Also, the work (mostly theoretical) by Vidmar [9-12], and others, at the Applied Research Laboratory, University of Texas, has delineated many effects of bottom interaction associated with shear waves in the sediment layers. Work on range dependent models with bottom interaction is represented by papers by Jensen and Kuperman [13] and McDaniel and Lee [14]. An equivalent bottom algorithm was developed by Bucker [15] so the very fast split-step parabolic equation of Tappert [16] can be used in cases with strong bottom interaction.

Finally, any review should include the proceedings of two workshops on bottom interaction. They are the Physics of Sound in Marine Sediments, ed. by Hampton [17] and Bottom Interacting Ocean Acoustics [18], ed. by Kuperman and Jensen.

SOUND PROPAGATION WITH 3-D BATHYMETRY

There are many ways to calculate the sound field with bottom interaction associated with 3-D bathymetry: coupled normal modes, adiabatic mode theory, 3-D parabolic equation models, finite difference or finite element algorithms, etc. In this section the Gaussian beam method will be presented because it generates accurate solutions using an algorithm of moderate complexity and reasonable computation times.

Beam tracing has recently received a great deal of attention for solving problems in seismic wave propagation. Many articles on beam tracing may be found in the October 1984 issue of the Geophysical Journal of the R. astr. Soc. In beam tracing, the radiated field of a source propagates as a fan of Gaussian beams. These beams move through the medium with the central ray of the beam following ray acoustics.

Each beam will have some influence at any receiver point. However, the influence falls off rapidly as the receiver becomes distant from the central ray of each beam, so only beams that pass "near" the receiver will have a noticeable affect.

Standard ray tracing often gives reasonable results, even at low frequencies, in bottom interaction problems. However, there are certain artifacts such as perfect shadows and caustics. These problems are greatly compounded when reflections from a three-dimensional ocean bottom must be considered.

The recent interest in beam tracing stems from the fact that it requires only slight modifications to standard ray tracing codes and because the field solutions are free of singularities at caustics. The modifications to codes that use dynamic ray tracing are so simple as to suggest a close relationship between ray tracing and beam tracing. In fact Deschamps [19] and Felsen [20] show that in some cases bounded beam equations can be obtained by displacing the rays into complex space and using their projection in real space. In the following sections the Gaussian beam solution for a case of unnaturally strong downward refraction (discussed by Pedersen and Gordon [21]) will be compared to standard ray theory and normal mode solutions.

SIMPLE GAUSSIAN BEAMS

An older semi-empirical method, simple Gaussian beams (SGB), used with the Ray-wave 2-D propagation model [22] will be presented first.

Consider one of a fan of beams radiating from a point source as shown in Fig. 1. The beam is traced from the source at 0 to a point P which is in the vicinity of a receiver at R. The contribution to the intensity at R due to the beam is

$$I = A \exp(-a\theta^2)/s^2, \tag{1}$$

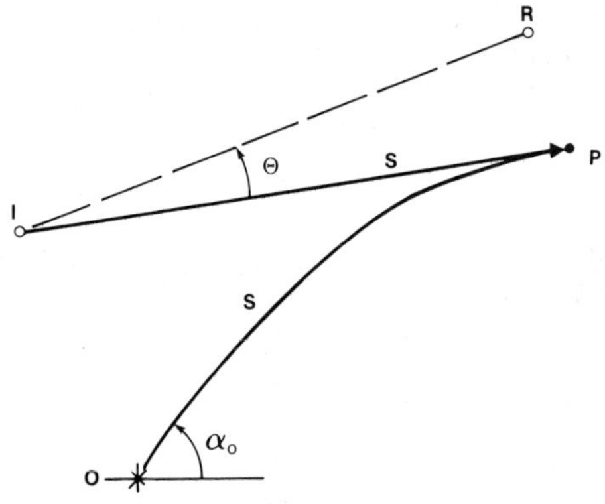

Figure 1. Simple Gaussian Beam.

where A represents the initial strength of the beam, s is the distance along the beam from 0 to P, θ is the angle between $\overline{IP}$ and $\overline{IR}$ (θ is not usually in the plane of the paper in the 3-D case), and "a" determines the fall-off of influence of the beam and is a function of frequency. The phase of the field at R is calculated by extension of the phase of the beam at P assuming a spherical wave front centered at image point I. Point I is located by extending a line, tangent to the beam at P, back a distance s. When the beam interacts with a boundary, there are usual changes to the strength, A, and to the phase.

Because the beam width parameter "a" does not change when the simple beam passes through the refracting ocean or sediments, it is necessary to first group together beam contributions of the same type. Then a complex pressure is calculated for each group and these are added to give the total field. For example, assume that there are contributions from three direct (D) beams with intensity I_1, I_2, I_3 and phase ϕ_1, ϕ_2, ϕ_3 and two surface reflecting (SR) beams (I_4, I_5) with phase ϕ_4, ϕ_5. Note that $\phi_1 \approx \phi_2 \approx \phi_3$ and $\phi_4 \approx \phi_5$. First calculate "complex group intensities"

$$I_D = \sum_{m=1}^{3} I_m \exp(i\phi_m) \text{ and } I_{SR} = \sum_{m=4}^{5} I_m \exp(i\phi_m). \tag{2}$$

Then the total complex pressure is

$$p = |I_D|^{\frac{1}{2}} \exp[i \arg(I_D)] + |I_{SR}|^{\frac{1}{2}} \exp[i \arg(I_{SR})]. \tag{3}$$

To determine the type of a beam, a group identification number (ID) is calculated as the beam propagates. An ID = 0 indicates that the beam has not yet been affected by any of the following: surface reflection, turn-over, turn-under, bottom reflection, touching a caustic. When the beam has a surface reflection, the number 2 is concatenated with the previous ID (a leading 0 can be omitted). A number 4, 6, or 8 is added for a turn-over, turn-under, or bottom reflection. When the central ray touches a caustic, the last number is increased by 1. For example, a ray ID = 294 would indicate that the beam had a surface reflection, a bottom reflection, touched a caustic, and then a turn-over. When beam splitting (e.g., part of a beam may reflect from an interface with part traveling through) occurs, the calculation must be done in several passes with only one type of action (e.g., the reflection) taking place during each pass.

GAUSSIAN BEAM EQUATIONS

A readable development of Gaussian beam (GB) equations is given by Červený et al. [23]. The GB equations are developed relative to a central ray that obeys standard ray equations. Consider first the 2-D Cartesian coordinate system (x, z) and let s represent a distance measured along the central ray.

The GB curvature and beam width are derived from auxilliary variables p(s) and q(s) which are obtained by integrating a pair or ordinary differential equations along the central ray,

$$dq/ds = c(s)p(s), \text{ and } dp/ds = -c_{2N}q(s)/c^2(s). \quad (4)$$

Here $c(s)$ is the sound speed at a point s units along the ray path and c_{2N} is the second derivative of the sound speed normal to the central ray.

In order to solve these equations, the initial conditions $p(0)$ and $q(0)$ must be specified. The final solution depends only on the ratio $p(s)/q(s)$. Therefore, the beam is characterized by a single complex constant $\Gamma = p(0)/q(0)$. This beam constant controls both the initial beam width and curvature and may be chosen in various ways. The optimal procedure for selecting Γ and other parameters of the beam solution (no. of beams, angular spread, etc.) is a matter of current research.

Once p and q are determined, the field due to the beam is given by

$$u(s,\Gamma) = A[(c(s)\, q(0))/(c(0)q(s))]^{\frac{1}{2}}$$

$$\cdot \exp\{i\omega[\tau + p(s)\rho^2/(2q(s))]\}, \quad (8)$$

where A is a source constant to be determined, $q(0)$ and $c(0)$ are values of q and c at the source, ρ is the normal distance from the central ray to the field point, ω is the angular frequency of the source and $\tau(s)$ is the phase delay (travel time) which satisfies the equation

$$d\tau/ds = 1/c(s) \quad (9)$$

EXPANSION OF POINT SOURCE INTO BEAMS

Some sources naturally give rise to Gaussian beams in a particular coordinate system. In other cases the appropriate superposition of beams must be determined. A canonical case for underwater acoustics is a point source in cylindrical coordinates. One approach is to match the high frequency asymptotic field of the Gaussian beam representation to the exact solution in a homogeneous medium. The Gaussian beam representation is [23]

$$u(x, z) = \int A(\alpha)u_\alpha(s,\rho)d\alpha, \quad (10)$$

where

$$u_\alpha = [q(0)c(s)/(q(s)c(0))]^{\frac{1}{2}}$$

$$\cdot \exp\{i\omega[\tau + p(s)\rho^2/(2q(s))]\}, \quad (11)$$

and α is the grazing angle of the beam at the source.

In the high-frequency limit ($\omega\to\infty$) the method of stationary phase yields the asymptotic representation.

$$u(x, z) \sim A(\alpha_o)[2\pi c(0)/(\omega r)]^{\frac{1}{2}} \exp[i(\omega r/c(0) - \pi/4)] \tag{12}$$

In the above, x is horizontal range, z is depth, and r is the slant distance between the source point and the receiver point. It follows directly that for point-source in two dimensions (line source in 3-D) that

$$A(\alpha_o) = 1/\pi \qquad \text{(point source in 2-D)} \tag{13}$$

and that, less formally,

$$A(\alpha_o) = [\omega/(2\pi r c(0))]^{\frac{1}{2}} \exp(i\pi/4) \tag{14}$$

for a point source in 3-D coordinate system with axial symmetry.

REFLECTION AND TRANSMISSION AT INTERFACES

The reflection and transmission of beams at interfaces has been extensively discussed in the literature. Reflection from an elastic medium is complicated and a matter of current research. Cervený and Pšenčík [25] provide complete equations for the case of a first-order curved acoustic interface but their results do not show the expected beam displacement. Felsen [20] discusses this point and suggests that the complex angle of the associated complex ray should be used in the equations. Until this matter is resolved, displacement of the central beam, as discussed by Tindle [26], will probably yield acceptable results.

COMPARISON OF ALGORITHMS

It is reasonably clear that there is a smooth extension of the Gaussian beam algorithms from two to three dimensions. Therefore, a useful check on the algorithms is a two dimensional problem with very strong gradient effects. The profile used has a single negative gradient layer ($c(z)^{-2} = a + bz$) with sound speed equal to 1677.33 yds/sec (1 yd = 0.9144 m) at the surface and equal to 1068.33 yds/sec at a depth of 1000 yds. Obviously, this gradient is much stronger than any that occurs naturally over a water layer with a few meters, or greater, thickness. This profile was studied by Pedersen and Gordon [21] in a comparison of ray and normal mode solutions.

Fig. 2 is a ray plot for a frequency = 2 kHz and a source depth at 66.7 yds. The level will be calculated for a receiver at the same depth and at various ranges. Fig. 3 is a comparison of the ray theory solution and a SGB solution with 0.2° beam spacing and a = 1.E+6, and Fig. 4 is a comparison of the normal mode solution (with 100 modes) and the SGB solution. At ranges less than ~570 yds the mode solution is not correct because of an insufficient number of modes. It is clear that for this extremely difficult test the SGB algorithm is the best of the three methods overall giving essentially exact results from 0 to 760 yds with a reasonable fit to the mode theory beyond 760 yds. Fig. 5 is a plot of the GB algorithm with a 0.225° beam spacing and the normal mode solution.

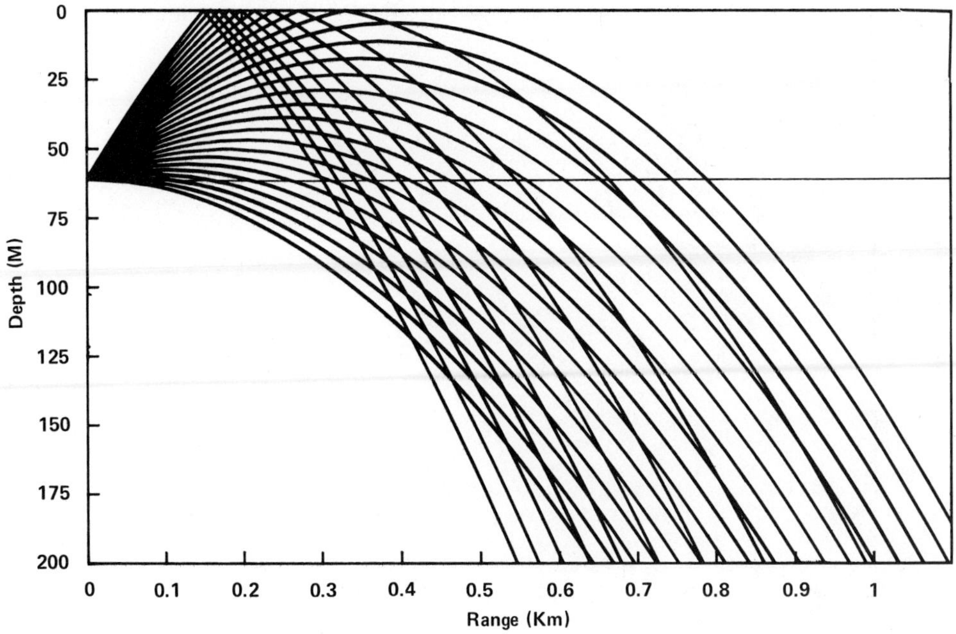

Figure 2. Ray Plot for Shallow Source.

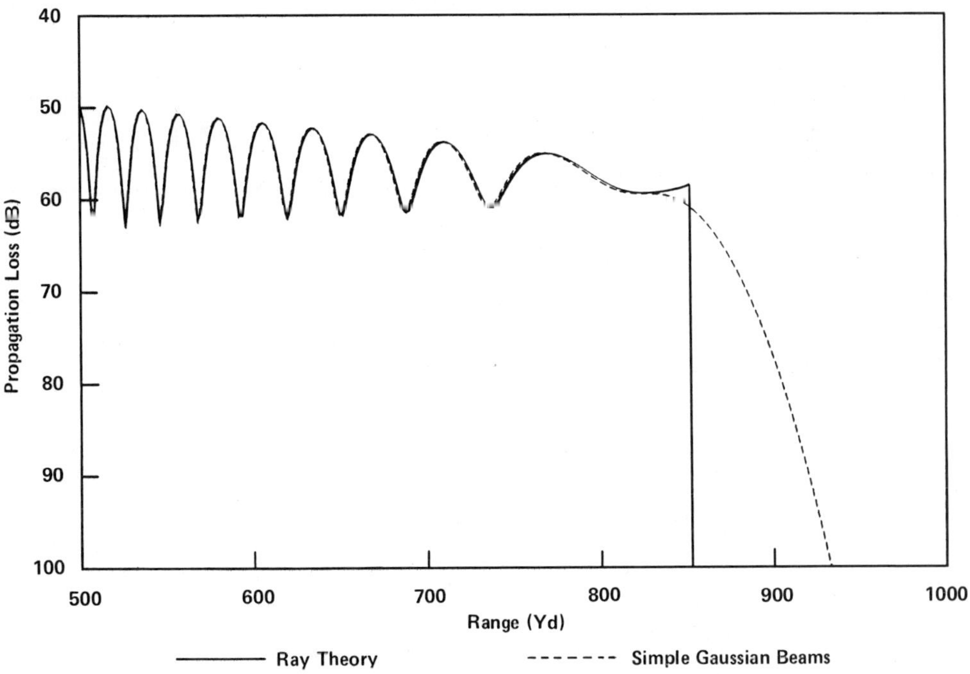

Figure 3. Comparison of Ray Theory and SGB Algorithm.

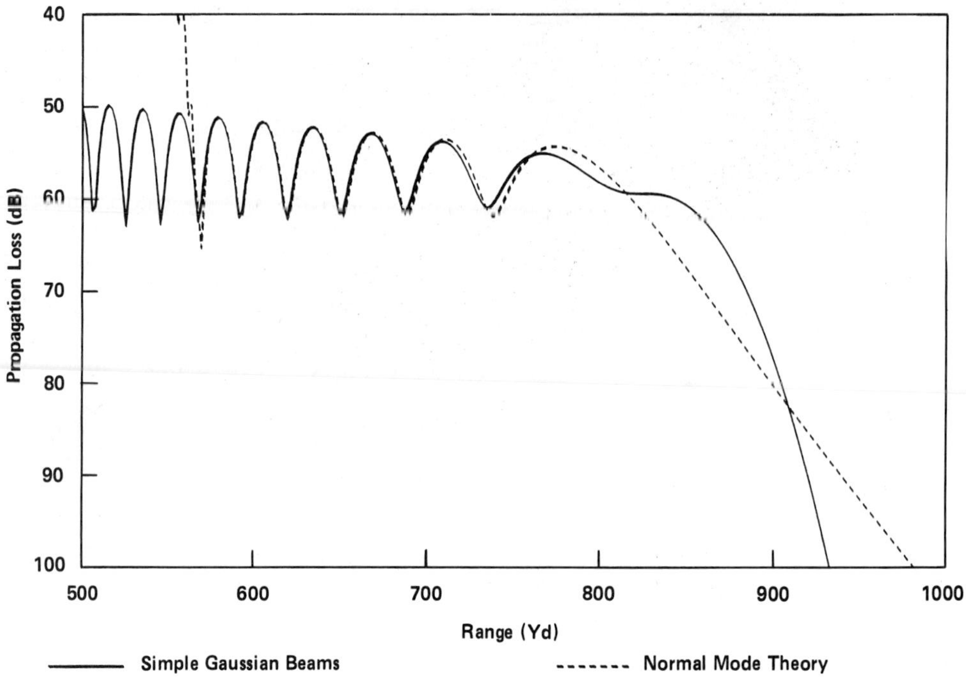

Figure 4. Comparison of Normal Mode Algorithm and SGB Algorithm.

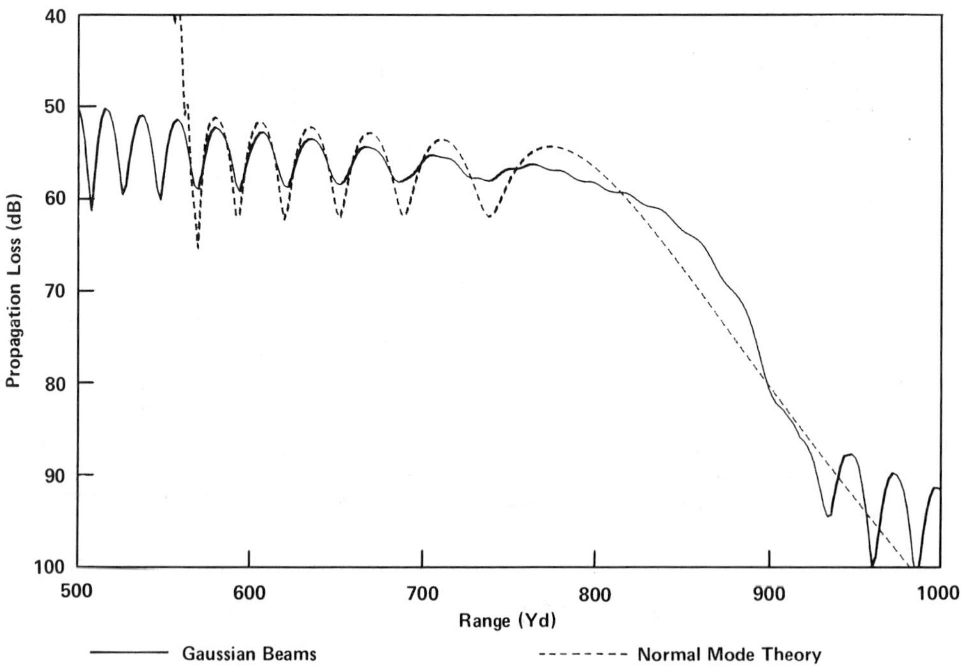

Figure 5. Comparison of Normal Mode Algorithm and GB Algorithm.

A second example is taken from Pedersen and Gordon where the source is at 1000 yds and the receiver at 800 yds. A range interval from 3100 to 3170 yds is examined. Over the interval the situation changes from interference between a downward refracted and a surface reflected ray, to interference between two downward refracting rays (one of which has touched a caustic), to a caustic (at 3158.8 yd), and to the shadow zone beyond the caustic. For this deep source it is not possible to examine the normal mode solution because of numerical problems. Fig. 6 is a comparison of a SGB solution with a 0.15° beam spacing and the ray theory solution. The small beam spacing is a consequence of the extreme gradient. Typically, beam spacings of 1° or so are suitable. Fig. 7 is a comparison of the SGB algorithm and the GB algorithm with a beam spacing of 0.18°.

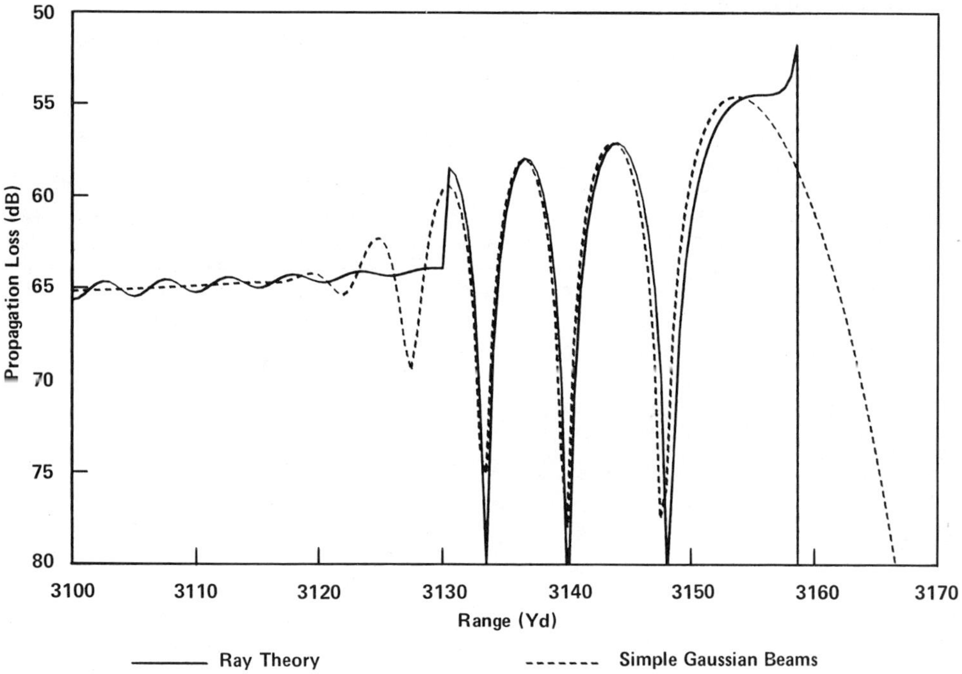

Figure 6. Comparison of Ray Theory and SGB Algorithm for Deep Source.

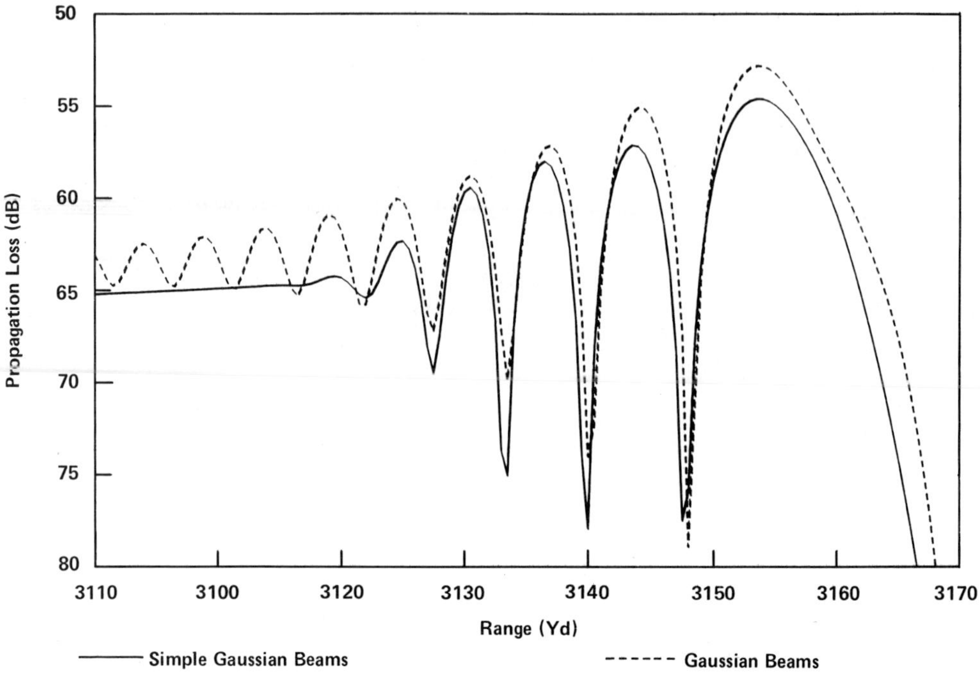

Figure 7. Comparison of SGB Algorithm and GB Algorithm for Deep Source.

GAUSSIAN BEAMS AND ACOUSTIC ARRAYS

To illustrate how bottom interaction affects an acoustic array, the bottom bathymetry shown in Fig. 8 is modeled. Fig. 9 is a blow-up of the bathymetry showing an 8 sensor line array (with 15 m separation) lying on the bottom. The 3-D Gaussian beam algorithm is used to calculate the field near the array and a normal mode (or parabolic equation) algorithm extends the field from the 3-D solution to a distant point.

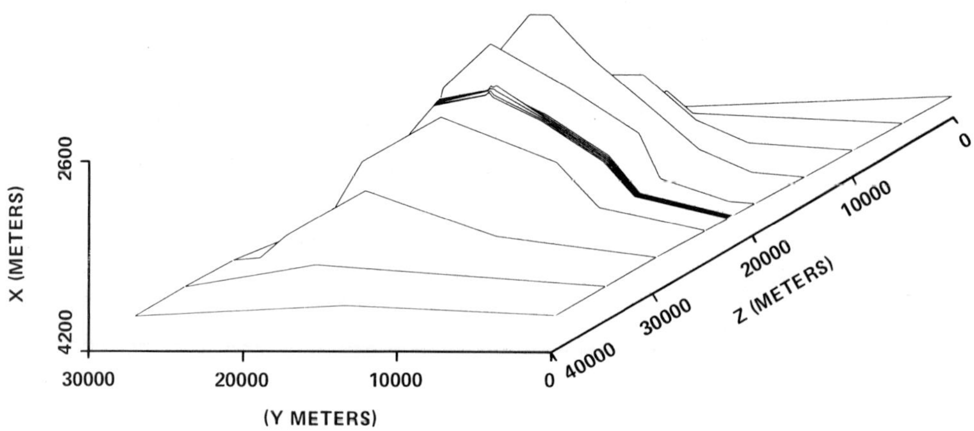

Figure 8. Model of Bottom Topology.

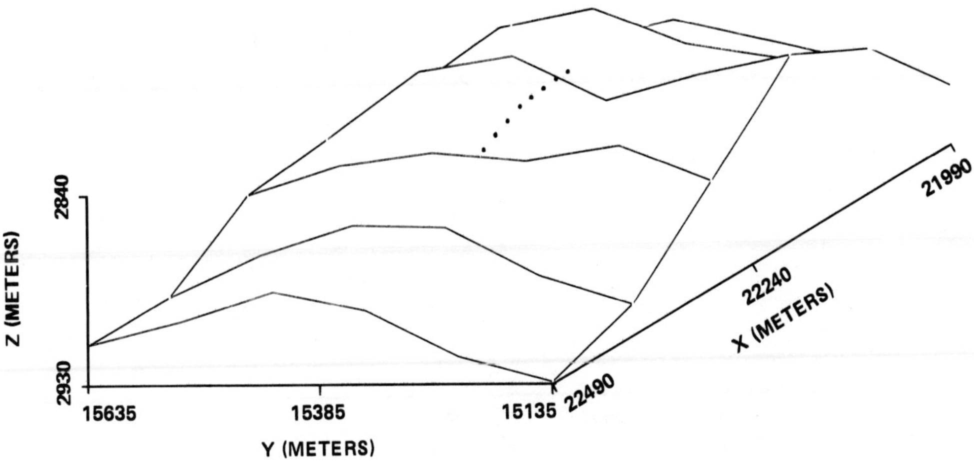

Figure 9. Close-up of Bottom Topology with Line Array.

In standard beam forming, a beam in look direction $\hat{\phi}$ is the dot product of the received signal $\vec{F}(F_1, F_2, \cdots F_N)$ and the expected signal $\vec{f_{\hat{\phi}}}(f_1, f_2, \cdots f_N)$, where $f_j = \exp[i(kx_j \cos\hat{\phi})]$, and x_j = location of the jth sensor in a line array. The beam output is squared and time averaged to stabilize the direction estimate $D_{\hat{\phi}}$. That is,

$$B_{\hat{\phi}} = \Sigma_j \; f_j^* \; F_j \text{ and } D_{\hat{\phi}} = \langle B \; B^* \rangle, \tag{15}$$

where * denotes complex conjugate and <·> a time average.

Fig. 10 shows the direction estimate $D_{\hat{\phi}}$ for a signal coming from 90° as a function of the range to the target. The actual range is 94 km and the source frequency is 46 Hz. The array is seen to have good performance, in general.

In matched field processing [27], the reference functions (f_j in Eq. 15) are calculated for an assumed range, bearing, and depth of the target. In this case, D becomes $D_{\hat{\phi},r,z}$ where r and z are the range and depth of the target. Fig. 11 is a plot of D as a function of $\hat{\phi}$ and r with z held fixed at the depth (91 m) of the target. Fig. 11 shows, as one might have deduced from the stable beam patterns of Fig. 10, that the array has very little range resolution. Range resolution can be improved by adding one or more sets of sensors above the set on the bottom. That is, each bottom sensor is replaced by a vertical line of sensors forming a billboard array. Fig. 12 shows the value of $D_{\hat{\phi},r,z}$ as a function of r for $\hat{\phi}$ and z equal to the bearing and depth of the target. The different curves correspond to different numbers of sensors (with 2 wavelength spacing) on the eight vertical lines.

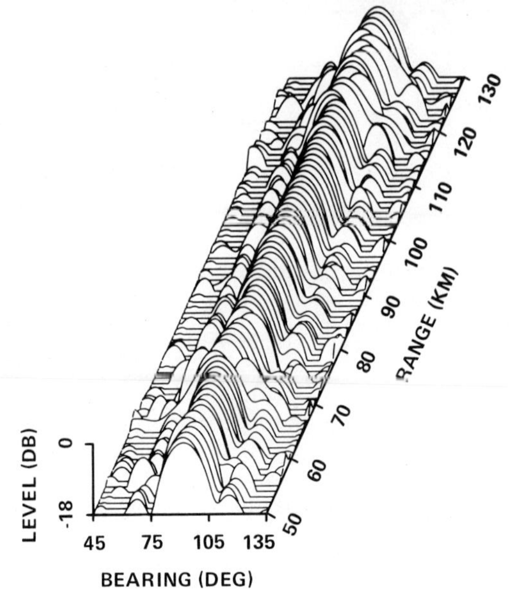

Figure 10. Direction Estimates as a Function of Range.

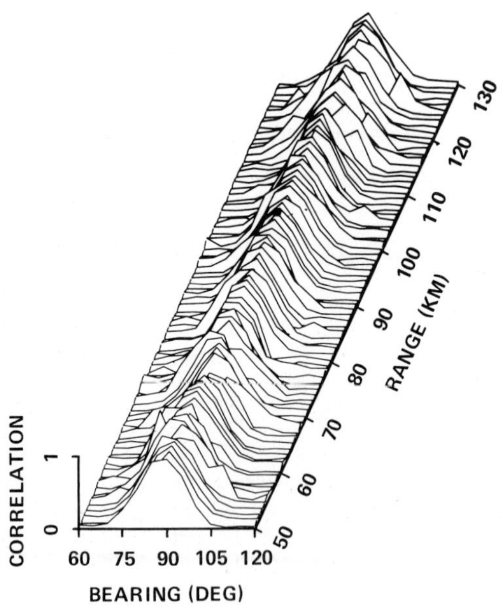

Figure 11. $D_{\hat{\varphi},r,z}$ as a Function of $\hat{\Phi}$ and r,z = const.

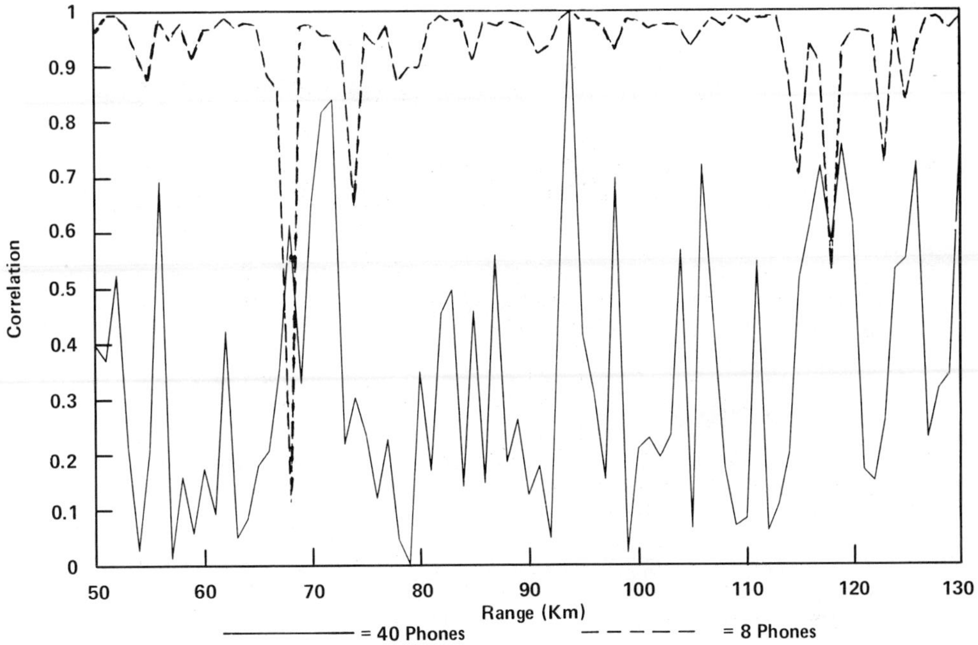

Figure 12. $D_{\hat{\Phi},r,z}$ as a Function of r for Different Number of Sensors on 8 Vertical Lines.

The above procedure applies to a one time look at the signal field. Significant processing gains should result from tracking the peaks of $D_{\hat{\phi},r,z}$ as a function of time while incorporating known characteristics of possible targets. This is called SMART processing (an acronym for Signals + Model of Acoustic Response + Tracking). There are myriad ways to organize a SMART processor, but the best performance will probably come from extending the time span of the coherent detection function (D) as long as possible.

CONCLUSION

Although the algorithms discussed in this paper are simple, they present a considerable computation burden when complex bathymetry is treated in three-dimensions. Certainly, at some frequencies, depending upon the location, there is too much scattering for the algorithms to work. Also it may not be feasible to gather the relevant environmental data. On the other hand, the ocean bottom is essentially constant, unlike the changing water column. Thus, bottom parameters can be collected over a long period of time and the propagation models can be improved by comparisons with experimental data.

For a long time, acousticians have wanted to incorporate their propagation models into signal processing. With the advent of efficient models and low cost computing machines, this may soon be done.

ACKNOWLEDGEMENTS

Support for this effort was provided by the NAVOCEANSYSCEN Independent Research Program and the Bottom Interaction Program at NORDA. Calculations for the matched field figures were done by Philip Schey, CSC Corp.

REFERENCES

1. J. L. Worzel and M. Ewing, Explosion sounds in shallow water, Section I, pp. 1-53 in Geological Society of America, Memoir 27, (1948).

2. C. L. Pekeris, Theory of propagation of explosive sound in shallow water, Sect. II, pp. 1-117 in Geological Society of America, Memoir 27 (1948).

3. H. P. Bucker, J. A. Whitney, G. S. Yee and R. S. Gardner, "Reflection of low frequency sonar signals from a smooth ocean bottom," J. Acoustic Soc. America, No. 37, pp. 1037-1051 (1965).

4. E. L. Hamilton, "Geoacoustic modeling of the sea floor," J. Acoustic. Soc. Am., No. 68, pp. 1313-1340 (1980).

5. E. L. Hamilton and R. T. Bachman, "Sound velocity and related properties of marine sediments," J. Acoustic. Soc. Am., No. 72, pp. 1891-1904 (1982).

6. E. L. Hamilton, R. T. Bachman, R. T. Berger, W. R. Johnson and L. A. Mayer, "Acoustic and related properties of calcareous deep-sea sediments," J. Sedimentary Petrology, No. 52, pp. 733-753 (1982).

7. O. F. Hastrup, "Some bottom-reflection loss anomalies near grazing and their affect on propagation in shallow water," pp. 135-152 in Ref. 18.

8. T. Akal, "Sea floor affects on shallow-water acoustic propagation," pp. 557-575 in Ref. 18.

9. K. E. Hawker, "The existence of Stoneley waves as a loss mechanism in plane wave reflection problems," J. Acoustic Soc. Am., No. 65, pp. 622-686 (1979).

10. P. J. Vidmar, "Ray path analysis of sediment shear wave affects in bottom reflection loss," J. Acoust. Soc. Am., No. 68, pp. 639-648 (1980).

11. H. Holthusen and P. J. Vidmar, "The affect of near-surface layering on the reflectivity of the ocean bottom," J. Acoust. Soc. Am., No. 72, pp. 226-234 (1982).

12. R. A. Koch, P. J. Vidmar, and J. B. Lindberg, "Normal mode identification for impedence boundary conditions," J. Acoust. Soc. Am., No. 73, pp. 1567-1570 (1983).

13. F. B. Jensen and W. A. Kuperman, "Sound propagation in a wedge shaped ocean with a penetrable bottom," J. Acoust. Soc. Am., No. 67, pp. 1564-1566 (1980).

14. S. T. McDaniel and D. Lee, "A finite-difference treatment of interface conditions for the parabolic equation," J. Acoust. Soc. Am., No. 71, pp. 855-858 (1982).

15. H. P. Bucker, "An equivalent bottom for use with the split-stop algorithm," J. Acoust. Soc. Am., No. 73, pp. 486-491 (1983).

16. F. D. Tappert, "The parabolic equation method," Sect. V in Lecture Notes in Physicis, No. 70, Ed. by J. B. Keller and J. S. Papadakis, Springer Verlag, Berlin, 1977.

17. Physics of Sound in Marine Sediments, ed. by Loyd Hampton, Plenum Press, New York, 1974.

18. Bottom Interacting Acoustics, ed. by W. Kuperman and F. Jensen, Plenum Press, New York, 1980.

19. G. A. Deschamps, "Gaussian beams as a bundle of complex rays," Electronics Letters, No 7, pp. 684-685 (1971).

20. L. B. Felsen, "Geometrical theory of diffraction, evanescent waves, complex rays and Gaussian beams," Geophys. J. R. astr. Soc., No. 79, pp. 77-88 (1984).

21. M. A. Pedersen and D. F. Gordon, "Normal-mode and ray theory applied to underwater acoustic conditions of extreme downward refraction," J. Acoust. Soc. Am., No. 51, pp. 323-368 (1971).

22. H. P. Bucker, "Some comments on ray theory with examples from current NUC ray trace models," pp. 32-36 in SACLANT Conf. Proceed., No. 5, (1971).

23. V. Červený, M. M. Popov, and I. Pšenčík, "Computation of wave fields in inhomogeneous media-Gaussian beam approach," Geophys. J. R. astr. Soc., No. 70, pp. 109-128 (1982).

24. L. Klimes, "Expansion of a high frequency time-harmonic wavefield given on an initial surface into Gaussian beams," Geophys. J. R. astr. Soc., No. 79, pp. 105-118 (1984).

25. V. Červený and I. Pšenčík, "Gaussian beams in elastic 2-D laterally varying structures," Geophys. J. R. astr. Soc., No. 78, pp. 65-91 (1984).

26. C. T. Tindle, "Ray calculations with beam displacements," J. Acoust. Soc. Am., No. 73, pp. 1581-1586 (1983).

27. H. P. Bucker, "Use of calculated fields and matched-field detection to locate sound sources in shallow water," J. Acoust. Soc. Am., No. 59, pp. 368-373 (1976).

DEVELOPMENT OF A PARABOLIC APPROXIMATION FOR THE COMPUTATION OF PROPAGATION LOSS IN A RANGE-DEPENDENT ENVIRONMENT

B. Grandvuillemin

GERDSM
Le Brusc, France

ABSTRACT

We propose to solve the Helmholtz equation by means of the parabolic approximation in a two-dimensional environmental propagation. We make clear the solution for a quadratic development of the propagation operator which allows us to take into account a larger angular aperture of the source. The numerical solution is given by an algorithm of finite differences using the Crank-Nicholson method. The problem of rectilinear, horizontal and oblique interfaces between two fluid media is solved. The problem of the medium where the sound velocity profile changes with range is treated by conditions of continuity between two media with the given sound speed profiles. The first results of numerical simulations executed on the VAX 11/780 and the array processor FPS 164 are presented. They show the ability of the model to describe the importance of environmental variation on propagation.

INTRODUCTION

The propagation of a monochromatic compression wave in an elastic medium is governed by the Helmholtz equation. When the environment is invariant with the range, approximate solutions are obtained by means of rays or normal mode. However, in the real case, these solutions are invalid. At a given depth, speed varies with range. This phenomenon is illustrated by currents or eddies. Moreover the utilization of very low frequencies to increase the range of sonars take into account the variability of the environment such as the transition from shallow to deep water. In this setting we use the method of the parabolic approximation which is related to the paraxial approximation of geometric optics. The field surrounds one determinate direction, the axial direction; a parabolic equation model allows us to analyze the variations along this direction. Physically the paraxial approximation omits the environmental backscattered waves. Only those which surround the axial directions are considered. The principal interest comes from the possibility of treating the propagation as a Cauchy problem (a problem with any initial conditions) with an easy numerical solution. The parabolic approximation was introduced for the first time by Fock and Leontovich in electro-magnetism. Tappert [1] introduced it in underwater acoustics in the 1970 s. His model (PAREQ) consists of executing an FFT of the field followed by the utilization of a propagation operator and an FFT reconstruction of the field. In the early 1980 s Lee, Botseas, and Papadakis [2], McDaniel and Lee [3] and Lee and McDaniel [4] proposed another algorithm, implicit finite difference (IFD). The numerical solution is found on finite differences with an algorithm of Crank-Nicholson. The computation of the propagation resolves a tridiagonal matrix. However the parabolic approximation fails because it generates errors when the aperture at the source is greater than about 15°. In this paper we review the formalism established by Thomson and Chapman [5] in order to extend the validity of parabolic approximation to an aperture of about 40°.

We show that the computation of propagation resolves a quindiagonal matrix. For treatment of rectilinear horizontal or oblique interfaces between two fluids we use the same procedure already established by Lee and McDaniel [3] to apply to the case of the quadratic development of the operator of the propagation. For environments where the sound velocity profile changes with the range we use a new method that uses the concept of a vertical rectilinear interface. In fact, as soon as the sound velocity profile changes we initialize again the parameters of the quindiagonal matrix according to the new environmental properties as well as the continuity conditions.

Finally we present some examples of modifications of the sound field. These use variations of the sound velocity profile, including the disappearance of the surface sound channel, collapse of the acoustic channel, and change in range of the convergence zones.

1 THE PARABOLIC EQUATION

For sinusoidal time-dependence the acoustic propagation satisfies the Helmholtz equation for a two-dimensional environment described by coordinates (r,z):

$$\left\{ \nabla^2 + k_o^2 n^2(r,z) \right\} p(r,z) = 0 \tag{1}$$

where k_o is the mean wave vector and $n(r,z) = c_o/c(r,z)$ is the reflectivity index. When we are far from the source ($k_o r \gg 1$) we usually write:

$$p(r,z) = u(r,z) \exp [ik_o r] \, r^{-1/2} \tag{2}$$

After substitution of the acoustic pressure given by expression (2) in equation (1), this can be written (neglecting the term which decreases as $1/r^2$)

$$\left\{ (P + ik_o Q)\,(P - ik_o Q) - ik_o [P,Q] \right\} u(r,z) = 0 \tag{3}$$

with

$$P = \frac{\partial}{\partial r} + ik_o \text{ and } Q^2 = n^2(r,z) \quad / k_o^2 \frac{\partial^2}{\partial z^2}$$

and where [P,Q] designates the commutator of the operators P and Q. When equal to zero (that is when the reflectivity index doesn't change with the range r) the backward and forward propagating waves separate.

The wave field is then represented by:

$$(P - ik_oQ)\ u(r,z) = 0$$

or

$$\frac{\partial}{\partial r} u(r,z) = (1+q_1+q_2)^{1/2}\ u(r,z) \tag{4}$$

with

$$q_1 = n^2(r,z) - 1 \text{ and } q_2 = \frac{1}{k_o^2}\frac{\partial^2}{\partial z^2}$$

The equation of propagation of the forward propagating wave is obtained from the development of the square root of the second term of (4) (see Tappert [1]). The idea is to choose a weakly inhomogeneous medium so that the norm of operator q_1 and q_2 is small. Therefore we have chosen a quadratic development of the square root of the second term of equation (4). This allows us to minimize the error coming from the truncation of the linear development and also to increase the angular aperture of the source with one small additional complication in the computations. The parabolic equation of the propagation of the progressive wave is written:

$$\frac{\partial}{\partial r} u(r,z) = a(r,z) + b(r,z)\frac{\partial^2}{\partial z^2} + c(r,z)\frac{\partial^4}{\partial z^4}\ u(r,z) \tag{5}$$

$$a(r,z) = \frac{ik_o}{2}\left\{[n^2(r,z) - 1] - 1/4\ [n^2(r,z) - 1]^2 - 1/4\ k_o^2 \frac{\partial^2}{\partial z^2}[n^2(r,z) - 1]\right\}$$

$$b(r,z) = \frac{i}{2k_o}\left\{1 - [n^2(r,z) - 1]/4\right\}$$

$$c(r,z) = -i \times (8k_o)^{-3}$$

The relationship (5) is computed numerically using a finite difference scheme. In this way, we go from continuous variables r and z to discrete variables nk and mh, where k is range increment and h the depth increment. We write:

$$u(r,z) = u(nk,mh) = u_m^n$$

We use the Crank-Nicholson scheme which consists of writing the second term of the equation as half of the sum of its expression taken at range steps n and n+1. In this expression the second and fourth derivates are given by:

$$\frac{\partial^2}{\partial z^2} u_m^n = (u_{m+1}^n - 2u_m^n + u_{m-1}^n)/h^2$$

$$\frac{\partial^4}{\partial z^4} u_m^n = (u_{m+2}^n - 4u_{m+1}^n + 6u_m^n - 4u_{m-1}^n + u_{m-2}^n)/h^4$$

Reorganizing these expressions in matrix format, and allowing for m to range from zero (surface) to M (bottom), we obtain:

$$
\begin{bmatrix}
X_1^{0+1} & S_1^{n+1} & \alpha_1^{n+1} & & & & & \\
S_2^{n+1} & X_2^{n+1} & S_2^{n+1} & \alpha_2^{n+1} & & & \bigcirc & \\
-\alpha_3^{n+1} & S_3^{n+1} & X_3^{n+1} & S_3^{n+1} & \alpha_3^{n+1} & & & \\
 & -\alpha_4^{n+1} & S_4^{n+1} & X_4^{n+1} & S_4^{n+1} & \alpha_4^{n+1} & & \\
 & & \ddots & \ddots & \ddots & \ddots & \ddots & \\
 & & -\alpha_m^{n+1} & S_m^{n+1} & X_m^{n+1} & S_m^{n+1} & \alpha_m^{n+1} & \\
 & \bigcirc & & \ddots & \ddots & \ddots & \ddots & \\
 & & & & -\alpha_{M-1}^{n+1} & S_{M-1}^{n+1} & X_{M-1}^{n+1} & S_{M-1}^{n+1} \\
 & & & & & -\alpha_M^{n+1} & S_M^{n+1} & X_M^{n+1}
\end{bmatrix}
\times
\begin{bmatrix}
u_1^{n+1} \\ u_2^{n+1} \\ u_3^{n+1} \\ \vdots \\ u_m^{n+1} \\ \vdots \\ u_{M-1}^{n+1} \\ u_M^{n+1}
\end{bmatrix}
+
\begin{bmatrix}
S_1^{n+1} u_0^{n+1} \\ -\alpha_2^{n+1} u_0^{n+1} \\ 0 \\ \vdots \\ 0 \\ \vdots \\ -\alpha_{M-1}^{n+1} u_{M+1}^{n+1} \\ S_M^{n+1} u_{M+1}^{n+1}
\end{bmatrix}
$$

$$
=
$$

$$
\begin{bmatrix}
Y_1^n & T_1^n & \alpha_1^n & & & & & \\
T_2^n & Y_2^n & T_2^n & \alpha_2^n & & & \bigcirc & \\
\alpha_3^n & T_3^n & Y_3^n & T_3^n & \alpha_3^n & & & \\
 & \ddots & \ddots & \ddots & \ddots & \ddots & & \\
 & & \alpha_m^n & T_m^n & Y_m^n & T_m^n & \alpha_m^n & \\
 & & & \ddots & \ddots & \ddots & \ddots & \alpha_{n-3}^n \\
 & \bigcirc & & & \alpha_{M-1}^n & T_{M-1}^n & Y_{M-1}^n & T_{M-1}^n \\
 & & & & & \alpha_M^n & T_M^n & Y_M^n
\end{bmatrix}
\times
\begin{bmatrix}
u_1^n \\ u_2^n \\ u_3^n \\ \vdots \\ u_m^n \\ u_{M-2}^n \\ u_{M-1}^n \\ u_M^n
\end{bmatrix}
+
\begin{bmatrix}
T_1^n u_0^n \\ \alpha_2^n u_0^n \\ 0 \\ \vdots \\ 0 \\ 0 \\ \alpha_{M-1}^n u_{M+1}^n \\ T_M^n u_{M+1}^n
\end{bmatrix}
\qquad (6)
$$

with

$$X_m^{n+1} = 1 - \frac{1}{2} k \, a_m^{n+1} + \frac{k}{h^2} b_m^{n+1} - 3 \frac{k}{h^4} c_m^{n+1}$$

$$Y_m^n = 1 + \frac{1}{2} k \, a_m^n + \frac{k}{h^2} b_m^n + \frac{3k}{h^4} c_m^n$$

$$\alpha_m^{n+i} = \frac{1}{2} \frac{k}{h^4} c_m^{n+i}$$

$$S_m^{n+1} = - \frac{1}{2} \frac{k}{h^2} b_m^{n+1} + \frac{2k}{h^4} c_m^{n+1}$$

$$T_m^n = \frac{1}{2} \frac{k}{h^2} b_m^n - \frac{2k}{h^4} c_m^n$$

Knowing the field at range step n, that is knowing u_m^n, the field at range step (n+1) is computed by means of the quindiagonal matrix relation for the previous range step. A linear development leads to the same model of the matrix equation but is tridiagonal instead of quindiagonal. Figure 1a, b shows the computed acoustic field using relationship (6) for frequencies of 200 Hz and 25 Hz, respectively.

The source is 600 m deep and the sound velocity profiles go through the points (0 m, 1480 m/s), (1200 m, 1432 m/s), (4800 m, 1500 m/s).

We notice that the field obtained by the parabolic approximation indicates caustics very well. However the caustics becomes poorly defined at 25 Hz. In effect, the parabolic approximation has, in contrast to ray theory (see Fig. 2), no low-frequency limitations.

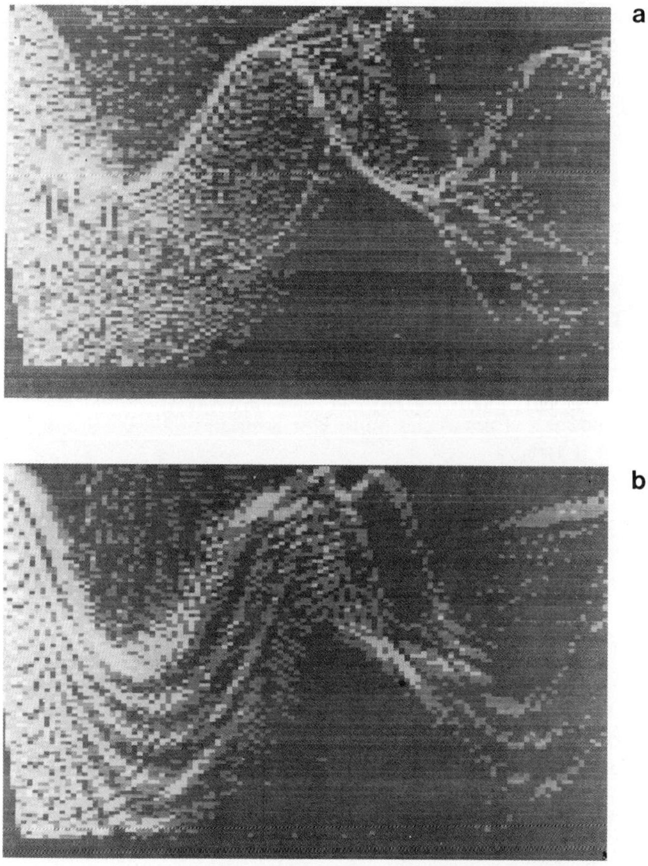

Fig. 1 Sound field from program "PARQUES". Source depth 600 m. a) 200 Hz, b) 25 Hz

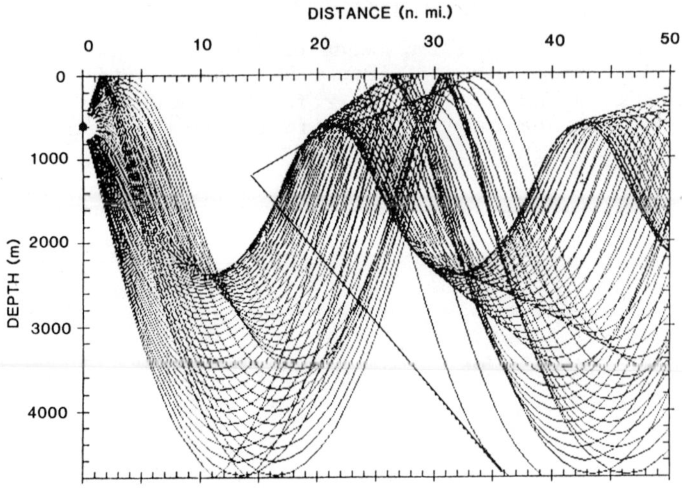

Fig. 2 Sound field from ray tracing. Source depth 600 m.

2 THE INTERFACES

A medium is characterized by its density, its sound attenuation properties, and its sound velocity profile. A large change in one of these parameters changes the character of the medium. Two different media meet at an interface, a distinct physical zone. At the interface, continuity of the pressure and of the normal component of particle velocity are required.

2.1 Horizontal Rectilinear Interface

The horizontal rectilinear interface coincides with the m_0th row of the matrix. For the calculation of the coefficients of the matrix of relation (6) for the interface ($m = m_0$) we adopt the method described by McDaniel and Lee [4] in which we take into account the quadratic development of the operator of relation (4). For the mathematical computation we consider also the mesh points ($n,m \pm 2$) and the expression of fourth derivative at the mesh points (n,m_0).

After some calculation, the treatment of the horizontal rectilinear interface changes the coefficients of the m_0th row (see reference [1] for the coefficients).

2.2 Oblique Rectilinear Interface

The interface makes an angle θ with the horizontal. All the mesh points on the interface are of the form ($n+i,m+i$). Going from range step n to range step n+1, the presence of the oblique rectilinear interface will alter the coefficients which join u_m^n (on the interface) at u_m^{n+1} (in medium one) and u_{m+1}^n (in medium 2) at u_{m+1}^{n+1} (on the interface).

We adopt the same method (Lee and McDaniel [3]) to describe the quadratic development [6].

The problem where the angle between the oblique rectilinear interface and the horizontal is $-\theta$ is treated similarly but the mesh points on the interface are of the form (n+i, m-i).

2.3 Vertical Rectilinear Interface

If the sound velocity profile changes with the range numerical processing can be done by two different methods. The usual method consists of associating the variation of coefficients α, X,Y,T... with the variation in sound speed. This method assumes that the variation in the sound velocity profile is small. However the matrix relation (6) is rigorously connected with the condition

$$[P,Q]\ u(r,z) = 0 \text{ or } \frac{\partial}{\partial r} n(r,z) = 0.$$

The second method supposes that the sound velocity profile is constant for the distance $N\Delta r$, changes quickly, and then is constant thereafter (Fig. 3).

In the first medium, $r \leqslant N\Delta r$, the propagation loss is calculated by the relation (6) where the coefficients α,X,Y,...are given by the parameters of Medium 1 $[\rho_1, c_1(z)]$. For $r \geqslant (N+1)\Delta r$ the propagation loss is calculated from relation (6) where the coefficients α,X,Y,...are given by the parameters of Medium 2 $[\rho_2, c_2(z)]$. We must begin the field in Medium 2 at the range $r = (N+1)\Delta r$ to find u_m^{N+1}. Therefore we assume the existence of a vertical rectilinear interface at the range $r = N\Delta r$ which separates the two media $[\rho_1,\ c_1(z)]$ and $[\rho_2,\ c_2(z)]$. We can write on the interface the equations of continuity of pressure and of the normal component of particle velocity. These equations give the new relations for the functions u(r,z). If we suppose that $\rho_1 = \rho_2 = \rho$ of sea water, we have

$$(u_1)_m^N = (u_2)_m^N \quad \text{and} \quad \frac{\partial}{\partial r}(u_1)_m^N = \frac{\partial}{\partial r}(u_2)_m^N$$

After expansion we use the equations of continuity to write,

$$(u_m^N) = 1/2\ [(u_2)_m^{N+1} + (u_1)_m^{N-1} + \frac{k}{4}\left(\frac{\partial}{\partial r}(u_1)_m^{N-1} - \frac{\partial}{\partial r}(u_2)_m^{N+1}\right)] \qquad (7)$$

In this relation $\frac{\partial}{\partial r}(u_1)_m^{N+1}$ and $\frac{\partial}{\partial r}(u_2)_m^{N+1}$ are calculated by the relation (5) where the coefficents a(r,z), b(r,z) and c(r,z) relate respectively to Medium 1 and Medium 2. Relation (7) permits us to calculate $(u_2)_m^{N+1}$ because u_m^{N-1}, u_m^N are known. For m going from 0 to M we obtain the following reinitialization in matrix format:

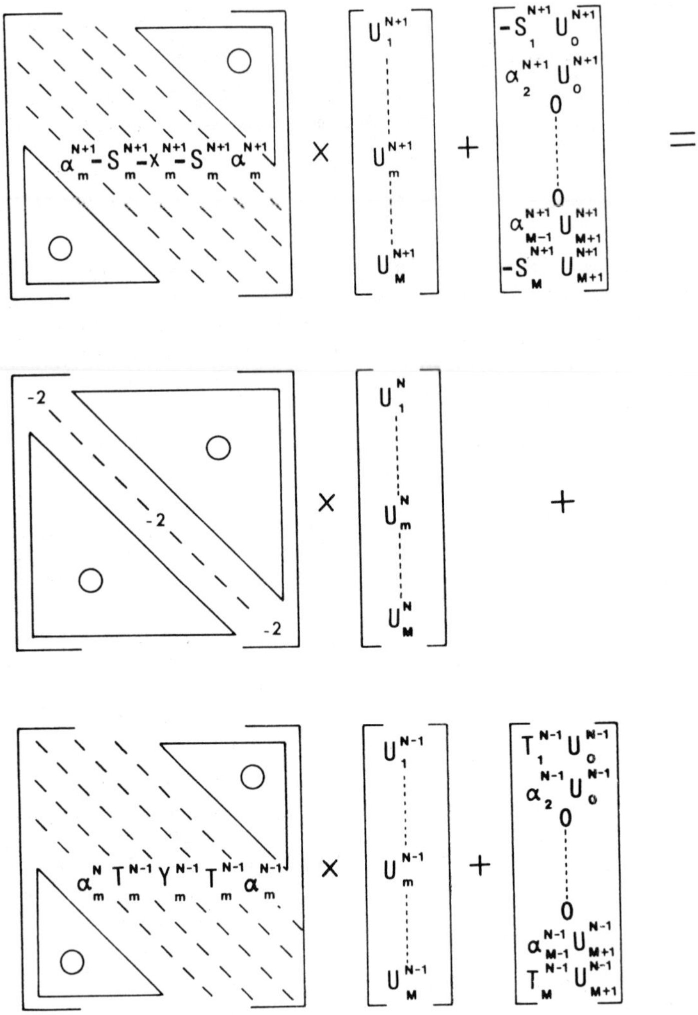

(8)

If we repeat this process many times we can follow gradually (but by successive discontinuities) the evolution of the sound velocity profile. Some examples given by this method are presented below.

4 NUMERICAL SIMULATION

The numerical simulation permits us to trace the sound field hereafter. The convention chosen to represent the sound field is light grey for low propagation loss to dark grey for high propagation loss. The sound velocity profile is traced on Fig. 4. In Fig. 5 the sound field is traced in accordance with ray theory. The name of the sound velocity profile is "JASA 1". The source depth is 100 m. Figure 6 presents the sound field obtained with the programme "PARQUES".

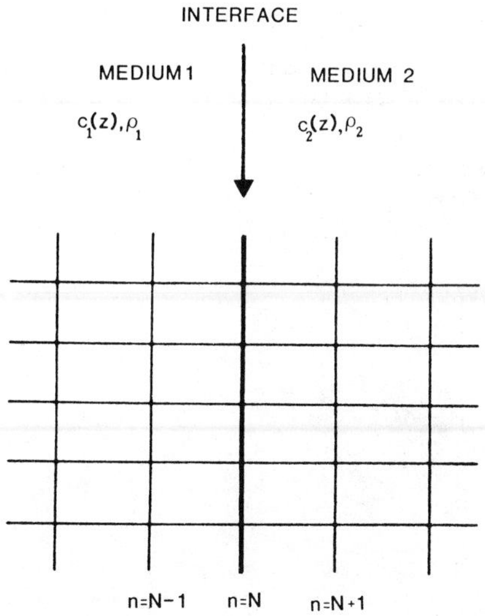

Fig. 3
Oblique rectilinear interface.

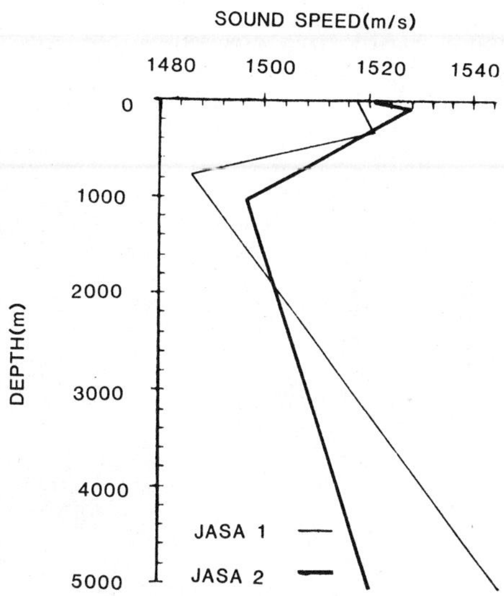

Fig. 4
Sound speed profiles.

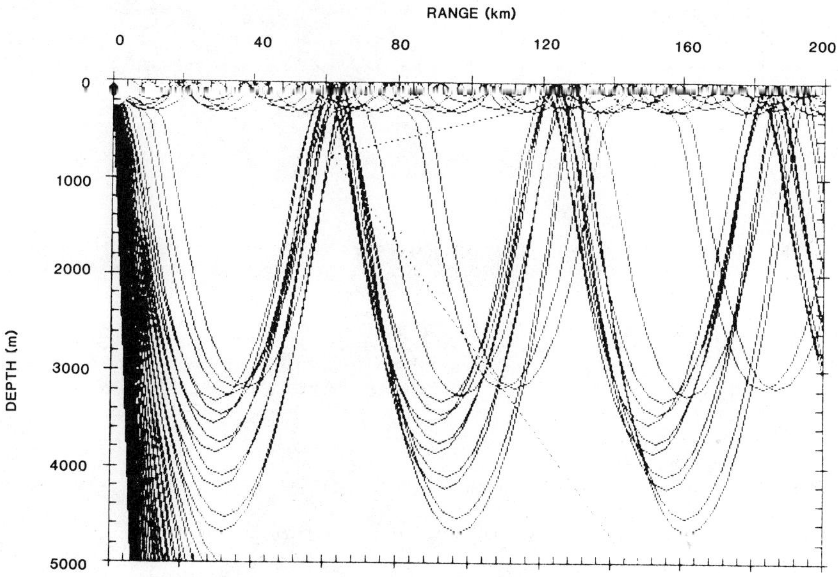

Fig. 5 Sound field from ray tracing.
Sound speed profile "JASA1". Source depth 100 m.

Figure 7 illustrates the modification of the sound field at 100 km from the source where the sound velocity profile changes abruptly from JASA 1 to JASA 2 (representing a strong oceanic front). We note the vanishing of the surface channel with a greater loss in transmission of about 25 dB, and a weakening of the second convergence zone. The third convergence zone is displaced an additional 22 km. Figure 8 presents the modifications of the sound field when the transition of "JASA 1" to "JASA 2" is accomplished in three stages at 110, 120 and 130 km from the source. The difference between Figs. 7 and 8 shows the sensitivity of the sound field to a change of the sound velocity.

Fig. 6
Sound field for single sound speed profile "JASA1". From program "PARQUES". Source depth 100 m, frequency 200 Hz

Fig. 7
Sound field for sound speed profile "JASA1" changing abruptly to "JASA2" at range 100 km. From program "PARQUES". Source depth 100 m, frequency 200 Hz

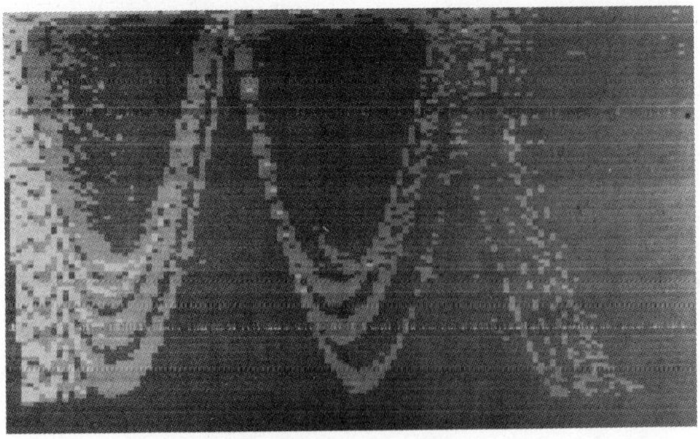

Fig. 8
Sound field for sound speed profile "JASA1" changing gradually to "JASA2" at ranges 110 to 130 km. From program "PARQUES". Source depth 100 m, frequency 200 Hz

Figure 9 presents the sound field obtained at 25 Hz for a source 760 m deep when after 100 km the sound speed changes from "JASA 1" to "JASA 2". On this same figure, we note the weakening and change of depth of the sound channel from 760 to more than 1000 m.

Figure 10 shows the influence of an oblique bottom. The source is 100 m deep and has a frequency of 10 Hz. Up to a range of 5 km from the source, the bottom is 3000 m deep. Then the bottom descends progressively until 5000 m depth at 15 km from the source, and remains flat up to a distance of 20 km. We note that a part of the energy is reflected by the oblique rectilinear interface, which greatly changes the sound field.

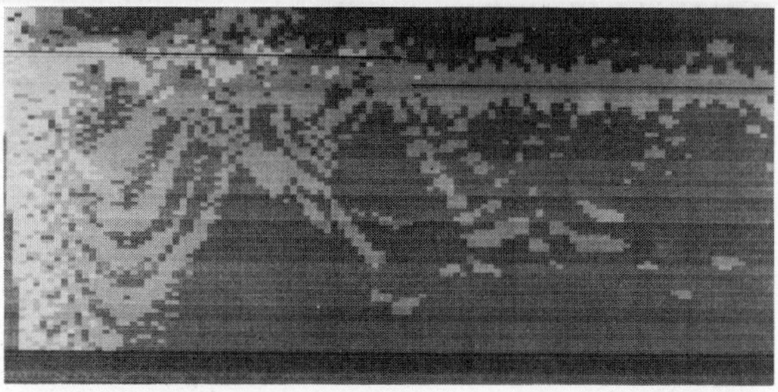

Fig. 9
Sound field for sound speed profile "JASA1" changing abruptly to "JASA2" at range 100 km. From program "PARQUES". Source depth 760 m, frequency 25 Hz

Fig. 10
Influence of a sloping interface on sound field from program "PARQUES". Source depth 100 m, frequency 10 Hz; slope is approximately 11°.

CONCLUSION

The first results given by the programme "PARQUES" are encouraging. The vertical rectilinear interface method seems to take into account the effects on the sound field made by a medium that is changing in range, e.g. the displacement of the convergence zone and the weakening of the sound channel.

In the future, we need to consider the effects of the sea surface and better represent the bottom effects.

REFERENCES

1. F.D. Tappert, The parabolic approximation method. In: J.B. Keller and J.S. Papadakis, eds. Wave propagation and underwater acoustic. Springer, Berlin, 1977.
2. D. Lee, G. Botseas, J.S. Papadakis, Finite-difference solution to the parabolic wave equation, J. Acoust. Soc. Amer. 70, 1981: 795-800.
3. D. Lee and S.I. McDaniel, A finite difference treatment of interface conditions for the parabolic wave equation: The irregular interface. J. Acoust. Soc. Amer. 73, 1983: 1441-1447.
4. S.T. McDaniel and D. Lee, A finite-difference treatment of interface conditions for the parabolic wave equation: The horizontal interface, J. Acoust. Soc. Amer. 71, 1982: 855-858.
5. D.J. Thomson and N.R. Chapman, A wide-angle split-step algorithm for the parabolic equation, J. Acoust. Soc. Amer. 74, 1848-1854.
6. B. Grandvuillemin, Application de l'approximation parabolique a l'acoustique sous-marine. These de 3eme cycle. Specialite en acoustique. Universite d'Aix, Marseille II (1985), France. 795-800.

LINEAR AND NONLINEAR OCEAN ACOUSTIC PROPAGATION MODELS

W.A. Kuperman*

Naval Ocean Research and Development Activity
NSTL STN, MS 39529

and

Naval Research Laboratory
Washington, DC 20375

B.E. McDonald

Naval Ocean Research and Development Activity
NSTL STN, MS 39529

ABSTRACT

A review of basic linear ocean sound propagation models as derived from fluid dynamics is presented. By retaining nonlinear terms in the equations of fluid dynamics a nonlinear time domain wave equation is then derived. A unidirectional approximation leads to a nonlinear time domain version of the parabolic equation, a nonlinear progressive wave equation (NPE), which can treat propagation of finite amplitude signals in a refracting medium in a wave theoretic sense and hence includes propagation through a caustic. In the limit of infinitesimal acoustics a linear time domain parabolic equation results which can be directly solved numerically without Fourier synthesis. Inserting a harmonic solution into this equation and taking the far field limit reduces this equation to the standard parabolic equation. Linear and nonlinear time domain examples are presented including the propagation of a shock wave through a caustic.

I. INTRODUCTION

Modern numerical physics and state-of-the-art computer technology have combined to provide an impetus for new developments in modeling ocean sound propagation. Some of these developments are presented in this short review, which is by no means complete, but is instead intended to indicate the direction of possible future avenues of research.

The study of sound propagation in the ocean dates back centuries; for example, Leonardo da Vinci noted that sound travels to great distances in the sea, and he even anticipated the passive-sonar fundamental consideration of avoiding self-noise by suggesting that the listener "...cause your ship to stop..."[1] before deploying a listening tube through the air-water interface. This listening tube was improved on by Colladen[2] who, on Lake Geneva in 1827 employed a trumpet-like device as an underwater sound receiver together with a submerged bell to, in essence, assemble an active-sonar (see Fig. 1) with which he measured the speed of sound to an accuracy of about two tenths of one percent. In addition, "Colladen marvelled that so little energy at the source could be transmitted so great a distance through the water medium and could still be detected by one trumpet receiver." It is still believed that only sound will propagate man-made signals to great distances in the sea.

*Present address Naval Research Laboratory

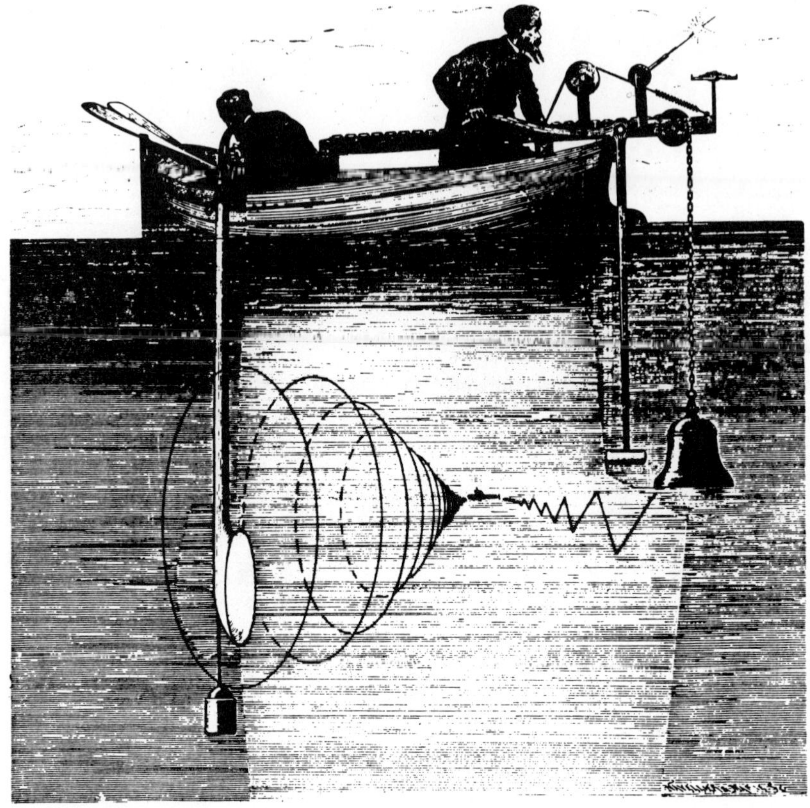

Figure 1. The experiment to measure the speed of sound was performed by Colladen and Sturm. They were actually in two boats ten miles apart using a light source and a stop watch to perform the time of propagation measurement. This figure is a composite of their two sketches of separate source and receiving boats put together to indicate the fundamental components of a sonar system. The source and listening device sonar system is shown here detecting a hypothetical submerged object in Lake Geneva. The fundamental acoustic phenomena that would be of interest in this scene are: sound propagation characteristics, noise, scattering/fluctuations and acoustic target properties.

The development of the theory of acoustics has followed a different historical path than optics/electromagnetic theory. While there had been a long running battle in optics between ray and wave theory, acoustics originally developed with wave and vibration concepts as far back as ancient Greek times. Thus, for example, musical sound is the result of air motion generated by a vibrating musical instrument. This sound is propagated in an analogous manner to water surface waves which propagate disturbances (signals) but not matter long distances. The bending of sound around corners is then easily understood in the context of diffraction in the water surface *wave* analogy. In acoustics, the ray picture evolved as a mathematical approximation to linear wave theory.[3] In underwater acoustics, in which the medium is inhomogeneous and bounded by complicated interfaces, ray theory has historically proved to be an indispensable tool for understanding and studying sound propagation. However, with the advent of computers, sound propagation in the ocean can now be studied and quantitatively described in greater and greater detail with the more exact wave theory.

Figure 2 schematically depicts the hierarchy of ocean sound propagation models which will be reviewed in part, in this paper. The models are all numerical solutions of some acoustic wave equation which in turn are derived from the equations of fluid mechanics. The "standard" sound propagation models, are shown directly linked to the "FREQ DOMAIN WAVE EQ" in Fig. 2. The derivation of these linear models is briefly reviewed in Section III.

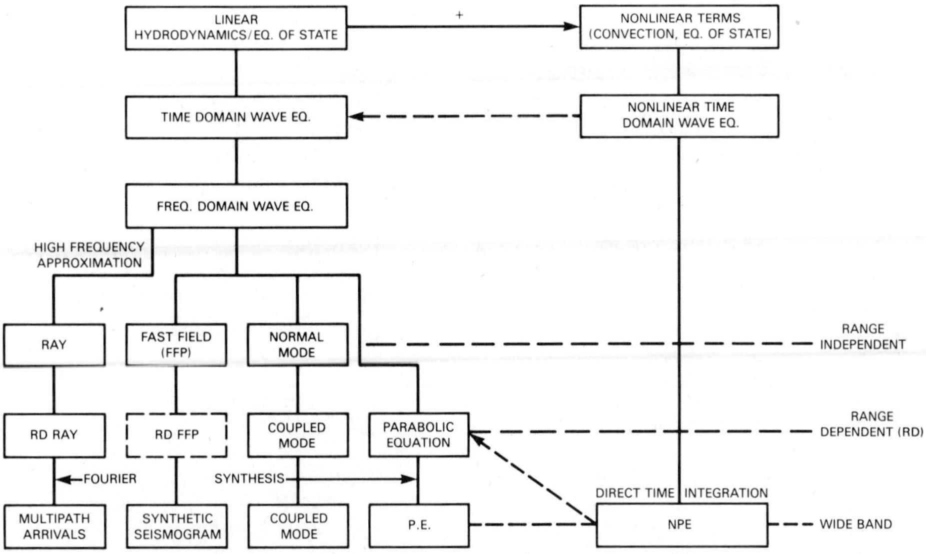

Figure 2. Hierarchy of ocean sound propagation models.

If the acoustic field is very high in intensity, sound propagation must be studied as a nonlinear phenomenon; the medium does not respond to a high amplitude disturbance in a simple harmonic oscillatory fashion as is the case for infinitesimal linear acoustics. With the acoustic field so disturbing the medium, the powerful tool of Fourier superposition cannot be rigorously applied to nonlinear acoustics which include such phenomena as the generation of new frequencies absent in the primary source of radiation and "self-refraction" which results from the local high amplitude field altering the acoustic (refractive) properties of the quiescent medium.

In this review, the more recent material will emphasize time-domain wave equations for two reasons. First, we are searching for direct numerical solutions to wideband ocean acoustic propagation problems. Second, the nonlinear propagation requires the time-amplitude history of the finite amplitude pulse. Thus, for example, Fig. 3 illustrates the steepening[4] of a finite amplitude pulse resulting from the amplitude dependent speed of the disturbance; here we illustrate the phenomenon that higher amplitudes travel with greater speeds. In the development of the work presented below, the development of the time domain solution of the nonlinear wave equation was first derived[5,6] afterwhich application to wideband linear acoustics was then studied omitting the nonlinear terms[7]. In the derivation below, the nonlinear terms are underlined. We will pause at certain steps of the derivation to retrieve conventional linear wave acoustic results. Hence, rather than follow the path directly below "LINEAR HYDRODYNAMIC/EQ OF STATE" in Fig. 2, we pick up the nonlinear terms on the right and derive a nonlinear progressive wave equation "NPE" (see Section V). Excluding the nonlinear terms we have a time domain version of the parabolic equation (PE) (see Section IV) a linear progressive wave equation which can be efficiently integrated directly without resort to Fourier synthesis as indicated by the horizontal dashed line from NPE to PE in Fig. 2. In addition, in the limit of linear acoustics, substitution of a harmonic solution in the NPE recovers the standard frequency domain PE. With new numerical techniques[8], the NPE can be efficiently integrated. Linear and nonlinear examples are given in Section VI. The final example is a nonlinear refraction problem: a shock wave passing through a "caustic." Previously, the problem could only be attacked using ray theory neglecting nonlinear effects in the high intensity caustic region (where ray theory diverges) which, because of the high intensity, is a region of increased nonlinearity.

II. THE LINEAR WAVE EQUATION

The wave equation in an ideal fluid is derived from hydrodynamics and the adiabatic relation between pressure and density. We include nonlinear effects by retaining the convective term $\vec{v} \cdot \vec{\nabla} \vec{v}$ in the Euler Equations and higher order terms in the equation of state (as underlined);

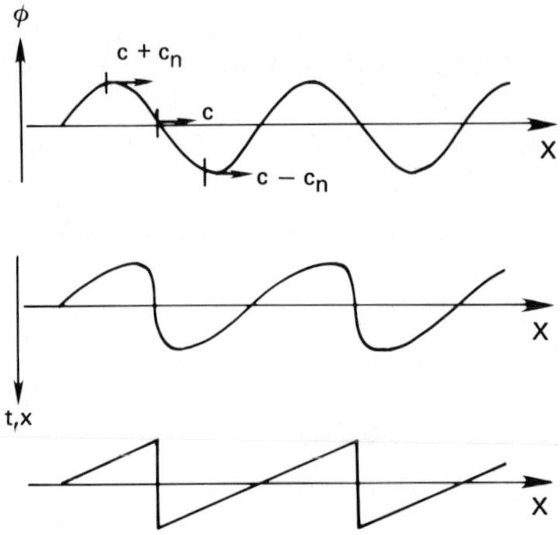

Figure 3. Propagation of a finite amplitude acoustic wave. Higher amplitudes (crests) propagate faster than lower amplitudes (such as rarefraction). As time progresses, fully developed shock wave is developed away from the source. Finally, shock wave ages with time and distance when attenuation of high frequency components is greater than shock growth rate.

later, we will use the underlined terms to derive a nonlinear wave equation. Hence, we start with the equation

$$\frac{\partial \rho}{\partial t} = -\vec{\nabla}\cdot\rho\vec{v}, \quad \text{[mass conservation]} \tag{1}$$

$$\frac{\partial \vec{v}}{\partial t} + \vec{v}\cdot\vec{\nabla}\vec{v} = -\frac{1}{\rho}\vec{\nabla}p(\rho), \quad \text{[Newton's 2nd law]} \tag{2}$$

$$p = p_0 + \rho'\left(\frac{\partial p}{\partial \rho}\right)_s + \frac{1}{2}\rho'^2\left(\frac{\partial^2 p}{\partial \rho^2}\right)_s + \cdots, \quad \text{[Eq. of State]} \tag{3}$$

$$\left(\frac{\partial p}{\partial \rho}\right)_s \equiv c^2, \quad \text{[speed of sound]} \tag{4}$$

where ρ is the density, v is the vector velocity field, p is the pressure, and the subscript s denotes the derivation with entropy constant evaluated at ambient density. Equations 1 and 2 can be manipulated to obtain

$$\frac{\partial^2 \rho}{\partial t^2} = \nabla^2 p + \partial_i\partial_j\,(\rho v_i v_j)\;. \tag{5}$$

Introducing small perturbations p' and ρ, we take $p = p_0 + p'$ and $\rho = \rho_0 + \rho'$ and we define the dimensionless density ratio as $\phi = \rho'/\rho_0$ where p_0 and ρ_0 are constant. Then, from the linear terms of Eqs. 3-5 (p has a nonlinear component), we obtain the linear wave equation in a source free region

$$\frac{\partial^2 \phi}{\partial t^2} = \nabla^2\, c^2\, \phi, \tag{6}$$

and by assuming a time dependence exp $(-i\omega t)$ we obtain the harmonic wave equation

$$\nabla^2\phi(x,y,z) + \frac{\omega^2}{c^2(x,y,z)}\,\phi(x,y,z) = -\delta(x-x_0)\,\delta(y-y_0)\delta(z-z_0), \tag{7}$$

where the right hand side (r.h.s.) of Eq. 7 indicates a point source at (x_0,y_0,z_0). We use a coordinate system for the ocean where the horizontal coordinates are (x,y) and the vertical coordinate, z, is taken positive downward.

III. STANDARD FREQUENCY DOMAIN UNDERWATER ACOUSTIC MODELS[9–22]

In linear theory, Eq. 7 is the jumping off point for Ray, Fast Field Program (FFP), Normal Mode (NM), and Parabolic Equation (PE) modeling. Direct time domain numerical solutions of Eq. 6 are still not computationally practical for realistic ocean environments without Fourier synthesizing solutions of Eq. 7. Equations 6 and 7 are also difficult to solve in practice in a range dependent environment, that is, when the sound speed varies in range such as when the ocean bottom has a slope. The PE[26–28] [and see Tappert and Nghiem-Phu in this book] makes range dependents frequency domain solution of Eq. 7 practical but still requires Fourier synthesis to construct solutions in the time domain. We can, however, obtain a time domain PE type solution of Eq. (6) directly as will be shown later in this section.

The goal of ocean acoustic modeling is to describe sound propagation phenomena in the ocean quantitatively. To clarify the complexity of the modeling problem, let us briefly review the environmental acoustics of the ocean. Figure 4 is a schematic of some propagation paths in the ocean; we show two possible locations of sources of sound on the left and sound is propagating to the right. The two dashed lines are sound-speed profiles that vary with both depth and range. Lines A, B, C, and D represent four possible propagation paths. The shapes of the paths are determined by the location of the source and the sound-speed structure over the extent of the propagation. Path A from the shallow source is "surface-duct" propagation, because the sound-speed profile is such that the sound is trapped near the surface of the ocean. Paths B, C, and D are from the deeper source. Ray B, leaving the source at a small angle from the horizontal, will tend to propagate in the "deep sound channel" without interacting with the boundaries (surface and bottom) of the ocean. At slightly steeper angles (path C) we have "convergence zone" propagation, which is a spatially periodic phenomenon of zones of high intensity near the surface. Here the path interacts with the ocean surface but not with the bottom. Path D is the "bottom-bounce path", which has a shorter cycle period than the convergence zone path. The right-hand side of Fig. 4 depicts propagation on the continental shelf (shallow waters) where a complicated bottom structure combined with variable sound-speed profiles result in rather complicated propagation conditions not always suited for a simplistic ray picture representation.

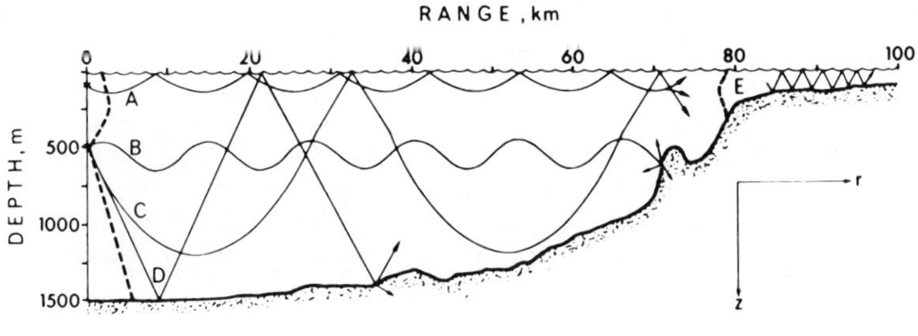

Figure 4. Schematic representation of types of sound propagation in the ocean.

We start with Eq. 7 where we take ϕ as the velocity potential. For the boundary conditions we begin with the ocean surface. The density of air is negligible compared with that of water; hence the pressure must vanish ("pressure-release surface") at the ocean surface. At a boundary between two media such as the ocean and the ocean bottom, the balancing of forces at the interface requires that both the normal particle velocity and the sound pressure be continuous across the boundary. Of course these boundary conditions are more complicated for elastic media which also support shear waves.

The standard frequency domain linear acoustic models are displayed in Fig. 2 directly below the "FREQ DOMAIN WAVE EQ" box.

A. Ray Theory

Ray theory is an infinite frequency approximation to the solution of Eq. 7. We assume as solution of the form

$$\phi(\vec{R}) = A(\vec{R}) \exp [iS(\vec{R})], \tag{8}$$

where the exponential term allows for rapid variations as a function of range and $A(\vec{R})$ is a more slowly varying "envelope" function which incorporates both geometrical spreading and loss mechanisms. Substituting Eq. 8 into the wave equation Eq. 7 within source free space and separating real and imaginary parts, we obtain

$$\frac{1}{A}\nabla^2 A - (\vec{\nabla} S)^2 + k^2 = 0; \tag{9}$$

$$\vec{\nabla} A \cdot \vec{\nabla} S + \frac{1}{2} A \nabla^2 S = 0.$$

The geometrical approximation assumes that the amplitude of the wave varies slowly in range. This can be expressed mathematically as follows:

$$\frac{1}{A}\nabla^2 A << k^2. \tag{10}$$

Thus, the first part of Eq. 9 may be expressed to good approximation as

$$(\vec{\nabla} S)^2 = k^2. \tag{11}$$

This is referred to as the eikonal equation in the literature. Here S is a phase function which includes rapid variations as a function of range. We note that the ray trajectories are perpendicular to surfaces of constant phase (wavefronts), S, and may be expressed mathematically as follows:

$$\frac{d}{dl}\left(k \frac{d\vec{R}}{dl} \right) = \vec{\nabla} k, \tag{12}$$

where l is the arc length along the direction of the ray and $\vec{R}$ is the displacement vector. One can determine from the above that the direction of average flux (energy flow) follows that of the trajectories and the amplitude of the field at any point can be obtained from the density of the rays. Once one solves for S it is possible to obtain the amplitude by employing the second part of Eq. (9). We also mention here that corrected ray theory assumes that A is a function of frequency and makes an expansion in powers of inverse frequency. The leading term is the infinite-frequency solution with the additional terms being corrections to the infinite-frequency solution.

The principle advantages of the ray theory are that computations are rapid and that ray traces give a very physical picture of the acoustic paths. The limitations of the approach are that it is inherently a high-frequency theory that ignores diffraction and other wave effects. Focal regions, caustics, where the density of rays become infinite occur in the infinite frequency approximation. Hence, in this approximation, finite frequency effects at caustic cannot be obtained without additional corrections. In addition, ray theory does not adequately treat such phenomena as bottom interaction and low-frequency duct propagation.

B. Fast Field Program (FFP) [22–24]

Range-independent wave theory solves the wave equation exactly when the ocean environment does not change in range. One of many possible derivations of this solution technique is to Fourier decompose the acoustic field into an infinite set of horizontal waves,

$$\phi(r,z) = \frac{1}{2\pi} \int d^2\vec{\eta}\, u(\vec{\eta},z) \exp(i\vec{\eta}\cdot\vec{r}) \tag{13}$$

which is substituted back into Eq. 7 to yield an equation for u,

$$d^2u/dz^2 + \left[k^2(z) - \eta^2\right] u = -\frac{1}{2\pi}\delta(z-z_0). \tag{14}$$

Assuming azimuthal symmetry, we can integrate Eq. 13 over the angular variable to obtain Hankel functions. The asymptotic form of the Hankel functions reduces Eq. 13 to

$$\phi(r,z) = \frac{\exp(-i\pi/4)}{(2\pi r)^{1/2}} \int_{-\infty}^{+\infty} d\eta (\eta)^{1/2} u(\eta,z) e^{i\vec{\eta}\cdot\vec{r}}. \tag{15}$$

Note that the factor $1/r^{1/2}$ arises from cylindrical spreading. We now discretize the above integral and thus transform it to a form amenable to the FFT technique. This can be done by setting $\eta_m = \eta_0 + m\Delta\eta$; $r_n = r_0 + n\Delta r$ where $n,m = 0, 1, 2, \ldots N{-}1$ with the additional condition:

$$\Delta r \Delta\eta = 2\pi/N$$

and N is an integral power of two. The discretization scheme will be seen to limit the solution to outgoing waves. Upon applying the above discretization to Eq. (15) and summing over all values, one arrives finally at the desired expression:

$$\phi(r_n,z) = \frac{\Delta\eta \exp\left[(\eta_0 r_n - \pi/4)\right]}{(2\pi r_n)^{1/2}} \sum_{m=0}^{N-1} X_m \exp(i2\pi mn/N), \tag{17}$$

$$X_m = (\eta_m)^{1/2} u(\eta_m,z) \exp(imr_0\Delta\eta). \tag{17a}$$

The above equation is now easily evaluated using the FFT algorithm, with the bulk of the effort going into evaluating u for expression (15) from Eq. 14. Although the method is labeled as "fast field" it is actually one of the slowest of the techniques because of the time required to calculate the u's. However, it has advantages when one wishes to evaluate the "nearfield" region or to include shear wave effects. The FFP method is often used as a benchmark for other less exact techniques. One such technique, not accurate in the "nearfield" limit but which is computationally faster, is the NM theory.

C. Normal Mode Model (NM)[9,10]

Rather than evaluate u in Eq. 14, one can utilize a normal mode expansion of the form

$$u(\vec{\eta},z) = \sum a_n(\vec{\eta}) u_n(z), \tag{18}$$

where the quantities u_n are eigenfunctions of the following eigenvalue problem:

$$d^2u_n/dz^2 + \left[k^2(z) - k_n^2\right] u_n(z) = 0. \tag{19}$$

The orthonormal u_n's are zero at $z = 0$ and satisfy a radiation condition for $z \to \infty$. Note that these eigenfunctions are elements of a Hilbert space. Hence, they can be represented as an orthonormal set with discrete eigenvalues. This last property distinguishes the NM from the FFP method, which includes eigenvalues in the continuum. The range of discrete eigenvalues is limited by the condition

$$\min[\omega/c(z)] < k_n < \max[\omega/c(z)]. \tag{20}$$

We now substitute the expression for u in Eq. 13 into Eq. 14 and multiply both sides by $\rho(z)u_n(z)$ and integrate over all z space, we obtain (using the orthonormality of the u's with weighting function ρ,

$$\phi(r,z) = \frac{\rho(z_0)}{(2\pi)^2} \int_{-\infty}^{+\infty} d\eta_x \int_{-\infty}^{\infty} d\eta_y \sum_n \frac{u_n(z_0)u_n(z)}{\eta^2 - k_n^2} \exp[i(n_x x + n_y y)].. \tag{21}$$

We choose a contour about the poles which leads to solutions with outgoing waves from the point source. Each integral in Eq. 21 is proportional to the two-dimensional plane-wave representation of the zero-order Hankel function of the first kind, and therefore ϕ can be expressed as:

$$\phi(r,z) = \frac{i}{4} \rho(z_0) \sum_n u_n(z_0) u_n(z) H_0^{(1)}(k_n r). \tag{22}$$

The asymptotic form of the Hankel function can be used in the above equation to obtain

$$\phi(r,z) = \frac{i\rho(z_0)}{(8\pi r)^{1/2}} \exp(-i\pi/4) \sum_n \frac{u_n(z_0)u_n(z)}{k_n^{1/2}} \exp(ik_n r). \tag{23}$$

Equation 23 is a "far-field" solution of the wave equation and neglects the "continuous spectrum" of modes; therefore we consider only a discrete number of modes. In effect, if we look at a ray picture of the modes we can associate a grazing angle at the ocean bottom with each of the modes: the continuous part of the mode spectrum corresponds to rays with grazing angles greater than the bottom critical angle and this sound suffers severe loss.

The advantages of the NM procedure are: that once the eigenvalue problem is solved one has the solution for all source and receiver configurations, that it is a numerically efficient technique in the farfield limit, and that it is easily extended to moderate range dependent conditions using the adiabatic approximation. Recently a fully coupled mode theory for range dependent environments has been developed.[25]

D. Parabolic Equation Model (PE)[26–28]

The Parabolic Equation Model is a far-field narrow-angle approximation to the wave equation. It is most accurate when propagation with respect to the horizontal is confined within a cone of about $\pm 20°$. Essentially we make the approximation that

$$\phi = \psi(r,z)\ S(r).$$

S will satisfy an average propagation equation and will be a rapidly varying function in range. The parabolic approximation is to assume that because the environment is changing slowly with range, $(\partial^2\psi/\partial r^2) << 2k_0(\partial\psi/\partial r)$, where k_0 is an average wave number. This leads to an equation for ψ with only the first derivative of ψ with respect to r. This approximation takes us from a boundary-value wave-equation problem to an initial-value parabolic equation that lends itself to marching-solution techniques. Rather than present one of the standard derivations, we show below that the standard PE can be recovered from Eq. 6 by obtaining a far field harmonic solution of a time domain progressive wave equation which we derive in the next section.

The advantages of the PE method are that it handles a range dependent environment and gives the acoustic field in the entire water column without additional computational effort. Wide angle PE solutions have also been developed.[29] Its disadvantages are that the procedure is not easily automated, and it is impractical for high-frequency propagation since computation times increase rapidly with frequency. In addition, there is no straightforward way of handling shear propagation in the bottom [see S.C. Wales in this book].

IV. LINEAR PROGRESSIVE WAVE EQUATION AND THE PARABOLIC EQUATION[5]

Motivated by the nonlinear problem, we are interested in time domain calculations to calculate the evolution of a propagating pulse. Equation 6 as derived refers to the rest frame of the medium. To follow the evolution of the pulse itself as it traverses large distances, it is convenient to transfer to a frame moving with some average speed c_0 such that the total sound speed is given by $c(r,z) = c_0 + c_1(r,z)$ with c_1 small. The time derivative in the moving frame is

$$D_t = \frac{\partial}{\partial t} + c_0 \frac{\partial}{\partial r}, \qquad (24)$$

where r is the range variable in the primary direction of propagation. (For a sinusoidal in homogenous infinite space, $D_t\phi = 0$.) Substituting Eq. 24 into Eq. 7 and assuming azimuthal symmetry, we obtain

$$\left(\underset{\uparrow\,a}{D_t} - c_0 \frac{\partial}{\partial t}\right)^2 \phi = \left(\frac{\partial^2}{\partial r^2} + \underset{\uparrow\,b}{\frac{1}{r}\frac{\partial}{\partial r}} + \underset{\uparrow\,c}{\frac{\partial^2}{\partial z^2}}\right)[(c_0 + \underset{\uparrow\,d}{c_1})^2\ \phi]. \qquad (25)$$

The arrows under Eq. 25 indicate terms within each bracket which are assumed to be small. Proceeding from left to right in Eq. 25, the smallness assumptions have the following physical interpretation: (a) the evolution of the pulse is gradual compared to its transit time over a given point; (b) the gradient scale size in range is much smaller than the total range; (c) the variation in the transverse direction is more gradual than that in range; (d) spatial variations in the sound speed are small. We now expand Eq. 25 keeping only the lowest order in small terms and then integrate with respect to r:

$$D_t\,\phi = -\frac{c_0}{2r}\,\phi - \frac{\partial}{\partial r}\,(c_1\phi) - \frac{c_0}{2}\int^r \frac{\partial^2\phi}{\phi z^2}\,dr. \qquad (26)$$

Each term on the r.h.s. of Eq. 26 describes a distinct physical process. Proceeding from left to right, these are radial spreading, refraction, and diffraction. Not only have the distinct physical processes been separated out, but we have also reduced the problem from second order in time to first order in time. Note that Eqs. 24-26 essentially involve a unidirectional or paraxial approximation. Hence, Eq. 26 should reduce to the frequency domain PE. Indeed, if we introduce ψ which varies slowly with range and let

$$\phi(r,z) = \frac{e^{i(kr - \omega t)}}{\sqrt{r}} \psi(r,z), \tag{27}$$

where $\omega \equiv kc_0$ and take the limit of large kr, we obtain

$$\frac{\partial \psi}{\partial r} = -ik \frac{c_1}{c_0} \psi + \frac{i}{2k} \frac{\partial^2 \psi}{\partial z^2}, \tag{28}$$

which is the usual frequency domain PE. In order to construct a wideband (or pulse) solution, a Fourier synthesis of solutions of Eq. 28 has been required. An alternative now is the direct solution of Eq. 26; an efficient numerical method to integrate Eq. 26 has been developed.[8] An example of a numerical result from Eq. 26 is presented in Section VI.

V. NONLINEAR PROGRESSIVE WAVE EQUATION (NPE)[5,6]

In addition to the procedure in the above section, we must now include the nonlinear terms in Eq. 2 and 3. From Eqs. 3 and 4, we have for the first term on the r.h.s. of Eq. 5

$$\nabla^2 p = \nabla^2 c^2 [\rho' + \frac{1}{c} \frac{\partial c\ (\rho_0)}{\partial \rho} \rho'^2] \tag{29}$$

We can obtain an expression for the convective term on the rhs of Eq. 5 by assuming propagation to be within a small angle θ to the x axis (same as paraxial approximation for the PE).

We then estimate v_i by using the linear impedance relation $v = p'/\rho_0 c$ together with Eq. 3.

$$v_i = \frac{c\rho'}{\rho_0} [\delta_{i,x} + O(\theta)] + O(\rho'^2), \tag{30}$$

and therefore,

$$\partial_i \partial_j (\rho v_i v_j) \approx \frac{1}{\rho_0} \nabla^2 c^2 \rho'^2. \tag{31}$$

Substituting Eqs. 29 and 31 into Eq. 5 and using the density ratio $\phi \equiv \frac{\rho'}{\rho_0}$ and the nonlinear parameter $\beta \equiv 1 + (\rho/c)(\partial c(\rho_0)/\partial \rho)$, we obtain

$$\frac{\partial^2 \phi}{\partial t^2} = \nabla^2 c^2 (\phi + \beta \phi^2) + O(\phi^2 \theta, \phi^3) + \ldots \tag{32}$$

We again use the time derivative in the moving frame, Eq. 24, to obtain from Eq. 32

$$\left(D_t - c_0 \frac{\partial}{\partial r} \right)^2 \phi = \left(\frac{\partial^2}{\partial r^2} + \frac{1}{r} \frac{\partial}{\partial r} + \frac{\partial^2}{\partial z^2} \right) (c_0 + c_1)^2 \ (\phi + \beta \phi^2) \tag{33}$$

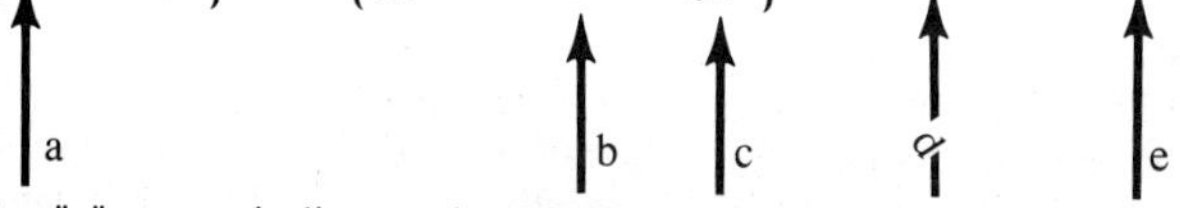

where the "e" arrow indicates the small nonlinear term. The same procedure leading to Eq. 26 now gives the NPE

$$D_t \phi = -\frac{c_0}{2r} - \frac{\partial}{\partial r} \left(c_1 \phi + \frac{c_0}{2} \beta \phi^2 \right) - \frac{c_0}{2} \int^r \frac{\partial^2 \phi}{\partial z^2} dr \tag{34}$$

where the β term includes nonlinear steepening effects. Thus, Eq. 34 illustrates how the presence of a finite amplitude disturbance affects the medium: the nonlinearity can be thought of as effective perturbation in the local sound speed. That is, the substitution $c_1 = c_1 + (1/2)\beta c_0 \phi$ achieves the same result. Where the disturbance is positive (negative) the local propagation speed is increased (decreased). As illustrated in Fig. 3, this leads to the well known steepening of initially smooth pulses into shock fronts.

VI. LINEAR AND NONLINEAR RESULTS FROM THE NONLINEAR PROGRESS WAVE EQUATIONS (NPE)

In this section we present numerical results for these problems using the NPE. New numerical methods were developed to construct our algorithm to solve the NPE.[8]

In order to test the velocity of the NPE approach, the first two problems solved (a linear and a nonlinear problem) have analytic solutions. The third problem: propagation in a refracting medium through a caustic is a new nonlinear wave solution result.

Problem 1. Linear pulse in an ideal waveguide.[7] The analytic solution for a pulse in a waveguide bounded above by a pressure release surface and below by a rigid bottom can be solved analytically and here we compare the analytic result (using the method of images) with the numerical solution of Eq. 26. The initial pulse used is of the form of a derivative of a Gaussian $x \exp(-x^2/L^2)$ for $|x| \leqslant 2L$ where $L = 11.7$m. The depth of the waveguide is 686 m, source depth is 266 m, $c_0 = 1520$ m/s, $c_1 = 0$, and the grid size for the numerics are $(\Delta r, \Delta z, \Delta t) = (.7, 14, .05)$ in mks units. The initial condition shown is that $ct = 6.86$km. The analytic and numerical results are shown in Fig. 5. The analytic results can be thought of as the wideband solution indicated below the FFP in the schematic of Fig. 3. Hence, the results of Fig. 5 indicate the applicability to direct time domain solutions of linear problems. We also note that the numerics developed to solve the NPE is not altered when the waveguide is bounded by penetrable media.

Problem 2. One dimensional nonlinear pulse developing into an "N" wave shock.[8] In this case we propagate an initial finite amplitude pulse in one dimension as indicated as step 1 in Fig. 6. With the antisymmetric pulse constructed as combining part of a parabola with itself inverted, the one dimensional analytic solution of the steepening to an "N" wave is known and is plotted in Fig. 6 together with the solution computed from the NPE. Again, the agreement is excellent. We note here that the shock conditions (the equivalent of the Rankine-Hugoniot condition) are built into the "higher order upwind flux numerics[8] developed for solving the NPE.

Problem 3. Linear and nonlinear propagation through a caustic.[5,6] Previously, much of shock propagation in a refracting medium has been approximated using empirically corrected ray theory since the history of the propagation can be followed along rays. For nonlinear spherical propagation, the empirical range factor[30] is $r^{-1,13}$ as opposed to linear spherical spreading, r^{-1}. (Actually, this nonlinear range dependence factor can be obtained from the NPE.[31]) At a caustic rays converge and the solution blows up. Hence, the empirical approximation used can only apply nonlinear corrections before and after the caustic and has to assume nothing special happened in the caustic region. However, the caustic is a region of high intensity and hence is the region of greatest nonlinearity. The NPE, a finite frequency formulation, contains no such artificial infinities and hence the nonlinear shock wave can be tracked right through a caustic. Consider the problem shown in Fig. 7 where an "N wave" is incident on a caustic. The classical linear solution[30] is that an unattenuated "U wave" emerges from the caustic (that is, there is only a phase change in the frequency domain components of the pulse.) Using the NPE with and without the nonlinear term, we compute the results for this problem as shown in Fig. 8. In the linear case, we see the N wave going to the unattenuated U wave within the numerical resolution of the computation. The contour of the linear acoustic field also shows the expected symmetry of the field on both sides of the caustic. The results are radically different in the nonlinear case. The N wave transforms to a dispersed and dissipated U wave because of the increased generation of the turbulence in the high intensity region of the caustic and the contour shows that the field is assymetric. That is, the shock does not follow the linear ray path of Fig. 7. The high intensity shock wave actually disturbs the medium modifying the local sound speed structure causing the shock wave to "self refract". This means, for example, that focusing of acoustic waves in the ocean such a convergence zone propagation will be different for shock waves than for the linear waves descriptive of sonar acoustics. This numerical solution indicates that we can calculate the nonlinear propagation of a pulse in an arbitrarily refracting medium as found in the ocean.

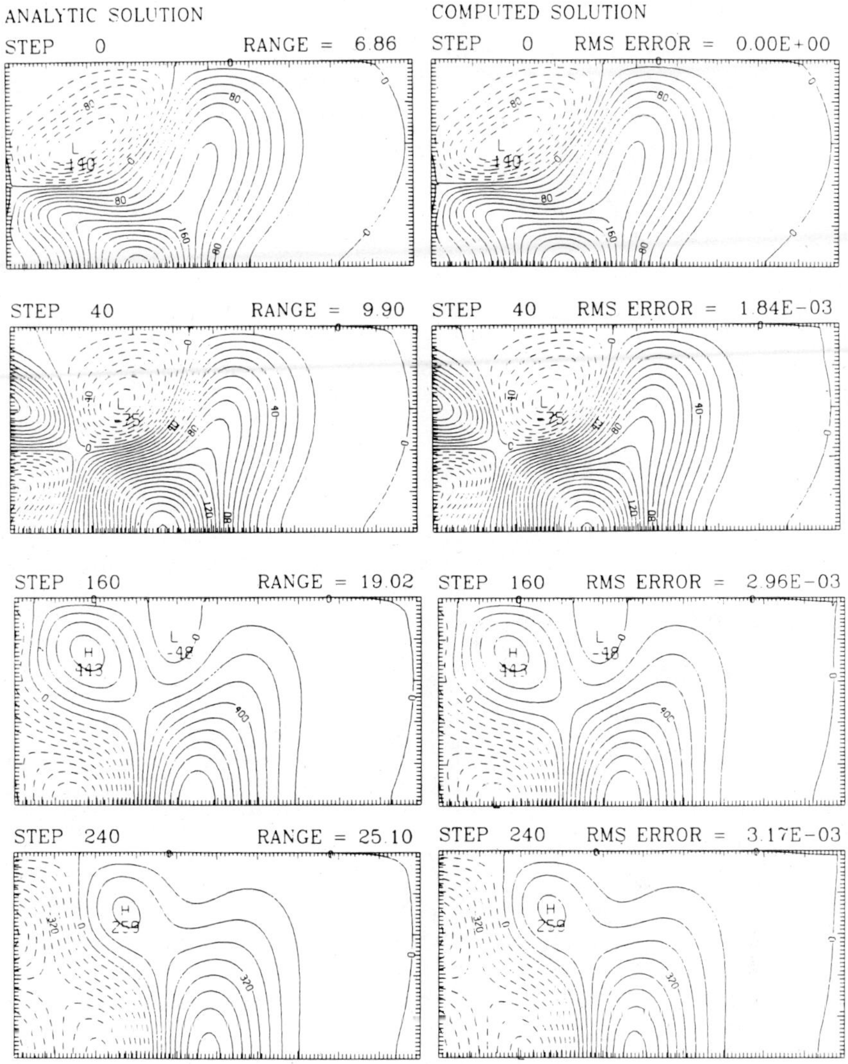

Figure 5. Analytic and computed solutions for Gaussian pulse in a waveguide. Contour values of acoustic density perturbation ϕ are in units of units of 10^{-5}. Range is in km. RMS ERROR is error relative to analytic solution, $\phi_a : ||\phi - \phi_a|| / ||\phi_a||$. Grid size is 686 meters vertical, 69.3 meters horizontal.

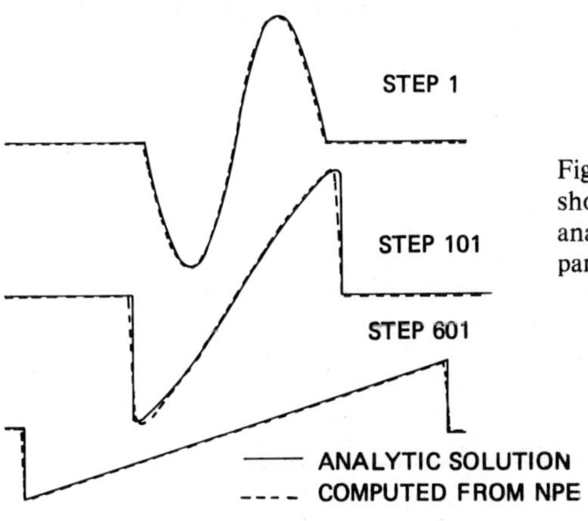

Figure 6. Development of one dimensional shock as computed from NPE and compared to analytic solution. Lobes of initial waveform are parabolas.

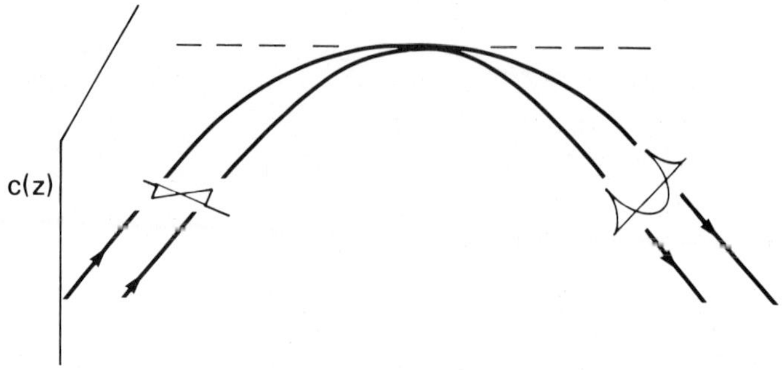

Figure 7. "N wave" nonlinear shock incident on a caustic resulting from refracting sound speed profile $c(z)$. Linear theory predicts emergence of unattenuated "U wave."

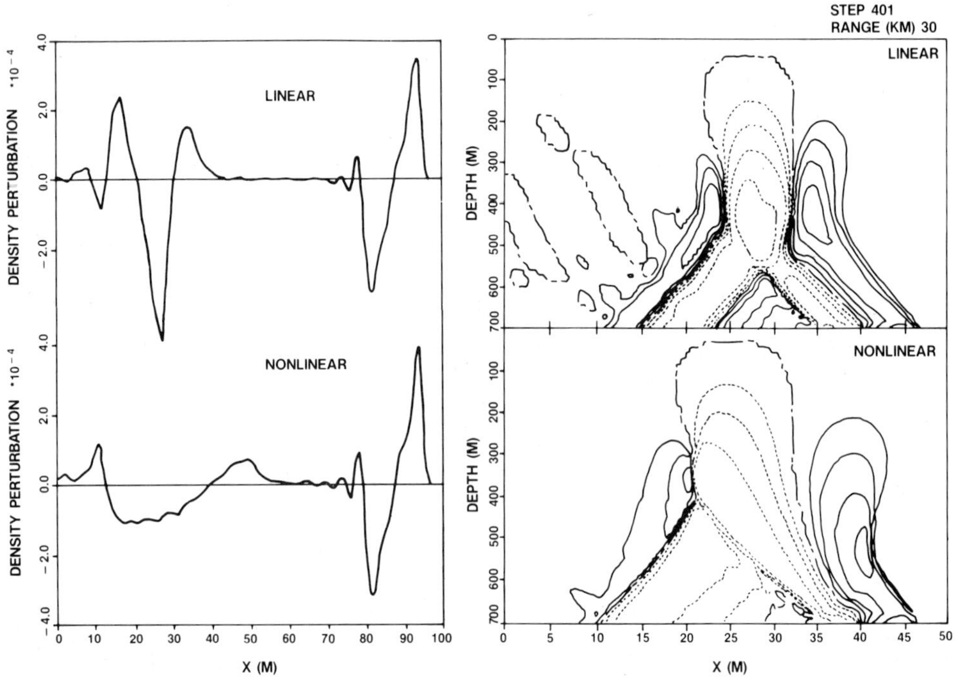

Figure 8. N wave incident on a caustic. The asymmetric nonlinear contour shows self-refraction. Nonlinear pulse form of outgoing wave indicates shock dissipation.

SUMMARY

We have shown that direct wideband calculations for sound propagation is complicated nonlinear refractive media can be performed. The hierarchy of underwater acoustic models, as schematically illustrated in Fig. 2 indicates the various levels where one can begin numerically evaluating complicated propagation phenomena. With the continuing advancement of computer technology, coupled with the better and more sophisticated numerical physics, solutions to more complicated problems will probably be implemented in the future by numerical schemes directed at the highest three levels of the right and left side levels of the block diagram in Fig. 2.

REFERENCES

1. Urick, R. J., Principles of Underwater Sound, 3rd Edition, McGrawHill Book Company, New York, 1983, p. 2.
2. Lasky, M., "Review of underwater acoustic to 1950," J. Acoust. Soc. Am. 61, 283-297 (1977).
3. Pierce, A.D., Acoustics: An Introduction to its Physical Principles and Applications, McGraw-Hill Book Company, New York, 1981, pp. 3-6.
4. Clay, C. S. and Medwin, H., Acoustical Oceanography, Wiley-Interscience, New York, 1977, pp. 395-404.
5. McDonald, B. E., and Kuperman, W. A., "Time Domain Solution of the Parabolic Equation Including Nonlinearity," Proc. Workshop on Comp. Ocean Acoust., Yale Univ. (eds. M. Schultz and D. Lee), August 1984.
6. McDonald, B. E., and Kuperman, W. A., "Time Domain Solution of Nonlinear Pulse Propagation," Proc. NATO Adv. Res. Workshop on Hybrid Formulation of Wave Propagation and Scattering, IAFE, Castel Gandolfo, Italy (ed. L. B. Felson), Sept. 1983.
7. McDonald, B. E., and Kuperman, W. A., "Broadband Pulse Propagation in the Time Domain," Proc. IMACS Symp. on Computing Methods in Ocean Acoustics, Oslo, August 1985.
8. McDonald, B. E., and Ambrosiano, J., "Higher Order Upwind Flux Methods for Scalar Hyperbolic Conservation Laws," J. Comp. Phys. 56, 448 (1984).
9. Kuperman, W. A., and Jensen, F. B., "Deterministic propagation model I Fundemental principles," in Underwater Acoustics and Signal Processing, edited by Leif Bjorno (D. Reidel Publishing Co., Dordrect, 1980).
10. Jensen, F. B., and Kuperman, W. A., "Consistency tests of acoustic propagation models," SACLANT ASW Research Centre, La Specia, Italy, SM-157 (1982).
11. Jensen, F. G., "Numerical Models in Underwater Acoustics," Proc. NATO Adv. Res. Workshop on Hybrid Formulation of Wave Propagation and Scattering, IAFE, Castel Gandolfo, Italy (ed. L. B. Felsen), Sept. 1983.
12. Officer, C. B., Introduction to the Theory of Sound Transmission. (McGraw-Hill, New York, 1958).
13. Tolstoy, I., and Clay, C. S., Ocean Acoustics: Theory and Experiment in Underwater Sound. (McGraw-Hill, New York, 1966).
14. Tucker, D. G., and Gazey, B. K., Applied Underwater Acoustics. (Pergamon Oxford, 1966).
15. Tolstoy, I., Wave Propagation. (McGraw-Hill, New York, 1973).
16. Urick, R. J., Principles of Underwater Sound. (McGraw-Hill, New York, 1975).
17. Clay, C. S., and Medwin, H., Acoustical Oceanography. (Wiley-Interscience, New York, 1977).
18. Keller, J. B., and Papadakis, J. S., (eds.) Wave Propagation and Underwater Acoustics. (Springer-Verlag, Berlin, 1977).
19. De Santo, J. A., (ed.), Ocean Acoustics. (Springer-Verlag, Berlin, 1979).
20. Brekhovskikh, L. M., Wave in Layered Media. (Academic, New York, 1980).
21. Brekhovskikh, L. M., and Lysanov, Yu, Fundamentals of Ocean Acoustics. (Springer-Verlag, Berlin, 1982).
22. DiNapoli, F. R., and Deavenport, R. L., "Theoretical and numerical Green's function field solution in plane multilayered medium," J. Acoust. Soc. Am. 67, 92-105, (1980).
23. Kutschale, H. W., "Rapid computation by wave theory of propagation loss in the Arctic Ocean," Rpt. CU-8-73, Columbia University, Palisades, N. Y., (1973).
24. Schmidt, H., and Jensen, F. B., "Efficient numerical solution technique for wave propagation in horizontally stratified ocean environments," Rpt. SM-173, SACLANT ASW Research Centre, La Specia, Italy (1984).
25. Evans, R. B., "A coupled mode solution for acoustic propagation in a waveguide with stepwise depth variations of a penetrable bottom," J. Acoust. Soc. Am., 74, 188-195 (1983).
26. Tappert, F. D., "The parabolic Approximation method," in J. B. Keller and J. S. Papadakis, Ed., Wave Propagation and Underwater Acoustics, Springer-Verlag, Berlin, 1977, pp. 224-287.
27. Jensen, F. B., and Krol, H. R., "The use of the parabolic equation method in sound propagation modelling," Rpt. SM-72, SACLANT ASW Research Centre, La Specia, Italy (1975).

28. Davis, J. A., White, D., and Cavanagh, R. C., "NORDA parabolic equation workshop," Rpt. TN-143, Naval Ocean Research and Development Activity, NSTL Stn, MS (1982).
29. Lee, D., and Gilbert, K. E., "Recent progress in modeling bottom interacting sound propagation with parabolic Equations," in Ocean 82 Conference Record (MTS-IEEE, Washington, D. C. 1982), pp. 172-177.
30. Rogers, P. H., and Gardner, J. H., "Propagation of Sonic Booms in the Thermosphere," J. Acoust. Soc. Am. 67, 78-91 (1980).
31. McDonald, B. E., and Ambrosiano, J., "Self-Similar Solution for Spherical Weak Shocks," J. Acoust. Soc. Am. Suppl. 1, Vol. 77 (1985).

MODELING OF PULSE RESPONSE FUNCTIONS OF BOTTOM INTERACTING SOUND USING THE PARABOLIC EQUATION METHOD

F. D. Tappert*, and L. Nghiem-Phu†

*University of Miami
Rosenstiel School of Marine and Atmospheric Science
Miami, Florida 33149, USA

†Daubin Systems Corporation
104 Crandon Blvd
Key Biscayne, Florida 33149, USA

ABSTRACT

A special-purpose microcomputer-based system with attached array processor has been developed for generating numerical solutions to the parabolic wave equation for on-board applications. We call this system PESOGEN (Parabolic Equation Solution Generator). We discuss the principles of geo-acoustic modeling as implemented on PESOGEN including: water-sediment interface with range-dependent bathymetry, range-dependent gradients in sound speed and loss in sediment layer, sediment-basement interface and artificial absorption, and especially modeling of rough-bottom scattering. We also discuss the principles of pulse response function modeling as implemented on PESOGEN including: band-limited signals and analytic signals, time-domain transmission loss and phase plots, and time-domain reciprocity. Results and examples include propagation studies and computation of pulse response functions in acoustically complex geographical areas. These studies concentrate on the related issues of sensitivity to bathymetry, and rough-bottom scattering and broad-band coherence.

INTRODUCTION

This article concerns sound propagation between sources and receivers located within the water column in bottom-limited shallow-water environments where propagation takes place to ranges of more than about ten water depths. The frequencies considered are above about 50 Hz. Under these conditions, a range-dependent acoustic model is needed for accurate predictions. Furthermore, the known characteristics of sea-bed sediments indicate that water-borne propagation paths dominate the received signals, and that significant angles of propagation are normally limited to less than the critical angle of about 20° from horizontal. These circumstances make the parabolic equation (PE) model of sound propagation the model of choice. In this article, we report preliminary results of using the PE model to study time-domain rough-bottom scattering in shallow-water environments.

PARABOLIC EQUATION MODEL FOR CW WAVES

We assume that propagation takes place within the vertical plane having horizontal coordiante x and depth coordinate z. Let the acoustic pressure p due to an omnidirecitonal source at $z = z_A$, $x = x_A$ having frequency components ω, be represented as

$$p = p_{Ao}(\omega)\left(\frac{R_o}{x-x_A}\right)^{1/2} \psi_A(\omega,z,x)\, e^{ik_o(x-x_A)} \tag{1}$$

where R_o is the reference range, $p_{Ao}(\omega)$ is the pressure at range R_o, $k_o = \omega/c_o$, and c_o is the reference sound speed. Then $\psi_A(\omega,z,x)$ satisfies the parabolic wave equation,[1,2,3]

$$i\frac{\partial\psi_A}{\partial x} + \frac{1}{2k_o}\frac{\partial^2\psi_A}{\partial z^2} - k_o\, U(z,x)\psi_A = (-2\pi i R_o/k_o)^{1/2}\,\delta(z-z_A)\,\delta(x-x_A) \tag{2}$$

The environmental function U(z,x) is given by

$$U(z,x) = u(z,x) + \beta(z,x) + v(z,x)\ , \tag{3}$$

$$u(z,x) = \frac{1}{2}\,[1-c_o^2/c^2(z,x)] - i\ \alpha(z,x)/k_o, \tag{4}$$

$$\beta(z,x) = \frac{1}{2k_o^2}\left[\frac{1}{\rho}\frac{\partial^2\rho}{\partial z^2} - \frac{3}{2}\left(\frac{1}{\rho}\frac{\partial\rho}{\partial z}\right)^2\right]. \tag{5}$$

This equation describes propagation in a fluid medium (including water, sediments, and substrate) having general range-dependent sound speed $c(z,x)$, volume loss $\alpha(z,x)$, density $\rho(z,x)$, and current $v(z,x)$.

At an interface between fluid layers where c, α, and ρ may be discontinuous, we employ a smooth "mixing" function F(z) having the property that F(z) tends rapidly to zero when $z \gg L$ and F(z) tends rapidly to unity when $z \ll -L$. We use the particular function (Fermi-Dirac),

$$F(z) = (1 + e^{z/L})^{-1} \tag{6}$$

The use of this mixing function is illustrated by the treatment of the water-sediment interface at depth $z = z_b(x)$. Let $u_w(z,x)$ be the function defined by Eq. (4) within the water column and let $u_b(z,x)$ be the corresponding function within the bottom sediments; both u_w and u_b are extrapolated to the other side of $z = z_b(x)$. Then the composite environmental function becomes

$$u(z,x) = u_w(z,x)\, F(z - z_b(x)) + u_b(z,x)[1 - F(z - z_b(x))]. \tag{7}$$

Density jumps are treated using the same mixing function and then $\beta(z,x)$ is calculated according to Eq. (5). By choosing $L = const/k_o$, one can assure that $|U(z,x)| \ll 1$ while retaining the physically correct small-angle reflection properties of the sea-bed.[1,2] In this manner, highly complex range-dependent geo-acoustic bottom models are easily included in the PE propagation model.

ROUGH BOTTOM MODEL AND RECIPROCITY

The bottom roughness model[4] used in this study is specified by allowing the water depth $z_b(x)$ to contain an additive zero-mean Gaussian random function,

$$z_b(x) = \bar{z}_b(x) + R(x) , \tag{8}$$

where $\bar{z}_b(x)$ is the deterministic water depth along the track between source and receiver as obtained from smooth bathymetric charts, and R(x) is a Gaussian random function with

$$\langle R(x) \rangle = 0 , \tag{9}$$

$$\langle R(x) R(x + x') \rangle = \int_{-\infty}^{\infty} S(k)\, e^{ikx'} dk . \tag{10}$$

The non-negative function S(k) is the power spectrum of bottom roughness, and the standard deviation of bottom roughness is

$$\sigma = \langle R^2(x) \rangle^{1/2} = [\int_{-\infty}^{\infty} S(k)dk]^{1/2} . \tag{11}$$

The random function R(x) may be regarded as representing either unchartable bottom roughness (random relief) or errors in the given bathymetric charts.

In this study, we have used the power spectrum,

$$S(k) = \frac{1}{2} \sigma^2 L_c (1 + k^2 L_c^2)^{-3/2} \tag{12}$$

where L_c is the horizontal correlation length of bottom roughness. Asymptotically for $kL_c >> 1$, this spectrum has the well-known self-similar fractal form,[5]

$$S(k) \sim \frac{1}{2} \gamma^2 k^{-3} \tag{13}$$

where $\gamma = \sigma/L_c$ is the slope parameter.

Realizations of R(x) are generated numerically using Fourier transform methods. For each realization, water depth given by Eq. (8) is put into the environmental function U(z,x) according to Eqs. (3), (4), (5) and (7). Then Eq. (2) is solved numerically using the split-step Fourier algorithm to obtain $\psi_A(\omega,z,x)$. Conceptually, this Monte Carlo procedure is similar to that used in Ref. 6 to study internal wave effects on sound propagation.

Let the receiver be located at $z = z_B$, $x = x_B$ (with $x_B > x_A$). The received pressure is given by

$$P_{AB}(\omega) = P_{Ao}(\omega) \left(\frac{R_o}{x_B - x_A} \right)^{1/2} \psi_A(\omega, z_B, x_B) e^{ik_o(x_B - x_A)} . \tag{14}$$

Similarly, propagation from B to A is governed by

$$-i\frac{\partial\psi_B}{\partial x}+\frac{1}{2k_o}\frac{\partial^2\psi_B}{\partial z^2}-k_oU(z,x)\psi_B=(-2\pi iR_o/k_o)^{1/2}\delta(z-z_B)\delta(x-x_B) \tag{15}$$

The environmental function U(z,x) is the same as used above. It is then easy to show[3] that, for each realization, transmission is exactly reciprocal, i.e.,

$$P_{AB}(\omega)/P_{AO}(\omega) = P_{BA}(\omega)/P_{BO}(\omega) \ . \tag{16}$$

Since this relation between stochastic fields holds for each realization of the random function R(x), it also holds for the ensemble average (coherent pressure),

$$\langle P_{AB}(\omega)\rangle / P_{AO}(\omega) = \langle P_{BA}(\omega)\rangle / P_{BO}(\omega) \ , \tag{17}$$

as well as for ensemble averages of moments such as the acoustic intensity,

$$\langle|P_{AB}(\omega)|^2\rangle/|P_{AO}(\omega)|^2 = \langle|P_{BA}(\omega)|^2\rangle/|P_{BO}(\omega)|^2. \tag{18}$$

TIME DOMAIN TRANSMISSION LOSS AND RECIPROCITY

Following Ref. 3, the complex-valued analytic signal is computed as

$$P_{AB}(t) = \left(\frac{R_o}{x_B-x_A}\right)^{1/2}\int_0^\infty P_{AO}(\omega)\psi_A(\omega,z_B,x_B)e^{i\omega[(x_B-x_A)/c_o-t)]}d\omega, \tag{19}$$

and the time domain transmission loss (dB re 1 yard) for a single realization is given by

$$TL_{AB}(t) = -10\ \log[|P_{AB}(t)|^2/|P_{AOP}|^2] \ , \tag{20}$$

where $|P_{AOP}|$ is the peak value of the transmitted signal at range R_o = 1 yard. Likewise, time domain transmission from B to A, $P_{BA}(t)$, is computed from solutions of Eq. (15) using a formula analogous to Eq. (19). Assuming that the transmitted pulses have the same shape, it was shown in Ref. (3) both theoretically and computationally that time domain reciprocity holds,

$$P_{AB}(t)/P_{AOP} = P_{BA}(t)/P_{BOp} \ . \tag{21}$$

Thus for each realization, we have

$$TL_{AB}(t) = TL_{BA}(t), \tag{22}$$

and it makes no difference whether the acoustic model propagates sound from A to B or from B to A. This is also true for ensemble averages.

It is always possible to separate the acoustic pressure into a coherent (specularly reflected) part and incoherent (scattered) part as follows:

$$P_{AB}(t) = \langle P_{AB}(t)\rangle + \delta P_{AB}(t), \tag{23}$$

$$< \delta P_{AB}(t) > = 0 \ . \tag{24}$$

The received intensity is a sum of the coherent intensity and incoherent intensity,

$$< |P_{AB}(t)|^2 > = |< P_{AB}(t) >|^2 + <| \delta P_{AB}(t)|^2 > \ . \tag{25}$$

We may define the coherent transmission loss,

$$TL_{AB}^{coh}(t) = - 10 \log [|<P_{AB}(t)>|^2/|P_{AOP}|^2], \tag{26}$$

the incoherent transmission loss,

$$TL_{AB}^{inc}(t) = - 10 \log [<| \delta P_{AB}(t)|^2>/|P_{AOP}|^2], \tag{27}$$

and the average total transmission loss,

$$TL_{AB}^{tot}(t) = - 10 \log [<|P_{AB}(t)|^2>/|P_{AOP}|^2] \ . \tag{28}$$

From Eq. (25) these three kinds of time domain transmission loss are related by logarithmic addition:

$$TL_{AB}^{tot}(t) = TL_{AB}^{coh}(t) \oplus TL_{AB}^{inc}(t). \tag{29}$$

Obviously we have $TL_{AB}^{coh}(t) \geq TL_{AB}^{tot}(t)$.

We have computed and displayed only the coherent and average total transmission loss since the third may readily be deduced from the other two. By time domain reciprocity, each of these quantities is also reciprocal:

$$TL_{AB}^{coh}(t) = TL_{BA}^{coh}(t) \ , \tag{30}$$

$$TL_{AB}^{tot}(t) = TL_{BA}^{tot}(t) \ . \tag{31}$$

Again, it makes no difference whether sound is propagated from the source to the receiver or from the receiver to the source.

NUMERICAL EXAMPLES

In order to explore rough bottom scattering in a simple situation, we first studied an idealized problem similar to that reported in Ref. 3, which is a simplified version of propagation in the Straits of Florida. The source transmitted a band-limited (100 Hz to 300 Hz) pulse of energy having duration (3dB full width) about 5 ms. The sound speed profile was linear and downward refracting:

$$c(z) = c_s + gz, \ c_s = 5070 \text{ ft/s}, \ g = -0.144s^{-1}.$$

The deterministic water depth was constant with z_b = 1538 ft. The bottom sound speed and loss were also constant: c_b = 5346 ft/s and α_b = 0.1 dB/nmi-Hz (thus the critical angle was about 25°). The bottom density was equal to the water density. The source depth was z_A = 1200 ft, and the receiver depth was z_B = 1172 ft. The range between them was 10.8 nmi.

Realizations of bottom roughness functions R(x) were generated using a 2048 point FFT on a mesh having Δx = 164 ft. We used L_c = 500 ft, and the following standard deviations: σ = 0, 25 ft, 50 ft, 100 ft. For each value of σ, we generated J = 8 independent realizations of R(x). For each of these, we solved Eq. (2) using the split-step Fourier algorithm for 128 equally-spaced frequencies within the 200 Hz band-width using depth mesh 3.75 ft and range step Δx = 164 ft. In this manner we obtained J independent functions

$$P_{AB}^{(j)}(t),\ j = 1,2,\ldots,\ J,$$

each having record length 640 ms (twice that used in Ref. 3). We then computed ensemble averages using

$$\langle P_{AB}(t) \rangle = \frac{1}{J} \sum P_{AB}^{(j)}(t)\ , \tag{32}$$

$$\langle |P_{AB}(t)|^2 \rangle = \frac{1}{J} \sum |P_{AB}^{(j)}(t)|^2\ . \tag{33}$$

Finally we computed $TL_{AB}^{coh}(t)$ and $TL_{AB}^{tot}(t)$ according to Eqs. (26) and (28).

The methodology developed in this article is clearly very general. In parallel studies,[4,7,9] we are applying the same methodology to determine the effects of rough bottom scattering in two geographical locations where much experimental data exists: the Straits of Florida, and a deep-ocean seamount case. In those cases, our environmental model is far more complicated: including range-dependent sound speed profiles and current profiles, range-dependent deterministic bathymetry, and more complex range-dependent geo-acoustic bottom models. Only the idealized example will be further discussed in this article. All numerical PE runs were actually made with the wide-angle version of the split-step Fourier PE model.[2,8]

RESULTS AND DISCUSSION

The results are shown in Fig. 1, (a) to (c). Transmission loss is displayed on an absolute scale between 70 dB and 130dB re 1 yard. The solid curves give the average transmission loss, and the dotted curves give the coherent transmission loss. The record length is 640 ms, and later arrivals are wrapped around.

When σ = 0 (flat bottom), we obtain a pulse response function that is dominated by about 6 strong RBR arrivals that are separated (although some have multiplicity greater than 1 due to caustics) and are spread over a time interval of about 100 ms. In addition there are many weaker SRBR arrivals, some of which are intermixed with the RBR arrivals and some of which trail out behind for several hundred milli-seconds. The peaks of both RBR and SRBR arrivals stand up about 20 dB above the temporal sidelobes. The expected ground wave arrival is much too weak to show on this plot. No sub-bottom arrivals are expected or seen, because the sediment layers were modeled as homogeneous and absorbing in this example.

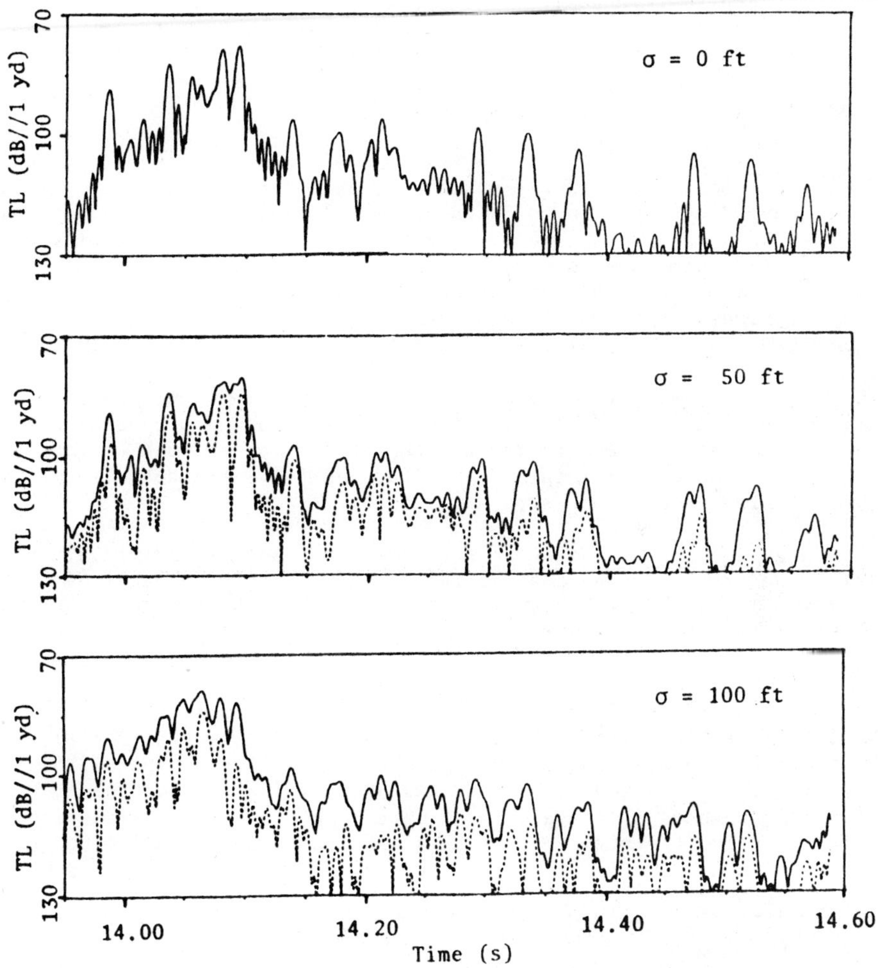

Fig. 1. Computed time domain transmission loss functions: Solid lines are $TL_{AB}^{tot}(t)$ and dotted lines are $TL_{AB}^{coh}(t)$. (a) $\sigma = 0$ ft, (b) $\sigma = 50$ ft, (c) $\sigma = 100$ ft.

When σ = 25 ft (not shown), we found that scattering is quite weak and $TL_{AB}^{coh} \approx TL_{AB}^{tot}(t)$. The loss of coherence is very small, less than about 3 dB.

When σ = 50 ft, we see that scattering causes the total energy arrivals to spread in time, and scattered energy tends to fill in the time intervals between the specularly reflected arrivals. Because of the sound speed profile, energy scattered from the specular paths may arrive either earlier or later than the unscattered waves, forming either precursers (typical for RBR arrivals) or tails (more typical for SRBR arrivals). Although the great dynamic range (60 dB) of our computations makes the effects of scattering appear to be large, it should be noted that the coherent intensity is reduced by only about 5 to 10 dB when compared to the flat bottom case. Further, the coherent part of the pulse response function is quite stable, even when the rms bottom roughness is 50 ft. Overlays show that the coherent arrivals are nearly in one-to-one correspondence with flat bottom arrivals.

Finally when σ = 100 ft we see that almost all of the received energy has been incoherently scattered, and that the coherent part has been reduced by about 10 to 20 dB or so. The total transmission loss of the RBR arrivals has become a nearly continuous smear lasting about 150 ms, with almost no separable arrivals. The last RBR arrival is very stable in time, but the earlier RBR arrivals are scattered toward even earlier times. At lower levels, some of the SRBR arrivals can still be separated, but are significantly spread in time and reduced in intensity.

Qualitatively these results are not surprising, and are what one would anticipate based on intuitive notions of sound propagation in a bottom-limited downward-refracting channel having a rough bottom. However, because of multiple scattering and full-wave near-field effects, quantitative predictions were not possible before now.

When looking at the pulse response functions of individual realizations, we have observed that arrival time "wander" dominates over time "spread" for the SRBR arrivals. This indicates that the size of the "footprints" of the higher-angle SRBR waves impinging on the bottom is smaller than the correlation length L_c = 500 ft. The opposite appears to be true for the smaller-angle RBR arrivals, for then time "spread" dominates over time "wander". One consequence of this is that the number of realizations used to compute ensemble averages (J = 8) was not sufficient to assure convergence of the coherent transmission loss, especially for σ = 100 ft and for the SRBR arrivals. For these arrivals, we have obtained only an upper bound on the coherent intensity. Unfortunately, the high cost of this massive computational effort (about 45 min. CPU time per realization) prevented additional runs.

CONCLUDING REMARKS

The idealized example studied in this article contains many parameters: the rms displacement and correlation length of the bottom roughness, the center frequency and bandwidth of the transmitted pulse, the water depth, the sound speed gradient in the water, the sound speed and volume loss of the bottom sediments, the source and receiver depth, and the range between them. A conventional theoretic approach to this problem of stochastic propagation would attempt to derive formulas relating measurable quantities to all of the above parameters. The theory would have to include near-field scattering effects, time-domain caustic effects, and other complicated but actually occurring effects. No such theoretical formulas are known, even for this idealized example. When other features of the real ocean environment are taken into account, such as range-dependent deterministic bathymetry and complex range-dependent sound speed profiles, then it is unlikely that such theoretical formulas will ever be known.

Although it is possible to exercise our acoustic model by systematically varying one parameter after another in such a way that approximate formulas could be extracted, we instead suggest that the conventional theoretic approach should be abandoned and that the acoustic model itself should be regarded as a theory. That is, the computer program that generated the results described in this article represents a theory, the correctness of which can be tested by comparing the predictions of the acoustic model (theory) directly with experimental measurements in the real ocean. If the predictions of the theory (acoustic model) agree with the experimental measurements, then the theory is correct and the physical phenomena associated sound propagation in shallow water having rough bottoms are understood. If the predictions are not correct, then the theory (acoustic model) needs improvement. We are presently engaged in a comparison of this kind using recently acquired experimental data of broad-band acoustic transmission in the Straits of Florida.[10]

ACKNOWLEDGEMENT

The PESOGEN system at DSC was graciously made available for all of the numerical computations. PESOGEN was developed at DSC under NORDA Contract No. N00014-81-C-0706.

REFERENCES

1. F. Tappert, "Selected applications of the parabolic equation method in underwater acoustics", in Intern. Workshop on Low-Frequency Propagation and Noise (Woods Hole, MA, Oct. 1974), Vol. 1, pp. 155-194.
2. F. Tappert, "The parabolic approximation method", in Wave Propagation and Underwater Acoustics, ed. J.B. Keller and J.S. Papadakis (Springer-Verlag Lecture Notes in Physics, Vol. 70, 1977), pp. 224-287.
3. L. Nghiem-Phu and F. Tappert, "Modelling of reciprocity in the time domain using the parabolic equation method", J. Acoust. Soc. Am. 78(1985).
4. L. Nghiem-Phu and F. Tappert, "Parabolic equation modelling of bathymetry: Sensitivity to bathymetric irregularity and rough bottom scattering", J. Acoust. Soc. Am. Suppl. 1, 77, S102(1985).
5. J.M. Berkson and J.E. Matthews, "Statistical properties of seafloor roughness", in Acoustics and the Sea-Bed, ed. N.G. Pace (Bath Univ. Press, Bath, UK, 1983), pp. 215-223.
6. S.M. Flatte and F.D. Tappert, "Calculation of the effect of internal waves on oceanic sound transmission", J. Acoust. Soc. Am. 58, 1151-1159(1975).
7. L. Nghiem-Phu and H.A. DeFerrari, "A comparison of parabolic equation predictions with reciprocal transmission experiments in the Straits of Florida", J. Acoust. Soc. Am. Suppl. 1: 75, S25 (1984).
8. D.J. Thomson and N.R. Chapman, "A wide-angle split-step algorithm for the parabolic equation", J. Acoust. Soc. Am. 74, 1948-1954 (1983).
9. L. Nghiem-Phu, "Numerical modeling of acoustic tomography in the Straits of Florida", Ph.D. Dissertation, Univ. Miami, May 1985.
10. H. A. DeFerrari and H. Nguyen "Acoustical reciprocal transmission experiments, Florida Straits", submitted to JASA, March 1985.

A STUDY OF PROPAGATION LOSS DEPENDENCE ON SEDIMENT LAYER THICKNESS USING THE FAST FIELD PROGRAM

C. H. Harrison and P. L. Cousins

CAP Scientific
40-44 Coombe Road
New Malden
Surrey KT3 4QF, UK

ABSTRACT

A propagation loss law of the form Alogr+Br+C is often used as an approximate expression for low frequencies and shallow water. At short ranges where there are contributions from many modes A=20, and at long ranges where there is only one surviving mode A=10. In the mode stripping region A has an intermediate value. It has been suggested that sediment layer thickness may often influence the value of the decay component B in shallow water because of its control over shear loss. If the bedrock has a lower shear speed than the compression velocity in water then downward propagating shear waves may be generated when the sediment layer is thin enough causing the waterborne modes and interface modes to decay. This mechanism produces decreasing loss as layer thickness increases since the layer acts as a screen preventing excitation of shear waves in the bedrock. An alternative mechanism where the intermediate layer is lossy produces an increasing loss with increasing layer thickness by straightforward absorption. This paper uses the Fast Field Program (FFP) to investigate these effects parametrically in a three layer simulation. It is shown that dependence of loss on layer thickness may take on a wide variety of laws.

INTRODUCTION

In submarine acoustics the main interest in the sea bed is in its influence on propagation loss measured between points in the water. From a theoretical point of view it is relatively easy to see mechanisms for losses involving compression and shear waves in the bed rock or sediment, absorption in, and tunnelling through sediment layers. From the practical point of view the large number of unknowns make it much more difficult to see what is going on, and one needs a more simplified approach. For instance, the accompanying paper by Lindop[1] suggests a relation between loss and sediment thickness. What is most useful in practice is to know the mean loss versus range. This can be obtained by fitting a function of the form Alogr+Br+C, and the most interesting term in the current context is the exponential fall-off, B.

In this paper we attempt to compare like with like and perform the same smoothing or curve fitting exercise on the output of a reliable wave treatment model - namely Fast Field Program [2,3]. This allows us to investigate the loss term B as a function of layer thickness L while varying other parameters of the sediment and bed rock. The plots of B versus L that are shown later display many different laws.

LOSS MECHANISMS

Several mechanisms for influencing the propagation loss are considered. These include absorption in one or more layers; generation of downward propagating shear waves at the rock boundary; tunnelling through a thin sediment layer, again to produce shear loss; upward refraction in the sediment layer; and resonance effects. In the three layer model used here we concentrate on varying the thickness and parameters of the intermediate sediment layer. Also, in order to distinguish the effects of shear loss we assume three bed rock types including shear speeds higher and lower than the sound speed in water.

Loss By Absorption In One Or More Layers

Absorption in any layer will cause modal attenuation. The importance of the absorption in any particular layer depends on the mode amplitude in that layer; a strong absorption in a lower layer would have little effect if the mode amplitude had decayed almost to zero in it.

Miller and Ingenito[4] give the modal attenuation coefficient δ (the imaginary part of the wavenumber) in a three layer medium as

$$\delta = \varepsilon_1 \gamma_1 + \varepsilon_{2s} \gamma_{2s} + \varepsilon_{2c} \gamma_{2c} + \varepsilon_{3s} \gamma_{3s} + \varepsilon_{3c} \gamma_{3c}$$

The absorption D in each layer in terms of dB per wavelength is related to the attenuation coefficients ε by

$$\varepsilon = \frac{2.303D}{10\,\lambda}$$

The γ terms are given in detail in Ref 4 and they are typically between 0 and 1. For instance the expression for γ_2 is proportional to the integral of mode amplitude squared across the layer. Since the mode amplitude is normalised over the whole depth each γ increases with the thickness of the appropriate layer. We therefore expect an increase in δ and hence the loss term B, (B= δ.10loge for a single mode), as layer thickness is increased.

Loss By Shear Wave Generation in the Sea Bed

With or without any absorption in the layers a solid sea bed with low shear velocity can result in a loss mechanism by production of downward propagating shear waves. The modal decay constant has been calculated by Ingenito and Wolf[5] (in the two layer case), and it agrees with the ray formula (grazing angle θ) for a uniform velocity duct of depth H with imperfect reflector of amplitude reflection coefficient R.

$$\delta = \frac{\tan\theta}{4H}\left(1 - |R|^2\right)$$

Brekhovskikh[6] derives a formula for reflection coefficient under these conditions which, for small grazing angles, reduces to a formula given by Weston[7].

$$1 - |R|^2 = 16\ m\ n_s{}^3\ (1 - n_s{}^2)^{\frac{1}{2}}\ \theta$$

where m is the ratio of rock to water density and n_s is the ratio of rock shear velocity to sound speed in water. The decay constant is now

$$\delta = \frac{4m\ n_s{}^3\ (1 - n_s{}^2)^{\frac{1}{2}}\ \theta^2}{H}$$

and this represents the relatively high losses associated with shear wave generation.

Loss Through a Thin Fluid Sediment Layer

If the bed rock has a low shear speed but is separated from the water column by a thick fluid sediment layer of relatively high velocity there is no longer a shear loss; the sediment traps modes in the water column. Clearly, there must be a transition as the sediment becomes thinner at which the shear loss returns.

The problem can be viewed as one of "tunnelling" out of a potential well as in quantum mechanics. The "well" is the water duct, the wall of the well to be tunnelled through is the the sediment layer, and the outside of the well is an infinite space with lower (shear) velocity than the inside of the well. The mode shape inside and outside the well is periodic, but in the intermediate layer it is composed of exponential or cosh and sinh functions. If the decay through the layer is small the mode carries on unimpeded in the vertical and produces a lossy mode. If the decay through the layer is large the mode is trapped, and there is no shear loss. There is therefore a slow transition from lossy mode to trapped mode as the sediment thickness is increased.

If the wavenumber in the sediment is k and the horizontal wavenumber is K then the mode decays vertically as exp $(-(K^2-k^2)^{\frac{1}{2}}z)$. For the first mode K is virtually the wavenumber in the water, and the change-over from lossy to trapped would be expected when

$$2\pi fL\ (1/c_1{}^2 - 1/c_2{}^2)^{\frac{1}{2}} = 1$$

where f, L, c_1, c_2 are frequency, sediment thickness, velocity in water and velocity in sediment. Putting f=25Hz, c_1=1500m/s, c_2=1600m/s, we obtain L=27 metres.

Upward Refraction in the Sediment Layer

Upward refraction in a sediment layer with positive velocity gradient has often been observed and has been reported in the literature as the cause of 'negative' reflection losses through the formation of a caustic in the water. An example is the experimental and theoretical investigation carried out by Frisk et al[8]. In the current context the upward curving ray in the layer provides a different (larger) transition depth between shear loss and no loss, but otherwise behaviour is similar to the isovelocity case considered above. The depth will be close to the point on the profile where the velocity in the sediment equals that in water.

Resonance in the Sediment Layer

If the sediment supports very low velocity shear waves then any ray in the water hitting the sediment boundary generates a shear wave that refracts almost vertically downwards. There is then a chance that interference between reflections from the two boundaries will produce significant resonance effects despite the quasi total internal reflection of the compression wave. One would expect peaks in resonance at spacings in layer thickness of one half (shear) wave length with the first null at about one quarter wavelength.

EXPECTED RANGE DEPENDENCE

Regardless of the detailed loss mechanisms the form of the loss law is predictable. The intensity due to an individual mode falls off as cylindrical spreading with a superimposed exponential decay, higher order modes decaying most rapidly. Expressing the pressure field as an incoherent sum of modes one can see that the decay rate of the total intensity is most rapid at short ranges where there are many modes. The change in decay rate in this 'mode stripping' region can be interpreted as a larger value of A in the expression Alogr+Br+C.

Weston[7] has put forward theoretical justification for A being 15 in this region for the two cases where the bottom is absorbing and where the bottom has a low shear velocity. Under these conditions the dB loss per bounce can be shown to be proportional to grazing angle (provided angles are small) so that there is indeed a higher loss for higher order modes. When only one mode remains then, of course, A reverts to 10.

The change-over range between mode stripping and single mode is expected to be[7]

$$r = 27\ H^3/\lambda^2\alpha$$

where H, λ , α are water depth, wavelength and dB loss per radian. As an example, setting H=100m, λ=60m, and α=2dB/rad. gives a change-over range of 3.7km.

APPROACH

The Fast Field Program

The approach in this study is to look at the dependence of B on sediment layer thickness L by performing numerous runs of FFP[2,3] with different environmental parameters. A smooth curve is then fitted to the loss versus range curve to obtain a value of B.

The FFP is a very powerful program for research because it gives an accurate solution of the wave equation for mixed liquids and solids (including interface waves) at any range including near field, above, below or near mode cut-offs. Its two shortcomings are that it can (so far) only deal with horizontally stratified media and its computation time is rather long.

In the context of curve fitting, an incoherent mode summation might have been more appropriate, but the benefits of FFP were thought to outweigh the losses.

Curve Fitting

Three types of curve fitting have been tried, using the last for the main investigation. The first uses a least squares matrix technique to fit Alogr+Br+C to the log of the intensity. The second uses a similar technique to fit 10logr+Br+C to the log of the intensity. The third uses an iterative NAG library routine [9] to fit 10logr+Br+C to the log of intensity but with least square deviation in linear intensity.

The first method tends to produce unstable values of A (sometimes much larger than 20) depending on the amount of modal interference and the range limits chosen for the fit. This is understandable in retrospect since at large r values the shape of a log curve is indistinguishable from a linear curve.

It was therefore decided to set A=10 regardless of the mode stripping-single mode change-over, and the third option with least square deviation in linear intensity was thought most desirable since this is the usual experimental technique. In the following examples range limits are about 5km to 50km which corresponds more or less to the single mode region.

Clearly, curve fitting is less straightforward than one would imagine, but the difficulties with fitting to the theoretical model simply highlight the hidden hazards when using real experimental data. They also make one ask the question, is it worth worrying about the true value of A?

LOSS FACTOR B VERSUS SEDIMENT THICKNESS L

Figures 1-6 show some examples of the many types of variation of loss with layer thickness that are possible. In all cases a three layer model is assumed with isovelocity water (1500m/s, 150m thick) and a frequency of 25Hz. Three types of bed rock are assumed as indicated by the names 'hard', 'intermediate' and 'soft'. In the 'hard' case the parameters are as follows:

c_p=3000m/s, c_s=2500m/s, a_p=.2dB/w, a_s=.5dB/w, ρ = 2.6

where c, a, ρ are sound speed, attenuation and density, and subscripts p and s indicate compression and shear. In the 'soft' cases all parameters are reduced in proportion so that the shear velocity is well below the sound speed in water.

c_p=2250m/s, c_s=1000m/s, a_p=.4dB/w, a_s=1.0dB/w, ρ = 2.2

Similarly the 'intermediate' case has

c_p=2600m/s, c_s=1550m/s, a_p =.3dB/w, a_s =.75dB/w, ρ = 2.4

The remaining variations are due to changing the parameters in the sediment, and these include c_p, c_s, a_p and a possible gradient in c_p. However, unless otherwise stated, the parameters in the sediment layer are

c_p = 1600m/s, c_s=0, a_p = 0, a_s = 0, ρ = 1.8

Sediment parameters are shown in each figure, and bottom type is indicated by name. Source and receiver depths in these examples are 61 and 80 metres respectively. Two other receiver depths 40 and 120 metres have been tried, but differences in B are usually less than 1%. Nevertheless, some cases involving a number of trapped modes result in variations of nearly a factor of 2 in B. This is caused by the varying emphasis given to modes of different decay rate by their own vertical shape.

Figure 1 shows variation with bedrock type when the sediment is fluid and loss-free. As expected there is a smooth transition between the shear loss for a thin layer and low loss for a thick layer. Increasing the shear speed in the bed rock reduces the shear loss effect, but an increase well past the water velocity is necessary to eliminate the loss.

Figure 2 again shows variation with bed rock type but with two cases of absorption, a_p=1 and 0.1 dB/w in the fluid sediment layer. Behaviour is similar to Fig 1 except that there is a significant residual loss due to absorption for very thick layers, and this forms the asymptote at the right hand side. Note that for the hard bottom B rises monotonically whereas for the soft bottom it falls monotonically.

The solid lines in Fig 3 show the effect of altering the offset of a constant gradient profile in the sediment. The compression velocity increases by 2 m/s per metre depth, and the velocities at the upper boundary are 1300, 1480, 1520 m/s. As expected, upward refraction allows deeper loss-free rays the lower the starting velocity at the upper boundary. This results in a shift of the curves to the right.

The dashed lines in Fig 3 show contrasting effects when the sediment is absorbing (a_p=1dB/w). The large velocity at the bottom of a thick sediment layer (caused by the gradient) implies an increasing number of modes. This makes little difference in the loss-free case, but as the sediment layer begins to form the greater part of the duct, absorption in the layer can produce large losses for thick layers.

The effect of increasing a uniform compression velocity above 1500 m/s is shown in Fig 4. The mode's exponential tail is shorter for high velocity sediment so the change over is expected at smaller L as shown.

In Fig 5 the solid line shows the effect of a finite shear velocity in the sediment layer of 200 m/s. There are strong resonance beats as predicted, although their exact spacing is complicated. The dashed line shows the fluid layer case (c_s=0) for comparison.

Figure 6 shows similar behaviour with absorption of a_p=1, a_s=2 dB/w introduced. There are still resonance effects but the amplitude is reduced.

CONCLUSIONS

Figures 1-6 show a number of dependences of the loss parameter B on sediment layer thickness L at one frequency (25Hz) in water of 150 metres depth. A qualitative frequency dependence may be seen through the rough inverse relation with layer thickness. Increasing, decreasing, U-shaped, and oscillatory behaviour of B are seen, and these result from the following significant effects (and combinations of them).

1) Absorption in the sediment.

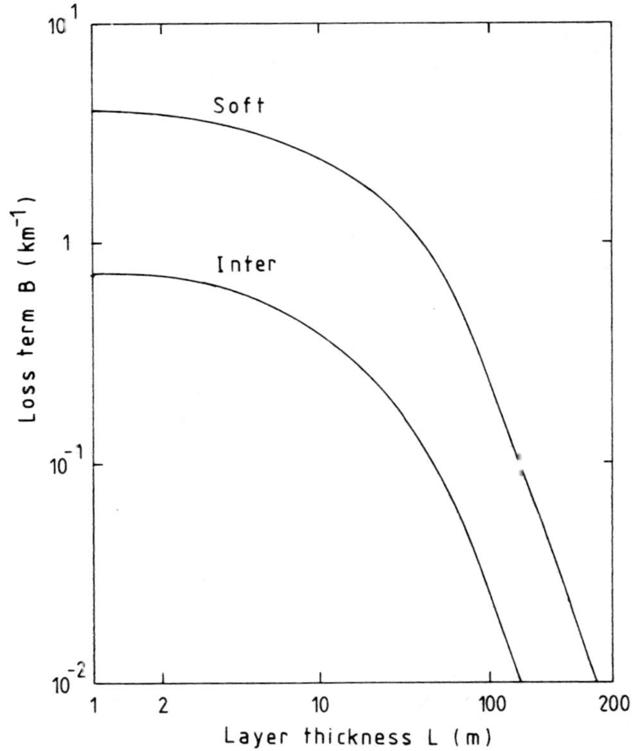

Fig. 1. Loss term B versus sediment layer thickness L showing shear loss and tunnelling effects. Layer parameters: c_p=1600, c_s=0, a_p=0, a_s=0; Bedrock: 'soft' and 'intermediate'.

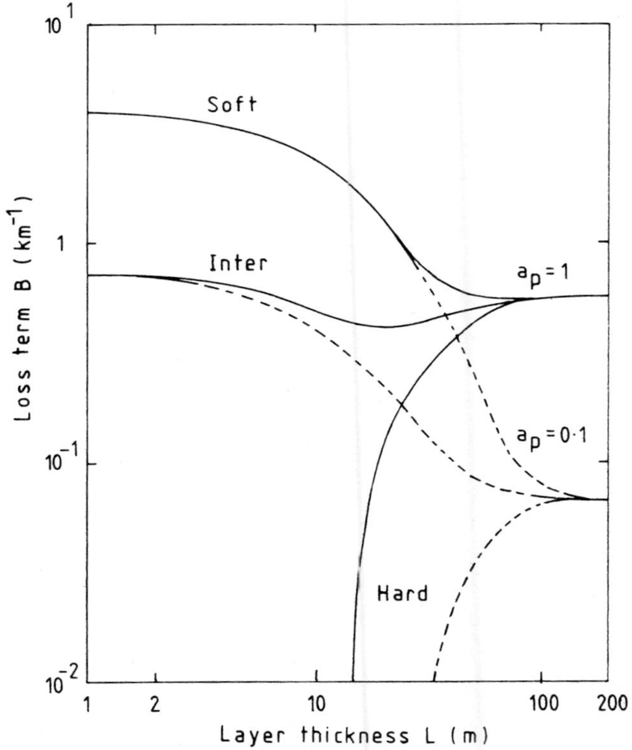

Fig. 2. Loss term B versus sediment layer thickness L showing tunnelling and absorption effects. Layer parameters: c_p=1600, c_s=0, a_p=1.0, 0.1, a_s=0; Bedrock: 'soft' 'intermediate' and 'hard'.

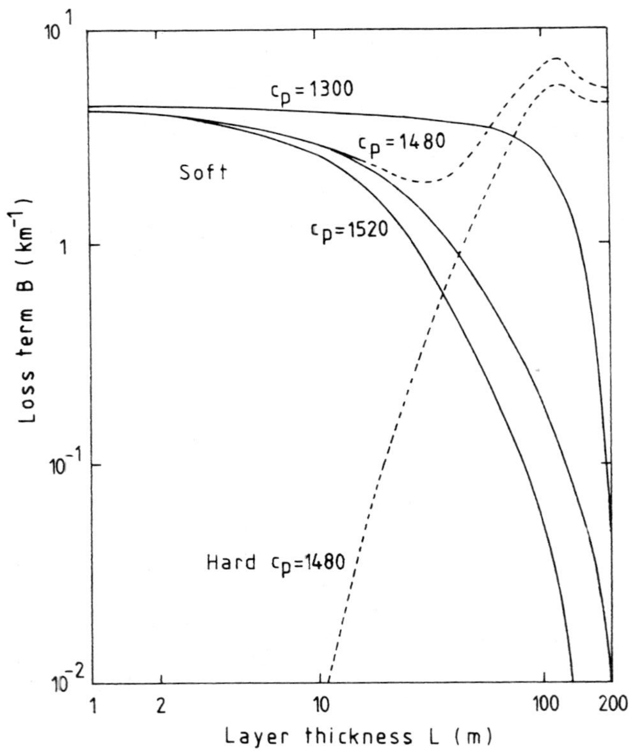

Fig. 3. Loss term B versus sediment layer thickness L showing upward refraction effects. Positive velocity gradient in layer =2m/s/metre, initial value of c_p shown on plot. Solid lines have a_p=0, dashed lines have a_p=1. Other layer parameters: c_s=0, a_s=0. Bedrock: 'soft' and 'hard'.

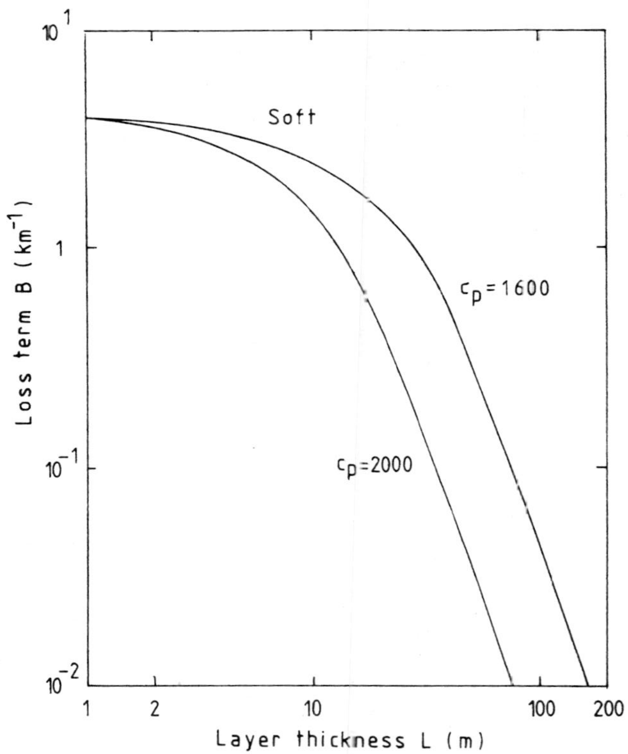

Fig. 4. Loss term B versus sediment layer thickness L showing sediment compression velocity effects. Layer parameters: c_p=1600, 2000, c_s=0, a_p=0, a_s=0; Bedrock: 'soft'.

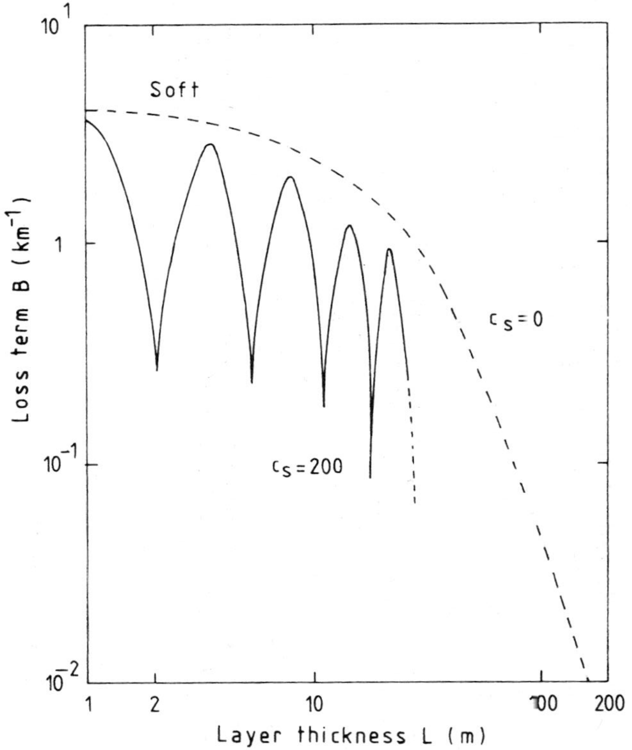

Fig. 5. Loss term B versus sediment layer thickness L showing resonance effects in loss-free sediment. Layer parameters: c_p=1600, c_s=200, 0, a_p=0, a_s=0; Bedrock: 'soft'.

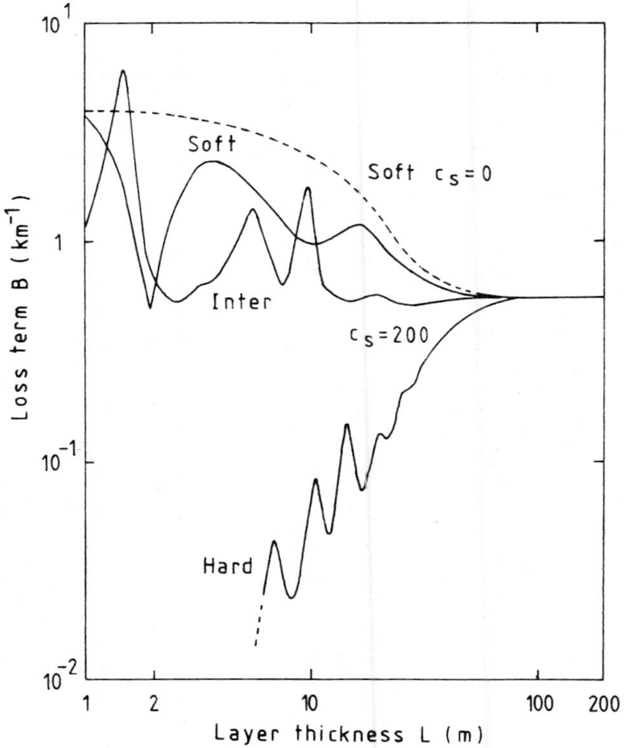

Fig. 6. Loss term B versus sediment layer thickness L showing damped resonance effects. Layer parameters: c_p=1600, c_s=200, 0, a_p=1, a_s=2; Bedrock: 'soft', 'intermediate' and 'hard'.

2) Tunnelling through the sediment layer when the bed rock has low shear velocity.

3) Shear wave resonance in the sediment layer.

4) Upward refraction in the sediment.

Although there may be a well defined relation between B and L in one locality there is certainly potential for a much wider variation on a global scale.

ACKNOWLEDGEMENT

The investigation reported here is an extension of some FFP validation work carried out under contract to the UK MOD. The authors would like to thank Mr P H Lindop for his assistance and cooperation, and they are indebted to him for drawing the subject to their attention.

REFERENCES

1. P. H. Lindop, "The deduction of approximate values of sediment depth from propagation loss measurements", (This Volume).

2. H. W. Kutschale, "Rapid computation by wave theory of propagation loss in the Arctic Ocean", Rpt. CU-8-73, Columbia University, Palisades, N.Y. (1973).

3. C. H. Harrison, "Modelling low frequency sound propagation in solid/fluid layers", in 'Acoustics and the Sea Bed', Conference Proceedings, Bath, UK (1983).

4. J. F. Miller and F. Ingenito, "Normal mode FORTRAN Programs for calculating sound propagation in the sea", NRL Memo Report 3071 (1975).

5. F. Ingenito and S. Wolf, "Acoustic propagation in shallow water overlying a consolidated bottom", J. Acoust. Soc. Am., 60:611 (1976).

6. L. M. Brekhovskikh, "Waves in Layered Media", Academic Press, N.Y. (1980).

7. D. E Weston, "Intensity-range relations in oceanographic acoustics", J. Sound and Vib., 18:271 (1971).

8. G. V. Frisk, J. A. Doutt and E. E. Hays, "Bottom interaction of low frequency acoustic signals at small grazing angles in the deep ocean", J. Acoust. Soc. Am., 69:84 (1981).

9. E04GDF "Minimising or Maximising a Function", Numerical Algorithms Group Ltd, NAG Central Office, Mayfield House, 256 Banbury Road, Oxford.

SHEAR WAVE EFFECTS ON PROPAGATION TO A RECEIVER IN THE SUBSTRATE

Paul J. Vidmar and Robert A. Koch

Applied Research Laboratories
The University of Texas at Austin
Austin, Texas 78713-8029

Some effects produced by shear wave processes in a layered seafloor are examined. Normal mode theory is used to calculate the displacement produced by a low frequency source in the water column at a receiver located inside the seafloor. The effects of the seafloor are represented by an impedance boundary condition obtained from the plane wave reflection coefficient. Shear wave processes are included by using the reflectivity of a solid, layered seafloor. A comparison of displacements obtained for solid and fluid descriptions of the seafloor is made to isolate the role of shear waves. Both vertical and horizontal displacements are found to increase significantly when shear wave processes are included. An analysis of the range dependence of displacement suggests that shear wave propagation is the major acoustical mechanism producing this increase.

I. INTRODUCTION

The relative importance of bottom interaction mechanisms influencing acoustic propagation depends on seafloor parameters such as sediment thickness, type, and layering, as well as on parameters such as frequency and source-receiver geometry. For the case of a single sediment layer overlying a substrate, the major acoustical mechanisms affecting the reflectivity of the seafloor are known and the geoacoustic parameters needed to predict their effects have been specified.[1] This knowledge forms the basis for routine calculation of plane wave reflectivity, and hence propagation.[2] One practical application of this knowledge is the reduction of computational and geoacoustic data collection requirements by specifying the conditions under which some mechanisms and geoacoustic parameters can be safely ignored. In particular, the conditions under which the sediment can be treated as a fluid, i.e., when shear wave effects are not important, can be accurately specified. Similar sophistication has not yet been developed for a layered seafloor.

This paper examines some of the effects of shear waves on the displacement field in a layered seafloor. A normal mode model[3] which treats the seafloor as an impedance boundary condition derived from the plane wave reflection coefficient[2] was used to calculate the acoustic field. This model includes the effects of shear waves propagating in a layered seafloor by using the appropriate plane wave reflection

Table 1. Geoacoustic Profile

	Depth into Bottom (m)	Density (g/cm^3)	Compressional Wave Velocity (m/s)	Compressional Wave Attenuation (dB/m-kHz)	Shear Wave Velocity (m/s)	Shear Wave Attenuation (dB/m-kHz)
Water	0	1.04	1545	0.000	0	0.00
Sediment	0	1.25	1540	0.010	100	4.00
	270	1.50	1775	0.037	483	4.00
Chert	270	2.60	5400	0.010	3150	0.01
	340	2.60	5400	0.010	3150	0.01
Substrate	340	2.50	4420	0.010	2330	0.01
	1640	2.50	5980	0.010	3370	0.01
Basement	1640	2.75	6500	0.001	3750	0.01

coefficient in evaluating the boundary condition. For this study we obtained the plane wave reflection coefficient using a generalization to multiple layers of the numerical approach previously developed for a single solid sediment layer over a semi-infinite solid basement.[4]

We expect shear wave effects to be important for our choice of source-receiver geometry, frequency, and geoacoustic profile. The geoacoustic profile (given in Table 1) has a sediment layer that is thin enough, and our frequency is low enough, for the compressional wave to generate shear waves at the chert layer beneath the sediment.[5] In addition, our receiver is located in the substrate where the compressional wave field is evanescent and the acoustic field should be dominated by shear waves.

The effects of shear waves in the seafloor were isolated by comparing displacements calculated for fluid and solid geoacoustic profiles. Including shear wave effects resulted in an increase in vertical displacement of about 20 dB relative to that obtained for a fluid geoacoustic profile. The horizontal displacement had a smaller increase of about 10 dB. Our analysis of the range dependence of horizontal and vertical displacements suggests that shear wave propagation in the seafloor is the major acoustical mechanism affecting the long range, low frequency displacement field near the top of the substrate.

Section II of this paper discusses the propagation model used to calculate the displacement field. Section III examines some effects of shear waves on the displacement field by comparing results for solid and fluid geoacoustic profiles. Section IV summarizes our results.

II. PROPAGATION MODEL

The normal mode model treats the seafloor as an impedance boundary condition.[2,3] The boundary condition is evaluated from the plane wave reflection coefficient as a function of horizontal wave number for a layered solid (or a layered fluid) geoacoustic profile. Six reflectivity calculations are used to obtain the normal modes: once with and once without attenuation at the frequency being analyzed, and without attenuation at two frequencies above and two frequencies below the frequency of interest. The reflectivities calculated without attenuation are used in a finite difference calculation of the frequency and wave number derivatives of the reflection coefficient required to normalize[2] the mode functions and evaluate their group velocities.[6] The reflectivity calculated with attenuation is used to obtain the normal mode attenuation coefficients.

The modal fields in the seafloor are used to evaluate the stress and displacement components at the receiver depth. For convenience, these stresses and displacements for the nth mode at depth z (in the seafloor) are written in four-component vector[4] Q_n as

$$Q_n = \left(u_z, \sigma_{zz}, \sigma_{zx}, u_x\right)_n^T \quad .$$

The first and fourth components give the vertical and horizontal displacements, respectively, and the second and third components give the vertical and horizontal stresses, respectively. For a fluid, $\sigma_{zx}=0$ and $\sigma_{zz}=\sigma$, where $\sigma(z)$ is the pressure. The superscript T denotes the transpose of the row vector.

The vector Q, defined as

$$Q(r,z;t) = -\sum_n \frac{p_o Q_n(z) r_o}{\omega_o^2 (1+R_n)} \left[\frac{2\pi}{k_n r}\right]^{1/2} \phi_n(z_o)\, \phi_n(H) \times \exp(-\delta_n r + i k_n r - i\omega_o t - i\pi/4) \quad , \qquad (1)$$

gives the complex displacements and stresses in the seafloor at a depth z for frequency $f=\omega_o/2\pi$ as a summation over contributions from each mode. The first and fourth components of Q are the vertical displacement d_v and the horizontal displacement d_h, respectively. In Eq. (1) the source depth is z_o (assumed to be in the water column), the water depth is H, and the range from the source is r. The plane wave reflection coefficient R_n is evaluated at the horizontal wave number eigenvalue k_n, δ_n is the modal attenuation, ϕ_n is the nth normalized eigenfunction in the water column,[2] and the asymptotic form of the Hankel functions has been used. The rms source strength is the pressure p_o at a reference range r_o from the source. For the calculations described in this paper r_o is taken to be 1 m.

For reference, the usual[2] (not impedance boundary condition) normal mode expression for pressure in a fluid is

$$\sigma(z) = -\sum_n p_o \rho(z) r_o \left[\frac{2\pi}{k_n r}\right]^{1/2} \phi_n(z_o)\, \phi_n(z) \times \exp(-\delta_n r + i k_n r - i\omega_o t - i\pi/4) \quad , \qquad (2)$$

where $\rho(z)$ is the fluid density at depth z. In this case, the expressions for the horizontal displacement d_h and the vertical displacement d_v are

$$d_v(z) = -\sum_n \frac{p_o r_o}{\omega_o^2} \left[\frac{2\pi}{k_n r}\right]^{1/2} \phi_n(z_o) \frac{d\phi_n}{dz}(z)$$

$$\times \exp(-\delta_n r + ik_n r - i\omega_o t - i\pi/4) \quad , \qquad (3)$$

and

$$d_h(z) = i \sum_n \frac{p_o r_o k_n}{\omega_o^2} \left[\frac{2\pi}{k_n r}\right]^{1/2} \phi_n(z_o)\, \phi_n(z)$$

$$\times \exp(-\delta_n r + ik_n r - i\omega_o t - i\pi/4) \quad . \qquad (4)$$

Our analysis uses the coherent and incoherent relative levels for pressure (for a fluid geoacoustic profile) and the components of displacement. The coherent relative levels are obtained directly from Q in Eq. (1): $20 \log_{10}|\sigma zz/p_o|$ for pressure, $20 \log_{10}|d_h/d_o|$ for horizontal displacement, and $20 \log_{10}|d_v/d_o|$ for vertical displacement, where d_o is a reference displacement. The incoherent relative levels use the same normalizations and are obtained by summing the magnitudes of the terms of Q in Eq. (1).

The reference displacement d_o can be chosen to be a convenient constant (e.g., 1.0 nm) or to be consistent with the reference pressure. For our analysis, d_o was chosen to be consistent with the nearfield produced by the pressure p_o. This yields $d_o = p_o/\omega_o^2\rho(z_o)r_o$ as the rms radial displacement. For $p_o = 1$ Pa and a frequency of 15 Hz, $d_o = 1.13 \times 10^{-7}$ m. For comparison, another choice consistent with p_o is the farfield expression resulting in $d_f = p_o/\omega_o\rho(z_o)c_o$, where c_o is the sound speed at the source depth. For $p_o = 1$ Pa, $c_o = 1500$ m/s, and a frequency of 15 Hz, $d_f = 7.07 \times 10^{-9}$m, which is about a factor of 10 smaller than d_o at 15 Hz. We note that the choice of normalization influences the offset between the relative levels of pressure and displacement and that d_o and d_f have different dependences on frequency.

For the normalizations given above and the expressions for pressure and displacement given in Eqs. (2) and (3), there is about a 40 dB offset between the relative level for pressure and that for horizontal displacement. Each term in d_h/d_o differs from the corresponding term in σ/p_o by the factor $k_n r_o \rho(z_o)/\rho(z)$. This factor can be estimated using the geoacoustic profile given in Table 1. The phase velocity of the normal modes is between the water sound speed and the shear speed of the basement (see discussion below). Taking an average of these velocities and the density at the top of the substrate, this factor is 1.42×10^{-2} at 15 Hz, yielding an estimated offset of about 37 dB.

III. SHEAR WAVE EFFECTS ON DISPLACEMENT

Relative levels for displacement were calculated for the geoacoustic profile in Table 1 and the sound velocity profile of Fig. 1. Table 1 describes a seafloor having a sediment (clay) layer 270 m thick over a 70 m chert layer on top of a 1300 m thick substrate layer and terminated by a semi-infinite basement. All parameters have constant gradients within a layer. Attenuation in the seafloor is assumed to be

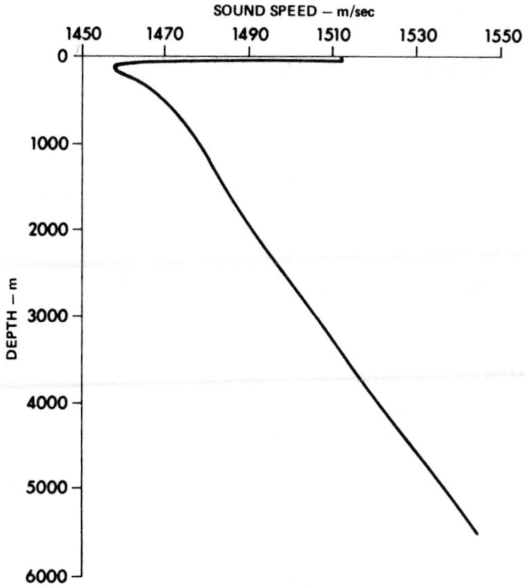

Figure 1. Water sound speed profile.

proportional to frequency.[7] For the frequency of 15 Hz used in our calculations, the attenuation in seawater is negligible and was not included. Relative levels were calculated at 1 km increments out to 200 km. The source depth was 60 m, and the receiver depth of 360 m was 20 m into the substrate.

Several observations can be made based on the geoacoustic profile. First, the shear and compressional speed profiles are upward refracting in the substrate, while both shear and compressional speeds decrease from the chert values to the substrate values. Thus, the upper portion of the substrate layer and the bottom of the chert layer form a solid acoustic waveguide. Second, the maximum horizontal phase velocity for any of the normal modes is the basement shear speed, because energy with higher phase velocity propagates into the basement. This energy is not described by the discrete spectrum of the normal modes, but by branch line integrals whose contributions are not important at long range.[8] Our analysis will concentrate on longer ranges where the normal modes dominate propagation. Finally, the compressional wave velocities in the chert and substrate layers are well above the basement shear speed, resulting in a compressional wave field in these layers which is evanescent and decays rapidly with increasing depth. On the other hand, shear waves are not evanescent, but propagate within the chert and substrate. We, then, expect shear wave contributions to the displacement at our receiver to be larger than those of the evanescent compressional waves. The displacement field will be larger for the solid geoacoustic profile than for the fluid profile.

Now we consider the results of the propagation calculations. For reference purposes we begin with calculations for a fluid geoacoustic profile. The fluid geoacoustic profile was obtained by setting all the shear velocity and attenuation values in Table 1, except those for the basement, to zero. Leaving the basement parameters unchanged keeps the same maximum horizontal phase velocity (the basement shear velocity) for the normal modes for both fluid and solid geoacoustic profiles.

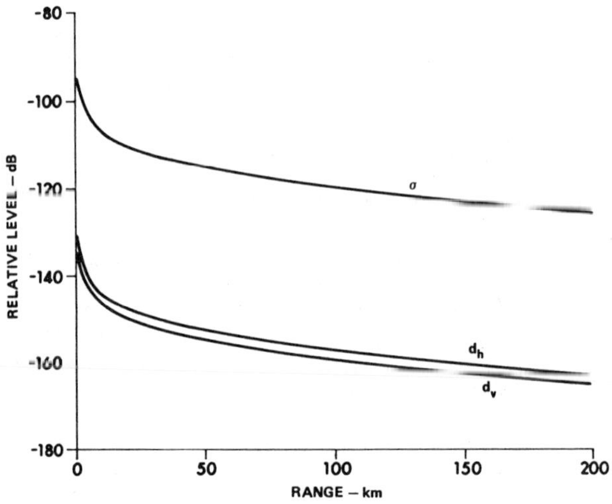

Figure 2. Incoherent relative levels of pressure (σ), horizontal displacement (d_h), and vertical displacement (d_v) calculated for the fluid geoacoustic profile.

Figure 2 shows the incoherent relative levels of σ, d_h, and d_v calculated for the fluid geoacoustic profile. Two features of these curves are noteworthy. First, as estimated above, there is an offset of 37 dB between the pressure and horizontal displacement curves. The second feature of note in Fig. 2 is that the range dependence of the three curves is nearly identical. (The coherent relative levels, not shown, are also nearly identical.) This is expected for the fluid geoacoustic profile since the range dependence is produced by cylindrical spreading and the attenuation of the modes. Since the relative contribution of a given mode is about the same for pressure and displacement, their relative levels will have about the same range dependence. Some difference is expected because of the dependence on mode number introduced by the factor k_n in Eq. (3) and the derivative in Eq. (4) that is not found in Eq. (2). The fact that the range dependence of the three curves is nearly identical means that this difference has a weak dependence on mode number for our case.

Figures 3 and 4 compare vertical and horizontal displacement, respectively, calculated for solid and fluid geoacoustic profiles. Both coherent and incoherent relative levels are shown. Figure 3 shows that the incoherent relative level for vertical displacement for the solid profile is nearly 20 dB higher (factor of 10 increase in displacement) than that for the fluid profile. This difference clearly shows that shear wave effects occurring in the geoacoustic profile in Table 1 can greatly increase the displacement near the top of the substrate. Figure 4 shows that the horizontal displacement is also significantly affected by shear wave effects, with the solid profile incoherent relative level about 10 dB more than the fluid profile level. Interference among the modes causes the coherent relative levels in Figs. 3 and 4 to oscillate about the incoherent levels.

Figure 5 shows the incoherent relative level for d_h and d_v for solid and fluid geoacoustic profiles. Some features of these curves suggest that the large difference in relative level between the solid and fluid

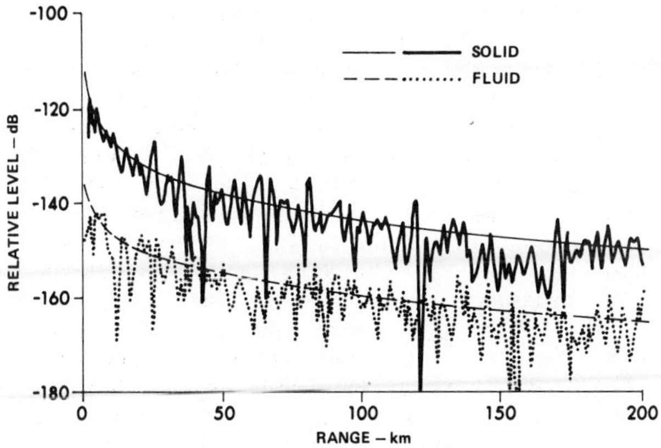

Figure 3. Coherent (oscillatory curves) and incoherent (smooth curves) relative levels of vertical displacement calculated for solid and fluid geoacoustic profiles.

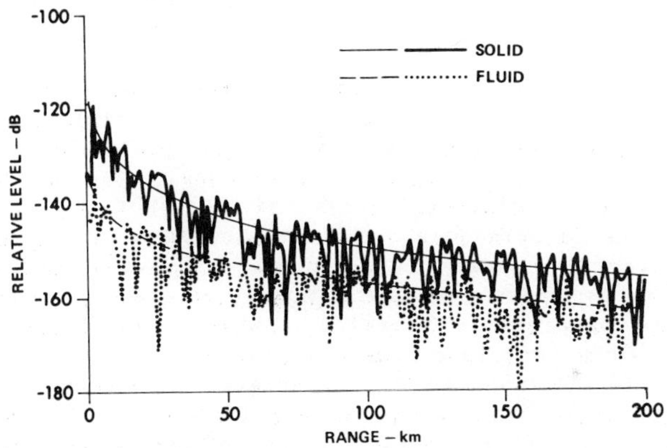

Figure 4. Coherent (oscillatory curves) and incoherent (smooth curves) relative levels of horizontal displacement calculated for solid and fluid geoacoustic profiles.

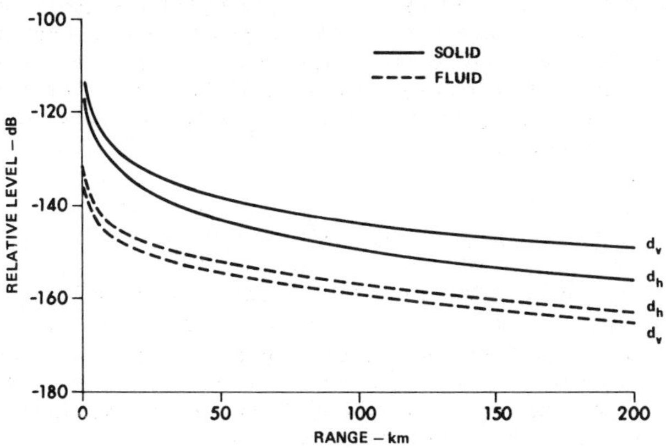

Figure 5. Incoherent relative levels of horizontal (d_h) and vertical (d_v) displacement calculated for solid and fluid geoacoustic profiles.

cases is due to shear wave propagation. One indication that shear waves are involved is found in the different range dependences in the solid and fluid cases over the first 50 km. From 10 to 50 km the relative level of d_h for the solid profile decreases 6 dB more than that for the fluid profile. For d_v the relative decrease is about 3 dB. The decreased level in the solid case is consistent with the attenuation of modes with shear components not found in the fluid case. A second indication that shear waves dominate propagation in the calculation for the solid profile comes from the comparison of d_h and d_v for the solid case in Fig. 5. The calculated incoherent level for the vertical displacement exceeds that of the horizontal displacement by about 4 dB near the source and by about 8 dB beyond 130 km, i.e., d_h is more rapidly attenuated than d_v. This can be understood in terms of the transverse nature of shear wave propagation. The more vertically propagating shear wave components, producing d_h, have longer, deeper paths in the substrate and hence are more heavily attenuated than the horizontally propagating components, producing d_v, which have turning points near the top of the substrate. The observed behavior of d_h and d_v is, then, consistent with shear waves producing the displacement field near the top of the substrate.

IV. SUMMARY

For the geoacoustic profile of the seafloor used in our calculations, shear wave propagation in the seafloor is the major acoustical mechanism affecting the long range, low frequency displacement field near the top of the substrate. This conclusion was reached through an analysis of the relative levels of the horizontal and vertical components of the displacement field calculated using a normal mode approach in which the seafloor is represented by an impedance boundary condition. The impedance boundary condition was calculated for both solid and fluid geoacoustic profiles. When the effects of shear wave propagation in the seafloor were included in the calculation, the relative level of the vertical displacement increased in magnitude by about 20 dB (10 dB for the horizontal displacement). Further analysis of the range dependence of incoherent relative levels for fluid and solid geoacoustic profiles suggested that shear wave propagation was the major acoustical mechanism producing the displacement field in the substrate for our geoacoustic profile.

The analysis presented here has concentrated on qualitative aspects of propagation to a receiver near the top of the substrate. Further research will concentrate on quantitative questions such as the frequency and depth dependence of the displacement field, the interface at which shear waves are being generated, and the depth of penetration of shear waves into the substrate.

ACKNOWLEDGMENTS

This work was supported by Naval Ocean Research and Development Activity. The normal mode computational model using the impedance boundary condition was developed with support from the Office of Naval Research. Typical parameters for chert, substrate, and basement were provided by Dr. F. Dennebier of the Hawaii Institute of Geophysics.

REFERENCES

1. P. J. Vidmar, "Linked Sets of Acoustical Processes and Geoacoustic Profiles Describing the Interaction of Sound with a Class of Sea Floor Structures", in Acoustics and the Sea-Bed, ed. by N. G. Pace (Bath University Press, Bath, UK, 1983) pp. 81-87.
2. H. P. Bucker, "Sound Propagation in a Channel with Lossy Boundaries", J. Acoust. Soc. Am. 48, 1187-1194 (1970).
3. R. A. Koch, P. J. Vidmar, and J. B. Lindberg, "Normal Mode Identification for Impedance Boundary Conditions", J. Acoust. Soc. Am. 73, 1567-1570 (1983).
4. P. J. Vidmar and T. L. Foreman, "A Plane-Wave Reflection Loss Model Including Sediment Rigidity", J. Acoust. Soc. Am. 66, 1830-1835 (1979).
5. P. J. Vidmar, "Ray Path Analysis of Sediment Shear Wave Effects on Bottom Reflection Loss", J. Acoust. Soc. Am. 68, 639-648 (1980).
6. R. A. Koch and J. B. Lindberg, "Normal Mode Cycle Distance and Beam Displacement, Time Delay, etc., for Low (Non-WKB) Frequencies", J. Acoustic. Soc. Am., to be published.
7. E. L. Hamilton, "Geoacoustic Modeling of the Sea Floor", J. Acoust. Soc. Am. 68, 1313-1340 (1980).
8. W. M. Ewing, W. S. Jardetzsky, and F. Press, Elastic Waves in Layered Media (McGraw-Hill, New York, 1957), pp. 156-164.

DESCRIPTION AND SOME RESULTS OF THE OPERATIONAL PROPAGATION LOSS MODEL ALMOST (= ACOUSTIC LOSS MODEL FOR OPERATIONAL STUDIES AND TASKS)

P. Schippers

Physics and Electronics Laboratory TNO
Oude Waalsdorperweg 63
2597 AK The Hague, The Netherlands

ABSTRACT

The propagation loss model ALMOST, which is primarily a deep water model, has been developed at the Physics and Electronics Laboratory TNO, The Hague between 1980 and 1984. It is based on ray-tracing solutions for the Convergence Zone Path and the Bottom Surface Path, taking the main features of the Sound Speed Profile (SSP) into account as well as several additional refinements. Moreover a number of channels (or ducts) are taken into account, which are derived by the program from the SSP. In this case averaged cylindrical spreading loss is calculated, to which adaptations are made afterwards in the vicinity of possible caustics and shadow zones. In particular for the lower frequencies diffraction loss is taken into account by frequency dependent averaging of the SSP, as well as by limitation of the effective channel width.

Broadband comparisons (25-5000 Hz) with MILOC-measurements show a good agreement for deep water. For shallow water the relevant acoustic properties of the bottom could only be derived from a core taken in the vicinity of the considered runs. Notwithstanding the poor description of the bottom and the initial stage of the shallow water application of ALMOST, the comparison with shallow water field data looks quite promising.

1. Introduction

The ALMOST model has been developed with the aim to obtain a practical propagation loss model for operational use in deep water. The model had to be small, fast and sufficiently accurate for the input data available at sea. A great deal of these goals has been achieved till now and this paper describes some of the basic features of the model.

The ALMOST model starts by distinguishing different propagation modes on the basis of the input Sound Speed Profile (SSP), see fig. 1. The contributions of the different propagation modes to the transmission of an acoustic signal from a source to a receiver depend on a number of factors. For channels and ducts for example, the contribution is primarily determined by the source and receiver depth. For sound paths (SRP, BSP, CZP) the contribution depends for a great deal on the shape of the SSP. All contributions are calculated individually and are finally added incoherently.

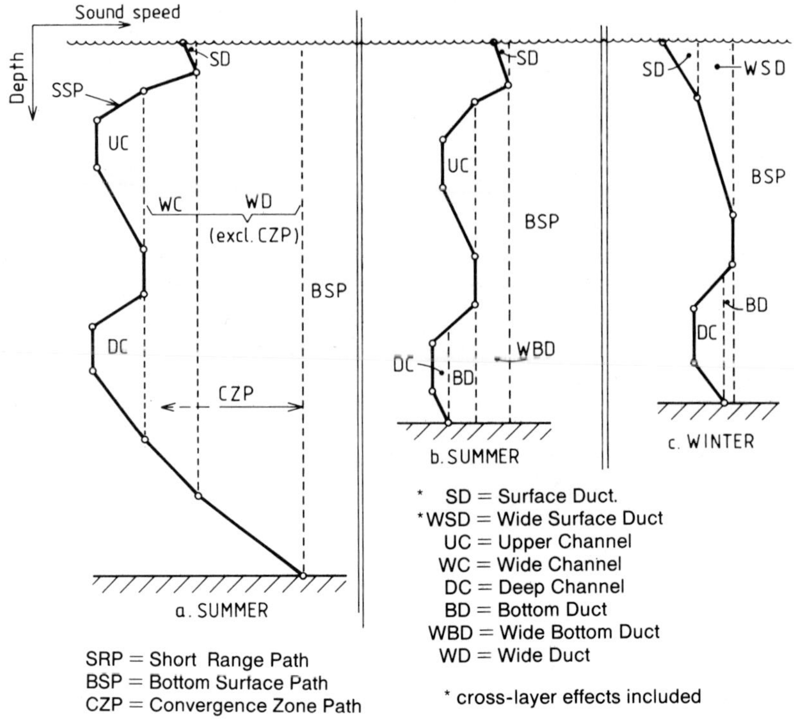

Fig. 1. Schematic representation of different propagation modes.

Because of the requirement of a short calculation time, the SSP is simplified without loosing the important characteristics which describe channels and ducts. This is achieved by the so-called "stylizing" procedure. Further linear and bilinear approximations are determined from the "stylized" SSP in order to obtain ray intensities and characteristic ranges for the propagation modes SRP, BSP and CZP.

For channels and ducts caustic like features are calculated, which particularly occur at the shorter ranges of these propagation modes. At the larger ranges the intensity in ducts and channels is determined by cylindrical spreading.

The surface loss in ducts is calculated on the basis of the Rayleigh scattering theory, combined with an approach which uses reflection by wave facets for the high frequencies. The surface loss is calculated from an integral over all possible grazing angles of the different intensity contributions. Also cross-layer effects are calculated on the basis of this surface scattering mechanism.

Bottom- and surface loss, which occurs in the paths BSP, CZP and SRP, is taken into account as a loss in dB per reflection. This Surface Loss is calculated according to Watson (1). The Bottom Loss (BL) per reflection is calculated with the formulas for the well-known MGS-provinces. It is also possible to calculate the BL on the basis of a physical model of the bottom. This possibility is of importance particularly for the condition of shallow water and low frequency.

Diffraction effects are accounted for in two stages: First a frequency dependent smoothing is applied to the SSP, before "Stylizing". Afterwards, the effective width of the channel or duct is adapted in such a way, that the width decreases for the lower frequencies. This diffraction description allows also for shallow water calculations by the primarily deep water model.

Important coherence effects like Surface Decoupling Loss in BSP, SD and/or WSD, and the Lloyd-Mirror-effect in SRP, are taken into account in a smoothed way.

In the above only a few of the main features of ALMOST are mentioned. In the following chapters some of these features will be described in more detail. Finally a comparison with field data will be made in order to give an impression of the reliability of the model.

2. Smoothing and Stylizing of the measured SSP

In the ALMOST model a frequency dependent smoothing is applied to the measured SSP, before any other calculation is made. With this kind of smoothing more reliable results are obtained for the lower frequencies and for the shallow water conditions in which often narrow channels and/or ducts are present.

The smoothing of the measured SSP is executed by calculating the following combinations of sound speed and depth:

$$C(i) = \frac{1}{\Delta D(i)} \cdot \int_{D(i) - \Delta D(i)/2}^{D(i) + \Delta D(i)/2} C(z) \, . \, dz$$

with: i = index for points of smoothed SSP
$D(i)$ = centre depth of interval $\Delta D(i)$
$\Delta D(i)$ = depth interval for point i
$C(z)$ = measured SSP

A smooth curve (D(i), C(i)) is obtained by using a sequence of overlapping depth intervals $\Delta D(i)$, see fig. 2. The size of $\Delta D(i)$ has been determined emperically. $\Delta D(i)$ increases with depth and has a minimum value of $7.\lambda$(acoustic). For this choice diffraction effects in the vicinity of a steep negative sound speed gradient (as caused by a thermocline) are described reasonably well.

The resulting smoothed SSP is subsequently "stylized". This procedure starts with searching for significant extrema in the SSP. This search procedure is able to find an eventual shallow Surface Duct, while on the other hand only the gross features are calculated at depths below such a duct. The second step in the stylizing procedure is the replacement of each extremum (except the one related to a shallow Surface Duct) by two points with equal sound speed, one just above the extremum, the other just below it. With these two points a better fit to the SSP is produced in the vicinity of extrema (fig. 2).

As a last step some characteristic points are generally added by the stylizing procedure in order to obtain a more complete SSP (fig. 2). In the case that a SSP does not show any extremum (apart from the surface- and bottom value), the stylizing procedure determines characteristic points which are close to the observed profile.

Finally the "class" of each stylized SSP is determined by comparing the result with each of 17 possible SSP shapes which have found to be relevant for the North Atlantic Ocean.

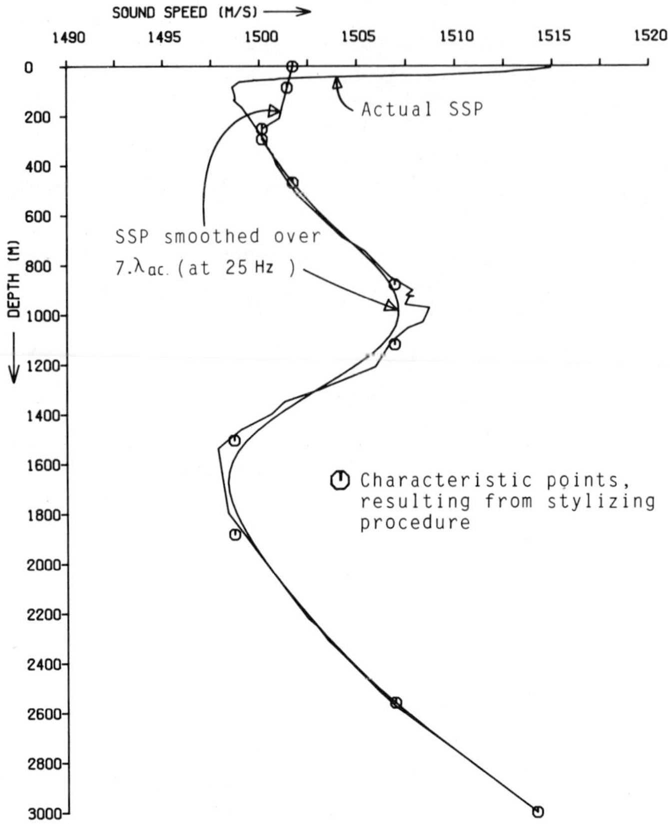

Fig. 2. Frequency dependent processing and stylizing of SSP.

3. Propagation Paths (SRP, BSP, CZP)

The losses for the propagation paths SRP, BSP and CZP are determined on the basis of the individual losses of all contributing rays. The loss per ray is the geometric loss combined with additional losses due to Bottom and/or Surface Reflections. Moreover the following coherence effects are taken into account as well: Surface Decoupling Loss (SDL) for BSP (4 rays), Lloyd-Mirror-effect for SRP (2 rays).

For the geometric spreading loss closed analytical formulations have been derived. This could only be achieved however for simple SSP's: linear or bilinear profiles. For this reason a procedure has been developed which determines such profiles from the stylized SSP. In general such simplified profiles are only acceptable for the steeper rays, as these occur in the propagation paths BSP and CZP.

No acceptable SSP-approximation could be found for the calculation of the geometric loss for the shallow rays of SRP. In this case spherical spreading loss is used for both direct and reflected rays. For the SRP also the Lloyd-Mirror-effect is taken into account, which describes a possible coherence effect between direct and surface reflected rays. The well-known limiting skip distances for both direct and reflected SRP-rays are calculated in a fast way, using a linear or bilinear SSP-approximation.

In the following paragraphs the propagation paths BSP and CZP are treated in more detail.

3.1. Bottom Surface Path (BSP)

The Bottom Surface Path calculation starts with the determination of the maximum skip distance per Bottom reflection. For this purpose a bilinear SSP-approximation is used (see fig. 3). The bilinear approximation is chosen in such a way that the mean sound speed is the same for both the original and approximated SSP-curve. Subsequently a linear SSP-approximation is chosen with such a gradient that the same maximum skip distance per Bottom reflection is achieved. Finally the geometric loss per ray is calculated based on this linear profile (see fig. 3).

For each ray the Bottom Loss is added per reflection. Always four rays with the same number of Bottom reflections are present, because of which also coherence effects occur: Surface Decoupling Loss. Further the total BSP intensity is calculated as the sum over all contributing rays corresponding to all possible numbers of Bottom reflections.

The Bottom Loss per reflection is usually calculated according to the given MGS-province number. However for shallow water the bottom will generally be acoustically flat below 1000 Hz, due to the mainly very small angles of incidence: In this case the theoretical Bottom Loss value is calculated on the basis of the physical parameters for the (semi-infinite) bottom. Above 1000 Hz the Bottom Loss value for the relevant province number is used partially, in order to account for bottom roughness.

3.2. Convergence Zone Path (CZP)

Because of the fact that rays of the propagation mode CZP do not reach the bottom, a bilinear SSP-approximation can be used for the calculation of the geometric loss per ray. The upper part of this bilinear SSP has a constant sound speed (see fig. 4). Again the mean sound speed values of both approximated and original SSP are the same, like for the BSP case.

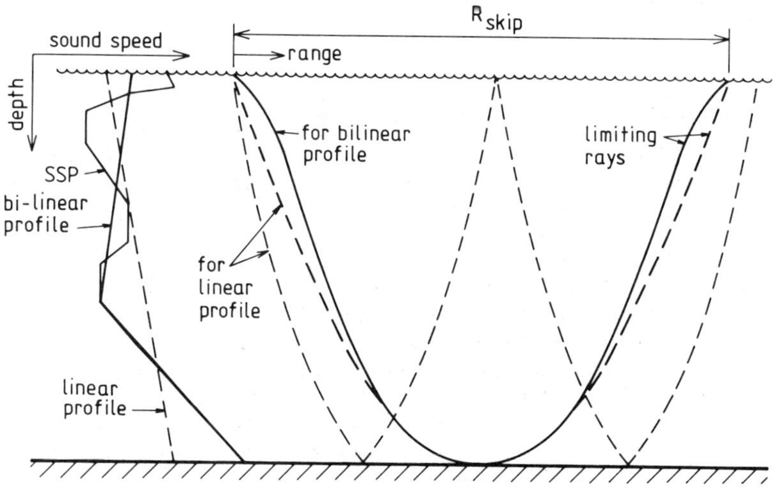

Fig. 3. Bottom Surface Path.

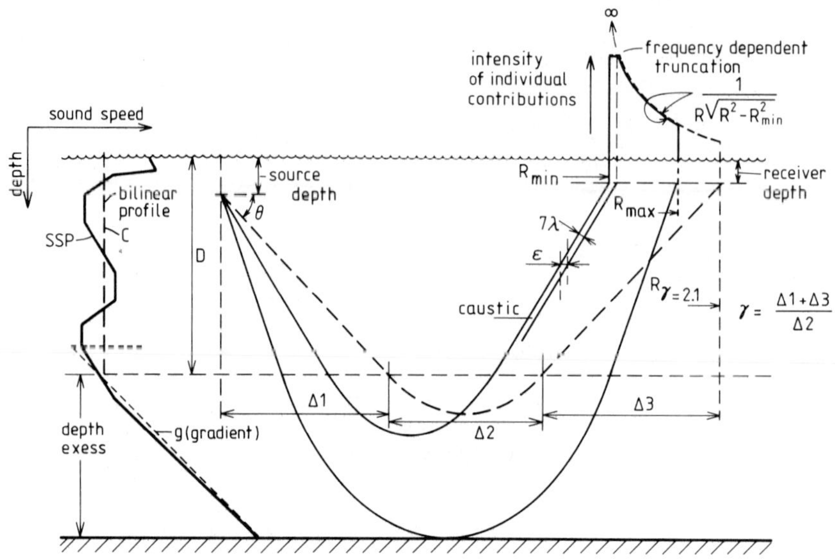

Fig. 4. Convergence Zone Path.

On the basis of this bilinear profile a minimum CZ-range is calculated with an error of less than 2%. The bottom depth sets a limit to the maximum range for the relatively deep rays (Rmax in fig. 4).

The intensity within the caustics (CZ's) is calculated on the basis of this bilinear profile. However for very small grazing angles this intensity calculation is not valid anymore due to the simple shape of the bilinear profile. The range until which the calculations are still valid is indicated by the limiting range R(γ). Beyond this range R(γ) the intensity is calculated according to Wide Channel-(WC-) and/or Wide Duct (WD) propagation (see ch. 4).

The intensity of the rays forming the CZ caustics itself are limited to a beam width of 7.λ(ac.) (fig. 4) in order to account for diffraction effects.

4. Channels and Ducts

Channels and Ducts are bounded by two sound speed gradients of different sign. The magnitudes of these two sound speed gradients differ often considerably (see fig. 1). This certainly applies to ducts, because one of the gradients is infinitely large in this case.

For a source or receiver situated outside the smaller gradient of the sound channel, a CZP-like propagation occurs. By applying the same SSP-approximations as for CZP, the various caustics which occur in channels, and in certain ducts, are calculated.

For source and/or receiver inside the smaller gradient, regions of higher intensity will also occur, with possibly shadow zones in between. This applies particularly to a source and/or receiver depth near the edge of the channel. However these caustics are generally less pronounced. For this reason cylindrical spreading loss is calculated within the range intervals of these caustics with the possible shadow zones in between.

This cylindrical loss follows from a calculation of the geometric spreading loss in the following way: The ray intensities of all possible rays are smoothed in range over their various skip distances in the channel. Only those rays that reach the given source and receiver depths do contribute to the total incoherent sum of smoothed ray intensities. This sum is transformed into an integral for which a closed analytical formulation has been found. This formula can be written as 1/(R.Rt), in which Rt represents the transition range.

For Surface Ducts coherence effects are taken into account as well, by applying the Surface Decoupling Loss calculation (of BSP) to the maximum grazing angle in the duct.

5. Comparison with measurements

As an example an ALMOST propagation loss calculation is made for the SSP of fig. 2. The result is given in fig. 5 for a frequency of 25 Hz. The Upper sound Channel (UC) in the original SSP is transformed into a Wide Surface Duct (WSD) for this low frequency of 25 Hz, as is shown by the "Stylizing" points in fig. 2. At about 25 km the first group of caustics of this WSD is shown. The first CZ together with the second group of caustics of the WSD is visible at about 60 km.

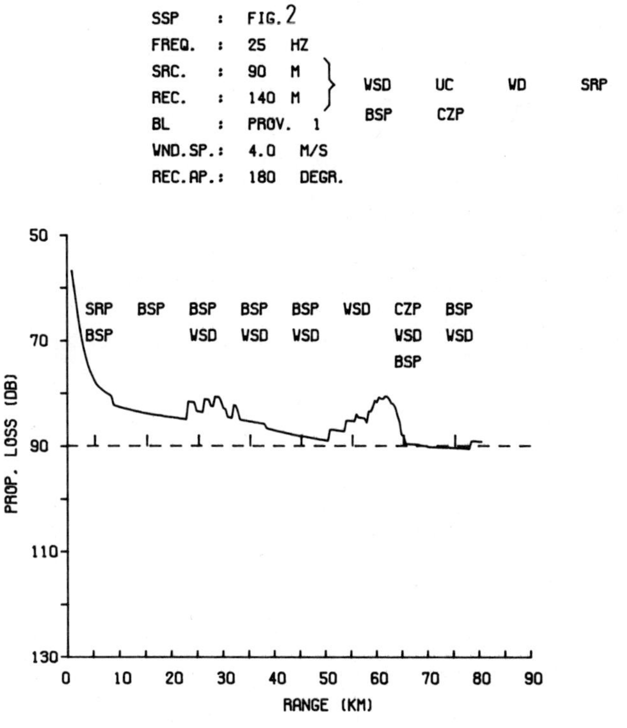

Fig. 5. ALMOST propagation loss versus range.

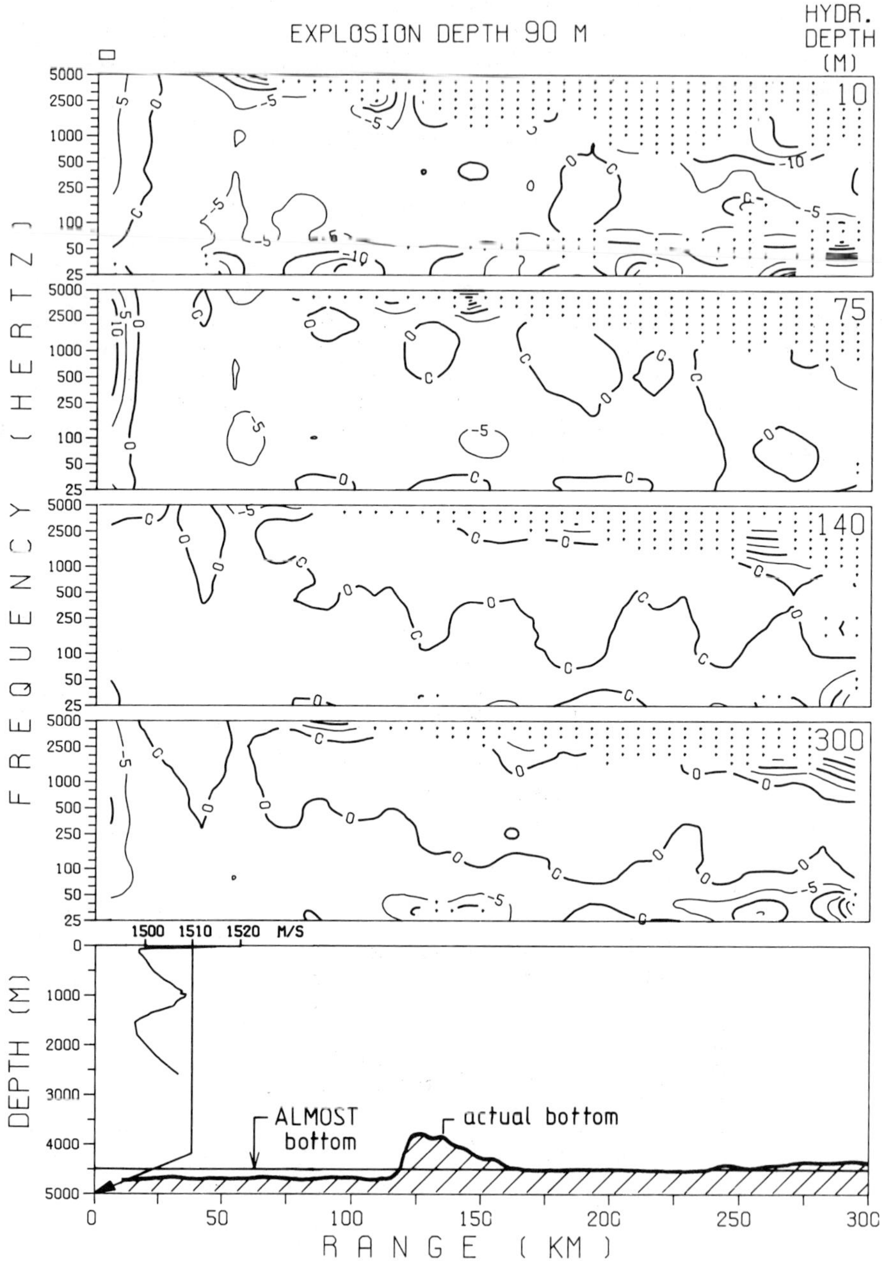

Fig. 6. Propagation loss difference (dB) between measured and ALMOST data (deep water).

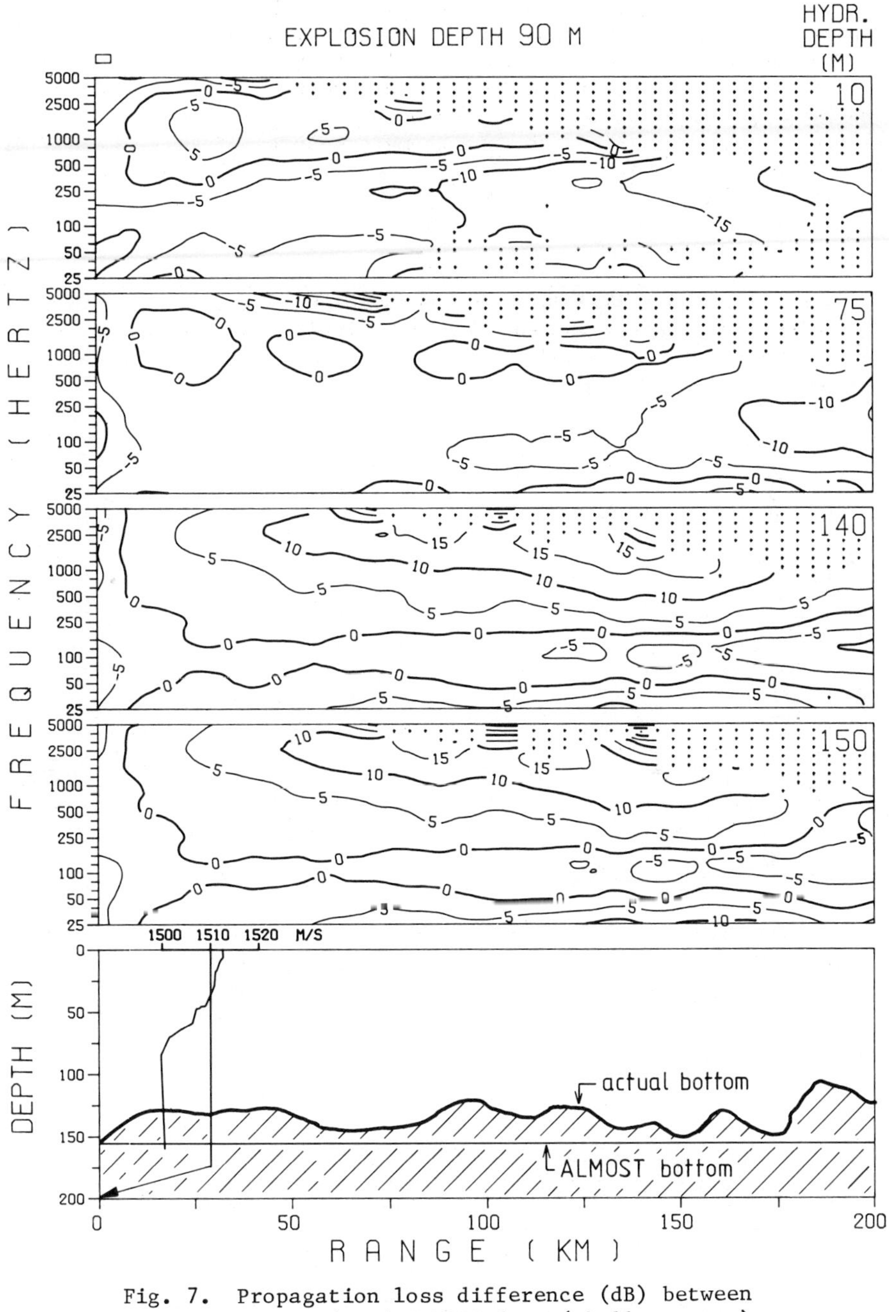

Fig. 7. Propagation loss difference (dB) between measured and ALMOST data (shallow water).

Extensive comparisons of ALMOST with measurements have been made for deep as well as shallow water. Examples are given in fig. 6 and 7. The shallow water results apply to an environment with a rather flat sandy bottom of considerable thickness. The bottom reflection loss is calculated here for the given physical parameters. Above 1000 Hz the results are combined with the values for MGS-province number 1, as described in para. 3.1..

The agreement is satisfactory for both comparisons, in spite of the fact that the shallow water adaptations of the ALMOST model are still in an early stage of development: Refinements concerned to the bottom interaction will be incorporated as soon as more results of shallow water measurements become available.

6. References

1. Watson, W.H.; Shallow water propagation loss model (I). Rep.nr. NUWC TN11, Naval Undersea Warfare Center, San Diego, California, July 1967.

EXPERIMENTAL CONFIRMATION OF HORIZONTAL REFRACTION OF SOUND PROPAGATION IN A WEDGE-LIKE OCEAN

R. Doolittle *, A. Tolstoy *, and M. Buckingham **

* Naval Research Laboratory, Washington, DC, 20375-5000, USA
** Royal Aircraft Establishment, Farnborough, Hampshire, GU146TD, UK

ABSTRACT

Analysis of experimental data obtained during the SET 1 exercise in the region of the East Australian Contentinental Slope has confirmed earlier theoretical predictions of energetic horizontal refraction in underwater acoustic propagation due to sound-slope interaction. The observed general modal behavior of the sound, the azimuthal characteristics of the field and the effects of bottom absorption were in accordance with the theory of sound transmission in wedge-like ocean acoustic environments. A experiment was conducted in which two ships, one towing a sound projector and the other a towed array, started at a 400 meter depth contour with an initial separation of 33 kilometers and proceeded out to deep water on diverging, constant line of bearing courses chosen so as to maintain the relative experimental geometry. Beamform data demonstrated the existence of horizontally refracted arrivals which showed significant bearing shifts due to repeated reflections from the ocean surface and bottom. The obliqueness of the azimuthal launch angle with respect to the shore line prevented the local grazing angle from becoming large enough to produce much bottom loss which is in accord with predictions. A combination of the particular source-receiver position and the critical angle effects serve to limit both the azimuthal extent of the sound reception and the total number of modes that arrive at any one position.

INTRODUCTION

The theory of acoustic propagation in wedge-like oceans was recently advanced with the publication of a normal mode solution by Buckingham[1] for the field due to a point source, in a three-dimensional wedge with perfectly reflecting, planar boundaries. Each mode contributing to the field can be interpreted in terms of equivalent rays, using arguments and concepts introduced by Weston[2].

The rays corresponding to a given mode are launched from the source at all azimuthal angles. At the source location, these rays (which may be upward or downward travelling) all show the same grazing angle with respect to the boundaries. This is the feature that identifies them with the mode. As a ray propagates away from the source, its grazing angle changes as it reflects from the inclined boundaries of the wedge. However, at any point along the ray's path, the grazing angle always satisfies the condition (which is familiar in the case of shallow water channels) that the upward and downward travelling wavefronts should interfere constructively. In fact, this condition gives rise to a set of discrete grazing angles which correspond directly with the modes at that point. At each reflection from either of the boundaries the modal rays are twisted in the horizontal, a phenomenon known as horizontal refraction. Thus, a horizontal projection of a ray onto the sea surface is not a straight line but follows a curved path. After smoothing out the kinks which occur at the points of reflection, the form of this curved path is hyperbolic. Thus, a ray which is launched obliquely up the slope approaches the apex of the wedge but, before reaching it, is turned around as it passes through the vertex of the hyperbola and then proceeds to propagate out into deep water. One consequence of this behavior is that the rays corresponding to a single mode can cross each others' paths, causing interference to occur within that mode, as well as creating caustics and shadow zones. None of these features of the acoustic field could be produced in a homogeneous channel with plane, parallel boundaries.

In late 1984, as part of the SET 1 exercise conducted jointly by New Zealand, the USA and Australia, an experiment was performed to investigate the properties of the sound field in a wedge-like ocean. The experiment was designed on the basis of the theory outlined above. The purpose of this paper is to present and interpret the results of an experimental run over the shallow, sloping bottom near the East Australian Rise. The interpretation of the experimental data discussed in the paper, although largely qualitative, suggests strongly that the effects of horizontal refraction are significant in the ocean, in accord with the predictions of the theory.

EXPERIMENT

The experiment to be discussed here to test for the effects of horizontal refraction was performed at a site on the East Australian continental shelf with a gradual bottom slope of between 0.4 and 1.2 degrees. A towed CW source was used at a frequency of 152 Hz, and the signal was received on a towed array. Over most of the acoustic path between the source and the receiver the bottom had a smooth sandy sediment about 200 meters thick overlying a rock basement (Fig. 1a). The direct line between the source and receiver was nominally North-South (approximately parallel to the shoreline) and remained so throughout the run. The initial source/receiver separation was approximately 33 kilometers (Fig. 1b). A southward current of about 2 knots had to be considered in determining the course and speed of the source and receiver ships during the experiment. In order to place the expected acoustic arrivals near broadside to the array the tow ship headed due east at about 1.5 knots, giving the array a true heading of 38 degrees. The speed and course of the ship towing the source were chosen so that both ships were equidistant from the shoreline at all times during the 4.5 hour run, and so that the North-South distance between the source and receiver increased throughout the run. A summary of these experimental details at the start of the run is given in Table 1. Navigation was kept by both SATNAV systems and by triangulation of known points on shore.

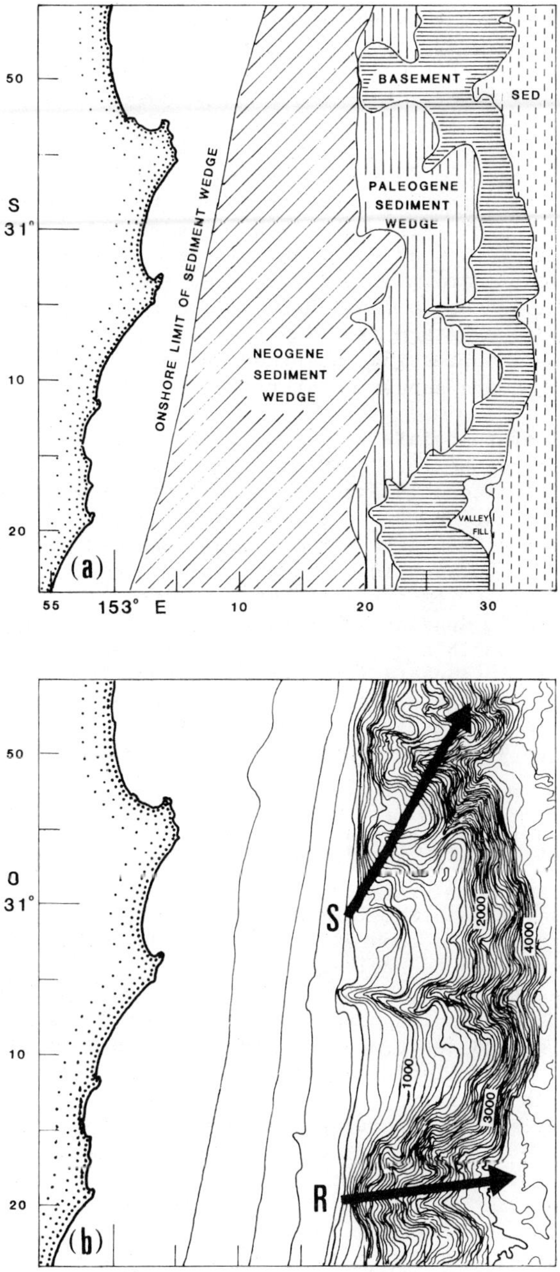

Fig. 1. Bottom geology (a) and bathymetry (b) in experimental area.

Table 1. Summary of Experimental Details

Source depth = 80 m
Receiver array depth = 80 m
Bottom depth at source and receiver = 400 m
Separation distance between source and receiver = 33 km
Bottom slope = 1.2 deg
Frequency = 152 Hz (wavelength = 9.87 m, wavenumber k = 0.64/m)
Array heading = 38 deg True

Narrow-band beamform data were recorded continually, from which a bearing-time record of acoustic arrivals was produced in the form of a gray-scale contour plot (Fig. 2). Each horizontal trace in the plot represents a time average of two minutes. Both conventional and maximum-likelihood beamforming algorithms were used to produce displays, although only the latter is shown here. The maximum-likelihood beamformer has the advantage of reducing sidelobe interference and, when spatially whitened, supressing any broadband noise in the background including the broadband acoustic interference from the towship[3].

The regions of high intensity (the light areas) in Fig. 2 represent the beam response to the 152 Hz CW signal incident at the array. The arrivals show a considerable angular spread, amounting to about 50 degrees between a bearing of appoximately 40 degrees and the broadside direction in the middle of the trace. Fig. 3 shows that the direct (great circle) line between receiver and the source is at an intial bearing of 38 degrees from array forward endfire. (The left edge of the trace in Fig. 2 is 30 degrees relative to array forward endfire). The main mechanism responsible for the observed angular spread in acoustic energy is horizontal refraction of the modal rays in the wedge-like ocean. Another contributing factor of lesser significance is the apparent bearing shift due to the response of the array (which has conical beam patterns) to rays which are inclined to the horizontal plane.

At the beginning of the run (the top of the trace in Fig. 2) the main arrivals extend between bearings of 38 to 62 degrees from forward endfire. These we shall refer to as the "principal arrivals". As the run continues, streaks appear intermittently across the trace between the principal arrival and broadside. The mechanism responsible for producing these streaks is unclear at present, but could be associated with sound propagation in the subbottom structure. The principal arrivals show a variation in bearing with time which is attributed to changes in heading and depth of the array due to a reduction in speed of the towship. In the following section, we concern ourselves with the principal arrivals and interpret the angular spread of these arrivals in terms of the horizontal refraction due to the sloping bottom.

INTERPRETATION OF THE RESULTS

Two theoretical models are used to help interpret the observed angular spread in principal arrivals. One is the normal mode solution of the acoustic field in a wedge with plane, perfectly reflecting boundaries[1], and the second is a ray model based on the concept of ray-mode equivalents. We begin with the ray model, since this provides insight into the nature of the field in the wedge when one of the boundaries is penetrable.

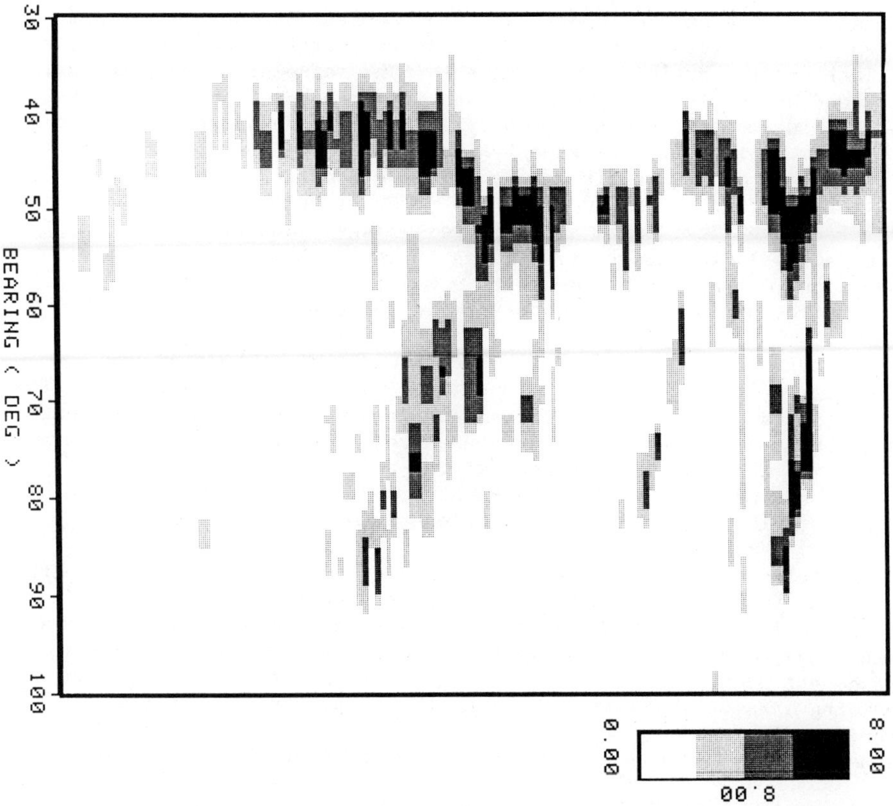

Fig. 2 Bearing-time trace for the 152 Hz CW signal using a maximum-likelihood beamformer, 4.5 hr of data shown.

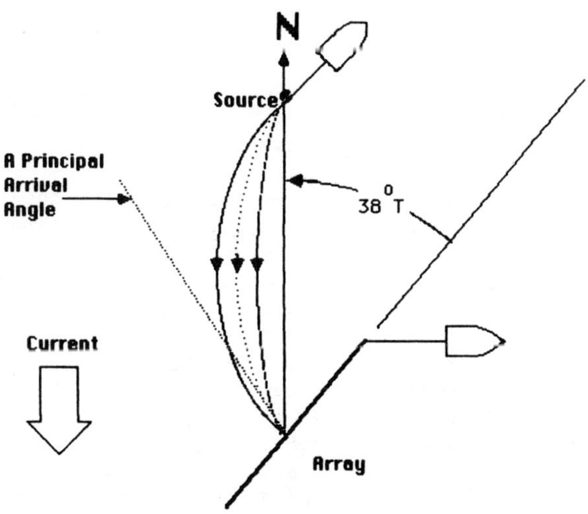

Fig. 3 Geometry of the experiment at the beginning of the run. "principal arrivals" defined.

At the site of the experiment, the compressional sound speed in the bottom sediment was greater than that in the water column, that is, a critical grazing angle, α_c, was evident. The value of α_c computed using the sound speed was 31 degrees and was in agreement with Ref. 3. Rays with grazing angles less than α_c are totally internally reflected from the bottom, while the more steeply inclined rays penetrate the interface in which case most of their energy radiates away into the sediment. For a given mode in the ocean wedge, this penetration occurs at a fixed range from the apex, known as the cutoff range. The effect of cutoff on the modal field is illustrated in Fig. 4 which shows two horizontal projections of rays in a wedge with a multi-faceted bottom, an idealization of the bottom topography in the experimental area.

In constructing Fig. 4a, it was assumed that the bottom was perfectly reflecting making it impenetrable to acoustic radiation. The horizontal ray paths in the diagram correspond to a single mode, and, as can be seen, no cutoff occurs even when the rays are launched up towards the apex of the wedge. (The head of the arrow corresponds to the position of the array). All the rays, irrespective of their launch angle, contribute to the field downslope from the source.

A similar set of ray paths is shown in Fig. 4b exept that here the bottom has been assigned a critical grazing angle of 31 degrees. In this case, it is evident that most of the rays launched towards the apex of the wedge are cut off before being turned around and hence these rays do not contribute to the acoustic field in the vicinity of the receiving array. Thus, we may expect bottom penetration at the experimental site to be a significant mechanism in determining the characteristics of the acoustic field observed at the receiving array.

For a given mode, only two of the hyperbolic rays (designated the "direct" and "indirect" rays) from the source can pass through the receiver position[1]. A simple calculation shows that, under the experimental conditions at the beginning of the run the indirect rays (i.e., all those launched upslope) associated with all propagating modes penetrate the bottom before reaching the receiver. Thus in the experiment, the field at the array consists only of the direct ray paths which themselves show a certain amount of horizontal curvature. The degree of curvature depends on the mode number, increasing as the order of the modes rises. A straghtforward calculation shows that, at the array, the total angular spread due to the direct rays associated with all the modes amounts to 18.2 degrees.

A further, minor angular spread is caused by the interaction of the beam pattern with the nonhorizontally travelling rays associated with the modes. The higher-order modal rays are affected most by this mechanism. A simple geometrical calculation shows that the expected angular spread due to this mechanism is 3.9 degrees.

When these two contributions are summed the total angular spread in energetic arrivals at the array is found to be 22.1 degrees, extending from a bearing of 38 degrees (lowest-order mode) to 60.1 degrees (highest-order mode) as measured from forward endfire. This compares remarkably closely with the angular spread of energy among the principal arrivals seen at the top of the trace in Fig. 2.

Evidently, the modal ray model provides a satisfactory explanation of the observed angular distribution of energy about the principal arrival at the array. However, it says nothing about the relative strengths of the modal arrivals. For this it is necessary to turn to the mode solution[1] for the field in the wedge.

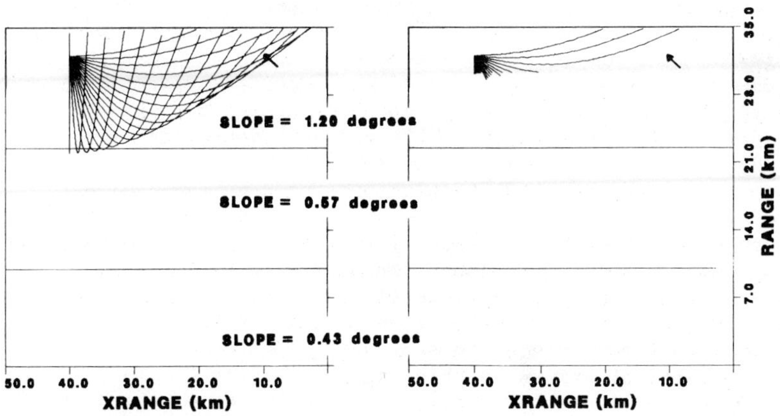

Fig. 4 Effect of critical grazing angle on the modal ray paths from source to receiver corresponding to mode 38 over a multi-faceted slope. (a) 90 degree critical grazing angle. (b) 31 degree critical grazing angle.

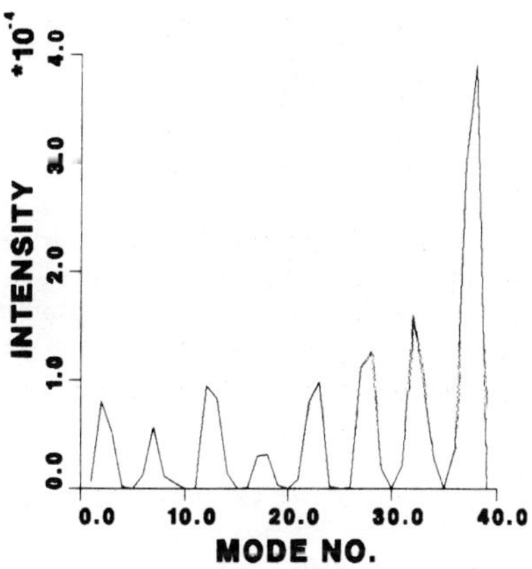

Fig. 5 Predicted mode intensities as a function of mode number for the source-receiver geometry at the beginning of the experimental run.

At the beginning of the experimental run, the source was one-fifth of the way down the water column at a depth of 80 meters. This suggests that ideally every fifth mode should not be excited. Fig. 5 shows the relative strengths of the modes at the first sensor of the array calculated from the normal mode solution taking into account the mode amplitude function (which depends on the range from the apex and the distance between the source and the receiver parallel to the apex) and the source excitation function. The oscillatory dependence of the mode intensities on mode number is apparent, with nulls occuring at every fifth mode as expected. Since many of the modes are effectively absent from the field, there should be perceptible angular gaps between the principal arrivals at the array. This can be investigated by using the normal mode solution to produce synthetic data which may be compared directly with the experimental results presented in Fig. 2.

By computing the field at each of the sensors in the array, using the normal mode solution in conjunction with parameter values appropriate to the experimental conditions and then beamforming, the time-bearing plot in Fig. 6 is obtained. (Time in this diagram actually has little significance since the same data appears all the way down the trace). The vertical arrows in this figure indicate the directions, calculated from the modal ray model, of some of the strongest arrivals which were identified from the mode intensity plot in Fig. 5. Note that the apparent bearing shift produced by the interaction of the conical beam pattern with modal rays, which are inclined to the horizontal, is automatically included in the calculations leading to Fig. 6.

On examining the synthetic trace in Fig. 6, gaps in bearing can be seen between acoustic arrivals, indicating that the modes can be resolved at the array. It has not yet been possible to separate the arrivals in the experimental data but this is currently being attempted. The angular spread of energy in the synthetic trace (Fig. 6) extends up through the vicinity of broadside. This is to be expected because the normal mode solution does not take into account the effect of cutoff. Hence, the computed field includes the contributions from the indirect modal rays which arrive at the array close to broadside as well as the direct rays which arrive at lower bearings.

If the arrivals near broadside are ignored it is clear that the synthetic trace of Fig. 6 strongly resembles the top of the experimental trace in Fig. 2, especially at bearings between 38 degrees and 62 degrees, where the principal arrivals occur. Thus, the normal mode solution also supports the conclusion that the observed angular spread among the principal arrivals in the experimental data is due to horizontal refraction in the wedge.

CONCLUSION

In a part of the international experiment (SET 1) reported here the objective was to investigate the theoretically predicted effects of horizontal refraction in the wedge-shaped ocean. A CW sound source and a towed array were located in the vicinity of the East Australian Continental Rise where the bottom slope is 1.2 degrees. Angular arrivals were received at the array which deviated initially by as much as 23 degrees from the great circle arrivals. These arrivals were interpreted, using a modal ray model and a normal mode solution for a field in a wedge-like ocean, as being largely due to horizontal refraction caused by repeated acoustic reflection from the sloping bottom. It was also possible to infer from the experimental data that the mode cutoff due to bottom penetration (occuring when the grazing

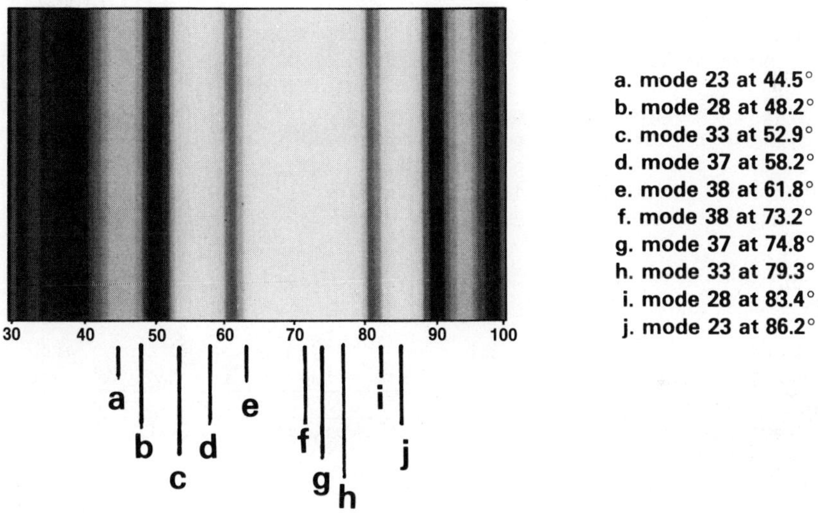

Fig. 6 Bearing plot of synthetic data computed from the normal mode solution for the source-receiver geometry at the beginning of the run. Principal arrivals indicated for strong modes by vertical arrows.

angle of the ray exceeds the critical grazing angle) is a significant mechanism in determining the characteristics of the sound field at the receiving array.

No satisfactory numerical model exists at present for computing three-dimensional sound fields. However, the two theoretical models used here to interpret the results obtained in the experiment, give, between them, a remarkably informative description of the acoustic field observed at the array. There is, nevertheless, a need for a more general modelling capability, which would allow the three-dimensional field in an ocean with a complicated bottom topography and a nonuniform sound velocity profile to be computed with relative ease. This is particularly true now that CW horizontal refraction in the ocean has not only been observed unequivocally but has also been shown to have a significant effect on array performance.

ACKNOWLEDGEMENTS

The able assistance and cooperation of the officers and men of HMNZS TUI, Derek Swann, Captain, and the HMNZS MONOWAI are gratefully acknowledged. Dr. R. Bannister and Dr. M. Guthrie of the Defence Scientific Establishment, Aukland, New Zealand aided the planning and conduct of the experiment as a portion of the SET 1 exercise. Geophysical data was provided by Dr. D. J. Kewley, The Defence Research Centre, Salisbury, Australia and by Dr. G. Packham, Univ. of Sydney. The data analysis and display system was engineered by Mr. B. Decina.

REFERENCES

1. M. J. Buckingham, Acoustic Propagation in a Wedge-shaped Ocean with Perfectly Reflecting Boundaries, Proceedings of the NATO Advanced Research Workshop on Hybrid Formulation of Wave Propagation and Scattering, IAFE, Castel Gandolpho (Rome), Italy, 77:105 (August 30 to September 3 1983).

2. D. E. Weston, Horizontal Refraction in a Three-dimensional Medium of Variable Stratification, Proc. Phys. Soc. 78, 46:52 (1961).

3. E. Hamilton, "Physics of Sound in Marine Sediments," L. Hampton, Editor, Plenum Press (1974).

ACOUSTIC PROPAGATION OVER LARGE-SCALE LINEAR OCEAN SLOPES

D. R. DelBalzo, J. E. Matthews, J. V. Soileau, and C. Feuillade

Naval Ocean Research and Development Activity
Ocean Acoustics and Technology Directorate
NSTL, Mississippi 39529-5004

ABSTRACT

This paper presents the results of a geoacoustic analysis of large-scale linear ocean slopes and their effect upon acoustic propagation. Large-scale linear ocean slopes are defined as features which are long relative to their width, have at least 2 km relief, and have an average slope angle greater than one degree. This classification includes most continental and island arc margins, and some intraocean ridges and fracture zones. These features probably represent the most range dependent of all marine acoustic propagation environments. A global geophysical analysis of these slopes and a geoacoustic assessment of their characteristics based upon 352 published studies and 613 actual slope profiles produced five generic classes. It was found that 79% of the slopes were covered with at least 200 m of sediment and that 55% of the slopes had an average angle of less than 4°.

The most prevalent generic class (passive margin) was chosen for detailed acoustic propagation calculations. The passive-margin slope is common in the North Atlantic, has a vertical angle of 2.8°, and is covered by 500 m of silt. Using a summer Sargasso Sea sound speed profile, numerous calculations of transmission loss at 100 Hz for various signal and noise geometries were performed. It was found that basin-generated noise is stripped away by the slope at some range and that submerged signals in the basin are stripped away at a range farther up the slope, thus creating a range interval wherein the signal-to-noise ratio (SNR) is enhanced by slope interaction. The amount of gain is depth dependent and is affected by downslope propagating noise that is generated over the continental shelf. The median values of slope-induced SNR gain are 16-23 dB, for several signal locations, in a range/depth box (dimensions of 10 km X 40 m) centered at an optimum range at a 20-m depth.

INTRODUCTION

One aspect of underwater acoustic propagation that has received only limited attention is the interaction of sound with a sloping sea floor, especially for signals propagating in the upslope direction. This is becoming an area of increased interest in naval applications. Now that advanced computational methods are available, accurate studies of the effect of slopes on acoustic energy are possible. This paper is concerned primarily with the combined acoustic fields generated by submerged signals in an ocean basin and surface noise originating both in the basin and over the continental shelf.

The vast majority of research in underwater acoustic propagation has traditionally centered around deep-water situations with little or no bottom interaction. Early acoustic propagation models (ray approach) were developed for (and, therefore, were useful in) these situations, especially at frequencies above 50 Hz. As naval needs in shallow water became more important, some researchers began developing the normal mode approach to propagation calculations. However, for efficient full-field computations at all frequencies and water depths in range-dependent environments, the development of the parabolic equation (PE) method is significant. Recent modifications in the PE approach allow for steep angles and discontinuous interfaces in the geoacoustic description of real-world cases, which are essential for the sloping environment of interest.

A precursor to this environmental-acoustic study was an effort to characterize all of the world's large-scale, linear bathymetric slopes into generic classes and to assign representative geoacoustic descriptions. This work produced five generic classes, and the most common of these was used for this paper. This class is representative of much of the North Atlantic Ocean, has an average angular slope of 2.8°, and is thickly sedimented. The geoacoustic description is believed to be accurate to handle frequencies down to approximately 30 Hz.

So often propagation studies are limited to either submerged signals or surface noise. In reality, the quantity of interest is always signal-to-noise ratio (SNR). Therefore, in addition to separately considering the effects of slopes on submerged signals and surface noise, this paper addresses the more important quantity of SNR. Calculations are presented for submerged signals at various deep-water positions in a variety of noise fields at 100 Hz.

This paper begins with a geophysical analysis of large-scale linear slopes. Then, the various models used for analysis are discussed—geoacoustic, oceanographic, and acoustic field. Next, the results of slope interaction on submerged signals, surface noise, and SNR are described.

GEOPHYSICAL PROPERTIES OF LARGE-SCALE LINEAR SLOPES

For this investigation, a large-scale linear slope is defined as a feature with at least 2 km of vertical relief, with a length many times its width, and an average slope in excess of 1°. By this definition most continental margins and island arc systems, as well as numerous submarine ridges and some fracture zones, are included. It is interesting to note, however, that some continental margins do not fall within this category. For example, the region in the Bay of Bengal off the mouth of the Ganges River consists of a large deposit of detrital sediment derived from erosion of the Himalayan mountains. The relief of this feature is adequate, but the fan extends so far out to sea that the average slope is less than 1°.

The identification and statistical characterization (such as average slope and height) of large-scale linear slopes were made from unpublished bathymetric maps (at an approximate scale of 1:4,400,000) produced by the U.S. Naval Oceanographic Office. After the slopes were delineated, 352 reference publications were assembled for a more detailed slope characterization and analysis of their shallow geology (<500 m). The details of this analysis, including maps showing the global distribution of slopes, are presented in Green and Matthews.[1] Table 1 gives the percentage occurrence of several "first-order" slope properties. This tabulation indicates that most large-scale slopes have relief of less than 4 km and average slope angles of less than 4°. It is not shown in this table but, by far, the dominant grain size of slope sediment corresponds to silt. The biogenic component of sediment is given by the percent occurrence of calcium carbonate (C), and silicious material (S) with the remainder being of terrigenous origin. The sediment thickness tabulation is somewhat misleading, because there are both sedimentary rocks and soft (unconsolidated) sediments. In geological discussions about sediment thickness this distinction is not always made; however, the geoacoustic implications are highly significant.

Table 1. Percentage occurrence of first-order slope geoacoustic properties. Under biogenic sediment, the abbreviations C and S denote calcium carbonate and silicious material, respectively.

RELIEF				SEDIMENT			
Height (km)	Occurrence (%)	Angle (deg)	Occurrence (%)	Biogenic (type)	Occurrence (%)	Thickness (m)	Occurrence (%)
2-3	49	1-2	24	C<30	38	<200	21
3-4	37	2-4	31	30<C<50	22	200-400	14
>4	12	4-8	35	C>50	34	>400	65
		>8	10	S>30	6		

A second method of comparison is to group the slopes according to their geophysical properties. Five generic classes of large-scale linear slopes were found. Four classes include various types of continental margins (passive, apparently passive, convergent, and active translation), while the fifth class (intraoceanic) includes features such as mid-ocean ridges and fracture zones. The nomenclature chosen has a plate tectonic connotation because the origin of these features has a distinct correlation with their physical characteristics. Although these generic classes are broad and contain considerable overlap, a natural grouping does appear. Previous comprehensive studies of continental margins, subduction zones and intraoceanic ridges took significantly different approaches. Only Emery[2] presented a global classification scheme, but this was for continental margins only. None of the previous studies included all types of large-scale linear features nor did they assess the parameters relevant to ocean acoustic modeling.

Table 2 summarizes the average properties of the five classes. Four of the five slope classes do not have relief comparable to that of full ocean basin depths due to two separate causes. The height of intraoceanic slopes is low because these features do not come to the ocean surface. The heights associated with passive, apparently passive, and active translation margins are low because the sedimentary rise provinces, which commonly occur at their bases, have not been included in the statistics. These rises should be treated in future studies. A strong correlation was found between height and average slope. Although 33% of the slopes consist of more than one-half outcropping rock, 24% have no rock outcrops at all. Most slopes have some outcropping rock with appreciable accumulations of sediment between the outcrops. Most rock outcrops consist of either terrigenous sedimentary or igneous rock. The strongest correlations between generic class and ocean basin are passive margin with the Atlantic Ocean, convergent margin with the Pacific Ocean, and apparently passive margin with the Mediterranean Sea.

Table 2. Generic classification for large-scale linear slopes. The following abbreviations are used: CCD-carbonate compensation depth, T-terrigenous sediment, C-calcium carbonate sediment, TS-terrigenous sedimentary rock, I-igneous rock, LS-limestone, DS-deformed sedimentary rock, NA-North Atlantic Ocean, SA-South Atlantic Ocean, NP-North Pacific Ocean, SP-South Pacific Ocean, I-Indian Ocean, M-Mediterranean Sea, ARC-Arctic Ocean, and AARC-Antarctic Ocean.

	PASSIVE MARGIN	APPARENTLY PASSIVE MARGIN	CONVERGENT MARGIN	ACTIVE TRANSLATION MARGIN	INTRAOCEANIC
% OCCURRENCE	35	12	20	10	23
AVERAGE HEIGHT (km)	2.6	2.7	4.5	3.2	2.7
AVERAGE SLOPE (deg)	2.8	3.6	5.5	5.1	3.6
SEDIMENT THICKNESS (m)	>400	highly variable	largely unknown	<400	<400 above CCD
SEDIMENT TYPE (%)	T(45) C(55)	T(40) C(60)	T(50) C(50)	T(50) C(50)	C(100)
% ROCK OUTCROP	40	38	60	50	50
ROCK TYPE	TS	TS,I,LS	I,DS	I,DS	I
DOMINANT OCEAN	NA, SA ARC,AARC	I,NP,M	I,NP,SP	NA,SA NP,SP	I,NA,SA ARC,AARC

GEOACOUSTIC, OCEANOGRAPHIC, AND ACOUSTIC FIELD MODELING

A. Geoacoustic: The passive-margin generic class of slope was chosen for the following analysis because it represents over one-third of all large-scale linear ocean slopes. This class is typical of much of the North Atlantic Ocean and is geoacoustically simple in that less than half of this class has outcropping rock or shallow acoustic basement. The geoacoustic model developed for the subsequent acoustic field calculations is shown in Figure 1. It consists of a sediment half-space of silt, a slope angle of 2.81°, and a height of 2600 m. At the top of the slope is a shelf-slope break at a water depth of 200 m. The shelf has a bottom slope of 0.26° and extends landward to a water depth of 100 m. At the slope base is a continental rise modeled as a clayey silt half-space with a slope of 0.5°. The rise province extends in the direction of the basin to the acoustic source location. In this report, all ranges are referenced to the 100-m bottom depth contour over the shelf and increase toward the deep-water basin, out to a maximum range of 150 km. Because the rigidity of these sediment types is low[3] they are treated as fluids. The remaining geoacoustic model parameters were estimated using the techniques outlined by Hamilton.[4]

Bottom loss, either calculated or measured, is highly dependent upon source-receiver geometry. It cannot be measured directly in shallow water, nor have these measurements been attempted over large-scale sea floor slopes. It is therefore preferable, in the current investigation, to input geoacoustic parameters directly into an acoustic field model rather than to perform the interim step of calculating theoretical bottom loss. The passive-margin slope class, especially when fully covered by thick sediment, has the least topographic irregularity of all slope classes. For this initial evaluation of slope propagation, no sea floor roughness has been included. A sediment layer of 500-m thickness is also adequate to preclude significant acoustic interaction with underlying rock for frequencies above 30 Hz. At frequencies above 300 Hz, thin sediment layering, not included in this model, will become increasingly important to acoustic bottom interaction. The present geoacoustic model is considered applicable to frequencies between 30 Hz and 300 Hz.

B. Oceanographic: Since the passive-margin class is representative of many sites along the east coast of the United States, a Sargasso Sea sound speed profile (SSP) was selected for the computer calculations. In order not to confuse oceanographic effects with bathymetric effects, a single range-independent SSP was used. In addition, a summer profile was chosen to accentuate the bottom interaction.

C. Acoustic Field: The results described in this work were obtained by the PE method first introduced by Tappert.[5] The study was performed using the Implicit Finite-Difference (IFD) formulation of the PE method developed by Lee and Botseas,[6] and modified by Botseas et al.[7] to include "wide-angle" capability.

The PE method has proven extremely important and successful in many underwater acoustic propagation problems. It may be applied in environments where propagation is predominantly radial, and where backscattering is negligible. This must be the case because the PE method computes the entire range and depth-dependent field by marching the solution forward in range and therefore neglecting backward-traveling waves. This limitation aside, however, the PE method includes other full-wave effects, like diffraction. It can also be used in many range-dependent environments. The IFD code is designed to solve the wave equation for compressional waves only: it does not include shear-wave effects. It is applicable in the work considered here since, as seen earlier, the rigidity of the sediment layers is low enough for them to be treated as fluids, which do not support shear-wave propagation.

The standard PE method is not valid for propagation in directions that make vertical angles of more than about 20° from the horizontal. The wide-angle capability, which is reviewed in detail in the NORDA PE Workshop Report,[8] extends the angular range of applicability to about 40° above or below the horizontal. Only with this capability is it possible to perform accurate calculations of propagation over steep slopes or over layers with discontinuities. This study deals with environments in which multiple reflections of sound between the sea surface and the sea bottom lead to acoustic ray paths greater than 30° above and below the horizontal, which makes the wide-angle capability essential.

In this study the inputs for sound-speed, density and acoustic attenuation, in the bottom layers, are average numbers for these materials, derived from direct geoacoustic measurements. Since these data are available, and since the IFD formulation can utilize them, the full geoacoustic description, vice bottom reflectivity, was chosen.

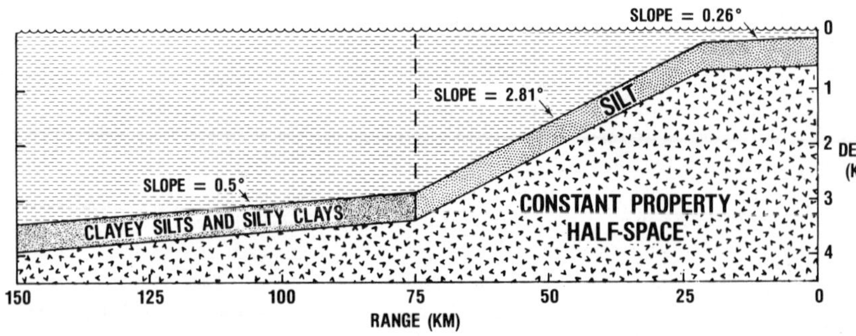

Figure 1. Geologic model for passive-margin geoacoustic situation.

ACOUSTIC FIELD CALCULATIONS

A. Philosophy: A suite of PE calculations was performed at 100 Hz with the passive-margin geoacoustic description for each situation under study; i.e., signal at 100-m depth in the basin at several ranges between 100 and 150 km from the 100-m bottom depth contour and noise at 7-m depth at similar ranges in the basin, plus several additional ranges over the shelf. The sequence of acoustic field calculations is shown in block diagram form in Figure 2 with a chronological order of execution given by the letters a-k.

The first quantity calculated was slope enhancement of the acoustic field for a 100-m deep signal (submerged target) by differencing the transmission loss obtained from two PE field calculations—one without a slope (a) and one with a slope (b). As will be shown later, the average level of enhancement (c) is only a few decibels for the cases under study.

Signal enhancement over the slope is interesting and important; however, if the noise field is similarly enhanced, then no real signal-to-noise ratio (SNR) gain is achieved. In a preliminary effort to incorporate noise into this study, two PE fields (again one without (d) and one with (e) the slope) were calculated for a 7-m deep source in the ocean basin to simulate the noise field produced by surface shipping. Obviously, this single-ship description of the entire ambient noise field is only partial, but it yields useful insight about the effect of the slope on surface noise. As before, the transmission losses for the two fields were subtracted to yield a plot of upslope-enhanced noise (f). There are many places in the world where the noise received over a slope comes predominantly from the deep water (e.g., selected sites in the Caribbean and Mediterranean Seas), so the results incorporating only upslope noise will apply at these locations.

A thorough treatment of noise is beyond the scope of this study and would require many detailed calculations for the appropriate horizontal distribution of noise sources and detailed geoacoustic and oceanographic descriptions in all directions. A very simple, but somewhat complete, noise model would involve noise coming from only three directions—the deep-water basin, along the continental slope/shelf, and directly landward of the receiver. In those situations where shipping over the shelf is extensive, one must include the coastal component of noise. However, those ships along the continental coast (but not directly landward) will probably be highly attenuated at a receiver positioned over the slope for two reasons. First, the entire propagation path will be in relatively shallow water so that the received level will be low due to bottom interaction. Second, this sound will propagate in a wedge and thus be refracted out to sea and away from the receiver after several bounces. This frequency-dependent (shadow zone) effect has been postulated[9] and observed.[10] The only situation left to consider is when the noise is directly landward of the receiver, which will happen only sporadically. This case was studied by calculating a PE field for a 7-m deep source over the shelf (i.e., downslope noise propagation) at three locations (over the 100, 150, and 200-m depth contours). One location (200-m depth contour) is at the shelf-slope break. The downslope noise field (h) was then subtracted from the noslope noise field (g = d) to produce a difference field (i) representing slope effects on noise generated over the shelf. The next step in the analysis

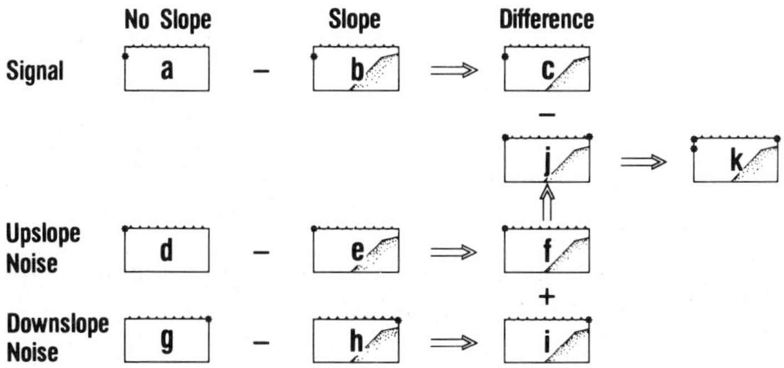

Figure 2. Schematic description of acoustic field calculations showing the chronological sequence of execution by the letters a-k.

is to combine the upslope-enhanced and downslope-enhanced noise fields to produce the enhancement of the total noise field (j).

At this point the effect of a slope on a submerged signal and on total surface-generated noise has been addressed. A final combination of signal enhancement over the slope with noise enhancement over the slope yields SNR enhancement (k). These three calculations (c, j, and k) represent absolute gains independent of signal and noise source levels. In other words, whatever the actual SNR is for the noslope case, the SNR enhancement is directly additive as real gain.

B. Upslope Propagation Results for Signals and Noise: The results for upslope propagation of signal (100-m depth) and noise (7-m depth), corresponding to inserts b and e of Figure 2, are given in Figure 3 as transmission loss (TL) vs. range at a receiver depth of 19 m for three initial source locations (150, 130, and 100 km). The first observation from Figure 3 is that the TL curves corresponding to the three source locations decay at approximately the same rate and position over the slope, despite the 50-km variation in initial source location. Since the upslope noise results are somewhat robust to source position, a single upslope noise source (150-km range) was chosen for the remainder of the calculations presented in this paper.

The second observation from Figure 3 is the severe attenuation of both signal and noise energy over the slope, which is due to penetration of sound into the bottom. Obviously, receivers placed well up the slope will measure drastically reduced levels from both signals and noise generated in the ocean basin. A significant observation is that noise is attenuated at greater ranges than is signal, thus opening the possibility for signal-to-noise ratio (SNR) gains through a restricted range interval over the slope. For example, the TL drops below 100 dB at ranges of 55 and 40 km for noise and signal, respectively, for the 150-km source location. Therefore, this 15-km interval is a potential location for SNR gains at the 19-m receiver depth.

The cause of the "limiting" noise range, R_{LN}, being different than the "limiting" signal range, R_{LS}, is apparent in Figure 4. The noise is generated near the surface at a relatively high sound speed, which corresponds to a deep conjugate depth and, consequently, to slope interaction at a deep depth.

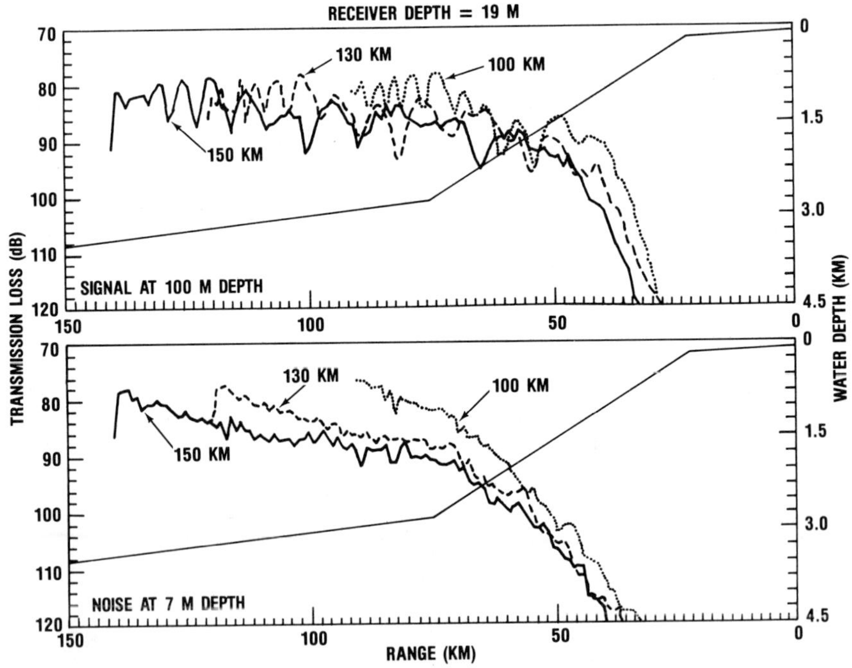

Figure 3. Slope effect on signal (upper) and on basin noise (lower) for the passive margin and a summer Sargasso Sea sound speed profile. The effect is shown by transmission loss at 100 Hz vs. range at a 19-m receiver depth for three initial source locations (150, 130, and 100 km) relative to the 100-m bottom depth contour.

Alternatively, the low sound speed at the signal depth creates a narrow sound channel and thus allows signal energy to propagate farther up the slope before conversion to steep vertical angles and ultimate loss in the bottom. The SSP and bathymetry in Figure 4 are meant to illustrate the concepts using a ray diagram. Actually, the summer Sargasso SSP has such a high near-surface sound speed that all cases studied in this report are bottom limited. Also, the differential range aspect of signal and noise stripping is accentuated by the summer Sargasso SSP chosen for this study.

The next quantity of interest is slope enhancement of basin-generated signals and noise (corresponding to inserts c and f of Figure 2). Figures 5 and 6 are contour plots of the enhanced acoustic fields for signal and noise at a 150-km range, respectively. For this frequency, geoacoustic model and geometry, it is apparent that no large gain in signal or noise level is created over large ocean areas by the 2.8° slope under study. Note that contours of "de-enhancement," or loss of energy, are not shown in these figures. The occasional 3-6 and 6-9 dB signal enhancements are interesting, but careful analysis indicates that these are often caused by a range-shift in the acoustic field; i.e., the slope causes some peaks to appear at ranges of valleys in the noslope case, and vice versa.

C. Downslope Noise Propagation Results: Although noise from the landward direction may not be important for many sites, for reasons mentioned above, it will certainly be important for those instances when a ship is nearby and directly landward. This situation is addressed in Figure 7 where a 7-m deep source is placed over the 100-m depth contour (zero range) and energy is propagated downslope. As the sound moves from the shelf out to the slope, steep-angled rays are converted into shallow-angle rays, as a result of bottom interaction, and then are trapped in a narrow sound channel.[11,12] Thus, it is apparent that noise from shelf shipping may fill in noise near the bottom over the slope and eliminate SNR gains in certain locations. However, noise levels at shallow depths over the slope will not be significantly affected by this shelf traffic. Of course, noise sources over the shelf on the opposite side of the ocean basin may contribute to the total noise field in the slope-enhanced region; however, it is felt that propagation loss over these long basin distances will reduce this component of noise.

The downslope noise calculation was repeated for noise source locations over the 150- and 200-m (shelf-slope break) depth contours. The results at two receiver depths (19 and 291 m) are shown in Figure 8 as TL vs. range for all three noise locations. At the shallow receiver depth, the TL for the two shelf noise sources is 100 dB at a range of about 30 km, while at the deeper receiver, a 100-dB TL occurs at about 50 km. This confirms the qualitative result of Figure 7 that downslope noise is more intense at deep depths than at shallow depths.

The TL is significantly lower for the noise source at the shelf-slope break (200-m water depth) than for the noise sources totally over the shelf. Thus, only in areas where local surface traffic tends to transit seaward of this break will shallow receivers over the slope be dominated by coastal shipping. For those sites where the local shipping is significantly landward of the break, downslope noise may not be a dominate noise factor for shallow receivers over the slope.

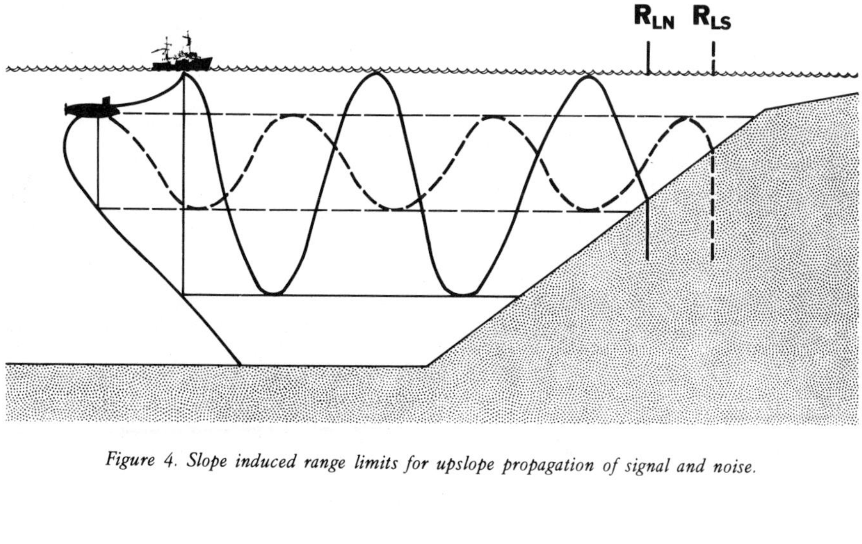

Figure 4. Slope induced range limits for upslope propagation of signal and noise.

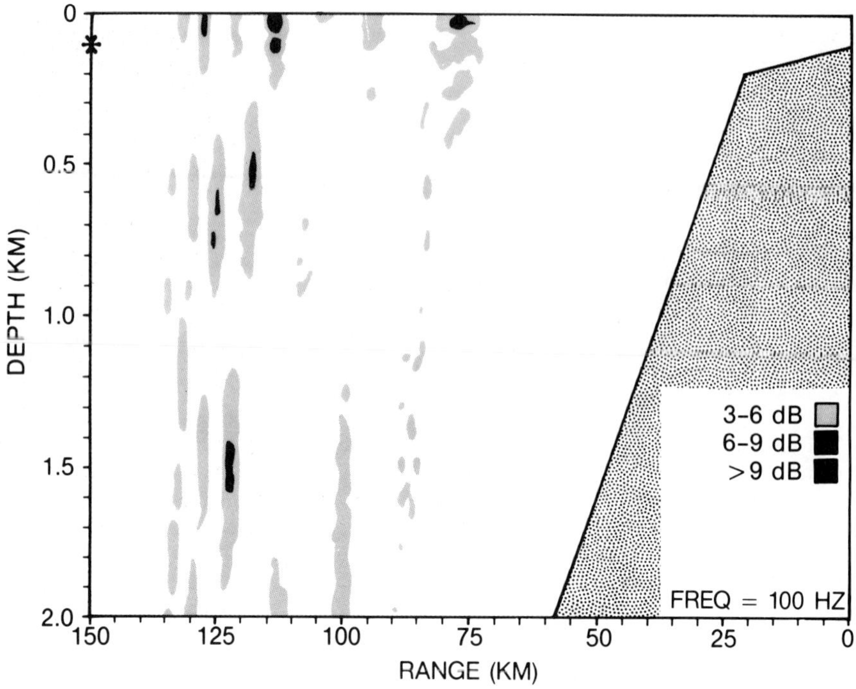

Figure 5. Signal enhancement contour plot at 100 Hz for the passive margin with a summer Sargasso SSP. The signal is 100 m deep at a range of 150 km in the basin.

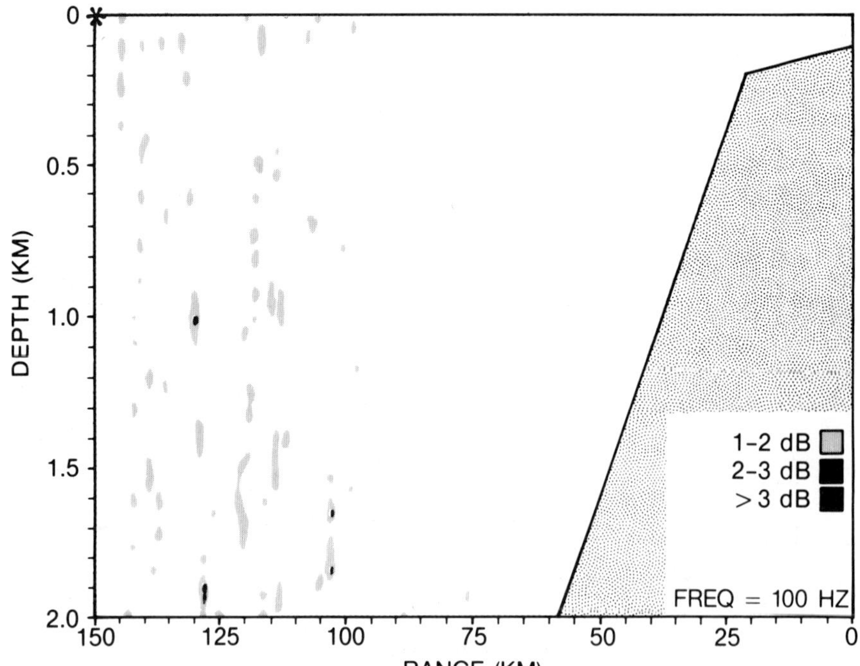

Figure 6. Upslope noise enhancement contour plot at 100 Hz for the passive margin with a summer Sargasso SSP. The noise is 7 m deep at a range of 150 km in the basin.

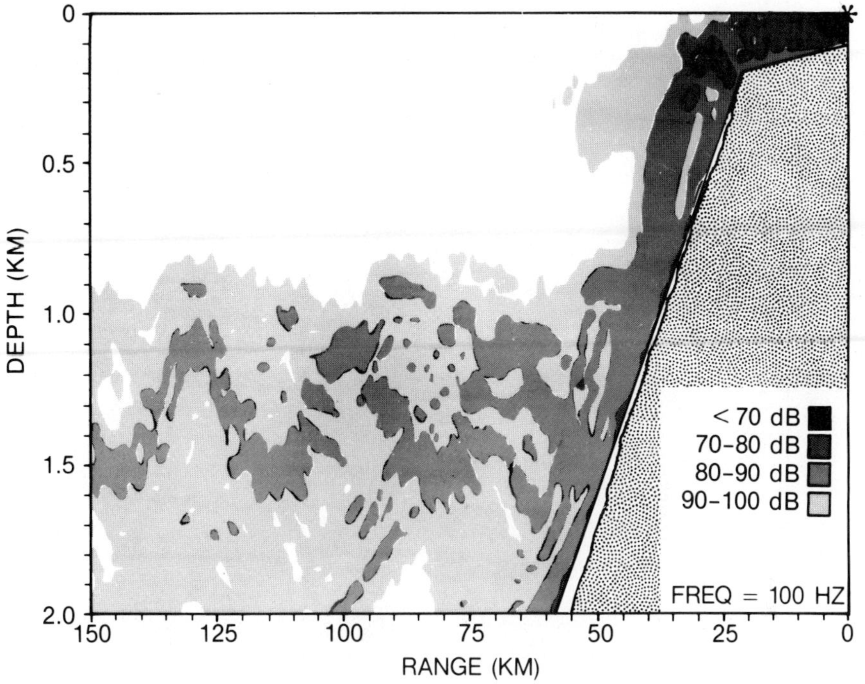

Figure 7. Downslope noise propagation at 100 Hz for the passive margin with a summer Sargasso SSP. The noise is 7 m deep at a range (0 km) where the water depth is 100 m.

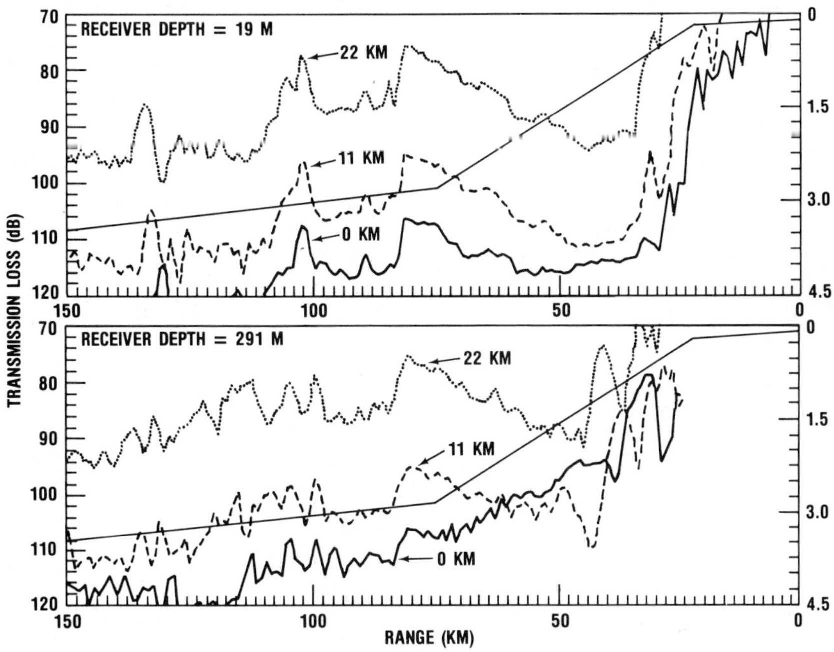

Figure 8. Slope effect on shelf-generated noise at 19 m (upper) and 291 m (lower) for the passive margin and a summer Sargasso SSP. The effect is shown by transmission loss at 100 Hz vs. range for three initial source locations (0, 11, and 22 km) relative to the 100 m bottom depth contour.

D. SNR Results Including Basin and Shelf Noise: The complete picture of signal combined with basin and shelf noise is given in Figure 9. This field results from subtracting the transmission loss values for noise enhancement (upslope from the 150-km range and downslope from the 0-km contour range) from signal enhancement for a submerged signal at a 150-km range. The significant result is that a large region in range and depth exists over the slope wherein SNR enhancements of 5-20 dB are calculated. At a 19-m depth, the range interval for a 5-10 dB enhancement is about 50 km, whereas 15-20 dB enhancements are calculated over a 25-km range interval. The lack of enhancement near the water/slope boundary is caused by the noise generated over the shelf.

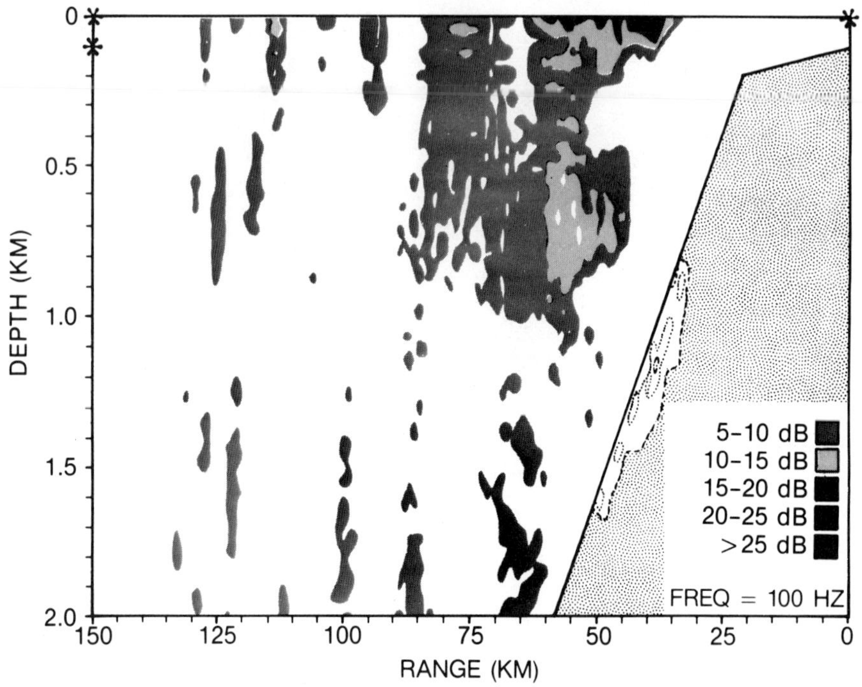

Figure 9. Signal-to-noise ratio enhancement for the passive margin with a summer SSP, including both basin and shelf noise.

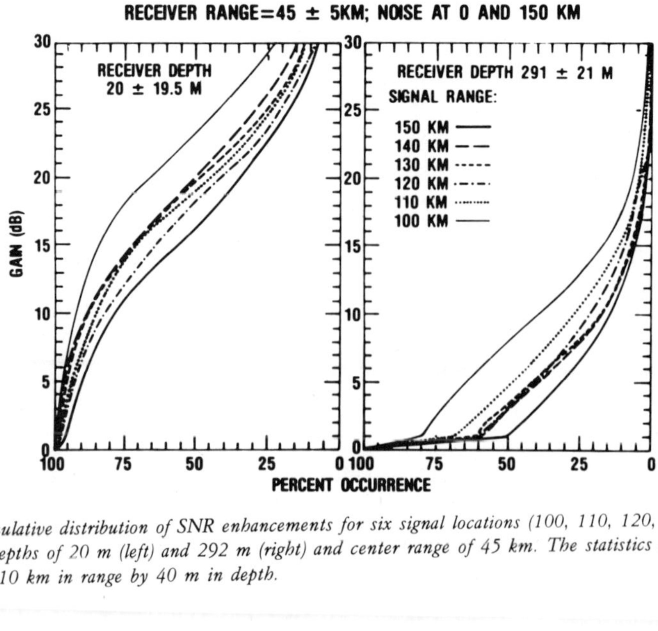

Figure 10. Cumulative distribution of SNR enhancements for six signal locations (100, 110, 120, 130, 140, and 150 km) at center depths of 20 m (left) and 292 m (right) and center range of 45 km. The statistics are based on values within an area 10 km in range by 40 m in depth.

A somewhat surprising ancillary result is that significant gains (up to 15 dB) are present in the thick sediment near the sound channel axis. This result is highly dependent on frequency (based on other calculations at 30 Hz) and on the position of shelf traffic. It is the subject of ongoing research at NORDA.

As a check on the robustness of this concept, the position of the submerged signal was moved in 10-km intervals from the most distant (150 km) range to a range of 100 km, while keeping the noise field constant. Then the cumulative distribution of SNR enhancements, in a range-depth box centered at one promising range (45 km) and 2 depths (20 and 292 m) was computed. The box covers a depth interval of 40 m and a range interval of 10 km, is centered over the 1328-m bottom depth contour, and contains 750 PE grid points (after some decimation to conserve computer memory). These results are given in Figure 10 where it is apparent that the concept is fairly robust against source location. The median enhancement is 16-23 dB and 25% of the values are 23-29 dB at the shallow depth. The results degrade with depth as expected.

CONCLUSIONS AND RECOMMENDATIONS

It has been determined that large-scale, linear slopes can be grouped into broad generic classes on a global basis and associated geoacoustic descriptions are feasible. Also, there now exist acoustic propagation models that can accurately predict acoustic fields over sloping seafloors with realistic geoacoustic descriptions. As a result of these descriptions and calculations, it was found that signal enhancement for upslope propagation amounts to several decibels. In addition, noise generated in the basin yields a few decibels of enhancement at different locations over the slope. These effects would appear to cancel and thus provide no appreciable SNR gain; however, there exists a region over the slope where severe bottom interaction greatly suppresses noise with only moderate signal suppression (for the summer Sargasso SSP) and thus yields significant SNR gains. Enhancements of 5-20 dB are common and robust to variations in signal range and occur mostly at shallow depths but over a large (25-50 km) range interval for the passive-margin case.

Additional work is needed in other types of environments to check the validity for various slope angles, subbottoms, and SSPs. The frequency dependence needs investigation and a more complete noise-field description is desired (especially to include locally generated ship and wind noise). Finally, measurements should be taken with a high spatial sampling density in search of decreased ambient noise over slopes, to confirm the conclusions of this paper.

REFERENCES

1. J. A. Green and J. E. Matthews, "Global Analysis of the Shallow Geology of Large-Scale Ocean Slopes," NORDA Tech. Note 197 (1983).
2. K. Emery, "Continental Margins of the World," in: *Geology of the East Atlantic Continental Margin*, ed., DeLaney, IC SU/SCOR Symp. 31, 7-29 (1970).
3. E. L. Hamilton, "Shear-Wave Velocity Versus Depth in Marine Sediments: A Review," *Geophysics* 41:985 (1976).
4. E. L. Hamilton, "Geoacoustic Modeling of the Sea Floor," *J. Acoust. Soc. Am.* 68:1313 (1980).
5. F. D. Tappert, "The Parabolic Approximation Method," in: *A Wave Propagation and Underwater Acoustics*, J. B. Keller and J. S. Papadokis, ed., Springer-Verlag, New York (1977).
6. D. Lee and G. Botseas, "IFD: An Implicit Finite Difference Computer Model for Solving the Parabolic Equation," NUSC Tech. Rep. 6659 (1982).
7. G. Botseas, D. Lee, and K. E. Gilbert, "IFD: Wide Angle Capability," NUSC Tech. Rep. 6905 (1983).
8. J. A. Davis, D. White, and R. C. Cavanagh, "NORDA Parabolic Equation Workshop," NORDA Tech. Note 143 (1982).
9. M. Buckingham, "Acoustic Propagation in a Wedge-Shaped Ocean with Perfectly Reflecting Boundaries," NRL Tech. Rep. 8793 (NTIS—ADA 139884) (1984).
10. R. Doolittle, A. Tolstoy, and M. Buckingham, "Experimental Confirmation of Horizontal Refraction of CW Sound Propagation in a Wedge-Like Ocean," in: *Ocean Seismo-Acoustics*, T. Akal and J. M. Berkson, ed., Plenum Press, New York (1985).
11. S. C. Wales and O. I. Diachok, "Ambient Noise Vertical Directionality in the Northwest Atlantic," *J. Acoust. Soc. Am.* 68:577 (1981).
12. R. Dashen and W. Munk, "Three Models of Global Ocean Noise," *J. Acoust. Soc. Am.* 76:540 (1984).

PREDICTED PARTITIONING OF VLF ACOUSTIC ENERGY IN A RANGE-DEPENDENT ENVIRONMENT

T. W. Tunnel and G. J. Tango

Naval Ocean Research and Development Activity
Ocean Acoustics and Technology Directorate
NSTL, Mississippi 39529-5004

ABSTRACT

The propagation of 10-Hz signals up the continental slope off the southeastern coast of the United States (Cape Fear, North Carolina) is modeled in support of the Cape Fear VLF Exercise to be conducted during June 1985. Because the first several hundred meters of subbottom in this area consist of sediments that can only weakly support shear, the effect of shear in the subbottom will be primarily to increase transmission loss. This increased transmission loss is modeled with a set of effective geoacoustic parameters derived from the fast field program models. The effective geoacoustic parameters are input to the parabolic equation model to predict the propagation of 10-Hz signals up the continental slope at Cape Fear. The details of deriving the effective subbottom parameters are discussed. The numerical results presented include the transmission loss as a function of depth and range and the relative fraction of acoustic energy that is bottom-interacting.

INTRODUCTION

This paper describes the acoustic propagation modeling of 10 Hz continuous wave (CW) signals up the continental slope off the southeastern coast of the United States near Cape Fear, North Carolina. This work was done to support a very low frequency (VLF, $f < 20$ Hz) exercise to be conducted during June 1985. The modeling effort focuses on identifying the dominant features that affect the acoustic field in the water column. The geology of the Cape Fear region is described in Section 2. Because the first several hundred meters of subbottom consist of thick sediments, it was felt that the effect of shear in the subbottom could be adequately modeled by effective subbottom parameters that do not include shear (where an increased P wave attenuation is used). The SACLANT FAst field program for Range Independent environments (*SAFARI*[1]) was used to derive the effective subbottom parameters; the details and results are given in Section 3. The upslope propagation is modeled with the implicit finite difference parabolic equation model (IFDPE). The transmission loss (TL) and simulated beamformed results for a vertical hydrophone array are presented in Section 4. A summary is given in Section 5.

MODEL GEOLOGY

The Cape Fear Sound was selected for the measurement exercise because of its strategic location with respect to coastal shipping lanes and because it encompasses a wide range of continental shelf bathymetry and geological conditions. In the region of interest (Figure 1), the geology consists of semiconsolidated hemipelagic muds extending seaward into the marginal geosyncline, overlying reefal limestones on a buried basement ridge. The availability of over 200 km of recent high-quality multichannel reflection seismic data[2] from the U.S. Geological Survey (USGS) for this area provided detailed measurements of subbottom and basement layer compressional wave interval velocities. On seismic records, a relatively uniform band of sediments appears as flat layering that gradually thickens seaward. Sediment thicknesses ranged from 100 m (rare) to over 800 m, with a median value of 500 m. Near the CW track (Shot Line 3 in Figure 1), as many as seven well-defined and continuous reflector horizons can be identified. These data provide an upper bound for probable velocity-depth gradients within the sediments using Hamilton's relations.[3]

Multichannel seismics reveal uniform fine-scale structure with range in post-rift sediments over the first 50 km of the track. However, at greater distances before the ridge dropoff, there are numerous evidences of lateral inhomogeneities, as well as diapiric intrusions and stratigraphic pinchouts (at ocean bottom depths greater than 1000 m). No simple velocity-range trends exist even in planar sections; only a weakly positive velocity trend for ranges greater than 100 km is inferred from refraction data. An original minimum number of 46 equispaced velocity/depth control points every 5 km were considered as candidates for shear effective modeling using the *SAFARI* FFP; further details of bottom fine-structure will be averaged out in the VLF regime. The shelf-to-abyssal plain slope was modeled using linear segments with range and depth increments of 25 m and 10 m, respectively ($\lambda/6$ and $\lambda/15$), of roughly 0.5^{0} inclination.

EFFECTIVE PARAMETER MODELING

A number of papers reporting numerical modeling and experimental results for seismic-acoustic propagation in the VLF regime over several environments have appeared in the recent literature.[4] Despite disagreements concerning the precise assignment of observed TL to shear-wave conversion and interface wave generation, it is generally acknowledged that overall propagation in shallow water (with respect to modal depth) allows modeling of shear rigidity effects as an increase in net P-wave attenuation in an acoustic-only model (without shear). As one recent option,[5] an equivalent bottom can be numerically derived that (in a least squares sense) minimizes the discrepancies between TL_{shear} and $TL_{no\ shear}$. While numerically efficient, however, the velocity/depth profiles so derived will oftentimes diverge grossly from realistic geology (which will be unacceptable where multifrequency pulse analysis is required). A simpler (but to date time-consuming and, hence, infeasible) procedure for deriving a shear-effective geoacoustic model is to simply neglect shear velocity and attenuation and, instead, to increase P-wave attenuation uniformly in all sediment-bottom layers.

If actual environmental conditions depart radically from those used here (compressional velocities derived from seismic refraction and multichannel reflection data), significant errors in predicted TL structure and levels might be incurred. Such departures of real from idealized environmental conditions could include strong lateral inhomogeneity in range, rough topography on the seabed, and strong localized perturbations in the water column sound-speed profile. Preliminary sensitivity studies based on available regional data have suggested that the use of more detailed and complete environmental models will not substantially improve predicted results.

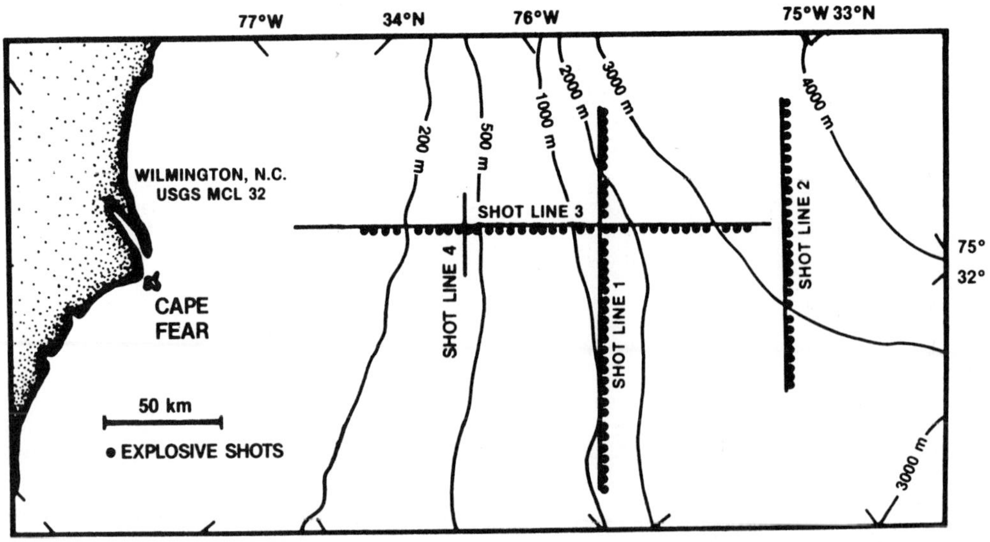

Figure 1. Geographic-shotline map of Cape Fear VLF field exercise.

In the following, an omnidirectional CW point source is located at a depth of 100 m. The water column (considered at varying depths from 150 to 4000 m) is characterized by a weak velocity gradient. The sediment and shallow crustal subbottom are divided into an arbitrary number of isovelocity layer steps representing a (assumed) positive velocity attenuation and density gradient in the sediments. A typical number of layers considered was 25. It was found that for sediment thicknesses greater than about 300 m, the approximate geometric ray turning depth, deeper high-shear velocity subbottom, and crustal layers are acoustically masked, thus, the net TL loss is effectively compressional. Comparison of the theoretical Green's impulse response functions for a full crustal and sediment-layers-only model reveals very little difference in the total number and amplitude of standing modes and seismic arrivals.

Some of the modeled results are presented below. The results show shear and no-shear computations by *SAFARI* FFP of plane wave reflection coefficient (RC) scaled as bottom loss in dB, over the

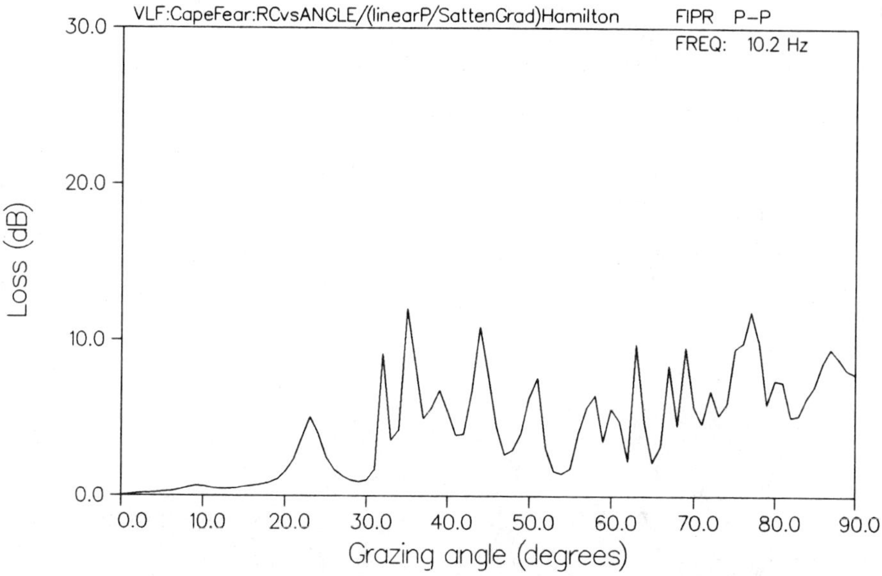

Figure 2a. Plane wave bottom reflection loss (full-shear) elastic bottom.

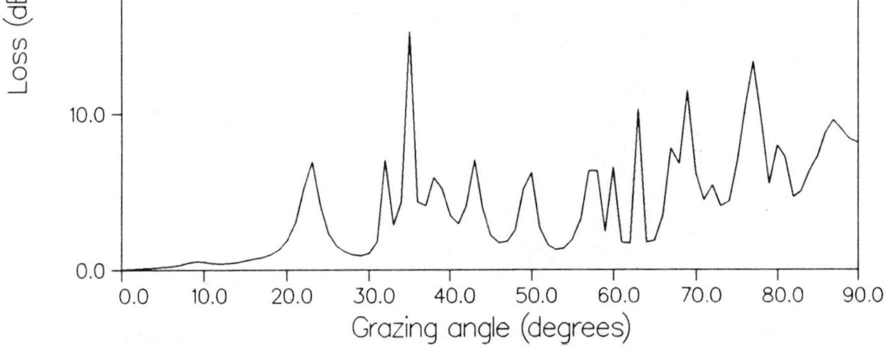

Figure 2b. Plane wave bottom reflection loss (acoustic-only) no shear, with effective attenuation = 1.25.

incident angles (0-90°) and VLF frequencies (Figure 2). The plane wave RC aids in identifying specifically bottom-interacting energy. As will be seen, the no-shear model underestimates both the low-loss trend and the correct, more complex pattern of peaks and dips for grazing angles greater than critical. This difference is a direct indication of the relative importance of shear conversion and interface wave effects (the latter only rarely seen as a relative peak in reflectivity loss at 0° grazing angle). This result agrees with earlier work.[6]

When the P-wave attenuation in all layers of the acoustic-only model is uniformly increased by a factor of 1.25, the agreement with bottom loss computed for the full shear model is almost exact for both trend and interference regions. This result is confirmed by comparing the agreement of CW transmission loss vs. range at 10 Hz, computed for shear, and for an acoustic-only model with a shear-effective P wave attenuation increase of 1.25. Here, both gross and almost all fine TL structure on the TL shear curve is completely tracked by the TL effective shear/acoustic only, confirming the previous RC predictions. Examination of the Green's function for shear and no-shear cases plainly shows that where the number of acoustic modes is small (as here) and seismic interface wave effects negligible, the above strategy of effectively matching shear-related loss via P attenuation succeeds well in reproducing observed TL behavior.

Although uncertainties in layer values were shown to have minor effects on predicted TL, in every case these geologic effects were dominated by the increased complexity of mode matching in greater water depths (for increased number of modes due to changes in bottom geometry). When the total number of propagating modes increases beyond 5, here for water depths greater than 1000 m, the optimal results attainable via TL shear/TL acoustic matching satisfactorily track the overall loss levels and trend without fully matching all phase features at mid- and farfield ranges. In a typical ocean acoustic setting, in the presence of background ambient noise and octave smoothing, further refinements to precise values of effective attenuation or the mode matching scheme are practically unnecessary.

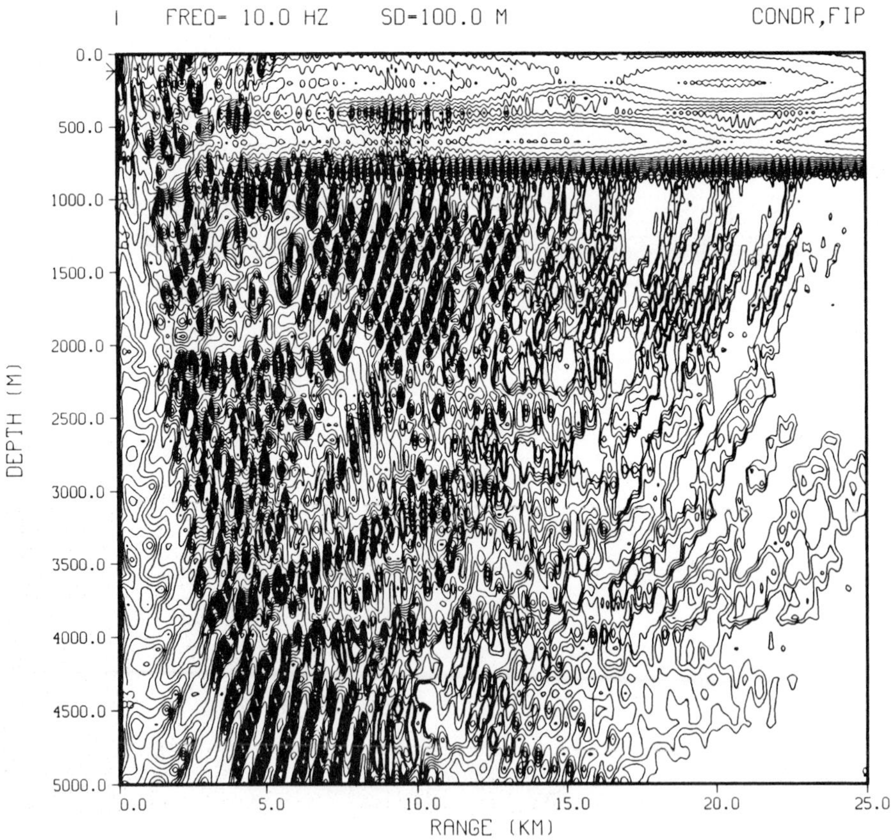

Figure 3. Range-depth contoured (full wave FFP) transmission loss field for range-independent shallow water segment of Cape Fear.

Examining net energy partitioning between water column and bottom layers (Figure 3), the *SAFARI* FFP range-depth contoured pressure loss field dramatically reveals the relatively slight effects of both shallow-ocean sound speed and water/bottom acoustic impedence contrast for this particular locality. Characteristically high loss into the bottom is seen for large grazing angles in the nearfield; overall mode structure is weak at 10 Hz. Notably, the suggested mechanism of energy return to the water column via reflection for a deep high-velocity layer or shear-velocity gradient was not observed in this model even at very large refraction-arrival distances.

CALCULATION OF UPSLOPE PROPAGATION VIA IFDPE

The depth-range contoured TL predicted for the range-dependent upslope propagation using the IFDPE with the effective subbottom parameters is presented in Figure 4. The dark line represents the bathymetry, which varies from 3600 to 400 m (left to right). In the near field (in the deep abyssal zone) a series of coherent zones originate at the source and disperse with range. The structure of the coherent zones shows that the water column is supporting a large number of modes ($N > 25$). In addition, there is considerable acoustic penetration into the subbottom that results from the bottom interaction of the higher order modes. As the sound propagates up the slope, the only change is that the penetration into the subbottom slightly increases. The water column is not able to support all the original modes. Approximately half-way up the slope, the structure of the coherent zones becomes more complicated and acoustic penetration into the subbottom significantly increases. This is the mode dropout zone where all but the lowest order modes dump into the subbottom. Above the mode dropout zone the structure of the coherent zones becomes simpler, and there is only slight penetration into the subbottom. Only the lowest order mode has survived the mode dropout zone. Over the entire range, no clear evidence of upslope enhancement in the water column is observed. The primary effect of the upslope propagation is to strip out all but the lowest order modes. When the calculation for the upslope propagation is repeated with no multiplication of the P attenuation, there is no noticeable difference in the results that are presented here.

To investigate predicted arrival structure, the IFDPE results were coupled into a maximum entropy method (MEM) beamformer to simulate the response of a vertical hydrophone array (of 16 receivers with 20-m separation at a depth of 80 m). A 10th-order MEM calculation was used, averaging over 100 of the 25-m range steps. The predicted beam response is shown in Figure 5 for source-receiver ranges of 100 to 170 km. The dark areas indicate a high-contour density and, therefore, a relatively well-defined beam. There are a large number of arrivals over the mode dropout zone, but none are dominant. Above the mode dropout zone, the two well-defined arrivals represent downward- and upward-propagating energy associated with the surviving mode. A conventional beamformer simulation was also used to determine degradation in the array signal gain, which was found to be 2 to 4 dB in deep water and slope regions, and 1 to 2 dB in the shallow sound (with few modes propagating)

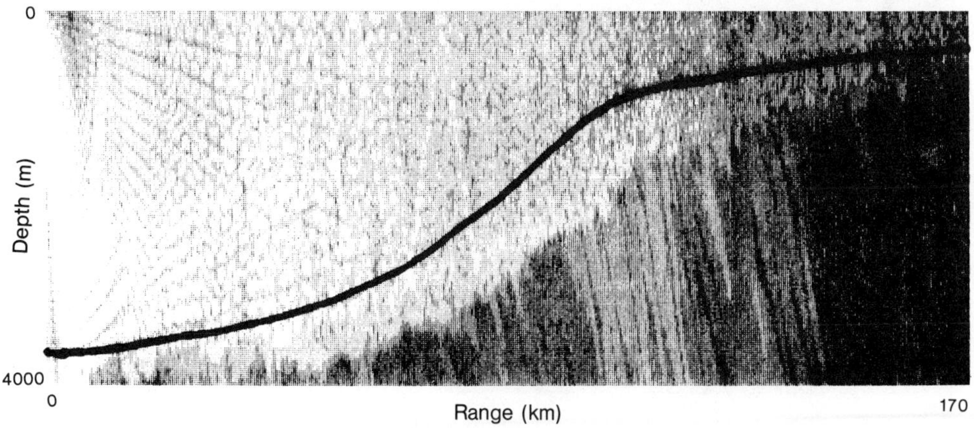

Figure 4. Range-depth isoloss plot (IFDPE) complex pressure field for range-dependent shallow-to-deep water Cape Fear profile.

The fraction of acoustic energy that would propagate through the water-column via combined water and earth-borne paths was also estimated with IFDPE. Here, bottom sound speed was held constant with depth and allowed to vary only in range. This procedure assumes that sound penetrating the subbottom is totally absorbed and would not re-enter the water column. This field was then compared with previous results computed with effective parameters. The results are displayed as a range-depth shade contour plot in Figure 6; contours are in 3-dB intervals and represent the fraction of acoustic energy

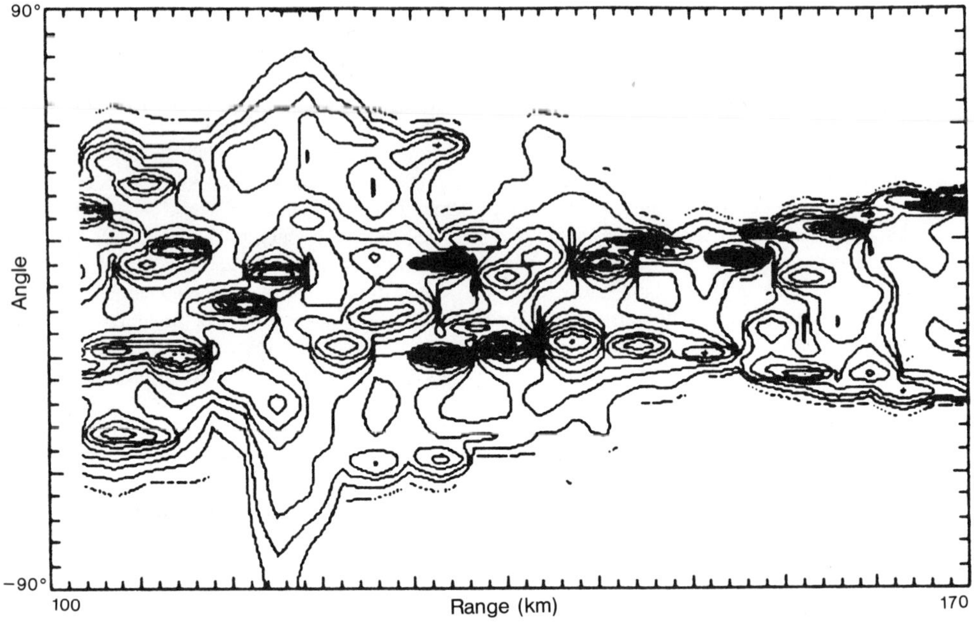

Figure 5. Simulated MEM array response in 3 dB contours (100-170 km range).

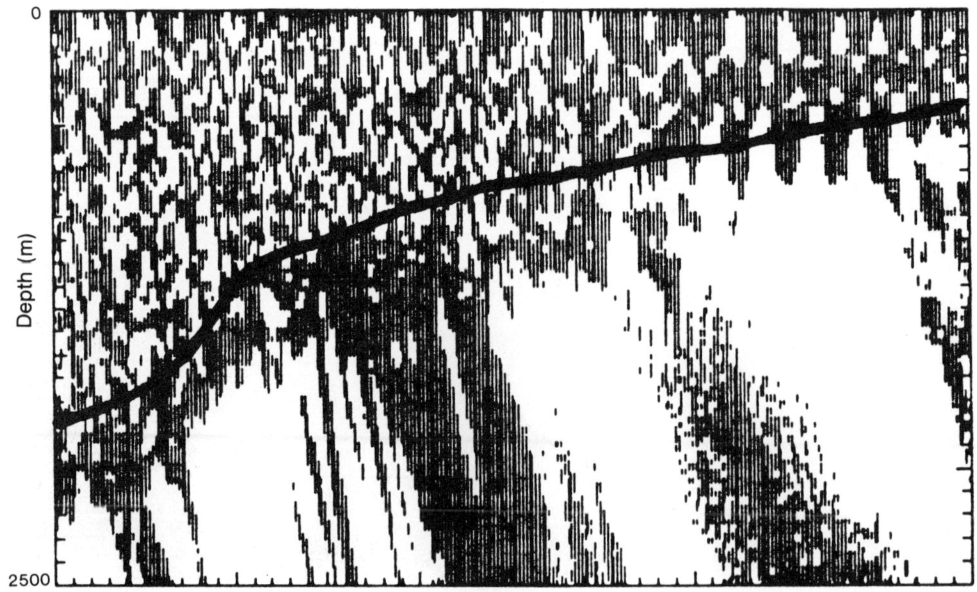

Figure 6. Contour of relative fraction of water/earth-borne energy (computed via IFDPE).

in the water column that is entirely water-borne. Looking only at shades in the water, the clear shade (0-3 dB fraction) represents acoustic energy that has only slightly penetrated into the subbottom. The dark shade identifies acoustic energy propagating primarily by combined water and earth-borne paths. An interesting result is that not until sound has propagated far upslope does any change appear in the relative fraction of acoustic energy that is primarily water-borne. Farther up the slope, combined propagation paths become more important, although penetration into the subbottom is decreasing.

The above results show that the range-dependent bottom geometry is the dominant feature affecting the acoustic field in the water column for upslope propagation at 10 Hz in the Cape Fear area. Decreasing water depth strips out all higher order modes, allowing only a single mode to survive in the shallow region. The surviving mode becomes increasingly bottom-interacting as water depth decreases further (with only slight penetration into the subbottom). Several conclusions for pre-experimental planning are immediate:

(1) Vertical array deployment is optimal after the mode dropout region (in shallow water). In this region the array spans nearly the entire water column.

(2) Beamforming on field array data allows direct and simple comparison with modeling predictions. Four ocean-bottom seismometers (OBS) will thus be emplaced along the CW tow profile, with two units above and two below the array. This configuration will provide field TL data to test modeling predictions through the mode dropout region and the shallow sound. Two OBS units are to be placed offside the CW profile near the array to study lateral variability of subbottom geology and the related seismo-acoustic field. Information concerning the relative contribution of combined water and earth-borne paths will be inferred from geophone/hydrophone comparisons.

SUMMARY

In late June 1985, a joint NORDA-USGS experiment will be carried out to investigate the characteristics of very low frequency seismo-acoustic sound propagation in the Cape Fear Atlantic Slope area using CW and explosive sources. Numerical results from the pre-experimental planning predict that the fullwave range-independent propagation modeling are here satisfactorily simulated through the use of effective geoacoustic layer parameters (where shear loss is effectively described by increasing P attenuation). Upslope propagation for this region is shown to be dominated nearly everywhere by geometric effects of bottom slope causing progressive mode cutoff and loss into the bottom. These effects (for Cape Fear) outweigh any specific geologic contributions to TL via seismic mode conversion and/or interface wave generation through the thick sedimentary cover. The instrumentation will be placed to test these predictions.

REFERENCES

1. H. Schmidt, "Excitation and Propagation of Marine Seismic Interface Waves (using a new Fast Field Program)," presented at Institute of Acoustics Underwater Sound Propagation Symposium, Bath University, UK. In: P.C. Pace (ed), *Acoustics and the Seabed*. Bath: University Press (1983).
2. D. R. Hutschinson, J. A. Grow, K. D. Klitgord, and B. A. Swift, "Deep Structure and Evolution of the Caroline Trough," In: *Geology and Geophysics of the Continental Margins*, Tulsa: Am. Assoc. Petrol. Geol. (1984).
3. E. Hamilton, "Prediction of deep-sea sediment properties: state-of-the-art," In: *Deep Sea Sediments, Physical & Mechanical Properties*, A.L. Inderblitzen, ed., New York, Plenum Press (1974).
4. P. J. Vidmar, "The dependence of bottom reflection loss on the geoacoustic parameters of deep sea (solid) sediments," *J. Acoust. Soc. Am.*, 68(5), pp. 1442ff (1980).
5. H. Bucker, "An equivalent bottom for use with the split-step algorithm," *J. Acoust. Soc. Am.*, 73(2), pp. 486ff (1983).
6. G. J. Fryer, "Compressional-shear wave coupling induced by velocity gradients in marine sediments," *J. Acoust. Soc. Am.*, 69(3), pp. 647ff (1981).

EXPERIMENTS ON SOUND PROPAGATION OVER SLOPED BOTTOMS

H. Hobaek*, C. T. Tindle+ and T. G. Muir

Applied Research Laboratories
The University of Texas at Austin
Austin, TX 78713, USA

ABSTRACT

Experiments to study sound propagation in a shallow water wedge have been conducted in a tank. Results confirm the assumption of adiabatic normal mode theory that the mode shapes adjust continuously to the local water depth. Mode coupling, if it exists, is negligible. For upslope propagation, the energy of a mode travelling beyond its cutoff depth is directed into the bottom with negligible coupling to lower modes. For downslope propagation, the normal modes in a wedge propagate with curved wavefronts and not with the straight vertical wavefronts of simple adiabatic normal mode theory. In addition, normal modes not present at the source can be "captured" when the water depth is sufficient to support them.

INTRODUCTION

Issues raised by recent theories of propagation into and out of a shallow water ocean wedge require precise and comprehensive measurements of the sound field in the water and the bottom for a variety of ranges and bottom slopes. These measurements are difficult in the ocean and so we have performed a scaled model experiment in an indoor tank. Typical operating parameters are a water depth at the source of 5 wavelengths, ranges up to 500 wavelengths and slopes up to 9°.

The objective of the experiment was to study propagation of normal modes in a smooth plane wedge. Early experimental work by Eby et al.[1] and theoretical work by Pierce[2] and Milder[3] suggested that normal modes far from cutoff behave adiabatically and adjust to the local water depth.

Recent numerical calculations by Jensen and Kuperman[4] suggested that as a normal mode passes upslope through its cutoff depth its energy is transferred to the bottom with negligible coupling to lower modes.

---

*On leave from Department of Physics, University of Bergen, Bergen, NORWAY.

+On leave from Department of Physics, University of Auckland, Auckland, NEW ZEALAND.

Our results for a wider range of slopes and a greater number of modes are in agreement with the earlier work. A selection of our results is presented here and two new effects associated with propagation on slopes are described.

EXPERIMENTAL ARRANGEMENT

The ocean bottom in the model experiment is formed by a tray of sand 10 m long x 1 m wide x 20 cm deep. The sides and bottom of the tray are lined with random scatterers to eliminate specular reflections. The tray is suspended in the water and the slope and depth of the bottom can be easily varied. The sand has a sound speed of 1787 m/sec, a density of 1.97 gm/cc and an attenuation of about 0.8 dB/wavelength.

The experiment was designed to operate with a line source in 10 cm of water at 80 kHz. There are six normal modes present in this case and the line source of seven elements can be shaded in phase and amplitude to generate individual normal modes. In practice there is always some slight contamination from other modes but this does not lead to any difficulties in interpreting the results.

In some cases it is convenient to use a single point source rather than a line source and this is easily achieved by exciting an individual element of the source array.

The acoustic signals are received by a small probe hydrophone which can be set at any given range and then moved accurately in a vertical direction. The vertical movement is connected via a servo-mechanism to an analog chart recorder.

Data can be taken by two complementary methods. Analog chart recordings of signal level (in dB) as a function of depth can be obtained by simply operating the vertical movement of the probe hydrophone. In this method we use a gating technique described by Eby et al.[1] An electronic gate is set to detect the peak value of the signal within a small time interval. The time interval is chosen to coincide with the arrival of the mode under investigation and this enables the contamination from other modes to be minimized.

The second method of taking data is to fix the range and depth of the probe hydrophone and to digitize the entire waveform of the received signal. It is a simple matter to take recordings at various depths and these can be treated as if they were simultaneous recordings on a vertical array of hydrophones because the start of the digitization interval is accurately controlled.

It is well known[5] that signals from a vertical hydrophone array can be combined in a suitably weighted summation to allow extraction of the waveforms present in each individual normal mode. In the following figures the normal mode waveforms were obtained using this technique.

ANALOG RESULTS

Analog results of signal level as a function of depth have been taken for a variety of cases. Figure 1 shows results for downslope propagation at 1° of bottom slope. The line source was shaded to generate modes 1, 2, 3 successively and the resulting analog traces are shown. It is clear that the modes have adjusted adiabatically to the local water depth of 22.9 cm. The small variation in the amplitude of

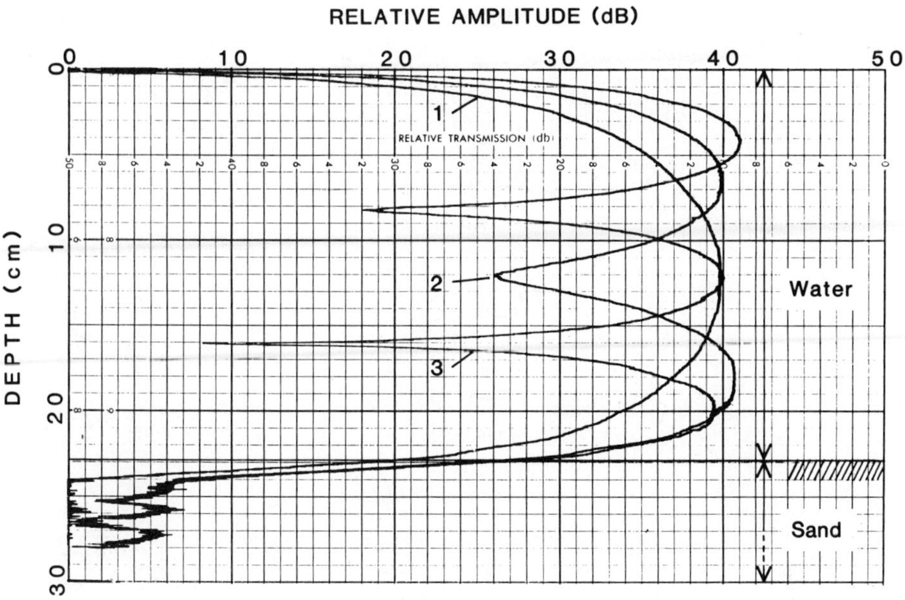

Fig. 1. Analog records at a range of 7.4 m for propagation downslope at 1°. The line source in 10 cm water depth was shaded to modes 1, 2, 3 successively. The records show that the modes have adjusted adiabatically to the local water depth of 22.9 cm.

modes 2 and 3 for different maxima of the same mode is consistent with contamination by other modes due to imperfect shading of the line source. It is not thought to be associated with coupling of energy between modes. The small oscillations in the bottom, some 35 dB below the maximum signal level in the water, indicate the noise level of our measurements and are possibly due to sound propagating down the hydrophone support.

Figure 2 shows results for propagation upslope at 0.6°. The line source in 10 cm water depth was shaded to preferentially excite mode 2. Figure 2(a) shows that mode 2 has adjusted to the local water depth of 4.8 cm. The field in the bottom falls off very rapidly. In figure 2(b) the water depth has decreased to 2.7 cm and the "tail" of the mode falls off much more slowly in the bottom. In figure 2(c) the water depth is 2.4 cm and mode 2 has passed beyond cutoff. There is now a substantial amount of energy propagating in the bottom. In figure 2(d) mode 2 has disappeared from the water column and the field in the water is due to the small amount of mode 1 generated by the line source. The remaining energy of mode 2 is now propagating as a broad, strongly attenuated beam of sound whose axis has moved deeper into the bottom. The result is consistent with the numerical calculations of Jensen and Kuperman.[4] The results shown in figure 2 should be regarded as preliminary as the origin of the small oscillations in the bottom is uncertain.

WAVEFORM RESULTS

In constant water depth each individual normal mode travels with its own group velocity. Thus if a point source emits a short pulse the signal received downrange will be a series of (possibly overlapping) pulses associated with each mode. The individual normal mode waveforms

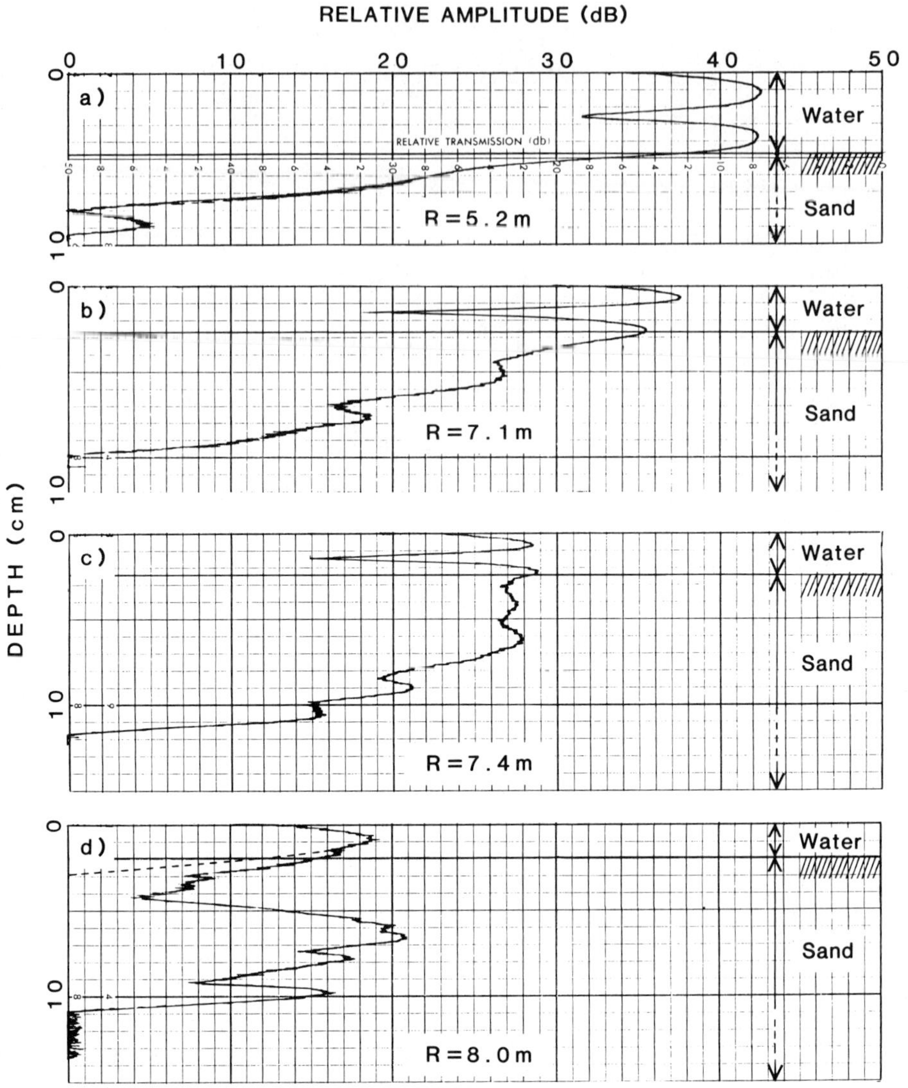

Fig. 2. Analog records for propagation upslope at 0.6°. The line source was shaded for mode 2. Ranges are (a) well before cutoff (b) just before cutoff (c) beyond cutoff (d) well beyond cutoff. The dashed curve in (d) indicates the separately measured level of mode 1. Water depth at each range is indicated on the figures.

can be extracted and an example is shown in figure 3(a). A point source at mid water depth was excited using a smoothly rising and falling pulse of approximately 4 cycles. The source element "rings" after the excitation is removed so the decay of the pulse in the water is slower than its rise time. The waveform of mode 1 in figure 3(a) is very similar to that of the pulse emitted by the source.

The other mode waveforms in figure 3(a) show the well known features of normal mode propagation in isovelocity layers. In particular

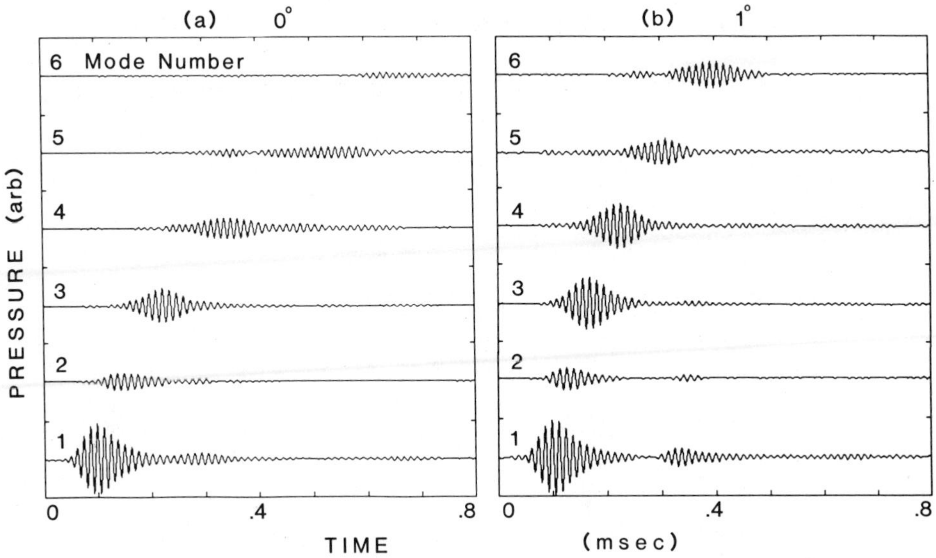

Fig. 3. Normal mode waveforms at bottom slopes of (a) 0° and (b) 1° at a range of 7 m using a single source at mid water depth. Water depth at the source is 10 cm and propagation is downslope. The small second arrivals present in mode 1 (and very slightly in mode 2) are reflections from the tank wall and should be ignored.

(i) higher modes arrive progressively later due to their lower group velocities,

(ii) the pulses of higher modes are more dispersed due to their greater variation of group velocity with frequency,

(iii) higher modes suffer progressively greater attenuation.

When the bottom is sloping the above features are still present but the extent of each feature is strongly affected by even a small slope as can be seen by comparing figures 3(a) and 3(b). Figure 3(b) shows the waveform obtained at the same range as figure 3(a) for downslope propagation at 1° of bottom slope. The relative arrival times of the higher modes are now much reduced. For example, the time interval between the peaks of modes 1 and 4 is reduced from 0.25 msec to 0.11 msec. The dispersion in the higher modes is also much reduced.

Another important effect of the slope is the severely reduced attenuation of the higher modes. Mode 6 is easily observed at 1° of bottom slope and yet for constant water depth, mode 6 is barely present at this range and could not be detected at greater ranges.

The above features of downslope propagation are all consistent with simple adiabatic normal mode theory. As the water depth increases the group velocity should decrease, the dispersion should be reduced and the attenuation should be reduced as the "exponential tail" of the mode penetrating into the bottom decreases.

We have taken data at bottom slopes of up to 9° and the above effects continue to be observed. Figure 4(c) shows normal mode waveforms at 9°. The source waveform used was a very sharply defined 4 cycle pulse

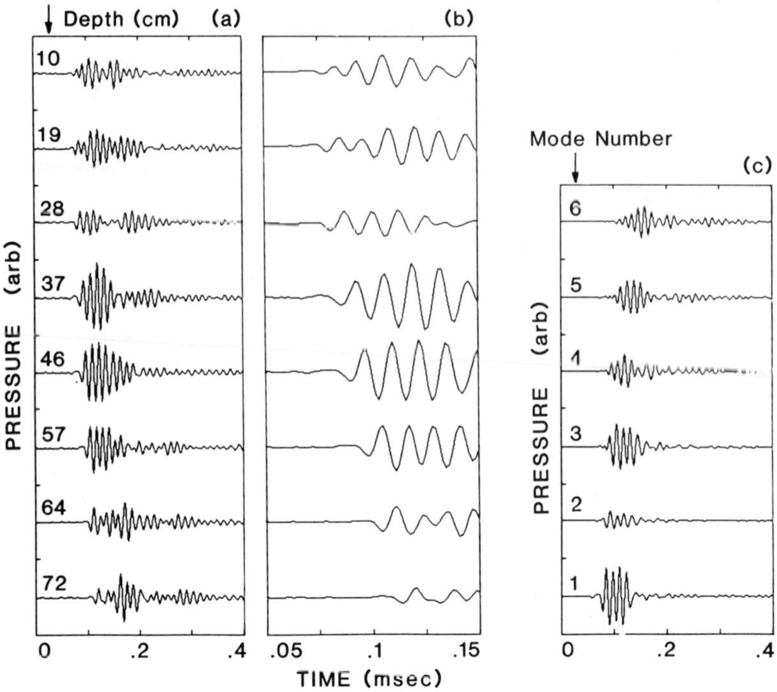

Fig. 4. Hydrophone waveforms and normal mode waveforms at a range of 4.5 m and a slope of 9°. Water depth at the source is 10 cm with the source at mid water depth and propagation is downslope. (a) Waveforms at successive depths as indicated, (b) as for (a) but with expanded time scale to show the delay between signals arriving at the top and successively lower hydrophones. (c) Mode waveforms extracted from the hydrophone waveforms of (a).

very similar to that shown for mode 1 which propagates with negligible dispersion. The time separation between successive modes is seen to be very small for large bottom slopes.

The mode waveforms of figure 4(c) were obtained from the hydrophone data of figure 4(a). The first part of the waveforms of figure 4(a) are shown on an expanded time scale in figure 4(b) to illustrate the time delay which occurs for the deeper hydrophones due to the sloping bottom. The time delay between top and bottom hydrophone waveforms for figures 4(a) and (b) is 0.035 msec and corresponds to 2.8 wavelengths.

The first arrival of the signals has a delay which is consistent with spherical spreading from the point source. In performing the extraction of the mode waveforms it was necessary to shift the hydrophone waveforms in time to synchronize them. The fact that the extraction of the mode waveforms is then successful shows that the normal modes do not propagate as wavefronts which are simultaneous at all depths. Instead they propagate as spherical wavefronts. This is in contrast to simple adiabatic normal mode theory which assumes the normal modes propagate as vertical wavefronts.

An interesting new effect is illustrated in figure 5 which shows mode waveforms at a range of 4 m and slopes of 1° and 3°. The source is placed near the surface in only 5 cm of water. This water depth can support only 3 normal modes but we see in figure 5(b) that 4 normal modes are present in the received signal. It is clear that energy is coupling into mode 4 even though mode 4 is not present at the source as a discrete mode. Therefore some of the energy which is usually associated with the continuum can be trapped when the water depth is sufficient to support another discrete mode. Thus a normal mode which is not present at the source can become "captured" as the water depth increases. This effect had been noted earlier in the theoretical calculations of Tindle and Deane.[6]

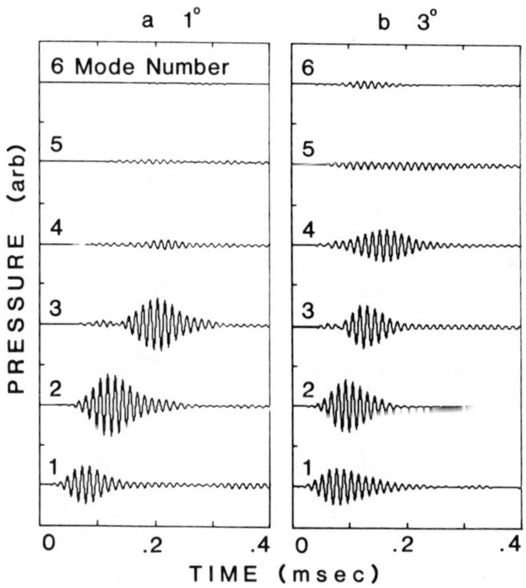

Fig 5. Normal mode waveforms at bottom slopes of (a) 1° and (b) 3° at range 4 m. The source is near the surface in 5 cm of water and propagation is downslope.

SUMMARY AND CONCLUSIONS

1. The experiment has confirmed the assumption of adiabatic normal mode theory that for a smooth plane wedge the mode shape adjusts continuously to the local water depth. The group velocities of the modes are also consistent with the adiabatic assumption.

2. The experiment has shown that as a mode passes through cutoff in upslope propagation, its energy enters the bottom as a beam and there is negligible coupling to lower modes.

2. Mode coupling effects, if they exist, are very small.

4. The normal modes propagate downslope as curved wavefronts and not as the vertical wavefronts of simple adiabatic normal mode theory.

5. Modes not trapped in the water depth at the source may be "captured" when the water depth increases.

ACKNOWLEDGEMENT

This work was supported by the U. S. Navy Office of Naval Research. H. Hobaek is grateful for partial support from the Norwegian Council for Science and the Humanities.

REFERENCES

1. R. K. Eby, A. O. Williams, Jr., R. P. Ryan and Paul Tamarkin, Study of acoustic propagation in a two-layered model, J. Acoust. Soc. Am. 32:88 (1960).

2. A. D. Pierce, Extension of the method of normal modes to sound propagation in an almost-stratified medium, J. Acoust. Soc. Am. 37:19 (1965).

3. D. M. Milder, Ray and wave invariants for SOFAR channel propagation, J. Acoust. Soc. Am. 46:1259 (1969).

4. F. B. Jensen and W. A. Kuperman, Sound propagation in a wedge shaped ocean with a penetrable bottom, J. Acoust. Soc. Am. 67:1564 (1980).

5. C. T. Tindle, K. M. Guthrie, G. E. J. Bold, M. D. Johns, D. Jones, K. O. Dixon and T. G. Birdsall, Measurements of the frequency dependence of normal modes, J. Acoust. Soc. Am. 64:1178 (1978).

6. C. T. Tindle and G. B. Deane, Sound propagation over a sloping bottom using rays with beam displacement, J. Acoust. Soc. Am. (accepted for publication).

THE EFFECT OF UNCONSOLIDATED SEDIMENT RIGIDITY ON LOW FREQUENCY ACOUSTIC PROPAGATION

J. H. Beebe and C.W. Holland*

Applied Research Laboratory
The Pennsylvania State University
Post Office Box 30
State College, PA 16804

*Planning Systems Inc.
7900 Westpark Drive, Suite 600
McLean, VA 22102

ABSTRACT

Most unconsolidated ocean sediments have been reported to possess small, but finite rigidity. In sandy sediments rigidity is due to sliding and rolling friction resulting from the interlocking grain structure; in silt clays rigidity is due to cohesion. For many marine environments acoustic propagation is not measurably affected by sediment rigidity. In some environments, however, the effect of sediment rigidity is the dominant mechanism in modal attenuation. Two coastal regions on the Scotian shelf will be contrasted by comparing the effects of sediment shear wave excitation on transmission loss at low frequency. The key environmental factor in unconsolidated sediment shear wave excitation is not the intrinsic sediment properties but the degree of elastic contrast found in the acoustic field.

INTRODUCTION

Sound propagation conditions in coastal regions are frequently constrained by the high losses encountered due to the interaction of the sound field with the seabed. Many coastal and shelf regions, while exhibiting relatively high transmission losses, can be treated by existing modeling capabilities. This is true for regions where the seabed is flat and has a relatively thick and homogeneous sediment cover over the basement structure. The two coastal regions to be discussed in this paper contrast acoustic propagation over such a thick, homogeneous sediment with propagation over a thin sediment that covers a rough granite basement. The modeling and predictive capability demonstrated for complex thinly-sedimented bottoms have not been nearly as successful as for the more simple bottom types, especially at low frequencies.

Shallow water transmission losses are attributable to the frequent encounters of the propagating sound field with the seabed. At each encounter, the sound field may be scattered, it may penetrate the bottom and undergo absorptive losses in the porous sediment, or it may be converted to shear waves at any interface, reducing the reflected component in the process. The bottom composition and structure determine the relative magnitude of these losses. Of particular interest in this paper are the sediment shear losses due to the bottom characteristics at the experimental sites. Other researchers have discussed losses due to sediment rigidity, including Fryer[1], Vidmar[2] and Akal and Jensen[3]. These studies have demonstrated that for certain environmental conditions the propagation of sediment shear waves influences transmission loss.

Two coastal regions on the Scotian shelf will be discussed in this paper. At one site (Site 1) the bottom consists of a relatively smooth, thick sediment layer over a lower velocity, semi-consolidated bedrock. The other site (Site 2) the bottom consists of a thin sediment cover over a granite bedrock structure which frequently projects through the sediment to provide a rough, complex bottom. The two bottom types will be contrasted by comparing the degree to which propagation of shear waves in the sediment effects the transmission loss.

Propagation over complex bottoms such as the second type has been discussed by Macpherson and Fothergill[4], Worley and Walker[5] and Beebe and McDaniel[6]. Measured one-third octave losses as high as 25 dB/km for frequencies from 50 Hz to 250 Hz were noted. This is in contrast to losses of 10 or 12 dB/km for shelf areas such as at the first site. It will be shown that sediment shear wave propagation accounts for the high losses at Site 2, but that sediment rigidity for Site 1 has little effect upon the transmission loss. It will be further argued that only for a thin sediment layer over a high velocity bottom, such as the granite, will shear wave propagation in the sediment have the dominant effect on modal attenuation as found at Site 2.

In the following sections the sites and experimental data are first discussed. A discussion of the models employed for studying propagation effects and for characterizing the bottom are then considered. The analysis proceeds by considering the effects of the sediment rigidity on propagation. This analysis focuses on the role of the sediment-basement interface as the location of shear wave excitation in the sediment.

EXPERIMENTAL DATA

Description of Experiment

The experimental data to be discussed was obtained off the coast of Nova Scotia during the fall of 1978 in a joint program with the Defense Research Establishment Atlantic (DREA) of Dartmouth, Nova Scotia. A vertical array of hydrophones was placed in the water column. The water depth was 57 m at Site 1 and 32 m at Site 2 . The data received on the bottom hydrophone of this array (1 m above the bottom) is discussed here. Explosive sources (one pound TNT equivalent) were detonated at depths of 18, 38 and 79 m, depending upon the water depth along the sloping track. While data was obtained out to a range of 123 km (Site 1) and 43 km (Site 2), only the first 8 km of the track, where the high losses occurred at Site 2 will be discussed. Sound velocity profiles along the Site 1 track are shown in Figure 1 and along the Site 2 track in Figure 2, where it can be seen that downward refraction of the wave front will ensure a high degree of bottom interaction at both sites.

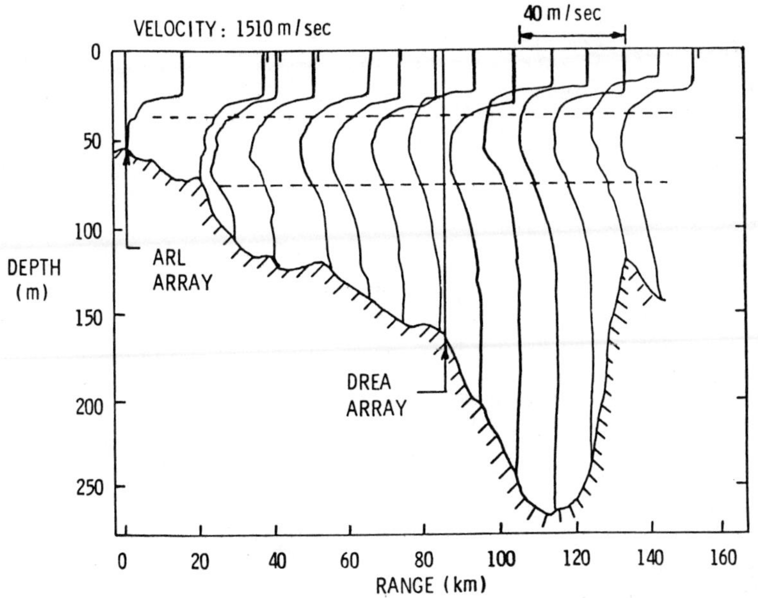

Figure 1. Site Sound Velocity Profiles and Bathymetry.

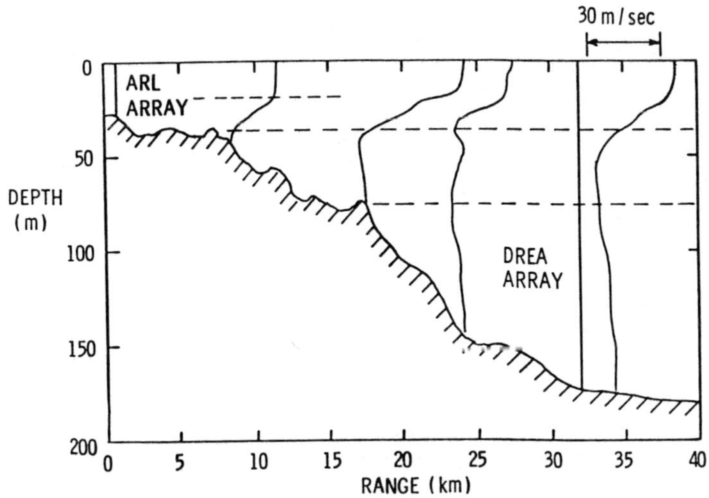

Figure 2. Site 2 Sound Velocity Profiles and Bathymetry.

Propagation Loss Data

Data received on the bottom hydrophone of the array were analyzed using one-third octave filters to determine the propagation loss. One-third octave center frequencies from 15 Hz to 500 Hz were used in processing the data. Source levels for the one pound explosive charges were generated by DREA using the Gaspin and Shuler[7] technique for frequencies above the bubble frequency and the depth-scaling technique of Hughes[8] for frequencies below the bubble frequencies.

Measured propagation loss is shown as a function of frequency in Figure 3 for frequencies from 25 - 500 Hz at Site 1 and from 125 - 500 Hz at Site 2. The Site 2 signals below 125 Hz were too low for recording.

The data was previously discussed by Beebe and McDaniel[6] where it was noted that attempts to model the propagation conditions for the thinly sedimented bottom found at Site 2 were unsuccessful, while efforts to model Site 1 were much more successful. The disparity in the low-frequency losses at the respective sites is obvious from the two curves. The high elastic contrast at the sediment-basement interface found at Site 2 is believed to control the transmission loss at low frequency.

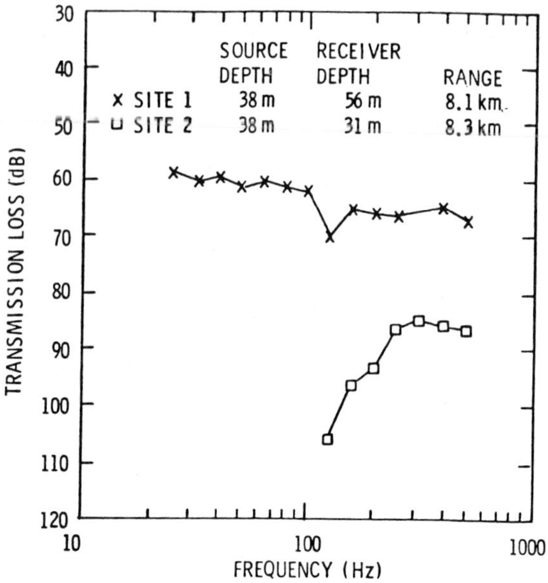

Figure 3. Transmission Loss vs. Frequency at Two Sites on the Scotian Shelf.

Geoacoustic Model

A geoacoustic model for the propagation track was generated using bottom and sub-bottom profiles obtained with the Huntec deep-tow system[9], surficial sediment and bedrock maps from the Bedford Institute of Oceanography (BIO), and dispersion and seismic refraction analysis. Tables 1 and 2 list the parameters of the geoacoustic model which describe the water and bottom properties along the tracks. Mean grain sizes for the surficial sediments were determined using the Huntec data for both sites plus grab sample data for Site 1. The Huntec data also provided the layer thicknesses and the depth of the bedrock layer. These values were confirmed using the sediment and bedrock maps from the BIO. Sediment compressional velocities were calculated from the mean grain sizes using the regression equations of Anderson[10] with corrections for the bottom temperature and depth. Velocity for the granite bedrock was obtained from seismic refraction data and bedrock attenuation values were obtained from handbook data[11]. Compressional attentuation values for the sediment were calculated using the Biot-Stoll sediment model[12] for Site 1; Hamilton's values[13] were used for shear velocities and attenuation at both sites and for compressional attenuation at Site 2.

Tables 1 and 2 represent only the gross features of the environmental properties. Unconsolidated sediments are known to exhibit significant vertical gradients associated with compressional and shear wave velocities and attenuations. Previous modeling efforts (Site 1) demonstrated, however, that transmission loss can be predicted with reasonable accuracy even when the gradients are neglected. The shear wave velocity was chosen to reflect the velocity at the bottom of the sediment column where excitation of the sediment shear wave is expected. Measured sound speed profiles in the water depicted in Figures 1 and 2 were used in the geoacoustic model; the speeds in Tables 1 and 2 represent values at the water surface.

Table 1. Site 1 Geoacoustic Model Parameters

Layer Type	Layer Depth (m)	Comp. Vel (m/s)	Shear Vel (m/s)	Comp. Attn (dB/km)	Shear Attn (dB/km)	Density (g/cm^3)
Water	57	1509	--	--	--	1.0
Sediment	50	1800	500	$.009f^{1.78}$	13f	2.0
Sub-bottom	∞	1950	600	$.007f^{1.3}$	13f	2.1

Table 2. Site 2 Geoacoustic Model Parameters

Layer Type	Layer Depth (m)	Comp. Vel (m/s)	Shear Vel (m/s)	Comp. Attn (dB/km)	Shear Attn (dB/km)	Density (g/cm^3)
Water	32	1498	--	--	--	1.0
Sediment	5	1780	170	0.7f	13f	2.0
Sub-bottom	∞	5500	2400	0.1f	0.06f	2.6

MODELING TECHNIQUES

The adiabatic normal-mode model of Ingenito et al.[14] was used to study the effects of the ocean bottom on sound propagation at the two sites in a previous analysis[15]. This model describes the ocean as a 3-layer structure which consists of the water layer, a fluid sediment layer, and a solid halfspace. The water and bottom acoustic properties are allowed to vary with range. The previous analysis showed that this model gave good results at the lower frequencies for Site 1, but not for Site 2.

The lack of agreement between predicted and experimental transmission loss at Site 2 suggested that the model did not account for all of the important loss mechanisms in the environment. In order to investigate the effect of sediment rigidity on transmission loss at Site 2 the Ingenito et al. model was modified. The result of treating the sediment layer as a solid instead of a fluid is to include losses from sediment shear waves in the mode attenuation calculation.

A normal-mode program developed by McDaniel and Caswell[16] calculates propagation through such a lossy-solid medium and was employed as a subroutine in the Ingenito et al. model. The routine is called from the model to calculate complex eigenvalues, the real part corresponding to the modal wavenumber and the imaginary part corresponding to the modal attenuation. The McDaniel-Caswell algorithm is an extension of Haskell's[17] formulation, allowing velocity and attenuation gradients through matrix descriptions of many thin homogeneous bottom layers.

The incorporation of the McDaniel-Caswell algorithm into the Ingenito et al. model gave rise to a model capable of predicting shear wave excitation in sediment layers, but only for range independent environments. Therefore, for the remainder of the study, only the first 8 km of the two tracks will be considered where range dependent effects should be small.

RESULTS

Site 1. - Thickly Sedimented Site

The environment at Site 1 is characterized by a thick, presumably homogeneous layer of unconsolidated sand and gravel over a semi-infinite (for the frequencies of interest) column of semi-consolidated glacial till. The sound velocity profile is downward refracting so that the acoustic field is expected to interact strongly with the bottom. The source depth for the explosive charges is 38 m and the receiver depth is 56 m, 1 meter above the bottom.

Predicted transmission loss as a function of frequency is shown in Figure 4 assuming (1) a fluid sediment and (2) a viscoelastic sediment; the two results are essentially identical. Measured one-third octave transmission loss as a function of frequency is also shown. The similarity of the two predicted curves suggest that the effect of sediment rigidity is small for this environment type. In other words coupling of energy from compressional to shear waves is negligible. Three mechanisms for shear wave excitation are expected. At low frequencies the acoustic field is expected to interact with the lower sediment layer. The sediment-sediment interface presents a boundary of elastic contrast so that some of the reflected energy at this boundary will be in the form of shear waves. At high frequencies the acoustic field will not interact with the lower sediment layer and conversion to shear waves occurs due to gradients in the elastic properties of the sediment. Some sediment shear wave excitation is also expected at the water-sediment interface. Comparison of the experimental data with measured data indicates that none of these mechanisms appreciably drove compressional to shear wave coupling in the sediment. The indication is that sediment rigidity does not influence the propagation of the acoustic field and may be neglected in predictive modeling for this environment type.

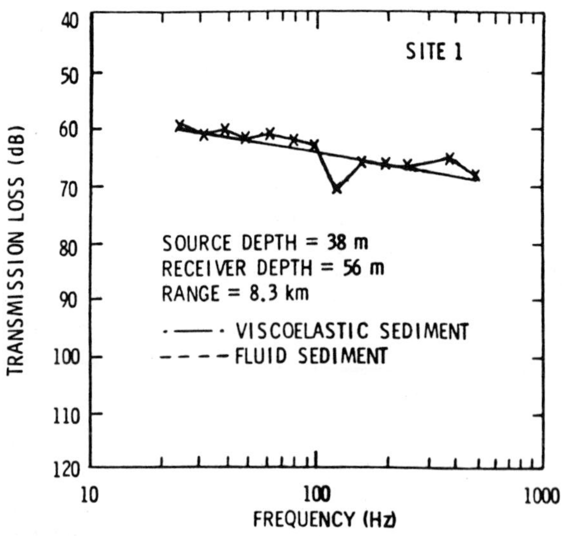

Figure 4. Experimental and Predicted Transmission Loss versus Frequency at Site 1. The x's represent experimental data.

Site 2 - Thinly Sedimented Site

The environment at Site 2 is characterized by a thin layer of unconsolidated coarse sand over a granite basement. The sound velocity profile is downward refracting so that the acoustic field is expected to interact strongly with the bottom. Data was taken for a source depth of 18.3 meters, and a receiver depth of 31 m, 1 meter above the bottom.

Predicted transmission loss as a function of frequency is shown in Figure 5 assuming: (1) a fluid sediment and (2) a viscoelastic sediment. Measured one-third octave transmission loss as a function of frequency is also given. The disparity of the two predicted curves suggests that the effects of sediment rigidity are large for this environment type. In fact, comparison of the predicted and measured data indicate that at low frequencies the generation of shear waves in the sediment is the dominant loss mechanism. The losses were so great at low frequencies that from 125 Hz down to modal cutoff at 15 Hz the signal could not be recorded above the small ambient and instrumentation noise.

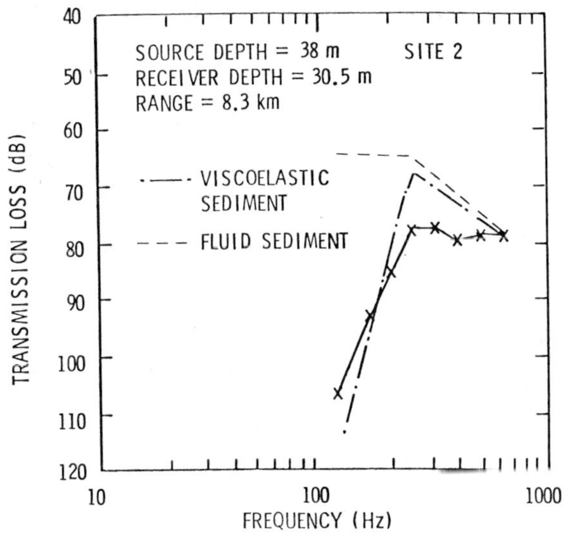

Figure 5. Experimental and Predicted Transmission Loss versus Frequency at at Site 2. The x's represent experimental data.

The frequency dependence of the losses due to sediment rigidity is not related to the properties of the sediment but rather can be shown to be dependent on sediment thickness. At 250 Hz the solid and fluid sediment models nearly converge implying that at this frequency and above the conversion of energy to shear waves in the sediment is no longer appreciable. The wavelength at 250 Hz is approximately 6 m and the acoustic field is evidently constrained to the sediment layer so that interaction with the basement is small. This result indicates that the sediment basement interface is the location of the large sediment shear wave excitation. At low frequency the acoustic field strongly interacts with the basement where a significant portion of energy is converted from compressional to shear waves that propagate in the sediment. The converted energy in the shear waves is largely trapped in the sediment column and therefore lost to waterborne acoustic field. Note that the measured data also shows a dramatic change of slope at about 250 Hz, lending credence to the suggested role of the sediment-basement interface at this site.

It may seem surprising that the small rigidity of an unconsolidated sediment can have such a drastic effect on the propagation of acoustic energy. The large sediment shear losses, however, are clearly not a function of sediment properties alone. This fact is evidenced by measured losses at Site 1 where the effects of sediment shear losses appeared to be negligible and sediment properties were comparable to Site 2. The important environmental difference between the two sites was the presence of the high velocity basement in the acoustic field at Site 2. The basement-sediment interface at this site provided a large elastic contrast so that at the low grazing angles of the propagating modes the energy in the compressional wave was largely reflected back into the sediment as a shear wave. These shear waves were subsequently trapped and lost to the waterborne acoustic field.

SUMMARY

It was shown that shear wave propagation in the sediment was responsible for the anomalous low-frequency losses at Site 2, but that sediment shear wave propagation at Site 1 had little effect on transmission loss. The influence of sediment rigidity on acoustic propagation is a function of the degree of excitation of sediment shear wave. The excitation of the sediment shear wave in turn is a function of the degree of elastic contrast in the acoustic field. At Site 1 the elastic contrast in the bottom was small, and hence the excitation of shear waves within the sediment was negligible. At Site 2 the unconsolidated sediment over the high velocity basement provided a large elastic contrast so that a significant portion of energy was converted to shear waves at the sediment-basement interface. It is necessary, therefore, to model sediment rigidity when an interface of large elastic contrast exists within the acoustic field.

ACKNOWLEDGEMENT

This work was supported by the Naval Sea Systems Command, Code 63R34, under the E/F program.

REFERENCES

1. G. J. Fryer, Reflectivity of the Ocean Bottom at Low Frequency, J. Acoust. Soc. Am., 63:35 (1978).
2. P. J. Vidmar, The Dependence of Bottom Reflection Loss on the Geoacoustic Parameters of Deep Sea (Solid) Sediments, J. Acoust. Soc. Am., 68:1442 (1980).
3. T. Akal and F. B. Jensen, Effects of the Sea-Bed on Acoustic Propagation, in: "Acoustics and the Sea-Bed," N. G. Pace, ed., Bath University Press, Bath, UK, (1983).
4. J. D. Macpherson and N. O. Fothergill, Study of Low Frequency Sound Progagation in the Hartlen Point Region of the Scotian Shelf, J. Acoust. Acoust. Soc. Am. 34:967 (1962).
5. R. D. Worley and R. A. Walker, Low-Frequency Ambient Ocean Noise and Sound Transmission Over a Thinly Sedimented Rock Bottom, J. Acoust. Soc. Am. 71:863 (1982).
6. J. H. Beebe and S. T. McDaniel, Geoacoustical Models of the Seabed to Support Range-Dependent Propagation Studies on the Scotian Shelf, in "Bottom Interacting Ocean Acoustics," W. Kuperman and F. Jensen, ed., Plenum Press, New York (1980).
7. J. B. Gaspin and V. K. Shuler, "Source Levels of Shallow Underwater Explosions," Naval Ordnance Laboratory Report, NOLTR 71-160 (1971).

8. R. D. Hughes, Low Frequency Underwater Explosive Source Levels, J. Acoust. Soc. Am., 60:572 (1976).
9. D. R. Parrot, D. J. Dodds, L. H. King, and R. G. Simpkins, "Measurement and Evaluation of the Acoustic Reflectivity of the Sea Floor," Huntec Report H7903-02/SB/DRP, presented at the First Canadian Conference on Marine Geotechnical Engineering, March 5, 1979.
10. R. S. Anderson, Statistical Correlation of Physical Properties and Sound Velocity in Sediments, in: "Physics of Sound in Marine Sediments," Loyd Hampton, ed., Plenum Press, New York (1974).
11. "Handbook of Physical Constants - Revised Edition," the Geological Society of America Memoir 97 (1966).
12. R. D. Stoll, Acoustic Waves in Saturated Sediments, in: "Physics of Sound in Marine Sediments," Loyd Hampton, ed., Plenum Press, New York (1974).
13. E. L. Hamilton, Geoacoustic Modeling of the Sea Floor, J. Acoust. Soc. Am. 68:1313 (1980).
14. F. Ingenito, R. H. Ferris, W. A. Kuperman, and S. N. Wolf, "Shallow Water Acoustics: Summary Report (First Phase)," Naval Research Laboratory, NRL Report 8179 (1978).
15. J. H. Beebe, S. T. McDaniel, and L. A. Rubano, Shallow-Water Transmission Loss Prediction Using the Biot Sediment Model, J. Acoust. Soc. Am., 71:1417 (1982).
16. Personal Communication
17. N. A. Haskell, The Dispersion of Surface Waves on Multilayered Media, Bull. Seism. Soc. Am., 43:17 (1953).

A PROPAGATION ANOMALY OBSERVED IN THE BARENTS SEA

E.H. Hug

Norwegian Defence Research Establishment
Box 115
N-3191 Horten, Norway

ABSTRACT

A low frequency sound propagation experiment was conducted in the Western Barents Sea from a towed continuous sound source and air-dropped explosive sound sources to a hydrophone lying on the sea bed. Anomalies in the propagation results were traced to sub-bottom features, in particular to a region of basement outcrop. This paper will present data suggesting the influence of a particular salt diapir occurring within this basement outcrop region upon the low frequency sound propagation. The results suggest alternative interpretations for the mechanism of influence of the sea-bed upon the sound field.

INTRODUCTION

In the spring of 1981 and 1982, experiments were carried out in the Western Barents Sea to determine the sound propagation properties in this largely shallow body of water. In the course of these experiments, propagation anomalies were encountered. Investigating the environmental causes of these anomalies led to a particular environmental effect that seemed worth reporting at this conference.

The effect was due to a salt diapir, an upwelling of salt from the deep basement containing an extensive layer of salt, termed by geophysicists "evaporites" and generally considered to be associated with passive continental margins. Evaporite layers have been formed in previous geologic epochs when seas or tidal pools have dried up, leaving behind their soluble but involatile remains as layers largely consisting of salt.

In later geologic epochs, as these layers have become covered and, still later, a new sea formed, the old salty layers remain as a remnant of the previous sea. As the eons pass, the salt, finding itself lighter than the sediments lying on top of it, tends to rise as slow upwellings until they reach almost to the sea-bed. These salt "diapirs" have generally been detected in one of two ways - as sharply defined, almost column-like features in geoseismic sections, and as salinity gradients in the interstitial water of the unconsolidated sediments arising from the rapid rate of ionic diffusion of the soluble salt. Evidence of evaporite layers and salt diapirs are also furnished by its magnetic properties, lower gravitational, density and higher thermal conductivity relative to the surrounding sediments.

These upwellings or columns of salt do not generally occur singly, but rather in groups, in a region. The group of salt diapirs of interest in this paper occurred in a region which we shall refer to as a "basement outcrop".

The purpose of this paper is to show evidence of the effect of one salt diapir upon the horizontal propagation of low-frequency underwater sound through a basement outcrop region peppered by such salt diapirs.

ENVIRONMENT AND EXPERIMENT

Figure 1 shows a rough outline of the salt diapir and a portion of the basement outcrop in which is situated. The track of a seismic profile through the diapir can be seen and to the right, the delineated seismic profile along this track with the salt diapirs is seen as distinct peaks or columns or upwellings through the surrounding sediments, which is generally Quaternary, glacial-glaciomarine. The salt has a density of 2.1 to 2.2 g/cm^3, p-wave velocity of 4200 to 5200 m/s and s-wave velocity of 2000 to 3000 m/s. The glaciomarine sediment has a density of 1.8 to 2.0 g/cm, a p-wave velocity of 1900 to 2100 m/s and s-wave velocity of 500 to 700 m/s. These values are estimates for a location directly over the salt diapir. The depth of the top of the diapir is estimated to be anywhere from a few to a few tens of meters beneath the sea-bed. It should be remarked here that such geological data are very sparse and it is not known with any accuracy how the bottom and sub-bottom parameters vary laterally. Also the shaded area in Fig. 1 must be taken as a very rough indication of salt diapirism.

Also shown on Fig. 1 are two other tracks. The "source track" is the track of the continuous source of 25 and 55 Hz acoustic energy while the "SUS Drop" track is the track of a series of explosive charge sources. The continuous source was at 120 m depth and received by a 285 m deep bottom hydrophone at point B. The explosive sources were at 18 m depth and received by a 270 m deep bottom hydrophone at point C. Both experiments were conducted in early April under expected similar sound-speed profile conditions.

Figure 2 shows a series of sound-speed profiles from STD casts at point B performed during the continuous source experiments. The profiles indicate a surface channel down to about 75-100 m and a bottom channel from about 200-225 m to the bottom at 285 m. It is probable that the same sound-speed profile conditions prevailed during the explosive source experiment, although no direct measurements were performed on that occasion. Figure 2 also shows the receiver's physical configuration at point B. The configuration at point C was functionally similar. In both cases it is expected that a near-source coupling into both channels occurs and with propagation to greater range, the channels propagate independently except possibly at the lowest frequency (25 Hz) where the channel separation (say, 177 m) was about 2 wavelengths. Hence the results to follow refer to the propagation in the bottom channel and its interaction with the sub-bottom features.

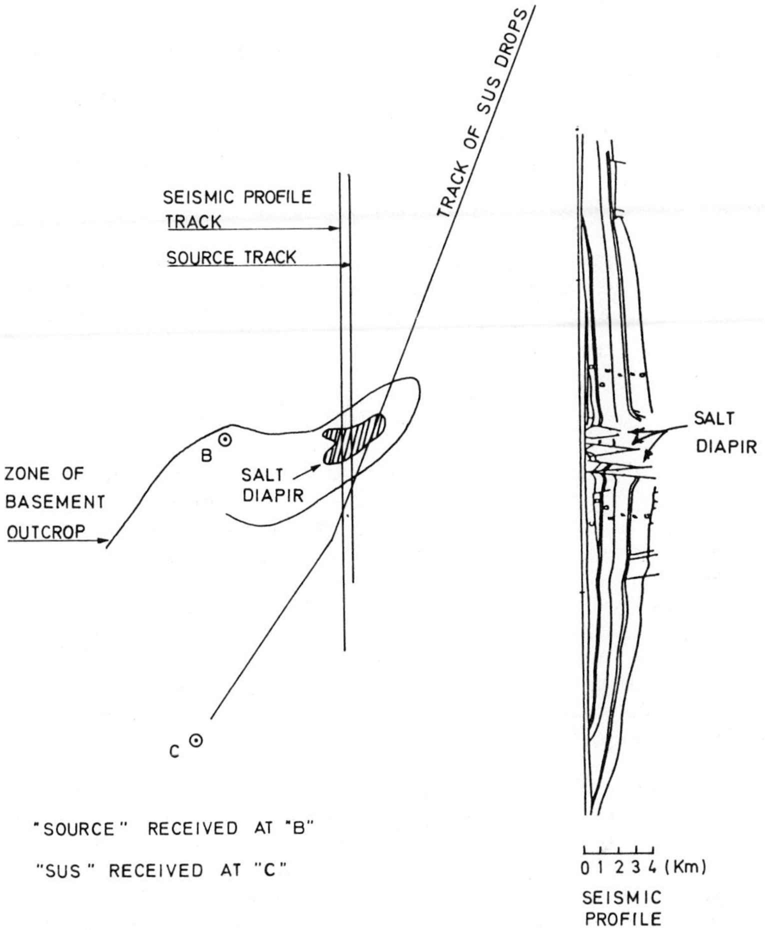

Fig. 1 Experimental geometry.

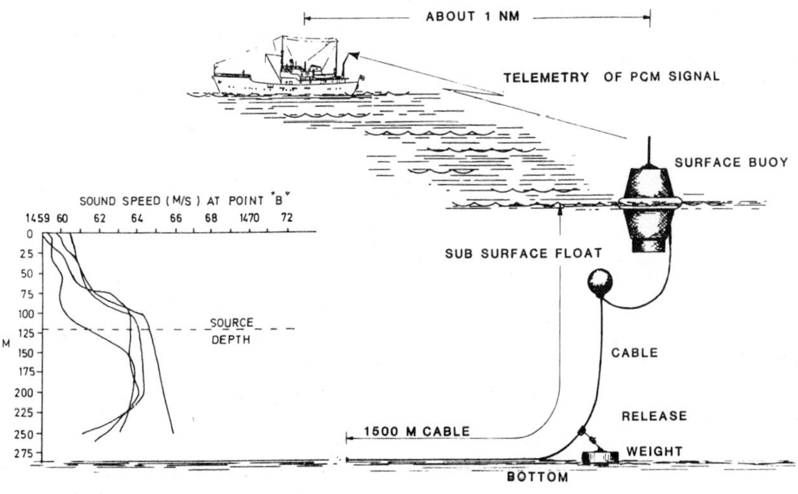

Fig. 2 Receiver system, physical configuration.

RESULTS

Figure 3 shows the transmission loss from the explosive shots to the bottom hydrophone at point C. This figure is made up of two data sets: one from 1978 covering from short range to about 100 nm and the other from 1981 from about 100 to about 300 nm. The frequency range 100-200 Hz is fairly well fit by the modified cylindrical spreading model shown.

In the lower octave bands there is a significant increase in transmission loss as the source enters the basement outcrop region. The increase in loss extends somewhat beyond the outcrop region and is then followed by decrease in transmission loss. This effect is more pronounced at the lower frequency octave bands in which sub-bottom penetration is also greater.

In searching for an explanation for this behaviour of the acoustic field, it must be remarked that very similar transmission loss vs range vs frequency features were observed in an experiment in Northeast Pacific Ocean deep water in acoustic propagation over the Dickins Sea Mount and reported by Chapman and Ebbeson [1]. Their explanation refers to the Medwin-Spaulding model [2] of seamount shadowing based on the theory of diffraction over a wedge. Another class of explanations of the behaviour depicted in Fig. 3 is based on the assumption that the acoustic energy is propagation largely as Scholte waves along the water-sediment interface and that this propagation channel is interrupted by changes in the shear properties of the sediment above the salt diapirs and their surrounding basement outcrop materials. These two classes of explanations cannot be resolved at present due to the sparseness of geophysical data in the area of the acoustic experiment.

Figure 4 shows the azimuthal dependance around point B, of the excess transmission loss in which the range-dependance of the transmission loss was removed by a simple loss vs range model not very different from that of Fig. 3. Each datum in this figure represents the excess transmission loss below which measured values fell ten percent of the time in a one-hour period. This effectively excluded fading due to interference among the multiple paths taken by the CW energy.

The main feature is a dip of about 5 dB (at 55 Hz) and 12 dB (at 25 Hz) in the direction of due east, which is where the source was passing approximately over the region of salt diapirism. Again, it is remarked that the region corresponding to the shaded area in Fig. 1 is mapped only by the single North-South seismic profile, and that the data is otherwise sparse. Hence the boundaries of the region shown must be considered very rough.

CONCLUSION

Some evidence has been shown linking an acoustic propagation anomaly observed in the Western Barents Sea with a known salt diapir occurring in a region of basement outcrop through which the acoustic energy was propagating. The data suggests that under the prevailing conditions of bottom channel and with the receiver hydrophone located on the bottom in shallow water, there appears to be a significant influence of the salt diapir and possibly other aspects of the basement outcrop region upon the transmission loss of low frequency sound propagating through the region.

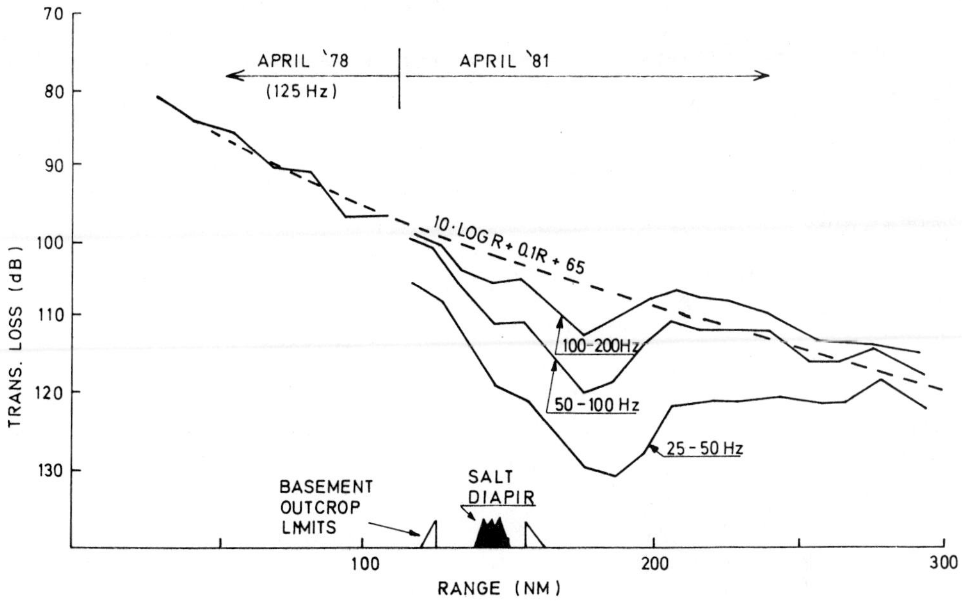

Fig. 3 Transmission loss from explosive shots to bottom hydrophone at point "C".

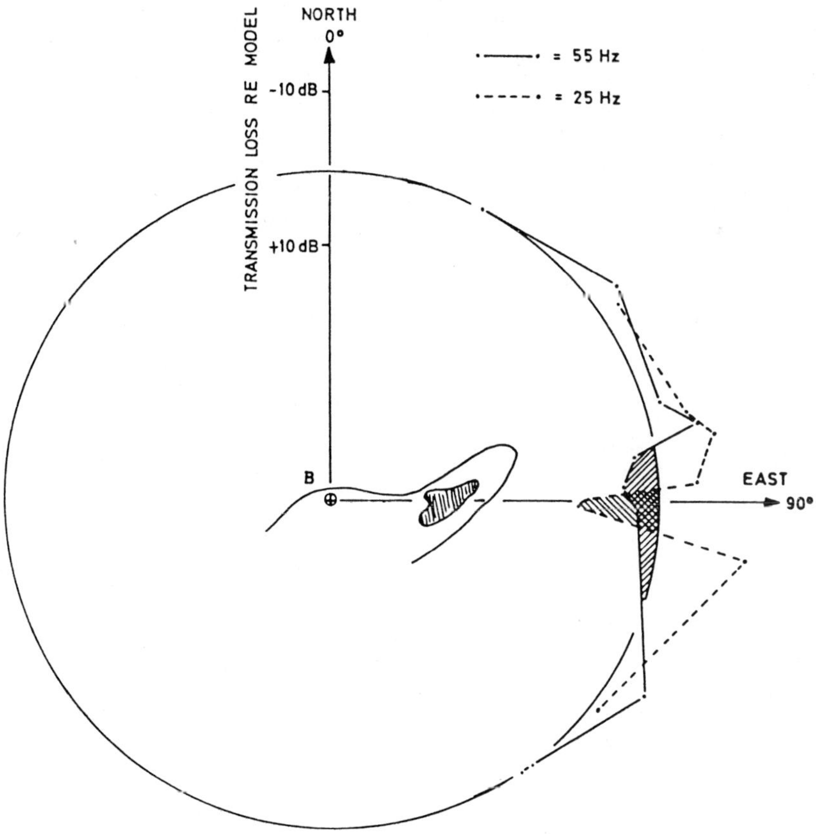

Fig. 4 Minimum excess transmission loss in 1 hour period as function of azimuth vs point "B" with range dependence removed.

The phenomena can be seen in terms of shadowing based on the theory of diffraction on the assumption that deeply penetrating compressional waves account for the acoustic energy transport, or it can be seen in terms of disturbances in the shear properties of the sediment at the water-sediment interface on the assumption that Scholte-like waves account for the acoustic energy transport. This issue is not resolved at the present time and awaits a future experiment in which the bottom compressional and shear properties are measured more densely along with water-column sound-speed measurements.

Salt diapirism and basement outcropping are quite prevalent in the shallow water regions of the world ocean as can be seen, for example in Rona [3]. Faleide and Gudlaugsson [4] give a more detailed description for the Western Barents Sea.

REFERENCES

1. R. Chapman and G. Ebbeson, Acoustic Shadowing by an Isolated Seamount, J. Acoust. Soc. Am. 73: 1979 (1983).
2. H. Medwin and R. Spaulding, The Seamount as a diffracting body. In: Bottom-Interacting Ocean Acoustics. W. KUPERMAN and F.B. JENSEN eds. Plenum Press, New York, NY p. 421 (1980).
3. P. Rona, Evaporities at Passive Margins, Atlantic Oceanographic and Meteorological Laboratories, NOA, Miami, Florida.
4. J. Faleide and S. Gudlaugsson, Geology of the Western Barents Sea, Cand Royal Thesis in Applied Geophysics, Dept. of Geology, University of Oslo 1981.

LONG RANGE LOW FREQUENCY PROPAGATION MEASUREMENTS IN DEEP WATER USING A SEISMIC TOWED ARRAY

Alain Plaisant

THOMSON-CSF ASM Division, 06801 Cagnes/Mer, France

B. De Raigniac

GERDSL Le Brusc, 83140 Six Four Les Plages, France

INTRODUCTION

A propagation experiment was carried out in March 1984 in Central Atlantic. A towed powerful source transmitted long single frequency waves in the band 30-120 Hz which were received on a 96 channels seismic towed array after travelling up to a distance of 2000 kilometers.
Results are presented concerning propagation loss, propagation loss fluctuations, frequency and space widening of the source.

DESCRIPTION OF THE EXPERIMENT

The experiment discussed in this paper took place in March 1984 in an area chosen for its regular deep bottom and low trafic in order to minimize propagation parameters variations and ambient noise at low frequencies.

Transmitted frequencies were in the band 30-120 Hz. Good signal to noise ratio were desired so we used a standard seismic towed array at reception in order to achieve processing gain against isotropic noise by classical beamforming, and a powerfull towed acoustic source transmitting LCW pulses. Both source and array were towed at low speed on straight divergent courses as shown in Fig. 1 ; speeds and headings were chosen to keep the source near the acoustic axis of the array. At the end of the experiment, distance between source and array was 2000 kilometers.

Transmission signals

Single frequency pulses of 30 minutes duration were transmitted continuously, switching from one frequency to the other. Four frequencies were used, the same one appearing every two hours. The results presented in this paper refer only to one frequency : 82 Hz for which the transmission level was 190 dB/μPa at 1 m. The source was towed at a depth of 60 meters.

Towed array configuration

A standard seismic towed array manufactured by "Ateliers Mécaniques de St. Gaudens - France" was used. The active acoustic length was 2375 meters, consisting of 96 equally spaced phase centers from which the acoustic signals were recorded on-board the receiving ship. Distance between

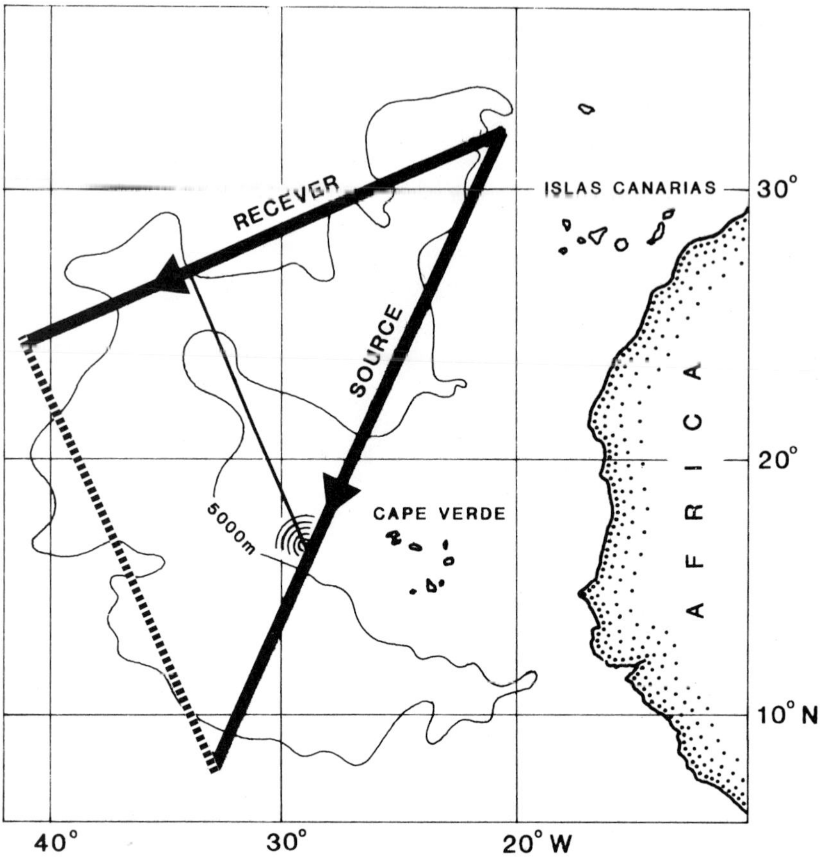

Fig. 1 - Ships courses during the experiment

two phase centers was 25 meters. The array was towed at a depth of 40 meters.

Ships courses

Both source and receiving ships started off the Madeira Island, source ship with constant heading 206 degrees and constant speed ; receiving ship with nominal constant heading 248 degrees but small and few corrections were made during the experiment in order to keep the source ship in a direction as close as possible to the acoustical axis of the array. The distance between source and receiver was then increasing almost linearly at a rate of 7 kilometers per hour. Good data was obtained from ranges 180 km to 2000 km.

Environmental data

Velocity profiles were regularly measured from both ships using XBT's. These profiles are displayed in Fig. 2. They indicate the presence of a surface duct of about 150 meters at the beginning of the experiment (northern latitudes), progressively decreasing in width as latitude decreases to about 55 meters at the lower latitudes reached by the source ship at the end of the experiment. The source was then just below the surface duct whereas the receiver remains in the duct during all the experiment.

Sea state was 6 to 8 at the beginning of the experiment, up to a distance of 400 km, then progressively decreasing to sea state 2. Trade wind were continuously blowing.

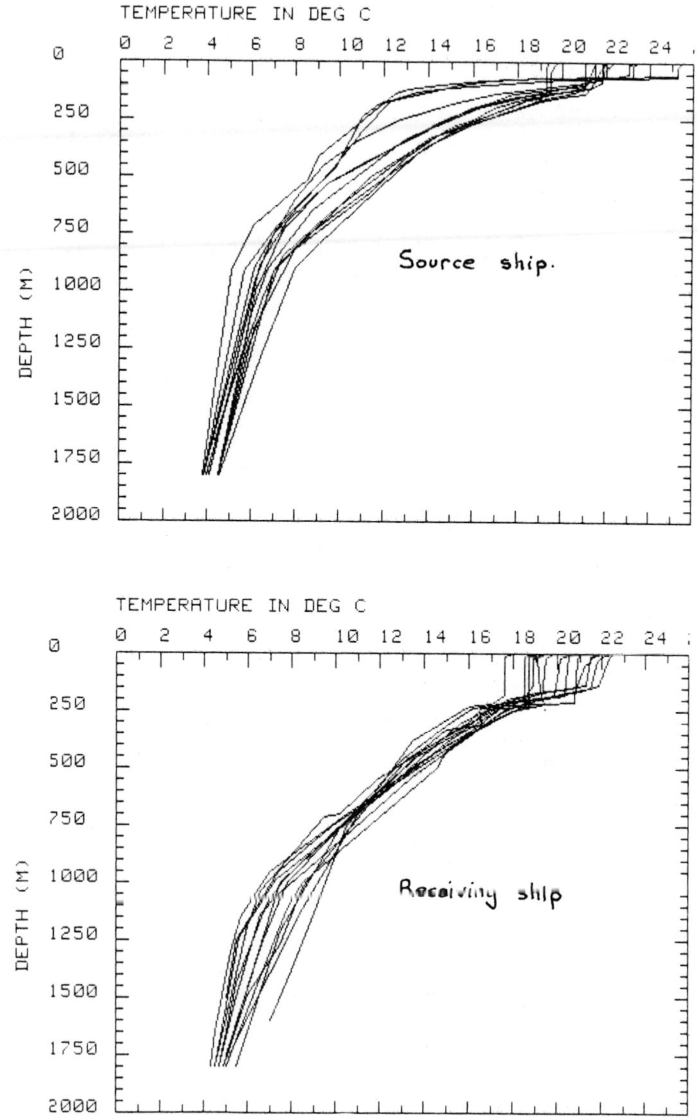

Fig. 2 - Velocity profiles measured from both ships

DATA PROCESSING

For all results presented in this paper signal to noise ratio higher than 30 dB, sometimes as much as 40 dB were achieved almost until the end of the experiment. The noise itself was decreasing regularly in the working bandwidth.

Classical beamforming was performed from the 96 channels outputs, using the double FFT technique in the space-frequency domain. Successive blocks of 16 seconds of data were Fourier transformed with no time shading so frequency resolution Δf is 1/16 Hz = 62.5 mHz.

The transmitted frequency lines had a much narrower frequency width (0.6 mHz corresponding to 30 minutes duration) and even after propagation we observed that the receiving frequency was covering only a few FFT lines, so interpolation with ratio 8 was made from the 16 FFT samples around the frequency 82 Hz of interest by adding zeros in the FFT of the signal to interpolate and returning to the original domain by FFT^{-1}. The frequency line observed around 82 Hz is then more smoothly defined ; transmission losses are computed from the maximum value of the interpolated spectrums and frequency spreadings are measured also from these interpolated spectrums.

Beams were formed by Fourier transforming the signals received at the same instant over the 96 aligned phase centers. 128 beams were formed, covering a sector of $\pm 21°$ from the acoustic axis of the array, stearing angle α being such as :

$$\sin\alpha = \frac{\lambda}{d} \cdot \frac{k-(N/2+1)}{N}$$

k : number of space FFT channel
N : block size of space FFT (128)
λ : wavelength (18.3 m)
d : hydrophones separation (25 m)

The 3 dB width of the central beam is about 0.4 degree at 82 Hz. Means and standard deviation of the interpolated peak spectrum level have been computed over 20 successive frequency spectrums in the beams where the maximum level is found ; in some cases, bad blocks of data have been removed thus reducing the number of spectrums over which the statistics were made.

PROPAGATION LOSS RESULTS

Fig. 3 displays propagation loss versus range. The small squares indicate the mean value obtained after averaging 20 spectrum or about 6 minutes, of signal and vertical bars represent the standard deviation.

In obtaining these propagation loss values, we implicitly supposed the signal from the source was perfectly coherent on the total acoustical length of the array and the array perfectly aligned. This point will be investigated later but preliminary results showed this hypothesis seem correct as far as the medium is concerned.

Transmission loss variations suggest the occurence of several convergence zones at short ranges. With velocity profiles such as those measured during the trial (see Fig. 2), the convergence zones should occur between 58 and 64 km. The 3 first large maximums in Fig. 3 are close to multiples of 63 km.

At larger ranges fluctuations remain important and it becomes difficult to relate observed fluctuation to any known environmental parameters. Almost all data remain within two limits which are :

$$PL_{up}(r) = 17 \log r + \alpha r$$

$$PL_{down}(r) = 20 \log r + \alpha r$$

α : volume absorption = 0,001 dB/km

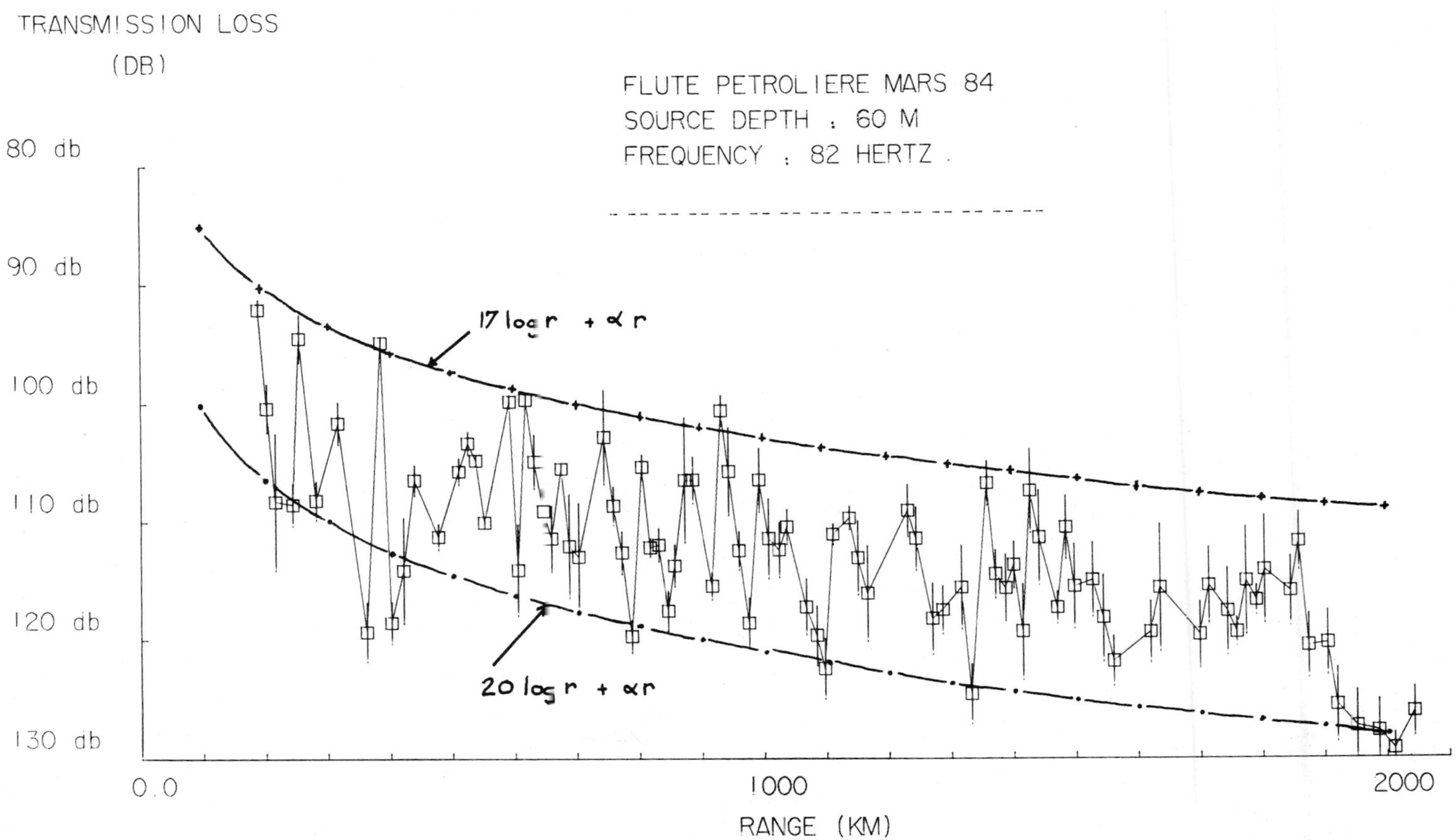

Fig. 3 - Propagation losses versus range

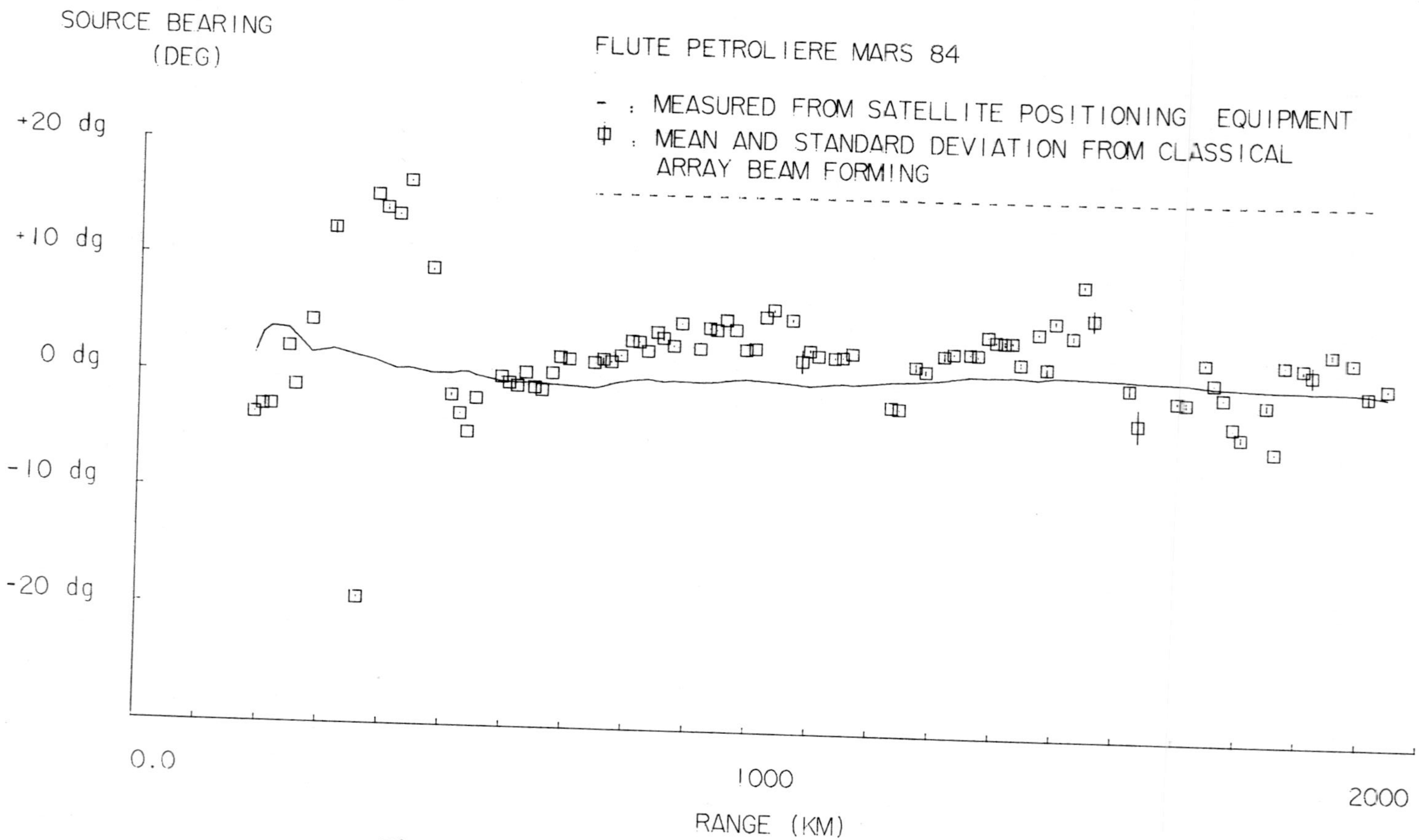

Fig. 4 - Source bearing assuming constant array heading

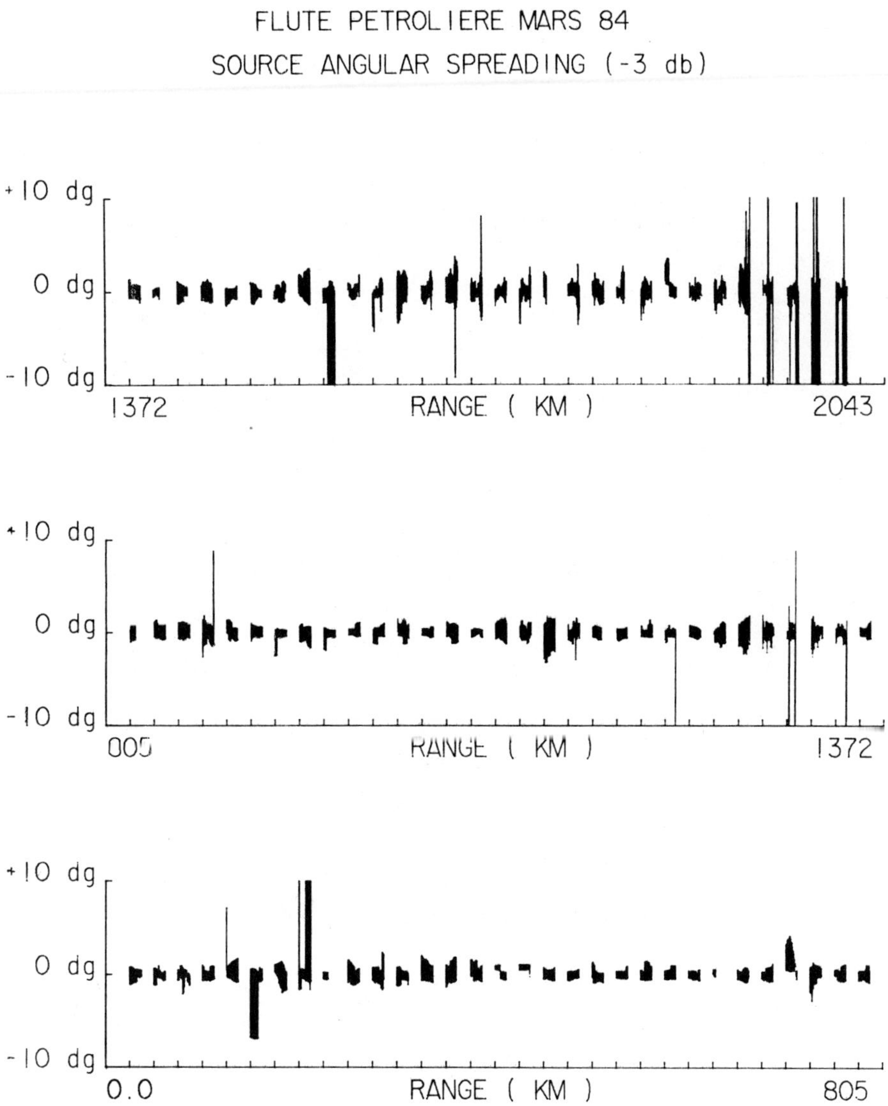

Fig. 5 - Source angular spreading

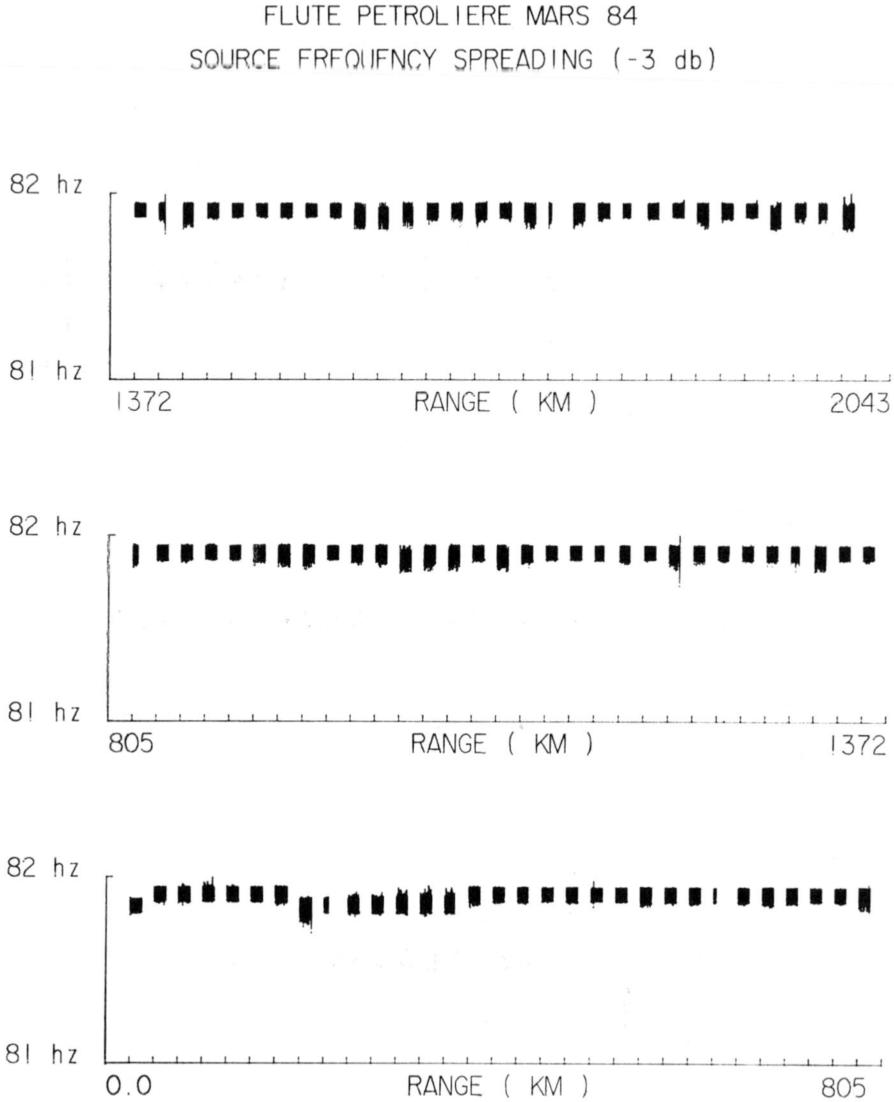

Fig. 6 - Source frequency spreading

High value of PL(r) observed around 400 km are probably due to defect of alignment of the array found to be -20° away from its nominal heading whereas the surrounding measurements indicate an alignment of +15 degrees. Fig. 4 shows the acoustical bearing of the source supposing the array is always aligned at heading 248°. At the end of the experiment we observe a sudden drop in the PL(r) curve which may be attributed to the presence of a sea mount rising up to 640 m in the middle of the propagation path (see Fig. 1). Another possible explanation is the fact that at the end of the experiment the source went just below the surface duct.

BEARING AND FREQUENCY SPREADING

If one plots the level of the different beams at the received frequency 81.9 Hz, the -3 dB points with respect to the maximum give an indication of the angular spreading of the energy coming from the source.

This angular spreading computed for all the 16 seconds spectrums in blocks of 20 are displayed on Fig. 5, centered around the mean value. Distance scale is not linear. Blocks of data have been packed in order to be able to reduce to range axis length. One can see the angular spreading is small : of the order of $\pm 1°$ at all distances except for blocks of data corresponding to low signal, particularly on the last 5 blocks.

Frequency spreading is shown in Fig. 6. One can notice the small negative doppler shift corresponding to a range rate of +7 km/hour. Frequency spreading looks quite small compared to the 1/16 Hz normal width due to troncation in the time domain. Spreads shown in Fig. 6 are about 0.1 Hz so one can say that spreading due to the medium and the possible movements of array and source, are at the most of the order of 0.1 to .4 Hz.

CONCLUSION

Results presented in this paper are preliminary. Further analysis remains to be done in the following fields : convergence zone structure and spatial coherence of wave fronts.

None of the well-known propagation loss models are suitable for handling such a situation where distances are several hundred of kilometers and therefore one cannot ignore variations of medium properties with range.

SOME CHARACTERISTICS OF VIRTUAL MODES IN SHALLOW WATER WITH HIGH SPEED BOTTOM

G.H. Brooke, D.J. Thomson and R.F. MacKinnon

Defence Research Establishment Pacific
FMO Victoria, British Columbia
Canada VOS 1BO

ABSTRACT

Seismic-acoustic properties of virtual (continuous spectrum) modes in an elastic waveguide are examined. Experimental observations made in a shallow water location in the Canadian Arctic Archipelago, where bottom compressional speeds are in excess of 4.5 km/sec, indicate that significant amounts of low frequency energy (<50 Hz) propagate with high speed to long range. The characteristics of this high speed energy are such that propagation occurs at preferred frequencies which can be associated with normal mode cutoffs. Moreover, there is a specific depth dependence of the energy in the water column at each frequency despite a high degree of time dispersion. Numerical results have been obtained using both a fast field program (FFP) technique and a normal mode approach applied to a layered elastic waveguide model of the location. The results are consistent with the experimental data and help define this unusual feature of shallow water propagation.

INTRODUCTION

In recent years[1,2] considerable interest has been focused on both the relationship and the interaction of water-borne acoustic energy with seismic-type energy carried in the ocean bottom. Very few, if any, investigations have been reported which involve high-speed bottom material in the absence of substantial sediment layering. In the shallow waters of Barrow Strait in the Canadian Arctic Archipelago precisely these conditions exist[3]; the uppermost bottom layers are characterized by compressional speeds in excess of 4.5 km/sec. Consequently shallow water propagation in this area exhibits several interesting features. Particular emphasis in this paper is given to low frequency energy which propagates with high speed to long range. This energy represents contributions from virtual (continuous spectrum) modes and has the property that it is propagated at preferred frequencies which can be associated with normal mode cutoffs. Furthermore, at each frequency, there is a specific depth dependence of this energy in the water column despite the fact that it is highly dispersed in time.

Two different numerical methods have been used to model the experimental observations. The first method is a normal mode technique applied to a simple Pekeris waveguide (two fluid layers). A feature of this model is the inclusion of a free surface bound at a depth well into the bottom; the mode spectrum of a bounded waveguide is entirely discrete and allows the contributions of the high speed energy (equivalent to the continuous spectrum in the unbounded waveguide case) to be included in a numerically efficient manner. The second method is a Fast-Field Program (FFP) approach applied to a more realistic waveguide model involving solid bottom layering. The SACLANTCEN FFP code SAFARI[4] is used for this purpose.

This paper consists of two parts: (i) a description of the features of the experimental data, and (ii) a presentation and a discussion of the numerical results obtained from the theoretical models. Both models yield results which are consistent with the experimental observations.

EXPERIMENTAL OBSERVATIONS

Several propagation experiments have been conducted by DREP in the Barrow Strait area. The bathymetry of one of the propagation tracks is sketched in Fig. 1. Roughly speaking, the water is between 150 and 180 m deep over a 15 km range. There is the characteristic upward refracting velocity profile[5] in the water column, however, this feature has little significance at the frequencies of interest in this paper. Similarly, the presence of ice cover (~2 m thick in later winter) has little effect at low frequencies over a range interval of 15 km; experiments have been conducted both in open water and under ice.

In all cases the receiver was located on the bottom at a depth of 150 m. Explosive sound sources (nominally 1 lb of TNT) were detonated throughout the water water column and at various ranges allowing for a thorough definition of the propagation over the track. A summary of shot locations is given in Fig. 1. Only data from the 15 km range will be discussed herein. At this range, shots were detonated at approximately 10 m depth intervals between 80 and 160 m.

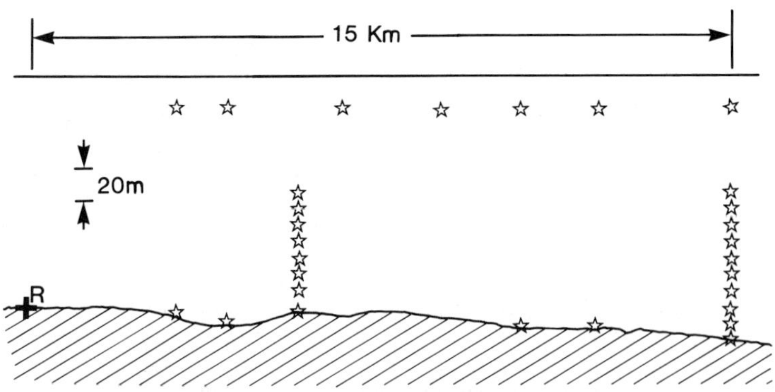

Fig. 1. Experiment: Source-Receiver Configurations.

A typical broadband time series and its corresponding time-frequency characteristic obtained from the 15 km range are shown in Figs. 2(a) and 2(b), respectively. The time series represents pressure fluctuations due to a one pound charge detonated at a depth of 89 m. It is apparent that the signal is dispersed over nearly twelve seconds allowing for time separation of several features. These features have been labelled in the figure and in order of arrival are as follows: (i) a group of well dispersed high speed arrivals travelling with speeds in the range 3.4 - 6.0 km/sec, (ii) a second group of well dispersed high speed arrivals travelling with speeds in the range 2.2 - 3.4 km/sec, (iii) a high frequency arrival which constitutes the water wave travelling with a speed of 1.44 km/sec, (iv) a low frequency wavetrain inversely dispersed over the frequency ranges of 5 - 10 Hz and travelling with speeds in the range 1.25 - 1.4 km/sec, and (v) a low frequency wavetrain of predominantly 10 Hz travelling with speeds in the range 1.0-1.1 km/sec. Thus, the time series represents energy travelling with a truly wide range of speeds (1.0 - 6.0 km/sec).

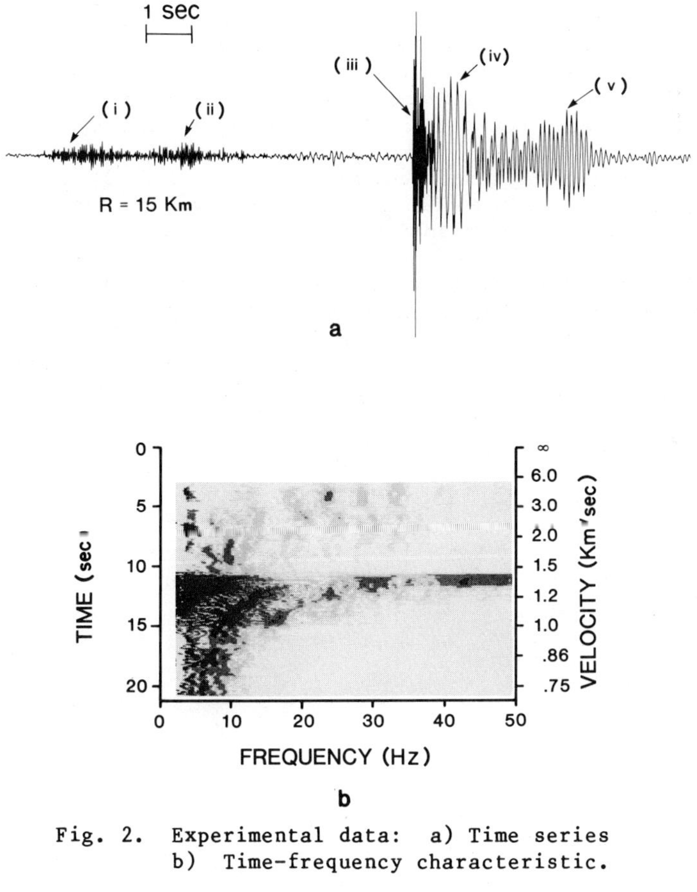

Fig. 2. Experimental data: a) Time series b) Time-frequency characteristic.

The time-frequency characteristic shown in Fig. 2(b) is a result of a frequency domain analysis undertaken to assist in the interpretation of the time series. A frequency resolution of 1 Hz was used for this purpose. Consequently, power spectral estimates were generated for overlapping (94%) one second intervals throughout each time series. These spectra were then stacked in time and converted to a grey scale representation. A dynamic range of 0 dB (white) to 48 dB (black) was used and effects due to filtering and to source spectrum were removed.

The frequency structure of each of the time series features is clearly illustrated in the time-frequency characteristic. The high speed groups have a marked frequency dependence with preferred frequencies occurring every 5 or 6 Hz starting at 13 Hz. Furthermore, within each group there is not much frequency dispersion. Also, there appears to be a frequency shift of about 1 Hz between the preferred frequencies of the two high speed groups ((i) and (ii) in Fig. 2(a)). This shift is consistent for all the time series and, hence is assumed to be significant. The water wave is shown as a general increase in energy level at all frequencies occurring just after ten seconds. An interesting property of the Barrow Strait data, however, is an obvious decrease in water speed energy between approximately 10 and 30 Hz. This phenomena has been tentatively ascribed to rough surface scattering (from the ocean bottom) peculiar to this area. The energy arriving after the water wave appears as three inversely dispersed branches occurring between frequencies 3 - 10 Hz, 10 - 17 Hz and 16 - 25 Hz. These branches can be associated with, respectively, the Scholte mode (mode 0) and modes 1 and 2 of the shallow water waveguide.

Identification of particular modes in the time series was made not only on the basis of the time-frequency characteristic, but also by observing the depth dependence of the energy in the water column. Examples of the depth dependence of the slow speed energy are shown on the left hand side of Fig. 3. The dashed line corresponds to the 6 Hz component of the inversely dispersed branch (see Fig. 2(b)) occurring between 3 and 10 Hz. A gradual decline in energy away from the ocean bottom is evident and, hence, the association with the interfacial Scholte mode. The solid line corresponds to the 10 Hz component of the inversely dispersed branch occurring between 10 and 17 Hz and has an obvious null at a depth of 130 m; this depth dependence is typical for mode 1.

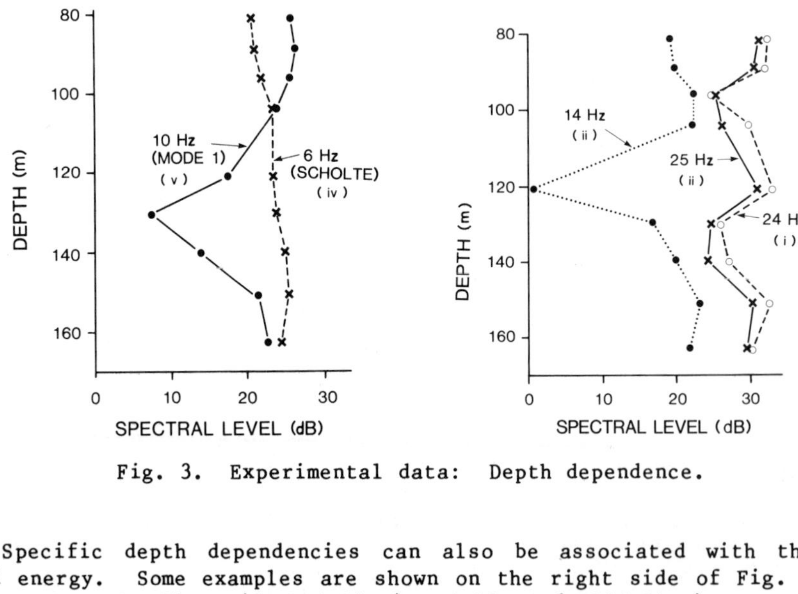

Fig. 3. Experimental data: Depth dependence.

Specific depth dependencies can also be associated with the high speed energy. Some examples are shown on the right side of Fig. 3. It appears that the 24 Hz (dashed line) and 25 Hz (solid line) components of arrival groups (i) and (ii), respectively, have very similar and distinct depth dependencies with nulls occurring at, approximately, 100 m and 140 m depths. Like slow speed normal modes, the amount of variation of the high speed energy with depth is frequency dependent. This is illustrated by the 14 Hz (dotted line) component of the arrival group (ii) which exhibits only one null which is situated at approximately 120 m depth.

In the preceding discussion the major characteristics of the Barrow Strait data have been established. The slow speed energy is readily identified with normal modes. A simple normal mode model consisting of 150 m of water over a homogeneous solid layer with shear and compressional velocities of roughly 3.0 km/sec and 5.2 km/sec, respectively, is sufficient to identify, in the data, the gross features of the normal modes. The high speed energy also possesses modal features (i.e. a frequency structure and a specific depth dependence). Because of its high speed, however, it cannot be associated with the normal mode spectrum and must, therefore, be associated with virtual mode energy. The following discussion relates to the modelling of this high speed energy.

THEORETICAL MODELLING

Theoretical models which can predict the major features of the Barrow Strait data must account for the contributions of the high speed virtual mode energy. Previous investigations[6,7] of virtual mode propagation have involved the approximation of the branch line integral[8,9] which expresses exactly, the contribution of these modes. In this paper, two models are employed which treat the virtual mode contributions in different ways: a normal mode model and a FFP model. A summary of the mathematical basis of the two models is shown in Fig. 4.

In general, the field, $\phi(r,z,t)$, in a waveguide may be represented in space and time by a double integral over the frequency-horizontal wavenumber (ω-k) domain. In the FFP model the k-integral is evaluated directly by numerical integration in k-space, as expressed in Fig. 4. Alternatively, in the normal mode model, the k-integral is evaluated using residue calculus - a procedure which results in a representation involving the complete spectrum of modes. Normally there is a branch line contribution from the continuous spectrum. Here, a free boundary is introduced at sufficient depth in the bottom (so as to not affect significantly the discrete normal modes) which makes the entire mode spectrum discrete. In this manner, the integral over the continuous (virtual) modes is replaced by a sum over an equivalent set of discrete modes.[10] Both the FFP model and the normal mode model employ an FFT to evaluate the ω-integral.

Frequency-Horizontal Wavenumber Integral

$$\Phi(r,z,t) = \int_{-\infty}^{\infty} d\omega \, \exp(j\omega t) \, S(\omega) \int_{0}^{\infty} dk \, k \, J_0(kr) \, G(k,z,z_s,\omega)$$

Normal Mode and FFP Approximations

$$\left\{ \begin{matrix} \Phi_{MODE} \\ \\ \Phi_{FFP} \end{matrix} \right\} \approx \sum_{m=0}^{M-1} S(m\Delta\omega) \left\{ \begin{matrix} \sum_{n=1}^{\infty} U_n^m(z) U_n^m(z_s) e^{-jk_n r} / \sqrt{r} \\ \\ \sum_{n=0}^{N-1} \sqrt{(2/\pi jr)} G_n^m(z,z_s) e^{-jn\Delta kr} \end{matrix} \right\} e^{j2\pi im/M}$$

Fig. 4. Basis of the Numerical Models.

Normal Mode Model

The Pekeris-type geoacoustic waveguide used in the normal mode calculations is shown in Fig. 5. In an attempt to model the Barrow Strait data, the sound velocity in the bottom (3.0 km/sec) was chosen to represent a shear velocity. A free surface bound is placed at a depth of 2000 m, well below the water depth of 150 m.

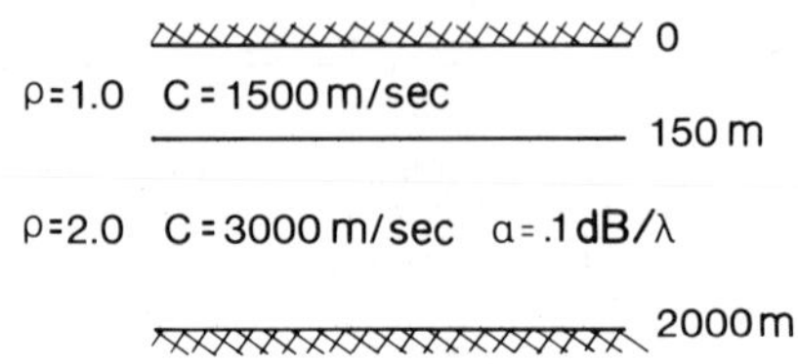

Fig. 5. Normal Mode Model Waveguide.

Phase velocity dispersion curves for the first ten modes on this structure are shown in Fig. 6(a). Briefly, the phase speed of each mode decreases rapidly from infinity (at cutoff) to approach the speed in the bottom layer. Over this part of the dispersion characteristic the mode is in a part of the spectrum, hereafter referred to as "continuous", which is associated with the true continuous spectrum in the absence of the free boundary. At regular transition frequencies (separated by five or six hertz) a low order mode becomes trapped in the water column; the phase speed of these modes drop rapidly to approach the water speed. Over this part of the dispersion characteristic the mode is in a part of the spectrum, hereafter referred to as "discrete", which can be associated with the familiar discrete mode spectrum in the absence of the lower free boundary. The transition frequencies are easily identified in Fig. 6(a). It is interesting and important to note that when a mode enters the "discrete" spectrum, the higher order "continuous" modes undergo a drop in phase velocity to compensate for the lost mode (i.e. a ripple effect is apparent in the "continuous" mode phase velocity curves at the transition frequencies).

Corresponding group velocity characteristics are shown in Fig. 6(b). Once again the transition frequencies are readily identified. Moreover, at these frequencies, each higher order "continuous" mode group velocity passes through a minimum. The group velocity minima have two implications, namely (i) there is an increase in the energy in the low speed medium (the water column) in order for the speed to drop, and (ii) the range dependence of these modes at the transition frequencies behaves in accordance with well-known Airy phase considerations.[9] Thus, at the transition frequencies, there should be significant high speed energy in the water column observable at long range.

A synthetic time series for the Pekeris-type waveguide is shown in Fig. 7(a). This time series represents a sum over twenty modes (including frequencies from 1 - 40 Hz) with a range of 15 km and with source and receiver depths of 90 m and 150 m, respectively. One cycle at 20 Hz is assummed for the source waveform. An attenuation of 0.1 dB/ for the bottom layer has been included in the calculations using standard perturbation theory[11]. There are at least three distinct features of the time series; they are (i) well dispersed high speed arrivals, (ii) a water speed arrival, and (iii) low speed wavetrains dispersed over several seconds. The corresponding time-frequency characteristic is shown in Fig 7(b) and should be compared with the group velocity curves shown in Fig

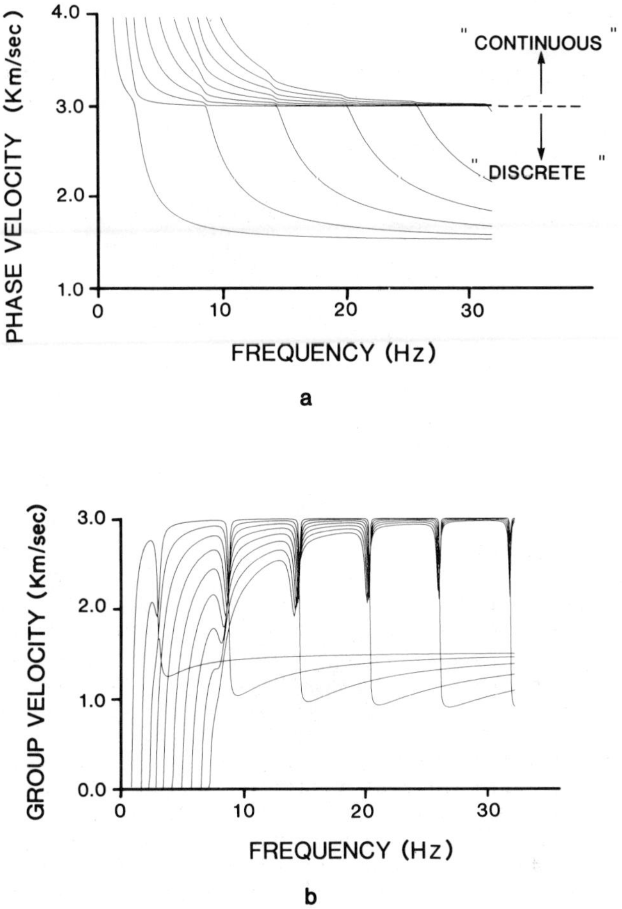

Fig. 6. a) Phase b) Group velocity dispersion.

6(b). In Fig 7(b) the grey scale has a similar dynamic range (0-48 dB) to that shown in Fig 3(b), and was generated from successive overlapping one second FFT's taken throughout the synthetic time series. The predominant low speed wavetrains in the time series may be identified as modes 3, 4, 5 and 6. Modes 1 and 2 are not well excited for the source-receiver configuration used but are still visible in the grey scale representation; in fact, the water speed arrival is seen to be predominantly mode 1. The high speed arrivals have a frequency structure which is very similar to that observed in the experimental data. In particular, frequencies which correspond to normal mode cutoffs are well excited as predicted in the discussion relating to the group velocity curves. Indeed time-frequency characteristics of individual "continuous" spectrum modes (not shown here) exhibit this frequency structure. Because these modes all have nearly identical depth dependencies at each frequency, it is not surprising that the total field also shows a specific dependence on depth as is observed experimentally.

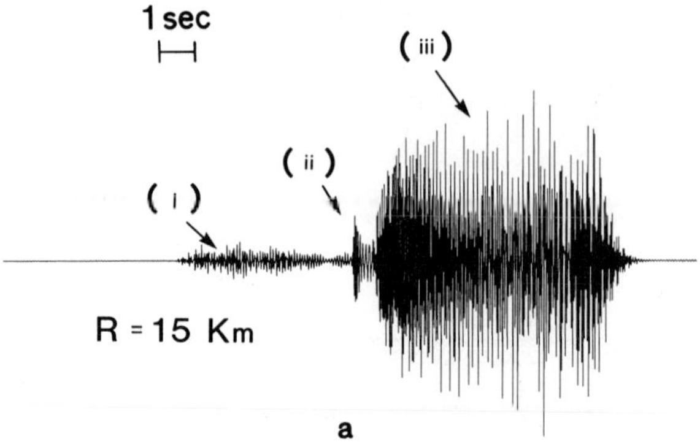

a

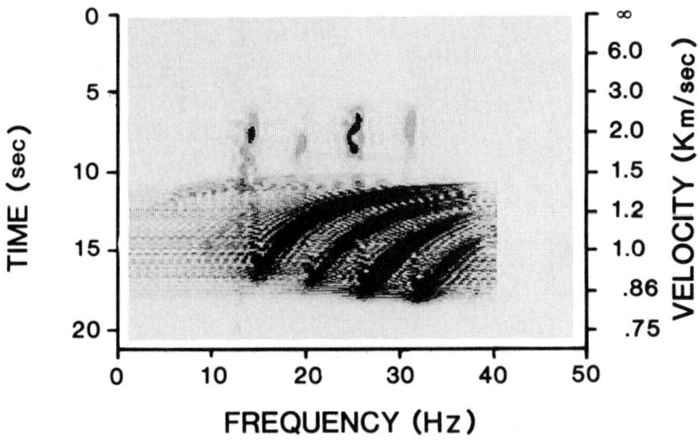

b

Fig. 7. Normal Mode Model a) Time series b) Time-frequency characteristic.

FFP Model

The geoacoustic waveguide used for the FFP calculations is shown in Fig 8. The parameters were chosen on the basis of the experimental time

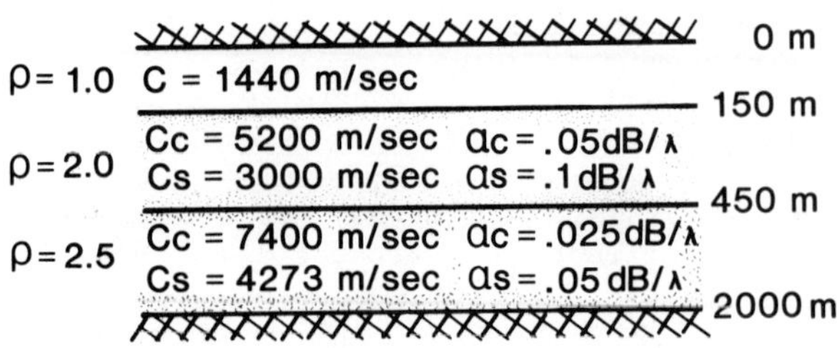

Fig. 8. FFP Model Waveguide.

series and the available geoacoustic data[3]. Attenuation values are comparable to other published data[1,2]. Note the presence of the free surface bound at 2000 m. Calculations were made with and without the bound, however, only with the bound did the virtual modes become well excited.

With the above parameters, SAFARI was used to generate the synthetic time series shown in Fig 9(a). This time series has the same basic features as those shown previously, namely (i) very high speed arrivals (5.0-6.0 km/sec) (ii) a water speed arrival followed closely by an inversely dispersed low frequency wavetrain, and (iii) low speed wavetrains travelling with speeds as low as 1 km/sec. The corresponding time frequency characteristic is shown in Fig 9(b). Once again the inversely dispersed branches of the Scholte mode and modes 1 and 2 are very prevalent. These modes are, however, much better excited in the model than in the experimental data (e.g. Fig 3(b), especially in the frequency range 10-30 Hz. At higher speed there is significantly less energy in the synthetic data than observed experimentally. An exception may be the high speed energy at frequencies less than 10 Hz; there appears to be a correspondence between theory and experiment and it is probably an indication that the attenuation values assummed for the bottom layers are reaso-

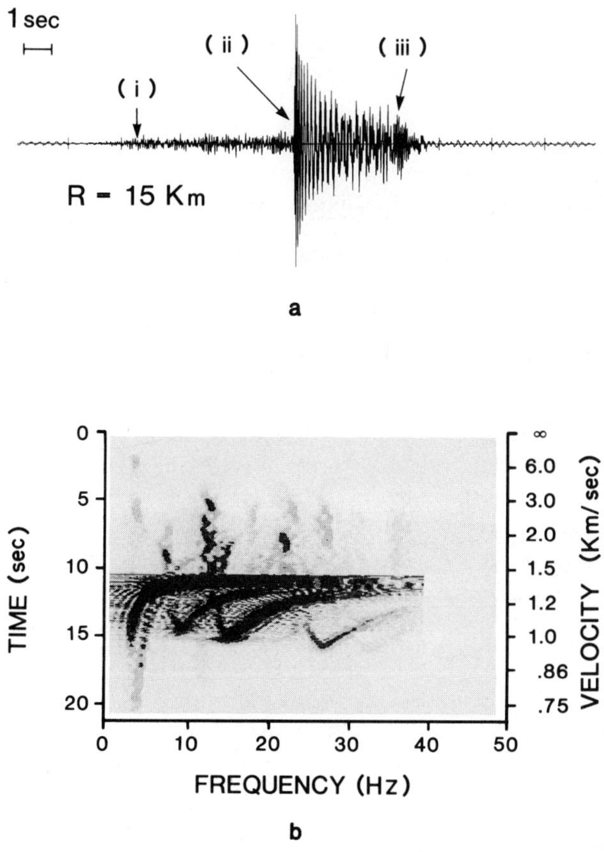

Fig. 9. FFP Model a) Time series b)Time-frequency characteristic.

nable. The frequency structure of the high speed synthetic data indicates that certain frequencies are better excited, however, there are discrepancies between theory and experiment as to which frequencies are preferred. This may, in part, be due to fact that the water depth varied between 180 m and 150 m over the experimental track.

SUMMARY

Shallow water propagation over a high speed bottom has been examined. Results generated from two theoretical models have been compared with experimental data collected in the Canadian Arctic. An outstanding feature of the data is the presence of strongly excited virtual mode arrivals. Both models yielded results consistent with the experimental data. Ironically, a free surface bound which was included in the simple normal mode model for numerical purposes, turned out to be necessary in the more realistic FFP model in order to achieve significant excitation of the virtual modes. The implication of these results is that simple geoacoustic models utilizing one or two homogenous solid bottom layers are not adequate to accurately predict the virtual mode properties, although the slow speed energy is well accounted for; strong sound speed gradients in the bottom may require consideration in order to duplicate the effect of the bound.

REFERENCES

1. "Acoustics and the Sea-Bed", Proceedings of an Institute of Acoustics Underwater Acoustics Group Conference edited by N.G. Pace, Bath University Press, Bath, UK (1983).
2. "Bottom-Interacting Ocean Acoustics", NATO Conference Series IV: Marine Sciences edited by W.A. Kuperman and F.B. Jensen, Plenum Press, New York (1980).
3. G.W. Sander and A. Overton, Deep Seismic Refraction Investigation in the Canadian Arctic Archipelago, Geophysics, 30:87 (1965).
4. H. Schmidt, Modelling of pulse propgation in layered media using a new Fast-Field program, in "Hybrid Formulation of Wave Propagation and Scattering" edited by L.B. Felsen, NATO ASI Series, Nijhoff, Dordrecht (1984).
5. A.R. Milne, Shallow water under-ice acoustics in Barrow Strait, J. Acoust. Soc. Amer., 32:1007 (1960).
6. A.O. Williams, Jr., Pseudoresonances and virtual modes in underwater sound propagation, J. Acoust. Soc. Amer., 64:1487 (1978).
7. C.T. Tindle, Virtual modes and mode amplitudes near cutoff, J. Acoust. Soc. Amer., 65:1423 (1979).
8. F.M. LaBianca, Normal modes, virtual modes, and alternate representations in the theory of surface-duct sound propagation, J. Acous. Soc. Amer., 53:1137 (1973).
9. C.B. Officer, "Introduction to the theory of sound transmission", McGraw-Hill, New York (1958).
10. G.H. Brooke and M.M.Z. Kharadly, Scattering by abrupt discontinuities on planar dielectric waveguides, IEEE Trans. Microwave Theory Tech., MTT-30:760 (1982).
11. C.T. Tindle, Attenuation parameters from normal mode measurements, J. Acoust. Soc. Amer., 71:1145 (1982).

THE DEDUCTION OF APPROXIMATE VALUES OF SEDIMENT DEPTH FROM PROPAGATION LOSS MEASUREMENTS

P. H. Lindop

Admiralty Research Establishment
Portland, Dorset
England

ABSTRACT

Sediment thickness is usually measured over shallow water areas by routing a ship with a vertical beam sonar along a lattice of tracks. This paper examines the possibility of deducing approximate values of sediment thickness from LF propagation loss measurements across the area. Evidence is presented to support an approximate frequency/sediment depth/propagation loss relationship which, at longer ranges, determines α in the expression T.L. = A + 10 log r + α r. An exact formulation of the dependence of loss on sediment thickness requires sea-bed parameters that are not always available. Parametric investigations can be carried out with wave theory models. It is concluded that loss measurements across a sediment covered area can give useful indications of shallow patches and bottom structure.

INTRODUCTION

This paper discusses a possible new method of determining approximate values for the depth of unconsolidated sediment on the Continental Shelf. It is assumed throughout the paper that the seabed structure can be represented by a layer of unconsolidated "sediment" (with a compressional sound velocity of less than 1.8 km/sec and low shear velocity) over a "rock" layer of infinite thickness. The depth of the upper layer is taken to be the variable that is to be determined over large areas in water depth of up to 200 metres.

This is not an accurate representation of the sub-bottom structure which is highly variable. Table 1 gives the findings from two bore-holes drilled by the British Geological Survey together with the depth of the dividing line in the simple model. Clearly the two regions are different and the simple model is a gross approximation. Other boreholes show sand as the upper layer. (It should be noted that current methods of shallow sub-bottom sounding have to assume sound velocities to obtain depths and this introduces inaccuracies if applied in the two regions in Table 1.)

At present, sediment depth is determined by routing a survey ship along a lattice of tracks to generate values beneath the keel using a hull mounted (or towed) vertical beam sonar. This gives sediment depth along the tracks but not between them. If a technique were available

Table 1. Borehole Information

Classification	Thickness (Metres)	Description
"Sediment"	66	Clay, dark grey, firmer at base, silty, gritty occasional pebbles.
"Rock"	5.75	Silt stone, hard, grey black. Thin limestone layers.
"Sediment"	5.48	Gravel
"Rock"	13.72	Mudstone greenish grey
"Rock"	1.86	Sandstone

that gave sediment depth using the horizontal transmission of sound then these types of surveys could be planned more economically. A possible method is discussed here.

THEORY

It is assumed, initially, that the sediment depth is of the order of tens of metres over an area of 50 miles x 50 miles containing a patch of exposed rock. What information can be deduced if measurements of propagation loss are obtained between a ship with a wide-band source on one side of the square and a ship (or sonobuoy) on the other side?

The contribution from one normal mode can be represented by

$$I = I_{n_1} \left(\frac{1}{r}\right) . \; 10^{-\alpha_n r . 10^{-3}}$$

where I = received intensity from range r

I_{n_1} = effective intensity at unit range (mode n) and the given source and receiver depths

α_n = attenuation

I is determined in part by the value of the term

$$\sin^2 \frac{X_n d_s}{D_s} \; \sin^2 \frac{X_n d_r}{D_r}$$

where D_s = water depth at the source

D_r = water depth at the receiver

d_s = depth of the source

d_r = depth of the receiver

X_n is the root of the equation in the normal mode formulation.

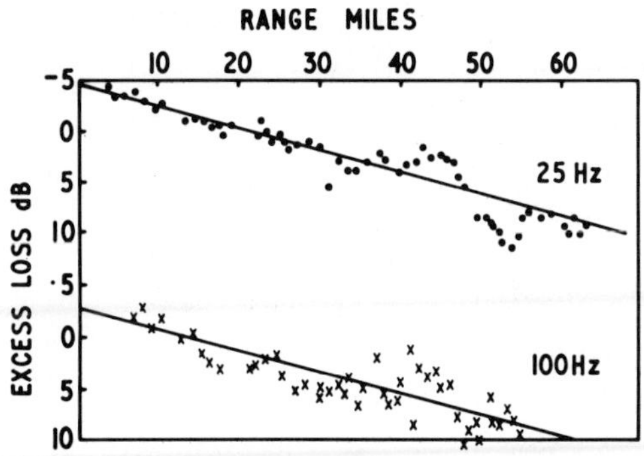

Fig 1 Straight line approximation to excess loss

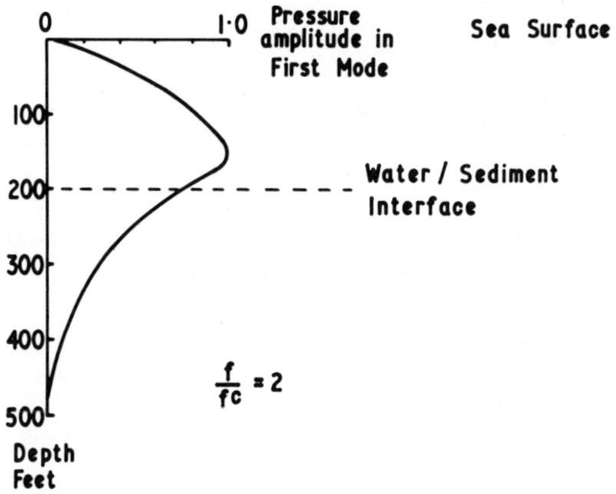

Fig 2 Representation of amplitude of first mode

d_s and d_r can be chosen so as to preferentially stimulate and receive one mode - say the first. Also since α_n increases with the number of the mode, at the longer range direct measurements will give an approximation to a single mode value. The "excess" loss over and above cylindrical spreading will then be a linear function of range.

Fig 1 gives measurements of excess loss (αr) as a function of range at 25 and 100 Hz over a region with a sea-bed expected to be uniform to approximately 35 miles range. The agreement with a straight line is good at 25 Hz and reasonable at 100 Hz over this range bracket.

A theoretical first mode depth dependence is shown in Fig 2 for a low velocity contrast across the water/sediment interface and a sound velocity minima between 100 and 200 ft depth. Measurements of $\int p^2 dt$ (where p = pressure) versus depth in a different environment are shown in Fig 3 for a source depth just above mid-water. It seems likely that the first mode only, is being observed at 25 Hz.

These results suggest that the excess loss can be represented as αr in dB notation; other results have shown this even when the experiments have not been set up to observe the first mode only.

The penetration into the sediment, (see Fig 2) is a function of frequency reaching 100 metres or so at the lowest frequencies considered.

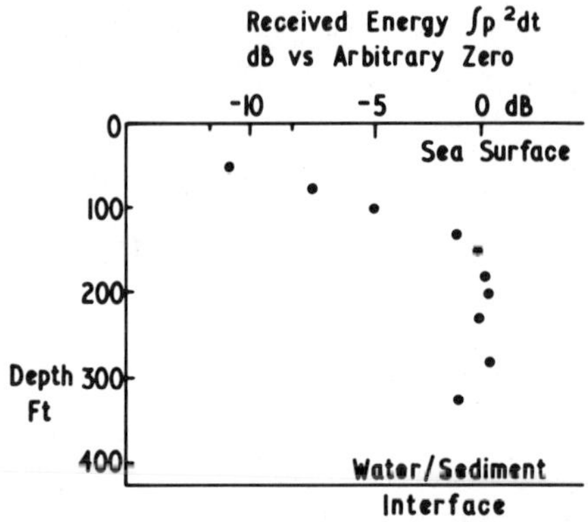

Fig 3 Measured energy vs depth (25 Hz)

Table 2 lists observed values of α at 25 Hz as a function of sediment depth showing a critical dependence on the depth of the overlay.

The theoretical depth above which there is 99% of the energy in the sediment is given in Table 3 for the first mode and assumed bottom parameters. The third column suggests that the dependence can be represented, very approximately by

$$E = e^{-xfd}$$

over a limited range of frequencies. E is the fraction of the energy travelling in the sediment which is above the depth d below the interface; f is the frequency and x is a constant for one type of uniform sediment and the water depth set at a constant value of 100 metres.

When the "rock" layer is introduced, some of the energy will interact with the rock to a degree depending on frequency as determined, roughly, by the above expression. (It is assumed for the purpose of this discussion that the presence of the rock does not alter this distribution.) Plane wave reflection losses at a rock boundary can be high for a wide range of combinations of compressional and shear wave velocities; for example over a rock sea-bed where the shear wave velocity is less than the compressional velocity in water. In general, the losses associated with horizontal propagation over rock are taken to be much greater than over sediment.

Table 2. Attenuation vs Sediment Depth and Mode Number (25 Hz)

Sediment Depth	Metres	Attenuation dB/kyd
60	Mode 1	0.06
20		0.35
10		0.6
0		1.5
Mode		
1		0.06
2		0.22

Table 3. Penetration into Sediment

Frequency Hz	Depth below water/ sediment interface for 99% of energy	From $E = e^{-xfd}$
25	58.5	60 (x=0.003)
63	20.8	23.8
100	13	15

The values of α observed are therefore a function of sediment depth. This is then the basis of the possible method; α is measured and the depth of sediment deduced.

The propagation loss can be expressed as

$$H_r = H_{r_o} + 10 \log \frac{r}{r} + \alpha (r - r_o)$$

where H_r = propagation loss at range r

H_{r_o} = propagation loss at range r_o

α = the loss to the sea-bed per unit range.

VERIFICATION

Some verification of the dependence of α on frequency and depth in the sediment can be obtained by measuring sediment depth below the tracks of propagation loss measurements. Figs 4-7 give comparisons for four areas with an assumed sound velocity in the sediment of 1.7 km/sec.

Fig 4 gives values of excess loss deduced from one-third octave measurements with explosive charges centred on 100 Hz above estimates of sediment depth over a distance of 40 miles. The losses scatter about an α value of approximately 0.05 dB/kyd to 27 miles and 0.4 from 27 to 40 miles. Sediment depths are greater than 10 metres out to ranges of 20 miles or so and zero beyond 28 miles. Thus these observations of α clearly differentiate between regions with greater than 10 metres of sediment and bare rock.

Fig 5 gives similar information but at 25 and 63 Hz to show frequency dependent effects. Here the apparent sediment depth is in the range 0 to 20 metres with a second set of depth measurements along a line 3-4 miles to the south of those terminating at a range of 30 miles. At 63 Hz α changes from 0.32 to 0.025 dB/kyd near the change from bare rock to some sediment cover at 23 miles. A similar though less distinct change in gradient of the excess loss line is apparent at 25 Hz. The depth of sediment is again very variable. The advantages of generating a spectrum of α in, say, the frequency range 25-200 Hz are demonstrated by this since the 25 Hz values alone do not define the change in the sea-bed with sufficient definition.

The full spectrum of α values is of particular significance when there is sediment cover over all of the area. When this is of uniform thickness a knowledge of the full spectrum should give an average depth across the region.

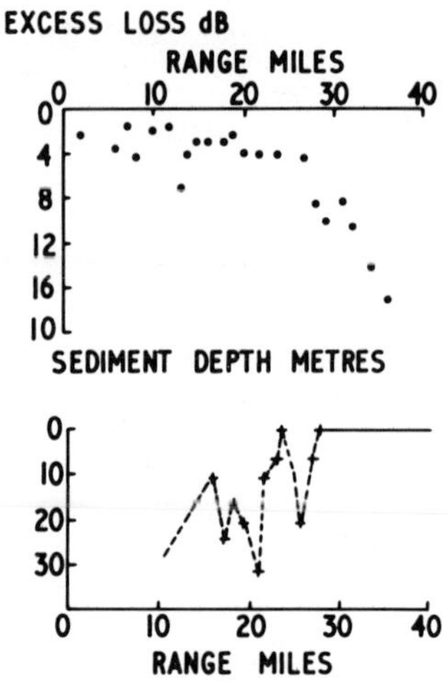

Fig 4 Excess loss/sediment depth 100 Hz (1)

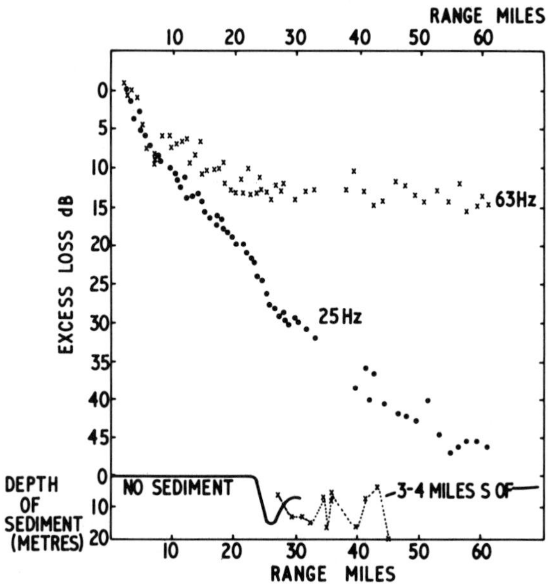

Fig 5 Excess loss/sediment depth 25,63 Hz (2)

Fig 6 compares measurements of excess loss at 25 and 1000 Hz across a change in measured sediment depth from zero to greater than 20 metres. Here the change is apparent at both frequencies with possibly a higher value of α at 25 Hz over depths in excess of 20 metres than at 1000 Hz.

The final example in Fig 7 is across a very variable region with sediment depths between zero and 13 metres. Here the 25 Hz losses are high throughout and the 500 Hz values do not reflect the changes very clearly.

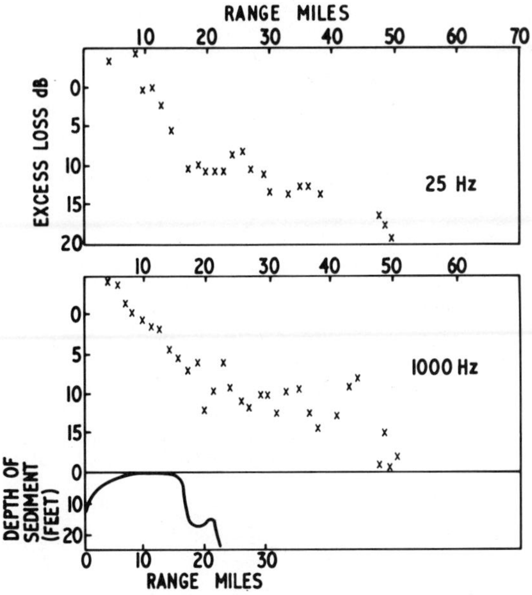

Fig 6 Excess loss/sediment depth 25 1000 Hz (3)

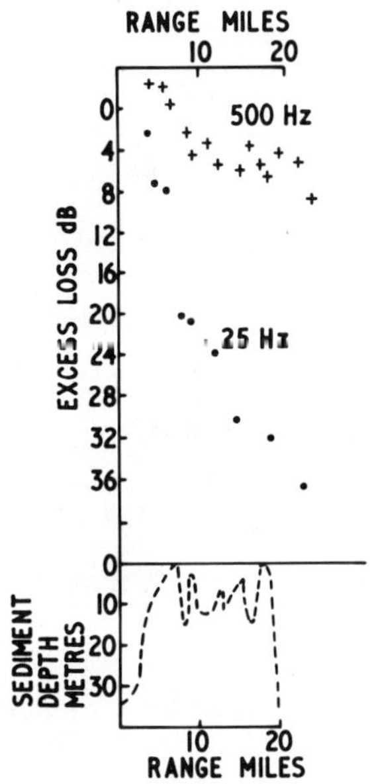

Fig 7 Excess loss/sediment depth (4)

The equation $H_r = H_{r_o} + 10 \log \frac{r}{r_o} + \alpha (r-r_o)$ can be used to predict losses across a sediment depth change; thus in Fig 8 changes from 0.1 to 1.5 dB per kyd at 30 miles range. The theory can be used to give representations of the effect in Figs 4-7.

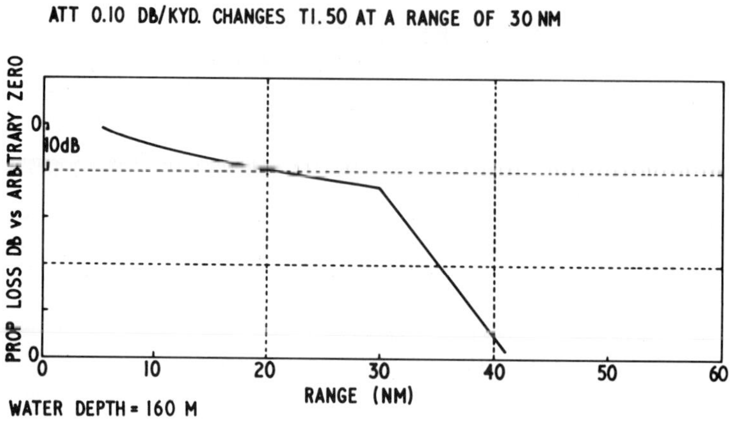

Fig 8 Predicted loss across sediment/rock boundary

A summary representation of the observed values of α is given in Fig 9 for rock, and sediment depths which are "deep"; that is E in the expression $E = e^{-xfd}$ is greater than 99%. The rock values show a considerable spread as is to be expected from the probable changes in type (although these have not been identified). However, at all times they can be clearly differentiated from the deep sediment values which show considerably less spread. This is especially so at 25 Hz.

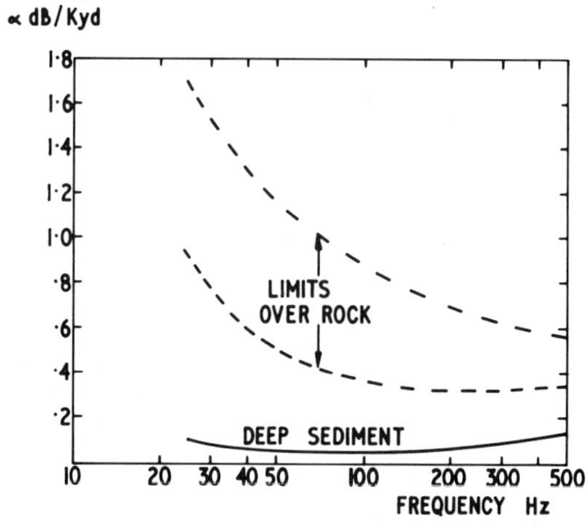

Fig 9 Limits of attenuation over rock and deep sediment

USE IN SURVEYS

A measurement of excess loss over, for example, 5 miles of sea-bed will determine whether the sea-bed is rock or deep sediment since the difference in propagation loss at 25 Hz will be at least 9 dB. Thus patches of rock in an otherwise deep sediment area should be distinguishable. How could this approach be used to survey an area? Two ships could measure the loss across the area along tracks that are at right angles as indicated in Fig 10. Lines along which there is loss over and above that for deep sediment could be recorded and an approximate area

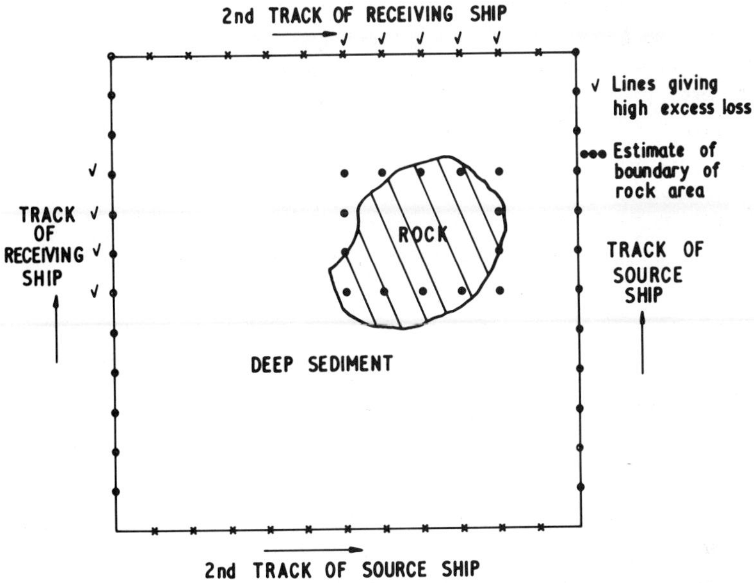

Fig 10 Location of rock region

of high loss noted. The figure suggests a representation for an irregular shaped area of rock which could be deduced in this way. A number of small areas would, of course, be more difficult to deduce.

If the sediment depth is less than "deep" in some region, the loss at some frequencies will exceed the deep sediment values and the spectrum between 25 and 100 Hz will give some clues to its depth. Thus nearly deep sediments will approximate the deep sediment line at 100-200 Hz but separate at 25 Hz. Very shallow sediment will only approach the deep sediment line at the highest frequency.

A number of small areas of rock or shallow sediment will be more difficult to map because of possible ambiguity; however, this is also ture of the standard technique where the tracks need to be closely spaced if some shallow (or deep) regions are not to be missed.

Another technique which might be employed is to lay a transmitting buoy (a long term sonabuoy) in the centre of the area to be surveyed. The ship could make good a course around this at an appropriate range measuring losses as frequently as is required to give the desired resolution.

PARAMETRIC STUDY

The simple model approach used here does not consider: the effect of losses in the sediment; the relationship of loss to water depth; the losses for various combinations of sediment and rock types. A parametric study with the Fast Field Model gives additional information.

CONCLUSION

The proposed inverse technique will provide useful information on sediment depths over long distances and could be used to locate areas for more detailed study with the conventional sounding method.

REFERENCE

1. C H Harrison, P L Cousins. A study of propagation loss dependence on sediment layer thickness using the fast field program. (This Volume)

RANGE AND TIME DEPENDENCE OF ACOUSTIC INTENSITY FLUCTUATIONS

B.J. Uscinski

Department of Applied Mathematics and Theoretical Physics
University of Cambridge, Silver Street
Cambridge CB3 9EW, U.K.

ABSTRACT

When an acoustic signal propagates in the body of the ocean it exhibits intensity fluctuations that vary randomly both in space and time. In these respects it differs from seismic type propagation in the bottom where the intensity fluctuations are time independent and related to a deterministic geological structure. The intensity fluctuation statistics for volume propagation are now fairly well understood. They have a strong dependence on the variance of the random intensity, and this quantity can be predicted theoretically by solutions of the parabolic fourth moment equation. Comparison of theory with data from the COBB and MATE experiments has shown satisfactory agreement. Fourth moment theory has now been extended to deal with the case when the medium contains a deterministic large scale profile.

INTRODUCTION

The propagation of acoustic waves in the ocean bears many similarities to the behaviour of such waves in the solid ocean floor. Just as large scale geological structures tend to guide and focus acoustic waves, so also do the large-scale sound channels that arise in the ocean because of the systematic variation of the acoustic refractive index with depth. In the same way as smaller geological features scatter and attenuate sound, so there are structures in the body of the ocean that do the same to acoustic waves. There is, however, a very important additional factor that operates in the ocean. Geological features are fixed, but ocean structures on all scales are continually changing with time. This means that the intensity fluctuations of an acoustic wave in the ocean have spatial structures that are also strongly time dependent. In recent years our understanding of these time varying intensity fluctuations has much improved. The aim of the present article is to give a clear physical picture of the space-time structure of the fluctuating intensity that arises when a sound wave propagates in a stretch of ocean and encounters the various temperature and salinity inhomogeneities that are always present.

When the frequency of the acoustic wave is less than five hertz the wavelength is over 300m and this is generally larger than the vertical scale of the most frequently encountered irregularities such as internal waves and turbulence. Thus, very low frequency acoustic waves are affected only by large scale features such as the surface, bottom, and overall sound channel, but the body of the ocean appears homogeneous at such large wavelengths. However, for frequencies greater than ten hertz the wavelength becomes smaller than the vertical scale of features like internal waves, and the sound wave is scattered in the body of the ocean. The result of such scattering is to produce random fluctuations of both intensity and phase in the acoustic field. These fluctuations have a complicated structure that varies with both space and time, and they are a dominant characteristic of acoustic signals of frequency greater than a few tens of hertz almost everywhere in the ocean.

Our understanding of such fluctuations has been greatly increased by the ability to model stochastic propagation numerically in a computer. Such simulations allow us to produce pictures of the fluctuating intensity in the ocean. We can see clearly the outstanding features and how they behave in both depth and range. By a suitable extension of these techniques it is also possible to represent their behaviour in time. The importance of numerical simulations cannot be stressed too much. By providing a picture of the full intensity field they allow the eye to detect features and patterns of growth and decay that may not be evident from particular statistical quantities such as spectra or correlation functions. The first part of the paper presents a picture of acoustic intensity fluctuations in the ocean based on the results of numerical simulations. The second part briefly recounts some recent theory that allows us to give a more quantitative description of the main features of the intensity fluctuations. Finally, mention is made of how our increased understanding has helped to interpret experimental results.

1. ACOUSTIC INTENSITY FLUCTUATIONS. SIMULATIONS

As a first step towards building up our picture of acoustic intensity fluctuations of the type outlined in the introduction we shall consider the numerical simulation of a sound wave propagating in a stretch of ocean containing internal waves. Naturally, both surface and bottom do play an important part but these, as well as the large scale sound channel, have long been studied by ocean acousticians and their overall effect, in the absence of statistical fluctuations, is well understood. Here we shall deal mainly with acoustic intensity fluctuations produced by the most common random structure present in the ocean, i.e. internal waves. There are, of course, other types of non-internal wave structures present at different times and places, and some mention of this will be made later in the paper.

(a) Internal Waves

The numerical simulations of propagation are done by sub-dividing the ocean into a set of slabs transverse to the direction of wave propagation and allowing the total scattering effect of such a layer to occur in a plane at the exit of the slab.[1] The ocean thus becomes a set of a large number of phase modulating screens whose space-time autocorrelation function is that of the internal waves summed within the slab. When standard models of the internal wave field are used this has the form[2]

$$\rho(\zeta,\tau) = \sigma(\zeta)\psi(\tau) , \tag{1}$$

where ζ is a separation in the vertical, and τ is a time separation. The spectrum corresponding to (2) is[2,3]

$$F(\beta,\omega) = \frac{C}{(\beta_o^2 + \beta^2)\beta} \frac{(n^2 - \omega^2)^{\frac{1}{2}}}{\omega^3} , \tag{2}$$

where β, ω are the vertical spatial frequency and time frequency respectively, β_o is the inverse of L_V the vertical scale size of the internal waves, n is the buoyancy frequency, and C is a normalizing constant.

(b) Spatial Intensity Fluctuations

The simulation technique allows us to produce a picture of the intensity fluctuations as they appear in space when sound propagates through an internal wave field. Time variation is suppressed for the moment. Each slab or screen introduces an additional phase modulation on the acoustic wavefront with an autocorrelation function $\sigma(\zeta)$. The wave is treated as composed of a spectrum of plane waves with spatial frequencies ν, and each component is given a phase advance of $\exp\{i\nu^2\Delta Z/2k\}$, where ΔZ is the distance to the next screen, and k is the acoustic wavenumber. Repetition of the process produces the complex acoustic field throughout the medium.

The spatial pattern of intensity fluctuations produced when a 4 KHz acoustic wave propagates through an ocean internal wave field at a depth of 1000m is shown on the left in Fig. 1 as given by numerical simulation. Only part of the ocean volume is shown. The dimensions are given, and we note that the most striking features are the strong spikes of intensity that arise and extend over large distances in the direction of sound propagation. Enclosing these high intensity regions by contours produces a new picture, that of the patches where the sound has been concentrated. A contour picture of this type corresponding to the simulation is given on the right in Fig. 1, with the dimension transverse to the direction of propagation added. Since the internal wave irregularities have horizontal and vertical scales that are in the ratio of about ten to one, so also the intensity fluctuations exhibit a similar anisotropy. However, the longitudinal scale of the sound patches in the direction of propagation is even greater.

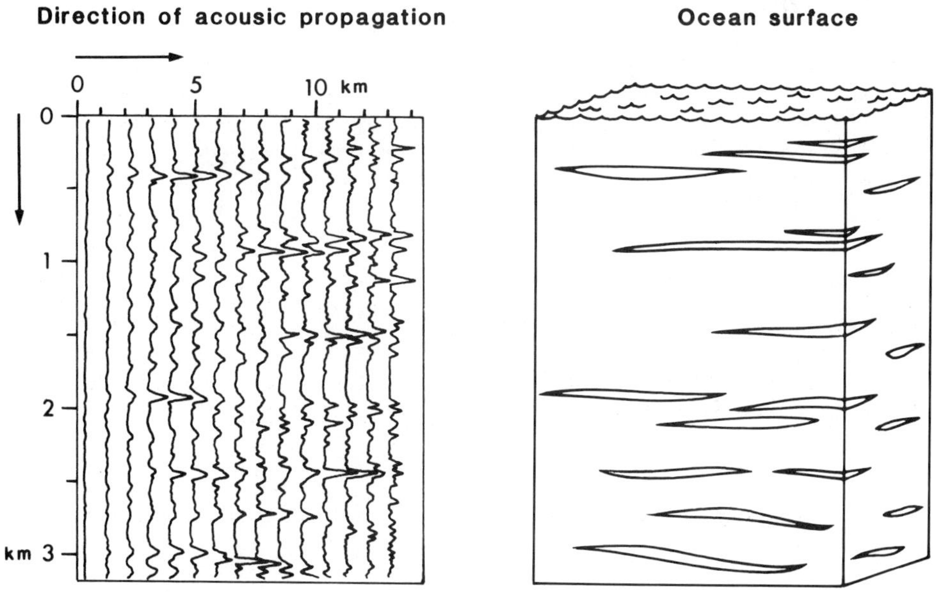

Fig. 1. Simulated acoustic intensity fluctuations and a contour diagram of the high intensity patches.

Simple analysis shows that in the region of well developed intensity fluctuations the vertical, transverse, and longitudinal scales of the high intensity sound patches are respectively

$$\zeta_o = \frac{L_V}{\sqrt{2\Gamma X}} , \quad \eta_o = \frac{L_H}{\sqrt{2\Gamma X}} , \quad \xi_o = \frac{kL_V^2}{2\Gamma X} , \qquad (3)$$

where L_V and L_H are the vertical and horizontal scales of the internal wave irregularities,

$$X = \frac{x}{kL_V^2} \qquad (4)$$

x being the distance in the direction of propagation, and Γ is a parameter characterizing the particular scattering regime involved. In the 4 KHz case illustrated in Fig. 1 the quantities turn out to be, when the range x is about 20 km,

$$\zeta_o \approx 20 \text{ m}$$

$$\eta_o \approx 200 \text{ m} \qquad (5)$$

$$\xi_o \approx 5 \text{ km} .$$

The high intensity patches appear as flat ribbons of concentrated sound, greatly extended in the direction of propagation. They tend to lie at small angles above and below the horizontal and to be separated one from the other by some characteristic distance in the vertical. A typical ribbon is shown in Fig. 2 for the 4 KHz example under discussion.

(c) Time Behaviour of the Intensity Fluctuations

It is a relatively simple matter to introduce time variation into the simulations, and when this is done the ribbons of high intensity tend to undulate irregularly, with some ribbons gradually disappearing and others appearing elsewhere, but they do not drift forward in the direction of sound propagation. An observer at a fixed depth will, from time to time, experience a high acoustic intensity as one of these ribbons passes over him. The time during which he experiences a high intensity depends both upon the rate at which the ribbon undulates and passes him by as well as the life-time of the particular ribbon. The time fluctuating intensity that he observes has a characteristic time scale

$$\tau_o = \tau_{Iw}/\sqrt{\Gamma X} , \qquad (6)$$

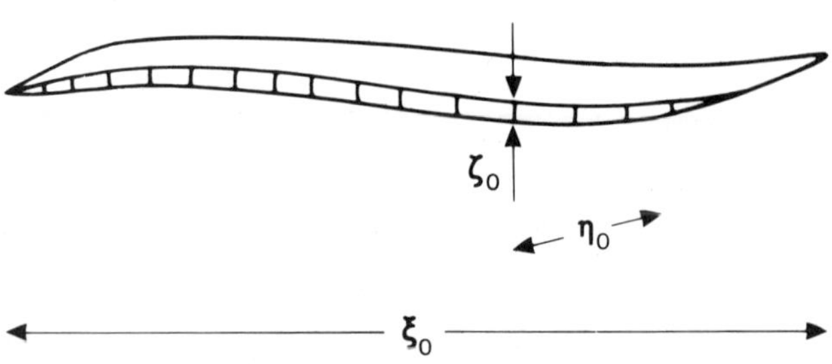

Fig. 2. Typical ribbon of high acoustic intensity.

where τ_{Iw} is the lifetime of internal wave irregularities. In the 4 KHz example

$$\tau_o \approx 10 \text{ mins.} \tag{7}$$

The actual lifetime of a ribbon as seen by itself is, of course, greater than τ_o.

(d) Spatial Frequency Spectrum of Intensity Fluctuations. Transverse Spectrum.

A quantity frequently employed to give a quantitative description of the intensity fluctuations in a randomly inhomogeneous medium is the spectrum of spatial frequencies ν present in the fluctuating intensity measured transverse to the direction of propagation at a fixed range x and time t. We shall denote this spectrum by $\Phi_I(\nu,x)$. In terms of Fig. 1 this would be the spatial frequency content of one of the irregular intensity traces at a fixed range x as the depth z changes.

The transverse spatial frequency spectrum $\Phi_I(\nu,x)$ is important because analytical expressions can be derived for it that depend only on the autocorrelation of the scattering irregularities and the wavelength of the acoustic wave.[4,5] One of these theoretical expressions will be discussed in detail below, and some examples of such spectra will be given. The Fourier transform of Φ_I gives the transverse autocorrelation function of the intensity fluctuations

$$\rho_I(\zeta,x) = \int \Phi_I(\nu,x)e^{i\nu\zeta}d\nu \ , \tag{8}$$

and it is ρ_I that has the characteristic scale ζ_o given by Eq.(3).[6,7]

Longitudinal Spectrum

The longitudinal spectrum of the intensity fluctuations is a less familiar quantity. It allows us to estimate the scale of the intensity fluctuations in the direction of propagation. If we return to Fig. 1 and replot the intensity on a line of constant depth we again obtain a fluctuating quantity and the spectral content of this is the longitudinal spectrum. An approximate but useful expression can be derived for the combined longitudinal-transverse autocorrelation function of the intensity fluctuations based on the fact that the acoustic field can be synthesized as a set or spectrum of plane waves. Thus in an acoustic field of wave-number k a single spatial frequency ν behaves like

$$\exp\{i(\nu z + \frac{\nu^2}{2k} x)\} \ , \tag{9}$$

when the phase reference e^{ikx} has been removed. Making the assumption that $\Phi_I(\nu)$ is slowly changing in the direction of propagation the obvious extension of (8) is

$$\rho_I(\zeta,\xi) = \int \Phi_I(\nu)e^{i(\nu\zeta + \nu^2\xi/4k)}d\nu \ . \tag{10}$$

The longitudinal correlation function is obtained by setting ζ equal to zero. Its characteristic scale is ξ_o as given by Eq.(3).

Two cross-sections of the correlation function $\rho_I(\zeta,\xi)$ obtained from simulated acoustic intensity fluctuations like those in Fig. 1 are shown in Fig. 3. The cross-sections $\rho_I(\zeta,0)$, the vertical transverse autocorrelation, and $\rho_I(0,\xi)$, the longitudinal autocorrelation, are shown in Fig. 3. The large longitudinal scale of the intensity fluctuations

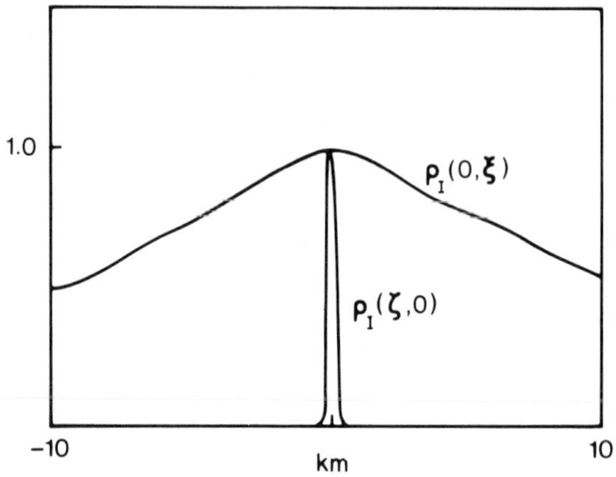

Fig. 3. Vertical and longitudinal autocorrelations of the acoustic intensity fluctuations at 4 KHz as in Fig. 1.

compared with the vertical transverse scale is clear. In this case the ratio of the two scales is of the order of 100:1.

(e) The Scintillation Index

Another quantity frequently used to characterise the intensity fluctuations is the scintillation index, or normalized variance of the random intensity I,

$$S_I^2 = (\langle I^2\rangle - \langle I\rangle^2)/\langle I\rangle^2 \ . \tag{11}$$

The intensity fluctuation spectrum Φ_I is defined so that S_I^2 at any point in the medium is the integral of Φ_I over both time and space frequencies. The scintillation index obtained from simulations like those in Fig. 1, is shown in Fig. 4 as a function of the scaled distance of propagation X. The peak in the S_I^2 curve and its eventual saturation to unity are the most important features of the quantity, and are a reflection of the way the probability distribution of the intensity fluctuations changes as the wave propagates and is multiply scattered.

(f) Inadequacy of the Statistical Quantities Currently in Use

Studying the intensity fluctuations of an acoustic wave by simulation techniques provides a rich supply of information about their structure and behaviour. Many of the features discussed in this section would not be apparent from the spectrum Φ_I and scintillation index S_I^2 alone. For example, the spectrum gives the mean transverse scale of the fluctuations, but does not reveal the existence and behaviour of the high intensity ribbons. However, knowing that they exist from the simulations, we are enabled to use Φ_I to confirm some of their characteristics. The simulations also indicate which new statistical quantities should be sought in order to give a better quantitative description of the features that are so clearly revealed by simulations.

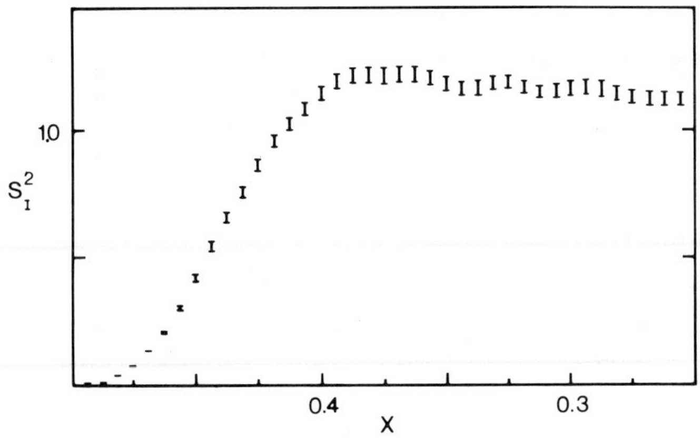

Fig. 4. Normalized variance of the intensity fluctuations (scintillation index) obtained from simulations for $\Gamma = 50$.

2. CONTEMPORARY THEORY OF INTENSITY FLUCTUATIONS

The most useful and extensively studied theoretical expression that describes quantitatively some aspects of the intensity fluctuation structure discussed above is the space-time spectrum of the intensity fluctuations. The most general form of this spectrum currently available is now given without derivation for a point source situated at the coordinate origin,[8]

$$\Phi(\underline{\nu},\omega,x) = \frac{1}{(2\pi)^3}\iint e^{-\int_0^x H(\underline{v}_1(x')\,;\,\underline{v}_2(x');\tau)\,dx'}$$

$$e^{-ik[\underline{v}_2'(x').\underline{v}_1(x)] - i\omega\tau}\, d\underline{v}_1(x)\,d\tau - \delta(\underline{\nu},\omega) \;, \tag{12}$$

where

$$H = 2k^2\mu^2[\rho(0) - \rho(\underline{v}_1) - \rho(\underline{v}_2) + \tfrac{1}{2}\rho(\underline{v}_2 + \underline{v}_1) - \tfrac{1}{2}\rho(\underline{v}_2 - \underline{v}_1)] \;, \tag{13}$$

μ^2 is the variance of the random variations of acoustic refractive index, and $\rho(\underline{v})$ is the transverse autocorrelation function of the irregularities defined as

$$\rho(\underline{v}) = \int_{-\infty}^{\infty} R(\underline{v},\xi)\,d\xi \;, \tag{14}$$

$R(\underline{v},\xi)$ being the three dimensional space-time autocorrelation function of the irregularities of acoustic refractive index. The time dependence of ρ, R has not been written explicitly but they are assumed to be functions of τ also.

Further, $\underline{v}_1$, $\underline{v}_2$ are the solutions of the equations

$$\underline{v}_1''(x') = \underline{v}_1(x')\underline{Q} \;, \tag{15}$$

$$\underline{v}_2''(x') = v_2(x')\underline{Q} + \nu k^{-1}\delta(x-x') \ . \tag{16}$$

where

$$Q = \begin{pmatrix} n_o(\underline{S})_{zz} & n_o(\underline{S})_{zy} \\ & \\ n_o(\underline{S})_{zy} & n_o(\underline{S})_{yy} \end{pmatrix} \tag{17}$$

where $n_o(x,y,z)$ is the smoothly varying part of the refractive index, sometimes referred to as the profile, and $S(x')$ is the ray path from the source to the point of observation given by the ray equation

$$\underline{S}''(x') = \nabla n_o(\underline{S}(x');x') \ . \tag{18}$$

The expression (12) has been derived by several independent methods, such as solving the parabolic fourth moment equation using the procedures of Uscinski,[4] Macaskill,[9] Frankenthal et al.,[10] or by path integral methods.[8]

Achievements of the Theory

Let us consider some examples of the theoretical spectra of (12) and the associated quantitative information that it yields about the structure of intensity fluctuations. For simplicity we take the case when n_o is constant, i.e. when the acoustic refractive index has no systematic profile. Since acoustic scattering by internal waves in the ocean occurs chiefly in the vertical plane Eq.(12) can be written for one transverse dimension only and becomes

$$\Phi(\nu,\omega,\tau) = \frac{1}{(2\pi)^2}\int_{-\infty}^{\infty} e^{-\int_0^X H(\frac{vx'}{X};\ \nu k^{-1}(x-x');\tau)dx'} e^{-i(v\nu+\omega\tau)} dv d\tau - \delta(\nu,\omega) \ . \tag{19}$$

The spatial frequency spectra corresponding to the simulations of Fig. 1 but obtained by evaluating Eq.(19) with the appropriate scattering cross-

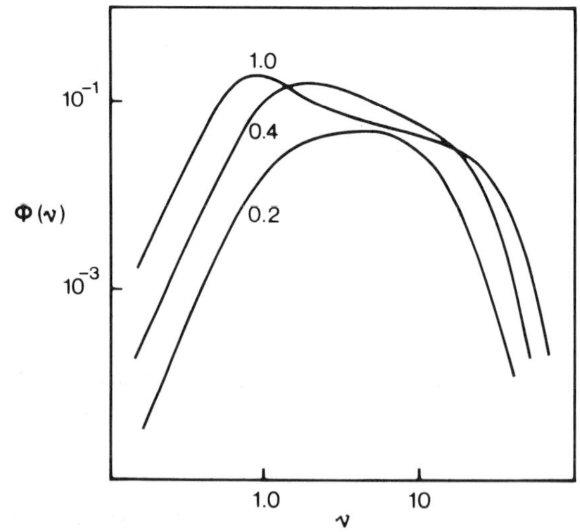

Fig. 5. Spatial frequency spectra of intensity fluctuations for $\Gamma = 1000$ at scaled distances $X = 0.2,\ 0.4,\ 1.0$.

section $\rho(v_1,\tau)$ are shown in Fig. 5. Theory gives the intensity fluctuation spectra at all distances in the scattering medium and is not restricted to asymptotic cases of small or large ranges. We note, in passing, that the theoretical spectra given by Eq.(19) agree very well with the corresponding spectra obtained from numerical simulations such as those in Fig. 1.

The theoretical spectra are defined in such a way that their integral over spatial and time frequencies gives the scintillation index

$$S_I^2 = (\langle I^2\rangle - \langle I\rangle^2)/\langle I\rangle^2 \tag{20}$$

Some typical examples of S_I^2 as a function of the scaled distance in the medium are shown in Fig. 6. These examples are chosen to illustrate the effect of varying the autocorrelation function of the scattering medium and of changing Γ the parameter characterizing the different scattering regimes.

Inspection of these variance or scintillation index curves reveals some interesting features not immediately apparent from the more complicated intensity plots such as in Fig. 1. The most important of these is the peak, which can be much larger than unity, in the variance and the eventual saturation to unity. As Γ becomes larger the peak becomes higher and occurs closer to the origin. The ability to predict the presence of this peak and give its height and position correctly is one of the most important achievements of the latest intensity fluctuation theory. The agreement between theory and the results of numerical simulations is very satisfactory. Finally, we note that the variance curves reflect the changes occurring in the probability distribution of intensity fluctuations as distance in the medium increases.

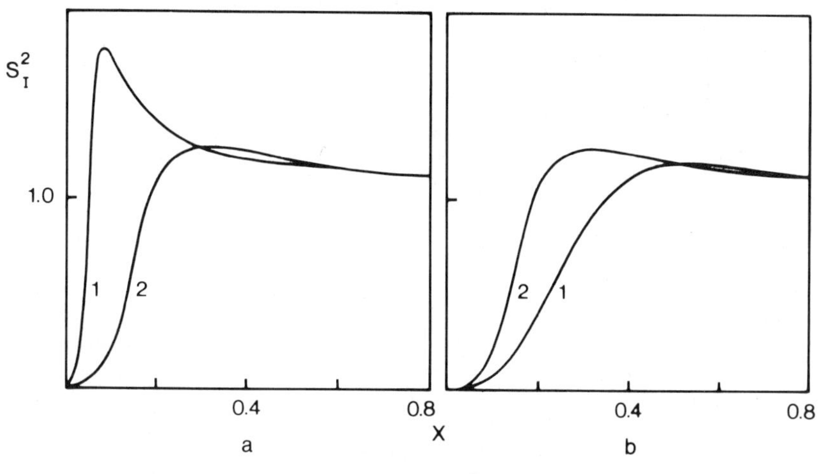

Fig. 6. Scintillation index as a function of X for
(a) medium with Gaussian autocorrelation function
1 - Γ = 1000, 2 - Γ = 50.
(b) medium with Γ = 50 and
1 - Power law autocorrelation function,
2 - Gaussian autocorrelation function

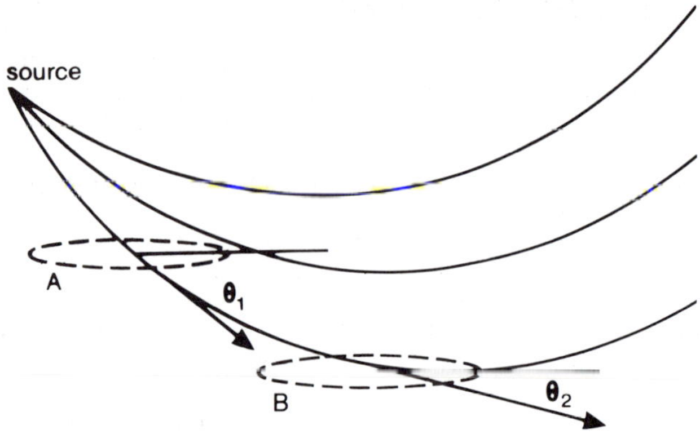

Fig. 7. Curved ray paths due to the linear profile(Eq.21).

Medium with a Refractive Index Profile

The theory outlined above can also deal with some aspects of the complications introduced when the medium has a refractive index profile n_o. The most obvious result of the presence of a profile is that the ray paths in the absence of scattering irregularities are now curves given by Eq.(18). This affects the form of v_1 and v_2 through Q , (17), and hence the form of the scattering cross-section, $\rho(v_1)$, (14) as well.[8] The appearance of the ray paths in the case of the linear profile

$$n_o = a_o + az \tag{21}$$

is shown in Fig. 7. The principal effect of curved ray paths occurs when the scattering irregularities are elongated in the direction of propagation. In this case the ray encounters different irregularities at different angles and the scattering cross-section $\rho(v)$ changes with this angle of incidence. The main drawback of the theoretical spectrum (12) is that this variation of the scattering cross-section with the ray angle does not emerge directly from any of the derivations so far produced. Although v_1 itself depends on n_o it simply describes the separation of adjacent rays due to geometrical spreading in the profile and does not allow for the dependence on angle of attack mentioned above.

This fact is clearly seen if we note that the spectrum (19) for a medium with no profile is the same as that obtained from (12) for a medium with the linear profile (21), since in both cases $Q = 0$. The separation of the rays v is just that due to cylindrical spreading and remains the same in both cases. The change in ρ with angle of attack is not present. This result is due to the fact that in the derivation the volume correlation of the scattering irregularities R has been approximated as depending only on the difference between angles of incidence and scatter and not on them separately, as should be the case. Of course the ad hoc remedy to this deficiency is simply to replace $\rho(v)$ by its value appropriate to the position on the ray path that it occupies.[11] This procedure will probably serve well enough when the ray angle does not change appreciably inside one irregularity (A of Fig. 7) but will fail when this condition is violated, (B of Fig. 7). This last deficiency is not easily remedied. It is not even physically clear what the effect of such a change in the ray angle will be and the problem requires more and detailed study.

Time Dependence of Intensity Fluctuations

The intensity spectrum (12), when integrated over all spatial frequencies is a function of the time frequency ω only. It is the spectrum of intensity fluctuations at a point in the fixed coordinate system. As remarked in section 1 this time fluctuation is due both to the undulation of the high intensity ribbons passing over the point of observation, as well as the intrinsic time decay of a particular ribbon. It is this time variation that is observed in many ocean acoustic experiments and so the theoretical spectrum (12) is useful for comparing with experimental data.[6,7] However, it does not answer the interesting question of the intrinsic life time of a high intensity ribbon which is clearly much longer than the time the ribbon takes to pass a fixed point.

The longer life of the high intensity ribbons is obvious from the numerical simulations. It is a good example of how such simulations provide a greater physical insight into the structure of intensity fluctuations than is immediately available from the usual theoretical expressions. Thus, in addition to providing a means of testing theory the simulations also indicate what new theory is required.

3. THE PROBABILITY DISTRIBUTION OF INTENSITY FLUCTUATIONS

Some limitations of the existing theory have just been pointed out and it is clear that new expressions will need to be derived in order to deal satisfactorily with the variety of interesting features that the simulations reveal. Some progress has already been made in understanding the probability distribution of the intensity fluctuations, but the picture is still far from being complete.

The theoretical spectrum discussed above is related to only one moment of the intensity fluctuations, namely the second. When integrated over all time and space frequencies it gives the normalized second moment at one point. The behaviour of the second moment with increasing distance in the medium is shown in Fig. 6. On comparing Fig. 6 with the corresponding distance behaviour of one realization of the intensity fluctuations (Fig. 1), we see that the change in the second moment is due to a systematic variation in the nature of the fluctuations with distance. At short ranges the peaks are small, they then become higher but remain relatively few in number, while at large ranges they are not so high but become densely packed as the sound field is broken up by the medium.

Clearly the probability distribution of the intensity fluctuations changes with distance of propagation. An example of such a changing probability distribution, obtained by taking ensemble averages of many numerical simulations like those in Fig. 1, is shown in Fig. 8.[12] So far it has not proved possible to derive, on a strictly theoretical basis, a single expression giving the intensity probability distribution at all distances in the medium. At short distances the log-normal distribution,[6] $P_{\ell n}(I)$,

$$\langle I\rangle P_{\ell n}(I) = \exp\{-(\ell n\, s\sqrt{y})^2/2\,\ell n\, y\}/s\sqrt{2\pi \ell n\, y}\,, \qquad (22)$$

where

$$s = I/\langle I\rangle \;,\; y = 2 + S_I^2 \;,$$

has a good theoretical justification and fits the data satisfactorily, while the same can be said of the K distribution,[6] $P_K(I)$,

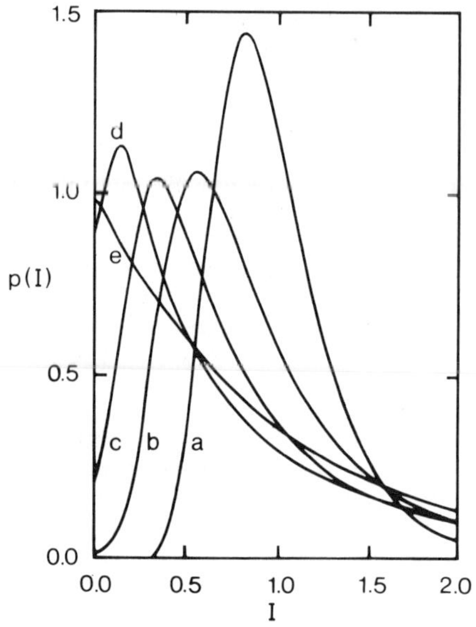

Fig. 8. Probability distribution of intensity fluctuations at increasing distances in the medium changing from (a) log-normal to (e) K-distribution

$$\langle I\rangle P_K(I) = \frac{2}{\Gamma(y)}\, y^{\frac{y+1}{2}}\, s^{\frac{y-1}{2}}\, K_{y-1}(2\sqrt{sy}) \;, \tag{23}$$

where

$$s = I/\langle I\rangle \;,\; y = 2/S_I^2 \;,$$

for ranges beyond the peak in the intensity variance curve of Fig. 6. The K distribution for intensity takes the form of an exponential distribution at very large distances in the medium, which reflects the fact that the acoustic field becomes completely randomised at large enough ranges and its amplitude is described by that of a two-dimensional random walk.

The probability distribution of intensity contains much more information than just the second moment, indeed it contains all the moments. However, it is interesting to note that the two distributions (22) and (23) are functions of the second moment alone when the intensity is scaled in terms of its mean value $\langle I\rangle$. This implies that if the moments are to be the parameters of the distribution then any general distribution that describes the whole range of scattering must depend on at least one more moment, perhaps $\langle I^3\rangle$. This becomes obvious when we consider that the same value of $\langle I^2\rangle$ can occur at two different ranges described by two different distributions in the same medium. For example, Fig. 6 shows that a value of $\langle I^2\rangle$ equal to 2, i.e. S_I^2 equal to unity, occurs on any of the curves both at a large range, $X \gg 1$, and at some short range, $X < 1$. The two quite different distributions (22), (23) are appropriate to these regions and so cannot depend on $\langle I^2\rangle$ alone.

Another useful way of describing the statistical properties of the intensity fluctuations is to display the higher moments as functions of the

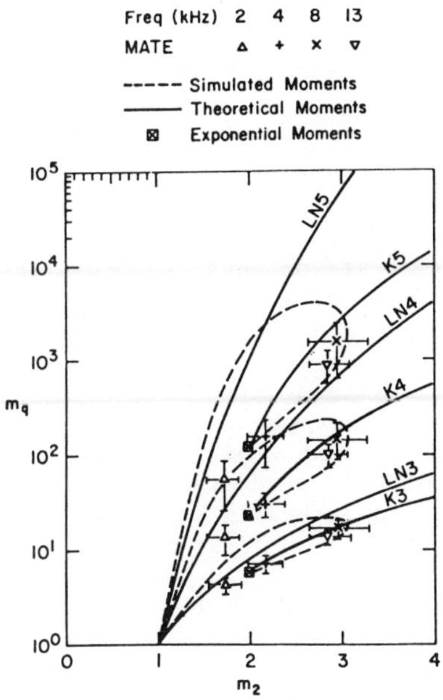

Fig. 9. Dependence of the intensity moments m_q on the second moment m_2.

second as in Fig. 9.[6,13] The intensities are normalized in terms of the mean $\langle I \rangle$ and so this does not appear. The behaviour of the moments

$$m_q = \langle I^q \rangle / \langle I \rangle$$

obtained by taking the ensemble of many numerical simulations is shown by the broken lines in Fig. 9 for $q = 3,4,5$. Some experimental results are also added and the behaviour of the corresponding moments, calculated theoretically from the log-normal and K distributions, is shown as well. Once again it is obvious that the m_q do not have a unique dependence on the second moment m_2, and so the general moment must be a function of at least one more parameter.[14] It can, in fact, be stated that the general probability distribution of intensity is a function of just two, and no more, fundamental parameters. These are Γ the parameter characterising the scattering regime, and X the scaled distance of propagation. This statement becomes obvious when it is realised that the parabolic equation for the general moment of intensity can be written in terms of these two quantities. Thus if all the moments depend on just Γ and X then so too does the probability distribution.

Finally it should be noted that the distribution discussed above $P(I)$ is the one point distribution and so has limitations on the description it can give of the intensity structure. For example, it cannot answer the question as to the positions of the high intensity ribbons relative to each other. For this purpose we would need at least the two point distribution $P(I_1,I_2)$, where I_1 and I_2 are the intensities at two different places. Indeed there are many interesting features of the intensity structures evident in the simulations that will necessitate the use of various higher order probability distributions in order to describe them quantitatively.

4. RESULTS OF OCEAN ACOUSTIC EXPERIMENTS

In the preceding discussion the intuition available from numerical simulations has played a dominating role. Surely the question should be asked, why rely on simulations, why not see what happens in the real ocean? Of course this is our ultimate aim and experiments are being done; some have been analysed and we should conclude this paper by talking about real experiments. However, the merits and function of simulations are obvious. Firstly they provide insight at low cost, secondly they are easily controlled, and thirdly they offer a sort of clean experiment that allows us to test our theory under known propagation conditions.

Of the ocean experiments that have yielded the most reliable and useful data concerning the time dependence of the acoustic intensity fluctuations, those carried out by Prof. T.E. Ewart at the COBB Seamount must certainly be mentioned.[6,7] Transmissions at acoustic frequencies of 2, 4, 8, 12 KHz between fixed towers 18 km apart at a depth of 1000 m yielded the received intensity as a function of time over several days. The features revealed by simulations, in particular the very high peaks corresponding to the ribbons, are present in the experimental intensity.

The time spectra of the fluctuating intensities observed in the MATE experiment at Cobb Seamount are shown in Fig. 10 for the four acoustic frequencies employed.[15] The curves are offset by a decade for purposes of clarity. On the same figure the theoretical time spectra predicted by Eq.(12) are also shown. They were calculated using a scattering cross section $\rho(v)$ based on the presence of the internal waves as well as a non-internal wave component revealed by the environmental measurements. The agreement between theory and experiment is good and the shortfall at the high frequency end of the curves is much reduced when a somewhat refined version of the theory is used.

The spatial structure of the intensity fluctuations, which could not be observed in MATE, was the focus of the recent Aiwex Acoustic Transmission

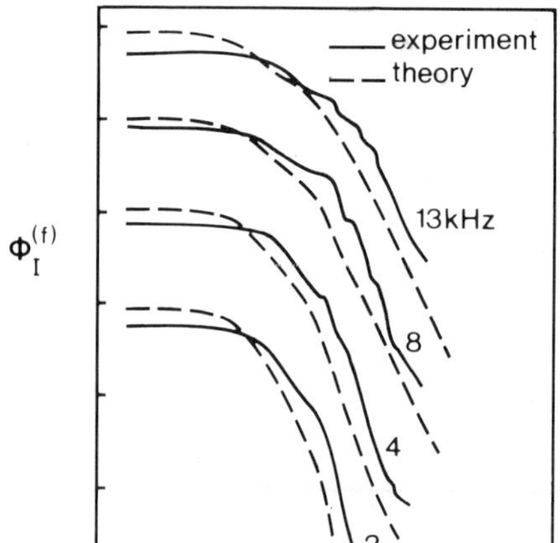

Fig. 10. Experimental intensity fluctuation spectra from MATE compared with theory. The time frequencies on the abscissa are in Hz.

Experiment (AATE) recently performed under the Arctic ice cap by Ewart and Uscinski. In AATE a vertical string of four receiving transducers was winched up and down periodically so as to give continuous reception over a depth of 200 m in the water column. The transmission was made at a range of 5 km at the four acoustic frequencies of 2, 4, 8 and 16 KHz. Processing of the data is now in progress and should provide experimental confirmation of the high intensity regions that we expect to encounter as we cross the ribbons with increasing depth.

ACKNOWLEDGEMENT

This work was supported by the Ministry of Defence, Procurement Executive.

REFERENCES

1. C. Macaskill and T.E. Ewart, Computer simulation of two-dimensional random wave propagation, IMA Journal Appl. Math. 33:1 (1984).
2. B.J. Uscinski, Parabolic moment equations and acoustic propagation through internal waves, Proc. R. Soc. Lond. A372:117 (1980).
3. W.H. Munk and F. Zachariasen, Sound propagation through a fluctuating stratified ocean, J. Acoust. Soc. Am. 59:819 (1976).
4. B.J. Uscinski, Intensity fluctuations in a multiply scattering medium. Solution of the fourth moment equation, Proc. R. Soc. Lond. A380: 137 (1982).
5. B.J. Uscinski, C. Macaskill and T.E. Ewart, Intensity fluctuations. Part I: Theory, J. Acoust. Soc. Am. 74:1474 (1983).
6. T.E. Ewart, C. Macaskill and B.J. Uscinski, Intensity fluctuations. Part II: Comparison with the Cobb Experiment, J. Acoust. Soc. Am. 74:1484 (1983).
7. T.E. Ewart and S. Reynolds, The mid-ocean acoustic transmission experiment, MATE., J. Acoust. Soc. Am. 75:785 (1984).
8. B.J. Uscinski, C. Macaskill and M. Spivack, Path integrals for wave intensity fluctuations in random media, J. Sound and Vib. (in press).
9. C. Macaskill, An improved solution of the fourth moment equation for intensity fluctuations, Proc. R. Soc. Lond., A386:461 (1983).
10. S. Frankenthal, A.M. Whitman and M.J. Beran, Two scale solutions for intensity fluctuations in strong scattering, J. Opt. Soc. Am. 1:585 (1984).
11. R. Dashen, Path integrals for waves in random media, J. Math. Phys. 20:894 (1979).
12. D. Roberts, Ph.D. Thesis, Cambridge University (1985).
13. C. Macaskill and T.E. Ewart, The probability distribution of intensity for acoustic propagation in a randomly varying ocean, J. Acoust. Soc. Am. 76:1466 (1984).
14. R. Dashen, The distribution of intensity in a multiply scattering medium, Opt. Lett., 10:110 (1984).
15. T.E. Ewart, C. Macaskill and B.J. Uscinski, The MATE acoustic frequency cross-correlations of intensity, J. Acoust. Soc. Am. (in press).

SHALLOW WATER TOMOGRAPHY

H. A. DeFerrari, D. S. Ko and C. L. Monjo

Applied Marine Physics Division
Rosenstiel School of Marine and Atmospheric Science
University of Miami
Miami, Florida 33149-1098

I. INTRODUCTION

Ocean acoustic tomography has been successfully demonstrated in deep ocean areas for transmission by wholly refracted paths. Other areas of the ocean such as straits, gulfs and the continental slope and rise are characterized as acoustically shallow having bottom limited transmission with sound paths that makes multiple interactions with the ocean floor. Since the vertical excursion of ray paths is confined in shallow water the differences in arrival time, amongst various ray paths, is usually small resulting in received pulses that overlap in time and cannot be cleanly separated. The forward scatter from bottom interactions transmits down the channel and is also received and serves to broaden pulses and produce a reverberation background which further complicates the tomographic interpretation of measured and channel pulse responses.

Yet, many shallow areas are of great oceanographic interest and potentially important applications of tomography. One such area is the Straits of Florida which contains the energetic Florida current that serves as the origin of the Gulf Stream. Recently, climate studies have renewed interest in the long-term variability of the Florida current. An acoustic monitoring approach has advantages over conventional oceanographic instrumentation in that spatial averages of temperature and current are obtained and the instruments need not be located in the areas of high velocity, but may be placed near the edges of the channel.

Reported here are the results of reciprocal transmission experiments in the Florida Straits to investigate the feasibility of shallow water tomography. Two 3-point reciprocal experiments were conducted with triangular geometries at ranges of approximately 25 and 45 km. Pseudo-random sequence signals were transmitted from a source having a 100 Hz bandwidth with a center frequency of 460 Hz. Thus, the pulse response of the propagation channel was measured with the resolution of 9 ms for reciprocal transmission between each source and receiver pair. The measurements were repeated every 12 minutes for a period of 30 days. The acoustic measurements accompanied with current and temperature measurements from moored instruments along the path of propagation.

Herein, both ray theory and split step parabolic equation method are used to analyze and invert experimental results to give depth averaged current and temperature measurements which are then compared with the direct measurements. The inversion problem for deep ocean tomography usually relies on ray models to predict the path of sound through the ocean and the arrival times of signals traveling along those paths. In shallow water ray theory must be used with caution. Complications involving the irregularities of the bottom, diffraction, phase interference, frequent caustics, leakage of energy into and out of the surface ducts and other effects are problematic for ray theory but can be handled correctly by the parabolic equation method. In order to interpret data, highly idealized ray models with smooth range dependent bottoms and sound speed profiles were used to estimate pulse responses of the channel and paths of eigenrays. The PE method is used to evaluate the effects of bottom rougness on paths of eigenrays and the order of arrival.

II. EXPERIMENTAL GEOMETRY

At least three source receiver pairs are necessary to resolve north-south, east-west components of current. By using an equilateral triangular array it is possible to cancel some errors which may result from clock drift or mooring motion. The first step is to measure the travel time along identified paths between source and receiver. We adopt the convention that t_{12} is a travel time for transmission from mooring 2 to 1.

The tomographic approach with reciprocal transmission requires measurements of travel time forward and back between a source-receiver pair. Three sound-receiver units are necessary to resolve north-south (v) and east-west (u) components of current. If the units are moored on the apexes of an equilateral triangle oriented 1 south, 2 north and 3 east, the contribution to the travel time from temperature and current along the sound path from unit 2 to 1 can be sorted out as follows:

$$t_{12} = t_{T12} + \varepsilon_{r12} + t_{V12} - \varepsilon_{N12}, \qquad (1)$$

and from unit 1 to unit 2

$$t_{21} = t_{T12} + \varepsilon_{R12} - t_{C12} + \varepsilon_{N12} \qquad (2)$$

where t_T the travel time associated with reciprocating factors of sound-speed along the path (mostly temperature, but also salinity and pressure).

ε_R = travel time errors that reciprocate (e.g., mooring displacement)

t_V = non-reciprocating effect of current assuming flow to the north is positive

$\varepsilon_{N12} = \varepsilon_{N2} - \varepsilon_{N1}$, non-reciprocating error (e.g. clock drift.

If clock of unit 2 becomes fast the signal s at unit 1 appears to arrive early; (-) minus sign in eq. 1; whereas the signal transmitted from unit 1 appears late to the fast clock of unit 2; (+) plus sign in eq. 2. Clock errors are indistinguishable from current effects.

For this experiment, short, very taut moorings were used so that mooring displacements are (reciprocating erros) negligible. The contribution from temperature and current can be computed by adding and subtracting eqs. 1 and 2.

$$t_{T12} = 1/2\ (t_{12} + t_{21}) \qquad (3)$$

$$t_{V12} = 1/2\ (t_{12} - t_{21}) + \varepsilon_{N12} \qquad (4)$$

The average northern component fow $\overline{V}$ passing between unit 3 and the north-south line of units 1 and 2 can be computed by

$$t_{\overline{V}} = 1/2\ (t_{13} - t_{31} + t_{32} - t_{23}) - \varepsilon_{N12} \tag{5}$$

Here the east-west components cancels as does the clock drift of unit 3. The vorticity can also be computed as

$$\text{vorticity} = 1/2\ (t_{12} - t_{21} + t_{23} - t_{32} + t_{31} - t_{13} \tag{6}$$

where all clock drifts cancel.

The following analysis and discussion is concerned with the geometry of reciprocal transmission path 1,2 the north-south leg of array 1 as shown in Figure 1. Sound speed profiles were computed from XBT casts during the first day of the experiment. At the current meter mooring locations, current speed and direction, and temperature were measured at three depths (courtesy of Fritz Schott). Model predictions of channel pulse response were computed for the three receiver depths shown. The results and conclusion are essentially similar for paths 2-3 and 3-1.

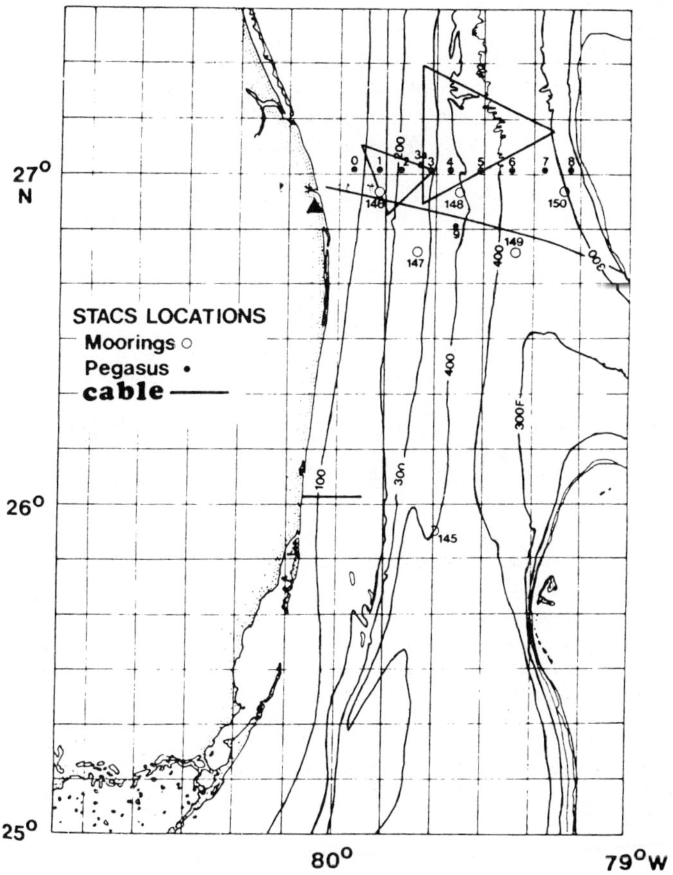

Fig. 1 Location of experiment.

III. MODELS OF PROPAGATION

1. Ray Models

Ray models are used to predict the pulse response of the channel and to identify paths for inversions. The mathematical and numerical methods of the Medusa Model[1] for range-dependent ocean environments have been employed here with some minor variation. Bi-cubic splines are used for the bathymetry. These splines are continuous in first and second derivatives, and therefore eliminate false caustics and shadow zones. The differential equations of rays are solved numerically with a Taylor series expansion and the Runge-Kutta method. Spreading loss, travel times and ray path geometry are computed for eigenrays.

Range-dependent sound speed profile, Figure 2, were computed from expandable bathythermographs (XBT) taken during the first day of the experiment. Paths of eigenrays illustrated in Figure 3 and the arrival

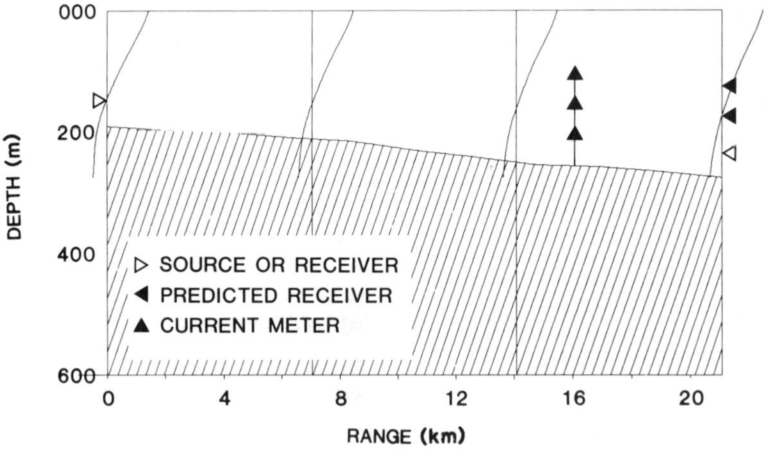

Fig. 2 Range-dependent sound speed profiles and experimental geometry.

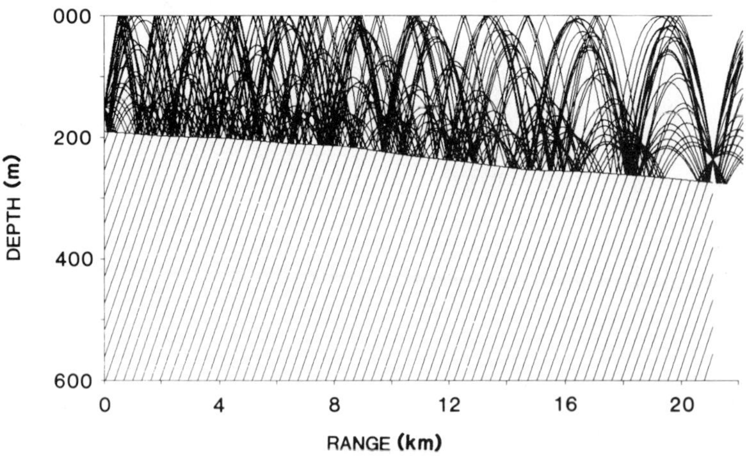

Fig. 3 Paths of eigenrays.

times of eigenrays are plotted in Figure 4a. The earliest arrivals are RBR paths which have upper turning points near the surface, or they are mixed paths which have made one surface interaction and then become trapped as RBR. The latst arrival of the RBR group travel deep in the channel and have upper turning points at the depth of the transmitter and receiver. Arriving later, are the mixed paths which make two or more surface reflections, and still later the SRBR group. The SRBR groups contain four arrivals and each group makes one more surface bounce than the earlier group.

There are two factors which determine the travel time and order of arrival for eigenrays. One is the length of path travelled between source and receiver and the second is the sound speed along that path. Flat rays have upper turning points near the bottom that is deep in the channel, whereas steep rays have upper turning points near the surface. Flat rays travel a shorter path length, but at a lower sound speed (assuming downward refracting profiles) compared with steep rays which travel in faster water but over a greater path length. If the sound speed gradient is constant (i.e., linear sound speed profile) as is nearly the case in the Florida Straits, the sound speed effect dominates and steep rays arrive earlier than flat. On the other hand, if the sound speed gradient decreases with depth being steep near the surface and gradual near the bottom then the path length effect can dominate, and flat rays arrive earlier than steep.

Ray theory predicts the amplitude, travel time, and phase of a signal traveling along paths between source and receiver. The model predictions can be used to formulate a system response function for the channel;

$$H(\omega) = \sum_{i=1}^{n} A_i \exp \{-j[\omega T_i + \operatorname{sgn}(\omega)\theta_i\} \tag{7}$$

where A = Amplitude

T = Travel time

and θ = Cumulative phase shift from multiple reflections.

The phase term θ is independent of frequency and can produce significant phase distortion of a pulse signal as well as an apparent shift in arrival time. The difference in energy between an undistorted 4-cycle pulse (any carrier frequency) and the phase distorted version is less than 20% for the worst case of $\theta = n\pi/1.$[2] Phase shift of $\theta = n\pi$ produce no distortion. If pulse arriving form multipaths overlap in time, as is the case under consideration, the pulse distortion will be insignificant compared with interference effects and can be neglected.

The average phase shift is retained if eq. (7) is written as

$$H(\omega = \sum_{i=1}^{n} A_i \exp \{-j[\omega T_i + \theta_i]\}, \tag{8}$$

and the corresponding channel impulse response function is computed by Fourier transform to be

$$h(t) = \sum_{i=1}^{n} A_i \exp \{-j\theta_i\} \delta(t - T_i) \tag{9}$$

The pulse response of the channel is computed by summing time delayed and phase shift replica of the transmitter pulse as represented by the convolution integral

$$p_o(t) = \int h(T - t)\ p_i(t)\ dt \tag{10}$$

Alternately, the complex amplitude spectrum of the received signal is computed using eq. (8) as

$$p_o(\omega = H(\omega)\ p_I(\omega) \tag{11}$$

Where $p_I(\omega$ is the spectrum of the transmitted signal. The pulse response $p_o(t)$ is then computed by Fourier transformer.

Figure 4b illustrates a model computation of the pulse responses from the arrival times shown in Figure 4a. Spreading and bottom loss are not included. The shape of the pulse response differs only slightly from the impulse response of the channel if losses are not included. There is a slight peak at about 6 dB in magnitude for the concentration of late arriving RBR paths.

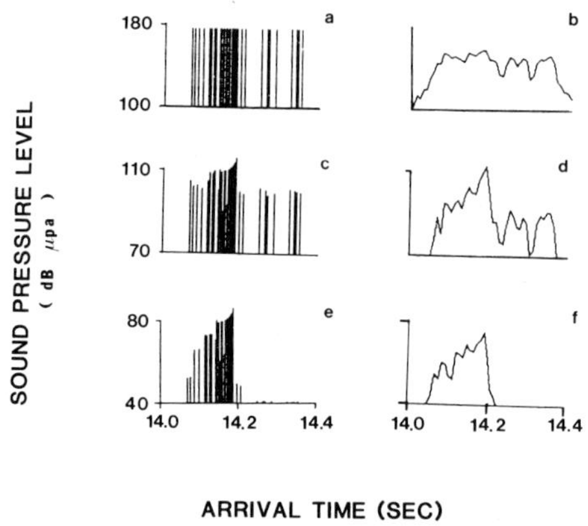

Fig. 4 Eigenray arrival times and pulse responses (see text).

The pulse response changes drastically when the effects of geometric spreading are included, Figures 4c and 4d. The mixed and SRBR paths suffer spreading loss in excess of 20 dB. The RBR paths undergo less spreading and tend to refocus at regular intervals in range. Steeper RBRs have greater loss than the flat rays which further accentuate the peak and the pulse response pattern associated with ray paths having upper turning points above and near the source receiver depth. There are a large number of such paths that have nearly equal travel time and are strongly refocused. The resulting pulse has a strong peak associated with the later arriving and more intense flat RBR paths. The lagging edge of the peak cuts off sharply owing to ray geometry which prohibits arrivals from rays which have turning points below the source receiver depth. The leading edge is more gradual and comes about from the combined effect of fewer overlapping arrivals with more losses from geometric spreading.

The inclusion of bottom loss further accentuates the pulse peak, Figure 4e and 4f. A McKenzie[3] bottom model gives losses as a function of grazing angle with the steeper mixed and SRBR rays suffering the greater losses. Also a nominal 1.5 dB loss per surface bounce is included. The net result is a pulse pattern which is entirely dominated by RBR paths. There is no indication of later SRBR paths which suffer losses greater than 30 dB below the peak RBRs. Thus, the model predicts pulse reesposes dominated by RBR arrivals, with mixed and SRBR paths making negligibly small contribution. This prediction agrees closely with all measured pulse responses.

The inversion process whereby observations of travel time can be interpreted as the average temperature and currents requires that the ray path through the ocean be known. The usual tomographic approach requires that arrivals from identifiable paths be clearly separable from other arrivals so that an arrival time associated with the path is directly measured. The predicted pulse response pattern shown in Figure 4f comes about from the super position of numerous multipaths which overlap in time and cannot be separated. However, the relationship between arrival time and heigh of the ray above the bottom is simple and unambiguous. Models are used to compute average height of the RBR ray lobes as a function of arrival time and results are plotted in Figures 5a and 5b. Figure 5a includes contributions from all eigenrays and curve b only those arrivals with signal levels no greater than 30 dB below peak level. The resulting nearly linear relationship between arrival time and ray height makes possible straightforward inversions.

Additional information about the depth dependence of current and temperature may be possible by employing a vertical array of hydrophones. A consistent feature of the pulse response is the sharp peak associated with the later arriving RBR paths. The average height above the bottom of the rays forming the late peak is determined by the receiver height. The purpose of the array is not to separate arrivals but to select average heights of the ray thus controlling depth of aevraging of current and temperature.

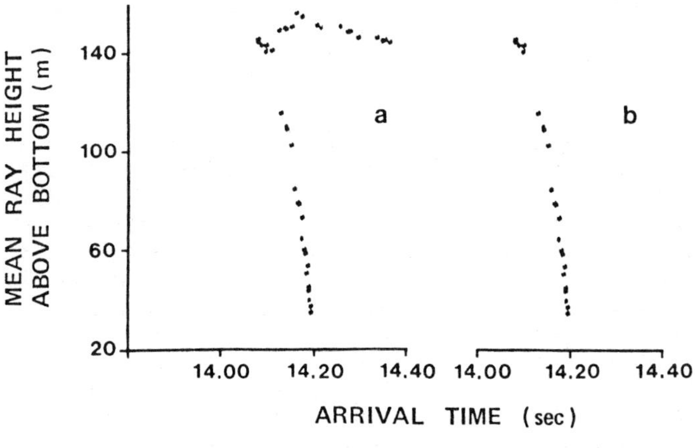

Fig. 5 Average height of the RBR ray lobes as a function of arrival time.

IV. PE MODEL PREDICTIONS

The parabolic equation method employing the split step Fourier alogorithm was used to study the effects of variability of bottom bathymetry on the paths of sound through the ocean. Color displays of tranmsmission loss predicted by the PE method exhibit ray-like patterns for a smooth representation of the bathymetry for a section between Miami and Bimini (the colored illustrations will be presented during the talk but not included in the written proceedings). Range dependence stochastic variability is then introduced to the bottom and the color contours of ray become blurred as the rms level of the fluctuations is increased. However, distinctive clear rays are observed for rms levels less than 5 m. In fact, the difference in intensity when comparing the smooth to the 5 m rms is generally only one or two dB for the steep SRBR rays and less than a half dB for the flat RBR rays. It follows that the bathymetry need only be known to within a few meters to allow for the use of ray theory for interpretation of results and tomographic inversion of travel time measurements.

V. DATA ANALYSIS

The log intensity (20) $\log_{10}[a(i,k)]$) is plotted in Figure 6 for the north-south transmission paths 12 and 21. Each plot contains 600 pulse responses, one every 72 minutes for 30 days. The pulse intensity is seen to spread over approximately 150 to 200 ms. The average arrival time of the entire pulse pattern fluctuates by more than 100 ms which corresponds to an average temperature change of about 2°C. The late arriving peak has a persistent high signal level and is more stable and temporally consistent than ealier arriving portions of the pulse which are associated with the steeper RBR paths.

In order to align pulses in time, the most distinctive and stable features was used. The peak formed by the late arriving RBR rays ("late peak") is a consistent and high signal level characteristic of the observed responses for all transmission paths. The time spread of the late pulse is about 40 ms. The alignment procedure consists of a series of computations including the lagging edge threshold cross, the centroid and phase tracking.[4] The resulting time series of an estimate of arrival time is stored and the pulses are then aligned by subtracting computed time shifts from each pulse as shown in Figure 7. These are the same pulse histories shown in Figure 6 except for the alignment and a two-hour smoothing which was only possible after alignment.

The resulting time series of travel time for transmisison paths 1,2 and 2,1 can be interpreted as fluctuations in temperature and current using eqs. (3) and (4) and the ray model prediction of Figure 5b. The computed average and difference converted to a temperature and current respectively are plotted in Figures 8 a and b along with temperature and current measured with conventional instruments at the lower and mid-depth of the current meter array shown in Figure 5. The inversions computation to give temperature provide only fluctuations around the mean because the navigation is not accurate enough to predict the range, but the current predictions are absolute. The predicted currents and temperatures correspond to the depth average of the rays forming the late peak. The average height above the bottom of these rays is approximately 65 m, thus inversions should agree more closely with the lower current meter at a depth of 50 m above the bottom. The strong correlation with the lower current meter is obvious for both the temperature and current. Low frequency temperature fluctuations agree well but the acoustic method predicts less variability at the tidal and higher frequencies which is probably the results of spatial averaging over the 21 km range. Both mean and fluctuations of the predicted current agree well with the lower current meter.

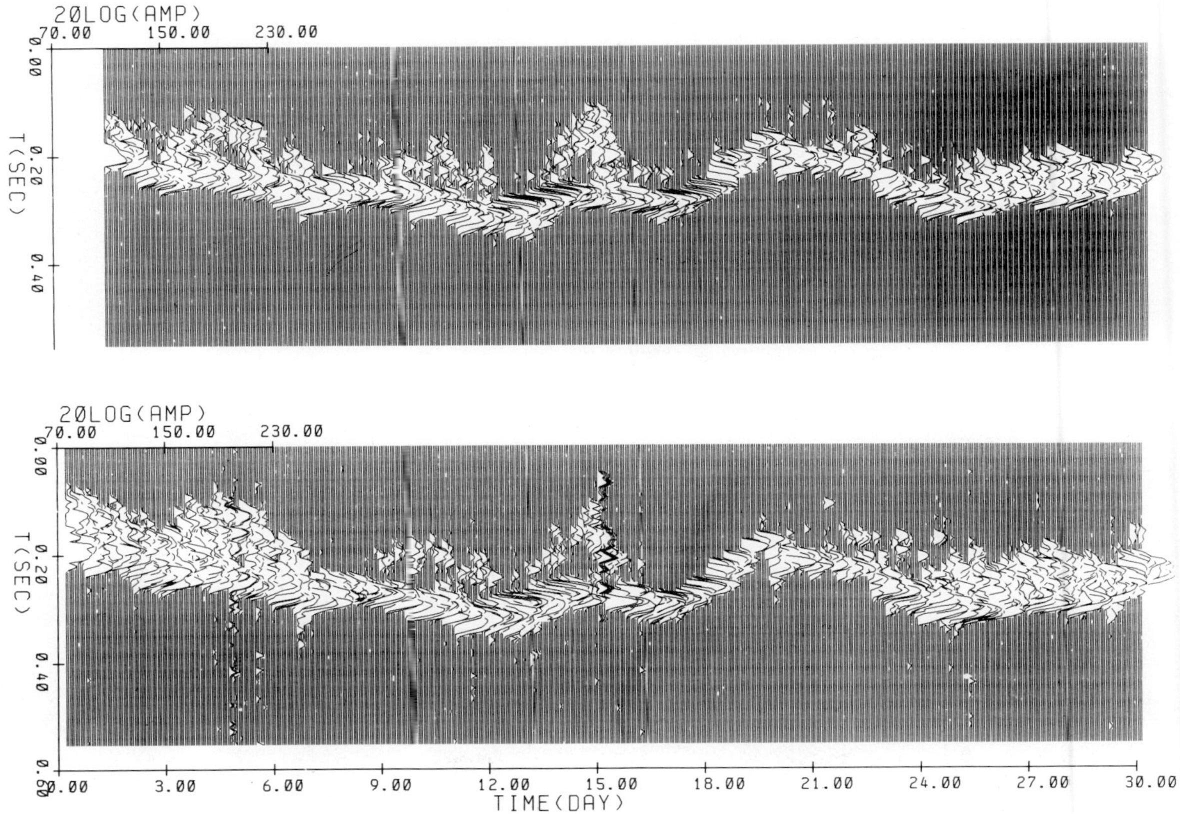

Fig. 6 Log intensity for north-south transmission paths 12 and 21.

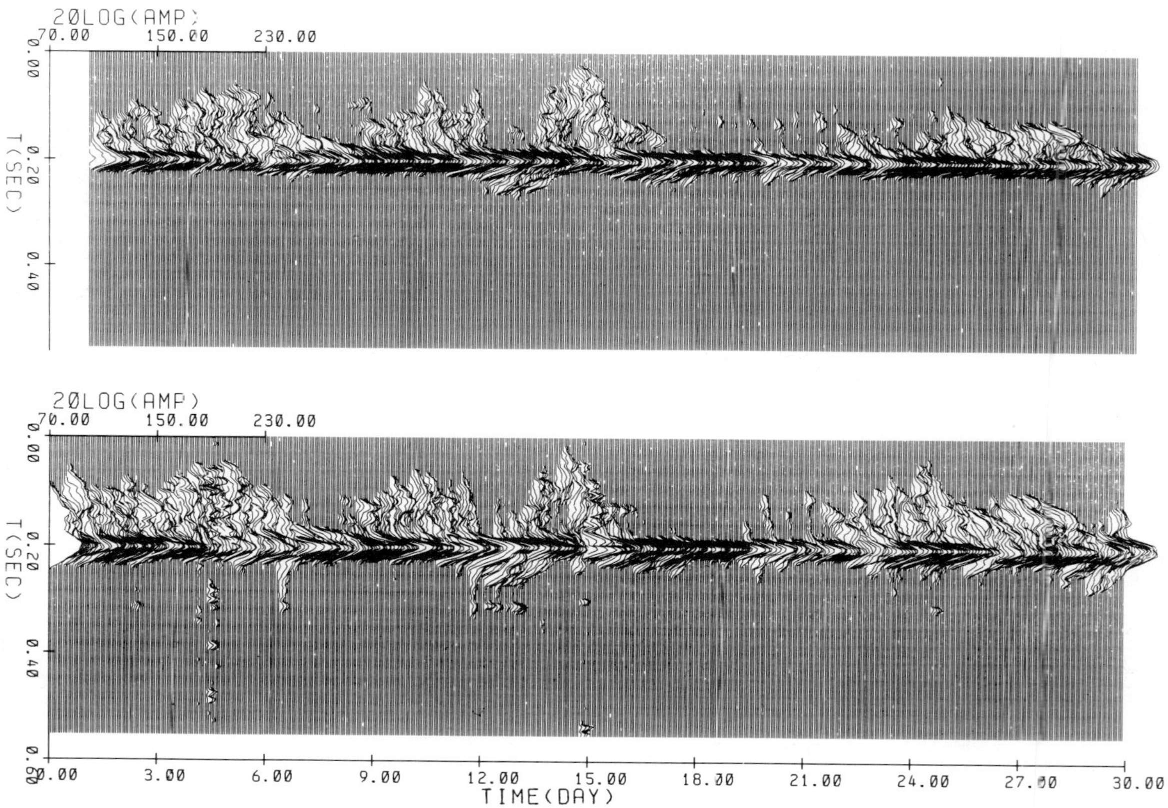

Fig. 7 Log intensity for transmission paths 12 and 21, aligned and smoothed.

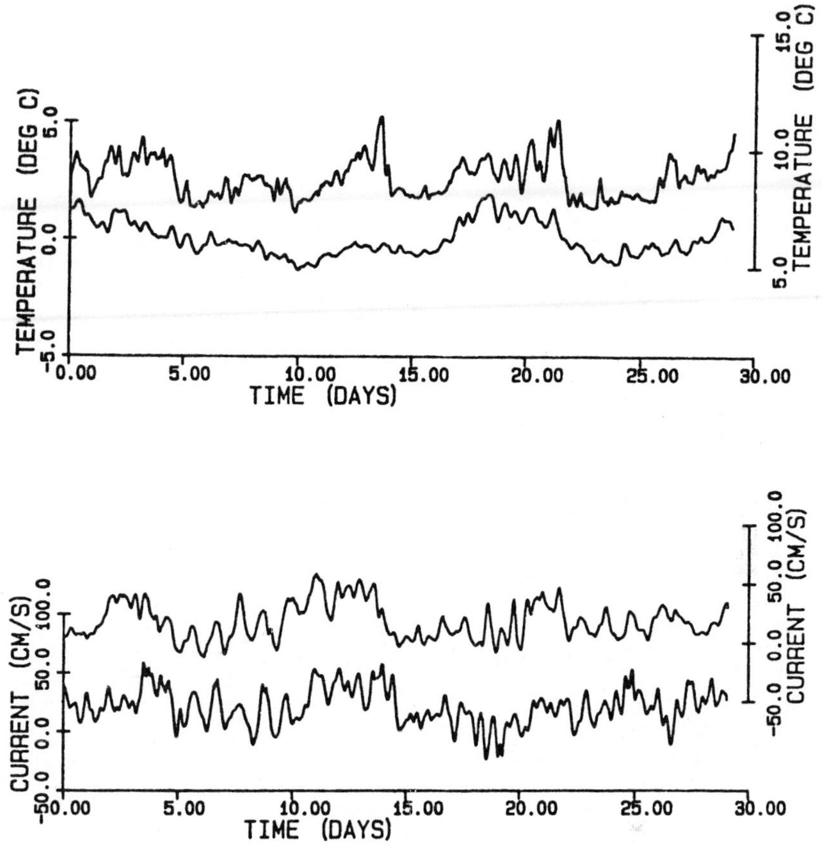

Fig. 8 Measured and computed fluctuations in temperature and current.

REFERENCES

1. T.L. Foreman, "Ray modelling methods for range-dependent ocean environment," ARL-TR-83-41. Univ. of Texas at Austin (December 1983).
2. B.F. Cron and A.H. Nuttall, J. Acoust. Soc. Am. 37, p 486-492 (1965).
3. K.V. MacKenzie, J. Acoust. Soc. Am. 32, p 221-231 (1960).
4. H.A. DeFerrari and H.B. Nguyen, "Acoustic reciprocal transmission experiment, Florida Straits," submittd to JASA, March 1985.

MEDIUM-INDUCED LOW-FREQUENCY FLUCTUATIONS IN ACOUSTIC TRANSMISSION LOSS: EXAMPLES FROM MEASUREMENTS IN SELECTED GEOGRAPHICAL AREAS

Hassan B. Ali, Melchiorre C. Ferla, and Sergio Fiori

SACLANT ASW Research Centre
Viale San Bartolomeo 400
19026 La Spezia, Italy

ABSTRACT

The ocean is a complex, highly variable acoustic medium. A propagating acoustic signal is affected by a host of phenomena, including the sea-surface and bottom, volume inhomogeneities, internal waves and tides, and non-stationary water masses. These effects cause fluctuations in the amplitude and phase of an acoustic signal and an accompanying loss in its coherence properties. The responsible mechanisms, and hence the acoustic effects, cover a wide range of temporal and spatial scales and, in general, can be understood only in terms of deterministic and random forces acting in concert. Although the listing of these mechanisms is generally easier than their isolation in realistic situations, it has been possible to correlate fluctuations in acoustic transmission loss with environmental variability, especially for very low acoustic frequencies. In particular, measurements conducted by SACLANTCEN in diverse geographical areas have identified semi-diurnal effects (tidal as well as heating), internal waves, inertial oscillations, and moving water masses as significant contributors to low-frequency acoustic fluctuations.

INTRODUCTION

The basic problem in using acoustics in the ocean is the complexity of the medium. The parameters controlling the propagation vary, usually unpredictably, both spatially, and, more significantly, temporally. Consequently, an acoustic signal propagating in such a medium is affected not only by interactions with the bottom and surface boundaries, but also by volume inhomogeneities caused by non-uniformities in temperature, density, and salinity distributions. The degradation of the acoustical signal is manifested by fluctuations in its amplitude and phase and by an accompanying loss in its coherence properties.

Generally, this degradation results in a decrease in performance of underwater acoustic systems, including sonar. In particular, the estimations of range and angular bearing of a sonar system can be adversely affected by temporal and spatial coherence losses. For this, and other reasons, an understanding of the causes of signal degradation is essential.

This paper will consider environmentally-induced signal fluctuations as they affect transmission loss.

Although the mechanisms leading to fluctuations in acoustic propagation are diverse, an essential common feature is an associated non-uniformity in the medium, either temporal or spatial or both. Depending on the temporal and spatial scales involved, the mechanisms can be considered either deterministic or random [1]. The general circulation of the ocean ("ocean climate") and its associated current systems (Gulf Stream, Kuroshio, etc.) are characterized by horizontal scales of variability limited only by the size of the basin, vertical scales of a few 100 m, and temporal scales from a few days to seasonal. These are deterministic structures. The intermediate scales of variability, including ocean motions such as fronts and eddies, can also be considered to be deterministic perturbations from the mean structure. The associated scales of variabilities are of the order of 100 to 1000 km in the horizontal, to ocean depth in the vertical, and days to months in time.

Internal waves, fine-structure, and microstructure comprise the smaller scales and are considered random features. The internal waves are characterized by scales from 100 m to 10 km or more in the horizontal, 1 to 100 m in the vertical, and from about 10 min to 1 day in time. Because they owe their existence to the restoring forces due to the density gradient and the Coriolis force, the frequency spectra of internal waves are bounded by the inertial frequency at the low end and by the buoyancy frequency (Brunt-Väisälä) at the high end. Variability induced by internal waves has been found to be a very significant source of sound scattering and has received considerable attention in recent years [2,3,4]. In essence, this increased activity arises from two complementary facts: on the one hand the spatial scales of internal waves match the acoustic wavelength over a broad frequency range, and thereby affect the acoustic field; on the other hand the internal wave is one of the few ocean features for which a reasonably effective statistical model (Garrett-Munk) is available. Fine- and microstructure variability involve horizontal scales from several metres to hundreds of metres, vertical scales from centimetres to about 10 m, and temporal scales of the order of milliseconds. Such variability generally affects sound propagation in the frequency range from approximately one to tens of kilohertz.

Table 1 [1] summarizes the temporal scales of environmentally-induced fluctuations often observed in acoustic propagation experiments. Of particular relevance to low frequency acoustic propagation is the information contained within the dotted area of Table 1. The remainder of the paper is devoted to a discussion of three SACLANTCEN measurements, the results of which show a clear correlation between transmission loss fluctuations and the environmental phenomena highlighted here.

In particular, long-period internal waves (inertial oscillations) and semi-diurnal, tide-related effects figure prominently in the measurements conducted in two areas of the North Atlantic Ocean and one in the Mediterranean. Moreover, although tidal forces are important at both sites, differences among the local oceanographic conditions lead to interesting contrasts between the nature and the degree to which these tidal forces affect the acoustic propagation. In particular, at the Mediterranean test site volume effects in the water column seemed to be of greater significance than in the North Atlantic test site, where acoustic interaction with the bottom was the predominant source of the fluctuations. Therefore, discussions of the results will be grouped under physical mechanisms, viz., under "volume effects" and "bottom effects", rather than under the less relevant geographical indicators.

Table 1 Temporal variation observed in acoustic data

PERIOD	PRINCIPAL ENVIRONMENTAL PARAMETERS	PRINCIPAL ACOUSTIC EFFECTS
ms	FINE AND MICROSTRUCTURE	AMPLITUDE, PHASE AND PROPAGATION PATH FLUCTUATIONS; HIGH FREQUENCY SCATTERING
5 s	SURFACE WAVE SIGNATURES	AMPLITUDE MODULATION; SCATTERING
10 min-24 h	INTERNAL WAVE SIGNATURE	AMPLITUDE AND PHASE FLUCTUATION; SIGNIFICANT REFRACTION OF NEARLY HORIZONTAL PATHS
4 h	SHALLOW WATER TIDES	PHASE AND AMPLITUDE FLUCTUATIONS OFTEN SIMPLY CORRELATED WITH TIDAL VARIATIONS.
12 h	SEMIDIURNAL TIDES	
24 h	DIURNAL TIDES	GREATEST EFFECTS IN SHALLOW WATER
4 day 15 day 1 month	MESOSCALE PHENOMENA (EDDIES,FRONTS)	FLUCTUATIONS IN PROPAGATION PATH, ACOUSTIC INTENSITY; DISTORTION OF SIGNAL SHAPE; SLOW MODULATION OF ACOUSTIC RAYS
1 month 1 year	OCEAN CLIMATE	FLUCTUATION IN AMPLITUDE, PHASE AND PROPAGATION PATH

1 MEASUREMENTS SHOWING VOLUME EFFECTS

1.1 Propagation Characteristics of the Test Environment

The acoustic propagation experiment was conducted in a shallow water region of the Mediterranean where the water depth varied from about 40 m to 85 m, as seen in Fig. 1. Although the bathymetry of the area is fairly complex, that along the propagation run is relatively simple. The water circulation in the region consists of a three-layer system: an eastward-flowing surface layer of Atlantic water, an intermediate layer in which turbulent mixing occurs, and a westward-flowing bottom layer of more saline (Levantine) water. A temperature/salinity profile of the measured data confirms this general behaviour.

For the experimental situation depicted in Fig. 1, broad-band (explosive) sources were dropped at approximately hourly intervals. A vertical array of hydrophones, 35 km distant, received the signals. Simultaneously, samplings were taken of the pertinent oceanographic parameters: sound speed, temperature, salinity and density (STDV casts). The test was conducted during summer conditions. Figure 2 presents typical depth profiles of the environmental parameters in this area. This type of profile tends to favour downward acoustic refraction and, therefore, results in greater bottom interaction than would occur in winter. Vertical stratification and some fine-structure are evident in the profiles. A closer analysis of the sound speed profile over shorter intervals in depth and sound speed reveals more clearly the presence of fine-structure, characterized by vertical dimensions of the order of centimetres to one or two metres. As already noted, such fine-structure will tend to scatter sound of frequency ranges higher than of interest herein.

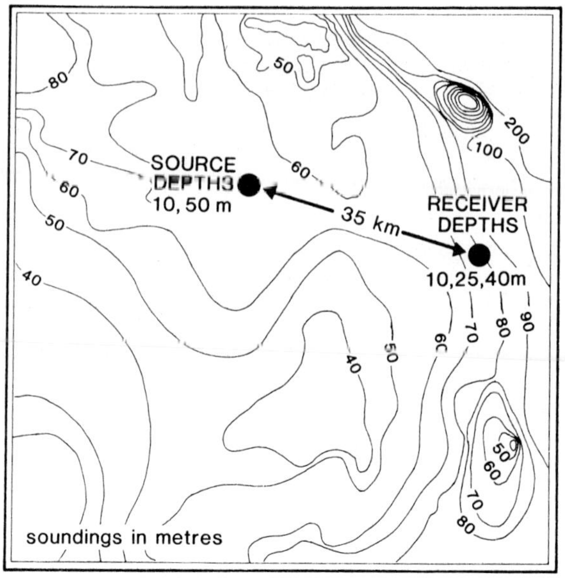

Fig. 1
Bathymetry of the experiment (Mediterranean site).

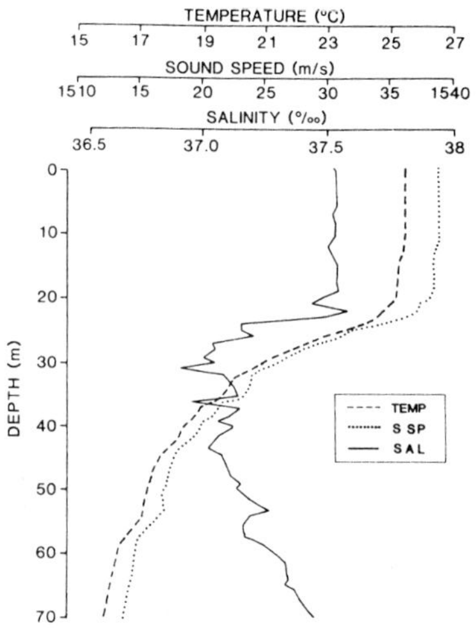

Fig. 2
Depth profiles of temperature, sound speed, and salinity (Mediterranean site).

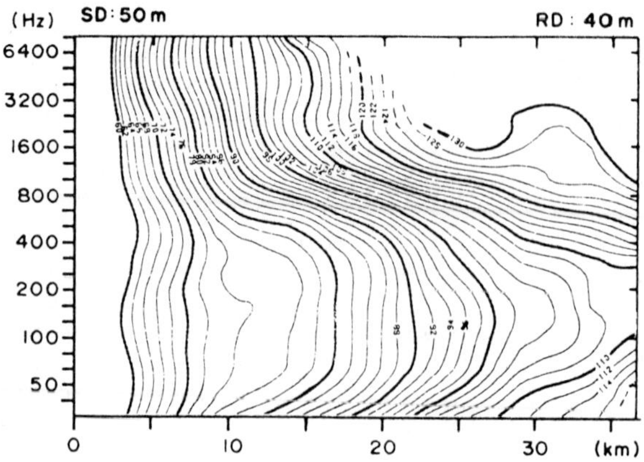

Fig. 3
Contours of measured transmission loss in 1/3 octave bands.

The effect on acoustic propagation is shown in Fig. 3 <5>, which presents contours of measured transmission loss, in 1/3 octave bands, in the frequency/range plane. The existence of an optimum frequency range (i.e. minimal transmission loss) for acoustic propagation is clearly evident. In this case, it lies between approximately 100 and 400 Hz. The explanation for this is that acoustic energy at very low frequencies suffers large attenuation as a result of bottom penetration which increases directly with wavelength. On the other hand the energy at the very high frequencies is greatly attenuated by absorption in the water column and, possibly, by scattering from fine-structure. Hence an optimum frequency range exists somewhere between the two extremes. In other words, for the conditions typefied by Fig. 3, the ocean behaves like a band-pass filter for propagating broadband acoustic signals.

1.2 Temporal Variability in the Environmental Data

Of primary interest for our purpose is the temporal variation of the relevant parameters at a fixed range, here 35 km. Figure 4 shows an example of the variation of sound speed with depth during a period of 25 hours.

The depth range from approximately 25 to 35 m comprises the steepest portion of the thermocline. The region down to 20 m or so is essentially isovelocity with a sound speed of approximately 1538 m/s. The fluctuations in sound speed are quite evident, particularly at a depth of 25 m or so within the thermocline. The plots clearly indicate an oscillation in the width of the mixed layer (surface duct). Figure 4b, in particular, indicates that typical features, e.g., "kinks" in the curve, appear to be migrating vertically with time - a behaviour often suggesting the presence of internal waves [Gregg, 6]. The frequency content of these oscillations is of particular interest because it provides clues to the responsible mechanisms. Examples of frequency spectra of the relevant environmental parameters, obtained from Fast Fourier Transforms of the corresponding normalized time series at 25 m depth, are shown in Fig. 5.

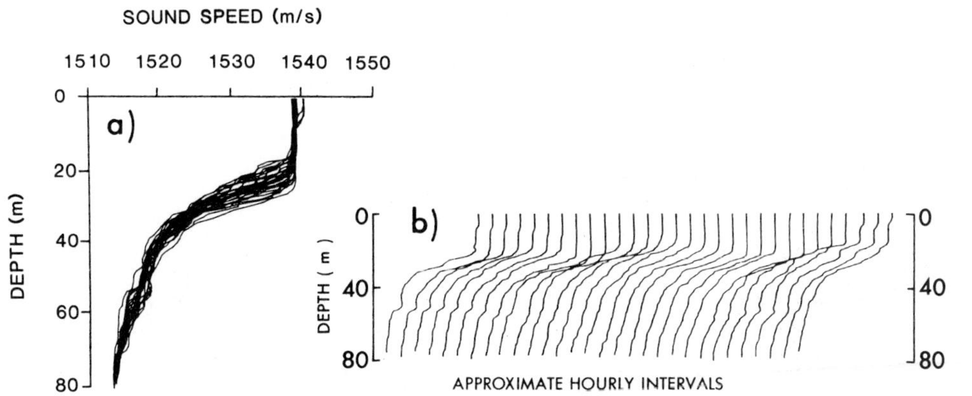

Fig. 4 Temporal variation of sound speed over approximately 25 hours.

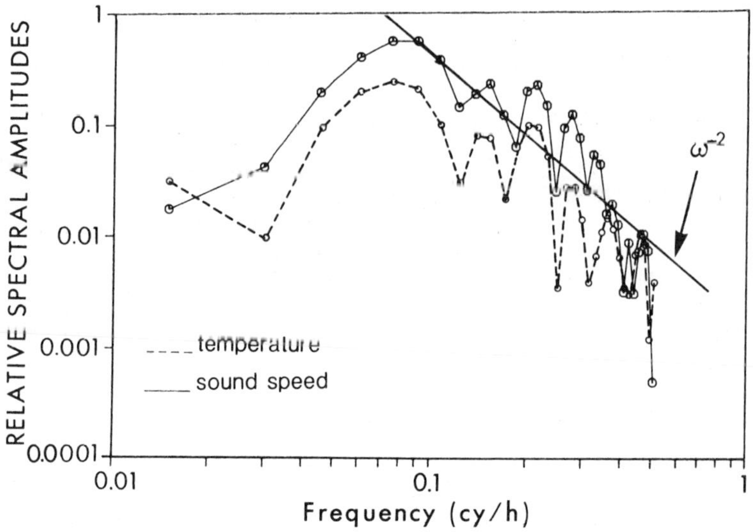

Fig. 5 Spectra of sound speed and temperature at 25 m.

The dominant fluctuations occur in the frequency range from 0.08 to 0.06 cycle/h, or for periods from 12 to 17 hours. This is a broad response and includes both the semi-diurnal period and inertial period (approximately 17 hours here). The time series was too short to allow for the spectral separation of the two responses. From the results obtained at other depths it does appear, however, that the relative significance of the semi-diurnal tidal and inertial effects depends on depth. In particular, from the analysis of the depth-dependent frequency spectra it appears that there are fluctuations of inertial period in the surface layers, and that as the depth increases the dominant fluctuations are semi-diurnal.

Inertial oscillations may occur in connection with a sudden change of the wind (i.e. a wind impulse of short duration) and changes in barometric pressure [7]. The whole of the Mediterranean is dominated by meteorological forcing; therefore it is likely that inertial oscillations are present in the study area. That they dominate the surface waters is probably related to their meteorological origin. The local water stratification retains most of this energy in the surface layers, so that at depth the semi-diurnal components are discernible.

In the spectra of Fig. 5 there also appear to be higher-frequency phenomena, with periodicities of about 2,4,5, and 7 h. These may be associated with internal wave activity. Note that the drop-off rate (sloping line) of the sound speed spectrum is approximately ω^{-2}, a slope consistent with that predicted by the Garrett-Munk model of internal waves.

1.3 Fluctuations in Acoustic Transmission Loss

Figure 6 shows the contours of measured transmission loss (1/3 octave bands) in the frequency/time plane for source and receiver depths of 50 m and 40 m, respectively. The higher frequencies (>1.6 kHz) exhibit far more pronounced fluctuations than the lower frequencies. This may indicate that the responsible environmental phenomena have physical dimensions that are comparable to the acoustic wavelengths of the higher frequencies. In order to demonstrate this selective frequency effect more clearly, the spectra of transmission loss at 200 Hz and 1600 Hz were compared (Fig. 7).

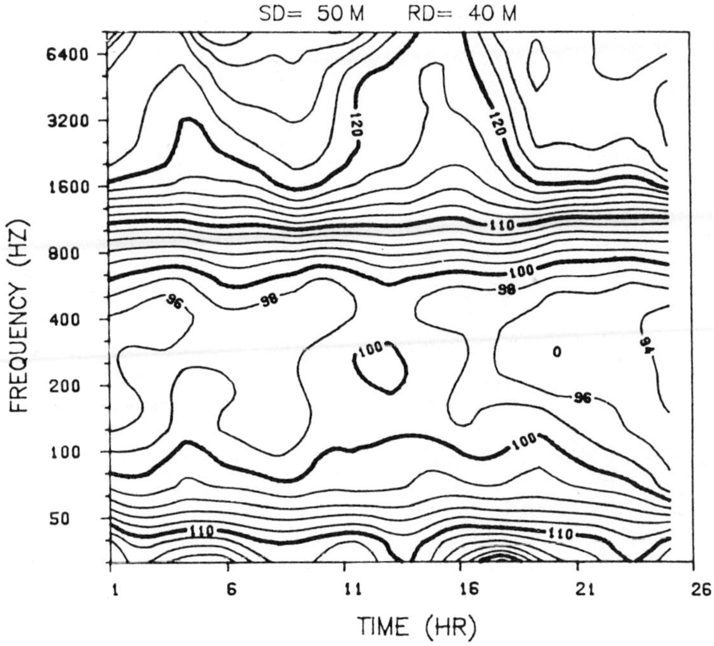

Fig. 6 Contours of measured transmission loss fluctuations at a fixed range (35 km).

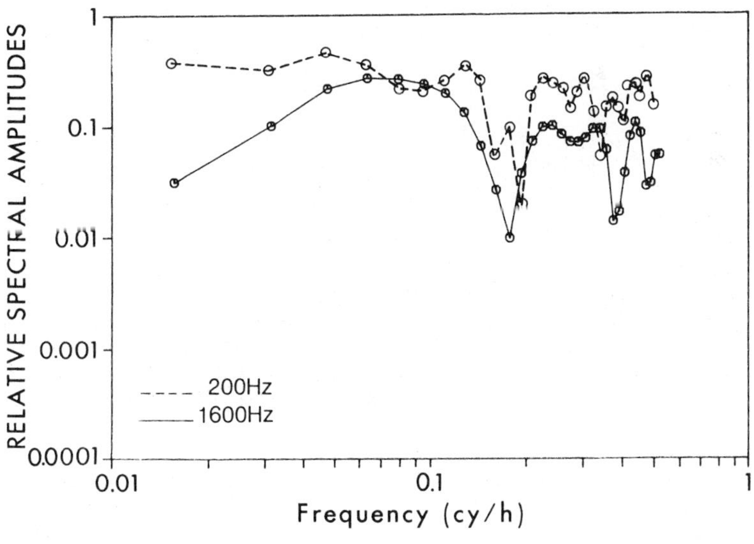

Fig. 7 Spectra of transmission loss for two selected frequencies.

The spectra, obtained for the same source/receiver as those in Fig. 6, emphasize the difference in the effect of variability. These results suggest that the optimum frequency range is less susceptible to environmental variability than other frequency ranges. A comparison of these spectra with those shown in Fig. 5 reveals a good correlation between the spectra of environmental variability and the higher frequency transmission loss spectrum.

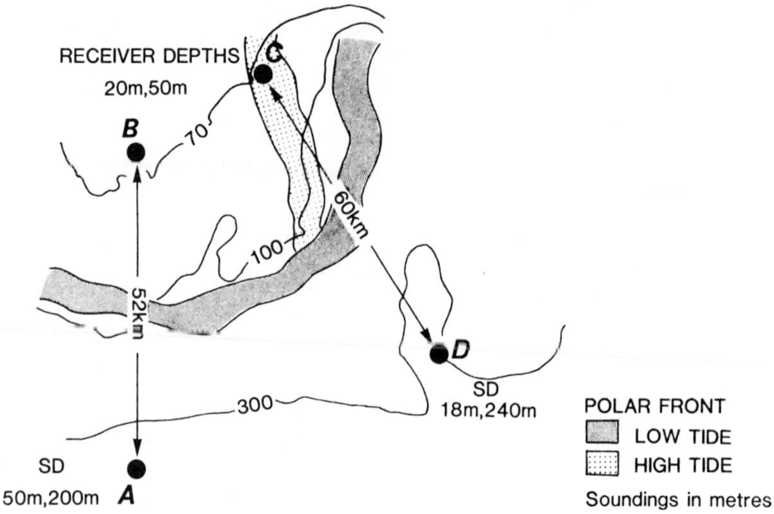

Fig. 8 Geometry of the North Atlantic measurements.

2 MEASUREMENTS SHOWING BOTTOM EFFECTS

2.1 Propagation Characteristics of the Test Environment

The North Atlantic test site is a very complex geological, oceanographic, and biological province and has been the object of many investigations [8]. The test site is located in an area affected by a large-scale permanent front resulting from the convergence of two water masses. In particular, at this location, cold, less saline Arctic water meets warmer Atlantic water to form the Polar oceanic front. The substantial differences in salinity and temperature of the two water-types (35‰ and 6° to 7°C for the Atlantic vs 34.6‰ and about 0°C for the Arctic) cause the Polar front to be characterized by steep gradients in these parameters, with significant implications for sound propagation. An added complication arises from the changing position of the front, which oscillates with semi-diurnal tidal periodicity.

The geometry and relevant details of the measurements are provided in Fig. 8. The two tracks, A-B and D-C, represent the propagation paths for measurements conducted one year apart. The environmental conditions along the two paths were somewhat different. Although both tracks A-B and D-C cross the Polar front, in water changing from deep to shallow, track A-B is over a generally hard bottom (sand) while track D-C is over a generally soft bottom (sand, silt, clay). From the point of view of temporal variability, however, a more fundamental difference between the two receiver locations arises from their relative proximity to the Polar front. In particular, station C is directly affected by movements of the front, the water above 20 m changing from Polar to Atlantic with an approximately 12-h periodicity, while the waters below 20 m remain of the Polar type. In comparison, station B remains in virtually isothermal Polar water, north of the front, as shown schematically in Fig. 8.

In effect, the Polar front serves as a demarcation line between two different propagation areas. To the north of the front, in isothermal water, an important part of the propagation will be in shallow water under upward-refracting conditions, resulting in small total transmission losses as a result of less bottom-interaction. On the other hand, in waters affected by the front, propagation will be under strong downward refracting conditions, as a result of the steep gradients, and thereby subject to higher losses, particularly at low frequencies. As noted, measurement site B was always in isothermal waters, whereas C was periodically subject to the effects of the front. From this one would expect the transmission losses measured at site B to be both lower in magnitude and subject to less temporal fluctuation than those at site C. Though generally valid, this conclusion is nevertheless based on an oversimplification of the propagation conditions. In particular, the propagation depends not only on the front, but also on frequency, source and receiver depths, bottom conditions (which are different for the two tracks), and, possibly, other features such as currents and inertial oscillations.

2.2 Selected Examples of Temporal Variability

In the following, typical results of measurements made over a 48-h period will be presented, emphasizing the data obtained at position C. The experimental procedure was essentially the same as that for the Mediterranean measurements, described previously.

Figure 9 provides a typical example of the fluctuations in measured transmission loss at site C for several frequencies, along with the predicted tidal curve for the area. It is clear that the fluctuations in transmission loss correlate well with tidal periodicity, particularly for the lower frequencies. This is consistent with the assumption that the lower frequencies are more affected by the tidally-induced change in downward-refracting conditions. Both diurnal and semi-diurnal periodicities are evident, the former seemingly the more significant. Further,

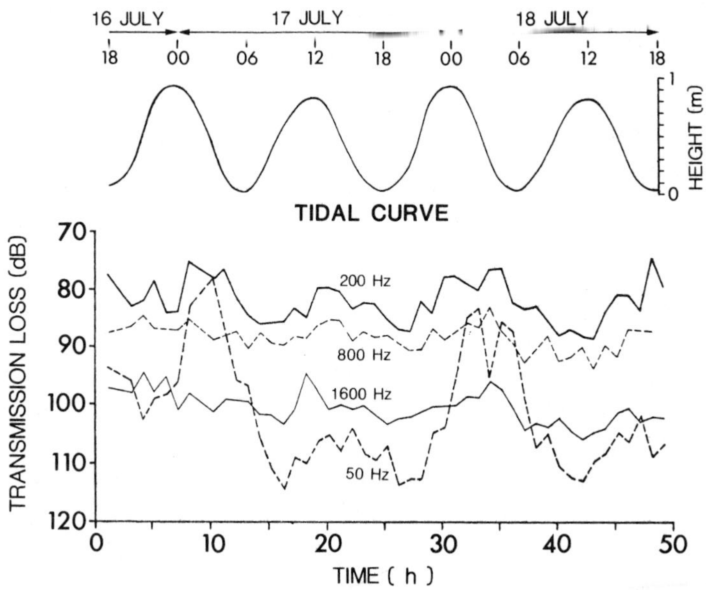

Fig. 9 Temporal variation of transmission losses and predicted tidal curve at site C.

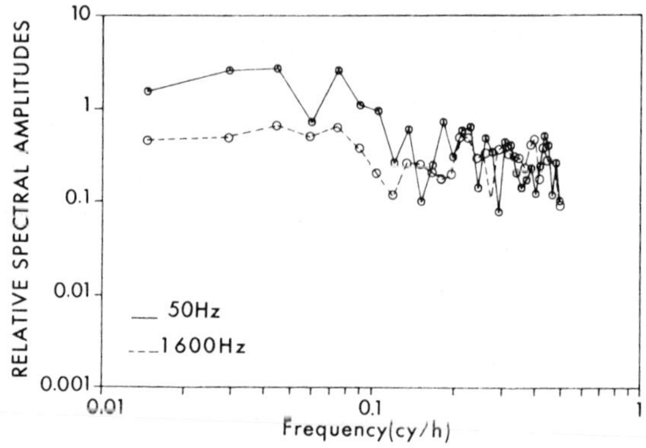

Fig. 10
Spectral composition of transmission loss at site C for two selected frequencies (SD 240 m / RD 50 m).

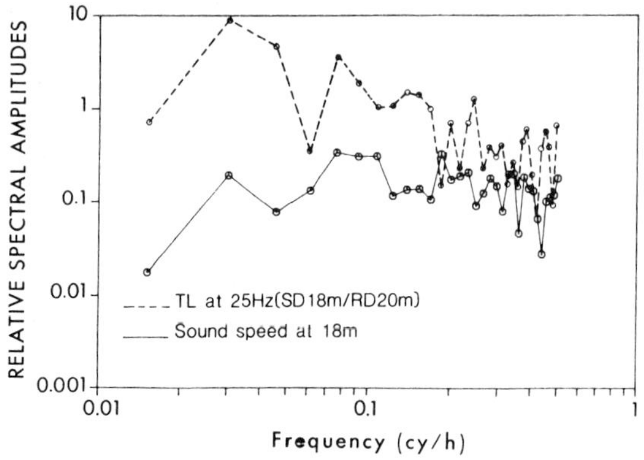

Fig. 11
Comparison of spectra of transmission loss at 25 Hz and sound speed.

there is a correspondence, albeit inexact, between high tide and high transmission loss, consistent with the fact that the front reaches site C during periods of high tide. Although the curves shown are for a particular source/receiver combination (240 m / 50 m) the observed trends are fairly representative of all the data obtained at this site. The spectral composition of the 50 Hz and 1600 Hz signals is shown in Fig. 10, again demonstrating the more marked fluctuations at the lower frequency, as well as the conspicuous presence of diurnal and semi-diurnal periodicities.

Figure 11 provides a comparison between the 25 Hz transmission loss spectrum at a depth of 20 m (source depth = 18 m) and the sound speed spectrum at a depth of 18 m. As seen the correlation between the two spectra is quite good.

It is instructive to compare the variability for the two positions B and C. An example of the fluctuations in temperature is shown in Fig. 12. Several features are immediately evident from the figure. First, it is clear that the amplitudes at C are greater than those at B and, in addition, show a definite dependence on depth. Further, although both data sets reveal tidal effects, they are more conspicuous at position C, particularly at 18 m depth. This behaviour no doubt reflects the position of the stations with respect to the Polar front, station C being directly affected by the movements of the front, while station B is constantly in isothermal waters, as discussed earlier. The significant differences in results between the two locations suggest that tidally advected changes in water masses, as at C, are more important than the indirect tidal effects (changes in water depth, currents, etc.) that are evident at B. The greater energy measured at 18 m (for site C) seems to confirm this conclusion since the transition zone between the Atlantic and Polar waters occurs at this depth.

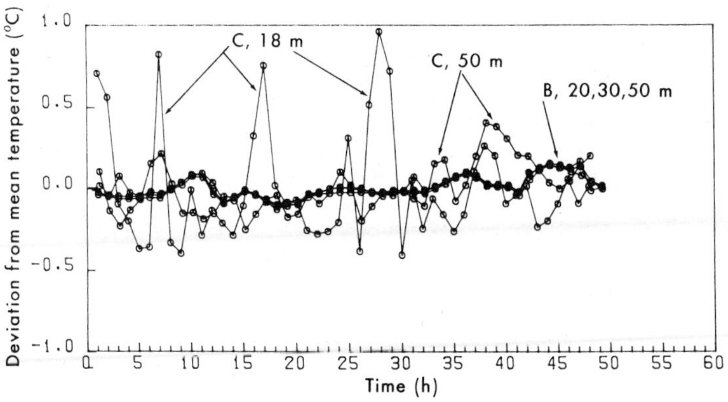

Fig. 12 Normalized temperature fluctuations at the two sites.

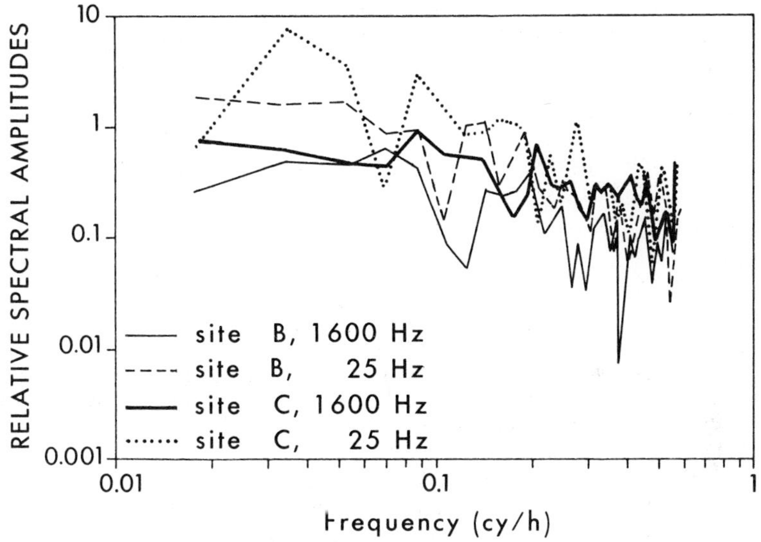

Fig. 13 Transmission loss spectra at receiver sites B and C (North Atlantic).

Finally, the comparison of the transmission loss spectra at the two sites is shown in Fig. 13. Note the fluctuations in low frequency transmission loss are greater at site C than at B; but there is little difference between the sites vis-à-vis the higher frequency transmission loss. This is consistent with the stronger downward refracting conditions at site C.

CONCLUSIONS

The measurements conducted in the Mediterranean and the North Atlantic indicate that fluctuations in acoustic transmission loss correlate well with the diurnal, semi-diurnal, and inertial periodic variability of the environmental parameters. However, the nature and extent of the causal phenomena differ for the two areas. At the Mediterranean test site volume effects appear to be the principal cause of the fluctuations, while at the N. Atlantic test sites bottom effects predominate. Both types of effects are described below:

Volume Effects

- Inertial effects dominate in the surface layers, whereas the semi-diurnal effects are of greater importance at greater depths. This suggests a meteorological forcing function.

- The fluctuations in the magnitude of acoustic transmission loss are greatest for the higher frequencies (1.6 kHz and above) and least for an optimum frequency range from approximately 100 to 400 Hz.

Bottom Effects

- At the N. Atlantic site the inertial period coincides with the semi-diurnal period and cannot be resolved by spectral analysis.

- The transmission loss oscillations correlate with tidal movements of the Polar front and are more marked at the lower frequencies. The effects on acoustic propagation are associated more closely with the tidally-advected changes in water masses than with the more subtle, indirect tidal effects.

ACKNOWLEDGMENT

The authors gratefully acknowledge the contribution made by Mr. S. Mari of SACLANTCEN.

REFERENCES

1. H.B. Ali. Spatial and temporal variabilities in underwater acoustic transmission: an analytical review, SACLANTCEN SM-166. La Spezia, Italy, SACLANT ASW Research Centre, 1983. [AD A 129 959].
2. Y. Desaubies. Statistical aspects of sound propagation in the ocean. In: Adaptive Methods in Underwater Acoustics, H. Urban, ed. Dordrecht, Reidel, 1985.
3. S.M. Flatte, R. Dashen, W.H. Munk, K.M. Watson and F. Zachariasen, eds. Sound Transmission through a Fluctuating Ocean. Cambridge, Cambridge University Press, 1979.
4. B.J. Uscinski, C. Macaskill and T.E. Ewart. Intensity fluctuations, part I: theory, part II: comparison with the Cobb experiment. J. Acoust. Soc. Am., Vol. 74, 1474-1499, (1983).
5. T. Akal. Sea floor effects on shallow-water acoustic propagation. In: W.A. Kuperman and F.B. Jensen, eds. Bottom Interacting Ocean Acoustics, New York, NY, Plenum, 1980.
6. M.C. Gregg. Finestructure: Oceanic "kinks and wiggles." Naval Research Reviews, Vol. 34, 52-72 (1982).
7. R.T. Pollard. On the generation of inertial waves in the ocean. Deep Sea Res., Vol. 17, 795-812, (1970).
8. J. Pietrzak, and H.B. Ali. Fluctuations in environmental parameters observed during acoustic measurements in the Barents Sea and the Strait of Sicily. La Spezia, Italy, SACLANT ASW Research Centre. (SM- in preparation).

WAVE PROPAGATION IN ANISOTROPIC MEDIA USING BORN SERIES THROUGH 4th ORDER

W.C. Meecham

School of Engineering and Applied Science
Department of Mechanical, Aerospace and Nuclear Engineering
University of California, Los Angeles, California 90024

ABSTRACT

The Born series is used to calculate a field propagating in a Gaussian medium which itself has a Gaussian-form correlation function. These specializations lead to integrals for the moments with integrands which are themselves mainly of Gaussian form. They can be integrated algebraically using multivariate Gaussian distribution formulas. Specifically for moments the, say, fourth order Born term involves a 12-fold integration of which eight can be done algebraically using the formulas. We are able, for a two-scale anisotropic medium, to sum the Born series for the average (the first moment) of the radiation field. The result reduce to familiar ones when we specialize to isotropic media. A novel characteristic is that for physically interesting parameter ranges diffraction effects can be substantially reduce the phase variance, thus reducing the randomising effect of the medium on the propagated field. Anisotropy also leads to an intensity dependance on angle. The fourth-order moment (the intensity variance) is considered through fourth-order Born terms; it is not possible to sum the Born series for this moment in terms of simple functions even for isotropic media. Previous results are in large part confirmed: when diffraction effects are slight, sound field fluctuation effects are confined to the phase, leading at sufficient range to a log-normal field distribution. When diffraction effects become appreciable, the sound amplitude fluctuates and the distribution becomes Gaussian. There is also an expected region of appreciable overshoot in the intensity variance, that variance exceeding its Gaussian value for intermediate ranges at lower frequencies.

INTRODUCTION

This paper is intended primarily to provide a post view of earlier work (see Ref. 1) in hopes of (among other things) improving access. Wave propagation in random media is an important and as yet not fully solved topic. There are many applications. Two that come to mind are: 1) degradation of laser beams when propagating through the atmosphere, 2) degradation of acoustic signals when propagating through the ocean. The latter is of primary interest to the present conference. We make no attempt to summarize the extensive literature on this subject and apologize to the numerous authors of excellent work who will not be directly cited here. For more comprehensive reviews of earlier work one can refer to Ref. 1 and to the book of which Flatté is editor (Ref. 2). Also it is important to cite the earlier extensive works of Tatarskii (Ref. 3) and Chernov (Ref. 4).

We discuss here wave propagation in random anisotropic, though homogeneous, media. Sound propagation in the ocean is influenced primarily by sound speed changes and these most importantly are caused by temperature fluctuations. Because of the ocean stratification (a similar stratification occurs in the atmosphere) the scales in the horizontal direction are typically considerably greater than those in the vertical direction. We shall assume that the process is statistically homogeneous in horizontal planes, as is approximately the case. We assume that the medium has Gaussian statistics. Additionally, we assume that the wavelength is small compared with both horizontal and vertical scales of the medium: we deal with a high frequency sound source. Rayleigh has shown that if on the contrary, the wavelength was large, the scattering would be isotropic with a dipole contribution. For small wavelengths, the scattering is mainly in the forward direction. This is a common assumption and will provide useful approximations later. For the purpose of this work, we suppose that the correlation functions in the medium are of Gaussian form. This is a restrictive assumption, but nevertheless we can discuss qualitatively the important physical characteristics of more general media using it. We shall expand the propagating field using a Born series approximation, which is an expansion in terms of a small parameter representing the index of refraction fluctuations. For our particular assumption of forward scattering, and Gaussian correlation functions, it is possible to use multi-variate Gaussian distribution relationships (between probability density functions and their characteristic functions). In this way two-thirds of the integrals in the Born series terms can be integrated using algebraic formulas of simple form. Finally, we assume that horizontal ranges are very large compared with medium scales.

It is emphasized that we treat the statistics of one ray. If macroscopic multipaths are present (say, many possible rays in the deep sound channel) then the field is further randomized - brought nearer to saturation. Clearly degradation is minimized, as would be desirable for information transmission, by processing for a single ray.

We treat here a single frequency as also is common, recognizing that pulses and the like can be obtained by Fourier transformation. We look for the Green's function solution of the resulting Helmholtz equation, that is for a simple source solution.

The equation to be solved is

$$[\vec{\nabla}^2+k_T^2(\vec{r})]G(\vec{r},\vec{r}_0) = \delta(\vec{r}-\vec{r}_0). \tag{1}$$

We assume that fluctuations in the quantity k_T^2 (due to index of refraction changes) are small compared with the mean. Rewrite the equation in the form

$$\{\vec{\nabla}^2+k^2[1+\varepsilon\nu(\vec{r})]\}G(\vec{r},\vec{r}_0) = \delta(\vec{r}-\vec{r}_0), \tag{2}$$

for ε small and

$$\overline{\nu^2(\vec{r})} = 1, \tag{3}$$

the fluctuation v is normalized. It is statistically homogeneous, but not necessarily isotropic, and is assumed to be isotropic in horizontal planes.

The approximation consists of expanding G in terms of a series of functions of successively higher order in ε. Thus

$$G(\vec{r},\vec{r}_o) = G_o(\left|\vec{r}-\vec{r}_o\right|) + G_1(\vec{r},\vec{r}_o)+G_2(\vec{r},\vec{r}_o)+\ldots\ldots\ldots \tag{4}$$

with

$$G_n = O(\varepsilon^n). \tag{5}$$

Substitute (2) to find

$$\sum_{n=0}^{\infty} \{\vec{\nabla}^2+k^2[1+\varepsilon\nu(\vec{r})]\}G_n=\delta(\vec{r}-\vec{r}_o). \tag{6}$$

The zeroth order term is known,

$$G_o(\left|\vec{r}-\vec{r}_o\right|) = -(1/4\pi)(e^{ik\left|\vec{r}-\vec{r}_o\right|}/\left|\vec{r}-\vec{r}_o\right|). \tag{7}$$

By iteration equating terms of the same order in ε we have $n \geqslant 0$,

$$(\vec{\nabla}^2+k^2)G_{n+1}(\vec{r},\vec{r}_o) = -k^2\varepsilon\nu(\vec{r})G_n(\vec{r},\vec{r}_o). \tag{8a}$$

Because of the forward scattering this can be simplified. Set the source position $\vec{r}_o = 0$ and $\vec{r} = x\hat{i}+\vec{\rho}$. The forward scattering characteristic leads to the result

$$\left|\vec{r}\right| = \left|x\hat{i}+\vec{\rho}\right| \simeq x+\rho^2/2x. \tag{8b}$$

We now solve by successive substitution. We find the nth order Born term

$$G_n(\vec{r},0) = C_g^{n+1}\, C_e^n\, e^{ikx} \int_0^x dx_n \int_0^{x_n} dx_{n-1} \cdots \int_0^{x_2} dx_1 \int\ldots\int d\vec{\rho}_1\ldots d\vec{\rho}_n$$

$$\exp \frac{ik}{2}\left[\frac{\vec{\rho}_1}{x_1} + \left(\frac{\vec{\rho}_2-\vec{\rho}_1}{x_2-x}\right)^2 +\ldots+\left(\frac{\vec{\rho}_n-\vec{\rho}_{n-1}}{x_n-x_{n-1}}\right)^2 + \frac{\rho_n^2}{x-x_n}\right]$$

$$\times\ [x_1(x_2-x_1)\ldots(x-x_n)]^{-1}\nu(\vec{r}_1)\ldots\nu(\vec{r}_n) \tag{9}$$

The only random quantity on the righthand side of the equation is the function $\nu(\vec{r})$. The coefficents are defined by

$$C_G = 1/(4\pi),\ C_e = -k^2\varepsilon. \tag{10}$$

1 THE FIRST MOMENT, THE AVERAGE FIELD

The methods of evaluating the various moments will be demonstrated for this, the simplest, the first moment. We want the average of (4),

$$\bar{G} = \sum \bar{G}_n \,. \tag{11}$$

Since the fluctuations v are Gaussian, we know that odd moments vanish and that even moments are the sum of products of the medium correlation function. For example, the 4th moment, the highest and most needed in this work, of the medium fluctuation is

$$\begin{aligned}\langle\nu(\vec{r}_1)\ldots\nu(\vec{r}_4)\rangle &= \langle\nu(\vec{r}_1)\nu(\vec{r}_2)\rangle\langle\nu(\vec{r}_3)\nu(\vec{r}_4)\rangle \\ &+ \langle\nu(\vec{r}_1)\nu(\vec{r}_3)\rangle\langle\nu(\vec{r}_2)\nu(\vec{r}_4)\rangle \\ &+ \langle\nu(\vec{r}_1)\nu(\vec{r}_4)\rangle\langle\nu(\vec{r}_2)\nu(\vec{r}_3)\rangle .\end{aligned} \tag{12}$$

Assume a Gaussiam-form of correlation function,

$$\langle\nu(\vec{r}_1)\nu(\vec{r}_2)\rangle = \exp\{-(1/2L_H^2)\,[(x_2-x_1)^2 + (y_2-y_1)^2]-(1/2L_V^2)(z_2-z_1)^2\} \,. \tag{13}$$

Because of the forward scattering we have nested integrals,

$$x_n > x_{n-1} >\ldots> x_1 \,. \tag{14}$$

It follows from this and the fact that the ranges are very large compared with the correlation lengths that only the first term, (e.g., in (12)), contributes. Then we find for the average of the nth order contribution to the Green's function

$$\begin{aligned}\bar{G}_n = C_G^{n+1}C_e^n e^{ikx} \int_0^x dx_n\ldots\int_0^{x_2} dx_1 \int\ldots\int d\vec{\rho}_1\ldots d\vec{\rho}_n[x_1(x_2-x_1)\ldots(x-x_n)]^{-1} \\ \times\exp\left[-\frac{k}{2i}\quad\frac{\vec{\rho}_1^{\,2}}{x_1}+\frac{(\vec{\rho}_2-\vec{\rho}_1)^2}{x_2-x_1}+\ldots+\frac{\vec{\rho}_n^{\,2}}{x-x_n}\quad-\frac{1}{2L_V^2}\,[(z_2-z_1)^2+\ldots+(z_n-z_{n-1})^2\,]\right] \\ \times\exp\left[-\frac{1}{2L_H^2}\,[(x_2-x_1)^2+\ldots+(x_n-x_{n-1})^2+(y_2-y_1)^2+\ldots+(y_n-y_{n-1})^2\,]\right] .\end{aligned} \tag{15}$$

Equation (15) shows the usefulness of the specialization to medium fluctuations with Gaussian-form correlation functions: the 2n integrals over the y and z directions, perpendicular to the direction of propagation, can be integrated using a formula. The multi-variate Gaussian formula needed is the following where we allow x_1 of the usual form to be complex here,

$$\int \ldots \int e^{\pm x_j y_j - 1/2 \; y_j m_{jk} y_k} dy_1 \ldots dy_N = (2\pi)^{N/2} M^{-1/2} e^{-1/2} \tag{16}$$

The M is the determinant of the matrix m (the entries of m are the ordinary application the covariances of the random variables x_i, and m^{-1} is the inverse of the matrix. After some manipulation involving the matrix and its inverse one finds (see Ref.1)

$$\overline{G}_n = G_o(x) \left(- \frac{(2\pi)^{1/2}}{8} \varepsilon^2 k^2 f(\gamma_H) L_H x \right)^{n/2} \left[\left(\frac{n}{2} \right) ! \right]^{-1} , \tag{17}$$

where the function f is defined

$$f(\gamma_H) = \left(\frac{2}{\pi} \right)^{1/2} \int_0^\infty e^{-\xi^2/2} (1 + i\gamma_H \xi)^{-1/2} d\xi . \tag{18}$$

The series of terms represented by (17) can be summed

$$\overline{G} = G_o \exp\{-[(2\pi)^{1/2}/8]\varepsilon^2 k^2 f(\gamma_H) L_H x\}, \tag{19}$$

$$= G_o(x) e^{-(\phi_o^2/2)} ,$$

where the parameter γ_h is defined by

$$\gamma_H = L_H / k L_v^2 . \tag{20}$$

For the isotropic case, where $L_h = L_V$, the forward scattering assumption means that γ_h can be neglected and f =1. Thus, for a statistically isotropic medium the average field is just a quantity determined by the variance of the phase where the phase fluctuation is given by geometrical optics. For anisotropioc media the interesting possibility arises that the field fluctuation in the forward direction of the average field may be reduced. Let us see physically how this can be.

In Fig. 1 we show a sketch of the model. Suppose that the contrast between the horizontal and vertical scales is large, then the field will be diffracted off vertical scale through an angle $\theta = (kL_v)^{-1}$. The distance which it travels passing through one eddy becomes L_V/θ. Evidently it will travel the distance to the field point marked II by passing through a much larger number of eddy regions (independent-fluctuation regions) than is the case for the ray proceeding in a horizontal direction to the field point I. This then, by the familiar random process central limit theorem, will reduce the amount of the fluctuation for the diffracted path. Figure 1 is drawn showing the diffracted ray as having traveled continuously at an angle θ. In fact the process (we are interested in the field in the horizontal direction) is one of a zigzag diffraction off successive eddy regions. Additionally we note that the ray with the lowest phase variance proceeding by zigzag path must travel a greater distance. This also can be seen from the algebra, the Im(f). For large contrasts between L_H and L_V it is easy to see that

$$\begin{pmatrix} \mathrm{Re}(f) \\ \mathrm{Im}(f) \end{pmatrix} \sim \gamma_H^{-1/2} \tag{21}$$

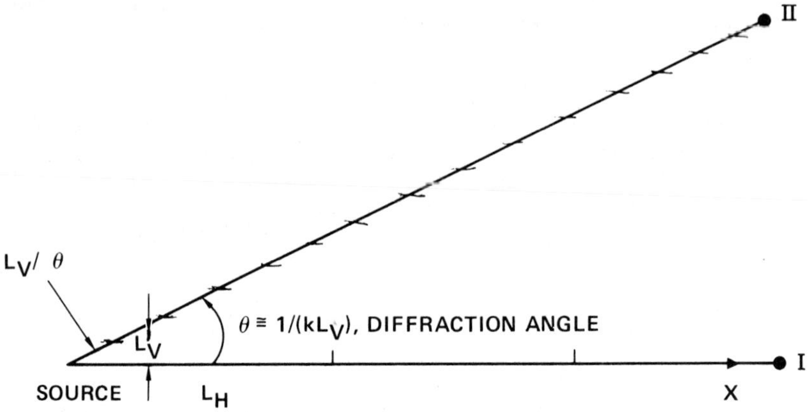

Fig. 1 Demonstration for markedly anisotropic media of the increased number of independent eddies through which the beam passes due to diffraction effects (see text).

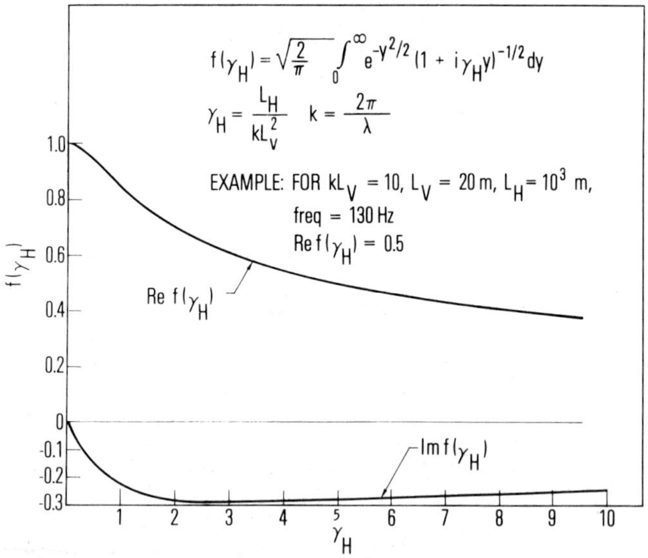

Fig. 2 Plot of the function $f(\gamma_H)$ giving the effect of anisotropy on the average field.

The phase variance then becomes

$$\phi^2 \simeq [(2\pi)^{1/2}/8]\varepsilon^2 k^2 L_V x(kL_H)^{1/2} , \qquad (22)$$

the ratio of the variances for the paths, II to I, is

$$(kL_V^2/L_H)^{1/2} \qquad (23)$$

This ratio cannot be fully, quantitatively obtained from the model because of the complications in the multiple scattering process, but the qualitative behavior is that described in connection with Fig. 1. Numerical computation gives the values for f (see Fig. 2). For small values of γ_H the propagation constant, because of changes in the imaginary part of f, becomes

$$k' = k(1 + \frac{1}{8} \gamma_H \varepsilon^2 kL_H). \qquad (24)$$

A slower wave; the slight decrease in the phase velociy comes (again) from the fact that the ray follows a zigzag path. This effect has been noted earlier (see Ref. 5). The result given in (19) has been given by a number of authors previously (see Ref. 1). Here it is obtained by a new method for the Gaussian-correlation medium. The physical model given the cause of the decrease in the field flutuation as a result of anisotropy has not yet been pointed out.

2. SECOND MOMENT (ACOUSTIC FIELD CORRELATION FUNCTION)

Define a "compensated phase" moment - one for which we calculate the average of the product of a random function by its corresponding complex conjugate. Because of this compensation, phase fluctuations are greatly reduced and the degradation of the average with range correspondingly reduced. Note that the average field itself, just discussed, is a not compensated phase quantity and thus shows a rapid decay with range. Define the correlation

$$\Gamma(f) = \langle G(x\hat{i}+f,0)\ G^*(x\hat{i},0)\rangle; \qquad (25)$$

using the Born series in the (9) this can be written (showing the correlaton measured at a range of x with x implicit),

$$\Gamma(f) = \sum_{m,n=0}^{\infty} \langle G_m(f)\ G_n^*(0)\rangle \qquad (26)$$

with the complex conjugate indicated by a star. (This conjugate we shorten to cc in the discussion). Proceed through 4th order Born, through terms of fourth order in ε. Use the multivariate Gaussian formula in a way similar to that described above and find for an isotropic medium that the terms in the correlation function can be summed to obtain the result (where erf is the error function)

$$\Gamma(\vec{\rho}) = G_o(\vec{\rho})G_o^*(0) \exp\left\{- \frac{(2\pi)^{1/2}}{4} \varepsilon^2 k^2 L_H x \times \left[1 - \frac{(2\pi)^{1/2}}{2} \frac{L_H}{\rho} \operatorname{erf}\left(\frac{\rho}{(2^{1/2}L_H)}\right)\right]\right\}. \qquad (27)$$

This has been shown by a number of other authors using other methods. For exemple see Tatarskii, Ref. 3. Indeed, his results are more general, being applicable to other than Gaussian-form correlation functions. It is noted in (27) that the correlation depends on the wavenumber k only through k^2 in the exponential, as a coefficent. This suggests that the result could be obtained from geometrical optics, allowing for phase fluctuations. In fact this is the case.

The anisotropic medium, however, presents complications. Following methods discussed here and a complicated argument (see Ref 1) it is shown that the fourth order Born term is not properly related to the second so that the Born series for the correlation function can be summed in exponential form. One might remain hopeful that the sum could be found in terms of some other elementary function, but the probability is low. Thus the surprising characteristic that certain of these low order moments can be summed break down here for the case of the anisotropic medium.

3. FOURTH MOMENT (THE INTENSITY VARIANCE)

Proceed now to substitute the Born series and calculate the normalized intensity variance (the correlations could be found) through fourth order terms. The moments needed are given as

$$
\begin{aligned}
[\langle I^2(0)\rangle - \langle I(0)\rangle^2]\langle I(0)\rangle^{-2} &= 4\varepsilon\langle(\mathrm{Re}G_oG_1^*)^2\rangle + \varepsilon^4[8\langle(\mathrm{Re}G_oG_1^*)(\mathrm{Re}G_oG_3^*)\rangle \\
&\quad + 4\langle|\mathrm{Re}G_oG_2^*|^2\rangle + 8\langle(\mathrm{Re}G_1G_2^*)(\mathrm{Re}G_oG_1^*)\rangle \\
&\quad + 4\langle|G_1|^2\,\mathrm{Re}G_oG^*\rangle + \langle|G_1|^4\rangle] = \sigma_I^2 , \qquad (28)
\end{aligned}
$$

We further specialize to isotropic media (the corresponding anisotropic result can be found with further effort). As is customary, having no other theortical guide, we use L_H in ϕ_o^2 for phase effects, and L_v in γ for diffraction effects. Further define a parameter J, the ratio of the eddy Fresnel length, kL_v^2 to the scattering-length-distance in which the average field is reduced to e^{-1}: x/ϕ_o^2. We have

$$
J = (2\pi)^{1/2}\,\varepsilon^2 k^3 L_H L_v^2/4 \qquad (29)
$$

Note J is small for relatively low frequencies and large for high frequencies. The comparison frequency for this determination of "low" and "high" frequencies is implicit in (29). We plot the intensity variance for J smaller and J larger respectively in Figs. 3 and 4. The definition of intensity variance given in (28) becomes unity for a fully Gaussian field. An interesting point is whether the variance will exceed this Gaussian value or not. The field according to the central limit theorem will become Gaussian at some (unspecified) large range, since the many multiply diffracted contributions will become statistically independent of one another at some such large range. In Fig. 3 we see that for frequencies relatively low it is believable that for small values of J the variance may approach the Gaussian value, unity, monotonically and thus not overshoot. (The shaded lines in Fig. 3 and 4 indicate the range-normalized-values for which the fourth order Born term is of the same order as the second. Thus, to the right of such shaded regions we cannot rely upon the calculation through fourth order shown in those figures). Concerning the overshoot it is clear that for values of J even as small as

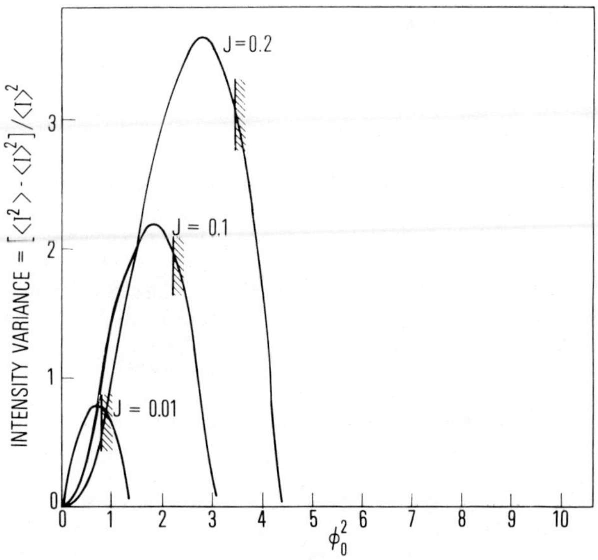

Fig. 3 *Plot of the normalized intensity variance as a function of the normalized range for smaller values of the quantity J (29). For nominal values* $L_{V_3} = 10^2$ *m,* $L_H = 10^3$ *m,* $k = 1\ m^{-1}$*,* $x = 10^4$ *m. We have* $J = 8 \times 10^{-3}$*. The curves are not valid to the right of the shaded lines.*

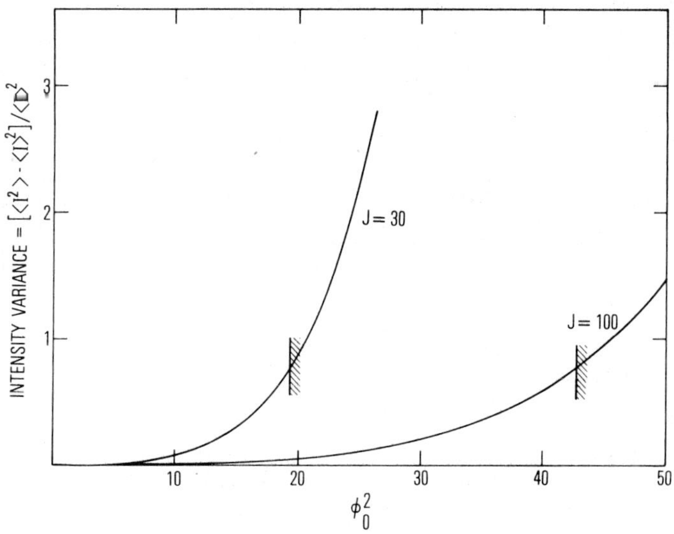

Fig. 4 *Plot of the intensity variance against normalized range for larger values of J (same as Fig. 3 with different values of J). The curves are not valid to the right of the shaded lines.*

0.1 there is substantial overshoot - of order a factor 2. As J increases in value this overshoot also increases. Indeed in Fig. 4, for large value of J, that is for high frequencies, we may reasonably expect a very large overshoot, since there is no tendency for the intensity variance to level out as a function of normalized range.

4. BRIEF DISCUSSION OF EXPERIMENTS

As yet, comparison with experimental results is somewhat difficult because of the necessity of dealing with single rays. In many experiments, until quite recently, it has been difficult to separate the data in this way. It is emphasized that for purposes of non-degraded signals, obtaining single rays is of prime importance. There is now some hope that platforms affixed to ice may successfully process the data (see Ref. 6).

It is interesting to examine some of the earlier experiments and see what the parametric values may be. Some of these are summarized in Table 1. We see that the values of J are uniformly large. Thus the intensity variance can be expected to greatly overshoot the Gaussian value, before settling down to a saturated field. It will be very interesting in future work to see if these intensity variances take on the characteristics described here. It is of course emphasized that we have posited a Gaussian-form correlation function. In fact power laws are more typical. When the appropriate data are available in large amount they can be examined with an eye to the effect of that correlation assumption.

Table 1 Parametric values from three sound experiments from Ref. 2.

Characteristic	Bermuda	Cobb	Azores
Range	550, 1250 km	17.2 km	35 km
No. of deterministic paths	14, 34	1	1
Frequencies	406 Hz	4000, 8000 Hz	400–5000 Hz
Sampling interval	300 s	15.7 s	10 s
Λ	0.12, 0.24	$10^{-3}, 5\times10^{-4}$	0.06–0.005
Φ	13, 20	5, 10	10–100
J	60, 69	1040, 7900	$69, 8.2\times10^{4}$

AKNOWLEDGMENT

This work has been supported by the Acoustics Branch of the Office of Naval Research.

REFERENCES

1. W.C. Meecham, M.T. Travis, and H.T. Yura, Born series sums for wave propagation in anisotropic random media, J. Acoust. Soc. Am., 70: 520-534 (1981).
2. S.M. Flatte, R. Dashen, W.H. Munk, K.M. Watson and F. Zachariasen, eds., "Sound Transmission through a Fluctuating Ocean", Cambridge University Press, Cambridge (1979).
3. V. Tatarskii, The Effects of the Turbulent Atmosphere on Wave Propagation. Springfield, VA, NTIS, 1971.
4. L. Chernov, "Wave Propagation in a Random Medium", McGraw-Hill, New York, NY (1960).
5. W.C. Meecham, On radiation in a randomly inhomogeneous medium, Space Technology lab. Rep.6120-0040-RU-000 (1961); Sound Propagation in a Randomly Inhomogeneous Medium. In: 4th International Congress on Acoustics, 1962, Paper 046.
6. B.J. Uscinski, private communication.

COUPLING OF ACOUSTIC MODES IN THE OCEAN, A COMPARISON OF APPROXIMATE SOLUTIONS

Yves Desaubies

Institut Francais de Recherche pour l'Exploitation de la Mer
B.P. 337 — 29273 Brest Cedex — France

INTRODUCTION

The representation of the acoustic field in a waveguide in terms of an infinite sum of normal modes is exact. However, when the properties of the waveguide are range-dependent, the modes (and their associated eigenvalues) can only be defined locally. The range variation of the amplitudes of the modes is a coupled set of ordinary second order differential equations. Under some conditions the coupling can be neglected and one obtains the adiabatic approximation to the solution[1,2]. We have discussed elsewhere[3,4] the conditions of validity of the adiabatic solution by carrying out a formal asymptotic series solution. It was found that for typical mesoscale eddies in the ocean coupling could be significant. Further modelling presented here confirms those findings.

Although an analytical approach is desirable to gain insight into the equations and the nature of the approximations, a useful solution can only be obtained numerically. There are various ways to derive the first order differential systems[5,6] required for numerical integration. We show here that they entail errors comparable to the asymptotic solution and that their range of validity (in terms of frequency, mode number, and range) is likewise limited.

THE COUPLED SET OF EQUATIONS

In a medium where the speed of sound $C = C(r,z)$ is a function of range and depth, one can, assuming cylindrical symmetry, seek a solution of the wave equation

$$[c^2\nabla^2 - \partial^2_{tt}]\ p = 0 \tag{1}$$

of the form

$$p = \mathrm{Re}\{r^{-1/2} \sum_m H_m(r)\ Z_m(r,z)e^{-i\omega t}]\} \tag{2}$$

Here (r, z, θ) represent cylindrical coordinates, $H_m(r)$ is a range dependent modal amplitude, and $Z_m(r,z)$ is a range dependent modal eigenfunction which satisfies:

$$[\partial^2_{zz} + k^2(r,z)]\, Z_m = \alpha^2_m(r)\, Z_m \tag{3}$$

$Z_m = 0$ at $z = 0$ (ocean surface); $\partial_z Z_m = 0$ at $z = h$ (rigid ocean bottom).

In the system (3), to be satisfied locally at each range, $\alpha^2_m = \alpha^2_m(r)$ are the eigenvalues, and the eigen modes can be orthonormalized according to:

$$\int_0^h Z_m(z')\, Z_n(z')dz' = \delta_{mn} \tag{4}$$

With these notations, the coupled system of equations governing the modal amplitudes can be shown to be[2,4]

$$[\partial^2_{rr} + \alpha^2_m(r)]H_m = \sum_{n\neq m} A_{nm}\, \partial_r\, H_n + \sum_n B_{nm}\, H_n \tag{5}$$

where

$$A_{nm} = 2(\alpha^2_m - \alpha^2_n)^{-1}\, U_{nm}, \qquad U_{nm} = \int_0^h (\partial_r k^2)\, Z_m Z_n dz \tag{6a}$$

$$B_{nm} = (\alpha^2_m - \alpha^2_n)^{-1}\, [V_{mn} + 2 \sum_s U_{sm}\, U_{sn}(\alpha^2_n - \alpha^2_s)^{-1} +$$

$$+ 2\, U_{mn}(U_{mm} - U_{nn})\, (\alpha^2_n - \alpha^2_m)^{-1}] \qquad m \neq n \tag{6b}$$

$$= 1/4 \sum_s A^2_{sn} \qquad n = m$$

$$V_{mn} \equiv \int_0^h (\partial^2_{rr}\, k^2)\, Z_m Z_n dz \tag{6c}$$

THE ADIABATIC APPROXIMATION

It is seen in Eq. 6 that the coupling terms involve range of derivatives of sound speed, through $\partial_r k^2$ (Eq. 6a) and $\partial^2_{rr} k^2$ (Eq. 6c). For a slowly changing environment these terms can be very small; if they are neglected Eq. 6 becomes decoupled

$$\partial^2_{rr}\, H_m + \alpha^2_m(r)\, H_m = 0 \tag{7}$$

with the approximate WKB solution

$$H_m(r) = \alpha_m^{-1/2}\, [C_+\, e^{i\theta_{om}} + C_-\, e^{-i\theta_{om}}] \tag{8}$$

where $C_\pm$ are constants and the WKB phase is given by

$$\theta_{om}(r) = \int_0^r \alpha_m(r')dr' \tag{9}$$

The constants must be determined from initial or boundary conditions. Usually an initial condition is given

$$H_m(0) = Z_m(0,\, z_o) \tag{10}$$

for a point source at z_o and a radiation condition of the form

$$\lim_{r\to\infty} (\partial_r H_m - i\alpha_m H_m) = 0 \tag{11}$$

is imposed, leading to the adiabatic solution

$$H_m(r) = Z_m(z_o)\ [\alpha_m(0)/\alpha_m(r)]^{1/2}\ e^{i\theta_{om}} \tag{12}$$

The effect of the radiation condition has been to discard the backscattered wave.

The main advantage of the adiabatic solution is that it decouples the system of Eq. 5; it still requires that the eigenvalue problem (3) be solved at a sufficient number of range steps to allow an accurate evaluation of the phase (9). The drawback of this solution is that it is not easy to assess the limits of its validity. Indeed while some of the effects of the range variability of the medium are accounted for by the use of local modes and the WKB approximation, the coupling effects of the medium have been neglected. It is not obvious that these are consistent approximations. Similarly the circumstances under which the backscattered field is negligible must be investigated.

Two approaches are available to assess the validity of the adiabatic approximation: one is to develop a systematic series expansion of the solution, taking advantage of the smallness of the rate of change of the medium along the direction of propagation, as was done by Desaubies et al[4]; the other is to implement an exact numerical solution of the coupled system (5). We discuss this latter procedure first.

NUMERICAL SOLUTION OF THE SYSTEM (5)

A numerical solution is most easily implemented if the problem is cast as a first order differential system. This can be achieved in a number of ways[5,6]. Expecting the solution to be essentially of the form (8), two functions, representing the forward and backward propagating waves, are defined by

$$H_n = \alpha_n^{-1/2}\ [F_n \exp(i\theta_n) + G_n \exp(-i\theta_n)] \tag{13}$$

$$\partial_r H_m = r\ \alpha_n^{1/2}\ [F_n \exp(i\theta_n) - G_n \exp(-i\theta_n)] \tag{14}$$

It follows that the system (5) is equivalent to

$$(\partial_r F_n)e^{i\theta_n} = \frac{1}{2}\frac{\alpha_n'}{\alpha_n} e^{-i\theta_n} G_n + \frac{1}{2}\sum_m A_{mn}\left(\frac{\alpha_m}{\alpha_n}\right)^{1/2} [F_m e^{i\theta_m} - G_m e^{-i\theta_m}]$$

$$-i/2 \sum_m B_{mn}(\alpha_m\alpha_n)^{-1/2}\ [F_m e^{i\theta_m} + G_m e^{-i\theta_m}] \tag{15a}$$

$$(\partial_r G_n)e^{-i\theta_n} = \frac{1}{2}\frac{\alpha_n'}{\alpha_n} e^{i\theta_n} F_n - \frac{1}{2}\sum_m A_{mn}\left(\frac{\alpha_m}{\alpha_n}\right)^{1/2} [F_m e^{i\theta_m} - G_m e^{-i\theta_m}]$$

$$+i/2 \sum_m B_{mn}(\alpha_m\alpha_n)^{-1/2}\ [F_m e^{i\theta_m} + G_m e^{-i\theta_m}] \tag{15b}$$

It is usual then to neglect the backscattered G_n on the right hand side of Eq. 15a which becomes

$$\partial_r F_m = \frac{1}{2}\sum_m A_{mn} \left(\frac{\alpha_m}{\alpha_n}\right)^{1/2} e^{i(\theta_m - \theta_n)} F_m \tag{16}$$
$$-1/2 \sum_m B_{mn}(\alpha_m\alpha_n)^{-1/2} e^{i(\theta_m - \theta_n)} F_m$$

Equation (13) is then

$$H_n = \alpha_n^{-1/2} F_n e^{i\theta_n} \tag{17}$$

while $\partial_r H_n$ is given by

$$\partial_r H_n = [i\alpha_n^{1/2} F_n - \frac{1}{2}\alpha'_n \alpha_n^{-3/2} F_n + \alpha_n^{-1/2} \partial_r F_n] e^{i\theta_n} \tag{18}$$

and not by Eq. 14 which would lead to an incompatibility.

It is hard to justify rejecting the G_n's while keeping the α_n's and A_{nm} terms in Eqs. 18 and 16, since it is precisely similar terms which generate the G_n's in Eq. 15b. In fact it is not entirely clear how the boundary value problem must be posed to determine uniquely the backscattered field: in an infinite, range-varying medium the radiation condition cannot be expressed simply; if the medium is assumed to become stratified beyond some range the amount of backscatter depends on the extent of the non uniform section. A possible solution would be to introduce a small dissipation; we have not attempted this yet and will not consider further the backscattering problem. Instead we focus on finding a solution according to the following criteria:

1. It must satisfy Eq. 5 and appropriate initial conditions (Eq. 10);

2. In the limit where the medium becomes stratified it must reduce to the usual outgoing solution $\sim \exp(i\alpha_n r)$;

3. The energy flux, integrated over depth, must be constant.

THE ENERGY FLUX

It follows from Eq. 2 and the basic acoustic equations, that the radial velocity component u is, in the far field, given by

$$u=(\omega\rho_o r^{1/2})^{-1} \mathcal{I}m\,[\sum_m Z_m \partial_r H_m + H_m \partial_r Z_m]\, e^{-i\omega t} \tag{19}$$

The energy flux through a cylinder of radius r and height h is then (* denoting complex conjugate)

$$\mathcal{F} = 2\pi r \int_o^h \langle pu\rangle dz = \frac{-\pi}{\omega\rho_o} \mathcal{I}m \{\sum_m H_m \partial_r H^*_m + \frac{1}{2} \sum_{m,n} A_{mn} H_m H^*_n \} \tag{20}$$

(the relation $\partial_r Z_n = \frac{1}{2}\sum_m A_{mn} Z_m$ has been used). Invoking Eqs. 13, 16, 18 and 20 we obtain the expression for the flux:

$$\mathcal{F} = \frac{\pi}{\omega\rho_o}\left[\sum_m |F_m|^2 - \frac{i}{4}\sum_{\ell m}\frac{A_{\ell m}}{(\alpha_\ell\alpha_m)^{3/2}}(\alpha_\ell - \alpha_m)^2 F_\ell F_m^* e^{i\theta_{\ell m}} - \frac{1}{4}\sum_{\ell m}(\alpha_m B_{m\ell} - \alpha_\ell B_{\ell m})(\alpha_\ell\alpha_m)^{-3/2} F_\ell F_m^* e^{i\theta_{\ell m}}\right] \tag{21}$$

where $\theta_{\ell m} \equiv \theta_\ell - \theta_m$

On the other hand one obtains immediately from Eq. 16

$$\partial_r \sum_m |F_m|^2 = \frac{1}{2}\sum_{m\ell}\frac{A_{\ell m}}{(\alpha_\ell\alpha_m)^{1/2}}(\alpha_\ell - \alpha_n) F_\ell F_m^* e^{i\theta_{\ell m}} - \frac{i}{2}\sum_{\ell m}(B_{\ell m} - B_{m\ell})(\alpha_\ell\alpha_m)^{-1/2} F_\ell F_m^* e^{i\theta_{\ell m}} \tag{22}$$

It follows therefore that the solution of Eq. 16 does not satisfy the conservation of energy flux. One can show similarly, by using Eqs. 16 and 18 to calculate $\partial^2_{rr}H_n$ and substituting into Eq. 5 that the original equation is not satisfied either. How serious is the mismatch and how self consistent are the approximations leading to Eq. 16 can only be assessed by introducing a formal asymptotic series expansion, as was done by Desaubies et al[4].

ASYMPTOTIC SOLUTION OF EQUATION 5

The asymptotic solution explicitely recognizes the slowness of the horizontal variations of the medium by introducing a "slow" variable $X = \varepsilon r$ where $\varepsilon \ll 1$, and considering that $C = C(X,z)$. Hence $\partial_r C = \varepsilon\partial_X c$ is small. We also note that in Eq. 5 $A_{m\ell} = O(\varepsilon)$, $B_{m\ell} = O(\varepsilon^2)$. The eigenfunctions and values are now functions of the slow variable: $Z_m = Z_m(X,z)$, $\alpha_m = \alpha_m(X)$ and the phase θ_{om} is redefined as

$$\theta_{om}(X) = \int_o^X \alpha_m(X')dx'$$

so that Eq. 8 becomes, for instance

$$\partial_r H_m = \varepsilon\partial_X H_m = i\alpha^{1/2}\exp[i\theta_{om}/\varepsilon] + O(\varepsilon)$$

The solution of Eq. 5 was found to be of the form

$$H_m(X) = H_m^{(0)} e^{i\theta_m/\varepsilon} + i\varepsilon\sum_n a_{nm}(x)\, e^{i\theta_n/\varepsilon} + O(\varepsilon^2)$$

$$H_m^{(0)}(X) = Z_m(z_o)\,[\alpha_m(0)/\alpha_m(x)]^{1/2}$$

$$a_{nm}(X) = H_m^{(0)} A_{nm}(x)\,\alpha_n / (\alpha_m^2 - \alpha_n^2) \tag{23}$$

$$\theta_m = \int_o^X \alpha_m \left\{1 + \frac{1}{2}\varepsilon^2\left[\sum_n A_{nm}^2(\alpha_m^2 - \alpha_n^2)^{-1} + \alpha_m^{-2}\left(\frac{3}{4}\left(\frac{\alpha_m'}{\alpha_m}\right)^2 - \frac{1}{2}\frac{\alpha_m''}{\alpha_m} - B_{mm}\right)\right]\right\}$$

This solution satisfies the three requirements listed above, to $O(\varepsilon^2)$*; its main features have been discussed elsewhere[3,4]. We note here the phase θ_m which is different from the adiabatic phase θ_{om} (Eq. 9), implying that the latter will give an error in the solution as the range increases; and the presence in the denominators of the correction terms of factors $\alpha_m - \alpha_n$ involving eigenvalue differences which can be very small. This contributes to making those terms large and invalidating the series solution. It was found that for propagation through ocean eddies in the NW Atlantic the validity of the adiabatic and series solutions was a sensitive function of range, frequency and mode number, the approximations being increasingly better as those parameters were smaller.

Similar results are obtained for acoustic modal propagation in the NE Atlantic. Sound speed profiles collected in 83 during the French campaign TOPOGULF (Fig. 1) were used as input. Ten modes at 200 Hz were propagated to ranges up to 500 km. Figure 2b shows at a range of 196 km the magnitude of the pressure as a function of depth, calculated numerically (from Eq. 16), and in the asymptotic form (Eq. 23); we conclude that for this eddy field coupling is significant and terms of $O(\varepsilon^2)$ cannot be neglected. In such a case neither of the three solutions presented here is accurate. In contrast, Fig. 2a shows the solutions (which are indistinguishable) calculated for a perturbation field δc reduced to 20% of its value in Fig. 2b. Now the solutions are nearly identical, but rather different from the fully perturbed field.

Another quantitative measure of the magnitude of coupling is provided by the correction terms in Eq. 23: a_{nm} and the ε^2 correction to the phase. Those are given in the table below for a few mode numbers and ranges, where $\delta\phi_n$ refers to the phase correction:

$$\delta\phi_n \simeq \alpha_n \times \varepsilon^2 \sum_n A^2_{nm} (\alpha^2_m - \alpha^2_n)^{-1} \tag{24}$$

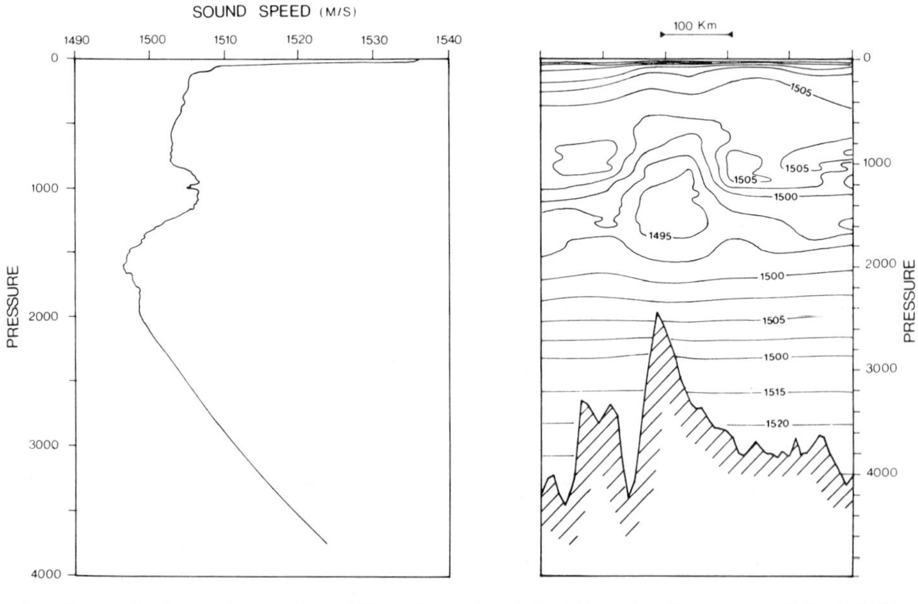

Fig. 1 Typical sound speed profiles from the N.E. Atlantic (Courtesy of M. ARHAN).

*When we state that two expressions are equal to $O(\varepsilon^2)$ we mean that their difference is $O(\varepsilon^2)$.

The values in the table, which are often close to or much larger than one, confirm the importance of coupling; they also show that in this case the environment does not vary slowly; and they indicate that the results of Ref. 4 on the dependence of those terms on mode number, frequency and range was a particular case related to the nearly bilinear form of the sound speed profile. Here we cannot detect any systematic variation.

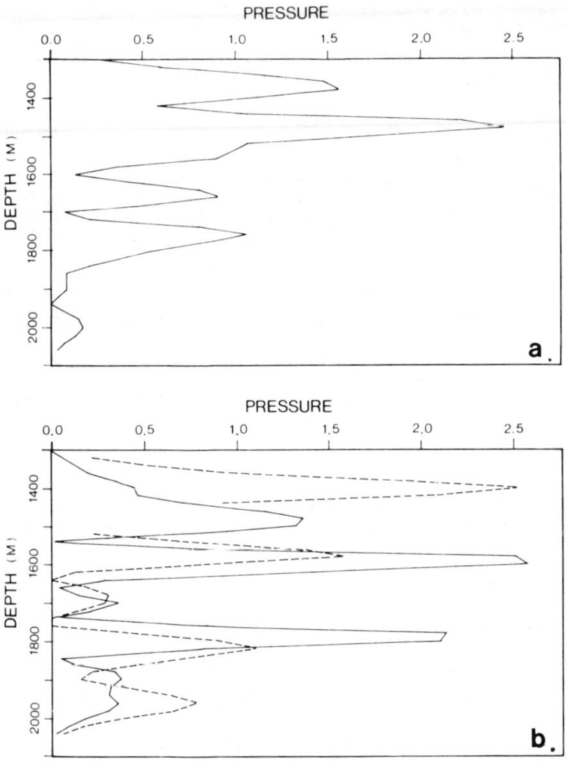

Fig. 2
Profiles of sound pressure as a function of depth.
Top, Fig. 2a: For very weak δc perturbations all solutions are identical.
Bottom, Fig. 2b: For the field of Fig. 1, the numerical and asymptotic solutions.

TABLE 1

MAGNITUDE OF A FEW CORRECTION TERMS

Upper entry: $\delta\phi_n$ (Eq. 24); lower entry $a_{n,n+1}$ (Eq. 23).
Approximate solutions are valid only when those terms are <<1.

Range (km)	Mode number 1	2	3	4	5
67	$2.8\ 10^{-2}$ 0.04	$5\ 10^{-2}$ 0.06	$9\ 10^{-2}$ 0.04	$6\ 10^{-2}$ 0.10	$8\ 10^{-2}$ 0.17
132	$8.8\ 10^{-3}$ 0.02	1.32 0.31	57.6 1.51	1.65 0.41	$5.57\ 10^{-1}$ 0.22
196	$1.2\ 10^{-1}$ 0.05	$6.0\ 10^{-2}$ 0.07	$8.6\ 10^{-2}$ 0.15	$6.6\ 10^{-2}$ 0.13	$1.2\ 10^{-1}$ 0.17
262	7.16 0.52	1.30 0.38	40.3 0.25	14.3 0.75	3.75 0.56

CONCLUSIONS

The results presented here confirm those of Ref. 4 that in presence of moderate ocean eddies, coupling between acoustic modes can be significant. In that case the three approximations discussed here (adiabatic, Eq. 12); numerical, Eq. 16); and asymptotic, Eq. 23) cease to be valid. The problem of finding a first order system for numerical solution is difficult; we have shown that the system (Eq. 16) yields solutions which entail two kinds of errors: those related to the backscattered field, and those attributable to the $0(\varepsilon^2)$ field which is not properly accounted for. It can be shown that the ratio of the backscattered to the forward coupled field is of the order

$$\Delta\alpha/\alpha = (\alpha_n - \alpha_m) / \alpha_n \simeq 0(10^{-3})$$

so that it can safely be ignored. On the other hand, the $0(\varepsilon^2)$ terms, including the phase correction in Eq. 23, can be significant and must be included.

It is well known that the decomposition of the field into forward and retrograde waves (Eq. 13) is not unique[6 7]; it depends on the subsidiary condition (Eq. 14). For instances Boyles[6] derives a first order system by introducing the modal amplitudes ($\bar{H}_m$ say) of the velocity field. With the condition

$$\bar{H}_n = i\alpha_n^{1/2} [F_n e^{i\theta_n} - G_n e^{-i\theta_n}]$$

in place of Eq. 14 the system

$$\partial_r F_n = \frac{1}{4} \sum_m A_{mn} (\alpha_m + \alpha_n) (\alpha_m \alpha_n)^{-1/2} F_m e^{i\theta_{mn}}$$

is obtained instead of Eq. 16. The two are seen to differ at $0(\varepsilon)$ by terms of $0(\Delta\alpha/\alpha)$, and to satisfy different equations at $0(\varepsilon^2)$; when $0(\varepsilon^2)$ terms cannot be neglected (as in Fig. 2b), both systems yield erroneous solutions.

We suggest that one way to improve the numerical solution is to incorporate the phase θ_m given in Eq. 23 into the system (13) and (16).

REFERENCES

1. A.D. Pierce, Extension of the method of normal modes to sound propagation in an almost-stratified medium. J. Acoust. Soc. Am., 37, 19-27 (1965).
2. D.M. Milder, Ray and wave invariants for SOFAR channel propagation, J. Acoust. Soc Am., 46, 1259-1263 (1969).
3. Y. Desaubies, A uniformly valid solution for acoustic normal mode propagation in a range varying ocean. J. Acoust. Soc. Am., 76, 624-626.
4. Y. Desaubies, C.S. Chiu, J. Miller, Acoustic mode propagation through mesoscale eddies in the ocean. J. Acoust. Soc. Am., submitted (1985).
5. S.T. McDaniel, Mode coupling due to interaction with the seabed. J. Acoust. Soc. Am., 72, 916-923 (1982).
6. C.A. Boyles, Acoustic waveguides. John Wiley and sons, 321 pp. (1984).
7. F.W. Sluijter, Arbitrariness of dividing the total field in an optically inhomogeneous medium into direct and reversed waves. J. Opt. Soc. Am., 60, 8-10.

LONG RANGE, LOW FREQUENCY ACOUSTIC BACKSCATTERING: A SURVEY

Arthur B. Baggeroer and Ira Dyer

Massachusetts Institute of Technology
Cambridge, Massachusetts 02139

ABSTRACT

At very low frequencies the signals observed from an impulsive source are dominated by scattering from the seafloor. Experiments demonstrating this have ranged from echoes locally reflected by seamounts to transoceanic reverberation from continental margins. Receivers have included omnidirectional sensors, towed line arrays, two dimensional horizontal arrays and vertical arrays.

Backscattering highlights have been observed from known topographic features as well as from tantalizing unknown ones. This suggests the use of backscattering signals for reconnaissance in poorly charted areas. One of the major uncertainties lies in the relation of the backscattering structure as a function of frequency to the topography (rms roughness and bathymetry). The experiments on acoustic backscattering and their data analyses are surveyed in this context.

1. INTRODUCTION

Acoustic backscattering at long ranges has been observed by many experimenters over a wide band of frequencies. As early as 1952, Luskin, et. al. conducted a series of experiments in the Caribbean which correlated backscattering with topographic features and even identified unsurveyed seamounts [1]. Later Mellen and Marsh noted that "reverberation in the Arctic is observable for no less than 30 minutes after an explosive detonation of moderate size" [2]. Side swipe echoes are often observed on real time monitors in the data acquisition for oil exploration. It should not be too surprising since the basic geometry of sidescan sonar is essentially the same. Scaling the frequency down by several orders of magnitude (100 kHz to 100 Hz) and the features of the targets up by the same (1 m to 1 km) should increase the ranges from a few hundred meters to few hundred kilometers. While this may be a very simplistic interpretation in that it ignores problems in the scaling, it does suggest that backscattering is more prevalent than commonly expected.

Low frequency backscattering experiments do present a unique set of problems not encountered at higher frequencies. The long wavelengths imply large array apertures for high angular resolution. Except for towed or ice drifting arrays this has been difficult to realize. Propagation analysis at low frequencies is more difficult because of the bottom interaction, so obtaining scattering strengths of targets is more difficult. Moreover, at very low frequencies absorption and ice cover losses are uncertain. While these are small, the ranges are large, and they do become a factor. Finally, explosives are still the only practical sources, and there is still some issues in signature of these, particularly for large ones when cavitation may be present.

While there have been several backscattering experiments which have produced echoes that can be correlated with topographic highs, the theoretical analysis for the target strength has received far less attention almost assuredly because it is a very difficult problem. The backscattering centers are associated with bathymetric relief either from the continental margins, seamounts or ridge escarpments. These features are seldom surveyed well enough to determine which of the several scattering theories are appropriate. Is the mechanism dominated by "highlights" off edges, specular scattering off locally smooth surfaces, or diffuse scattering? Much remains to be done theoretically if the data are to be interpreted correctly.

Our intent here is to survey low frequency, long range acoustic backscattering. Specifically, our attention is on long range, therefore low grazing angle, experiments and not on high frequency nearly vertical backscattering from the seafloor where there is a very extensive literature. Our discussion is divided into four parts: i) previous experiments; ii) the signal processing methods used to resolve the backscattering; iii) propagation and scattering processes; and sonar system implications. Results from the CANBARX, Fram II and Fram IV experiments in the Arctic are used as examples. While somewhat parochial, these data are the most readily available to the authors.

2. BACKSCATTERING EXPERIMENTS

Low frequency backscattering of explosive signals echoing from topographic features were observed and reported in several early experiments. Luskin, et al, reported on the effects of topographic echoes as early as 1952 and did an extensive travel time and amplitude analysis correlating returns with many features in the Caribbean [1]. Mellen and Marsh, [2] and Kutschale [3] noted the reverberation produced by explosive sources and their correlation with topography. These early experiments used small explosives, less than 25 kg, and omnidirectional receivers. The observations were simply correlated with the travel times expected from known bathymetric features.

Using omnidirectional sources and receivers on a grander scale, reverberation studies were made from several shots of the ARTEMIS experiment in the North Atlantic and the CHASE experiments in both the Atlantic and Pacific. These shots used very large charges ranging 10 tons to 1 kilotons and bistatic receivers. Northrop [4] identified the margins and seamount chains as the primary reverberation mechanism at long ranges. Goertner and Blatstein [5,6,7] analyzed the same data, and by computer modeling of the bathymetry and travel times demonstrated in a convincing manner that the reverberation levels were correlated with continental margin and seamount chains suggesting transoceanic backscattering and decay durations up to several hours. For the very high yield shots they attributed the late arrivals in terms of multiple scattering. Their primary conclusion was that variations in reverberation levels could be accounted for by geometric considerations.

The first experiment using directional receivers was done by Spindel and Heirtzler [8]. Over 800 SUS charges, 1.8 lbs of TNT, were detonated at 243 m and 457 m on an 800 km line in the vicinity of the Bahamas and the Blake Plateau. The receiver was a 30 m towed array with directly summed hydrophones, i.e. a broadside beam, towed at 600 m. An analysis band of 30 - 200 Hz was used. Strong echoes 3.5 - 8.5 min corresponding to ranges of 158 - 385 km were received. The range time spread was approximately 24 s. Prominent features of topographic relief could be correlated with the echoes. The topography was well charted and known to be steep which led to the strong echoes. In addition, the topographic relief was confined primarily on one side so the port/starboard ambiguity intrinsic with towed arrays was not present.

In a series of ice camps in the Arctic Ocean, Dyer and colleagues have performed several experiments on low frequency, transoceanic backscattering [9-11]. The unique aspect of these experiments has been use of the ice canopy for deploying two dimensional hydrophone arrays with 11 to 24 channels with apertures up to 1.5 km by 1.5 km. Fig. 1 illustrates one of the arrays used during the Fram II experiment in 1980 [12]. This permitted angular resolution of approximately 6 degrees in azimuth using recently developed array processing algorithms of the reverberation returns. Explosive charges up to 800 kg were used. Topographic highs associated with continental margins, mid-oceanic ridges and known escarpments were repeatably resolved up to ranges of 2700 km. Several scattering highs were identified which do not correlate with known bathymetry but do with magnetic anomalies. Fig. 2a indicates one of the processed returns which can be well correlated with the known bathymetry indicated in Fig. 2b for ranges up to 1000 km. These unknown scattering locations are currently being examined. These experiments demonstrated that extremely large explosive yields (they were a factor of 10 to 1000 less than the ARTEMIS and CHASE shots) are not necesary to excite basin wide reverberation if one has enough array gain and resolution to resolve the spatial struture of the field. One of the more remarkable observations was that backscattering from features known to be very deep and well below the sound channel was evident.

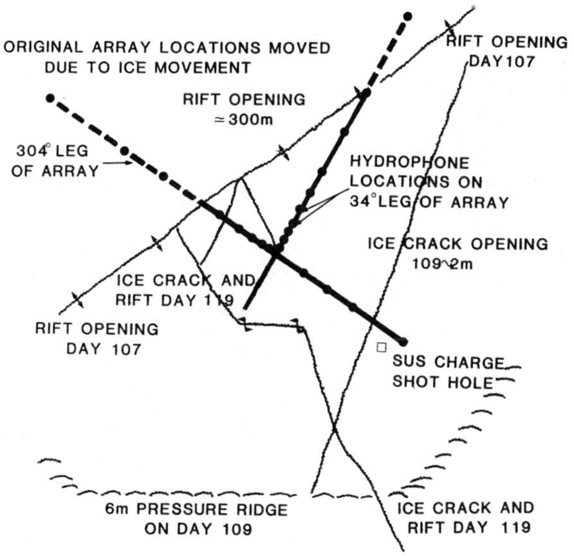

Fig. 1: Two dimensional array used for backscattering experiments at the FRAM II ice station. The shot was 5 km south and set off at 243m.

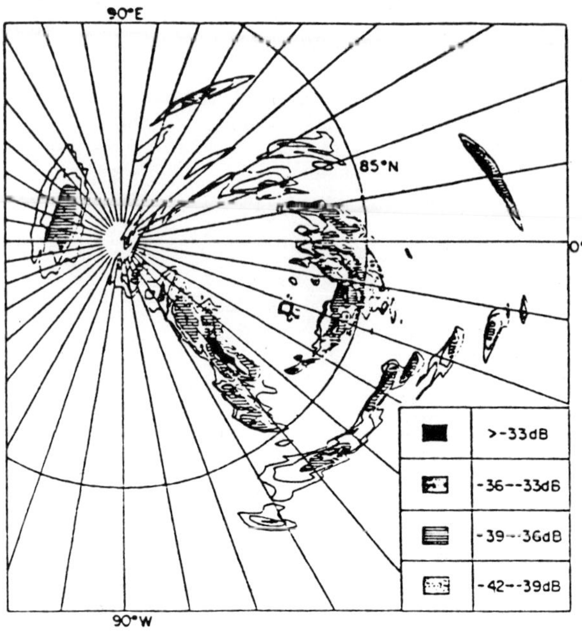

Fig. 2a: Backscattering levels measured at FRAM II from a 400 kg shot

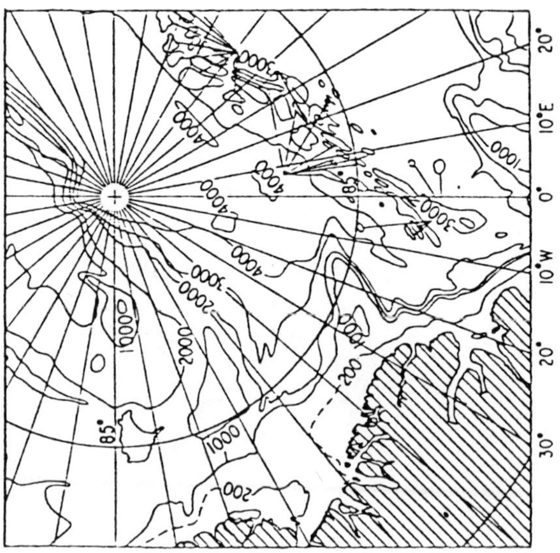

Fig. 2b: Bathymetry of the Arctic Ocean near FRAM II

Multichannel towed array experiments have recently been reported by two groups. The Naval Research Laboratory has performed two experiments, one northeast of New Zealand and one west of Gibraltar, which mapped both known and uncharted topographic features in 2000 km square regions [13,14]. The array forms a set of broadband beams which are stabilized to account for array crabbing. Port/starboard ambiguity is resolved by incoherent averaging over several azimuths, in effect producing a backscattered tomographic section of the basins with feature resolution of 2 - 6 km in azimuth and 2 km in range. Moreover, algorithms for converting backscattered levels to bathymetry are developed which compare well with other remote sensing methods.

SACLANT Center has also recently conducted towed array experiments in the Tyrrhenian Sea [15]. Data from several shots were processed in eight frequency bands from 20 to 750 Hz using both time domain and adaptive beamforming. Scattering from topographic features from seamounts and the coastline for ranges up to 100 km again dominated the processed data. Scattering strengths were measured and reported to be consistent with previously published data.

2. SIGNAL PROCESSING FOR LOW FREQUENCY BACKSCATTERING EXPERIMENTS

The early backscattering experimenters used omnidirectional receivers to process their data or bistatic observations combined incoherently, i.e. by mean square levels. Most of the recent backscattering experiments have used arrays to extract directionality information. The signal processing used in the analysis of low frequency, backscattering data is especially important because the array almost certainly has limited apertures compared to the wavelengths. The objectives are both challenging and different than much of the emphases in the current literature. The basic problem is to estimate the directional spectrum of a signal field that is only "short term stationary" because of the transient nature of the explosive shot and the range dependence of the reverberation. "High resolution" in both bearing and range is not as important as in target tracking problems. In contrast, stability, sidelobe suppression and biases are far more important in measuring scattering cross sections reliably. These parameters, however, are limited by the small "time-bandwidth" product of the signals scattered from a particular feature.

The array processing methods used to date can be classified into three categories: i) broadband, time domain beamforming [8,13,14]; ii) frequency domain beamforming [9-11]; and iii) parametric methods [15]. These are illustrated in Fig. 3. In the time domain processing the signals are simply delayed, weighted and summed as in conventional sonar beamforming. If implemented digitally, the major constraint is that the sampling rate must be five to ten times the signal bandwidth to avoid beam distortion due to interpolation errors. Typically, the relatively short arrays lead to poor sidelobe patterns; however, with wideband signals the sidelobes do not interfere constructively across a wide bandwidth since their separation from the main lobe scales inversely with frequency so this problem is mitigated to a degree. The major advantage of the time domain approach is its simplicity and robustness; the disadvantage is that it produces a noisy response and is most useful with strong, directional signals and not the diffuse, low level signals encountered in backscattered signals.

Frequency domain methods are based upon the methods used for ambient noise studies in acoustics and seismic arrays [17-19]. In its simplest form it is a direct dual of the time domain beamforming. The data are transformed with an FFT, steered with phase shifters for the

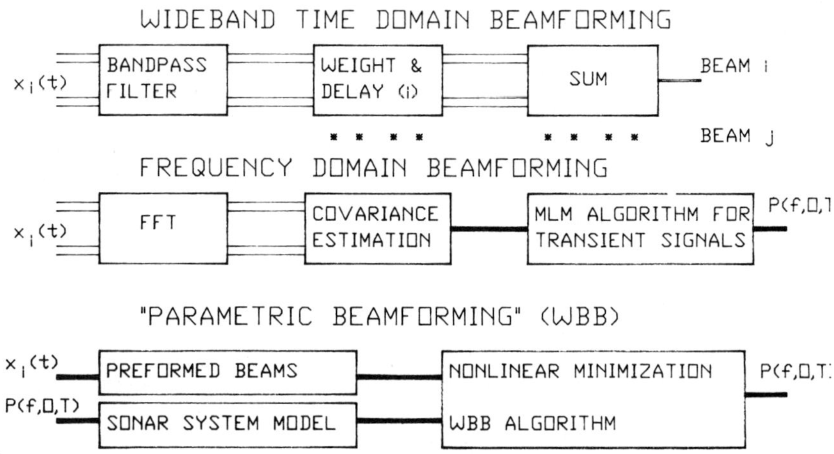

Fig. 3: Array processing methods used for backscattering experiments

dual operation of the time delays, and energy detected. In this form it has all the same advantages and disadvantages as the time domain approach, so the relative merits are based on computational efficiencies. Formulation in the frequency domain does permit the use of adaptive array processing algorithms which have some useful properties in terms of sidelobe reduction. Most algorithms developed for time series analysis can be used for towed arrays with equally spaced elements; however, the choice is far more restrictive with two dimensional arrays and unequally spaced elements. Currently, the maximum likelihood method is the only one which has been used because of its applicability to arbitrary array geometries such as those used in [7,17]. In addition, virtually all the signal processing literature focuses on resolution while the quantification of the biases in virtually all the adaptive algorithms remains essentially an open issue because of their nonlinear nature, so one must use some degree of care in computing scattering cross sections.

The "parametric" approach is based upon an algorithm introduced by Wagstaff [20]. Essentially, it assumes the frequency- wave number function of the ambient field can be parameterized by sectors. The effect of these sectors upon the output of a set of preformed beams is modeled in terms of the response of the processing system. When converted to decibels the formulation leads to a nonlinear minimization of the field entropy subject to a least squares constraint. The formulation is appealing because of its direct coupling of the field parameters through the processing system; however, some of the mathematical issues associated with the parameterization such as the number of degrees of freedom permitted, continuity constraint and bounds, as well as the minimization algorithm are still under investigation.

The issue that makes the backscattering array processing different from ambient noise analysis is the time varying nature of the reverberation. Since stability in spectral estimation is inversely proportional to the time-bandwidth product of the available data, one must compromise the length of the analysis windows and the stationarity of the signal field. The duration of the scattering process from a particular scattering complex appears to be approximately 5 - 20 s, corresponding to range resolution 7 - 15 km, so window lengths have been chosen accordingly. At long ranges the signal bandwidth is necessarily low due to high propagation loss, so backscattering energy above 20 Hz from features at ranges beyond 500 km is quite small. This places an upperbound on the degrees of freedom available for achieving stable estimates. Moreover, the frequency selective properties of the propagation usually require using bandwidths of just a few Hertz.

Several experiments have "imaged" scattering centers from different locations. In towed array applications this is necessary to eliminate the port/starboard ambiguity implicit with the linear geometry. In the two dimensional ice camp experiments this was done to resolve regions of unusually high backscattering compared to the known topography. Fig. 4 indicates backscattering maps generated separately from the Fram II and Fram IV camps in the Arctic Ocean. Correlation with the map of the current bathymetry, as demonstrated in Fig. 2b, indicates that most of the backscattering highs are associated with the continental margins and the Lomonosov and Nansen midocean ridges. There are also some regions, e.g. one at (87 N, 62 E) along a region of low relief of the Nansen ridge and another between the Morris Jessup Rise and the Yermak Plateau, which corresponds to a fracture zone offset of the mid oceanic ridge, which both indicate high and unexpected scattering.

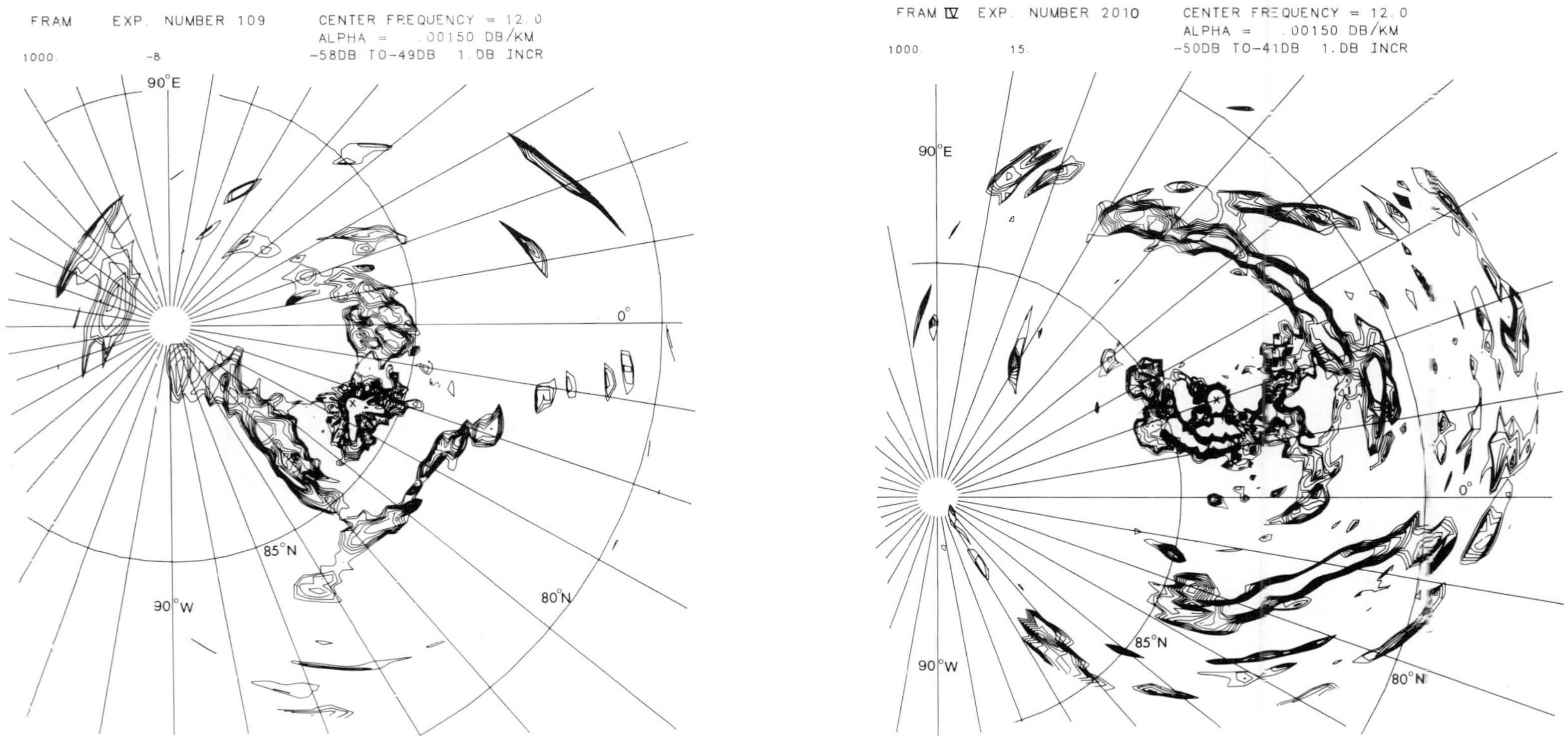

Fig. 4: Backscattering from FRAM II on the left and from FRAM IV on the right processed in identical manner. These should be viewed in context with the bathymetric map of this part of the Arctic in Fig. 2b. Note the coincident highs along the Lomonosov and Nansen ridges. The Nansen ridge is approximately 2500 m deep while the Lomonosov rises to 900 m in some regions, yet the backscattering intensity is comparable. Also note the region of coincident highs not correlated with previously published bathymetric maps.

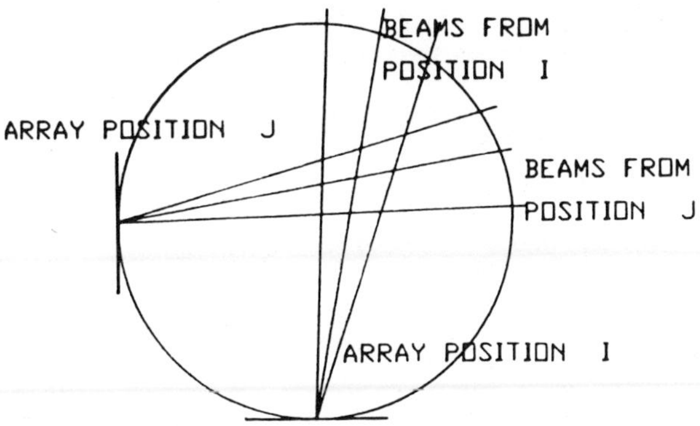

Fig. 5: Backscattering measurement using towed arrays positioned around a scattering complex. Conceptually, the scattering levels as a function of source/receiver location, target range and bearing can be inverted to obtain an estimate of scattering strength.

When this concept is extended to include many perspectives, it becomes a backscattering tomography experiment, so many of the same signal processing issues previously encountered in other fields should be appropriate. In fact, it is very close to tomography as used in CAT scanners for medical tomography with the major differences being the use of backscattered energy instead of forward scattered energy and the use of several beams from one source-receiver location. Fig. 5 indicates the concept. Many perspectives are needed, however, to make this concept work since the processing gain depends upon the energy combination which grows roughly inversely with the number of independent images.

3. ESTIMATING BACKSCATTERING STRENGTH

Converting an estimate of the time varying directional spectrum into a map of backscattering strength involves a number of compensations for the explosive sources used, the propagation to and from the target as well as receiver effects. Moreover, the target scattering physics are very complicated and not well understood.

Any estimate of scattering strength requires and accurate model of the source signature. Several extensively cited authors, e.g. Weston, [21] and Wakeley [22}, have published formulae characterizing the signature. Since large, explosive sources are used, sound pressure can exceed the cavitation threshold of the water especially at shallow depths. Measurements of source signatures and cavitation effects have been made extensively, e.g. [21] and [23]. Weston suggests the use of a cavitation cone within which the sound pressure levels exceed 245 db re 1 uPa, the maximum pressure that the water surface can instantaneously support before cavitation onset. In more detail Gaspin, et al.[23] indicate a cavitation region where the surface reflected pulse induces cavitation. For long range propagation the cavitation issue needs further refinement. First, the bubble pulse and not the shock wave is the primary source of energy, and it cavitates well below the shock pulse. Secondly, only the energy radiated at angles which do not have significant bottom interaction are of concern. Fig. 6 schematically

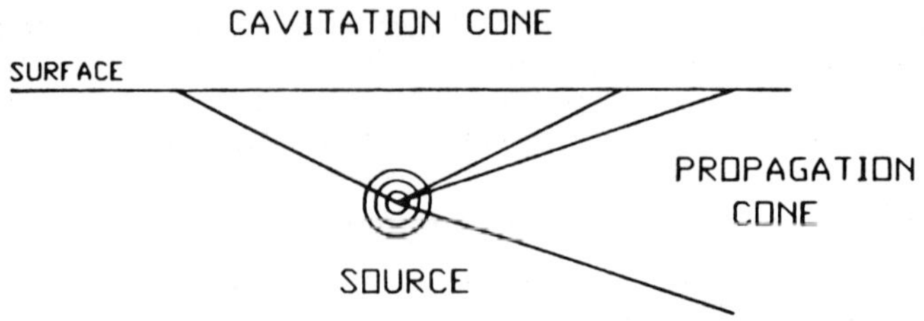

Fig. 6: Cavitation and propagation cones indicating situation when cavitation does not affect source signature at long range.

summarizes the issue. There is a cavitation cone of a width defined by where the bubble pulse cavitates at the surface; there is also a propagation cone encompassing the angular sector important for long range propagation. When these two intersect, the source signature at long ranges may not be consistent with that predicted by the formulae in [22-24]. For all but the very highest explosive yields, this suggests that the net effect of the cavitation at long ranges is not consequential.

The frequencies needed for long range imply propagation that interacts with the surface and more importantly the bottom. The sound that interacts with the boundaries at high angles is attenuated rapidly and is not a factor in long range propagation; however, the useful part of the field that is within the sound channel must be compensated for boundary and absorption losses. This has been done empirically by trying to apply a compensation which trades off propagation loss with the ambient noise field at long ranges. Factors of .001 to .003 dB/km have been found to work in practice over ranges of 2000 km in the Arctic [9].

To obtain physically useful quantities such as backscattering strength, appropriate models and theories need to be applied to the data. Graphs of low frequency bottom backscattering strength, typically contains significant variability. Numerous propagation studies have resulted in many models. It should not be surprising that backscattering experiments and theories are not well defined. Generally, scattering theory is among the more complicated intellectual endeavors in wave physics, but a tradition of creative work over more than a century has yielded a rich base of understanding including underwater acoustics. At long ranges, however, backscattering has recently presented new challenges principally because such long ranges usually entail nearly horizontal incidence upon nearly horizontal rough surfaces. At near grazing incidence, effects such as shadowing become important. Of even greater concern is the realization that traditional approaches to backscattering theory may have to be abandoned or significantly modified.

For example, one of the issues that has arisen in interpreting backscattering data from the Arctic is indicated in Fig. 4. Examination of both the FRAM II and FRAM IV maps of backscattering strength suggests that both the Lomonosov Ridge and the Nansen Ridge produce echoes of comparable level. The Lomonosov Ridge, however, is known to be much shallower with depths approaching 900 m, while the Nansen Ridge is a typical midocean spreading center with a nominal depth of 2500 m at its shallowest points. Because of its upperward refracting sound speed gradient most of the propagating long range acoustic energy in the Arctic is contained near the surface. Table I from [10] demonstrates this as summary of the average relative level for a 10 Hz signal at a depth of 500 m. The unresolved issue is why does the deeper and presumably less intensely illuminated Nansen Ridge produce the relatively high returns?

Traditional backscattering theory, e.g. Bass and Fuks,[24], relies upon the Kirchoff approximation in which one takes the scattered field as the incident one, multiplied by a plane wave reflection coefficient for every point on the rough surface. Evidence has now accumulated to render this as an inappropriate approximation at low angles in forward scatter, e.g. [25]. In forward scatter the low angle, coherent loss amplitude is classically [26]

$$R = \exp(-2k^2\xi^2\sin^2\theta) \qquad \text{Eq. 1}$$

where k is the wavenumber;

ξ the rms roughness height;

θ the grazing angle.

Recent forward scattering theories [27] and some experimental evidence [29] suggest that the classical forward scatter Kirchoff results to be incorrect at very low angles. Low angle backscatter is also unlikely to be well predicted by the classical Kirchoff theory. Jebsen and Medwin, [26], have shown the classical theory to be in error for the backscatter from a semi-infinite plate and from a wedge at low angles while a theory based upon boundary wave theory did lead to satisfactory agreement between experiment and theory [29].

Research on rough surface, low angle backscatter is ongoing. It is not unreasonable to assume that as in the case of forward scatter improvements upon the classical theories will become available. Meanwhile, interpretations such as ours, [10], of long range backscatter measurements in terms of classical theory must be viewed with a degree of skepticism.

Table 1: Relative distribution of energy at long range for a 10 Hz signal propagating in an Arctic sound speed profile (approximate)

Depth (m)	Relative level dB re level @ 500m	Gradient of Energy Level
500	0.	-1. dB/100m
800	-2.	-.5 dB/100m
1200	-4.	-.33 dB/100m
1900	-6.	-.25 dB/100m

SONAR SYSTEM IMPLICATIONS

The persistence of low frequency reverberation after a shot has important consequences for sonar systems operating in the same band since it raises the interferenc level. The effect upon omnidirectional systems is straightforward and can be computed using the methods indicated in [4-6]. As an example, fitting an exponential to the data in [6] from a 500 ton explosion suggests time constants of the decay rate in dB/s of approximately 1000 s at 20 Hz and 500 s at 50 Hz. It is straightforward to work the sonar equation for any source/receiver geometry to determine how long it takes to return to within 3 dB of the ambient level. Typically, shots of modest yield result in excess noise for periods of 10 to 30 minutes.

The effects of the reverberation upon array based systems is much more difficult to assess. The fundamental result from all the experiments with directional capability cited above is that the reverberation is geometrically controlled. Fig. 7 is an indication of the reverberation as a function of time after shot and direction for a 200 kg shot before compensation to measure the backscattering level. The directionality of the interfering reverberation can be exploited in order to minimize its effects. Quantifying this, however, is a complicated analysis which depends very intimately upon the array, the spatial distribution of the reverberation and the array processing algorithms

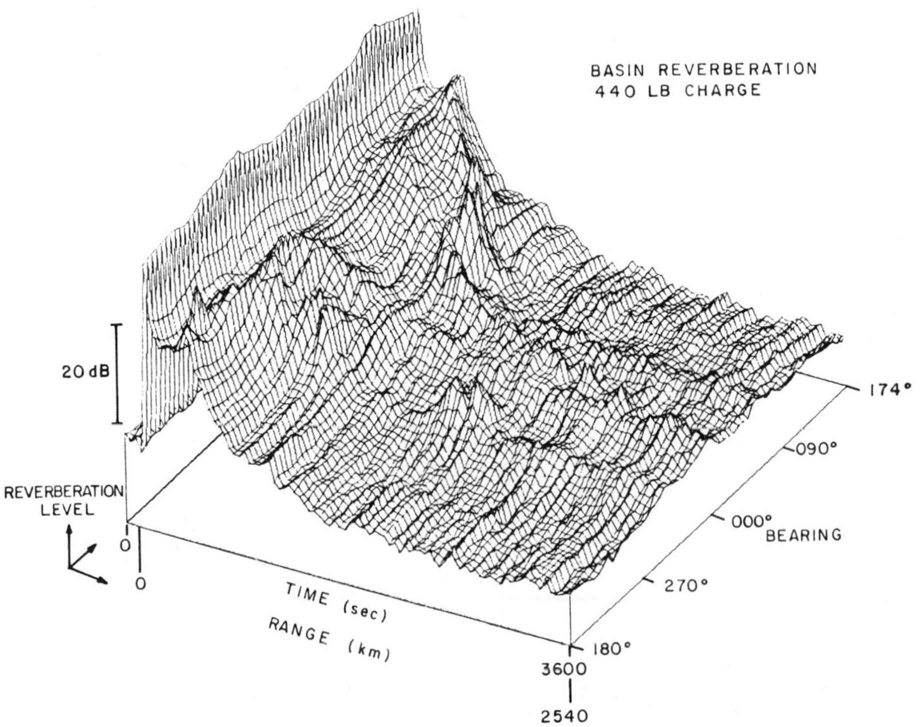

Fig. 7: Reverberation level as a function of time after shot for a 200 kg shot at the CANBARX ice camp in the Beaufort Sea [10]. Note the spatial structure generated by the topography of the basin.

used to resolve the target signature. Given some assumptions about these quantities it is straightforward to compute the array gains attainable either with direct sum beamforming or with some of the high resolution algorithms [16,19]. Generally one should be able to obtain a gain of at least 10 log (# of elements) unless the target is near the same bearing as the reverberation maxima. It is evident, however, that it does not take many explosives to have a strong influence upon the performance of a low frequency sonar system especially if the shots are sited to "focus" their effects upon a particular location by exploiting the geometry of the topography.

SUMMARY

The state of long range, low frequency backscattering has been reviewed. The dominant effect of bottom topography has been noted since the earliest experiments. In spite of early observation and a modest literature the theoretical understanding of the backscattering is still not well understood because of the extremely low grazing angles. The backscattering can be exploited both to locate unknown topographic features and to impair the performance of low frequency surveillance systems. The signal processing required for the topographic mapping is now within the capability of low cost computers; however, performance of a sonar system to such "jamming" remains an open question until more is known about the spatial structure of the reverberation.

ACKNOWLEDGEMENT

The authors thank our students, past and present, who have participated with us on the several Arctic experiments on long range backscattering. We also thank our colleagues at Woods Hole Oceanographic Institution who assembled the equipment required for these studies. This work was supported by the Arctic Program of the Office of Naval Research under contract no. N 00014-77-C-0266.

REFERENCES

1. Luskin, B. M., Landisman, M., Tirey, G. B. and Hamilton, G. R., "Submarine topographic echoes from explosive sound," Bulletin Geological Soc.Am.,63, 1053-1068, [1952].
2. Mellen, R. H. and Marsh, H. W., "Underwater acoustic reverberation in the Arctic Ocean," J. Acoust. Soc. Am., 56, 80-82, [1963].
3. Kutschale, H., "Arctic hydroacoustics," Arctic, 22, 246-264, [1969]
4. Northrop, J., "Submarine topographic echoes from CHASE V," J. Geophys. Res., 73, 12, [1968].
5. Goertner, J. A. and Blatstein, I. M., "Computer model for predicting acoustic reverberation from underwater explosions," Proc. OCEANS '75,, 227-232, IEEE Publications, [1975].
6. Blatstein, I. M., "Ocean basin reverberation at very long frequencies," J.Acoust.Soc.Am., 64, S44-S45, [1978].
7. Goertner, J. A., "Computer model predictions of ocean basin reverberation for large underwater explosions," Bottom Interacting Ocean Acoustics, ed. W. A. Kuperman and Jensen, F. P., 593-608, Plenum Press, New York [1980].
8. Spindel, R. and Heirtzler, J. H., "Long range echo rangings," J. Geophys. Res., 77, 35, 7073-7088, [1972].
9. Dyer, I., Baggeroer, A. B., Zittel, J. D. and Williams, R. J., "Acoustic backscattering from the basin and margins of the Arctic Ocean," J. Geophys. Res., 87, C12, 9477-9488, [1982].
10. Zittel, J. D., "Ocean basin reverberation," Engineer Thesis, Mass. Inst. of Tech., Cambridge, Mass. and Woods Hole Oceanographic Inst., Woods Hole, Mass., [1979].

11. Williams, R. J., "Backscattering of low frequency sound from the topographic features of the Arctic Ocean basin," S.M. Thesis, Mass. Inst. of Tech., Cambridge, Mass. [1981].
12. Baggeroer, A. B. and Dyer, I., "Fram II in the Eastern Arctic," EOS Trans. Am. Geophys. Union, 61, 217-222, [1980].
13. Schifter, D., Franchi, E., Griffin, J. and Adams, B., "Hydrographic reconaissance of large undersea topography using scattered acoustic energy," J. Acoust. Soc. Am., S66, pp. S25, [1979].
14. Erskine, F. T. and Franchi, E. R., "Rapid, basin-wide surveying of large undersea topography using scattered acoustic energy," Proc. OCEANS '84,, Vol.2, pp. 990-995, IEEE Publications, [1984].
15. Berkson, J. M., Akal, T., and Berrou, "Techniques for measuring backscattering from the sea floor with an array," Adaptive Methods in Underwater Acoustics, ed. H. Urban, [1985].
16. Baggeroer, A. B., "Sonar signal processing," in Applications of Digital Signal Processing, ed. A. V. Oppenheim, Prentice-Hall, Inc., Englewood Cliffs, N. J., 1978.
17. Capon, J., "High resolution frequency wavenumber spectral analysis," Proc. IEEE, 57, 1408-1418, [1969].
18. Kay, S. M. and Marple, S. L., "Spectrum analysis - a modern perspective," Proc. IEEE, 69, 1380-1419, [1981].
19. Proc. IEEE, special issue on spectral estimation, 70, [Sept. 1980]
20. Berrou, J. L. and Wagstaff, R. A., "A robust iterative technique for high-resolution spatial processing and spectral estimation," inHigh Resolution Spatial Processing, ed. R. Wagstaff and A. Baggeroer, 215-239, [1985].
21. Weston, D. E., "Underwater explosions as acoustic sources," Proc. Phys. Soc., 76, 233-249, [1960].
22. Wakeley, J., Jr., "Pressure signature model for an underwater explosive charge," U. S. Navy Journal of Underwater Acoustics, 27, 445-451, [1977].
23. Gaspin, J. B., Goertner, J. A., and Blatstein, I. M., "The determination of acoustic source levels for shallow underwater explosions, J. Acoust. Soc. Am., 66, 1446-1452, [1979].
24. Bass, F. G. and Fuks, I. M., Wave Scattering from Statistically Rough Surfaces, Pergamon Press, New York, [1979].
25. Tolstoy, I. and Clay, C. S., Ocean Acoustics: Theory and Experiment in Underwater Sound, McGraw-Hill, New York, [1966].
26. Chapman, D. M. F., "An improved Kirchoff formula for reflection loss at a rough ocean surface at low grazing angles," J. Acoust. Soc. Am., 73, 520-527, [1983].
27. Duckworth, G. L. and Baggeroer, A. B., "Estimation of ice surface scattering and acoustic attenuation in Arctic sediments from long range propagation data," (this volume), [1985].
28. Jebsen, G. M. and Medwin, H., "On the failure of the Kirchoff assumption in backscatter," J. Acoust. Soc. Am., 72, 1607-1611, [1982].
29. Biot, M. A. and Tolstoy, I., "Formulation of wave propagation in infinite media by normal coordinates with an application to diffraction," J. Acoust. Soc. Am., 29, 381-391, [1957].

DIRECTIONAL MEASUREMENTS OF LOW-FREQUENCY ACOUSTIC BACKSCATTERING FROM THE SEAFLOOR

J.M. Berkson, T. Akal, H.J. Kloosterman*, and J.L. Berrou+

SACLANT ASW Research Centre
Viale San Bartolomeo 400
I-19026 La Spezia, Italy

*Present Address:
Laboratory of Seismics and Acoustics
Delft University of Technology
Delft, The Netherlands

+Present Address:
Thomson-CSF
ASM Division
06802 Cagnes-sur-Mer, France

ABSTRACT

Long-range, low-frequency directional measurements of acoustic backscattering from the seafloor were made in the Tyrrhenian Sea using explosive sound sources. The signals were received by a horizontal towed array of hydrophones and were processed by a beamforming procedure to obtain the directional distribution of the scattered field as a function of time. These data are used to form images of scatterers and to estimate the backscattering strength of specific physiographic features. From the data obtained the scattering strengths were estimated to range from -25 dB to -35 dB and did not exhibit strong dependence on frequency.

INTRODUCTION

Low-frequency acoustic backscattering from the seafloor has often been studied using omnidirectional measurement techniques [1-4]. Generally, omnidirectional sources and receivers are used. The calculation of scattering strength assumes uniform scattering in a ring-shaped area on the seafloor for a given element of travel time. Such measurements can be contaminated by non-bottom returns having the same travel time as the scattering ring [5] or by non-uniform roughness within the scattering ring [6]. Although careful measurements using omnidirectional geometries will yield proper estimates of scattering strength, these techniques are more limited than directional methods in application. Directional methods make it possible to obtain images of the scattering features and to estimate the scattering strength of specific physiographic features such as seamounts and continental slopes. Images of seafloor topography from long-range directional measurements of acoustic backscattering have been obtained by various methods: broad-side beam, side-scan mode with towed array [7], full beamforming with a towed array [8,9], and full beamforming with two-dimensional arctic arrays [10]. Backscattering values have been reported for some of these measurements [9,10].

Previously the authors have reported beamforming methods applied to sensors of a towed array receiving backscattered sound [11] and aspects of resolution of seafloor images that were obtained [12]. In this paper we further describe the technique used for processing broadband data received from an explosive source and report additional measurements of bottom backscattering strength for the Tyrrhenian Sea.

EXPERIMENTAL METHODS

An omnidirectional explosive source (broadband) is deployed near a hydrophone array. The sounds reflected and scattered back to elements of the array are processed to form beams, as in Fig. 1. For processing, the hydrophone signals are split into time increments. For each time increment, a beamforming procedure is used to obtain the angular distribution of energy over a given frequency band of the source.

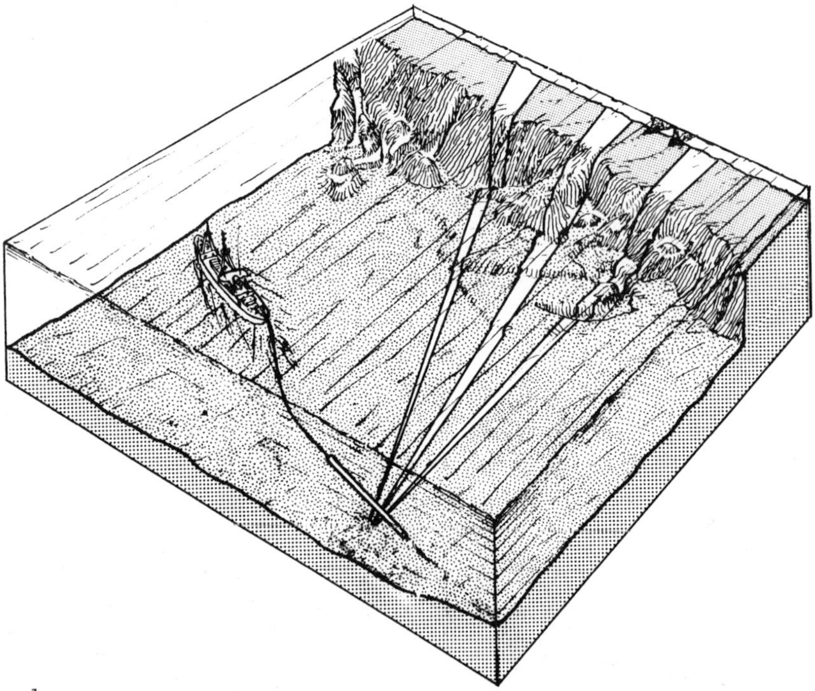

Fig. 1
Experimental geometry for backscattering measurement (not to scale): returns from an omnidirectional sound source (not shown) are received on directional receiving beams. The beams, which actually scan a vertical section of the volume, are indicated by their wedge-shaped areas on the seafloor. For simplicity only three beams on the right-hand side of the array are shown in their far field and the complementary (ambiguity) beams on the left-hand side of the array are omitted.

The beamforming procedure has been described in Refs. [11] and [12] and is based on the assumption of plane-wave sound signals of uniform velocity and is an application of frequency-domain beamforming based on the Fast Fourier Transform [13]. The procedure applied to sensors of a line array results in the estimation of beam power values equally spaced in spatial wavenumber, k_x, for the different acoustic frequencies, f, of the analysis band. With the assumption of uniform sound velocity, we can discriminate plane waves with different beam-pointing angles, α, where

$$k_x = (\omega/c) \sin \alpha. \tag{1}$$

in which c is the speed of sound and ω is the angular acoustic frequency, $\omega = 2\pi f$. The method of processing broadband data is shown graphically in Fig. 2. For each k_x value at the central frequency of the band, a set of interpolated beam powers is determined for the other frequencies. This is

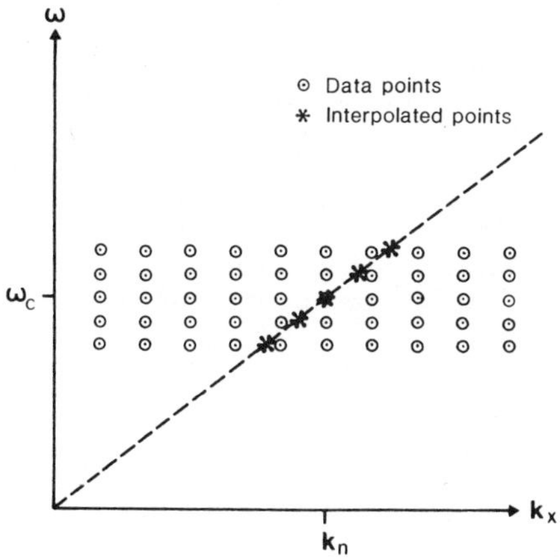

Fig. 2
Beam-power interpolation for wavenumber k_n: beam-power levels are interpolated for the points indicated by stars from the data at the points indicated by circles. The interpolated points fall along the azimuth line (dotted) of the central frequency of the band ω_c.

done by linearly interpolating between the appropriate beam-power values in such a way that the beam azimuth equals the beam azimuth that corresponds to the central frequency of the band. An average beam-power value over a given frequency band is obtained at each k_x value by averaging the interpolated powers. This procedure corresponds in $k_x-\omega$ space to averaging data points along a line.

The scattering area is defined by the intersection of the beam and the scattering ring boundaries for the particular travel time and processing-time window, as in Fig. 3. Assuming a process of totally incoherent scattering, the scattering strength S, for a given frequency band may be

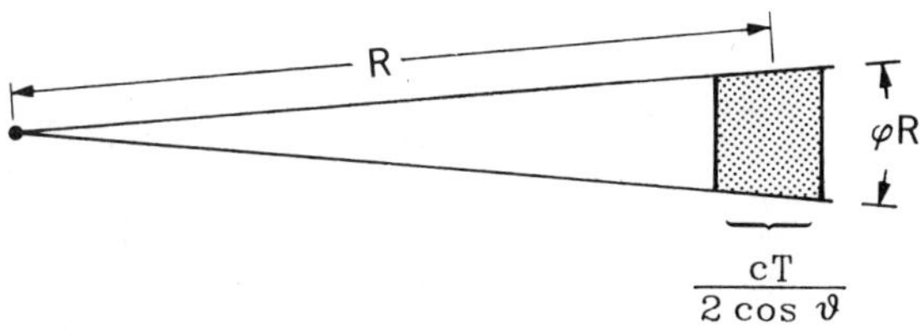

Fig. 3 Plan view of scattering area at seafloor: approximate dimensions of scattering area for small grazing angles;
c is the sound speed in seawater;
T the processing time;
ϕ the 3-dB beamwidth (rad);
θ the grazing angle at the bottom; and
R the range.

expressed in terms of the sonar equation as

$$S = RL - SL + TL1 + TL2 - 10 \log A , \tag{2}$$

where S is the scattering strength in dB/m^2; RL and SL are respectively the equivalent reverberation level and source level in dB re 1 μPa; TL1 and TL2 are the one-way transmission losses in dB; and A is the effective scattering area of the bottom in m^2. If the process involves a combination of scattering, reflection, and diffraction, then the estimated scattering strength may have errors or may be inappropriate. However, if the process is entirely that of specular reflection for an area of the seafloor greater than a Fresnel zone, a reflection loss parameter would be more appropriate.

MEASUREMENTS

Measurements were made in the Tyrrhenian Sea at the locations shown in Fig. 4; the same figure also indicates the major scattering features. Explosive SUS sources (0.8 kg TNT) were set at a depth of 245 m near a receiving array towed at a depth of 100 m. The hydrophone signals were split into increments that were processed by the beamforming technique described above and implemented in the processing system presented schematically in Fig. 5. Repetition of the beamforming procedure for adjacent

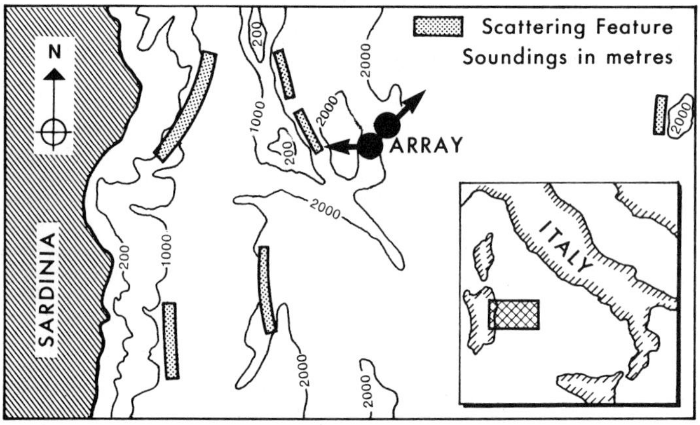

Fig. 4 Location of experiment: estimated positions of major scattering features are shown in shading.

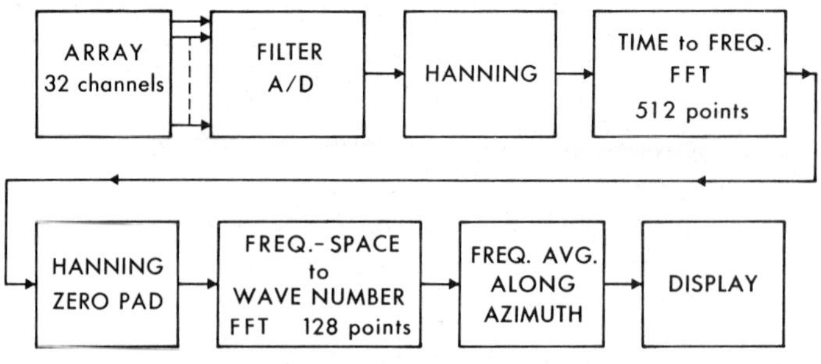

Fig. 5 Acquisition and processing system.

time segments gave a display of the beam-time history of features of the seafloor that scattered sound back to the array. Examples of beam-time histories are shown in Figs. 6 and 7. It can be seen that the direct arrival is followed by arrivals due to multiple reflection from the seafloor at high grazing angles. These appear strong on all beams because of sidelobe reception of the overloaded signals. Later arrivals are primarily due to backscattering from the seafloor at low grazing angles. The records also

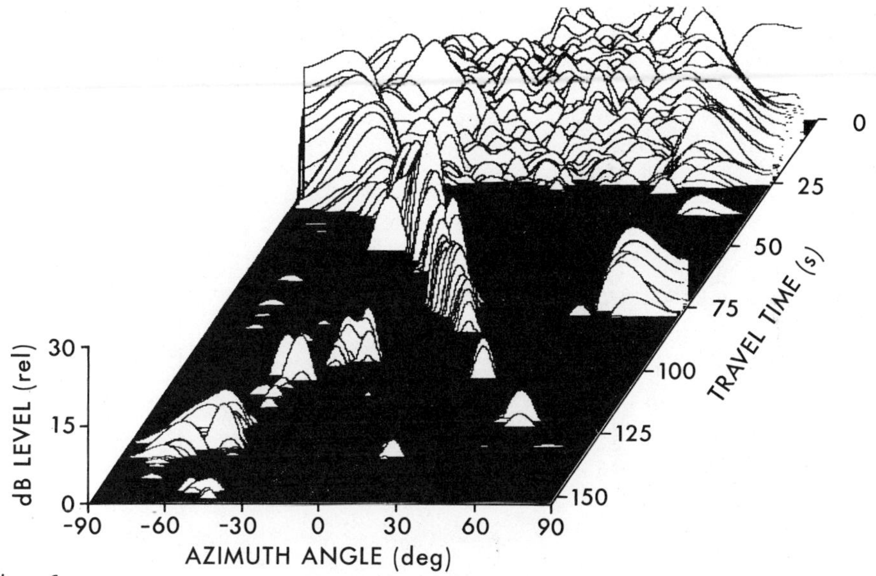

Fig. 6
Beam-time history at 600 Hz: 50-Hz bandwidth and array orientation 45°; positive azimuth toward ship; explosive source occurs at t = 6 s.

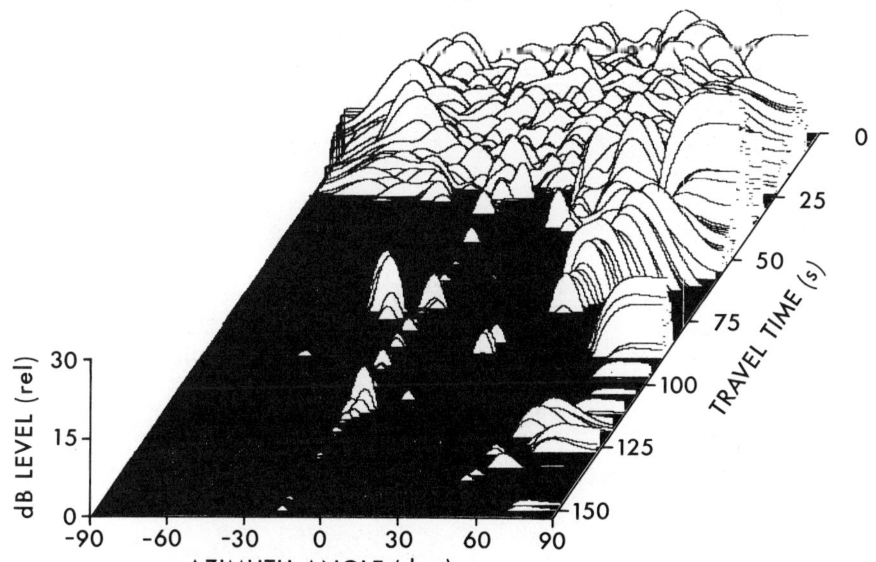

Fig. 7
Beam-time history at 700 Hz: 50-Hz bandwidth and array orientation 270°; positive azimuth toward ship; explosive source occurs at t = 6 s.

show continuous arrivals (artifacts) that are unrelated to returns from the explosive source and are comprised of towship and shipping noise. The major features are caused by backscattering from major physiographic features, e.g. island slope of Sardina and the Baconi Seamounts.

Scattering strengths were calculated using Eq. (1) and are shown in Fig. 8 as a function of frequency. Also shown are other scattering strength values calculated from directional measurements at low frequency [9,10]. The frequency dependence of the scattering strengths of these data is apparently weak. This is in agreement with most measurements of scattering from the seafloor at frequencies of 2 kHz to 100 kHz [14]. More data are required before the relationship at low frequencies can be established. However, at this point these limited data certainly do not suggest the very strong frequency dependence (about 1.6 power dependence, or 4.8 dB per octave) found in some of the high-frequency (13 to 290 kHz) data of McKinney and Anderson [15].

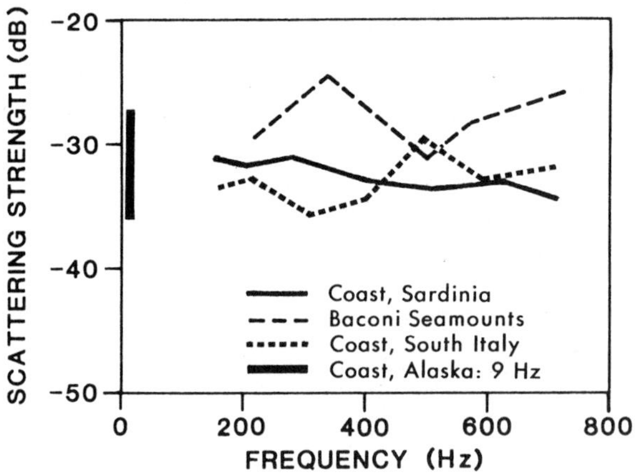

Fig. 8
Scattering strength versus frequency: data for coasts of South Italy and Alaska taken from Refs. 9 and 10 respectively.

CONCLUSION

Directional measurements of backscattering were made with an omnidirectional explosive source and a linear towed receiving array. The hydrophone signals were processed to form receiving beams and the data are displayed as images of the scattering features of the seafloor. The scattering strengths obtained exhibit weak frequency dependence, but more data are required before the frequency dependence of the seafloor features can be established.

REFERENCES

1. K.V. Mackenzie, Bottom reverberation for 530- and 1030 cps sound in deep water, J. Acoust. Soc. Am. 33:1498-1503 (1961).

2. R.J. Urick and D. S. Saling, Backscattering of explosive sound from the deep sea bed, J. Acoust. Soc. Am. 34:1721-1724 (1962).

3. P.B. Schmidt, Monostatic and bistatic backscattering measurements from the deep ocean, J. Acoust. Soc. Am. 50:326-331 (1971).

4. H.M. Merklinger, Bottom reverberation measured with explosive charges fired deep in the ocean, J. Acoust. Soc. Am. 44:508-513 (1968).

5. R.R. Goodman, Comment on two papers concerning reverberation from the ocean floor, J. Acoust. Soc. Am. 37: 174 (1965).

6. J.M. Berkson, C.S. Clay, and T.K. Kan, Mapping the underside of Arctic sea ice by backscattered sound, J. Acoust. Soc. Am. 53:778-781 (1973).

7. R.C. Spindel and J.R. Heirtzler, Long-range echo ranging, J. Geoph. Res. 77:7073-7088 (1972).

8. D.E. Schifter, E.R. Franchi, J.M. Griffin, and B.B. Adams, Hydrographic reconnaissance of large undersea topography using scattered acoustic energy, Rec. IEEE Electronics and Aerospace Systems Conventions 270-274 (1980).

9. J.M. Berkson, T. Akal, and J.L. Berrou, Techniques for measuring backscattering from the seafloor with an array, In: Adaptive Methods in Underwater Acoustics, H. Urban, ed., Reidel, Dordrecht, (1985).

10. I. Dyer, A.B. Baggeroer, J.D. Zittel, and R.J. Williams, Acoustic backscattering from the basin and margins of the Arctic Ocean, J. Geoph. Res. 87:9477-9488 (1982).

11. J.M. Berkson, H.J. Kloosterman, T. Akal, and J.L. Berrou, Application of beamforming techniques to measurements of acoustic scattering from the ocean bottom, IEEE J. Oceanic Engineering, in press (1985).

12. J.M. Berkson, H.J. Kloosterman, T. Akal, and J.L. Berrou, Aspects of spatial resolution for long-range, low-frequency imaging of the seafloor, In: "Acoustic Imaging, Vol. 14," A.J. Berkhout, J. Ridder, and L. VanderWal, eds., in press, (1985).

13. J.R. Williams, Fast beamforming algorithm, J. Acoustic. Soc. Am. 44:1454-1455 (1968).

14. A.V. Bunchuk and Y.Y. Zhitkovskii, Sound scattering by the ocean bottom in shallow-water regions (review), Sov. Phys. Acoustic 26:363-370 (1980).

15. C.M. McKinney and C.D. Anderson, Measurements of backscattering of sound from the ocean bottom, J. Acoust. Soc. Am. 36:158-163 (1964).

SEAMOUNT HEIGHT ESTIMATION FROM LONG-RANGE, LOW FREQUENCY ACOUSTIC BACKSCATTER

F.T. Erskine, E.R. Franchi, and B.B. Adams

Naval Research Laboratory
Washington, D.C. 20375-5000, USA

INTRODUCTION

We are developing a long-range acoustic backscatter method for mapping major undersea topography in an ocean basin (approximately 2000 km x 2000 km) and for determining the sizes of topography (seamounts and ridges). In this paper we discuss our results on estimating sizes of ocean topography using data from the Bathymetric Hazard Survey Test (BHST). BHST was a proof of concept experiment conducted by the Naval Research Laboratory in the eastern Atlantic Ocean in September 1980[1,2].

In BHST we deployed explosive sound sources at 4 hour intervals and digitally recorded the topographic backscattered signals using a directive receiver array. These signals are processed to yield received levels as a function of time and azimuth. The test geometry is shown schematically in Fig. 1. By assuming a mean sound speed to relate arrival time to range, and by using information about receiver orientation and shot location, we are able to transform from time-azimuth format to latitude-longitude format, to make a "single-shot" backscatter map for each shot detonation[3-5]. It is these maps which we use for seamount height estimation.

We have previously reported[3-5] that this active acoustic technique provides a relatively inexpensive method for rapid (about 1 week) survey of an entire ocean basin for the presence of major bathymetric features and is well suited for use in ocean basin areas which at present have been mapped only to a limited extent. In this paper we show that with this technique it is possible to obtain first-order estimates of the heights of all major topographic features in a basin to better than +/- 1.3 km.

The low frequency reverberation mapping method[3-6] which we use complements the only other techniques currently under development for large-scale ocean bathymetry mapping, i.e. spacecraft altimetry[7], and airborne magnetics/gravimetry[8,9]. BHST followed two earlier experiments performed by NRL in 1976 and 1979 called TOPO I and TOPO II respectively[10,11]. These earlier experiments were conducted in a deep basin approximately 1000 km northeast of New Zealand. The TOPO experiments used explosives and a somewhat low resolution towed receiver array, but they showed that this acoustic mapping technique works and led to the conduct of BHST. Historical references to earlier topographic ranging tests with very limited receiver directionality and range capability, as well as tests in an arctic environment, were previously reported[4,5,11].

SEAMOUNT HEIGHT ESTIMATION FROM REVERBERATION MAPS

BHST was conducted in a well-surveyed ocean basin to allow relationships between seamount size and acoustic scattering level to be developed using known topography. Such relationships would then provide estimates of feature size for uncharted topography in a basin. We develop an analysis procedure for estimating the sizes of seamounts and other major ocean topography from our single-shot reverberation power spectral density maps[3-5], which are 21 deg 20 min x 21 deg 20 min (about 2000 km on a side). Each map pixel is about 1 arc-min square. We search digitally along selected great circle radials from the position of a shot through the positions of undersea topographic features, as guided by an existing bathymetry map of the Canary Basin area. We use the NRL Reverberation Ray Trace Model[12] to make estimates of propagation losses along these radials. We calculate esimates of "flat-bottom" backscattering area, and use our known source level to compute a measure, "SS", related to backscattering strength.

The basic procedural steps in our seamount height estimation method are:

(a) Prepare a SYNBAPS (Synthesized Bathymetry[13]) overlay (Fig. 2) for our single-shot reverberation power spectral density maps[3-5]. Fig. 2 shows bathymetry contours for the Canary Basin with 9.3 km resolution as well as the positions of individual shot detonations (at 4 hour intervals).

(b) Examine all single-shot reverberation power spectral density maps for candidate known seamounts, using the overlay of (a).

(c) For each reverberation map, select great circle radials from shot position through candidate seamounts. Each radial is specified by shot position and initial bearing (as computed from positions of shot and seamount of interest).

(d) Choose a manageable subset of 22 radials of interest from 12 different shots (Fig. 2). These radials are chosen to have seamounts at a variety of ranges (less than 200 km to about 1100 km), and bearings (relative to the receiver array heading). It is not possible to always choose a bearing so that the reflected or conjugate receiver beam direction[3-5] is looking down relatively flat bathymetry, although this is desirable and is the case for about a third of the radials.

(e) Run the NRL Reverberation Ray-Trace Model[12] for each of the 42 radials to compute two-way propagation loss. Inputs for the model are archival sound speed profiles and SYNBAPS bathymetry vs. range as well as range dependent bottom loss vs. grazing angle estimates. Output is a data file for each radial containing the propagation loss vs. range.

(f) Compute the following along each selected radial:

$$SS = RLNET - SL + 2TL - AREA$$

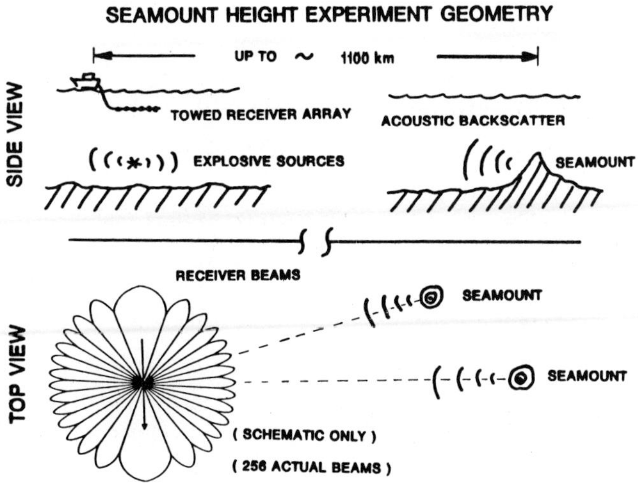

Fig. 1 Schematic Diagram of BHST experiment geometry.

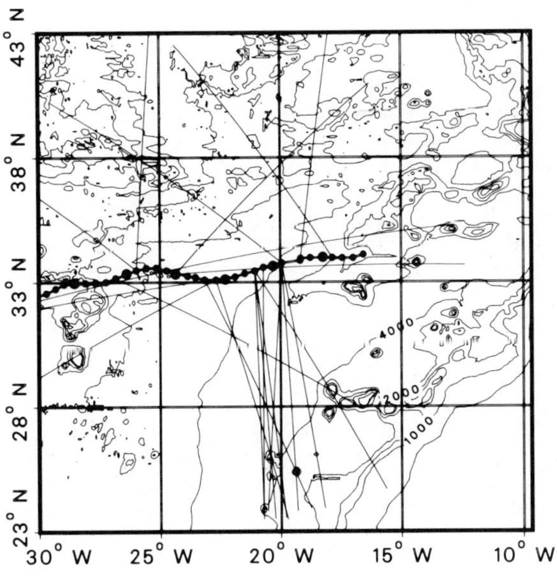

Fig. 2 Contour map of SYNBAPS bathymetry for BHST op-area. Map is 21 deg 20 min x 21 deg 20 min (about 2000 km on each side). Map shows track of ship (from east to west) and positions of shots (dots) (one every 4 hours). Also shown are great circle primary radials for seamount height study.

Where:

RLNET = 10 log (reverb level - ambient noise), in (dB re 1 $\mu Pa^2/Hz$)

SL = source level, in (dB re 1 $\mu Pa^2/Hz$)

2TL = round-trip propagation loss estimate from (e), in (dB)

AREA = geometric scattering area of an assumed flat bottom piece of ocean (no attempt to compute true seamount area), in (dB re 1 m^2)

So SS is a scattering strength measure, but not true scattering strength, because of the way we calculate the AREA term.

(g) Compile a list of the following for as many seamounts or other bathymetric features as possible from the subset of radials run in (f):

1. Scattering Strength measure, SS; (record "peak" and "mean") (dB)
2. Range of SS "peak" (km)
3. SYNBAPS bathymetry feature height (km)
4. Range of SYNBAPS feature peak (km)
5. SYNBAPS bathymetry feature slope (deg)

(h) Perform a least squares correlation analysis to look for correlation between SS and SYNBAPS bathymetric feature heights and slopes. Derive the "functional dependence" of SS on seamount height and seamount slope.

(i) Use the functional dependence from (h) to plot "reverberation estimated bathymetry" for selected radials.

Fig. 3 shows received level vs. range (solid) in a receiver beam looking along the primary great circle radial in a digital single-shot reverberation map; as well as SYNBAPS archival bathymetry along this same radial (dashed) and along the conjugate radial (dotted). Fig. 4 shows SS, "scattering strength measure" vs. range, as well as SYNBAPS bathymetry as in Fig. 3. After computing SS for numerous radials, we correlate this backscattering strength measure with seamount heights and slopes from archival bathymetry. Fig. 5 shows the results of our correlation analysis for approximately 32 topographic features along 22 separate radials from 12 different shots. The topography upon which Fig. 5 is based varied from small features only 0.8 km to 1.0 km above the nominal 5.5 km basin depth to islands reaching the surface. These topographic features were at a variety of ranges from the source and receiver, from about 200 km at the nearest to about 1100 km at the

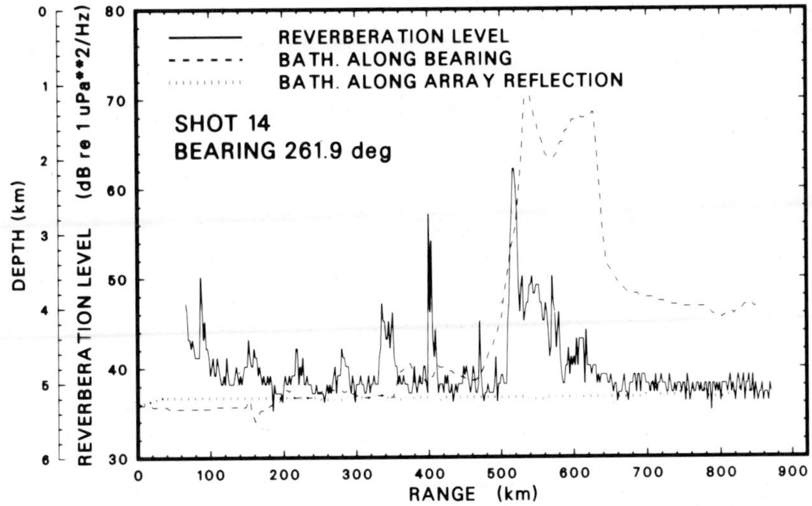

Fig. 3 Reverberation power spectral density (dB re 1 $\mu Pa^2/Hz$) vs. range along a primary radial for shot 12. Near-range bottom reverberation (inside of about 60 km) is omitted from our single-shot maps, hence the short gap at left. Also superimposed to same range scale is SYNBAPS bathymetry for primary radial (dashed) and conjugate radial (dotted).

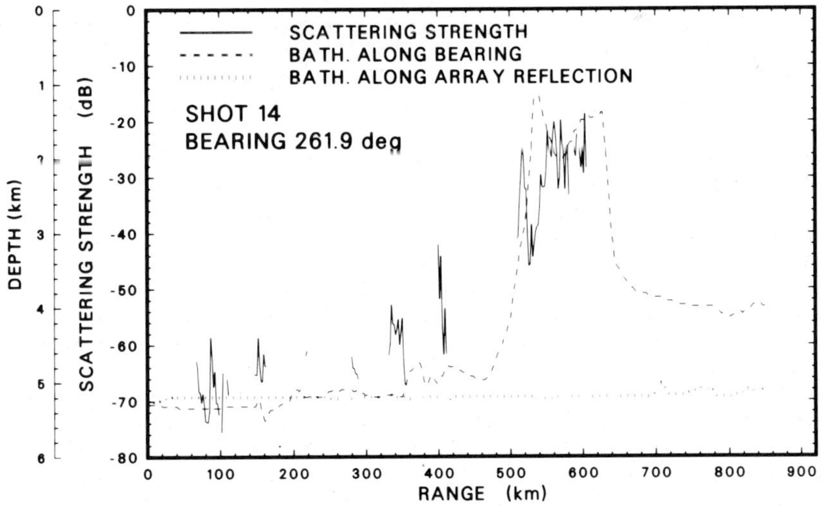

Fig. 4 "Scattering strength measure" (see text) vs. range for a selected radial. Also shown is SYNBAPS bathymetry for primary radial (dashed) and conjugate radial (dotted).

farthest. Fig. 5 shows that it is possible to develop relationships to remotely measure the heights of seamounts in an ocean basin by long range acoustic backscatter. The scattering strength measure "SS" is linearly correlated with bathymetry height relative to a nominal 5.5 km basin depth (linear correlation coefficient = 0.8). The 90 % confidence curves associated with this fit imply that we can predict individual seamount heights from long range reverberation to within about +/- 1.3 km. In similar fashion, we correlate scattering strength measure with mean seamount slopes from archival bathymetry. The results of this correlation (Fig. 6) show that SS is linearly correlated with seamount slopes as well as heights. The linear correlation coefficient for slopes is about 0.5, and the implied uncertainty in prediction of slopes from long range reverberation is about +/- 3 deg.

We are also interested in our ability to predict seamount position or range along a given radial from the reverberation data. The acoustic backscatter will in general originate from the facing slopes of seamounts or bathymetric features, and thus we expect a position offset between the reverberation signatures and the centroids of seamounts. Fig. 7 shows the range of SS peaks vs. range of seamount peaks (centroids) for our sample radials. Fig. 7 shows that we are able to predict seamount positions quite well, but with a slight systematic offset (generally less than 15 km). Since we can remotely estimate seamount positions, heights, and slopes, we can show "estimated" bathymetry along a selected radial, compared to archival bathymetry[13]. Fig. 8 shows such a bathymetry "estimation" for a typical radial and represents our initial effort to remotely estimate bathymetry in a deep ocean basin from long range reverberation returns.

We may compare our bathymetry height estimations to those from remote sensing techniques currently under development for wide area estimates of bathymetry. These techniques include: a) spacecraft altimetry[7], b) aeromagnetics[8,14] and c) airborne gravimetry[9,15]. Briefly, each of these techniques depends critically on a physical model relating the physical quantity measured to topographic size. The physical quantity measured in each case is: a) sea surface height relative to a nominal mean, b) deviation of magnetic flux density from local mean, c) deviation of earth's gravitational acceleration from local mean. In each case, the relationship between the measured quantity and the actual bathymetry height is not uniquely given by a physical or mathematical model. At present the typical uncertainties in estimation of seamount heights by these techniques, when applied to a deep ocean basin under optimal conditions are approximately as follows: a) +/- 500m to several km; b) +/- 500m to several km; c) +/- hundreds of m to several km depending on type of basement. Thus, our seamount height estimation technique using long range reverberation compares favorably against these other techniques. Further refinements in the analysis procedures may reduce the uncertainty and merit investigation.

SUMMARY AND CONCLUSIONS

We have described a remote sensing acoustic backscatter method for estimating seamount heights and locations at ranges up to 1100 km. The method uses reverberation data from explosive shots in a deep ocean basin received on a horizontally directional receiver. We compute a scattering strength measure along selected radials in our digital reverberation power

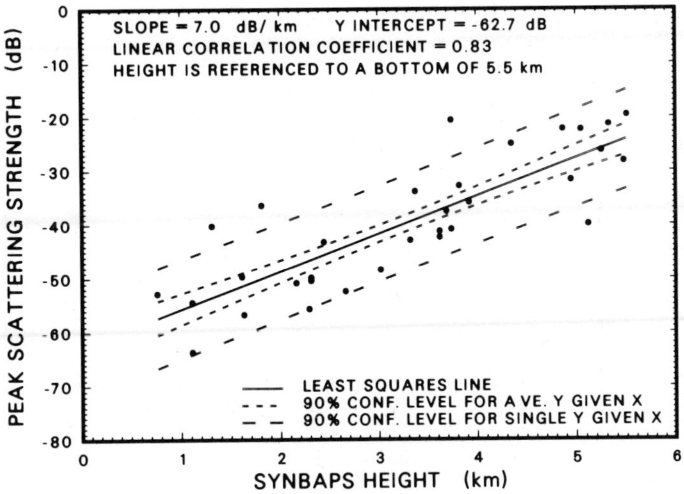

Fig. 5 Peak "scattering strength measure" vs. SYNBAPS bathymetry height (re 5.5 km depth).

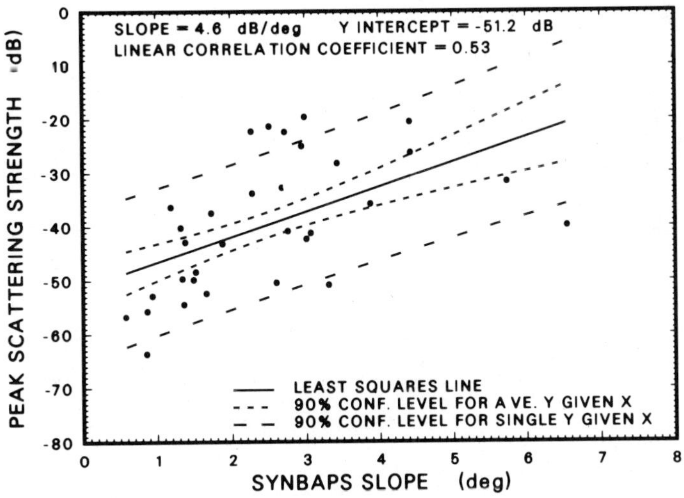

Fig. 6 Peak "scattering strength measure" vs. SYNBAPS bathymetry slope.

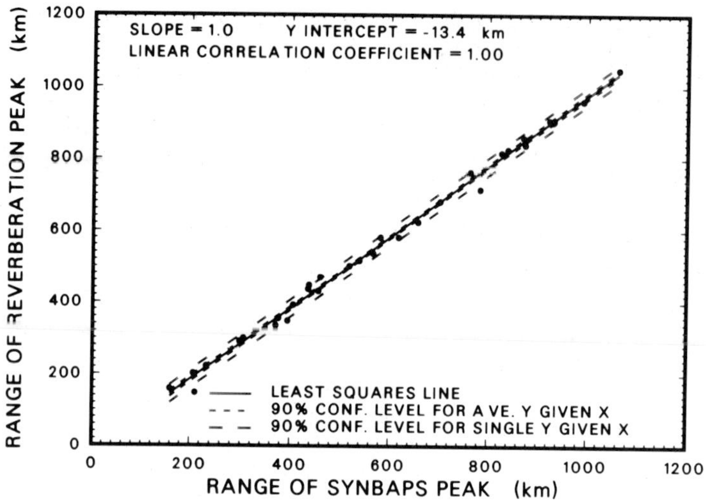

Fig. 7 Range of scattering strength peak vs. range of SYNBAPS bathymetry.

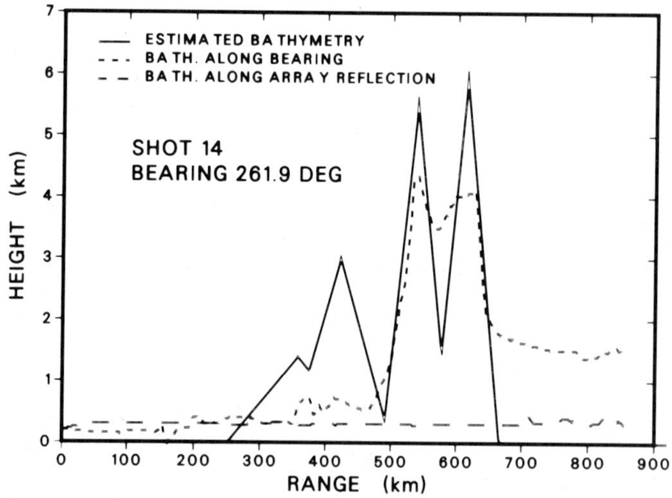

Fig. 8 Reverberation estimated bathymetry compared to SYNBAPS bathymetry for typical radial.

spectral density maps. We show that this scattering strength measure is linearly correlated with seamount heights (correlation coefficient = 0.8), and seamount slopes (correlation coefficient = 0.5). Also, we show that that reverberation comes from facing slopes of seamounts (typical range offset = 13 km). Using our measured functional dependence of scattering strength on seamount height and slope, we can draw estimated bathymetry along a selected radial in a deep basin to within about +/1.3 km in height and +/- 3 deg in slope. These results compare favorably with those using other available techniques for seamount height estimation.

ACKNOWLEDGEMENTS

Funding for this analysis was provided by the Defense Mapping Agency (DMA) and U.S. Navy/OP-952.

REFERENCES

1. E. R. Franchi, and D. E. Schifter, Bathymetric Hazard Survey Test: Test Results, BHST Report No. 1, Acoustics Division, Naval Research Laboratory, Washington, D.C. (1980).

2. D. E. Schifter and E. R. Franchi, Bathymetric Hazard Survey Test: FY-81 Processing, BHST Report No. 2, Acoustics Division, Naval Research Laboratory, Washington, D.C. (1981).

3. F. T. Erskine, G. Bernstein, and E. R. Franchi, Imaging Techniques for Analysis of Long Range Backscatter Data from Ocean Basin Regions, IEEE Proc. Third Working Symposium on Oceanographic Data Systems, Woods Hole, MA, 160-164 (1983).

4. F. T. Erskine, G. M. Bernstein, E. R. Franchi, and B. B. Adams, Imaging of Long Range Reverberation from Ocean Basin Topography, Marine Geodesy, to be published (1985).

5. F. T. Erskine, G. M. Bernstein, S. M. Brylow, W. T. Newbold, R. C. Gauss, E. R. Franchi, and B. B. Adams, Bathymetric Hazard Survey Test: Scientific Results and FY 82-84 Processing, BHST Report No. 3, Acoustics Division, Naval Research Laboratory, Washington, D.C. (1985).

6. F. T. Erskine, and E. R. Franchi, Rapid, Basin-Wide Surveying of Large Undersea Topography Using Scattered Acoustic Energy, IEEE OCEANS 84 Proc., Washington, D.C., 2:990-995 (1984).

7. T. H. Dixon, M. Naraghi, M. K. McNutt, and S. M. Smith, Bathymetric Prediction from SEASAT Altimeter Data, J. Geophys. Res., 88(C3):1563-1571 (1983).

8. R. H. Feden, P. R. Vogt, and H. S. Fleming, Magnetic and Bathymetric Evidence for the 'Yermak Hot Spot' Northwest of Svalbard in the Arctic Basin, Earth and Planet. Sci. Lett., 44:18-38 (1979).

9. J.M. Brozena, , A Preliminary Analysis of the NRL Airborne Gravimetry System, Geophysics, 49(7):1060-1069 (1984).

10. D. Schifter, E. Franchi, J. Griffin, and B. Adams, Hydrographic Reconnaissance of Large Undersea Topography Using Scattered Acoustic Energy, J. Acoust. Soc. Amer., Suppl. 1, 66:S25 (1979).

11. D. E. Schifter, E. R. Franchi, J. M. Griffin and B. B. Adams, Reverberation Mapping for Basin-Wide Bathymetric Surveys, Marine Geodesy, to be published (1985).

12. E. R. Franchi, J. M. Griffin, and B. J. King, NRL Reverberation Model: A Computer Program for the Prediction and Analysis of Medium to Long Range Boundary Reverberation, Naval Research Laboratory Report 8721, Washington, D.C. (1984).

13. R. J. VanWyckhouse, Synthetic Bathymetric Profiling System (SYNBAPS), Technical Report TR 233, U.S. Naval Oceanographic Office, Washington, D. C. (1973).

14. M. Czarnecki, personal communication, 1984.

15. J.M. Brozena, personal communication, 1984.

AN ACOUSTIC MODEL FOR BATHYMETRIC SCATTERING WITH LOW-FREQUENCY APPLICATIONS

R. N. Baer, J. S. Perkins, E. B. Wright, and B. B. Adams

Naval Research Laboratory
Code 5160
Washington, DC 20375-5000, USA

ABSTRACT

We have developed a computer model to estimate the modification of the directional acoustic field caused by scattering from rough ocean bathymetry. The model is applied to a situation with a 30-Hz source, source, selected as the highest frequency commensurate with the available bathymetric statistics for a seamount. The model is then used to compute predictions of horizontal and vertical energy spreading. We discuss these redistribution effects of rough bathymetry, manifested by observed high side-lobe levels, and consider the possibility of estimating bottom statistics by way of the scattered component of the energy.

INTRODUCTION

When acoustic energy strikes a rough, hard ocean floor, some of the energy is specularly reflected, and the remainder is scattered in all directions in three-space. This scattering can be measured with directional receivers operating in the vicinity of the ocean floor. When conducting bathymetric and sub-bottom surveys with acoustic apertures, it is judged valuable to obtain these additional roughness estimates. The Ocean-Refraction and Bathymetric-Scattering (ORBS) model[1] has been developed as a first step to predict these effects for hard boundaries. It calculates the directional pattern of received energy at a finite-sized receiver, incorporating both the effect of a varying sound-speed structure in the ocean volume (refraction) and scattering from rough bathymetry. The acoustic field for receivers sufficiently distant from the source is computed by a combination of the split-step parabolic equation method, propagation of the coherence function along characteristics, and rough-surface scattering. For short-range applications the coherence propagation alone is used. The ORBS model especially includes scattering out of the vertical plane, and thus it is a three-dimensional model. The model has been implemented on a VAX 11/780 at NRL.[2]

In this paper we discuss the application of the ORBS model to the propagation of low-frequency sound scattered from seamount bathymetry from a continuous source at a range long enough to columnate the energy in order to simplify the applications and the interpretation of results. In the first section we describe briefly the components of the ORBS model. In the next section we discuss the influence of rough bathymetry on a

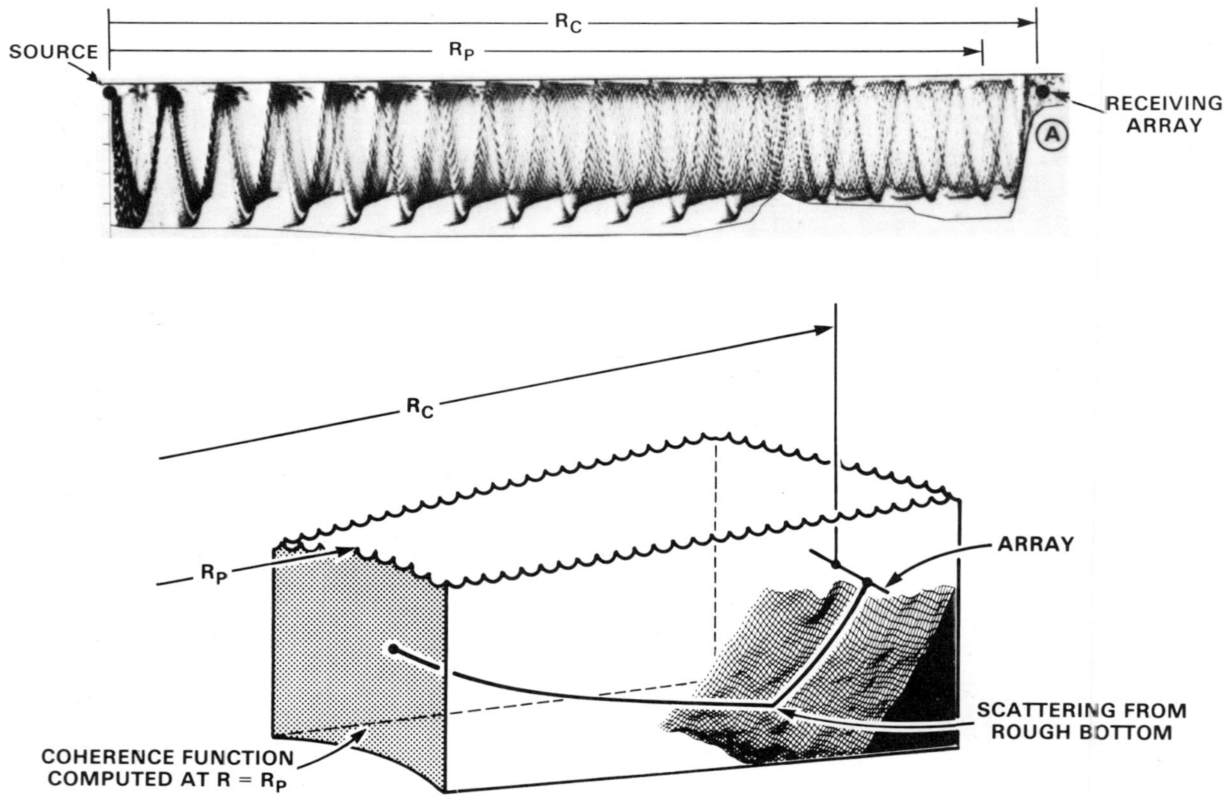

Fig. 1. ORBS model description. The upper diagram shows the long-range (parabolic equation) propagation region. The lower diagram is a three-dimensional detail showing the coherence function propagation in the vicinity of the boundary scatter and the acoustic array (region A in the upper diagram).

horizontal array. We then describe the influence of rough bathymetry on a vertical array and discuss the implications for estimating roughness based on acoustic measurements. The final section is a short summary of the paper.

THE ORBS MODEL

Since the ORBS model is discussed in detail elsewhere[1,2] this section will summarize its main features. Figure 1 illustrates the basic structure of the two-part ORBS model. The upper part of the figure shows the long-range, deep-water, non-boundary-controlled region where two-dimensional, parabolic equation (PE) model is used[3]. The range R_p is chosen as a range up to which the bathymetric influence is minimal, the cross-range variation in the acoustic field is small, and the propagation is consequently describable by two-dimensional analysis. At R_p the second moment of the acoustic field is formed from the acoustic pressure, p, calculated from the PE. A transform of this quantity results in the mutual coherence function, a function which propagates along the characteristics of the Helmholtz equation (i.e., the rays) and is directly related to the acoustic energy. As stated earlier, for operations of the source close to the receiver, the first part of the model can be omitted. A fuller discussion of coherence-function propagation is contained in Ref. 4.

In order to determine the distribution of received energy at the array located at R_c from the known distribution at R_p, we use a ray tracing algorithm on each side of the bathymetric scattering point. (See bottom part of Fig. 1.) For computational efficiency, we trace backwards from R_c to R_p so that no effort is wasted on paths which miss the receiver. It is, however, necessary to consider paths from each element on the sensor and in all directions from each element. Furthermore, at the scattering point another set of paths is introduced by the scattering. At this point a

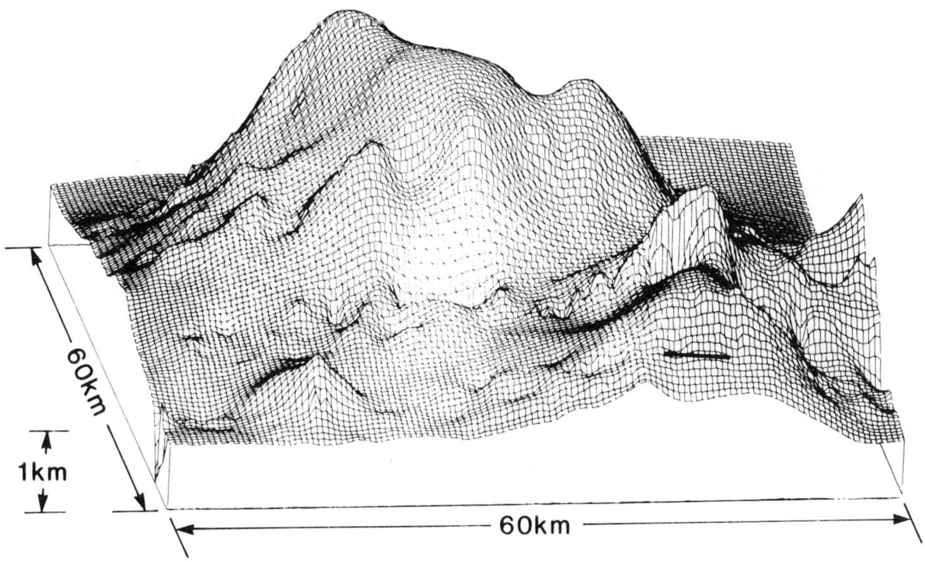

Fig. 2 Seamount used in examples in text. The 5000-m horizontal array location is indicated by the horizontal line in the lower right.

scattering cross section based upon a combination of the Kirchhoff approximation and small perturbation theory is used.[5] This formulation is valid in a region at least equal to the union of the regions where its component parts is valid. The scattered paths introduce at this point potentially strong cross-range components and consequently three-dimensional variation into the acoustic field. ORBS uses the two-dimensional bathymetric roughness spectrum in the scattering region and the range varying ocean sound-speed structure in the long-range propagation region as environmental input.

In general, a total of tens and even hundreds of thousands of paths need to be considered. For this reason an approximation to estimate the Kirchhoff integral has been developed which greatly increases computational efficiency.[6] Specularly reflected energy and paths that do not interact with the bathymetry are also incorporated in ORBS. As the ranges R_p and R_c are taken to be close together, the ray approximation used between R_p and R_c does not lead to significant error as compared to the Helmholtz equation.

HORIZONTAL DIRECTIONAL SENSOR RESULTS

In this section, we consider the application of the ORBS model to low-frequency propagation in the vicinity of a rough slope.

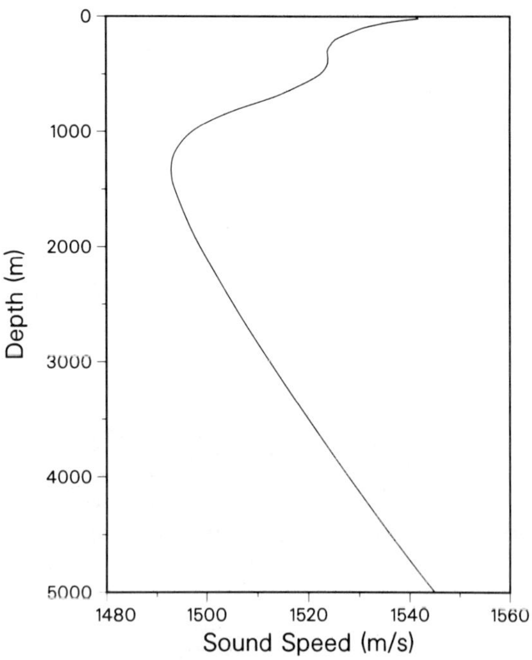

Fig. 3 Canonical summer sound-speed profile used for illustrative examples.

Detailed bathymetric height data has been obtained from the U.S. Naval Oceanographic Office for a typical seamount in the North Atlantic Ocean. In Fig. 2, we show a computer-generated view of this seamount. Czarnecki and Bergin of NRL have developed a sequence of programs[7] which have been used to process and analyze this bathymetric data. From these programs an ensemble of realizations of the spectum of the data is obtained. The ensemble is then used to compute the bathymetric information needed by ORBS.

A 30-Hz omnidirectional sound source is assumed at a depth of 150 m. The receiver is a sparse 5000-m (100 wavelengths) horizontal array of 26 equally spaced modules. It is towed at a depth of 3750 m, 150 m above the slope of the seamount, and broadside to the source. For illustrative purposes the canonical sound-speed profile of Fig. 3 is assumed valid in the entire propagation channel. The water depth in the deep-water region is 4800 m, and we take the source-receiver separation to be 93 km.

Figure 4 is a gray-scale plot of acoustic intensity (with respect to cylindrical spreading) produced by the PE algorithm. After considerable high-angle energy is stripped away in the first bottom-bounce region, we note an essentially periodic-with-range intensity distribution which is typical for deep-water acoustic propagation.

Figure 5 is the result of another PE run showing the energy being intercepted by the array at 93 km. The seamount is diagrammatically illustrated by its average slope of 15°. At the range R_p = 90 km, the coherence function is computed by ORBS, and the ray-path computation portion of ORBS starts. Thus, the remainder of Fig. 5 is for illustrative purposes only as it does not include the bathymetric scattering.

In Fig. 6 we show the primary output of ORBS, the output beam intensity as a function of beam direction (i.e., an angular power distribution or spectrum). Hanning shading has been applied and since the source is in the near-field of the array, the array is focused at a distance of 93 km to eliminate the spreading due to extra Fresnel zones. The dashed curve is the ORBS output without scattering (the rms surface roughness being taken as zero). The solid curve incorporates scattering as computed from the ensemble of bathymetric spectra appropriate for the region -- a k^{-4} power law, an rms roughness of 20 m, and a horizontal correlation distance of 400 m. The bathymetric statistics were taken as isotropic to not needlessly complicate the problem. The curves were normalized so that 0 dB

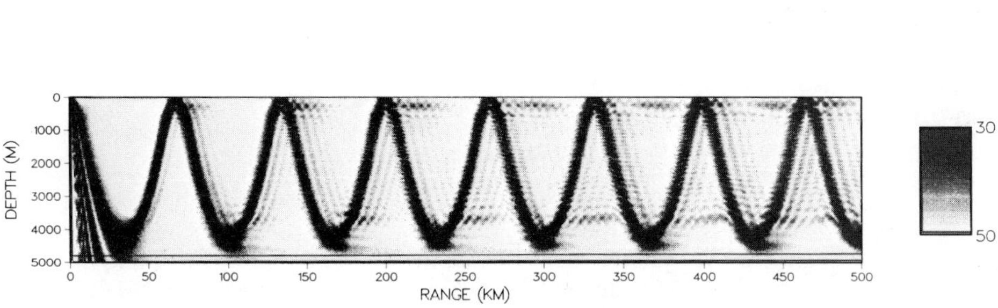

Fig. 4 Parabolic-equation gray-scale plot for 30-Hz sound source at depth of 150 m. Darker regions correspond to higher acoustic intensity than lighter regions, with black to white representing a 20-dB dynamic range. (Cylindrical spreading has been removed before plotting.)

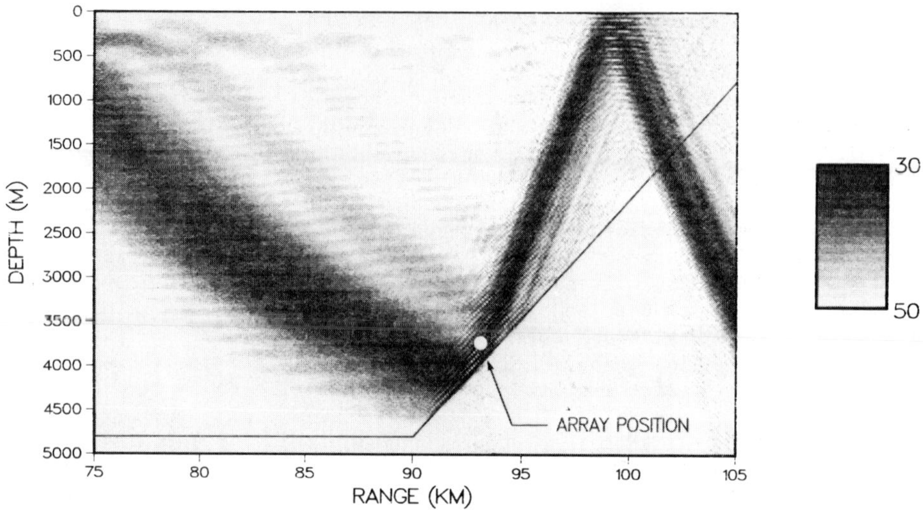

Fig. 5. Gray-scale plot for source-receiver separation of 93 km. Seamount with 15° average slope is shown. The horizontal array is perpendicular to the plane of the plot. The vertical array of the next section is in the plane.

is the value at broadside (0°) for the no-scattering case. Had array signal gain been plotted instead of beam intensity, a 2.3-dB degradation due to the scattering would have been observed in the broadside direction for the spectrum that includes scattering.

The scattering causes approximately 94% of the energy that reaches the seamount to be scattered and the scattered spectrum shows this redistribution. The remaining 6% is specularly reflected. The major difference between the two curves, observed in Fig. 6, is the great increase in the side-lobe level due to the scattering. Thus, it is quite important to incorporate the effect of bathymetric scattering in propagation models where the bottom statistics have spatial wave components short compared to the aperture size and variance comparable to or greater than the acoustic wavelength.

We also consider the inverse problem of determining the rms roughness from acoustic measurements. The great difference between the two curves in Fig. 6 would allow us to estimate the rms roughness from acoustic measurements using a sequence of curves calculated for various values of the roughness.

VERTICAL DIRECTIONAL SENSOR RESULTS

We now consider the beam output of a 150-m (3-wavelength) vertical array, also 93 km from the 30-Hz sound source and centered at a depth of about 3750 m. In Fig. 7, we have graphed the array beam intensity as a function of beam angle. There are two main arrivals; one slightly above 0°, and the other slightly above 30°. The 0° arrival corresponds to directly arriving energy, i.e. energy that has not interacted with the bathymetry, while the 30° arrival corresponds to the scattered energy. Comparing the solid curve with the dashed curve in Fig. 7, we see that the direct arriving energy is, as expected, not influenced by the bathymetry. However, there is about a 3-dB difference between the solid (rough bathymetry) and dashed (smooth bathymetry) curves in the vicinity of 30°.

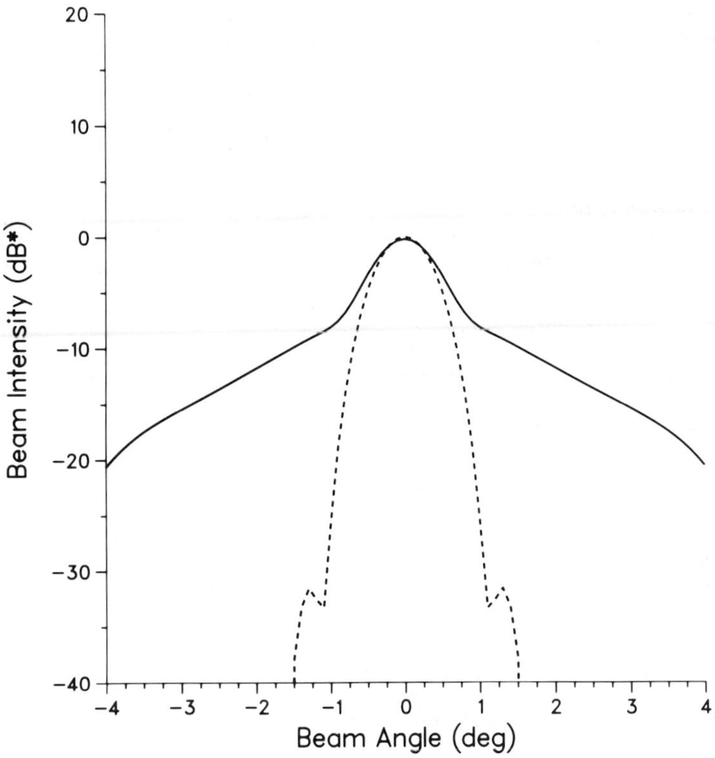

Fig. 6. Beam intensity computed by ORBS as a function of beam angle for a 5000-m horizontal array at a source-receiver separation of 93 km. The dashed curve is for smooth bathymetry (no scattering) while the solid includes the scattering produced by bathymetry with a 20-m rms roughness. The array has been focused for 93 km.

Figure 7 shows that while we can discriminate between energy that scatters from a smooth (0-m rms roughness) seamount surface and a rough seamount surface, the effective difference at low frequency is small. This would make the vertical array not nearly as suitable for use in determining roughness from acoustics as the horizontal array discussed in the previous section.

CONCLUSIONS

We have shown how the ORBS computer model can predict the effect of rough-surface scattering on horizontal and vertical receiver output. The low frequency (30 Hz) was chosen to have a length scale comparable to that of the available bathymetric statistics. These bottom scattering effects can be estimated because of the very significant increase in out-of-plane energy manifested in the side-lobe levels for a horizontal array, because of incoherent scattering from the rough ocean floor, for which the seamount slope is a good example. For a 100-wavelength horizontal array, we observed

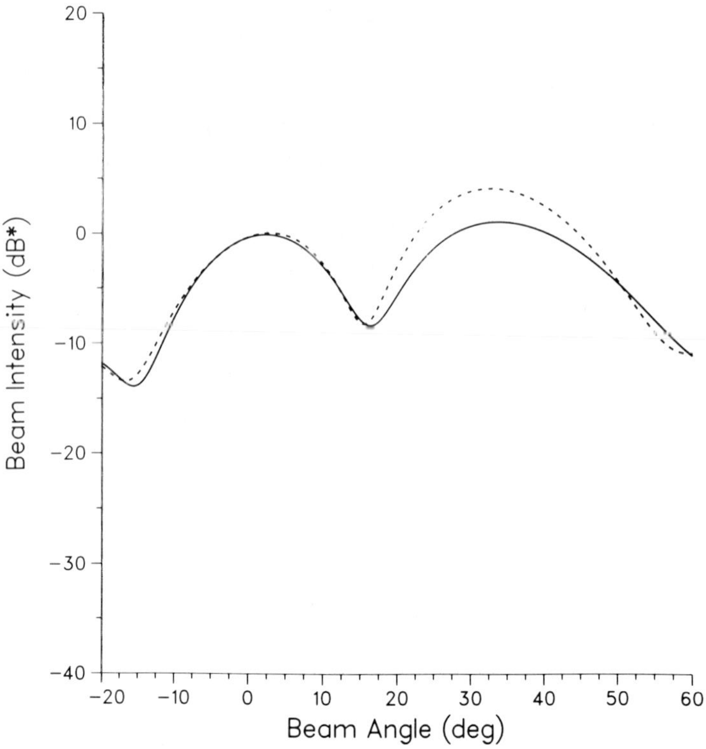

Fig. 7. Beam intensity computed by ORBS as a function of beam angle for a 150-m vertical array at a source-receiver separation of 93 km. The dashed curve is for smooth bathymetry (no scattering) while the solid includes the scattering produced by bathymetry with a 20-m rms roughness.

only a 2.3 dB reduction in the coherent signal measured by array signal gain due to bathymetric roughness, while the off-specular levels rose by 40 dB or more. We have also used results from the ORBS model to illustrate that vertical arrays, which might be thought effective, are much less suitable for determining roughness values from acoustic measurements due to the difficulty in rejecting direct and specular components.

ACKNOWLEDGMENT

This work was funded by the U. S. Naval Research Laboratory.

REFERENCES

1. R. N. Baer, D. H. Berman, J. S. Perkins, and E. B. Wright, A Three-dimensional Model for Acoustic Scattering from Rough Ocean Bathymetry, to be published in J. Comp. and Math. with App. (1985).

2. E. B. Wright, D. H. Berman, J. S. Perkins, and R. N. Baer, The ORBS Bathymetric Scattering Model, Naval Research Laboratory Report, in preparation (1985).

3. J. S. Perkins, R. N. Baer, E. B. Wright, and L. F. Roche, Solving the Parabolic Equation for Underwater Acoustic Propagation by the Split-Step Algorithm, Naval Research Laboratory Report 8607 (1982).

4. J. J. McCoy and M. J. Beran, Directional Spectral Spreading in Randomly Inhomogeneous Media, _J. Acoust. Soc. Am._ 66:1468 (1979).

5. D. H. Berman and J. S. Perkins, The Kirchhoff Approximation and First-Order Perturbation Theory for Rough Surface Scattering, to be published in _J. Acoust. Soc. Am._ (1985).

6. D. H. Berman and J. S. Perkins, Exponential Substitution for Kirchhoff Scattering from Gaussian Rough Surfaces, to be published in _J. Acoust. Soc. Am._ (1985).

7. M. F. Czarnecki and J. M. Bergin, Spectral Characterization of Multibeam Bathymetric Data, in preparation (1985).

EVALUATION OF LOW-FREQUENCY BOTTOM BACKSCATTERING STRENGTH VS GRAZING ANGLE BY MEANS OF MULTIPLE BEAMFORMING

D. Marandino and T.G. Goldsberry

SACLANT ASW Research Centre
Viale San Bartolomeo 400
I-19026 La Spezia, Italy

ABSTRACT

Bottom backscattering strength was measured in deep water as a function of grazing angle in the Balearic Abyssal Plain of the Mediterranean. The measurements were made with a towed, narrowband low-frequency omnidirectional source and a towed, horizontal linear array. The method takes advantage of the multiple beamforming capability of the receiving array and processor to accurately discriminate returns at given grazing angles from interfering returns. Approximations for the evaluation of the backscattering area are described and discussed. The results are presented and compared with previously reported measurements. At the higher grazing angles, specular reflection from normal incidence dominates the returns. However at medium and small grazing angles data have a large plateau region, with small variations. A Lambert's rule relationship, with a coefficient of -33 dB, can approximate results for grazing angles as low as about 20°.

INTRODUCTION

Low-frequency bottom backscattering strength values have been previously measured and reported by several authors [1 to 5]. These results refer to experiments with broad-band sources (explosives) and omnidirectional receivers. However the data presented herein were collected using narrowband sources and directional receivers. By taking advantage, on reception, of the directional properties of a towed linear array, it was possible to accurately discriminate returns at various grazing angles. Indeed, it is well known that beams formed by a linear array have conical symmetry around the array's axis. This property was exploited to discriminate returns in the vertical plane and to estimate arrivals from progressively longer times and thus smaller grazing angles.

Measurements were made in a deep water area in the Balearic Abyssal Plain. While the results are pertinent only to a similar environment the evaluation criteria can be generally applied to horizontal towed arrays.

The next section provides a description of the experiment and is followed by a discussion on the measurement criteria that emphasizes the evaluation of the backscattering area. Finally, the results are presented and discussed.

1 DESCRIPTION OF THE EXPERIMENT

The experiment was conducted in an area of the Balearic Abyssal Plain which is typical of Mediterranean-type abyssal plains. The water depth at the site is 2800 m. Cores, 6 m long that were previously collected in the same area, show layers of sand interspersed with thin layers of clayey and silty deposits. An isothermal water column was present, giving rise to typical Mediterranean winter propagation conditions characterized by a totally upward refracting profile. Wind speed was less than 7 kn and sea state was 1 to 2, with a moderate swell.

The experimental set-up for the collection of monostatic reverberation used the SACLANTCEN R/V Maria Paolina to tow a sound source and a linear array receiver at constant speed, course and depth. During the run, the source generated pulses at regular intervals, and the bottom echoes were received by the towed linear array and subsequently processed on board. Both the source and array depth were 100 m.

Figure 1 shows a simplified block diagram of the data acquisition and processing system. The array has 32 hydrophones with half wavelength spacing. Hamming shading is applied to the hydrophone data. The filtered and digitized returns are beamformed via a time-domain, programmable multiple beamformer, and subsequently match-filtered in the array processor. Figure 2 depicts the beamformer file that was used during the experiment and indicates both a plan and side view of the beam-pointing directions,' as determined by conical symmetry and assuming a constant sound speed profile.

The transmitted waveform was a linear frequency modulated (LFM) pulse with a one second duration and 10 Hz swept bandwidth at a centre frequency of 365 Hz. The received signals were processed through a quadrature replica correlator with Hanning shading; the final equivalent slant range resolution is approximately 120 m.

2 BOTTOM SCATTERING STRENGTH EVALUATION METHOD

2.1 The Data Base

Returns from a number of pings were ensemble-averaged to determine a mean reverberation envelope. Reverberation at initial ranges is dominated by bottom backscattering that affects returns in two distinct ways:

- The first is referred to as the "fathometer effect", and is caused by the energy that is radiated by the omnidirectional source in the vertical direction and that propagates vertically in the water column, repetitively bouncing off the bottom and the surface. This is evidenced in the processed beam output time series as a sequence of peaks of the received power, which are regularly spaced at a distance equal to twice the water depth.

- The second effect is the progressive decay of the received power which is caused by backscattering from the bottom at progressively longer ranges, and thus smaller grazing angles.

Both effects can be seen in Fig. 3 where the ping-to-ping (ensemble) average of the returns received at the broadside beam over a limited set of ranges is shown on a logarithmic scale in an A-scan format. Also shown is the measured ambient noise level in the processed band (noise floor). The lower half shows the standard deviation of the averaging procedure of

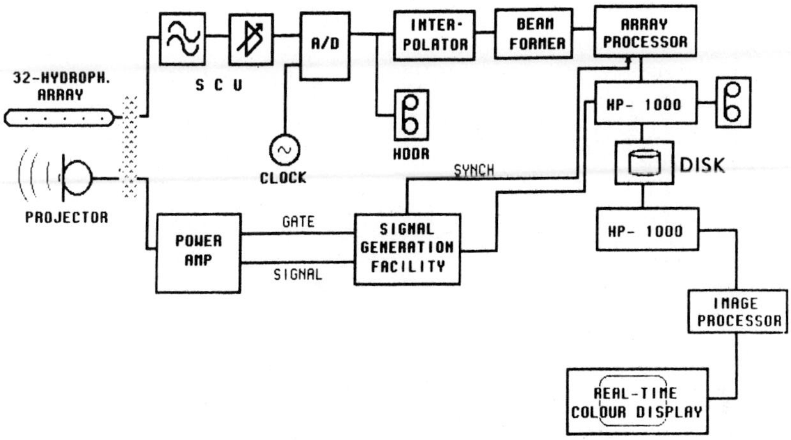

Fig. 1 Block diagram of SACLANT's data acquisition and processing system for reverberation measurements.

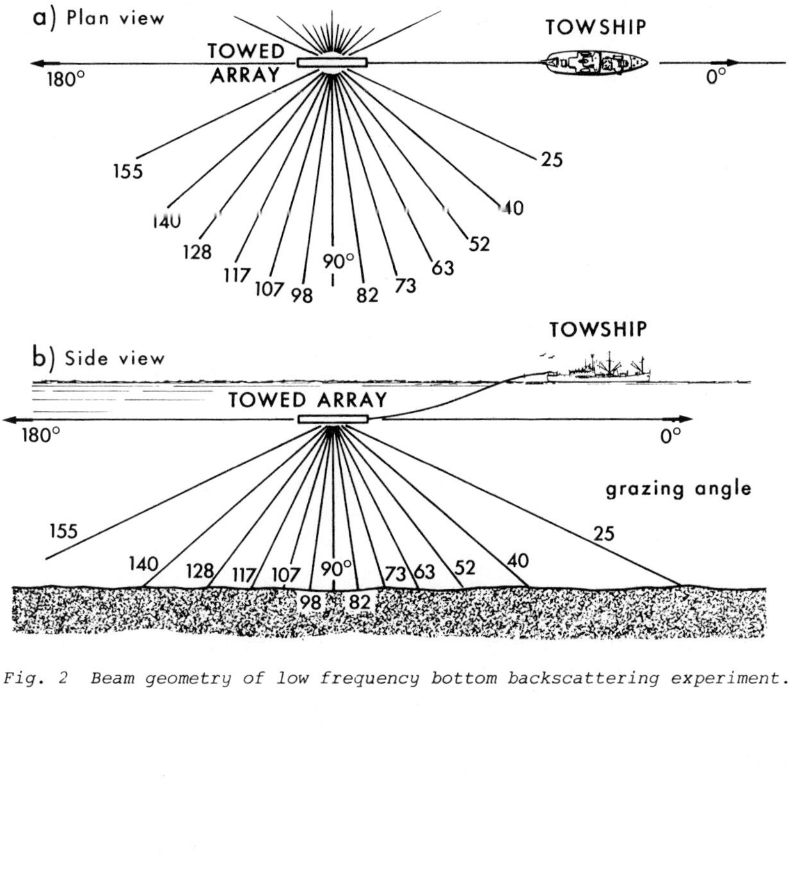

Fig. 2 Beam geometry of low frequency bottom backscattering experiment.

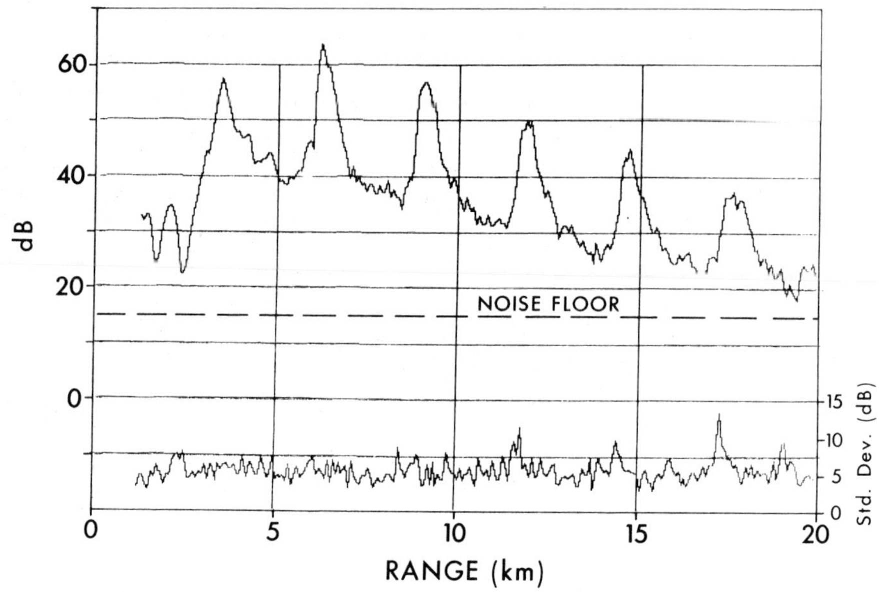

Fig.3 Average of reverberation returns received at the 90° beam, ambient noise floor, and standard deviation. (water depth: 2800 m; centre frequency: 365 Hz).

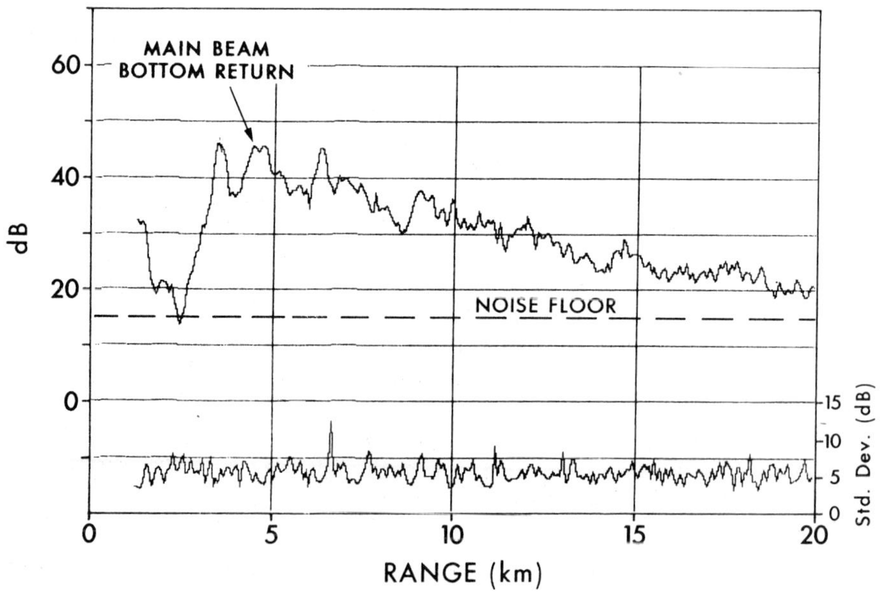

Fig. 4 Average of reverberation returns at the 38° from broaside beam, ambient noise floor, and standard deviation. (water depth: 2800 m; centre frequency: 365 Hz).

the figure which has a mean value of 5.5 dB, as expected from the averaging of Rayleigh distributed log normalized data. The fathometer peaks and the regular decay of bottom reverberation vs range are evident. However, broadside beam data cannot be used to accurately determine bottom backscattering strength because of the high sidelobes associated with the fathometer returns. To overcome that, the data from the off-broadside beams have been used. In this case, the fathometer effect is attenuated because vertically propagating energy is picked up through the angular sidelobes of the various beam patterns. Also, the beam power time series shows a local maximum corresponding to the onset of the bottom returns at the grazing angle related to the main beam pointing direction. This correspondence confirms the hypothesis of the predominance of bottom reverberation.

This is illustrated in Fig. 4 which show the time series associated with the beam pointing at 38° from broadside, which corresponds to a 52° grazing angle. An arrow marks the main beam bottom return. The advantages of evaluating backscattering strength in this manner are that: a) a (local) maximum of the signal-to-noise ratio is used; b) the influence of the fathometer effect can be considered negligible; c) the propagation loss can easily be estimated by assuming spherical spreading.

Bottom backscattering strength vs grazing angles can then be determined by direct application of the sonar equation, using the power output time series of the various beams.

2.2 Evaluation of the Backscattering Area

In order to determine scattering strength by means of the sonar equation, an estimate of the backscattering area (A_b) must be obtained. The evaluation of A_b is determined by three factors: propagation mechanism, the characteristics of the waveform and the beam shape at reception.

Figure 5 provides a plan and side view of the geometry of the experiment which is applicable to the evaluation of A_b and defines the pertinent axes and parameters. The y-axis is orientated along the array axis direction. In the figure,

A_{bM} is backscattering area at medium to small grazing angles

A_{bH} is backscattering area at high grazing angles

d_w is water depth

h is water depth less source depth

γ is the grazing angle (angle between bottom and propagation path)

γ_{eq} is the "equivalent" grazing angle, at small grazing angles

d_r is the annular width related to the signal range resolution

d_b is the annular width related to the mainbeam intersection and

d_{cr} is the "cross-range" dimension of A_b.

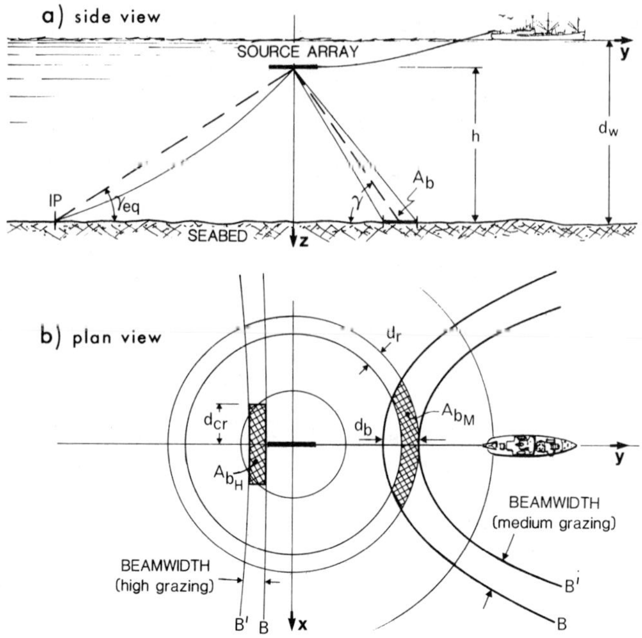

Fig. 5 Geometry for the backscattering area.

The energy radiated by the source propagates spherically and intersects the (flat) bottom along circles, whereas the various beams, which define cones around the array axis, intersect the bottom along hyperbolic curves B and B' in Fig. 5. However, such representation is only applicable to high and medium grazing angles (> 20°) where ray bending caused by the true sound speed profile are negligible. In this case the grazing angle is also equal to the complementary angle of the ray at the source/receiver. For smaller grazing angles, say below 20°, it has been assumed that the geometry remains approximately valid, provided that the actual grazing angle is replaced by an "equivalent" angle obtained by the straight-line approximation to the actual intercept point (IP in the figure) as determined by the actual sound speed profile. For this experiment, the ray at 13.86° at the source grazed the bottom at 0° with an equivalent grazing angle of 6.93° and slant range of 22370 m.

For an omnidirectional receiver, power comes from a circular annulus with a width (d_r) expressed by:

$$d_r = d_s \,/\, \cos(\gamma) \tag{1}$$

where d_s = slant-range signal resolution (here, 120 m).

Since the receiver is directional, only that part of the annulus which is included in the mainbeam contributes to backscattering, under the assumption that contributions from the beam's sidelobes are negligible. The intersection of the mainbeam with the bottom (hyperbolic curves B and B' in Fig. 5) is expressed by:

$$y_i^2 = \frac{x^2+h^2}{\tan^2\gamma_i} \quad ; \; i = 1, 2 \tag{2}$$

where γ_1 and γ_2 are the grazing angles corresponding to the 3 dB beamwidth.

This area has a minimum radial width, d_b, along the trace of the array axis on the bottom, expressed by

$$d_b = h * b_w(90°) \, / \, (\sin(\gamma))^3 \tag{3}$$

where $b_w(90°)$ = array beamwidth at broadside, with the assumption that the beamwidth for any beam pointing direction is given by

$$b_w(\gamma) = b_w(90°) \, / \, \sin(\gamma).$$

The backscattering area can then be estimated through the rectangular approximation:

$$A_b = D_r * D_{cr} \tag{4}$$

where D_r is the radial dimension, and

D_{cr} is the cross-range dimension.

Two cases may thus be distinguished:

- $d_r > d_b$. This is valid at high grazing angles (A_{bH} in Fig. 5). In this case the width of the transmission annulus is larger than the width of the hyperbolic annulus generated by the mainbeam intersection; therefore, in Eq. 4 it is:

$$D_r = d_b \tag{5}$$

- and, for the cross-range dimension, simple geometrical considerations on returns included between R and R+d_s, where R is the slant range in the mainbeam give:

$$D_{cr} = 2 \, d_s \sqrt{\left(1 + \frac{2h}{d_s \sin \gamma}\right)} \tag{6}$$

- $d_r < d_b$. This is valid at medium to small grazing angles. The backscattering area (A_{bM} in Fig. 5) is that part of the circular annulus that is delimited by the intersection with the beam hyperbola. For a given range, the area can then be expressed by the rectangular approximation (4), where the radial dimension is

$$D_r = d_r \, , \tag{7}$$

and the cross-range dimension D_{cr} is the arc of circle generated by the intersection of the lowermost hyperbola and the circle related to that range. Then:

$$D_{cr} = \frac{2h}{\tan \gamma_R} \cdot \arcsin \sqrt{\frac{\tan^2 \gamma_L - \tan^2 \gamma_R}{1 + \tan^2 \gamma_L}}$$

where γ_L = higher grazing angle, which corresponds to the 3 dB beamwidth boundary

γ_R = grazing angle corresponding to the given range.

At small grazing angles, however, ray bending increases the distance at which the rays intercept the bottom and the beam intersections with the seabed are not expressed as hyperbolas. However the evaluation of A_b is still based on formulas (7) and (8), in which equivalent grazing angles that correspond to the linear approximation to the actual rays are used.

The relationship between the actual and the equivalent grazing angle can easily be established since the experiment was carried out in isothermal water and thus with a constant velocity gradient. In this case the rays are arcs of circles whose centers lay at a fixed distance

$$R_o = C_{REF} / \zeta \tag{9}$$

where

C_{REF} is the reference sound velocity (1500 m/s), and

ζ is the velocity gradient (0.017 s^{-1})

It is then:

$$\cos \gamma = \left(1 + \frac{h}{R_o}\right) \cos \gamma_{RAY} \tag{10}$$

and

$$\gamma_{eq} = \frac{1}{2} (\gamma_{RAY} + \gamma) \tag{11}$$

where

γ is the angle of the grazing ray

γ_{RAY} is the angle of the (monostatic) ray at the source/array and

γ_{eq} is the equivalent grazing angle.

The approximations used in the evaluation of A_b at small grazing angles can be considered satisfactory because d_r is much smaller than the other dimensions.

The behaviour of the bottom backscattering area for beams off broadside, as a function of the true grazing angle, is indicated in Fig. 6 along with the two-way propagation loss at the pertinent ranges. A_b is expressed in dB relative to one square meter and is evaluated with the parameter values applicable to the experiment.

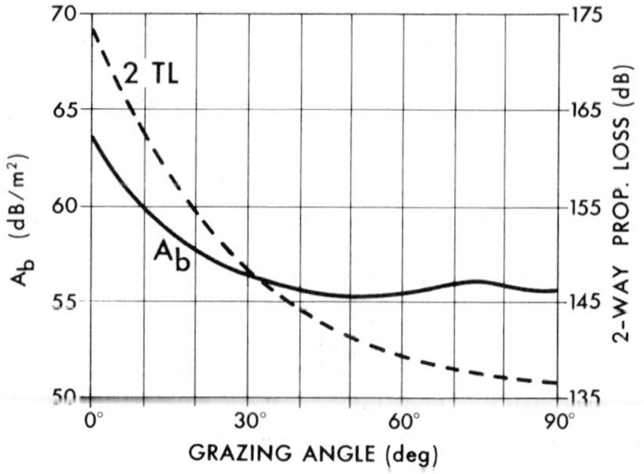

Fig. 6
Bottom backscattering area and two-way propagation loss as a function of the true grazing angle.

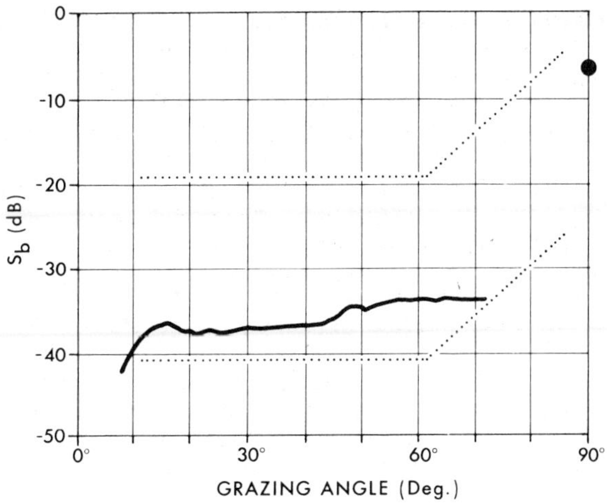

Fig. 7
Backscattering strength coefficient s_b, versus grazing angle, at a centre frequency of 365 Hz. The dotted lines show the limits of reported values for low frequency bottom backscattering strength.

3 RESULTS AND CONCLUSIONS

By applying the sonar equation [6] and using the appropriate curves for the backscattering area and the propagation loss, a point estimation of the backscattering strength coefficient versus grazing angles is obtained (see solid line in Fig. 7).

Two cases must be distinguished:

- At high grazing angles, because of the signal slant-range resolution returns from 73° to 90° cannot be separated and are merged into the fathometer peak and sidelobes. Nevertheless, the fathometer peak itself can be regarded as high grazing angle backscatter. Therefore, an estimation of S_b at near incidence can be obtained by measuring the power at the peak of the fathometer return. However, since the first-bounce fathometer return saturates the hydrophones, it is impossible to determine S_b reliably. Instead the second-bounce fathometer return was used with the addition of the estimated bottom loss incurred at the first bounce (6 dB). The applicable backscattering area here is that of the broadside beam at 90°. The result is given in Fig. 7 by the value at the 90° grazing angle.

- The evaluation of S_b for medium to small grazing angles is outlined in Sect. 2.2. Both rear and forward beam pointing directions were used; however forward beams have a higher noise level due to the towship contributions and are not suitable for estimates at small grazing angles. The estimate is quite accurate for angles above 30° because the backscattering area has small variations and, more importantly, because the measurements are made before the onset of the second fathometer return and are thereby free of multiple bounce scattering. At smaller grazing angles, the estimate of S_b becomes more difficult because the output ratio of signal-to-ambient noise is lower, small variations of the array tilt cause large variations of the intercept area and contributions from multiple-bounce scattering are more likely. However measurements can still be made and an approximate, upper bound for S_b can be obtained for these smaller grazing angles.

The overall behaviour of the bottom backscattering strength indicated in Fig. 7 shows that S_b has high values in that region where backscattering is dominated by specular reflection. The range of angles of high backscattering is determined by the system geometry and the type of waveform such that the coherent (specular) component dominates incoherent (diffuse) scattering.

For decreasing angles, S_b quickly decays by about 30 dB. Then, for medium and small grazing angles, it has a plateau region where the variations are small. Values of -33 to -37 dB are measured for medium angles (60° to 30°) and of -40 to -42 dB for smaller angles (around 10°).

A comparison with Lambert's rule for diffuse backscattering shows that a value of -33 dB for the Lambert coefficient is reasonable for medium grazing angles. The limits of reported data for low-frequency bottom backscattering strength range from -20 to -40 dB, and are shown in Fig. 7 by the dotted lines (see, for instance, Ref. 6, page 246). These include results at various frequencies and for different types of bottom. As can be seen, the range of previously reported data spans the present measurements, although similar types of seafloor are usually assumed to have slightly higher S_b values.

In conclusion, a method has been presented to evaluate low frequency bottom backscattering strength which takes advantage of the multiple beamforming capability of a towed linear array. The intrinsic system directivity confines backscattering to small, homogeneous seafloor patches and suppresses interference from other directions.

REFERENCES

1. K.V. Mackenzie, Long range shallow-water bottom reverberation, J. Acoust. Soc. Amer. 34: 62-66 (1962).
2. R.J. Urick and D.S. Saling, Backscattering of explosive sound from the deep-sea bed, J. Acoust. Soc. Amer. 34: 1721-1724 (1962).
3. A.W. Burstein and J.J. Keane, Backscattering of explosive sound from ocean bottoms, J. Acoust. Soc. Amer. 36: 1596-1597 (1964).
4. P.B. Schmidt, Monostatic and bistatic backscattering measurements from the deep ocean bottom, J. Acoust. Soc. Amer. 50: 327-331 (1971).
5. H.M. Merklinger, Bottom reverberation measured with explosive charges fired deep in the ocean, J. Acoust. Soc. Amer. 44: 508-513 (1968).
6. R.J. Urick, Principles of Underwater Sound, 2nd ed. New York, NY, McGraw-Hill (1975).

NEAR GRAZING PROPAGATION OVER A LOW ROUGHNESS HARD OCEAN BOTTOM

Herman Medwin

Physics Department
Naval Postgraduate School
Monterey, CA 93943, USA

ABSTRACT

When low frequency sound interacts with a low roughness rigid surface at grazing or near-grazing angles, a coherent boundary wave is generated by repeated forward scatter. The boundary wave extends several wavelengths into the fluid and affects both the phase velocity and the amplitude of the forward propagating sound near the surface. The propagation is subsonic and slightly dispersive and, under certain conditions, the reflection coefficient can be greater than unity. Laboratory model results for periodic roughness elements have now been extended to demonstrate that these boundary wave phenomena are significant for steep sloped randomly rough surfaces. Predictions are made of the consequences of boundary wave propagation at a rough rigid ocean bottom.

INTRODUCTION

In the 1950s Twersky[1] presented a series of papers describing scattering by arrays of bosses. More recently, Tolstoy[2,3] specialized to the case of grazing propagation from a low frequency point source near a low roughness boundary and predicted the existence of a coherent boundary wave in a physically intuitive way. By "low frequency" we understand $kh<1$ where k is the wavenumber and h is the separation between roughness elements. By "low roughness" we mean $kH<1$, where H is the height of the roughness element. Most recently Lucas and Twersky[4] have extended the rigorous boss analysis to the same case of a point source near a low roughness boundary and new, more detailed, aspects of the coherent boundary wave have been described. These results are for steep-sloped roughness elements; there are no limitations to small slopes and no avoidance of the important effect of secondary scatter between roughness elements. Indeed those factors are crucially important in near grazing propagation.

The theoretical predictions have been supported by extensive controlled laboratory experiments[5,6] which have demonstrated the growth, attenuation, dispersion and diffraction of the boundary wave. Both two dimensional and three dimensional roughness elements have been tested in periodic assemblages and the effect of three dimensional randomly rough natural elements has now been evaluated for grazing and near-grazing forward propagation. Among the most interesting of the predictions of these theories is that, at certain ranges and frequencies, the reflection coefficient will be greater than unity; this has been verified in the laboratory experiments. Also predicted[7] is a modified refraction behavior for bottom-limited rays due to the subsonic propagation of the near-grazing waves at the rough bottom. This presentation will consider the results of those experiments and their application to propagation at a rough hard ocean bottom.

THE COHERENT SCATTER WAVE: THEORY

Any good textbook on acoustics will show that low frequency Rayleigh scatter from a single fixed sphere can be represented by monopole and dipole components.[8] Together these components show a strong forward scattered beam of magnitude that is dependent on the relative elasticity and density of the sphere, compared to that of the medium. This description for a single sphere suggests the method of synthesis of the boundary coherent scatter caused by an array of small scatterers. Following ideas of Biot and Lighthill, Tolstoy replaces the array of small protuberances on the rough surface by a continuous distribution of equivalent monopole and dipole mini-sources which, for a real point source on or near the rough surface, generate a cylindrically divergent wave in the boundary region. The strengths of the equivalent sources on the rough plane depend also on whether the elements are fixed and how closely they are packed. The reader is referred to the Tolstoy review paper[3] for the mathematical details of the technique. The coherent scatter wave is significant only at near-grazing propagation. There the spectral amplitude is predicted to be

$$P_S = \frac{\varepsilon}{2\pi} k^2 \Big[\theta^2 + \frac{2\pi}{kr} \exp\left[-2\varepsilon k^2(z+z_0)\right] \tag{1}$$

$$+ 2\left(\frac{2\pi}{kr}\right)^{1/2}\theta \sin\left(kr - \frac{\pi}{4}\right) \exp\left[-\varepsilon k^2(z+z_0)\right]\Big]^{1/2},$$

where k is the free-space wavenumber, r is the range, and $\theta=\tan^{-1}[(z+z_0)/r]$ with z and z_0 the receiver and source heights.

The critical quantity ε, with units of length, is proportional to the volume of the protuberances per unit area and is a function of the shape and packing density of the roughness elements. ε is called the scattering parameter. For close-packed hemispheres[2], $\varepsilon = 4.44 \times 10^{-2}\, h$ where h is the separation between elements.

When z_0 and z approach zero so that $\theta \to 0$ the scatter wave is simply the boundary wave with amplitude

$$P_B = \varepsilon(2\pi r)^{-1/2}k^{3/2} \;. \tag{2}$$

Since the normalized spectral amplitude for the delta function source is the frequency-independent direct wave amplitude in half space

$$P_D = (2\pi r)^{-1}, \tag{3}$$

this leads to the ratio of the boundary wave amplitude to the direct wave amplitude at grazing incidence (ignoring attenuation)

$$(P_B/P_D)_0 = \varepsilon(2\pi r)^{1/2}k^{3/2} \;. \tag{4}$$

The boundary wave is also predicted to be dispersive. The theoretical dispersion of the grazing specularly scattered sound is rewritten from Eq. (2.5) of Ref. 2.

$$(c-v)/c = 1-k_S/k_R \cong \Lambda \varepsilon^2 k^2, \tag{5}$$

where $k_R \simeq k_S(=k)$ are propagation constants over the rough and smooth surfaces, (and free space), c is the speed in free space, $v = v(k_R)$ is the phase velocity of the wave propagating over the rough surface, and $A \simeq 0.5$ for close-packed hemispheres. The experimental value of c is obtained by making the measurement on the smooth surface.

The grazing propagation of a low frequency wave over a low roughness surface is therefore described as

$$P_{rough} = P_{smooth} + P_{boundary},$$

where

$$P_{smooth} = P_D \exp[i(\omega t - k_S r)], \quad (6)$$

and

$$P_{boundary} = P_B \exp[i(\omega t - k_b r)],$$

where k_B is the propagation constant for the boundary wave.

The dream of an intermidable increase in the grazing boundary wave amplitude relative to the direct wave amplitude, Eq. (4), is brought to a rude termination by the realities of incoherent boundary wave scatter and destructive self-interference at larger ranges and frequencies. Including these effects[6] leads to

$$P_B/P_D = (P_B/P_D)_0 \, e^{-\delta r - \beta r^2} \quad (7)$$

The δ, the attenuation rate due to scatter, has been given by Tolstoy[9], and is proportional to f^6. The β, due to self-interference of the various out-of-phase contributions to the boundary wave was given by Medwin et al[6] as

$$\beta = (A^2/2)\varepsilon^4 k^6 \quad (8)$$

The consequence of the attenuation processes is a range-dependent, low-frequency pass band within which the boundary wave amplitude may indeed surpass the direct wave amplitude and thereby yield reflection coefficients that are greater than unity.

EXPERIMENTAL RESULTS

The laboratory experimental technique uses a model rough surface and a reference smooth surface. A pulse from a small source is received by a small receiver and the relative spectra are calculated. Ref. 6 has a detailed description of the method. How well the theoretical expectations are fulfilled, in the test case of close-packed rigid spheres on a rigid plate, is shown in Fig. 1, for the amplitude, and Fig. 2 for the dispersion.

Of greater interest to the seismo-acoustician are the questions: Is there a boundary wave for a randomly rough surface? Can a boundary wave follow the vagaries and the topographical meanderings of a real rough bottom? Does the subsonic dispersion have any practical effect on wave propagation near a rough bottom? We believe that the answer is "yes" to all of these questions.

EXCESS REFLECTION AT A RANDOMLY ROUGH PLANE

When a layer of gravel is spread over a rigid plate the relative boundary wave amplitude again (see Eq. 4) increases approximately as $f^{3/2}$, peaks at $P_B/P_D>1$, and is ultimately destroyed by attenuation. The peak values are found at lower frequencies for larger ranges (Fig. 3). The behavior can be reformulated in dimensionless form, as shown in Fig. 4. The ordinate here is the excess beyond perfect reflection, the quantity $(P_R/P_D - 1)$, expressed in percent. P_R is the pressure amplitude at the rough surface. The two gravels, with their different rms heights and correlation functions, yield different excess reflections relative to a smooth perfectly reflecting surface. A generalized empirical law is being developed which will predict this excess reflection.[10]

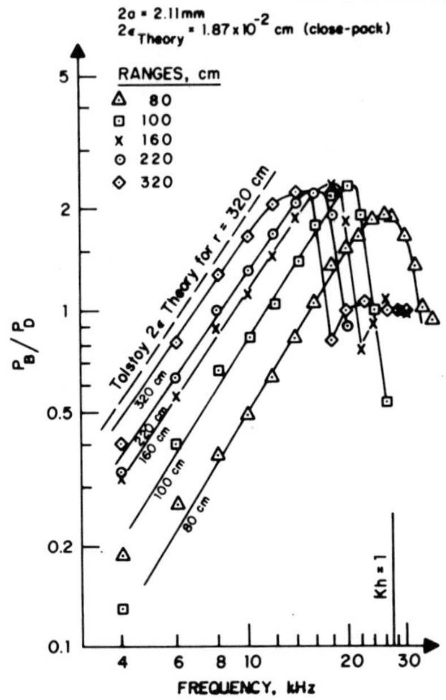

Fig. 1

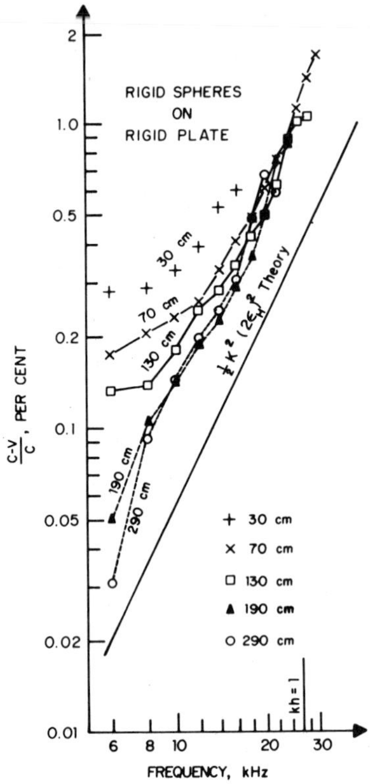

Fig. 2

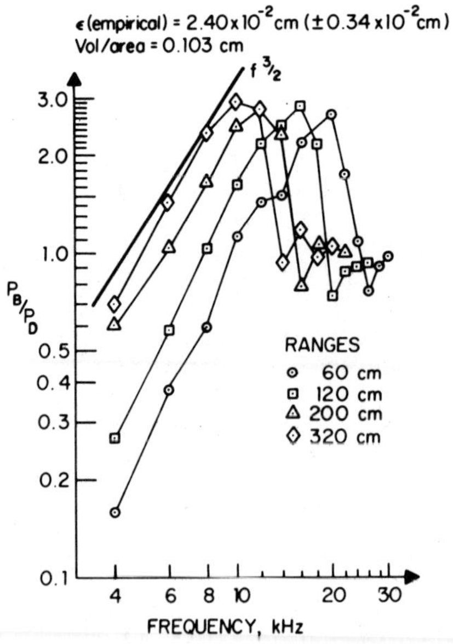

Fig. 3

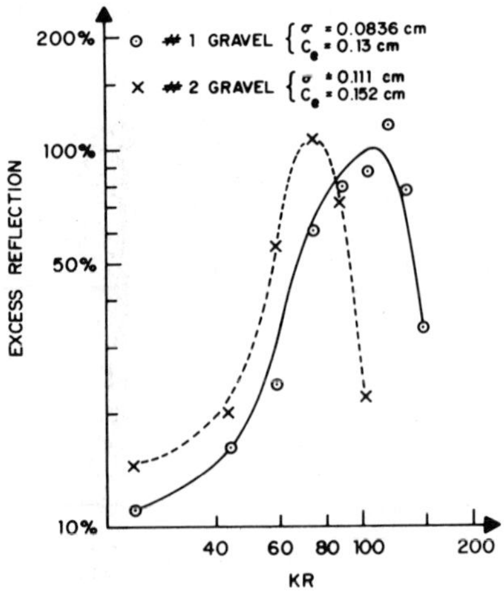

Fig. 4

Ref. 6 describes laboratory experiments which showed that a boundary wave will diffract over a ridge (a 14° upslope and 14° downslope were studied) which has been roughened by close-packed, hemispherical bosses. The diffraction caused the spectrum of the boundary wave which is proportional to $f^{3/2}$ to change by the factor $f^{1/2}$ due to diffraction loss over a wedge; the spectrum then had the slope P_B/P_{DD} proportional to f^2, where P_{DD} is the diffracted direct wave.

Another proof of the diffracting capability of the boundary wave was obtained by using a scale model of DICKINS seamount.[11] The upper ocean shadowing by this seamount had previously been explained by comparison of model results with ocean experiments from 50 to 500 Hz. When a thin layer of #20 sand (rms height 0.016cm, correlation length to e^{-1}, 0.031cm) is used to roughen the model seamount we get boundary wave amplitudes at the seamount bottom as for a rough plane surface, but at the expected spectral slope for the diffracted boundary wave, P_B/P_{DD} approximately proportional to f^2. This is shown in Fig. 5. A photograph of the model surface is in Fig. 6. The scale model ranges of 35, 45, 65 cm correspond to ocean ranges of 2.8, 3.5 and 5.1 km from the crest of the seamount.

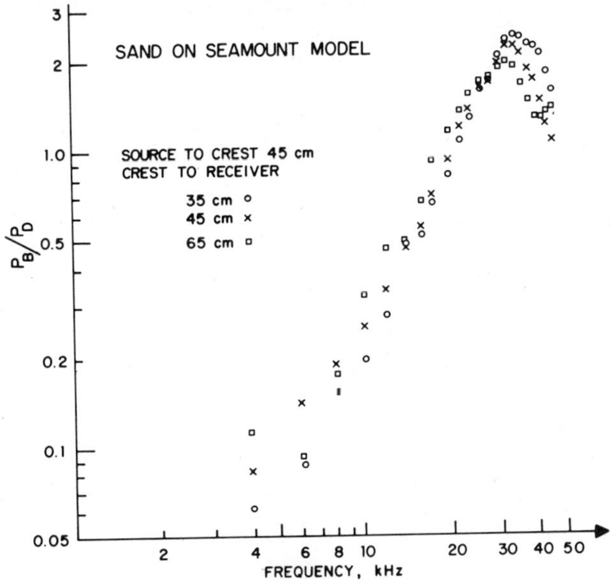

Fig. 5

Fig. 6

On the surface the dispersion at grazing incidence is described by Eq. 5. Above the surface, the phase velocity of the rough surface wave, v, continues to be subsonic until it fairs into the free space value, c, at several wavelengths above the surface. Fig. 7 shows this effect as obtained in scale model experiments.[10] The vertical bars indicate the e^{-1} limit of the boundary wave amplitude at two frequencies. Z is the distance from the rough surface. The consequence of the behavior shown in Fig. 7 is that, due to the roughness, there is an intrinsic gradient given by

$$b_i \simeq -(\omega/2)\varepsilon^3 k_o{}^3 \tag{9}$$

which is superimposed on the extrinsic gradient which is accounted for by gradients of salinity, temperature and pressure. In the presence of an upward refracting, isothermal, isohaline, velocity gradient, b, such as is common for the deep ocean, the bottom limited ray propagates in a region characterized by the gradient

$$b_s = |\, b \,| - |\, bi \,| \quad . \tag{10}$$

Fig. 8 shows that the effect of the weakened gradient will be to increase the radius of curvature and thereby to increase the skip distance of the bottom-limited ray. The range added to the classical skip distance, EX, will be a function of the parameters of the ocean and of the rough surface.

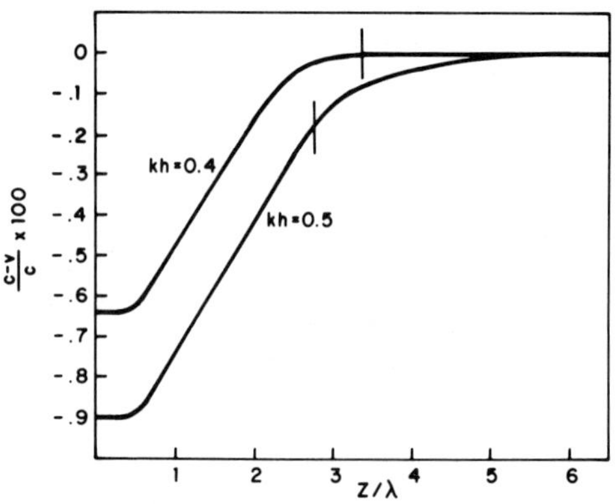

Fig. 7

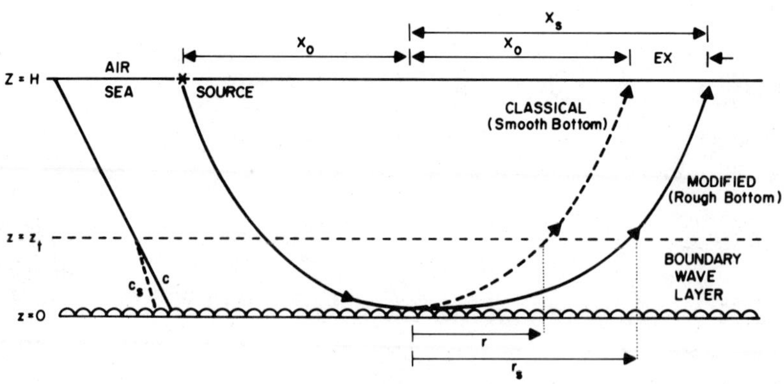

Fig. 8

A prediction of the amount of the excess skip distance for a particular case is shown in Fig. 9 as a function of the surface roughness parameter ε and the sound frequency. Although the predicted effect is severely limited in extent, the magnitude can be quite significant and would be easily observable under the specified conditions.

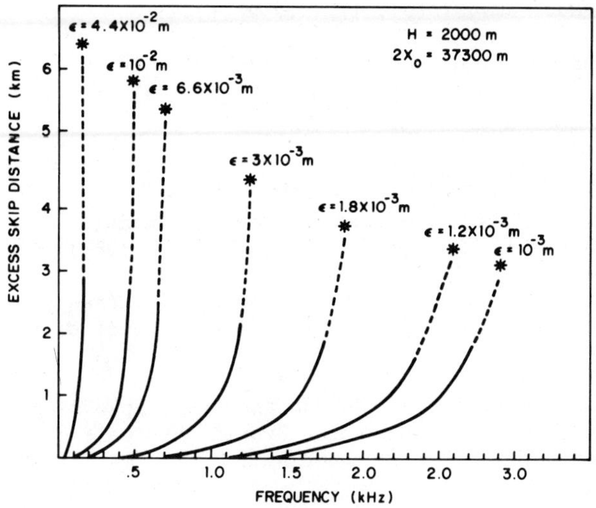

Fig. 9

CONCLUSION

The presence of low roughness on a hard sea floor can have significant effects on both the amplitude and refraction of low frequency sounds at near-grazing incidence.

ACKNOWLEDGEMENT

I am indebted to Gerald L. D'Spain, Jorge C. Novarini and Emily H. Childs of Ocean Acoustics Associates who competently performed the sensitive experiments described in this paper. The financial support provided by the U.S. Office of Naval Research is appreciated.

REFERENCES

1. V. Twersky, On scattering and reflection of sound by rough surfaces, J. Acoust. Soc. Am., 29(2), 209-225 (1957)

2. I. Tolstoy, The scattering of spherical pulses by slightly rough surfaces, J. Acoust. Soc. Am., 66(4), 1135-1144 (1979)

3. I. Tolstoy, Smoothed boundary conditions, coherent low-frequency scatter, and boundary modes, J. Acoust. Soc. Am., 75(1), 1-22 (1984)

4. R. J. Lucas and V. Twersky, Coherent response to a point source irradiating a rough plane, J. Acoust. Soc. Am., 76(6), 1847-1863 (1984)

5. H. Medwin, J. Bailie, J. Bremhorst, B. J. Savage and I. Tolstoy, The scattered acoustic boundary wave generated by grazing incidence at a slightly rough rigid surface, J. Acoust. Soc. Am., 66(4), 1131-1134 (1979)

6. H. Medwin, G. L. D'Spain, E. Childs and S. J. Hollis, Low-frequency grazing propagation over periodic steep-sloped rigid roughness elements, J. Acoust. Soc. Am., 76(6), 1774-1790 (1984)

7. H. Medwin and J. C. Novarini, Modified sound refraction near a rough ocean bottom, J. Acoust. Soc. Am., 76(6), 1791-1796 (1984)

8. C. S. Clay and H. Medwin "Acoustical Oceanography" Wiley, New York (1977)

9. I. Tolstoy, Rough surface boundary wave attenuation due to incoherent scatter, J. Acoust. Soc. Am., 77(2), 482-488 (1985)

10. H. Medwin and G. L. D'Spain, Near grazing, low frequency propagation over a randomly rough rigid surface, To be submitted to J. Acoust. Soc. Am.

11. H. Medwin, E. Childs, E. A. Jordan and R. P. Spaulding, Jr., Sound scatter and shadowing at a seamount: Hybrid physical solutions in two and three dimensions, J. Acoust. Soc. Am., 75(5), 1478-1490 (1984).

ESTIMATION OF ICE SURFACE SCATTERING AND ACOUSTIC ATTENUATION IN ARCTIC SEDIMENTS FROM LONG-RANGE PROPAGATION DATA

G.L. Duckworth and
A.B. Baggeroer

Massachusetts Institute of Technology
Cambridge, MA 02139
U.S.A.

ABSTRACT

A single explosive shot at a range of 341.3 km in the Pole Abyssal Plain of the Arctic Ocean is used to assess the components of propagation loss for this region. The acoustic energy propagated between a satellite ice camp and the Fram II ice station in a water column of nearly uniform depth. Much of the observed energy interacted with the upper 200 meters of sediments along a path which was nearly parallel to the Arctic Mid-Ocean Ridge. In addition, the upward refracting sound channel of the Arctic Ocean also caused the observed energy to interact extensively with the ice canopy, which was contiguous over the entire path. The deterministic lateral homogeneity of the bathymetry and sediments, and the statistical lateral homogeneity of the ice canopy in this region allow us to attempt to separate the effects of geometrical spreading, ice surface scattering, and effective sediment compressional wave attenuation. The primary data for this work are observations of the signal from a 25.8 kg explosive charge received on a 24 channel two-dimensional hydrophone array with a 1 km aperture. The data are inadequate to resolve attenuation at depth, but provide an estimate for frequency independent Q of 200 to 300 for the upper 200 meters of the sediments and a dependency of surface scattering on frequency over the 5-50 Hz band. It is shown that partially coherent summation of the surface multipaths is required to predict the observations.

INTRODUCTION

Discussion

The goal of this paper is to assess the effects of sediment attenuation and surface scattering on acoustic propagation in the Arctic Ocean. A model relating observations, source characteristics, surface scattering, and attenuation for a series of multiple arrivals observed at a constant offset from a single explosion is given in decibels by:

$$P(f,n) = S(f) + R_2(f,n) + G(n) + R_1(f,n) + A(f,n) \tag{1}$$

where

f	is the temporal frequency in Hz;
n	is the order of the multiple observed;
$P(f,n)$	is the observed data energy spectral density;
$S(f)$	is the source explosion energy spectral density;
$R_2(f,n)$	is the factor accounting for the near source and near receiver reflections and the random travel-time component due to the rough under-ice surface;
$G(n)$	is the geometrical spreading factor;

$R_1(f,n)$ is the amplitude component of surface scattering and ocean bottom transmissions for the linear surface and bottom interactions;

$A(f,n)$ is the attenuation factor due to effective Q along the travel path.

In this equation, the sound speed model is assumed known, so the multiple order, n, completely specifies the turning point depth, z, the ray slowness, p, and the length and time of the travel path in the sediments and water column.

Knowledge of the sound speed in the water column and sediment also specifies the geometrical spreading factor, $G(n)$, for a constant offset. The sound-speed profile is computed from the data separately. We compute $S(f)$ from the well known and regarded empirical spectra of TNT explosions at known depths, and estimate the scattering terms, $R_i(f,n)$, and the attenuation, $A(f,n)$, from the data.

We examine both the Fraunhofer and Fresnel approximations in the identification of under-ice roughness parameters. It is seen that the combination of travel-time fluctuation measurements through $R_2(f,n)$ and amplitude scattering from $R_1(f,n)$ allows characterization of both the rms surface roughness averaged over a Fresnel zone, $\bar{\sigma}_z$, and the instantaneous rms roughness, σ_z.

One assumption that is made, and one that has come under intense scrutiny in recent work [1,2], is an effective sediment quality factor, Q, which is constant over the frequency band from 5 to 50 Hz. For this work, this assumption is necessary in order to separate ice surface scattering effects from attenuation. Our assumptions and experimental geometry also do not allow estimation of Q as a function of multiple, n, and thus of the path integral which a ray follows to its maximum depth. Even if this had been possible, we would not have attempted to invert these estimates for "instantaneous" depth because it is unclear what the result of this inversion may mean. Due to the lack of separation of intrinsic Q from the geometrical effects of layered bedding the Q encountered by a ray at a given depth depends on the direction of the ray with respect to the bedding[2]. It is our feeling that a value of constant Q averaged over all multiples with a maximum depth of penetration of 200 meters is a more reliable quantity, and is consistent with the dominant energy wavelengths of 60-150 meters.

One very important point that will be made is that for propagation distances of 340 km in sediments and water column it is important to retain the coherency of near source and receiver multipaths to predict the spectral shape and amplitude of the received arrivals in the 5-50 Hz band. Though they will not be dealt with in this paper, the incoherent scattering aspects of the propagation may also be treated by analysing the scattered energy observed on our records between the strong arrivals. In this work, however, we will quantify random scattering only through expression as losses from the primarily coherent data. We are able to do this because of our fortuitous location in the nearly laterally invariant Pole Abyssal Plain, and because of the great temporal and spatial stability of the Arctic Ocean[3].

Background

Spectral ratio methods such as those used for this work are quite popular for estimation of attenuation in marine sediments. Stoll and Houtz[2] provide a good bibliography of recent work. This work differs from many of those efforts in that it attempts to separate source directivity effects, surface scattering, bottom transmission, and effective Q in the sediments using both travel-time and amplitude information. In addition, estimates for these parameters in the Arctic are quite sparse. This work should also compliment recent work by Lee and Kutschale[4], which used full-wave forward modeling to determine sediment Q in the Arctic. Their work required the inclusion of a frequency independent Q of 190-220 to obtain the observed dispersion and amplitude characteristics of later arriving sediment interaction multiples in experiments identical to ours. They also assumed that surface scattering was negligible in their frequency band from 5 to 30 Hz.

Experiment Location and Receiver Characteristics

The data used for this work are from the Fram II experiment in the Pole Abyssal Plain of the Arctic Ocean. The shot was initiated at a remote ice camp (Camp 1) at 88.850°N, 18.75°W on 26 April, 1980 at 1427 GMT. The data were received at the Fram II base camp drifting with the ice pack at 85.786°N and 24.41°W. This is a great circle distance of 341.3 km at 2° true from the Fram II camp. Bathymetric charts show a water depth of 4-4.3 km along most of the path[4]. This path was nearly parallel to the spreading center at the Arctic Mid-ocean Ridge and followed magnetic anomaly 24. The constant age and bathymetry along the path predispose us to believe that the sediments are laterally homogeneous. Few samples of the sediment are available, and we assumed the sediments to be turbidites. Other experimental evidence suggests that the assumption of a constant Q with frequency is good for this type of sediment[5,6].

The logistics of this experiment, including the receiving array configuration and recording capabilities are detailed by Duckworth et al.,[7]. For the shot analyzed in this paper, 22 hydrophone recording channels were used in a two-dimensional array configuration over a 1 km aperture. The hydrophone depth was 93 meters. The data are good from 5 Hz to 80 Hz. Below 5 Hz hydrophone strum became a problem, and the digital floating point acquisition system had anti-alias filters at 80 Hz.

Paper Outline

We first discuss the models that will be used to interpret the data. These include the source spectrum, geometrical spreading, near surface reflection characteristics, surface and bottom scattering effects, and the attenuation model. The data processing sequence is outlined, and some of the observational aspects of the data requiring little interpretation are presented. We then present the procedure used to obtain the Q estimate and the model parameters for the scattering loss. Finally, we comment on the reliability, resolution, and interpretation of our results, compare them to previous work, and indicate what further analysis would benefit the goals of this paper.

MODELING

Source

The source was 25 kg of TNT detonated by a depth activated Mk 82 SUS charge (0.8 kg TNT). The TNT explosive charge has been studied extensively as a pressure source in the water column. We implemented the equations of Wakeley[8] and computed a source signature for a depth of 93 meters and received at a range of 1 km. This range introduces some additional nonlinear attenuation of high frequencies over the calculation at 1 meter. This was corrected by +60 dB to obtain the signature at 1 meter, and then by -160 dB to obtain the voltage waveform we would obtain from our -160 dB re $1V/\mu Pa$ hydrophones. Voltage spectral levels, not pressure, will be used throughout the paper assuming phones of this sensitivity. The waveform was sampled at 10 kHz to catch the high frequency components of the shock wave and passed through a digital filter with an 80 Hz cutoff to mimic our analog anti-alias filter, decimated to a 4 msec sample interval, and its spectrum computed by a padded, unwindowed FFT with appropriate corrections for one-sided energy spectral density re $1V^2/Hz$. The spectrum was then smeared over a 4 Hz band by a running frequency average to give it the same bandwidth averaging characteristics as our data analysis procedure. The spectrum is shown in Fig. 1 and agrees well with that published by Urick[9].

Geometrical Spreading

Because we are attempting to estimate the scattering losses in absolute terms, rather than just the slope of the spectral ratio to obtain Q[10], it is necessary to compute the geometrical spreading loss. In our model, this is assumed to be independent of frequency and will be computed by ray theory from a laterally invariant sound speed profile for the water column and sediments.

The water column profile was obtained from a combination of direct sound speed measurements, theory, and inversion of the observed water column modal dispersion curves for this shot. The procedure is presented by Duckworth[11 (Chapter 6)]. The

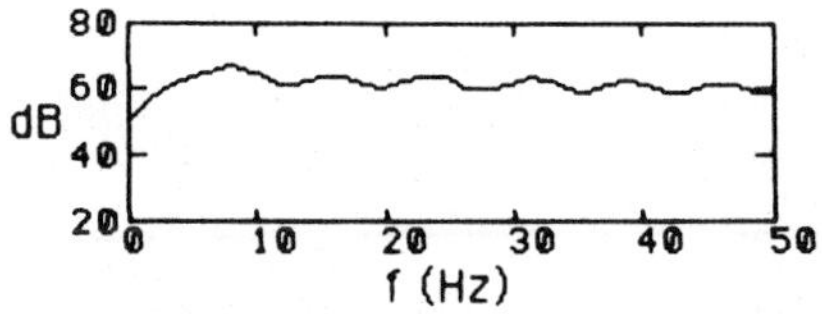

Fig. 1. Explosion Energy Spectrum

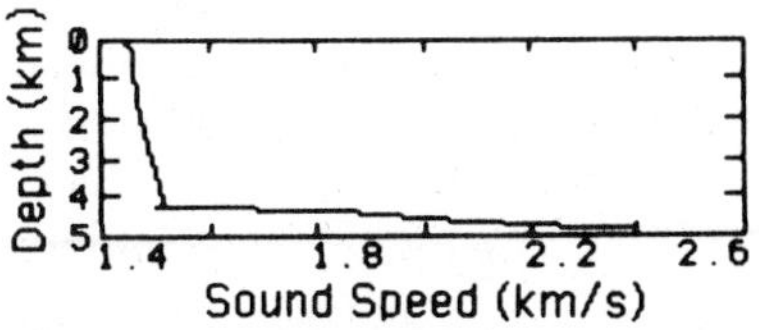

Fig. 2. Sound speed model.

sediment profile was determined by travel-time inversion of $\tau(p)$ and $x(p)$ curves estimated from the data by the methods described by Duckworth et al.[7]. The inversion methods of Diebold and Stoffa[12] and Dorman and Jacobson[13] were used after correcting the array estimated slownesses for slight bathymetric dip local to the receiving array. A preliminary inversion of this type is given by Duckworth[11 (Chapter 6)]. This profile is nearly identical to that of Lee[4]. Lee also found that shear wave conversion was not important for these paths.

The resulting sound speed profile is shown in Fig. 2. All travel times, phase velocities, and geometrical spreading factors were computed using linear velocity gradient segments between speed/depth points specified by this model. The geometric spreading loss assumes a spherically symmetric pressure source and no free surface interaction. The travel times of the data were measured at 8 Hz, the peak energy frequency of the source signature, using travel time detection on the maximum-likelihood beamformed output of the array[11 (Chapter 3)].

Bottom Transmission and Surface Scattering Losses

Bottom Transmission Coefficient: First, and seemingly least important to propagation loss in this area, is the transmission/reflection coefficient at the ocean bottom. Due to the long spatial wavelengths of bottom features in abyssal plains, we model this interaction independent of frequency in our band. Since we are dealing strictly with refracted arrivals, the transmission effects are the only ones to be considered. We feel that this frequency independent term in $R_1(f,n)$ should be very small due to gradual matching between the impedance of the water and the bottom by a very unconsolidated layer of about 35 meters thickness. In this layer the sound speed in the sediments is modeled to drop below that of the water column. Samples of the bottom obtained from OBS units dropped from the Fram II camp support this hypothesis [H.R. Jackson, personal communication, 1984]. Because of the extremely small losses at low frequencies where surface scatter is less important, we find that the bottom reflection coefficient is less than 0.2. This would give a maximum transmission loss of 0.35 dB/multiple, which will be seen to be consistent with our data. Helmberger, et al.[14] found the same bound for deep sediments in the Bering Sea.

Surface Scattering: The pressure of the shock wave at the surface above a 25.8 kg TNT explosion at a slant range of 186 meters is 233 dB re $1\mu Pa$. This is greater than the 222 dB pressure due to the atmosphere and 2 meter water surface suppression by the ice, thus indicating that there may be clipping of the rarefaction peak. However, our observations with low sensitivity monitor phones indicate that no cavitation takes place, and that all losses must be due to scattering from the uneven upper surface of the ice. We could not resolve any reflection from the bottom of the ice with a positive reflection coefficient in observations up to 1 kHz, indicating that a slushy matching layer is in effect here, as well as at the sea bottom. Experiments by Weston[15] indicate that the cavitation threshold in seawater is 246 dB re $1\mu Pa$ for short duration stresses, such as the explosive shock wave, supporting our observations.

If the multipaths shown in Fig. 3 are considered, where any equal number, $(n-1)$, of surface interactions may be present at the breaks in the rays between the source and receiver, then the transfer function due to surface scattering and bottom transmission from the explosion source to that received for the multiple of order n is given by:

$$P(f,n) - S(f) = H(f,n) \tag{2}$$

$$= 10 \log_{10} \left|\left| \alpha_1^{n-1} e^{-i2\pi f \tau_a} + \alpha_2 \alpha_1^{n-1} e^{-i2\pi f \tau_b} + \alpha_1^n e^{-i2\pi f \tau_c} + \alpha_2 \alpha_1^n e^{-i2\pi f \tau_d} \right|\right|^2$$

where

$\tau_{a,b,d}$	are the travel times for paths a, b, and d shown in Fig. 3;
τ_c	is the same as τ_b with reflection instead at receiver;
$\alpha_1 = \alpha_1(f,n)$	is the effective reflection coefficient for a single bounce assuming linear interaction with the surface;
$\alpha_2 = \alpha_2(f,n)$	is the reflection coefficient for a single bounce which may interact nonlinearly with the surface due to cavitation.

Note that α_2, the "nonlinear" reflection coefficient, appears only in two of the terms.

To determine if α_1 and α_2 were equal we applied the source, geometrical loss, and transfer function, to the data to compute the residual spectrum:

$$\Delta P(f,n) = P(f,n) - S(f) - G(n) - H(f,n) \tag{3}$$

for many realistic combinations of α_1 and α_2. Only when $\alpha_1 = \alpha_2$ would the ripples in the residual observed spectrum due to the near source/receiver interferences be removed correctly. This also lends credence to the assumption that the nonlinearity of the near source surface interaction is low. We also found that for all multiples the combination $\alpha_1 = \alpha_2 = -.65$ and $n = 1$ worked the best, although not perfectly. The reason for this is that for large n, say 9-24, the coefficients in front of all the exponentials in eqn. 2 are approximately equal, thus yielding a filter with very deep notches which over-corrected the observed spectra, and introduced sharp peaks in the residual spectra at the nulls of $H(f,n)$. One way to reduce the depths of the nulls was to set $n = 1$ and the α_i small so that only very incomplete cancellation could occur. This has, however, no physical justification, and it was found that the resulting uniform depth of the nulls for all frequencies at which nulls occurred was unrealistic. The observations indicated deeper nulls at low frequencies than high frequencies. In addition, the observed "residual" spectra after correction still showed quite a bit of variability and were not well "flattened" as we expected, leaving only the smooth effects of attenuation and statistical surface scattering.

The explanation for the variability of the corrected spectrum, and the changing null depth is quite simple, however, and lies in the time delays of the exponents in eqn. 2. First we approach the null width. The time delays, τ_a, τ_b, τ_c, τ_d, in eqn. 2 have the following form when the free surface is smooth and source and receiver depths are equal; $\tau \equiv (\tau_b - \tau_a) = (\tau_c - \tau_a) \approx (\tau_d - \tau_a)/2$. This leads to deep nulls in $H(f,n)$ at $f = m/\tau$, $m = \{0,1,2,...\}$. However, if the four different multipaths interact n times with a surface with rms roughness σ_z, then the density function of each path length, s is approximately Gaussian, with mean given by the smooth surface value and standard deviation given in the Fraunhofer approximation [16, 17] by $\sigma_s = \sqrt{n}\, 2\sigma_z \gamma \cos\vartheta(n)$. In this relation γ is a factor which accounts for the averaging over a Fresnel zone during the interaction and is given by $\gamma = \sqrt{2}l / r(f,n)$ where l is the correlation length of the ice roughness and $r(f,n)$ is the Fresnel zone radius as a function of frequency and multiple. This approximation should be adequate since the Rayleigh parameter, $2k\sigma_z \cos\vartheta$, is much less than 1 for our experiment. In general, $\bar{\sigma}_z \equiv \sigma_z \gamma$ is less than σ_z due to statistical averaging over the interacting area.

If the correlation length of the surface is less than the distance between interaction points of the four multipaths, then the path lengths are statistically independent. This makes the travel times, τ_a, τ_b, τ_c, τ_d, independent, with mean given by ray theory for a smooth surface, and standard deviation, $\sigma_\tau = \sigma_s / c_0$, where where c_0 is the water sound speed at the surface. Finally, we compute the width over which the mth "null" of $H(f,n)$ should be spread by the variations in τ; $\sigma_f^m = (m/\tau^2)\,\sigma_\tau$. Since τ depends on n, and $f(n) \equiv 1/\tau(n)$, then the null width is:

$$\sigma_f^m(n) = mf^2(n)\sigma_\tau = \frac{mf^2(n) 2\sqrt{n}\,\bar{\sigma}_z}{c_0} \cos\vartheta(n). \tag{4}$$

The concept of nulls remains valid as long as σ_τ is a small fraction of the periods of the waves we are studying.

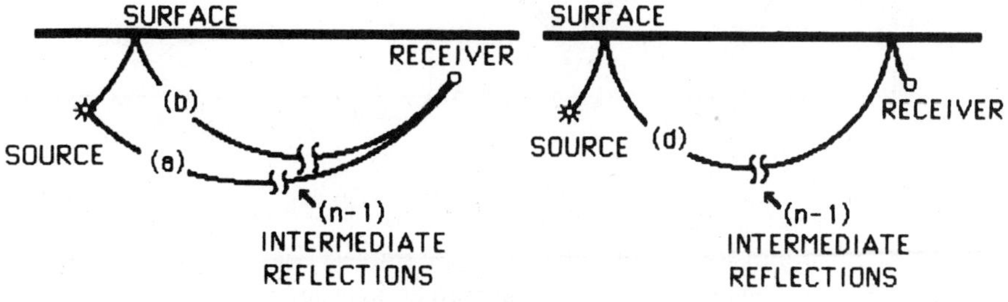

Fig. 3. Near-surface multipaths.

As the null width increases, the null depth is decreased. Our observations show an increase in null width with f as dictated by eqn. 4. The dependence of the null width on frequency may be even stronger than f^2 since the Fresnel zone size decreases with increasing frequency. The amount of averaging depends on the size of the Fresnel zone and the correlation length of the surface.

The variability of the residual spectra after correction with a filter computed using a particular set of $\alpha_1 = \alpha_2$ and $\tau_a, \tau_b, \tau_c, \tau_d$ values is a result of the fact that the observations are only one realization of a stochastic process. We find that variations of 1 msec in the τ values cause 15 dB changes in the null depth, and 5 msec variations cause lumps to come and go in different regions of the spectra and nulls to disappear and appear spontaneously. These correspond to the "fades" commonly observed in sonar work due to differential pathing in an unsaturated environment.

Fig. 4 indicates the various stages in the calculation of a residual spectrum. Fig. 4a shows the measured spectrum for multiple 11. Fig. 4b gives an ensemble of possible source spectra after correction by ten different Monte Carlo realizations of the surface reflection filter with $n = 1$, $\alpha_1 = \alpha_2 = -1$ and $\sigma_\tau = 1$ msec. Fig. 4c shows the residual spectrum obtained by subtracting the geometrical spreading loss and the energy averaged filtered source spectra of Fig. 4b (reconverted to dB) from the measured spectrum (eqn. 3). Fig. 4d shows the residual spectrum computed using the same procedure, but using the deterministic filter in eqn. 2 with $\alpha_1 = \alpha_2 = -.65$, $n = 1$, and $\sigma_\tau = 0$ msec. Note that the residuals are very similar, with 4c being slightly better smoothed. In addition, Fig. 4c uses a physically reasonable model and Fig. 4d an *ad hoc* one. This shows the equivalence between changing the coefficients of the filter and the delays in controlling the depth and breadth of the nulls. We note that $\sigma_\tau = 1$ msec predicts a γ of 1/9 to 1/16 for an rms roughness of 2.3 meters, a value generally accepted for the Arctic[18]. This figure for σ_τ gave very good null depth and breadth results. Applying the average $H(f,n)$ computed by Monte Carlo techniques using these values gave the most "pleasing" residual spectra for all multiples and all frequencies. The prospects of formal inversion of these spectra for σ_τ will be discussed later in this paper.

These results lead us to a revised model for the combination of bottom transmissions and surface scattering given by:

$$H(f,n) = 10 \log_{10}\left[|\alpha(f,n)|^{2n} < |e^{-i2\pi f\tau_a} + e^{-i2\pi f\tau_b} + e^{-i2\pi f\tau_c} + e^{-i2\pi f\tau_d}|^2 >\right] \tag{5}$$

where the < > indicate averaging over an ensemble of τ values with $\sigma_\tau = 1$ msec. For $n > 5$ and $\alpha(f,n) > 0.9$ the differences between the coefficients of the exponentials in eqn. 2 become negligible and they may be lumped outside the average. For small travel time variance the averaged component has little effect on the maximum height of the filter, but governs only the width and depth of the nulls. The frequency dependency of α controls the height and trend of the filter maxima and is governed by the scattering loss in the ice interaction. There may also be a frequency independent part of α due to the ocean bottom transmission coefficient.

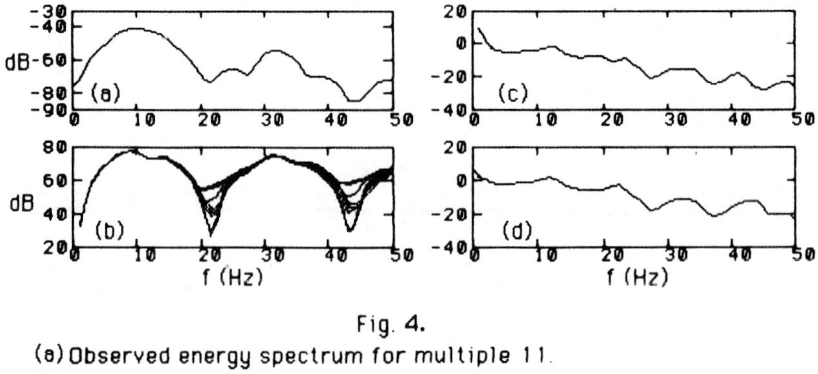

Fig. 4.

(a) Observed energy spectrum for multiple 11.

(b) Ensemble of $R_2(f,11)$ filters for $\sigma_\tau = 1$ msec.

(c) Residual Spectrum after correction by average of filters in (b)

(d) Residual Spectrum after correction by deterministic filter.

Equations 1 and 5 and the discussion above combine to give

$$P(f,n) = S(f) + G(n) + H(f,n) + A(f,n) \tag{6}$$

where

$$H(f,n) = n\cdot 10\log_{10}[\alpha^2(f,n)] + 10\log_{10} < |F(f,n)|^2 > \tag{7}$$
$$= R_1(f,n) + R_2(f,n)$$

and $<|F(f,n)|^2>$ is the averaged quantity in eqn. 5. The $R_2(f,n)$ of eqn. 1 depends on the multiple order n through the τ values. $R_1(f,n)$ depends only on the scattering coefficient, $\alpha(f,n)$, thus we have separated random travel-time and scattering amplitude effects. We have parameterized the R_2 term by a single statistical quantity, σ_τ, and through it $\bar{\sigma}_z$, since travel-time variations due to sound speed effects are quite small in the Arctic[3]. It now remains to find a simple parameterization for the amplitude scattering component, $R_1(f,n)$.

The frequency dependence of the scattering term, $R_1(f,n)$ is the subject of great debate. The primary argument centers around whether Fraunhofer theories, which give the loss an f^2 dependence, or Fresnel theories, which behave as f^1[16, 17] are more valid for scattering from the pack ice. For a Rayleigh parameter of $g = 2k\sigma_z \cos\vartheta \ll 1$, where k is the total wavenumber, $2\pi f / c_0$, of the insonification at the scattering boundary and ϑ is the angle of incidence, Fraunhofer theory predicts:

$$R_1(f,n) = -n\left[10\log_{10}(e)\, g^2(f,n) + T\right]. \qquad \text{Fraunhofer} \tag{8}$$

T is the 2-way transmission gain at the seabottom. When g is similarly small, the first terms of the expansion for Fresnel theory predict:

$$R_1(f,n) = -n\left[5/2\, g(f,n) + T\right]. \qquad \text{Fresnel} \tag{9}$$

Dyer [unpublished manuscript, 1985] indicates that observations of water column transmission loss in the Arctic are best predicted by the first term of eqn. 9. The 5/2 coefficient is close to the slope of the theoretical loss shown in Horton and Melton[16] for small Rayleigh parameter. With eqns. 8 and 9 we have a simple parameterization for the scattering loss in terms of σ_z and T.

Attenuation

In the previous section we found simple parameterizations for the transmission and scattering effects. This is also easily done for attenuation if frequency independent Q is assumed. If we are to determine the attenuation in the sediments, then loss must first be compared to that governed by dissipation in the water column. For the 9th multiple, the travel time in the water column is 233.32 seconds, and only 2.35 seconds are spent in the sediments. The attenuation coefficient for seawater is about 10^{-5} dB/km at 10 Hz and 10^{-4} dB/km at 50 Hz. The water column path length is 343 km, giving total water column losses of only 0.03 to 0.3 dB. The expected losses in the sediment are given by:

$$A(f,n) = 20\log_{10}\left[e^{\frac{-\pi f t(n)}{Q(n)}}\right] = -27.29\, t(n) f\ Q^{-1}(n) \tag{10}$$

where $t(n)$ is the travel-time of multiple n in the sediments and Q is the effective quality factor. We choose to solve for $Q^{-1}(n)$, the integrated effective Q^{-1} along the path for multiple n, rather than the attenuation coefficient, because there is no ambiguity in the definition of path and local slowness along the path. For multiple 9, and a Q of 200, $\alpha(10,9)$ is 0.9 dB/km, and $\alpha(50,9)$ is 4.5 dB/km. The total loss for multiple 9 is 3.2 dB at 10 Hz and 16 dB at 50 Hz. This signature is certainly observable compared to the attenuation along the water path.

Complete Model

The final modeling equation is given by:

$$\Delta P(f,n) \equiv P(f,n) - S(f) - G(n) - R_2(f,n) \tag{11}$$

$$= n \left\{ \begin{array}{ll} \dfrac{-686 \, f^2 \cos^2\vartheta(n)}{c_0^2} \sigma_z^2 \\ \dfrac{-31.4 \, f \, \cos\vartheta(n)}{c_0} \sigma_z \end{array} \right\} + nT - 27.29 \, t(n) \, f \; Q^{-1}(n). \quad \begin{array}{l} \text{Fraunhofer} \\ \\ \text{Fresnel} \end{array} \tag{11}$$

These equations will be inverted for σ_z or σ_z^2, T, and Q^{-1}. They are formally invertible for Q^{-1} as a function of n, however the conditioning for this problem is quite poor, and the issues outlined earlier make a constant Q^{-1} a more reasonable parameter to solve for.

DATA PROCESSING

Spectral Analysis

To obtain the arrival times and phase velocities of the different multiples the data were processed by a velocity analysis algorithm using the maximum-likelihood method[7] in 4 Hz bands. The results of the analysis for the first few seconds of the data are given in Fig. 5 for a center frequency of 8 Hz using a 0.25 second analysis window. The resulting stacking velocity/travel-time functions were quite similar for all frequencies in the band for the sediment arrivals. The modally dispersed water column arrivals obviously do not have this feature. The average velocity/travel-time function was used to dynamically move-out correct and stack the 22 channel data to obtain an increase in the signal to noise ratio of the bottom interacting pulses of 13.5 dB.

The stacked trace was then spectral analyzed with a tapered 0.5 second window advanced every 0.1 seconds to obtain an instantaneous energy density spectrogram. The 0.5 second window was chosen to include most of an arrival at a specific frequency in a single window for coherent addition. The combination of length and window taper gave a spectral resolution of 4 Hz, the same as the resolution of the source energy spectra computed earlier. A few seconds of the spectrogram are given in Fig. 6. The energy density of each multiple was computed by integrating in power over a small time window (1-1.5 seconds) on this spectrogram to include all the slightly dispersed arrivals which appeared in different time windows. A few examples of the observed energy density spectra $P(f,n)$ are shown in Fig. 7.

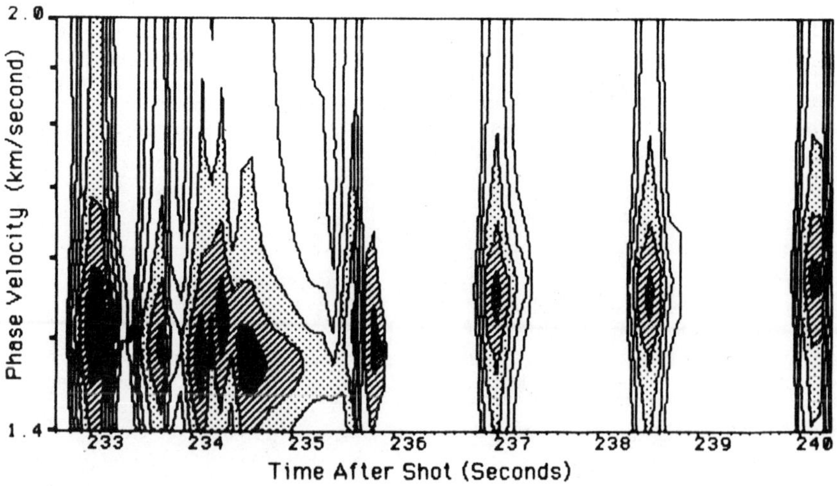

Fig. 5. Output of MLM Beamformer at 8 Hz center frequency. Contour intervals are 5 dB and are shaded with respect to local peak, not absolute level.

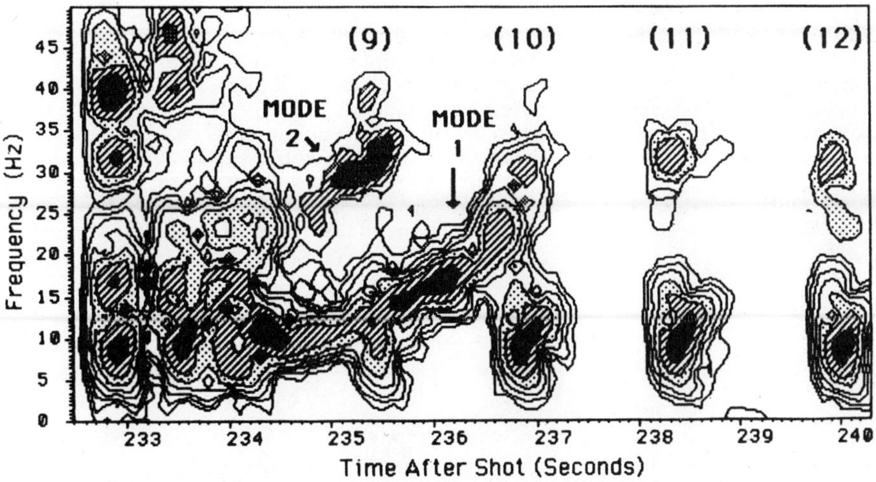

Fig. 6. Spectrogram of the stacked arrivals. Contour intervals are 5 dB, and are shaded with respect to the local peak, not absolute level. Numbers indicate order of multiple.

The ambient noise power spectrum level averaged over channels was -80 to -85 dB re 1 V^2/Hz over the 10-50 Hz band when measured on a signal free section of data before the shot. This would predict a noise energy spectral density across the band of -93 to -98 dB for multiples 9-20 with 1 second observations and -91 to -96 dB for multiples 21-24 with 1.5 second observations. These figures are a result of the array gain afforded by the 22 sensors. Note on Fig. 7 that at high frequencies and high multiple orders, the measurements are very close to the ambient noise level.

Residual Spectra

We can now compare the observations, $P(f,n)$, and the predictions given by $S(f) + G(n) + R_2(f,n)$. These are contour plotted in Fig. 8. These figures show that the locus and width of the peaks and nulls are well predicted by the statistically derived $H(f,n)$ used for R_2. We have not predicted the variations in geometrical spreading very well, as indicated by the focusing and defocusing of the power as a function of multiple in the data, but not in the predictions. These effects will, however, be diminished by the averaging over multiple order in the estimation of the attenuation and scattering. In addition, multiples 9 and 10 have some contribution from the modes that were not completely suppressed by stacking.

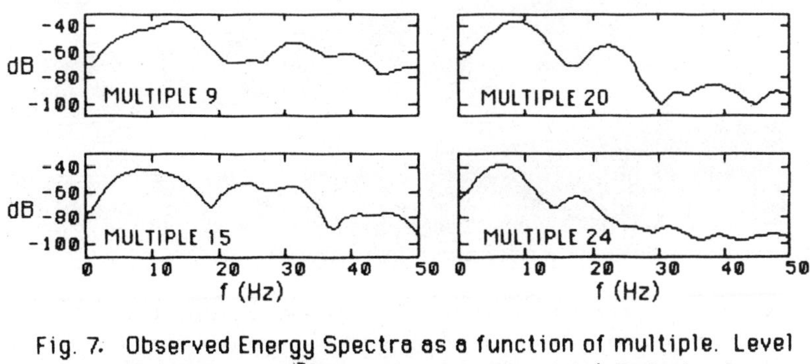

Fig. 7. Observed Energy Spectra as a function of multiple. Level is in dB re 1 V^2/Hz with -160 dB re 1 V/uPa hydrophones.

The difference between Fig. 8a and 8b should show smooth trends; i) a linear fall-off with increasing frequency for attenuation due to frequency independent Q losses, ii) a level shift downward due to frequency independent reflection or ocean bottom transmission losses, and iii) an additional frequency dependence due to surface scattering. The third term is expected to be smooth if the under-ice roughness spectrum is smooth. The variability between nearby frequencies is due to the coherent interference effects described earlier, and is unavoidable unless many realizations with different surface interaction locations are available and averaged. In addition, for high multiple orders and high frequencies, the observed spectra of Fig. 7 are very close to the ambient noise levels, even on their peaks. The "filling in" of the interfer-

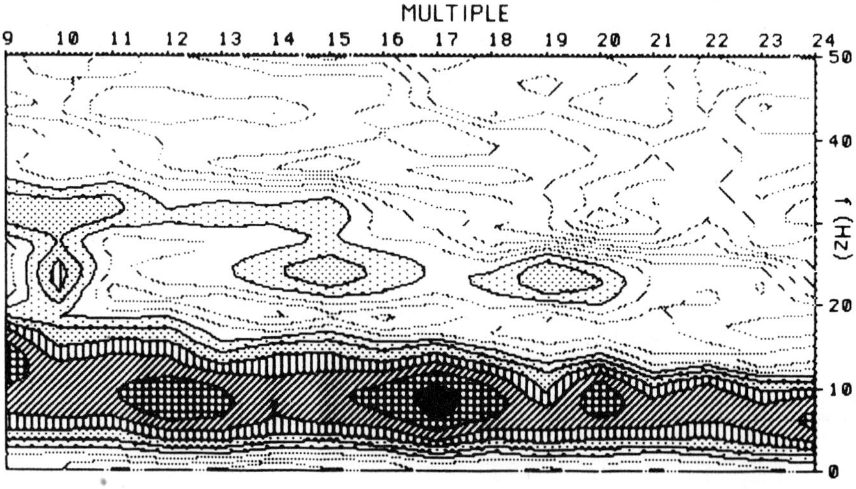

Fig. 8a. Observed energy spectrum as a function of multiple and frequency. Contour intervals are 5 dB and are shaded in absolute level.

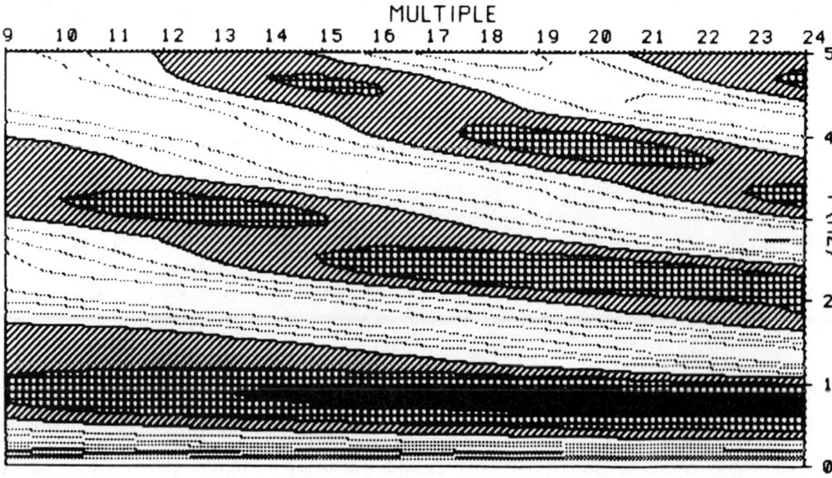

Fig. 8b. Predicted energy spectrum as a function of multiple and frequency. Contour intervals and shading are the same levels as Fig. 8a. No attenuation or scattering loss has been included in the prediction.

ence nulls by ambient noise are seen as artificial peaks of the residual spectra when corrected by the $R_2(f,n)$ filters. Finally, the strong modal contribution introduces an extraneous peak at 22 Hz on the residual spectrum for multiple 10.

Because errors in the R_2 spectral correction, noise, and extraneous arrivals tend to artificially increase the residual spectra, the data were edited from 832 raw spectral values to 628 and 545 values that were not felt to be corrupted The two edited datasets were used to check the sensitivity of the parameter estimates to this possible data abuse.

INVERSION AND RESULTS

Unconstrained Inversion

The massively overdetermined systems given by eqn. 11 were solved by double-precision singular value decomposition to obtain both the estimates and the resolution and statistical properties of the inversion. The successful separation of σ_z or σ_z^2 and Q^{-1} relies on the fact that the coefficients of these terms in eqn. 11 behave differently with n, especially in the Fresnel case where linear dependence on f is expected for both terms. Unfortunately, $n \cos\vartheta(n)$ and $t(n)$ both have similar behavior. While the resolution was perfect with condition numbers of 10^5 in the Fraunhofer case and 10^3 in the Fresnel case, the correlation coefficient between σ_z^2 and Q^{-1} was -.962. In the Fresnel case, the value was -.986. For the 545 point dataset, the parameters which give the minimum rms error fit to the data are given below.

Theory	T (dB)	σ_z (meters)	Q	rms error (dB)
Fraunhofer	$-.318 \pm .008\ \sigma_P$	$4.09 \pm .06\ \sigma_P$	$2865 \pm 614\ \sigma_P$	4.32
Fresnel	$.21 \pm .005\ \sigma_P$	$11.0 \pm .13\ \sigma_P$	$-208 \pm 5.4\ \sigma_P$	3.89

Although the spectral data are certainly non-Gaussian, we still deal only with 2nd moment statistics. σ_P is the standard deviation of the data spectral values, and may be approximated by the rms error. Both of these results are clearly unrealistic, the Fraunhofer yielding an rms roughness that is too large, and a Q that is too high, and the Fresnel yielding a negative Q (energy gain!) and a roughness that is too large in order to compensate for the negative Q. Allowing T to vary with n to compensate for geometrical spreading correction errors did not change the results significantly.

Constrained Inversion

Because Q and σ_z co-vary so strongly, it was felt that $\overline{\sigma}_z$ might be constrained by the travel-time variance measurements inferred from the $R_2(f,n)$ filters given earlier, and σ_z calculated from it and the Fresnel zone averaging term, $\gamma = \sqrt{2}\, l/r(f,n)$. We have not formally inverted the spectra for $\sigma_T(f,n)$ since this would require much more stable data spectral estimates than we have for this single shot. However, it is possible to use the theory developed for $R_2(f,n)$ and geometrical optics to calculate $\gamma(f,n)$ to obtain σ_z from the spectral shapes. To illustrate this procedure using the measured value of σ_T = 1 msec, and σ_z = 2.5 meters, we obtain $\gamma(f,9) = 1/4$, $\gamma(f,16) = 1/6$, and $\gamma(f,24) = 1/10$ assuming a correlation length of 20 meters [18]. This gives Fresnel radii of 100, 170, and 280 meters, respectively for multiples 9, 16, and 24. These are quite comparable to the normal incidence Fresnel radius for a homogeneous water column at 20 Hz with a depth of h=4 km of $r=\sqrt{\lambda h/2}$ = 400 meters. We use 20 Hz because the nulls near this frequency were modeled most accurately by σ_T=1 msec. Note that as the higher multiples approach normal incidence, the agreement is better

Because the single shot data do not allow an unconstrained separation of the Q and scattering contributions using the current formulation, we plot the tradeoff curve for σ_z and Q in which Q is varied as a constraint and σ_z and T are chosen to minimize the rms data fitting error. This gives a region of acceptable results to which one can apply their personal bias, Q implying σ_z and vice versa. This is illustrated in Fig. 9a. We feel that these results constrain Q to lie from 200 to 300. The rms error curves are quite flat for these estimates, and are shown in Fig. 9b. We feel that the current inversion is inadequate to distinguish between the Fresnel and Fraunhofer theories. The Fraunhofer result is attractive because its rms error is closer to the unconstrained value at the physically feasible solutions. However, the constrained solutions for both are quite similar, and the constrained Fraunhofer predictions fall off nearly as linearly with frequency as the Fresnel since they are strongly controlled by the attenuation term. The 545 point data set and predictions for selected multiples are given in Fig.

10. It should also be noted that a cubic polynomial in place of the scattering theories had almost the same minimum rms error as the best fit values of Fig. 11, indicating that a model with more degrees of freedom, but still smooth, is not the solution to the problem. The Q predicted by this method was 393.

CONCLUSIONS

The disparity between the best fit parameter standard deviations and their feasible values indicates that the loss model in the inversion has not yet been parameterized correctly. This is also indicated by the extremely flat rms fitting error curve with large minimum value at the best solution (Fig. 9b). It is felt that the correct parameterization will include the $R_2(f,n)$ term directly in the inverse problem. Since this is a nonlinear term in the parameter $\bar{\sigma}_z(f,n)$, it will require a more complicated inversion procedure. This was not carried out for this work because it does not make sense to

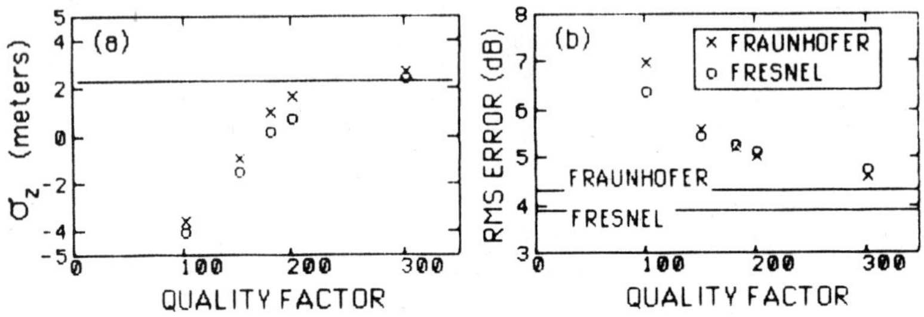

Fig. 9. (a) Estimated rms surface roughness as a function of Q. Line at 2.3 meters shows accepted value for Arctic.
(b) RMS fitting error as a function of Q constraint value. Lines give minimum errors of unconstrained problem.

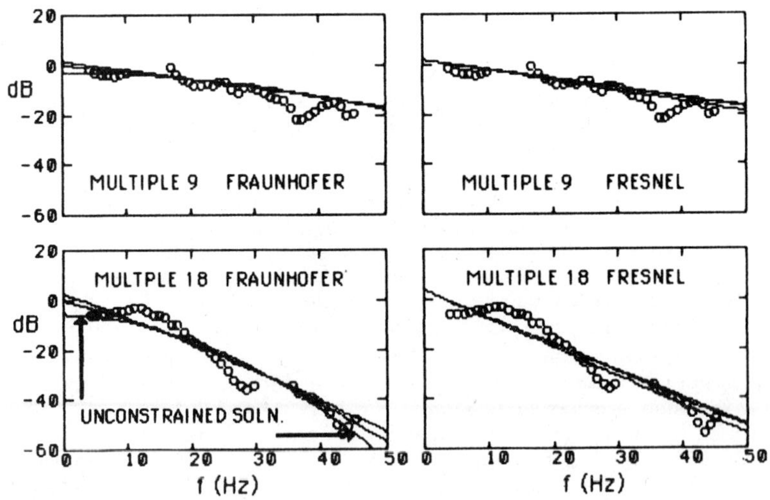

Fig. 10. Sample spectra from edited dataset (o labels) and fits of minimum RMS error, Q=200 and Q=300 constrained solutions.

invert data which are not ergodic, and are only one realization of the process, for this statistical parameter. For this work we have chosen the most smoothed parameters that would yield scientifically interesting answers, but still found the data somewhat inadequate. Future work will include improving the dataset by averaging, and inclusion of the coherent part of the scattering process directly in the inversion procedure. The separation of scattering and attenuation will also be greatly improved by utilizing the modal arrivals which do not interact with the ocean bottom.

We feel that the constrained inversion results indicate a very high sediment Q of 200 to 300 for the upper 200 meters. This value is consistent with the work of Lee [4] which reported values ranging around 200 when scattering was neglected, thus biasing his results downward. Helmberger, et al., [14] report Q values of 500 for similar sediments in the Bering Sea. We note that Q estimates are very sensitive to surface roughness assumptions in the range of roughnesses reported for the Arctic.

Finally, we have shown that partially coherent summation of travel paths is required to predict the magnitude and shape of the observed spectral functions. For this region, we also note that about half of the energy observed after 340 km of propagation has interacted with the ocean bottom, and all of it has interacted with the ice, making the separation of these effects very important to understanding propagation in this region.

ACKNOWLEDGEMENTS

The authors wish to thank Ira Dyer and Wafic Beydoun for their helpful discussions. This work was supported by the Office of Naval Research, Arctic Programs Office under contract #N 00014-77-C-0266.

REFERENCES

1. R.D. Stoll, "Marine sediment acoustics," *J. Acoust. Soc. Am.* **77**(5) pp. 1789-1799 (1985).
2. R.D. Stoll and Houtz, R.E., "Attenuation measurement with sonobuoys," *J. Acoust. Soc. Am.* **73**(1) pp. 163-172 (1983).
3. Mikhalevsky, P.N., "Characteristics of CW signals propagated under the ice in the Arctic," *Journal of the Acoustical Society of America* **70**(6) pp. 1717-1722 (1981).
4. Lee, T., "Long-range sound transmission of bottom-interacting signals in the Arctic channel: Experiment and theory," Ph.D. Thesis, Columbia University, University Microfilms International, Ann Arbor, MI (1983).
5. Hamilton, E.L., "Prediction of in situ acoustic and elastic properties of marine sediments, North Pacific," *Geophysics* **36** pp. 266-284 (1971).
6. Hamilton, E.L., "Geoacoustic modeling of the sea floor," *J. Acoust. Soc. Am.* **68** pp. 1313-1340 (1980).
7. Duckworth, G.L., Baggeroer, A.B., and Jackson, H.R., "Crustal Structure Measurements near FRAM II in the Pole Abyssal Plain," *Tectonophysics* **89** pp. 172-215 (1982).
8. Wakeley, J., "Pressure-signature model for an underwater explosive charge," *U.S. Navy Journal of Underwater Acoustics* **27**(2) pp. 445-449 (1977).
9. Urick, R.J., *Principles of Underwater Sound*, McGraw-Hill Book Co., New York, NY (1975).
10. R.S. Jacobson, G.G.Shor, Jr., and Dorman, L.M., "Linear inversion of body wave data - part II: Attenuation versus depth using spectral ratios," *Geophysics* **46** pp. 84-94 (1981).
11. Duckworth, G.L., *Processing and Inversion of Arctic Ocean Refraction Data*, Sc.D. Thesis, Massachusetts Institute of Technology; Department of Electrical Engineering and Computer Science; Woods Hole Oceanographic Institution; Department of Oceanographic Engineering (September, 1983).
12. Diebold, J.B. and Stoffa, P.L., "The traveltime equation, tau-p mapping, and inversion of common midpoint data," *Geophysics* **46**(3) pp. 238-254 (1981).
13. Dorman, L. and Jacobson, R.S., "Linear Inversion of Body Wave Data; Part I: Velocity Structure From Travel Times and Ranges," *Geophysics* **46**(2) pp. 138-151 (1981).
14. D.V. Helmberger, G. Engen, and Scott, P., "A note on velocity, density and attenuation models for marine sediments determined from multi-bounce phases," *J. Geophys. Res.* **84** pp. 667-671 (1979).
15. Weston, D.E., "Underwater Explosions as Acoustic Sources," *Proceedings of the Physical Society* **76** p. 233 (1960).

16. Horton, C.W. Sr. and Melton, D.R., "Importance of the Fresnel Correction in Scattering from a Rough Surface. II. Scattering Coefficient," *Journal of the Acoustical Society of America* **47**(1) pp. 299-303 (1970).
17. Melton, D.R. and Horton, C.W. Sr., "Importance of the Fresnel Correction in Scattering from a Rough Surface. I. Phase and Amplitude Fluctuations," *Journal of the Acoustical Society of America* **47**(1) pp. 290-298 (1970).
18. Mellen, R.H. and Dinapoli, F.R., "Underwater Acoustics in the Arctic Ocean," *NATO Advanced Study Institute on Adaptive Methods in Underwater Acoustics*, (1984).

LOW FREQUENCY ATTENUATION IN THE ARCTIC OCEAN

Frederick R. DiNapoli

Naval Underwater Systems
Center, New London Lab
New London, CT 06320

Robert H. Mellen

Planning Systems Inc.
Marine Sciences
New London, CT 06320

The deep water abyssal plain regions of the Arctic Ocean are almost text book examples of acoustic propagation in a range independent ocean at low frequencies, if one could ignore the influence of the unique rough ice canopy. This can be seen by examining the empirically derived expression and plots of propagation loss versus range at several low acoustic frequencies shown in Fig 1. The term, α R, which is predominantly determined by the scattering of energy from the rough surfaces of the ice canopy was found to have an $f^{(3/2)}$ dependence on frequency and a magnitude which exceeds sea water absorption by two orders of magnitude.

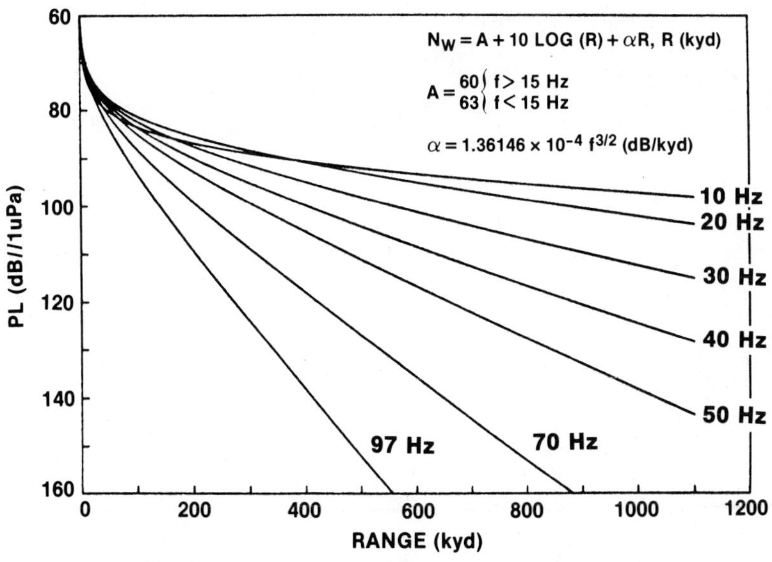

Fig. 1. TRISTEN PACK ICE EMPIRICAL PROPAGATION LOSS EQUATIONS

Presented below is a synopsis of our progress in a continuing effort to gain some insight into this problem which hopefully will lead to providing an accurate propagation loss prediction capability.

The application of the method of small perturbations (MSP) has received considerable attention[1-10] in the past, and since we believe it to be applicable with minor, but rather tedious algebraic modifications, shall employ it to find expressions for the coherent* scattered plane wave reflection coefficient represented by $\tilde{R}$. The unperturbed reflection coefficient, R , used in the Fast Field Program[11] (FFP) and referenced to the mean rough surface is then replaced with $\tilde{R}$ and predictions of propagation loss using a measured SVP are compared with the empirical curves of Fig. 1.

The availability of concurrent measures of both the acoustic field and the statistics of the rough surface from which it was scattered is a rare occurence, which depending on your outlook is either fortunate or frustrating. The theoretician may view the absence of concurrent measurements as fortunate since he can make approximations whose validity cannot be easily evaluated. Our frustration stems from convincing ourselves that obtaining agreement with experimental data in the absence of concurrent measurements and thus based upon the use of free parameters, is not a good criteria for judging the validity of mathematical approximations. In short. we have not solved the problem to date, but based upon the concurrent measurements we can indicate what the answer cannot be.

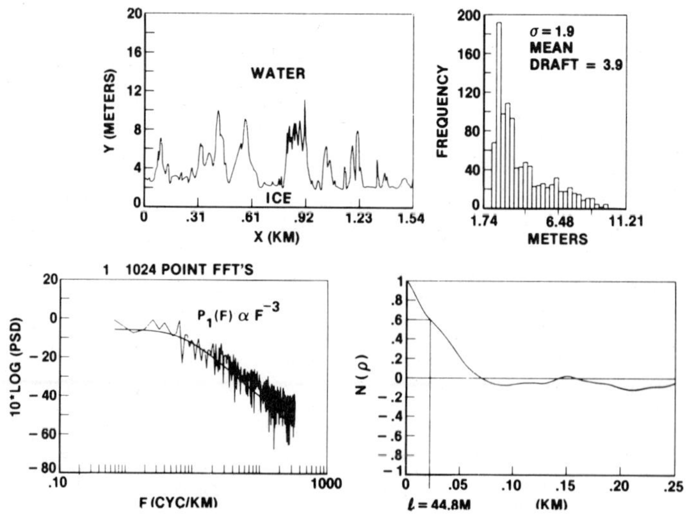

Fig. 2. MEASURED ICE PROFILE STATISTICS

*Examination of the measurements reveals that the incoherent scattered field is nugatory at frequencies below 100 Hz.

Typical results for the statistical measures of the bottom side roughness are provided in Fig 2. Shown in the upper left hand corner is a segment of the ice canopy. The air-ice interface is the flat line at the bottom. (Concurrent measurements of the topside roughness were not available.) The rough ice-water boundary is above that. This is a greatly exaggerated view since the horizontal scale is in km while the vertical scale is in m. It is not uncommon to find ice keels of 40–60 m in depth. The probability distribution of ice keels taken about the mean depth which is 3.9m is shown to the right. The standard deviation about the mean was found to be 1.9m. The normalized auto correlation function is shown immediately below this. We define the correlation length, which for this case was found to be 44.8m as that point for which the auto correlation function has a value of .6.

The one dimensional density function (PDF) is shown in the lower left along with a smooth curve provided by an analytic function which will be described shortly.

Mathematical models for the ice canopy under consideration are shown in Fig. 3. Model I is a full elastic layer with roughness at both boundaries, represented by σ_T, ℓ_T, σ_B, ℓ_B, both shear and compressional constant speeds of sound and absorption coefficients. Owing to the sparseness of information regarding sea ice at the frequencies of interest, all of these parameters are free except for the density, compressional speed of sound, $(C_1)_P$ σ_B and ℓ_B. If the elastic properties of ice are ignored one is left with the liquid layer given as Model II. The free parameters here being σ_T, ℓ_T, and α_1. Finally, if all properties of the ice are ignored except for roughness one has Model III with no free parameters if the roughness is associated with both the free parameters of sea-ice and the validity of theoretical approximations often made to arrive at tractable scattering formulas we choose to start with Model III and bottom roughness so that no free parameters exist.

Utilizing the validity conditions given by Bass and Fuks[3] with the average experimental values of σ_B=2m and ℓ_B=44m one finds that MSP should certainly be valid below 100 Hz considering the relatively shallow grazing angles which predominate at low frequencies for the typical deep Arctic profile. We also make the usual assumption that the roughness statistics are stationary (although our data suggest that this is not the case) and isotropic. If the topside statistics are even loosely correlated with those of the bottom side then the isotropic assumption is also inaccurate.

We have not determined the impact of these inaccurate assumptions in our results to date.

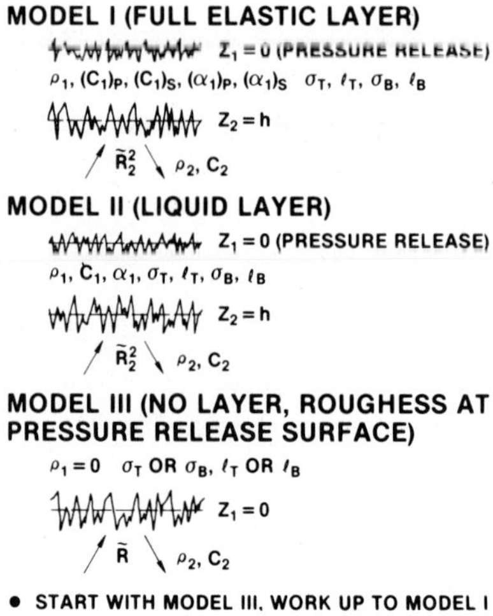

Fig. 3. MATHEMATICAL MODELS FOR SCATTERED PLANE WAVE REFLECTION COEFFICIENT

The MSP result for the plane wave scattered reflection coefficient associated with Model III is given, by the double integral of Fig. 4 with the incident angle and the two dimensional PSD, $\tilde{P}_2(F)$, which is related to the measured one dimensional PSD, $\tilde{P}_1(F)$, using the isotropic assumption. Using the analytic fit to $\tilde{P}_1(F)$ shown in Fig. 2 we find that the associated one dimensional auto correlation function is given in terms of the modified Bessel function of the first kind, which for $\rho=.6$ has the value of (e^{-1}). After some straight forward but intricate manipulation the result can be reduced to a single integration which is evaluated numerically.

$$\widetilde{R}(f,\theta_0,\sigma,\ell) = -1+k\cos\theta_0 \frac{\sigma^2}{\pi}\int d\vec{k}_\perp \widetilde{P}_2(\vec{K}_\perp)\left\{k_2^2-(\vec{k}_\perp-\vec{K}_\perp)\cdot(\vec{k}_\perp-\vec{K}_\perp)\right\}^{1/2}$$

$\vec{k}_\perp$; HORIZONTAL WAVENUMBER COMPONENT OF INCIDENT PLANE WAVE

$\vec{K}_\perp$; HORIZONTAL WAVENUMBER COMPONENT OF ROUGH SURFACE

— ASSUMING ISOTROPIC ROUGHNESS

$\widetilde{P}_2 = \widetilde{P}_{21}$ WHICH IS RELATED TO MEASURED $\widetilde{P}_1$

— USING ANALYTIC FIT TO $\widetilde{P}_1(F)$, F SPATIAL FREQUENCY FIND

$$N_1(\rho) = (2/\ell)\, K_1\,[(2/\ell)\rho] \qquad \widetilde{P}_1(F) = \frac{\pi\,(2\sigma/\ell)^2}{[(2/\ell)^2+(2\pi F)^2]^{3/2}} \simeq F^{-3} \text{ FOR LARGE F}$$

AND

$$\widetilde{P}_{21}(F) = \frac{4\pi\,(2\sigma/\ell)^2}{[(2/\ell)^2+(2\pi F)^2]^2} \simeq F^{-4} \text{ FOR LARGE F}$$

$$\widetilde{R} = -1+k^4\cos\theta_0\,(\sigma\ell)^2\int_0^{\pi/2} \frac{d\theta\,\cos^2\theta\,\sin\theta\left\{1+\left(\frac{k\ell}{2}\right)^2(\sin^2\theta+\sin^2\theta_0)\right\}}{\left\{\left[1+\left(\frac{k\ell}{2}\right)^2(\sin^2\theta+\sin^2\theta_0)\right]^2-\left[\left(\frac{k\ell}{2}\right)^2\sin\theta\,\sin\theta_0\right]^2\right\}^{3/2}}$$

Fig. 4 MODEL III (NO LAYER, ROUGHNESS AT PRESSURE RELEASE SURFACE)

One can also evaluate the scattering integral using either the Kirchhoff or Marsh approximations shown in Fig. 5. The Kirchhoff approximation, which ignores the slopes of the rough surfaces yields a reflection coefficient having a squared dependency on both frequency and cos θ_0. The Marsh approximation, which is apparently appropriate for shallow grazing angles and incorporates the correlation length has a 3/2 dependency on frequency and a linear dependency on cos θ_0.

KIRCHHOFF APPROXIMATION

$k\ell \gg 1,\ k\ell \cos^2\theta_0 \gg 1$ (APPROPRIATE FOR STEEP GRAZING ANGLES)

$$\widetilde{P}_2(\dot{K}_\perp) \cong 2\pi\,\delta^2(\dot{K}_\perp) \qquad N_2(\rho) \cong 1 \qquad \frac{dN_2}{d\rho}(\rho) \cong 0$$

$$\widetilde{R}(f, \theta_0, \sigma) \cong -1 + 2\sigma^2 k^2 \cos^2\theta_0$$

MARSH APPROXIMATION

$k\ell \gg 1,\ k\ell \cos^2\theta_0 \ll 1$ (APPROPRIATE FOR SHALLOW GRAZING ANGLES)

$$\widetilde{R} \cong -1 + 2\sqrt{2}\,k^{3/2}\cos\theta_0\,\sigma^2 \int d|K_\perp|\,\widetilde{P}_2(|K_\perp|)\,|K_\perp|^{3/2}$$

USING MEASURED $\widetilde{P}_1$ FIND

$$\widetilde{R}(f, \theta_0, \sigma, \ell) \cong -1 + 2.396\,\frac{k^{3/2}}{\ell^{1/2}}\,\cos\theta_0\,\sigma^2\,e^{i\pi/4}$$

Fig. 5. MODEL III (NO LAYER, ROUGHNESS AT PRESSURE RELEASE SURFACE)

The reflection loss associated with the three results are compared in Fig. 6 at 47 Hz. For incident angles from about 77 to 90 which is the region of interest to low frequency Arctic acoustics, the Marsh approximation yields a greater loss and the Kirchhoff approximation less loss that the full evaluation of the MSP scattering integral. The juxtaposition of the three results in this region remains the same for all frequencies below 100 Hz.

An indication of how the propagation loss predictions from all three results compare with the experimental data can be obtained by comparing FFP results using only the Marsh (which yields the greatest loss over the important angles of incidence) approximation for the reflection coefficient. This comparison presented at the top of Fig. 7 for 47 Hz and σ_B=2m, ℓ_B =44m which were the average statistical measures of the rough surface shows that the Marsh approximation gives too little loss. This discrepancy persists for all frequencies spanned by the data and thus all there results for Model III fail to provide agreement.

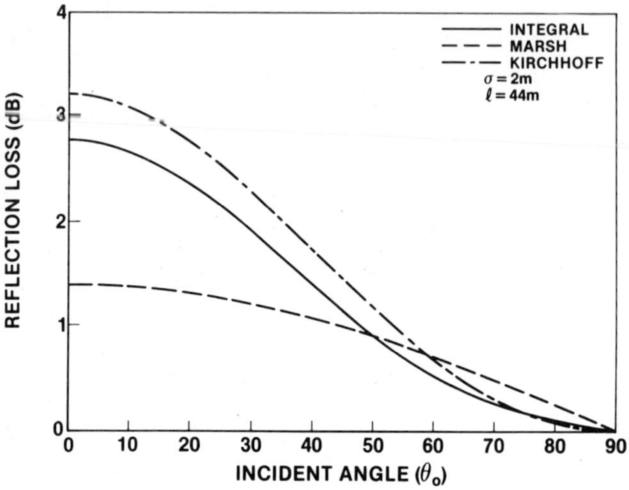

Fig. 6. MODEL III REFLECTION COEFFICIENT

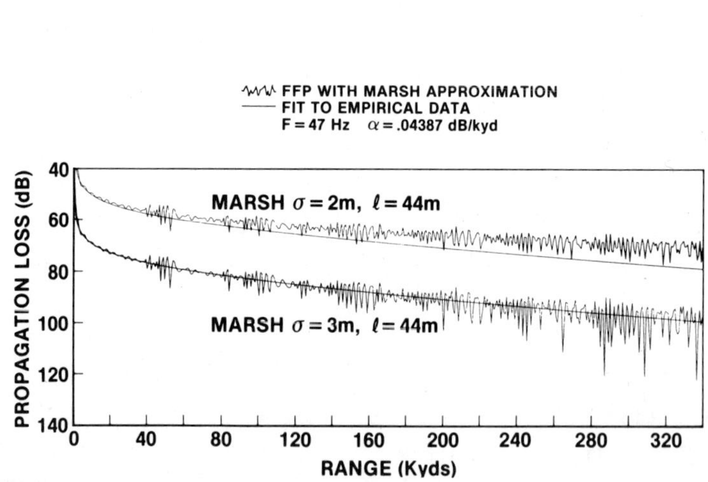

Fig. 7. COMPARISON OF MARSH APPROXIMATION WITH EXPERIMENTAL DATA

If concurrent acoustical and statistical measures were not available and values for free parameters could be assigned arbitrarily one finds that almost perfect agreement with the data can be obtained from the Marsh approximation at all frequencies merely by increasing σ_B to 3m. The result (displaced by 20dB) at 47 Hz is shown at the bottom of Fig. 7. Excellent agreement can also be obtained with the Kirchhoff approximation at all frequencies if σ_B is taken as 5m which of course is unrealistic for the deep water Arctic Abyssal Plains.

If the scattering surface were taken as the ice-air interface instead of the ice-water interface we suspect the comparison using the average values for σ and ℓ would be even worse since the σ of the topside roughness should be smaller owing to melting and the effects of wind erosion. We thus conclude that Model III is inaccurate.

The scattered plane wave reflection coefficient for the liquid layer of Model II with no roughness at the top and no absorption loss within the layer is given, Fig. 8, in terms of the unperturbed reflection coefficient reference to the mean level of the rough surface and the double integrals I_1 and I_2 where the $\tilde{M}$ functions involved expression of the acoustic field evaluated at the mean surface.

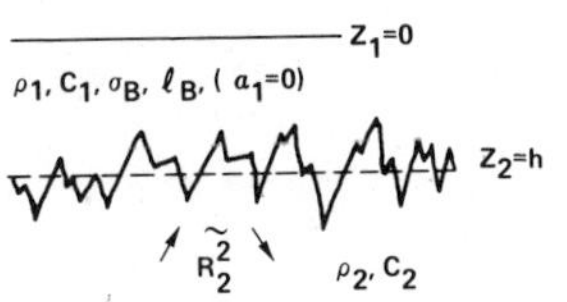

$$\tilde{R}_2^2 = R_2^2 + \frac{\sigma^{\ell}}{2\rho_2 k_{2z}} \left\{ -k_{2z}(1-R_2^2) I_1 + \rho_2 (1+R_2^2) I_2 \right\}$$

$$I_1 = \frac{1}{2\pi}\int d\vec{\xi}_\perp \, \Delta_c^{-1}(\xi_{1z}, \xi_{2z}) \tilde{P}_2(\vec{k}_\perp - \vec{\xi}_\perp) \tilde{M}(\vec{\xi}_\perp, \vec{k}_\perp)$$

$$I_2 = \frac{1}{2\pi}\int d\vec{\xi}_\perp \, \Delta_c^{-1}(\xi_{1z}, \xi_{2z}) \tilde{P}_2(\vec{k}_\perp - \vec{\xi}_\perp) \left\{ \tilde{M}_2(\vec{k}_\perp \vec{\xi}_\perp) - \tilde{M}_3(\vec{\xi}_\perp, \vec{k}_\perp)\, \vec{\xi}_\perp \cdot (\vec{\xi}_\perp - \vec{k}_\perp) \right\}$$

R_2^2 – **UNPERTURBED PLANE WAVE REFLECTION COEFFICIENT**

$c_1 > c_2$ **RESULTING IN CRITICAL ANGLE**

$|R_2^2| \equiv 1$ **FOR ALL ANGLES**

$\rho_1 \cong \rho_2$

Fig. 8. MODEL II (LIQUID LAYER, WITH NO ATTENUATION, NO ROUGHNESS AT TOP)

We make two observations which hopefully provide some insight into the results for this problem. First, the unperturbed reflection coefficient has unit magnitude for all angles of incidence. This results because the absorption in the liquid layer is zero and the pressure release surface behind the boundary perfectly reflects any transmitted energy back into the water. Thus, the presence of a critical angle has no special significance to the problem. The second point to note is that the difference in densities between the two media are almost zero.

The result for the scattered reflection coefficient under the Kirchhoff approximation is shown in Fig. 9. Recalling that the unperturbed reflection coefficient has unit magnitude and thus no loss, the influence of the roughness at the bottom of the liquid layer is determined by the deviation of the curve from zero. For comparison purposes the Kirchhoff result for Model III is also plotted in Fig 9. As frequency increases the deviation for Model II does increase slightly but, in all cases, is considerably less than the Kirchhoff result for Model III. We believe the factor controlling the scattering for Model II is the density difference which is practically zero and, thus, the result would not substantially change if the full MSP scattering integrals were to be evaluated. We thus conclude that Model II is also inappropriate.

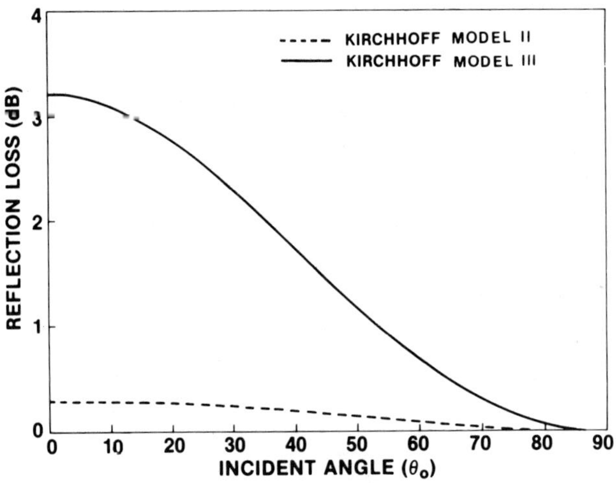

Fig 9. MODEL III REFLECTION COEFFICIENT FREQUENCY 47 HZ

Our future investigations will include an examination of the impact of the elastic properties within the layer as well as the influence of non-stationary rough surface statistics.

REFERENCES

1. H. Wysor Marsh, M. Schulkin, and S. G. Kneale, Scattering of Underwater Sound by the Sea Surface, J. Acoust. Soc. Am., Vol. 33, Number 3, (March 1961).
2. E. Y. T. Kuo, Wave Scattering and Transmission at Irregular Surfaces, J. Acoust. Soc. Am., Vol. 3, Number 11, (Nov., 1964).
3. F. G. Bass, I. M. Fuks, "Wave Scatteringfrom Statistically Rough Surfaces", Pergamon Press, New York, (1979).

4. Yu. P. Lysanov, Part IV, Scattering of Sound by Irregular Surfaces, from "Acoustics of the Ocean", USSR Academy of Sciences, Acoustics Institute, Nauka Press, Moscow, 1974. NISC Translation No. 4048, (Oct., 1978).
5. J. A. DeSanto, Chapt. II, in: "Ocean Acoustics", Springer-Verlag, New York, (1979).
6. William A. Kuperman, Coherent Component of Specular Reflection and Transmission at a Randomly Rough Two-fluid Interface, J. Acoust, Soc. Am., Vol. 58, No. 2, (Aug., 1975).
7. L. M. Brekhovskikh, Yu, P. Lysanov, Chapt. IX in: Fundamentals of Ocean Acoustics", Springer-Verlag, New York, (1982).
8. A. D. Lapin, Scattering of Sound by a Solid Layer with Rough Boundaries, translated form Akusticheskii Zhurnal, Vol. 12, No. 1.,(Jan.-Mar., 1966).
9. F. I. Kryazhev, V. M. Kudryashov, and N. A. Perov, Propagation of Low-frequency Sound Waves in a Wavequide with Irregular Boundaries, Sov. Phys. Acoust.,Vol. 22, No. 3, (May-Jun., 1979).
10. F. I. Kryazhev, and V. M. Kudryashov, Sound Field in a Waveguide with a Statistically Rough Admittance Boundary", Sov. Phys. Acoust., Vol 30, No. 5, (Spt.-Oct., 1984).
11. F. R. DiNapoli, R. L. Deavenport, Theoretical and Numerical Green's Function Field Solution in a Plane Multilayered Medium, J. Acoust. Soc. Am., vol 67(1), (Jan., 1980).

EFFECT OF SUB-BOTTOM INHOMOGENIETIES ON SHALLOW WATER SPATIAL COHERENCE

S. T. McDaniel and D. F. McCammon

Applied Research Laboratory
The Pennsylvania State University
Post Office Box 30
State College, Pennsylvania 16804 USA

ABSTRACT

A coupled-mode approach is applied to examine the spatial coherence of the acoustic field in an inhomogeneous waveguide. The formulation derived is a matrix analog of the equations resulting from application of strong fluctuation theory to propagation in an inhomogeneous medium. The expression that results is evaluated to determine the effect of sub-bottom roughness on horizontal spatial coherence in shallow water. Results are presented for low-frequency propagation in an environment in which a homogeneous clay-silt sediment overlies irregular bedrock. Numerical examples demonstrate the effects of frequency, water depth and source-receiver geometry on coherence. For the cases examined, the field is coherent over many wavelengths in qualitative agreement with existing data.

INTRODUCTION

Low-frequency acoustic propagation in shallow water takes place through multiple interactions with the surface and bottom. If these boundaries are irregular, the transmitted field becomes incoherent due to the large number of interactions. Mode-coupling theory[1] provides a means of treating the effects of boundary roughness on shallow-water propagation. The major advantage of this approach is that it permits a three-dimensional treatment of scattering as opposed to the more usual assumption of cylindrical symmetry.

In this paper, a coupled-mode approach is applied to treat the effects of boundary irregularities on horizontal spatial coherence in a shallow-water waveguide. The theoretical development makes use of the forward scattering approximation and is valid provided the surface correlation length is large compared to the acoustic wavelength and that the scattering is small over a surface correlation length. The result obtained is an extension of the theory of strong fluctuations in inhomogeneous media[2] to the case of an inhomogeneous waveguide, yielding a matrix equation for a mutual coherence vector.

The theory is applied to investigate the effects of sub-bottom roughness on transverse horizontal spatial coherence. The environment addressed is that considered by McCammon and McDaniel[3] in which irregular bedrock is overlain by a clay-silt sediment. Numerical results predict a strong dependence on acoustic frequency, source to receiver range and water depth. In addition, it is demonstrated that the coherence loss is dominated by mode-coupling effects, not intramodal phase interference.

THEORETICAL DEVELOPMENT

We consider the transverse horizontal coherence $C(x,d)$ of the signals received by two hydrophones located at a range x from the source and having the transverse coordinates $y = \pm d/2$ as shown in Fig. 1.

$$C(x,d) = \langle p(x,d/2)p^*(x,-d/2)\rangle |p^2(x,0)|,^{-1} \tag{1}$$

where p is the acoustic pressure. The pressure will be expanded in terms of the mean depth eigenfunctions $Z_n(z)$ of the shallow-water waveguide

$$p = \sum_n u_n(x,y)Z_n(z)\exp(ik_n x), \tag{2}$$

where z is the depth, and k_n is the mean propagation constant of the nth mode. Note that the expansion in terms of mean eigenfunctions limits this development to cases where the density discontinuities at irregular interfaces may be neglected. Substituting Eq. (2) into Eq. (1), and assuming that

$$\langle u_n(x,d/2)u_j^*(x,-d/2)\rangle = 0, \ n \neq j,$$

we obtain

$$C(x,d) = \sum_n \Gamma_n(x,d)Z_n^2\exp(-2\alpha_n x)\left[\sum_n \Gamma_n(x,0)Z_n^2\exp(-2\alpha_n x)\right]^{-1}, \tag{3}$$

where the mutual coherence vector Γ_n is defined by

$$\Gamma_n(x,d) = \langle u_n(x,d/2)u_n^*(x,-d/2)\rangle,$$

Z_n is evaluated at the receiver depth, and losses are accounted for by the factor $2\alpha_n = i(k_n - k_n^*)$.

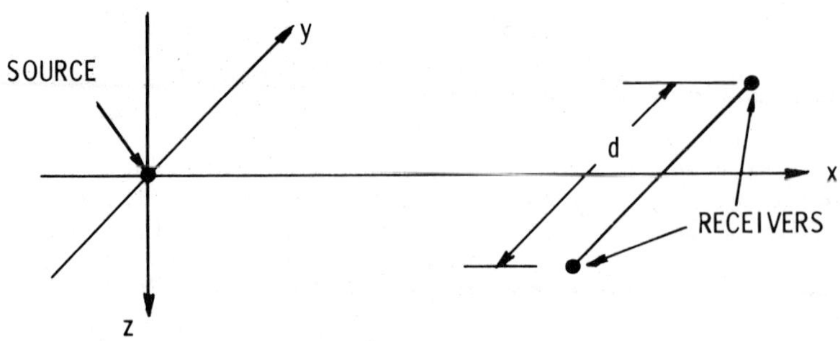

Fig. 1. Geometry for coherence modeling.

To compute the spatial coherence, we need to relate the mutual coherence to the statistical properties of the bottom. The initial step is to substitute Eq. (2) into the Helmholtz equation and invoke the forward scattering approximation, whence

$$\sum_n \{[k^2(x,y,z)-k_o^2(z) + 2ik_n \frac{\partial}{\partial x} + \frac{\partial^2}{\partial y^2}]u_n\}Z_n \exp(ik_n x) = 0, \tag{4}$$

with k(x,y,z) and $k_o(z)$ being the local and mean wave number profiles, respectively. To obtain Eq. (4), use was made of the relation

$$\frac{d^2 Z_n}{dz^2} + [k_o^2(z)-k_n^2]Z_n = 0,$$

obeyed by the mean eigenfunctions.

Next, we use the orthonormality of the eigenfunctions

$$\int_0^\infty dz\; Z_n(z)Z_m(z) = \delta_{nm},$$

to obtain from Eq. (4) the result

$$2ik_n \frac{\partial u_n}{\partial x} + \frac{\partial^2 u_n}{\partial y^2} = - \sum_m M_{nm}u_m, \tag{5}$$

where

$$M_{nm}(x,y) = \int_0^\infty dz\; Z_n(z)Z_m(z)(k^2-k_o^2)\exp[i(k_m-k_n)x]. \tag{6}$$

Before continuing, we note that two types of terms occur on the right hand side of Eq. (5). The term with m=n produces random perturbations in the modal propagation constant and causes phase interference. For n ≠ m, the right hand side of Eq. (5) couples the propagating modes. We will examine the relative magnitude of these two effects on coherence later in this paper.

To proceed, we write the complex conjugate of Eq. (5) at the position $(x,\tilde{y})$

$$\left(-2ik_n \frac{\partial}{\partial x} + \frac{\partial^2}{\partial \tilde{y}^2}\right) u_n^*(x,\tilde{y}) = - \sum_m M_{nm}^*(x,\tilde{y})u_m^*(x,\tilde{y}). \tag{7}$$

Multiplying Eq. (5) by $u_n^*(x,\tilde{y})$ and Eq. (7) by $u_n(x,y)$, the expectation of their difference is given by

$$\left(2ik_n \frac{\partial}{\partial x} + \frac{\partial^2}{\partial y^2} - \frac{\partial^2}{\partial \tilde{y}^2}\right) \Gamma_n(x,y,\tilde{y}) = - \sum_m \{\langle u_n^*(x,\tilde{y})M_{nm}(x,y)u_m(x,y)\rangle - \langle u_n(x,y)M_{nm}^*(x,\tilde{y})u_m^*(x,\tilde{y})\rangle\} . \tag{8}$$

Equation (8) is the matrix analog of the result obtained in Reference 2,

To solve this equation, we must first evaluate the expectation appearing on the right hand side.

If the sub-bottom roughness is assumed to be Gaussian[4], the Navikov-Furutsu formula may be used to express the expectations appearing in Eq. (8) as

$$\langle M_{nm}(x,y)U(x,y,\tilde{y})\rangle =$$

$$\int_{-\infty}^{+\infty}\int_{-\infty}^{+\infty} dx'dy'\langle M_{nm}(x,y)M_{nm}(x',y')\rangle\left\langle\frac{\delta U(x,y,\tilde{y})}{\delta M_{nm}(x',y')}\right\rangle, \tag{9}$$

where δ denotes the functional derivative. We make the following assumptions[5] to evaluate the functional derivative:

1. The surface correlation length Λ is large compared to the acoustic wavelength.

2. Coupling is small over a distance Λ.

3. The range x is much larger than Λ.

4. The functional derivative $\delta M_{nj}(x,y)/\delta M_{mn}(x',y')=0$, for $j\neq m$.

5. For the purposes of computing the coupling coefficients, k and k_o may be treated as real, and coupling takes place only between modes such that $(k_m-k_n)\Lambda \ll \pi$.

With these assumptions, Eq. (8) may be cast in the form

$$\left(2ik_n\frac{\partial}{\partial x}+\frac{\partial^2}{\partial y^2}+\frac{\partial^2}{\partial \tilde{y}^2}\right)\Gamma_n(x,y,\tilde{y}) =$$

$$-2ik_n\sum_m\left[A_{nm}(0)\Gamma_n(x,y,\tilde{y}) - A_{nm}(y-\tilde{y})\Gamma_m(x,y,\tilde{y})\right], \tag{10}$$

where A_{nm} is given by

$$A_{nm}(y-\tilde{y}) = (2k_n)^{-2}\int_{-\infty}^{+\infty} dx\,\langle M_{nm}(x,y)M_{nm}(x,\tilde{y})\rangle. \tag{11}$$

For cylindrically spreading waves, the solution to Eq. (10) is

$$\Gamma_n(x,d) = \frac{\exp(-\beta_n x)}{x}\sum_m \exp\left[\int_o^x dx'\, A_{nm}\left(\frac{x'd}{x}\right)\right]\Gamma_m(0,d), \tag{12}$$

where $\beta_n = \sum_m A_{nm}(0)$, and we have set $y = \tilde{y} = d/2$.

Equation (12) has a simple physical interpretation. The term β_n is the rate at which completely coherent energy is scattered from the nth mode. The second exponential describes the partial coherence of the scattered energy. To solve Eq. (12), initial values for $\Gamma_n(0,d)$ are needed. Substituting a point source into Eq. (2) yields

$$\Gamma_n(0,d) = AZ_n^2(z_s)k_n^{-1},$$

where z_s is the source depth and A is an arbitrary constant.

Table 1. Acoustic Environment

Layer	Thickness (m)	Sound Speed (m/sec)	Density (gm/cm^3)
Water	35	1517	1.0
Sediment	25[a]	1530	1.5
Basement	∞	2300	2.3

[a] mean thickness

NUMERICAL EXAMPLES

To illustrate the effect of sub-bottom inhomogenieties on transverse spatial coherence, we consider the environment of Ref. 3 in which the bottom consists of fine sediment over irregular rock. To simplify the problem, the surficial sediment will be treated as homogeneous and the water column as isovelocity. Table I presents the mean acoustical parameters used in this three layer model. The rms roughness of the sub-bottom feature of interest is 1.8 m and its correlation function is roughly Gaussian with a correlation length of 600 m.

For the case of a single rough sub-bottom interface between media having propagation constants K_1 and K_2, Eq. (6) may be approximated as

$$M_{nm}(x,y) \doteq Z_n(z_i)Z_m(z_i)(K_2^2-K_1^2)h(x,y)\exp i(k_m-k_n)x, \qquad (13)$$

where z_i is the mean depth of the interface and $h(x,y)$ is the departure of the interface from its mean depth. Equation (11) may then be written

$$A_{nm}(y-\tilde{y}) = Q_{nm} \int_{-\infty}^{+\infty} dx' \langle h(x',y)h(x',\tilde{y}) \rangle, \qquad (14)$$

where

$$Q_{nm} = [Z_n(z_i)Z_m(z_i)(K_2^2-K_1^2)/2k_n]^2,$$

and use has been made of the fact that coupling occurs only for modes such that $(k_m-k_n)\Lambda \ll \pi$.

The first case considered is the variation with range of the coherence between two receivers having a horizontal separation of 50 acoustic wavelengths as shown in Fig. 2. In this example, the source depth is 10 m and the receiver is located at the water-sediment interface. (The receiver will be bottomed in all the cases considered.) As is evident in Fig. 2, the acoustic field is highly coherent over 50 wavelengths even at source-to-receiver ranges of 10 km. This prediction is thus in accord with experimentally observed[6] coherences in shallow water.

A more traditional manner of presenting coherence is as a function of receiver separation at fixed range as shown in Fig. 3. In this case the ordinate is the normalized hydrophone spacing $2\pi d/\lambda$, the range is 1 km and the remaining parameters are the same as in Fig. 2. We see in this example that the coherence falls with frequency for frequencies between 50 and 200 Hz, but that this trend is reversed at 400 Hz. This occurs because for the source depth considered, 10 m, the primary modes excited at 400 Hz do not penetrate the bottom sufficiently to interact strongly with the sub-bottom interface.

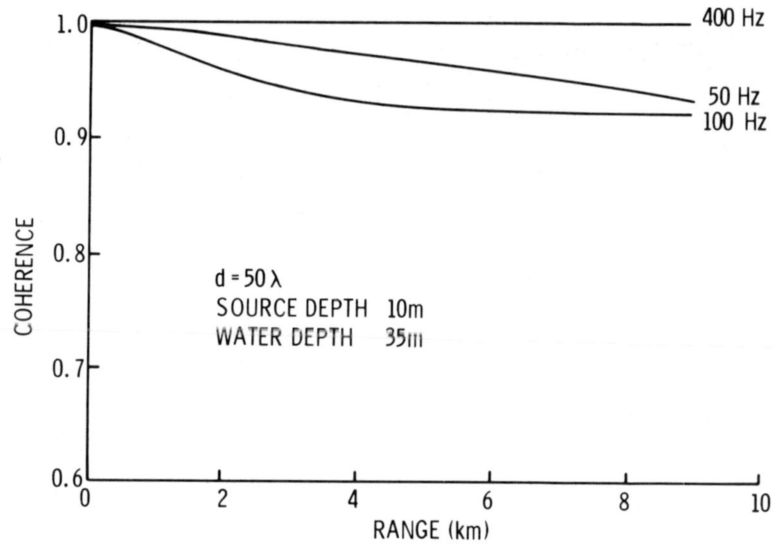

Fig. 2. Horizontal spatial coherence as a function of range for fixed receiver separation of 50 wavelengths.

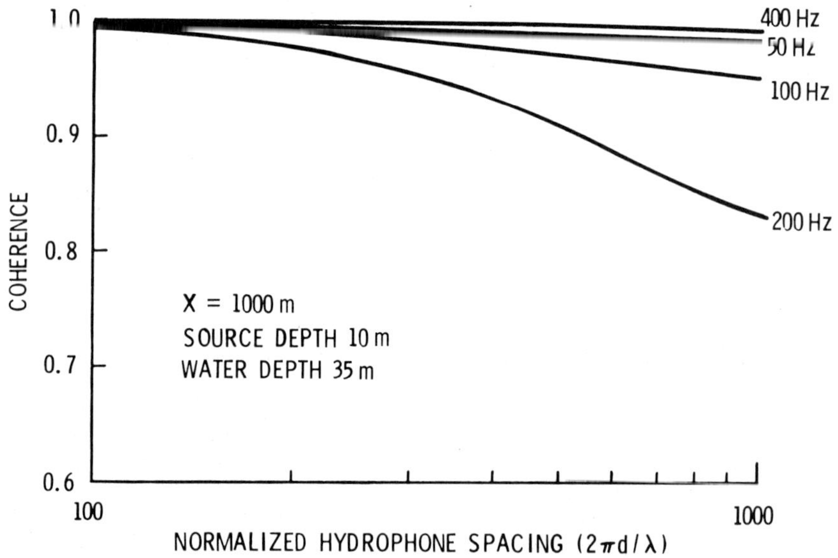

Fig. 3. Spatial coherence as a function of normalized hydrophone spacing at a range of 1 km.

For the isovelocity water conditions considered here, if we move the source closer to the surface, higher order modes will be predominantly excited. These modes penetrate deeper into the bottom than do the lower order modes, and hence should suffer more scattering at the rough sub-bottom interface with a resulting greater loss of coherence. This is borne out by the results shown in Fig. 4 which depict the variation of coherence with source depth for a frequency of 200 Hz at a range of 1 km.

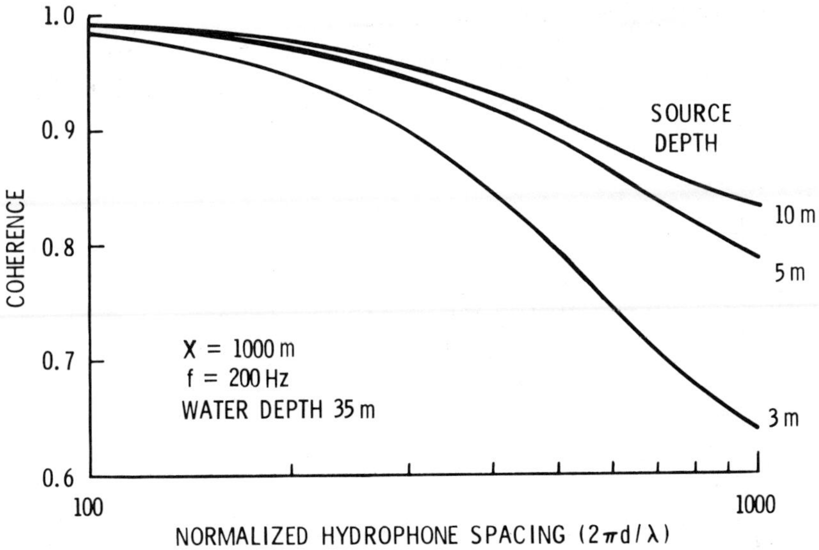

Fig. 4. Effect of source depth variation on 200 Hz spatial coherence.

Studies also revealed that the spatial coherence is strongly dependent on water depth. Figure 5 shows coherence at a frequency of 200 Hz and a range of 1 km for water depths between 25 and 50 m. In this example, only the depth of the isovelocity water has been changed, the bottom is treated as in the earlier examples. Here we see that little change in the coherence results from reducing the water depth from 35 to 25 m. However, a deeper water layer produces a highly coherent field for this source and receiver geometry. This occurs from a ray theoretic point of view, because there are fewer interactions with the bottom over the propagation range, and from the viewpoint of normal-mode theory, because there is less penetration into the bottom for the deeper water.

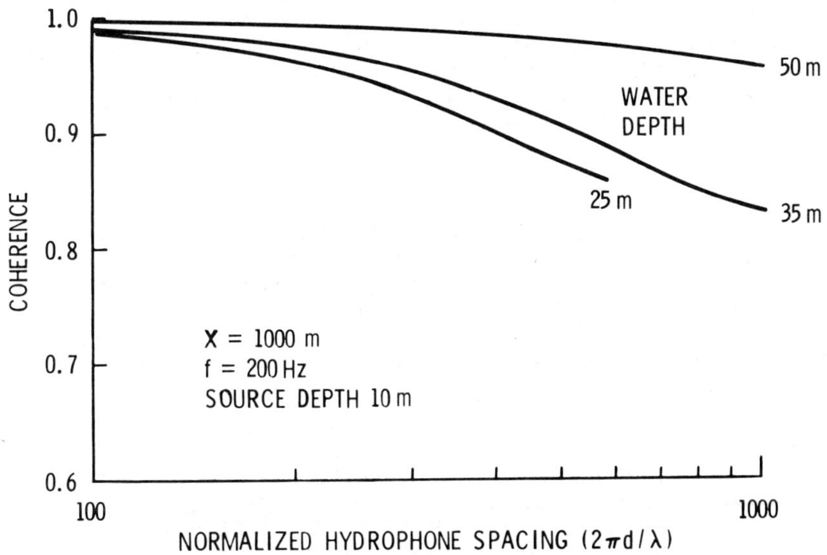

Fig. 5. Water depth dependence of spatial coherence at 200 Hz.

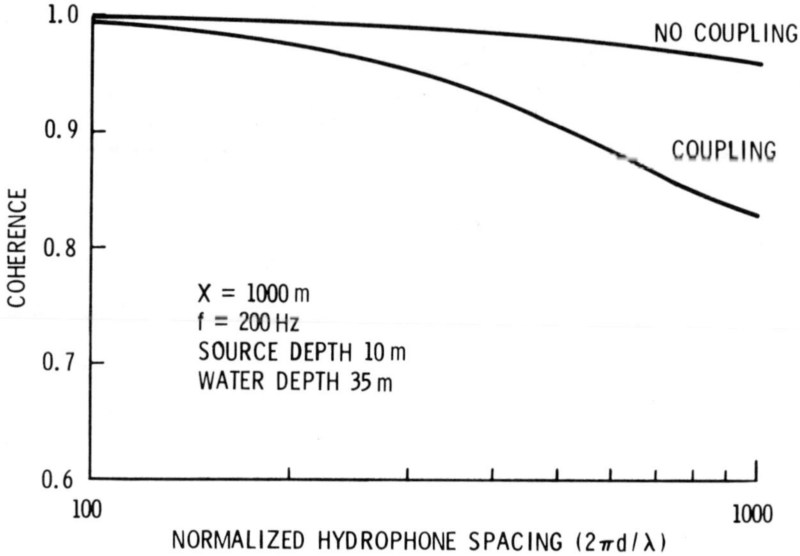

Fig. 6. Relative effect of mode coupling and intramodal scattering on coherence.

The last example that we consider is the relative effect of mode coupling versus intramodal scattering on the coherence. The coherence in the absence of mode coupling may be obtained simply be retaining only those terms for which n=m in Eq. (13). In this case, no energy is exchanged between modes, and the loss of coherence is governed solely by changes in the model propagation constant. Figure 6 compares the predicted coherence including coupling with that obtained in the absence of coupling. It is evident that the exchange of energy between modes plays a dominant role in coherence loss.

SUMMARY AND CONCLUSIONS

Coupled-mode theory has been applied to examine the effects of environmental variations on the transverse spatial coherence of the acoustic field in a shallow water waveguide. For the case considered, that of sub-bottom irregularities, the coherence was found to depend strongly on frequency, water depth and geometry. Generally, high coherences were predicted for receiver separations of several tens of wavelengths yielding qualitative agreement with field data.

One important finding of this study was that the exchange of energy between propagating modes was the dominant factor governing coherence loss, i.e., that adiabatic mode theory was not sufficient to treat scattering from features having large coherence lengths. The roughness scales and coherence lengths of the ocean surface and water-sediment interface[4] are generally smaller than those associated with sub-bottom features. The effect of these inhomogenieties on shallow-water spatial coherence remains to be investigated.

ACKNOWLEDGEMENT

This research was performance at the Applied Research Laboratory under the sponsorship of the Office of Naval Research.

REFERENCES

1. S. T. McDaniel, Mode coupling due to interaction with the seabed, J. Acoust. Soc. Am., 72:916 (1982).
2. A. Ishimaru, "Wave Propagation and Scattering in Random Media," Academic Press, New York (1978).
3. D. F. McCammon and S. T. McDaniel, Mode Coupling from Subbottom Roughness, in this volume.
4. J. M. Berkson and J. E. Matthews, Statistical Properties of Sea-Floor Roughness, in: "Acoustics and the Sea-Bed," N. G. Pace, ed., Bath University Press, Bath (1983).
5. C. C. Yang and K. C. Yeh, Electromagnetic Scattering from a Strongly Turbulent Medium, TR-69, Ionosphere Radio Laboratory, University of Illinois, Urbana, Ill. (1984).
6. P. Wille and R. Thiele, Transverse horizontal coherence of explosive signals in shallow water, J. Acoust. Soc. Am., 50:348 (1971).

MODE COUPLING FROM SUBBOTTOM ROUGHNESS

Diana F. McCammon and Suzanne T. McDaniel

Applied Research Laboratory
The Pennsylvania State University
Post Office Box 30
State College, Pennsylvania 16804 USA

ABSTRACT

Low frequency sound propagation in shallow water can entail extensive bottom penetration if the sediments that overlay the basement are very fine grained. In this case, the effective scattering from the roughness of the water-sediment interface may be of secondary importance compared to the effective scattering of the basement irregularities. In addition, at low frequencies, this scattered energy is not lost, rather it is transferred between the normal modes of propagation through a process called mode coupling. The object of this paper is to explore the relationship of subbottom roughness to the propagated modal structure. We will show the effects of frequency, source depth, water depth and subbottom layer depth. A dimensionless parameter $\gamma = fh/c$ has been suggested as a measure of coupling influence. For large γ, significant coupling has been shown to occur between nearest neighbor modes. In this paper, we will examine more fully the relationship between coupling and γ, particularly in the intermediate region $1 < \gamma < 10$.

INTRODUCTION

The coupling of acoustc energy between normal modes is a phenomenon that occurs whenever the properties of the oceanic wave guide have a range dependence. However a complete computation of this coupling effect is seldom feasable. Instead, approximations such as the adiabatic theory or the parabolic approximation are commonly used when obvious range dependent conditions exist. McDaniel[1] has examined the importance of mode coupling that occurs for a smooth sloping bottom, and has concluded that, for an isovelocity water over an isovelocity halfspace, mode coupling is important in two regimes of the dimensionless parameter $\gamma = fh/c$. In the first of these regimes, for large γ, coupling occurs between nearest neighbor modes, while in the second regime, $\gamma \rightarrow \gamma_{cutoff}$, energy is transferred to the continuous spectrum.

In this paper, we will examine the importance of mode coupling between nearest neighbor modes as a function of subbottom roughness. We have selected the sea bed in the Block Island Fishers Island (BIFI) Corridor as our environmental model because it contains many interesting geological formations and because extensive subbottom surveys have been conducted at this region.

We will show the relationship between the rate of transfer of energy among modes and the frequency, source depth, water depth and layer depth. Multiple layers are also considered. Section I will present the environmental model and the theory of mode coupling applied to subbottom roughness. Section II will discuss the numerical computations and draw conclusions concerning the factors that influence mode coupling.

ENVIRONMENTAL MODEL AND THEORETICAL APPROACH

During the period October 8 to October 27, 1971, the Marine Sciences Institute of the University of Connecticut conducted a marine seismic reflection survey over portions of Block Island Sound between Fishers Island and Block Island. An example of the seismic profiles they obtained is shown in Figure 1. In this figure, a strong scattering layer can be detected below the surficial sediment layer. Many overlaying reflectors that closely conform to the bedrock reflector are also apparent. A careful core study was done of this region and the reflectors were found to consist of couplets of clay and silt called varved clay[2]. This type of deposit was formed by glacial action upon a fresh water lake, that is, clay and silt were carried into the lake by spring melts, whereupon the silt precipitated out while the smaller clay particles remained in suspension. In the winter, as the waters cooled, the clay was deposited, thus forming alternating layers of silt and clay. The number of these layers indicates how deep the lake was and the thickness of each layer gives some information about rates of glacial melting. The required suspension of clay could not take place in salt water, thus proving that the region must have been a fresh water lake at one time[3]. Similar geological records have been found in parts of Sweden. Another unusual formation found in the Block Island Sound are moraines, also associated with glaciers. These formations provide a very rough boundary for the wave guide from which the acoustic energy can be scattered, since the overlaying varved clay allows considerable acoustic penetration.

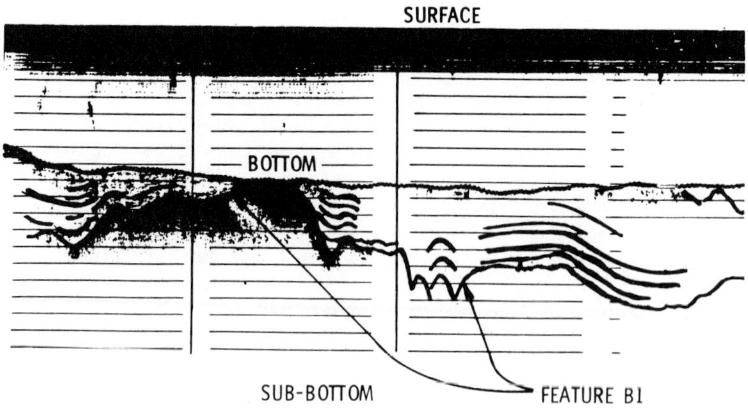

Fig. 1. Seismic profiles obtained in the Block Island sound, New York. Horizontal lines represent 5 msec and the two center vertical lines mark 5 minutes with a constant ship speed of 5 kts. This profile spans approximately 2.7 km and reveals water depth of approximately 70 m.

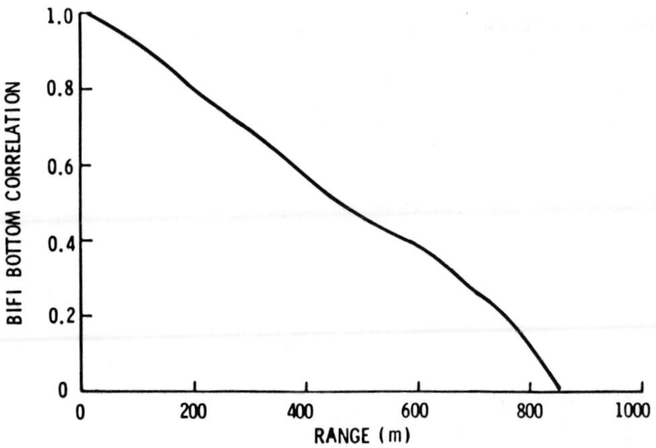

Fig. 2. The coherence of the sub-bottom feature B1. The coherence length is shown to be 616 m.

For this study a section of seismic profile has been sampled at 7 m spacing for statistical analysis. Figure 2 shows the coherence of the section shown in the previous figure. The coherence length is 616 m and the RMS roughness is 1.75 m. Table I contains the basic environmental parameters that we have associated with this region. The sound velocity of the varved clay is an average velocity associated with an average grain size of $\phi = 7.5$.

Mode coupling terms can be derived from the wave equation[4] if the depth eigenfunctions are assumed to have range dependence.

$$[\nabla^2 + k^2(r,z)]\ \psi = 0, \tag{1}$$

$$\psi = \sum_n \phi_n(r)\ Z_n(r,z).$$

The depth functions obey the equation

$$\frac{\partial^2 Z_n}{\partial z^2} + [k^2(r,z) - k_n^2]\ Z_n = 0, \tag{2}$$

with k_n the local modal wavenumber. The mode amplitude functions ϕ_n obey the equations

Table 1. BIFI Environmental data set

water column	isovelocity 1517 m/sec water depth 35 m
sediment (varved clay)	isovelocity 1530 m/sec mean depth 25 m density 1.50 attenuation .056f dB/km/hz
basement (granite)	C_p = 2300 m/sec C_s = 1626 m/sec density 2.3 attenuation .009f dB/km/hz

$$(\nabla_\perp^2 + k_n^2)\,\phi_n = -\sum_{m\neq n} (B_{mn} + \vec{C}_{mn}\cdot \vec{\nabla}_\perp)\phi_m, \tag{3}$$

where the coupling coefficients are given by

$$B_{mn} = \int_{-\infty}^{0} Z_m \nabla_\perp^2 Z_n \, dz,$$

and $$\vec{C}_{mn} = 2\int_{-\infty}^{0} Z_m \vec{\nabla}_\perp Z_n \, dz.$$

For ocean bottoms of slow range dependence B_{mn} can be neglected with respect to $\vec{C}_{mn}$. With the restriction to small slopes, modal power transfer coefficients can be derived. In the forward scattering limit,[1] the rate Q_{mn} at which energy is transfered from the nth mode is

$$Q_{mn} = 1/2 \left(\frac{k_m}{k_n}\right)^{1/2} \int_0^\infty d\rho \, \langle C_{mn}(r) C_{mn}(r+\rho)\rangle \cos[\rho(k_m - k_n)]. \tag{4}$$

We define

$$Q_{mm} = \sum_n Q_{mn} - \alpha_m,$$

with α_m equal to the mode attenuation. The mode coupling coefficients C_{mn} are found to be

$$C_{mn} = \frac{2Z_{n1}Z_{m1}}{(k_n^2 - k_m^2)}\left[\gamma_{1n}^2 + \frac{\rho_1}{\rho_2} g_{2n}^2 + g_{2m} g_{2n} \frac{\rho_1}{\rho_2}\left(\frac{\rho_1}{\rho_2} - 1\right)\right]\left(\frac{dz_B}{dr}\right), \tag{5}$$

for an isovelocity sediment. In this equation,

$\gamma_{1n} = [(2\pi f/c_{water})^2 - k_n^2]^{1/2}$, $g_{2n} = [(2\pi f/c_{sediment})^2 - k_n^2]^{1/2}$

z_B=sediment depth, Z_{n1} = eigenfunction at subbottom interface.

On substitution into Equation (4) we obtain an integral that is the Fourier integral transform of the bottom slope correlation function

$$\int_0^\infty \left\langle \frac{\partial z_B(r)}{\partial r} \frac{\partial z_B(r+\rho)}{\partial r} \right\rangle \cos[\rho(k_m - k_n)]\, d\rho, \tag{6}$$

that is, the bottom slope spectrum. Thus the rate of transfer of energy between modes is proportional to the bottom slope spectrum and inversely proportional to the modal eigenvalue difference. Mode coupling can be thought of as a type of resonance phenomenon of the rough surface slope spectrum and the modal wavenumber difference.

We will use this representation of mode coupling together with the bottom slope spectra measured from a section of the BIFI corridor to examine the properties and effects of subbottom roughness on modal propagation. For propagated effects, we solve the equation

$$\frac{du_m}{dr} = Q_{mn} u_n, \tag{7}$$

where u is the forward propagated portion of the mode amplitude and has been corrected for spreading loss. The technique employs a stable implicit finite difference solution

$$u_m(r+\Delta r) = \left[I + \frac{Q_{mn}\Delta r}{2}\right] \left[I - \frac{Q_{mn}\Delta r}{2}\right]^{-1} u_n(r). \qquad (8)$$

NUMERICAL COMPARISONS

The propagation loss of the first three normal modes at 200 Hz for the BIFI data set outlined in Table I is shown in Figure 3. The graph on the left shows the absence of coupling where the first mode dominates at all ranges, and swiftly becomes the principle propagating mode. The even numbered higher order modes, not shown here, contain the energy equivalent of the second mode and the odd numbered modes are grouped around the third mode. The introduction of a subbottom roughness such as the BIFI profile shown in the graph on the right, causes a transferring of energy primarily from the first mode into the second, and from the second into the third, etc. A dramatic change in propagation of the higher order modes occurs.

Let us examine the energy transfer as a function of frequency. Figure 4 contains graphs of energy gain or loss referenced to the flat bottom case as a function of range from 50 to 400 Hz. In the 50 Hz case, a minute amount of energy is radiated into the second mode, the gain only becoming apparent beyond 10 km. At 100 Hz, the second mode is the primary beneficiary. Little loss is evident in the first mode. For 200 Hz, the first mode shows a small but steady energy loss with range, while the third mode shows the greatest gain. Up to 8k m the second mode loses more energy to higher modes than it gains from below. At 400 Hz, both first

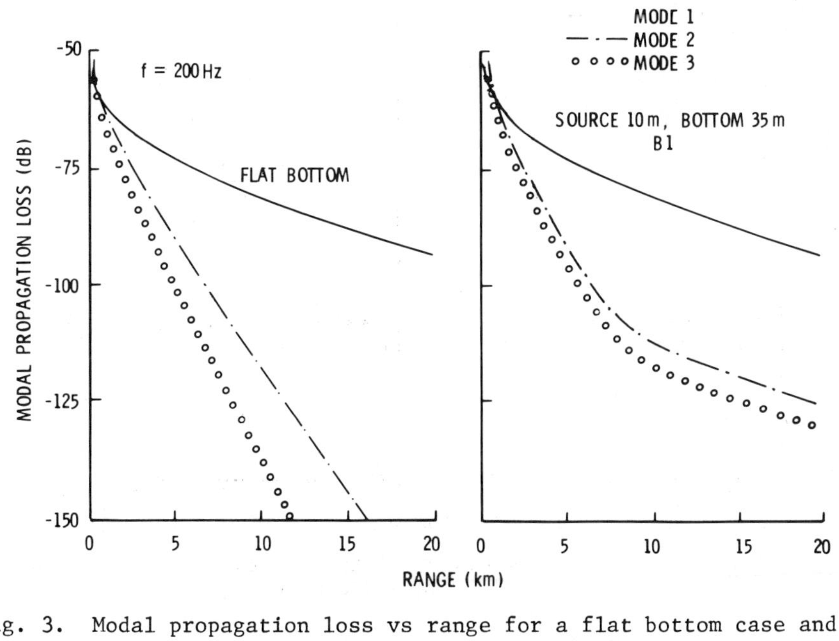

Fig. 3. Modal propagation loss vs range for a flat bottom case and the roughness measured from BIFI sub-bottom feature B1.

and second modes are losing energy to the higher modes. The mode coupling is significantly greater with increasing frequency because the modal wave-number difference (k_m-k_n) that occurs in the denominator of the coupling coefficient equation decreases with increasing frequency. That is, the modes increase in number and become more closely spaced.

The value of the dimensionless parameter $\gamma = fh/c$ ranges from 1 to 9 in these figures, with noticeable coupling occurring above $\gamma = 4.6$.

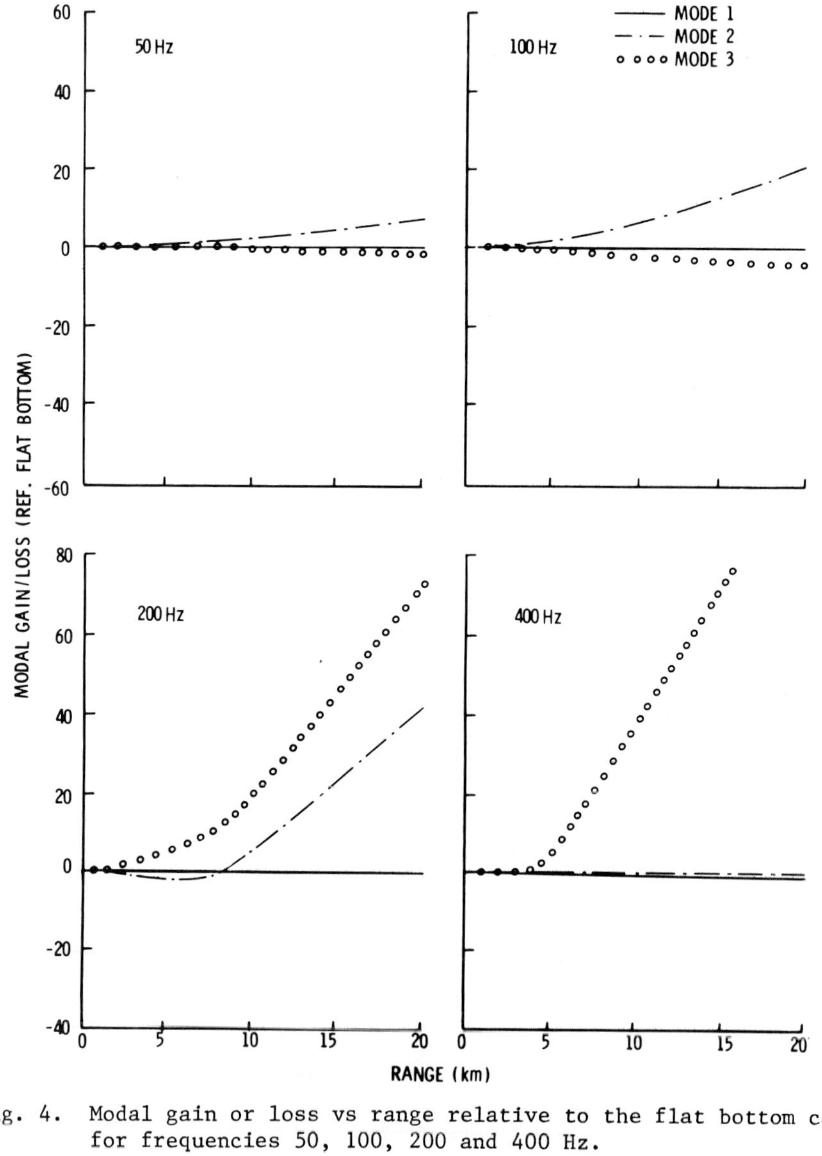

Fig. 4. Modal gain or loss vs range relative to the flat bottom case for frequencies 50, 100, 200 and 400 Hz.

Table II. Mode Coupling as a Function of Water Depth (200 Hz Frequency)

Modal Propagation Differences (relative to the flat bottom case) at 20 km

(dB)

water depth (m)	Total PL	Mode 1	Mode 2	Mode 3	Mode 4	γ
25	-2.30	-2.50	+107.00	+26.00	+20.	3.3
35	-0.32	-0.34	+ 39.00	+78.00	+40.	4.6
50	-0.04	-0.04	- 0.60	+64.00	+35.	6.6
70	-0.02	-0.01	- 0.06	- 0.31	+52.	9.2

While the increase in modal loss with frequency is clear, the suggested increase with depth requires some study. Table II details the findings of a study of water depth variations. The maximum loss from the first mode to the second mode occurs at the shallowest water depth of 25 m. As the water deepens, the amount of coupling increases but shifts to the higher order modes. This occurs because the deeper water supports more modes while at the same time the eigenfunctions of the lower modes approach zero at the rough interface. At the deepest depth, energy is being lost from the first three modes into the higher modes, while at the shallowest depth, only the first mode is shedding energy.

The conclusion drawn from this analysis seems to be that mode coupling can be reliably predicted when γ is large, with a viable definition of "large" as 3 or greater.

Next we examine the effects of the scattering layer depth and the mode excitation structure. An increase in the subbottom layer depth decreases the observed coupling between the lower modes because it decreases the effective energy available for interaction. The loss from the first three modes is shown in Table III as a function of subbottom layer depth. Note the dramatic reduction in mode loss when the layer thickness is increased from 10 m to 25 m compared to the reduction achieved when going from 25 to 50 m. In the case of the first layer depth, sufficient energy penetrated to the basement to insure good couping with the subbottom roughness for both first and second modes, however at the deeper layer depth, little energy is available for coupling, hence only the first mode is shedding energy.

Table III. Mode Coupling as a Function of Subbottom Layer Depth (200 Hz Frequency)

Subbottom layer depth (m)	Total PL	mode 1	mode 2	mode 3
10	-5.63	-6.20	- 4.80	+30.
25	-0.32	-0.34	+ 39.00	+78.
50	-0.02	-0.02	+101.00	+43.

Source depth location is important in the determination of the extent to which each mode is excited, however the amount of mode coupling that occurs in propagation out to 20 km is only weakly dependent on a particular mode excitation structure. The rate of transfer of energy, Q_{mn}, is a function of the eigenvalue differences and the eigenfunctions at the rough interface and these two quantities are properties only of the waveguide. Source depth is important in the computation for the mode amplitude, whose gains and losses are considered here, however since we are comparing losses relative to the flat bottom case, mode level excitation differences are generally subtracted out. Figure 5 displays modal gain/loss relative to a flat bottom for four source depths, 3, 10, 20 and 30 m in a water depth of 35 m. The loss from the first mode is identical in all cases. Minor differences in the higher modes occur, probably due to the presence or absence of available modes for coupling.

In the previous examples, near sediment layering was neglected. To examine this effect, five cascaded layers of clay and silt, each with a velocity ratio v_1/v_2 = 1498/1530 and a thickness of 5 m, are examined. Propagation is weakly affected by the five layers, with the energy loss from the first mode about 3 dB greater at 20 km than the loss from a single scattering layer. While these five layers do not represent the majority of oceanic environments, this example does serve to demonstrate that in cases of deep modal penetration, all subbottom structures will cause mode coupling to some extent.

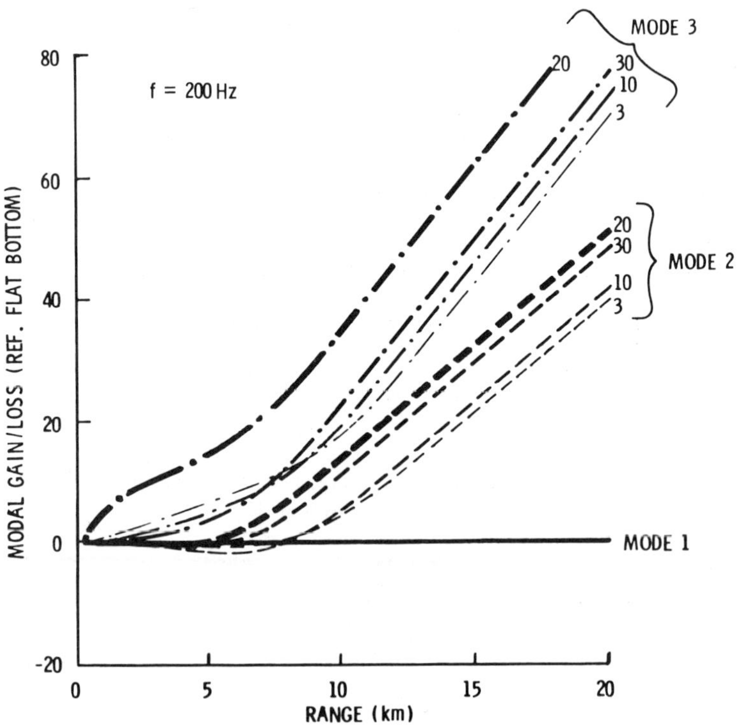

Fig. 5. Modal gain or loss vs range relative to the flat bottom case for source depths 3, 10, 20 and 30 m.

In summary, we have seen that energy propagating through a fine grained sediment to the rough basement interface can undergo significant loss via the mechanism of mode coupling. When these factors are present, the amount of mode coupling can be predicted by the simple parameter $\gamma = fh/c$ because both frequency and water depth directly affect the number of propagating modes and the modal wave number spacing.

We have found that in addition to frequency and water depth the sediment layer thickness also controls the extent of mode coupling. Thick overlaying sediments lessen the degree of mode coupling that occurs because they prevent modal penetration to the rough rock interface. In contrast, source geometry has almost no effect on mode coupling. This is because the amplitude of modal excitation is controlled by the source depth, but the modal wave number spectrum and the mode shapes are unchanged, and these are the factors that affect the rate of transfer of energy.

ACKNOWLEDGMENT

This work was performed at the Applied Research Laboratory under the sponsorship of the Office of Naval Research.

REFERENCES

1. S. T. McDaniel, Mode Coupling due to interaction with the seabed, J. Acoust. Soc. Am. 72, 916-923 (1982).
2. J. J. Dowling, R. Bertoni and R. Miller, Seismic Reflection Profiling in the BIFI Corridor, BIFI Corridor Block Island Sound, Rhode Island-New York, University of Connecticut (1971).
3. Personal communications with Dr. J. J. Dowling, Marine Sciences Institute, University of Connecticut, April 20, 1983 and Mr. Ralph Lewis, U.S. Geological Service, Avery Point, Conn. April 20, 1983.
4. A. D. Pierce, Extension of the method of normal modes to sound propagation in an almost-stratified medium, J. Acoust. Soc. Am. 37 19-27 (1965).

ACOUSTIC WAVES IN MARINE SEDIMENTS

Robert D. Stoll

School of Engineering and Applied Science
Columbia University
New York, New York 10027

and

Lamont-Doherty Geological Observatory
Palisades, New York 10964

ABSTRACT

Over the past fifteen years a mathematical model to describe acoustic wave propagation in water-saturated sediments has been developed on the basis of the classical Biot theory. As a result of extending the basic theory to allow for energy dissipation owing to various viscous and frictional mechanisms in the skeletal frame of the sediment, a model has evolved that may be used as a practical research tool for studying the acoustic behavior of unconsolidated marine sediments. The development of the model is reviewed in this paper and various physical parameters that influence wave velocity and attenuation, such as grain size, porosity, permeability and effective stress, are discussed. Comparisons are made between the predictions of the model and experimental data for different kinds of sediment.

INTRODUCTION

Acoustic waves propagating in marine sediments may be thought of as a special case of a more general class of mechanical waves that can propagate in fluid-saturated porous media. Wave motion in this kind of material has been extensively studied by acousticians, geophysicists and engineers for many diverse reasons. Low frequency waves of rather large amplitude are of interest when analyzing conditions near the source of an earthquake or near any facility housing heavy vibrating machinery. High frequency waves, that can resolve thin layering, are used to study near-bottom ocean sediments whereas in deep geophysical exploration or seismological studies of the earth's structure, interest is often centered on waves of low frequency and amplitude that have traversed long distances. As a result of this wide spectrum of interest, geological materials ranging from soft, consolidated sediments to rock have been studied extensively in an effort to understand and exploit the different modes of wave propagation that are observed.

The interaction of waterborne acoustic waves with the sediments in the ocean bottom becomes increasingly important when the focus shifts to low frequencies and shallow water. Deep accumulations of unconsolidated, terrigeneous sediments underlie many areas of shallow water and in many of these locations the velocity-depth profile has a form which results in a significant amount of energy being returned to the water column, far from the source, after deep penetration into the bottom. Because of the many combinations of different kinds of sediment, stratification and structure, accurate mathematical description of the wave field produced by a particular source depends on being able to specify the acoustic properties of each of the individual components in the sediment column. In any one strata of sediment, there are a number of parameters which play a principal role in controlling the acoustic response. These include:

A. Porosity.
B. Static, intergranular stress (overburden).
C. Degree and kind of lithification.
D. Gradation and grain shape.
E. Dynamic strain amplitude.
F. Material properties of individual grains.
G. Structure as determined by mode of deposition.

In this paper we review the details of a mathematical model for wave propagation that includes the influence of a number of these parameters in an explicit manner. The model is based on the classic work of Maurice Biot who developed a general theory for the mechanics of fluid-filled, porous media in a number of papers spanning the time period from 1941 to 1973.[1-7] Biot started with a paper on the quasistatic consolidation of a porous, linearly elastic medium and then extended the theory to cover the dynamic response for a number of different cases which include considering the solid to be viscoelastic, anisotropic and nonlinear in its response. The efficacy of the theory is due largely to the fact that the average motion of both the fluid and the solid are followed separately so tht the phenomenological properties of each of these two interpenetrating media may be specified separately. Moreover it allows local interaction between components that is not specifically included in the general theory, to be approximated by changing the model of either or both of the components. For example, the phenomenological model of the porous solid may be defined by viscoelastic operators which contain the relaxation constants of the "hidden degrees of freedom" corresponding to local fluid motion (i.e. "squeeze film" motion near points of intergranular contact or "squirt flow" in cracks). Alternatively the compressibility of the fluid and the apparent permeability may be changed to reflect the presence of gas bubbles in the fluid. Thus a number of different factors that are important in geophysical applications can be built into the model without changing the basic formulation.

Over the past 15 years, the author and several colleagues have adapted the Biot model to describe acoustic propagation in unconsolidated marine sediments. Since our objective was to develop a practical working model that could be used as a guide in planning and interpreting experimental work, we began with the simplest possible form and added various complexities only as they were needed to expand our work to cover different kinds of sediment. Thus the number of material parameters that had to be measured or assumed at any stage in the development of the model was kept to a minimum. Since the first version of the model was introduced in 1970, we have published a number of technical papers covering various stages of its development and application[8-14] and over twenty papers have been published by colleagues who have used our work in various ways. As a result, the theory has evolved into a practical research tool for studying unconsolidated sediments.

BASIC THEORY

In order to generate the equations of motion that describe propagation in a fluid-filled porous medium, we start with a set of constitutive equations of the form

$$\begin{aligned}
\tau_{xx} &= He - 2\mu(e_y + e_z) - C\,\zeta \\
\tau_{yy} &= He - 2\mu(e_z + e_x) - C\,\zeta \\
\tau_{zz} &= He - 2\mu(e_x + e_y) - C\,\zeta \\
\tau_{xy} &= \mu\,\gamma_z \\
\tau_{yz} &= \mu\,\gamma_x \\
\tau_{zx} &= \mu\,\gamma_y \\
p_f &= M\zeta - Ce
\end{aligned} \tag{1}$$

where H, C, M and μ are operators (either differential or integral) that relate the components of stress to the components of strain and strain rate. The τ_{ij} are total stresses on an element of volume attached to the skeletal frame, p_f is the pressure in the pore-water, and e_x, e_y, e_z, γ_x, γ_y, γ_z, are components of small strain. The volumetric strain of an element attached to the frame e, is given by

$$e = e_x + e_y + e_z = \text{div}\ \underset{\sim}{\mathbf{u}} \tag{2}$$

and the volume of fluid that has flowed in or out of the element, ζ, is

$$\zeta = \beta\ \text{div}\ (\underset{\sim}{\mathbf{u}} - \underset{\sim}{\mathbf{U}}) \tag{3}$$

where β is porosity, $\underset{\sim}{\mathbf{u}}$ is frame displacement and $\underset{\sim}{\mathbf{U}}$ is fluid displacement.

In Eq. 1, the operator μ, which relates shear stress to shear strain, is clearly analogous to the shear modulus used in conventional solid mechanics. However, the other three operators H, M, and C are not directly comparable to the usual moduli of solid and fluid mechanics and therefore require special attention. To understand their significance we consider two experiments designed to measure the "jacketed" and unjacketed" compressibility of the sediment.

In the first case, we consider a specimen of sediment that is completely encased in a thin flexible membrane, Fig. 1a. The fluid filling the interstices of the sediment is allowed to flow into or out of the specimen via a tube passing through the membrane. When there is a change in the fluid pressure -p', on the outside of the jacket, the specimen will change volume and some water will be expelled or drawn into the specimen via the tube. If the test is performed slowly enough to prevent any change in the pore-water pressure within the specimen, or the "excess" pore pressure is allowed to dissipate after a more rapid change of pressure, the ratio of the volume change of the specimen to the applied pressure is termed the "jacketed" compressibility. Geotechnical engineers will recognize the procedure just described as similar to the first step (isotropic consolidation) in a triaxial compression test. Adding the first, second and third of Eqs. 1 and taking the reciprocal of the "jacketed" compressibility we obtain the drained bulk modulus of the sediment in terms of H, M, C and μ

$$K_b = - p'/e = H - 4\mu/3 - C^2/M \tag{4}$$

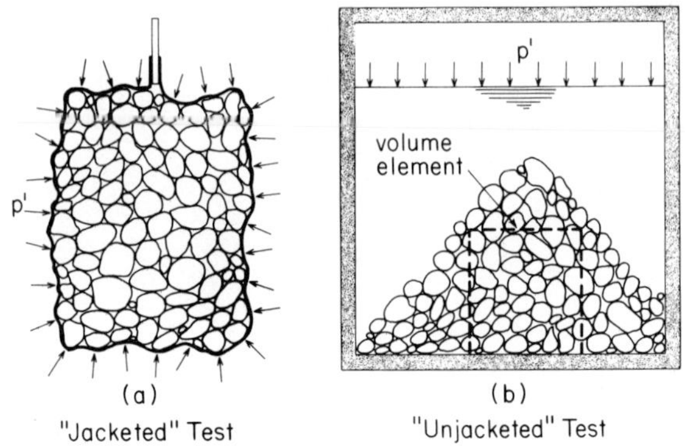

Fig. 1. Isotropic tests to determine moduli.

In the second case we consider a hypothetical test where the pore-water pressure $-p'$, in an unjacketed specimen is increased as shown in Fig. 1b. We measure the dilatation e, of an element of volume attached to the skeletal frame and also the increment of fluid content ζ, for the same volume. The "unjacketed" compressibility δ, and the "coefficient of fluid content" γ, are then related to H, M, C and μ with the help of Eq. 1.

$$\delta = - e/p' = \frac{1 - C/M}{H - 4\mu/3 - C^2/M} \tag{5}$$

$$\gamma = \zeta/p' = \frac{H - 4\mu/3 - C}{(H - 4\mu/3 - C^2/M)M} \tag{6}$$

Eqs. 4, 5 and 6 define the basic relationship between the operators H, M, C and μ from the Biot theory and potentially measureable physical quantities. Unfortunately it is not practicable to measure the unjacketed compressibility or the coefficient of fluid content in any "real" test so that we are still left without a connection to familiar quantities that can be measured or estimated. In order to make this connection, a simplifying assumption is now made that is crucial to the practical application of the Biot theory. We assume that the volumetric strain of the pore volume within the sediment is the same as the volumetric strain of the total element of volume attached to the skeletal frame during an unjacketed test. Stated another way, the porosity of the sediment is assumed to remain constant during unjacketed loading. This permits us to write the following expression relating δ and γ to more familiar quantities such as the bulk modulus of the fluid K_f, the bulk modulus of the individual mineral grains, K_r.

$$\gamma = \beta(1/K_f - 1/K_r) \tag{7}$$

$$\delta = 1/K_r \tag{8}$$

Finally with the help of Eqs. 4 through 8 we may derive explicit expressions for H, C and M in terms of the quantities K_b, K_f, K_r and μ which can be measured or estimated in a variety of different ways.

$$H = \frac{(K_r - K_b)^2}{D - K_b} + K_b + 4\mu/3 \tag{9}$$

$$C = \frac{K_r(K_r - K_b)}{D - K_b} \tag{10}$$

$$M = \frac{K_r^2}{D - K_b} \tag{11}$$

where

$$D = K_r(1 + \beta(K_r/K_f - 1)).$$

The assumption made to obtain Eqs. 7 through 11 is a truism in the case of sediments composed of isotropic particles and is generally considered to be reasonable for many other sediments even though the individual particles may be anisotropic. However it is an important step in the practical application of the theory and by overlooking it, some authors have gotten into difficulty in trying to determine H, C and M in terms of other material properties.

In most of our work to date we have considered the bulk moduli of the fluid and the individual grains, K_f and K_r, to be elastic constants whereas the bulk and shear moduli of the skeletal frame, K_b and μ have been considered to be operators that take into account energy dissipation during oscillatory motion (i.e. viscoelastic operators in the time domain or complex moduli in the frequency domain). The energy dissipation is the result of intergranular friction and "local" fluid motion between particles. The local fluid motion occurs in regions of close proximity between the surfaces of adjacent particles and is visualized as a "pumping" or "squeeze film" motion of the type which is quite familiar in the theory of lubrication. The viscous damping from "local" fluid motion depends on the particle size distribution of the sediment and, in general, cannot be characterized by a single relaxation time.

EQUATIONS OF MOTION

If Eqs. 9, 10 and 11 are combined with the stress equations of motion and a generalized form of Darcy's law, two coupled equations of motions result:

$$\begin{aligned} \nabla^2(He - C\zeta) &= \frac{\partial^2}{\partial t^2}(\rho e - \rho_f\zeta) \\ \nabla^2(Ce - M\zeta) &= \frac{\partial^2}{\partial t^2}(\rho_f e - m\zeta) - \frac{\eta}{k}\frac{\partial \zeta}{\partial t} \end{aligned} \tag{12}$$

in which ρ is the total mass density, ρ_f is the fluid mass density, η is fluid viscosity and k is the coefficient of permeability of the porous frame with units (L^2). The parameter m, is an apparent mass density given by the expression

$$m = c\rho_f/\beta \ , \ c > 1 \tag{13}$$

and accounts for the fact that not all of the fluid moves in the direction of the macroscopic pressure gradient because of the shape and orientation of the interconnected void spaces. Because of the complex geometry of the pore spaces in real sediments, c must be determined experimentally. In order for the second of Eqs. 12 to remain valid at high frequencies, the viscous resistance to fluid flow given by η/k must be made frequency dependent. Biot studied the actual microvelocity field within the pore channels for certain idealized shapes and deduced a complex correction factor to be multiplied times the ratio η/k in the frequency domain. His factor F, which is a function of angular frequency ω, is given by

$$F(\kappa) = F_r(\kappa) + i\, F_i(\kappa) = \frac{1}{4}\,\frac{\kappa T(\kappa)}{1-2T(\kappa)/i\kappa}, \tag{14}$$

where

$$T(\kappa) = \frac{\mathrm{ber}'(\kappa) + i\ \mathrm{bei}'(\kappa)}{\mathrm{ber}(\kappa) + i\ \mathrm{bei}(\kappa)},$$

and

$$\kappa = a(\omega\rho_f/\eta)^{1/2}$$

The function ber(κ) and bei(κ) are the real and imaginary parts of the Kelvin function and a is a parameter with the dimension of length that depends on the size and shape of the pores. Recent studies of the microvelocity field for pore geometries that more nearly replicate those in real sediments, suggest that both m and η/k should be frequency dependent.[15]

Eqs. 12 are satisfied by solutions of the form

$$e = A_1 \exp(i(\omega t - \ell x))$$

and

$$\zeta = A_2 \exp(i(\omega t - \ell x))$$

provided the following frequency equation is satisfied:

$$\begin{vmatrix} \bar{H}\,\ell^2 - \rho\omega^2 & \rho_f\omega^2 - \bar{C}\,\ell^2 \\ \bar{C}\,\ell^2 - \rho_f\omega^2 & m\omega^2 - \bar{M}\,\ell^2 - \dfrac{i\omega F\eta}{k} \end{vmatrix} = 0 \tag{15}$$

The two complex roots of Eq. 15, of the form $\ell = \ell_r + i\ell_i$, yield the attenuation ℓ_i, and the phase velocity ω/ℓ_r, for two kinds of body waves. Biot called these waves dilatational waves of the "first" and "second" kinds. The dilatational wave of the first kind is similar to the ordinary compressional wave in solid mechanics. The fluid and solid components are moving only slightly out of phase so that the wave attenuates slowly and there is very little dispersion. In contrast, the compressional wave of the second kind attenuates very rapidly since the motions of the fluid and solid components are essentially out of phase.

It is only recently that convincing laboratory documentation of the existence of this kind of body has been obtained.[16,17]

A frequency equation for shear waves may also be obtained by combining Eqs. 1 with the stress equations of motion and the generalized Darcy's law. The resulting equations of motion are

$$\mu\nabla^2\underset{\sim}{\boldsymbol{\omega}} = \frac{\partial^2}{\partial t^2}(\rho\underset{\sim}{\boldsymbol{\omega}} - \rho_f\underset{\sim}{\boldsymbol{\theta}})$$

$$\frac{\eta}{k}\frac{\partial\underset{\sim}{\theta}}{\partial t} = \frac{\partial^2}{\partial t^2}(\rho_f\underset{\sim}{\omega} - m\underset{\sim}{\boldsymbol{\theta}}) \tag{16}$$

where

$$\underset{\sim}{\boldsymbol{\omega}} = \text{curl } \underset{\sim}{\mathbf{u}}$$

and

$$\underset{\sim}{\boldsymbol{\theta}} = \beta \quad \text{curl } (\underset{\sim}{\mathbf{u}} - \underset{\sim}{\mathbf{U}})$$

The frequency equation for Eqs. 16 takes the form

$$\begin{vmatrix} \rho\omega^2 - \mu\ell^2 & \rho_f\omega^2 \\ \rho_f\omega^2 & m\omega^2 - \frac{i\omega F\eta}{k} \end{vmatrix} = 0 \tag{17}$$

Eq. 17 indicates that only one kind of shear wave can propagate, however it also shows that this wave may suffer some dispersion because the solid and fluid components can move out of phase with one another causing viscous dissipation.

When the permeability of the sediment is low enough to justify the assumption of a "closed system (i.e. $\zeta \to 0$) then the Biot equations reduce to the equivalent of the Gassmann equation.[18] In this case ($H-4\mu/3$) may be considered as the undrained bulk modulus of the sediment and a little algebra will show this is the reciprocal of the "closed system" compressibility used by Gassmann. Finally in very soft sediments or suspensions, where both K_b and μ become vanishingly small, we can recover the Wood equation for compressibility of a water-sediment mixture[19] by letting K_b=μ=0 in the expression ($H-4\mu/3$).

$$c_m = \frac{1}{K_r}(1-\beta) + \frac{1}{K_f}\beta \tag{18}$$

where c_m is the compressibility of the mixture.

CHOOSING MODEL PARAMETERS

In order to use Eqs. 15 and 17 to determine the velocity and attenuation in real sediments, it is necessary to choose realistic values for 13 different parameters. Several of the choices are straight forward and reasonable values may be obtained from standard references however a few of the parameters need careful consideration in order to produce meaningful predictions. We will consider these parameters in two categories - the complex moduli of the skeletal frame K_b and μ and the three parameters (a,m, and k) that control the overall mobility of the fluid component and its interaction with the skeletal frame.

In the examples to follow we assume that the wave amplitude and the logarithmic decrement are small so that the different measures of damping may be related in a particularly simple way. For example, the quality factor Q, the logarithmic decrement δ_G (for both traveling waves and stationary vibrations), the attenuation coefficient $\alpha = \ell_i$, and the ratio of the imaginary and real parts of the appropriate complex modulus, $M^* = M_r + iM_i$, are related approximately by the expression[12]

$$\frac{1}{Q} = \frac{\delta_G}{\pi} = \alpha(\frac{2V}{\pi f}) = \frac{M_i}{M_r} = \tan \theta_L \tag{19}$$

where V is phase velocity, f is frequency and θ_L is the phase difference between stress and strain under harmonic loading.

In the case of dry granular materials such as sands, silts and sandstones, the principal mechanism of energy loss is friction which governs the minute amount of slip occurring within the contact area between particles. The slip is confined to small, annular regions of the contact area and does not result in any gross sliding between particles. As a result, the damping observed in a dry granular material is usually very small and essentially independent of frequency. Typically, 1/Q will by of the order of 10^{-2} or slightly larger in sands and sandstones. However, when these granular materials are located in a water environment, 1/Q increases and becomes frequency dependent at all but the lowest frequencies because of the additional energy losses attributable to the viscosity of the fluid component. The effects of viscosity manifest themselves in two different ways. There is an overall motion of the fluid field relative to the skeletal frame of the sediment and a local motion in the vicinity of intergranular contacts. The effect of the overall motion of the fluid is built into the basic Biot theory and is controlled by the parameters (a, m and k) or any equivalent set that is devised to couple the motion of the fluid and solid components. The effects of viscosity in local fluid flow must be built into the complex moduli K_b and μ. The frequency dependent damping produced by local fluid motion becomes larger as the particle size and the relative distance between contiguous surfaces near contacts become smaller. Thus this mechanism of loss tends to become more important in fine-grained sediments such as silt and clay where the particle size is small and the overall mobility of the fluid is low.

In modeling the coarser granular materials, we have generally chosen the moduli K_b and μ to be complex constants with the ratio of the imaginary to real parts chosen on the basis of experimental measurements on dry specimens or water-saturated specimens tested quasistatically. Thus in this case, the damping built into K_b and μ is due to friction. The real parts of K_b and μ may be determined from the resonant frequencies in resonant column experiments or from measurements of wave velocity in the lab or field. In choosing values for K_b and μ it is important to remember that these must be determined for drained conditions and for very small amplitudes of motion (strains $< 10^{-6}$).

We will discuss the choice of complex moduli and other parameters in more detail however at this point it is instructive to see what Eqs. 16 and 17 predict when one or more of the parameters is varied in a systematic way. In the first group of figures we have assigned a single set of values to all parameters except a (the pore size parameter) and k (the coefficient of permeability). In Figs. 2 and 3, k and a have been varied systematically in such a way that k/a^2 = constant (in analogy with the predictions of the Kozeny-Carman equation[20]). Fig. 2 shows the logarithmic decrement predicted for shear waves and Fig. 3 the decrement predicted for p-waves of the first kind. The low frequency end of all of the curves in Fig. 2 approaches asymptotically to the value of δ_G that was input to the model. Thus as the frequency approaches zero, the log decrement is determined by the expression $\delta_G = \pi G_i/G_r$ which is a measure of frictional losses due to shearing motion. As the frequency is increased, the decrement increases to some peak value after which it decays in a manner very similar to a typical viscoelastic relaxation curve. As the permeability decreases the peaks shift to higher and higher frequencies. The curves for p-waves of the first kind shown in Fig. 3 are similar to those of Fig. 2 except for one major difference - the asymptotic value of the log decrement at the low frequency end of each curve is less than either of the values of log decrement input to the model by a factor of about 30! Thus by saturating the sediment with water, we end up with a material with much lower compressibility but with the same basic mechanism for dissipating energy by friction; the net result is a much lower overall logarithmic decrement at low frequencies for any motion involving volume change.

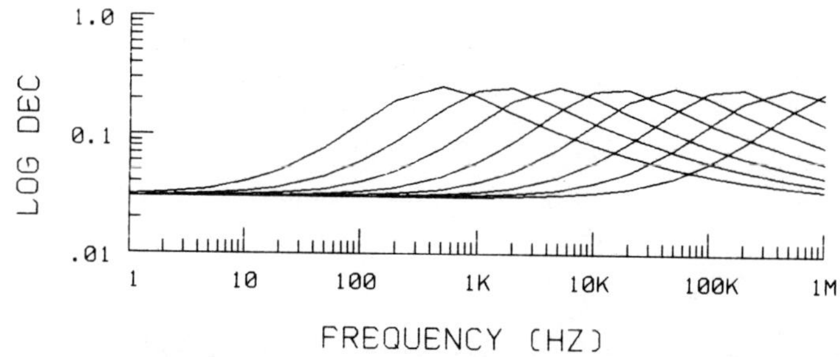

Fig. 2. Logarithmic decrement for shear waves in water-saturated sand. (k/a^2= constant - permeability decreasing from left to right).

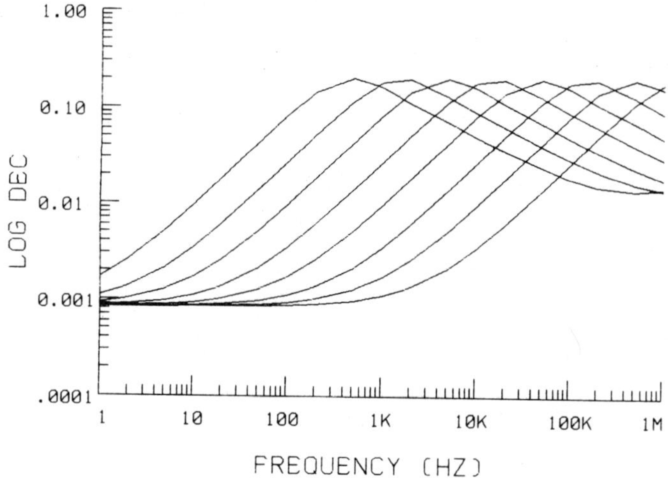

Fig. 3. Logarithmic decrement for p-waves in water-saturated sand (k/a^2=constant).

In order to calculate the attenuation exponent a, it is necessary to know both the logarithmic decrement and the phase velocity. Because of damping, there is a small amount of dispersion in p-waves of the first kind; the pattern of this dispersion is shown by the curves of Fig. 4 where it can be seen that the velocity increase occurs over a rather narrow band of frequency for each curve. As the fluid mobility decreases because of lower permeability, this band shifts to higher and higher frequencies in the same manner as the peaks of the damping curves in Fig. 3.

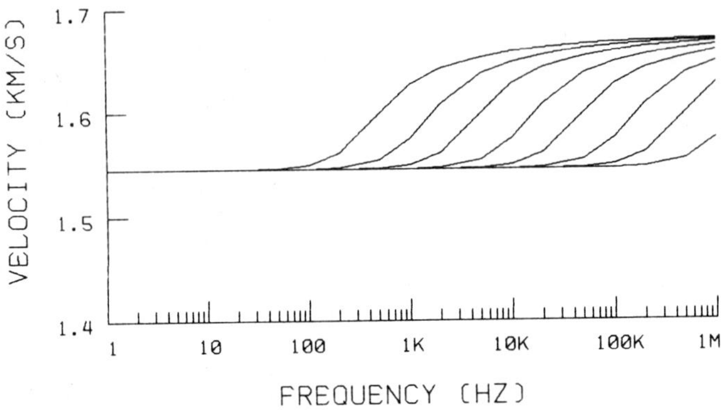

Fig. 4. Velocity of p-waves as a function of frequency.

Fig. 5. shows the attenuation of p-waves of the first kind that corresponds to the logarithmic decrements shown in Fig. 3 and the velocities shown in Fig. 4. The highly nonlinear form of this family of curves suggests that there is no simple power law of the form $\alpha = Kf^n$ (f is frequency; K and n are constants) that lends itself to a fit of the data over an extended range of frequencies. This kind of power law has often been fitted to experimental data obtained over a narrow range of frequencies and the values of K and n cataloged for future use. It can be seen from this figure that once the values of K and n are taken out of the context of a complete description of the original experiment, they may be very misleading.

The shape of the curves shown in Fig. 5 depends on both the pore size parameter a, and the permeability k. As an example of the effect of different combinations of these parameters, the value of k has been fixed and four different values of a were assumed in order to generate the curves shown in Fig. 6. From this figure it is clear that a and k are related in a rather subtle way so they cannot be chosen independently in an arbitrary manner if one hopes to obtain a good fit to a particular set of experimental data.

It must be emphasized that the curves given in Figs. 2 through 6 are parametric studies designed to highlight the effect of one or two of the 13 input parameters. In real sediments a number of the input parameters would be changing simultaneously so that not all of the curves shown in these figures represent realistic combinations corresponding to real sediments.

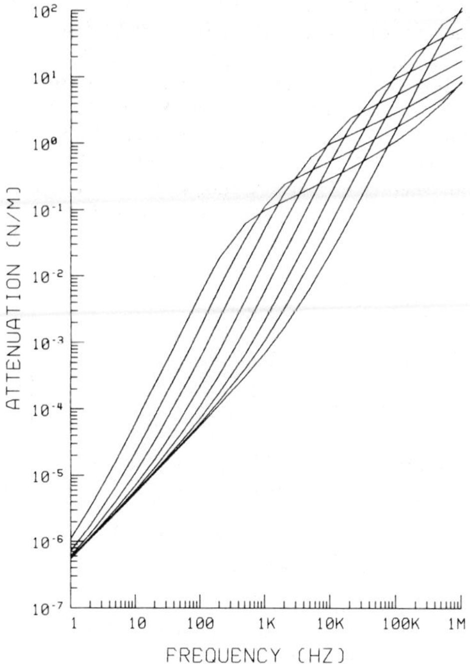

Fig. 5. Attenuation of p-waves corresponding to logarithmic decrements shown in Fig. 3 and velocities shown in Fig. 4.

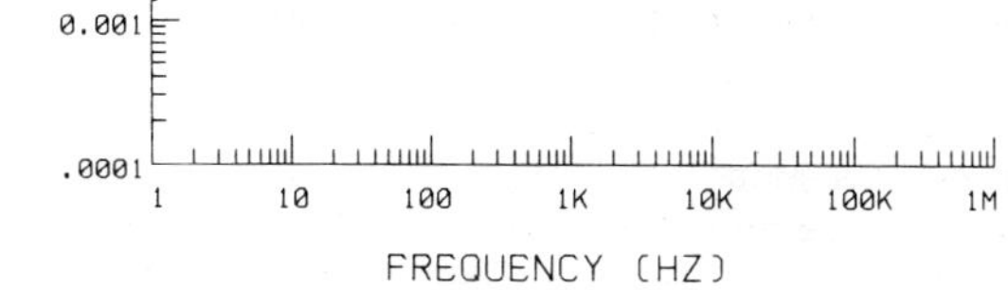

Fig. 6. Logarithmic decrement for p-waves - k fixed, a varying.

In order to predict the p-wave response in finer sediments where the permeability is very low, it is necessary to cast the moduli K_b and μ as frequency dependent couplex quantities either directly on the basis of experimental results or on the basis of some viscoelastic model. We have recently completed a set of torsional resonant column experiments on a micaceous silt material of fairly low permeability (approximately $2.5\text{x}10^{-10}$ cm^2) covering the frequency range from 1 Hz to about 1.5 kHz. The decrement measured in these experiments is shown in Fig. 7. To determine the logarithmic decrement at low frequencies, we measured the difference in phase between the driving torque and the resulting torsional motion using a very sensitive capacitive probe. At high frequencies the decrement was measured by measuring the decay of motion after the specimen was driven to one of several different resonant frequencies.[10,12]

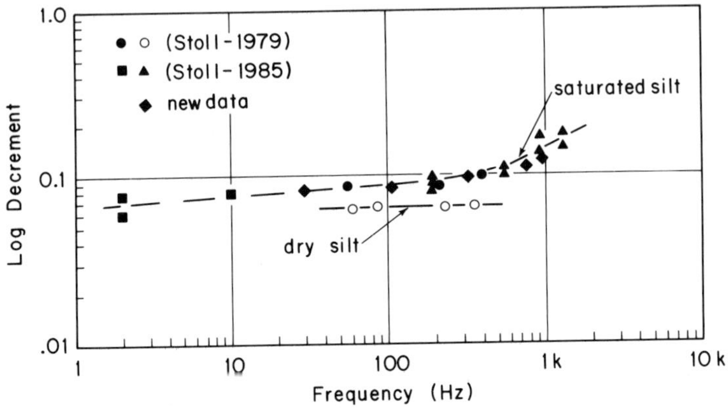

Fig. 7 Experimentally determined logarithmic decrement for silt.

If we take the low frequency asymptote of Fig. 7 to determine a value of δ_G representing frictional damping under quasistatic motion and then choose K_b and μ to be complex constants, as was done previously for the example involving sands, (Figs. 2 through 6) we obtain the curve labeled "constant complex moduli" shown in Fig. 8. Clearly the permeability is so low that the form of viscous damping included in the basic Biot theory only begins to show a marked effect at the extreme high frequency end of the curve. In order to incorporate the frequency dependence of damping that is obvious in Fig. 7 we now determine the imaginary parts of μ and K_b on the basis of the experimental data in this figure. Since we do not yet have a complete set of direct experimental values for the log decrement corresponding to drained, bulk motion, we have assumed that the log decrement for extensional motion, $\delta_E = \pi E_i/E_r$ (where $E^* = E_r + iE_r$ is the drained Young's modulus) undergoes a variation with frequency similar to that of δ_G and falls somewhere in the range

$$0.9 < \delta_E/\delta_G < 1.5 \qquad (20)$$

Eq. 20 is based on new experimental and theoretical work described in Ref. 12. When the variation of δ_G with frequency from Fig. 7 and the spread of δ_E/δ_G from Eq. 20 are used to pick a range of values to be introduced into the frequency equation (Eq. 15), we obtain the speckled band shown in Fig. 8. Thus we anticipate small but significant changes in the logarithmic decrement for p-waves to occur at much lower frequencies than predicted by the basic Biot theory.

In order to see how the response depicted by Figs. 5 and 8 compare with field measurements, a composite plot with data from a number of sources is shown in Fig. 9. Most of the older data in this figure are based on compilations of field data given by Hamilton.[21-23] Almost all

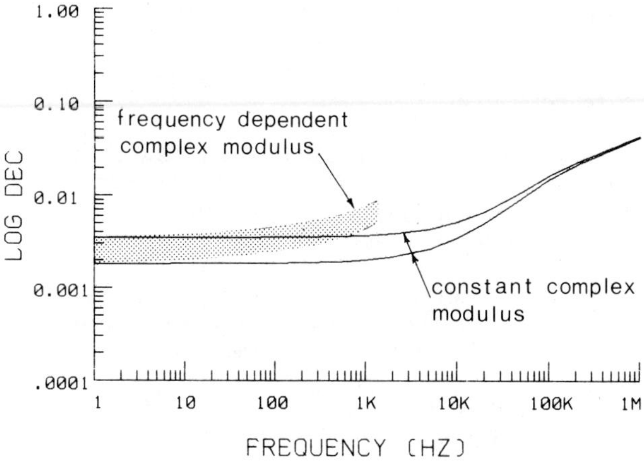

Fig. 8. Logarithmic decrement for p-waves in silt

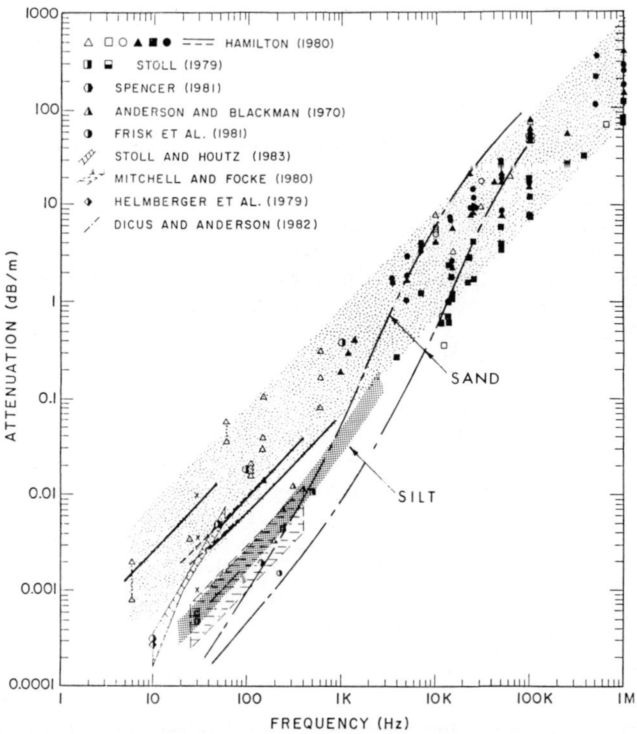

Fig. 9. Laboratory and field data for attenuation versus frequency with curves showing predictions of the Biot model superimposed.

of the data that he plotted falls within the lightly speckled band shown on the figure and on the basis of the general trend, he suggested that one may estimate propagation loss by assuming the attenuation varies as the first power of frequency. In recent years there have been a number of new studies of velocity and attenuation in the frequency range from 10 to 250 Hz[24-29] and many of these indicate an overall attenuation that is significantly lower than had previously been measured. These data have been included in Fig. 9 and the result is a spread of over two decades in the value of α in the low frequency range. Finally, if typical values of attenuation for sands and for finer sediments are calculated on the basis of the model and superimposed on the measured data, the figure is complete.

In Fig. 9, the theoretical curves labeled "SAND" start well below the speckled band, straddling the field data point from Frisk et al.[28] (from experiments over a sand botom) and end up near the upper part of the band in the high frequency range. A careful study of the data within the speckled band shows that most of the data points for sands (open or solid circles) are clustered near the upper edge of the band in the high frequency range. Thus the theoretical curves provide the correct link between high and low frequency measurements.

The heavily shaded band in Fig. 9 labeled "SILT", is based on the experimentally determined variation of logarithmic decrement shown in Fig. 8. If the data in the lightly speckled band are carefully examined, it can be seen that the plotting symbols for finer sediments (open and solid squares) tend to fall near the bottom of the band in the high frequency range. Again the theoretical curves predict the correct transition between the older high frequency data and the new data obtained at lower frequencies.

MODEL PREDICTIONS

In addition to the effects on attenuation and velocity, illustrated by Figs. 2 through 9, the Biot model predicts many important differences in the acoustic field near boundaries between water and sediment or between sediments of different impedance. For example, the reflection coefficients at a plane interface between two sediments become frequency dependent and each incident wave is partitioned into three refracted and three reflected waves. However, in a typical water saturated sediment, the amount of energy converted to p-wave of the second kind is very small.[9,14] Thus the bulk of the energy is refracted and reflected as p-waves of the first kind or shear waves.

Fig. 11 shows the effect of frequency on the reflection coefficient of a p-wave of the first kind at different angles of incidence for the case of an incident wave in water over clean sand. The reflection coefficient is the ratio of the amplitudes of the incident wave to the reflected p-wave (i.e. the amplitudes are those of the displacement potentials). The curve labeled "Elastic" corresponds to the conventional Rayleigh reflection coefficient for water over an elastic solid and the familiar critical angle occurs at about 73°. On the other hand the curves generated by the model suggest that the notion of a critical angle is not really valid in a lossy material. The frequency dependence that is obvious in the different reflection curves, shows that the sand boundary will act as a frequency selective filter over a certain range of grazing angles. Thus the predictions of the model are important in studies of bottom loss and the propagation of normal modes. The Biot model has recently been used to study certain aspects of both of these problems.[30-32]

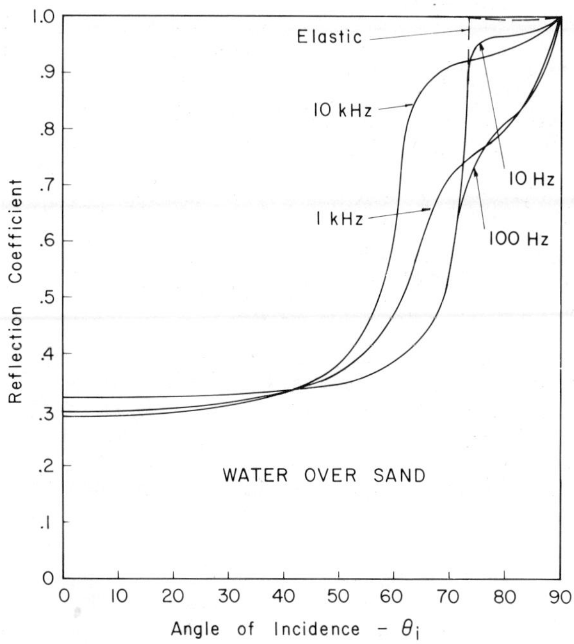

Fig. 10. Reflection coefficients for a wave incident in water over sand at four different frequencies. (Ref. 14)

While the amount of energy converted to p-waves of the second kind may be very small for a wave crossing a single interface, it becomes increasingly significant in the case where many layers are involved; moreover the amount of energy conversion increases as the pore fluid becomes more compressible. Thus in situations where the presence of dissolved or free gas justifies modelling the fluid component as being more compressible than pure water, one expects to see a more significant fraction of the total energy converted to this kind of wave and then rapidly dissipated. There are many geological strata that contain significant amounts of gas in one form or another so that the model may prove very useful in studying these cases.

Finally, in the thick layers of unconsolidated sediment that cover large areas of the ocean bottom, we are interested in studying the variation of velocity and attenuation with depth. In addition to the important effects of particle density and porosity, which depend on the geological history of a particular area, the velocity is strongly influenced by changes in the stiffness and compressibility of the skeletal frame as expressed by μ and K_b. Both of these quantities are inputs to the model and they both depend on the static effective stress (overburden stress) at any depth. Many experimental and theoretical studies have shown that both of these moduli tend to vary in proportion to the effective stress raised to a power in the range of 1/6 to 1/4, as long as the individual particles remain unbonded to one another. When diagenesis begins to cause cementation between particles, prediction of these moduli becomes more difficult. Several recent papers have addressed the problems of specifying K_b, and μ as functions of depth in real sediments however, much work still needs to be done in this area.[33,34]

SUMMARY AND CONCLUSIONS

The objective of this paper has been to present a summary and tutorial covering the development of a mathematical model for acoustic wave propagation in marine sediments based on the Biot theory. When frictional losses and various forms of viscous damping are included, the Biot model is able to match much of the observed response in real sediments. In the low frequency range it correctly predicts lower attenuations than had previously been predicted on the basis of linear extrapolation from higher frequency data. Moreover, it appears to provide the correct nonlinear link between high and low frequencies once the input parameters are properly evaluated.

While we have covered the basic theory in some detail, only a few of the most important practical applications have been mentioned. Because of limitations on space, only the most recent of the references describing these applications have been cited. Many other important contributions have appeared in the literature over the past 10 years and we apologize to our colleagues for not having the chance to discuss each one.

ACKNOWLEDGEMENTS

The author and his colleagues thank the Office of Naval Research, code 425, for continued support of the work described in this paper. Lamont-Doherty Geological Observatory Contribution No. 3828.

APPENDIX

Table 1. Parameters used to generate the theoretical curves shown in Fig. 2-8.

Input Parameters	Figs. 2-6	Fig. 8
Bulk Modulus of grains, K_r(Pa)	$3.6\ 10^{10}$	3.6×10^{10}
Bulk modulus of fluid, K_f(Pa)	2.0×10^{9}	2.0×10^{9}
Specific gravity of grains	2.65	2.65
Specific gravity of fluids	1.0	1.0
Shear modulus of frame, Reμ^*(Pa)	5.7×10^{7}	5.7×10^{7}
Bulk modulus of frame, Re K_b^*(pa)	7.6×10^{7}	5.2×10^{7}
Logarithmic decrement of frame (shear)	.03	.07 to .15
Ratio of extensional to shear decrement, δ_E/δ_G	1.0	.90 to 1.5
Porosity	0.4	0.5
Permeability, k(cm^2)	10^{-6} to 10^{-8}	2.5×10^{-10}

REFERENCES

1. M. A. Biot, General theory of three-dimensional consolidation, J. Applied Phys. 12:155 (1941).
2. M. A. Biot, Theory of elastic waves in a fluid-saturated porous solid. I. Low-frequency range, J. Acoust. Soc. Am. 28:168 (1956).
3. M. A. Biot, Theory of elastic waves in a fluid-saturated porous solid. II. Higher frequency range, J. Acoust. Soc. Am. 28:179 (1956).
4. M. A. Biot and D. G. Willis, The elastic coefficients of the theory of consolidation, J. Appl. Mech. 24:594 (1957).
5. M. A. Biot, Mechanics of deformation and acoustic propagation in porous media, J. Appl. Phys. 33:1482 (1962).
6. M. A. Biot, Generalized theory of acoustic propagation in porous dissipative media, J. Acoust. Soc. Am. 34:1254 (1962).
7. M. A. Biot, Nonlinear and semilinear rheology of porous solids, J. Geophys. Res. 78:4924 (1973).
8. R. D. Stoll, Acoustic waves in saturated sediment, in: "Physics of Sound in Marine Sediments," L. Hampton, ed., Plenum, New York (1974).
9. R. D. Stoll, Acoustic waves in ocean sediments, Geophysics 42:715 (1977).
10. R. D. Stoll, Experimental studies of attenuation in sediments, J. Acoust. Soc.Am. 66:1152 (1979).
11. R. D. Stoll, Theoretical aspects of sound transmission in sediments, J. Acoust. Soc. Am. 68:1341 (1980).
12. R. D. Stoll, Marine sediment acoustics, J. Acoust. Soc. Am. 77:1789 (1985).
13. R. D. Stoll and G. M. Bryan, Wave attenuation in saturated sediments, J. Acoust. Soc. Am. 47:1440 (1970).
14. R. D. Stoll and T. K. Kan, Reflection of acoustic waves at a water-sediment interface, J. Acoust. Soc. Am. 70:149 (1981).
15. A. Bedford, R. D. Costley and M. Stern, On the drag in virtual mass coefficients in Biot's equations, J. Acoust. Soc. Am. 76:1804 (1984).
16. T. J. Plona, Observation of a second bulk compressional wave in a porous medium at ultrasonic frequencies, Appl. Phys. Lett. 36:259 (1980).
17. D. L. Johnson and T. J. Plona, Acoustic slow waves and the consolidation transition, J. Acoust. Soc. Am. 72:556 (1982).
18. F. Gassmann, Uber die elastizitat poroser medien, Vierteljahrschr. Naturjfosch. Ges., Zurich 96:1 (1951).

19. A. B. Wood, "A Textbook of Sound," G. Bell and Sons, London (1911).

20. P. C. Carman, "Flow of Gases Through Porous Media," Academic Press, New York (1956).

21. E. L. Hamilton, Compressional wave attenuation in marine sediments, Geophysics 37:620 (1972).

22. E. L. Hamilton, Sound attenuation as a function of depth in the sea, J. Acoust. Soc. Am. 59:528 (1976).

23. E. L. Hamilton, Geoacoustic modeling of the sea floor, J. Acoust. Soc. Am. 68:1313 (1980).

24. R. S. Anderson and A. Blackman, Attenuation of low frequency sound waves in sediments, J. Acoust. Soc. Am. 49:786 (1971).

25. D. V. Helmberger, G. Engan and P. Scott, A note on velocity, density and attenuation models for marine sediments determined from multibounce phases, J. Geophys. Res. 84:667 (1979).

26. S. K. Mitchell and K. C. Focke, New measurements of compressional wave attenuation, J. Acoust. Soc. Am. 67:1582 (1980).

27. J. W. Spencer, Stress relaxations at low frequencies in fluid saturated rocks: Attenuaton and modulus dispersion, J. Geophys Res. 86:1803 (1981).

28. G. V. Frisk, J. A. Doutt and E. E. Hays, Bottom interaction of low frequency acoustic signals at small grazing angles in the deep ocean, J. Acoust. Soc. Am. 69:84 (1981).

29. R. L. Dicus and R. S. Anderson, Effective low frequency geoacoustic properties inferred from measurements in the northeast Atlantic, Naval Ocean Research and Development Activity Report No. 21 (1982).

30. T. Yamamoto, Acoustic propagation in the ocean with a poro-elastic bottom, J. Acoust. Soc. Am. 73:1587 (1983).

31. M. Badiey and T. Yamamoto, Propagation of acoustic normal modes in a homogeneous ocean overlaying layered anisotropic porous beds, J. Acoust. Soc. Am. 77:954 (1985).

32. M. Stern, A. Bedford and H. R. Millwater, Wave reflection from a sediment layer with depth-dependent properties, J. Acoust. Soc. Am. 77:1781 (1985).

33. P. J. Digby, The effective moduli of porous granular rocks, J. Appl. Mech. 48:803 (1981).

34. P. R. Ogushwitz, Applicability of the Biot theory, I. Low-porosity materials; II. Suspensions; III. Waves speeds versus depth in marine sediments, J. Acoust. Soc. Am. 77:429 (1985).

PORE SHAPE AND THE BIOT-STOLL MODEL FOR SATURATED SEDIMENTS

Keith Attenborough

Faculty of Technology
The Open University
Milton Keynes UK

ABSTRACT

The Biot-Stoll model for acoustic propagation in poroelastic saturated sediments includes the frequency-dependent effects of viscous drag within the pores by means of a viscosity correction function and a pore shape parameter. An alternative approach for the description of the acoustical characteristics of rigid-framed porous media allows for viscous action through a frequency-dependent effective density of air in the pores. This approach introduces a pore shape factor ratio as a parameter. The Biot-Stoll pore shape parameter may be re-interpreted in terms of the shape factor ratio. The two approaches should be reconciled if the solid displacement in the Biot-Stoll theory is set equal to zero. It is shown that reconciliation is possible only if the Biot-Stoll viscosity correction function is multiplied by the square of shape factor ratio. Support for this modification is provided by acoustical data for dry soils and sands. The consequent sensitivity of the modified Biot-Stoll theory to assumed pore shape in saturated sediments is explored.

INTRODUCTION

The Biot-Stoll model for wave propagation in saturated sediments is finding increasing application in studies of ocean bottom interaction. More recently it has been used to explain features of the interaction of airborne sound waves with the surface of porous soils[1-3]. In his original formulation[4], Biot suggested that extremes for the microvelocity field within pores, whose walls were able to move to and fro parallel to their axes, would correspond to those inside a circular capillary and a parallel-sided slit. This suggests a scaling factor, for the argument (λ) of the frequency-dependent viscosity correction function $F(\lambda)$, which should lie between 1 and 4/3 fpr pores of arbitrary shape.
The function $F(\lambda)$ is given by

$$F(\lambda) = -(1/4)\frac{\lambda\sqrt{i}\ T(\lambda\sqrt{i})}{1-[2/(\lambda\sqrt{i})]T(\lambda\sqrt{i})} \qquad (1)$$

$$\text{where } T(\lambda\sqrt{i}) = J_1(\lambda\sqrt{i})/J_0(\lambda\sqrt{i})$$

and the usual Kelvin Functions have been replaced by cylindrical Bessel functions of the first kind. It should be noted that the convention for time dependence understood here is $\exp(-i\omega t)$. Biot went on to combine this shape factor with tortuosity within the F function and referred to this combination as the structural factor. In his application of the Biot theory, Stoll[5], introduced a pore size parameter, a_s, which has the dimension of length and depends upon both the size and the shape of the pores, which may, in turn, be expected to be related to grain size, permeability and porosity. a_s and λ are related through

$$\lambda = a_s(\omega/\nu)^{\frac{1}{2}} \tag{2}$$

ω being angular frequency and ν being kinematic viscosity. a_s is assumed constant and, consequently, requires the frequency correction for viscous resistance to be the same for all pores in a given sediment. This assumption may break down in materials where pore size is distributed over a wide range causing a redistribution of viscous friction between the pores as the frequency is varied[6]. In the remainder of this paper, however, it will be assumed that the effects of pore size may be represented by a constant parameter.

The pore size parameter has often been considered as an empirical or adjustable parameter, as first suggested by Biot[4]. Stoll[5] has used values between one-sixth and one-seventh of grain size (diameter) in the absence of acoustic data. Hoven and Ingram have identified the pore size with twice the hydraulic radius (r_h)[7], where

$$r_h = \frac{\text{volume filled with fluid}}{\text{wetted surface}} = \frac{\Omega}{(1-\Omega)S_o} \tag{3}$$

Ω being the volume porosity of interconnected pores and S_o being the specific surface. This gives a direct method of calculating pore size parameters from grain diameter if the grains are assumed to be spherical. Alternatively, irregular shape and non-uniform size distribution of grains may be included by means of the Kozeny-Carman relationship[8]. Brunson[6] has investigated these three methods of calculating pore size parameter, (empirical, hyraulic radius assuming spheres, hyraulic radius with non-spherical particles), and has compared their predictions with laboratory data for water-saturated beads, sorted angular sand and unsieved angular sand. He has concluded that none of these methods was adequate for prediction of viscous attenuation loss in the absence of acoustical data.

In this paper an interpretation of the influence of pore shape is offered based upon analysis of sound propagation in a rigid-framed porous medium. It is demonstrated that this propagation is equivalent to that of the slow wave in the corresponding poroelastic medium. The associated theoretical development introduces a shape factor ratio. It is shown that for the Biot viscosity correction function (equation (1)) to be consistent with the governing equation of motion, in the limit of zero solid displacement, it is necessary for it to be defined differently. The use of the shape factor ratio in fitting data on attenuation and phase velocity in air-filled sands and soils is demonstrated. Some of these data and predictions have been published elsewhere[8,9]. Finally the resulting sensitivity of viscous attenuation in water-filled sediments to the shape factor ratio is explored.

PROPAGATION IN AN AIR-FILLED RIGID-FRAMED POROUS MEDIUM

The large density contrast between the imbedding fluid and mineral grains enables the assumption of a rigid frame to be made for frequencies above 100 Hz. Consequently intergranular friction may be ignored. It is usual to treat both viscous and thermal effects separately since the results are reasonable approximations to those of a combined analysis. The viscous effects may be incorporated into a complex or frequency-dependent density of air in the pores and the thermal effects incorporated into a complex bulk modulus. In the following outline only the results for viscous attenuation are presented. They are found to be the dominant effects anyway in soils and sands below 1000 Hz, and the complex bulk modulus may be replaced by the isothermal bulk modulus with sufficient accuracy. Moreover the thermal effects are relatively unimportant in water saturated sediments.

For pores of arbitrary shape the complex bulk density of air in the pores may be written[10]

$$\rho_b(\omega) = q^2\rho_f[1-2(\lambda\sqrt{i})^{-1}\,T(\lambda\sqrt{i})]^{-1} \qquad (4)$$

where $\lambda = (1/n)\ (8\rho_f q^2 s\,/\,\Omega\sigma)^{\frac{1}{2}}$ (5)

ρ_f = equilibrium density of air

$q^2 = \Omega^{-n'}$

n' is the grain shape factor

σ is the flow resistivity in MKS rayls per metre = μ/K where K is the permeability in m^2 and μ is the dynamic viscosity

s is the steady flow pore shape factor, such that $2q^2s = K_0$, the Kozeny constant,

and n is the dynamic pore shape factor.

The expected range of n is $0.5 \leqslant n \leqslant 1.0$ corresponding to extreme pore shapes of parallel-sided slit and circular capillary respectively. The corresponding extreme values of s are 1.5 and 1.0 respectively, leading to the purely phenomenological hypothesis that

$$n = 2 - s. \qquad (6)$$

From (2), (5) and (6) it may be deduced that

$$\lambda = (\sqrt{s}/n)\ (8q^2s\mu/\sigma)^{\frac{1}{2}}(\omega/\nu)^{\frac{1}{2}} \qquad (7)$$

Since the shape factors always appear in the ratio $n/\sqrt{s}$, this is replaced by a single parameter S_f called the <u>shape factor ratio</u>. Consequently it is possible to write

$$a_s = (2/S_f)\ (2\mu q^2 s/\Omega\emptyset)^{\frac{1}{2}} = (2/S_f)\,(KK_0/\Omega)^{\frac{1}{2}} \qquad (8)$$

The differential equation governing propagation in the rigid-framed medium may be written

$$\nabla^2\emptyset_f = \Omega\rho_b(\omega)\,C(\omega)\ \frac{\partial^2\emptyset_f}{\partial t^2} \qquad (9)$$

where $C(\omega)$ represents the frequency-dependent compressibility of air in the pores and $\emptyset_f$ represents the velocity potential within the constrained fluid (air). With the notation used in one of Biot's later papers[11], the equations governing propagation in the corresponding poroelastic medium may be written

$$\nabla^2(H\emptyset_s + \alpha M\emptyset_f) = \frac{\partial^2}{\partial t^2}(\rho\emptyset_s - \rho_f\emptyset_f) \tag{10}$$

and $$\nabla^2(\alpha M\emptyset_s + M\emptyset_f) = \frac{\partial^2}{\partial t^2}(\rho_s\emptyset_s + \rho_f{}^1\emptyset_f) \tag{11}$$

where $$\rho_f{}^1 = q^2\rho_f/\Omega + i\eta F(\lambda)/\omega K \tag{12}$$

and where H and M are elastic constants having the standard definitions[5] K is the permeability (= η/σ), η is the dynamic coefficient of viscosity and α may be shown to be given by

$$\alpha = 1 - \frac{K_b}{K_r} \tag{13}$$

where K_b is the unjacketed bulk modulus of the poroelastic frame and K_r is the bulk modulus of the constituent grains. $\emptyset_s$ and $\emptyset_f$ are scalar potentials related to solid and fluid displacement respectively, and ρ is the average density of the porous medium.
When $\emptyset_s = 0$, the rigid-frame limit, then equation (11) becomes

$$\nabla^2\emptyset_f - \frac{\rho^1{}_f}{M}\frac{\partial^2}{\partial t^2}\emptyset_f \tag{14}$$

Comparing equations (9) and (14) it is possible to deduce that

$$M = [\Omega\, C(\omega)]^{\frac{1}{2}} \tag{15}$$

and that

$$F(\lambda) = \frac{-S_f{}^2}{4}\;\frac{\lambda\sqrt{i}\;T(\lambda\sqrt{i})}{1-[2/(\lambda\sqrt{i})]\;T(\lambda\sqrt{i})} \tag{16}$$

where use has been made of equations (4) and (12).
Equation (16) for $F(\lambda)$ differs from that deduced orginally by Biot (equation (1)) by a factor of $S_f{}^2$. This factor is required to ensure consistency between the poroelastic theory and that for a rigid-framed porous medium. In addition, the expression for a_s derived as equation (8) differs from that obtained by assuming that the pore size parameter is twice the hydraulic radius[7]. The difference is a factor of S_f.

Table 1 shows propagation constants calculated according (i) to rigid-framed porous theory and (ii) to poroelastic theory, for the slow wave, incorporating the modifications described hitherto and using the properties listed in Table II which correspond to the physical properties of a dry sandy soil[1]. It is clear that the two sets of predictions are practically identical. Consequently it may be deduced (a) that measurements of sound pressure within a dry granular medium due to an external source will correspond to the induced slow wave and (b) that propagation of this induced wave may be approximated, reasonably, by the only wave that would propagate in a rigid-frame medium. The large density contrast between the solid and constrained fluid means that the wave propagation is largely independent of the frame moduli which may be inferred reasonably from measurements of p- and s-wave speeds.

TABLE I Predicted propagation contents in sandy soil

Model	Rigid-framed porous		Poroelastic (slow wave)	
Freq/Hz	Phase Constant rad/m	Attenuation constant m^{-1}	Phase Constant rad/m	Attenuation m^{-1}
100	13.01	12.61	12.89	12.49
200	18.7	17.56	18.53	17.38
300	23.26	21.18	23.06	20.96
400	27.29	24.09	27.06	23.83
500	31.0	26.53	30.74	26.24
600	34.5	28.64	34.22	28.32
700	37.85	30.49	37.56	30.15
800	41.09	32.14	40.79	31.77
900	44.26	33.62	43.95	33.23
1000	47.38	34.96	47.05	34.55
2000	77.21	43.91	76.84	43.38

COMPARISON BETWEEN THEORY AND EXPERIMENT FOR SLOW WAVE ATTENUATION IN DRY GRANULAR MEDIA

Ferraro and Sacerdote[12] deduced the phase and attenuation constant of the slow wave as a function of frequency in a dry medium consisting of spherical lead shot by reflection measurements. Figure 1 shows a comparison between their data and predictions obtained from equation (9), with $C(\omega)$ equal to the inverse isothermal bulk modulus using the measured porosity and flow resistivity and using S_f as an adjustable parameter. For extreme pore shapes of circular capillary and parallel-sided slit, the value of S_f ranges between 1.0 and 0.41 respectively. The minimum value is found to enable a tolerable fit to the data for stacked spheres. Although it should be noted in this and subsequent examples that no attempt has been made to discover the optimum value of S_f for each data set. Figure 2 shows data and theoretical predictions for various depths in a dry washed sand of flow resistivity 210 000 MKS units and porosity, 0.4. In this case an intermediate value of S_f (0.75) is found to give reasonable predictions[9]. The data shown for this example were obtained using a specially constructed probe microphone[2]. The final example (Figure 3) is for another sand. Several data sets are shown corresponding to several probe microphone positions at various depths. Here the mean flow resistivity was measured to be 450 000 MKS units. A value of $S_f = 1$ is found to give good matching between data and predictions. Other measurements have been made in less homogeneous sands and soils where conventional invasive techniques of measuring flow resistivity and porosity do not give values that are related to acoustic measurements in a straightforward manner[8,13].

TABLE II Measured Physical properties of sandy soil

Property	Value
Upper layer depth	2 m
P-wave speed in upper layer	270 ms^{-1}
S-wave speed in upper layer	190 ms^{-1}
P-wave speed in lower medium	530 ms^{-1}
S-wave speed in lower medium	325 ms^{-1}
Flow resistivity near surface of upper layer	366000 ± 108 Nsm^{-4}
Air volume % in upper layer	26.9
Water volume % in upper layer	9.3
Solid matter volume % in upper layer	63.8
Bulk density in upper layer	1700.3 kg/m^3

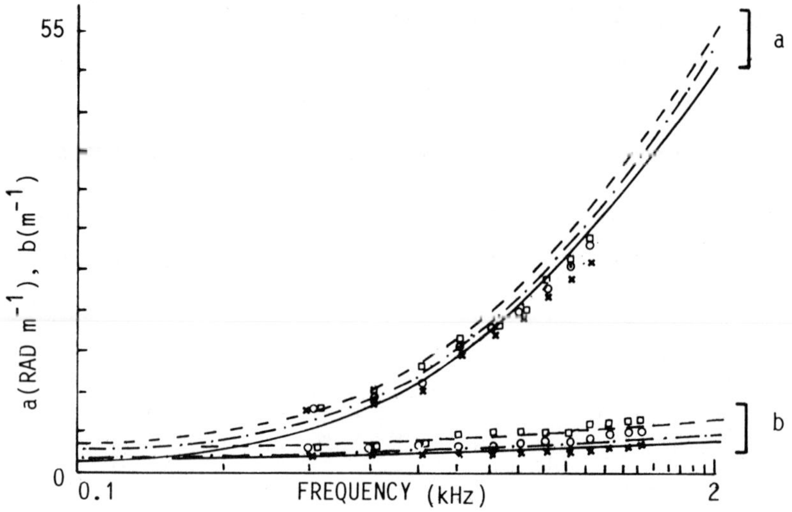

Fig 1 Slow wave propagation constant in spherical lead shot

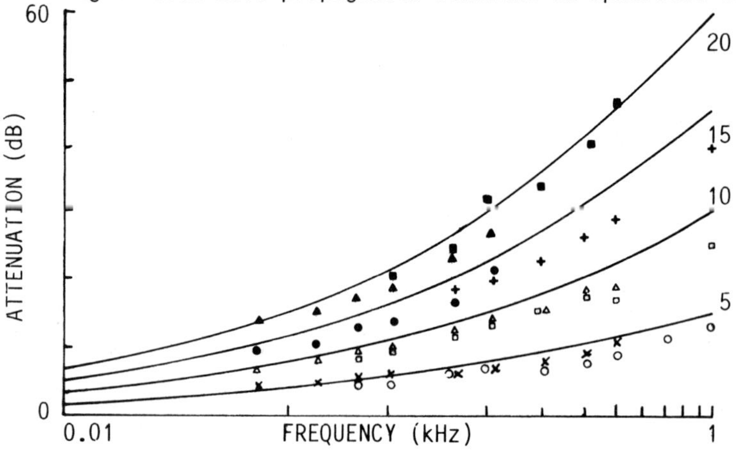

Fig 2 Attenuation in washed sand: σ = 210000(MKS); depths in cm

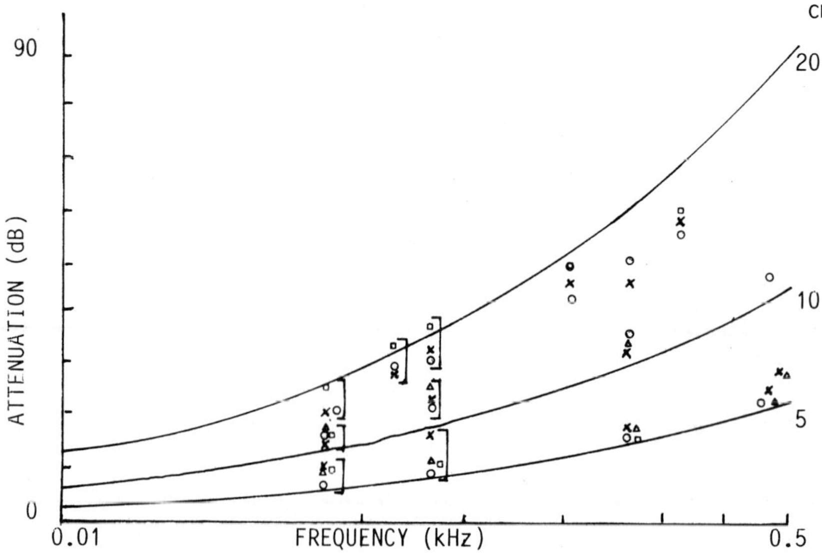

Fig 3 Attenuation in sand: σ = 150000(MKS); depths in cm

TABLE III Parameter values for water-saturated sand

Parameter	Value (cgs units)
Volume porosity	0.4
Permeability	1×10^{-7}
Grain shape factor	1.25
Frame rigidity modulus	8.27×10^{9}
Frame bulk modulus	1.38×10^{10}
Grain bulk modulus	3.6×10^{11}
Grain density	2.65
Fluid bulk modulus	2.2×10^{10}
Fluid density	1.0

PREDICTED INFLUENCE OF SHAPE FACTOR RATIO ON PROPAGATION IN FULLY-SATURATED SEDIMENTS

To study the predicted sensitivity of fast and slow wave characteristics in water-saturated sand to variation in the shape factor ratio the base values listed in Table III have been chosen. Results are shown in Figures 4 to 7 where the slotted lines correspond to $S_f = 0.41$ and the continuous lines to $S_f = 1.0$.

CONCLUSIONS

The shape factor ratio used in theories of propagation in rigid-frame media is equivalent to the pore-size parameter used in the Biot-Stoll model for propagation in a fluid saturated poroelastic material. Measurements of reflection and transmission of sound at the surfaces of spherical lead shot and of various dry sands where the transmission of sound is measured by a purpose-built probe microphone clearly show

(i) the dominance of the slow wave in air-saturated particulate materials

and

(ii) the importance of the shape factor ratio in distinguishing between the acoustic behaviour of of the different media.

When modified to agree with theory for a rigid-frame porous material in the limit of zero solid displacement, the Biot-Stoll model predicts slow-wave characteristics for a water-saturated sand that also are highly-sensitive to values of shape factor ratio. This influence may be significant in calculating penetration loss in layered sediments. Furthermore the frequency range for which the fast-wave shows dispersion is affected considerably by the value of the shape factor ratio. When comparing laboratory measurements of shear wave attenuations in water-filled sands and spherical glass beads with values predicted by the Biot-Stoll model, Brunson[6] computes pore size parameter both using the Kozeny-Carman (hydraulic radius) formulation,

$$a_s = 2(KK_0/\Omega)^{\frac{1}{2}} \tag{17}$$

and by adjusting its value to give best fit to the measured data. He finds that the best-fit or optimum values are larger than the values computed from equation (17) for all three of the water-filled samples tested. These results are summarised in Table IV.

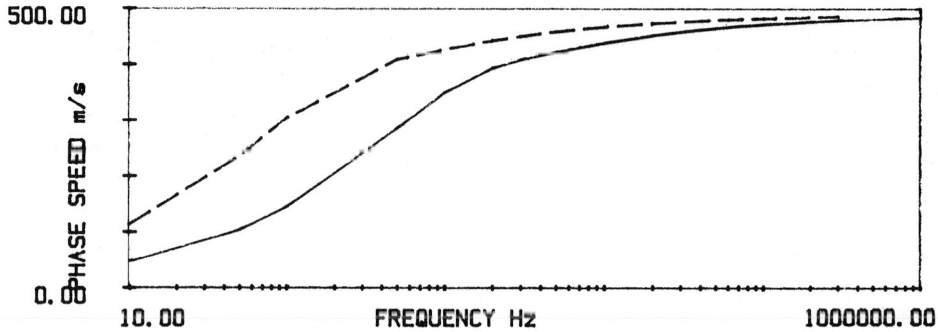

FIG.4 INFLUENCE OF S_f ON SLOW WAVE SPEED

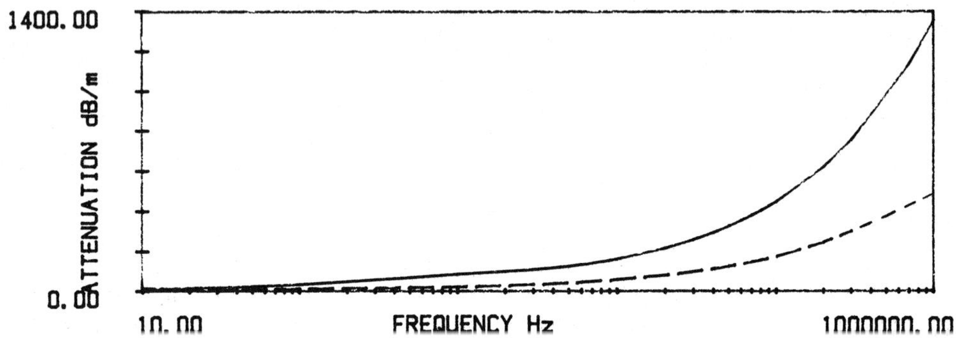

FIG.5 INFLUENCE OF S_f ON SLOW WAVE ATTENUATION

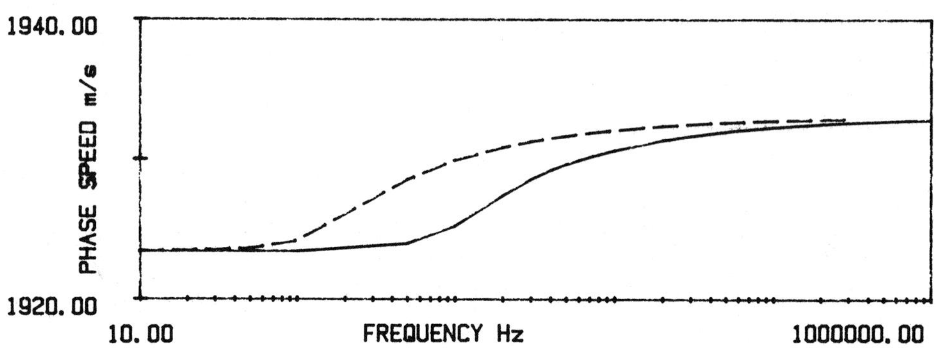

FIG.6 INFLUENCE OF S_f ON FAST WAVE SPEED

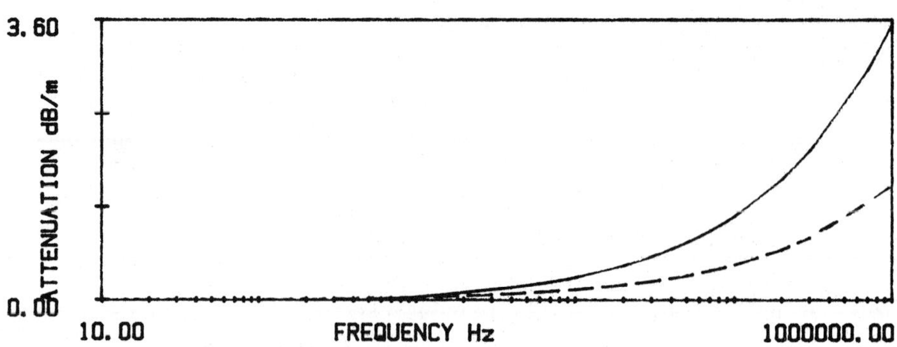

FIG.7 INFLUENCE OF S_f ON FAST WAVE ATTENUATION

TABLE IV Values of pore-size parameter obtained by Brunson[6]

Material	a_s from equation (17) (cm)	Optimum a_s (cm)	ratio
Spherical glass beads (40/45) mean diameter 0.385 mm	0.0071	0.03	0.237
Angular sand (40/45)	0.0066	0.012	0.55
Angular sand (unsieved)	0.0059	0.01	0.59

The difference between equations (8) and (17) suggests that the ratio of the value of a_s predicted by equation (17) to the best-fit value should correspond to S_f. Although the values of this ratio are not those expected from measurements on air-filled materials (0.41 for spheres and higher values up to 1.0 for sands) they do show the expected trend ie the smallest value for spherical beads and the greatest value for the unsorted sand containing the greatest proportion of non-spherical grains. It should be noted that the values for S_f obtained from Brunson's data are subject to uncertainties in the value of K_0, assumed equal to 5.0, for each medium.

Although analysis has not yet been advanced sufficiently to enable calculation of shape factor ratio from knowledge of grain shape and size distribution, it seems that much empirical evidence of appropriate values may be gained from studies of slow wave propagation in air-saturated materials where effectively the solid frame may be regarded as rigid. Indeed, recent progress in using finite element techniques to deduce the influence of pore shape on propagation in rigid (air-filled) porous media[14] may prove useful for acoustic modelling of water-filled sediments.

ACKNOWLEDGEMENTS

This work was supported in part by the US Army through the European Research Office and was carried out, in part, while the author was on study leave at the USAE Waterways Experiment Station, Vicksburg, Mississippi. He would like to thank Mr J. Lundien, Chief, Battlefield Environment Group, and colleagues Drs. Sabatier, Bass and Bolen (University of Mississippi, Physical Acoustics Research Laboratory) for their contributions to a joint effort on acoustic-to-seismic coupling.

REFERENCES

1. J. M. Sabatier, H.E. Bass, L.N. Bolen, K. Attenborough and V.V.S.S. Sastry, The acoustic surface impedance of layered poroelastic soil. J Acoust Soc Amer (submitted for publication 1985).
2. K. Attenborough, J. M. Sabatier, H.E. Bass and L. N. Bolen, The acoustic transfer function at the surface of layered poroelastic soil. J Acoust Soc Amer (submitted for publication 1985).
3. J. M. Sabatier, H. E. Bass, L. N. Bolen and K. Attenborough, Acoustically-induced seismic waves, J Acoust Soc Amer, (submitted for publication 1985).
4. M. A. Biot, Theory of propagation of elastic waves in a fluid-saturated porous solid II. Higher frequency range J Acoust Soc Amer 28 179-191 (1956).

5. R. D. Stoll, Acoustic waves in saturated sediments in Physics of Sound in Marine Sediments edited by L. Hampton (Plenum, New York 1974).
6. B. A. Brunson, Shear wave attenuation in unconsolidated laboratory sediments, NORDA Technical Note 159 (1983).
7. J. M. Hovem and G. D. Ingram, Viscous attenuation of sound in saturated sand, J Acoust Soc Am 66 1807-1812 (1979).
8. P. C. Carman, Flow of Gases through Porous Media Academic Press, New York, 1956.
9. K. Attenborough, H. E. Bass, and L. N. Bolen, Sound transmission through plane porous ground surfaces Acoustics Letters 6 87-100 (1982).
10. K. Attenborough, Acoustical characteristics of rigid fibrous absorbents and granular materials. J Acoust Soc Am 73 3 785-799 (1983).
11. M. A. Biot, Mechanics of deformation and acoustic propagation in porous media, J Appl Phys, 33 (4) (1961).
12. M. A. Ferraro and G. G. Sacerdote, Parameters of sound propagation in granular absorbent materials, Acoustic 1, 137-142 (1951).
13. K. Attenborough and H. M. Hess, Acoustical surveying of porous soils, Proceedings of the 14th Symposium on Acoustical Imaging, The Hague (1985).
14. A. Craggs and J. G. Hildebrandt, Effective densities and resistivities for acoustic propagation in narrow tubes. J Sound Vib 92 321-331 (1984).

LOW-FREQUENCY ANOMALIES IN THE REFLECTION BEHAVIOUR OF MARINE SEDIMENTS

Markus von Haumeder

SACLANT ASW Research Centre
Viale San Bartolomeo 400
I-19026 La Spezia, Italy

ABSTRACT

It is frequently found that the reflectivity of the sea bottom deviates considerably from the values that are expected by considering its impedance. A type of sea bottom in which anomalies in the reflectivity are particularly strong is a coarse marine sediment consisting of two components: particles that form a stiff frame and a liquid that fills the interstices. Such sediments can be considered porous materials. A theory that analyses the response of a porous material to an elastic wave is the Biot theory. It treats the two components as two interpenetrating elastic continua and predicts some remarkable anomalies in the acoustic behaviour of porous materials. One important prediction of the Biot theory is that acoustic energy of a certain frequency band can penetrate the sediment even in the range of total reflection. The porous material behaves here like a frequency filter with respect to broadband acoustic energy. This energy transfer into the porous material is associated with a relaxation process. We give here a derivation of the pertinent relaxation times and demonstrate how the relaxation time can be evaluated from the reflectivity anomaly. The relaxation time is expressed in terms of parameters descriptive of the internal flow. Use of this interrelation permits us to estimate the flow-permeability of a sediment on the basis of a bottom reflection loss measurement. The method is applied to a real marine sediment.

INTRODUCTION

A large fraction of the naturally occurring marine sediments can be considered porous materials. Being immersed in a fluid, such a porous material consists of two clearly separated components: a porous skeleton or frame and a fluid that fills the pore space. Each of the two components forms a large, interconnected cluster and the two clusters fit exactly into each other. This situation can be viewed as two interpenetrating elastic continua. A pressure wave impinging on such a composite material results not only in a deformation of the single components but also in a relative motion of the two continua with respect to each other. The fluid is free to move through the frame from areas of high local pressure to areas of low local pressure.

The theory that describes the propagation of elastic waves through a porous, fluid-filled material was developed by Maurice Biot [1-3]. The Biot theory is a microscopic theory that takes the particular geometry of the porous skeleton into consideration. It introduces two particle-displacement vectors, one for the solid frame and one for the pore fluid, and expresses the internal flow in terms of a difference vector. The composite nature of the porous material is reflected in the theory by a pair of wave equations that are coupled by a flow term. Thus, the Biot theory is capable of describing the underlying microscopic processes in some detail.

According to the Biot theory, porous materials exhibit some anomalies in their response to a pressure wave. One of these anomalies pertains to the reflection process at the interface between a liquid and a porous material. In an angle-of-incidence range close to grazing angle where total reflection is expected for a solid of equivalent acoustic impedance, a fluid-filled porous material absorbs energy of a certain frequency band. The porous material acts here as a band-pass filter with respect to broad-band acoustic energy, and the fraction of energy that is transferred into the interior of the porous material is used to drive the internal flow. For many frequencies this internal flow will follow the excitation by the pressure wave with a delay. Such a delay results in energy dissipation due to relaxation. Relaxation can be described best in terms of a characteristic response time, the so-called relaxation time, τ.

This paper gives a derivation of the relaxation times for a porous, fluid-filled material on the basis of the Biot theory. The relaxation times are expressed entirely in terms of flow parameters, which gives evidence that the underlying microscopic process is a flow process.

Further we present a measurement of the bottom-reflection loss that shows the reflectivity anomaly and permits the relaxation time to be determined experimentally. This measurement is used in combination with the given formula to calculate the permeability/porosity ratio of the sea bottom. The procedure used can be considered a new technique for remote bottom-permeability determination.

1 THEORY

The Biot theory treats an isotropic, homogeneous, porous material and its pore-filling fluid as two interpenetrating elastic continua [1-5]. Each of the two continua shows its own characteristic response to a pressure wave, and the composite of the two materials requires the use of two particle displacement vectors, one for the skeletal frame, u, and one for the pore fluid, U. The difference between the two displacements is a measure of the internal fluid flow, which is expressed in terms of the vector quantity, w:

$$w = \beta \left[u - U \right] . \tag{1}$$

The quantity w represents the average fluid displacement relative to the frame and β is the porosity defined as the fraction of volume occupied by the pore fluid. By expressing the vectors u and w in terms of scalar potentials, ϕ, and vector potentials, Ψ,

$$u = \nabla \Phi_s + \mathrm{curl}\, \Psi_s \quad , \quad w = \nabla \Phi_f + \mathrm{curl}\, \Psi_f \tag{2}$$

the Biot equations of motion [3] can be seperated [5] into a pair of equations involving only dilatational modes and another pair of equations for the rotational mode. The subscript s refers to 'solid' and the subscript f to 'fluid'. We start our consideration with the simpler set of wave equations for the shear-wave:

$$\mu \nabla^2 \Psi_s = \rho \ddot{\Psi}_s - \rho_f \ddot{\Psi}_f \quad , \quad \eta/k \ \dot{\Psi}_f = \rho_f \ddot{\Psi}_s - \rho_f/\beta \ \ddot{\Psi}_f \tag{3}$$

where μ is the shear modulus, ρ and ρ_f are the densities of the fluid and of the fluid-filled porous material, respectively, η stands for the fluid viscosity, and k designates the flow permeability. The composite nature of the porous, fluid-filled material is reflected in the theoretical description by a set of two coupled wave equations. The term on the left-hand side of the second equation is a flow term that has to be compared to the lag term in relaxation theory. This term contains the particular physics of porous materials. Its prefactor η/k is the viscous resistance to fluid flow. For high frequencies the fluid viscosity, η , is not constant and the viscous resistance to fluid flow must be made frequency dependent. Considering the friction forces within pore channels of simplified geometry, Biot derived [1] a complex correction factor, $F(\kappa) = F_r(\kappa) + iF_i(\kappa)$, for oscillatory motion. The factor η/k may be replaced by $F\eta/k$ in the frequency domain, where

$$F(\kappa) = \frac{1}{4}\,\frac{\kappa\, T\{\kappa\}}{1 - T\{\kappa\}/i\kappa} \quad , \quad T\{\kappa\} = \frac{ber'[\kappa] + i\, bei'[\kappa]}{ber[\kappa] + i\, bei[\kappa]} \tag{4}$$

The functions ber(κ) and bei(κ) are the real and imaginary parts of the Kelvin function with ber'(κ), bei'(κ) being their derivatives. The argument, κ , of the correction factor, F,

$$\kappa = a\sqrt{\rho_f\, \omega/\eta} \tag{5}$$

is non dimensional and depends on the pore-size parameter a, the fluid density ρ_f , the viscosity, η , and the angular frequency, ω. Since we are going to make use of this correction factor $F(\kappa)$ in a quantitative way, we show the variation of its real and imaginary part with frequency in Fig. 1.

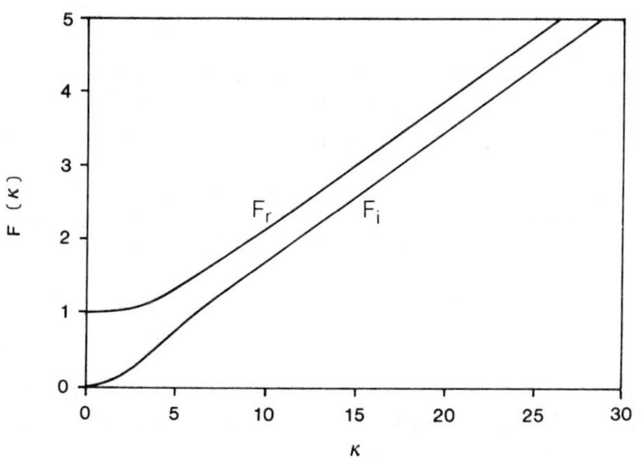

Fig. 1
Real part (F_r) and imaginary part (F_i) of the frequency correction function for three-dimensional, oscillatory flow.

For very low frequencies $F(\kappa)$ approaches unity, resulting in the same equation as when Poiseuille flow is assumed. For high frequencies, and hence large values of κ , the curves become straight lines with a slope of $1/4\sqrt{2}$= 0.177. The derivation of the complex correction factor is valid only for frequencies where the wave length is large compared with the pore size. For sands, this puts the upper limit on frequencies at about 10^5 to 10^6 Hz.

Incorporating the complex correction factor into Eq. 3 by replacing η by $\eta\, F(\kappa)$, we derive now an explicit relationship between attenuation and frequency.

By Fourier transforming Eq. 3 into the frequency domain and assuming plane-wave type potentials, the two equations can be combined to give one homogeneous wave equation of the form:

$$\nabla^2 \hat{\Psi} + \frac{\omega^2}{V^2}\hat{\Psi} - \frac{\rho_f^2\,\omega^4/\mu}{\rho_f\,\omega^2/\beta - i\omega\eta F/k}\,\hat{\Psi} = 0 \;. \tag{6}$$

which permits solutions of the type

$$\hat{\Psi}[x,\omega] = A \exp i[\omega t + kx] + B \exp i[\omega t - kx] \;. \tag{7}$$

Equation 6 can be cast into the form of a lossy Helmholtz equation of the general type $(\Delta^2 + \vec{k}^2) = 0$, with $\vec{k}^2 = (k - i\alpha)^2$. In this form it appears as

$$\left[\nabla^2 + \left\{\frac{\rho\,\omega^2}{\mu} - \frac{\rho_f^2\omega^4/\mu}{\rho_f\omega^2/\beta - i\omega\eta F/k}\right\}\right]\hat{\Psi} = 0 \;. \tag{8}$$

The complex k-vector, $\vec{k}$, is to be identified with the expression in the brackets. Separating the real and imaginary parts of the wave vector renders two equations from which we derive the attenuation coefficient:

$$\alpha = \frac{\omega}{V\sqrt{2}}\sqrt{-1 + \frac{1}{1+[\omega\tau + F_i/F_r]^2}\left\{\left[K\omega^2\tau^2 + F_i/F_r\right] \mp \sqrt{\left[1 + \left\{\omega\tau + F_i/F_r\right\}^2\right]\left[1 + \left\{\left[\omega\tau + F_i/F_r\right] - K\omega\tau\right\}^2\right]}\right\}} \tag{9}$$

$$K = \frac{\beta\,\rho_f}{\rho} \qquad \tau = \frac{\rho}{\beta}\,\frac{k}{\eta F_r} \tag{10}$$

in which τ has the dimension of a time and can be identified as a relaxation time. This typical response time of the porous material is entirely expressed in terms of parameters that are descriptive of the internal flow of the pore fluid through the frame, i.e. permeability, k, fluid density, viscosity, and the real part of the frequency correction factor, F_r . K is a dimensionless factor that is determined by the density ratio of the fluid and the porous material ($\rho = \beta\rho_f + (1-\beta)\rho_r$). This factor introduces an altered frequency into the attenuation formula, which largely affects the reflection coefficient.

If we now assume the resistance to fluid flow to be frequency independent by setting the correction factor, $F(\kappa)=1$, we obtain for the attenuation the simpler formula:

$$\alpha = \frac{\omega}{V\sqrt{2}}\sqrt{-1 + \frac{1}{1+\omega^2\tau^2}\left\{K\omega^2\tau^2 \mp \sqrt{\left[1+\omega^2\tau^2\right]\left[1 + \omega^2\tau^2\left\{1-K\right\}^2\right]}\right\}} \;. \tag{11}$$

Equation 9, which is valid for all frequencies, takes the form of Eq. 11 when $f \to 0$, $F_i/F_r \to 0$. For the case of a pure liquid ($\beta = 1$), K equals one and Eq. 11 takes the form of the well-known formula for the relaxation attenuation in a viscous fluid:

$$\alpha = \frac{\omega}{V\sqrt{2}}\sqrt{-1 + \frac{1}{1+\omega^2\tau^2}\left[\omega^2\tau^2 \pm \sqrt{1+\omega^2\tau^2}\right]} \;. \tag{12}$$

This shows that the derived formula is capable of producing the proper results in the limit $\beta = 1$. We now want to test in how far Eq. 11 reproduces the overall variation of the attenuation as obtained by the Biot computations.

For this purpose the attenuation according to Eq. 11 is plotted versus the frequency for a porosity of 0.5 (curve 'b' of Fig. 2). In this case of a frequency-independent flow resistance (Poiseuille flow) the attenuation tends towards a constant value at higher frequencies. Curve 'a', however, represents a plot of Eq. 9 in which the flow resistance is made frequency dependent. The important effect of a frequency-dependent flow resistance is apparent, mainly at high frequencies. These curves are in very good agreement with corresponding curves shown by Stoll [5,6], which were obtained as a result of a Biot model computation.

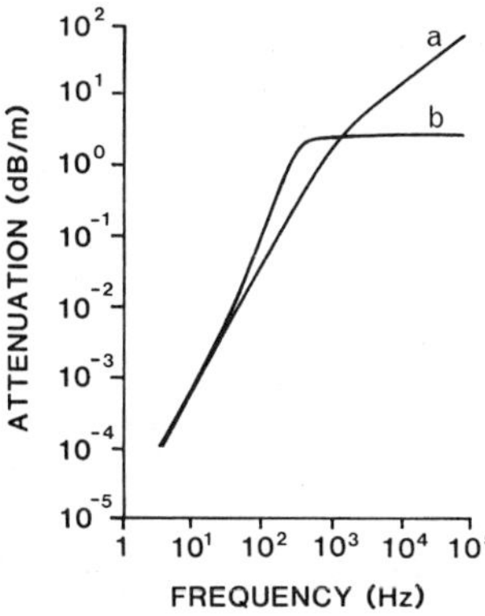

The agreement gives evidence that the derived formula reflects the attenuation in porous materials as considered by the Biot theory. In the following chapters we make use of the bottom-reflection loss anomaly in order to determine τ experimentally.

Fig. 2
Frequency dependence of the attenuation for a frequency dependent flow resistance (a) and a constant flow resistance (b).

2 BIOT PREDICTIONS OF BOTTOM-REFLECTION LOSS

The frequency dependence of the bottom-reflection loss exhibits one of the most significant anomalies in porous materials. In an angle-of-incidence range where total reflection would normally be expected, a liquid/porous-composite interface acts as a band-pass filter with respect to broadband energy. Figure 3 shows the Biot predictions for the frequency dependence of the reflection coefficient for different grazing angles. At high grazing angles the energy is reflected almost independently of frequency. At grazing angles below about 10°, however, energy of a certain frequency band penetrates the porous material, whereas the energy of higher and of lower frequencies is reflected. The line shape of the minimum resembles that of an inverted relaxation curve. In order to verify that the internal flow-

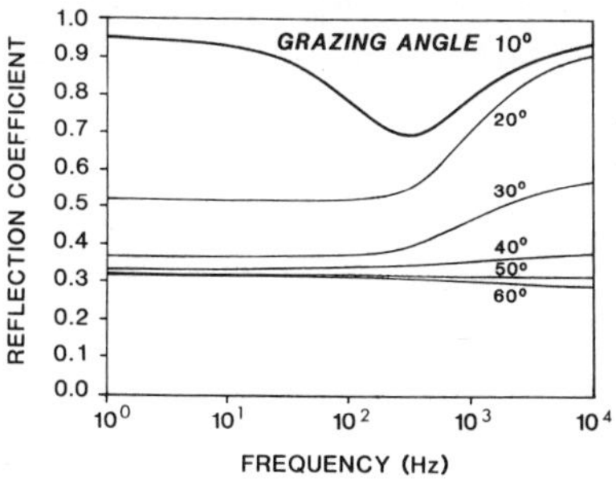

Fig. 3
Frequency dependence of the reflection coefficient for different grazing angles as predicted by the Biot theory.

relaxation creates such a minimum in the reflectivity we display the relaxation and reflectivity curves together in Fig. 4. A maximum of the relaxation manifests itself most clearly in the logarithmic decrement of the viscous losses, which are plotted in the lower part of Fig. 4 for different permeability/pore-size combinations. The appearance of the maximum in the relaxation is closely correlated to the appearance of the minimum in the reflection loss.

By varying the permeability, the curve is strongly shifted within the frequency spectrum; for comparison we display on top of Fig. 4 the corresponding minima in the bottom-reflection loss. It turns out that the corresponding minima and maxima do not show up at the same frequency. Instead they are offset by a constant ratio: the frequency at which the minimum in the bottom-reflection loss occurs is 0.53 times the frequency at which the maximum of the logarithmic decrement is observed. This is a direct result of the Biot theory and will be used later for the evaluation of τ.

The location of the maximum of the logarithmic decrement in the frequency spectrum permits τ to be determined via the peak-condition $\omega\tau = 1.1 - \frac{F_i}{F_r}$.

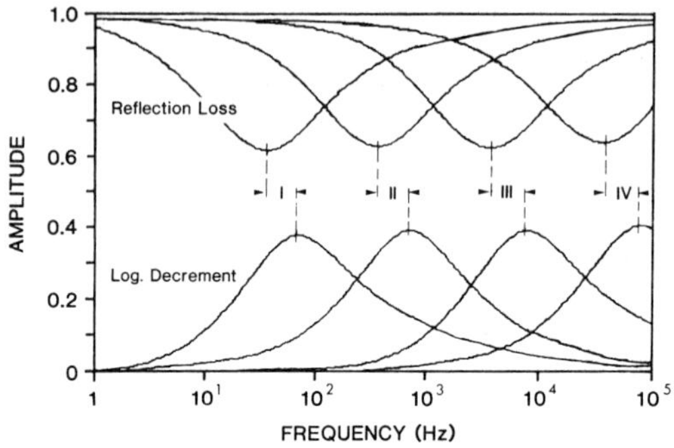

Fig. 4
Comparison of the viscous loss per wave-length (log.decrement) with corresponding reflection loss for four different permeability(k)/pore-size(a) combinations:

I : $a = 1.5 \cdot 10^{-4}$ m, $k = 10^{-9}$ m^2
II : $a = 4.0 \cdot 10^{-5}$ m, $k = 10^{-10}$ m^2
III : $a = 1.0 \cdot 10^{-5}$ m, $k = 10^{-11}$ m^2
IV : $a = 4.0 \cdot 10^{-6}$ m, $k = 10^{-12}$ m^2

We are now in the position to use the minimum in the reflection loss to find the maximum of the corresponding logarithmic decrement, and from the frequency where this maximum occurs, to find the relaxation time. In principle, the minimum in the bottom-reflection loss can be measured, although it is not clear whether the Biot effect is large enough to be measurable in a real marine sediment.

3 MEASUREMENT OF BOTTOM-REFLECTION LOSS

In 1978 the SACLANT ASW Research Centre had used a vertical array of twelve hydrophones to measure the bottom-reflection loss in certain deep basins of the central Mediterranean at frequencies from 50 Hz to 7 kHz [7]. To measure the reflection losses over this wide frequency band, explosive sources of 500 g TNT were used. The experimental set-up is sketched in Fig. 5. The receiving ship suspended a 750-m long vertical hydrophone string,

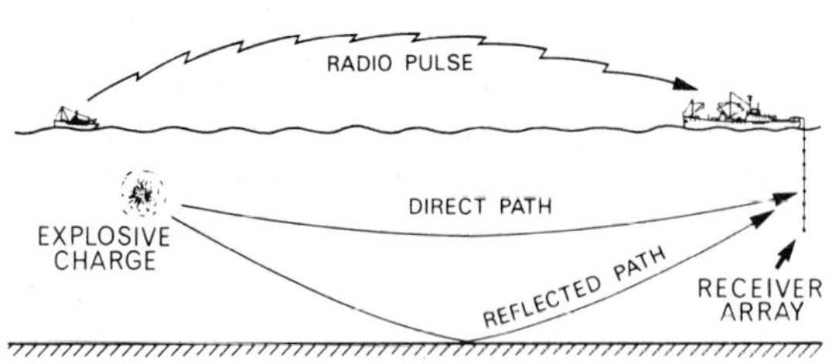

Fig. 5
The experimental set-up for bottom-reflection loss measurements [7].

while the source ship moved on a predetermined fixed course, launching charges set to explode at 550 m depth. The direct and reflected acoustic signals were recorded in digital form by the receiving ship together with a radio pulse transmitted from the source ship at the moment when it received the direct acoustic pulse from the charge. Some of the data collected in those studies will be used here.

Because the sea bottom often exhibits surface roughness and/or layering, the reflection anomaly studied in this paper is frequently masked by distortion of the acoustic signals. To isolate the reflection features related to the Biot effects from those related to the bottom morphology we choose a measuring site that shows a negligible degree of geomorphological discontinuities. This choice was made by examining underwater photographs and profiles of relative sound speed (0.98), relative density (1.5), and porosity (0.5), deduced from core samples. The area selected for the measurements used in this paper was on the Messina abyssal plain in comparatively deep water (3700 m). The morphology of the sea floor over the abyssal plain is extremely flat, with a very soft top layer of fine clay. The area is homogeneous enough to permit the measurement of the Biot reflectivity anomaly.

Accordingly the signals themselves are undistorted, almost textbook-like, which easily permits Fourier analysis. As one result of this analysis, we obtain the impulse response for different angles of incidence, as shown in Fig. 6.

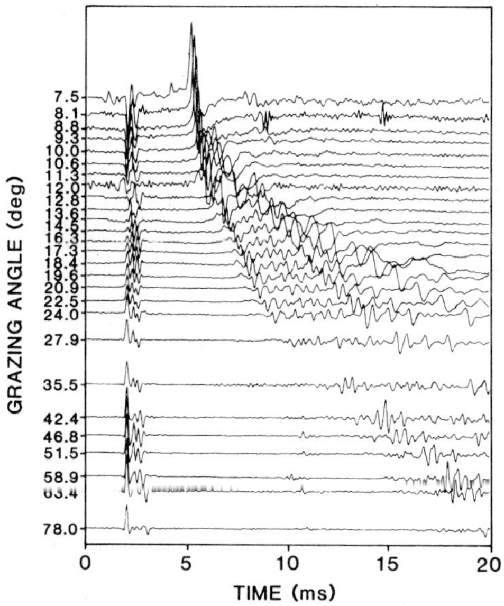

Fig. 6
Detailed display of the first 20 ms of impulse responses for different grazing angles [7].

An analysis of the signal pattern indicates a very soft sea-bottom. This bottom type, with high permeability and high porosity, implies that the sound speed in the bottom is less than that in sea water, which represents an intromission-angle situation for the frequencies discussed here. Accordingly, the first reflection remains small for all grazing angles. However, the first layer is terminated at a depth of 14 m by another, high-velocity layer, as can be deduced from the signal pattern in Fig. 6 (extrapolation to vertical incidence). As long as the grazing angle is less than 10°, all reflections/refractions merge here to create a single, large-amplitude, phase-shifted pulse. It is mainly at this 14-m interface that a reflectivity anomaly similar to the one predicted by the Biot theory was measured.

4 COMPARISON BETWEEN MODEL PREDICTIONS AND MEASUREMENTS

By Fourier analyzing the measured signal we obtain the bottom-reflection loss at different frequencies. If we use the whole signal for the computation, i.e. the full-length signal of Fig. 6 comprising the contributions of all layers, we obtain the oscillating curve depicted in Fig. 7. The bottom-reflection loss is here shown for a very small grazing angle of 10°.

It is striking that the overall tendency of this curve resembles very much the one predicted by the Biot theory for liquid/porous-composite interfaces, as shown in Fig. 3. Some features are different, due to the layering and mixing of different materials in a real sea floor and due to the fact that the signal is also effected by the refracted arrival. Nevertheless, an appropriate choice of the input parameters to the Biot model permits the measured curve to be fitted in a way that reproduces the overall tendency; this is particularly true for the position of the minimum which is strongly shifted in frequency by small changes of the permeability (see Fig. 4). The superimposed solid line in Fig. 7 represents the fit by the Biot theory. The values of the parameters used for this computation are given in Table 1 and were chosen to obtain agreement between the experimental and the theoretical curves in Fig. 7. However, in choosing their values, constraints were imposed in order to keep them within physically realistic limits. Accordingly they are of the same order of magnitude as commonly used parameter sets given by Stoll [8].

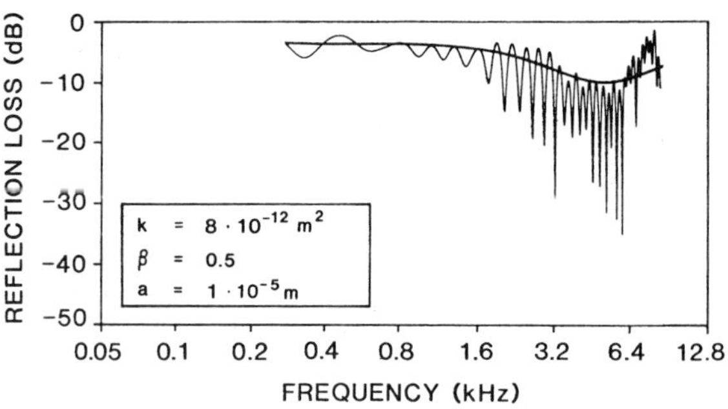

Fig. 7
Bottom-reflection loss versus frequency for a 10° grazing angle. The smooth line represents the prediction by the Biot theory.

TABLE 1

POROSITY	$\beta = 0.5$	ISOTROPY FACTOR	$\alpha = 3$
PERMEABILITY	$k = 8 \cdot 10^{-12}\ m^2$	BULK MODULUS OF FRAME	$K_b = 4.2 \cdot 10^7$ Pa
PORE SIZE	$a = 1 \cdot 10^{-5}$ m	SHEAR MODULUS OF FRAME	$\mu = 2.5 \cdot 10^7$ Pa
DENSITY OF GRAIN MATERIAL	$\varrho_s = 2.5 \cdot 10^3$ kg/m³	LOG DECREMENT FOR SHEAR VIBRATIONS OF THE FRAME	$\delta_\mu = 0.015$
DENSITY OF FLUID	$\varrho_f = 10^3$ kg/m³	LOG DECREMENT FOR BULK VIBRATIONS OF THE FRAME	$\delta_K = 0.015$
VISCOSITY	$\eta = 10^{-3}$ Pa·s		

Regarding the position of the miniumum in the frequency spectrum, which is critically influenced by the input parameters, the fit can be considered accurate. From this fit we obtain values for the bottom parameters, in particular for those determining the above-derived relaxation time, τ . This allows us to apply our concept of a relaxation time directly to a measurement. On the one hand the relaxation time is known by the frequency at which the measured minimum occurs according to the peak condition: $\tau = (1.1 - F_i/F_r)/2\pi(1.8f)$, where the factor 1.8 is determined by the shift factor of Eq. 10. On the other hand we can easily calculate τ by introducing the fit parameters into our derived formula. A comparison of the two values of τ shows close agreement and proves the applicability of the derived formula to real marine sediments. Experimentally we obtain $\tau = 1.5\times10^{-5}$s; the formula gives $\tau = 1.6\times10^{-5}$s. The close agreement convinces us that we are dealing with a measured Biot effect. It opens up the possibility of determining bottom parameters from a measurement of the bottom-reflection loss.

CONCLUSIONS

The derived expression for the relaxation time — which correlates an overall response time with the parameters of the microscopic structure— can be used for determining bottom parameters remotely. The idea is first to use the minimum in a measured bottom-reflection-loss curve to evaluate the relaxation time and then to employ Eq. 10 to deduce values for the correlated structure parameters. The relaxation time itself is expressed in terms of the permeability/porosity ratio and of additional parameters descriptive of the liquid. With a simple assumption concerning the relation between flow permeability and porosity we are thus in a position to determine the flow permeability of the bottom under examination. This method of remotely sensing the permeability may not only be useful for characterizing the sea bottom, but may also· have applications to the sensing of liquid-carrying layers. By analyzing bottom-reflection losses in the central Mediterranean we have demonstrated the applicability of this method to an actual sea-floor measurement.

ACKNOWLEDGMENTS

The author wishes to thank O. Hastrup and T. Akal of SACLANTCEN for the permission to use their bottom-reflectivity measurements.

REFERENCES

1. M.A. Biot, 'Theory of propagation of elastic waves in a fluid-saturated porous solid. I. Low-frequency range, II. Higher frequency range', J. Acoust. Soc. Am., 28: 168-191 (1956).
2. M.A. Biot, 'Mechanics of deformation and acoustic propagation in porous media', J. Appl. Phys., 33: 1482-1489 (1962).
3. M.A. Biot, 'Generalized theory of acoustic propagation in porous dissipative media', J. Acoust. Soc. Am., 34: 1254-1264 (1962).
4. R.D. Stoll and T.K. Kan, 'Reflection of Acoustic Waves at a water-sediment interface', J. Acoust. Soc. Am., 70: 149-156 (1981).
5. R.D. Stoll, 'Acoustic waves in saturated sediment'. In: L. Hampton, ed. Physics of Sound in Marine Sediments, New York, Plenum, 19-39 (1974).
6. R.D. Stoll and G.M. Bryan, 'Wave attenuation in saturated sediments', J. Acoust. Soc. Am., 47: 1440-1447 (1970).
7. T. Akal, 'Acoustical characteristics of the sea floor: experimental techniques and some examples from the Mediterranean Sea'. In: L. Hampton, ed. "Physics of Sound in Marine Sediments", New York, Plenum, p. 447 (1974).
8. R.D. Stoll, 'Experimental studies of attenuation in sediments', J. Acoust. Soc. Am., 66: 1152-1160 (1979).

THE INTERACTION OF LOW-FREQUENCY ACOUSTIC WAVES

WITH A ONE-DIMENSIONAL RANDOM SEDIMENT MODEL

Ioannis M. Besieris and Werner E. Kohler

Virginia Tech
Blacksburg, Virginia 24061
U.S.A.

ABSTRACT

The onset of stochastic effects at low frequencies is studied for a thick sediment layer resting upon a rock basement. Reasonable values of density and sound speed fluctuations and refractive gradients are incorporated; the model, however, is one-dimensional, i.e., lateral homogeneity is assumed. Appropriate scales are introduced and the asymptotic theory of White and Franklin[1] is applied to approximate the fields within the sediment as the sum of a deterministic component and a small, zero-mean Gaussian fluctuation. All aspects of this approximation - in particular the covariance matrix of the fluctuation process - can be explicitly determined.

I. INTRODUCTION

Consider an acoustic plane wave incident upon a thick sediment layer lying on a rock basement. We shall assume the sediment to be a randomly laminated, refracting medium in the depth direction; however, we assume transverse or lateral homogeneity. The rock basement is assumed to be a homogeneous semi-infinite medium. With these idealizations, the problem reduces to a (one-dimensional), stochastic, two-point boundary value problem. At sufficiently low frequencies, when the acoustic wavelength greatly exceeds the length scales of the random density and sound speed fluctuations within the sediment, virtually no multiple scattering is generated by these fluctuations. In this regime, the acoustic response of the sediment layer is basically a deterministic one, obtained by analyzing the geoacoustic model. As frequency increases, multiple scattering effects similarly increase; a small stochastic component begins to become apparent in the acoustic response of the sediment layer.

We shall analyze the problem, i.e., the stochastic two-point boundary value problem, in the frequency regime where stochastic effects are small. Using the asymptotic theory of White and Franklin,[1] we shall show that a first-order stochastic correction to the deterministic geoacoustic response can be asymptotically approximated by a zero-mean Gaussian process. The covariance matrix of this process can be explicitly determined from the geophysical parameters of the problem.

In Section II we shall formulate the problem of interest as a stochastic two-point boundary value problem. Parameter values and the appropriate scaling are addressed in Section III. The asymptotic Gaussian characterization of the fluctuations is considered in Section IV, while in Section V we shall present our conclusions.

For purposes of exposition, a number of simplifications will be adopted. In Section V we comment on extensions to more realistic models.

II. PROBLEM FORMULATION

Sediments are complicated porous media; Burridge and Keller[2] have used homogenization theory to derive a macroscopic Biot theory from a microscopic one. For our purposes, however, we shall adopt the following simplified description:

$$\begin{aligned} &\partial_t p + \rho c^2 \nabla \cdot \underline{v} = 0, \\ &\partial_t \underline{v} + \rho^{-1} \nabla p = \underline{0}, \end{aligned} \tag{1}$$

where p, $\underline{v}$, ρ and c denote pressure, particle velocity, density, and sound speed, respectively. In the water column ($z < 0$), the density and sound speed are assumed to have constant values ρ_o and c_o, respectively; the rock basement ($z > L$) will likewise have constant constitutive parameters. Within the sediment layer ($0 \leqslant z \leqslant L$), density and sound speeds will vary with depth as functions of a single, underlying stochastic process; the core samples of Tucholke[3] support this assumption. We shall assume that density and sound speeds both have linearly increasing mean values and zero-mean random fluctuations.

Assume that a unit-amplitude, time-harmonic plane wave is incident upon the sediment layer from the water column at incidence angle Θ_o (measured from the vertical). Within the layer, the pressure and velocity can be represented as

$$\begin{aligned} p(\underline{r},t) &= e^{i(kx\sin\Theta_o - \omega t)}\hat{p}(z), \\ \underline{v}(\underline{r},t) &= e^{i(kx\sin\Theta_o - \omega t)}[\hat{v}_x(z)\underline{x}_o + \hat{v}_z(z)\underline{z}_o], \end{aligned} \tag{2}$$

where $k \equiv \omega/c_o$ is a wave number, and $\underline{x}_o$, $\underline{z}_o$ denote unit vectors. Substitution of equations (2) into (1) leads to a first order system of ordinary differential equations for $\hat{p}$, $\hat{v}_z$. It becomes convenient, however, to introduce a new pair of dependent (scattering) variables as follows:

$$\begin{aligned} a &\equiv \frac{1}{2}[\hat{p} + \hat{v}_z(\rho_o c_o \sec\Theta_o)], \\ b &\equiv \frac{1}{2}[\hat{p} - \hat{v}_z(\rho_o c_o \sec\Theta_o)]. \end{aligned} \tag{3}$$

For brevity, let $\tilde{\rho} \equiv \rho/\rho_o$ and $\tilde{c} \equiv c/c_o$. The stochastic two-point boundary value problem of interest becomes then

$$\frac{d}{dz}\begin{bmatrix} a \\ b \end{bmatrix} = \frac{ik}{2}\begin{bmatrix} A_{11} & A_{12} \\ -A_{12} & -A_{11} \end{bmatrix}\begin{bmatrix} a \\ b \end{bmatrix}, \quad 0 < z < L;$$

$$A_{\substack{11 \\ 12}} \equiv \pm \tilde{\rho}\cos\Theta_o + \tilde{\rho}^{-1}\sec\Theta_o(\tilde{c}^{-2} - \sin^2\Theta_o); \tag{4}$$

$$a(0) = 1, \quad b(L) = \Gamma a(L),$$

where Γ is the (known) sediment-basement interface reflection coefficient. The continuity of pressure and the normal component of velocity across $z = 0$ and $z = L$ implies a similar continuity of the variables a and b across these interfaces.

The dependent variable "a" represents an incident or downward propagating wave, while the variable "b" corresponds to a reflected or upward propagating wave. In the absence of dissipation, one can show directly from (4) that energy flux is invariant within the layer, i.e., $\frac{d}{dz}(|a|^2 - |b|^2) = 0$.

III. PARAMETER VALUES AND SCALING

Within the sediment layer, we assume that the density and the sound speed are given respectively by

$$\rho(z) = \rho_1(1+\eta z) + \delta\rho(z/\ell),$$

$$c(z) = c_1(1+\eta z) + \delta c(z/\ell), \tag{5}$$

where ρ_1 and c_1 are constants, while $\delta\rho(\cdot)$ and $\delta c(\cdot)$ are zero-mean, unit-correlation length random fluctuations. Thus, ℓ represents the actual correlation length; η represents the gradient parameter for the linearly increasing mean values. We shall adopt the following (order of magnitude) values:

$$\rho_o = 1 \text{ gm/cm}^3, \qquad \rho_1 = 1.5\text{gm/cm}^3$$

$$c_o = c_1 = 1500\text{m/sec}, \tag{6}$$

$$\eta^{-1} = 1500\text{m}, \qquad \ell = 0.15\text{m}.$$

With these values, the mean sound speed $\langle c(z)\rangle$ has slope $c_1\eta = 1 \text{ sec}^{-1}$; the 15cm correlation length corresponds roughly to values measured at the Hatteras abyssal plane.

We shall be interested in the frequency range 1-100Hz. For purposes of scaling, let $f_o \equiv 10\text{Hz}$ denote a reference frequency ($\lambda_o \equiv c_o/f_o = 150\text{m}$). In terms of this reference, we introduce the following dimensionless quantities:

$$\kappa \equiv \pi(f/f_o) = \pi(\lambda_o/\lambda),$$

$$\tau \equiv L/\lambda_o, \tag{7}$$

$$\mu \equiv \eta\lambda_o,$$

$$\varepsilon \equiv \ell/\lambda_o.$$

For the frequencies of interest, and for a sediment layer several hundred meters in thickness (say $L = 500m$), κ, τ, and μ are roughly $O(1)$ constants. The parameter $\varepsilon \tilde{=} 10^{-3}$ will serve as the small parameter underlying the asymptotic analysis.

Let $\xi \equiv z/\lambda_o$ and $\underline{v} \equiv (a\ b)^T$. In terms of the parameters introduced in (7), differential equation (4) becomes

$$\frac{d}{d\xi}\underline{v} = i\kappa A(\mu\xi,\xi/\varepsilon)\underline{v}, \qquad 0 < \xi < \tau;$$

$$A \equiv \begin{bmatrix} A_{11} & A_{12} \\ -A_{12} & -A_{11} \end{bmatrix}, \tag{8}$$

where the dependence of the matrix A upon $\eta z = \mu\xi$ and $z/\ell = \xi/\varepsilon$ has been explicitly noted.

For brevity, let the mean coefficient matrix and the fluctuation matrix be denoted by $\bar{A}$ and $\mathscr{A}$, respectively; that is,

$$\langle A(\mu\xi,\xi/\varepsilon)\rangle \equiv \bar{A}(\mu\xi), \qquad \mathscr{A} \equiv A - \bar{A}. \tag{9}$$

The scaled boundary value problem, consisting of equation (8) and the boundary conditions in (4), will be studied in the context of the asymptotic theory of White and Franklin. The averaged matrix $\bar{A}$ basically defines the geoacoustic model; it determines the low frequency deterministic behavior about which the small random fluctuations develop.

IV. APPLICATION OF THE ASYMPTOTIC THEORY

We first convert the boundary value problem of interest from a 2-dimensional complex-valued system to a 4-dimensional real-valued system. Let

$$a \equiv a_1 + ia_2, \quad b \equiv b_1 + ib_2, \quad \underline{u} \equiv (a_1\ a_2\ b_1\ b_2)^T,$$

$$\Gamma \equiv \Gamma_1 + i\Gamma_2,$$

$$B \equiv \begin{bmatrix} 0 & -A_{11} & 0 & -A_{12} \\ A_{11} & 0 & A_{12} & 0 \\ 0 & A_{12} & 0 & A_{11} \\ -A_{12} & 0 & -A_{11} & 0 \end{bmatrix}, \quad \begin{matrix} \Lambda_0 \equiv \begin{bmatrix} 1 & 0 & 0 & 0 \\ 0 & 1 & 0 & 0 \end{bmatrix}, \\ \Lambda_\tau \equiv \begin{bmatrix} \Gamma_1 & -\Gamma_2 & -1 & 0 \\ \Gamma_2 & \Gamma_1 & 0 & -1 \end{bmatrix}. \end{matrix} \tag{10}$$

The new, real-valued, boundary value problem becomes then

$$\frac{d}{d\xi}\underline{u} = \kappa B\underline{u}\ , \ 0<\xi<\tau; \quad \Lambda_o\underline{u}(0) = \begin{bmatrix} 1 \\ 0 \end{bmatrix}, \quad \Lambda_\tau\underline{u}(\tau) = \begin{bmatrix} 0 \\ 0 \end{bmatrix}. \tag{11}$$

As before (cf. (9)), let $\bar{B} \equiv \langle B \rangle$ and $\mathscr{B} \equiv B - \bar{B}$. We first consider the centering trajectory, i.e., the deterministic solution to (11) when the fluctuation matrix $\mathscr{B}$ is neglected. Let $\underline{u}^o(\xi)$ denote the unique solution to the boundary value problem

$$\frac{d}{d\xi}\underline{u}^o = \kappa\bar{B}\underline{u}^o; \quad \Lambda_o\underline{u}^o(0) = \begin{bmatrix} 1 \\ 0 \end{bmatrix}, \ \Lambda_\tau\underline{u}^o(\tau) = \begin{bmatrix} 0 \\ 0 \end{bmatrix}. \tag{12}$$

Let $\Phi(\xi)$ denote the 4 x 4 fundamental matrix for the first order system in (12) that reduces to the identity matrix at $\xi = 0$; let $\underline{\phi}_j$, $j = 1,\ldots,4$, denote the 4 columns of Φ. Define, also, the 2 x 2 constant matrix

$$J \equiv \Lambda_\tau[\underline{\phi}_3(\tau),\underline{\phi}_4(\tau)]. \tag{13}$$

We shall assume that the fluctuation matrix $\mathscr{B}$ is an almost surely bounded, strongly mixing, matrix-valued process. This strong mixing hypothesis is tantamount to the assumption that the sediment density and sound speed fluctuations tend toward statistical independence as vertical separation distance increases. A precise statement of this physically reasonable assumption can be found, for example, in refs. 1 and 4.

With little effort, the hypotheses of ref. 1 can be verified. The basic theorem of Khasminskii[5] can be modified to accomodate the dependence of B upon the slow scale ξ (cf. ref. 4). Since the solution trajectories of (11) are, in fact, uniformly bounded, the ostensibly unbounded dependence of the right side of (11) upon $\underline{u}$ in reality presents no problem. Remark 1 of ref. 6 addresses this point. The existence of a unique solution of (12) and the invertibility of J can easily be demonstrated if $\kappa\tau$ is small enough.[7] These conditions can be directly checked in specific cases.

The White-Franklin result states basically that the acoustic fields within the sediment can be approximated by the superposition of a deterministic trajectory $\underline{u}^o$ and a small $O(\varepsilon^{1/2})$ Gaussian fluctuation. In order to define this fluctuation process, we must first define several intermediate processes.

Let $\underline{W}^o(\xi)$ be a Gaussian process with mean zero, independent increments, and covariance matrix

$$C(\xi) \equiv \kappa^2 \int_0^\xi \lim_{\zeta\to\infty} \left[\zeta^{-1} \int_0^\zeta \int_0^\zeta \langle\ \mathscr{B}(\sigma,s_1)\underline{u}^o(\sigma)\underline{u}^{oT}(\sigma)\mathscr{B}^T(\sigma,s_2)\ \rangle \right.$$
$$\left. \times\ ds_1 ds_2 \right] d\sigma. \tag{14}$$

Let, furthermore, $\underline{y}^o(\xi)$ denote the Gaussian Markov process which is the solution of

$$\underline{y}^o(\xi) = \underline{W}^o(\xi) + \kappa \int_0^\xi \bar{B}(\sigma)\underline{y}^o(\sigma)d\sigma. \tag{15}$$

Lastly, define

$$\underline{z}^o(\xi) \equiv \underline{y}^o(\xi) - [\underline{\phi}_3(\xi),\phi_4(\xi)]J^{-1}\Lambda_\tau \underline{y}^o(\tau). \tag{16}$$

A very crude statement of the White-Franklin result then states that in the asymptotic limit as $\varepsilon \to 0$, one can approximate the solution process $\underline{u}$ as

$$\underline{u}(\xi) \tilde{=} \underline{u}^o(\xi) + \varepsilon^{1/2}\underline{z}^o(\xi). \tag{17}$$

Reference 2 establishes approximation (17) in terms of weak convergence, while the ideas of references 4 and 8 can be used to obtain convergence of moments. The solution process is thus approximated as the superposition of the deterministic trajectory $\underline{u}^o$ and the Gaussian fluctuation process $\varepsilon^{1/2}\underline{z}^o$. The approximation is explicit since (15) can be solved to obtain

$$\underline{y}^o(\xi) = \underline{W}^o(\xi) + \kappa \int_0^\xi \Phi(\xi)\Phi^{-1}(s)\bar{B}(s)\underline{W}^o(s)ds. \tag{18}$$

V. CONCLUSIONS

The importance of (17) lies in the fact that the zero-mean fluctuation process $\underline{z}^o$ is totally characterized by its covariance matrix, which, in turn, is determined by the underlying deterministic geoacoustic model (12) and the statistics of the density-sound speed fluctuation process. If the fluctuations are small, so that a linearization of the matrix elements A_{11} and A_{12} in $\delta\tilde{\rho}$ and $\delta\tilde{c}$ is possible, only the correlation function of the geophysical fluctuation process is required.

Although equations (1) were adopted as reasonable model equations, the ideas presented are not inherently limited to these equations. These ideas extend very naturally to any (similarly scaled) linear stochastic two-point boundary value problem; more realistic equations, arising for example from Biot theory, can, in principle, be accomodated. Dissipation, although not explicitly discussed, can be introduced into (1) by making the sound speed complex. This simply alters the structure of the coefficient matrix B. The ideas presented also can be extended to low frequency pulse propagation. In these problems, the use of Fourier transforms reduces the problem to that of determining the multifrequency statistics of the solution process. We can obtain these multifrequency statistics by adjoining multifrequency copies of (11), the basic boundary value problem of interest. For example, if two-frequency statistics of the solution process are needed to ultimately determine two-point, two-time correlations of the acoustic fields, we would consider the (8-dimensional) process formed by the direct sum of solutions to (11) at two generic scaled wavenumbers, say κ and κ'.

The asymptotic description (17) is relatively explicit and complete. As frequency increases and multiple scattering becomes more pronounced, the asumptotic characterization becomes more difficult; some results along these lines are given in ref. 9.

REFERENCES

1. B. S. White and J. N. Franklin, A Limit Theorem for Stochastic Two-Point Boundary-Value Problems of Ordinary Differential Equations, *Comm. Pure Appl. Math.* 32:253 (1979).
2. R. Burridge and J. B. Keller, Poroelasticity Equations Derived From Microstructure, *J. Acoust. Soc. Am.* 70:1140 (1981).
3. B. E. Tucholke, Acoustic Environment of the Hatteras and Nares Abyssal Plains, Western North Atlantic Ocean, Determined From Velocities and Physical Properties of Sediment Cores, *J. Acoust. Soc. Am.* 68:1376 (1980).
4. G. C. Papanicolaou and W. Kohler, Asymptotic Theory of Mixing Stochastic Ordinary Differential Equations, *Comm. Pure Appl. Math.* 27:641 (1974).
5. R. Z. Khasminskii, On Stochastic Processes Defined by Differential Equations With a Small Parameter, *Th. Prob. and Applic.* 11:211 (1966).
6. B. S. White, Some Limit Theorems for Stochastic Delay-Differential Equations, *Comm. Pure Appl. Math.* 29:113 (1976).
7. H. B. Keller, "Numerical Methods for Two-Point Boundary-Value Problems," Blaisdell, Waltham, Mass. (1968).
8. W. Kohler and G. C. Papanicolaou, Fluctuation Phenomena in Underwater Sound Propagation, I and II, in: "Stochastic Differential Equations and Applications," J. D. Mason, ed., Academic Press, New York (1977).
9. W. Kohler, Reflection From a One-Dimensional, Totally Refracting Random Multilayer, *SIAM J. Appl. Math.*, to appear.

PROPAGATOR MATRIX FOR ACOUSTIC WAVE PROPAGATION THROUGH ANISOTROPIC POROUS MEDIA

Tokuo Yamamoto and Mohsen Badiey

Applied Marine Physics Division
Rosenstiel School of Marine and Atmospheric Science
University of Miami
Miami, Florida 33149-1098

INTRODUCTION

The body of oceanic water is bounded by sediments and rocks at the bottom and by ice at the surface. Sediments, rocks and sea ices are porous and their pores are filled with sea water or mixtures of air and sea water. It is well known that marine sediments and rocks are anisotropic both elastically (Anderson, 1961; Saada, 1978) as well as hydrodynamically (Davis, 1969). Elastic and hydrodynamic structures of sea ice are also very anisotropic (Anderson, 1961). Furthermore, marine sediments, rocks and sea ices have layered structures. Therefore, ocean bottoms and ocean ice sheets may be considered as layered anisotropic poro-elastic media.

In a fluid-filled poro-elastic medium, a complicated interaction between the pore fluid and the solid skeletal frame takes place as a result of the inertia coupling and the viscous coupling when the medium is excited. In a series of papers, Biot (1956, 1962) proposed a simple phenomenological theory of acoustic propagation in porous, fluid-filled, macroscopically homogeneous and isotropic media. According to the Biot theory, when excited in rotational mode, the pore fluid and the solid frame move relative to each other but there is only one phase velocity. On the other hand, under a compressional excitation the pore fluid and the solid frame not only move relative to each other but this mode of fluid-solid interaction generates two compressional waves of different phase velocities simply because the pore fluid and the skeletal frame have different compressibilities and densities. Characteristically for the viscous coupling, the phase velocities and attenuations of these three elastic waves are strongly dependent of the excitation frequency, thus they are very dispersive. The viscous coupling is governed by the relaxation frequency, f_r, of the porous medium which is the inverse of the time required for the viscous vorticity to travel the pore space. In terms of the bulk properties of a porous media this relaxation frequency is given by Yamamoto (1983) and Badiey and Yamamoto (1985) as

$$f_r = \frac{1}{3\pi} \frac{\beta\nu}{k_s} \tag{1}$$

in which β = porosity, ν = kinematic viscosity of pore fluid and k_s = coefficient of permeability. At a frequency much lower than the relaxation frequency, the pore fluid and the skeletal frame is completely coupled due to the fluid viscosity. At a frequency much higher than the relaxation frequency, the viscosity of the pore fluid can no longer keep the pore fluid coupled to the solid frame. Here, the coupling between the pore fluid and the solid frame is governed by the hydrodynamic pressure induced on the accelerating solid in the inviscid fluid - an inertia coupling. At a high frequency as the effective density of the solid is reduced considerably the compressional wave and the shear wave propagate faster than at a low frequency.

In an anisotropic porous medium, the relaxation frequency of the system or the viscous coupling becomes dependent of the direction of wave propagation. In addition, because a volumetric change is induced by a shear deformation, shear waves are coupled with the compressional waves in an anisotropic medium. Although Biot (1962) has extended the theory for the anisotropic poro-elastic media, the theory has very seldom been applied to the problem of acoustic wave propagation. Recently, Badiey and Yamamoto (1985) has examined the effects of hydraulic and the elastic anisotropies on propagation of the acoustic normal modes in the shallow ocean overlaying a layered sediment bottom. A more complete derivation of the propagator matrix for a layered poro-elastic bed of transverse isotropy is given in this paper. Example calculations made for the acoustic reflection by an anisotropic sediment bed and an anisotropic sea ice sheet are also presented.

THEORY

Elastic Equations

The stress-strain relation of a transversely isotropic poro-elastic medium is given by Biot and Willis (1957) as

$$\begin{Bmatrix} \sigma_{xx} \\ \sigma_{yy} \\ \sigma_{zz} \\ \sigma_{yz} \\ \sigma_{yx} \\ \sigma_{xy} \\ \sigma \end{Bmatrix} = \begin{bmatrix} P & A & F & 0 & 0 & 0 & M \\ A & P & F & 0 & 0 & 0 & M \\ F & F & C & 0 & 0 & 0 & Q \\ 0 & 0 & 0 & L & 0 & 0 & 0 \\ 0 & 0 & 0 & 0 & L & 0 & 0 \\ 0 & 0 & 0 & 0 & 0 & N & 0 \\ M & M & Q & 0 & 0 & 0 & R \end{bmatrix} \begin{Bmatrix} e_{xx} \\ e_{yy} \\ e_{zz} \\ e_{yz} \\ e_{yx} \\ e_{xy} \\ \varepsilon \end{Bmatrix} \qquad (2)$$

in which $P = A + 2N$.

Here σ_{ij} is the effective stress tensor, σ is fluid stress, e_{ij} is the strain tensor of solid frame and ε is the volumetric strain of the pore fluid. Note that eight independent elastic coefficients are needed to describe the stress-strain relation of a transversely isotropic poro-elastic medium.

The inverse of the stiffness matrix of Eq. (2) will define the softness matrix. Let us designate the corresponding elements of the softness matrix by the corresponding lower case letters.

The non-zero elements of the softness matrix are given by Biot-Willis (1957) as

$$
\begin{aligned}
p &= 1/E_1 , \qquad a = -\nu_1/E_1 \\
f &= -\nu_3/E_1 , \qquad c = 1/E_2 \\
m &= \left[\frac{\delta_1}{\beta} - \left(\frac{1-\beta}{\beta}\right)\left(\frac{1-\nu_1-\nu_3}{E_1}\right)\right] \\
q &= \left[\frac{\delta_2}{\beta} - \left(\frac{1-\beta}{\beta}\right)\left(\frac{1-2\nu_2}{E_2}\right)\right] \\
r &= \left[\frac{\gamma}{\beta^2} - \left(\frac{1-2\beta}{\beta}\right)\left(\frac{2\delta_1+\delta_2}{\beta}\right) + \left(\frac{1-\beta}{\beta}\right)^2\left(\frac{1-2\nu_2}{E_2} + \frac{2(1-\nu_1-\nu_3)}{E_1}\right)\right] \\
n &= 1/N, \quad l = 1/L
\end{aligned}
\tag{3}
$$

in which E_1, E_2 are horizontal and vertical Young's modulus respectively ν_1, ν_2, ν_3 are Poisson's ratios (see Fig. 1), δ_1, δ_2 are the coefficient of compressibility of grain material and γ is the coefficient of fluid content.

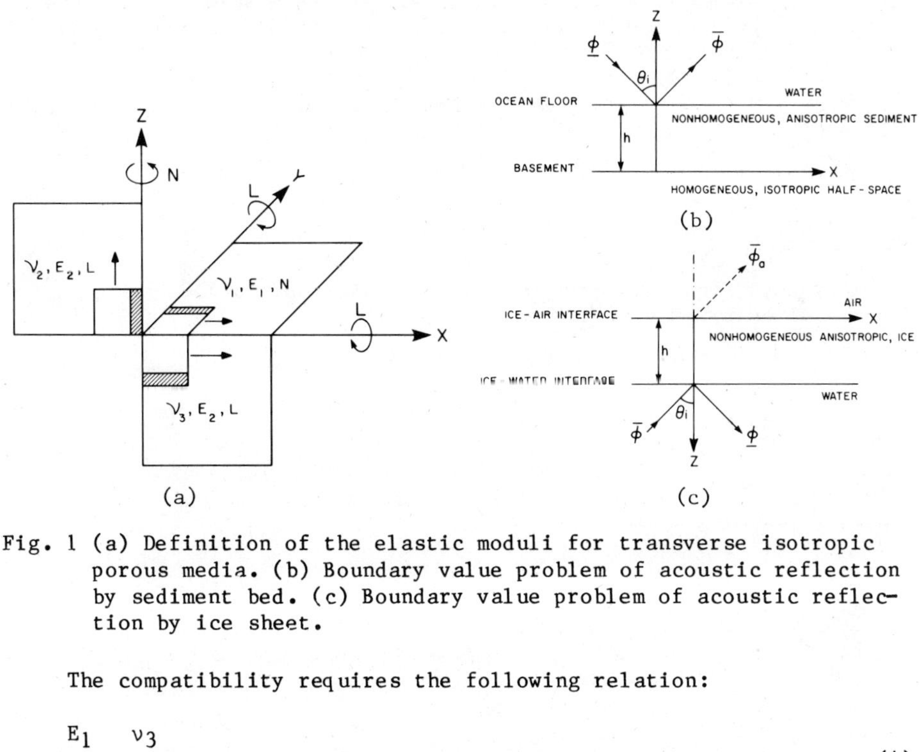

Fig. 1 (a) Definition of the elastic moduli for transverse isotropic porous media. (b) Boundary value problem of acoustic reflection by sediment bed. (c) Boundary value problem of acoustic reflection by ice sheet.

The compatibility requires the following relation:

$$
\frac{E_1}{E_2} = \frac{\nu_3}{\nu_2} = n_e \tag{4}
$$

We will call this ratio n_e the elastic anisotropy. The two shear moduli L and N are related to the Young's moduli and Poisson's ratios as

$$
L = \frac{E_2\nu_3}{\nu_3+\nu_2+2\nu_2\nu_3} , \quad N = \frac{E_1}{2(1+\nu_1)} \tag{5}
$$

In this paper, the Poisson's ratio ν_3 (relating the vertical strain due to the horizontal stress) is assumed to be equal to ν_1 (relating the horizontal strain due to the horizontal stress). ν_2 is the Poisson's ratio relating the horizontal strain due to the vertical stress. In addition, it is assumed that the grain material itself is isotropic, therefore $\delta_1 = \delta_2$. δ_1 and γ are given in terms of the bulk modulus of solid K_r and the bulk modulus of fluid K_f as

$$\delta_1 = \frac{1}{3K_r} \quad , \quad \gamma = \frac{1}{K_f} \tag{6}$$

Therefore, the following six independent parameters are needed to describe the elastic property of a transverse-isotropic poro-elastic medium, they are ν_1, L, n_e, β, K_r, K_f. The losses due to the grain to grain friction are considered to be independent of frequency. They may be introduced by replacing the elastic coefficient by complex coefficients

$$L = L(1+i\delta) \tag{7}$$

here, δ is the loss coefficient.

The total stress tensor τ_{ij} and the pore pressure p are related to the strain tensor e_{ij} and the fluid strain $\zeta = -\beta(\varepsilon-e_{xx}-e_{yy}-e_{zz})$. Let us designate the stiffness matrix for the total stress system by putting a prime to the stiffness matrix given in Eq. (2). Then, the primed matrix is related to the non-prime matrix as:

$$\begin{aligned} P' &= P + 2M + R \;, & A' &= A + 2M + R \;, & L' &= L \\ F' &= F + Q + M + R \;, & M' &= -(M + R)/\beta \;, & N' &= N \\ Q' &= -(Q + R)/\beta \;, & R' &= R/\beta^2 \;, & C' &= C + 2Q + R \end{aligned} \tag{8}$$

Dynamic Equation

The dynamic equations of a transversely isotropic poro-elastic medium are given by Biot (1962) as

$$\begin{aligned} \frac{\partial \tau_{xx}}{\partial x} + \frac{\partial \tau_{xz}}{\partial z} &= \rho \frac{\partial^2 u_x}{\partial t^2} + \rho_f \frac{\partial^2 w_x}{\partial t^2} \\ \frac{\partial \tau_{zz}}{\partial z} + \frac{\partial \tau_{xz}}{\partial x} &= \rho \frac{\partial^2 u_z}{\partial t^2} + \rho_f \frac{\partial^2 w_z}{\partial t^2} \\ -\frac{\partial p}{\partial x} &= \rho_f \frac{\partial^2 u_x}{\partial t^2} + m_x \frac{\partial^2 w_x}{\partial t^2} + \frac{\eta_x}{k_x} \frac{\partial w_x}{\partial t} \\ -\frac{\partial p}{\partial z} &= \rho_f \frac{\partial^2 u_z}{\partial t^2} + m_z \frac{\partial^2 w_z}{\partial t^2} + \frac{\eta_z}{k_z} \frac{\partial w_z}{\partial t} \end{aligned} \tag{9}$$

where ρ is the bulk density of medium ρ_f is the density of water, m_x and m_z are the virtual mass density of the medium in x- and z-direction. They are given in tems of the added mass coefficients by

$$m_i = \rho_f (1+\alpha_i)/\beta \;, \; (i = x,z) \tag{10}$$

We define the anisotropy in the inertia coupling by

$$n_a = \alpha_x/\alpha_z \qquad (11)$$

η_x and η_z are the oscillatory viscosity in x- and z-direction defined by Yamamoto (1985)

$$\eta_j = \eta_f\, F(\kappa_j); \quad F(\kappa) = \frac{1}{3}\frac{\sqrt{i\kappa_j}\tanh\sqrt{i\kappa_j}}{1-(1/\sqrt{i\kappa_j})\tanh\sqrt{i\kappa_j}}, \quad (j = x,z)$$

$$\kappa_j = \left(\frac{\omega}{2\nu}\,\frac{3k_{sj}}{\beta}\right)^{1/2} \qquad (12)$$

k_{sx} and k_{sz} are the permeability in x- and z-direction. The anisotropy in viscous coupling may be defined by

$$n_p = k_{sx}/k_{sz} \qquad (13)$$

An exact treatment of wave propagation through layered transversely isotropic poro-eleastic media may be made deriving a 6 x 6 propagator matrix for the 6-component displacement-stress vector $(u_x, u_z, w_z, \tau_{xx}, \tau_{zz}, p)$ from Eqs. (8) and (9). The propagator matrix solution for the case of isotropy is given by Yamamoto (1983b). Application of the exact 6 x 6 propagator matrix is, however, limited to low frequency waves due to a numerical restriction.

An approximation method which is numerically efficient yet accurate has been derived by Badiey and Yamamoto (1985). In this method, the Biot's slow compressional wave is treated only as an energy sink while the fast compressional wave and the shear wave are treated as propagating body waves.

In this paper, the problem is treated in a similar manner. However, we derive the phase velocities of the three waves exactly. In addition the slow waves are treated as surface waves at the interfaces.

From the Biot-Willis elalstic equation (Eq. 8) and the dynamic equation (9), one can derive the characteristic velocities of the fast and slow compressial waves in x- and z-direction exactly as

$$\left.\begin{matrix} V_{1x} \\ V_{2x} \end{matrix}\right\} = \left[\frac{2(P'R'-M'^2)}{\rho R'+m_x'P'+2\rho_f M' \mp [(m_x'P'-\rho R')^2+4(\rho_f R'-m_x'M')(\rho_f P'-\rho M')]^{1/2}}\right]^{1/2} \qquad (14)$$

$$V_{3x} = \left(\frac{L}{\rho}\right)^{1/2}(1-\rho_f/\rho m_x')^{-1/2}; \quad m_x' = \rho_f(1+\alpha_x)/\beta + i\left(\frac{\eta_{osx}}{k_x\omega}\right) \qquad (15)$$

For V_{1z}, V_{2z}, V_{3z}, replace P', M', L, m_x' by C', Q', N, m_z' respectively.

The wave propagation in layered transversely isotropic media is represented by a complex 4 x 4 propagator matrix as

$$\frac{d}{dz}\begin{Bmatrix} u_x \\ u_z \\ \tau_{xz} \\ \tau_{zz} \end{Bmatrix} = \begin{bmatrix} 0 & -ik & 1/L^* & 0 \\ -ik\frac{F^*}{C^*} & 0 & 0 & \frac{1}{C^*} \\ \zeta & 0 & 0 & -ik\frac{F^*}{C_*} \\ 0 & -\rho\omega^2 & -ik & 0 \end{bmatrix} \begin{Bmatrix} u_x \\ u_z \\ \tau_{xz} \\ \tau_{zz} \end{Bmatrix} \qquad (16)$$

in which $\zeta = -\rho\omega^2 + P^*k^2 - k^2\frac{F^{*2}}{C_*}$

The poro-elastic coefficients in the propagator matrix are given in terms of the characteristic velocities as

$$P^* = \rho(V_{1x})^2 \quad , \quad C^* = \rho(V_{1z})^2$$
$$L^* = \rho(V_{3x}{}^2 + V_{3z}{}^2)/2 \quad , \quad F^* = \alpha(P^* - 2L^*) \tag{17}$$

where, $\alpha = \dfrac{F'}{A'-2L'}$

At the interface, the seepage displacement w_z is recovered and accounted for the matching boundary conditions as

$$w_z = \Delta P / iM'\lambda_2 \tag{18}$$

where ΔP is pore pressure jump at the boundary and λ_2 is the vertical wave number of the slow wave at the boundary.

The propagator matrix method described above is implemented for two boundary value problems. One is the acoustic reflection by a sediment bed (Fig. 1b) and the other is the acoustic reflection by an ice sheet (Fig. 1c).

Acoustic Reflection by Anisotropic Sediment Beds

Acoustic reflection by a homogeneous isotropic poro-elastic half-space was studied analytically by Stoll and Kan (1981). It was found that the acoustic reflection by a poro-elastic sediment becomes strongly frequency dependent compared to an elastic bed which shows a frequency-independent reflectivity. Furthermore, there is no critical angle of incidence beyond which a total reflection takes place. We investigate the effects of poro-elastic anisotropies on acoustic reflection through comparisons with this isotropic sediment case.

The values of the physical properties of the sand used by Stoll and Kan (1981) are reproduced in Table 1. Using these values, the reflection coefficient versus the angle of incidence is calculated by the present method. The result at frequency f = 1 kHz, is shown by the curve designated by n_p = 1. in Figure 2. A good agreement with the analytical result by Stoll and Kan (1981) is obtained. The phase difference between the incident and the reflected wave is also shown in Figure 2. For the isotropic sand (n_p = 1), the phase is zero for the range of incident angle of 0 to 50° and gradually moves to −180° as the incident angle tends

Table 1. Physical Properties of Example Models

Property	Sand	Sea Ice
Poisson's ratio	0.25	0.32
Porosity	0.47	[0.01+0.09 Exp(−Z/0.144)]
Shear Mod. of frame, P_a	2.6E+7	[2.8−1.8Exp(−Z/0.144)]*1.0E+9
Permeability, m^2	1.00E−10	[0.02+1.98 Exp(−Z/0.144)]*1.0E−11
Density of frame, kg/m^3	2.65E+3	0.90E+3
Density of pore fluid, kg/m^3	1.025E+3	1.025E+3
Bulk Mod. of frame, P_a	3.60E+10	3.60E+10
Bulk Mod. of fluid, P_a	2.00E+9	2.25E+9
Kinematic Viscosity of pore fluid, m^2/s	1.00E−6	1.00E−6
Loss coefficient	0.02	0.02

to 90°. This indicates that the isotropic sand bed behaves like a lossy rigid wall at a small incident angle while it behaves as a pressure relief surface as the incident angle tends to 90°.

Effects of Hydrodynamic Anisotropy

To investigate the effect of anisotropy in the sediment permeability, the upper 1.5 m of sediment bed is replaced with an anisotropic sand with a various permeability ratio $n_p = k_{sx}/k_{sz}$. Other parameters are the same as the isotropic sand given in Table 1. Below depth 1.5 m, is a halfspace of the isotropic sand. It is clearly shown in Figure 2 that the anisotropy in the permeability has significant effects on the acoustic reflection. for n_p = 10 and 100, a dramatic increase in reflection takes place at incident angles larger than 60° while the phase is practically the same as the isotropic sand. Therefore, it is important that the anisotropy in sediment permeability is taken into account when a propagation model is constructed. The broken line in Figure 2 shows the result when the slow wave generation at the water-sediment interface is ignored for the case of n_p = 10. Very small difference due to the slow wave is shown.

Similar comparisons are shown in Figure 3 at frequency f = 500 Hz. Again a significant effect of the anisotropy in permeability on the reflection at the incident angle above 60° is shown.

Although not shown here, effect of the anisotropy in the added mass coefficient is found to be relatively small on the acoustic reflectivity of a sediment bed.

Effects of Elastic Anisotropy and Inhomogeneity

In order to investigate the effect of elastic anisotropoy of sediment on the acoustic reflection, the elastic anisotropy ratio $n_e = E_1/E_2$ alone is varied from 0.5 to 1.0 while other parameters are kept the same as the isotropic sand of Table 1. Comparisons of the reflection diagrams so calculated indicate that the effect of elastic anisotropy of a sediment is relatively small on the acoustic reflection by a sediment bed. Figure 4 shows the effect of inhomogeneity of the bed. As can be seen, layered structures introduce finer dependence of reflection coefficient on incident angle as compared to homogeneous bed.

Acoustic Reflection by an Anisotropic Poro-Elastic Ice

As the second example of the acoustic reflection by an anisotropic poro-elastic medium, we consider a 1 m thick layered ice defined in Table 1. The elastic and hydrodynamic characteristics of the ice sheet have been estimated from the sea ice crystals measured by Mellor et al, (1984).

The reflection diagrams computed at frequencies of 500 Hz, 1000 Hz and 3000 Hz are shown in Figure 5. A virtually total reflection at all angles take place for low frequency waves at 500 Hz. At a higher frequency, significant energy is lost in the ice due to the viscosity of pore fluid. Comparative calculations indicate that the anisotropies in both elasticity and permeability are important on the acoustic reflection by an ice sheet at frequency higher than 500 Hz. It is also determined that the slow wave generation at the ice-water interface is significant and thus should be included in calculations.

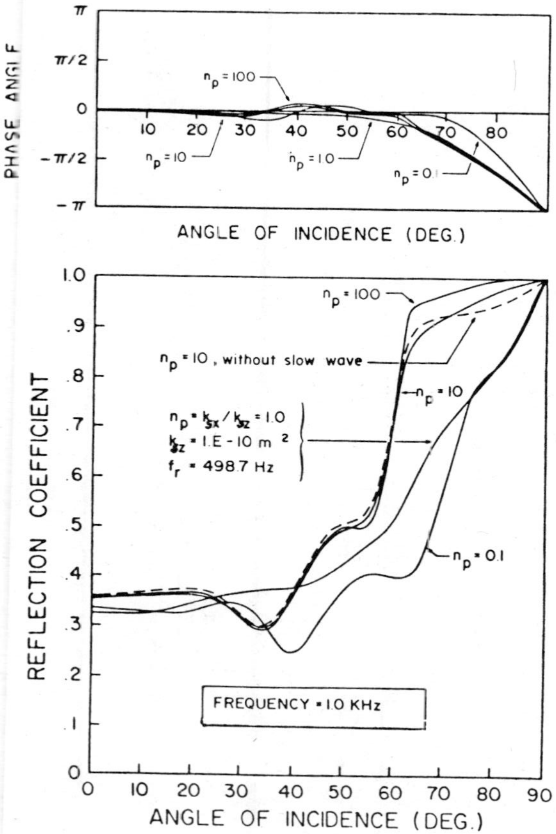

Fig. 2 Reflection diagrams for transversely isotropic bed at 1. KHz.

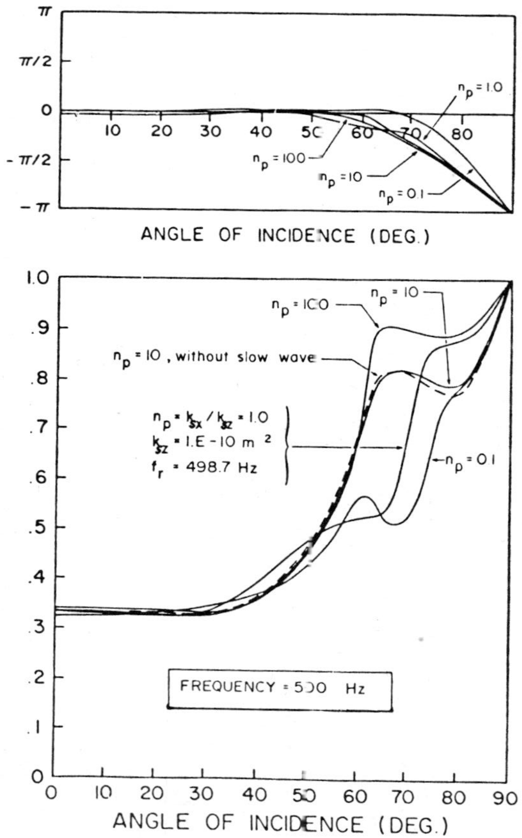

Fig. 3 Reflection diagrams for transversely isotropic bed at 0.5 KHz.

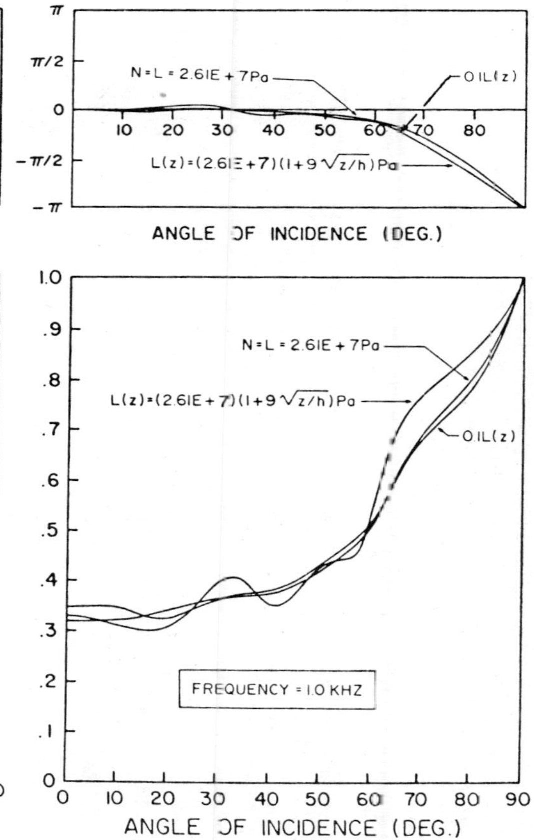

Fig. 4 Reflection diagrams for inhomogeneous sediment bed at 1. KHz.

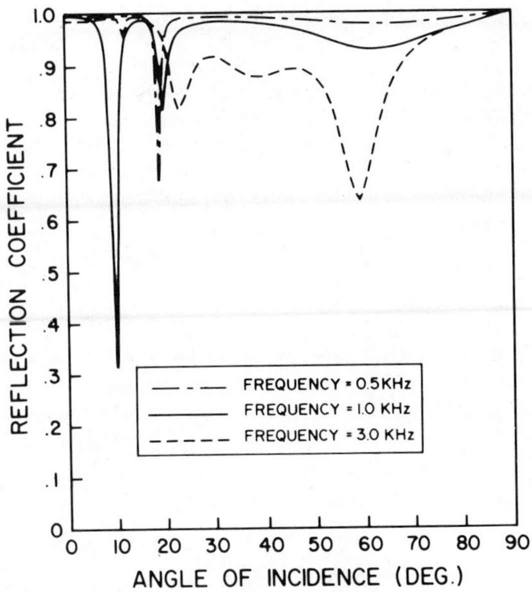

Fig. 5 Reflection diagrams for inhomogeneous transversely isotropic porous sea ice at 0.5, 1.0 and 3. KHz.

CONCLUSIONS

Exact expressions for the complex velocities of the fast compressional waves, slow compressional waves and shear waves in transversely isotropic poro-elastic media have been obtained from the Biot's equation of motion and the Biot-Willis elastic coefficients. A complex 4 x 4 propagator matrix for the acoustic propagation in transversely isotropic poro-elastic media has been formulated. Example calculations are made for acoustic reflections by sediment beds and sea ice layers. It is found that the anisotropy in the permeability is important for acoustic reflection by a sediment bed whereas the elastic anisotropy is not so significant. For modeling of acoustic reflection by sea ice, however, the effects of both elastic and hydrodynamic (permeability) anisotropies are found to be important.

ACKNOWLEDGEMENT

This research was totally sponsored by ONR Code 425UA (Dr. Raymond Fitzgerald), through contract N00014-85-C-0020. The authors are grateful for this support.

REFERENCES

Anderson, D. A., 1961, Elastic wave propagation in layered anisotropic media, J. Geophys. Res., Vol. 66, No. 9, pp. 2953-2963.

Badiey, M. and T. Yamamoto, 1985, Acoustic normal modes in a homogeneous ocean overlaying layered anisotropic porous beds, JASA 77(3), 954-961.

Biot, M. A., 1956, Theory of elastic waves in a fluid-saturated porous solid. II. Higher frequency range, JASA (28), 179-191.

Biot, M. A., 1962, Mechanics of deformation and acoustic propagation in porous media, J. Applied Phys. Vol. 33, 1482-1596.

Biot, M. A. and D. G. Willis, 1957, The elastic coefficient of the theory of consolidation, J. of Applied Mech. 24, 594-601.

Davis, S. H., 1969, Porosity and permeability of natural materials in flow through porous media, edited by R.J.M. Dewiest, Academic Press pp. 53-86.

Mellor, M., G. F. N. Cox and H. Bosworth, 1984, Mechanical properties of multi-year sea ice, CRREL, Repport 84-8.

Saada, A. S., G. F. Bianchimi, and L. P. Shook, 1978, The dynamic response of anisotropic clay, Proc. Geotechnical Engineering Div., Specialty Conference on Earthquake Engineering and Soil Dynamics, Vol. II, ASCE, pp. 777-801.

Stoll, R. D., 1977, Acoustic waves in ocean sediments, Geophysics 42, 715-725.

Stoll, R. D. and T. K. Kan, 1981, Reflection of acoustic waves at a water-sedimemt interface, JASA 70(1), 149-156.

Yamamoto, T., 1983a, Acoustic propagation in the ocean with a poro-elastic bottom, JASA 73(5), 1587-1596.

Yamamoto, T., 1983b, Propagator matrix for continuously layered porous sea beds, Bull. Seismol. Soc. Am. 73, 1599-1620.

ACOUSTIC-TO-SEISMIC COUPLING AT POROUS GROUND SURFACES

Keith Attenborough

Faculty of Technology
The Open University
Milton Keynes MK7 6AA UK

ABSTRACT

Theoretical models and data for seismic activity induced by airborne acoustic sources above porous ground surfaces are presented. Solid particle motion in the ground is shown to result from viscous action associated with fluid motion in the pores. An analysis for grazing spherical wave incidence on a poro-elastic half-space, using a modified Biot-Stoll model, has been carried out. It shows that the motion directly-induced by the air wave incident above the buried sensor gives the dominating contribution except at low frequencies for incidence on a high P-wave, low damping ground where the Rayleigh-type wave contribution may be comparable. An analysis for plane waves incident on a poro-elastic layer resting upon a non-porous elastic substrate is shown to predict ratios of surface vertical particle velocity to surface pressure that correspond closely to measured values for several different ground surfaces. Finally a prediction for a collocated hydrophone-seismometer pair at a similarly structured sea-bed is presented and discussed.

INTRODUCTION

Air-to-ground coupling has been of considerable interest in geophysics both because of the measured ground roll caused by intense acoustic sources and because of the possibility of carrying out shallow seismic refraction surveys using air-shots. Theoretical analyses have been carried out for spherical wave incidence on a ground consisting either of a fluid or solid layer above a fluid or solid half-space.[1,2] Interesting predictions are those of (i) an induced non-despersive Rayleigh wave corresponding to a third branch in the velocity dispersion curve[2] and (ii) maxima in the induced solid particle velocity. Hastrup[3] has investigated cases of shallow water propagation where the sea bottom consists of a layer of soft unconsolidated sediments above a high velocity layer. He shows that under certain conditions, near to grazing incidence, very high reflection losses may occur.

Excess attenuation representing additional frequency dependent attenuation over and above that due to atmospheric absorption and spherical spreading and resulting from the interaction between airborne sound and the ground, has been studied extensively.[4] A typical excess attenuation spectrum over acoustically-soft ground shows a low-frequency peak in the frequency range 200 Hz to 800 Hz that is fairly insensitive to geometry, depending, as it does, on the phase change on reflection at the boundary, which may be approximated, usually, by a locally-reacting one. The existence of this peak has been extensively verified by experiment and may be predicted, adequately, by assuming the soil matrix to be rigid.

Recent efforts at explaining the buried geophone signal induced acoustically by sources above ground, have adopted a model for typical soil surfaces of a poro-elastic layer above a non-porous substrate.[5,7] A modified Biot-Stoll analysis of wave propagation in a poro-elastic medium predicts four compressional (two down going and two up going) waves and two shear waves within the poro-elastic layer. The compressional waves of the first and second kind are decoupled in the frequency range of interest 50 Hz to 1500 Hz. The waves of the second kind, or slow waves, are highly attenuated and dispersive and may be approximated closely by the acoustic pore-wave within the corresponding rigid-framed porous medium.[8] This explains why the reflection properties of many soil surfaces may be modelled by assuming a rigid solid matrix and local reaction. On the other hand, the rapid attenuation of the slow wave is associated with a transfer of momentum to the pore walls inducing seismic motion which is detected by a buried geophone. The induced seismic motion is affected greatly by the near-surface layering.

In this paper the results of a theory of spherical wave transmission at a rigid porous interface are summarised. For typical flow resistivities and porosities of air-filled granular media it is shown that the transmitted field is represented adequately by plane waves travelling at the refracted angle. A brief outline of a theory for spherical wave transmission into a semi-infinite poroelastic medium is given. This shows that, despite the possibility of an acoustically-induced Rayleigh wave, the direct coupled air-to-ground wave is dominant for a large range of soil properties. Finally some implications of the analysis of plane wave transmission at the surface of a poro-elastic layer above a non-porous elastic substrate are presented and related to the analysis by Hastrup[3] of propagation above a layered ocean bottom.

SPHERICAL WAVE TRANSMISSION AT A RIGID POROUS INTERFACE

Near grazing-incidence on a boundary with a high refractive index, the ratio (T) of the field detected at some depth z_2 to that at the boundary is given by

$$T = \exp[ik_1 z_2 (n^2 - \sin^2\theta)^{\frac{1}{2}}] \qquad (1)$$

where

$$\sin\theta = r/(r^2 + z_1{}^2)^{\frac{1}{2}}$$

r being the horizontal source receiver separation, z_1 being the height of the source, n being the complex refractive index = k_2/k_1, k_i being the propagation constant in the upper (i = 1) and porous (i = 2) medium. For very high values of $|n|$ the term in $\sin\theta$ may be neglected and T represents the acoustic transfer function from spherical waves travelling above the surface. Hence equation (1) provides the theoretical justification for deduction of k_2 directly from probe microphone measurements made in air-filled soils and sands.[5,7]

SPHERICAL WAVE TRANSMISSION AT A PORO-ELASTIC INTERFACE

The problem of spherical wave incidence on a poro-elastic half-space is approached in the same manner as that for the more well-known case of fluid-fluid, fluid-solid and solid-solid interfaces, which are described by, for example, Ewing et al.[9] The axisymmetric displacement potentials are expressed in terms of inverse Hankel transforms, the transformed potentials being determined by the boundary conditions at the interface. In this case there are four potentials corresponding to the reflected compressional wave in the upper medium, the transmitted slow and fast compressional waves, and the transmitted shear wave. The complexity of the resulting integrals prohibits a thorough analysis using the saddle point method as in the fluid-fluid case used to model spherical wave interaction with a rigid porous ground surface. Some simplification is possible by considering the limiting case when the bulk modulus and density of the saturating fluid are much smaller than the elastic modulus and density of the solid. A consequence of this 'light-fluid' limit, which should be applicable to air-filled ground surfaces, is that the propagation constants of the fast and shear waves are relatively unaffected by the presence of the fluid, while that for the slow wave is the same as for the rigid-porous case.[8] Details of the resulting analysis are given elsewhere.[10]

With the light-fluid approximation, the restriction to zero receiver height, and the further approximation of large Kr it is possible to make numerical comparisons of the contributions of various possible wave types to the vertical soil particle motion at the porous ground surface. Since at least one wave number is complex in the applications of interest it is necessary to search for the zeros of the determinant of boundary conditions by numerical means and to establish criteria as to whether they lie within the closed integration contour being evaluated. The search reveals, not unexpectedly, that with the 'light-fluid' approximation there are two possible surface waves; one being a Rayleigh-type wave and the other corresponds to the surface wave identified in many analyses of propagation above a rigid-porous boundary.[11] Figures 1 and 2 computed for an air-filled soil with parameters listed in Table 1 show that the predicted relative contributions of the various wave types are highly dependent upon source height and frequency. P-wave speed and the damping of the body waves within the poro-elastic medium are found also to have an important influence on the relative strengths of the wave contributions. The Rayleigh-type wave may give the dominant contribution at low frequencies and grazing incidence on a high P-wave material with low internal damping. Once the source is sufficiently elevated however, or in a material with high internal damping the dominant contribution is from the direct-coupled airwave. This is consistent with many experiments that show times-of-flight from airborne impulsive sources to buried geophones, that relate directly to the speed of sound in air.

PLANE WAVE INCIDENCE ON A PORO-ELASTIC LAYER WITH A NON-POROUS ELASTIC BACKING

This analysis takes account of a total of nine waves. In the air incident and reflected compressional waves are allowed. Fast and slow compressional waves and a shear wave both down going and up going are allowed within the poro-elastic layer. In the substrate a single compressional and a shear wave are transmitted. Nine boundary conditions are applied. These are continuity of normal particle velocity, pressure, total normal stress and tangential stress at the upper interface; and continuity of normal frame velocity, tangential frame velocity, fluid velocity, total normal stress and tangential stress, at the lower-interface.

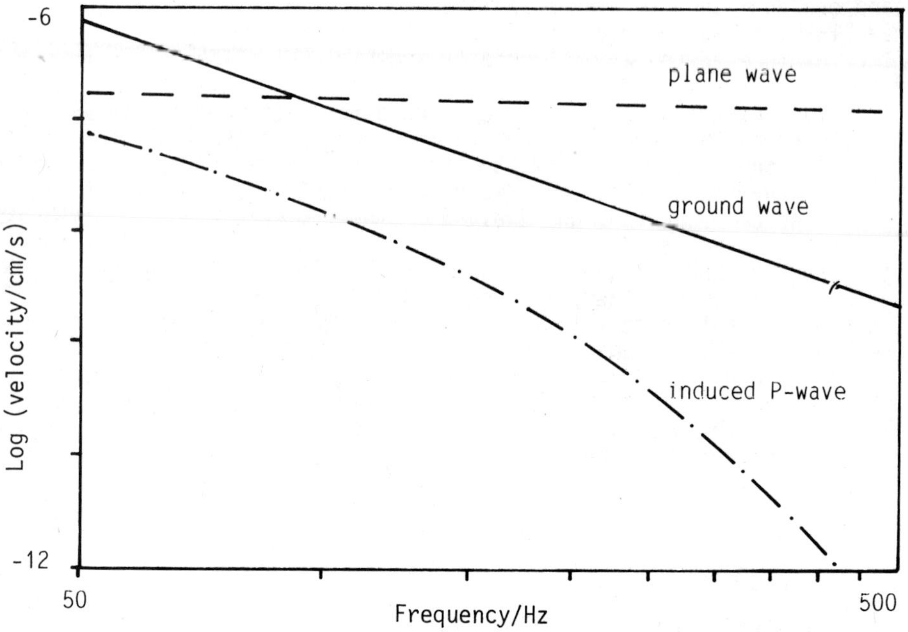

Figure 1 Wave contributions to surface vertical particle velocity at 500 m from a unit source at a height of 100 m.

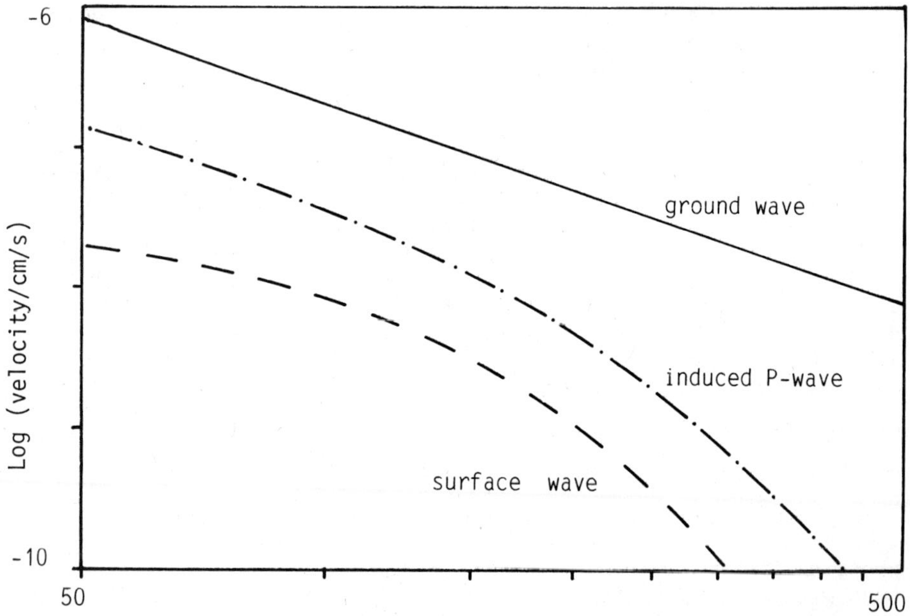

Figure 2 Wave contributions to surface vertical partical velocity at 500 m from a unit sound source at the surface.

For oblique incidence, these conditions have been solved numerically to yield wave amplitudes, normal surface impedance, defined as the ratio of the pressure at the surface to the vertical component of the air particle velocity at the surface and the seismic transfer function, defined as the ratio of the vertical component of the soil particle velocity at the surface to the incident fluid pressure.[5,7] For normal incidence the nine boundary conditions may be reduced to four and solved analytically. Figure 3 shows the predictions of normal surface impedance at 0°, 68° and 80° incidence on a sandy soil (see Table 2). Also shown are measurements made using an inclined track method.[12]

The peaks and dips in the impedance spectrum that are predicted at oblique incidence may be shown to be emphasised by increasing the seismic wave speeds in the elastic substrate. Furthermore the frequencies at which they occur are influenced by the assumed thickness of the poro-elastic layer. The dips in the impedance spectrum are associated with maxima in the spectrum of the seismic transfer function. Figure 4 shows measured and predicted seismic transfer spectra at the surface of a soya bean field characterised by the parameters listed in Table 1.

Following the argument used by Hastrup,[3] the maxima in the seismic transfer function may be associated with trapped modes within the poro-elastic layer. Near grazing-incidence, the condition for these modes may be written

$$\left(\frac{H}{\lambda}\right)_n = \frac{(2n+1)\pi - S\mu_o}{4\pi\sin\mu_o} \qquad n = 0, 1, 2, 3 \ldots \qquad (2)$$

where H is the thickness of the poro-elastic layer,

$$S = \frac{2(\rho_3/\rho_2)}{[1-(c_2/c_3)^2]^{\frac{1}{2}}}$$

$$\cos\mu_o = c_2/c_1$$

c_i and ρ_i are the sound speeds and bulk densities in medium i, where i = 1 represents air, i = 2 represents poro-elastic medium and i = 3 represents non-porous elastic substrate.

TABLE 1 Measured and assumed parameter values at test sites

Parameter	Sandy Soil	Soyabean field
Flow resistivity (Ns m^{-4})	366000	300 000
Porosity (dry)	0.269	0.15
Upper layer density (kg m^{-3})	1700.3	1700
Upper P-wave speed (m/s)	270	207
Lower P-wave speed (m/s)	530	1594
Upper S-wave speed (m/s)	190	85
Lower S-wave speed (m/s)	325	116
Layer depth (m)	2	1.98
Assumed substrate density (kg m^{-3})	2650	2650
Grain shape factor	0.5	0.5
Shape factor ratio	0.725	0.725

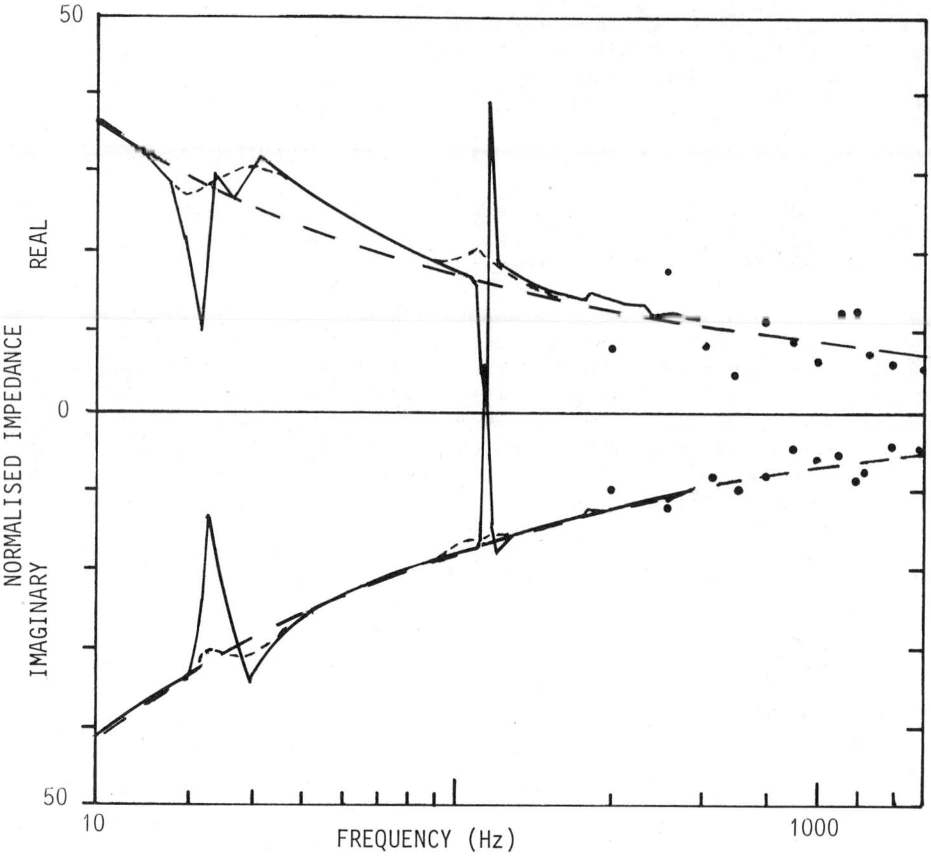

Fig 3 Surface normal impedance predicted for sandy soil (Table 1) at normal incidence (slotted), 68° and 80° (continuous) incidence.

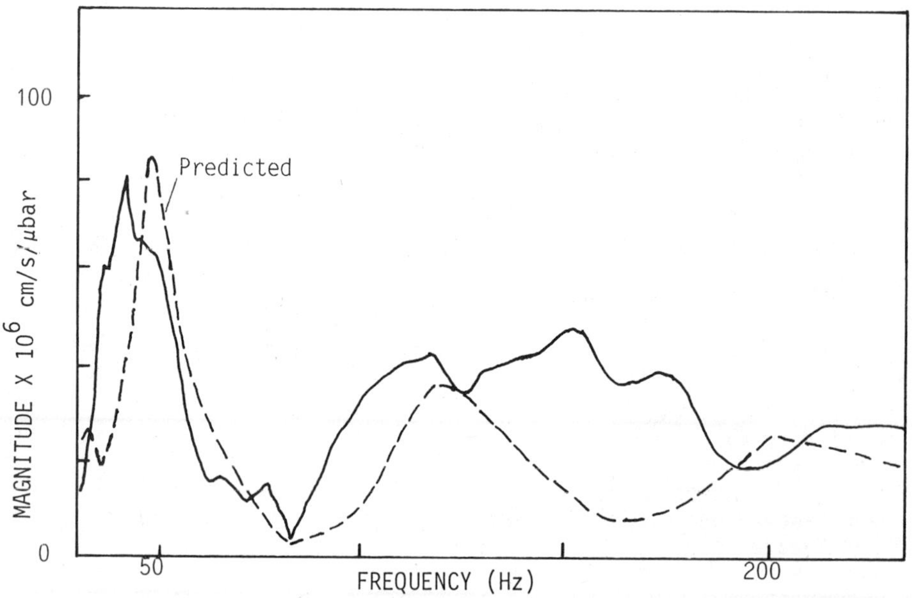

Fig 4 Acoustic-to-seismic transfer function spectrum measured at the surface of a soya bean field and predicted from values in Table 1.

Alternatively if the substrate is assumed rigid then for any angle of incidence θ_i, the condition is

$$\left(\frac{H}{\lambda}\right)_n = \frac{(2n+1)}{4[1-(c_2/c_1)^2\sin^2\theta_i]^{\frac{1}{2}}} \tag{3}$$

Equation (3) reduces to the classical quarter wavelength resonance condition for normal incidence. Table 2 shows the frequencies at which peaks occur in the measured seismic transfer function over sandy soil with parameters listed in Table 1. The asterisked values correspond to major peaks. These are compared with peak frequencies calculated from equations (2) and (3). The softness of the substrate reduces the frequencies of the peaks compared with a rigid substrate. Inclusion of the elastic properties of the substrate produces tolerably good agreement with the measured spectrum shape.

TABLE 2 Frequencies (Hz) of seismic transfer maxima in sandy soil

Measured	Predicted			
	Hard-backing [equ (3)]		Soft-backing [equ (2)]	
	P	S	P	S
	54.7 (f_o)	28.5 (f_o)	45.6 (f_o)	51.25 (f_1)
		85.5 (f_1)		
*105				108.3 (f_2)
*116			122.9 (f_1)	
155		142.5 (f_2)		
	164.1 (f_1)			165.35 (f_3
*174				
		199.5 (f_3)		
*220				222.4 (f_4)
235			231.6 (f_2)	
250		256.6 (f_4)		
	273.5 (f_2)			
275				279.5 (f_5)
303				
*330		313.5 (f_5)		
335				
			340.95 (f_3)	

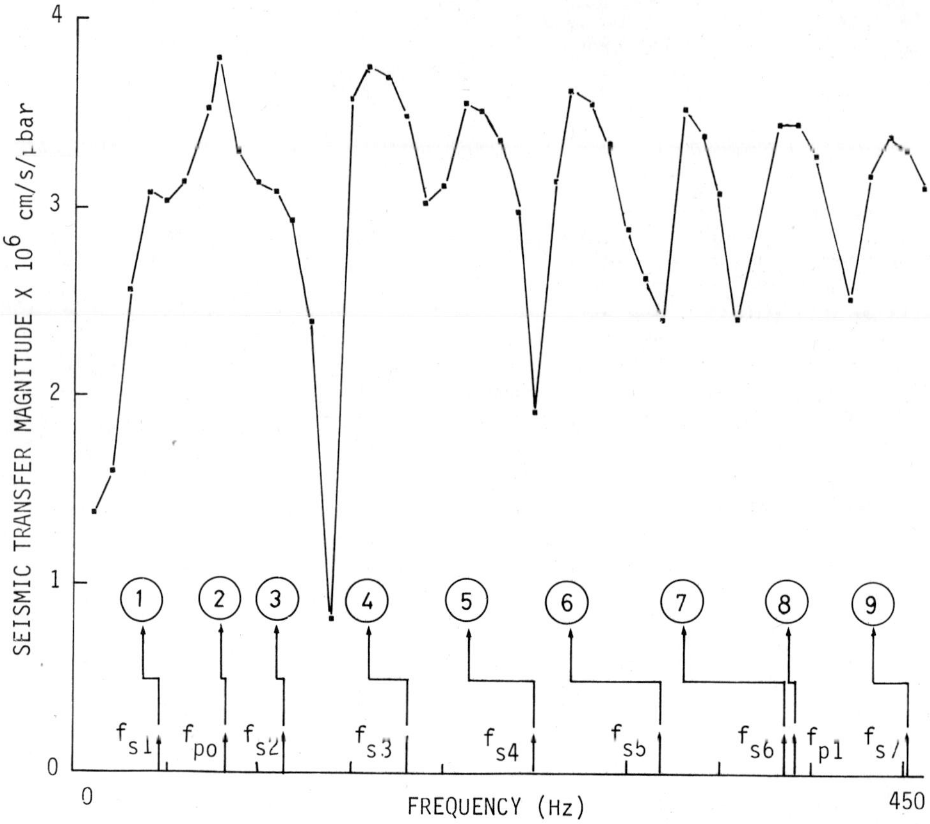

Fig 5 Seismic transfer function spectrum predicted at the surface of an ocean sediment. f_{sn} and f_{pn} are frequencies of shear and P-wave interferences calculated from equation (2).

In Table 2 the S-wave interference frequencies are calculated by replacing the sound or P-wave speeds in equations (2) and (3) by the respective S-wave speeds. Finally results of a calculation of the seismic transfer function spectrum for plane waves in water at 85° incidence on an idealised ocean bottom, with the structural model outlined above, and having the assumed characteristics listed in Table 3, are shown in Figure 5. Also shown in this figure are the frequency locations of P- and S-wave interference peaks predicted by equation (2) for grazing incidence. Although the correct number of peaks is predicted by equation (2), this simple model seems to give a systematic overprediction of the S-wave interference frequencies. The difference is not attributable to the assumption of grazing incidence implicit in equation (2).

CONCLUSIONS

Theoretical and experimental studies of the transfer of energy from an airborne sound wave into porous ground surfaces have shown the usefulness of the Biot-Stoll model in explaining the measured data. This has been incorporated into models of the ground surface both as that of a semi-infinite poro-elastic medium and as that of a poro-elastic layer above a non-porous elastic substrate. The large difference in densities and stiffness between air and soil grains allow the reflection coefficient and normal surface impedance to be calculated usefully from a simple rigid-porous model of the ground. The high attenuation and low wave speeds of the Biot-Stoll slow wave within air-filled media enable porous ground surfaces to be modelled as locally-reacting. The acoustic-seismic transfer spectrum measured by a collocated microphone - buried geophone pair is dominated at low frequencies by interferences between directly-induced and substrate-reflected P- and S-waves. The frequencies at which these lead to maxima in the seismic transfer function near-grazing incidence may be calculated from a plane wave reflection model which ignores the slow wave in the porous layer. The model of a poro-elastic layer above a non-porous elastic substrate may be useful also for the prediction of the seismic transfer function measured by a collocated hydrophone-seismometer pair at a layered ocean bottom. An example calculation for an idealised ocean bottom predicts the kind of low frequency structure that is often measured at air-soil surfaces. However, the simple calculation method for the seismic transfer function maxima turns out to be not as accurate. In particular the predicted location of shear wave interference maxima is in error by as much as 16% near 300 Hz. This may be due to the effects of the induced slow wave that has relatively low damping in an ocean sediment compared with a dry soil.

TABLE 3 Characteristics of idealised layered ocean bottom

Parameter	Value
Upper layer depth	3.7 m
P-wave speed in porous layer (c_2)	1455 m/s
S-wave speed in porous layer	400 m/s
P-wave speed in non-porous substrate	1900 m/s
S-wave speed in non-porous substrate	665 m/s
Flow resistivity of porous layer	100,000,000 MKS rayls
Porosity	0.305
Bulk density of porous layer	1450 kg m^{-3}
Bulk density of substrate	2650 kg m^{-3}
Bulk modulus of grams	3.6 x 10^{11} cgs
Fluid bulk modulus	2.0 x 10^{10} cgs
Pore shape factor ratio	1.0
Grain shape factor in porous layer	0.5

ACKNOWLEDGEMENTS

Much of this work has been supported by the US Army through its European Research Office and carried out in collaboration with the Physical Acoustics Research Laboratory at the University of Mississippi, directed by Professor H E Bass, and with the Battlefield Environment Group, Chief, Mr Jerry Lundien, at the USA Waterways Experiment Station, Vicksburg, MS. The author would like to thank Dr J M Sabatier for his assistance with the field work and the calculations.

REFERENCES

1 F Press, and M Ewing. 'Theory of Air-coupled Flexural Waves' J Appl Phys (1951).

2 W S Jardetzky and F Press. 'Rayleigh-wave coupling to atmospheric compression waves'. Bulletin of the Seismological Society of America 42 135-144 (1952).

3 O F Hastrup. 'Some bottom-reflection loss anomolies near grazing and their effect on propagation in shallow water'. Bottom-Interacting Ocean Acoustics Papers presented by Saclanten staff at a conference organised by W A Kuperman and F B Jensen, 9-13 June, (1980).

4 J E Piercy, T F W Embleton, and L C Sutherland. 'Review of noise propagation in the atmosphere'. J Acoust Soc Amer 61 1403-1418. (1977).

5 J M Sabatier, H E Bass, L N Bolen, K Attenborough and V V S S Sastry. 'The Acoustic Surface Impedance of a Layered Poro-elastic Soil'. J Acoust Soc Amer. (submitted for publication 1985).

6 K Attenborough, J M Sabatier, H E Bass and L N Bolen. 'The Acoustic Transfer Function at the Surface of a Layered Poro-elastic Soil'. J Acoust Soc Amer. (submitted for publication 1985).

7 J M Sabatier, H E Bass, L N Bolen and K Attenborough. 'Acoustically-induced Seismic Waves'. J Acoust Soc Amer. (submitted for publication 1985).

8 K Attenborough. 'Pore shape and the Biot-Stoll model for saturated sediments'. Proc symposium on Ocean Seismo-Acoustics, La Spezia. June (1985).

9 W M Ewing, W S Jardetzky and F Press. Elastic Waves in layered Media. McGraw-Hill. (1957).

10 K Attenborough, N W Heap, T L Richards and V V S S Sastry. 'Acoustic-to-Seismic Coupling at Porous Ground Surfaces'. Final Report on Contract DATA US Army, European Research Office, January 1984 (unpublished).

11 K Attenborough, S I Hayek and J M Lawther. 'Propagation of sound above a porous half-space'. J Acoust Soc Amer. 68 1493-1501 (1980).

12 L van der Heijden. 'The influence of vegetation on acoustic properties of the ground'. Doctoral thesis, The Catholic University of Nijmegen, Department of Botany (1984).

GEOTECHNICAL CHARACTERISTICS OF THE SEA BED RELATED TO SEISMO-ACOUSTICS

Denzil Taylor Smith

School of Ocean Sciences
University College of North Wales
Bangor, U.K.

ABSTRACT

The pattern of seismo-acoustic phenomena in marine sediments is often paralleled by such sediments' geotechnical behavior. Thus a compressional wave velocity variation with depth is closely akin to the void ratio change with pressure seen in the consolidation test, and can be utilised to calculate the uni-directional compressibility of the sediment. Further, by use of a Biot model, permeability can be determined from the velocity-porosity data which allow the rate of settlement of any engineering structure on the sea floor to be derived. The Biot calculation requires an assessment of both frame and shear modulus, achieved by the in situ measurement of the shear wave velocity. This quantity is a definitive geotechnical parameter; it particularly allows the degree of anisotropy of a soil fabric to be postulated of benefit to acoustics and geotechnics alike. The full interdependence of the two can be obtained by the strain rate and strain level correction curves produced by the resonant column apparatus.

INTRODUCTION

Clearly, there are close similarities between the needs of the design engineer and the acoustician in the underwater environment. Apart from anything else both are in the prediction game. From a series of mechanical measurements, which hopefully seek to classify and identify the foundation material, the engineer attempts to predict the material behavior under the levels of stress and strain to be expected through impulsive, cyclic or even long-time loading. The acoustician, operating with the same material properties, aims at predicting the seismo-acoustic propagation phenomena through the use of some porous medium model of which that based on Biot's differential equations seems to be the most favored .

Unfortunately for the predictors the surface materials are often highly variable in distribution and in their relevant properties. Even in areas of apparent homogeneity, variations in saturation - the replacement of the pore water by gas - and anisotropic effects on a large or small scale can lead to complex behavior under load, whether static or dynamic. The presence of thin bands of clay, in otherwise competent foundations, can be potential slip surfaces; and intercalated sands and silts in clay formations, whose drainage is consequently inhibited, can lead to the build-up of excess pore pressures producing a foundation failure[2]. Such anisotropy is also a nightmare to the geoacoustician giving spurious bottom reflections and transmission responses through interference from multiple reflections.

Much of the information, both qualitative and quantitative, required by the geotechnical engineer can be obtained by the use of geophysical techniques, particularly through the measurement of seismic wave velocities and electrical formation factor, although acoustic impedance is being increasingly favored. In seismo-acoustics the now common practise of digital data collection, with the range of processing techniques thus made available, allow some considerable definition to be made in the mapping of the various sediment boundaries and their lateral distribution. Pockets of gas clearly show on the record and thin clay layers are often identifiable on the basis of the phase relationships of the returned signals. The processing technqiue can also provide vertical "logs" of compressional wave velocity and acoustic impedance[3,4,5,6]. Similar data can also be obtained from downhole, crosshole, and inhole borehole measurements; importantly such techniques can provide direct assessments of density and shear wave velocity[7]. While not yet standard commercial practise, some organisations (such as the British Geological Survey) offer the sea floor measurement of seismic wave velocities and electrical formation factor by a range of devices[8]. Other authors[9,10], describe techniques measuring shear wave velocity on the sea floor, and Kiel University seems to have had some success with a shear wave profiler[11].

However, a dilemma arises: a complete interpretation of the geophysically-collected quantities, to provide the geotechnical design parameters, requires the use of a viable seismo-acoustic model (such as Biot); and the viable seismo-acoustic model needs the same geotechnical parameters in order to solve the differential equations. Probably the parameters of most importance to both geotechnics and seismo-acoustics are the in situ values of sediment compressibility and permeability. The need for these are best seen in considering separately the Biot model and consolidation theory, and then the usefulness of the interrelations.

THE BIOT MODEL

The Biot theory has been considered by many authors (most recently by Ogushwitz[1]) with some mathematical rigor. It will only be cursorily examined here largely to define the parameters involved. For an isotropic porous solid it is based on stress-strain relationships of the following sort:

$$\sigma_x = X\Delta - 2G(\varepsilon_y + \varepsilon_z) - Y\zeta \quad (1)$$

$$\sigma_y = X\Delta - 2G(\varepsilon_z + \varepsilon_x) - Y\zeta \quad (2)$$

$$\sigma_z = X\Delta - 2G(\varepsilon_x + \varepsilon_y) - Y\zeta \quad (3)$$

$$\tau_{xy} = G\gamma_{xy} \quad (4)$$

$$\tau_{yz} = G\gamma_{yz} \quad (5)$$

$$\tau_{xz} = G\gamma_{xz} \quad (6)$$

$$\lambda = Z\zeta - Y\Delta \quad (7)$$

where σ and τ are normal and shear stresses on an element of the solid frame, G is the shear modulus, λ is the pore-water pressure, and ε and γ

are strain components appropriate to the stress applied. The volumetric strain, or dilatation, is given by

$$\Delta = \varepsilon_x + \varepsilon_y + \varepsilon_z \tag{8}$$

ζ is the "relative dilatation" representing the change in the volume of fluid caused by flow in or out of the elemental volume and expressed as

$$\zeta = n(\Delta - \xi) \tag{9}$$

where **n** is the porosity and ξ the absolute dilatation of the fluid.

The parameters X, Y, and Z, which have units of pressure, are each a form of sediment elastic modulus and can be related to the compressibilities of the constituents by use of a more detailed theory relating stress and strain in a fluid-saturated porous medium[12,13,14,15,16,17,18]. All authors arrive at relationships which are fundamentally identical:

$$X = \frac{(1-R)^2}{(1-n-R)\,C_s + nC_w} + \frac{1}{C_f} + \frac{4G}{3} \tag{10}$$

$$Y = \frac{(1-R)}{(1-n-R)\,C_s + nC_w} \tag{11}$$

and

$$Z = \frac{1}{(1-n-R)\,C_s + nC_w} \tag{12}$$

where C_f is the compressibility of the skeletal frame, C_s the compressibility of the solid material constituting the frame, R the ratio C_s/C_f, and C_w the compressibility of the fluid (usually sea water) contained in the pore spaces of the frame.

If the stress-strain equations are combined with Darcy's equation for flow through a porous medium, and with the stress equations of motion derived from Newton's Second law, two partial differential equations are obtained which govern the propagation of a dilatational (compressional) wave through the medium:

$$\nabla^2 (X\Delta - Y\zeta) = \frac{\partial^2}{\partial t^2} (\rho\Delta - \rho_w\zeta) \tag{13}$$

and

$$\nabla^2 (Y\Delta - Z\zeta) = \frac{\partial^2}{\partial t^2} (\rho_w\Delta - m\zeta) - \frac{\mu}{k}\frac{\partial\zeta}{\partial t} \tag{14}$$

In this set of equations ρ is the bulk density of the system given by

$$\rho = (1-n)\rho_s + n\rho_w \tag{15}$$

ρ_w is the liquid density, μ is the liquid viscosity and **k** is the coefficient of permeability in units of area. The parameter **m** has the dimensions of density and accounts for the fact that not all of the liquid moves in the general direction of the pressure gradient because of the shape of the grains, the way they interlock, and the consequent form of the cavities. This parameter may be written

$$m = \frac{b\rho_w}{n} \tag{16}$$

where **b** is a mass-coupling factor which must have a numerical value greater than unity. Stoll[17] states that the value for **b** must be determined experimentally; in general modelling he used a value of 1.25 for coarse-grained materials such as sands and one of 3.0 for clays. Berryman claims that **b** can be defined theoretically, though the components of his formula require a microscopic study of the frame-fluid relationships. As the structure factor was introduced by Biot to account for the tortuosity of the pores, Brown[20] draws an analogy to electrical tortuosity in a porous medium through formation factor F-porosity-permeability considerations:

$$b = Fn \tag{17}$$

The differential equations (13) and (14) allow solutions of the form

$$\Delta = A_1 \exp\{i(\omega t - lx)\} \tag{18}$$

and

$$\zeta = A_2 \exp\{i(\omega t - lx)\} \tag{19}$$

where ω is the angular frequency of the applied seismic wave and **l** is the wave number. The velocity of these waves is

$$V = \omega/l \tag{20}$$

Biot's analysis considers solutions to the equations within a low frequency range where Poiseuille flow is valid, and also extends it to a high frequency range where such flow breaks down. The latter analysis requires the introduction of a complex viscosity correction for frequency through the dimensionless ratio ω/ω_c where ω_c is a characteristic angular frequency for the material under consideration. The full assessment of this effect involves the use of a complex Kelvin function which demands a specification of the dimensions of the pore space[1]. As the rigorous mathematical treatment requires information about the sediment sample which is difficult or even impossible to obtain, there is much to be said for the approximate solution favoured by Geerstma[21] or Ishihara[15], particularly for the former's handling simplicity and accordance with the experimental data for most sea floor sediments, as illustrated in a later section. The approximate solution makes use of two velocities: the "zero frequency velocity" V_o for applied frequencies very much lower than the characteristic frequency of the material; and the "infinite frequency velocity" V_∞ where the applied frequency is very much higher than the characteristic frequency. Thus the sediment compressional wave velocity V_p is

$$V_p^2 = \frac{V_\infty^4 + V_o^4(\omega_c/\omega)^2}{V_\infty^2 + V_o^2(\omega_c/\omega)^2} \tag{21}$$

where

$$V_o^2 = X/\rho \tag{22}$$

$$V_\infty^2 = \frac{Z\rho + Xm - 2Y\rho_w}{m\rho - \rho_w^2} \tag{23}$$

and

$$\omega_c = \mu\rho/k(m\rho - \rho_w^2) \tag{24}$$

If the permeability coefficient **k** is rewritten in terms of units of velocity (ϕ) rather than units of area, then

$$k = \phi\mu/g\rho_w \tag{25}$$

and

$$\omega_c = gn\rho/\phi\,(b\rho - n\rho_w) \tag{26}$$

Given the range of permeabilities found in sediments, the characteristic frequency range is from about 100kHz (sands) to over 100MHz (clays).

In summary, therefore, to predict a compressional wave velocity in a sea-floor sediment the following (mostly geotechnical) quantities are required:

(a) frame, solid and liquid compressibilities C_f, C_s, and C_w,
(b) shear modulus G,
(c) sediment porosity n,
(d) solid and liquid densities ρ_s and ρ_w,
(e) permeability ϕ, and
(f) mass coupling factor b.

CONSOLIDATION THEORY

Consolidation in the geotechnical sense is the reduction in volume of a sea-floor sediment under a compressive load. The load may be either produced by some engineering construction, or by the overburden pressure from the accumulating superincumbent material itself. In either case the end result is settlement. There are two types of settlement (Fig. 1):

(1) Initial settlement which occurs as soon as the load is applied to be carried by the mineral framework and the pore fluid. This causes the framework to deform slightly but, because the fluid phase is virtually incompressible, no real deformation can occur, the total load being supported by the fluid.

(2) Consolidation settlement (time dependent):

(a) Primary, or hydrodynamic consolidation, which is controlled by the escape of water from the soil voids resulting in the change of volume.

(b) Secondary consolidation as a result of deformation and rearrangement of the solid grains, usually occurring after the primary consolidation has ceased and may continue for some time.

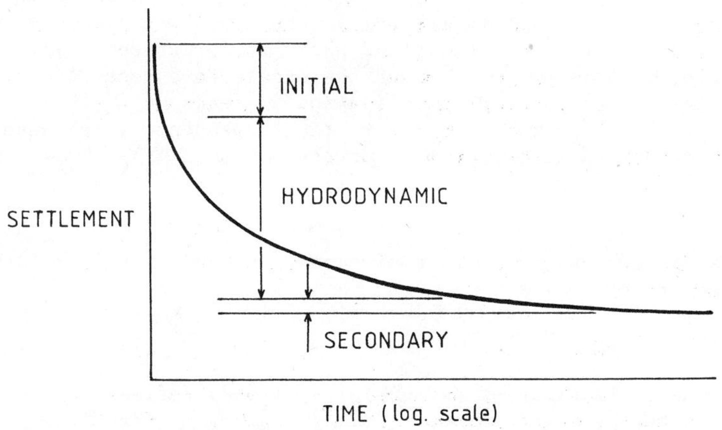

Fig. 1. Time-settlement in consolidation

A typical example of settlement is given by the Ekofisk oil storage platform resting on the floor of the North Sea (Fig. 2). The total settlement of the sediment column caused by the applied load can be derived from the theory of elasticity by assuming that the medium is elastic, isotropic and homogeneous.

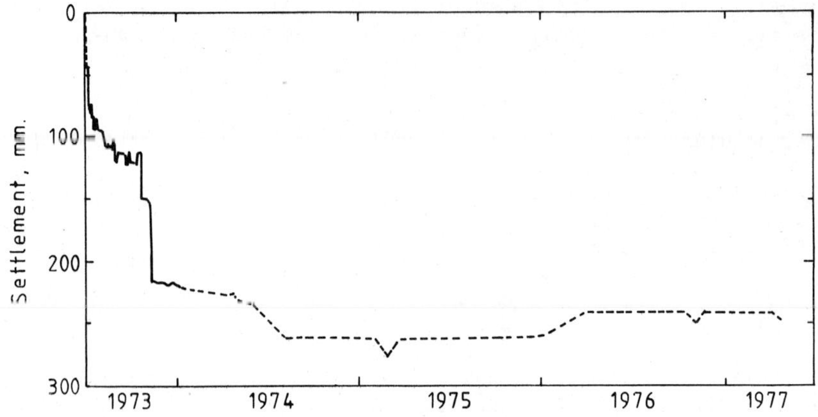

Fig. 2. Time-settlement for the Ekofisk tank

An approximation seemingly valid for most foundation calculations gives the settlement **S** as

$$S = pRA/E \tag{27}$$

where **p** is the load applied, **R** is the radius of the loaded area, **E** is Young's modulus, and **A** is a constant directly related to Poisson's ratio ν.

While settlement can be determined from equation (27) through in situ measurement of **E** and ν, its proper time-related behavior is usually estimated in the laboratory from a uniaxial consolidation test where a cylinder of material is strained in one dimension only under an axial stress. The load is carried initially by the water in the pores and is then gradually transferred to the solid skeleton by the water draining out of the sample. There is, therefore, a change in the excess pore pressure with time as drainage proceeds; this results in a change in volume of the sample, seen as a reduction in its axial length for the cylindrical surface is constrained. Each load is applied until the pressure is the total effective stress, this occurring when the excess pore pressure is reduced to zero. This "100% primary consolidation" depends on the soil thickness, compressibility, and permeability; its time dependence is defined by a differential equation in excess pore pressure:

$$c_v \, d^2u/dz^2 = du/dt \tag{28}$$

where c_v is called the coefficient of consolidation. It is related to other parameters by

$$c_v = \phi / m_v g \rho_w \tag{29}$$

where ϕ is the coefficient of permeability already defined in the Biot theory, $\mathbf{m_v}$ is the uniaxial compressibility, and ρ_w is the density of the pore-water. Of particular interest to seismo-acoustics is the uniaxial compressibility: this is the reciprocal of an elastic modulus - the constrained modulus **D** - derived from the slope of the stress-strain curve, usually plotted as the logarithm of the applied pressure against void ratio (Fig. 3). The constrained modulus is theoretically identical to the combination of elastic coefficients defining the compressional wave velocity obtained from a solution of the standard wave equation. Of interest also to Biot modellers is the decompression portion of the stress-strain curve.

This is achieved by unloading the sample at certain pressures once the "100% primary consolidation" stage has been reached. As no excess pore pressure exists at this stage, then the decompression curve relates to the elastic rebound of the skeletal frame and its slope must provide a value for the frame compressibility at this pressure. This point will be referred to in the section on shear waves.

Another method of determining the consolidation behavior of a sediment is to construct the void ratio-pressure curve through considerations of the submerged weight of the superincumbent material at selected depths in the sediment column. This is the sediment-compression curve of Terzaghi[22], and may be derived by a procedure discussed by Skempton[23]. To illustrate this, Fig. 3 has the equivalent depths for the DSDP red clay sample being tested by calculation from

$$p = \rho_b g z \tag{30}$$

where

$$\rho_b = \text{submerged density} = \rho - \rho_w$$

$$= (\rho_s - \rho_w)/(1 + e)$$

Thus, whichever technique is used, the consolidation curve expresses the variation of void ratio **e** (or porosity) with depth in the sediment column. As there is a time-honored relationship between compressional wave velocity and porosity it is to be expected that the gradients shown in consolidation will equally express themselves in velocity terms.

VELOCITY-CONSOLIDATION RELATIONSHIPS

The literature on compressional wave velocity gradients in marine sediments is considerable (for example, Hamilton[24]). These data have been

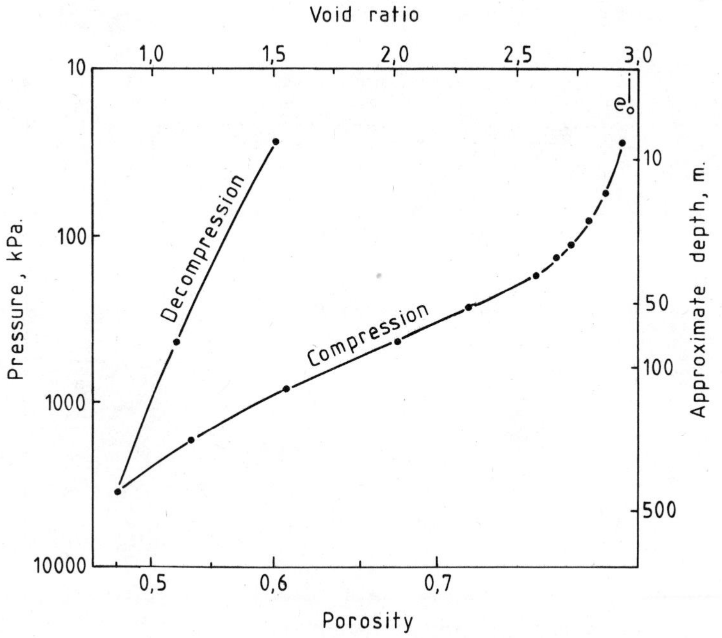

Fig. 3. Consolidation stress-strain curve for a DSDP red clay sample

obtained by a variety of techniques both in situ, from seismic refraction and reflection experiments, and by laboratory or shipboard measurements on long cores. Ogushwitz[1] in his detailed study of the applicability of the Biot model compares these experimental values to predictions based on the model. The estimating technique requires information concerning the variation of porosity with depth; for this purpose Hamilton's regression equations[23] based on laboratory analyses of DSDP samples, and where five distinct classes emerge, are used:

$$n = 0.720 - 0.816z + 0.361z^2 \quad \text{(terrigeneous)}$$

$$n = 0.720 - 0.987z + 0.830z^2 \quad \text{(calcareous)}$$

$$n = 0.814 - 0.813z - 0.164z^2 \quad \text{(pelagic clay)}$$

$$n = 0.900 - 0.016z - 3.854z^2 \quad \text{(radiolarian ooze)}$$

$$n = 0.861 - 0.549z - 0.492z^2 \quad \text{(diatomaceous ooze)}$$

where the depth **z** is in kilometers. It is interesting to note that the DSDP red clay sample, used in Fig. 3 as an example of the consolidation experiment, has a porosity depth relationship of the following form:

$$n = 0.765 - 1.236z + 1.229z^2$$

Hamdi[26], in a comprehensive examination of consolidation-velocity correlations (obtained by continuously measuring compressional and shear wave velocities during the consolidation test) shows that the velocity variation with effective pressure can also be divided up into five clsses (Fig. 4), and that the variation of velocity with void ratio for all samples is a function of the effective stress (Fig. 5). The latter correlations show that the rate of increase of velocity with decreasing void ratio decreases with overburden pressure. Consequently, for any

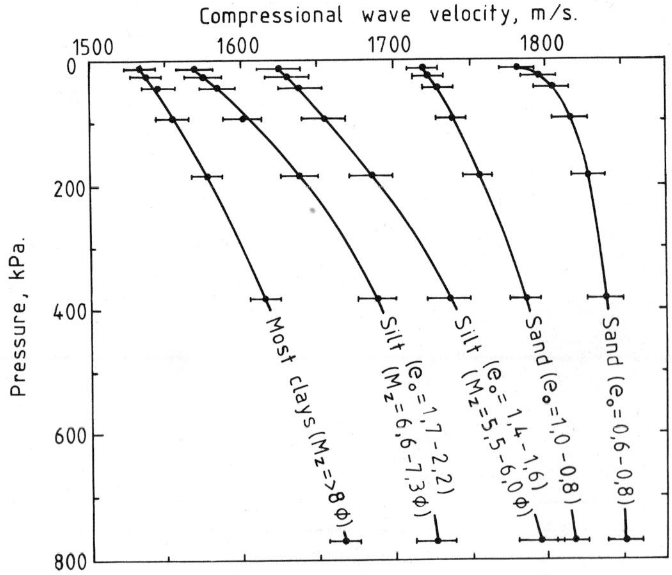

Fig. 4. Compressional wave velocity variations with effective stress

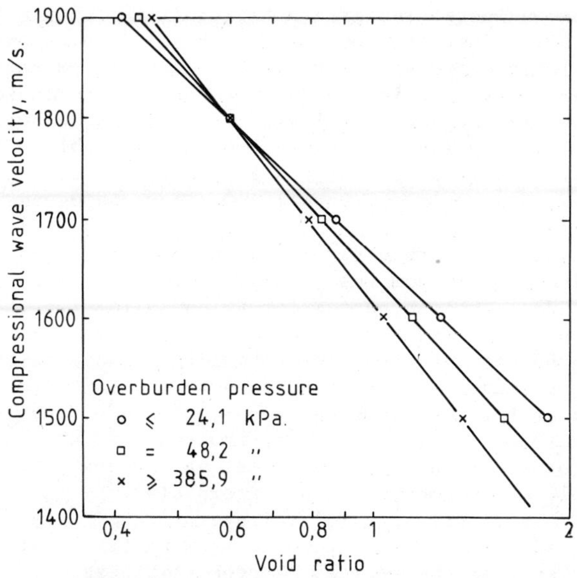

Fig. 5. Compressional wave velocity variations with void ratio for different effective stresses

homogeneous sediment column there is an increase in velocity with depth, the gradient decreasing with depth. This rather contradicts the findings of Ogushwitz[1] where the gradients are predicted to increase with depth: for a calcareous clay from 0.4 s^{-1} at the top of the sediment column to 1.1 s^{-1} at 300 meters decreasing thereafter. Ogushwitz's sand predictions depend on what value of porosity is assumed: for a constant porosity of 0.4 there is a gradient decrease from about 5.0s^{-1} at the immediate sea bottom to 1.4 s^{-1} at 20 m to 0.3 s^{-1} at 400 m; but a porosity decreasing with

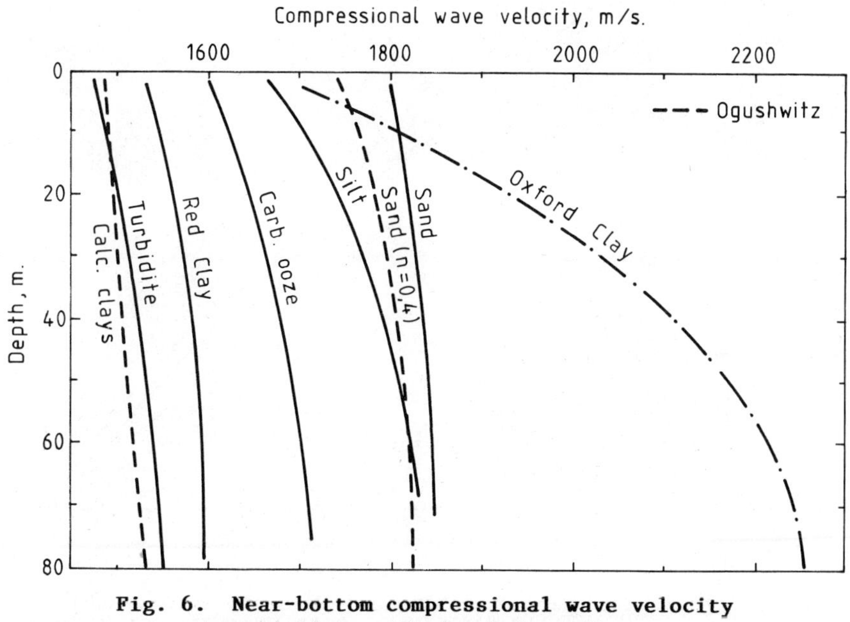

Fig. 6. Near-bottom compressional wave velocity variations with depth in sediment

depth produces an increasing gradient virtually identical to that of the calcareous clay. There is some observational support for the idea of velocity gradients decreasing with depth: apart from Hamilton's comment[24], Lovell & Ogden[27] have gradient magnitudes similar to those of Ogushwitz but decreasing with depth; and Davis'[28] downhole measurements at a saturated Oxford clay land site give exceptionally high initial gradients which fall off rapidly at depth. Fig. 6 represents a selection of the data from Hamdi[26], Davis[28], Lovell & Ogden[27] and Ogushwitz[1].

Whichever author's results or predictions are taken it is clear that there is an increase in compressional wave velocity with depth and that this increase cannot be accounted for by consolidation (that is, by a reduction in porosity or void ratio) alone. It may well be the case that the immediate bottom deposits have very large gradients through a rapid consolidation process, but for sand-size sediments the overall change in porosity with increasing depth, as a consequence of the increase in the overburden pressure, would seem to be too small to account for the gradients involved. The rapid gradient at the near surface, shown in Davis' borehole measurements on the Oxford clay[28], is probably a result of a rebound release of pressure following erosion of this highly overconsolidated clay. Thus it would seem that the main cause of the velocity gradients is the increase in the stiffness of the sediment frame with depth in the sediment column; such a stiffness increase can be attributed to many factors among which effective stress, chemical bonding and, even, geological time itself loom large.

While in any prediction process relating sediment consolidation and compressional wave velocity it is difficult to account for full or partial cementation, and certainly for geological time, the velocity correlations with effective stress (Fig. 4) and with void ratio (Fig. 5) do allow an attempt to be made on a prediction of the void ratio-effective stress curve.

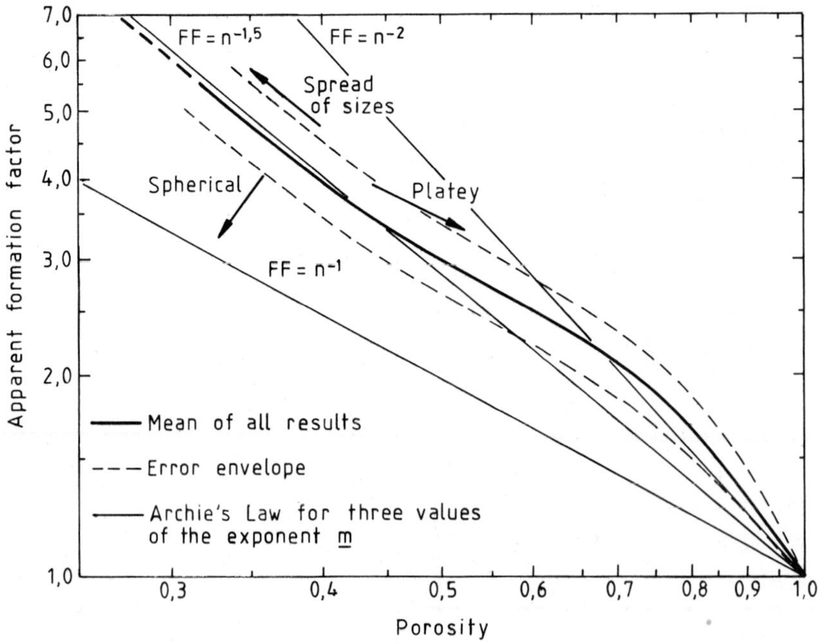

Fig. 7. Electrical formation factor - porosity relationships for full range of sediments (sands to clays)

Table 1 Predicted and measured uniaxial compressibility for different sediment samples (10^{-7} m^2/N)

Soil type	e_o	Predicted	Measured
Sand	0.83	2.1	1.8
Clay	1.65	3.9	3.7
Clayey sand	1.34	4.1	3.4
Clay	1.71	4.1	3.6

Such a prediction[29] requires (a) the in situ measurement of the compressional wave velocity on the sea-bottom, either by careful seismic refraction or reflection methods or by probe techniques, (b) the in situ determination of the initial void ratio perhaps by a measurement of electrical resistivity and the use of a curve as given in Fig. 7, and (c) a determination of the graphic mean size of the sea-floor sediment using a grab sample. With this information it is possible to construct the required curve. Using Cernock's data[31] an example of the predicted and measured stress-strain curves is given in Fig. 8. Table 1 shows a comparison between predicted and measured uniaxial compressibility for several different samples using this procedure. To complete the study of the consolidation process the settlement-time relationship needs to be established through a calculation of the coefficient of consolidation, and this requires knowledge of the permeability. This quantity may be guessed at through well-established empirical formulae or may be calculated from the seismo-acoustic model. The latter approach needs a measurement of the sediment shear wave velocity.

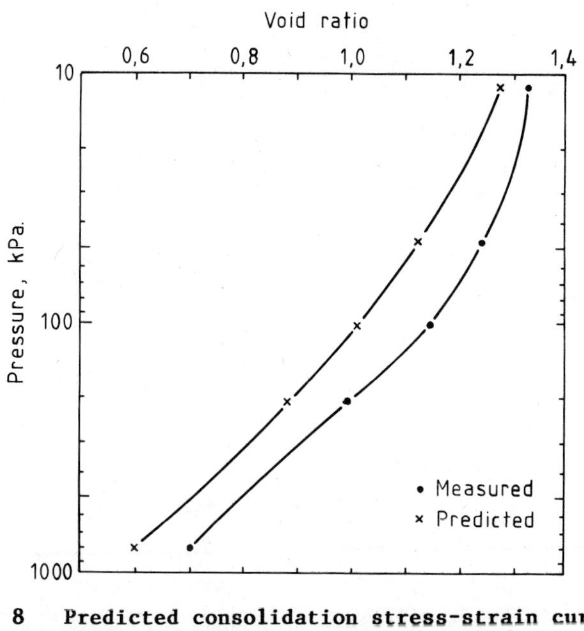

Fig. 8 Predicted consolidation stress-strain curve compared to Cernock's measured data

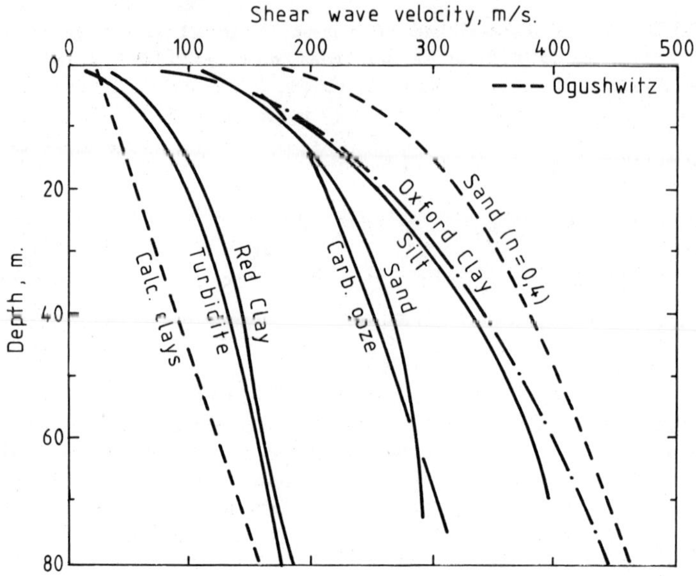

Fig. 9 Shear wave velocity variations with depth in near-bottom sediments

While a close affinity exists between consolidation and compressional wave velocity, probably the most definitive seismic parameter from a geotechnical point of view is a material's shear wave velocity. Stating this is not to ignore the considerable importance of seismic damping, particularly in earthquake areas, but so much success has been achieved by the use of the shear wave velocity in geotechnical site investigations that it has now become one of the required parameters for the design of large structures such as dams and nuclear power stations.

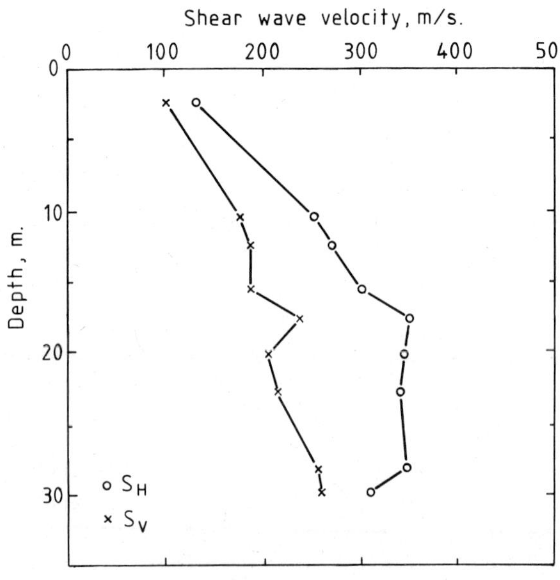

Fig. 10. Shear wave velocity anisotropy with depth in overconsolidated London clay

The usefulness of the shear wave in civil engineering is basically to define the shear modulus **G** and its directional variations, particularly with depth, so that the performance of any proposed structure can be predicted. Figs. 9, 10 and 11 are examples of such depth variations; all indicate that for a near-surface material - whether sand, ooze or overconsolidated clay - there is a very rapid increase in the velocity over the first few tens of meters of depth in the (supposedly) homogeneous media. The pressuremeter data of Fig. 11 in a uniform sand show the same trend. The sensitivity of the polarisation direction to anisotropy is clearly indicated in Fig. 10[28].

As the magnitude of the shear wave velocity is fundamentally related to the stiffness of the sediment frame, and to very little else, the cause of the gradients must somehow be wrapped up in the factors which control the bonding of the frame. Undoubtedly effective stress is the dominant factor in the near-surface sediments providing such large gradients at such depths. An interesting example of such control is given by a study of liquefaction phenomena[32]. Fig. 12 shows the effect of cyclic loading where the permeability of the medium is such that the excess pore pressure developed at each stress application has not had enough time to dissipate before the load is again applied. This leads to a build up of pore pressure with a consequent reduction in effective stress until ultimately this becomes zero; at this position the medium has no resistance to shear, the liquid-solid mixture becoming a suspension and behaving like a liquid. When this occurs, as it can do frequently on the sea floor, the foundation has no capacity to support the structure resting upon it and (often) catastrophic failure results. The shear wave at the point of liquefaction disappears, the sediment frame having collapsed completely.

In contrast to the compressional wave there is no clear division of the various sediments into groups obeying the same sort of velocity gradient. For most sediments, some power law obtains but the exponent can be quite variable from about 0.25 to 0.50; this seems to have no relation to the type of material. For example, Lovell & Ogden[27] have the following gradients for two DSDP red clay samples:

$$V_s = 38.1127 \; z^{0.2484}$$

$$V_s = 20.4994 \; z^{0.4496}$$

Ogushwitz's predictions[1] give linear gradient ranges from 2.8 s^{-1} at the sediment-water interface to 0.9 s^{-1} for sands at 500 meters and 2.4 s^{-1} to 0.7 s^{-1} for clays over the same depth range. However, Hamdi[26] and Lovell & Ogden[27] have much higher linear gradients for surficial sediments, the former measuring gradients for surficial sands up to 20 s^{-1}. Interestingly Davis[28] obtained directional linear gradients for overconsolidated clays:

$$V_s = 122 + 9.21z \quad \text{(parallel to bedding)}$$

$$V_s = 84 + 16z \quad \text{(perpendicular to bedding)}$$

The first of these is similar to Hamilton's equation for shallow depth silt-clays and turbidites :

$$V_s = 116 + 4.65z$$

Clearly, the empirical equations for shear wave velocity variations with depth are extremely variable, particularly for the near-bottom sediments where it is most important for both geotechnical and seismo-acoustic propagation studies. At depth, as Fig. 9 shows, most sediments have similar gradients.

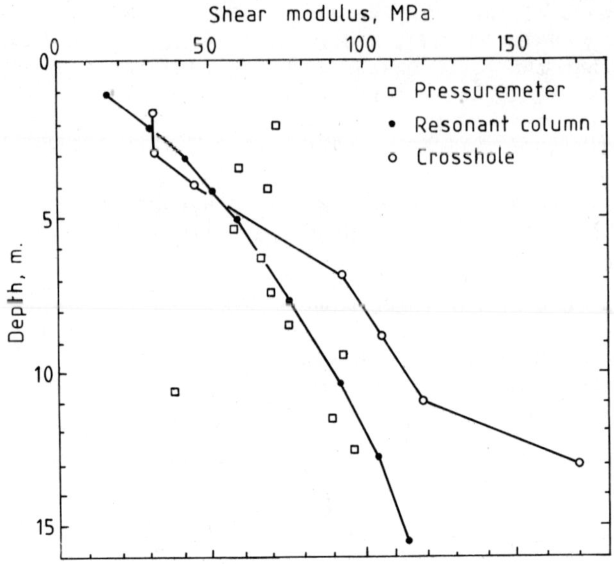

Fig. 11. Resonant column predicted shear modulus depth variations compared to in situ pressuremeter and seismic measurements

Having established the importance of the shear wave velocity in geotechnical design, the roles of each can be inverted to examine the Biot model predictions of compressional wave velocity. As indicated earlier, important parameters in the calculation are the sediment frame modulus and shear modulus. Both of these quantities, being directly dependent on the sediment structure and having only a very small association with the contained fluid, are interconnected through

$$B_f = 2(1 + \nu)G/3(1 - 2\nu)$$

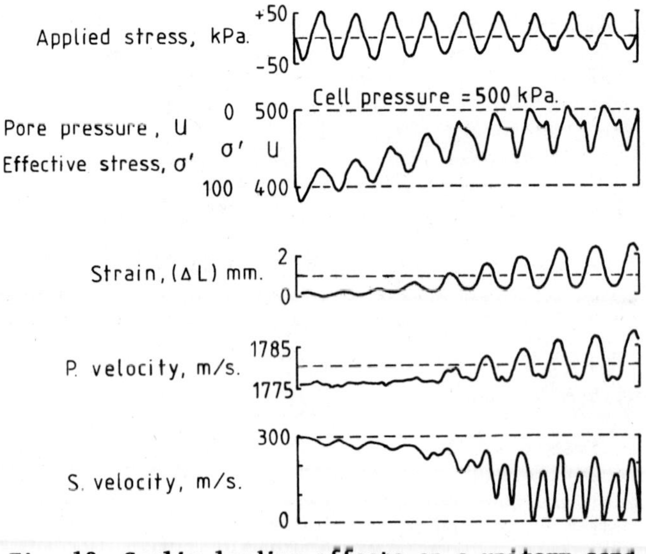

Fig. 12. Cyclic loading effects on a uniform sand

Stoll[34] suggests that as Poisson's ratio ν has a limited range from 0.1 to 0.2, $\mathbf{B_f}$ can be approximated as **G**. While there are various ways to obtain G[34], a satisfactory method is by a measurement of the shear wave velocity which, incidentally, can also provide an estimate of the frame modulus. Fig. 13 shows the variation of the shear wave velocity with frame modulus, obtained from the decompression portion of the void ratio-pressure curve (Fig. 3). The best fit to the data shows that

$$B_f = 0.0005 \ V_s^{2.1}$$

but

$$B_f = 0.001 \ V_s^2$$

would fit almost as well. The importance of this in geotechnical studies can not be overestimated as may be seen when permeability predictions are required.

PERMEABILITY

From all the foregoing it is clear that sediment permeability is an important parameter in any geotechnical design equation. But the parameter also plays an essential part in the Biot model. This quantity, more than any other, highlights the dilemma of the acoustic modeller and the design engineer alike: the former needs the quantity to solve the Biot equations, while the latter needs a (seismo-acoustic?) technique to measure the magnitude of the in situ permeability and its variations: in modern parlance, a Catch-22 situation. However, there is a way out using Eqs. 21-26 and the in situ measurement of the seismic wave velocities and electrical resistivity.

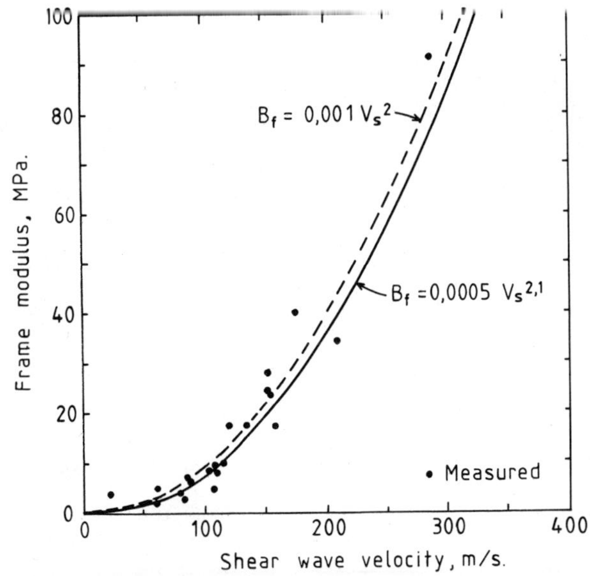

Fig. 13. Frame modulus derived from the consolidation decompression curve (see Fig. 3) correlated to sediment shear wave velocity

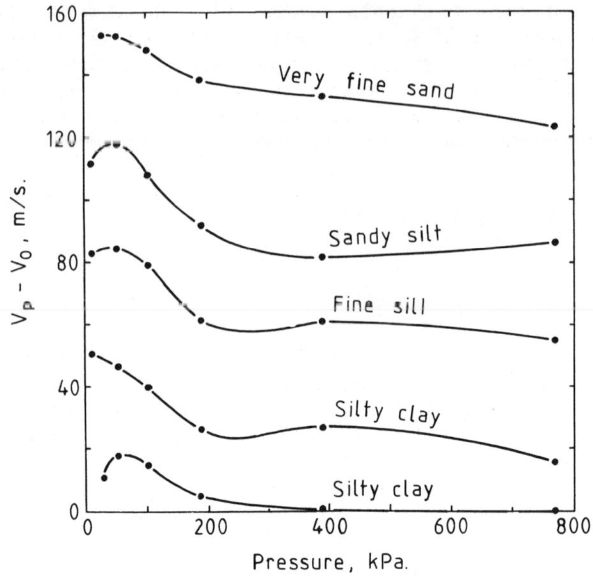

Fig. 14. "Zero frequency" velocity subtracted from the measured compressional wave velocity for various effective stresses and sediments

The electrical measurement, suitably normalised for sea water resistivity, will provide a value of the formation factor which in turn provides a value of the porosity from Fig. 7. Porosity gives a value for the density from Eq. 15 and, with formation factor, an estimate of the mass coupling factor from Eq. 17. Both the shear modulus and the frame modulus can be calculated from the shear wave velocity which leads to a solution of Eq. 22. For low permeability clays, where the characteristic

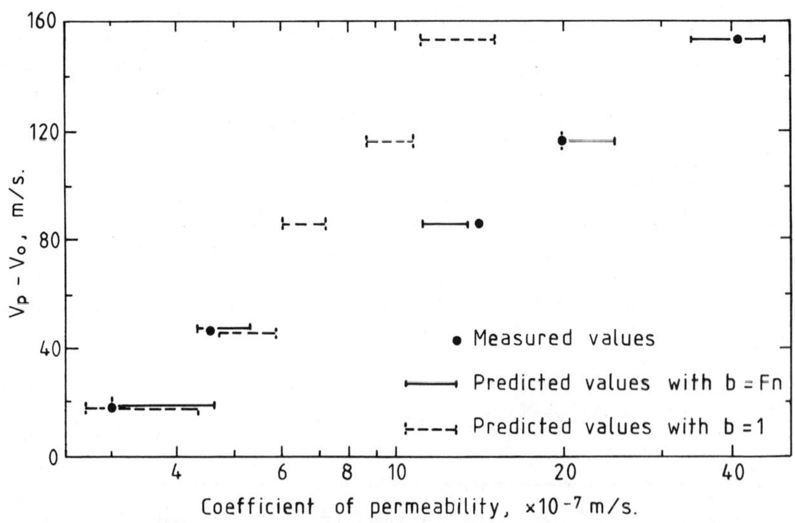

Fig. 15. Samples of Fig. 14 shown against measured permeability and against permeability predicted from Eq. 31

frequency ω_c is always very much greater than the applied frequency ω, the "zero frequency velocity" V_o will approximate to the measured value of the compressional wave velocity V_p. The difference between the two velocities (that is, between the in situ measured compressional wave velocity and the predicted "zero frequency" velocity) will increase as the permeability increases, the relationship between them being the same at all depths (Figs. 14 and 15). Fig. 15 also shows the predicted values of permeability, with error bars created by the uncertainties of the input data, obtained from a combination of Eq. 21 and Eq. 26

$$\phi = \frac{ng}{\omega}\left\{\frac{\rho}{b\rho - n\rho_w}\right\} \frac{Vo^2}{V\infty^2} \left\{\frac{(V_p/V_o)^2 - 1}{1 - (V_p/V\infty)^2}\right\}^{\frac{1}{2}}$$

Permeability, of course, could be assessed empirically through the electrical measurement itself, or through the Kozeny-Carman equation, or even directly from Fig. 15[27].

CONCLUSIONS

Thus the geotechnical characteristics of the sea-bed and seismo-acoustics are intertwined and indivisible. Each needs the other. Attempts to utilise seismic measurements to predict geotechnical quantities have not found universal favor among design engineers through the lack of acceptance of the strain amplitudes and rates employed. However, these can be easily corrected through the use of resonant column measurements[35] to the strain required. But that's another story.

REFERENCES

1 P.R. Ogushwitz, Applicability of the Biot theory Parts I,II & III, J.Acoust.Soc.Am., 77; 429 (1985).

2 P.W. Rowe and W.H. Craig, Application of models to the prediction of offshore gravity platform performance, in: "Offshore Site Investigation", D.A. Ardus ed., Graham & Trotman, London (1980).

3 D.D. Caulfield and Yung-Chang Yim, Prediction of shallow sub-bottom sediment acoustic impedance while estimating absorption and other losses, J.Can.Soc.Exp.Geophys., 19: 44 (1983).

4 R.W. Hutchins, Development of new geophysical methods for site investigation studies, in "Offshore Site Investigation", D.A.Ardus ed., Graham & Trotman, London (1980).

5 R.E. Sheriff, "Seismic Stratigraphy", I.H.R.D.C., Boston (1980).

6 A.M. Berge, G. Drivenes, R. Kanestrom and B. Berkow, Acoustic modelling of the sea floor, Geophysical Prospecting (in press).

7 A.M. Davis and P.J. Schultheiss, Seismic signal processing in engineering site investigation: A case history, Ground Engineering, 13: 44 (1980).

8 J.D. Bennell, P.D. Jackson and P.J. Schultheiss, Further development of sea-floor geophysical probing, Oceanology International 82, 4: 8 (1982).

9 S.D. Schwarz and F.R. Conwell, A technique for the in situ measurement of shear wave velocities for deep marine foundations, Offshore Technology Conference, Paper OTC 20: 755 (1974).

10 R.B. Whitmarsh and R.C. Lilwall, A new method for the determination of in situ shear wave velocity in deep-sea sediments, Oceanology International 82 (1982).

11 T. Gehrmann, P. Gimpel and F. Theilen, Marine shear wave profiling, Geophysics, 50: 336 (1985) (Abstract only).
12 F. Gassmann, Uber die elastizitat poroser medien, Vierteljahrsshr. Naturforsch.ges.Zurich, 96:1 (1951).
13 M.A. Biot and D.G. Willis, The elastic coefficients of the theory of consolidation, J.App.Mech., 79:594 (1957).
14 J. Geertsma and D.C. Smit, Some aspects of elastic wave propagation in fluid-saturated porous solids, Geophysics, 26: 169 (1961).
15 K. Ishihara, Approximate forms of wave equations for water-saturated porous materials and related dynamic modulus, Soils and Foundations (Japan), 10: 10 (1970).
16 D. Taylor Smith, Acoustic and mechanical loading of marine sediments, in: "Physics of sound in marine sediments", L. Hampton ed., Plenum, New York (1974).
17 R.D. Stoll, Acoustic waves in saturated sediments, in: "Physics of sound in marine sediments", L. Hampton ed., Plenum, New York (1974).
18 A.W. Bishop and D.W. Hight, The value of Poisson's ratio in saturated soils and rocks stressed under undrained conditions, Geotechnique 27: 369 (1977).
19 J.G. Berryman, Long wavelength propagation in composite elastic media Parts I & II, J.Acoust.Soc.Am., 68: 1809 (1980).
20 R.J.S. Brown, Connection between formation factor for electrical resistivity and fluid-solid coupling factor in Biot's equations for acoustic waves in fluid-filled porous media, Geophysics, 45: 1269 (1980).
21 J. Geertsma, Velocity-log interpretation: the effect of rock bulk compressibility, J.Soc.Pet.Eng., 1: 235 (1961).
22 K. Terzaghi, Undisturbed clay samples and undisturbed clays, J.Boston Soc.Civ.Eng., 28: 211 (1941)
23 A.W. Skempton, Notes on the compressibility of clays, Q.J.Geol.Soc. Lon., 100: 119 (1944).
24 E.L. Hamilton, Sound velocity gradients in marine sediments, J.Acoust. Soc.Am., 65: 909 (1979).
25 E.L. Hamilton, Acoustic and related properties of the sea floor: density and porosity profiles and gradients, NUC Tech.Pap., No. 459 (1975).
26 F.A.I. Hamdi, Variations of seismic velocity with depth in near-surface marine sediments and their engineering significance, Ph.D. thesis, University of Wales (1981).
27 M.A. Lovell and P. Ogden, Remote assessment of permeability/thermal diffusivity of consolidated clay sediments, EEC Rpt., EUR 9206En (Brussels) (1984).
28 A.M. Davis, Dynamic-static properties of sedimentary materials, Ph.D. thesis, University of Wales (1982).
29 F.A.I. Hamdi and D. Taylor Smith, Soil consolidation behavior assessed by seismic velocity measurements, Geophysical Prospecting, 29: 715 (1981).
30 P.D. Jackson, D. Taylor Smith and P.N. Stanford, Resistivity-porosity-particle shape relationships for marine sands, Geophysics, 43: 1250 (19
31 P.J. Cernock, Sound velocities in Gulf of Mexico sediments as related to physical properties and simulated overburden pressures, Texas A & M University Tech.Rept. 70-5T (1970).
32 P.J. Schultheiss, The influence of packing structure on seismic wave velocities in sediments, Ph.D. thesis, University of Wales (1983).
33 E.L. Hamilton, Shear-wave velocity versus depth in marine sediments: A review, Geophysics, 41: 985 (1976).
34 R.D. Stoll, Theoretical aspects of sound transmission in sediments, J.Acoust.Soc.Am., 68: 1341 (1980).
35 A.M. Davis and J.D. Bennell, Dynamical properties of marine sediments, in: "Ocean seismo-acoustics", T. Akal and J. Berkson eds., Plenum, New York (this volume) (1985).

DYNAMIC PROPERTIES OF MARINE SEDIMENTS

A. M. Davis and J. D. Bennell

School of Ocean Sciences
University College of North Wales
Bangor, U.K.

ABSTRACT

In the marine environment dynamic stresses can be generated in a variety of ways, for example by water wave motion, earthquake activity, foundation vibrations etc., and once translated to the seafloor, transmission of the stress wave is dependent on the dynamic properties of the sediment itself. An understanding of the response of seafloor sediments to dynamic stresses is essential to several fields of marine investigation - seismic hazard evaluation, radioactive waste disposal, slope stability, foundation investigation - and this can best be achieved by studying the various parameters which control elastic wave propagation in porous media. Such studies have been undertaken at the Marine Science Laboratories where the dynamic elastic parameters and damping characteristics of sediment samples have been obtained via testing in a modified resonant column apparatus. In this device, concurrent acoustic measurements using pulse propagation techniques and continuous sinusoidal wave techniques have allowed the dynamic properties to be obtained over a range of strain levels, strain rates and effective stress conditions. The typical strain amplitudes vary between 0.0001% and 1% with the duration of a single stress reversal ranging from 0.00001 second up to 100 seconds. The results from tests carried out on a range of marine sediment samples have allowed the production of a catalogue of strain level and strain rate correction curves. Thus in situ results from low strain, short duration dynamic loadings, as imposed by a seismo-acoustic pulse, can be corrected via these curves to produce information applicable to dynamic loading under a wide range of strain levels and strain rates.

INTRODUCTION

The dynamic stress-strain behaviour of marine sediments can only be described as purely elastic for small amplitude cyclic loading. Strictly the amplitude of these loadings should approach zero, then the material damping would also approach zero and there would be no slippage at the sediment particle contacts. In practice an upper limit for small amplitude strains can be taken as about 0.001%[1]. One of the best ways to study this type of loading in sediments is to measure the propagation velocity of small amplitude stress waves. For larger strain amplitudes greater than 0.001% considerable slippage at the particle contacts will occur and the

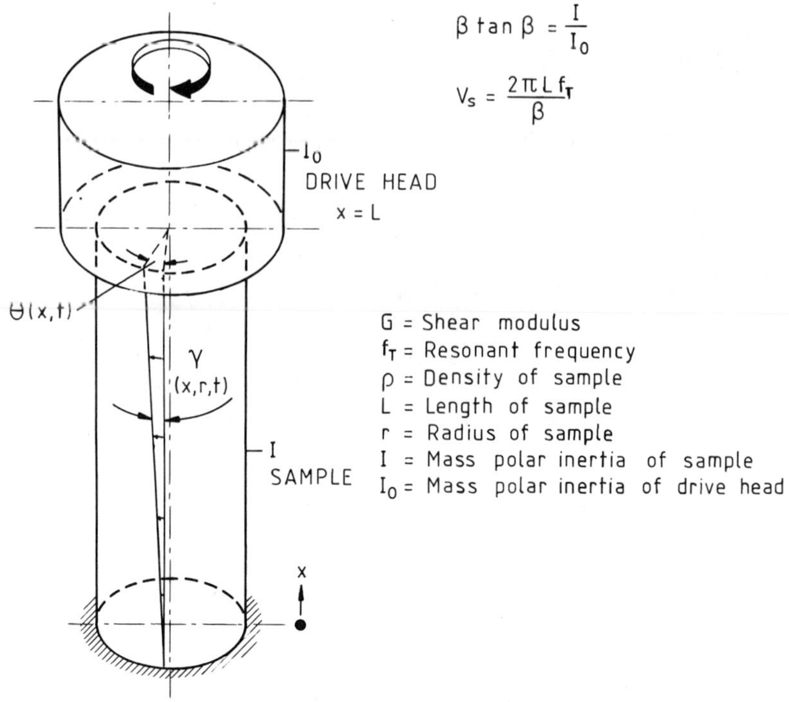

Fig. 1. Principle of the resonant column test.

sediment no longer behaves nearly elastically. Thus a knowledge of these strain dependent dynamic properties is essential for determining the dynamic response of marine sediments to large stresses induced by for example earthquake activity, water wave motion and foundation vibration.

APPARATUS

The modified resonant column device developed at the Marine Science Laboratories, and described by Bennell et al.[2] provides an ideal means of assessing the dynamic properties of marine sediments over a large range of stress-strain conditions in the laboratory. The equipment is based on the Drnevich Resonant Column Apparatus in which the sediment sample is fixed at the bottom and free to oscillate at the top as shown schematically in Fig. 1. Both torsional and longitudinal vibrations can be applied to the top of the sample by means of coil magnet drive mechanisms. The resulting motion is monitored by means of horizontal and vertical accelerometers attached to the top of the specimen. A linear variable differential transformer is used to detect vertical height changes of the specimen during testing. The specimen consists of a vertical, solid or hollow, cylindrical sediment sample contained in a rubber membrane to which a confining pressure is applied.

A signal generator supplies a sinusoidal voltage to the drive coils. The frequency of excitation is varied between 15 and 300 Hz. until the specimen vibrates at the 1st. mode resonant frequency of the oscillator specimen system. This occurs when the voltage applied to the drive coils is in phase with the velocity of vibration. The quantities measured during a test include: resonant frequency, vibration amplitude, force or torque

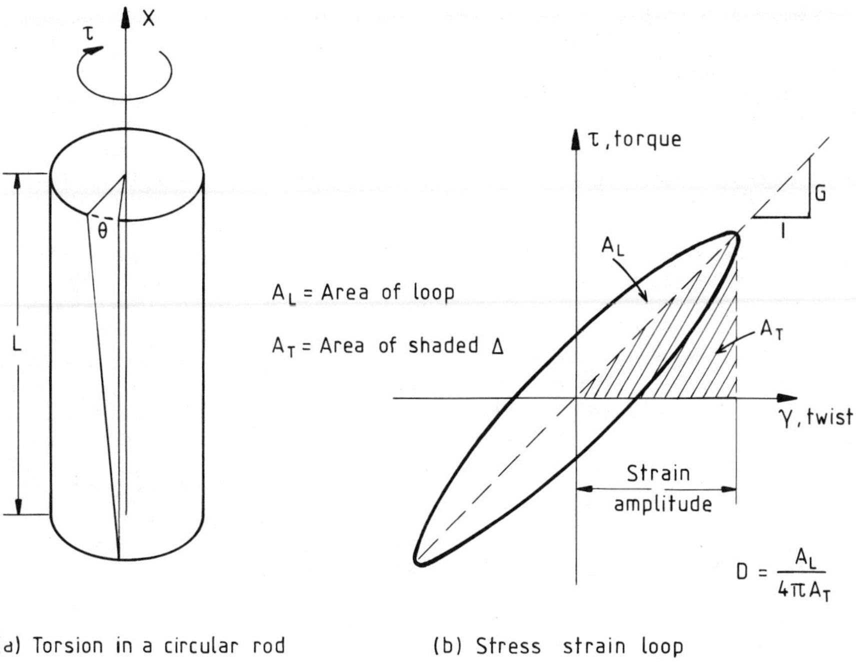

Fig. 2. Torsional simple shear.

produced by the oscillator, the length and volume changes of the sample and the pore and confining pressures. From these data together with the system constants and initial specimen properties the velocity of propagation, modulus, strain amplitude and damping can be determined for both longitudinal and torsional vibrations. The damping may also be determined from the free vibration decay curve which is generated by shutting off the driving power to the coils at resonance. The strain amplitude can be increased from 0.0001% up to 1%, depending on the type of material and effective stress, by increasing the power to the drive coils until either the capacity of the apparatus is reached or the sample fails.

This apparatus has been adapted to perform as a torsional simple shear device by the installation of displacement transducers which measure the twist or shear strain developed at the top of the sample in response to a simple shear stress reversal. Thus instead of measuring the resonant response of the sample at frequencies greater than 15 Hz. the actual load deformation behaviour of the specimen is measured at frequencies below 1Hz. by monitoring the torque-twist curves with an X - Y recorder (Fig. 2). For a complete stress reversal the resulting plot is a closed stress-strain loop as shown. The shear modulus is determined from the slope of the lines joining the end points of the loop, the shear damping from the area or width of the loop and the shear strain amplitude from the average shear strain developed in the sample. For a solid sample the shear strain varies from zero in the centre to a maximum at the edge as shown in Fig.2. The average shear strain is therefore taken as the strain developed at two thirds the radius of the sample. For a hollow sample the strain developed is almost the same throughout the sample.

The test uses the same drive coils as with a resonance test but the frequency is controlled so that a single stress reversal can take from between 1 to 100 seconds. The strain amplitude can also be increased but the maximum attainable is much less than is obtainable from the magnification of the response which occurs at resonance. The effect of the number of cycles of loading can also be studied more easily than in the resonance test since in the latter case some 500 to 3000 cycles of loading occur during the period required to perform a measurement. This modification also allows an assessment of the effect of strain rate on the measurement of shear modulus by performing tests at a number of different frequencies.

Piezoelectric transducers have also been incorporated into the end platens of the resonant column apparatus in order to propagate low strain shear and compressional waves through the length of the sample by the pulse technique. Compressional waves are generated by PZT 4 disc transducers mounted behind the porous stones which facilitate the drainage of pore fluid from the sample during consolidation. Shear waves are generated by bimorph transducers which protrude through the porous stones and penetrate some 3 mm. into the sample[3]. The bimorph transducer is energised with a 10 volt, 50 Hz. square wave from a pulse generator whilst the disc transducer is driven by a 240 volt. spike impulse. Thus knowing the path length and fixed delays of the system the compressional and shear wave velocities can be computed allowing a direct comparison to be made between the propagation of isolated seismic events (pulse propagation) and continous vibrations (resonant column and torsional simple shear) in marine sediments.

TEST PROCEDURE

Specimen construction for both cohesive and cohesionless soils is identical to that for conventional triaxial testing and is adequately covered in the soil mechanics literature[4]. When the test apparatus has been constructed about the sample the sediment is then fully saturated either by applying a back pressure or a combination of a vacuum and then back pressure. When the sample is 100% saturated the confining pressure is increased to the required effective stress level, and the sample is left to consolidate until all the excess pore pressure developed within the specimen has been dissipated. In cohesionless sediments this is usually fairly rapid but can be as long as 24 hours for cohesive sediments. During this period the sample dimensions and low amplitude dynamic properties are monitored. Once primary consolidation has been achieved, and the state of effective stress has been established, torsional simple shear tests are performed at several frequencies and strain amplitudes.

High amplitude torsion and longitudinal resonance measurements can then be performed by increasing the amplitude of vibration in steps. After each step the low amplitude measurements are repeated; these serve as reference values to determine if any change in the specimen has occurred as a result of the high amplitude vibrations, and they are also used to correct the high amplitude values so as to be representative of the initial "undisturbed" state of the sample (i.e. before any high amplitude vibrations were applied). At each strain level about 30 seconds of high amplitude vibration are required to establish resonance and a rest interval of at least one minute is allowed between all measurements. Drainage valves are closed during each high amplitude measurement and opened during the rest interval and low amplitude measurements. After completetion of a high amplitude test sequence a period of at least one day is required to allow the sample to recover to its "undisturbed" state before another test sequence at a subsequent effective stress is initiated. If the low amplitude moduli do not return to their original values or the sample is failed then any subsequent measurements will no longer be representative of the initial specimen characteristics.

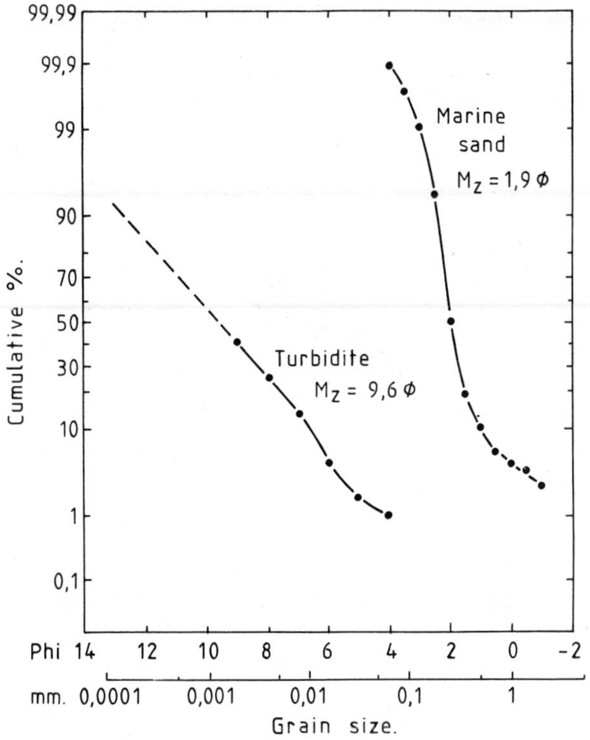

Fig. 3. Grain size distributions of the marine sand and turbidite samples.

TEST RESULTS

A number of sediment samples have been tested in this apparatus ranging from stiff overconsolidated clays, continental shelf muds, silts and sands, and deep sea sediments. It is impossible to summarise the results from all these tests in the confines of this paper so representative results from just two samples will be presented to indicate the typical response of sediments to dynamic loading. These samples are a marine sand of mean particle diameter 1.9ϕ and a deep sea turbidite of mean particle diameter 9.6ϕ. The grain size distributions of these samples is shown in Fig.3.

The dynamic properties obtained from torsional or shear vibations are the shear wave velocity V_s, shear modulus G and shear damping D_T. Those from longitudinal vibrations are the rod wave velocity V_R, Young's modulus E, and rod damping D_L. The logarithmic decrement from the free vibration decay curve is equal to 2π times the damping ratio. While the responses of both torsional and longitudinal vibrations are essentially similiar, the discusion will concentrate on the torsional properties since they are the parameters more frequently required[5].

In general the effect of increasing the effective stress on the sediment sample (effectively simulating depth) leads to an increase in propagation velocities, density and moduli. A corresponding decrease in the damping also occurs. Fig. 4 illustrates the variation of shear wave velocity both from low amplitude resonance and pulse measurements, as a function of effective stress. Woods[5] gives an empirical formula for sands

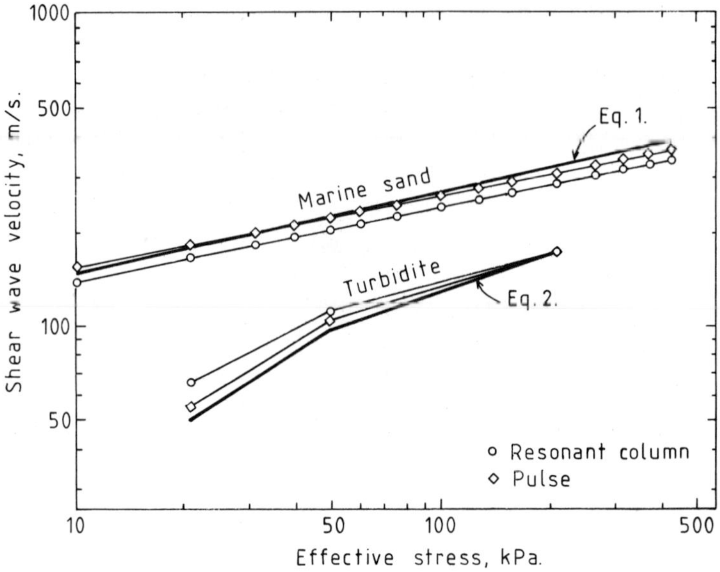

Fig. 4. Effect of effective stress on the shear wave velocity of a marine sand and a turbidite.

with rounded grains, relating the shear wave velocity (in m/sec) to void ratio e and effective stress (in Pa.)

$$V_s = (19.7 - 9.06).(\sigma_o)^{0.25} \quad \text{-----------} \quad (1)$$

This equation is also plotted in Fig.4, with e=0.584 and is seen to give a reasonable fit to both the pulse and resonance results for the marine sand sample, having a void ratio between 0.586 and 0.582. The fit would be improved if the exponent of the effective stress were slightly less than 0.25. In this case the values of shear wave velocity from the pulse measurements are slighty greater than for the resonance measurements. This is probably a frequency effect with the pulse measurements being made in the frequency range 10 to 50 kHz. compared to the resonance frequencies of between 20 to 60 Hz.

Fig.4 also shows the shear wave velocities measured for the turbidite sample after 24 hours consolidation, at confining pressures of 20, 60 and 200 kPa. and having corresponding void ratios of 2.29, 1.99 and 1.67. These velocities would increase slightly with increasing consolidation time because of creep and viscous effects[6]. Woods[5] also gives an equation which applies for angular grained sand and which Hardin and Black[7] have confirmed for clays with low surface activity

$$V_s = (18.4 - 6.20e).(\sigma_o)^{0.25} \quad \text{-----------} \quad (2)$$

This equation is also plotted in Fig.4 and shows good agreement to the experimental values. In this case the pulse velocities are slightly lower than the resonance values.

Fig.5 shows the variation in shear modulus with shear strain amplitude

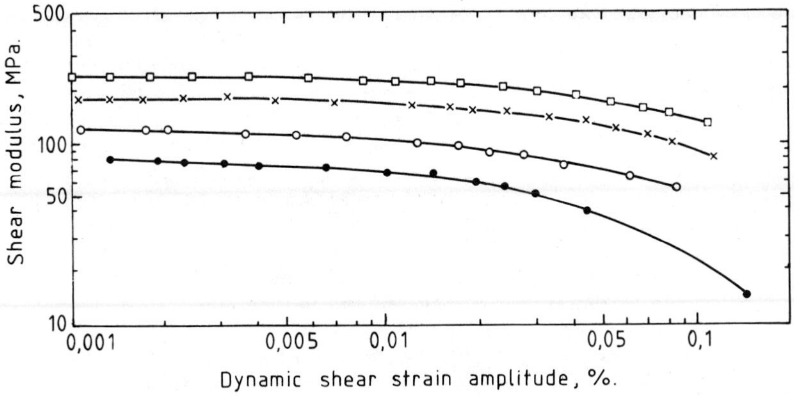

Fig. 5. Variation of shear modulus with dynamic shear strain amplitude and effective stress for the marine sand sample.

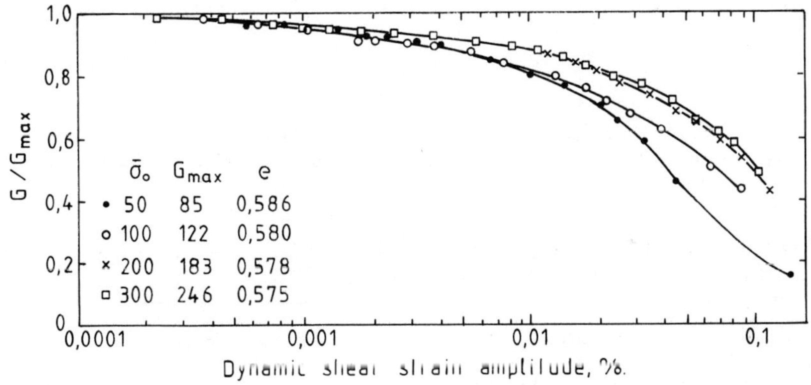

Fig. 6. Variation of normalised shear modulus G/G_{max} with dynamic shear strain amplitude and effective stress for the marine sand.

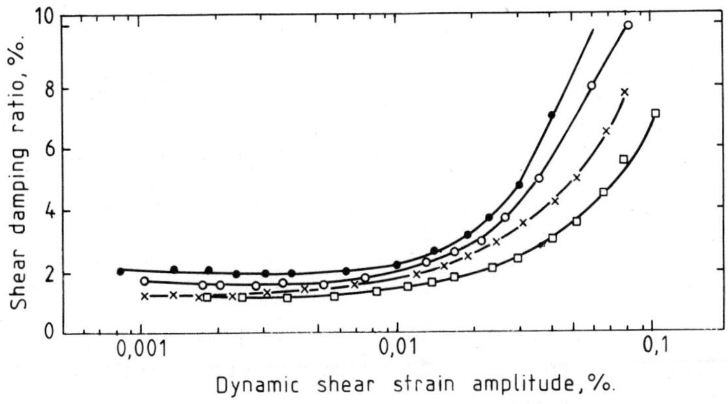

Fig. 7. Variation of shear damping ratio with dynamic shear strain amplitude and effective stress for the marine sand sample.

for the sample of marine sand derived from high amplitude torsional resonant column results. Each of the high amplitude moduli have been adjusted using the low amplitude moduli measured immediately afterwards to take into account any change in the sample as a result of the high amplitude vibrations. This figure illustrates that the shear modulus or the strength of the sand decreases dramatically for shear strains in exess of 0.01% indicating that considerable slippage is occurring at the interparticle contacts. This high amplitude modulus corresponds to a hysteresis type of modulus according to Drnevich et al.[8]. Similiar behaviour also occurs for all the sediments examined so far. Young's modulus is also observed to decrease in a similiar fashion with increasing longitudinal strain amplitude. Fig. 6 shows the effect of normalising the data in fig. 5 with respect to the shear moduli extrapolated back to zero strain amplitude, termed G_{max}, for each effective stress. Here the decay in the normalised shear modulus G/G_{max} for shear strains in excess of 0.01% is seen to be less the greater the effective stress. This is attributable to the increase in shear strength of the sediment at higher effective stresses requiring a correspondingly larger shear strain amplitude to promote slippage at the interparticle contacts. Fig. 7 shows the effect of shear strain amplitude on the shear damping ratio for the marine sand sample. As can be seen the values of D_T are fairly constant below strain amplitudes of 0.01% but increase rapidly as interparticle slippage occurs at higher strain amplitudes. The pattern is similiar to Fig.6: i.e. the higher the effective stress on the sample the higher the strain amplitude required to produce slippage at the grain contacts which induces increased material damping.

The resonant frequency of the turbidite sample confined at an effective stress of 20 kPa. was less than 10 Hz; this is too low a

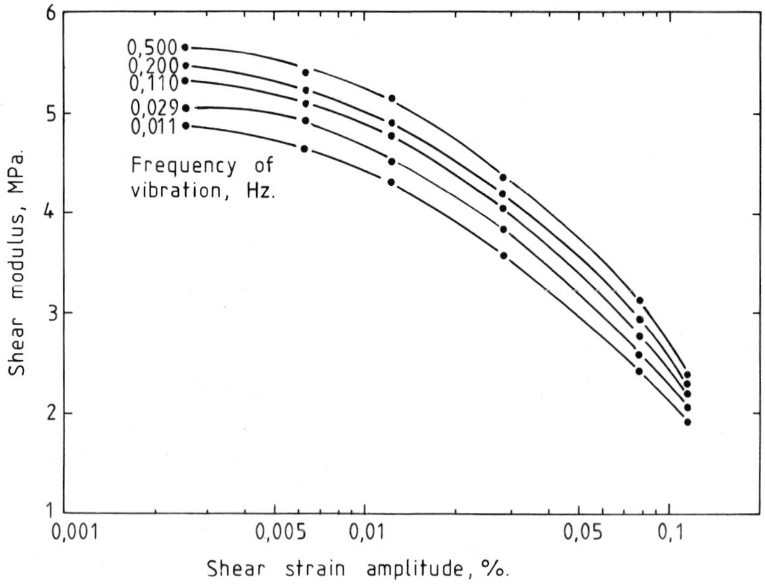

Fig. 8. Variation of shear modulus with shear strain amplitude and frequency of vibration for the turbidite sample at σ_o=20 kPa.

frequency to perform a full high amplitude resonant column test. It was possible, however, to use the torsional simple shear method to determine moduli at high strain amplitudes and the results are shown in Fig. 8. This method has the advantage over the resonant column method in that the frequency of excitation is predefined and the testing can be confined to a single stress reversal. In the resonant method, however, the frequency of vibration depends on the material properties and usually decreases with increasing strain amplitude. In addition between 500 and 3000 stress reversals occur during the measurement which can lead to considerable sample disturbance or even sample failure at high amplitude cycling. Fig.8 also shows the decrease in shear modulus with increasing shear strain amplitude. In addition the modulus is shown to consistently decrease with decreasing frequency of vibration (or increasing period of a single stress reversal). A similiar effect has been found in all the sediments investigated so far and can be attributed to a creep or viscous phenomena resulting in an increase in the measured strain with respect to time of stress application (i.e. a strain rate effect).

Abbiss[9] has shown from viscoelastic theory that the shear modulus determined from seismo-acoustic measurements can be used to predict shear moduli applicable to high strain levels and long time periods using the relation

$$G_2 = G_1 . (f_2 / f_1)^{(4.D / \pi)} \quad \text{---------} \quad (3)$$

where G_1 is the seismic modulus measured at frequency f_1 and G_2 is the predicted modulus at vibration frequency f_2. D is the damping ratio

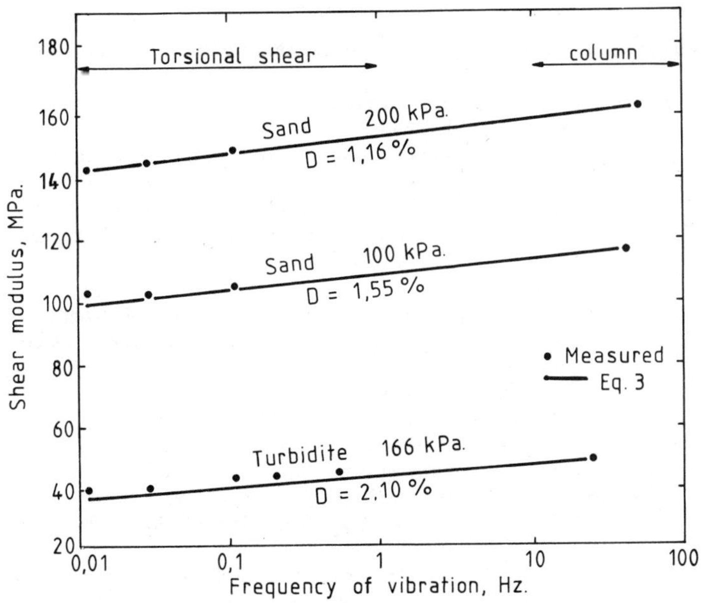

Fig. 9. The effect of frequency of vibration on the shear modulus of the marine sand and turbidite samples.

corresponding to the strain level of the required modulus. This relationship is shown in Fig. 9 with G_1 equal to the shear modulus determined from the resonant column test at a strain level of 0.004 % for the marine sand at effective stresses of 100 and 200 kPa., and the turbidite at 166kPa. The predicted range of values is seen to give a reasonable fit to the data derived from the torsional simple shear measurements at frequencies of vibration below 1 Hz. If the damping ratio corresponding to a higher strain level is used in Equation 3 then, the corresponding low frequency moduli applicable to that strain level can be predicted.

CONCLUSIONS

The modified resonant column apparatus has proved to be well suited to determining the dynamic properties of marine sediment samples over a range of effective stresses from 0 to 700 kPa., frequencies from 0.01 Hz to 100 kHz., and strain amplitudes from 0.0001 to 1 %. A range of different samples have been tested allowing the accumulation of a catalogue of strain level and strain rate correction curves, for both shear and Young´s moduli, similiar to the normalised curves in Fig. 6. Thus low amplitude in situ measurements of the dynamic properties of marine sediments can be corrected via these curves to produce information applicable to dynamic loading under a wide range of strain levels and strain rates.

ACKNOWLEDGEMENTS

This work was carried out under the Marine Technogy, (Sea-bed Studies) programme of the Science and Engineering Research Council.

REFERENCES

1. B. O. Hardin, The nature of stress-strain behaviour for soils, in "Proceedings of American Society of Civil Engineers, Specialty Conference on Earthquake Engineering and Soil Dynamics", Pasadena, Cal., 1, pp. 3 - 90. (1978).
2. J. D. Bennell, D. Taylor Smith, and A. M. Davis, Resonant column testing of marine sediments, Oceanology Int. 84. (1984).
3. P. J. Schultheiss, A note on simultaneous measurement of P and S wave velocities during conventional laboratory soil testing procedures, Marine Geotechnology, 4, pp. 343 - 367. (1981).
4. A. W. Bishop, and D. J. Henkel, "The measurement of soil properties in the triaxial test", 2nd Ed., Edward Arnold, London, 228p. (1962).
5. R. D. Woods, Parameters affecting elastic properties, in "Proceedings Dynamical Methods in Soil and Rock Mechanics", 1, B. Prange, ed., Balkema, Rotterdam. (1978).
6. D. G. Anderson, and K. H. Stokoe, II, Shear modulus: A time dependent soil property, in: "Dynamic Geotechnical Testing", ASTM, STP 654 American Society for Testing and Materials, pp. 66 - 90. (1978).
7. B. O. Hardin, and W. L. Black, Vibration modulus of normally consolidated clay, J. Soil Mech. and Found. Div., ASCE, 94, pp. 353 - 369. (1968).
8. V. P. Drnevich, J. R. Hall, Jr., and F. E. Richart, Jr., Effect of amplitude of vibration on the shear modulus of sand, in "Proceedings of the International Symposium on Wave Propagation and Dynamic Properties of Earth Materials", Albuquerque, N.M., pp.189-199,(1967).
9. C. P. Abbiss, Shear wave measurements of the elasticity of the ground, Geotechnique, 31, pp. 91 - 104. (1981).

RELATIONSHIP BETWEEN THE ACOUSTICAL CHARACTERISTICS OF DEEP SEA SEDIMENTS AND THEIR PHYSICAL ENVIRONMENT

C.C. Leroy*, J.M. Daupleix** and P.J. Longuemard***

* SINTRA Marine Depart.
Z.I. des Paluds
13400 Aubagne, France

** L.M.A.I
University of Perpignan
66000 Perpignan, France

*** Ecole Centrale des Arts et Manufactures
Grande coie des Vignes
92290 Chatenay Malabry, France

INTRODUCTION

The propagation of sound at long ranges in the deep ocean is influenced by the sea bottom characteristics especially when both source and receiver are located outside the sound channel created by the velocity profile in the sea: under these conditions, bottom reflected paths largely contribute to the total energy transmitted. As long as medium frequencies (say, from 2 to 12 Hz) are concerned, the calculation of the total sound field can be carried out by entering, in ray theory, a so called "bottom loss" that depends on incidence angle, frequency and type of bottom, and is mainly controlled by the acoustic contrast between sea water and the very first sediment layers, plus some consideration on bottom roughness. At low frequencies (say below 500 Hz) this simplified approach is no longer valid: the sound energy that penetrates into the sediment is not rapidly dissipated by absorption and can travel far enough to be refracted by the sound speed gradient and partly returned to the ocean. The existence of a systematical increase of sound speed has been recognized as a universal feature of deep sea sediments. Sound speed values and profiles, however, will depend on the type of sediments.

Various methods can be used to calculate sound propagation losses, taking into account the above described effect. All of them require, as far as the sediments are concerned, the following data as a function of depth ζ below the sea floor interface:

- Sound speed $V_p(\zeta)$ of the compressional waves.
- Attenuation coefficient $\alpha_p(\zeta)$ associated to $V_p(\zeta)$.
- Density $\rho(\zeta)$.

and to a minor degree:

- Sound speed $V_s(\zeta)$ of the transverse waves.
- Attenuation coefficient $\alpha_s(\zeta)$ associated to $V_s(\zeta)$.

Finally, an underlaying semi-definite substratum has to be defined to provide boundary conditions.

The above values cannot usually be obtained from maps of in situ measurements, and one is faced with the problem of making best estimates of these physical parameters from the available geophysical data. For this reason, a research program has been undertaken under the sponsorship of the French underwater laboratory of LE BRUSC (GERDSM). The research is organized around four main subjects:

a) Review the principal relationships that permit to estimate acoustic parameters from geophysics and geology.

b) Investigate into greater details the "depth effect" by laboratory measurements.

c) Examine the mathematical models that can take into account, without too much computer time, the "depth effect" for low frequencies sonar applications.

d) Divide areas of interest into zones, as large as possible, that can be considered as homogeneous and properly characterized.

The present paper is only concern with subjects (a) and (b). However the importance of the sound speed gradient effect needs illustration. Figure 1 shows an example of the equivalent plane wave bottom loss computed (i) with an isovelocity fluid bottom overlying an elastic substratum and (ii) under the same conditions with a 1.3 S^{-1} gradient in the sediments.

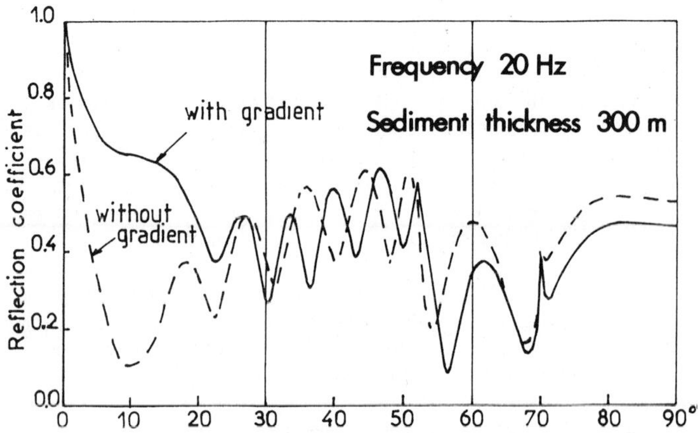

Fig. 1 Reflection coefficient vs grazing angle.

RELATIONSHIP CONCERNING THE ACOUSTICAL PROPERTIES OF DEEP SEA SEDIMENTS

The general theoretical formulations for the compressional and shear waves velocities V_p and V_s in a medium are:

$$V_p = [(\lambda + 2\mu)/\rho]^{1/2}$$

$$V_s = [\mu/\rho]^{1/2}$$

where μ is the shear rigidity modulus, λ the first Lame constant and ρ the specific mass. The deep sea sediments are complex media, where the various above parameters depend on such factors as porosity or mean grain size, distribution of sizes and shapes in the grains, chemical composition, in situ conditions like pressure and temperature, etc. A great number of relationships between various parameters have been proposed since practically two decades and studies are still continuing. This paper will just select a few number of equations and examples, all related to the evaluation of velocities, attenuations and densities.

The study of sediments has been for a long time concentrated upon porosity and grain size. The shear rigidity modulus μ and the specific mass ρ are both strongly dependant on the porosity n (water volume to total volume ratio). It seems that the present trend would be, to describe the acoustical effects, to consider ρ in priority for (i) it needs to be known anyway (ii) it is easier to measure than porosity and (iii) there exist well established relationships between ρ and n. As examples we can quote from Hamilton and Bachman [1].

$$n = 150.1 - 51.2\ \rho \qquad \text{in pelagic sediments}$$
$$n = 155.9 - 56.0\ \rho \qquad \text{in turbidites}$$

and from Akal [2]

$$\rho = 2.604 - 1.606\ n \qquad \text{(from 15124 samples)}$$

It is more common, however, to find values of V_p as a function of n than as a function of ρ. Figure 2 illustrates the results from several authors as collected by Anderson [3]. Values are generally above those given by Wood's equation that does not take rigidity into account (Anderson [1], Akal [2], Hamilton [3]) but are still often lower than the sea water sound speed for the porosities encountered in deep sea sediments. A study carried out by Longuemard and Moussiessie [4] in 1983 from 130 combinations of V_p and ρ led to a modelisation of the form:

$$V_p = a + b(\rho - d)^m$$

where the a, b, d and n are constants depending upon the nature of the bottom. For small values of ρ, V_p tends to 1500 m/s and one find $V_p < V$ sea water when $\rho \sim 1300\ kg/m^3$.

Sediments that are especially rich in silicium or carbonates present some peculiarities. For carbonated sediments samples analyzed by Horn and Horn and Delach [5] in 1968 did not arrive to a relationship between V_p and the $CaCO_3$ percentage. For this same type of sediments, Morton [10] considers that the usual methods to predict sound speed from physical properties can no longer be employed. Velocity/porosity dependance cannot be utilized when sands are constituted by hollow foraminifera tests and it would be preferable to use the rigidity modulus that does not take into account the origins and transport mechanism of the sediment. Hamilton et al [6] however are less restrictive, as they indicate that specific mass is a good approach to the calculation of sound speed in the absence of tests of intact calcarous organism. The following formula is then proposed:

$$V_p = 1\ 828 - 480.1\ \rho + 193.4\ \rho^2$$

To take into account the carbonate content, Moussiessie [7] considers an equation as follows:

$$V_p = a_1 + a_2\Psi + (b_1 + b_2\Psi)\ (\rho - 1.2)^m$$

where Ψ is the percentage of $CaCO_3$, the a_1, a_2, b_1, b_2 are constants and ρ is in g/cm^3. The exponent m is a function of location and type of bottom and is of the order of 2.5 for deep sediments.

Other sets of equations can be found that take into account the percentage of sound or silt the grain size, or the sorting index, for example (Hamilton et al [6]):

$$V_p = 1\ 576 - 5.7\ Mz \qquad (Mz = \text{mean grain size})$$

$$V_p = 1\ 526 - 0.54\ S \qquad (S = \text{sand content in }\%)$$

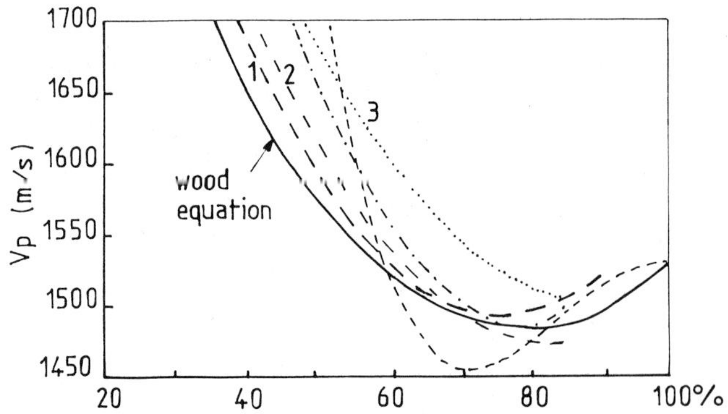

Fig. 2 Variation of V_p with porosity from various authors.

The influence of depth ζ inside the sediments is clearly marked by a systematical increase of both ρ and V_p. Figures 3 and 4 illustrate these effects. Figure 3, after Hamilton [8] gives curves of $\rho(\zeta)$ which all correspond to formulas of the type

$$\rho(\zeta) = \rho_0 + a\zeta - b\zeta^2$$

with $a > 0$ and $B > 0$ except one case. For the most common pelagic clays one would get:

$$\rho = 1.357 + 1.338\ \zeta - 0.284\ \zeta^2 \qquad (\rho \text{ in } g/m^3,\ \zeta \text{ in km})$$

The curves of Fig. 4, after Matthews [9], give $V_p\ (\zeta)$ from theoretical consideration introducing a porosity decay constant S. Experimental data show curves that are somewhat less bent. In all cases, and from the various authors, the compressional sound speed $V_p(\zeta)$ can be expressed by formulas of the form:

$$V_p(\zeta) = V_{p0} + \alpha\zeta - \beta\zeta^2 \qquad (\alpha \text{ and } \beta > 0)$$

As for the attenuation coefficient αp, usually taken as 1 dB per wave length around the tens of kHz, it reaches lower values (0.1 to 0.4 dB/λ) at low frequencies but measurements are difficult and necesserily include the depth effect due to a variation $\alpha_p(\zeta)$. The trend of this variation seems due to a decrease in absorption with ζ for sand/silt sediments but an increase for silt/clay.

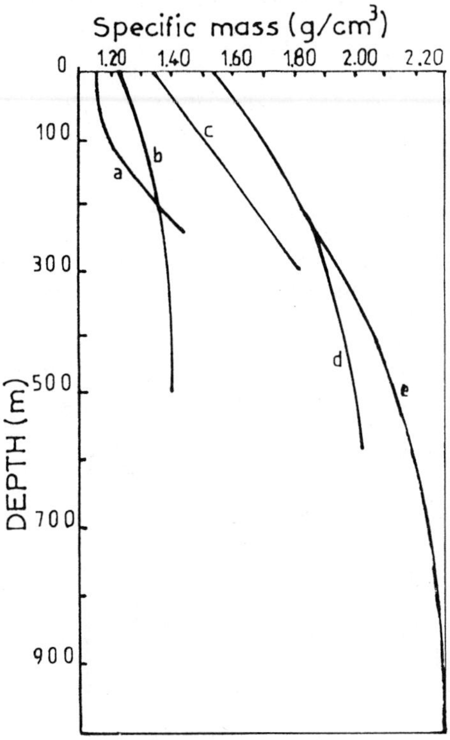

Fig. 3
Variation of specific mass with depth ζ in sediment.

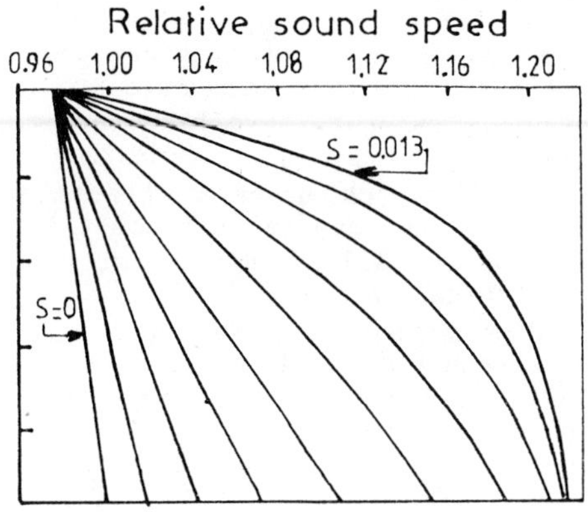

Fig. 4
Variation of sound speed with depth ζ in sediment.

EXPERIMENTAL EQUIPMENT

The increase of specific mass with depth ζ is obviously due to the compaction of sediments by pressure and aging. The variation of V_p, knowledge of which is most important to propagation loss measurements need some more investigation to understand the various phenomena that can be involved. In fact the sediments at a certain depth ζ withsand the weight of the column of sediment above them plus the pressure. The interstical pressure of contained water is also increased above that at the sea floor interface. Finally, due to the earth internal heat the temperature at depth ζ is higher than at the sea bottom. In all, there must exist a relationship of the form:

$$V_p(\zeta) = f(P_t, P_i, T, \rho)$$

where P_t and P_i are respectively the total and the intersticial pressures, T and ρ the temperature and the specific mass, all of them depending on ζ. The coefficients in function f should also depend upon the type of sediment.

In order to investigate the relative importance of the various parameters, an experimental equipment has been built. It consits (Fig. 5) of a cylindrical enclosure capable of supporting up to 6000 bars. Sediments are introduced from one end (left on the figure) by removing a closing piston. The physical force on the sediments is applied from this piston actuated by a hydraulic jack. The interstitial pressure can be varied by a separate water pressure circuit and the temperature by an appropriate heating system. Two acoustic probes are located on oposite sides of the cylinder and are used to measure sound speed from a few kHz to 50 kHz. Calibration techniques including signal processing to separate multiphats and resonance effects will be used. At this time the equipment versus temperature have been performed. An example is given in Fig. 6.

The research program will be carried out with local, characteristic sediments and deep sea cores obtained in the Atlantic during a cruise last september. All sediments will be examined for their geological and geotechnical properties (water content, specif mass, Attenberg limit, grain size, shear modulus, etc.). Systematic measurements of sound speed and attenuation will be carried out with changes of the parameters P_t, P_i and T. All in all it is expected to arrive at a better understanding of the physics of the "depth effect" and formulate equations of the above type.

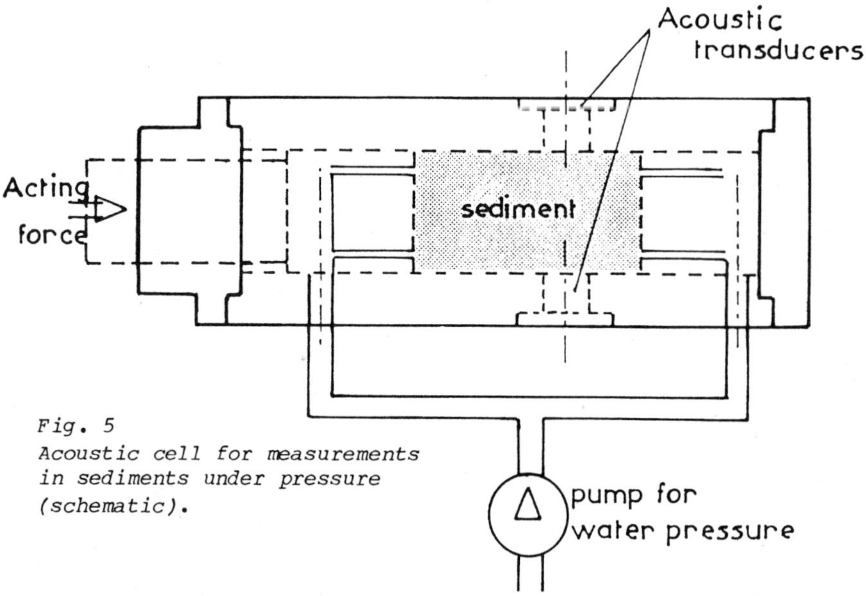

Fig. 5
Acoustic cell for measurements in sediments under pressure (schematic).

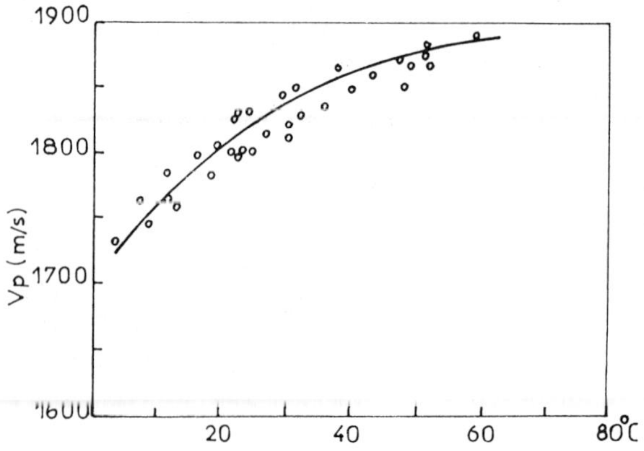

Fig. 6
Measured values of sound with temperature inside saturated sediments.

CONCLUSION

The calculation of long range, low frequency propagation in the deep ocean requires the knowledge of the acoustic properties of the sediments as a function of depth below the sea floor. Those properties can be evaluated through a set of relationships with order parameters that are more generally obtainable. The main formulas have been presented and it is expected that definite acoustic zones can be defined in the near future. A better understanding of the effect of the depth is however desirable and a specific equipment has been constructed and will be used for systematic measurements on real deep sea sediment samples. Propagation modelling taking the depth effect into account has been started and the studies are under way.

REFERENCES

1. E.L. Hamilton, R.T. Bachman, Sound velocity and related properties of marine sediments, J. Acoust. Soc. Am., 72: 1981 (1982).
2. T. Akal, The relationship between the physical properties of underwater sediments that affect bottom reflection, Marine Geology, 13: 251 (1972).
3. R.S. Anderson, Statistical correlation of physical properties and sound velocity in sediments, in: "Physics of sound in Marine Sediments". L. Hampton, ed, Plenum Press, New York (1974).
4. J.P. Longuemard and J. Moussiessie, Relation celerite des ondes de compression - masse volumique de sediments marins satures, Revue du Cethedec, n° 76: 183 (1983).
5. D.R. Horn, B.M. Horn and M.N. Delach, Correlation between acoustical and other physical properties of deep sea cores, J. Geophys. Research, 73: 1939 (1968).
6. E.L. Hamilton, R.T. Buchman, W.H. Berger, T.L. Johnson and L.A. Mayer, Acoustic and related properties of calcareous deep sea sediments, J. Sedim. Petrology, 52: 733 (1982).
7. J. Moussiessie, Contribution a l'etude des relations entre la acoustique et les qualites geotechniques des sediments marins, Thesis n° 2944, University of Toulouse (1984).
8. E.L. Hamilton, Variations of density and porosity with depth in deep-sea sediments, J. Sedim. Petrology, 46:280 (1976).
9. J.E. Matthews, Heuristic physical property model for marine sediments, J. Acoust. Soc. Am., 68:1361 (1980).
10. R.W. Morton, Sound velocity in carbonate sediments from the Whiting Basin, Puerto-Rico, Marine Geology, 19: 1 (1975).

LOW FREQUENCY MEASUREMENTS OF THE ACOUSTIC PROPERTIES OF MARINE SEDIMENTS

J.I. Dunlop

School of Physics
University of New South Wales
P.O. Box 1, Kensington, Australia 2033

INTRODUCTION

In many applications of underwater acoustics e.g. shallow water propagation or sonar depth sounding, the sea floor plays an important role. There is an acoustic mismatch at the water sea-floor interface and there is therefore a need for accurate measurement of the acoustic properties of the marine sediment - the sound speed, attenuation constant and density. These properties may be determined indirectly by several methods but their direct in-situ measurement is difficult, the most useful method being based on the measurement of high frequency pulses propagating between transducers inserted directly in the medium.

The required acoustic properties can also be derived from measurement of the acoustic impedance of the medium. Shirley (1977) developed a technique for in-situ measurement by determining the effects of acoustic loading on the measured electrical impedance of an acoustic transducer immersed in the sediment. This technique operated at the resonant frequency of the transducer (65 kHz). The acoustic impedance of a material may also be measured using the acoustic impedance tube, a technique widely used for acoustic measurements in air. This technique has been operated in a water medium (Fay et al 1947) and applied to marine sediments (Dunlop and Whichello 1980) but its use has been restricted to laboratory sample measurements. In this paper is discussed the development of modified impedance tube techniques (Dunlop 1985) for in-situ measurement of marine sediments at low sonic frequencies

MEASUREMENTS BY IMPEDANCE TUBE

The acoustic impedance tube consists of a rigid walled tube or duct along which acoustic waves propagate to impinge on a sample fixed at the end of the tube. Some fraction of the acoustic waves is reflected back along the tube and combine with the incident waves. Measurements of the acoustic field in front of the sample will yield information of the reflectivity of the sample and hence its acoustic properties.

In its normal mode of operation the impedance tube is terminated at its end by a rigid end plate on which the sample to be measured is placed. This mode of operation offers severe limitations in its application to marine sediments in that a sample of the sediment must be removed and reconstituted in the tube, thereby irrecoverably altering its acoustic properties. There are however two other possible configurations of deployment of the tube, namely the open flanged end and the open protruding

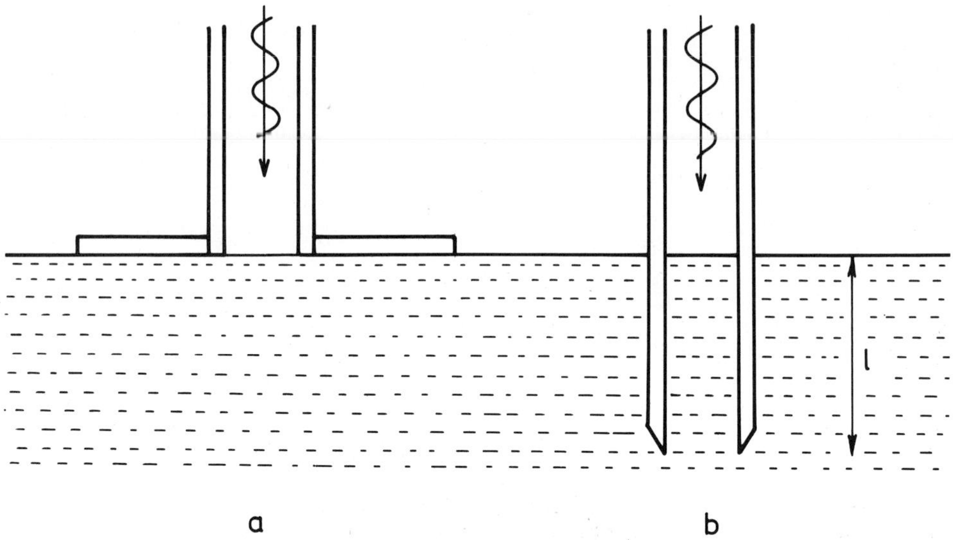

Fig. 1. (a) Open flanged tube (b) Open protruding tube

end (Fig. 1). These methods have recently been evaluated for measuring the acoustic properties of air foam mixtures (Dunlop 1985).

In the case of the open flanged tube, the acoustic impedance measured at the end of the tube is the radiation impedance into the sample medium, in the case of interest, marine sediment. The acoustic radiation impedance Z_r at the end of the tube at frequencies (f) such that f << 0.1 C/a (C being the sound speed and a the tube radius), may be expressed as

$$Z_r/Z_0 = R(2\gamma a) + j\,X(2\gamma a) \qquad \text{.............. (1)}$$

where Z_0 is the characteristic impedance of the fluid in the tube, γ the propagation constant, and the functions of R and X given by

$$R(y) = y^2/2.4 - y^4/2.4^2.6 + y^6/2.4^2.6^2.8 - \ldots$$

$$X(y) = 4\pi[y/3 - y^3/3^2.5 + y^5/3^2.5^2.7 - \ldots]$$

A possible limitation of this technique to the measurement of marine sediments is that the effective penetration may be limited, to the surface layers of the medium whereas the properties of marine sediments are expected to vary with depth due to changes in composition and to increasing pressures of overburden. The open protruding end configuration overcomes this difficulty to some extent but, at the expense of disturbing the sample as the protruding end is pushed into it. However provided the tube walls are thin enough this disturbance may be minimal. In this configuration the measured sample impedance (Z) is the impedance of a length (ℓ) of sample termination by the radiation impedance Z_u at the end of the tube in the sample medium. This may be expressed as (Zwikker and Kosten 1949)

$$Z/Z_0 = \frac{Z_0 \cosh \gamma\ell - Z_u \sinh \gamma\ell}{Z_0 \sinh \gamma\ell - Z_u \cosh \gamma\ell} \qquad \text{............ (2)}$$

where Z_0 is the characteristic impedance of the medium. The tube termination impedance Z_u is the radiation impedance at the end of an unflanged pipe. An expression for this was first derived by Levine and Schwinger (1948). A much simpler approximation (Kinsler & Frey 1962) has proved to be of sufficient accuracy for these measurements, viz

$$Z_u/Z_0 = \gamma^2 a^2/4 + j\ 0.4\ \gamma a$$

PROPAGATION IN MARINE SEDIMENTS

The theory of elastic wave propagation in saturated granular media has been formulated by several researchers (e.g. Stoll 1974). In its simplest terms the theory assumes a wave propagating in a fluid modified by the presence of solid grains. A major mechanism for dissipation is viscous flow of the fluid in the pores between the particles, so that the system might be compared to fluid propagation in a rigid porous material as a first approximation. There are however other mechanisms involved e.g. the elastic response of grains and skeletal frame, and losses due to grain to grain contact, which will limit the accuracy of this simple model.

Acoustic propagation of a fluid within a rigid porous frame is characterised by the presence of energy dissipation and a change in sound speed. The dissipation may be expressed (Zwikker and Kosten 1949) in terms of a loss angle denoting the phase difference (δ) between acoustic pressure and particle velocity, or density variation, and is related to the attenuation constant α (via $\alpha = \omega\delta/2C$). A structure factor (k) is introduced to account for change in the sound speed, the sound speed being related to the free medium sound speed via $1/\sqrt{k}$. The acoustic characteristics of the material, γ the propagation constant and Z_0 the characteristic acoustic impedance then become a function of frequency ω viz

$$\gamma = j\omega\sqrt{k}/C\ (1 - j\ \delta/2), \text{ and}$$

$$Z_0 = \rho C\ \sqrt{k}\ (1 - j\ \delta/2)$$

Substitution of these expressions for γ and Z_0 in the impedance equations 1 and 2 will then produce impedance plots as a function of the frequency. Various values may be substituted for the parameters δ and k until good agreement is obtained between these impedance plots and the measured impedance values.

THE WATER FILLED IMPEDANCE TUBE

The impedance tube technique was originally developed to make absorption measurements in an air medium. In this instance the walls of the tube may be easily made rigid enough that there is no significant interaction between the air in the tube and the tube walls and the acoustic properties of the air in the tube is the same as that of free air. The characteristic impedance of water however is very much greater so that the walls of the tube can no longer be considered rigid and there is coupling of the water waves to the walls of the tube. The chief effects of this coupling are a reduction in the speed of sound in the tube (C_t) and an increase in the attenuation coefficient of sound in the tube (α_t). This problem has been investigated previously (Kuhl 1953, Fay et al 1947) and it has been shown that with a proper choice of material and wall thickness a tube could be made which satisfactorily measured acoustic impedance in a water medium.

The tubes used in these experiments were of cadmium plated steel, internal diameter 89 mm and wall thickness 7 mm. The sound field in front of the measurement end of the tube was measured using Bruel and Kjaer type 8103 hydraphones inserted in the tube through grommets in the wall. The water in the tube was excited by a sound source positioned at the end of the tube. This was driven by limited band noise, and the signals from the hydraphones processed by computer to yield the pressure ratio (P_{12}) and the phase difference (ϕ_{12}) between the hydraphone positions at frequency intervals within the bandwidth of the noise signal employed. The impedance (Z) at the end of the tube was then derived from P_{12} and ϕ_{12} by the method of Seybert and Ross (1977).

The acoustic characteristics of the tubes were first investigated to ensure the behaviour was as predicted by theory. The sound speed in the water filled tube (C_t) is given by

$$1/C_t^2 = 1/C_w^2 + (2a/b \times \rho_w/\rho_s)/C_s^2$$

where C_w is the open water sound speed

a and b are the tube internal radius and wall thickness

ρ_w and ρ_s the densities of water and steel and

C_s the sound speed in steel.

Acoustic measurements were therefore made to determine the sound speed and also the attentuation in the empty tube immersed in water. These involved measurement of the acoustic field inside the tube with either a closed or open end termination. In the case of the closed end mode it is difficult to ensure a rigid tube termination and the very large impedance values involved precluded its use as a measurement parameter. However the measured values of P_{12} and ϕ_{12} can be compared with expected values following substitution of suitable values for C_t and α_t yielding estimates of the required parameters.

The open ended tube proved a more sensitive method for determining C_t and α_t. For example, in Fig. 2 is shown the measured impedance values of a length of water filled tube terminated by the open unflanged end. The calculated impedance values derived using equation (3), the known value of C, the sound speed in open water, and suitable adjusted values of C_t and α_t are shown superimposed on the measured values. The good agreement between measured and calculated values is evident from the impedance plots and the results are summarised in Table 1.

Table 1: Sound Speed of Water in Tube

Method	C	δ
Calculated values	1390	-
Closed Tube Determination	1335 ± 10	0.05
Open Tube Determination	1325 ± 5	0.01

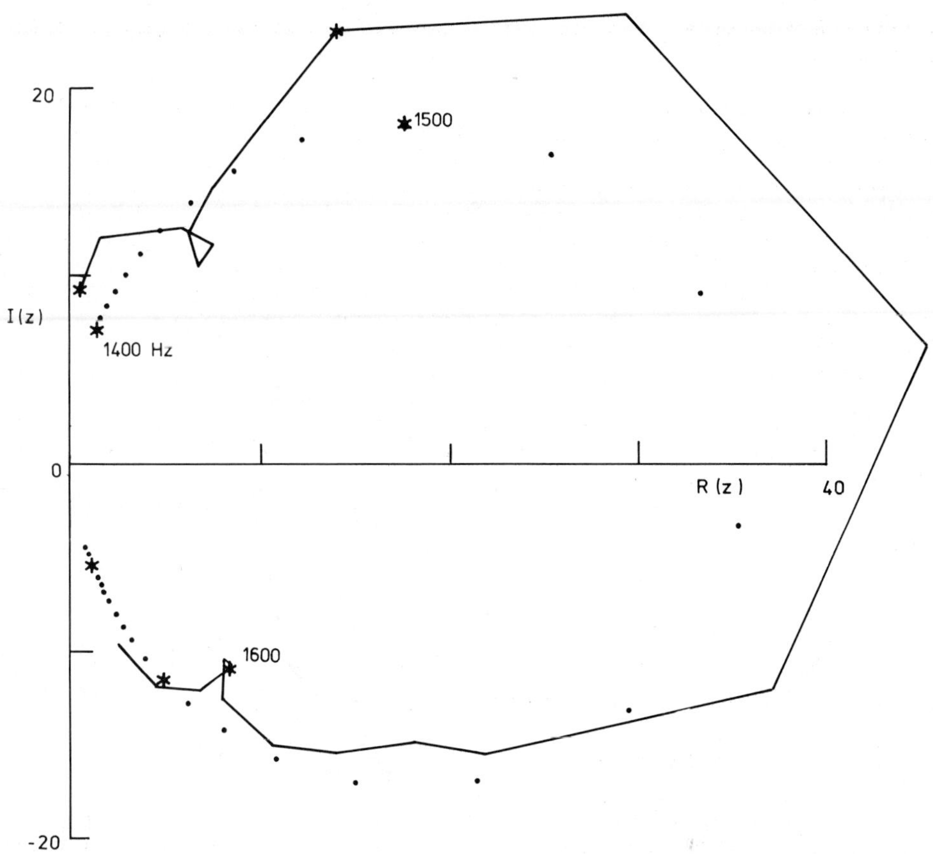

Fig. 2. Acoustic impedance values of a 191 mm length of water filled pipe terminated by the open end.
Measured values (-). Values calculated by equation 2 (....)

MEASUREMENTS ON MARINE SEDIMENTS

Acoustic impedance measurements were made using the two open end tube methods on sand in a laboratory tank filled with fresh water and on a sea floor at a water depth of approximately 10 m, the apparatus being sited by Scuba divers. The sediment in this case was a fine silty sand, density 1950 kg/m^3 and sound speed 1.04 C (as measured by the pulse transit method at 30 kHz).

When the open flanged tube was placed on the sea floor very little change was observed in the acoustic impedance at the end of the tube. This suggested small acoustic mismatch at the sea floor interface, and the relative insensitivity of the method to making measurements into the sediment. The protruding tube results were more promising, an example of the measured impedance of a 127 mm protrusion into the sea floor sediment being shown in Fig. 3. The theoretical impedance curve with optimal values substituted for the parameters δ (0.125 ± 0.01) and k (0.9 ± 0.05) is also shown in Fig. 3. The values of sediment sound speed obtained (1.05 C) agrees with that measured by the pulse transit method and the attenuation constant (α = 4.5 dB/m) is within accepted limits.

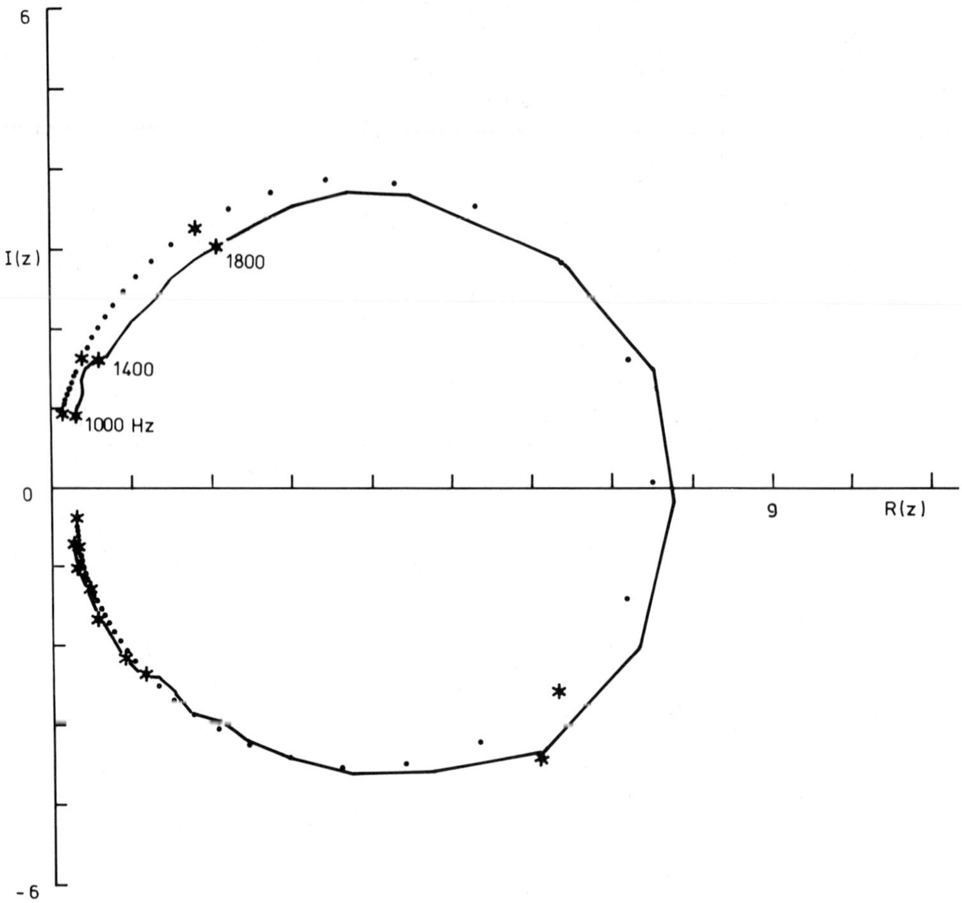

Fig. 3. Acoustic impedance values of length of pipe protruding 127 mm into marine sediment.
Measured values (-). Values calculated by equation 2 (.....) with δ = 0.125 and k = 0.9.

CONCLUSION

A new method for the direct in-situ measurement of the acoustic properties of marine sediments is presented. The method, based on the use of an acoustic impedance tube, produces opportunity for making measurements of sound speed C and attenuation constant α at frequencies much lower than traditional methods and with considerable precision in α. The technique is restricted to the accessible upper layer of the sediment.

ACKNOWLEDGMENT

The author wishes to acknowledge the Australian Marine Science and Technologies Grants Scheme fortheir support of this project.

REFERENCES

Dunlop, J.I., 1985, An Open Tube Technique for the Measurement of Acoustic Properties of Porous Sound Absorbing Materials, J. Acoust. Soc. Am. (in press).

Dunlop, J.I. and Whichello, A., 1981, Acoustic Impedance of Marine Sediments, Marine Geophys. Res., 5, 157.

Fay, R.D., Brown, R.L. and Fortier, O.V., 1947, Measurement of Acoustic Impedances of Surfaces in Water, J. Acoust. Soc. Am., 19(5), 850.

Kinsler, L.E. and Frey, A.R., 1962, Fundamentals of Acoustics, John Wiley 2nd. Ed., N.Y.

Kuhl, W., 1953, The Properties of Water-Filled Tubes for the Measurement of Impedances and Sound Velocity, Acustica, 3, 111.

Levine, H. and Schwinger, J., 1948, On the Radiation of Sound from an Unflanged Circular Pipe, Phys. Rev., 73(4), 383.

Seybert, A.F. and Ross, D.F., 1977, Experimental Determination of Acoustic Properties Using a Two Microphone Random Excitation Technique, J. Acoust. Soc. Am., 61(5), 1362.

Shirley, D.J., 1977, Method for Measuring In-situ Acoustic Impedance of Marine Sediments, J. Acoust. Soc. Am., 62(4), 1028.

Stoll, R.D., 1974, Physics of Sound in Marine Sediments, Ed. L. Hampton, Plenum Press, N.Y.

Zwikker, C. and Kosten, C.W., 1948, Sound Absorbing Materials, Elsevier, Amsterdam.

SPATIAL VARIABILITY OF SURFICIAL SHALLOW WATER SEDIMENT GEOACOUSTIC PROPERTIES

Michael D. Richardson

SACLANT ASW Research Centre
Viale San Bartolomeo 400
I-19026 La Spezia, Italy

ABSTRACT

Variability of surficial sediment geoacoustic properties was determined from cores collected at eight shallow-water continental shelf regions in the U.S., Italy and Australia. Highly porous muds found in low energy environments exhibited the lowest range of values in physical and acoustic properties; mixtures of sand and shell found in higher energy environments exhibited the highest range of values. Compressional wave attenuation consistently exhibited the highest variability followed by mean grain size, porosity and compressional wave velocity. Vertical variability was generally greater than horizontal variability for all properties measured. Sediment geoacoustic properties of most coastal marine sediments are controlled by the interaction of biological and hydrodynamic processes. Biological processes tend to dominate in finer sediments, whereas hydrodynamic processes control sediment geoacoustic properties in sandy substrates. Understanding these processes in various environments not only explains the spatial distribution of sediment geoacoustic properties but leads to improvement of predictive geoacoustic models.

INTRODUCTION

Knowledge of the spatial variability of surficial sediment geoacoustic properties is important to the prediction of acoustic scattering from the sediment-water interface and to the prediction of propagation of acoustic energy through the sediment[1]. It is within this surficial zone (upper 50 cm of sediment) that the most active and rapid diagenic changes in sediment properties are found. Gradients and variability of geoacoustic properties result from dewatering caused by overburden pressure, as well as active chemical, sedimentological, biological and hydrodynamic processes which mix and alter sediment properties[2,3].

For high frequency (>10 kHz) acoustic applications, geoacoustic properties of the upper tens of centimeters must be known, whereas for low frequency applications surficial geoacoustic properties provide the initial conditions used for prediction of depth gradients of sediment physical properties[1]. Accurate values, including variability, of surficial sediment geoacoustic properties are therefore required for geoacoustic models covering the wide range of frequencies of interest to those studying underwater acoustics, marine sedimentology, geophysics and marine geotechnique.

In this paper I compare data on the spatial variability of sediment geoacoustic properties among eight shallow-water continental shelf regions in the U.S, Italy and Australia. I also discuss the relative importance of biological, hydrodynamic and sedimentological processes in determining spatial distribution of these geoacoustic properties. Results are then compared to a similar study of geoacoustic properties of three sedimentary provinces in the Venezuela Basin (3500 to 5050 m water depth).

MATERIALS AND METHODS

Replicate sediment samples were collected from eight sites along the continental shelves of the United States, Italy and Australia (Table 1). Sediment types ranged from clayey-silts found in Long Island Sound and the Arafura Sea (off northern Australia) to coarse sands off San Diego, California. All sediments were collected with 6.1 cm (inside diameter) cylindrical cores either in situ by scuba divers or from relatively undisturbed 0.25 m^2 USNEL box core samples collected from deeper sites (Washington coast and Arafura Sea). Collection, measurement and handling procedures were designed to minimize sampling disturbance and maintain an intact sediment water interface with overlying sea water. Sediment compressional wave velocity and attenuation were measured after sediments equilibrated with laboratory temperature, usually one-half to two days after collection. Temperature and salinity of the overlying water were measured with a YSI model 43TD temperature probe and an AO Goldberg temperature-compensated salinity refractometer. Compressional wave velocity and attenuation were measured at 1 cm intervals on sediments in 129 subcores using a pulse technique.

*Table 1. Collection sites, including mean values of porosity, mean grain size, compressional wave velocity (V_p-ratio) and attenuation (dB/m at 400 kHz) (*unpublished data).*

Site	Date	Water Depth (m)	Substrate type	Porosity (%)	Grain Size (ϕ)	V_p-ratio	Attenuation (dB/m)	References
Long Island Sound	XIII 80							3,4
FOAM		10	clayey-silt	73.2	7.4	0.99	-	
NWC		16	clayey-silt	77.2	8.4	0.99	-	
San Diego,CA	IV - V 82							5
fine sand		18	very fine sand	-	3.5	1.10	188	
coarse sand		18	coarse sand	-	1.0	1.15	116	
Montauk Point,NY	V 82	35	fine sand	36.6	2.1	1.14	88	6
Quinault, Washington	IV 83	49	fine sand	41.2	2.9	1.11	160	7
Charleston S.C.	VI 83	18	medium sand	37.9	1.6	1.12	292	8,*
La Spezia, Italy	XI 83							8,9
MT		17	gravel	-	-1.6	1.20	-	
ST		8	silty-clay	69.8	8.9	0.98	104	
PV		13	silty-clay	67.6	9.4	0.98	64	
AV		5	very fine sand	43.3	3.6	1.10	136	
Arafura Sea	V 84	47	clayey-sand	69.2	5.3	0.99	336	*
Panama City, Florida	IX 84							10
pre-site		18-33	fine, medium sand	39.3	1.9	1.14	216	
experimental		33	fine sand	39.0	2.6	1.13	228	

Time delay measurements were made through sediments and a distilled water reference with a Hewlett-Packard 1743A dual time interval Oscilloscope. Signals were generated by driving a Underwater Systems, Inc. (Model USI-103) transducer-receiver head with a 400 kHz, 20 volts p-p sine wave triggered for 25μs duration every 10 ms with a Tektronic PG 501 Pulse Generator and FG 504 Function Generator. Differences in time delay between distilled water and sediment samples were used to calculate sediment compressional wave velocity (Vp). Compressional wave velocity was expressed as the dimensionless ratio of measured sediment velocity divided by the velocity in the overlying water that was calculated for the same temperature, salinity and depth[11]. This ratio is independent of sediment temperature, salinity and depth and therefore ideal for comparison to other geoacoustic properties. Attenuation measurements were calculated as 20 log of the ratio of received voltage through distilled water to received voltage through sediment[12]. Values of attenuation were extrapolated to a 1 m pathlength and expressed as dB/m.

After completion of the acoustic measurements, core samples were extruded and sectioned at 2 cm intervals for determination of sediment porosity and grain size distribution. Porosities were determined by weight loss of sediment dried at 105°C for 24 hours. Values were not corrected for pore water salinity. Salt-free porosity may be obtained by multiplying values by 1.012. Sediment grain size distribution was determined on disagregated samples by dry sieving for sand-sized particles and with a Micromeritics Sedigraph and/or pipette for silt and clay-sized particles. Grain size statistics were determined using the graphic formula of Folk and Ward[13].

RESULTS

Sediment geoacoustic data were measured on 129 cores that were collected for this study at eight different sites. I have restricted the data presentation to a summary of sediment geoacoustic properties from each site together with selected graphical representations of the spatial variability of sediment geoacoustic properties. Detailed data presentation can be found elsewhere[3-10].

Sediment samples were collected from two sites in Long Island Sound. Both sites were characterized by high porosity, fine grained, low velocity sediments. At the deeper (18 m) NWC site sediments, were characterized by a uniform distribution of geoacoustic properties (Table 2) on scales of centimeters to meters with no apparent depth gradients in the upper 20 cm. Richardson et al.[3] attributed this low variability to sediment mixing by macrofaunal animals that feed on bottom deposits (bioturbation). At the shallower FOAM site (10 m) sediments exhibited a much higher variability in sediment geoacoustic properties (Table 2). This higher variability was a result of storm induced erosional and depositional events creating laminations within the upper 35 cm of sediment. These laminations, each with different values of geoacoustic properties, were preserved because sediment mixing by macrofaunal animals at the FOAM site rarely extended below the upper few centimeters of sediment[3].

The shallow water samples collected off San Diego came from two distinct sediment types: a fine sand that had lower compressional wave velocity and higher attenuation than the coarser sand over which it was migrating. Of particular interest were the significant positive gradients in compressional wave velocity and attenuation with depth (Fig. 1) for both substrate types without a concordant change in mean grain size. These gradients probably resulted from increased packing and compaction of sands with depth in the sediment.

*Table 2. Coefficient of variation (CV) of porosity, mean grain size, compressional wave velocity ratio (Vp-ratio) and attenuation, calculated for eight shallow water and one deep-sea location (*data from the upper 2 cm of sediment only).*

Site	Porosity	Grain Size	V_p-ratio	Attenuation
Long Island Sound				
FOAM	7.33	11.53	0.82	-
NWC	1.50	1.91	0.35	-
San Diego				
fine sand	-	11.75	1.16	16.41
coarse sand	-	7.42	0.97	25.73
Montauk Point, NY	3.36	6.45	0.93	15.72
Quinault, Washington	4.98	3.63	1.19	33.08
Charleston S.C.	6.28	18.84	1.03	37.76
La Spezia, Italy				
ST	6.07	3.89	0.31	36.63
PV	6.86	1.80	0.34	36.11
AV	3.11	5.42	0.85	10.05
Arafura Sea	5.81	14.56	0.51	47.26
Panama City				
presite	5.60*	36.51*	1.07	33.03
experimental	3.72	6.14	0.87	15.59
Venezuela Basin				
carbonate	2.82	11.23	0.45	19.75
turbidite	9.88	16.22	1.67	105.88
hemipelagic	3.41	4.51	0.25	23.68

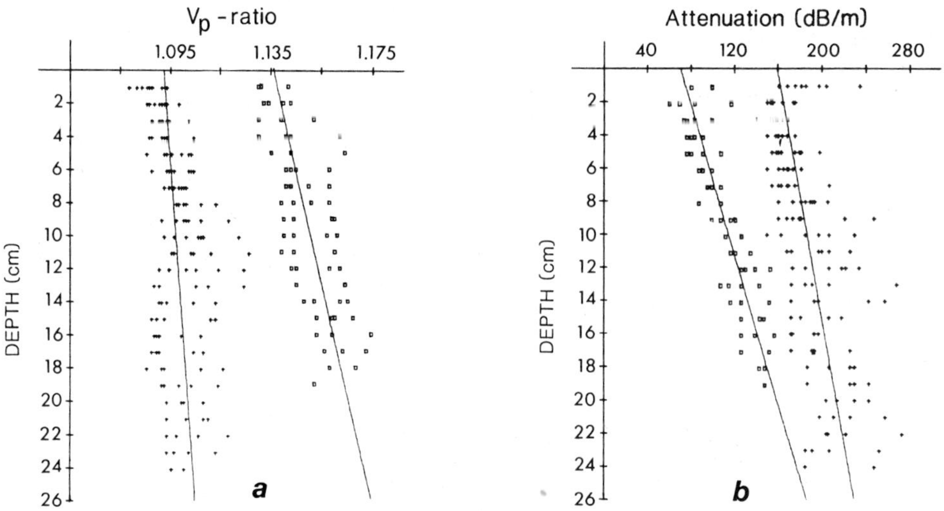

Fig. 1 Regressions of compressional wave velocity ratio (V_p-ratio) and attenuation (dB/m at 400 kHz) with depth for fine (+) and coarse (□) sand substrates found off San Diego, California.

The fine sand sediments collected in 35 m of water 25 km east of Montauk Point, Long Island, New York exhibited relatively low variability in sediment geoacoustic properties, especially attenuation. This low variability was attributed to mixing of sediments by an abundant population of sand dollars[6]. The positive gradient of compressional wave velocity with depth with little change in mean grain size was similar to gradients found in sediments from San Diego. Porosity generally decreased with depth supporting the argument that, for sand, positive gradients in compressional wave velocities can result from increased compaction. Surficial sediment samples collected within a 0.5 km^2 area around the prime collection site exhibited considerable variability in mean grain size (CV=108) because the study site was located at the base of a drowned barrier spit. At this site coarse grained lag deposits formed during the last glacial advance were covered intermittently by fine-grained sediments which were in equilibrium with current hydrodynamic conditions.

Fine sand surficial sediments collected in 50 m water depth on the continental shelf off Washington State exhibited little variation in values of geoacoustic properties from the surface to 12 cm depth (Fig. 2). Below 12 cm, the sediments had higher percentages of silt-sized particles that resulted in higher values of porosity, mean grain size, and attenuation as well as lower values of compressional wave velocity. Variability of geoacoustic properties was also highest below 12 cm depth. A high abundance of molluscan shells below 12 cm also contributed to higher variability of geoacoustic properties, especially attenuation.

Scuba divers observed two surficial substrate types in a one km^2 study area off the Charleston, S.C. coast. Fine to medium sand (mean grain size

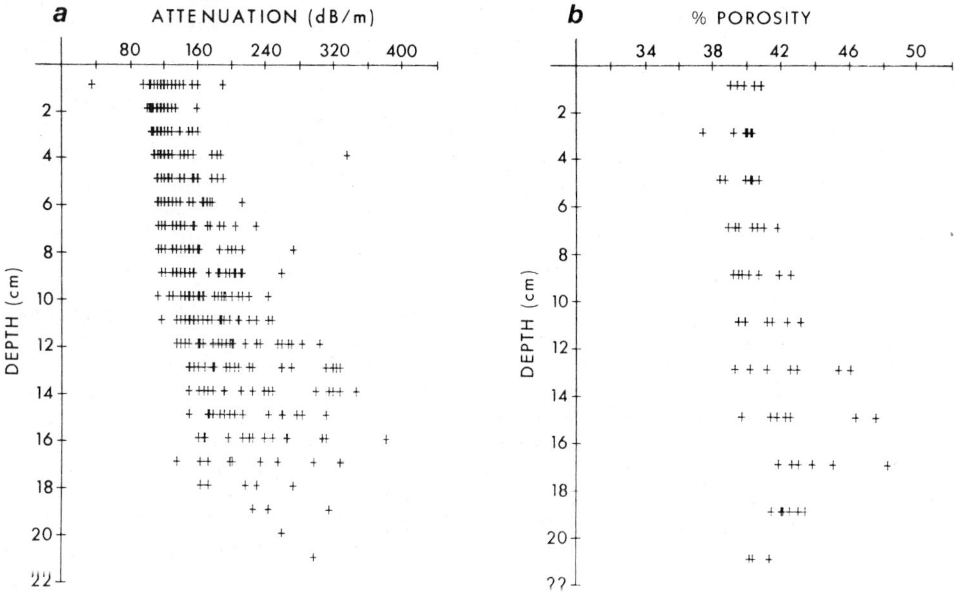

Fig. 2 Depth distribution of values of compressional wave attenuation (dB/m at 400 kHz) and porosity (%) for sediments collected in the Quinault acoustic tracking range - Washington.

1.8 to 2.2ϕ) with porosity values of 37 to 42% were found in troughs of shallow sand waves. The coarser (0.9 to 1.7ϕ) sediments on crests contained abundant shell material and generally had lower values of porosity (30 to 40%). Below 6 to 10 cm, shell material was abundant in all cores. Coefficients of variation for geoacoustic properties of the Charleston samples were higher than at most other locations (Table 2). Most of the variability at the Charleston site was associated with samples that had greater than 2% gravel-sized particles. If these samples are excluded, coefficients of variation are reduced to 0.72% for compressional wave velocity; 23.25% for attenuation; 5.19% for grain-size; and 2.20% for porosity.

Sediment samples were collected from four locations in the vicinity of La Spezia, Italy as part of an international program to compare various acoustical, geotechnical and geophysical measurement techniques[9]. Locations were chosen to include a wide spectrum of sediment types (Table 1). The Monasteroli site (MT) was located in a high energy environment next to the Ligurian coast. Sediments were composed of gravel-sized particles (75% by weight) making it difficult to collect adequate samples.

The coarser sediments (mean grain-size -1.57ϕ) probably originated from erosion of the nearby steep coastline[9]. Frequent storms winnowed finer material from the gravel sediments leaving only 0.17% silt-and clay-sized particles. Compressional wave velocity ratio (mean 1.20) and its coefficient variation (3.45%) were higher than at any other location. Porosity, mean grain size and attenuation were not accurately measured to permit the calculation of coefficients of variation. Sediments from the PV site located at the eastern entrance of the Portovenere channel were protected from physical disturbance, e.g. storms. Sediments were composed of silt and clay mixed with up to 5% gravel-sized shell particles. Compressional wave velocity and mean grain size values varied little (Table 2). However, the high coefficient of variation for porosity reflected 10 to 15% higher values of porosity found in the uppermost 4 cm of sediment. Both biological (bioturbation) and hydrodynamic processes probably maintained this higher surficial porosity. The Santa Teresa site (ST) was located in a small protected harbor on the eastern side of the Gulf of La Spezia. Samples consisted of very soft muds. Compaction of the upper 6 cm produced a negative gradient of porosity with depth without apparent change in sediment acoustic properties. The lack of vertical gradients in compressional wave velocity resulted in a low coefficient of variation. The Venere Azzurra site (AV) was located on a shallow (6 m), flat, hard-packed sandy bottom off Lerici in the Gulf of La Spezia. These well sorted and hard packed sediments had the lowest coefficient of variation for values of attenuation of any site listed in Table 2.

Twelve sediment cores were collected from a one km^2 area in the central Arafura Sea. The clayey-sand sediments contained an abundant quantity of both whole and broken gravel sized molluscan shells (2 to 23%). The lack of orientation of shells suggested considerable biological mixing had occurred. Variability of mean grain size was a function of the patchy distribution of shells whereas variability of porosity was related to vertical gradients caused by bioturbation and by strong currents which resuspend sediments. Very high values of attenuation measured at 400 kHz (mean 336 dB/m) were a direct result of scattering of the high frequency signal from shell material. Even at 125 kHz values of attenuation were high (mean 65 dB/m) and highly variable (45.7%). Compressional wave velocity ratio appeared to be much less affected by shell material, and was controlled primarily by the clayey-sand matrix (0.989 at 125 kHz vs 0.988 at 400 kHz).

As part of a pre-site survey, 27 sediment cores were collected from a variety of sandy sites (18 to 33 m water depth) off Panama City, Florida. The upper 2 cm of sediments ranged from coarse to fine sand (0.42ϕ to 2.75ϕ) with a relatively narrow range of porosities (36.3 to 43.9%). The coefficient of variation for attenuation and compressional wave velocity were approximately the same as other sandy sites (see Table 2). As part of the acoustic experiment 15 cores were collected from a one km^2 area that had uniform surficial sediment properties[10]. As expected the coefficient of variation for all geoacoustic properties was lower than that calculated from the pre-site survey (Table 2, Fig. 3).

DISCUSSION

The data in Table 2 demonstrate that shallow-water geoacoustic properties of sediment can be quite variable on scales of a kilometer or less. Not evident from Table 2 are the spatial sources of this variation (either vertical or horizontal), the scales (cm to km) or the processes that create this variation. Understanding these relationships can increase the predictability of sediment geoacoustic properties in time and space.

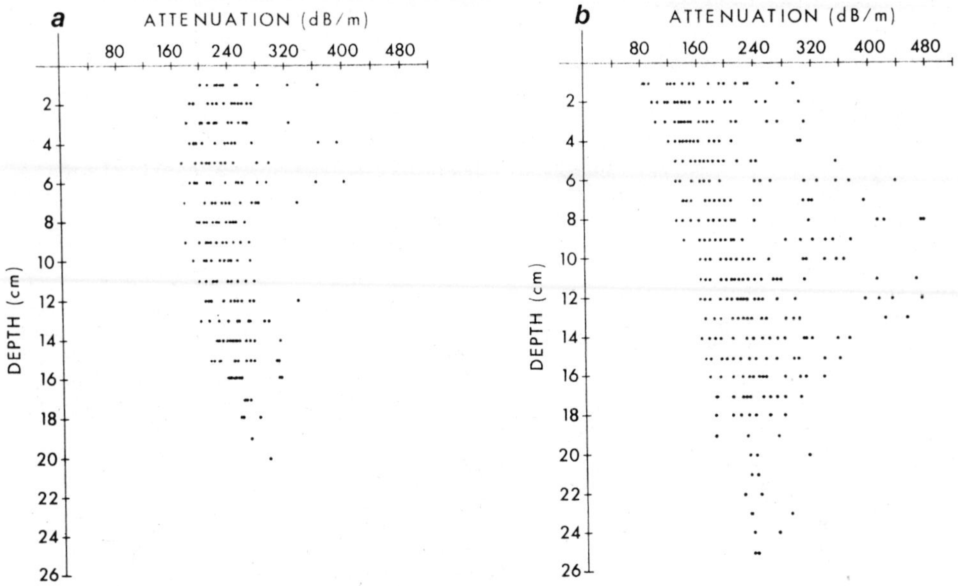

Fig. 3 Comparison of the variability and depth distribution of attenuation (dB/m at 400 kHz) for a one km² experimental site (a) and a larger presite survey area (b) off Panama City, Florida.

Several general observations can be made from the data presented in Table 2. Compressional wave attenuation consistently had the highest values of variation, and velocity the lowest. These results were expected given the previously reported range of values for compressional wave velocity and attenuation[12,14,15]. The coefficient of variation for mean grain size was generally higher than for porosity. The lowest coefficients of variation for mean grain size and compressional wave velocity were found at muddy sites, whereas attenuation and porosity exhibited the lowest coefficients of variation in fine hard packed sands. The highest variability of attenuation was found at sites with an high percentage of shell material.

Depth gradients in geoacoustic properties at several sites (FOAM, San Diego, Montauk Point, Washington and Charleston) accounted for a considerable percentage of variation of sediment geoacoustic properties. At the San Diego site compressional wave velocity and attenuation significantly correlated with depth (F-test, >0.001) for both fine and coarse sand substrates (see Fig. 1). The coefficients of variation for these properties after correction for depth gradients were 11.59 to 11.79 for attenuation and 0.72 to 0.79 for velocity. Depth gradients in geoacoustic properties at the San Diego and Montauk Point sites were the result of increased packing of sand-sized particles with depth[5,6]. At the FOAM site sediment laminations produced by storm induced erosional and depositional events resulted in the higher values of mean grain size and porosity and lowest values of compressional wave velocity between 5 to 15 cm depth in the cores[3,4]. Increased percentages of silt with depth at the Washington continental shelf site resulted in higher values of porosity and lower values of compressional wave velocity below 12 cm depth in the cores[7]. At both the Washington and Charleston sites the higher abundance of shell material with depth resulted in higher values of attenuation with depth[7,8]. Knowledge of the depth distribution of values of sediment geoacoustic properties can greatly increase predictability of these properties.

The magnitude of horizontal variation in geoacoustic properties varied considerably among sites. At the FOAM site in Long Island Sound horizontal patchy distribution of animals resulted in considerable variation in geoacoustic properties on scales of 10 to 100 meters[4]. At San Diego, sediments were classified as fine and coarse sands based on a mosiac created from overlapping tracks of side-scan sonar records[5]. Knowledge of the distribution of these two substrate types greatly reduced coefficients of variation of geoacoustic properties used in correlations with the values of acoustic bottom backscatter[16]. Side-scan mosaics from the Charleston, S.C. and Panama City sites also indicated considerable spatial variability in sediment types. The improved predictability of sediment geoacoustic properties for the Panama City site is clearly demonstrated in Table 2. At the Charleston site both horizontal and vertical variations in sediment geoacoustic properties must be taken into account to improve the predictability of sediment properties. Large scale horizontal variation, such as found in data collected from the Gulf of La Spezia, can usually be accounted for by utilizing detailed sediment distribution maps which are available for many continental shelf regions.

Table 3. Dominant factors affecting geoacoustic properties of sediment (xx = very important; x = important).

Experiment	Hydrodynamic Processes	Biological Processes	Relict (Inherited) Features
Long Island Sound	x	xx	
San Diego CA	xx		x
Montauk Point	x	x	xx
Quinault, Wash	xx		
Charleston S.C.	xx	x	x
La Spezia, Italy	xx		xx
Arafura Sea	x	xx	x
Panama City, Fl.	xx	x	

Sediment geoacoustic properties of many coastal marine sediments are controlled by the interaction of biological and hydrodynamic processes[3]. Table 3 presents the relative importance of these processes as well as that of relict (inherited) features in determining values and variability of sediment geoacoustic properties. Bioturbation (mixing of sediment by deposit-feeding animals) tends to reduce horizontal variation in sediment physical properties especially in fine grained sediments (NWC, Arafura Sea). The vertical gradients created by bioturbation are usually restricted to the upper 10 cm and are predictable on a seasonal basis given knowledge of biological processes present. A high abundance of tube dwelling species tends to stabilize the sediment surface, thus preserving the spatial variability of geoacoustic properties (FOAM). In fine sediment where the dominance of hydrodynamic and biological processes alternate horizontal and vertical variability is great and predictability poor (FOAM).

In sandy substrates, surface deposit feeding species, such as sand dollars, tend to reduce horizontal variability[5]. Feeding also reduces packing in the upper few centimeters increasing porosity and decreasing compressional wave velocity (Montauk Point, Panama City). At Montauk Point sand dollars were so abundant that they acted as surface point scatterers of acoustic energy[17]. The presence of surface and buried shell material contributes to considerable fine scale variation of geoacoustic properties

(Quinault, Charleston, Arafura Sea and Panama City sites). The high values and variability of compressional wave attenuation associated with shell material are due to scattering from the shells as opposed to the intrinsic absorption described by Hamilton[12]. The distribution of shells in the sediment is controlled by both biological and hydrodynamic processes. The random distribution and lack of preferred orientation of shells at the NWC and Arafura Sea sites suggests biological mixing processes were dominant, whereas hydrodynamic mixing processes control the distribution of shells at sandy sites (Charleston, Quinault and Panama City).

Large scale variability is generally controlled by hydrodynamic processes acting on given relict features. At the San Diego site a recently deposited fine-grained sand was found to be migrating over a coarser-grained offshore Pleistocene deposit[5]. At the Montauk Point site a light colored fine grained sand of modern origin discontinously covered a reddish granular sediment which was a lag deposit formed by the erosion of a drowned Pleistocene barrier spit[6]. At both locations the sand of modern origin was in dynamic equilibrium with recent storm events, whereas the coarser Pleistocene deposits were in equilibrium with winter storm events.

Hamilton and Bachman[14] presented data on geoacoustic properties from 340 sediment samples collected on the continental shelf and slope. They classified sediments into nine groups based on grain size. The coefficient of variation for porosity (range 5.0 to 13.8%; mean 10.3%) and compressional wave velocity ratio (0.7 to 3.2%; mean 2.3%) were much greater within those nine sediment classes than for the individual sites included in this study (Table 2). Direct measurements of geoacoustic properties are therefore preferred to predictions based on sediment type despite considerable fine-scale variability of geoacoustic properties demonstrated in this paper.

Briggs et al[18], using the same techniques employed here, characterized sediment geoacoustic property variability for three sedimentary provinces in the Venezuela Basin (3450 to 5050 m water depth) (see Table 2). Most of the variability was associated with vertical gradients of geoacoustic properties (cm) as opposed to horizontal variability (on scales of cm to km)[10]. As would be expected in the deep sea coefficients of variation for the carbonate and hemipelagic sites were lower than at most shallow-water sites, whereas the turbidite site had very high coefficients of variation. This variability resulted from the presence of alternating layers of pelagic and terrigeneous sediments. This study indicates the importance of considering the variability of geoacoustic properties for geoacoustic modelling of both deep-sea and shallow water sediments.

ACKNOWLEDGMENTS

I wish to thank Kevin Briggs and Ricky Ray of NORDA for their assistance in data collection, laboratory analysis and data preparation. I also thank Dave Young for his encouragement and support during field exercises and report preparation. Data collection and analysis were conducted while the author was at the Naval Ocean Research and Development Activity and was supported by NAVSEA Program Element 62759N (Bob Martin, Program Mgr) and NORDA Program Element 61153N (Herb Eppert, Program Mgr). The paper was written at the SACLANT ASW Research Centre (R.R. Goodman, Technical Director).

REFERENCES

1. E.L. Hamilton, Geoacoustic modelling of the sea floor, J. Acoust. Soc. Am. 68: 1313-1340 (1980).

2. M.D. Richardson and D.K. Young, Geoacoustic models and bioturbation, Mar. Geol. 38: 205-218 (1980).
3. M.D. Richardson, D.K. Young, and K.B. Briggs, Effects of hydrodynamic and biological processes on sediment geoacoustic properties in Long Island Sound, USA, Mar. Geol. 52: 201-226 (1983).
4. M.D. Richardson, D.K. Young, and K.B. Briggs, Acoustical, physical and biologial properties of surface sediment cores collected from Long Island Sound, August 27-28, 1980, NORDA TN 150. NSTL, MS, Naval Ocean Research and Development Activity, (1982). [AD A 118 183]
5. M.D. Richardson, D.K. Young, and R.I. Ray, Environmental support for high frequency acoustic measurements at NOSC Oceanographic Tower, 26 April - 7 May 1982; Part I: Sediment geoacoustic properties, NORDA TN 219. NSTL, MS, Naval Ocean Research and Development Activity, (1983).
6. M.D. Richardson, J.H. Tietjen, and R.I. Ray, Environmental support for Project Weapons Environmental Acoustic Program (WEAP) East of Montauk Point, New York, 7-28 May 1982, NORDA Report 40. NSTL, MS, Naval Ocean Research and Development Activity, (1983). [AD A 139 800]
7. M.D. Richardson, K.B. Briggs, R.I. Ray, and W.I. John, Environmental support for high frequency acoustic experiments conducted at the Quinault Range off the Washington Coast, 28 April - 1 May 1983. NORDA Report. NSTL, MS, Naval Ocean Research and Development Activity, (1985).
8. M.D. Richardson, Environmental bottom characterization required for modeling and prediction of high-frequency acoustic bottom scattering, J. Acoust. Soc. Am. 75: S50-S51 (1984).
9. T. Akal, P. Curzi and E. Michelozzi, Geoacoustic measurements: physical property variations and sedimentary processes. In: Convegno sul tema: La Geologia Marina — aspetti di ricerca pura ed applicata. S. Benedetto del Tronto, 16-18 aprile 1984, Ascoli Piceno, Societa' Geologica Italiana (to be published).
10. S. Stanic, M.D. Richardson, P. Fleischer and B.E. Eckstein, High frequency acoustic bottom scattering experiments conducted off Panama City, Florida. NORDA Report. NSTL, MS, Naval Ocean Research and Development Activity, (1985).
11. E.L. Hamilton, Prediction of in-situ acoustic and elastic properties of marine sediments, Geophys. 36: 266-284 (1971).
12. E.L. Hamilton, Compressional-wave attenuation in marine sediments, Geophys. 37: 620-645 (1972).
13. R.L. Folk and W.C. Ward, Brazos River bar, a study in the significance of grain size parameters, J. Sediment. Petrol. 27: 3-26 (1957).
14. E.L. Hamilton and R.T. Bachman, Sound velocity and related properties of marine sediments, J. Acoust. Soc. Am., 72: 1891-1904 (1982).
15. T. Akal, The relationship between those physical properties of underwater sediments that affect bottom reflection, Mar. Geol., 13: 251-266 (1972).
16. H. Boehme, N.P. Chotiros, L.D. Rolleigh, S.P. Pitt, A.L. Garcia, T.G. Goldsberry, and R.A. Lamb, Acoustic backscattering at low grazing angles from the ocean bottom. Part I. Bottom backscattering strength, J. Acoust. Soc. Am., 77: 962-974 (1985).
17. W.I. Roderick and R.K. Dullea, High resolution bottom backscatter measurements, NUSC Tech. Document 7181. New London, CT, Naval Underwater Systems Center, (1984).
18. K.B. Briggs, M.D. Richardson, and D.K. Young, Variability in geoacoustic and related properties of surface sediments from the Venezuela Basin, Caribbean Sea. Mar. Geol. (in press).

A BOTTOM SHEAR MODULUS PROFILER: INVERSE ANALYSIS OF THE FIELD DATA ON WAVE-INDUCED BOTTOM MOTION

Tokuo Yamamoto and Tsuyoshi Torii

Applied Marine Physics Division
Rosenstiel School of Marine and Atmospheric Science
University of Miami
Miami, Florida 33149-1098

INTRODUCTION

Accurate information on the shear modulus profile within the seabed is essential for modeling bottom interactions of acoustic wave propagation in the shallow ocean. Such information is also important for design of offshore structures and for modeling bottom interactions of gravity water waves. Conventional methods using penetrometers, vanes and bore holes are costly and time consuming. They are often impossible in severe environments such as in the Arctic and are impractical for covering a large area of the seafloor. To help ease this situation, a concept of a bottom shear modulus profiler has been developed.

A preliminary attempt was recently made to determine the shear modulus profile of a clay bed by a linear inverse theory (Figueroa et al., 1985). This idea has been refined and advanced to be much more practical and efficient. This paper presents the concept of the bottom shear modulus profiler (BSMP) as an alternative to the conventional methods. The BSMP is consisted of the software which is based on the Yamamoto (1983b) theory of wave-soil interaction combined with the linear inverse theory and the hardware for the instrumentation to measure wave-induced bottom motion. The BSMP software has been developed and applied to a set of field data on the wave-induced mudline motion obtained by Suhayda et al., (1976). The BSMP software and the example application are presented in this paper.

THEORY

The Forward Theory of Wave-Soil Interaction

An extensive review of the theories of wave-soil interaction is given by Yamamoto et al., (1983). Because the permeability of sediments is very small and because the frame structure of sediments is soft, the solid frame and the pore fluid of the sediments are completely coupled in their motion under gravity water waves (Yamamoto et al., 1978). In addition, the stiffness of pore water is much larger compared to the rigidity of sediment frame, the wave-induced motion of the water-saturated sediments is entirely due to shear deformation. Furthermore, because water waves have very low frequencies, inertia forces are negligibly small compared with elastic restoring forces in the equations of motion of water-saturated sediments (Yamamoto, 1983a). In other words, a water-saturated sediment behaves as a massless incompressible solid under water waves. Therefore, the wave-induced motion of a sediment bed is independent of the density of the sediment but only dependent of the shear modulus of the sediment. This simple behavior of a sediment bed under water waves is the basis for the bottom shear modulus profiler proposed in this paper.

The dispersion relation of the gravity water waves propagating over a sediment bed is given by Yamamoto (1982) as

$$\tanh kh = \frac{\omega^2}{gk}\left(1 - \frac{W_o + V_o}{a_o}\,\text{sech}\,kh\right) \tag{1}$$

Here, k = wave number; h = water depth; $\omega = 2\pi/T$ angular frequency; T = wave period; g = gravitational acceleration; W_o = amplitude of vertical displacement at the bed surface; V_o = amplitude of vertical seepage displacement of pore water at the bed surface and a_o = amplitude of surface water waves. Seepage displacement V_o is usually negligible. As the bed admittance W_o/a_o is of the order of one percent or so at most according to Suhayda et al., (1976) and Forristal and Reece (1985), the dispersion relation is reduced to the familiar one given by

$$gk \tanh kh = \omega^2 \tag{2}$$

The wave induced motion of an arbitrarily layered bed may be given by a propagator matrix representation (Yamamoto, 1983b).

Let the displacement-stress vector $\underset{\sim}{r} = (r_1, r_2, r_3, r_4)$ represent the depth functions of the horizontal displacement U, the vertical displacement W, the shear stress T_{xz} and the vertical normal stress T_{zz}, as

$$\begin{aligned} U &= i\, r_1\, \theta, & W &= r_2\, \theta \\ T_{xz} &= i\, r_3\, \theta, & T_{zz} &= r_4\, \theta \end{aligned} \tag{3}$$

where $\theta = \exp[i(\omega t - kx)]$

Then, the wave-induced motion of a layered massless incompressible elastic bed is given as

$$\frac{d}{dz}\begin{bmatrix} r_1 \\ r_2 \\ r_3 \\ r_4 \end{bmatrix} = \begin{bmatrix} 0 & k & 1/G & 0 \\ -k & 0 & 0 & 0 \\ 4Gk^2 & 0 & 0 & k \\ 0 & 0 & -k & 0 \end{bmatrix} \begin{bmatrix} r_1 \\ r_2 \\ r_3 \\ r_4 \end{bmatrix} \tag{4}$$

in which G(z) = shear modulus which may be an arbitrary function of depth. Note that the propagator matrix is made up by wave number k and shear modulus G(z) alone. This simple structure of the propagator matrix provides stability and precision for the inversion.

The linear differential equation (4) may be integrated numerically as an initial value problem. A fourth order Runge-Kutta method is adopted as our numerical integration technique. Two independent sets of initial values may be defined at the basement (z = − d). z is the vertical distance from the mudline and is positive in the upward direction.

If the basement (at z = − d) is a rigid boundary, the two initial vectors may be given as

$$\underset{\sim}{r}_s\ (-d) = (0,0,1,0)$$

and (5)

$$\underset{\sim}{r}_p\ (-d) = (0,0,0,1)$$

On the other hand, if the basement at z = -d is an elastic half space, the two initial vectors may be given as

$$\underset{\sim}{r}_s\ (-d) = (1,0,-2kG(-d),\ 0)$$

and (6)

$$\underset{\sim}{r}_p\ (-d) = (0,1,0,\ -2kG(-d))$$

The displacement-stress vector in the sediment bed is given as a linear combination of the two particular vectors as

$$\underset{\sim}{r}(z) = A\underset{\sim}{r}_s\ (z) + B\underset{\sim}{r}_p\ (z) \tag{7}$$

The proportionality constants A and B can be determined from the boundary conditions at the sea-bed interface. At the interface (z = 0) shear stress vanishes;

$$r_3(0) = 0 = Ar_{s3}(0) + Br_{p3}(0) \tag{8}$$

and vertical stress must be balanced by wave-induced bottom pressure P(0) θ as

$$r_4(0) = -\ P(0) = Ar_{s4}(0) + Br_{p4}(0) \tag{9}$$

in which P(z) is the wave induced-pressure amplitude at z = z as

$$P(z) = \rho g a_o \text{ sech } kh \cosh kz \tag{10}$$

in which ρ = density of water. In the experiments, bottom pressure may be measured directly or deduced from measured wave amplitude through Eq. (10).

The Linear Inverse Problem

The linear inverse problem most often encountered in geophysics can be written in the discrete form (Aki and Richards, 1980) as

$$e_i = \sum_{j=1}^{M} C_{ij}\ m_j\ (i=1,2,\ldots,\ N)$$

or in matrix notation

$$\underset{\sim}{e} = \underset{\sim}{C}\ \underset{\sim}{m} \tag{11}$$

where e_i is one of N observations, C_{ij} is a kernel derived from a forward theory, and m_j is an unknown function for which we solve. Actual problems may be nonlinear. However, the solution for a nonlinear problem may be obtained iteratively through a succession of the piecewise linear steps to the true solution.

Our problem discussed as the forward theory in the previous section can be regarded as the same inverse problem. For our case, $\underset{\sim}{e}$, $\underset{\sim}{m}$ and $\underset{\sim}{C}$ are approximated by the first term of the Taylor expansions as,

$$\underset{\sim}{e} = (\underset{\sim}{U}_D - \underset{\sim}{U}_I, \underset{\sim}{W}_D - \underset{\sim}{W}_Z)^T = (e_1, e_2, \ldots, e_N)^T \tag{12}$$

$$\underset{\sim}{m} = (\cdots, (\frac{\Delta G}{\alpha G})_i, \ldots) = (m_1, m_2, \ldots, m_M)^T \tag{13}$$

$$\underset{\sim}{C} = \begin{pmatrix} C_{11} & C_{12} & \cdots & C_{1M} \\ C_{21} & C_{22} & \cdots & C_{2M} \\ \vdots & \vdots & & \vdots \\ C_{N1} & C_{N2} & \cdots & C_{NM} \end{pmatrix} \tag{14}$$

$$C_{ij} = (\underset{\sim}{U}_j' - \underset{\sim}{U}_I)_i \text{ and } (\underset{\sim}{W}_j' - \underset{\sim}{W}_I)_i \tag{15}$$

where

$\underset{\sim}{U}_D$, $\underset{\sim}{W}_D$ = the measured horizontal and vertical displacement spectra, respectively

$\underset{\sim}{U}_I$, $\underset{\sim}{W}_I$ = the horizontal and vertical displacement spectra calculated by the forward theory for the initial shear modulus profile,

$\underset{\sim}{G}(z) = (G_1, G_2, \ldots, G_j, \ldots, G_M)^T$

$\underset{\sim}{U}_j'$, $\underset{\sim}{W}_j'$ = the horizontal and vertical displacement spectra calculated by the forward theory for the modified initial shear modulus profile $G_j'(z) = (G_1, G_2, \ldots, (1+\alpha)\, G_j, \ldots G_M)^T$.

To solve Eq. (11) for $\underset{\sim}{m}$, we employ the singular value decomposition of the matrix $\underset{\sim}{C}$ (e.g. Aki and Richards, 1980);

$$\underset{\sim}{C} = \underset{\sim}{H}\ \underset{\sim}{\Lambda}\ \underset{\sim}{V}^T \tag{16}$$

and the construciton of the solution is

$$\begin{aligned} \hat{\underset{\sim}{m}} &= \underset{\sim}{C}^{-1}\ \underset{\sim}{e} \\ &= \underset{\sim}{V}\ \underset{\sim}{\Lambda}^{-1}\ \underset{\sim}{H}^T\ \underset{\sim}{e} \\ &= \sum_{i=1}^{N} \frac{\underset{\sim}{H}_i\ \underset{\sim}{e}}{\lambda_i}\ \underset{\sim}{V}_i \end{aligned} \tag{17}$$

where

$\underset{\sim}{H}$; the eigenvectors of a coupled eigenvector-eigenvalue problem, N-dimensional vectors of the data space.

$\underset{\sim}{V}$; the eigenvectors of a coupled eigenvector-eigenvalue problem, M-dimensional vectors of the solution space.

$\underset{\sim}{\Lambda}$; the diagonal matrix with the corresponding eigenvalues,

λ_i ; the eigenvalues. Zero valued eigenvalues are not included.

For the data with noise such as actual experimetnal data, we have constructed the solution using the truncated singular value expansion;

$$\hat{\underset{\sim}{m}} = \sum_{i=1}^{L} \frac{\underset{\sim}{H}_i\ \underset{\sim}{e}}{\lambda_i}\ \underset{\sim}{V}_i \quad (L < N) \tag{18}$$

We have, also, adopted the resolution matrix for the quality of our solution.

$$\underset{\sim}{C}^{-1}\ \underset{\sim}{C} = \sum_{k=1}^{L} V_{ik}\ V_{kj} \tag{19}$$

In the ideal case $\underset{\sim}{C}^{-1}\ \underset{\sim}{C}$ is the identity matrix and resolution is perfect. When $\underset{\sim}{C}^{-1}\ \underset{\sim}{C}$ has near unity values on the diagonal and small values elsewhere, the computed solution represents a smoothed version of the true solution.

In order to compute Eqs. (17) through (19) a Fortran computer program which was developed by M. G. Brown (1983) has been used.

<u>Numerical Tests</u>

The consistency and uniqueness of the inversion procedure has been ascertained by synthetic data without noise which are produced through the forward theory using prescribed shear modulus profiles.

It was found that the forward theory is quite linear and appropriate for linear inversion in the range of increments $-1.0 < \alpha < 8.0$ in Eq. (13). Using a reasonable criterion for the truncation for the solution given by Eq. (18) we have been able to determine consistently and uniquely the shear modulus profile(s) starting from various initial profiles (Torii, 1985).

FIELD MEASUREMENTS

During periods in March, 1976, Suhayda et al.(1976) made continuous simultaneous measurements of bottom oscillations and wave characteristics at the East Bay of the Mississippi Delta (see Fig. 1). Wave staffs, pressure sensors and a 3-component bottom accelerometers were placed in about 19 m of water at an oil platform in an area having a fine-grained clay bottom. The accelorometer package was placed about 0.3 m below the mudline (see Fig. 2).

Table 1 - Field data of the amplitudes of wave-induced pressure (at z = 13.2 m) and bottom accelerations. (from Dawson and Suhayda, 1981)

Wave	Wave-induced Pressure (lbs/ft^2)	Horizontal Acceleration (ft/sec^2)	Vertical Acceleration (ft/sec^2)	Wave Period (sec)
1	94.7	.0110	.0156	6.2
2	73.6	.0086	.0115	6.8
3	67.2	.0069	.0108	7.0
4	50.4	.0065	.0069	6.1
5	77.7	.0079	.0140	8.0
6	52.5	.0047	.0069	5.2
7	71.4	.0086	.0082	6.0
8	43.1	.0056	.0069	6.7
9	62.0	.0069	.0125	7.6
10	47.2	.0042	.0069	5.7

The results of the experiments indicated that the bottom undergoes an elastic response to wave-induced bottom pressure in a quite linear manner. The amplitudes of the wave-induced pressure at z = 13.2 m, the bottom horizontal acceleration and the bottom vertical acceleration under ten waves of nearly sinusoidal shapes are depicted in Table 1. (Dawson and Suhyada, 1981).

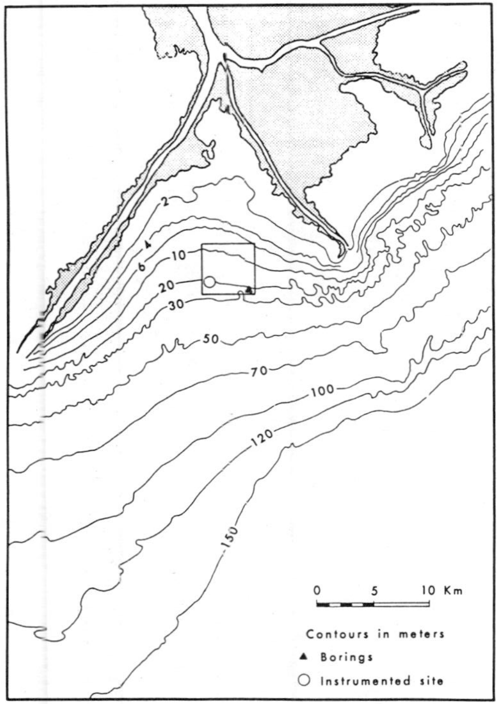

Fig. 1. Location of test site and boring site in East Bay, Louisiana. (from Suhayda et al., 1976)

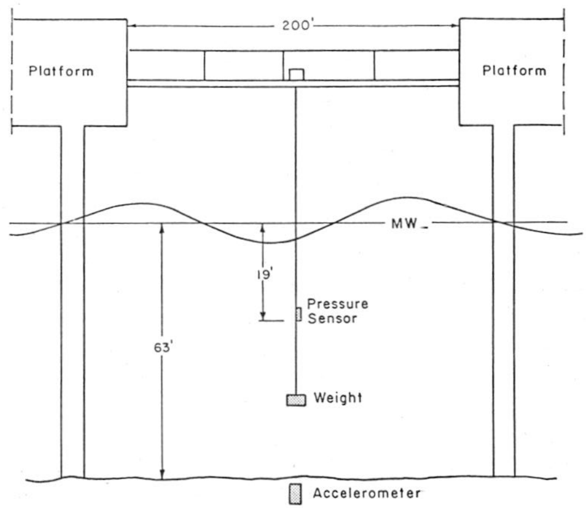

Fig. 2. Experiemtntal setup of the field measurements (from Suhayda et al., 1976)

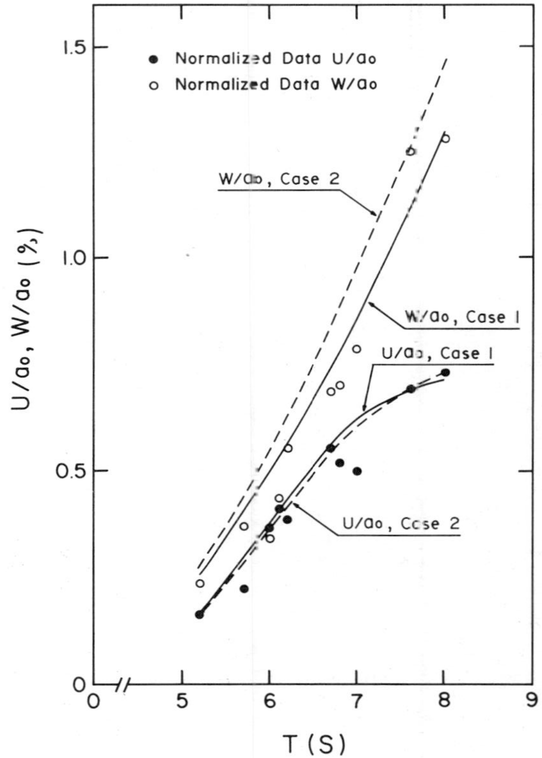

Fig. 3. Measured bottom admittance spectra and theoretical spectra for the two final shear modulus profiles obtained through inversion.

Using this set of data, the horizontal bed admittance U/a_0 and the vertical bed admittance W/a_0 are calculated for the ten waves and are plotted in Fig. 3. These values of bottom admittance are used as input data for the inversion analysis.

The geological characteristics of the cores taken in the boring from this area were presented by Roberts, et al., (1976). Figure 4 shows the shear strength profile in the bottom taken at Boring 5 which is about 5 km east of the experimental site as shown on the map (Fig. 1). Note that an approximately linear shear strength profile exists above a thick sublayer of soft mud below about -25 m.

INVERSION RESULTS

Iterative Inversion Procedure

The iterative linear inversion procedure outlined in the previous section is implemented for the bed admittance data and illustrated in Fig. 5. The grid system of the bed used for the numerical integration is also shown in Fig. 5. The upper 50 m of the bed is represented at a constant integration step of 1 m. The basement below -50 m is assumed to be an elastic half space whose shear modulus is equal to that of the grid point at z = -50 m. In the inversion, the values of shear modulus at twelve elevations (-z = 1, 3, 5, 8, 11, 14, 18, 22, 26, 30 40 and 50 m) are varied as independent variables. The shear modulus at the grid points between these points is linearly interpolated.

Figure 5 shows how the initial shear modulus profile corrects itself through eight successive iterative inversions and finally converges to the final profile.

Tests for Uniqueness and Consistency

In order to evaluate the uniqueness and the consistency of the inverted shear modulus profile, another independent iterative inversion procedure is conducted using a completely different initial shear modulus profile as shown in Fig. 6. Comparison between Fig. 5 and Fig. 6 indicates that the overall featurs of two final profiles are very similar except near the peaks at about z = -18 m. Both profiles clearly reveal the existence of a very soft sublayer below about -25 m. This is an indication of the consistency of the inversion. The fact that the two final profiles differ from each other weakens the uniqueness of the inverted solution to some degree. This is largely due to the noise or the scatter in the measured data as may be observed in Fig. 3. The calculated bottom admittance spectra for the two final shear modulus profiles are compared with the measured data in Fig. 3. It may be concluded that the iterative inversion procedure tends to smooth the data and converges to one of many possible averaged lines through data. This indicates some degree of uncertainty introduced by the noise in the real data which is absent in the inversion using the synthetic data without noise (Torii, 1985).

Comparison between the Inverted Shear Modulus Profile and the Boring Data

The two final shear modulus profiles obtained from the inversion analysis are compared with the shear strength profile obtained from the boring in Fig. 4. A conversion factor C = 200 is used to convert shear strength into shear modulus. It must be noted that the boring site and the test site is separated by about 5 km so that the comparisons are only very approximate. Nonetheless, good overall agreement between the inverted

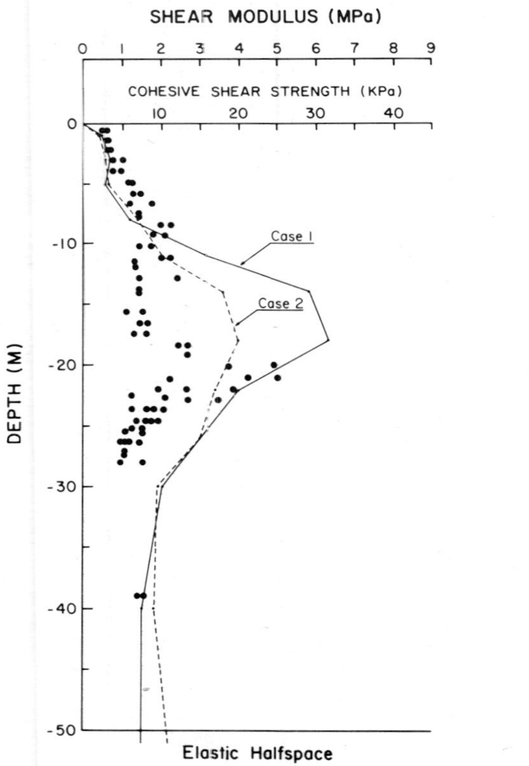

Fig. 4. Shear strength profile taken at Boring 5 site (from Roberts, et al., 1976). Two curves are the shear modulus profiles at the test site (5 km west of Boring 5) obtained through inversion.

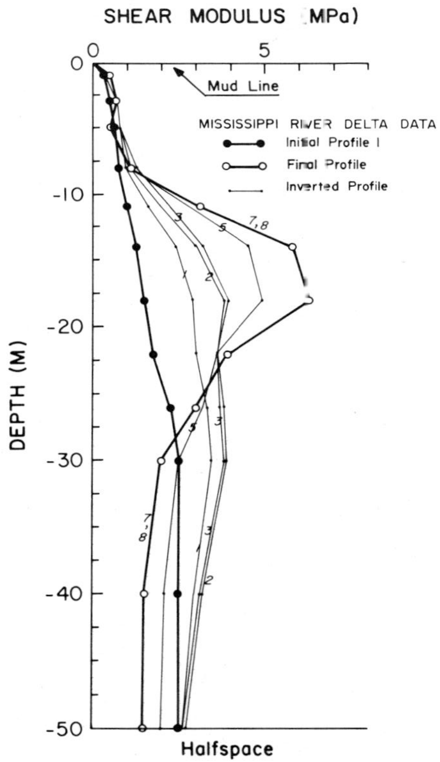

Fig. 5. Inverted shear modulus profile, Case 1, the initial profile 1, the intermediate profiles (the number of iterations indicated) and the final profile.

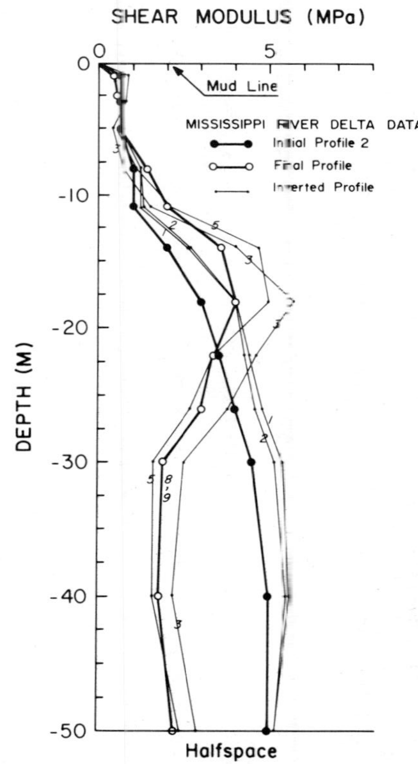

Fig. 6. Inverted shear modulus profile, Case 2, the initial profile 2, the intermediate profiles, and the final profile.

profiles and profile by direct measurements is obvious. In particular, the very soft sublayer below about -25 m revealed by the inverse analysis is most remarkable.

CONCLUSIONS

A software of a bottom shear modulus profiler (BSMP) has been developed based on the Yamamoto theory of wave-soil interaction combined with a linear inversion technique. The BSMP software has been successfully applied to determine the shear modulus profile of the clay bed at the East Bay of the Mississippi River Delta from the field data of wave-induced mudline motion.

ACKNOWLEDGEMENT

This research was solely supported by ONR Code 425UA (Dr. Raymond Fitzgerald) through a contract N00014-85-C-0020. The authors are grateful for this support. The authors wish to thank Dr. Peter Rogers for his encouragement at the initial stage of the BSMP development and Dr. Michael Brown for his extensive consultation during the investigation and kindly allowing us to use his computer subroutine for linear inversion. The two-year scholarship given to the junior author by Toa-Kensetsu Co is appreciated.

REFERENCES

Aki, K. and P. G. Richards, 1980, "Quantitative Seismology; Theory and Methods", Vol. 2, W. H. Freeman and Co., New York, N.Y.

Brown, M. G., 1984, Linearized travel time, intensity and waveform inversions in the ocean sound channel - A comparison, J. Acoust. Soc. Amer., Vol. 75, No. 5, pp. 1451-1461.

Dawson, T. H., J.N. Suhayda and J. M. Coleman, 1981, Correlation of field measurements with elastic theory of sea-floor response to surface waves, Preprint of 13th Offshore Tech. Conf., OTC 3973, Houston, TX.

Figueroa, L., T. Yamamoto and T. Nagai, 1985, Inverted shear modulus from wave-induced soil motion, J. Geotech. Engr., ASCE, Vol. 111, No. 1, Jan., pp. 115-132.

Forristall, G. Z. and A. M. Reece, 1985, Measurements of wave attenuation due to a soft bottom: The SWAMP Experiment, J. Geophys. Res., Vol. 90, No C2, pp. 3367-3380.

Roberts, H. H., D. W. Cratsley and T. Whelan, III, 1976, Stability of Mississippi Delta sediments as evaluated by analysis of structural features in sediment borings, Preprints of 8th Offshore Tech. Conf., OTC 2425, Vol. 1, pp. 1-28, Houston, TX.

Suhayda, J. N., T. Whelan, III, J. M. Coleman, J. S. Booth, L. E. Garrison, 1976, Marine sediment instability: Interaction of hydrodynamic forces and bottom sediments, Preprints of 8th Offshore Tech. Conf., OTC 2426 Vol. 1, pp. 29-40, Houston, TX.

Torii, T., 1985, Determination of shear modulus profile by inverse analysis, M.S. Thesis submitted to University of Miami, May.

Yamamoto, T., 1982, Nonlinear mechanics of wave interactions with sediment beds, App. Ocn. Res., Vol. 4, No. 2, pp. 99-106.

Yamamoto, T., 1983a, On the response of a coulomb-damped poro-elastic bed to water waves, Mar. Geotechn., Vol. 5, No. 2, pp. 93-130.

Yamamoto, T., 1983b, Numerical integration method for seabed response to water waves, J. Soil Dynamics and Earthq. Engr., Vol. 2, No. 2, pp. 92-100.

Yamamoto, T., H. L. Koning, H. Sellmeijer and E. Van Higum, 1978, On the response of the poro-elastic bed to water wave, J. Fluid Mech., Vol. 87, Part 1, pp. 193-206.

Yamamoto, T., S. Takahashi and B. Schuckman, 1983, Physical modeling of sea-bed interactions, J. Engr. Mech., ASCE, v. 109, pp. 54-72.

ESTIMATION OF SUBSURFACE PARAMETERS IN LOSSY MEDIA

Philip M. Carrion

Henry Krumb School of Mines
Columbia University
New York, N.Y. 10027 USA

INTRODUCTION

In recent years a number of inversion techniques which can estimate the velocity of compressional waves in acoustic media by inversion of reflection reponses have been presented. The majority of these methods ignore the presence of attenuation and estimate the sub-surface parameters by inversion of the reflection data. Recently Carrion (1985) showed that ignoring the effects of attenuation can lead to erroneous results especially when the attenuation is relatively large. The presence of attenuation can lead to ambiguities in the recovery of the acoustic impedance. Jaulent (1976) pointed out that in general a solution of the inverse problem is nonunique as long as the attenuation in a medium can not be neglected.

In this paper I treat the problem of estimation of Q-profiles by inversion of reflection data. The following problem is considered: the upper region ($z\leq 0$) in a one-dimensional model is loss-free with a constant velocity. An impulsive point source generates incident waves which are reflected back from the vertical inhomogeneities of the velocity of viscoacoustic waves and from the vertical inhomogeneities of the absorption in the medium ($z>0$). The goal is to recover separately the attenuation and variations in the velocity from the impulse response. It is shown that indeed the variations in the velocity of viscoacoustic waves (we call it the "quasivelocity") as well as the attenuation can be recovered separately if the quality paramater Q and the quasivelocity do not depend on the frequency.

BASIC EQUATIONS

Let us consider a set of integrodifferential equations which describe the propagation of compressional waves in an acoustic medium with absorption:

$$\rho(x)\,\frac{\partial V(t,x)}{\partial t} = -\nabla P(x,t), \qquad x\varepsilon R^3 \tag{1}$$

$$\frac{\partial P(x,t)}{\partial t} = -\rho(x)c^2(x)\nabla\cdot V(x,t) + P(x,t)*A(t,x) \tag{2}$$

where $*$ denotes a convolution in time, $x = (x,y,z)^T$ is a coordinate vector in R^3, $P(x,t)$ is the pressure, $V(x,t)$ is the particle velocity, $A(x,t)$ is the attenuation, $\rho(x)$ is the density of the medium, $c(x)$ is the velocity of compressional waves in a loss-free medium. Applying the operator $\nabla\cdot$ to both sides of equation (1) we obtain:

$$\nabla\rho(x)\ \frac{\partial V(t,x)}{\partial t} + \rho(x)\ \frac{\partial}{\partial t}\ \nabla\cdot\ V(t,x) = -\nabla^2\ P(t,x) \tag{3}$$

Then using (1) and (2), equation (3) can be reduced to:

$$\nabla^2 P(t,x) - \frac{1}{c^2(x)}\ \frac{\partial^2 P(t,x)}{\partial t^2} - (\nabla \ln\ \rho(x))\cdot\nabla P(t,x) +$$
$$+ \frac{1}{c^2(x)}\ \frac{\partial}{\partial t}\ P(t,x) * A(t,x) = 0 \tag{4}$$

Let us apply a Fourier transform to equation (4):

$$\nabla^2\ \hat{P}(\omega,x) + \frac{\omega^2}{c^2(x)}\ \hat{P}(x,\omega) - (\nabla \ln\rho(x))\cdot\nabla\ \hat{P}\ (\omega,x) +$$
$$+ \frac{1}{c^2}(x)\ i\omega\ \hat{P}(\omega,x)\ \hat{A}\ (\omega,x) = 0 \tag{5}$$

where:

$$(\hat{P},\hat{A})^T\ (\omega\ ,x) = {}^{\infty}\!\int_0\ (P,A)^T(t,x)\exp(-i\omega t)dt \tag{6}$$

represents the pressure and the damping function in the frequency domain. Equation (5) can be rewritten in terms of the complex velocity $\alpha(\omega,x)$:

$$\nabla^2\hat{P}(\omega,x) + \frac{\omega^2}{\alpha^2(\omega,x)}\ \hat{P}(\omega,t) - \nabla \ln\rho(x)\cdot\nabla\hat{P}(\omega,x) = 0 \tag{7}$$

where:

$$\alpha(\omega,x) = \frac{c(x)}{\sqrt{1 + \frac{i\hat{A}(\omega,x)}{\omega}}} \tag{8}$$

Equation (7) is similar to the acoustic wave equation in a loss-free medium. This means that the propagation of viscoacoustic waves in a medium with absorption is equivalent to propagation of compressional waves in a loss-free medium but the real velocity should be changed on the complex velocity defined by (8). Assuming that:

$$\left|\frac{\hat{A}(\omega,x)}{\omega}\right| << 1 \tag{9}$$

(for all frequencies within the bandwith of the data

$$\omega\varepsilon[\omega_1,\ \omega_h]$$

where ω_1 is the low cutoff frequency and ω_h is the high cutoff frequency), formula (8) can be replaced by its approximation:

$$\alpha(\omega,x) = c(x)\ (1 - \frac{i}{2}\ \frac{\hat{A}(\omega,x)}{\omega}\ ) \tag{10}$$

or

$$\alpha(\omega,x) = \beta(\omega,x) + \frac{ic(x)}{2Q(\omega)} \tag{11}$$

where the real component of the particle velocity:

$$\beta(\omega,x) = c(x)\ (1 + \frac{\hat{A}_i(\omega,x)}{2\omega}\ ) \tag{12a}$$

and the quality parameter Q is:

$$Q(\omega,x) = \frac{\omega}{\hat{A}_r(\omega,x)} \qquad (12b)$$

The quality parameter Q is usualy defined as the ratio of the peak strain energy stored in the volume and the energy lost in each cycle (see for example Aki and Richards (1980)). In (12a-12b) $\hat{A}_r$ and A_i are the real and the imaginary parts of the damping function $\hat{A}(\omega, x)$ respectively.

$$\hat{A}(\omega,x) = \hat{A}_r(\omega,x) + i\hat{A}_i(\omega,x)$$

In this paper I will consider only the one-dinensional case and I will start with the one-dimensional version of equation (5) assuming that the density is constant:

$$\frac{\partial^2\hat{P}(\omega,z)}{\partial z^2} + \frac{\omega^2\hat{P}(\omega,z)}{c^2(z)} + \frac{i\omega\,\hat{P}(\omega,z)\,\hat{A}(\omega,z)}{c^2(z)} = 0 \qquad (13)$$

This equation is the starting point for our analysis, which should answer the following question: can the Q-profiles be uniquely recovered from seismic data?

INVERSE PROBLEM AND COMPUTATION OF Q-PROFILES

Our objective in this part of the paper is to determine c(z) and $\hat{A}(\omega,z)$ under the following conditions:

1) c(z) is a real positive function determined at each depth z; it is constant for z<0. Following Cohen and Bleinstein (1979) c(z), for z>0 can be presented as:

$$\frac{1}{c^2(z)} = \frac{1 + \mu(z)}{c_0^2}$$

where is assumed that

$$|\mu(z)| \ll 1$$

2) $\hat{A}(\omega,z)$ is uniquely defined for any ω and z and is zero for z< 0. For z>0 we assume that $|\hat{A}(\omega)/\omega|$ is small (of the order $O(\Delta), |\Delta| \ll 1$)

The source is impulsive and is located at z = -0. The data are recorded at z = 0 in the time domain or in the frequency domain. Under the above mentioned conditions, equation (13) can be rewritten in the following form:

$$\left[\frac{\partial^2}{\partial z^2} + \frac{\omega^2}{c_0^2}\right] \hat{P}(z,\omega) = -\frac{\omega^2}{c_0^2}\mu(z)\,\hat{P}(z,\omega) - \frac{i\omega^2}{c_0^2}\Big[$$

$$\left[(1 + \mu(z))\,\frac{\hat{A}(\omega,z)}{\omega}\,\hat{P}(z,\omega)\right] + \delta(z + 0)\hat{W}(\omega) = 0 \qquad (14)$$

where $\hat{W}(\omega)$ is the spectrum of the source function or in short:

$$L_0\hat{P}(z,\omega) = -\frac{\omega^2}{c_0^2}\,v(\omega,z)\hat{P}(z,\omega) + \delta(z+0)\hat{W}(\omega) \qquad (15)$$

where:

$$v(\omega,z) = \mu(z) + \frac{i\hat{A}(\omega,z)}{\omega}$$

where:

$$v(\omega,z) = \mu(z) + \frac{i\,\hat{A}(\omega,z)}{\omega} \qquad (16)$$

is the scattering potential for the medium with absorption. (Terms smaller than $O(\Delta)$ where neglected). It is important to note that the scattering potential is a function of frequency. The unperturbed operator L_0 in equation (15) is:

$$L_0 = \frac{\partial^2}{\partial z^2} + \frac{\omega^2}{c_0^2} \qquad (17)$$

This operator describes the propagation of compressional waves in the upper region ($z\leq 0$) where the velocity of propagation of compressional waves $c(z)$ is constant and there is no absorption. Let us consider now equation (15). A solution of this equation can be expressed in terms of the Green's function of this operator. If the problem has any boundary conditions than the Green's function which corresponds to (15) should satisfy them. For scalar equations (for the acoustic wave equation for example a combination of the free-space Green's functions can mimic these boundary conditions). In our particular case when the scattering potential is small we can always present the Green's function as a combination of the unperturbed Green's function and the scattering potential. This combination usually can be presented via the Born series. Let us first consider the free-space Green's function for the unperturbed operator L_0:

$$G_0(z,\xi) = \frac{\exp\,[2ik|z-\xi|\,]}{2ik} \qquad (18)$$

where k is the wave number. Then solution of (14) can be presented in terms of a Lippman-Schwinger integral:

$$\hat{P}(z,\omega) = \hat{W}(\omega)\; G_0(k,z) - \int_0^\infty \frac{\omega^2}{c_0^2}\; v(\omega,\xi)P(\omega,\xi)G_0(z-\xi)d\xi \qquad (19)$$

Setting z = 0 and defining:

$$D_s(\omega,0) = \hat{P}(0,\omega) - \hat{W}(\omega)\; G_0(\omega,0) \qquad (20)$$

where $D_s(0,\omega)$ is back scattered data. Equation (19) can be presented as:

$$D_s(0,\omega) = -\int_0^\infty \frac{\omega^2}{c_0^2}\; \frac{e^{ik\xi}}{2ik}\; v(\omega,\xi)\; \hat{P}(\xi,\omega)\; d\xi \qquad (21)$$

Thus, the total recorded wave field minus the direct wave from the source to receiver is related to the scattering potential v and the total pressure field P(equation (21)) via the Lippmann-Schwinger integral equation.

REMARK

It is interesting to observe that when $\mu(z) = 0$ for all $z>0$ the scattering potential $V(\omega,z) = 0$ and the reflection response will be found from the damping function $\hat{A}(\omega,z)$ only.

Let us now substitute in (21) instead of the total pressure field, only the incident field, (this approximation is equivalent to the first term of the Born approximation). Then intergral (21) becomes:

$$D_s(0,\omega) = -\frac{W(\omega)}{c_0^2}\int_0^\infty \frac{e^{2ik}}{(2ik)^2}\; v(\omega,\xi)\; d\xi \qquad (22)$$

This integral operator corresponds to a Fredholm integral equation of the kind which operator does not always have an inverse. Moreover we can not guarantee that in general a solution of (22) exists or that it is unique. However if $v(\omega,\xi)$ did not depend on frequency, equation (22) would be a Fourier intergral and under our assumption that $v(\omega,\xi)$ is a square integrable function with respect to ξ a Fourier integral operator has an inverse in accordance with a Plansherel's theorem. Assuming that:

$$\frac{\hat{A}(\omega,z)}{\omega} = B(z) \tag{23}$$

(B(z) is in general complex), which simply means that the damping function $\hat{A}(\omega,z)$ depends linearly on the frequency, equation (22) can be presented as:

$$D_s(0,\omega) = \frac{W(\omega)}{4} \int_{-\infty}^{\infty} e^{2ik\xi} [\mu(\xi) + i\, B(\xi)]\, d\xi \tag{24}$$

where we have extended the limits of integration because $B(z) = 0 = \mu(z)$ for $z<0$. A Fourier inverse of (24) yields:

$$\frac{4}{\pi} \int_{-\infty}^{\infty} \frac{D_s(0,\omega)}{\hat{W}(\omega)} \exp(-2ikz)\, dk = \mu(z) + iB(z) \tag{25}$$

which represents a general solution to the problem. The fact that W(w) is in the denominator means that certain problems can be caused by the bandlimitation. If we assume that the source time function is close to the Dirac delta function we can easily separate the real and complex parts of (25) obtaining a unique representation:

$$\frac{4}{\pi} \int_{-\infty}^{\infty} [D_{s,r}(0,k)\cos(1kz) + D_{s,i}\sin(2kz)]\, dk = \mu(z) - B_i(z) \tag{26}$$

$$\frac{4}{\pi} \int_{-\infty}^{\infty} [D_{s,i}(0,k)\cos(2kz) - D_{s,r}(0,k)\sin(2kz)]\, dk = B_r(z) = Q^{-1}(z) \tag{27}$$

where $D_r(o,k)$ and $D_i(0,k)$ are the real part and the imaginary part of the recorded data. Our results show that the quantity

$$\mu(z) - B_i(z)$$

as well as the Q-profiles can be uniquely recovered from the data in the one-dimensional case.

REMARK

Let us consider formula (12a) which can be rewritten as:

$$\beta(z) = \left(1 + \frac{B_i(z)}{2}\right) \frac{c_0}{\sqrt{(1 + \mu(z))}} = (1 + B_i(z)/2)\, c_0\, (1-\tfrac{1}{2}\mu(z)) =$$

$$= \frac{c_0}{2} (B_i(z) - \mu(z)) \tag{28}$$

Since we know the quantity $\mu(z) - B_i(z)$ and the reference velocity are known, we can estimate the real part of the complex velocity $\beta(z)$ and the attenuation by inverting the observed data.

CONCLUSION

A method presented in this paper shows that for the one-dimensional case, the Q-profiles as well as the real component of the complex velocity can be uniquely determined from the impulse response if neither the real part of the complex velocity nor the Q-parameter depend on the frequency.

This work was sponsored by the NSF grant EAR-8407511. The author would like to thank Dr. K-H. Chen, Dr. C. Frasier and Dr. A. Trorey for their valuable discussion and comments.

REFERENCES

Aki, K., and Richards, P., 1980, Quantitative seismology, W. H. Freeman and Co. San Francisco

Carrion, P., 1985, Recovery of the acoustic impedance in media with absorption, Submitted to Geohysical Prospecting

Cohen, J., and Bleinstein, N., 1979, Velocity inversion procedure for acoustic waves, Geophysics v 44 pp 1077-1087

Jaulent, M., 1976, Inverse scattering problems in absorbing media, Journal of Math. Physics v 17 pp 1351-1360

ANALYSIS OF OCEAN-SUBBOTTOM SEISMOGRAPH (OSS) DATA

Jerry A. Carter, George H. Sutton, and Anne Suteau-Henson

Rondout Associates, Incorporated
P.O. Box 224
Stone Ridge, New York 12484 USA

Fred K. Duennebier

Hawaii Institute of Geophysics
2525 Correa Road
Honolulu, Hawaii 96822 USA

ABSTRACT

Data from OSS-IV, a deep-sea borehole seismic instrument installed approximately 750 kilometers southeast of the Central Kuril Islands, have been analyzed for signal directionality and the importance of scattering. Azimuths from instrument to source for a series of small SUS charge explosions determined by polarization analysis agree well with those obtained by navigation and travel time ranging. Synthetics generated for receivers at the ocean bottom and at the OSS location for the SUS line demonstrate the differences between the two sensor locations. Comparison with the data is good except for the large amplitude water wave arrivals at small offsets in the radial data. Noise from a passing ship is used to evaluate the ability of the polarization method to track continuous moving sources.

INTRODUCTION

On September 11, 1982 the fourth Ocean Subbottom Seismometer (OSS-IV), a deep-sea three component seismic borehole package developed by the University of Hawaii (Duennebier and Blackinton, 1983), was installed off the coast of Kamchatka at 43.924°N, 159.797°E. The hole penetrated 357.6 meters of sediment and 20.6 meters of basalt at an ocean depth of 5467 meters (Duennebier et al., 1985). The location of the site, 750 kilometers southeast of the Central Kuril Islands, was ideal for studying the seismic activity of the Kuril trench and Kamchatka Peninsula.

Long term data were recorded from September 12, 1982 to November 16, 1982 during which time about 1500 regional earthquakes, many teleseismic events, and several semi-continuous signals produced by sources such as ships and whales were recorded.

During the data recovery phase of the OSS-IV in late May of 1983, a refraction line of 21, 57-gram SUS charges was recorded in real time at 100 samples/second by Hawaii Institute of Geophysics scientists on the R/V Kana Keoki. The charges, set to detonate at 18 meters depth, were dropped by plane at ranges extending from directly over the hole to nearly 68 kilometers. Shot locations were determined by aircraft navigation and travel time ranging.

Both the SUS refraction line and the signals from one of the passing ships have been analyzed for signal directionality and the importance of scattering using three component polarization methods.

POLARIZATION ANALYSIS TECHNIQUES

Polarization filtering and polarization analysis make use of the vector particle motion characteristics associated with different wave types to identify, and emphasize or eliminate certain waves or particle motions coming from certain directions (Phinney and Smith, 1963; Sutton and Pomeroy, 1963; White, 1964; Archambeau et al., 1965). If the wave type is properly identified, the particle motion indicates the direction of propagation (sometimes with π ambiguity) and, for some wave types, the phase velocity of the signal. Conversely, the apparent direction of approach and phase velocity can be used to identify the wave type.

Two polarization techniques are used to analyze the OSS-IV data; adaptive polarization analysis (Carter et al., 1984) and polarization state detection (Samson and Olson, 1977). These methods are outlined below.

Adaptive polarization analysis, instead of constraining the rectilinearity of motion to a particular azimuth, computes the azimuth and apparent angle of incidence as functions of time by finding the direction of maximum signal strength.

To determine the direction of maximum signal, a small time window is chosen; one that is at least as long as one or two cycles of the predominant frequency. The azimuth within the window is determined by examining the zero lag cross covariance between the two horizontal components as a function of azimuth angle. The angle at which this function is zero is the direction of maximum signal strength within the window. This azimuth is used to rotate the horizontal components to the radial and transverse directions at the center of the window. The window is then advanced one sample and the process repeated.

The apparent angle of incidence is calculated in much the same manner as the azimuth. For this measurement, the adapted radial and vertical components are used as the orthogonal components from which the angle of incidence is determined.

An estimate of the relative accuracy of the azimuth and apparent angle of incidence measurements is obtained through the maximum value of the zero lag cross correlation function. A large value is indicative of strong rectilinear motion, and thus a small error, over the window and a small value indicates the lack thereof. Error determined in this manner is not an absolute quantity but rather a relative measure of the accuracy of the value.

The signed square root of the absolute value of the product of the radial and vertical traces is included as one of the outputs of adaptive polarization analysis. Its purpose is to aid in phase identification through increased amplitude and sign. Where the signal motion is compressional, r and z have the same sign and $\mathrm{sign}(rz)(|rz|)^{\frac{1}{2}}$ (hereafter $\underline{rz}$) is positive. When the motion is vertically polarized shear, r and z have opposite signs and $\underline{rz}$ is negative. $\underline{rz}$ alternates for elliptical motion. Because $\underline{rz}$ emphasizes coherent peaks between r and z, it also aids in the identification of different phases.

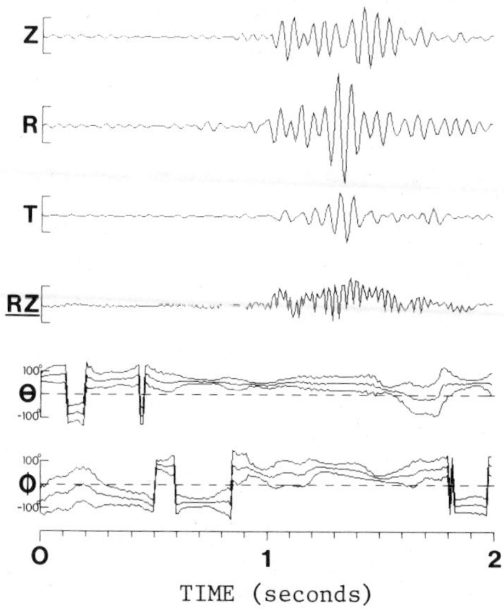

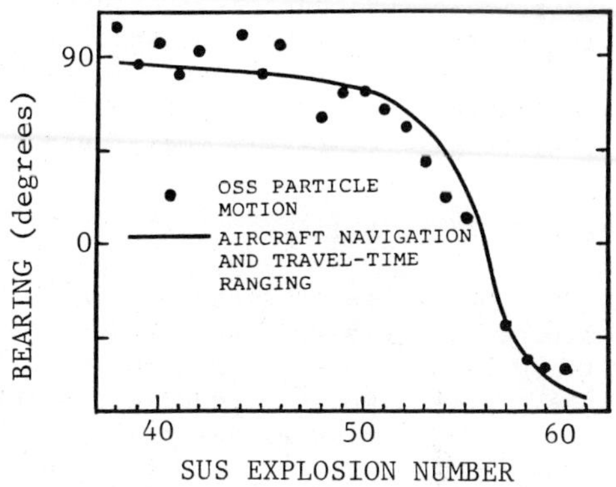

Figure 1 (left). Adaptive polarization analysis of the second water wave arrival for shot 52. Z, R, and T are the vertical, adaptive radial, and adaptive transverse components, respectively; $\underline{RZ} = \mathrm{sign}(RZ)(|RZ|)^{1/2}$; θ is the azimuth; and ϕ is the apparent angle of incidence. The two outer traces surrounding the estimates of θ and ϕ are relative error bars.

Figure 2 (right). Azimuth to the SUS explosions from OSS-IV. Solid line determined from aircraft navigation and water wave travel time data; dots are the values obtained from adaptive polarization analysis. Shot 38 is 67 kilometers west of the hole and shot 60 is 16 kilometers east of the hole.

Polarization state detectors are based on a theory used in optics and quantum mechanics that describes vector processes in terms of their polarization states (Samson, 1973). When a pure polarization state is present in a multichannel signal, information about this state can be extracted from the spectral matrix, S. This applies to three component recordings of seismic waves propagating in simple modes, such as p, s, and Love waves, which are in pure rectilinear states, and Rayleigh waves, which are in pure elliptical states (Samson, 1977).

Samson and Olson (1980) show that a measure of the degree of polarization P of S can be found where 0 <= P <= 1. P=0 corresponds to totally unpolarized data and P=1 to a pure polarized state. The measure of degree of polarization, P, enables us to discriminate between unpolarized noise and polarized signal as a function of frequency and time. When three-component data are available, the spectral matrix S also contains information about any specific polarization state that the data are in, including the azimuth and apparent angle of incidence of the traveling waves. These too are given as functions of frequency and time.

POLARIZATION ANALYSIS OF SUS DATA

Twenty shots from the SUS refraction line have been analyzed using the adaptive polarization method to determine the azimuth of energy propagation. The records are characterized by numerous water layer multiples with the later multiples becoming dominant at larger ranges (Figure 3). An example of the analysis for the second water wave multiple of shot 52 is shown in Figure 1.

The later water waves give the most reliable azimuth estimates even though azimuth resolution decreases for this method at steeper angles. This indicates that the steeper arrivals are less distorted by scattering, lateral refraction and irregular boundaries. A comparison of the azimuth results from adaptive polarization analysis and the aircraft navigation and travel time data is shown in Figure 2. The scatter of azimuths is relatively small, about +/-10°.

SUS DATA/SYNTHETICS COMPARISON

Comparisons between OSS data from the SUS charges and full waveform synthetic seismograms for both OBS and OSS locations at ranges from 10 to 30 km are shown in Figures 3-5. The synthetics were generated using the "locked mode" procedure developed by Harvey (1981) with the preliminary velocity model determined by HIG scientists given in Table I. Frequencies range from 10 to 25 Hz and phase velocities were filtered to pass velocities below 10 km/sec.

Inspection of the Figures shows some interesting relationships. The water waves of both the data (Figure 3) and the OSS synthetics, (Figure 4) are largest for the second multiple and nearly nonexistent for the direct wave. The vertical component traces have the best overall match as the radial data have very large amplitudes at small range compared to the synthetics. Between water wave arrivals on the synthetics are other large arrivals that we attribute to s wave reflections from the base of layer 2. These arrivals are not clear on the data indicating that the true s wave structure is a gradient rather than the constant velocity used here. The OBS synthetics (Figure 5) show strong water wave arrivals with the later arrivals predictably becoming larger than the first water wave at greater ranges. The large arrivals on the OBS radial component synthetics cause severe wrap around problems. Compared to the OSS synthetics, the OBS synthetics show that radial motion is much larger on the sea floor relative to the vertical motion than in the borehole. Clearly, much of the signal recorded on the OBS is reflected at the ocean bottom and is thus never recorded by the OSS. Looking to the OSS-IV data, the radial component water wave arrivals have approximately the same amplitudes as the vertical component data; nearly twice that predicted by the synthetics.

The largest body wave arrivals in Figures 3 through 5 are marked with solid lines at 6.0 and 3.4 km/sec corresponding to the p and s velocities of layer 5. Both components of the OSS synthetics show clear arrivals at these velocities. On the OSS-IV data, however, the 6.0 km/sec arrival dominates the vertical component while the 3.4 km/sec arrival is seen primarily on the radial component.

SHIP DATA

During the two months of continuous recording by OSS-IV, the noise from several ships was recorded as vessels passed close to the site. Ships act as steady state moving sources and provide an excellent test for the polarization analysis technique. For transient sources, travel times along the various wave paths differ and scattered energy generally arrives later in the wavetrain; but for a steady state source all wave paths will be present at any given instant, thus complicating the arrivals and making analysis more difficult.

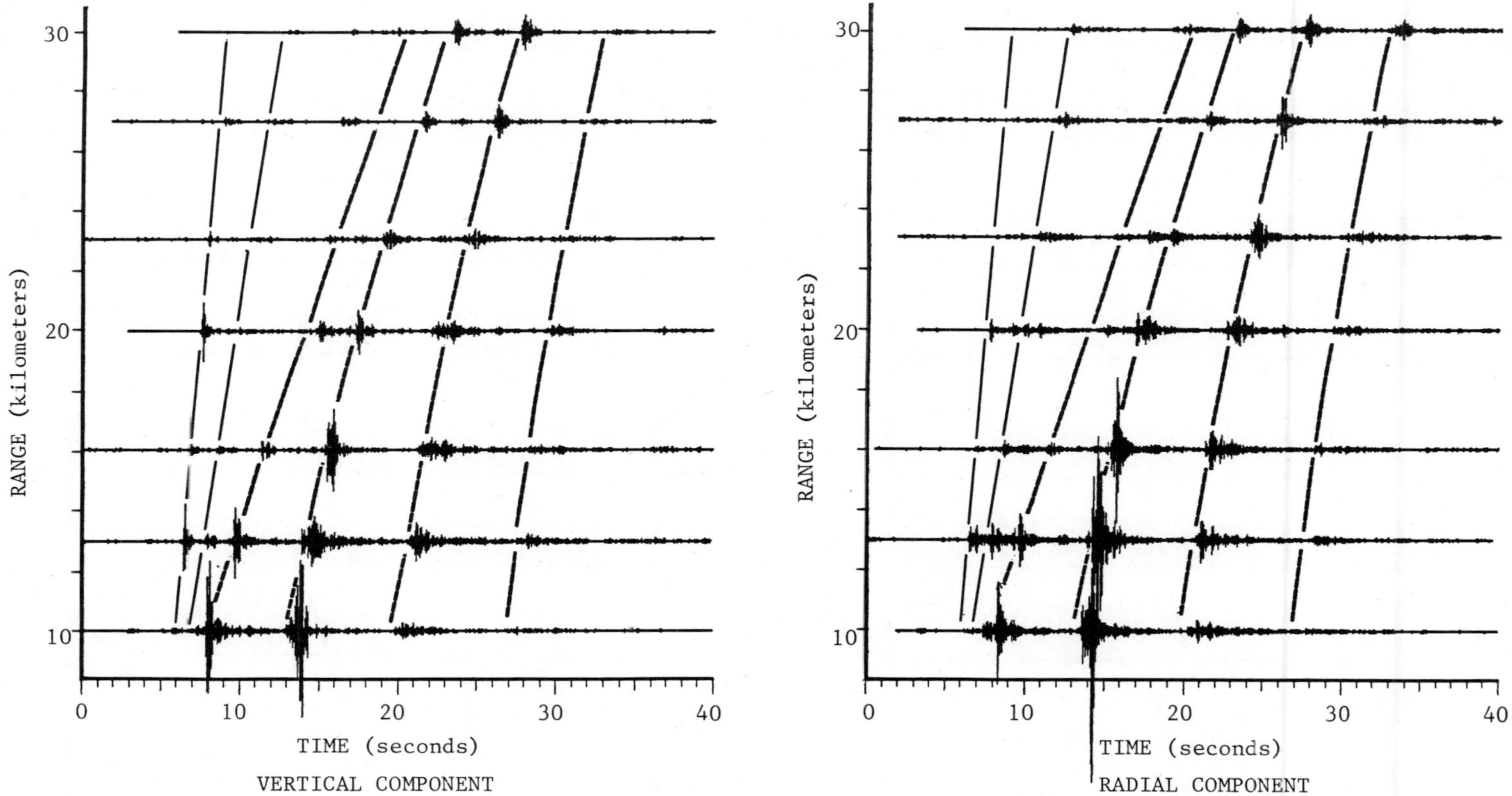

Figure 3. OSS-IV data for six explosions between 10 and 30 km range. Solid lines are 6.0 and 3.4 km/sec. Dashed lines are the theoretical travel times to the top of the hole for the first four water waves. Vertical and radial amplitude scales are the same.

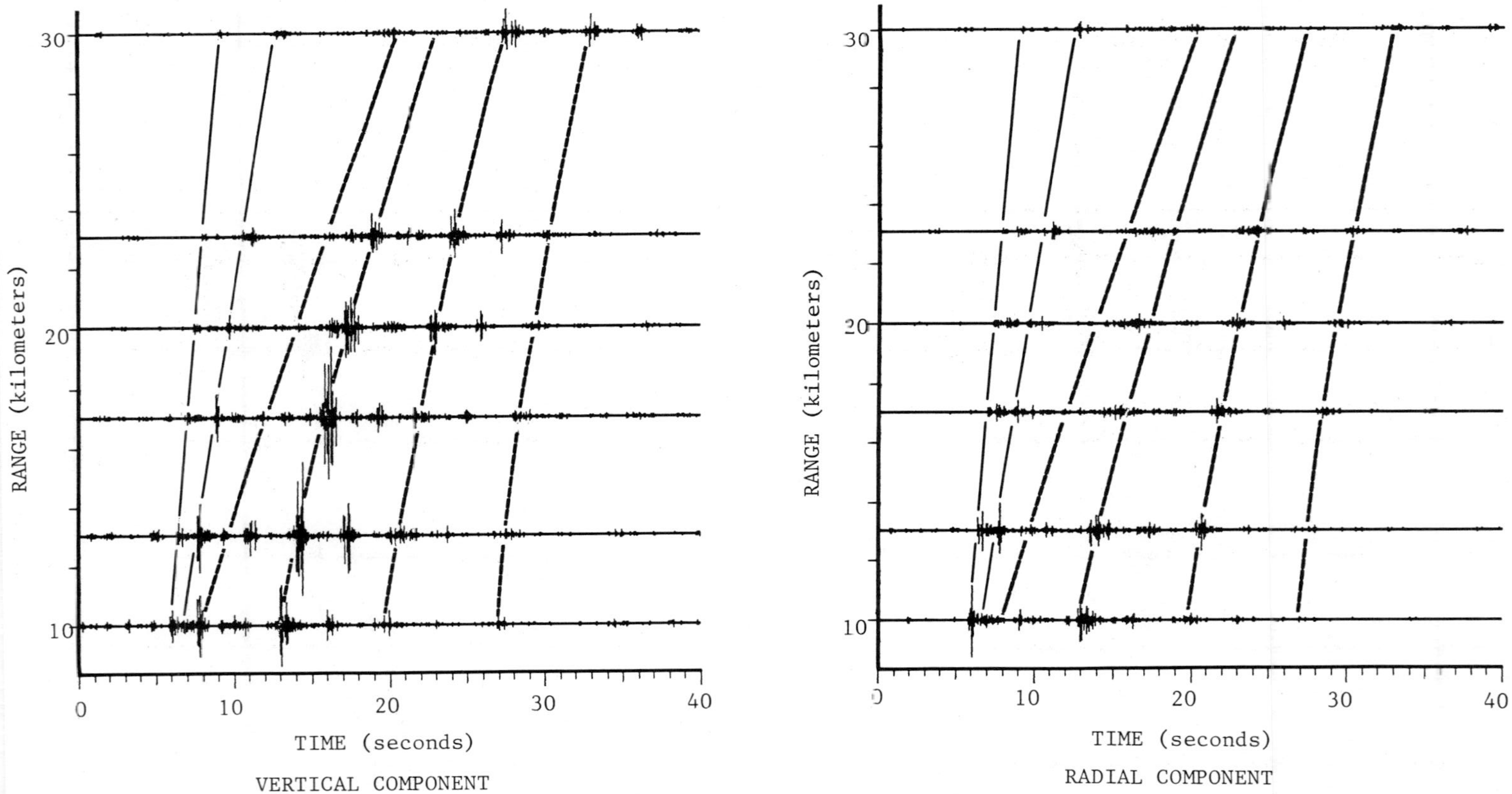

Figure 4. OSS synthetics for six explosions between 10 and 30 km range. Solid lines are 6.0 and 3.4 km/sec. Dashed lines are the theoretical travel times to the top of the hole for the first four water waves. Vertical and radial amplitude scales are the same.

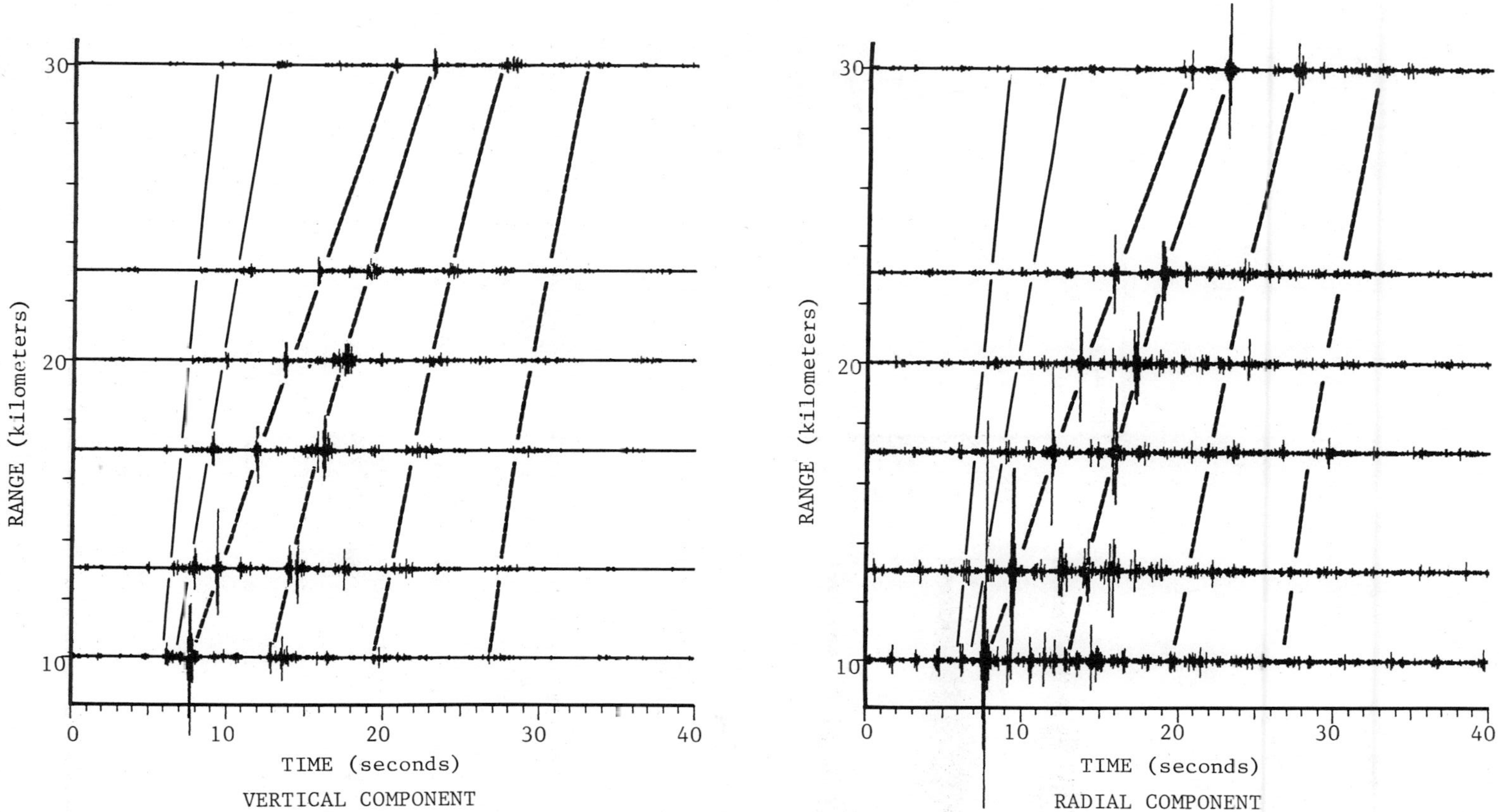

Figure 5. OBS synthetics for six explosions between 10 and 30 km range. Solid lines are 6.0 and 3.4 km/sec. Dashed lines are the theoretical travel times to the top of the hole for the first four water waves. Vertical and radial amplitude scales are the same.

TABLE I
OSS VELOCITY/ATTENUATION STRUCTURE

LAYER	Vp(km/sec)	Vs(km/sec)	(g/cm3)	Thk(km)	Qp	Qs
1	1.5	0.00	1.0	5.47	10,000	
2	1.8	0.24	1.8	0.36	200	100
3	4.0	2.70	2.5	0.18	5,000	5,000
4	4.4	3.20	2.6	1.46	5,000	5,000
5	6.0	3.40	2.7	1.00	5,000	5,000
6	6.6	3.90	2.8	1.50	5,000	5,000
7	7.0	3.90	3.0	3.50	5,000	5,000
8	8.1	4.60	3.3	51.53	5,000	5,000
9	20.0	20.00	4.0			

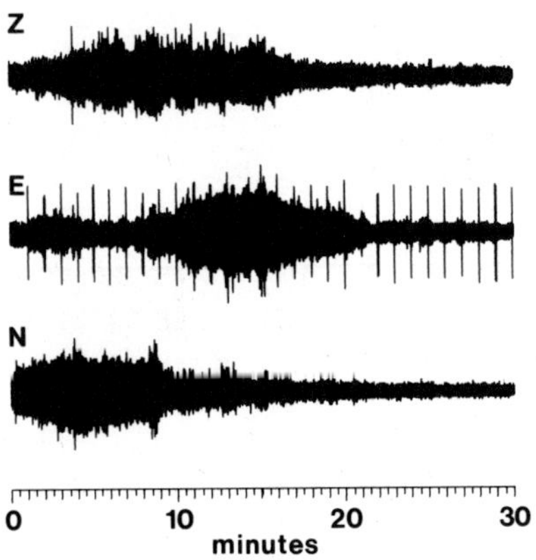

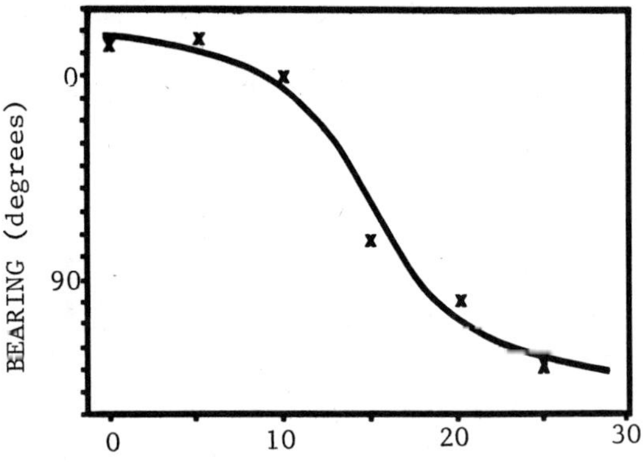

Figure 6 (left). OSS-IV three component data during the passing of a ship. The periodic bursts on the E component are minute marks.

Figure 7 (right). Azimuth of the ship relative to OSS-IV determined at 5 minute intervals using adaptive polarization analysis (X's). The solid line is the theoretical curve for a ship passing within 2.1 km of the hole on a heading of N 34° E with a speed of 17 knots.

One half hour of data during which a ship passed the hole are shown in Figure 6. Six 30 second samples of the data taken at 5 minute intervals were analyzed for azimuth using the adaptive polarization method. Averages of the azimuth angles determined by the analysis over the sample windows are plotted versus time as x's in Figure 7. These are compared with the theoretical curve for a surface ship traveling at 17 knots on a constant heading. The fit is reasonably good (standard deviation from theoretical of 8.9°) and indicates a ship's track of either S 34° W passing 2.1 km west of OSS-IV or N 34° E passing 2.1 km east of OSS-IV.

The ship data in Figure 6 were also processed using polarization state detectors for degree of polarization and azimuth. The azimuth estimates are valid only at those times and frequencies when the degree of polarization is high. Because both upcoming and downgoing arrivals are coincident for a steady state source, two different configurations for the three component data were tried and compared: using all three components of motion, and using only the two horizontal components of motion.

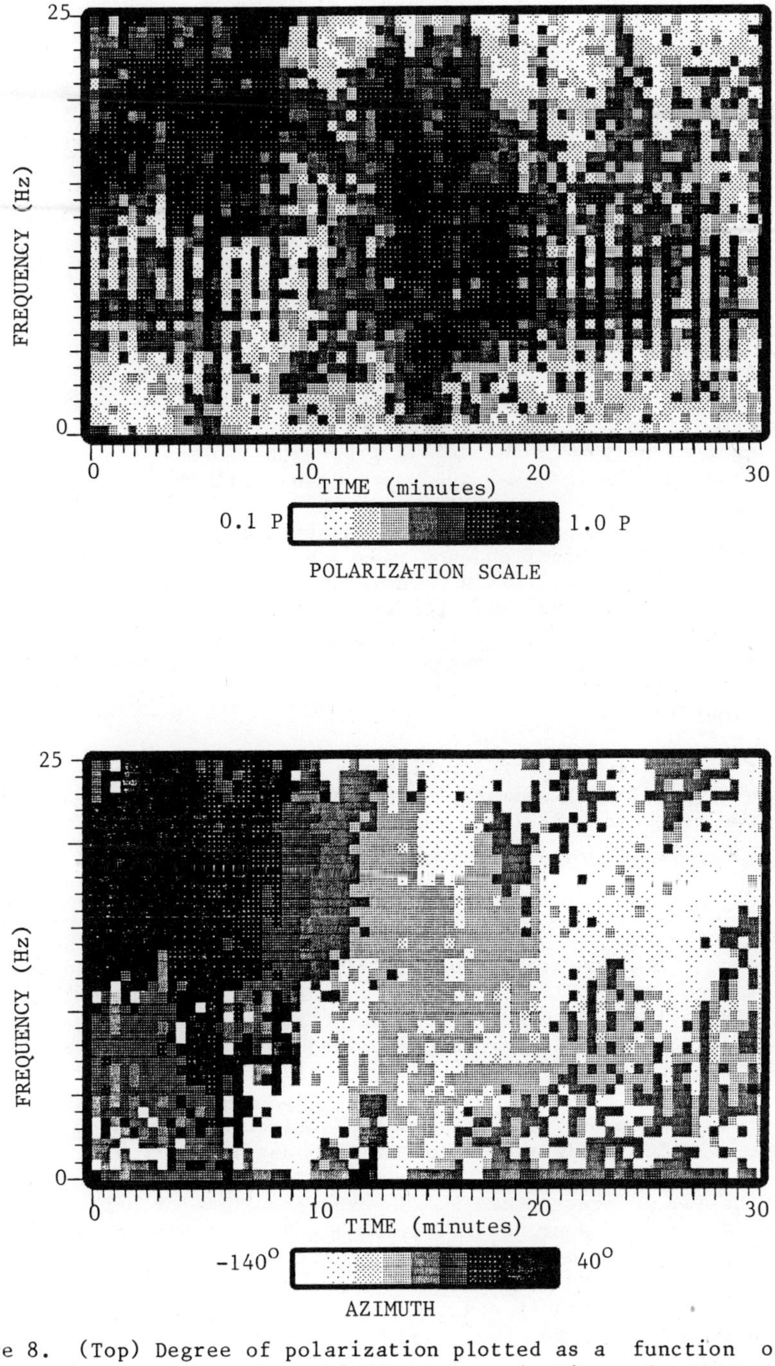

Figure 8. (Top) Degree of polarization plotted as a function of frequency and time for the ship data shown in Figure 6. Darker shades represent a higher degree of polarization. (Bottom) Azimuth as a function of frequency and time for the ship data. Between 15 and 22 Hz (where the degree of polarization is high) the azimuth changes smoothly as the ship passes.

The degree of rectilinear polarization and azimuth for the three component data give rather poor results whereas the degree of polarization and azimuth for the two component data exhibit a high degree of polarization and stable azimuth determinations (Figure 8). This is most likely the result of signals with varying angles of incidence reducing the correlation between the vertical and the two horizontal components. In principle the three dimensional data provides the direction of a line in space along which the signal oscillates. If signals were arriving only from above (or below), the azimuth of approach would be determined uniquely. Conversely, if the azimuth were known within +/- 90°, the angle of incidence could be determined uniquely. The addition of data from a pressure sensor would remove this ambiguity and the direction of energy propagation would be completely determined in three dimensions.

The degree of polarization plot shows frequency-time regions of strong linear polarization that should produce the most reliable and stable azimuth determination. Random noise gives a value of 0.18. The highly polarized energy indicated by vertical lines around 10 Hz is due to minute marks in the E-W channel. For the plots shown, E-W polarization is equivalent to +/- 90° and will be either black or white within a given frequency-time interval depending upon small amounts of noise.

Energy between about 15 and 22 Hz appears to be the most highly polarized. Within this frequency range the azimuths vary smoothly with relatively little noise. This frequency band gives results nearly identical to those obtained from the adaptive polarization method. At the beginning of the analysis, the azimuth is close to 20° and as the ship passes, the angle decreases passing through the N-S line and into the negative azimuths. By the end of the half hour of data, the azimuth is approximately -130°.

A strong narrow band of polarization is observed at 7 Hz. The signal shows a different azimuthal progression than the 15-22 Hz band and may be from a second ship in the area.

CONCLUSIONS

The two methods of polarization analysis applied to the OSS-IV data have shown that accurate azimuth information can be obtained from both transient and steady state sources using three component deep-sea borehole data. It is interesting to note that only two of the components were useful for azimuth determination; when the vertical component was used to track the passing ship, poor results were obtained. Simultaneous arrivals from both above and below were the probable cause. The addition of a hydrophone channel to the horizontal component data would eliminate the 180° ambiguity in the azimuth determinations.

Synthetics generated for the SUS explosion line at both the ocean bottom and in the upper basalt layer have demonstrated a large difference in the character of the records between a borehole instrument and an OBS. The amplitudes at the sediment/water interface are much larger than they are in the basalt and the radial component of the OBS synthetics is larger than the vertical component while the opposite is true for the OSS synthetics.

It is also observed that the direct water wave dominates the OBS synthetics while on the downhole synthetics, little energy is seen beyond 12 km for the direct water wave; refracted and reflected arrivals are more prevalent. The OSS data, however, does show a much stronger water wave contribution than predicted by the synthetics, especially for the second water wave on the radial component.

ACKNOWLEDGEMENTS

This research was supported in part by contract with the United States Navy.

REFERENCES

Archambeau, C.B., J.C. Bradford, P.W. Broome, W.C. Dean, E.A. Flinn, and R.L. Sax, 1965, Data Processing Techniques for the Detection and Interpretation of Teleseismic Signals, Proc. IEEE, 53, 1860-1884.

Carter, J.A., L.L. Peseckis, P.W. Pomeroy, and G.H. Sutton, 1984, Regional Seismic Wave Propagation, Final Technical Report, 1 October 1982-30 September 1984, DARPA/AFOSR Contract No. F49620-83-C-0017, Rondout Associates, Incorporated, Stone Ridge, New York, 185 p. plus four appendices.

Duennebier, F.K. and G. Blackinton, 1983, The Ocean Subbottom Seismometer, in Geophys. Exploration at Sea, R. Geyer, Ed., CRC Press, Boca Raton, Florida, 317-332.

Duennebier, F.K., R. Cessaro, and P. Anderson, 1985, Deep-Ocean Bore-Hole Seismo-Acoustic Results from the North-West Pacific Ocean, this volume.

Harvey, D., 1981, Seismogram Synthesis Using Normal Mode Superposition: the Locked Mode Approximation Method, Geophys. J. Roy. Astr. Soc., 66, 37-69.

Phinney, R.A. and S.W. Smith, 1963, Processing of Seismic Data from an Automatic Digital Recorder, Bull. Seis. Soc. Am., 53, 3, 549-562.

Samson, J.C., 1973, Descriptions of the Polarization States of Vector Processes: Applications to ULF Magnetic Fields, Geophys. J. Roy. Astr. Soc., 34, 403-419.

Samson, J.C., 1977, Matrix and Stokes Vector Representations of Detectors for Polarized Waveforms: Theory with some Applications to Teleseismic Waves, Geophys. J. Roy. Astr. Soc., 51, 583-603.

Samson, J.C. and J.V. Olson, 1980, Some Comments on the Descriptions of the Polarization States of Waves, Geophys. J. Roy. Astr. Soc., 61, 115-130.

Sutton, G.H. and P.W. Pomeroy, 1963, Analog Analyses of Seismograms Recorded on Magnetic Tape, J. Geophys. Res., 68, 2791-2815.

White, J.E., 1964, Motion Product Seismograms, Geophys., 29, 288-298.

THE RELATIVE AMPLITUDES OF PRIMARY AND MULTIPLE SIGNALS REFRACTED IN THE OCEAN CRUST

Arthur B. Baggeroer,
Gregory L. Duckworth, and
Karl J. Ellefsen

Massachusetts Institute of Technology
Cambridge, Massachusetts 02139

Edward K. Scheer

Woods Hole Oceanographic Institution
Woods Hole, Massachusetts 02543

ABSTRACT

In deep ocean seismic refraction experiments the energy in the primary signal is often less than that in the multiple arriving at approximately twice the travel time. This has perplexed interpreters since the multiple has a greater path length and interacted with the seafloor once more, so its amplitude should presumably be less. High resolution slowness travel time spectral analysis of an expanding spread data set suggests an explanation in terms of the differences in the sound speed gradient in the oceanic crust. The gradient in the upper part of the crust is typically much greater than that in the deeper crust; therefore, energy is focused much more and has a decreased geometrical offset. The multiple then should have a lower phase speed and a higher amplitude when it is refracting in the higher gradient part of the upper crust. Synthetic modeling using the WKBJ seismogram method and the slowness-travel time spectra of an ESP data set support this hypothesis.

INTRODUCTION

Seismic refraction experiments have been done throughout the oceans of the world since the 1940's. In fact our knowledge of the deep crust "has been based largely on the results of marine refraction work and correlation of the seismic velocities in these experiments with possible rock types." [1] The principal analysis technique was based upon the travel time of the first arrivals and fitting a model assuming constant speed layers, and numerous algorithms now exist for model inversion based upon travel times, e.g. [2] and [3]. Recently, other attributes of the refracted signal have been used to further constrain the inversion procedures. For example, much refraction data is now inverted by forward modeling using waveform synthesis. Synthetic algorithms have been based upon the approximate methods based upon WKBJ approaches of Wiggins [4] or Chapman [5] or full wavefield synthesis using reflectivity layers.[6] (The latter is essentially equivalent to the FFP algorithm in ocean acoustics. [7]) This has led to the incorporation of amplitude contraints and the use of velocity gradients within the seismic section to match the amplitude behavior as well as travel times. [8-10] In addition, multiple signals, especially those that have traversed the water column more that twice, have been used to provide additional measurements of travel times and phase velocities to enhance the offset sampling of the data. [11] The significant point is that much more of the waveform beyond the first arrival time is now being used in the inversion algorithms.

One of the common and intriguing observations about the refracted signal is that the multiple signals are often more energetic than the primary one. Furthermore, this is often observed for the second and third multiples as well as the first one. This is contrary to intuition because the multiples have longer travel paths and have gone through more bottom interactions than the primary. Numerous hypotheses have been proposed to explain this phenomenon. The two most often cited and their counter argument include the following.

i) Bathymetric focusing: The topography of the seafloor can act as lens due to the difference in sound speed between the seafloor and that of the water. This focusing can certainly occur; however, the phenomenon is often observed over regions where the seafloor is flat and horizontal layering is a good model.

ii) Constructive interference of head waves: Refracted signals were usually modeled in terms of head waves. [12] Since the head wave leaks off the interface continuously, one can argue that the multiple must be a superposition of all the possible travel paths and that constructive interference leads to a more energetic signal. The phenomenon, however, is observed over regions were the bathymetric variations produce travel time anomalies much larger than one half the dominant period of signal. Furthermore, it is considered that even though the travel times are approximately correct, the decay rate of the head wave amplitude versus range is far too high low to be the mechanism leading to the refracted signal.

There are other arguments, including leakage to shear of the primary and the transmission/reflection coefficients across the seafloor or sediment-basement interfaces; however, neither provide a robust explanation.

The argument advanced in this paper is that the controlling mechanism is the sound speed gradient versus depth in the earth crust. The reason that the observation is so common is that the oceanic lower crust is statistically very uniform. In addition, the multiple signal is more complex than the models commonly employed for it consisting in the situation examined here is composed of at least two signals with very close travel times and phase velocities which are seldom resolved unless high resolution slowness-travel time spectra are used.

OCEANIC CRUST MODELS

Statistical summaries of ocean sediment and basement sound speeds indicate that the crust may be crudely divided into three layers. [13-15] More recent summaries suggest that layers 2 and 3 may be subdivided, i.e. into layers 2A & 2B, 3A & 3B. [16] The data of the most recent summary are tabulated below.

TABLE I

Layer 1: (sediments)	sound speed: from approximately 1.5 km/s for silty clays to 4.0 km/s for highly consolidated, carbonates thickness: usually less than 0.5 km, but may be virtually none e.g at very young crust or several kms in basins.
Layer 2:	sound speed: 5.04 +/- 0.69 km/s thickness: 1.39 +/- 0.50 km
Layer 3:	sound speed: 6.73 +/- 0.19 km/s thickness: 4.97 +/- 1.25 km
Moho	sound speed: 8.15 +/- 0.31 km/s

The "average" profile is plotted on the left in Fig. 1. The more contemporary view of the crustal layering based upon amplitude modeling cited above is that it consists of layers with a continuous sound speed increase in depth. [9-10] Layer 2 has a steep gradient between 1 to 2 /s, while layer 3 has a much lower gradient. For the purposes of modeling the multiple phenomenon an "average" crustal sound speed profile on the right in Fig. 1 is used. It has a linear gradient from the top of layer 2 with its nominal speed of 5.04 km/s to the top of layer 3 with its nominal speed of 6.73 km/s. Layer 3 is terminated with a speed of 7.50 km/s at its nominal thickness by a 1 km transition to the Moho with a gradient of .6 /s. This leads to a sound speed gradient of 1.2 /s for layer 2 and .15 /s for layer 3. This "average" model agrees essentially with the current literature regarding the structure of the lower crust. In addition, signals based upon this model would lead to measurements of the refraction sound speeds close to those that would be estimated by the analysis methods used in the data base of the above table. While these are only approximate models, since the detailed structure of the lower crust is still a subject of ongoing research, they do indicate the typical oceanic crust measured at many sites in all oceans has sound speed gradients in layer 2 that are significantly larger than in layer 3. This difference in gradients leads to the common occurence of more energetic multiples in the seismic refraction experiments.

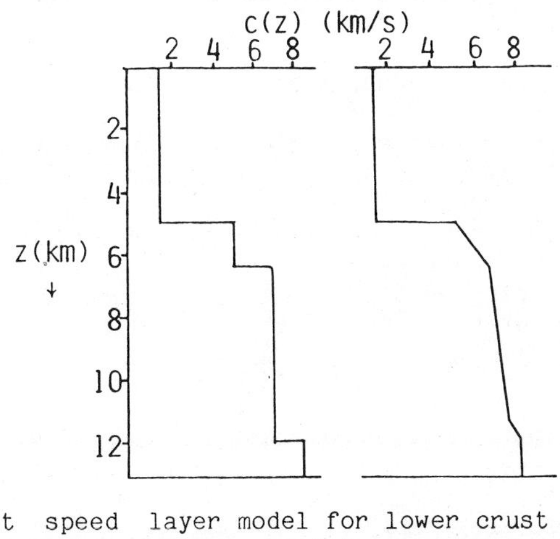

Fig. 1: Constant speed layer model for lower crust based upon data summary of [14] on left; gradient model for synthetics on right.

SOUND SPEED GRADIENTS AND SIGNAL AMPLITUDES

The major effect of the gradient is to control the amplitude of the refracted signal. More specifically, the gradient at the depth where the ray turns if one uses WKBJ models or where the transmission-reflection matrix becomes critical is the most important parameter in the determination of the amplitude of the refracted signal. This can be seen from both a transmission loss/WKBJ approximation or from the more exact reflectivity models. If one examines the factor controlling the amplitude in the WKBJ or ray disk synthetics, one finds, not surprisingly, that after appropriate normalization for the different source models it is the same formula as that used for transmission loss calculations via the ray bundle spreading approach. [3,4,18] The transmission losses as a function of ray parameter, TL(p), are given by

$$TL(p) = -10.\cdot \log\left[\left(p/(X(p)\,|dX(p)/dp|)\right)\cdot \left(c_o^2/(1. - p^2 c_o^2)\right)\right] \qquad \text{Eq. 1a}$$

where p is the ray parameter,

$X(p)$ is the offset function,

c_o is reference sound speed at the surface.

The offset function can be computed by integrating along the ray path. The dominant term in transmission loss is the derivative of the offset, $dX(p)/dp$, which incorporates the ray bundle spread. In general this is a complicated function of the sound speed profile however, the case of piecewise linear profiles where analytic results are available can be used to suggest the important role of the gradient. In this case one obtains

$$\frac{dX(p)}{dp} = -\frac{2}{p^2}\left[\sum_{\substack{K-1 \\ \text{layers}}} \left(\frac{1}{b_k}\right)\left([1 - p^2 c_{i,k}^2]^{-\frac{1}{2}} - [1 - p^2 c_{b,k}^2]^{-\frac{1}{2}}\right) + \frac{1}{b_K}[1 - p^2 c_{i,K}^2]^{-\frac{1}{2}}\right] \qquad \text{Eq. 1b}$$

where $c_{i,j}$; $c_{b,k}$ are the sound speeds at the top and bottom of each in each layer,

and b_k are the gradients in each layer.

A few calculations for sample profiles shows that the dominant term is the last one, or the layer at which the refracted signal turns.

Alternatively, one can use reflectivity modeling. Most algorithms approximate the gradient by a sequence of constant velocity layers incrementally increasing in depth. A simple model of a purely acoustic medium indicates that the reflectivity at one of these layers is given by

$$\Gamma(p,z) = \frac{1}{2}[1 - p^2 c^2(z)]^{-\frac{1}{2}}(\Delta \ell n c(z)/\Delta z) \qquad \text{Eq. 2}$$

where $\Gamma(p,z)$ is the reflectivity per unit depth.

The reflection coefficient at each interface is proportional to the local gradient; however, at the critical depth when p*c(z) equals unity, the coefficient is amplified by a factor theoretically becoming singular (Examine the slope of any test box for the Rayleigh reflection coefficient going through the critical angle.)

A simplified ray path geometry for the "average crust" given in Table 1 is shown in Fig. 2. The sediment column is absent in order to simplify the discussion; its presence complicates but does not alter it significantly. The diagram has no vertical exaggeration so true path lengths are indicated. There are five rays of concern: i) the primary refracted path which enters both layers 2 and 3 and penetrates the deepest; ii) the primary seafloor reflection; iii) the 1st refraction multiple which enters layer 2 twice but does not penetrate as deep as the primary; iv) the 1st reflection multiple; and v) the two "peg legs" which enter layer 2 once and reflect once off the seafloor interface.

A hypothesis which explains the more energetic multiple signals is that the typical gradients encountered in the Earth's lower crust cause the primary signal to turn in layer 3 and the multiple in layer 2. As a result the steeper gradient in layer 2 focuses the energy enough to overcome the extra spreading incurred by making two loops through the crust.

There is one additional complication to this argument. The impedance contrast at the seafloor and at the base of the sediment column is often large. This is especially true when young and unconsolidated sediments are not present or have been eroded away leaving exposed consolidated ones or basement rock. Consequently, at the seafloor and at the base of the sediments there is often a high contrast in sound speed and density. This leads to "peg leg" multiples which have travel times and ray parameters which for typical crustal models are very close those of the twice refracted signals.

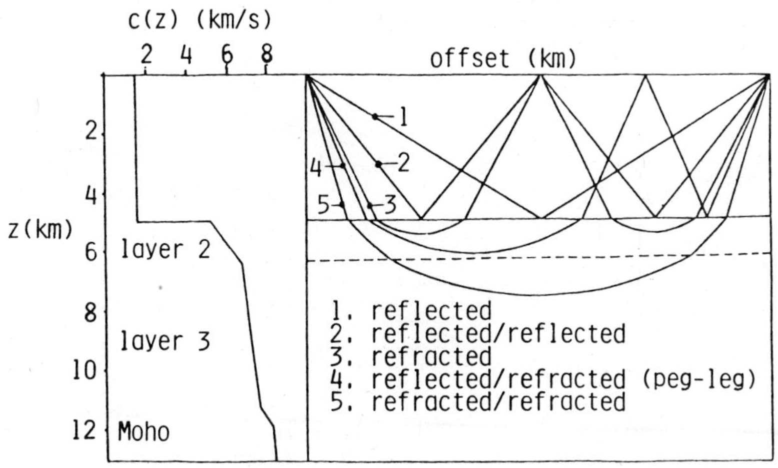

Fig. 2: Ray paths for primary and multiple refracted and reflected signals.

The travel time versus offset for the average crustal model is illustrated in Fig. 3. Each of the five events are also labeled. The apparent slowness, or ray parameter, for each event can be inferred by the derivative of travel time function. One can observe that the arrival sequence of the events at short offsets consist of the primary refraction at low slowness, the primary reflection at high slowness, the "peg leg" multiples and the 1st refraction multiple at low slowness and finally the 1st reflection multiple at intermediate slowness. At longer offsets the "peg legs" and the refraction multiples overtake the primary reflection. The signal arriving at the time commonly associated with the multiple actually consists of three signals, the two "peg legs" and the 1st refraction multiples arriving nearly simultaneously. (The "peg legs" may separate if there is a small amount of bathymetric topography.) Fig. 2 indicates that the rays "peg legs" turn deeper in the section than the 1st refraction and their ray parameters are correspondingly higher. Both the difference in the ray parameters and the effects of the gradients should be observable with an array of sufficient resolution and high enough signal to noise ratio.

AMPLITUDE ANALYSIS

The effects of the gradients are assessed by examining the ray bundle amplitude versus offset and the synthetic seismograms generated using the WKBJ algorithm [5]. The effects of the seafloor and possibly the sediment/basement interfaces must be included. For the acoustic model this can be done using the Rayleigh reflection coefficient, for the elastic case by the Knott or Zoeppritz equations. [16] The solution of the Knott equations for transmission and reflection coefficients for the "average" crustal model and a Poisson's ratio of .25 are illustrated in Fig. 4. (The discussion here is simplified by assuming the elastic case, but all energy converting to shear is not reconverted back to compressional energy.) The most important effect on the amplitude is the null in the transmission coefficient at the compressional speed of the basement. The synthetic seismograms for the acoustic case are illustrated in Fig. 5. The important effect is to note the increase in the amplitude of the multiple in the range of 8 to 16 km. Over this offset the primary and the multiples are refracting in the different layers. A similar phenomenon also occurs when the rays interact with the Moho interface.

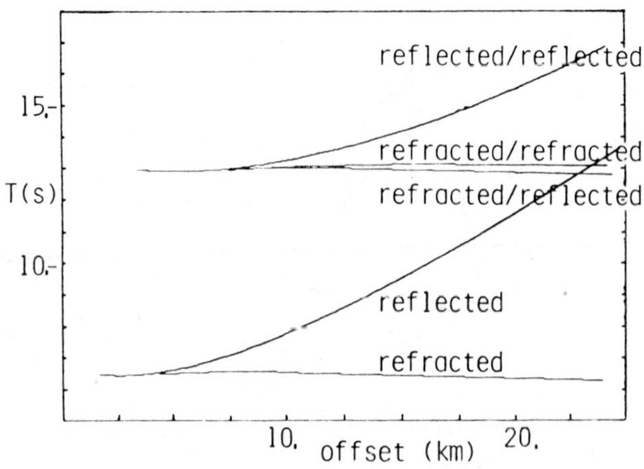

Fig. 3: Travel time function (reduced by 6. km/s) for the primary and multiple signal and the "average" crustal model.

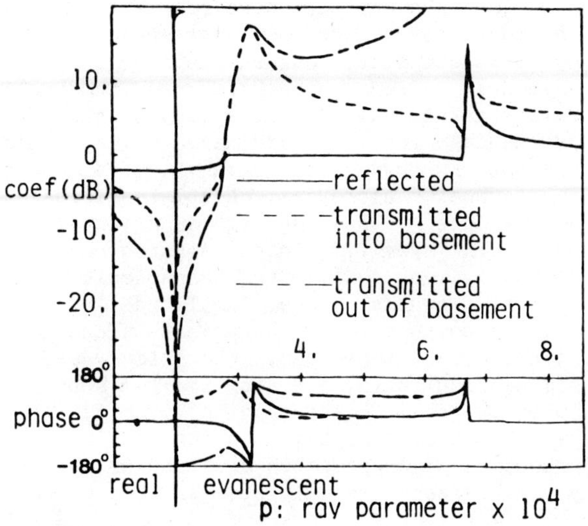

Fig. 4: Transmission and reflection coefficients for the seafloor-basement interface of the "average" crustal in Fig. 1. A Poisson's ratio of .25 and a basement density of 2.3 g/cc are assumed.

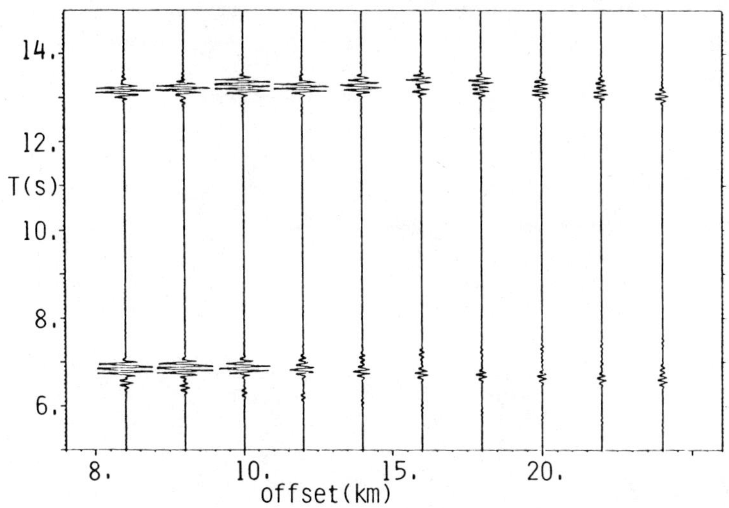

Fig. 5: Synthetic seismograms (travel time reduced by 6. km/s) generated using the WKBJ method of [5] for the acoustic case. The reflected signals with their high slownesses have been suppressed by not including them in the synthesis sumation. Note the very energetic primary refraction which decays after it turns in layer 3 and the amplitude of the two later multiples arriving almost at the same time.

The amplitude pattern can be interpreted using the offset versus ray parameter function illustrated in Fig. 6 in conjunction with the transmission-reflection coefficients. At close offsets just the reflection events exist. Both events have relatively large amplitudes for all offsets since the reflection coefficients are high and their offset function derivatives are low. The primary refraction first appears at the critical distance when the ray parameter equals the compressional slowness. For the acoustic case the amplitude is high, but for the full elastic case it is low because the transmission coefficients through the basement interface has a null at this slowness. As the offset increases further, the primary refracted signal becomes larger as the ray parameter decreases. Next, the critical distances for the multiples are reached. The "peg leg" is slightly closer because of the steep gradient in layer 2. In the elastic case initially these signals are relatively weak because the transmission coefficient is low near the critical ray slowness. Their amplitude increases rapidly, and they soon have levels comparable to the primary since all are refracting in layer 2. As the offset increases, the primary amplitude decreases since it starts to turn in layer 3 with its low sound speed gradient. Since the multiples are still refracting in layer 2, they are stronger than the primary until they also begin to turn in layer 3. The primary refraction then becomes stronger again as it turns at the Moho interface and at the very long offsets the multiples as they encounter it.

EXPERIMENTAL RESULTS

The data from one of the expanding spread profile (ESP) can be used to substantiate the multiple ampltitude phenomenon. The data are from a 24 channel, towed array data from ESP 5 gathered by Stoffa and Buhl. [18] In this data aquisition method source and receiving ships separate from

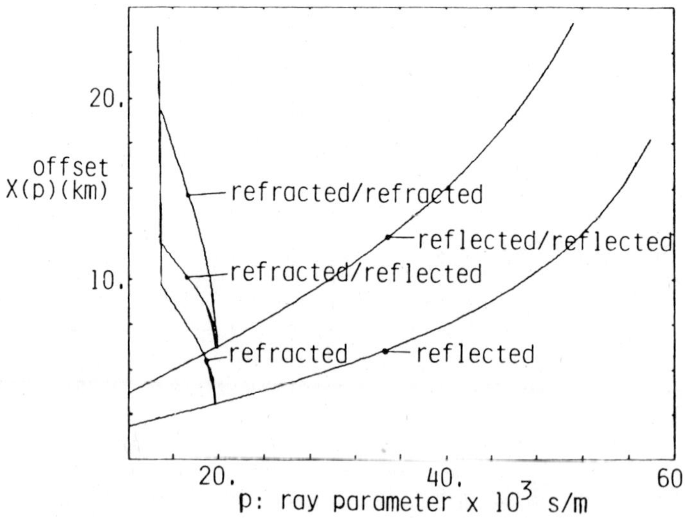

Fig. 6. Offset function for the primary and multiple events for the "average" crustal model of Fig. 1.

a common point. Except for the extremely long offsets the shooting geometry is similar to the common depth point shooting procedure for seismic exploration for oil. The geological region is (145 W, 30 N) in the east Pacific. The region is approximately 60 mybp and has very thin sediments less than 200m thick. [19] The crustal age and thin sediments lead to "old" basement rock at or near the seafloor with sound speeds of layer 2. (The inversion in [18] indicates speeds of 6. km/s at the seafloor.) The thin layer of sediments are characterized as "silty clays" which typically have densities of 1.5 - 2.0 gm/cc and sound speeds close to 1.5 km/s, so the impedance contrast of the sediments is relatively low and one can concentrate on the basement interface.

The multichannel data were analyzed for each shot offset using the slowness-travel time spectra methods described in [10]. Fig. 7a illustrates the signals from 12 of the 24 channels at an offset of 15.5 km where the multiple signal is approximately equal to the primary. The time series start at 11 s after the shot instant. The primary refraction is the first event in the time series and the multiples occur roughly 7 sec later. Fig. 7b is the slowness-travel time spectra for these data. The events of interest are indicated on the spectra. Note that the multiple actually consists of two events. There is higher speed (slowness decreases vertically) event which is interpreted as the "peg leg" and this is followed by a lower speed event event interpreted as the twice refracted multiple. Fig. 8a is an even more compelling example. At this offset the primary and peg are encountering the Moho and a series of "micro events" are present. The twice refracted multiple still is turning in layer 2 so it also is strong. The slowness-travel time spectrum indicates the complex of events with a high speed primary, peg leg at lower speed and finally the refracted one at the slowest speed.

The amplitude dependence can be analyzed using the offset versus ray parameter slowness shown in Fig. 10. The estimated slownesses for the primary signal at each offset are indicated by the "dots." The sound speed versus depth model in Fig. 10 is a fit to these data. The offset versus slowness functions of the model for the primary and the multiples are superimposed in Fig. 10. Since amplitude is inversely proportional to the slope of these functions, one can rapidly identify the regions where each signal should be strong. Note that for offsets less than 14 km the slope of offset function for the primary is low leading to relatively high amplitudes which is consistent with observation in [18]. After that there should be a region where the "peg leg" is stronger. At longer offsets the primary should increase and then decrease as it dives through the Moho transition.

Finally, the estimated amplitudes of the primary and the "peg leg" are plotted versus offset in Fig. 12. These data are subject to the absolute uncertainty in the source amplitude. While the shot weight and depth were nominally constant, partial detonations and variations in "sink rate" which influences the bubble pulse do occur leading to changes of a few dB in source amplitude. Since a source monitor was not used, it is difficult to be assured of a constant source signature. Nevertheless, a pattern in the region of 14-18 km does emerge, which is consistent with the model predictions based upon the travel time and slowness data. It is certainly, not as intense as the predictions of the synthetics; however, these ignored several effects such as absorption and the full elastic model of the seafloor and basement interfaces.

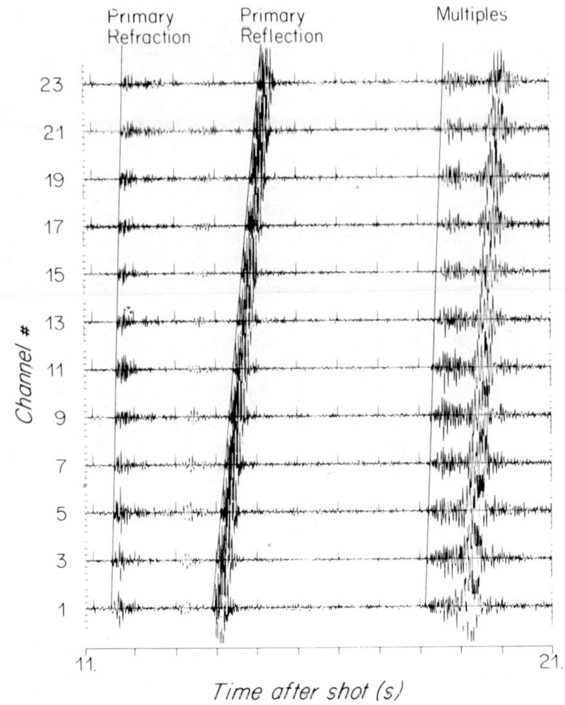

Fig. 7a: Multichannel ESP data at 15.5 km primary refracting in layer 3; multiple just emerging at critical offset and refracting in layer 2.

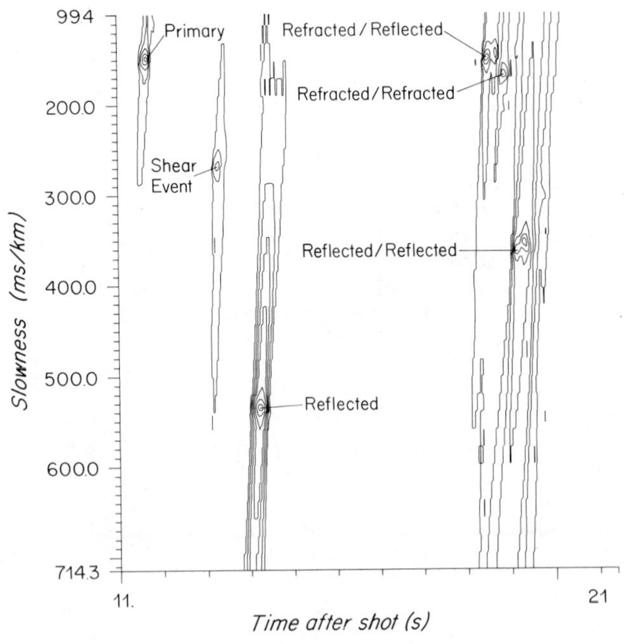

Fig. 7b: Slowness-travel spectra at 15 km (5 db increments) note multiple is actually two events.

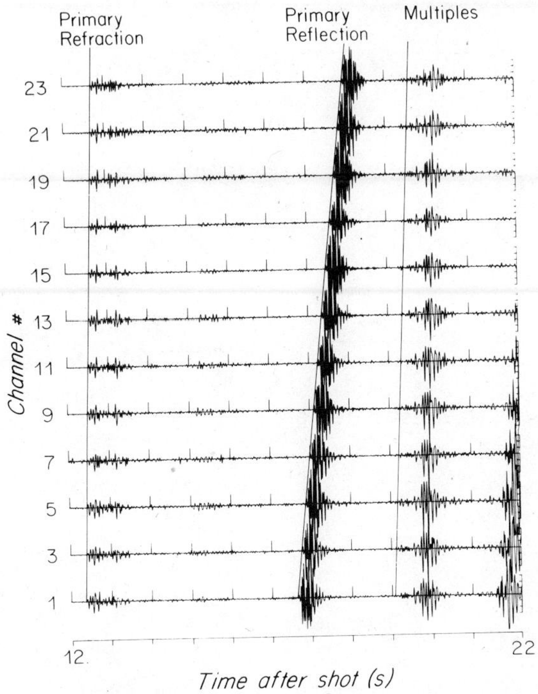

Fig. 8a: Multichannel ESP data at 22. km. Primary refracting near Moho, multiples peg leg in layer 3 and twice refracted in layer 2.

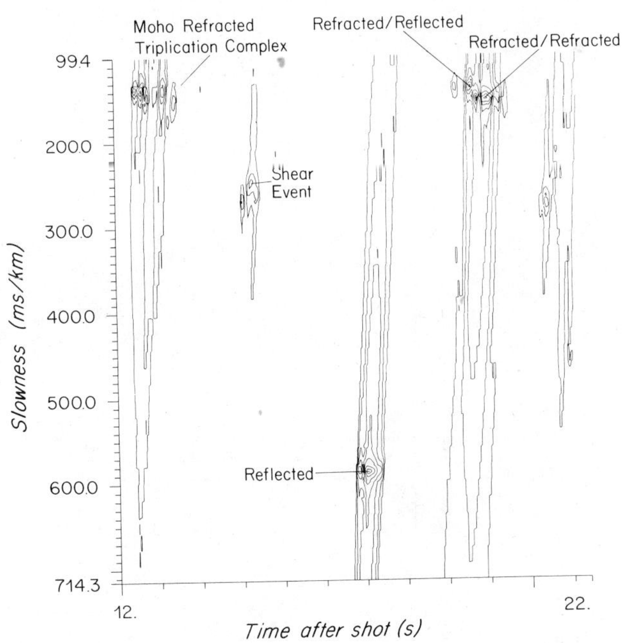

Fig. 8b: Slowness-travel time spectra at 22 km (5 db increments) primary complex is associated with Moho boundary triplication.

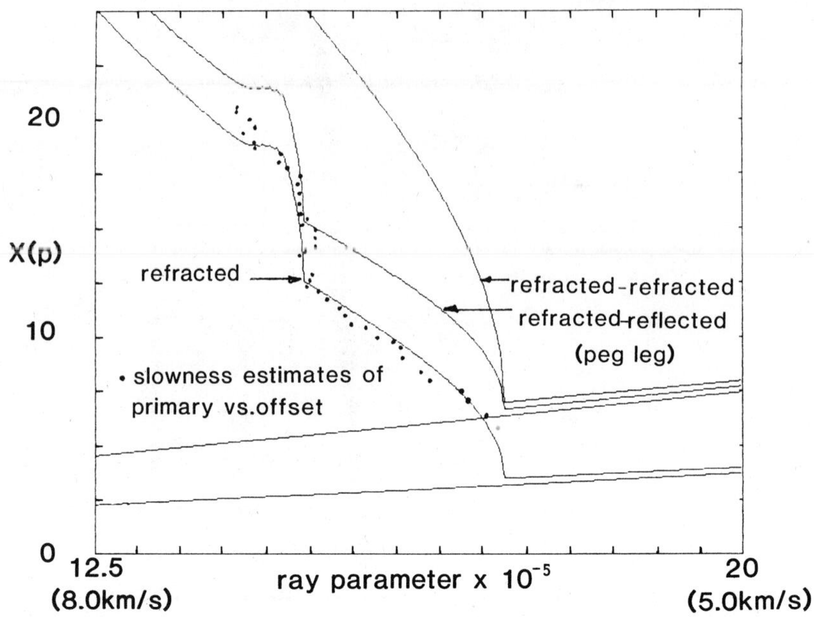

Fig. 9: Estimate of offset function, X(p), and model offset functions for refracted primary, refracted/reflected peg leg multiple and refracted/refracted multiple.

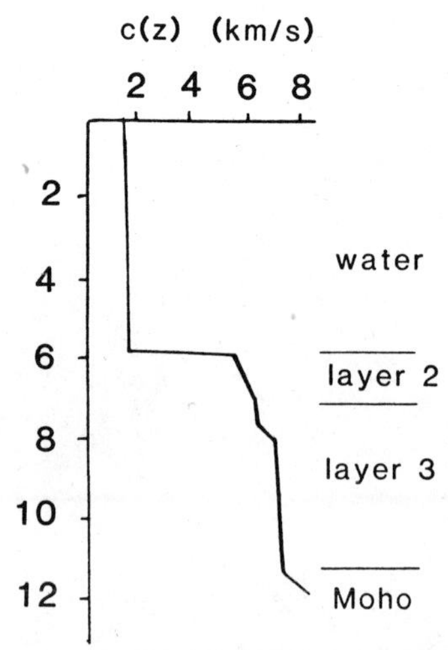

Fig. 10: Estimated sound speed profile for data of ESP 5. The estimate essentially agrees with 18 except in the transition region from layer 2 to layer 3 at approximately 7.5 km.

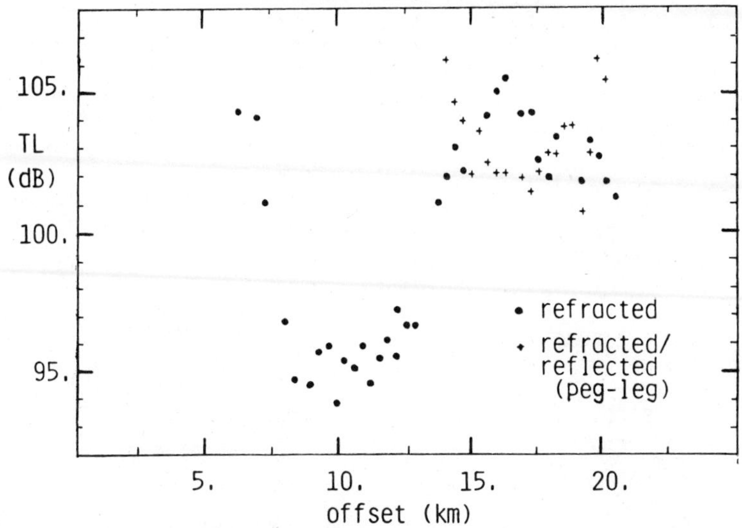

Fig. 11: Relative amplitude estimates of the primary refracted event and the refracted/reflected peg leg multiple. Note that low level, i.e. high transmission loss, TL, is up on the figure. At short offsets the primary emerges from the null at critical refraction and is very strong whhile refracting in layer 2. The multiple undergoes the same behavior at almost twice the offset, and for a range of offsets from 14 to 19 km is stronger thaı the primary.

SUMMARY

The gradients present in typical oceanic crust provide a robust explanation to the amplitude behavior of the multiples observed in refraction profiling. High resolution slowness-travel time spectra can separate the slownesses of the individual events and provide a consistent explanation of the amplitude pattern of both the primary and muultiple events.

ACKNOWLEDGEMENTS

The authors thank J. Ewing, R. Stephen and I. Dyer for their many useful commnents and discussion. This work was supported by the Office of Naval Research, Contract #N 00014-77-C-0266.

REFERENCES

1. Kennett, J. P., Marine Geology, pp 47-50, Prentice-Hall, Inc., Englewood Cliffs, N. J. [1982].

2. Dobrin, M. B., Introduction to Geophysical Prospecting, 3rd edition, Chap. 9, McGraw-Hill, Inc, New York, N.Y., [1976].

3. Telford, W. M., Geldhart, L. P., Sheriff, R. E. and Keys, D. A., Applied Geophysics, Chap. 4, Cambridge University Press, Cambridge, England, [1976].

4. Wiggins, R. A., "Body wave amplitude calculations - II," Geophysical J. of the Royal Astronomical Society, 46, 1-10, [1976]

5. Chapman, C. H., "A new method for computing synthetic seismograms," Geophysical J. of the Royal Astronomical Society, 54, 481-518, [1978]

6. Fuchs, K. and Muller, G., "Computation of synthetic seismograms with the reflectivity method and comparison with observation," Geophysical J. the Royal Astronomical Society, 23, 417-433, [1971]

7. Di Napoli, F. R. and Deavenport, R., Computer Models of Underwater Acoustic Propagation, NUSC Technical Report 5867, Naval Underwater Systems Center, New London, Conn., [1980]

8. Kennett, B. L. N., "Towards a more detailed seismic picture of the oceanic crust and mantle," Marine Geophysics Research, 3, 7-42, [1977]

9. Lewis, B. T. R., "Evolution of ocean crust and seismic velocities," Annual Rev. of Earth and Planetary Science, 6, 377-404, [1978]

10. Spudich, P. and Orcutt, J., "A new look at the seismic velocity structure of the oceanic crust," Review of Geophysics and Space Physics, 18, 627645, [1980]

11. Duckworth, G. L., Baggeroer, A. B ,and Jackson, H. R., "Crustal structure measurements near FRAM II in the Pole Abyssal Plain," Tectonphysics, 89, 172-215, [1982]

12. Grant, F. S. and West, G. F., Interpretation Theory in Applied Geophysics, McGrawHill, New York, N. Y., [1965]

13. Raitt, R. W., [1963]. "The Crustal Rocks", in The Sea, Vol 3, ed. by M. N. Hill, pp. 85-102, Wiley Interscience, New York, [1963]

14. Christensen, N. I. and Salisbury, M. H., "Structure and constitution of lower oceanic crust," Review of Geophysics and Space Physics, 13, pp. 57-86, [1975]

15. Bott, M. H. P., The Interior of the Earth: its structure, constitution and evolution, Elvesier Science Publishing, Great Britain, [1982]

16. Houtz, R. and Ewing, J. I., "Upper crustal structure as a function of plate age," J. of Geophysical Research, 81, 2490-2498,[1976]

17. Clay, C. S. and Medwin, H., Acoustical Oceanography: Principles and Applications, Wiley-Interscience, New York, N.Y., [1977]

18. Stoffa, P. L. and Buhl, P., "Two-ship multichannel seismic experiments for deep crustal studies: expanded spread and constant offset profiles," Journal of Geophysical Research, 84, B13, 7645-7660, [1979]

19. Ewing, J. I., Ewing, M., Ludwig, and Aiken, "The crust and upper mantle of the Pacific Ocean," The Sea, Vol. 2, ed. by A.Maxwell, Wiley Interscience, New York, [1962]

THE PROPAGATION OF P_n

Thomas Sereno and John Orcutt

Institute of Geophysics and Planetary Physics
Scripps Institution of Oceanography, A-025
La Jolla, California 92093

ABSTRACT

A propagation model for the oceanic P_n phase is presented in which the onset consists of a refraction from the lower oceanic lithosphere while the subsequent coda is a result of near-receiver sediment and water reverberation. The first arrival of the oceanic P_n phase remains entirely within the lithosphere and for epicentral ranges in excess of 25° to 30° contains a free surface reflection, similar to the seismic phase PP. At large epicentral distances, or similarly for deep events, the P_n wavetrain is preceded by a low frequency forerunner which has traversed the low Q asthenosphere. The coda of the P_n phase was analyzed through the computation of synthetic seismograms using wavenumber integration. Spectral characteristics associated with both sediment and water reverberation are clearly observed in both the synthetics and data collected in the southwest Pacific during the 1983 Ngendei Seismic Experiment.

INTRODUCTION

The oceanic P_n phase is characterized by low spatial attenuation, a high frequency wavetrain of extremely long duration and an onset velocity near 8.0 km/s[1]. Through the computation of synthetic seismograms in a laterally homogeneous, vertically stratified medium we will show that these properties can be achieved in the absence of scattering, low velocity channels or high Q waveguides.

SYNTHETIC SEISMOGRAMS

Wavenumber integration, in which a quadrature over wavenumber is performed at fixed frequency points, was used to generate the synthetic seismograms[2]. The time domain response was obtained by applying a Fourier transform to the resulting complex spectrum. The seismograms contain all reverberations, including those off the free surface, as well as all P to S and S to P mode conversions. Attenuation is included by allowing the material velocities to be complex thereby allowing the integration to be performed along the real wavenumber axis. The most serious limitations of the method include the restriction to models consisting of a stack of

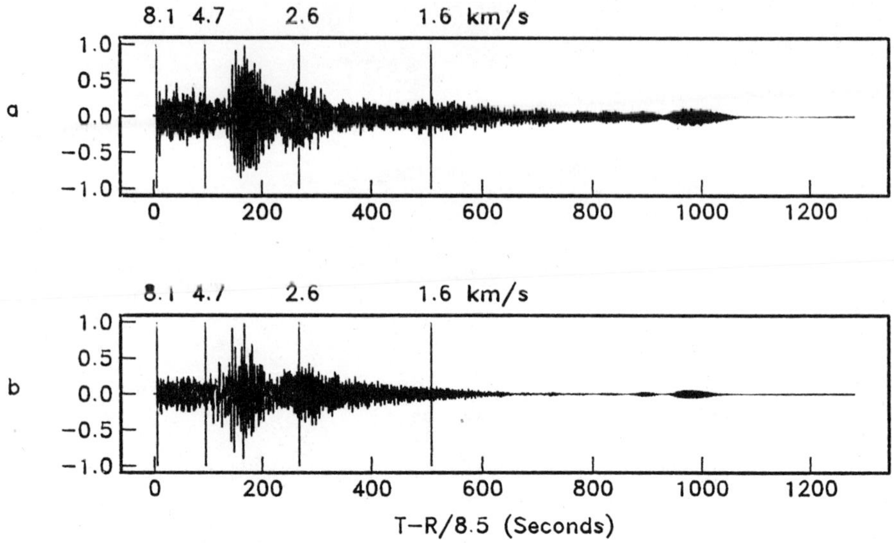

Figure 1. Complete synthetic seismograms at an epicentral range of 1000 km for a thrust fault source at 14.0 km depth. The fault had a dip of 45°, a rake of 270° and an azimuth relative to the receiver of 60°. (a) Vertical component and (b) horizontal component rotated 25° clockwise from the radial direction. Group velocities of 8.1, 4.7 2.6 and 1.6 km/s are indicated. Phase velocities between 3.85 and 25.0 km/s were included in the calculation.

homogeneous layers and the high computational costs involved for models with more than 10 to 15 layers.

As an exploratory model we have adopted that of Gettrust and Frazer[3] which includes a 4 km thick water layer and a 500 m thick sediment layer overlying a vertically inhomogeneous oceanic crust and upper mantle. Compressional and shear wave velocities are monotonically increasing functions of depth. The values of the quality factor, Q, were less than 500 in the crust and upper mantle and 50,000 in the oceanic water column.

Figure 1 is an example of complete synthetic seismograms generated for a sub-MOHO thrust fault at an epicentral range of 1000 km. The late arriving energy on the vertical component (figure 1a) with a group velocity of 1.6 km/s is emergent in nature and has a substantial duration characteristic of oceanic abyssal T-phases[4]. Through a comparison of delay time wavefields between Green's functions that contain this late arriving energy and Green's functions that do not, it was observed that the ray parameter range comprising this group corresponds to that of post-critical reflections off the MOHO. The synthetic T-phase is thus explained as a reverberation between the free surface and the MOHO with a substantial number of water and crustal reverberations acting to extend the wavetrain. The 4.7 km/s arrival on the horizontal component (figure 1b) is grossly identified as the oceanic S_n phase. The first arrivals of this group are horizontally polarized shear waves and are thus absent from the vertical component. Work is continuing in an effort to synthesize more realistic S_n phases.

The synthetic P_n wavetrain has an onset velocity of 8.1 km/s, a duration of over 100 seconds and extends into the onset of S_n. The amplitude of the synthetic P_n wavetrain was within a factor of 2 or 3 of data amplitudes obtained during the 1983 Ngendei Seismic Experiment in the southwest Pacific (23° 49′ S, 165° 32′ W). An event located near the Tonga Islands by the National Earthquake Information Service (NEIS) with a magnitude of 5.8 and a moment of 2.6×10^{24} dyne-cm was recorded by three ocean bottom seismometers (OBS's) at the Ngendei site. The maximum amplitudes of these recordings ranged between 0.76 and 1.44 volts. Convolution of the synthetic P_n phase with the OBS instrument response and scaling by the NEIS published moment yielded a maximum amplitude of 0.42 volts. Differences of this order can be explained simply in terms of differences in model parameters, fault orientation and focal depth.

REVERBERATIONS

The importance of reverberations within the oceanic water column and sediment layer is most effectively demonstrated in the frequency domain. It is simple to show that the power spectrum of a wavelet reverberating in a single layer has eigenfrequencies given by

$$f_n=(2n+1)\frac{df}{2} \; ; \; (n=0,1,2,...) \qquad (1)$$

where $df=\frac{1}{T}$ and T is the time between multiples. In the case of near-vertical incidence T can be approximated by the two-way travel time in the layer and (1) represents the eigenfrequencies for an organ-pipe mode[5]. Arrivals within the P_n wavetrain must penetrate the mantle in order to have group velocities between 6 and 8 km/s and must therefore have phase velocities greater than 8 km/s. A result of this is that the rays are quite steep at the receiver, justifying the use of the two-way travel time in equation (1). The $\frac{df}{2}$ offset of the spectral peaks is due to a π-phase shift between multiples. The width of the spectral peaks at frequencies given by (1) is governed by the impedance contrast at the boundaries of the layer. Thus leaky modes, which suffer a continuous loss of energy along the propagation path due to leakage into the surrounding medium, will attenuate more rapidly than locked modes and will thus have substantially wider spectral peaks. The decay of the reverberation peaks with increasing mode number, n, is a function of Q in the layer.

The computation of the full synthetic seismograms pictured in figure 1 required extremely fine frequency sampling to avoid time series wrap-around. The calculation extended to a Nyquist frequency of only 6.4 Hz which involved obtaining the response at 8193 frequency points. While the wavenumber integration algorithm is not strictly bandlimited, the costs of extending the calculation to higher frequencies are formidable. Synthetic power spectra do not, however, require this fine frequency sampling and can be computed rather efficiently up to a Nyquist frequency of 64.0 Hz. Thus direct comparison between data collected at the Ngendei site and synthetics is easily achieved in the frequency domain.

The Ngendei site was the location of hole 595B of Deep Sea Drilling Project (DSDP), Leg 91. The depth of the oceanic water column at the site is 5.5 km and, from core sample measurements, the sediment thickness is 70 m with a compressional velocity of 1.6 km/s[6]. For compressional wave sediment reverberation the two-way travel time in the layer is 0.0875 s yielding an eigenfrequency spacing of df=11.43 Hz. The impedance contrast at the ocean/sediment interface is quite small for compressional waves. Thus the sediment layer acts as an extremely leaky waveguide for compressional wave propagation and the peaks associated with sediment reverberation have a large characteristic width. The reverberations within the water column actually include the sediment layer and take place between the free surface and the base of the sediments. The resulting eigenfrequency spacing

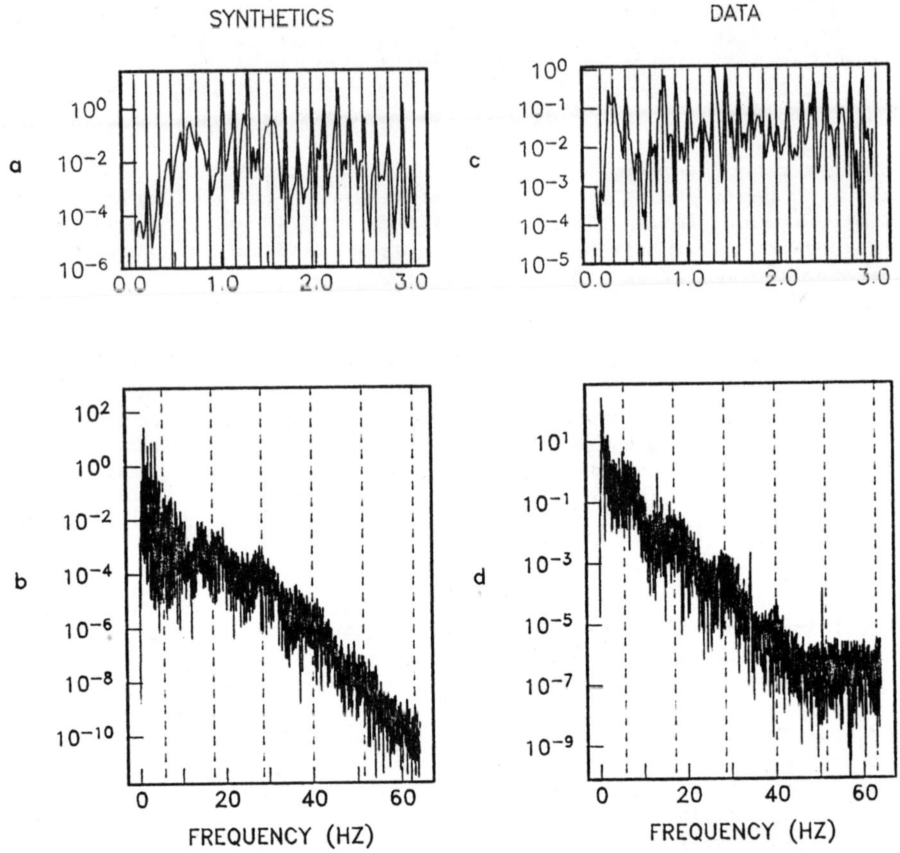

Figure 2. Power spectra illustrating the presence of reverberation within the P_n coda. (a) First 3.0 Hz of the synthetic vertical component spectrum depicted in (b). (c) First 3.0 Hz of a hydrophone spectrum and (d) a complete OBS vertical component spectrum. In figures (a) and (c) vertical lines indicate predicted eigenfrequencies associated with water reverberation (equation (1)). In figures (b) and (d) vertical lines indicate predicted eigenfrequencies for P-wave sediment reverberation.

for water reverberations is therefore df=.1375 Hz. Both interfaces in this case have very strong impedance contrasts and the associated spectral peaks are quite narrow. Thus the power spectrum of a wavelet reverberating in the oceanic water column and sediment layer consists of widely spaced spectral "humps" due to P-wave sediment reverberation superimposed on finely spaced spectral "peaks" due to water reverberation.

Figure 2 is an illustration of this behavior. Figure 2b is a synthetic power spectrum computed using wavenumber integration in which we have used the sediment and water properties of the Ngendei site. In an effort to reduce computational costs we have included only phase velocities between 6.9 and 25.0 km/s. Figure 2a depicts the first 3.0 Hz of the spectrum with vertical lines indicating the eigenfrequencies of water reverberation computed using equation (1). Figure 2b contains the full spectrum[5] after correcting for the OBS instrument response and a source spectrum of the form

$$|S(f)|^2 = S(0)^2\left[1+\frac{f}{f_c}\right]^{-4} \qquad (2)$$

where f_c is the corner frequency and was given a value of 0.6 Hz. In this case the vertical lines indicate positions of spectral "humps" associated with P-wave sediment reverberation. To verify our interpretation of these features we doubled the layer thicknesses and recomputed the spectra. The effect was to decrease the spacing of the spectral peaks by a factor of 2 thereby validating the interpretation.

Figures 2c and 2d contain power spectra of data collected at the Ngendei site. Figure 2c is the first 3.0 Hz of a hydrophone spectrum illustrating the presence of water reverberations and figure 2d is an OBS vertical component spectrum illustrating the presence of sediment reverberation. These spectral features were quite common in the data and were independent of the source characteristics.

The presence of S-wave sediment reverberation is most easily detected in horizontal component spectra due to the steep incidence of the rays. For shear waves the impedance contrasts at both the ocean/sediment and sediment/crust interfaces are quite large yielding relatively narrow spectral peaks. The low quality factor for shear waves in the sediments, however, results in a rapid decay of the spectral peaks with increasing mode

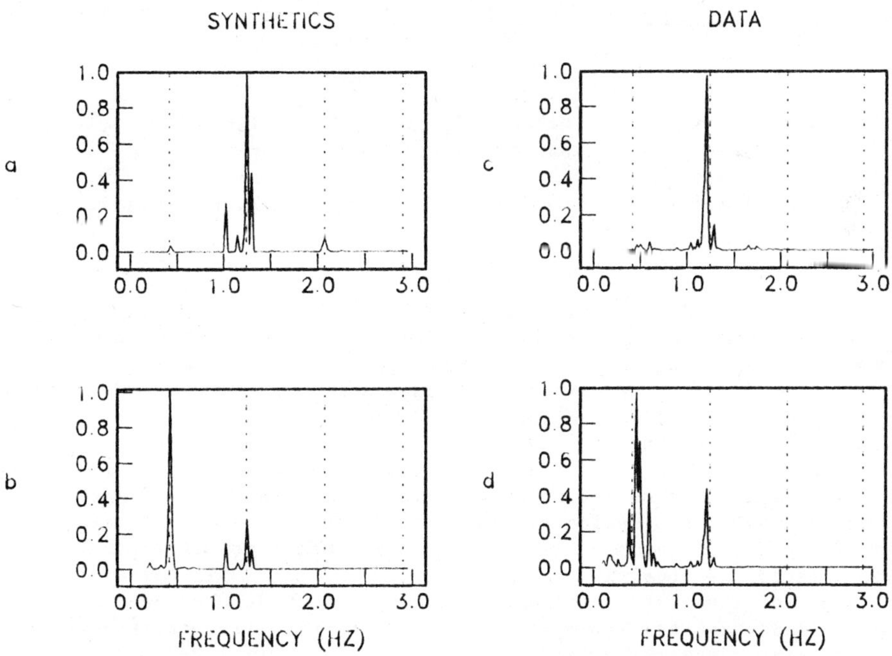

Figure 3. Power spectra illustrating shear wave sediment reverberations within the P_n coda. (a) First 3.0 Hz of the synthetic horizontal spectrum after including the instrument response and source spectrum. (b) Synthetic spectrum of (a) without including the instrument response. Power spectra of horizontal component data taken at the Ngendei site (c) before removing the instrument response and (d) after removing the instrument response. In (a) (d), dashed vertical lines indicate predicted eigenfrequencies of S-wave sediment reverberation. The peak amplitudes in each figure were scaled to a value of one.

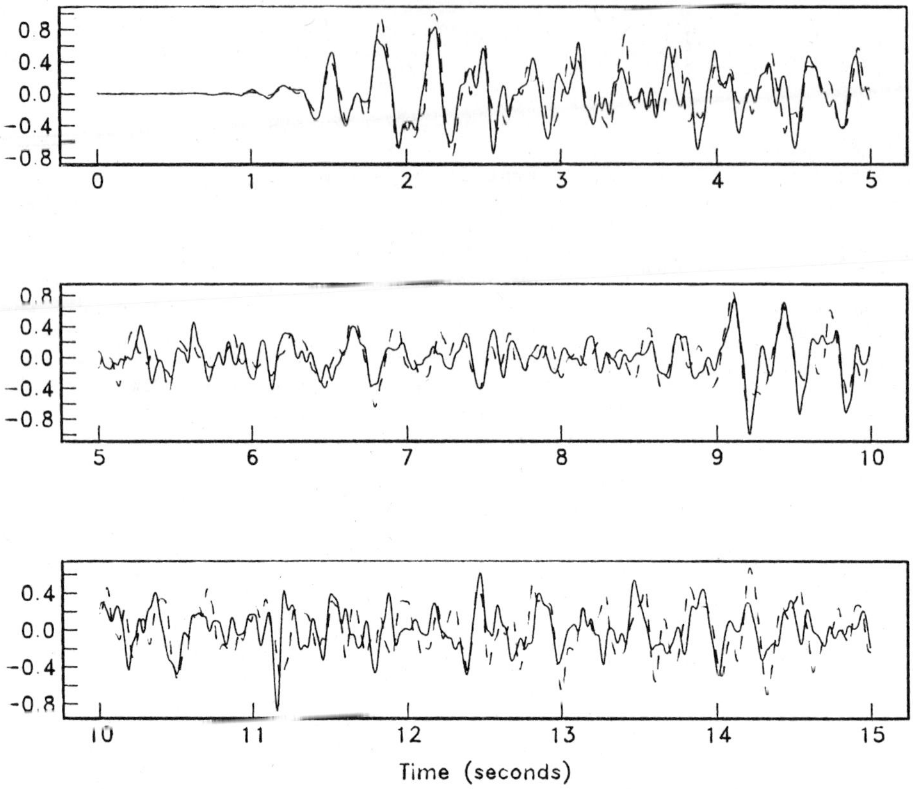

Figure 4. Coherence of the oceanic P_n phase. Comparison of recordings of the same event taken by OBS's separated by a distance of 1.9 km. OBS Karen (solid) vs. OBS Janice (dashed). The records were bandpassed between 2.0 and 25.0 Hz prior to comparison.

number. Figure 3a depicts the first 3.0 Hz of the synthetic horizontal component power spectrum after including the OBS instrument response and source spectrum given by (2). The shear wave sediment velocity used in the computation of the synthetic spectrum was 116 m/s yielding an eigenfrequency spacing of 0.83 Hz. Figure 3b is the synthetic spectrum without including the instrument response. Figure 3c is the first 3.0 Hz of the power spectrum of horizontal component data taken at the Ngendei site and 3d was obtained after removing the instrument response. In figures 3a-3d, dashed vertical lines indicate predicted frequencies (equation (1)) of spectral peaks associated with S-wave sediment reverberation. The positions, widths and spectral decay rates of these peaks all agree with predictions for shear wave sediment reverberation. Once again we altered the sediment thickness and recomputed the spectra to verify our interpretation of the peaks in the synthetic spectra.

COHERENCE

Strong coherence of the P_n wavetrain between OBS's separated by large distances is not expected unless the water and sediment properties at the two sites are nearly equal. It was found in the Ngendei data that far greater coherence could be achieved by bandpassing the data between 2.0 and 25.0 Hz prior to comparison. The result of even a small change in either S-wave sediment velocity or sediment thickness would result in a

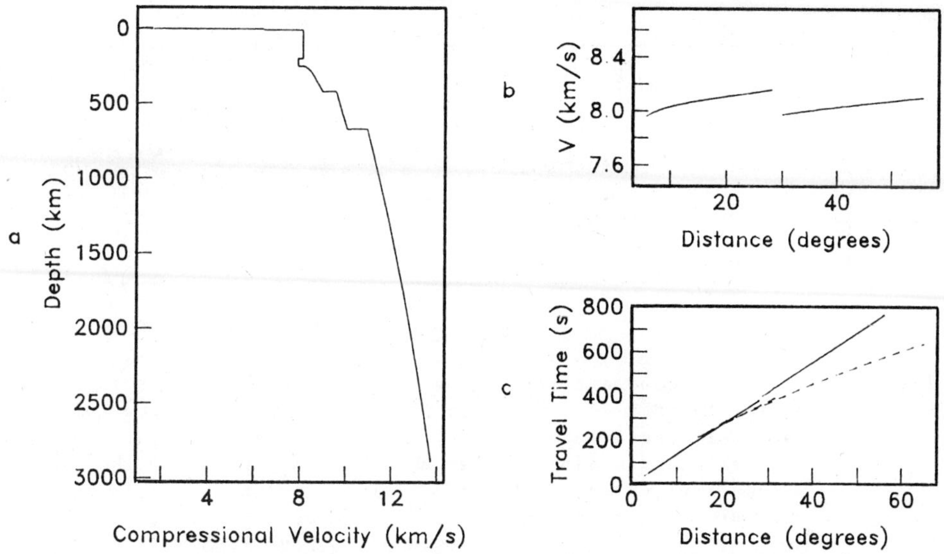

Figure 5. (a) Velocity model used to compute group velocities and travel times of P_n and normally refracted P. The results are shown for a surface source and receiver. (b) Group velocity of P_n as a function of epicentral distance for the model in (a). Group velocity has been defined as epicentral distance divided by total travel time. The discontinuous break at 30° represents the crossover between a ray normally refracted in the lithosphere and one that contains a free surface reflection. (c) Travel time curve of P_n (solid) verses normal refracted P with turning points below the base of the lithosphere (dashed) for the model in (a).

measurable shift in the dominant frequency of the wavetrain (figure 3). The frequency band between 0.0 and 2.0 Hz is the most sensitive to these changes resulting in a reduction of coherence at these lower frequencies. Figure 4 is a comparison of data collected by two different OBS's separated by a distance of 1.9 km for the same event after bandpassing between 2.0 and 25.0 Hz. The first water reverberation is evident at about 9 seconds and the excellent coherence well into the P_n wavetrain is evidence for reverberation, as opposed to scattering, as the dominant mechanism of P_n coda generation. Work is continuing on the problem of coherence as well as that of polarization within the P_n coda.

PROPAGATION PATH

The observation that the group velocity of P_n is generally higher at epicentral ranges between 12° and 30° than at distances outside this range[7] can be explained by the confinement of P_n to the oceanic lithosphere. At ranges less than 30° the onset of P_n is associated with a normal refracted P-wave remaining in the lithosphere, however at distances in excess of 30° the first arrival that remains in the lithosphere contains a free surface reflection, similar to the seismic phase PP, thereby decreasing the group velocity. This is demonstrated in figure 5b which displays the group velocity of P_n as a function of epicentral distance for the velocity model shown in figure 5a. In order to achieve a crossover distance of 30°, a lithosphere thickness of nearly 200 km is required. The decrease in group velocity at ranges less than 12° is the result of

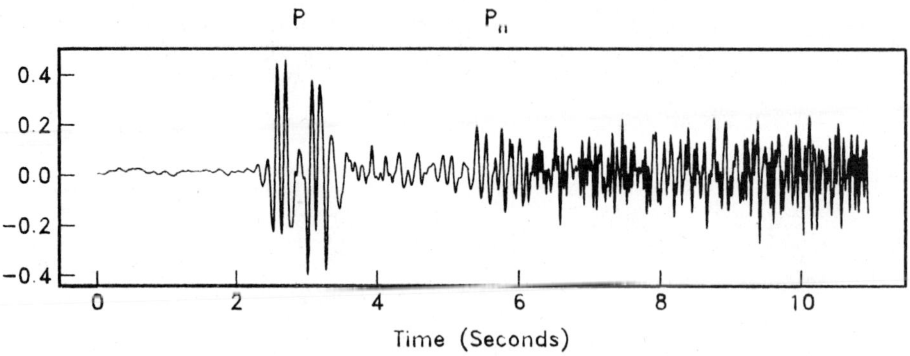

Figure 6. Vertical component seismogram recorded by a borehole seismometer at the Ngendei site. The focal depth of the event was 293 km in the Kermadec trench (m_b=5.0, Δ=12.91^o). The complex P waveform between 2 and 4 seconds is a triplication from the 400 km discontinuity. Arrivals following 5.5 seconds constitute the P_n arrival. The enrichment in high frequency content of P_n relative to P is indicative of the low Q of the sub-lithosphere mantle.

shallower turning points in the lithosphere. A similar interpretation was presented by Menke and Richards[8] in terms of whispering gallery modes at the MOHO.

Figure 5c is a travel time curve illustrating the arrival times of normal refracted P verses P_n for the model in figure 5a. The curves diverge at an approximate range of 25° where the normal refracted arrival is a ray that has crossed the low Q asthenosphere constituting the commonly observed low frequency forerunner of the P_n wavetrain[9]. Figure 6 is an example illustrating the normal refracted P forerunner of oceanic P_n. For deep events the separation of normal refracted P and P_n occurs for smaller ranges as P_n must propagate up and around the corner of the descending slab. During the Ngendei Experiment the deployment of a continuously recording borehole seismometer enabled the recording of strong P_n phases for events with focal depths in excess of 450 km. The suggestion that a descending slab can act as a waveguide to propagate the high frequency oceanic S_n phase was made by Isacks and Barazangi[10]. Their model consists of a transformation of the basaltic oceanic crust to a thin, high velocity, eclogite layer as the slab descends to greater depths. Thus the upper flank of the slab provides an effective reflecting surface and, in combination with a velocity gradient within the slab, could provide an efficient waveguide for the propagation of high frequency seismic energy.

SUMMARY

The model of the propagation of P_n presented consists of an arrival confined to the oceanic lithosphere with subsequent water and sediment reverberations comprising the ensuing wavetrain. The low spatial attenuation observed for P_n does not require an intrinsically high Q lithosphere waveguide but rather can be explained in terms of the extremely high Q in the oceanic water column. The depletion of high frequencies in the normal refracted arrival relative to P_n is due to propagation through the low Q asthenosphere.

Acknowledgements. The data used for this study were collected with the support of DARPA Grant No. AFOSR-84-0043 while the synthetic calculations were supported by ONR Contract No. N00014-80-C-0440.

REFERENCES

1. D. Walker, High-frequency P_n and S_n phases recorded in the western Pacific, _J. of Geophys. Res._ 82:3350 (1977).
2. R. Apsel, Ph. D. thesis, University of California, San Diego, La Jolla, 349 pp. (1979).
3. J. F. Gettrust and L. N. Frazer, A computer model study of the propagation of the long-range PN phase, _Geophys. Res. Lett._ 8:749 (1981).
4. R. H. Johnson and R. A. Norris, T-wave generation mechanisms, _Hawaii Institute of Geophysics Rept. HIG-70-7_, 16 pp. (1970).
5. K. Aki and P. G. Richards, "Quantitative Seismology: Theory and Methods," W. H. Freeman, San Francisco (1980).
6. I. I. Kim, D. K. Smith, J. A. Orcutt, T. H. Jordan and H. W. Menard, Seismic reflection site survey: correlation with physical properties, leg 91, Deep Sea Drilling Project, _Initial Rep. Deep Sea Drill. Proj., Leg_ 91 (1985).
7. D. Walker, High-frequency PN, SN velocities: some comparisons for the western, central and south Pacific, _Geophys. Res. Lett._ 8:207 (1981).
8. W. H. Menke and P. G. Richards, Crust-mantle whispering gallery phases: a deterministic model of teleseismic P_n wave propagation, _J. of Geophys. Res._ 85:5416 (1980).
9. G. V. Latham and G. H. Sutton, Seismic measurements of the ocean floor, _J. of Geophys. Res._ 71:2545 (1966).
10. B. L. Isacks and M. Barazangi, High frequency shear waves guided by a continuous lithosphere descending beneath western South America, _Geophys. J. R. astr. Soc._ 33:129 (1973).

EFFECTS AND CONSEQUENCES OF TRANSVERSE ISOTROPY IN THE SEAFLOOR

Gerard J. Fryer and Daniel J. Miller

Hawaii Institute of Geophysics
University of Hawaii at Manoa
Honolulu, HI 96822, USA

ABSTRACT

Transverse isotropy, a form of anisotropy with a single vertical axis of symmetry, is extremely widespread in marine sediments, but its effects on seismic interpretations have been inadequately quantified. This anisotropy increases with depth, just as propagation velocities themselves increase. Indeed, it is impossible to separate the effects of velocity gradients and anisotropy. Using elastic parameters appropriate for a carbonate sequence, we have used ray-tracing and wavefield modeling codes to investigate the consequences of ignoring anisotropy. If isotropy is erroneously assumed, sound-speed gradients will be underestimated, the thickness of the sedimentary sequence overestimated (often seriously) and the shear velocity overestimated (hence estimates of Poisson's ratio will also be incorrect). Transverse isotropy may also affect the bottom reflection loss, but from our preliminary studies such effects appear to be minimal. From compressional-wave traveltime data alone the identification of transverse isotropy appears almost impossible. Shear-wave information is vital if the phenomenon is to be adequately characterized.

INTRODUCTION

Transverse isotropy, a variation of seismic velocity with angle from the vertical, has been widely observed in Deep Sea Drilling Project cores and is probably ubiquitous in marine sediments. The anisotropy may be caused by subhorizontal cracks or pores, by recrystallization of anisotropic minerals, or by compositional layering causing very fine interleaved beds of high and low velocity material.[1] The latter two mechanisms can explain the observed variation: an increase in anisotropy with depth of burial. Typically, P-wave anisotropy increases from zero at the seafloor to about 10% at a depth of 1000 m, with horizontal waves travelling faster than vertical.

The property of marine sediments which seems most important in determining their reflection properties, especially at small grazing angles, is the rate of increase in velocity with depth. Marine sediments are inferred to have extremely high velocity gradients because of the rapid decrease of porosity with depth of burial,[2] but estimates of the gradients are made ignoring anisotropy. Behaviour of a ray near its turning point is controlled by the horizontal propagation velocity, which in transversely isotropic sediments is higher than the vertical. Thus, anisotropic sediments with only a mild velocity gradient could conceivably be mis-identified as being isotropic but with a strong gradient. In reality, the effects of anisotropy can be safely ignored at shallow depths where gradients are the strongest, but with increasing depth it becomes impossible to separate the two phenomena. Gradients and anisotropy must be considered together if the reflectivity of a sedimentary sequence is to be adequately described. This is especially true at low frequencies, when even grazing rays may be affected by the structure at some depth. To date, no complete descriptions of the combined effects of gradients and anisotropy have been presented. Most work on transverse isotropy has concentrated purely on traveltime analyses for homogeneous layers,[3,4] with the effects of gradients only cursorily considered.[5,6] In this paper we describe a preliminary investigation of the combined effects of transverse isotropy and gradients.

ENERGY PROPAGATION AND RAY TRACING

There are fundamental differences between wave propagation in isotropic and transversely isotropic media. In a homogeneous isotropic medium, a wavefront expanding away from a source is spherical, so that the group velocity direction (the direction along which energy moves) and the phase velocity direction (the normal to the wavefront) coincide. In an anisotropic medium, since velocities vary with direction, the expanding wavefront is not spherical, and group and phase velocity directions will, in general, diverge. Because of this difference, a compressional disturbance in an anisotropic medium is denoted qP for ′quasi-P′. In a transversely isotropic medium symmetry demands that phase and group velocities diverge only in the vertical plane, with the result that, as in isotropy, P-SV motion decouples from SH.

The phase velocity in a particular direction is the velocity of a plane wave in that direction. The propagation characteristics of plane waves in an anisotropic solid may be found from the Christoffel equation,[7] from which

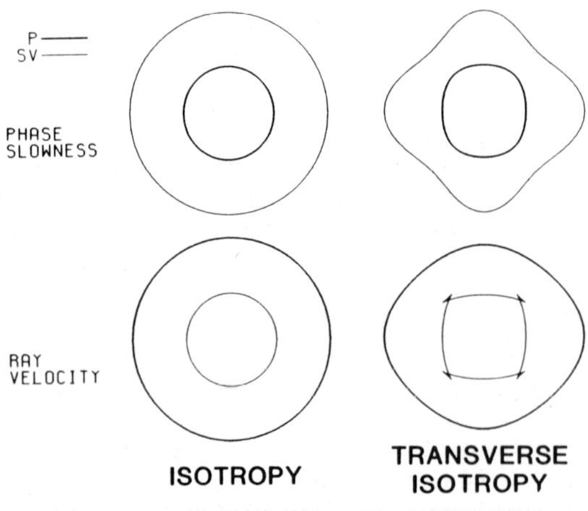

Fig. 1. Slowness and ray (group) velocity surfaces for isotropic (left) and transversely isotropic (right) media. The heavy curve represents P or qP motion, the lighter curve SV or qSV.

Table 1. Carbonate Model

Subbottom Depth (m)	Elastic Parameters (x 10^9 Newtons/m^2) C_{11}	C_{13}	C_{33}	C_{44}	Density	Velocities (km/s)[a] V_{Pz}	V_{Px}	$V_{Sz}=V_{Sx}$
Water	2.25	2.25	2.25	0.0	1.00	1.50	1.50	0.0
0[b]	4.10	3.536	4.10	0.282	1.60	1.60	1.60	0.42
800	12.81	4.50	10.59	2.233	1.95	2.33	2.56	1.07
Basement	21.60	11.508	21.60	5.046	2.40	3.0	3.0	1.45

[a] V_{Pz} is the vertical qP-velocity, V_{Px} the horizontal qP-velocity; V_{Sz} and V_{Sx} are vertical and horizontal qSV velocities. V_{Sz} and V_{Sx} are the same in a transversely isotropic medium, but velocities are different at other angles to the vertical, as shown in Figure 1.

[b] Within the region from 0 to 800m depth, elastic parameters vary such that <u>velocity</u> increases linearly. Density also varies linearly in this range.

it is possible to identify the slowness surface, the surface in phase slowness space that gives the permissible values of slowness as a function of propagation direction. Slowness surfaces for isotropy and transverse isotropy are shown in Figure 1.

The direction of a seismic ray path is, of course, the direction of energy propagation, and is defined by the group velocity vector. This vector is normal to the associated slowness surface,[7] so once the slowness surface has been found the energy propagation directions may be determined. Figure 1 shows group velocity surfaces for isotropy and transverse isotropy; these figures reflect the shape of an expanding wavefront from a point source. Note that in transverse isotropy it is possible for triplications to form on the qSV wavefront (with associated concentration of energy) even in a homogeneous medium.

With the group velocity vectors known, it is possible to trace rays. For transverse isotropy, performing a complicated mapping of phase to group velocity direction is unnecessary as closed-form integral expressions for distance and travel time directly in terms of elastic parameters have been provided by Cerveny et al.[8] Numerical integration of these expressions is straightforward.

TRAVELTIME ANALYSIS OF A SIMPLE CARBONATE STRUCTURE

To demonstrate anisotropic effects, we sought the simplest possible structure with both velocity gradients and transverse isotropy. A transverse isotropic material is described by five elastic constants, but we wish to limit ourselves to qP-qSV motion only, so only four constants need be specified.[10] Our final choice is shown in Table 1. Basically, the structure is the carbonate sequence of Milholland et al.[9] but excluding the uppermost ooze and imposing a constant velocity gradient. The values of the elastic parameter C_{13} were arbitrarily chosen, following the guidelines of Bachman.[6] We admit that the model is unrealistically simple, however, our purpose is to demonstrate only that ignorance of transverse isotropy can lead to identifiable errors; in future when we try to quantify these effects in real data, we shall consider more realistic ocean-bottom structures.

Figure 2 shows traveltime results for qP and qSV refractions within the carbonate sequence and reflections from basement. The water layer has been ignored so that source and receiver are located at the ocean bottom. Note that there is a triplication in the qSV refraction. Figure 2 also shows the traveltime predictions of a tau-sum inversion[11] of qP wide-angle reflections and refractions, and the shear travel times which would be inferred from this inversion using the Milholland et al. values of Poisson's ratio.[9]

Details of the tau-sum inversion are shown in Figure 3. P and S travel times were inverted independently, assuming that the structure is isotropic. Considering first the P inversion, we find that the tau-sum method consistently picks a velocity between the vertical and horizontal velocities, but the depth estimate for the bottom of the carbonates is seriously in error. This result makes intuitive sense: the underlying assumption of the inversion is that at any depth the propagation velocity is the same as the apparent velocity of a ray reaching a turning point at that depth. In our structure this assumption is not true; the velocity along any ray penetrating a particular depth will always be less than the apparent velocity of a ray turning at that depth (i.e., less than the horizontal velocity). To reconcile its correct estimate of the horizontal velocity with too-large travel times, the tau-sum method is forced to overestimate depth. Note that when the basement reflector is sensed, at 925m, the velocity estimate matches perfectly the horizontal velocity at the base of the carbonates. The non-physical reversal of the velocity-depth curve occurs because wide-angle reflections have been included in the inversion. Since reflections everywhere travel at a slower speed than refractions, the interface must be drawn up to satisfy the travel times.

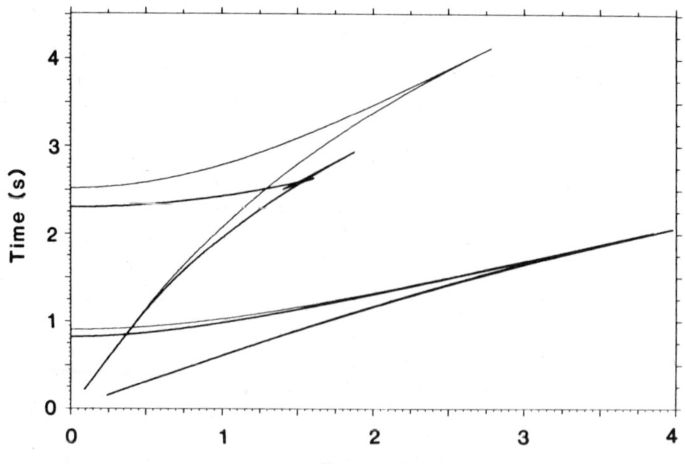

Fig. 2. Travel times for qP and qSV reflections and refractions for the structure of Table 1 (bold). Note that qSV has a triplication. Shown as lighter curves are the travel times predicted by an inversion of the P refractions, and the shear traveltime curve for the same inversion obtained by assuming Poisson's ratio.

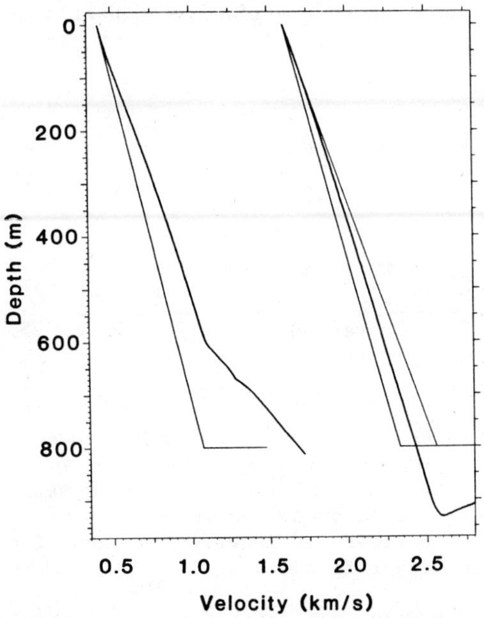

Fig. 3. Tau-sum inversions of the traveltime curves of Fig. 2. The true vertical and horizontal velocities are shown as lighter lines.

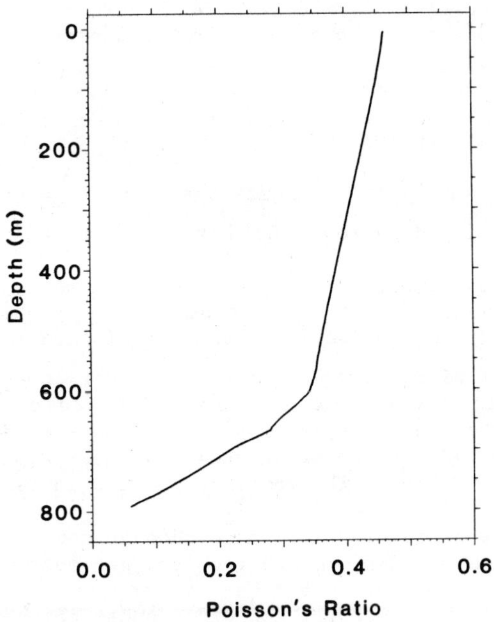

Fig. 4. Poisson's ratio obtained from the inversions shown in Fig. 3.

Table 2. Isotropic Model

Subbottom Depth (m)	Compressional Velocity km/s	Shear Velocity km/s	Density
Water	1.50	0.0	1.0
0	1.60	0.42	1.60
925	2.56	1.18	1.95
Basement	3.00	1.45	2.40

Velocities and density vary linearly from 0 to 925m.

The shear inversion shown in Figure 3, performed independently of the P inversion, has a similar explanation. Now the horizontal velocity is a minimum velocity (see Fig. 1), so the tau-sum method consistently underestimates depth. The reflector is incorrectly placed at 600m depth, and (from wide-angle reflections) is incorrectly identified as region of higher gradient rather than as a reflector.

Using these P and S inversions to compute Poisson's ratio would confirm that the traveltime data have unusual properties. Poisson's ratio inferred from these inversions is shown in Figure 4. It has a sensible variation from the top of the sediments down to 600m, but then rapidly proceeds to unlikely values. The true Poisson's ratio, when measured horizontally, varies from 0.46 at the top of the carbonates to 0.39 at the bottom; when measured vertically, it varies from 0.46 to 0.36 (vertical and horizontal Poisson's ratios are different because the response of a transversely isotropic medium to stretching is different in vertical and horizontal directions). The Poisson's ratio computed from the inversions is everywhere less than the true values.

WAVEFIELD MODELING AND BOTTOM LOSS

Transverse isotropy in sediments has an unknown effect on the reflection loss of a stratified bottom. Oakley and Vidmar[10] have investigated reflection from a homogeneous transversely isotropic bottom, but have not considered the modifying effects of gradients. To see if anisotropy might have any identifiable effects on bottom loss for the carbonate model of Table 1, we have computed reflectivities and performed a wavefield modeling using the techniques of Fryer and Frazer.[12,13] For comparison, we have also considered an "equivalent isotropic model" based on the P traveltime inversion. Parameters of this isotropic model have been chosen to match as closely as possible both the compressional and shear wide-angle reflection and refraction travel times. The model is defined in Table 2.

Synthetic pressure records for the carbonate and isotropic models are displayed in Figure 5. Source and receiver are both located at the ocean bottom and the direct water wave has been supressed. The very large amplification was chosen to enhance the differences between the two responses (the largest amplitude arrival, the P-refraction, is identical for the two models). Apart from the anticipated traveltime differences for the P-reflection from the base of the sediments, the two responses are essentially identical. In particular, note that there is no appreciable burst of energy where the traveltime plot (Fig. 2) suggests we might see shear waves (between 2.5 and 3.0 seconds at about 1.5km range). The reason for the lack of shear energy is that the gradient has acted as an impedance matcher and no shear waves have been excited (note also that there are quite small impedance contrasts across the sediment-basement interface, so shear waves are not excited there either). What little shear energy there is is more noticable on horizontal displacement records.

From the synthetic pressure records we might anticipate that transverse isotropy has negligible effect on the bottom loss, at least at small grazing angles. This is indeed the case. Bottom loss computed from the reflectivities used to produce Figure 5 are displayed in Figure 6. For grazing angles of less than 60° (the region in which P-energy is critically reflected at basement), the bottom loss is everywhere small. The differences between bottom losses for the carbonate and isotropic models for grazing angles greater than 60° are primarily due to the differences in the P-velocity structure of the two models.

We must emphasize here that the claim that transverse isotropy has negligible effect on the bottom loss is valid only for the carbonate model. Similar investigations involving more realistic structures must be performed before the effects of anisotropy on bottom loss can be dismissed. Certainly, the claim of Oakley and Vidmar[10] that transverse isotropy has a definite effect on bottom loss remains valid if the immediate subbottom has appreciable anisotropy. Gradients, however, seem to reduce its effects. We shall investigate this matter further.

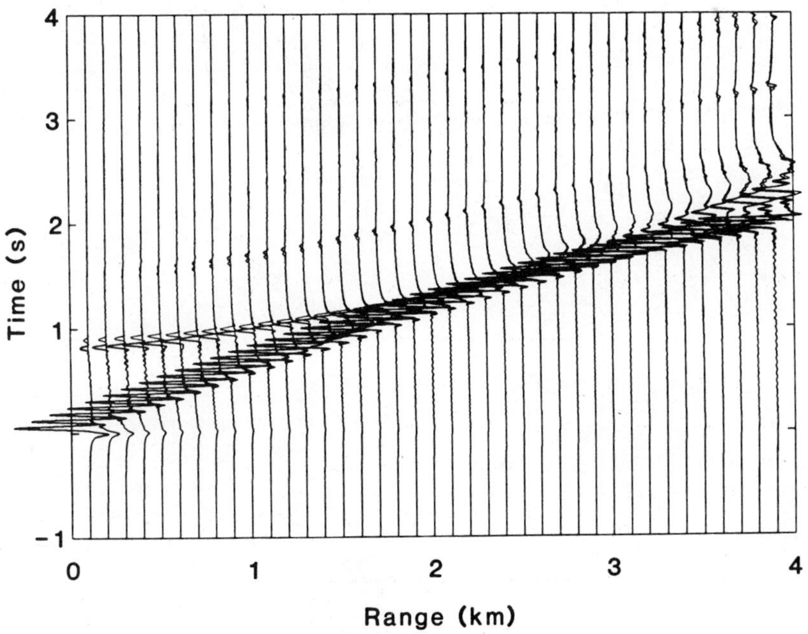

Fig. 5. Pressure response at the ocean-bottom for the carbonate and isotropic models. The source is a pulse of pressure at the ocean bottom. Each trace is actually a doublet, with the carbonate model response shown by a bolder line. The direct ray has been suppressed.

CONCLUSIONS AND DISCUSSION

It is clear from the ray-tracing experiments that shear information is critically important in identifying transverse isotropy; there is little in the compressional wave travel times alone which might indicate a departure from isotropy. If shear wave data were available, a joint inversion of P and S should be possible and should be adequate to quantify the anisotropy. Because of the necessity to record shear arrivals, the use of ocean-bottom seismometers should be encouraged. The results of an investigation of transverse isotropy in sediments using ocean-bottom seismometer data will be reported later.

Regarding bottom loss, it appears that velocity gradients might reduce the effects of transverse isotropy to the point where it can be ignored in marine acoustics. The work reported here can hardly be regarded as definitive, however; further investigations, including analyses of field data, are essential before the effects of transverse isotropy can be dismissed.

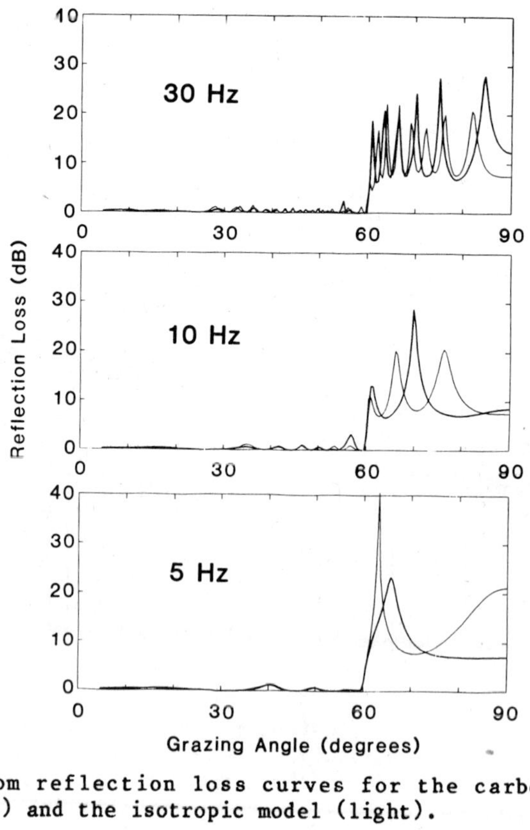

Fig. 6. Bottom reflection loss curves for the carbonate model (bold) and the isotropic model (light).

ACKNOWLEDGMENTS

This research was supported by the ocean acoustics program of the US Office of Naval Research.

Hawaii Institute of Geophysics Contribution no. 1611.

REFERENCES

1. R. L. Carlson, C. H. Schaftenaar and R. P. Moore, Causes of compressional-wave anisotropy in carbonate-bearing, deep-sea sediments, Geophysics, 49:525 (1984).
2. E. L. Hamilton, Sound velocity gradients in marine sediments, J. Acoust. Soc. Am., 65:909 (1979).
3. B. J. Radovich and F. K. Levin, Instantaneous velocities and reflection times for transversely isotropic solids, Geophysics, 47:316 (1982).
4. S. Crampin and B. J. Radovich, Interpretation of synthetic common-depth-point gathers for a single anisotropic layer, Geophysics, 47:323 (1982).
5. R. T. Bachman, Acoustic anisotropy in marine sediments and sedimentary rocks, J. Geophys. Res., 84:7661 (1979).
6. R. T. Bachman, Elastic anisotropy in marine sedimentary rocks, J. Geophys. Res., 88:539 (1983).
7. B. A. Auld, "Acoustic Fields and Waves in Solids," Vol. 1, John Wiley & Sons, New York (1973).
8. V. Cerveny, I. A. Molotkov and I. Psencik, "Ray Method in Seismology," Univerzita Karlova, Praha (1977).
9. P. Milholland, M. H. Manghnani, S. O. Schlanger and G. H. Sutton, Geoacoustic modeling of deep-sea carbonate sediments, J. Acoust. Soc. Am., 68:1351 (1980).
10. D. W. Oakley and P. J. Vidmar, Acoustic reflection from transversely isotropic consolidated sediments, J. Acoust. Soc. Am., 73:513 (1983).
11. J. B. Diebold and P. L. Stoffa, The traveltime equation, tau-p mapping and inversion of common midpoint data, Geophysics, 46:238 (1981).
12. G. J. Fryer and L. N. Frazer, Seismic waves in stratified anisotropic media, Geophys. J. Royal Astron. Soc., 78:691 (1984).
13. G. J. Fryer and L. N. Frazer, Seismic waves in stratified anisotropic media II -- Elastodynamic eigensolutions for some anisotropic systems, in preparation.

SEISMIC ANISOTROPY IN THE UPPER CRUST AT DSDP SITE 504B

Ralph A. Stephen

Woods Hole Oceanographic Institution
Woods Hole, MA 02543

ABSTRACT

Seismic anisotropy in the upper oceanic crust is observed in borehole data obtained at Deep Sea Drilling Project (DSDP) Site 504B on DSDP Leg 92. Particle motion analysis of converted shear wave arrivals from explosive sources at various azimuths reveals a set of patterns which is indicative of hexagonally isotropic structure with a horizontal symmetry axis. There are four diagnostic patterns: 1) Along symmetry axes, where vertically polarized shear waves (SV) are generated, but horizontally polarized shear waves (SH) are not generated, the particle motions are purely vertical; 2) For azimuths at which both SV and SH are generated and the SH velocity is significantly faster than SV, a cruciform pattern with horizontal first motion is observed; 3) For azimuths at which both are generated and the SV velocity is significantly faster than SH, a cruciform pattern with vertical first motion is observed; and 4) for azimuths at which both are generated and SV and SH velocities are similar elliptical particle motions are observed. The shear wave particle motions and compressional wave travel times (from a 2 km radius circle) are consistent with an anisotropic model with hexagonal symmetry. The compressional wave velocity has a two theta variation between 4.0 and 4.8 km/sec. The symmetry axis is horizontal with an azimuth of N20°W ± 10°. The spreading direction at the site (6 my age) is north-south. The observed seismic anisotropy is most probably caused by the preferred orientation of large scale fractures and fissures in upper layer 2 which were created in the early stages of crustal development by near axis extensional processes and normal block faulting.

INTRODUCTION

In this paper, we present shear wave particle motion data which specifically require anisotropic structure in the upper crust. The anisotropy indicated by the shear wave observations is consistent with the anisotropy indicated by compressional wave arrival times for a short range (2 km) circle which also have azimuthal dependence. The observed seismic anisotropy is most probably caused by the preferred orientation of large scale fractures and fissures in upper layer 2 of the oceanic crust. These fractures and fissures were created in the early stages of crustal development by near axis extensional processes and normal block faulting.

Stephen (1981) reported the first observations of seismic anisotropy in upper oceanic crust from borehole data at DSDP Site 417 in the West Atlantic. The two phenomena observed were deviations of compressional wave particle motions from the shot azimuths and shear wave splitting (horizontal polarizations arriving after vertical polarizations). Unfortunately data was only obtained at azimuths parallel and perpendicular to the predicted spreading direction. Continuous variation of compressional wave particle motion deviations and shear wave particle motions with azimuth could not be demonstrated. Crampin et al. (1982) and Shearer and Orcutt (1985) have pointed out that the compressional wave particle motion deviations described in Stephen (1981) were too large to be explained by reasonable anisotropic models, however, the shear wave splitting observations remain as evidence for anisotropy. White and Whitmarsh (1984), Shearer and Orcutt (1985) and Little and Stephen (1985) have subsequently reported observations of anisotropy in the upper oceanic crust based on compressional wave arrival times.

As pointed out by Backus (1965) it is important to distinguish between anisotropy and lateral heterogeneity when studying arrival time data. Typically data sets for the upper oceanic crust have too much scatter in arrival time to unambiguously demonstrate anisotropy. Furthermore, the observation of Stephen (1981) of SH arriving after SV could be attributed to a scattering phenomena rather than anisotropy. Consequently, although seismic anisotropy has been a preferred explanation for a number of observations of this type, it has not previously been specifically required by the data.

The Data Set

The data was acquired at DSDP Site 504B in 6 my old crust on the Nazca Plate in the eastern equatorial Pacific. The receiver was a clamped borehole seismometer which ensured good coupling to true ground motion. Small explosive charges (15 pounds) were fired along radial lines and in circles around the receiver (Figure 1). Data was obtained with the geophone clamped at 42 m, 272 m, 452 m and 667 m into basement. Shot spacing was approximately 500 m.

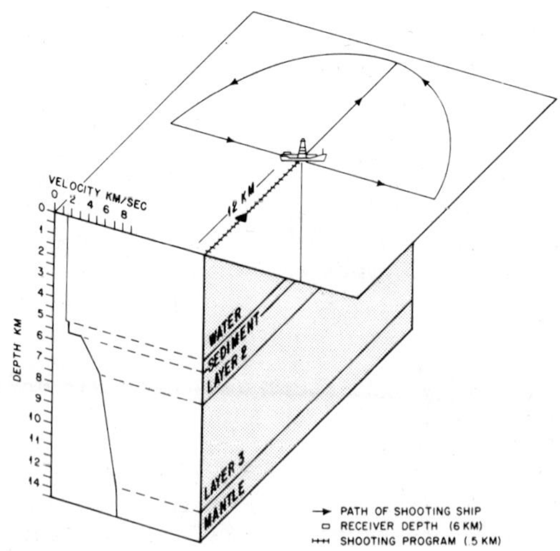

Figure 1. Schematic diagram of a borehole seismic experiment over oceanic crust. Excellent quality three-component seismograms are obtained by clamping a borehole seismometer in the basaltic upper crust. Shot patterns consisting of circles and lines are fired out to a maximum range of 12.0 km.

Ranges were obtained from radar transponders on both the drilling and shooting vessels. Azimuths of the shooting ship were obtained from the drilling ship's radar. Shot instants were transmitted to the drilling vessel from the shooting ship and travel-time corrections were made in the conventional manner. Little and Stephen (1985) describe the experiment in more detail and present the initial interpretation of the data. The site has anomalously high upper crustal velocities of 4.3 k/s for its age (6 my). Bathymetry at the site shows less than 50 meters of relief and is lineated perpendicular to the nominal spreading direction (0°N). Reflection profiling shows that the basement topography has similar relief to the bathymetry with a mean sediment thickness of 275 m.

Figure 2 shows a typical vertical component record section with the three prominent arrivals (P, S and direct wave) indicated. The direct wave is an evanescent wave below the sediment/basement interface (Stephen and Bolmer, 1985) and is only observed at the shallowmost geophone location (42 m). The compressional and converted shear waves follow 'diving', Snell's Law ray paths in the velocity gradient of the upper crust. The evidence for seismic anisotropy discussed in this paper is 1) compressional wave arrival times from the two kilometer radius circle shot to the shallowest geophone position and 2) the particle motions of the converted shear waves.

Compressional wave arrival times versus azimuth

Figure 3 shows the compressional wave arrival times as a function of azimuth for circles of 2, 4, 6 and 8 km radius. The arrival times have been corrected for small changes in range. The two kilometer circle shows an obvious two theta dependence indicative of anisotropy with a slow direction north-south and a fast direction east-west. The larger radius circles do not show any azimuthal dependence but have greater scatter. The greater scatter in these circles can be attributed to: i) greater uncertainty in picking arrival times, (these arrivals correspond to diving energy in the crust, have more emergent wave forms and lower signal-to-noise ratios than the direct wave observed at 2 km range) and ii) more sensitivity to lateral heterogeneities in the crust (the arrivals from larger radius travel farther and deeper in the crust and are hence more effected by lateral velocity changes).

As recommended by Backus (1965), a Fourier decomposition of the arrival times for the two kilometer circle was performed (Table 1). The largest component is two theta which supports the anisotropy hypothesis. The azimuth of the largest component is N20.4°W. The magnitude of the arrival time variations can be explained by a north-south velocity of 4 km/s and an E-W velocity of 4.8 km/sec. The next largest component is one theta which can be attributed to a sloping sea floor immediately above the receiver. The remaining components are of similar magnitude and represent the scatter in the data.

Table 1. Fourier Coefficients for the Compressional Wave Arrival Times for the 2 km Circle

Amplitude (in seconds)		Phase (in degrees)
one theta	0.013	-70.2
two theta	0.016	-20.4
three theta	0.007	-20.7
four theta	0.009	26.6
five theta	0.005	13.8
six theta	0.008	-3.8

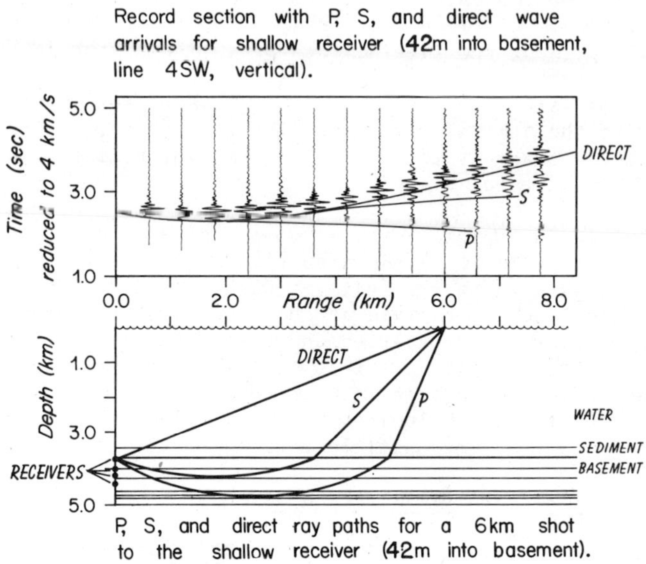

Figure 2. The principle arrivals which occur on the borehole seismograms are shown on this vertical component section. The direct wave root (Stephen and Bolmer, 1985) decays rapidly with depth and is not seen at deeper receivers. Evidence for seismic anisotropy comes from the travel times of P waves and the particle motions of S waves.

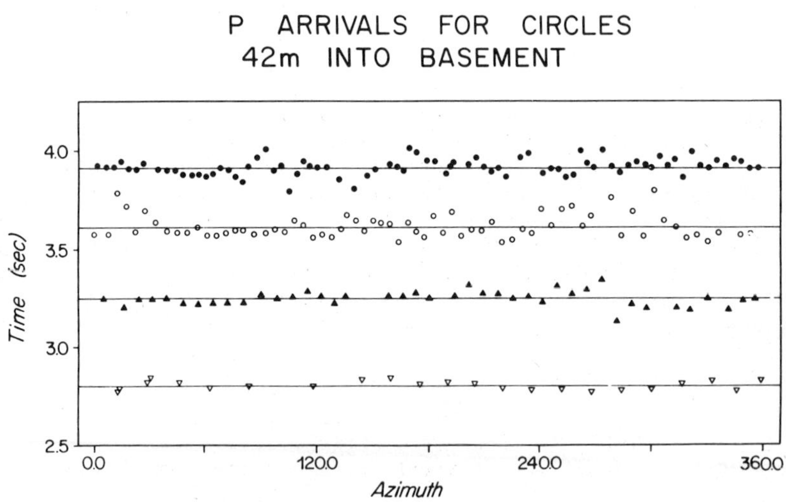

Figure 3. The P wave arrival times versus azimuth at 2 km range (▽) show a clear two theta dependence indicative of anisotropy. At larger ranges (4, 6 and 8 km, ▲, ○, ●) the scatter in the travel times is too large to see any anisotropic effect.

Shear Wave Particle Motions versus Azimuth

The particle motions for converted shear waves generated at the interface between a homogeneous liquid and a hexagonally isotropic solid (with the symmetry axis parallel to the interface) have a distinctive variation with azimuth. In Figure 4 we show compressional and horizontally and vertically polarized shear wave velocities versus azimuth for an isotropic model with hexagonal symmetry. The bottom diagram shows the transmission coefficients for horizontally and vertically polarized shear waves computed by the method of Keith and Crampin (1977). The compressional wave velocity is consistent with the arrival time observations. Along the fast compressional wave axis the compressional to horizontally polarized shear wave transmission coefficient is zero. Shear wave particle motions at this azimuth will be purely vertical. As one moves away from the principle axis horizontally polarized shear waves are generated and they travel faster than vertically polarized shear waves. At these azimuths, shear wave particle motions will be initially horizontal and subsequently break into vertical motion. Between about 20° and 60° from the principle axis the two shear waves have similar conversion coefficients and velocities. Particle motions here will be elliptical or obliquely linear depending on the phase shift between the two waves. Still further around in azimuth, the vertically polarized shear wave will have a faster velocity. First motions will be vertical with subsequent horizontal motions. At the cusp in the SH curve

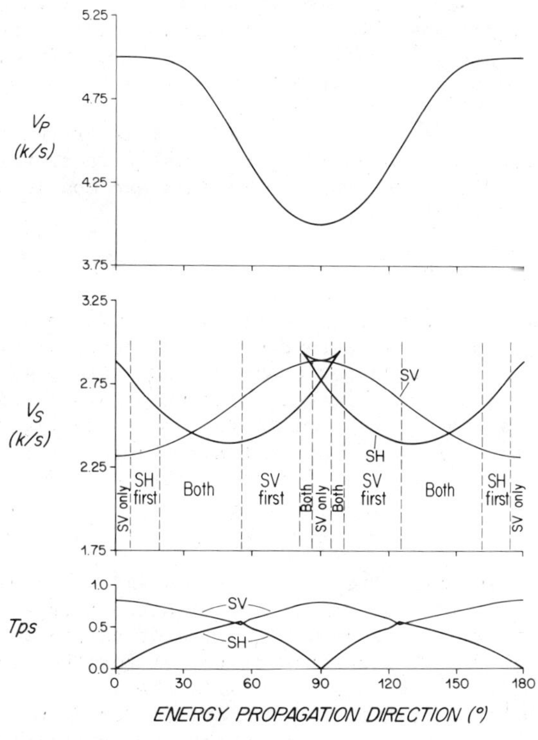

Figure 4. The anisotropic model shown above explains the compressional wave arrival times and shear wave particle motions observed in the data. Curves representing velocities for compressional (Vp), horizontally polarized shear (SH) and vertically polarized shear (SV) are shown in the two frames. In the bottom frame the compressional to shear wave (SV and SH) transmission coefficients for a representative angle of incidence are shown. Horizontally polarized shear waves are not generated along the principle axes.

there may be a small region where both waves arrive together but right at the second principle axis (90°), no horizontally polarized shear waves are generated and one would obtain only vertical particle motions.

All of the above types of shear wave particle motions are observed in the Site 504B data. Figure 5, shows particle motions for compressional and shear waves for the geophone clamped at 452 m in basement. At this depth the geophone is one wavelength below the interface and the effects of basement roughness on particle motions are reduced. The compressional waves are predominantly linear, as expected, but the shear waves show the above azimuthal patterns. The various types of particle motions can be identified on these plots.

The 'SH first' diagram at this depth is convincing evidence for anisotropy as opposed to scattering. If SH is being generated by scattering but is propagating in an isotropic medium it will always arrive at the same time or later than SV. However, if SH is being generated by conversion into an anisotropic medium and it propagates in an anisotropic medium there will be some azimuths at which it travels faster and arrives before SV.

The shear wave arrivals for the circle of shots at six kilometer radius were identified as belonging to one of the four groups (SV only, SH first, SV first, SV and SH together) and the polarization type is shown as a function of shot location in Figure 7. The outer circle of symbols in Figure 7 indicates the observed particle motion. The inner circle shows the predicted particle motions based on Figure 4 but rotated to fit the observations as well as possible. The correspondence of the predicted and observed shear wave particle motions is extremely convincing evidence for the presence of seismically anisotropic structure in the upper crust.

Also indicated on Figure 7 are the directions of the fast and slow compressional wave axes based on the compressional wave travel times of Figure 3. The shear wave and compressional wave data are consistent with the same anisotropic model to within 10°.

Discussion and Conclusions

Compressional wave arrival times at short ranges and shear wave particle motions as functions of azimuth demonstrate conclusively that seismic anisotropy is present in the upper crust at DSDP Site 504. The observation of horizontally polarized shear waves occuring before vertically polarized shear waves requires some form of anisotropy. Scattering effects can generate SH, but they cannot account for SH traveling faster than SV. This data set resolves the earlier question (raised by Stephen, 1981) as to whether SH in the oceanic crust is primarily a scattering effect or an anisotropic effect.

Both the compressional wave arrival times and the shear wave particle motions are consistent with a transversely isotropic upper crust with a horizontal symmetry axis. The slow direction of compressional waves is within 20° of the predicted spreading direction for the crust at this site. Compressional wave velocities are 4.0 k/s in the slow direction (spreading direction) and 4.8 k/s in the fast direction (perpendicular to the spreading direction). The compressional and shear wave velocities for the anisotropic model (4.0-4.8 k/s, 2.3-2.8 k/s) bracket the velocities obtained for upper crust under the isotropic assumption using travel-time analysis and reflectivity method synthetic seismogram analysis (4.56 k/s, 2.63 k/s). Both sets of observations suggest that the anisotropy is restricted to the upper 500 m of crust.

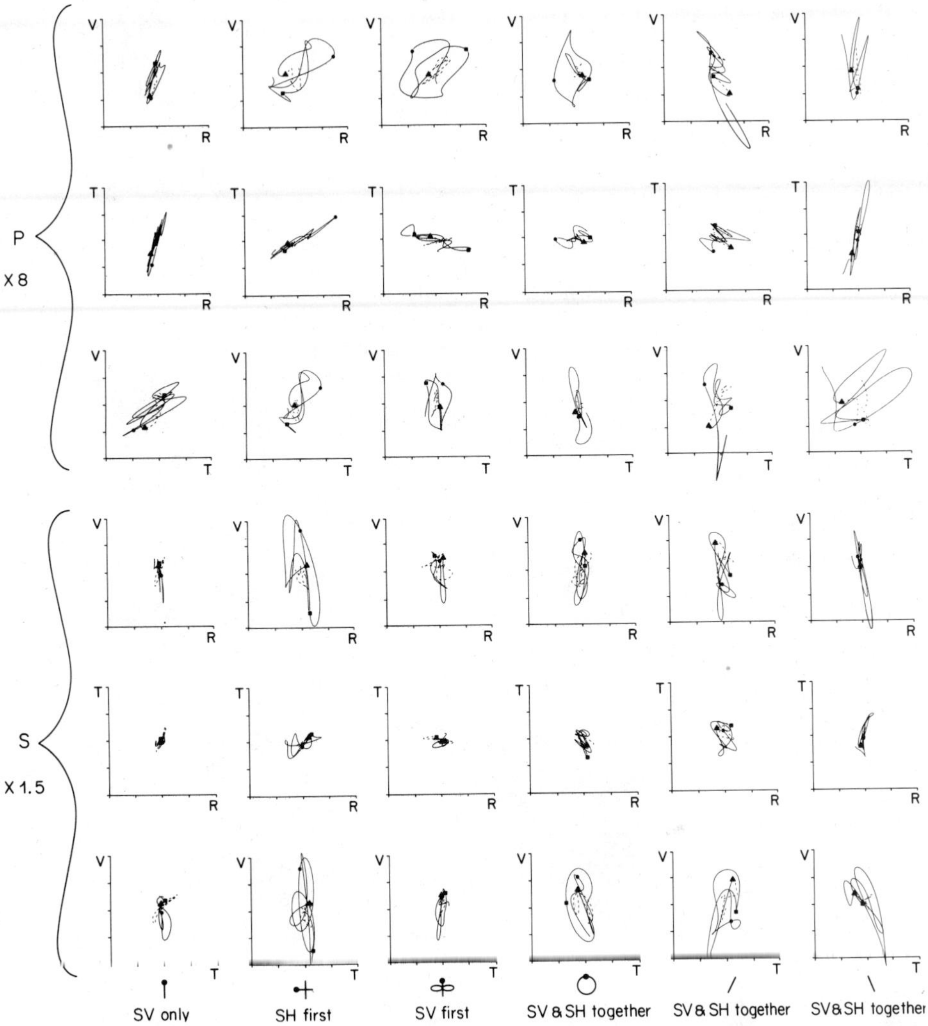

Figure 5. Complete sets of particle motion diagrams are shown for compressional and shear arrivals. (Axes are vertical (V), radial (R) and transverse (T).) The compressional waves are predominantly linear, however the shear wave arrivals can be linear or cruciform depending on azimuth. The key for the plots is shown in Figure 6.

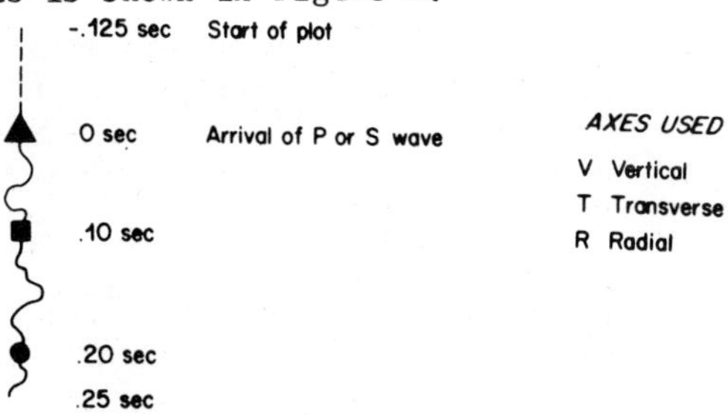

Figure 6. The key for the particle motion plots in Figure 5. The dashed line starts 0.125 seconds before the arrival and gives an indication of background noise levels and motions.

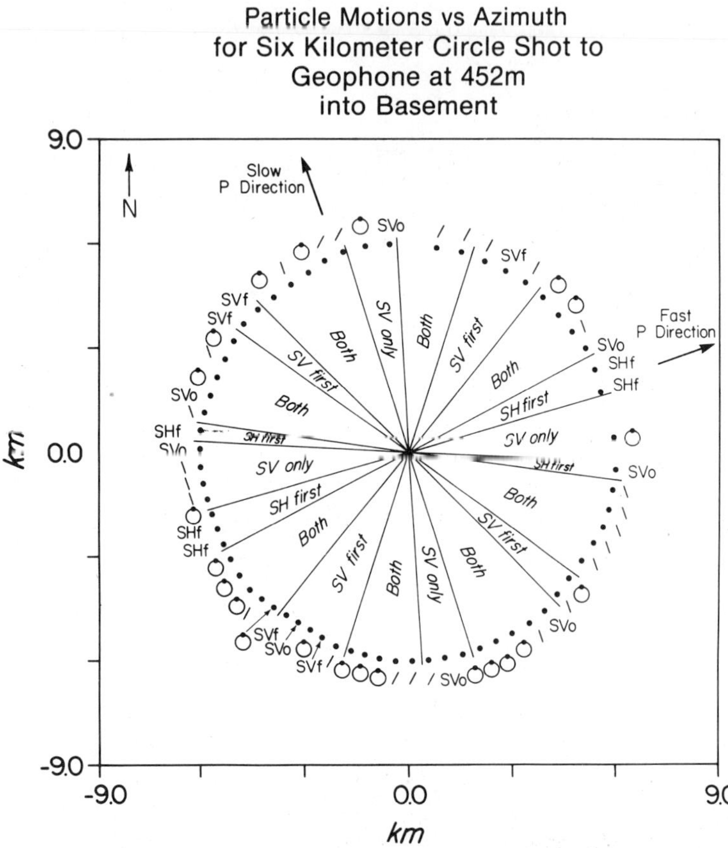

Figure 7. The type of shear wave particle motion is shown at each shot location for a circle of shots at 6 km range. The inner circle shows the particle motions which would be expected for the anisotropic model of Figure 4. Excellent agreement between observed and predicted particle motions is obtained. The directions of fast and slow P wave velocities (Figure 3) are also shown. They agree with the shear wave particle motion data to within 10°.

What can cause anisotropy in the upper crust? The upper crust is identified by drilling to be pillow basalts and recovered samples are isotropic. A promising explanation is that remanents of the faults and fissures which are observed in young crust persist in older crust and provide well-oriented planes of elastic weakness. These faults and fissures would not influence the elastic properties of the small scale hand samples recovered from the hole. However, they would have a significant effect on the larger scale seismic properties of the crust. The anisotropy observed at Site 504 is qualitatively in agreement with this explanation. Apparently the effects of the remanent fissures and faults diminishes with depth. As yet there is no observable anisotropy deeper than 500 m. Although one would expect the sheeted dike sequence below 600-700 m to also be anisotropic there is no indication of this in the data.

Acknowledgements

I would like to thank Tom Bolmer for his invaluable assistance in data reduction and Pam Foster for laying out and typing the manuscript. This work was supported by the U.S. National Science Foundation (Contract No. OCE81-17571 and OCE81-09155).

References

Backus, G.E., 1965, Possible forms of seismic anisotropy of the uppermost mantle under oceans, J. Geophys. Res., 70, 3429-3439.

Crampin, S., R.A. Stephen and R. McGonigle, 1982, The polarizations of P-waves in anisotropic media, Geophys. J.R. Astr. Soc., 68, 477-485.

Keith, C.M. and S. Crampin, 1977, Seismic body waves in anisotropic media: reflection and refraction at a plane interface, Geophys. J.R. Astr. Soc., 49, 181-208.

Little, S.A. and R.A. Stephen, 1985, Costa Rica Rift borehole seismic experiment, Deep Sea Drilling Project Hole 504B, Leg 92, In: Anderson, R.N., Honnorez, J., Becker, K., et al., Init. Repts. DSDP, 83: Washington (U.S. Govt. Printing Office), 517-528.

Shearer, P. and J. Orcutt, 1985, Anisotropy in the oceanic lithosphere-theory and observations from the Ngendie seismic refraction experiment in the south-west Pacific, Geophys J.R. Astr. Soc., 80, 493-526.

Stephen, R.A., 1977, Synthetic seismograms for the case of the receiver within the reflectivity zone, Geophys. J.R. Astr. Soc., 51, 169-181, 1977.

Stephen, R.A., 1981, Seismic anisotropy observed in upper oceanic crust, Geophys. Res. Lett., 8, 865-868, 1981.

Stephen, R.A. and S.T. Bolmer, 1985, The direct wave root in marine seismology, Bull. Seism. Soc. Am., 75, 57-67.

White, R.S. and R.B. Whitmarsh, 1984, An investigation of seismic anisotropy due to cracks in the upper oceanic crust at 45°N, Mid-Atlantic Ridge, Geophys. J.R. Astr. Soc., 79, 439-468.

A BROAD-BAND STUDY OF ATTENUATION IN OCEAN BOTTOM SEDIMENTS

Jeffrey S. Barker

Woodward-Clyde Consultants
566 El Dorado St.
Pasadena, CA 91101

Donald V. Helmberger

Seismological Laboratory
California Institute of Technology
Pasadena, CA 91125

ABSTRACT

Seismic pulses trapped between the ocean surface, bottom, and the subbottom structure commonly observed in refraction profiling may be modeled synthetically. The relative amplitudes of single- and multi-bounce phases are used to study reflection coefficients and effective attenuation. This technique was used to investigate a thick section of mature sediments located in the Bering Sea. Attenuation is quite low (Q > 500) for pulses dominated by 10 Hz energy, but becomes increasingly important at higher frequencies.

The broad-band properties of effective attenuation were modeled by applying a relative waveform analysis procedure for PP and PPP phases. An absorption-band attenuation model with frequency-dependent Q, parameterized by t^* and τ_M, yields Q of 70 above 32 Hz with no appreciable absorption at lower frequencies. Since absolute amplitudes are not considered, this analysis is somewhat insensitive to the specific frequency dependence of Q. However, by interpreting the broad-band results along with the long-period results, for which absolute amplitudes were modeled, a frequency-dependent Q model is required.

INTRODUCTION

In this study, we investigated the effective attenuation properties of the subbottom using refraction data from the Leapfrog expedition in the Bering Sea[1]. The thick, sedimentary basin structure of the Bering Sea is optimal for this type of study, since the various subbottom phases may be isolated in time. A sample of the data is shown in Figure 1. These responses are the output of three hydrophones at slighltly different ranges and depths. In addition to the broad band signals, the three data channels are low-passed (LP) and high-passed (HP). We began with a foreward synthetic modeling approach to determine the

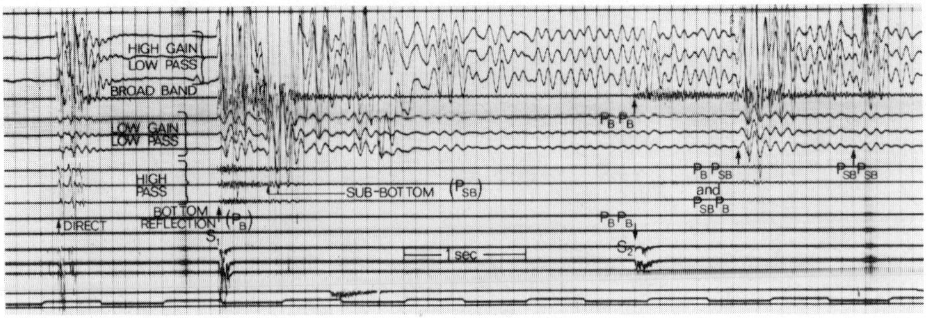

Fig. 1. An example a refraction record taken in the Bering Sea at a range of 10 km. The traces labeled S_1 and S_2 are rectified output used for identifying waterbound arrivals. The other labels are identified in the text and are displayed geometrically in Figure 2.

characteristics of propagation and attenuation for the long-period data. This was followed by a relative waveform analysis of broad-band, multiply-reflected phases that differ by one traverse of the subbottom, in order to obtain a differential attenuation operator for the subbottom.

MODELING THE LONG-PERIOD BEHAVIOR

The objective of an earlier study by Helmberger, et al.[2] was to model the first two bottom reflections in order to determine the density

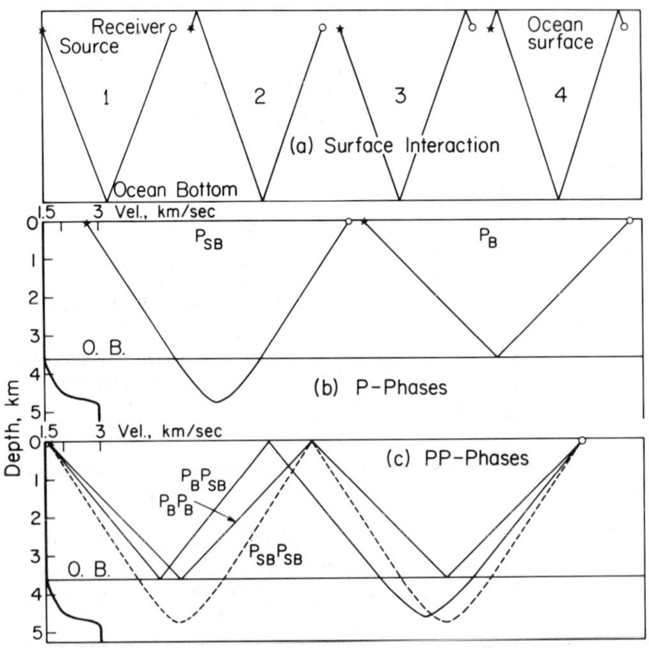

Fig. 2. (a) Ray paths used in the construction of the source transfer function. (b) Paths followed by the phases P_B and P_{SB} appropriate for a range of 7 km in the velocity model displayed on the left. (c) Similar paths for the PP phases at a range of 14.3 km.

structure of the Bering Sea. Synthetic seismograms were computed using generalized ray theory, in conjunction with the Cagniard-deHoop method[3], and a layered model was obtained such that the synthetics agreed with the long-period observations. Following the same technique, we have modeled the behavior of the second and third bounces as a check of the effectiveness of the model. As illustrated in Figure 2, we denote compressional waves that reflect from the water-sediment interface as P_B, and those that traverse the subbottom as P_{SB}. The multiple reflections include phases such as P_BP_B, P_BP_{SB}, $P_{SB}P_{SB}$, etc.

A record section constructed from one of the low-gain, LP channels is presented in Figure 3, where the traces are aligned with respect to the ocean bottom response (P_B). The synthetic record section, plotted on the same scale, is shown in Figure 4. Due to the subbottom triplication, P_{SB} becomes strong near 7 km, $P_{SB}P_{SB}$ near 14 km, and $P_{SB}P_{SB}P_{SB}$ near 21 km. The PP phase contains P_BP_B, (P_BP_{SB} + $P_{SB}P_B$) and $P_{SB}P_{SB}$, while the PPP phase contains $P_BP_BP_B$, ($P_BP_BP_{SB}$ + $P_BP_{SB}P_B$ + $P_{SB}P_BP_B$), (P_BP_{SB} + $P_{SB}P_BP_{SB}$ + $P_{SB}P_{SB}P_B$) and $P_{SB}P_{SB}P_{SB}$. In the synthetics, the phases with mixed interactions are assumed to have equivalent paths, but in the data, these are probably slighltly different. The P_BP_B and $P_BP_BP_B$ phases are quite small due to the low value of the reflection coefficient at the water-mud interface (R = .2). The phases with mixed interactions reach critical angle before the subbottom reflections, but are only slightly smaller because the redundancy factors of 2 and 3 due to the multiple arrivals for PP and PPP, respectively, offset the low water-mud reflection coefficient. In general, the synthetics predict the timing and relative amplitudes of the observed data quite accurately. Although we will not discuss PPPP in this paper, the observations of this phase and higher multiples are quite apparent in the data.

A comparison of the P, PP and PPP observations with synthetics computed for various assumptions of Q is presented in Figure 5. The

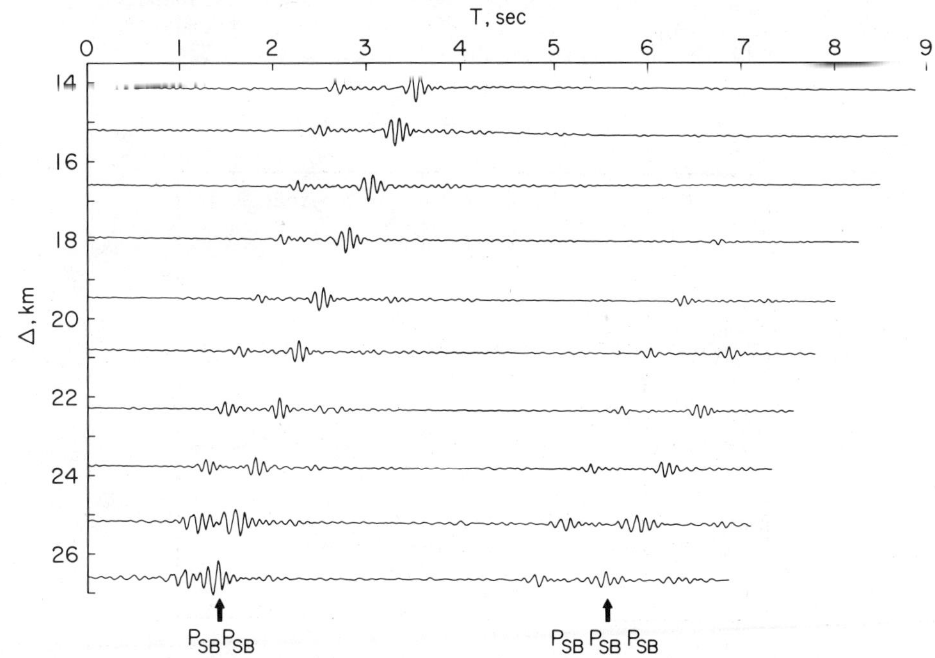

Fig. 3. Record section constructed from the low-gain, long-period channels at ranges appropriate for the PP and PPP phases. The traces have been aligned with repect to the P_B arrival.

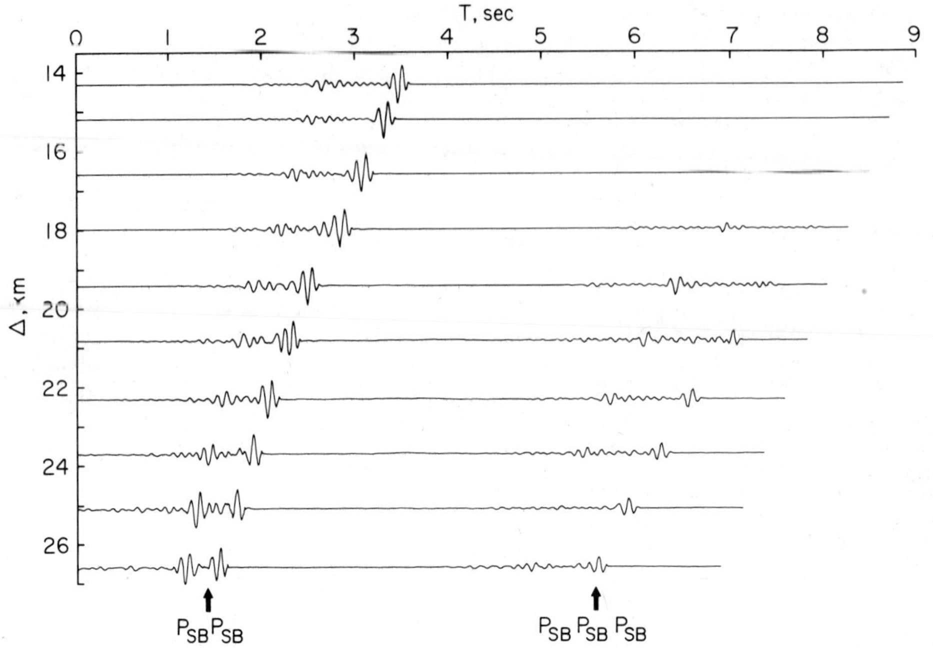

Fig. 4. Record section constructed from long-period synthetics. The amplitude and time scales are the same as those in Figure 3.

traces are plotted on common time and amplitude scales. In these figures we assumed a uniform, frequency-independent Q throughout the sedimentary column and used the Futterman[4] Q operator. The P_B phase does not change since we assume no attenuation in the water at these frequencies. The phases P_{SB} and $(P_BP_{SB} + P_{SB}P_B)$ have nearly the same attenuation because of the similarity in the length of the subbottom path. The PPP phases follow similar patterns, with $P_{SB}P_{SB}P_{SB}$ being the most sensitive to Q.

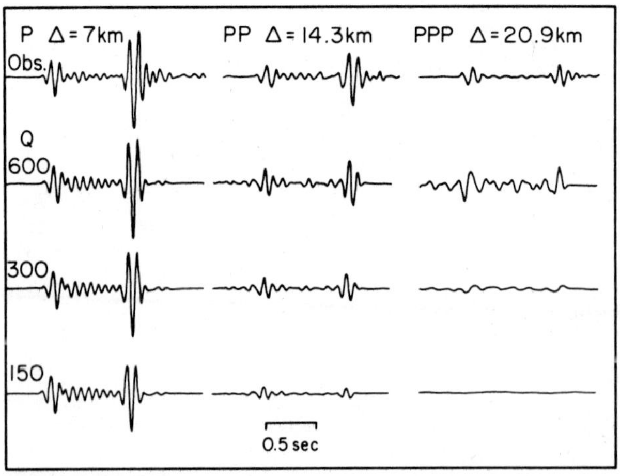

Fig. 5. Comparison of the observed long-period P_{SB}, $P_{SB}P_{SB}$ and $P_{SB}P_{SB}P_{SB}$ phases with synthetics for various degrees of attenuation.

After comparing the results of Figure 5 and the many ratios of possible redundancies, we conclude that the 10 Hz energy is propagating essentially elastically (Q > 500), with relatively little energy lost due to attenuation. However, the relative frequency content of the broad-band, multiple-bounce signals shows considerable change, suggesting significant attenuation at higher frequencies.

BROAD-BAND ANALYSIS

The broad-band traces for the PP and PPP observations (as displayed in Figure 2) were photographically enlarged and hand digitized. An example of the PP data is displayed in Figure 6 along with Fourier spectral estimates of the three phases P_BP_B, $(P_BP_{SB} + P_{SB}P_B)$ and $P_{SB}P_{SB}$ (Barker and Helmberger[5] includes similar plots for all of the observed PP and PPP data sets). The P_BP_B phase is obviously of much higher relative frequency content than the other two. The phase $P_{SB}P_{SB}$ seems to be the most depleted in high frequencies. These features are expected in the

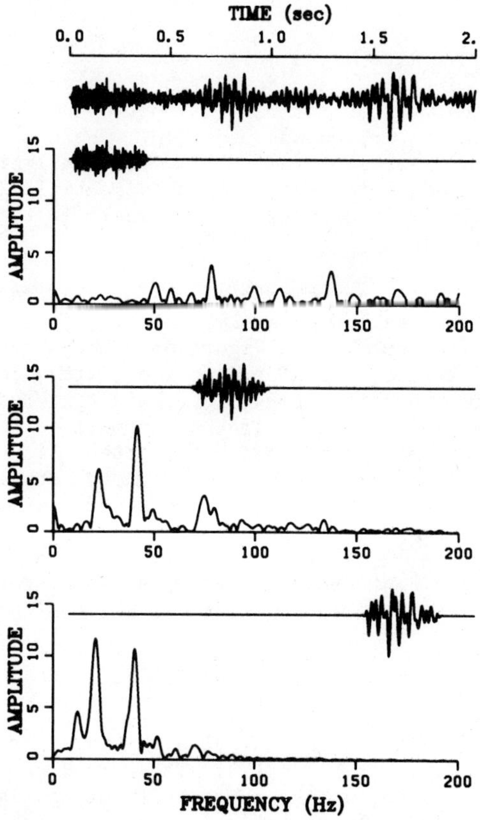

Fig. 6. An example of Fourier spectral estimates for the PP phases. Shown at the top is the broad-band signal for shot 766 (14.3 km). Below this are the windowed seismograms and amplitude spectra for the P_BP_B, $(P_BP_{SB} + P_{SB}P_B)$ and $P_{SB}P_{SB}$ phases.

presence of absorption and/or scattering effects within the subbottom. However, making useful estimates of the spectral fall-off and corresponding Q values proves difficult given the strong periodicities in the data.

Another difficulty in the spectral analysis for Q is the fact that the spectral peaks change from shot to shot. Some of this variability is probably caused by lateral variations of the bottom structure itself (asymmetries in P_BP_{SB} and $P_{SB}P_B$ for example), but much of it is caused by the recording practice. Refraction shots are conventionally detonated at depths appropriate to enhance 20 Hz energy by constructive interference with the ocean-surface reflections. In addition, the hydrophones are placed roughly at the proper depths for a similar enhancement. The source excitation may be computed from the equations given by Weston[6], including the shock wave and bubble pulses. The response computed with the various surface reflections can account for much of the periodicity observed in the data. This idealized response is probably appropriate for describing the longer periods, but since the source-receiver geometry varies from shot to shot in an unpredictable manner, a forward synthetic calculation was deemed unfeasible. Instead, we decided to measure differential attenuation in the subbottom using a relative waveform method which would implicitly correct for the unknown source and receiver characteristics.

MEASUREMENT PROCEDURE

The P_BP_B and P_BP_{SB} data sets discussed in the previous section present some clear difficulties to further analysis. Identifying the exact onset of the arrivals of interest is not always possible, which makes time windowing an ambiguous operation. The coda arriving between these two major pulses represents noise in our analysis. A relative waveform analysis procedure for measuring differential Q, which is stable in the presence of timing errors and noisy signals is required. Burdick[7] developed such a procedure.

The method relies on a normalized cross-correlation coefficient for comparing waveforms to each other analytically. First we window out the phases in question, as displayed in Figure 6. The phase (P_BP_{SB} + $P_{SB}P_B$) is filtered by a suite of transfer functions, T(t), corresponding to a reasonable range of attenuation operators. This filtered phase is then compared to the phase $P_{SB}P_{SB}$ using a normalized cross-correlation procedure to determine the optimal effective attenuation operator for the subbottom. The procedure is repeated in order to obtain estimates of attenuation for the PPP phases by cross-correlating the filtered ($P_BP_{SB}P_{SB}$ + $P_{SB}P_BP_{SB}$ + $P_{SB}P_{SB}P_B$) phase with the $P_{SB}P_{SB}P_{SB}$ phase.

The expression for the cross-correlation norm for comparing two waveforms, a(t) and b(t), defines a best fit between them as the maximum of

$$N[a,b] = a(t)\ \underline{cc}\ b(t)$$

where $\underline{cc}$ represents cross-correlation. The signals are normalized so that Max {N[a,a]} = Max {N[b,b]} = 1.0. The norm can be recognized as a least-squares norm evaluated at the phase shift of maximum correlation. When applied to the signals directly, the norm is unstable for complex (multi-peaked) data because the wrong peaks may begin to correlate with each other. However, when a(t) and b(t) are auto-correlations of the signals rather than the signals themselves, the central auto-correlation peaks are always compared to each other. In this way, errors due to incorrect alignment of the signals are avoided. Similarly, by normalizing the signals, frequency-independent amplitude differences between the signals are removed. It is important to emphasize that the normalized cross-correlation coefficient being used defines a best fit in the time and frequency domains, simultaneously. Therefore, though we display our results in the time domain, our conclusions would not change if we had chosen to display their spectra instead.

ATTENUATION OPERATORS

The Futterman[4] attenuation operator has been used extensively in seismology. It assumes a constant Q within the frequency band of the data. Aside from a phase shift designed to retain causality, the operator consists of an exponential decay in amplitude of the form $\exp(-\omega T/2Q)$, where ω is circular frequency ($=2\pi f$), T is the travel time, and Q is the seismic quality factor. The operator is commonly parameterized by a single parameter, $t^* = T/Q$, and may be related to Hamilton's[8] k parameter by $8.686\ \pi t^* = kL$, where L is the path length in m, t^* has units of s, and k has units of dB/Hz/m.

The existence of multiple reflections such as PPP and PPPP, and the long-period modeling results discussed above indicate that Q is very high (very little attenuation) at low frequencies. However, the broad-band analysis indicates a significant degree of absorption of high frequencies each time the subbottom is traversed. This suggests that a differential attenuation operator with frequency-dependent Q is necessary for the subbottom. One such operator is the standard linear solid operator of Minster[9], for which Q is constant within a limited frequency band and increases for frequencies outside of the band. This operator is parameterized by t^* for the constant-Q band, τ_M ($=1/\omega_M$) defining the low-frequency limit of this band, and τ_m ($=1/\omega_m$) for the high-frequency limit. We require high Q for low frequencies, and a constant low-Q for frequencies above some limit. Therefore, we will solve for t^* and τ_M, assuming τ_m is outside of the high-frequency range of the data. Note that as τ_M approaches the low-frequency end of our data, the standard linear solid operator approaches the Futterman operator.

In testing the possible differential Q operators, we allow a range in t^* of 0.000 - 0.020 s, in increments of 0.001 s. Assuming an average two-way travel time in the subbottom of 0.70 s, this corresponds to a range of from 35 to essentially infinite Q. We fix $\tau_m = 0.0003$ s and allow τ_M to range from 0.000 to 0.012 s in increments of 0.001 s.

RESULTS

Figure 7 provides an example of the measurement procedure and the results. Shown as trace A is the ($P_BP_{SB} + P_{SB}P_B$) phase for shot 766, located 14 km from the receiver. Trace B is the $P_{SB}P_{SB}$ phase for the same shot. The trace labeled A*Q is the convolution of trace A with the best-fit standard linear solid attenuation operator. For this case, $t^* = 0.018$ s and $\tau_M = 0.003$ s. The time lag apparent in the third trace is due to an arbitrary time shift in the attenuation operator, but has no effect on the final result. The fourth trace is the auto-correlation of the attenuated trace, A*Q, and the fifth trace is the auto-correlation of trace B. The best-fit solution is obtained by stepping through the $t^*-\tau_M$ parameter space and choosing the result that maximizes the normalized cross-correlation coefficient between the bottom two traces. For this example, the maximum of the cross-correlation coefficients was 0.784. In an ideal situation, the perfect solution would give a cross-correlation coefficient of 1.000.

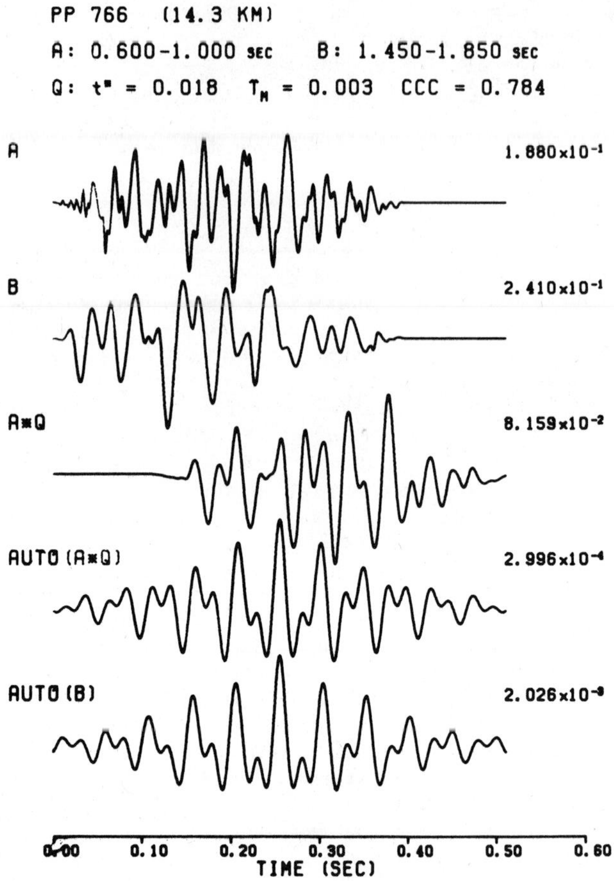

Fig. 7. Example of the measurement procedure and the results. See text for explanation.

There is significant trade-off between the parameters t^* and τ_M. The resolution of the solution may be investigated by inspecting the measure of the quality-of-fit (the cross-correlation coefficient) as a function of the model parameters. Figure 8 shows such an analysis for the solution obtained in the previous figure. τ_M increases to the right and out from the paper, so the front-right face of the figure is essentially the quality of fit of the constant-Q Futterman operator as a function of t^*. The solution space is fairy broad, but is peaked at about $t^* = 0.010$ s. As τ_M decreases, or as we move to the back and left, the corner of the attenuation filter moves to higher frequencies, so t^* increases to compensate. That is, high frequencies may be attenuated by a suite of operators ranging from a relatively gentle decay across the entire frequency band to a more severe decay that begins only at higher frequencies. The best-fit attenuation operator determined for this case actually corresponds to the highest point on the ridge in the back-right portion of the plottted parameter space. However, any of the solutions

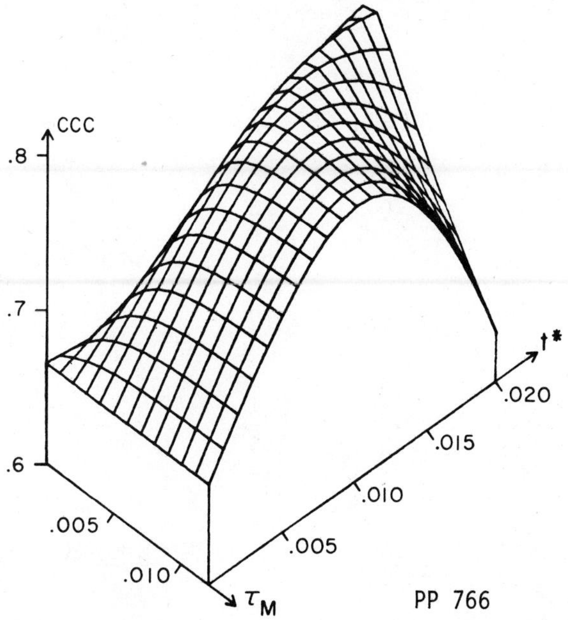

Fig. 8. The normalized cross-correlation coefficient for the relative waveform analysis on the PP observation from shot 766, plotted as a function of the effective attenuation parameters t^* and τ_M.

on the curving ridge might be considered equally valid. For this reason, we present results for the best-fit standard linear solid operator and also for the approximate Futterman operator, determined as the best-fitting t^* for the maximum value of τ_M, in Table 1. Plots of the waveforms and their auto-correlations, along with the plots of the solution as a function of t^* and τ_M, are included in Barker and Helmberger[5] for all of the PP observations analyzed as part of this study.

For the most part, the scatter in the results may be attributed to the trade-off between t^* and τ_M, but some data pairs yield poor cross-correlations or anomalous results. This may reflect laterally varying reflection or transmission properties within the subbottom or anomalous interference of the multiple arrivals. Such cases, if not apparent by an inspection of the waveforms, may be identified by the unususal shape of the solution plot in $t^*-\tau_M$ space. Excluding these anomalous results, average differential attenuation operators for the subbottom may be obtained from the results of Table 1. For the Futterman operator, this gives $t^* = 0.005$, or $Q = 140$, independent of frequency. Similarly, the standard linear solid operator gives $t^* = 0.010$ s and $\tau_M = 0.005$ s, or $Q = 70$ above 32 Hz. For path lengths in the subbottom of the order of 3000 m, these values of t^* would correspond to k values of .045 and .091 dB/kHz/m. Although these values were obtained assuming a constant Q with depth, they are slightly smaller than the average of the depth-dependent values found by Hamilton[10] and by Jacobson, et al[11], but are significantly larger than the results of Mitchell and Focke[12].

If the frequency-dependent attenuation operator obtained by this procedure sufficiently accounts for the absorption and scattering properties of the subbottom, a similar solution should be obtained for PPP phases as for PP phases. Figure 9 shows the cross-correlation

Table 1 - Average Differential Attenuation Results

			Futterman[4]		Minster[9]		
	Shot	R (km)	t^* (s)	Norm	t^* (s)	τ_M (s)	Norm
PP							
	766	14.3	.010	.779	.018	.003	.784
	767	15.2	.003	.763	.003	.011	.764
	768	16.6	.020**	.714	.020**	.010**	.716
	769	18.0	.007*	.584	.020*	.002*	.603
	770	19.5	.003	.745	.010	.001	.747
	771	20.9	.012*	.358	.020*	.003*	.414
	average		.005		.010	.005	
PPP							
	770	19.5	.001	.642	.005	.001	.643
	771	20.9	.003	.855	.003	.012	.855
	772	22.4	.020**	.729	.020**	.010**	.731
	773	23.8	.002	.751	.002	.012	.751
	774	25.2	.000**	.742	.009**	.000**	.758
	775	26.7	.005	.672	.005	.012	.672
	average		.003		.004	.009	

* Poor correlation between phases; not included in average.
** Anomalous solution; not included in average.

results for the filtered ($P_BP_{SB}P_{SB}$ + $P_{SB}P_BP_{SB}$ + $P_{SB}P_{SB}P_B$) phase compared to the $P_{SB}P_{SB}P_{SB}$ phase for shot 771, located 21 km from the receiver. The first two bottom reflections in these phases follow essentially the same paths as the PP phases for shot 766 (14 km), shown in Figure 7. The additional traverse of the subbottom, however, means that high frequencies have already been substantially attenuated, and there is little difference in frequency content between the PPP pairs. As such, the PPP phases are somewhat less sensitive to variations in the parameters of the differential attenuation operator. The plot of the cross-correlation coefficient as a function of t^* and τ_M for this case is shown in Figure 10. Although the general shape of the solution in the parameter space is similar to that of the PP phase (Figure 8), it is quite broad, indicating poor resolution of the parameters. Plots of the PPP waveforms, their auto-correlations, and the plots of the solution in $t^*-\tau_M$ space are included in Barker and Helmberger[5]. The best-fit attenuation operator parameters are listed in Table 1. The solutions are generally similar to those of the PP phases, although the scatter in the parameters is greater. Once again, anomalous results were obtained for two of the PPP observations. Excluding these results, the average Futterman operator for the PPP phases has $t^* = 0.003$, or $Q = 233$, independent of frequency. Similarly, the average standard linear solid operator for the PPP phases has $t^* = 0.004$ s and $\tau_M = 0.009$ s, or Q = 175 above 18 Hz in the subbottom layer. These t^*s translate into k values of .027 and .036 dB/kHz/m.

SUMMARY AND CONCLUSIONS

Long-period, multiple reflections from the ocean bottom in the Bering Sea can be accurately modeled with synthetic seismograms. A comparison of P, PP and PPP phases with synthetics indicates that the 10 Hz energy propagates essentially elastically ($Q > 500$) in the subbottom. However, changes in frequency content in the broad-band data with

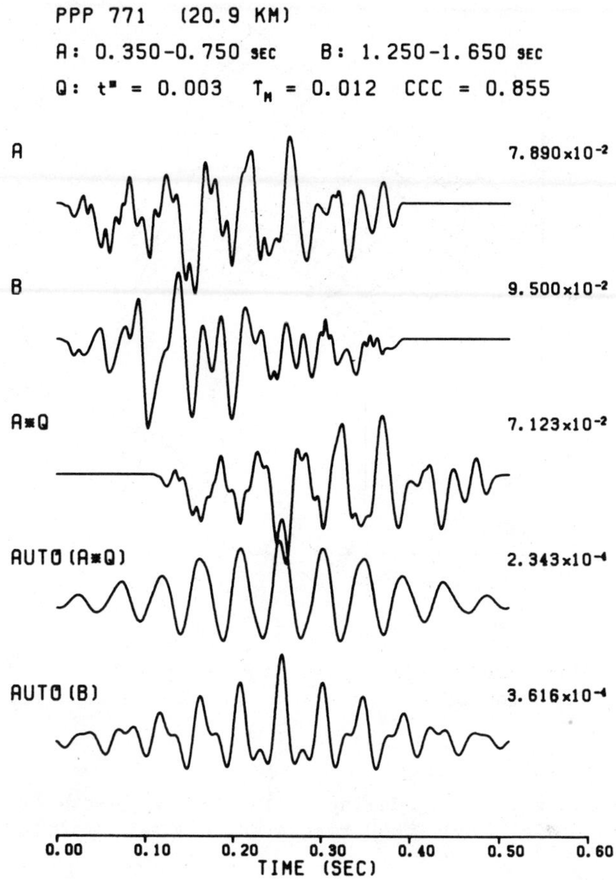

Fig. 9. The relative waveform analysis results for PPP phases from shot 771.

successive traverses of the subbottom suggests that Q in the subbottom must be frequency dependent. Foreward modeling of the broad-band data is quite difficult due to periodicities in the data which vary from shot to shot. We have, therefore, applied a relative waveform analysis procedure to determine an average differential attenuation operator for the subbottom. For PP phases, we obtain Q = 70 for frequencies above 32 Hz, and for PPP phases, we obtain Q = 175 above 18 Hz. Due to the insensitivity of the relative waveform analysis to absolute amplitudes, the frequency dependence of Q is not well resolved from the broad-band data. However, the long-period data, for which absolute amplitude was modeled, demand a frequency-dependent Q.

ACKNOWLEDGEMENTS

This work was supported by NORDA under contract N00014-82-C-0192. We gratefully ackowledge the efforts of David Cole and Stephen Hartzell

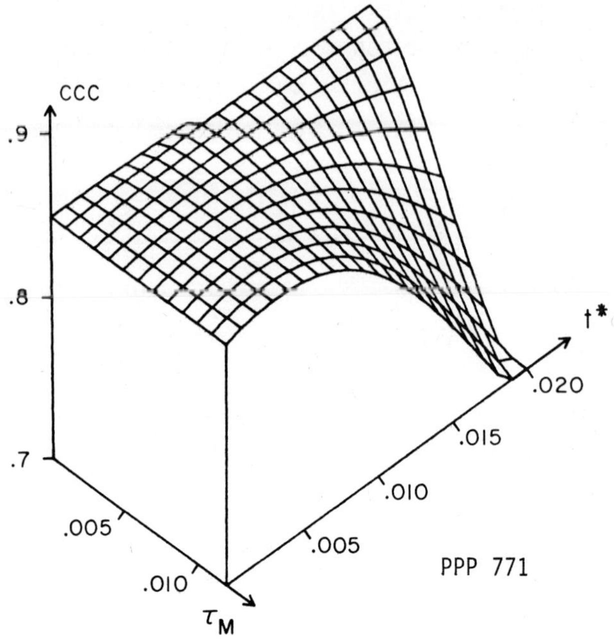

Fig. 10. The normalized cross-correlation coefficient as a function of t^* and τ_M for the PPP phases from shot 771.

on long-period waveform modeling during the early stages of this study. Cindy Arvesen digitized the broad-band data and drafted several of the figures.

REFERENCES

1. G. G. Shor, Jr., Structure of the Bering Sea and the Aleutian Ridge, Marine Geol., 1:213-219 (1964).

2. D. V. Helmberger, G. Engen and P. Scott, A note on velocity, density, and attenuation models for marine sediments determined from multibounce phases, J. Geophys. Res., 84:667-671 (1979).

3. D. V. Helmberger, The crust-mantle transition in the Bering Sea, Bull. Seismol. Soc. Am., 58:179-214 (1968).

4. W. I. Futterman, Dispersive body waves, J. Geophys. Res., 67:5279-5291 (1962).

5. J. S. Barker and D. V. Helmberger, A broad-band study of attenuation in ocean bottom sediments, Woodward-Clyde Consultants, Final Technical Report WCCP-R-85-04 (1985).

6. D. Weston, Underwater explosives as acoustic sources, Proc. Phys. Soc., 76:233-249 (1960).

7. L. J. Burdick, Estimation of t^* for Asian travel paths using sP and sS phases, Geophys. J. Roy. astr. Soc., in press (1985), see also L. J. Burdick, Estimation of the frequency dependence of Q from ScP and ScS phases, Geophys. J. Roy. astr. Soc., 80:35-47 (1984).

8. E. L. Hamilton, Compressional-wave attenuation in marine sediments, Geophysics, 37:620-646 (1972).

9. J. B. Minster, Transient and impulse responses of a one-dimensional linearly attenuating medium - II. A parametric study, Geophys. J. Roy. astr. Soc., 52:503-524 (1978).

10. E. L. Hamilton, Sound attenuation as a function of depth in the sea floor, J. Acoust. Soc. Am., 59:528-535 (1976).

11. R. S. Jacobson, G. G. Shor, Jr. and L. M. Dorman, Linear inversion of body wave data - Part II: Attenuation versus depth using spectral ratios, Geophysics, 46:152-162 (1981).

12. S. K. Mitchell and K. C. Focke, New measurements of compressional wave attenuation in deep ocean sediments, J. Acoust. Soc. Am., 67:1582-1589 (1980).

ON THE ROLE OF BOTTOM INTERFACE WAVES IN OCEAN SEISMO-ACOUSTICS: A REVIEW

Dieter Rauch

Istituto di Acustica "O.M. Corbino"
Via Cassia 1216, I-00189 Rome, Italy

ABSTRACT

Ocean bottom interface waves of the Scholte type (sometimes also called oceanic Rayleigh or Stoneley waves) are guided by the acoustically most significant interface separating the water column (including water-sediment mixtures) from the underlying consolidated strata of the basement. The steadily growing efforts of underwater acousticians, marine geophysicists, and offshore engineers to improve the surveillance, exploration, and exploitation of shallow water areas have created a strong interest in bottom-interacting acoustics in general and these boundary waves in particular. The physics is discussed, theoretical and experimental work is reviewed, and recent field data are given. The usefulness of these low-velocity waves for seismic sensing of infrasound in coastal waters (below the hydro-acoustic cut-off) is demonstrated by means of deterministic events (i.e. ground-roll from small explosive charges and sweep signals from a towed low-frequency projector) and ambient noise records. Some particular aspects of these low-frequency waves are mentioned to introduce topics in the advanced research papers, e.g. their promising application for in-situ microzonation of the upper seabed with regard to its extremely important shear profiles.

INTRODUCTION

This paper intends to enlarge and consolidate knowledge of free or genuine seismo-acoustic interface waves that propagate along the sea floor. The underlying physics will be emphasized to encourage further reading, especially of the increasing number of advanced research papers in this field [1]. Due partly to real-world dispersion and attenuation effects, and partly to the progress in underwater infrasonics, the frequency range from about 1 to 30 Hz has become extremely important. Because of the growing interest of acousticians in waterborne sound-bottom interactions and especially the need for low-frequency propagation studies in coastal waters (below the hydro-acoustic cut-off) seismo-acoustic studies have been carried out more systematically during the last twenty years than previously. To some extent these recent efforts were sustained by the development of more rugged, high-fidelity ocean bottom seismometers and by the search for offshore hydrocarbons.

In seismics about 0.2 Hz is traditionally the frequency that separates medium- and long-period earth movements from short-period (SP) tremors; in acoustics 30 Hz divide infrasonics (IS) from sonics. Thus SP-IS identifies the frequencies of interest in this paper. Pure (non-radiating) interface waves are classified according to the names of their discoverers or investigators. Thus, we have Rayleigh waves, at the free surface of a solid (solid-vacuum interface); Scholte waves, at the interface of a solid and a liquid; and Stoneley waves (for highly restricted parameter combinations) at the interface of two solids. Here we are mainly interested in oceanic Scholte waves and especially the modified or generalized versions that result from realistic environmental conditions (especially regular stratification). Scholte waves are guided by the most significant acoustic interface, the one that separates the water column, which includes unconsolidated sediments and suspensions, from the cemented or even rocky substratum. Modified Rayleigh and Stoneley waves define, respectively, the possible low- and high-frequency limits of real-world Scholte wave modes. Thus, we assume that the reader is at least familiar with the phase-velocity curves, particle motions (restricted to the "sagittal plane") and penetration-depth of short-period Rayleigh wave modes on land.

1 REVIEW OF CLASSICAL ANALYTICAL APPROACHES AND IMPORTANT ADVANCED NUMERICAL MODELLING TECHNIQUES AND EXPERIMENTS

In the late 1930's Cagniard [2] solved the problem of an impulsive source located close to a planar solid/solid or liquid/solid interface. However, because of the sophistication of the mathematics, these results never became a practical tool to solve geoacoustic problems. In the early 1960's, some efforts were devoted to the derivation of simpler mathematical descriptions and to testing them in model experiments. Although the results of those small-scale experiments can reflect only some very pronounced features of real-world data they provide a helpful cross-check on the theoretical set-up and give some confidence in applying it under more realistic conditions.

Before discussing briefly the theoretical aspects of the simplest case of two half-spaces, we would like to review the underlying mathematics in general. The wave propagation of interest along an interface is characterized by up to four branch points of the pertinent integrands that correspond to a maximum of four undisturbed bulk waves. To restore the uniqueness of the complex plane of integration we therefore chose up to four branch lines (or cuts) that divide the ambiguous Riemann surface of the integrals into a maximum of five sheets. Due to the different ways of making these cuts, and to the resulting definition of those sheets, several ways to calculate the different contributions to the integrals are possible. The final choice is usually determined by the requirement for a fast convergence and by a need to facilitate the evaluation. For a fixed source/receiver configuration those joined sheets, separated by the arrivals of the bulk waves, can be reasonably interpreted as the subsequent "time windows" of the synthetic seismogram. Thus the contribution of the sheets can be interpreted as the pertinent sections of the calculated signal. The only sheet without any contribution (i.e. without physical meaning) is usually classified as the lowest one corresponding to the period between the onset of the source and the first arrival at the receiver position. The genuine interface waves of interest are characterized by poles on the top sheet that contribute to the tail of the signal, while the leaky pseudo-interface waves result from poles on lower sheets between the extremes.

Despite the fact that the integration has to be performed on the top sheet, those lower sheet poles may have an influence on the synthetic seismograms by their proximity to the cuts. With respect to the more practical approaches that reveal this correlation between the mathematical and the physical aspects, papers by Roever, Vining and Strick [3] and by Phinney [4] are milestones in the field of elastic interface waves. To simplify the mathematics, Strick reduced the problem to two dimensions, assuming a delta function excited horizontal line source instead of a point source [3]. The resulting field terms of these sources differ at large ranges by essentially an amplitude factor and a half-order differentiation; such a line source may be regarded as a point source with an excitation between a delta- and a step-function. The chosen line source is therefore a good simulation of a detonating chemical charge or of an electrical-spark discharge, the most common source devices for both field studies and laboratory tests. The only substantial adulteration results from the characteristic tail-effect of a line source, which can be easily understood from Huygen's principle. (The source cannot be blanketed off and thus cannot be weighted by an amplitude assignment). Assuming two adjacent half-spaces, Strick calculated the pressure response and the particle velocities at a water/solid interface corresponding to a soft bottom (SB: $c_s < c_w$) and hard bottom (HB: $c_s > c_w$) sea floor. In both cases he proved the excitation of a relatively strong Scholte-wave pulse, and demonstrated impressively that the suppression of the pseudo-Rayleigh wave at a soft ocean bottom is due to the off-axis position of the pertinent pole. In addition, Strick has demonstrated that the real part of this pseudo-Rayleigh pole is practically identical with the true, real-Rayleigh pole and thus with the zero-crossing of the pertinent signals propagating with the same velocity [3].

The agreement of the model experiments with Strick's predictions is outstanding. As expected, the Scholte-wave pulse is more compressed under HB bottom conditions and also more difficult to separate from the direct P-wave arrival (water wave) in the water column.

In the early 1960's, Spitznogle, McLeroy et al extended Strick's theoretical work and the experimental studies by Roever and Vining [3] to a two-layered, solid half-space covered by a homogeneous liquid counterpart [5] or by a liquid layer with an adjacent gaseous half-space [6]. To compose the theoretical seismograms, their calculations took into account up to two subsequent interactions of the primarily-radiated P-wave front with the crucial liquid/solid interface. Using four different small-scale models and a spark gap they recorded numerous pressure responses at the liquid/solid interface. In almost all cases the measured signals again showed very good agreement with the synthesized pressure/time curves. Usually the Scholte-wave pulse turned out to be the most prominent portion of the signal after the above mentioned direct arrival in the liquid.

Since these earlier studies there have been many attempts to adapt the mathematics to real-world conditions, especially by proceeding to computer-oriented numerical methods. Two milestones in the development of these modern modelling techniques need description. Based on the well-known Thomson-Haskell matrix method a range-independent Green's function approach has been elaborated for depth-dependent arbitrary horizontal stratifications and a single source-receiver combination. This approach is very successful but computationally is quite slow especially concerning broadband signals. This approach, widely known as the Fast Field Program (FFP) [7,8], was recently replaced by a much more versatile computer-code, called SAFARI [9], which allows multiple source-receiver combinations and is faster (at least one order of magnitude). With regard to most seismic propagation studies, and especially the interesting interface waves, this modern numerical tool (using a global matrix approach similar to the well-known finite element method) has already had a significant impact [10].

2 IMPORTANT FIELD-TESTS TO INVESTIGATE SEISMIC SENSING OF WATERBORNE SOUND

Generally, the approach to most realistic sources (e.g. explosive charges, sparkers, boomers, air-guns, CW-projectors) is point source excitation which generates annular wave-fronts with cylindrical spreading and thus an intrinsic geometrical amplitude-decay in radial direction (range) according to $r^{-1/2}$. Rarely used source patterns or extended generators/targets may be easily synthesized from arrays or clusters of those point-sources. Attenuation coefficients are given below and provide true medium losses that are extracted from field data after having corrected for this basic geometrical decay. Some examples of these losses are electrochemical processes, fluid-flow phenomena that depend on pore geometry and fluid viscosity, skeleton-friction, and internal scattering.

Since World-War II a series of small- and large-scale field experiments has been carried out by seismologists and underwater acousticians. The most significant of these tests will be briefly summarized below. Many of the large-scale experiments gave contradictory results both internally and externally because most of these experiments, lacking clear concepts, were performed in a non-systematic way. The answers to the following fundamental questions are needed before the initiation of elaborate and expensive field-tests:

a) Which wave type should be investigated for the given objective?

b) What are the characteristics of this wave type under different environmental conditions?

c) What sensors are needed and what are their requirements?

SACLANTCEN's success in the field of seismic sensing of waterborne sound is due to the timely consideration of these questions and the appropriate answers.

During 1943 and 1944 Worzel and Ewing [11] installed a sensor station with two hydrophones and one vertical geophone on the ocean bottom in five widely separated shallow-water areas. At water depths between 15 m and 180 m they sensed the pressure and displacement signals from explosive charges dropped at distances between 0.4 and 50 km. Their main interest was devoted to the so-called "ground-wave", which is composed of the high-speed shallow-refracted LF arrivals having dominant frequencies in the 10 to 30 Hz band.

In the late 1950's Worley and Walker [12] performed a similar study in coastal water 27 m deep with a more versatile receiver package. Despite the failure of the horizontal geophone they were able to show that the signals from the hydrophone and the vertical geophone had equal range-dependent fall-offs at 15 to 17 Hz.

Approximately at the same time, Blaik and Clay [13] tried to sense waterborne sound on land. They towed a powerful low-frequency transducer in shallow water and placed vertical geophones in a well on shore at depths between about 120 and 600 m. Profiting from the refracted field at shallow and medium depth they succeeded in detecting CW signals (at 10, 24, 88 and 148 Hz), with losses consistent with spherical spreading, over ranges up to about 9 km.

Since 1958 McLeroy and co-workers have carried out different theoretical studies, model experiments, and extensive field tests on the use of seismic paths for detection and communication in coastal waters. Using a shallow, buried, triaxial ocean bottom seismometer station and a nearby hydrophone they recorded the signals from high-power broadband CW sources (10 to 700 Hz) and explosives at distances up to a few kilometers. Although their model experiments successfully synthesized the different contributions to the sound field at the interface, they often couldn't correlate the complex real-world data with well-defined wave types of their model. In general, they merely distinguished between the ensembles of the "ground-wave" and the "water wave". However, they noticed at all frequencies a much higher signal-to-noise ratio on the horizontal geophones (up to 20 dB) than on the hydrophone and the vertical geophone. In one field experiment they observed a distinct interface wave train (6-7 Hz) that they interpreted as a Scholte wave despite some deviation from the expected characteristics [14].

From 1960 to 1963 Shorthouse (unpublished PhD thesis, referenced in [15]) developed and successfully applied a seismic system to the observation of P-head waves from velocity discontinuities both in the oceanic crust and in the sedimentary layers. Using a geophone and several hydrophones on the deep sea floor he recorded the signals from 2.3 kg (5 lb) charges exploded on the ocean bottom at different distances. As a by-product, late-arriving, long wave trains were observed at ranges up to 2.5 km. These wave trains showed a substantial normal dispersion and quite strong amplitude decay proportional to $r^{-1.4}$. Analyzing the pertinent data, Davies [15] clearly identified these tails of the records as Scholte waves with phase velocities between 130 and 60 m/s in the frequency band of 2 to 10 Hz. His measured dispersion curves for the fundamental and first higher mode propagating along a regularly layered sea bed possibly are the first of their kind.

In 1964 Bucker and co-workers [16] performed two small-scale propagation experiments (water-depth, 1 and 20 m). They used an array of 4 vertical geophones spaced 5 m apart as sensors and blasting caps as broadband sources. These experiments demonstrated the potential use of Scholte wave trains (center frequencies 20 to 25 Hz) for in situ determination of sediment shear velocity ($c_s \approx 110$ and 215 m/s) and attenuation coefficient ($\alpha_s \approx 1.4$ dB/λ at those frequencies).

In 1968 Urick [17] tried to track small craft (controlled runs) and large freighters (uncontrolled runs) with a triaxial seismic sensor package and a hydrophone (at 18 m depth) on a steeply sloping sea floor at about 18 m depth. Below frequencies of 100 Hz the signal-to-noise ratio of the vertical geophone was less than that of the hydrophone and attenuated rapidly with range; that of the horizontal sensors was always 5 to 10 dB and decreased more slowly with range. Probably because infrasonic frequencies were not considered, none of these experiments indicated interface waves. The unexpectedly high levels of the cross-deflection (up to 50% of the radial displacement) confused the records. These levels were probably caused by a relatively high energy transfer to Love waves caused by inhomogeneities and irregular layering.

In 1976 McLeroy and his co-workers [18] again performed a very extensive real-world experiment in coastal waters. Using a receiving array of nine triaxial OBS stations and six bottom-mounted hydrophones at 20 to 30 m depth, they conducted numerous acoustical runs with towed, powerful CW sources that operated at distances up to about 50 km. They used about 200 explosive charges at 20 m depth and/or 1 m above the sea-floor which were detonated at distances of up to about 16 km. This time numerous environmental parameters were carefully recorded, and even preliminary results indicated a much higher signal-to-noise ratio for both the radial and the vertical geophone than for the transverse sensor and the nearby hydrophone. These results, contradicting those of earlier field-tests by the same investigators and others, may indirectly indicate the presence of Scholte waves.

In 1976 Schirmer [19] conducted interface wave experiments using explosive charges that were detonated at a depth of 130 m at distances from 0.8 to 1.7 km. A weak Scholte wave train was observed only in a few cases at frequencies in the 3.5 to 5.5 Hz range. The hodographs were vertically aligned, retrograde, and elliptically circumscribed. The results were unsatisfactory partly because the 10 Hz geophones used were apparently inappropriate and partly because of the sensitivity (confirmed by several SACLANT experiments) of the poorly understood excitation mechanism to local bottom conditions. Nevertheless Schirmer was able to infer a shear-speed of about 120 m/s and a shear attenuation coefficient, α_s, of approximately 0.2 dB/λ at those very low frequencies.

In 1977 Essen et al [20] carried out a follow-up experiment along the shore at flood-tide. Strong interface wave trains were recorded, dominated by a 4 Hz wave train displaying normal dispersion and a group velocity of about 200 m/s (first higher mode, horizontally aligned, retrograde circumscribed elliptical hodographs) and a 4 to 6 Hz train with a group velocity around 100 m/s (fundamental mode, vertically aligned, prograde elliptical hodographs) showing no dispersion.

In 1978 Sutton et al [21] performed the first important OBS-intercomparison experiment at 7 m depth off Lopez Island, British Columbia. The main purpose was to determine the effects of bottom-coupling, sensor cross-coupling, and bottom currents on the performance of twelve OBS with completely different mechanical configurations. However, a small part of the study was dedicated to Scholte waves generated by small detonators, air-guns and ambient microseismic noise fields. The particular features of these interface waves and their influence on marine seismic work were identified unambiguously and verified the results of earlier experiments.

During 1978 and 1979, Northrop [22] buried arrays of horizontal geophones and triaxial sensor units together with hydrophones in the seabed in coastal waters (about 15 m depth). The objective was to investigate Urick's [17] questions regarding the gain in S/N ratio that is possible with a geophone compared with that from a hydrophone (the seismic gain). Despite the large number of explosives and towed LF-projectors (50 and 100 Hz) on many different tracks (ranges up to 80 km) and proximity patterns, the outcome was not much different from Urick's results, which dated back to 1968, because of conceptual shortcomings. About 1982 Holt and co-workers [23] repeated Bucker's 1964 small scale propagation study [16] with a tunable CW-source (10 to 130 Hz) and more versatile receivers. Both on land and on the sea floor they recorded pronounced Rayleigh and Scholte wave pulses at ranges of 5 to 40 m, verified the most prominent characteristics, and determined the shear attenuation coefficient, α_s (= 2.3 dB/λ at 35 Hz).

3 SCHOLTE WAVE PROPAGATION RESULTS IN COASTAL WATERS

After review of background studies the writer undertook field-work in 1979 using SACLANTCEN's newly-developed advanced digital OBS system. The OBS, a rugged, compact and lightweight sensor-package, consists of three orthogonal, active 1 Hz geophones (Teledyne Geotech S-500) and one external, variable depth hydrophone. The package connects to a surface buoy that houses the power supply and the VHF telemetry system (see [1], [24] and [25]) for details). From 1979 to 1982, the system acquired a large body of Scholte wave data near La Spezia (see [26 to 29]).

Scholte-wave trains from small explosive charges

Figure 1 illustrates the experimental technique used to record explosion-generated Scholte waves. The receiving ship was anchored approximately 1 km from the radio buoy. A source ship deployed the

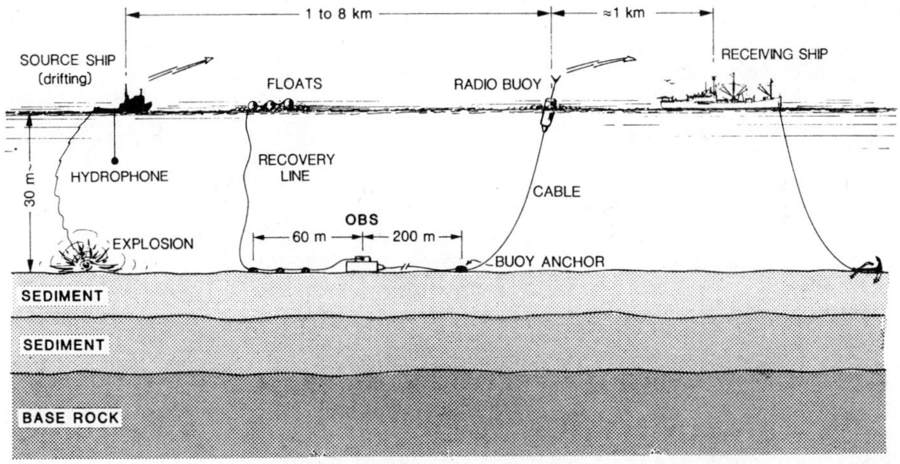

Fig. 1 Deployment of OBS system and experimental procedure to record explosion-generated Scholte waves.

explosive charges at successive, pre-selected ranges from the OBS. For each range the explosive charge (45 g to 1450 g TNT) was lowered to the sea floor and fired electrically. The use of explosive charges as broadband sources is a standard technique in underwater acoustic measurements; they furnish extremely low energy per Hertz within the very low frequency band (1 to 6 Hz) of interest in these experiments. A hydrophone below the source ship received the pressure pulse from each explosion. The pulse was then transmitted to the receiving ship, where it triggered data acquisition.

A pre-processing software package furnished real time plots (Fig. 2) of the outputs from all four geoacoustic sensors. The following three figures present time-distance plots and corresponding dispersion analyses for the vertical and radial geophone responses obtained from two different bottom types. In particular, Fig. 3 and the left column of Fig. 5 result from a soft bottom area, while Fig. 4 and the right column of Fig. 5 are from a harder bottom area. Figure 3 shows the time-distance plots of the vertical and radial particle velocities recorded on a soft sea floor (6 n.mi off Marina di Carrara) for six bottom charges (water depth 20 m; hydroacoustic cut-off about 7 Hz). The first arrivals with HF-oscillations (dotted line) are associated with the direct water wave that has a speed of 1534 m/s. The LF-interface wave train is delineated by the two thick solid lines corresponding to a group velocity of 235 m/s at the front and of 78 m/s at the rear of the seismo-acoustic tremor. The vertical components consist merely of single, normally dispersed, wave-packets with 1.5 to 6 Hz oscillations in the rear section and group velocities roughly equal to 70 to 110 m/s. In addition to this contribu-

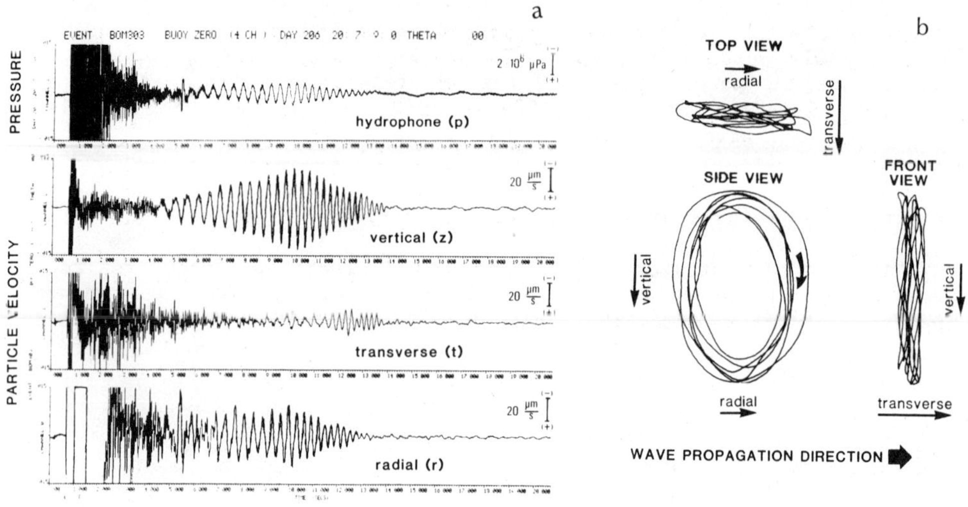

Fig. 2 Seismograms and resulting hodographs
a) Seismograms from the three geophones and one hydrophone of an ocean-bottom seismometer
b) Resulting hodograph plotted in the radial, vertical and transverse planes, for the indicated 2 s time window.

tion (fundamental mode), the radial component is composed of two normally dispersed wave-packets with similar frequencies running ahead. The centre of the more powerful second packet (first higher mode) is indicated by the dotted line (133 m/s). The corresponding amplitude-and phase-contours from a high-resolution dispersion analysis (multiple-filter technique with Δf = 0.05 Hz; see e.g. [25] or [28]) offer more insight into the modal structure of the seismograms. These contour plots (Event A, with range = 1.7 km; see arrows in Fig. 3) are reproduced in the left column of Fig. 5. The amplitude contours for the vertical particle velocity component reveal a strongly dominating fundamental mode with almost no dispersion at a frequency higher than 2 Hz. Thus even the deeper penetrating large wavelength oscillations down to about 2 Hz do not indicate any drastic change in the crucial bottom parameters. In other words, the depth gradients of the shear parameters should be quite moderate and smooth down to the critical depth of about 90 m at the top of the highly consolidated (hard) subbottom. In addition to this fundamental mode the radial particle velocity demonstrates a dominant first mode with weak normal dispersion and a slightly more dispersive second mode. Superimposing the radial amplitude contours (lighter shade) on the pertinent phase-plots we realize that a frequency-dependent inversion of the rotation is attributable to the fundamental mode (prograde circumscribed, vertically aligned ellipses up to 3.5 Hz; retrograde particle motion for higher frequencies) and the first higher mode while the second mode is characterized by horizontal ellipses with retrograde motion below 3.5 Hz. These phase-inversions of all modes around 3.5 Hz reveal a minor but still significant change of the shear parameter gradients at about mid-depth (approximately 45 m) of the local sediment cover.

Combining Fig. 4 with the right column of Fig. 5 (Event B, source-receiver distance 1.9 km) provides a test on a harder ocean bottom (8 n.mi off Viareggio) at a water-depth of 30 m (hydroacoustic cut-off at about 6 Hz). The maximum range was three times that of the previous experiment at Marina di Carrara. The water arrival speed was 1520 m/s , the direct P-wave speed in the upper sea-bed was 1715 m/s and the speed of its subbottom refracted counterpart was 2850 m/s (estimated depth, 750 m). In this case the front of the Scholte wave train was slightly faster (250 m/s; 1.5 to 2 Hz oscillations) and its rear slightly slower (65 m/s; 2.5 to 3 Hz oscillations). The superimposed wave packet of the first higher mode (around 4 Hz) was also shifted towards higher group velocities (145 m/s). The strong dispersion of the fundamental mode below 3 Hz (prograde) and the almost negligible dispersion of the first higher mode above 3 Hz (retrograde) indicate that at this new position (about 10 n.mi SE of the former one) the sediment cover is thinner (rough estimate: 75 m). The greater consolidation at this site is not only manifested by the broader velocity spectrum but also by the greatly diminished attenuation of the wiggle traces in Fig. 4 as compared to the signal amplitudes in Fig. 3 obtained at different ranges for the same charge size. Much more comprehensive data sets (also for a hard bottom type) may be found in [25]; however the attenuation coefficients calculated there are broadband results and thus only suggestive rather than specifically comparable. A rough estimate of velocity and attenuation may be inferred from the (depth-averaged) shear velocity and attenuation-coefficient at about 3 Hz for these thick and relatively soft sediment covers; thus $c_s \approx 100$ to 150 m/s, and $\alpha_s \approx 0.1$ to 0.3 dB/λ. These results for the Mediterranean are comparable to those recently published for the North Sea [19]. Outstanding examples of regular elliptical hodographs may be found in references [25,26,28,29]. The left column of Fig. 6 provides a typical example of the well-defined correlation between pressure p_w at the sea floor ($\pi/2$ ahead) and vertical particle velocity u_z, resulting from an explosion. The corresponding results for the ambient noise, to be discussed later, are shown in the right column of Fig. 6.

Figure 7 shows the normalized amplitude contours of the vertical and radial particle velocity of a sweep signal (11 to 18 Hz in about 18 seconds) from a powerful mine sweeping source (type AMK VI; source level, 190 dB re 1μPa at 1 m). The source operated 12 m below the sea surface in coastal waters of 16 m depth; the OBS recorded at a range of 1.05 km from the source on a soft sea bed. Even at ranges up to several kilometers the peak levels of the vertical and radial particle velocity exceeded the corresponding relatively high microseismic background (noisy area due to heavy shipping and industry on shore) by 10 to 20 dB.

Ambient noise background

Based on much evidence, seismic noise propagates essentially as interface waves, although Love wave modes often contribute considerably. As early as 1960 Bradner and co-workers [30] confirmed the role of Scholte waves as one of the most important and sometimes highly directive carriers of microseismic energy (besides hydroacoustic organ-pipe modes and converted Love waves) by demonstrating some of their characteristics. The right column of Fig. 6 provides a verification of the dominant part played by Scholte waves at frequencies up to 4 Hz (very high coherence in the interval from 1.5 to 3 Hz and a stable phase shift of about 80° from 1.5 to 4 Hz). Thus the underlying transmission characteristics are indeed the same as those of the shot-generated signal on the left. Similar results concerning ambient noise can also be found in [29] which demonstrates that the coherence becomes low above 4 Hz, where the phase shifts are random. This fact, and the often observed directionality [30] of narrow-band seafloor seismicity, is demonstrated by the normalized azimuthal power plots (see also [26, 27, 28, 29]) of Fig. 8. This data set stems from recent ambient noise measurements on the continental shelf off the Cinque Terre coast northwest of La Spezia at a water depth of about 100 m. Approximately half, or more, of the Scholte wave energy propagates in the liquid overburden. Sea water is almost lossless, which explains the

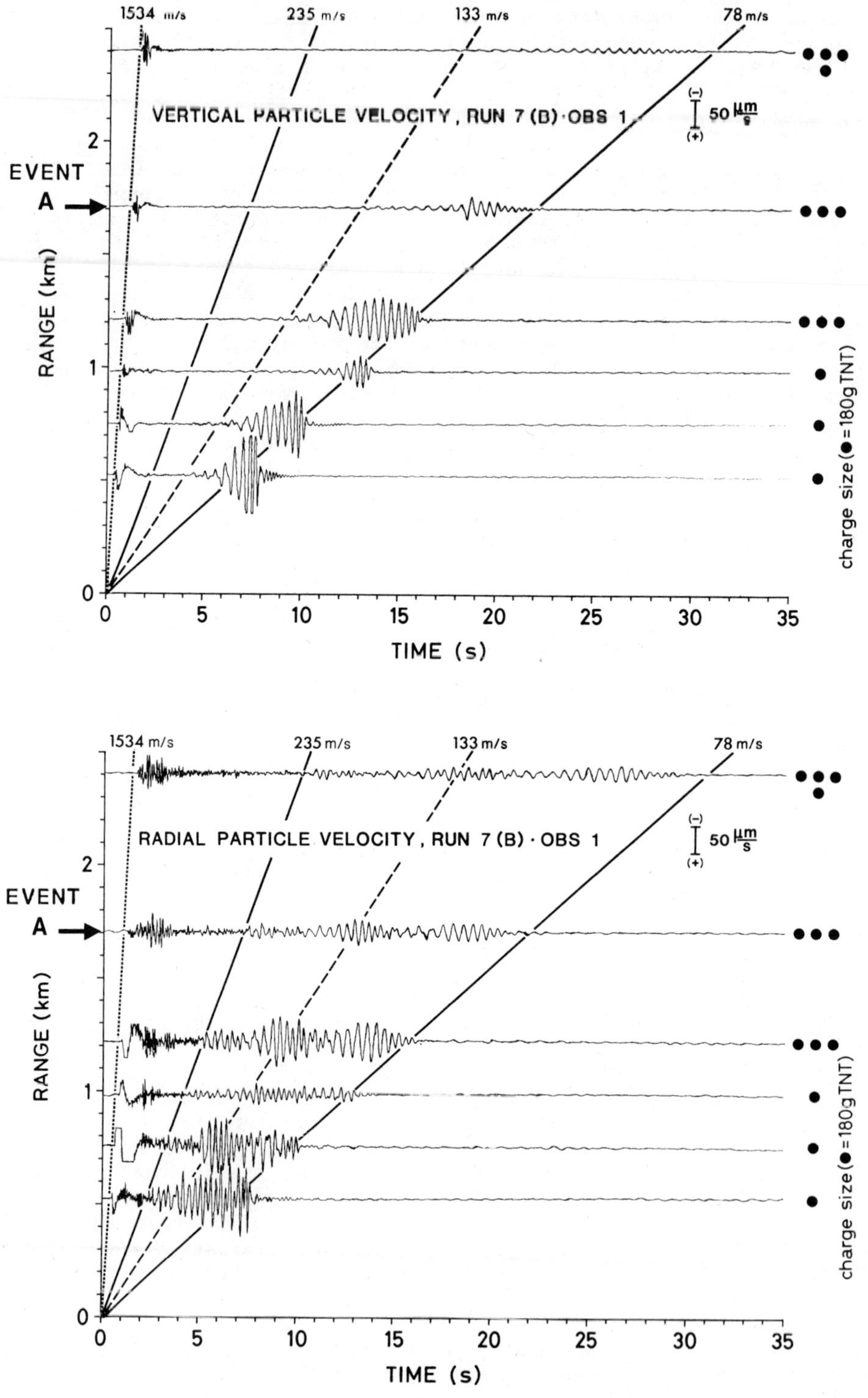

Fig. 3 Time-distance plots of the vertical and radial particle velocities recorded on a soft sea floor at a water depth of 20 m.

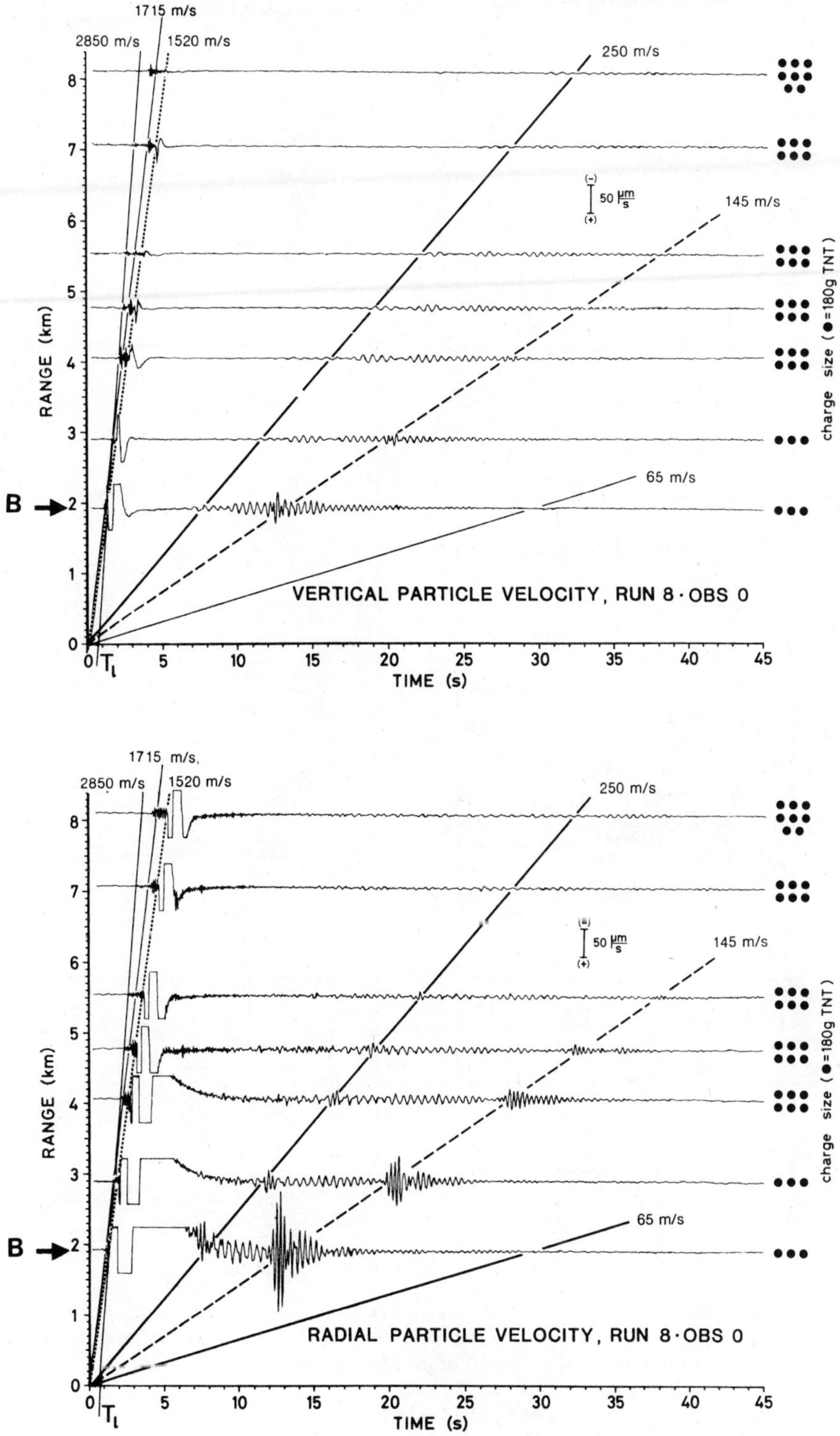

Fig. 4 Time-distance plots of the vertical and radial particle velocities recorded on a harder sea-floor at a water depth of 30 m.

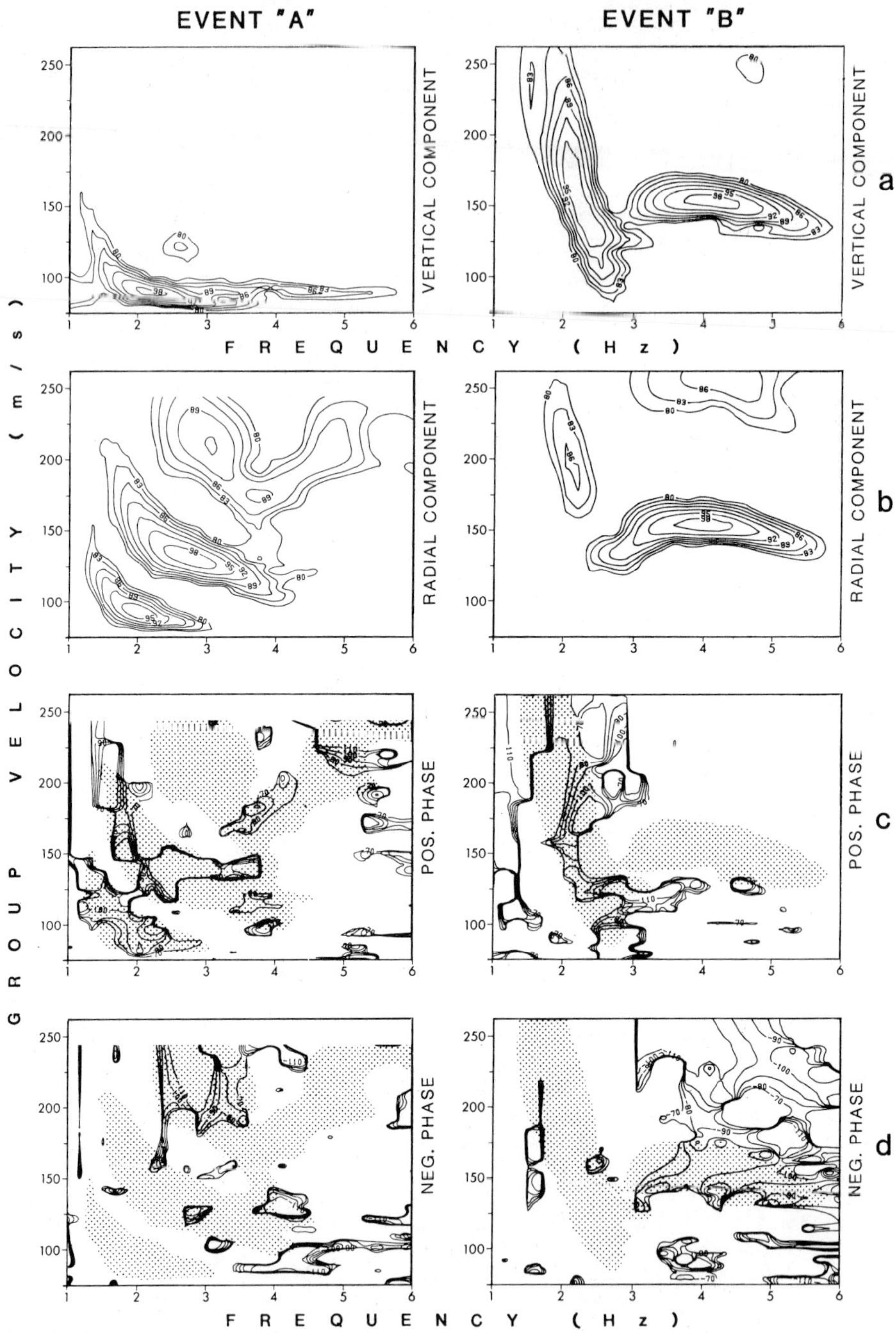

Fig. 5 Results of a dispersion analysis (group-velocity vs frequency) for interface waves generated by Events "A" (soft bottom) and "B" (hard bottom); these consist of normalized amplitude contours in the interval from 80 to 98 dB for

a) the vertical particle velocity,

b) the radial particle velocity

and contours (contour interval = 10°) of the phase between the radial and vertical velocity component

c) in the interval from 70° to 110°, prograde and

d) in the interval from 70° to -110°, retrograde.

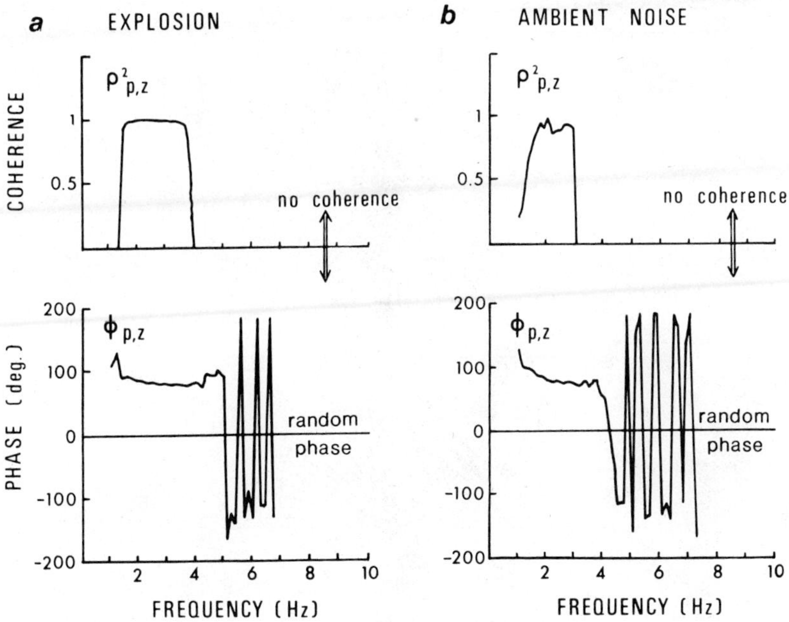

Fig. 6 *Coherence- and phase-spectra between bottom pressure p_W (hydrophone) and vertical particle velocity u_z (z-component geophone) for*
a) a shot-generated signal, b) an ambient noise sample.

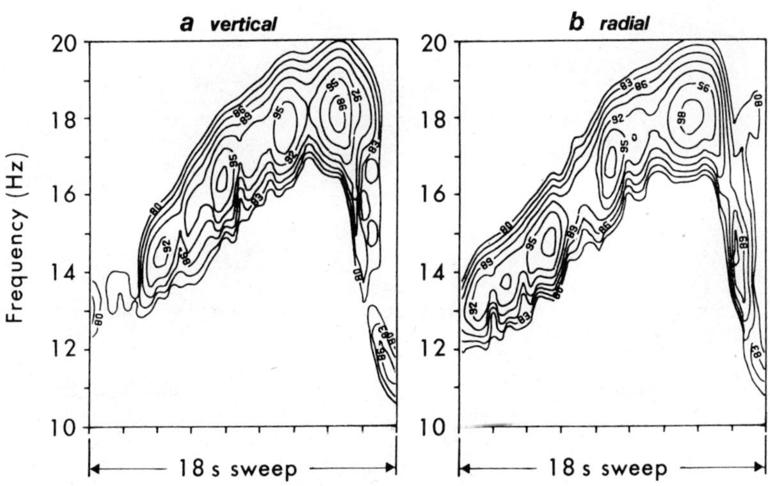

Fig. 7 *Normalized amplitude contours in the interval from 80 to 98 dB of the*
a) vertical particle velocity, b) radial particle velocity generated by a LF-sweep signal from a powerful minesweeping projector.

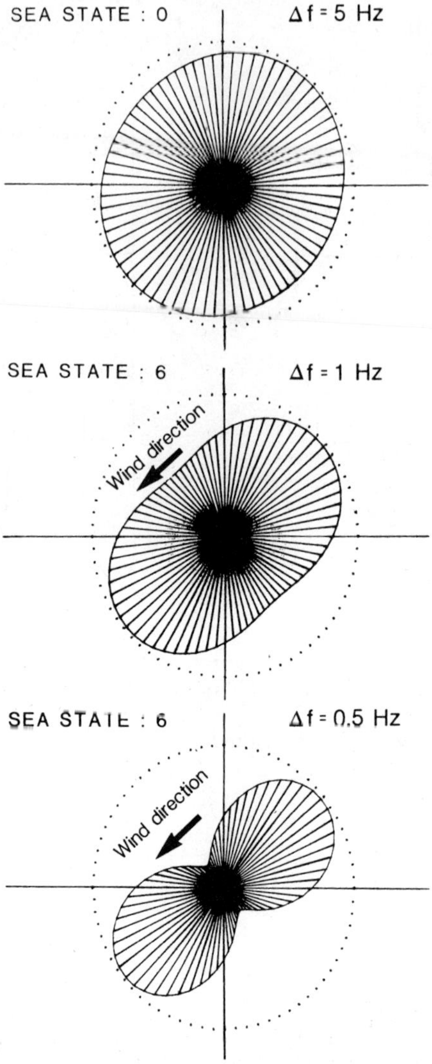

Fig. 8 Normalized azimuthal power-distributions of the horizontal particle velocities for sea state 0 and 6 for the frequency range of 1 to 6 Hz. (Δf : Bandwidth of the analysis)

reduced attenuation of short-period oceanic microseismic noise as compared to its counterpart on land.

Figure 9 reproduces two sets of typical background spectra from the geographical area of Fig. 8; the results are blanketed out below 1 Hz due to the high drop-off rate (-20 dB/oct) of the geophone response below the natural frequency of our sensors. A detailed discussion of these data is given by Akal et al [this volume].

Independently Bradner [31] and Toksöz [32] predicted the potential exploitation of this ambient noise background for passive bottom probing.

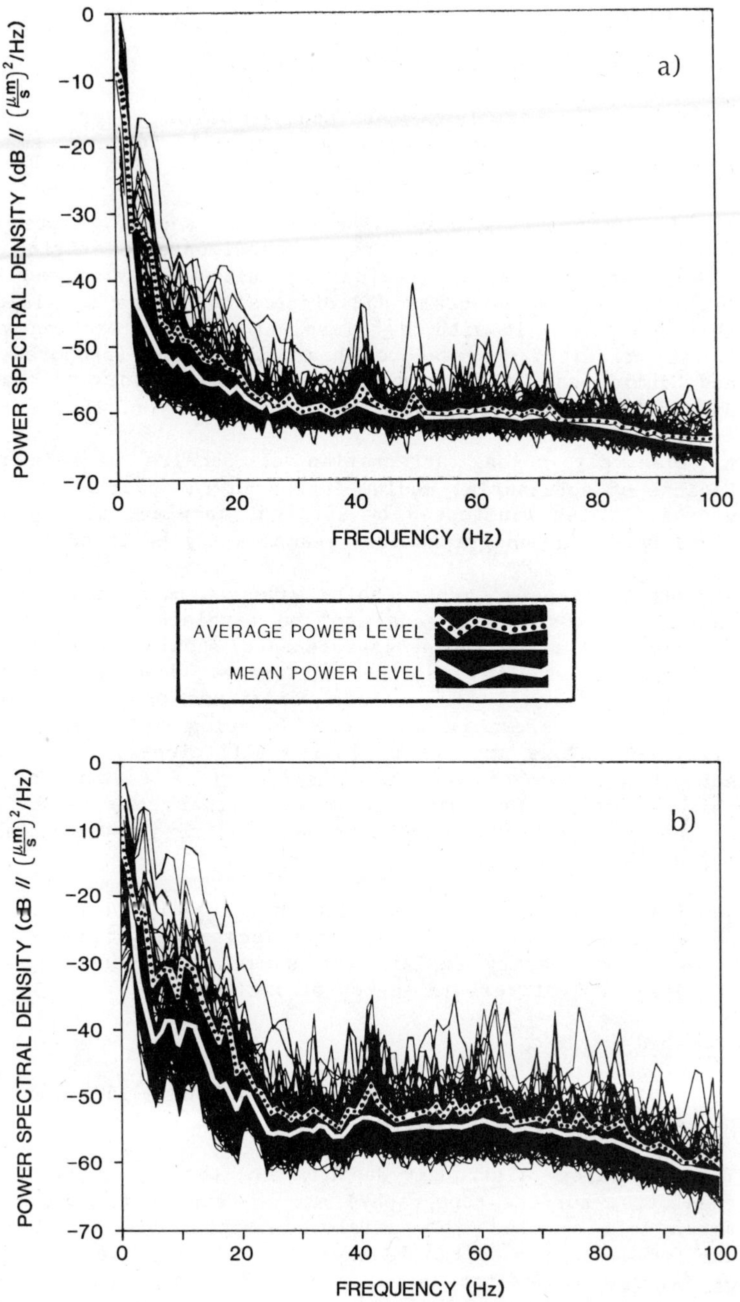

Fig. 9 Overlay of a set of 105 power-spectra of ambient noise (one 5 min - sample per day) with the resulting average (APL) and mean (MPL) power level estimates for
a) the vertical component
b) the horizontal component.

4 BOTTOM INTERFACE WAVES: SUMMARY, CONCLUSIONS AND RECOMMENDATIONS

These final statements apply to Scholte wave propagation in both deep and shallow water.

Soft Seabed ($c_s \ll c_w$)

a) Scholte wave speeds (c_{Sch}) are usually of the order of some tens to a few hundreds of metres per second (always smaller than $c_R \approx 0.9\ c_s$).

b) A relatively strong dispersion (several modes) occurs in Scholte wave propagation because of regular sedimentary layering with relatively large parameter gradients. This dispersion can be used to fit synthetic seismograms and dispersion curves to field data and thereby achieve in situ high resolution bottom probing with regard to the difficult-to-measure shear parameters (proposed by Jensen and Schmidt and by Szelwis in this volume). In very shallow water dispersion can be influenced by the proximity of the sea surface.

c) A relatively strong attenuation of Scholte wave energy exists because of substantial medium losses in the sediments. The attenuation is also manifested by a slight in-plane tilt of the elliptical particle hodographs (two examples may be found in [29]).

d) The penetration depth of Scholte waves on both sides of the interface is approximately equal to the wavelength. Therefore microseismic background noise measurements should be performed at or very close to the water-bottom interface. However, the undisturbed detection of small LF pressure variations or weak seismic phases should rely preferably on sensors floating sufficiently high above (i.e. hydrophone arrays) or buried sufficiently below (i.e. ocean subbottom seismometers) the interface (see Carter and Sutton in this volume). In very shallow water there may be no chance for undisturbed LF-hydroacoustic propagation measurements (also due to cut-off).

Hitting soft surficial sediment strata with obliquely incident sound beams generated at much higher frequencies the local excitation of Scholte or Stoneley waves may easily explain the sometimes observed peaks of the reflection loss [33] at certain angles of incidence.

Hard Seabed ($c_s > c_w$)

a) c_{Sch} is roughly in the range from 1250 to 1500 m/s but always slightly smaller than c_w (see Brooke et al in this volume).

b) The dispersion is strongly dependent on the actual layering (and on the bottom surface topography) but is usually less pronounced than for a soft sea bed. On an almost homogeneous rock basement only the fundamental mode can be expected. It may even exhibit anormal or inverse dispersion (e.g., see [25], [26], and Brooke et al in this volume). Even in deeper waters the sea surface may be sensed and thus slightly influence the dispersion.

c) The medium losses are usually much lower and all gradients more gentle than in the soft bottom case but substantial scattering losses may occur due to large-scale bottom roughness features such as grooves and outcrops or pronounced Bragg reflection losses because of resonant scattering at periodic small-scale bottom structures [34].

d) The penetration depth in the ocean floor is approximately equal to the Scholte wavelength and thus is slightly smaller than the corresponding water wavelength; however it may exceed this value many times in the water column because of the much reduced decay there (confirmed by Brooke et al in this volume) which may also produce coupling with hydroacoustic modes and the above mentioned sensing of the sea-surface. Thus there is almost no chance to operate hydrophones or geophones (even far from the water-bottom interface) without recording superimposed LF-Scholte waves (microseisms and man-made LF-noise).

e) The Scholte wave itself causes a strong differential horizontal motion between sea-floor and water mass (particle deflection may be 5 to 6 times bigger in the water column than in the bottom); thus serious problems exist for high fidelity sensing by bulky ocean bottom seismometers. This detrimental influence is usually attributed merely to local currents.

If attenuation is not considered, a Scholte wave propagating along the ocean floor provides an energy-flux density that is many times higher for a soft sea bed than for a hard sea bed. Therefore a soft bottom is much more favourable for the seismic sensing of waterborne LF-sources than a hard sea bed. However the soft bottom is also much more lossy and noisy. Thus some intrinsic difficulties exist for all measurements with bottom mounted geophones.

For many different technical and operational reasons the ocean bottom seismometer is still the most convenient and easiest instrument to use (see [21]) and Sutton in this volume). Because of the differential horizontal motion at the interface (discussed above) the sensor unit should be smooth and flat with a small cross-section. On soft sediments the sensor unit should be as light as possible (low bearing pressure) or almost neutrally buoyant within the top-layer so as to minimize the distortion of the free-field motion.

Additional comprehensive, long-term ambient noise measurements are needed in the frequency range from 1 to about 100 Hz to learn more about the interplay between LF-hydroacoustics and HF-seismics especially in shallow waters.

At higher frequencies (flat sections of the dispersion curves) the phase velocity $c_{Sch,o}$ of the fundamental mode may be calculated from the simple formula

$$c_{Sch,o} = c_w \Big/ \sqrt{(\rho_w c_w \,/\, r_o)^2 + 1}$$

where ρ_w stands for the water density and r_o for the amplitude-ratio of bottom pressure and vertical particle velocity. Thus $c_{Sch,o}$ may be determined with a single sensor unit in contrast to standard phase velocity measurement techniques which require the deployment of bottom-mounted hydrophone-or geophone-arrays (in suitably chosen geometric patterns and sizes) to apply cross-power spectral methods.

ACKNOWLEDGMENTS

The author is indebted to his former colleagues at SACLANTCEN, especially to those working in the Environmental Acoustics Group (EAG), the Environmental Modelling Group (EMG), and the Scientific and Technical Information Department (STI), for the generous technical support they offered to him, and the many useful scientific discussions which took place during the preparation of this paper.

REFERENCES

1. RAUCH, D. Seismic interface waves in coastal waters: a review, SACLANTCEN SR-42. La Spezia, Italy, SACLANT ASW Research Centre, 1980, [AD A 095 018].
2. CAGNIARD, L. Reflection and Refraction of Progressive Seismic Waves, 2nd edn. New York, N.Y., McGraw-Hill, 1962.
3. ROEVER, W.L., VINING, T.F. and STRICK, E. Propagation of elastic wave motion from an impulsive source along a fluid/solid interface. Phil. Trans. Roy. Soc., 251A, London, 1959: pp. 465-523.
4. PHINNEY, R.A. Propagation of leaking interface waves. Bull. Seism. Soc. Am., 51, 1961: pp. 527-555.
5. SPITZNOGLE, F.R. and McLEROY, E.G. Computed pressure-time waveforms for an impulsive source near a fluid-two-layer-solid interface. J. Acoust. Soc. Am. 36, 1964: pp. 972-973.
6. VICTOR, A.S., SPITZNOGLE, F.R. and McLEROY, E.G. Propagation at short ranges of elastic waves from an impulsive source in a shallow fluid overlying a layered elastic solid. J. Acoust. Soc. Am. 37, 1965: pp. 894-898.
7. KUTSCHALE, H.W. Rapid computation by wave theory of propagation loss in the Arctic ocean, Tech. Rep. CU-8-73. Palisades, N.Y., Lamont-Doherty-Geological Observatory, Columbia University, 1973.
8. JENSEN, F.B. Numerical models in underwater acoustics. In: Felsen, L.B. ed. Hybrid Formulation of Wave Propagation and Scattering. Dordrecht, The Netherlands, Nijhoff, 1984: pp. 295-335.
9. SCHMIDT, H. Modelling of pulse propagation in layered media using a new fast field program. In: FELSEN, L.B., ed. Hybrid Formulation of Wave Propagation and Scattering. Dordrecht, The Netherlands, Nijhoff, 1984.
10. SCHMIDT, H. and TANGO, G. Efficient global matrix approach to the computation of synthetic seismograms, J. Roy. Astr. Soc. (in press).
11. WORZEL, J.L. and EWING, M. Explosion sounds in shallow water. In: Propagation of Sound in the Ocean. Mem. Geol. Soc. Am. 27, 1948.
12. WORLEY, R.D. and WALKER, R.A. Low frequency measurements in the Gulf of Maine. J. Under. Acoust. 9, 1959: pp. 435.
13. BLAIK, M. and CLAY, C.S. Detection in the ground of sound from a source in shallow water, TR-76. Dobbs Ferry, N.Y., Hudson Labs., Columbia University, 1959.
14. McLEROY, E.G. and DELOACH, A. Measurements of sea bottom elastic waves from underwater explosions. Rep. No. U2727, U.S. Navy Mine Defense Laboratory, Panama City, Fla, 1968.
15. DAVIES, D. Dispersed Stoneley waves on the ocean bottom. Bull. Seism. Soc. Am. 55, 1965: pp. 903-918.
16. BUCKER, H.P., WHITNEY, J.A., and KEIR, D.L. Use of Stoneley waves to determine the shear velocity in ocean sediments. J. Acoust. Soc. Am., 36, 1964: pp. 1595-1596.
17. URICK, R.J. Underwater sound transmission through the ocean floor. In: HAMPTON, L., ed. Physics of Sound in Marine Sediments. New York, N.Y., Plenum, 1974: pp. 161-180.

18. McLEROY, G. The NCSC shallow water seismic propagation study. In: U.S. DEPARTMENT OF THE NAVY, OFFICE OF NAVAL RESEARCH. Proceedings, workshop on seismic propagation in shallow water, 6-7 July, 1978. Arlington, Va, U.S. Office of Naval Research, 1978: pp. 4/1-4/28.
19. SCHIRMER, F. Experimental determination of properties of the Scholte wave in the bottom of the North Sea. In: KUPERMAN, W.A. and JENSEN, F.B., eds. Bottom Interacting Ocean Acoustics. New York, N.Y., Plenum Press, 1980: pp. 285-298.
20. ESSEN, H.H., JANLE, H., SCHRIMER, F. and SIEBERT, J. Propagation of surface waves in marine sediments. J. Geophysics 49, 1981, pp. 115-122.
21. SUTTON, G.H., LEWIS, B.T.R., EWING G.J., DUENNEBIER, F.K., IWATAKE, B. and TUTHILL, J.O. Lopez Island ocean bottom seismometer intercomparison experiment. Final report, HIG-80-4. Honolulu, Hawaii, Institute of Geophysics, University of Hawaii, 1980.
22. NORTHROP, J. Some underwater sound propagation studies in the vicinity of the NOSC oceanographic tower, NOSC TR 424, San Diego, CA, Naval Ocean Systems Center, 1979. [AD A 073 492].
23. HOLT, R.M., HOVEM, J.M. and SYRSTAD, J. Shear modulus profiling of near bottom sediments using boundary waves. In: PACE, N.G. ed. Acoustics and the sea-bed. Bath, U.K. Bath University Press, 1983: pp. 317-325.
24. BARBAGELATA, A., MICHELOZZI, E., RAUCH, D. and SCHMALFELDT, B. Seismic sensing of extremely-low-frequenxy sounds in coastal waters. In: Proceedings, IEEE Conference on Acoustics, Speech and Signal Proc., (ICASSP 82) Paris, 1982. Piscataway N.J., IEEE Service Center, 1982: pp. 1878-1881.
25. SCHMALFELDT, B. and RAUCH, D. Explosion-generated seismic interface waves in shallow water: experimental results, SACLANTCEN SR-71. La Spezia, Italy, SACLANT ASW Research Centre, 1983, [AD A 134 551].
26. RAUCH, D. Experimental and theoretical studies of seismic interface waves in coastal waters. In: KUPERMAN, W.A. and JENSEN, F.B., eds. Bottom-Interacting Ocean Acoustics. New York, N.Y., Plenum Press, 1980: pp. 307-327.
27. SCHMALFELDT, B. and RAUCH, D. Seismic and hydroacoustic sensing of infrasonic noise in coastal waters. In: WAGSTAFF, R.A. and BLUY, O.Z. eds. Underwater ambient noise, SACLANTCEN CP-32, La Spezia, Italy, SACLANT ASW Research Centre, 1982: pp. 5-1 to 5-12.
28. RAUCH, D. and SCHMALFELDT, B. Ocean-bottom interface waves of the Stoneley/Scholte-type: Properties, observations and possible use. In: PACE, N.G., ed. Acoustics and the Sea-Bed. Bath, UK, Bath University, 1983: pp. 307-316.
29. ALI, H.B., and SCHMALFELDT, B. Unpublished report, 1984.
30. BRADNER, H. and LATHAM, G.V. Prograde Rayleigh waves and microseism sources. J. Geophys. Res. 77, No. 32, 1972: pp. 6422-26.
31. BRADNER, H. Probing Sea-bottom sediments with microseismic noise. J. Geophys. Res. 68, No. 6, 1963: pp. 1788-91.
32. TOKSÖZ, M.N. Microseisms and an attempted application to exploration. Geophysics 24, No. 2, 1964: pp. 154-177.
33. HAWKER, K.E. The existence of Stoneley waves as a loss-mechanism in plane wave reflection problems. J. Acoust. Soc. Am. 65, 1979: pp. 682.
34. BROOKE, G.H. Low frequency acoustic propagation in Canadian arctic waters. J. Acoust. Soc. Am. 74, Suppl. 1, 1983: pp. 520.

OBSERVATIONS OF INTERFACE WAVES AND LOW-FREQUENCY ACOUSTIC PROPAGATION OVER A ROUGH GRANITE SEABED

Philip R. Staal and David M.F. Chapman

Defence Research Establishment Atlantic
P.O. Box 1012, Dartmouth, Nova Scotia
Canada B2Y 3Z7

ABSTRACT

Experimental measurements of hydrophone signals from explosive sources in shallow water over a rough granite seabed have revealed several interesting seismo-acoustic phenomena. The pressure versus time waveforms from sources at 2 km to 40 km range show strong interface waves in the vicinity of 4 Hz arriving at a group speed of about 1200 m/s. The frequency-time dispersion of these arrivals is consistent with a simple dispersion model, including shear wave effects, based on a uniform water layer overlying a granite half-space. The propagation loss data show a notch of high loss in a band between 10 Hz and 100 Hz, centred at 30 Hz. A propagation loss model using the same geo-acoustic input does not reproduce these high losses but agrees well with the data in the range 2 Hz to 6 Hz. There are two hypotheses that may explain the discrepancy: either the seabed is made up of two or more layers having different acoustic properties giving rise to the observed propagation loss, or scattering at the rough interface (not included in the model) causes the extra loss. We favour the latter interpretation.

INTRODUCTION

The study of low-frequency underwater acoustic propagation in the shallow ocean has begun to overlap with the study of seismic propagation in marine sediments and rock. Acousticians have realised that shear wave excitation in the seabed can have a profound effect on the acoustic field in the water: in soft seabeds such as unconsolidated sediments or sedimentary rock, in which the shear speed is less than the speed of sound in the water, energy is robbed from the water layer and radiated into the seabed via bulk shear waves; in hard seabeds such as crystalline rock, in which the shear speed is higher than the speed of sound in the water, this loss mechanism is not significant. In both cases, the ability of the seabed to support shear waves contributes additional modes of propagation, variously called interface waves, Scholte waves or Stoneley waves, whose energy is concentrated in the vicinity of the ocean bottom (or any other interface involving a solid). These waves have come under considerable scrutiny by acousticians in recent years,[1] both as a propagation mechanism and as a means of probing the acoustic and seismic properties of the seabed for acoustic modelling purposes.

Previously, DREA has reported[2,3] the experimental and theoretical study of shallow water acoustic propagation over a chalk seabed, in which shear wave losses played a dominant role at low frequencies. We observed a wave at the water/chalk interface which allowed us to estimate a shear speed in the chalk of about 1000 m/s, lower than the sound speed in water. A simple two-layer normal mode propagation model including shear wave effects reproduced the very high losses observed experimentally below 300 Hz. The interface wave itself was weak and did not contribute significantly to acoustic propagation, having decayed after a few kilometres.

More recently, during a similar experiment over a rough granite seabed, we observed a strong interface wave in the vicinity of 4 Hz that propagated well for several tens of kilometres. This wave accounted for the very low propagation loss observed at this frequency, the frequency of optimum propagation. We also observed very high propagation loss in the band 10 Hz to 100 Hz, centred at 30 Hz; this was unexpected. In this paper we describe the experiment, the observed hydrophone signals, our dispersion analysis of the interface wave, the measured propagation loss from 2 Hz to 813 Hz and our initial, unsuccessful, modelling attempts. We also suggest a loss mechanism, scattering at the rough seabed, which may explain the high losses and we describe our current investigation of the theory governing this mechanism.

THE EXPERIMENT

In June 1983, DREA conducted a shallow water acoustic experiment in an area of the eastern Canadian continental shelf for which the seabed was rough granite with little or no sediment cover. As shown schematically in Fig. 1, we deployed the Hydra array[4] with a horizontal leg on the bottom and a vertical leg in the water. Our quiet research ship, CFAV Quest, stood by to receive telemetered acoustic data. The source ship travelled on a radial course from the array, dropping 0.45 kg explosive charges set to explode at 61 m depth at roughly 3 km intervals. Although Hydra consisted of several hydrophones in the vertical and the horizontal, we will only report here the

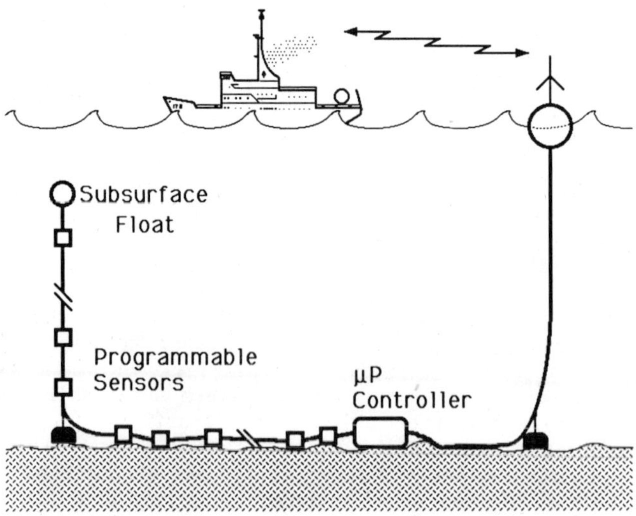

Fig. 1. Typical deployment of the Hydra array on the seabed.

results of a single hydrophone on the bottom. A bottomed hydrophone is more sensitive to interface wave signals and is less susceptible to potential self-noise problems.

The bathymetry and sound speed profiles under this experimental track are shown in Fig. 2, along with the charge depth. The water depth varied from 177 m at Hydra, decreasing to 125 m at 30 km and increasing to 200 m and more beyond 40 km. The scale for the sound speed profiles appears at the top of the figure; vertical ticks at the ocean surface indicate a reference sound speed value of 1480 m/s for each profile. The point at which each profile intersects the bottom indicates its spatial position. The sound speed profile was not uniform: there was a downward refracting thermocline for the first 50 metres depth and a sound channel existed at Hydra, gradually weakening with decreasing water depth.

We know little about the geology of the area other than that the seabed is one large body of granite or granodiorite of pre-Cambrian age and it has several metres of relief, sometimes as much as 10 or 20 metres. Significant outcroppings occur irregularly over length scales of several hundred metres. Small areas have been filled in by gravel and are relatively smooth, but hard. We are currently having detailed seismic profile records of the area analysed to produce a more comprehensive map of the roughness.

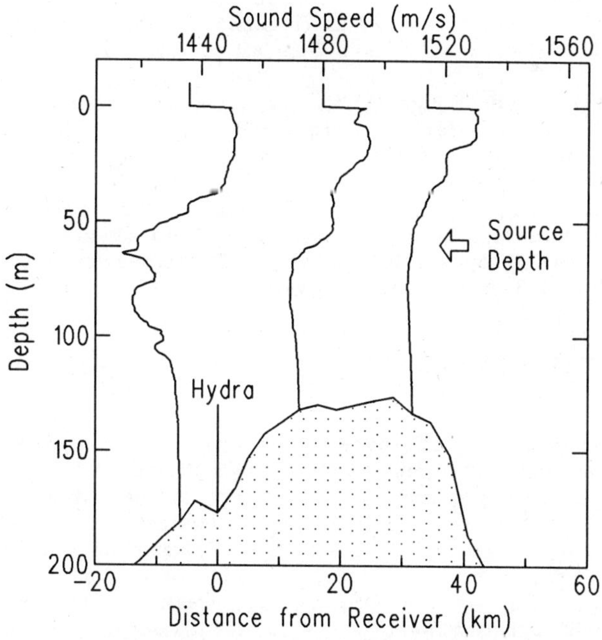

Fig. 2. Bathymetry and sound speed profiles for the experiment. The sound speed scale is at the top; vertical ticks at the sea surface are reference values of 1480 m/s. Profiles join the seabed at the appropriate range.

DATA ACQUISITION

In the Hydra array, the hydrophone signals of acoustic pressure are filtered, amplified, sampled and digitized by electronics inside the hydrophones. These digitized signals are sent in a parallel form to a micro-processor controller, then in a serial form to a surface buoy and then by radio link to the ship. The microprocessor controller also accepts radio commands from the ship and sets the hydrophone electronics in the desired operating state. This includes setting individual hydrophone gains, bandpass filters, calibration signals and the selection of the hydrophone signals desired from the total number of hydrophones available.

The acoustic data received at the ship are recorded directly on high density digital recorders for later playback and analysis, are converted to parallel data streams for direct recording by a PDP 11/34 minicomputer and some are converted to analogue form for monitoring. The spectral bandwidth of the system for this experiment was 1 Hz to 1 kHz and the dynamic range was 12 bits or 72 dB.

DATA ANALYSIS

Dispersion analysis of the time series data was carried out by filtering the signals in a band a few Hertz wide around the frequency of maximum amplitude of the interface wave and examining the filtered signal for zero-crossings. We used the time span of three consecutive zero-crossings to calculate an instantaneous frequency of the wave at the time of the middle zero-crossing; we calculated the group speed of that wavelet by dividing the shot-to-array distance by the time-of-flight of the wave. For each shot, we produced a plot of group speed versus frequency; these were later combined to produce a single plot containing data from all shots.

To calculate received signal levels, spectra of the received signals were computed using a fast Fourier transform (FFT) algorithm. The resulting power spectra were grouped into 1/3 octave bands, then added. Then the levels were corrected for system frequency response, gain and ambient noise levels. The noise levels were sampled before the receipt of the shot; only 1/3 octave measurements with a signal to noise ratio of 3 dB or greater were used in subsequent analysis. Propagation loss was calculated from these signal levels by subtracting the 1/3 octave source levels for the explosion.

DISPERSION OF THE INTERFACE WAVE

Fig. 3 shows the time history of the signals received at a bottomed hydrophone from shots at 11 successive ranges. The time axis is reduced travel time, that is, travel time minus range divided by a reference sound speed. In this case, the reference sound speed is the average sound speed in the water, so that the water arrivals align vertically. A vertical line thus represents a speed of 1480 m/s; the sloping line drawn on the figure represents a speed of 1000 m/s. The water wave arrival is identified by a burst of high frequency energy. This is preceded by very weak low frequency waves that are probably bulk compressional waves in the granite. The interface wave appears as a strongly dispersed low frequency arrival following the water wave at a group speed of about 1200 m/s. This wave is so dispersed at the longer ranges that its time of arrival is unclear. It is possible that there are several different interface modes present, corresponding to a multi-layered seabed, or that some interface waves are travelling by other than the direct path, that is, they may have been scattered in azimuth, arriving later than the direct arrival.

The simplest ocean acoustic environment admitting interface waves consists of a finite, homogeneous, water layer over a semi-infinite, homogenous, solid layer. Essen[6] has presented the characteristic equation governing the dispersion of the trapped modes in this environment:

$$4\sqrt{1-\left[\frac{C}{C_S}\right]^2}\sqrt{1-\left[\frac{C}{C_P}\right]^2}-\left[2-\left[\frac{C}{C_S}\right]^2\right]^2=r\left[\frac{C}{C_S}\right]^4\sqrt{\frac{1-\left[\frac{C}{C_P}\right]^2}{1-\left[\frac{C}{C_W}\right]^2}}\tanh\left[\frac{2\pi fH}{C}\sqrt{1-\left[\frac{C}{C_W}\right]^2}\right]$$

In this equation, C is the phase speed of the mode, Cs is the shear speed in the solid, Cp is the compressional speed in the solid, r is the ratio of water density to solid density, Cw is the sound speed in the water, f is the source frequency, and H is the water depth. The solution of this equation for C, regarding the other parameters as given, provides at least one but perhaps several real solutions: the phase speeds of the trapped modes. There are two special limiting cases contained within this equation: (1) if r=0 or fH/C<<1, the single real root is the Rayleigh speed, Cr, which is always just less than Cs; (2) if fH/C>>1, the tanh function approaches unity and the single real root is the Scholte speed, Csch, which is always less than the smaller of Cw and Cr. Physically, the first limiting case corresponds to the Rayleigh wave at a vacuum/solid interface; the second case corresponds to a Scholte wave at a fluid/solid interface. These two interface waves are dispersionless: their phase and group speeds are the same and independent of frequency.

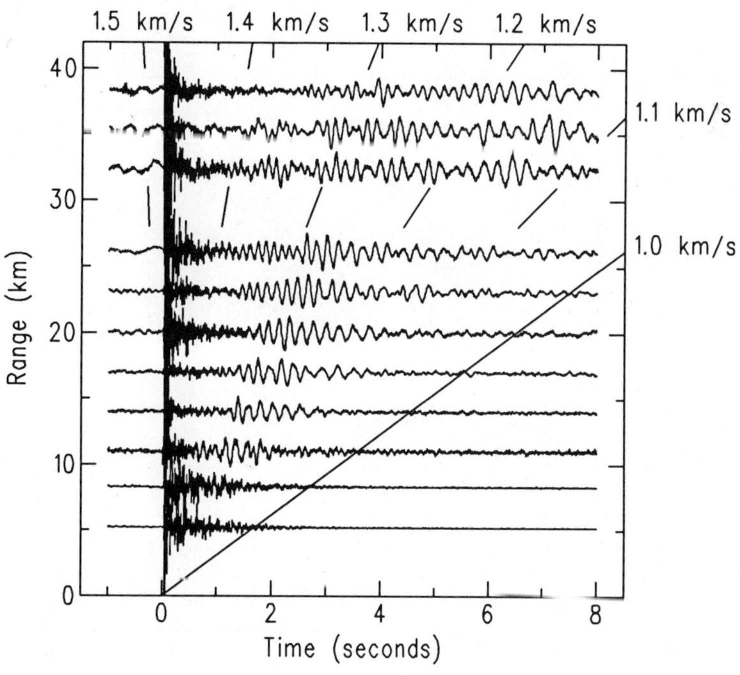

Fig. 3. Acoustic pressure versus time for several shots at different ranges. The water wave arrivals are aligned vertically; the straight line represents a group speed of 1000 m/s.

Returning to the general case, for which the water layer is finite, the phase speed C of the interface wave monotonically decreases from Cr at low frequency to Csch at high frequency. It is possible that other trapped modes exist. If Cw>Cs, the interface wave is the only trapped mode, the other modes are leaky. This condition leads to poor propagation at low frequencies. If Cw<Cs, other trapped modes may exist; if Cs is sufficiently high, there will be several trapped modes and the low-frequency propagation will be good.

Since the interface wave is dispersive, the group speed V is different from C and is also dependent upon frequency, but it may be calculated from the relation $V = C/[1-(f/C)(dC/df)]$. The quantity dC/df can be derived by differentiating the characteristic equation. The phase speed C must be calculated numerically for each frequency of interest, and then the group speed V may be calculated. In this way, a curve of group speed versus frequency can be generated for a specific acoustic environment for comparison with experimentally measured group speeds. To model our data, we used the parameter sets: Cw = 1468 m/s, Cp = 5500 m/s, Cs = 3300 m/s, r = 0.38, and H = 125 m and 177 m. Since the water depth was not constant, we have selected these two values to define limiting curves.

The plot of group speed versus frequency in Fig. 4 illustrates the dispersion of the interface waves, both measured experimentally and calculated from theory. The observed group speeds agree reasonably well with the model curves. We felt that the agreement between theory and experiment was sufficiently good to apply the same geo-acoustic model of the seabed to a propagation loss model.

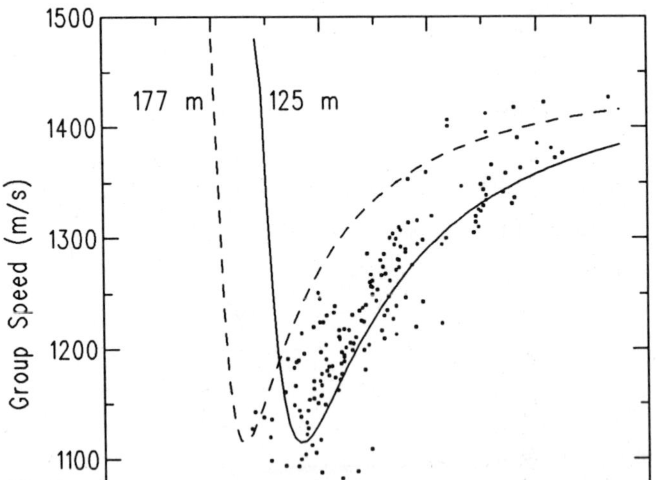

Fig. 4. Group speed of the interface wave versus frequency: experiment and theory. The two curves represent model results using the minimum and maximum water depths.

PROPAGATION LOSS

Fig. 5 shows the measured 1/3 octave propagation loss from 2 Hz to 813 Hz for several different ranges, and illustrates the interesting frequency dependence of propagation in this environment. If it was not clear from Fig. 3, Fig. 5 shows that the frequency of optimum propagation is associated with the interface wave, about 4 Hz. The propagation loss increases with increasing frequency dramatically. There is a band of very high propagation loss between 10 Hz and 100 Hz, centred at 30 Hz. At higher frequencies, the propagation loss reduces to a moderate level. Since Cs for granite is well above Cw, we did not expect to see the large losses at low frequencies that we saw for chalk, for which Cs<Cw; there ought to be several trapped modes. We were not comforted by the fact that our subsequent modelling efforts agreed with our expectations but not with the data.

We chose the fast-field program (FFP) developed at SACLANT ASW Research Centre[7] to attempt to model the acoustic propagation. This model seemed appropriate because it permits a multi-layered medium (necessary to include the depth-dependent sound speed profile) and because any of the layers may be a solid. Like other models of its type, which employ an FFT algorithm in the calculation, it assumes a range-independent medium; for our initial attempts, we did not consider this to be troublesome.

The FFP codes are based on transforming the range part of the acoustic wave equation from spatial coordinates into wavenumber coordinates using a Hankel transform. In this way the acoustic field is imagined to be the sum of a continuum of fields G(k) each propagating horizontally as a cylindrical wave with wavenumber k, that is, each travels at a different horizontal phase speed. For each value of k, the wave equation for the acoustic field reduces to a one-dimensional differential equation for G as a function of depth. For a

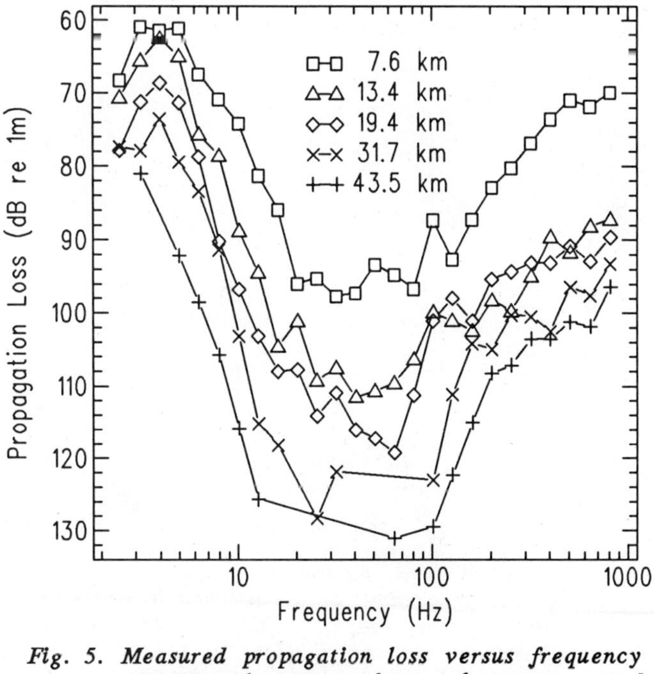

Fig. 5. Measured propagation loss versus frequency at several ranges for a bottom-mounted hydrophone. Source depth was 61 m.

medium made up of several homogeneous layers, it is straightforward to calculate G for any given value of k. The rapidity of the FFP models follows from recognizing that the transformation from wavenumber coordinates back to spatial coordinates can be accomplished via the FFT algorithm, provided that the Hankel function can be approximated by its asymptotic form and provided that the function G is sufficiently described by discrete samples calculated at evenly spaced values of k.

In order to calculate propagation loss for our experiment, we used the same geo-acoustic model of the seabed used in the dispersion analysis. Rather than assuming a homogeneous water layer, we used the sound speed profile at Hydra. We also introduced typical compressional and shear attenuations of 0.55 and 1.55 dB per wavelength, respectively, in the granite layer. To investigate the contribution of the interface wave to the propagation, we performed two sets of calculations: one including only the interface wave and one including all waves other than the interface wave. This is easily accomplished at an intermediate stage within the FFP model by equating G to zero at undesired wavenumber values before transforming from wavenumber coordinates to spatial coordinates. In a span of frequencies for which both calculations contributed significant energy, we computed their sum. The calculations were carried out from 2 Hz to 512 Hz and the propagation loss values at 20 km were selected for comparison with the experimental data.

Fig. 6 shows the comparison between experiment and theory. Note that the model agrees acceptably with the experiment only at those frequencies for which it predicts that the interface wave is solely responsible for the propagation. The disagreement at higher frequencies admits two hypotheses: either our geo-acoustic model of the seabed is inappropriate for higher

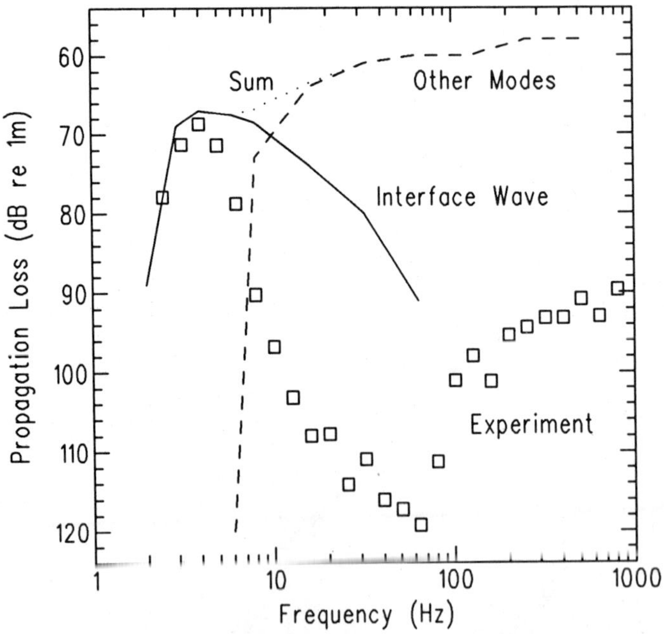

Fig. 6. Modelled and measured propagation loss versus frequency for a bottomed hydrophone at 20 km range. The three curves represent the contributions of the interface mode, all other modes and their sum.

frequencies or there is a loss mechanism active in the experiment which has not been included in the propagation model. Increasing the compressional or shear attenuations in the granite simply increases the propagation loss over all frequencies.

Addressing the first hypothesis, the agreement at 2 Hz to 6 Hz only verifies that our geo-acoustic model of the seabed is a reasonable description of the average seabed properties over a depth range of the order of an acoustic wavelength, that is, over several hundred metres. Thinner layers having quite different acoustic properties would not significantly alter the results at these frequencies since their influence is proportional to the amount of energy propagating within the layers relative to the total energy in the wave. On the other hand, such layers would dominate the propagation at wavelengths of the order of their thickness if they were near the ocean floor. It is possible that such depth variation of old rocks is the result of cracking and weathering over long periods of time. Although better agreement can be produced by introducing a multi-layered geo-acoustic model, the results are not satisfying since we have no evidence to guide the construction of such a model and, given enough adjustable parameters, we could generate any desired propagation loss curve.

We favour the second hypothesis, that scattering at the rough granite surface is responsible for the high losses. The geological evidence shows that the surface is quite rough and it would be unreasonable to propose that this has no observable effect. If this is the explanation, it is clear that a theory describing the scattering of the interface wave is necessary to explain the rapid increase of propagation loss above 4 Hz. We can see from the model predictions that, even if all other modes were highly damped due to scattering, the propagation loss due to an unattenuated interface wave alone is insufficient. Consequently, scattering of the interface wave itself should be important. In fact, below 8 Hz, a theory describing only the attenuation of the interface wave may account for the observed propagation loss. Of course, at higher frequencies, a scattering theory would have to apply to all modes to fit the data.

FUTURE WORK

The theory of the scattering of Rayleigh waves[8] and Scholte waves[9] has shown that the attenuation of these waves increases as frequency raised to the 5th power when the wavelength is large compared with the correlation length of the rough surface and the average height of the irregularities. Also, laboratory modelling of the transmission of ultrasonic Scholte waves over irregular surfaces[10] has produced wave trains and attenuation characteristics remarkably similar to our seismo-acoustic data.

Accordingly, we are beginning to investigate the theory of scattering of interface waves travelling along a rough solid surface under a finite fluid layer. This would be an extension of the work just mentioned, but would have novel features, since the interface wave is dispersive and there are other modes of propagation.

We also plan to repeat the experiment in nearby areas of higher and lower roughness.

CONCLUSIONS

We have performed an acoustic propagation experiment in a shallow ocean over a rough granite seabed which revealed a strong interface wave in the vicinity of 4 Hz and very high propagation losses in the band 10 Hz to 100 Hz. A simple dispersion model based on a water layer over a semi-infinite granite layer agrees reasonably well with the measured group speed of the interface wave. A propagation loss model using the same geo-acoustic model as an input only agrees at the lowest frequencies, and does not reproduce the high losses in the intermediate frequency band. Although another hypothesis exists, we propose that scattering at the rough granite surface is responsible for the observed losses.

ACKNOWLEDGEMENTS

We are indebted to Dr. H. Schmidt of SACLANT ASW Research Centre for providing DREA with a copy of their FFP computer code for our modelling efforts.

REFERENCES

1. D. Rauch, and B. Schmalfeldt, Ocean-bottom interface waves of the Stoneley/Scholte type: properties, observations and possible use, in: "Acoustics and the Sea-Bed," N.G. Pace, ed., Bath University Press, Bath, UK, p. 307 (1983).

2. P. R. Staal, Acoustic propagation measurements with a bottom mounted array, in: "Acoustics and the Sea-Bed," N.G. Pace, ed., Bath University Press, Bath, UK, p. 289 (1983).

3. D. D. Ellis and D. M. F. Chapman, Modelling of shear-wave related acoustic propagation on the U.K. continental shelf, DREA Technical Memorandum **84/P** (1984).

4. P. R. Staal, R. C. Hughes and J. H. Olsen, Modular Digital Hydrophone Array, in: "Proceedings of IEEE Oceans '81", Boston, p. 518 (1981).

5. P. R. Staal, Computer control of the underwater acoustic array Hydra, DREA Technical Memorandum **84/S** (1984).

6. H. -H. Essen, Model computations for low-velocity surface waves on marine sediments, in: "Bottom-Interacting Ocean Acoustics," W. A. Kuperman and F. B. Jensen, eds, Plenum Press,NewYork, p. 299 (1980).

7. H. Schmidt, Excitation and propagation of interface waves in a stratified sea-bed, in: "Acoustics and the Sea-Bed," N.G. Pace, ed., Bath University Press, Bath, UK, p. 327 (1983).

8. A. G. Eguiluz, and A. A. Maradudin, Frequency shift and attenuation length of a Rayleigh wave due to surface roughness, Phys. Rev. B **28**:728 (1983).

9. A. D. Lapin, Scattering of surface waves propagating over an uneven liquid-solid interface, Sov. Phys. Acoust. **15**:336 (1970).

10. J. R. Chamuel, "Transient Scholte and Rayleigh waves in periodic structures/ Experimental studies", DREP contract # 04SB.97708-3-3842, Sonoquest Advanced Ultrasonics Research, P.O. Box 584, Sudbury, Massachusetts 01776, USA (1984).

A COMPARISON OF SEISMIC AND HYDROACOUSTIC MEASUREMENTS AT VERY LOW FREQUENCIES IN DIFFERENT SHALLOW WATER AREAS

B. Schmalfeldt*

SACLANT ASW Research Centre
viale S. Bartolomeo 400
I-19026 La Spezia, Italy

ABSTRACT

Numerous sea-trials concerning the detection and propagation of very low frequency sound (2 to 24 Hz) have been conducted by SACLANTCEN, mostly in the Mediterranean Sea. These field-experiments were carried out by using triaxial Ocean Bottom Seismometers with additional bottom mounted hydrophones. The sound sources involved were explosions, CW and low-frequency sweep sources and very low frequency monochromatic sources. The paper will present and discuss some of the results in terms of seismic interface waves (Scholte waves). Some properties of these waves, such as phase-relationships, dispersion, attenuation and group velocity, are demonstrated exemplarily by means of spectral analysis methods. By comparing the results of different areas, it is demonstrated how different bottom types may affect these interface wave properties, making evident the strong correlation between bottom parameters and the interface wave propagation. Thus these experiments offer important input knowledge for the deduction of sediment properties by means of interface wave modelling.

INTRODUCTION

In recent years there has been a growing interest in bottom-borne propagation at low frequencies in shallow water. The study of interface waves, in particular, has resulted in the increase of knowledge concerning the shear velocity of the upper sea-floor since interface waves are closely related to shear velocity. On the other hand, shear velocity is one of the most important parameters for modelling the sound propagation at low frequencies in shallow water.

This paper deals with the presentation of some measurements of the interface wave in which the author was involved while working at the SACLANT ASW Research Centre from 1979 to 1983. In particular, the results of 3 field-experiments will be presented, 2 from the Mediterranean Sea and 1 from Danish coastal waters.

----

* Present address: MTG Marinetechnik GmbH, P.O.Box 70 12 49
D-2000 Hamburg 70, F.R.G.

It will be shown how different bottom types affect the interface wave characteristics such as group velocity, phase-relationships, attenuation and the S/N ratio of the seismic sensors and bottom-mounted hydrophones. How the shear velocity can be determined from this field-experimental data will be shown in two other papers at this conference[1,2].

PHYSICAL BACKGROUND

To make the paper self-explanatory we shall, very briefly, summarize the main properties of these interface waves as well as describe the main features of the SACLANTCEN Ocean Bottom Seismometer (OBS) and the data acquisition and processing facilities. For more details concerning the theoretical background refer to Ref. 3 and for more technical information refer to Ref. 4 .

Seismic interface waves are those waves that are guided by the water/sediment interface or by other sediment/sediment interfaces. The main properties are the following:

1. the energy is confined to the radial/vertical plane
2. there exists a phase-shift of about ±90° between the radial and the vertical component, resulting in an elliptical particle motion
3. there also exists a 90° phase-shift between the bottom pressure and the vertical seismic component
4. the energy decreases exponentially in a direction perpendicular to the interface
5. the phase velocity is of the order of some 100 m/s and is always less than the minimum of the sound speed in water and the shear velocity in the bottom
6. the propagation is only slightly affected by the water column, but very strongly by the upper sediment layers.

THE SACLANTCEN OBS AND THE DATA HANDLING

The OBS discussed here was developed at SACLANTCEN. It is a versatile and rugged tool consisting of three orthogonal lightweight geophones (Teledyne S 500), a hydrophone, orientation sensors and a compass. The system is operable up to a water depth of 200 m. The data is digitized in the OBS and then telemetered to the research vessel and stored on a disc of a 21MX Hewlett Packard computer. Together with this a comprehensive preprocessing package is available. In particular, a conceptual rotation of the geophone coordinate system can be applied to the seismic channels, thereby aligning one of the horizontal components into the propagation direction (radial component). The off-line processing includes spectral analysis packages as well as a module for energy calculations within selectable frequency bands.

RESULTS AND DISCUSSION

Table 1 gives an overview of the 3 locations which are considered in more detail later. Throughout this paper the three locations are classified by the letters "C", "R" and "K". The test site R is situated off Viareggio in the Ligurian coastal waters about 25 km southeast of site C. The location K is situated off Frederikshavn in the Danish Kattegat. At all 3 locations the water depth is less than 30 m.

The knowledge of the sediment layering of the upper sea-floor is very poor because normally no special investigations were conducted for this purpose. The sea-floor data of the Kattegat stems from the

Table 1. Overview of the 3 considered field-experiments

ID-label	C	R	K
location	Gulf of La Spezia	off Viareggio	Kattegat
water depth	16 m	28 m	30 m
sea-floor characte-ristics	silt (?)	consolidated, layered sediments	10 m clayey silts 200 m silty clays chalk basement
sources involved	explosives, LF-sweep	explosives, monochrom. VLF source	explosives, LF-sweep

literature[5]. The sources involved were, as indicated, explosives at all sites, a LF sweep source (11 to 24 Hz) at site C and K and a very low frequency monochromatic source (2 Hz) at R.

Seismic Ambient Noise

Before discussing the results of interface waves generated by deterministic sources, let us have a quick look at the seismic ambient noise which results from various local and distant sources. Figure 1 summarizes the standard deviation of ambient noise spectra (see the shaded area) of 9 shallow locations in the Mediterranean Sea, recorded at sea state less than 1. It shows the characteristic enormous increase of spectral energy below about 5 Hz. The solid line within the shaded area characterizes one representative spectrum from the Kattegat area. This demonstrates that the seismic background noise is consistent with the average noise levels found in the Mediterranean Sea.

Although the ambient noise characteristics are similar in the Danish and the Mediterranean sites, the propagation properties are quite different, as will be shown in the following.

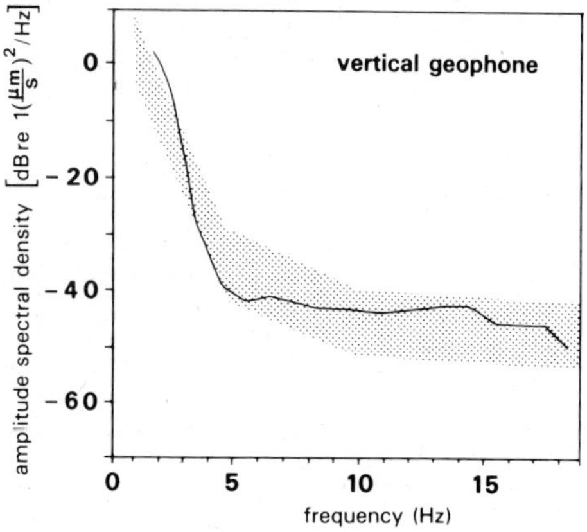

Fig. 1. Seismic ambient noise. Shaded: ±σ range of the Mediterranean Sea (9 different locations). Solid line: representative spectrum from site K.

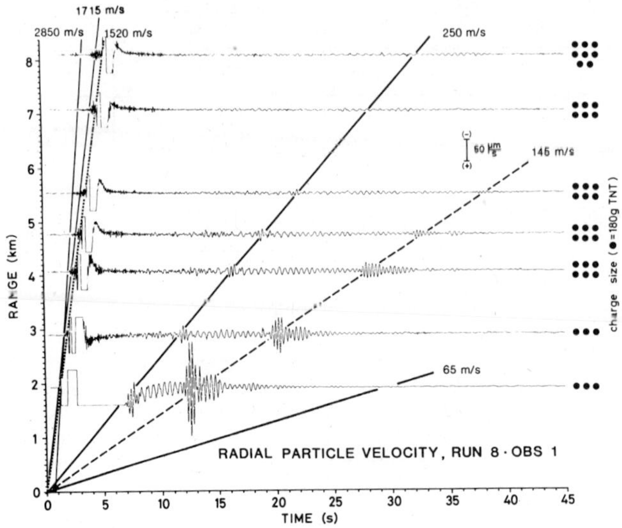

Fig. 2. Time/distance plot of the radial geophone component, site R.

Explosion-Generated Interface Waves

We shall now consider some results of explosion-generated interface waves. The explosives were TNT charges of less than 1.5 kg, which were fired on the sea-bottom.

Figure 2 shows a representative time/distance plot of a profile measured at site R. As one can see, the radial geophone component clearly picks up the direct water wave (1520 m/s) and even 2 forerunners (1715 and 2850 m/s) as well as pronounced interface wavelets with their low velocities (65 to 250 m/s) and low frequencies (2 to 5 Hz). Since all the seismograms have the same post-amplification factor, one can easily read the particle velocity of the wavetrains in absolute units (look at the 50μm/s bar) and hence draw some qualitative conclusions about the interface wave attenuation.

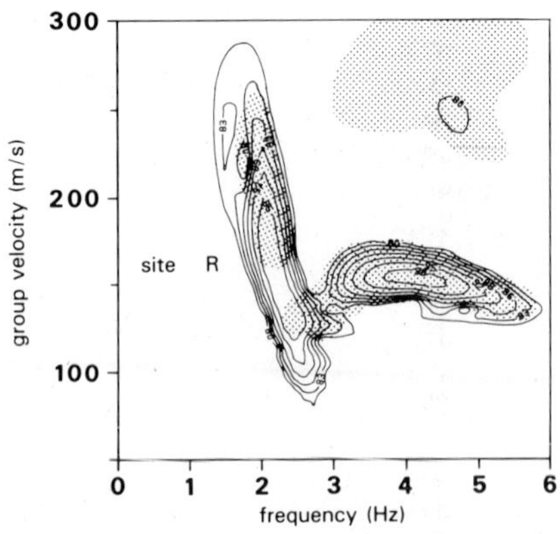

Fig. 3. Dispersion curves of the vertical and radial (shaded) geophone components. Shot distance: 1850 m.

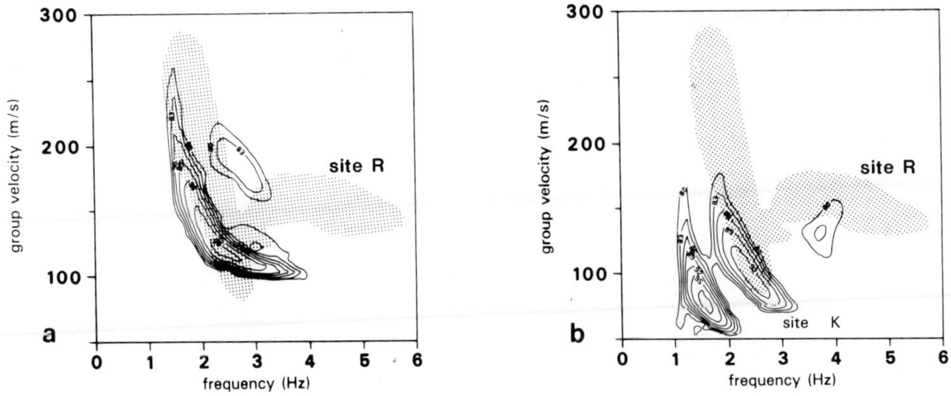

Fig. 4. Dispersion curve of the vertical geophone component.
a) site C at a distance of 2150 m
b) site K at a distance of 1040 m
along with the (shaded) dispersion curve from site R.

One of these wavetrains (the one at a distance of 1850 m) has been subjected to a dispersion analysis by means of the multiple-filter technique[6]. This technique results in a contouring of the iso-amplitude levels over a range of 20 dB (with arbitrary 99 dB as a maximum) in the group velocity/frequency domain.

Figure 3 represents the result of the dispersion analysis for the vertical and radial (shaded) geophone component. Both components reveal 2 dispersed branches (modes) centered at 2.2 Hz with 150 m/s (fundamental mode) and at 4.0 Hz with 145 m/s (1st mode). The fundamental mode is highly dispersed with drastically increasing group velocity with decreasing frequency. From this one can already qualitatively see that the shear velocity increases with depth by keeping in mind that the penetration depth of the interface waves is proportional to its wavelength.

It is often found that the dispersion of the corresponding radial component reveals an additional mode, as can also be seen in this particular case by looking at the shaded area in Fig.3 which clearly depicts a 3rd mode of the radial component.

Let us now have a look at the two other locations, beginning with site C which is about 25 km from the previous site. The dispersion curve of the vertical component is shown in Fig.4a (from an explosion at a distance of 2150 m) along with the corresponding (shaded) curve from site R, which has been discussed before. We notice that the dispersion properties are quite similar in this case, with a highly dispersed fundamental mode which, however, is slightly shifted towards lower frequencies. One might expect such a similarity since the test sites are close to each other.

A completely different behaviour is revealed by the dispersion of the vertical component from test site K, as can be seen in Fig. 4b, again along with the (shaded) dispersion curve from site R. We notice a conspicuous very low frequency/low velocity fundamental mode. From this alone we can conclude -qualitatively- that the bottom at this location must be much slower and softer. It should be mentioned that at this site, K, interface waves were obtained only infrequently. The example shown here stems from a shot at a distance of 1040 m from the OBS.

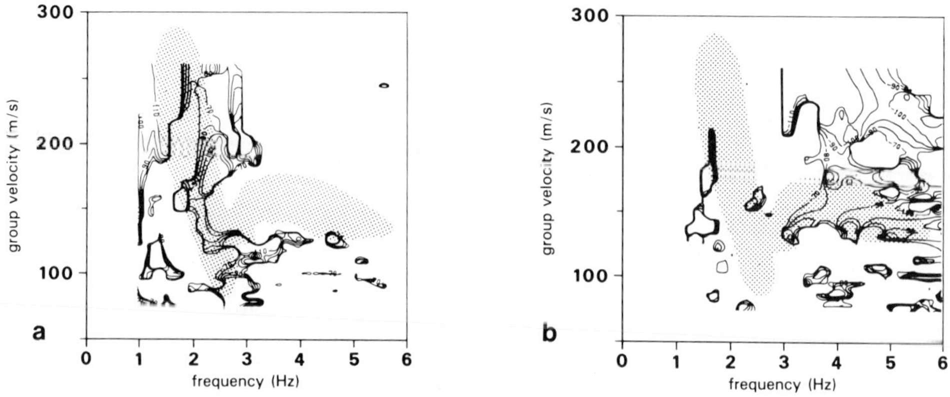

Fig. 5. Contours of the phase between the radial and the vertical component, along with the amplitude contours of the vertical component (shaded). a) positive phase, b) negative phase.

It is by fitting model computations to this kind of measured amplitude dispersion curves that reasonable information on the upper sediment structure can be derived. In addition to the amplitude dispersion curves one could benefit from the phase-relationship between the radial and vertical geophone sensors or, more strictly, between the corresponding particle velocities. The Figs. 5a,b represent these phase-relationships, from site R, along with the appropriate amplitude dispersion curve (shaded) of the vertical geophone component. In particular, the figures show the phase contours in the interval from 70° to 110° and from -70° to -110°, corresponding to prograde and retrograde particle motions, respectively. In this particular case the phase changes from prograde in the fundamental mode (below 3 Hz) to retrograde in the first mode above that frequency. Thus the consideration of the radial/vertical phases may offer additional help in checking the sediment layer models by fitting the experimental interface wave data.

At site C we were also able to derive the phase velocity of explosion-generated interface waves, by exploiting the phase spectrum (in the highly coherent frequency regime) between two OBS stations which were 17.5 m apart, see Fig. 6. From that we obtain the wavelength with its corresponding frequency, for example

$$\lambda = 135 \text{ m} \quad \text{at } 2.0 \text{ Hz}$$
$$\lambda = 45 \text{ m} \quad \text{at } 4.0 \text{ Hz}.$$

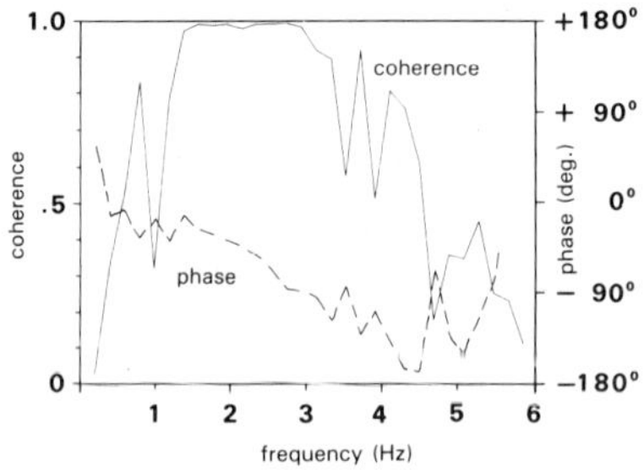

Fig. 6. (Quadratic) coherence and phase-shift between the vertical components of 2 OBS.

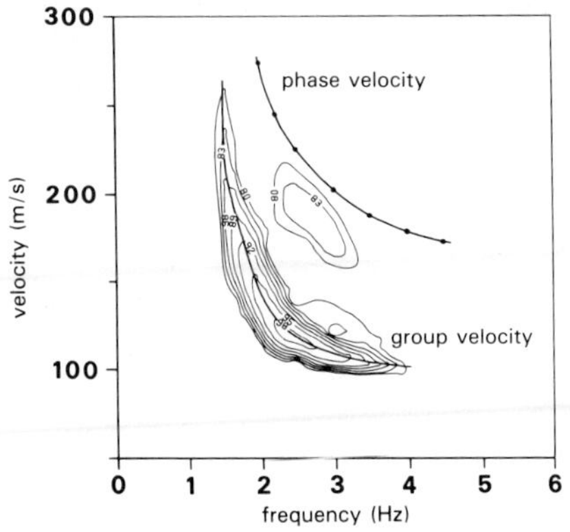

Fig. 7. Phase and group velocity of explosion-generated interface waves.

Finally, Fig. 7 represents, for site C, the calculated phase and group velocities along with the measured group velocity.

Measured phase and group velocities can in like manner be exploited for sea-floor modelling studies. There is one point to be kept in mind, however. From phase velocity measurements we can model the bottom parameters between the OBS stations only, which means that the method is limited to distances in the range of about 100 m or less; greater distances result in ambiguity of the phase-shift. On the other hand, from the group velocity, average shear parameters can be derived which are valid up to several miles.

<u>LF-sweep source</u>

In the following we shall present some results which were obtained by operating a low frequency (LF) source at the locations C and K. At both sites the projector was used in the "sweep mode" which means its frequency was cyclically increased from about 10 to 20 Hz at a repetition rate of about 40 s. At site K this source was towed 12 m below the surface, whereas at site C the source ship was anchored when operating the projector.

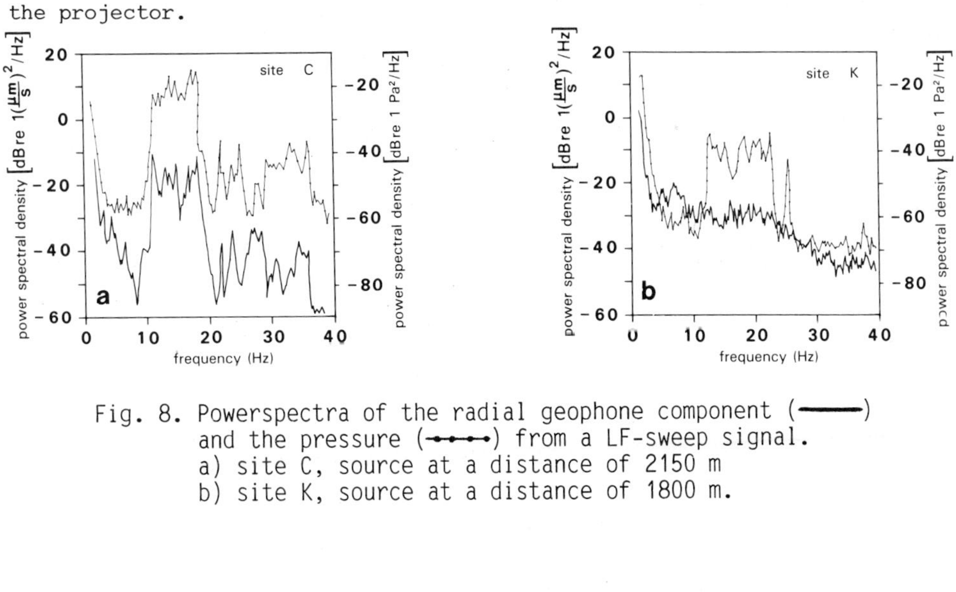

Fig. 8. Powerspectra of the radial geophone component (——) and the pressure (-•-•-•-) from a LF-sweep signal.
a) site C, source at a distance of 2150 m
b) site K, source at a distance of 1800 m.

Figure 8a shows the power spectra of the response of the radial geophone component and the (bottom mounted) hydrophone to a sweep signal at a distance of 1800 m, taken from location C. It reveals a high energy content in the 10 to 20 Hz interval, in both of the sensors, which exceeds the background noise by about 25 dB. In contrast to this, the propagation in the Kattegat area (site K) is much poorer, at least with respect to the seismic path. As shown in Fig. 8b the radial geophone hardly responds to the radiated sweep signal, whereas the performance of the hydrophone is much better. However, comparing the absolute spectral levels sensed by the hydrophones at the two sites, we realize that they are 15 to 20 dB higher at site C, even though the source at this site is about 350 m further away.

These findings are summarized in Fig. 9 which depicts the attenuation of the seismic and hydroacoustic energy along the track on which the projector was towed. From this we learn that for site K the overall attenuation (no corrections applied for spreading loss) within the first 1.5 km is about

17 dB/km for the radial geophone and
21 dB/km for the bottom mounted hydrophone.

This can be compared with the corresponding values from site C, see the crosses in Fig. 9. If we connect the (only) two points by a straight line we obtain the following attenuation coefficients:

about 4 dB/km for the radial geophone and
about 5 dB/km for the hydrophone.

The latter result compares quite well with our previous findings for the attenuation of explosion-generated interface waves (frequency band from 2 to 5 Hz) in an area close to the one under consideration.

Very Low Frequency Monochromatic Source

Finally, we shall present one result of probing the sea-floor by means of a very low frequency (VLF) source which radiated a stable 2 Hz frequency. Figure 10 shows the power spectral density, in the form of shaded dB-levels, of the radial geophone response, when the source was

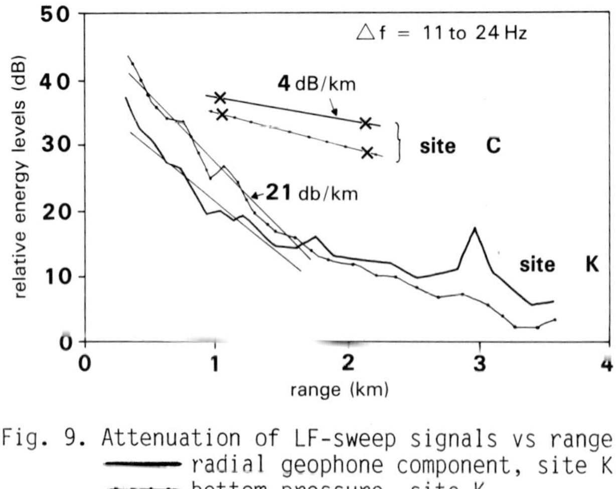

Fig. 9. Attenuation of LF-sweep signals vs range.
——— radial geophone component, site K.
-•-•-•- bottom pressure, site K.
——— radial geophone component, site C.
x···x bottom pressure, site C.

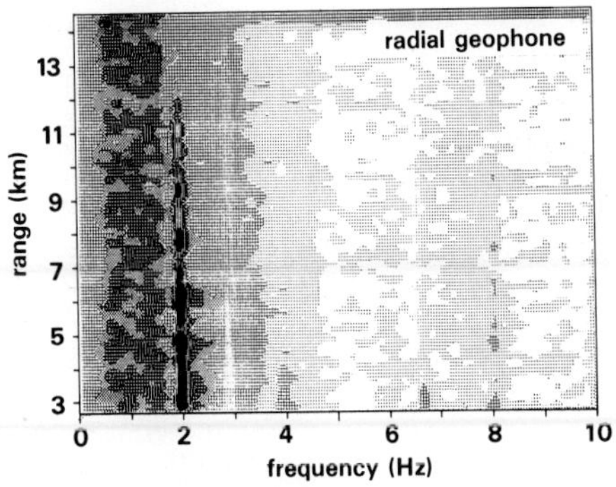

Fig. 10. Power spectral density in the form of shaded dB-levels vs range (time) from a VLF source radiating at 2 Hz, site R.

moving along a track away from the OBS. It reveals the strong 2 Hz line, which is discernible up to a distance of about 12 km. That interface waves are involved in this propagation phenomenon is proved by the fact that the 2 Hz frequency is well below the cut-off frequency of the 28 m water duct. Furthermore, there exists a quite stable phase-shift of about + 90° between the radial and the vertical geophone components (not shown here).

Along the track of the VLF source we calculated the attenuation of the energy within the narrow frequency band from 1.6 to 2.4 Hz including the 2 Hz of the sound source. Figure 11 shows the result for one seismic component and also for the hydrophone[7]. From this we determine an average value of the (overall) attenuation coefficient of about 2 dB/km. This value is consistent with that used by H. Schmidt[8] in successfully modelling the sea-floor by fitting experimental dispersion curves.

In Fig. 11 we also notice an energy fluctuation of 1 to 2 cycles per km. We believe that these fluctuations stem from the interaction of the interface wave with irregularities in the subbottom structure. UNIBOOM profiling along the track provides evidence for this assumption[7,9].

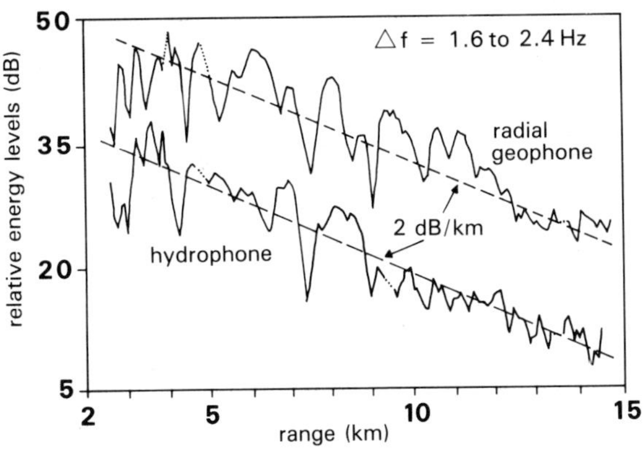

Fig. 11. Attenuation of a VLF signal vs range for the radial geophone component and the bottom pressure.

Additional details on the measurements discussed in this paper can be found Ref. 10 .

CONCLUSIONS

It has been demonstrated how different bottom types may affect the properties of interface waves. In particular, the splitting into different dispersion modes, the energy attenuation, and the relative S/N ratios of seismic sensors and hydrophones have been investigated. Explosion-generated interface waves reveal main frequencies from 1.5 Hz to about 4 Hz and group velocities from about 80 m/s to 185 m/s. The attenuation coefficients of the seismic components range from 2 dB/km to about 21 dB/km and the S/N ratio of the hydrophones can sometimes exceed that of the seismic sensors. Together with the phase-relationships between the radial and the vertical components all these characteristics may be used in sea-floor modelling studies, which are based on experimental field data.

ACKNOWLEDGEMENTS

This paper is largely based on material prepared in collaboration with Dr. D. Rauch who worked with me at SACLANTCEN. The data from the Danish Kattegat stems from a joint experiment with the NDRE in Horton, Norway in 1982. Part of the presented results has been analyzed at the SACLANTCEN after my departure from that research centre. In these activities I was kindly aided by the Computer Department (COM).

REFERENCES

1. F. B. Jensen and H. Schmidt, Shear properties of ocean sediments determined from numerical modelling of Scholte wave data, (this volume).
2. R. Szelwis, Near-surface shear wave velocity from multimodal Rayleigh waves. To be published.
3. D. Rauch, Seismic interface waves in coastal waters: A review, SACLANTCEN Report SR-42, La Spezia, Italy (1980).
4. B. Schmalfeldt and D. Rauch, Explosion-generated seismic interface waves in shallow water: Experimental results, SACLANTCEN Report SR-71, La Spezia, Italy (1983). [AD A 134551].
5. J. I. Faleide, private communication, University of Oslo, Norway.
6. A. Dziewonski, S. Bloch and M. Landisman, A technique for the analysis of transient seismic signals, Bull. Seism. Soc. Am. 59:427 (1969).
7. H. B. Ali and B. Schmalfeldt, unpublished report (1984).
8. H. Schmidt, Determination of bottom properties from seismic interface waves, SACLANTCEN, CP-34, La Spezia, Italy (1984).
9. E. Michelozzi, private communication, SACLANTCEN, La Spezia, Italy.
10. B. Schmalfeldt and H. B. Ali, Low-frequency seismic and hydroacoustic studies in the Oslo Fjord and the Kattegat, SACLANTCEN Report (in preparation).

INTERFACE WAVE STUDIES ON THE LIGURIAN SHELF USING AN OBS ARRAY: EXPERIMENTAL RESULTS AND PROPAGATION MODELS

Michael Snoek, Guido Guidi, and Enzo Michelozzi

SACLANT ASW Research Centre
Viale San Bartolomeo 400
I-19026 La Spezia, Italy

ABSTRACT

Seismic waves travelling in the water/sediment or subbottom sediment/sediment interface have been the subject of considerable interest in recent years. Experiments have confirmed the existence of interface waves in different geological environments. Some progress has been made in understanding the propagation and attenuation characteristics of these waves. However, the generating mechanisms are poorly understood. In particular, what are the roles of geometry and bottom characteristics in coupling waterborne sound into the sea bottom? What is the acoustic-to-seismic energy conversion process? Using explosives as sound sources, an experiment was performed in which the propagation along different interface paths was measured by an array of three tri-axial ocean bottom seismometers. To obtain realistic input data for the propagation models, continuous seismic profiles were run and cores were taken along the propagation path. Results of these experiments and a comparison of measured and modelled data are presented.

INTRODUCTION

Results from SACLANTCEN experiments and other published data show that the propagation and attenuation characteristics for seismic interface waves are still poorly understood and are ofttimes ambiguous (Ali & Schmalfeldt [1] and Jensen & Schmidt [2]).

Rauch [3,4] has outlined the theoretical background for the mathematical treatment of interface waves. Jensen & Schmidt [2] have also presented a modelling approach to reconstruct shear speed and shear attenuation from interface waves for the upper relevant sediment layers.

Unclear, however, is how and under which conditions acoustic energy converts into seismic energy. It has been frequently observed that in changing the bearing of a profile or the frequency spectrum of the source propagation, interface wave characteristics can change dramatically. The changes can range from signal enhancement to disappearance. Before a reasonable prediction of attenuation and propagation characteristics can be made, the driving mechanism for interface wave conversion and propagation have to be carefully studied.

To describe this conversion, one must determine the sensitivity of the process to differences in the physical properties and geometry of the surface layers, to geological structure, to the frequency content of the seismic source and to the bearing and position of the source with respect to the receiving sensors. The interrelations of these parameters affect the complexity of the conversion process. The interpretation of the data has to be supported by a modelling phase based on input data which should be as complete as possible. This requires a multi-parameter survey involving methods and systems schematically shown in Fig. 1.

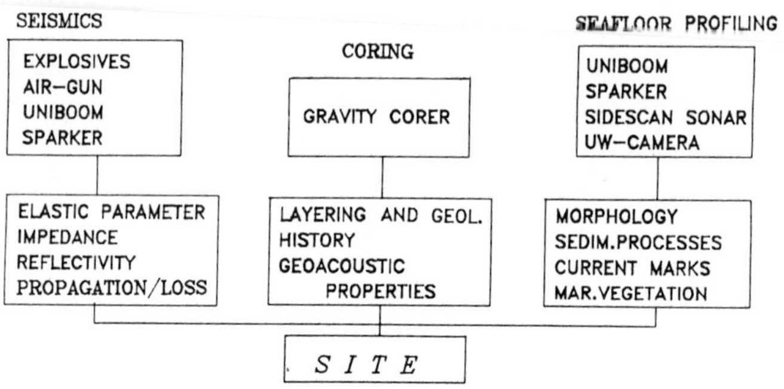

Fig. 1
Direct and indirect methods of investigations for a detailed ocean bottom/subbottom survey

INTERFACE WAVE STUDIES

We have conducted a set of experiments on the Ligurian shelf off La Spezia, Italy, to study the interface waves along different propagation paths (Fig. 2) in a region where we could expect silty clay and clayey silt layers over a more sandy subbottom.

The instrumentation and the setup of SACLANTCEN interface wave experiments have been described in detail by Rauch & Schmalfeldt [5] and Ali & Schmalfeldt [1]. We operated with two ships and deployed three tri-axial ocean bottom seismometers in a triangular configuration, with distances of 537 m, 352 m and 315 m between them. All distances were relative to a marker buoy. Positions were controlled by Loran C, satellite navigation, radar, and terrestrial navigation.

Charges were fired electrically on the sea bottom, their sizes ranging from 180 g (standard size) to 1080 g. The data were available a short time after each shot, thus allowing for correction of charge size if necessary. After the shot instant, the distances to the receivers were controlled by using the unprocessed recordings; all data were bandpass-filtered, 2-14 Hz, and stacked in time-distance plots. The response of the four sensors of seismometer No. 2, for different runs B and C, are displayed in Figs. 3 and 4. In Fig. 5 we see the response of the z-component (the vertical seismometer) for profiles of different azimuth. We can trace interface wave phases up to a range of 4 km; beyond that the signal-to-noise ratio becomes too small. However, even for the close ranges different modes cannot be identified easily. This is a further indication of a rather complex subbottom structure. We marked events of the interface wave train and correlated them with group velocities, as shown in Table 1. The values obtained are in the expected range and typical for the environment encountered. They show a certain scatter, amounting to about 6% for the slow end and 35% for the higher end of the interface wave train for profiles A-B and C-D. On profile F, however, velocities are considerably higher (see Table 1), remarkable because profiles B and F vary in direction by only 9°. The data also contains p-wave phases travelling as headwaves or refracted waves in deeper interfaces with velocities of 1.8 to 2.4 km/s. These data will be included when modelling the full wave field.

We calculated the interface wavelet energy generated by standard charges for different runs and varying ranges. Having only a few profiles, we obtained information on the distribution of interface wave energy for selected azimuthal angles, rather than a continuous azimuthal distribution of energy. We compared two pronounced frequency bands of 4-5 Hz and 8-9 Hz for three different ranges (Fig. 6). There are considerable differences in energy levels, ranging as high as 18 dB. At this stage, however, the results apparently are contradictory, with inconsistency in range and frequency.

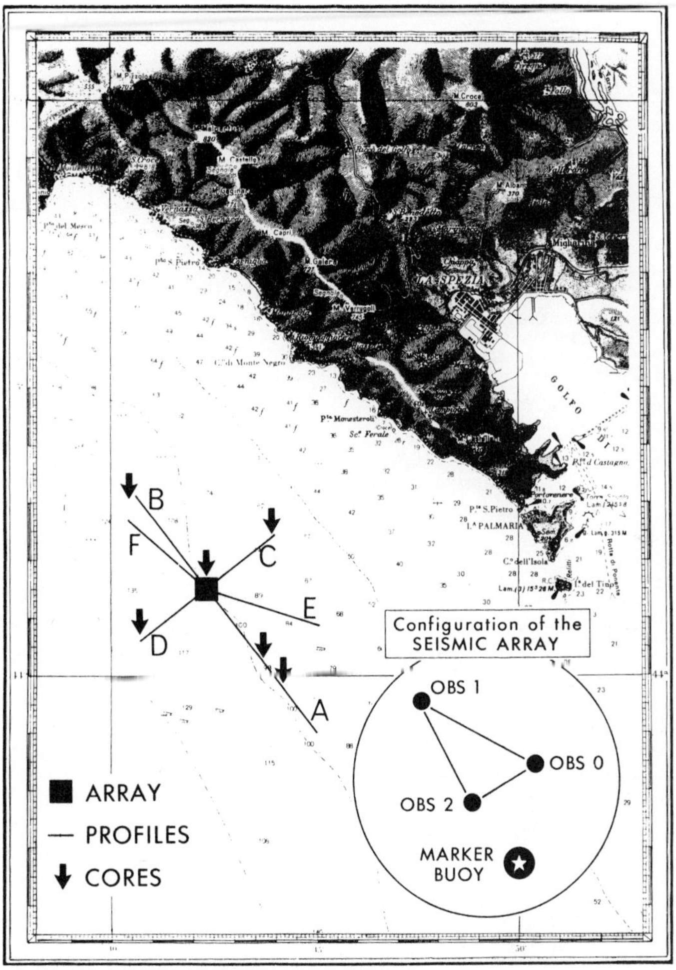

Fig. 2 Area of operation: Ligurian Sea off the coast of Italy

Whereas we can observe a trend in the behaviour for the 8-9 Hz frequency band - lowest attenuation in direction of profile D and highest attenuation on profile B (data were not available at all ranges for profile F) - no general trends can be observed for the 4-5 Hz frequency band. As with the group velocity analysis, the most pronounced effects can be observed on profile F, where we find extremal values for the attenuation, highest for the 8-9 Hz frequency band and, at the same range, the lowest for the 4-5 Hz frequency band. This analysis was based on only three ranges and two frequency groups; further analysis should reveal a more coherent picture of spatial energy distribution when extended to full range and all frequencies.

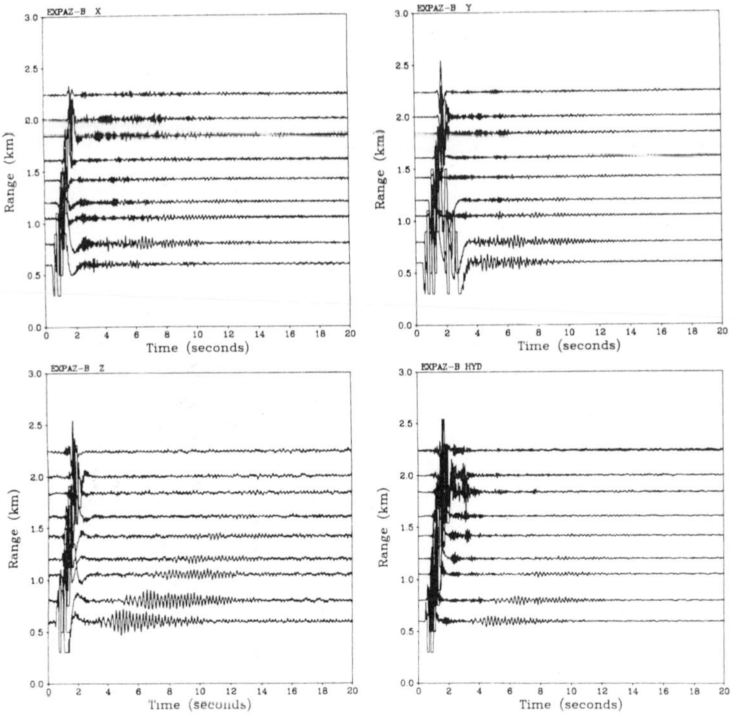

Fig. 3 Stacks of seismograms for profile B, X = radial, Y = transversal, Z = vertical and HYD = hydrophone data

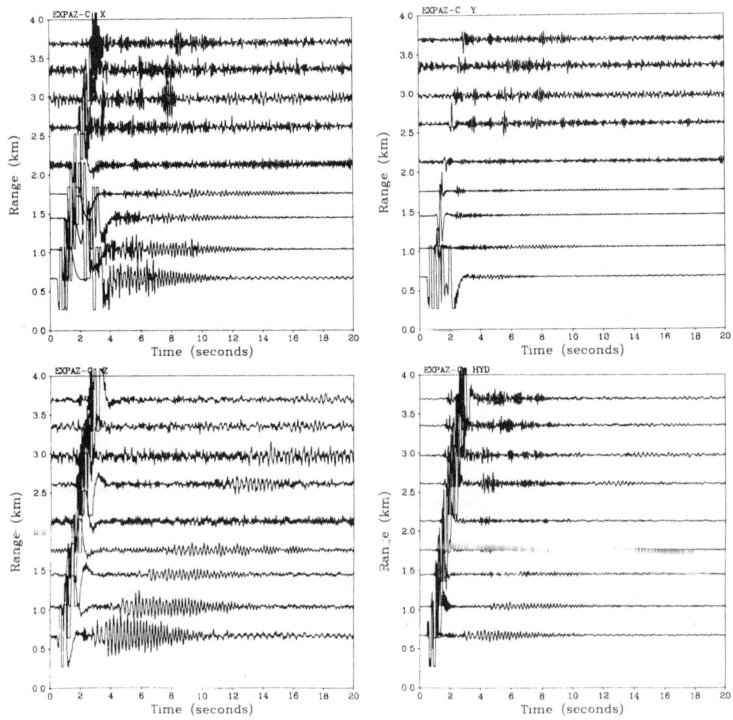

Fig. 4 Stacks of seismograms for profile C, X = radial, Y = transversal, Z = vertical and HYD = hydrophone data

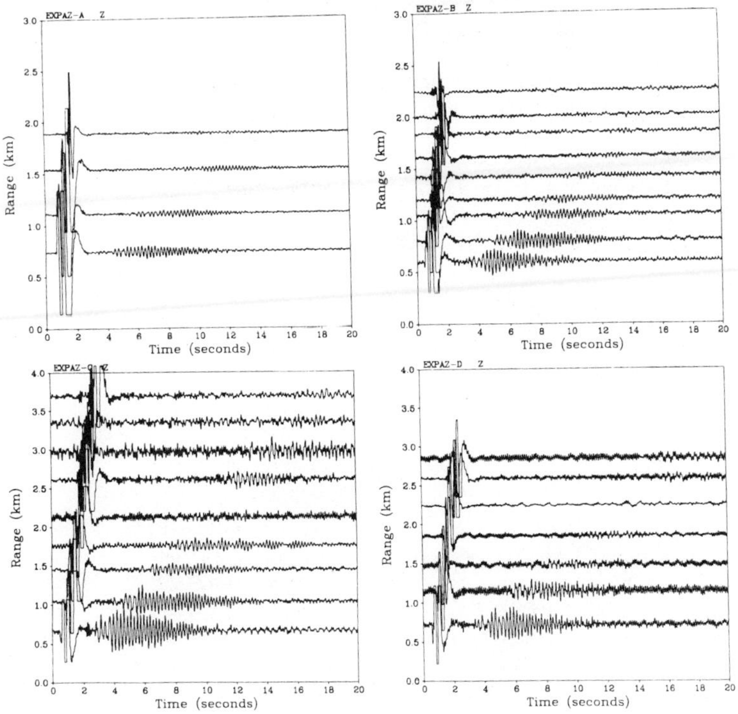

Fig. 5 Stacks of seismograms for profiles A, B, C, D. Only the Z-component is displayed.

Table 1
Interface mode velocities

PROF.A	65	122	170	260
PROF.B	60	121	154	190
PROF.C	62	132	228	296
PROF.D	66	143	192	242
PROF.F	88	210	235	325
	Group velocities (m/s)			

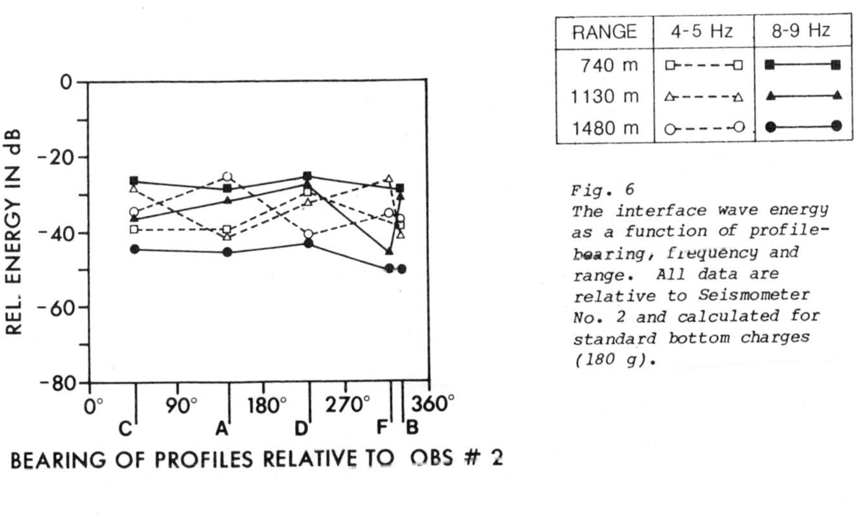

Fig. 6
The interface wave energy as a function of profile-bearing, frequency and range. All data are relative to Seismometer No. 2 and calculated for standard bottom charges (180 g).

A further approach in analyzing the interface wave trains was to apply the multiple filter technique or dispersion analysis suggested by Dziewonski [8]. This technique provides a tool to separate modes of different order in a seismogram. However, this procedure has limitations in this case because the interface wave trains display only a poor modal structure.

We will present herein a characteristic case, which will also be used in modelling the data by applying an inverse method. The dispersion analysis was run and displayed in a Gabor matrix of the data from profile C. Prior to the analysis a conceptual rotation of the geophone orientation was performed, i.e. an operation was applied to obtain maximum response of the radial sensor in the direction of the profile. We determine the radial (R) and transverse (T) components from the actual X and Y components of the seismometer for a given source bearing θ by:

$$R = x \cos\theta + y \sin\theta$$

$$T = -x \sin\theta + y \cos\theta$$

Hence, only Gabor matrices for the radial and vertical components are displayed, (Fig. 7). The most striking feature is the high velocity gradient effective over a very narrow frequency band of 2 Hz. We find a reduction in group velocity from about 260 m/s to 74 m/s over a frequency range from 3.5 to 5.5 Hz. In addition to this we find some further features of interest on the radial geophone data, marked as local maxima, which at this stage are taken as indicators of the complexity of the subbottom structure.

Earlier experiments by Essen et al [6], Rauch & Schmalfeldt [5] and Schmalfeldt & Rauch [7] have clearly demonstrated a correlation between the physical properties of the seafloor and subbottom and the propagation characteristics of interface waves. Therefore in this area, with its pronounced lateral variations in sedimentation rates and processes, the effects depend on the bearing (direction) of the profiles. Further, there is always a certain amount of scatter involved in arrival times of seismic waves travelling near the surface of the crust, or, in this case, in the top sediment layers. Because of this, one cannot attribute with certainty the observed effects to anisotropy or to lateral heterogeneities.

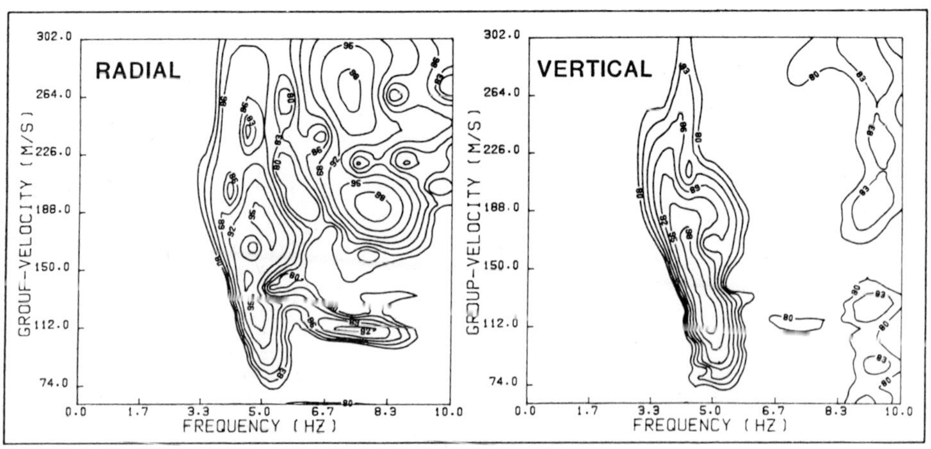

Fig. 7 Gabor matrix for shot No. 2 on profile C (see Fig. 2) for range 0.550 km Displayed are the vertical and radial components.

Prior to the actual interface wave experiment a set of high resolution seismic UNIBOOM profiles were run. The experiment had three goals: to determine experimentally relevant coring locations, to enable later correlation of geoacoustical parameters and geological features and to map the topography of the top layers.

Two typical profiles are shown: profile A-B with a bearing of 320° (Fig. 8), and profile C-D with a bearing of 25°, representing a sloping

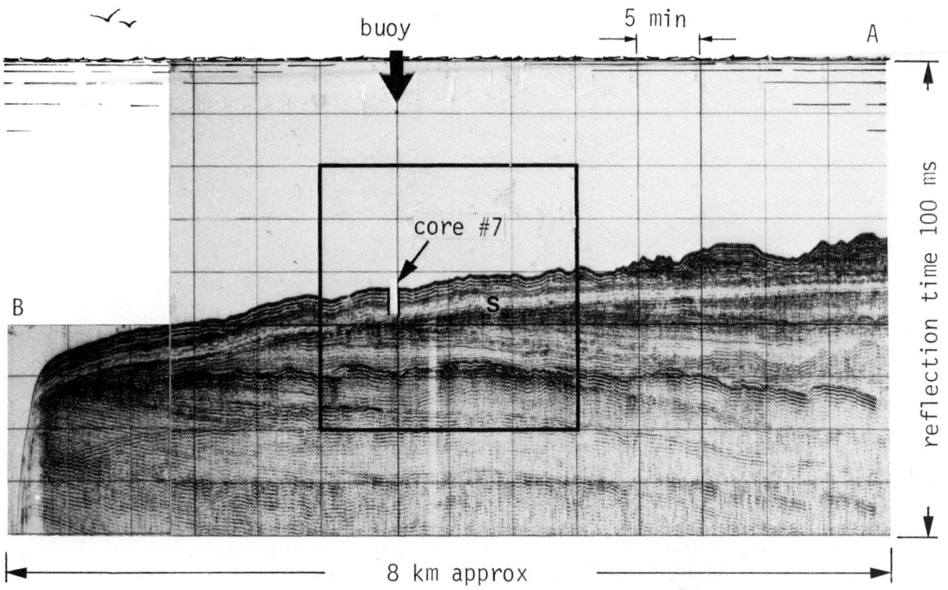

Fig. 8 Reflection seismic UNIBOOM profile A-B

seabottom (Fig. 9). Profile A-B has a uniform water depth (it follows the 100 m contour; all other profiles crossed contour lines).

The major structural feature recorded on profile A-B is a slightly tilted reflecting surface that outcrops on the slope. This reflector, which is interpreted as the bedrock surface during the last glacial period, shows fracturing as well as erosion. It is overlain by sediment wedges that thin out towards the slope and consist of some weaker reflectors, a dark band of sediments with some surface roughness and a laminated top layer. In general we can distinguish two different regions: the western part with a more regular and uniform sedimentation and an eastern part that displays strong lateral variations in sediment accumulation. This coincides with results from the seismic runs, which indicate that at a range greater than 2.5 km eastwards from seismometer No. 2 either poor or no interface wave trains were recorded. On profile C-D the most interesting features are cross bedding and the slumps, indicators of recent sediment dynamics. Marks of earlier gravitational sliding as well as the stratification of the upper sediments can be clearly identified.

Eleven cores were taken with a recovery range from 2.0 to 3.95 m and an average recovery of 3.80 m. Three of the cores were taken along profile A-B: One was taken at the site of the seismometer, one on the slope, and one in the region of anomalous sediment accumulation (see Fig. 2). Eight cores were taken along the other profiles to allow for spatial interpretation of sediment distribution. The No. 7 core, closest to the array, has been analyzed and the others have been inspected.

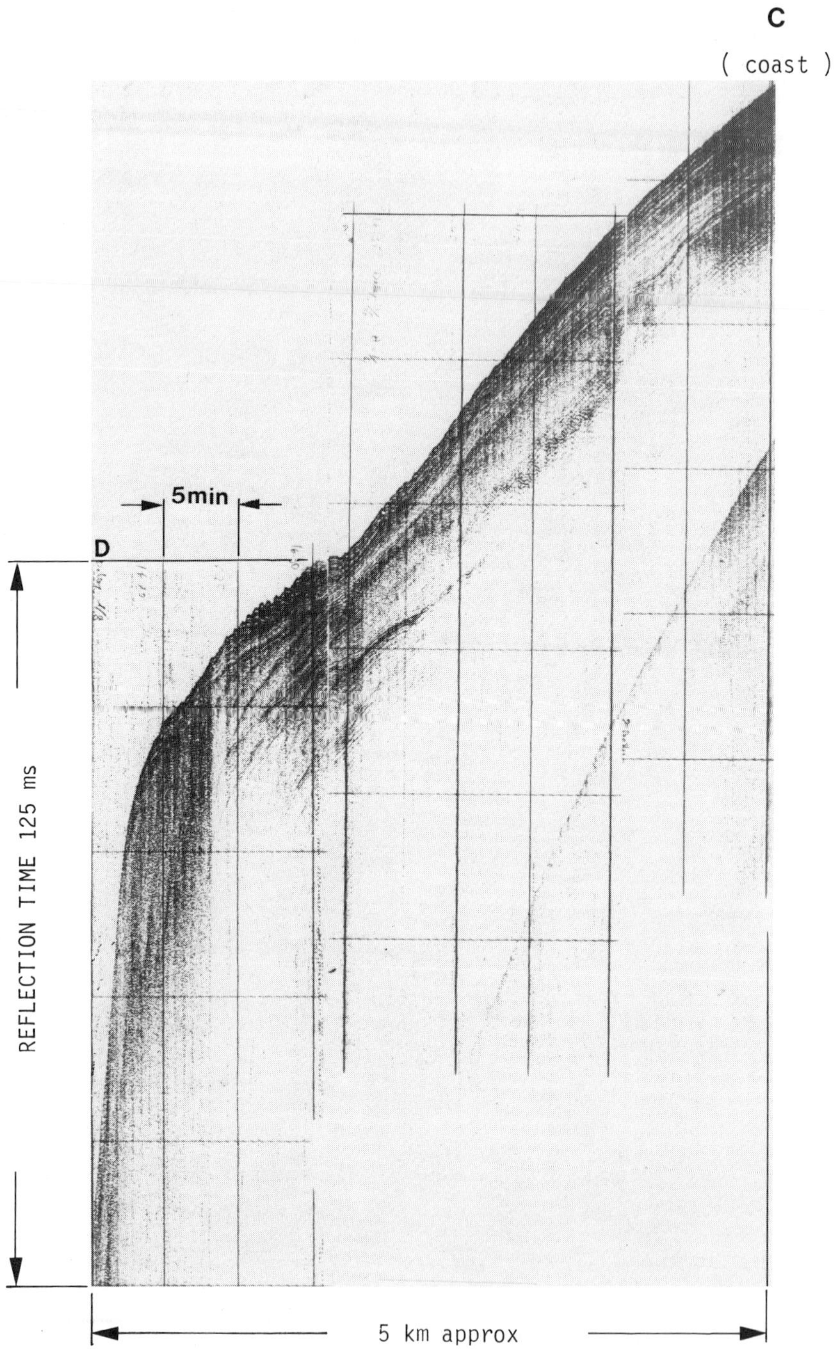

Fig. 9 Reflection seismic UNIBOOM profile C D

The core analysis shows that the lower end (3.70-3.85 m) has a sandy composition (46.8% sand, 27.6% silt, 25.5% clay) that may correlate with sand at 3.95 m at a nearby core site. We find some dark laminations that correspond to regions of higher p-wave velocities (see Fig. 10). From the available data we can conclude that there exists a sandy interface at about 4 m of depth. We can correlate this event to the reflector S marked in Fig. 7.

The continuous seismic reflection profiles indicate strong lateral heterogeneities. It therefore can be assumed that the observed dependencies of interface velocities and energy on the direction of propagation are caused primarily by lateral inhomogeneities. However, this does not rule out the effects of anisotropy. The further analysis of data from all three ocean bottom seismometers (OBS) should allow more precise discrimination of the various effects involved.

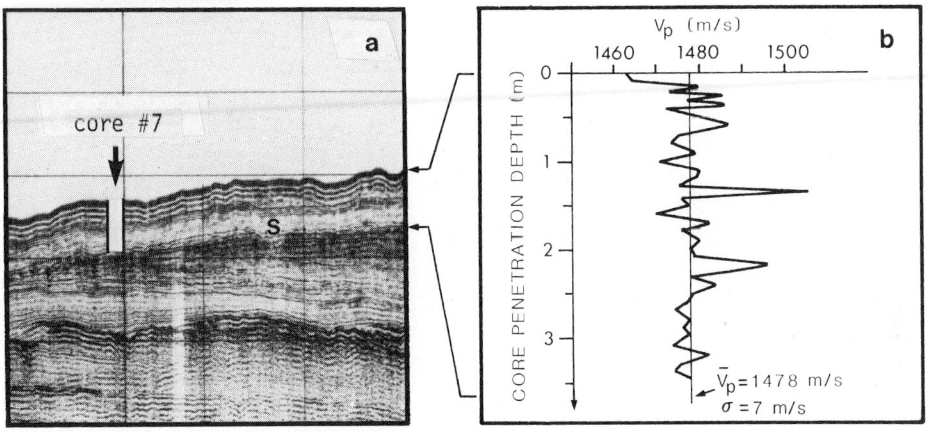

Fig. 10 a) Site of the seismometer array and core No. 7 with indicated analyzed column.

b) P-wave analysis of core No. 7

MODELLING

The inverse method for calculating the dispersion of interface waves trains needs realistic input data, such as measured velocity-depth profiles for P and S waves. We have at this stage the following kind of data and information:

- Sound speed profiles in the water column.
- A velocity depth profile for P waves in the core.
- The actual thickness of the uppermost soft layer.
- Evidence for a sandy substratum.
- Group velocities from seismogram stacks.
- Geoacoustic properties from the core analysis.

Using these data we applied modelling codes from which the group and phase velocities for different modes were calculated as a function of frequency (Fig. 11).

Fig. 11
Calculated dispersion curve from input model based on experimental results.

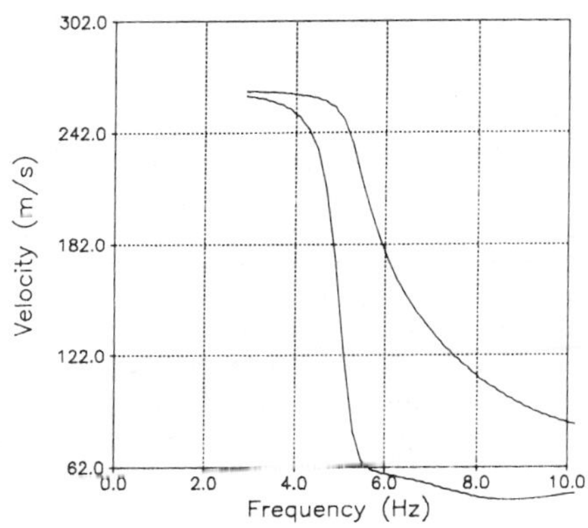

The first or fundamental mode has been successfully modelled. The results fit the data reasonably well. However, modes of higher order could not be matched to the data, although several models and variations were tried. The final model used was a sandy halfspace subbottom with a shear velocity of V_s=300 m/s, overlain by a 4 m thick layer of soft sediments with a shear velocity profile of 78 m/s on top and 96 m/s on the bottom. A shear velocity of 300 m/s was used for the halfspace, since this has proved to be a realistic value for other measurements in this region.

Not all features seen in the Gabor matrix display can be modelled. This is because the local maxima revealed in the dispersion analysis of the field data may be caused by structural effects of the subbottom rather than higher order modes. Also, the input model needs considerable adjustment to the real world in order to allow the modelling of higher order modes. Nevertheless, if theoretical results agree reasonably with field data for at least the fundamental mode, then the geoacoustic properties of the uppermost layer, that is the dispersion behaviour of interface waves, can be predicted to a reasonable degree of accuracy. With the sensors sitting on the top layer of the seafloor, in general only conversion processes at the seafloor or near seafloor interfaces can be recorded because interface wave energy decreases exponentially perpendicular to the propagation path.

CONCLUSION

By evaluating all of the measured data, we can begin to resolve the pertinent problems that are related to the effects and the role of the geological environment on interface wave propagation. The preliminary data analysis has shown that a multiparameter survey combined with a powerful analysis and modelling phase can provide a deeper understanding of the relevant conversion processes. This will lead to a greater accuracy for predictions of interface wave characteristics.

REFERENCES

1. Ali, H., Schmalfeldt, B. Unpublished report, 1984.
2. Jensen, F., Schmidt, H. Shear properties of ocean sediments determined from numerical modelling of Scholte wave data (this volume).
3. Rauch, D. Seismic interface waves in coastal water: a review, SACLANTCEN SR-42, La Spezia, Italy, SACLANT ASW Research Centre, 1980.
4. Rauch, D. On the role of bottom interface waves in ocean seismo-acoustics: a review (this volume).
5. Rauch, D., Schmalfeldt, B. Ocean-bottom interface waves of the Stonley/Scholte type: properties, observations and possible use. In: Pace, N.G. In: Acoustics and the sea-bed. Bath, U.K., Bath University Press, 1983.
6. Essen, H.-H, Janle, H., Schirmer, F. and Siebert, J. Propagation of surface waves in marine sediments. Journal of Geophysics, 49, 1981.
7. Schmalfeldt, B., Rauch, D. Explosion-generated seismic interface waves in shallow water: experimental results, SACLANTCEN SR-71, La Spezia, Italy, SACLANT ASW Research Centre, 1983.
8. Dziewonski, A., Bloch, S. and Landisman, M. A technique for the analysis of transient seismic signals. Bulletin. Seis. Soc. Am., 59, 1959.

A STUDY OF SEA FLOOR STRUCTURE USING OCEAN BOTTOM SHOTS AND RECEIVERS

A.W. Sauter, L.M. Dorman, and A.E. Schreiner

Scripps Institution of Oceanography
La Jolla, California 92093, U.S.A.

INTRODUCTION

Much of the ocean floor is covered with layers of soft sediment that cannot withstand large shearing forces without tearing. As a consequence, shear-wave slowness is an order of magnitude larger in ocean sediments than in crustal rock. The soft sea floor is a low velocity zone that focuses seismic energy, causing interface waves (or Stoneley waves) to travel along the ocean bottom in the same fashion that Rayleigh waves travel along the surface of the earth. In this paper we present Stoneley wave data from sea-floor sources recorded on ocean bottom seismometers (OBS's) at deep water sites off the California coast. To interpret the data, we generate synthetic seismograms based on thinly layered, low shear velocity models.

EXPERIMENT

The goal of this experiment is to independently determine the elastic sediment properties for use in studying OBS-sea floor coupling. We can infer sediment densities and velocities at the OBS-source site by measuring Stoneley waves traveling in the oceanic sediments. In March 1984, 3 OBS's and 8 bottom shots were deployed in 3,800 meters of water 400 km west of San Diego, California near DSDP Site #469. Deep water Stoneley waves are not often recorded because of the problems involved with deploying an energy source directly on the sea floor. Using a safe method of deploying explosives in pressure cases (D. Davies, 1965), we dropped 5-pound TNT charges to the ocean floor. Electronic timers preset from 2 to 10 hours after deployment detonated the TNT, allowing us to leave the site, eliminating the ship as a major source of noise. Although we collected airgun data at the same location, Stoneley waves were not noticeable on those seismograms because airgun sources in the water column do not excite the Stoneley modes efficiently.

Stoneley waves from the bottom shots were recorded by OBS's closer than 1.25 km from the source, but because of the high attenuation in the sediments they were not recorded at larger ranges. Figure 1 shows a vertical displacement record section with Stoneley waves visible on the closest three seismograms (denoted by arrivals 1,2, and 3). Note the late arrival times in Figure 1, and the dispersive nature of the Stoneley arrivals.

We decompose the data into slowness-frequency plots using the multiple filter technique described in Dziewonski (1969). The Fourier transform of the data is multiplied by a set of Gaussian windows centered at a discrete set of frequencies and the output from each product is transformed back to the time domain. The resulting series are arranged by window center frequency, and the amplitude envelopes of all the time series are contoured producing a 2-dimensional map in frequency and slowness (arrival time/range). Figure 2 shows the results of multiple filtering the closest Stoneley wave (arrival 1). The contoured ridges reveal at least two individual modes, differentiated from one another by the number of horizontal planes of no motion (nodes) they have. The fundamental mode (#0) has no nodes.

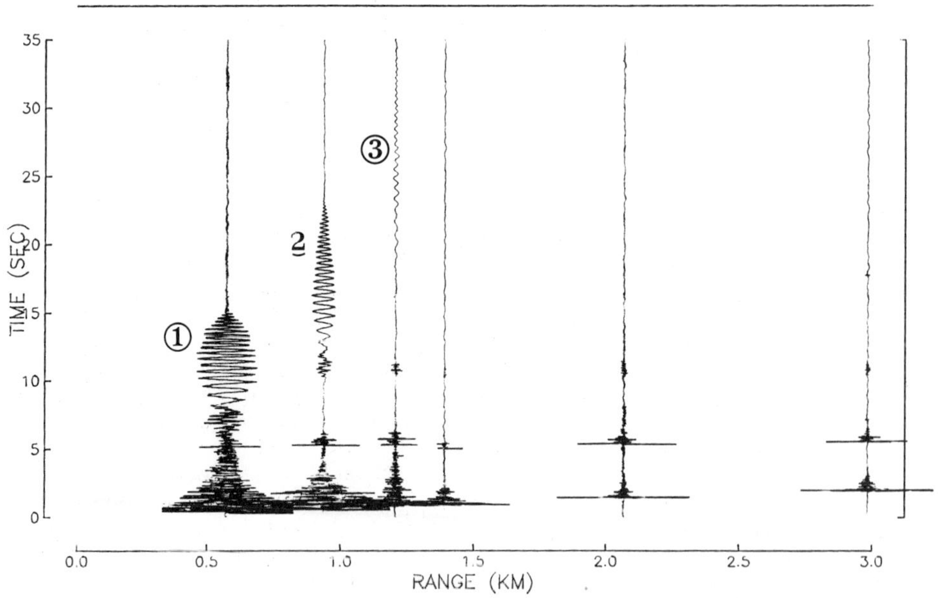

Fig. 1. Record section of the bottom shot arrivals. The OBS Suzy recorded the 1st and 3rd shots and the OBS Karen (1.5 km south of Suzy) recorded the 2nd Stoneley arrival.

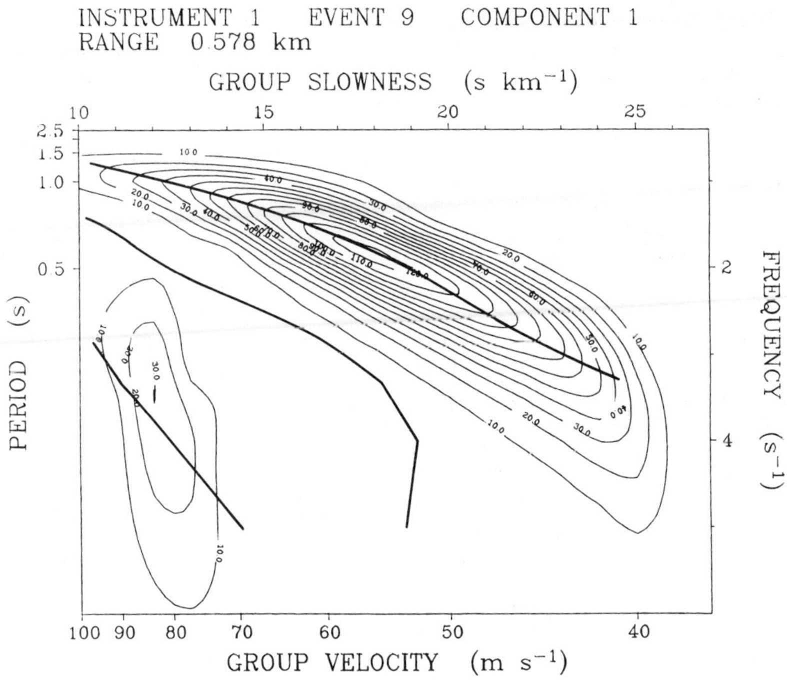

Fig. 2. Slowness-frequency contour of the closest Stoneley arrival. The dark lines show the dispersion curves calculated from a layered sediment model.

The darker curves in Figure 2 are computed dispersion curves fit to the contour data by trial-and-error The earth model used to generate the 3 curves consists of 3800 meters of water over layers with steadily increasing shear wave velocities starting at 35 meters/second. A shear velocity gradient of 3.2/second fits our data, and this compares with Davies value for a location in the Indian Ocean of 15/second in the upper 4 meters and 6.2/second in the next 16 meters. There is no data peak corresponding to the synthetic middle mode, whereas the fundamental mode matches the data closely and the 2nd mode passes through a definite peak. The multiple-filter map of the Stoneley arrival #2 is similar to figure 2 for arrival #1: modal ridges are located in the same slowness-frequency regions of the plane. However, the 3rd arrival map differs significantly. The fundamental mode contour for #3 is shifted to the right in slowness almost 40 sec/km. Using the wrong shot-receiver range would cause the same effect, but in order to make #3 contour match the others, the shot range would have to be shifted by 650 meters, and the water arrival constraint does not allow this. Local variations in sediment velocities could explain the difference in observed slowness for #3.

By examining the spectral ratios of the 3 Stoneley arrivals, we can determine the attenuation of the fundamental mode in the sediments. From Jacobson et al (1981):

$$P_s(\omega) = |S(\omega)\cdot E(\omega)\cdot I(\omega)|^2$$

P_s is the Stoneley power spectrum, S is the source spectrum, I is the instrument response, and E is the earth response:

$$E(\omega) = Y(\omega)R(x)G(\omega);$$

Y is the vertical mode function which depends on the earth properties $\rho(z), \alpha(z), \beta(z)$, and ω, R is the geometrical spreading loss function, and $G(\omega)$ is the attenuation term:

$$G(\omega) = e^{-\frac{\omega t}{2Q_t}}$$

The spectral ratio of two Stoneley arrivals is then:

$$\frac{P_i(\omega)}{P_j(\omega)} = \frac{R^2(x_i)\, e^{-(\frac{\omega t_i}{Q_t})}}{R^2(x_j)\, e^{-(\frac{\omega t_j}{Q_t})}} = \frac{R_i^2}{R_j^2}\, e^{-\omega(\frac{t_i - t_j}{Q_t})}$$

Figure 3 shows the spectral ratios of the 3 Stoneley arrivals: the slope of the log spectral ratio is:

$$\frac{t_j - t_i}{Q_t} .$$

Given the error bars in choosing the source time and OBS time, we number the arrivals from nearest to furthest shot 1 to 3 and get temporal Q values ranging from:

$$Q_{12} = 27 - 40 \;,\quad Q_{13} = 11 - 15 \;,\quad Q_{23} = 14 - 17 \;.$$

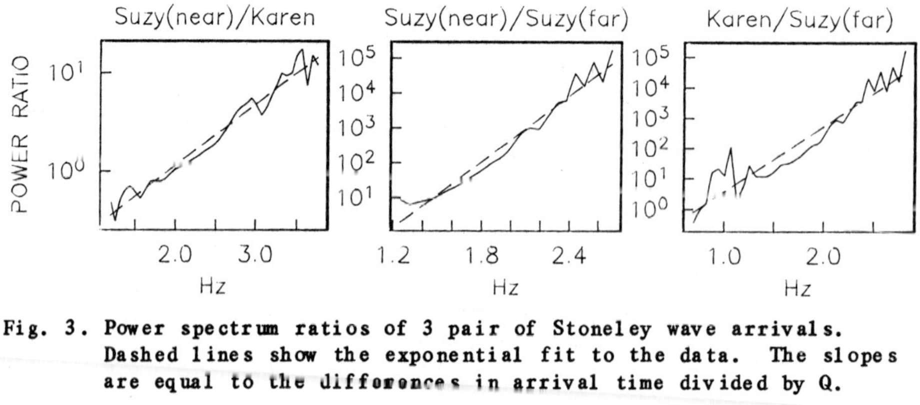

Fig. 3. Power spectrum ratios of 3 pair of Stoneley wave arrivals. Dashed lines show the exponential fit to the data. The slopes are equal to the differences in arrival time divided by Q.

Q_{12} is approximately twice as large as Q_{13} or Q_{23}. The large spectral slopes for ratios 13 and 23 (see figure 3) overwhelm the increased arrival differences ($t_3 - t_1$ and $t_3 - t_2$) producing low values for Q_{13} and Q_{23}. The lower Q and slower group velocities calculated for arrival #3 may also be explained by sediments having a lower shear velocity to the north of OBS Suzy than to the east or south (arrival 1 traveled from the east, 2 from the south, and 3 from the north of Suzy).

SYNTHETIC GENERATION

In this section we will give a brief explanation of modal synthetic seismogram generation. We assume an elastic, layered earth below the ocean floor, and since the wave equation is separable, displacement can be written (following Aki and Richards) as:

$$\underline{u}_{ST}(r,\theta,z,t) = \frac{1}{2\pi}\int_{-\infty}^{\infty} e^{-i\omega t}\int_{0}^{\infty}\sum_{m=0}^{\infty}\Big[y_1(z,k,\omega,m)\cdot\underline{S}_k^m(r,\theta) + y_2(z,k,\omega,m)\cdot\underline{R}_k^m(r,\theta)\Big]\,k\,dk\,d\omega; \qquad (1).$$

$$\underline{S}_k^m \simeq i\,\frac{1}{\sqrt{2\pi kr}}\,e^{i(kr - \frac{2m+1}{4}\pi + m\theta)}\,\hat{r};$$

$$\underline{R}_k^m \simeq -\frac{1}{\sqrt{2\pi kr}}\,e^{i(kr - \frac{2m+1}{4}\pi + m\theta)}\,\hat{z}.$$

$\underline{R}_k^m$ and $\underline{S}_k^m$ are the asymptotic form of the horizontal functions valid for waves that have traveled several wavelengths from the source. $\underline{u}_{ST}$ is the displacement vector, and is evaluated at $z = z_f$, the sea floor. y_1 and y_2 are the depth-dependent displacement components transformed to the (k,ω) space. Eqn. (1) is valid for the inhomogeneous case, where a vector point source is applied at $(0,0,z_s)$. The solution is the Green's function, which we use to find the displacement caused by a more generalized source represented by a moment tensor. Transformed to (k,w) space, the point source is a stress discontinuity, constant for all k at $z = z_s$.

To account for the source discontinuity, we build y_1 and y_2 from a combination of a discontinuous and a homogeneous solution:

$$\underline{y}(z,k,\omega,m) = \underline{y}_d + \frac{\underline{y}_h}{\Delta(\omega)}, \qquad (2);$$

$\underline{y} = (u_x, u_z, \sigma_{xz}, \sigma_{zz})$ is the displacement-stress vector, y_{3d} and y_{4d} are the 2 stress discontinuity components at $z = z_s$.

The k integral of eqn. (1) is evaluated by residue theory since the trapped Stoneley modes correspond to poles where $\Delta(\omega) = 0$.

$$\underline{u}_{ST} = \int_{-\infty}^{\infty} e^{-i\omega t}\sum_m\sum_n ik_n\left|\frac{y_{1h}(k_n,m,z,\omega)}{\partial\Delta/\partial k}\cdot\underline{S}_k^m + \frac{y_{2h}(k_n,m,z,\omega)}{\partial\Delta/\partial k}\cdot\underline{R}_k^m\right|$$

Using eqn. (2), the equation of motion integrated over z, the boundary conditions of vanishing displacement at $z = -\infty$, and vanishing stress at the water surface, we find $(\partial\Delta/\partial k)^{-1}$ at $z = z_f$, then we transfer its evaluation to $z = z_s$ using

$$\frac{d}{dz}\left|y_{1h}y_{3d}+y_{2h}y_{4d}-y_{1d}y_{3h}-y_{2d}y_{4h}\right| = 0$$

derived from the equations of motion for the result:

$$\left|\frac{\partial\Delta}{\partial k}\right|^{-1} = \frac{-(y_{1h}y_{3d} + y_{2h}y_{4d})_{z_s}}{4kUCI_1} ; \quad U = \frac{d\omega}{dk} ; \quad C = \frac{w}{k} ;$$

$$I_1 = \frac{1}{2}\int_{-\infty}^{\infty} \rho \, (y_{1h}^2 + y_{2h}^2) \, dz ;$$

At z = s, the transformed force components y_{3d} and y_{4d} correspond to stress discontinuities and can be expressed in terms of the original point force components using the orthogonality of the basis set $\underline{S}_k^m$ and $\underline{R}_k^m$, and in the process, the sum over m is accomplished since y_{3d} and y_{4d} vanish except for m = 0, ± 1.

$$\underline{u}_{ST}(r,\omega) = \sum_n \frac{1}{\sqrt{2\pi k_n r}} e^{i(k_n r + \pi/4)} \frac{1}{4cUI_1} \cdot$$

$$\left[F_z(\omega)\, y_{2h}(z_s) + i(F_x(\omega) \cos\phi + F_y(\omega) \sin\phi) y_{1h}(z_s)\right] \cdot$$

$$\left[-iy_{1h}(z_f) \cos\phi \, \hat{x} - iy_{1h}(z_f) \sin\phi \, \hat{y} + y_{2h}(z_f) \, \hat{z}\right] \qquad (3)$$

for a source at the origin, separated from the receiver a distance r; ϕ is the angle between the x axis and the source-receiver line. It is easy to build the Green's tensor (G_{ij}) out of (3) since $u_i(\omega) = F_j G_{ij}$ (the Green tensor's defining equation) where the F_j's are the vector force components. u_i can also be expressed in terms of a moment tensor.

Any source that obeys the law of conservation of linear and angular momentum (eg. earthquakes, explosions, or phase changes) and is applied at a volume small compared to the wave lengths of interest, can be expressed as a symmetric moment tensor. Given a force component - $F_j(\omega)$ at the origin, we add another force in the opposite direction separated from the first force by ds_k in the kth direction, then

$$u_i(r,z,\omega) = F_j(\omega)\, G_{ij}(r,z,\omega,\underline{s}+\underline{ds}_k) - F_j(\omega)\, G_{ij}(r,z,\omega,\underline{s})$$

By Taylor's theorem,

$$u_i(r,z,\omega) = F_j \, |ds_k| \, \frac{\partial}{\partial x_k} G_{ij}(r,z,\omega,\underline{s}) .$$

$F_j \, |dx_k| = M_{jk}$ has dimensions of force-length and is a moment component. By summing over all nine components, we get the displacement caused by the moment tensor description of the source:

$$u_i(r,z,\omega) = M_{jk}(\omega) \frac{\partial}{\partial x_k} G_{ij}(r,z,\omega,\underline{s}) . \qquad (4).$$

The partials of the Green's tensor with respect to changing the source location affect only the multiplicative terms containing source-receivers distance r, and the source-depth dependent functions y_i's. The terms containing ϕ only vary to the second order. In addition, the partial of $1/\sqrt{2\pi k_n r}$ is of order $r^{1.5}$ and can be neglected since the distance separating the shot from the receiver is large compared to the wavelengths. So, for example, the first few Green's partials are:

$$\frac{\partial}{\partial x} G_{xx} = - \sum_n A_n \, ik_n \, y_{1h}(z_s) \, y_{1h}(z_f) \cdot \cos^3\phi \, ,$$

$$\frac{\partial}{\partial y} G_{xx} = - \sum_n A_n \, ik_n \, y_{1h}(z_s) \, y_{1h}(z_f) \cdot \cos^2\phi \, \sin\phi \, ,$$

$$\frac{\partial}{\partial z} G_{xx} = \sum_n A_n \, y_{1h}(z_s) \, \frac{dy_{1h}(z_f)}{dz} \cdot \cos^2\phi; \quad \text{where}$$

$$A_n = \frac{1}{4CUI_1} \cdot \frac{1}{\sqrt{2\pi k_n r}} \, e^{i(k_n r + \pi/4)}$$

The depth dependent terms $\underline{y}(z,k_n,\omega,m)$ are found numerically using the Haskell-Thompson method of propagating minor vectors up to the free surface and checking for roots where $\Delta_n(\omega)$ vanish. Then for the $k_n(\omega)$ eigenvalues, the eigenvectors $\underline{y}$ are evaluated at all layer interfaces. By knowing the eigenvalue locations, we can calculate the phase dispersion curves directly using $k = \omega c$. The group dispersion curves (shown in figure 2) are found when the group velocity, U, is determined using variational results.

Transformation of eqn. 4 to the time domain provides us with synthetic seismograms. We start with the earth model that generated the modal curves in figure 2 and a moment tensor representation for an isotropic explosive source buried 1 meter in the sediments. The Stoneley wave arrival 1 is plotted at the top of figure 4 followed by the first 3 synthetic modes calculated at a range of .578 km.

DISCUSSION

The synthetic fundamental mode and the actual data match in phase, however, more high frequency energy is present in the unfiltered synthetic than in the data. This may have several causes: 1) the source, originally modeled as an impulse in time, may be less energetic at high frequencies; 2) the source depth may be more than 1 meter; 3) the Q in the upper layers may be smaller than we modeled. The direct water arrival spectra show that the explosions release a great deal of high frequency energy that is not showing up on the Stoneley arrivals. Perhaps the source-sediment coupling efficiency is frequency dependent and at higher frequencies the sediment becomes thixotropic and is unable to support shear energy. To test the effect of source depth, we generated a fundamental mode synthetic with the source buried 10 meters within the sediments. The amount of energy in the frequency band above 4 Hertz. decreases for the deeper source, but not enough to equal that of our data. In any case, it is unlikely that the explosives in the pressure cases could free fall 10 meters into the sediments. Using Q values lower than they are presently modeled is inappropriate in the synthetic mode program because the attenuation effect is calculated using perturbation theory, and very small values of Q correspond to large perturbations. In addition, low Q's change the dispersion, and a more suitable synthetic program would have to include this effect. As a first approximation, we find that by lowpass filtering the synthetic seismogram using a filter

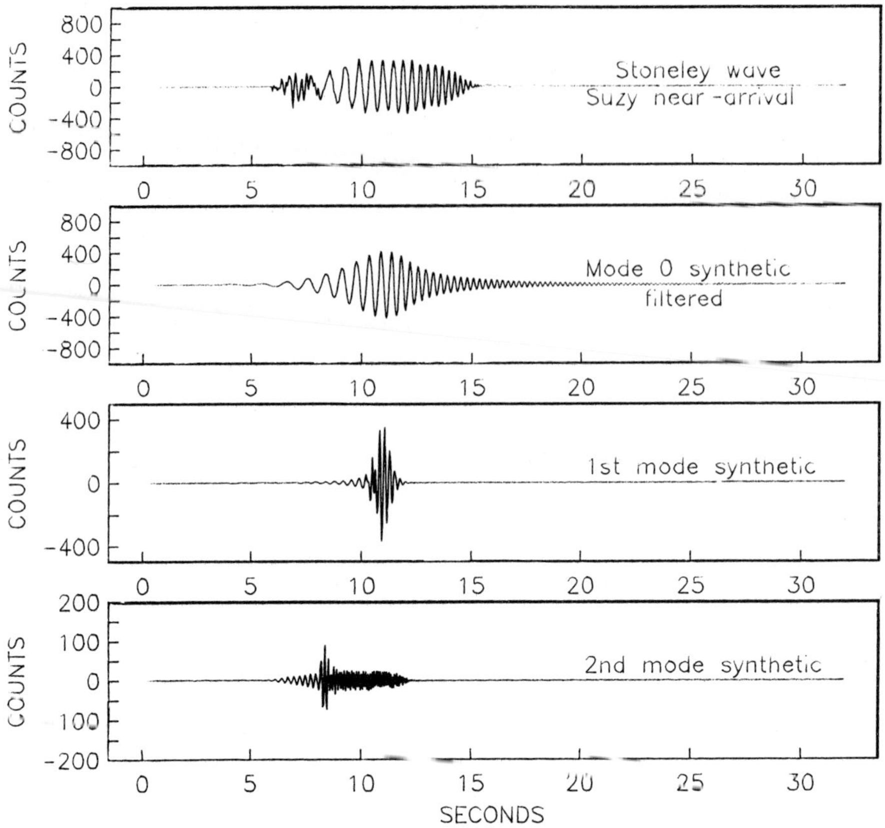

**Fig. 4. Top: OBS recording of Stoneley arrival # 1.
Below: the first 3 synthetic modes.**

with the corner frequency at 2 Hertz we can match the spectral ratios of the actual to synthetic data. The 2nd trace in figure 4 shows the filtered synthetic fundamental mode.

The fundamental synthetic matches the 1st arrival well at the onset, but there is a long tail on the synthetic whereas the real Stoneley arrival is soon lost in the noise. We can shorten the synthetic tail by increasing the velocity of the uppermost layer, but at the cost of poorer phase matching at the start of the arrival. A method of inverting the surface modes for earth structure has been carried out by A. Lerner-Lam (1983) and we are presently at work applying it to our data.

The second mode arrives at approximately the same time for synthetics and actual data, but the first mode is completely missing from the 2 closest OBS records. One explanation for the lack of the first mode is that the source may be located at its nodal depth and so can not excite it. Various synthetics were generated with different source depths to test for a nodal plane for mode 1. None were found at reasonable depths (0 to 10 meters in the sediments), and the synthetic-OBS data mismatch for mode 1 is still unexplained.

SUMMARY

We have recorded deep oceanic Stoneley waves and inferred the shear velocities and attenuation factors for the site area. A layered model with a shear velocity gradient of 3.2/second, an upper-layer shear velocity of 35 meters/second, and a temporal Q ranging from 11 to 40 fits the data. Synthetic Stoneley waves have been generated as a first step to a more formal data inversion.

ACKNOWLEDGEMENTS

Thanks to ONR for funding, T.G. Masters for the explanation and use of his mode program, J.A. Hildebrand for improving this draft, and to M. Olsson for suggesting the use of electronic wrist watches to perform tasks on the sea floor.

REFERENCES

Aki, K., and Richards, P.G., 1980, "Quantitative Seismology Theory and Methods", Freeman, San Francisco.

Davies, D., 1965, Dispersed Stoneley waves on the ocean bottom, Bull. Seism. Soc. Amer., 55:903-918.

Dziewonski, A., Bloch, S., and Landisman, M., 1969, A technique for the analysis of transient seismic signals, Bull. Seism. Soc. Amer., 59:427-444.

Jacobson, R.S., Shor, G.G.Jr., and Dorman, L.M., 1981, Linear inversion of body-wave data - Part II: Attenuation versus depth using spectral ratios, Geophysics, 46:152-162.

Lerner-Lam, A.L., and Jordan, T.H., 1983, Earth structure from fundamental and higher-mode waveform analysis, Geophys. J. R. astr. Soc., 75:759-797.

Mendiguren, J.A., 1977, Inversion of surface wave data in source mechanism studies, J. Geophys. Res., 82:889-894.

Saito, M., 1967, Excitation of free oscillations and surface waves by a point source in a vertically heterogeneous earth, J. Geophys. Res., 72:3689-3699.

Yeats, R.S., Haq, B.U., et al., 1981, Site 469: Base of Patton escarpment, Init. Repts. DSDP, 63:173-226.

SHEAR PROPERTIES OF OCEAN SEDIMENTS DETERMINED

FROM NUMERICAL MODELLING OF SCHOLTE WAVE DATA

Finn B. Jensen and Henrik Schmidt

SACLANT ASW Research Centre
Viale San Bartolomeo 400
19026 La Spezia, Italy

ABSTRACT

Recently it has become apparent that shear properties of ocean sediments are most easily determined from measured propagation characteristics of seismic interface waves. We use a newly developed numerical model of wave propagation in horizontally stratified viscoelastic media to reconstruct from Scholte-wave records the shear-speed and shear-attenuation profiles in the upper sediment layers of a shallow-water test area. For an unconsolidated sand-silt bottom we find the shear speed to increase with depth from 85 m/s at the sea floor to 300 m/s at 60 m depth. On the other hand, the shear attenuation at 3 Hz decreases with depth from 0.45 dB/λ at the sea floor to 0.15 dB/λ at 60 m depth, with an approximately quadratic frequency dependence.

INTRODUCTION

It is well established that shear rigidity of the ocean bottom affects propagation of waterborne sound through the conversion of acoustic energy into shear waves. This energy conversion is of particular importance in low-frequency shallow-water acoustics, in which the excitation of shear waves in the bottom often becomes the dominant loss mechanism for waterborne sound [1]. Under these circumstances a realistic physical model of the ocean bottom is a viscoelastic solid described by compressional and shear-wave velocities, by the attenuation factors associated with these waves, and by the material density.

While density and the compressional-wave properties (speed and attenuation) in sediments can be determined by direct methods, the shear-wave properties are difficult to measure. This is because of the usually high attenuation of these waves and because it is difficult to generate a wave that consists of predominantly transverse particle motion. The shear speed and attenuation can be indirectly determined, however, through the measured propagation characteristics of the ocean-bottom interface wave. The existence of this wave is intrinsically related to the shear properties of the sediments.

We shall review the characteristics of seismic interface waves and give some simple formulas that relate their speed and attenuation to those of shear bulk waves. We shall also provide examples from the literature of indirectly determined shear speeds in sediments. Finally, we will demonstrate how the use of a sophisticated numerical model of seismic-wave propagation allows us to reconstruct from seismic records the shear speed and shear attenuation profiles in the upper sediment layers.

SEISMIC INTERFACE WAVES

This particular wave type - sometimes named a boundary, surface, or interface wave - is a guided wave propagating along the interface between two media with different shear speeds [2,3]. The wave is generally given different names according to the media of propagation involved [4]. Hence, if propagating on a free surface of a solid, it is called a Rayleigh wave, if propagating along a liquid/solid interface, it is called a Scholte wave and if associated with a solid/solid boundary, it is called a Stoneley wave. Note that at least one of the media must be a solid for the interface wave to exist. In the case of a water/sediment interface, the pertinent wave type is a Scholte wave with the following characteristics:

- The wave propagates along the sea floor with exponentially decaying amplitude away from the guiding interface (the wave is evanescent in both media).
- Its particle motion is elliptical in the depth/range plane.
- There is no low-frequency cutoff.
- Its propagation speed and attenuation are closely related to the shear properties of the sediment.

For a simplified environment consisting of two homogeneous media in contact, the Scholte wave is non-dispersive. Its propagation speed (V_{sch}) and attenuation (Δ_{sch}) then relate in an extremely simple manner to the sediment shear properties as:

$$V_{sch} \simeq 0.9\ V_s \tag{1}$$

and

$$\Delta_{sch} \simeq 1.1\ \Delta_s, \tag{2}$$

where the attenuation is in decibels per unit length. Using numerical simulation these formulas were determined to be accurate to within a few percent for unconsolidated sediments (clay, silt, sand) with low shear speeds. The formulas indicate that the Scholte wave properties are virtually independent of densities and compressional wave properties in the two media [5]. Consequently the Scholte interface wave is ideal for determining the shear properties of ocean sediments.

In a realistic environment with sediment layering, the Scholte wave becomes dispersive. In addition, a finite number of ducted shear modes will be present. Figure 1 shows dispersion curves for a solid layer of low shear speed on a high-speed solid substratum. The fundamental mode (M_0) is an interface wave that in the low-frequency limit, $kH \to 0$, propagates as a Rayleigh wave on the substratum, and in the high frequency limit, $kH \to \infty$, propagates as a Rayleigh wave on the surface of the upper layer. The higher-order modes are ducted shear modes in the upper layer. They all have a cutoff frequency at which they propagate with the shear speed of the substratum. In the high-frequency limit these modes propagate with the shear speed of the upper layer. In this case there is no

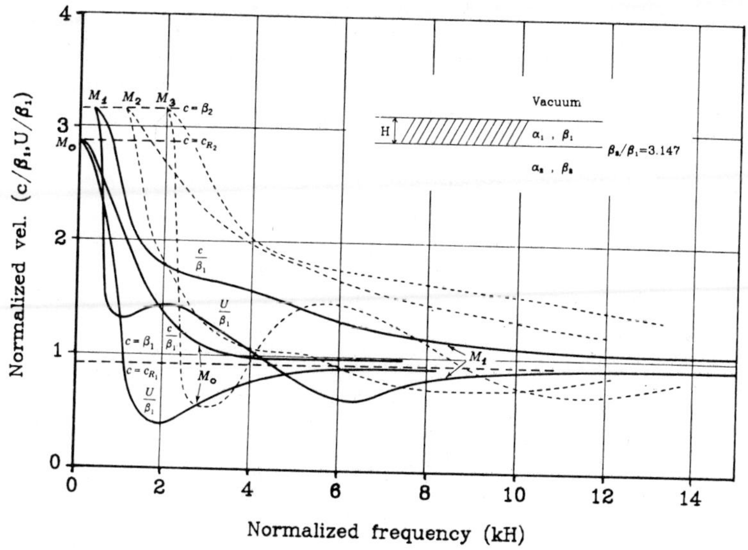

Fig. 1 Dispersion curves for elastic layer on a semi-infinite elastic substratum (Tolstoy and Usdin [2]).

Stoneley wave due to the stringent conditions necessary for the existence of this type of wave. A Stonely wave exists only when the change in shear speed across a solid/solid boundary is less than 10%.

In the following section we show how the complex information contained in experimentally determined dispersion curves similar to those of Fig. 1 can be used to extract shear information about the ocean bottom. Such an analysis generally requires the use of a numerical model. However, if experimental results show negligible dispersion, we can use simple formulas (Eqs. 1 and 2) to determine shear properties for the upper sediment layer. No dispersion means that the sediment is homogeneous to a depth of 1 to 2 wavelengths, the approximate penetration depth of the Scholte wave.

SEDIMENT SHEAR PROPERTIES

Over the past 20 years various investigators [6-17] have used Scholte waves to determine the shear properties of unconsolidated sediments. Their experimental results are summarized in Table 1. Note that experiments were done in water depths ranging from 1 to 5260 m, over sandy or silty bottoms, and for frequencies between 3 and 35 Hz. The measured Scholte wave attenuations (with geometrical spreading removed) vary more than three orders of magnitude, with the highest attenuation being associated with the highest centre frequency. The inferred shear speeds are between 25 and 260 m/s and in excellent agreement with the values quoted by Hamilton [18].

Assuming the shear attenuation to be linear with frequency [18], we can conveniently express it in decibels per wavelength. This has been done in the last column of Table 1, using the simple formulas of Eqs. (1) and (2) to relate measured attenuation of the Scholte wave with the bulk shear-wave attenuation. We obtain values that still vary more than two orders of magnitude (0.02 - 2.3 dB/λ_s), whereas Hamilton [18], based on

Table 1 Interface wave experiments

Investigators	Year	Water depth (m)	Bottom type	Centre freq. (Hz)	Measured att. (dB/km)	Inferred shear speed (m/s)	Inferred shear att. (dB/λ_s)
Bucker, Whitney, Keir[6]	1964	1 20	sand sand	20 25	300 200	100 195	1.4 1.4
Davies[7]	1965	4410	-	6	-	50-190	-
Herron, Dorman, Drake[8]	1968	5	silt	5	-	40-115	
Hamilton et al.[9]	1970	390 985	silt silt	- -	- -	100 90	- -
Schirmer[10]	1980	130	sand	4.5	7	120	0.2
McDaniel, Beebe[11]	1980	32	sand	10	-	200	-
Essen et al.[12]	1981	1	silt	4	-	75-250	-
Tuthill et al.[13]	1981	7	mud	4.5	-	25-50	-
Whitmarsh, Lilwall[14]	1982	5260	-	4.5	-	25-170	-
Holt, Hovem, Syrstad[15]	1983	-	sand	35	600	135-195	2.3
Brocher et al.[16]	1983	67	sand	5	0.43	260	0.02
Schmalfeldt, Rauch[17]	1983	20 30	- -	3 3	10 2	100 150	0.3 0.1

sparse in situ measurements for silt and sand sediments, quotes attenuation values around 0.5 - 2 dB/λ_s. Note, however, that the very low attenuation values are all associated with low-frequency experiments (3 to 5 Hz).

There are two possible explanations for the observed large variation in shear attenuation with frequency. First, that the shear attenuation in sediments decreases with depth as reported by Hamilton [18]. This leads to lower attenuation values for the low-frequency experiments, since the Scholte wave here penetrates deeper into the bottom. The second explanation is that the attenuation does not depend on the first power of frequency as assumed above, but rather follows a power law (f^n) where the exponent is between 1 and 2. Some experimental evidence for this latter assumption does exist [18]. In particular, Stoll in a recent paper [19] has strongly advocated the hypothesis of a marked frequency dependence of attenuation in fluid-saturated sediments, based both on recent field and laboratory data, and on theoretical results from the Biot model. However, more experimental work is needed in order to definitely resolve the depth and frequency dependence of shear-wave attenuation in marine sediments.

NUMERICAL MODELLING OF SCHOLTE WAVES

Figure 2 displays the Scholte wave data from which shear speed and shear attenuation profiles can be determined using a sophisticated propagation model. Stacked time signals for both the vertical and horizontal (radial) particle velocities are shown as recorded by a geophone on the ocean bottom [17]. The source was an explosive charge detonated near the sea floor in 20 m of water. The charge size was increased with range as indicated by the black dots in Fig. 2. We clearly see the dispersed low-frequency Scholte-wave arrivals with group velocities between 78 and 235 m/s.

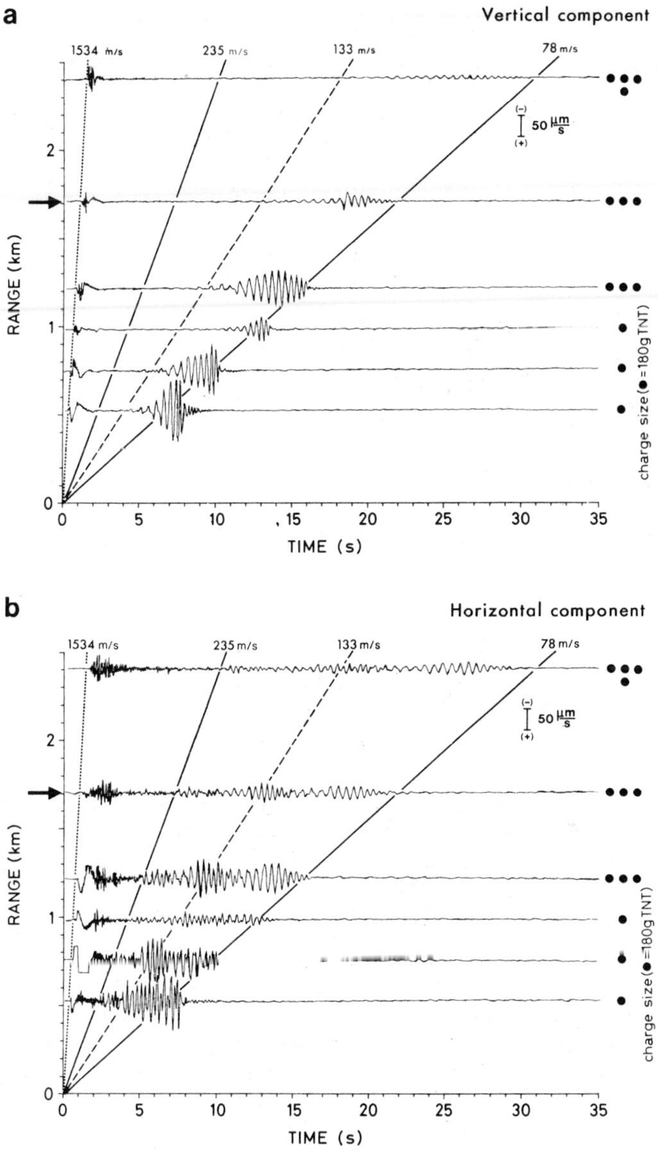

Fig. 2 Stacked time signals for vertical (a) and horizontal (b) particle velocities as recorded by a geophone on the sea floor.

Before performing detailed modelling of this propagation situation, it is convenient to carry out a dispersion analysis of the experimental data. By applying the multiple filter technique [17] to the recorded time series at range 1.7 km, we obtain the dispersion diagrams shown in Fig. 3. The contours indicate energy levels in arbitrary decibels (max level = 99 dB). It is evident that the radial component of the particle velocity gives more information about the propagation situation than does the vertical component. Thus Fig. 3b shows that energy is arriving in three discrete modes, of which the slowest arrival is the Scholte mode (M_0) with its energy centred around 2 Hz. The first shear mode (M_1) is strongly excited, with maximum energy around 2.8 Hz, while the second shear mode (M_2) is only weakly excited.

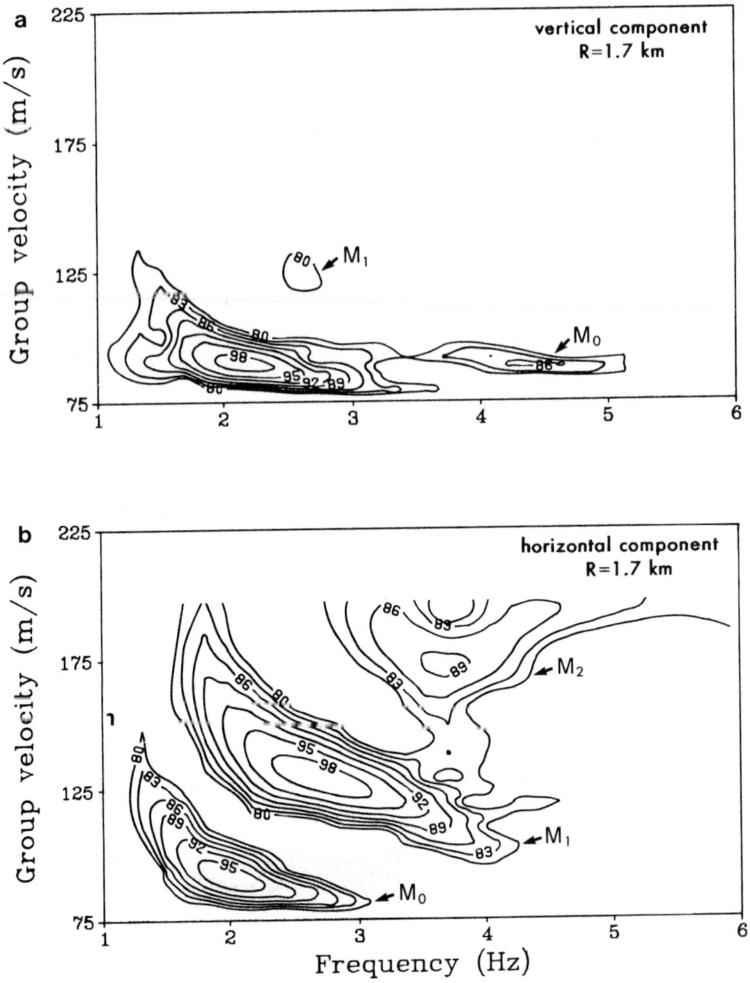

Fig. 3 Dispersion curves obtained by applying a multiple filter technique to the experimental records at range 1.7 km. Energy is seen to be propagating in three discrete modes: M_0 *,* M_1 *, and* M_2 *.*

The modelling is done with a newly developed numerical model of seismic wave propagation in horizontally stratified media [20,21]. The aim is to construct a model environment leading to computed dispersion characteristics that agree with the experimental results of Fig. 3. The modelling is done in a trial-and-error fashion, in which environmental parameters are changed in a systematic way until acceptable agreement is obtained between theory and experiment. Since we are interested in determining the shear properties for an unconsolidated sediment, we can fix, a priori, the compressional-wave properties and densities, which are known to have negligible effect on the propagation characteristics of bottom interface waves [5]. The compressional-wave properties (C_c,β_c) and densities (ρ) used as input to the seismic model are shown in Fig. 4. Also shown are the final choices of shear-speed (C_s) and shear-attenuation (β_s) profiles for the bottom.

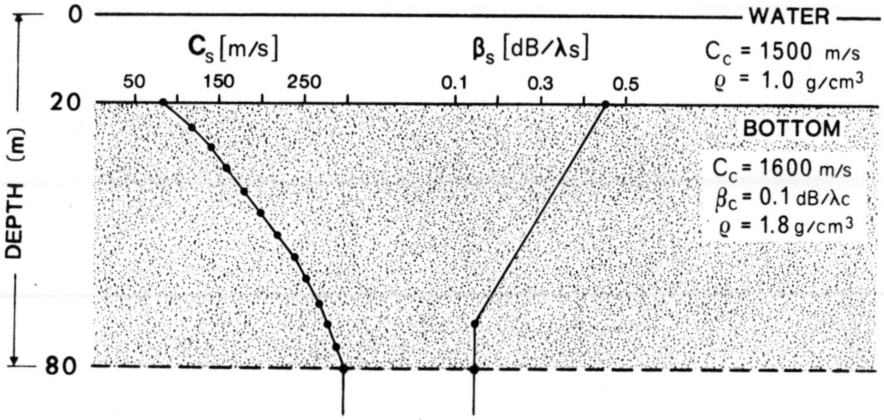

Fig. 4 Model environment used for generating synthetic seismograms.

The computed low-frequency dispersion curves for the model environment are shown in Fig. 5. In Fig. 5a we have superimposed the theoretical dispersion curves (dashed lines) on the experimental data. Clearly there is good agreement between theory and experiment for the modal arrival structure, particularly for the two lower-order modes. This, in turn, means that we have chosen an appropriate shear-speed profile. Figure 5b shows the relative energy distribution in the first three modes as determined from the numerical model. Again there is good agreement with the experimental results in Fig. 5a, indicating that the choice of shear-attenuation profile is also appropriate. It should be pointed out, however, that it was necessary to assume a quadratic frequency dependence of shear attenuation in order to get agreement on energy levels over the entire frequency band.

The inferred shear speeds (85 to 300 m/s) and shear-speed gradients ($<$ 6 m/s/m) agree well with values given by Hamilton [18] for sand-silt bottoms. Reported data concerning the shear-attenuation profiles are so sparse that no comparison with those in the literature can be made.

As a final confirmation of the validity of this modelling exercise, we have created synthetic seismograms for both the vertical and horizontal (radial) particle velocities (Fig. 6). We notice a good overall agreement with the experimental results in Fig. 2, the main difference being fast, low-frequency arrivals present in the synthetic seismograms but filtered out in the experimental data.

CONCLUSIONS

Direct measurements of shear properties of ocean sediments are difficult to perform. However, both shear-speed and shear-attenuation profiles can be inferred from measured dispersion characteristics of bottom interface waves through the use of a sophisticated propagation model. The modelling is considerably simplified because the propagation characteristics of Scholte waves depend entirely on the sediment shear properties and have only negligible dependence on the compressional-wave properties and the density. For a shallow water test area we find sediment shear speeds (85 to 300 m/s) and shear attenuations (0.15 to 0.45 dB/λ at 3 Hz) that are in good agreement with values reported in the literature. Moreover, in accordance with recent sediment studies, we find the shear attenuation to have an approximately quadratic frequency dependence.

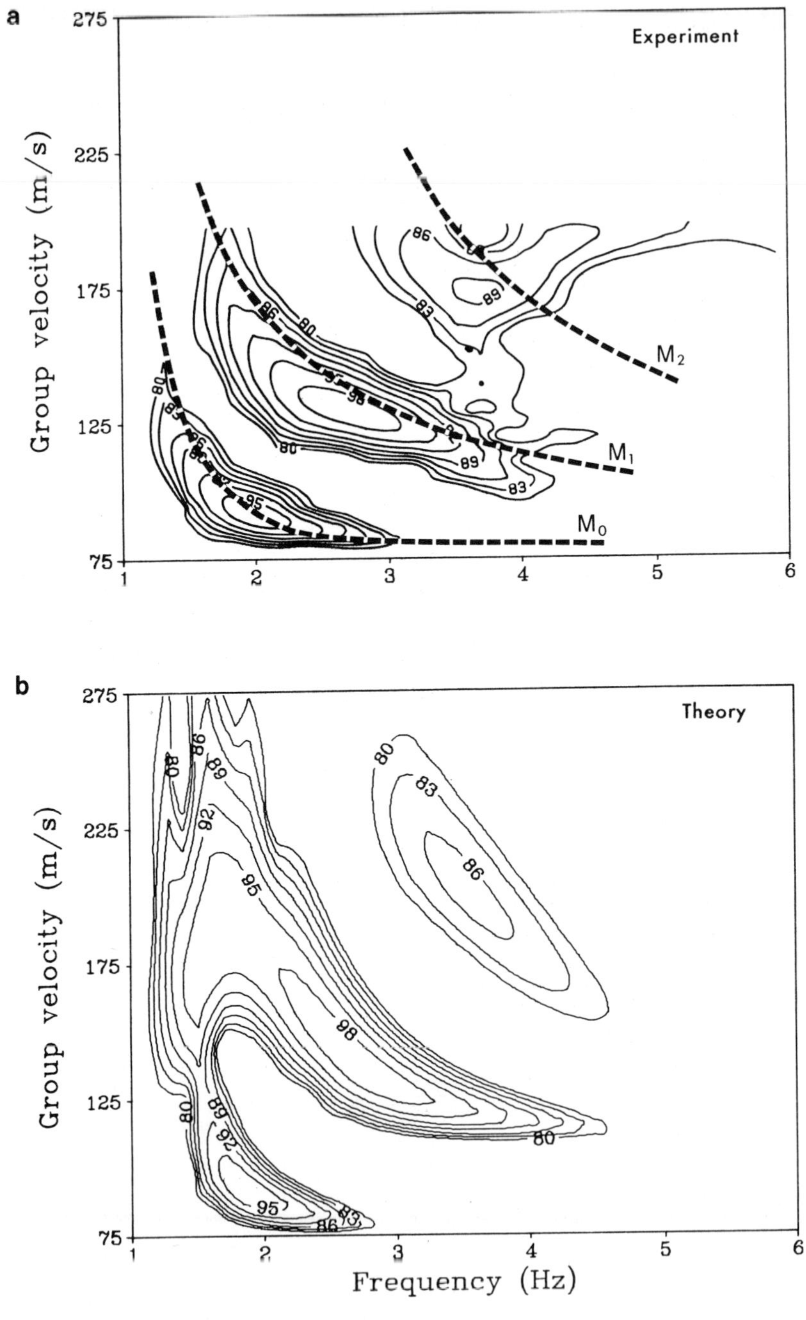

Fig. 5 Experimental (a) and theoretical (b) dispersion curves associated with the horizontal particle velocity of the sea floor at range 1.7 km.

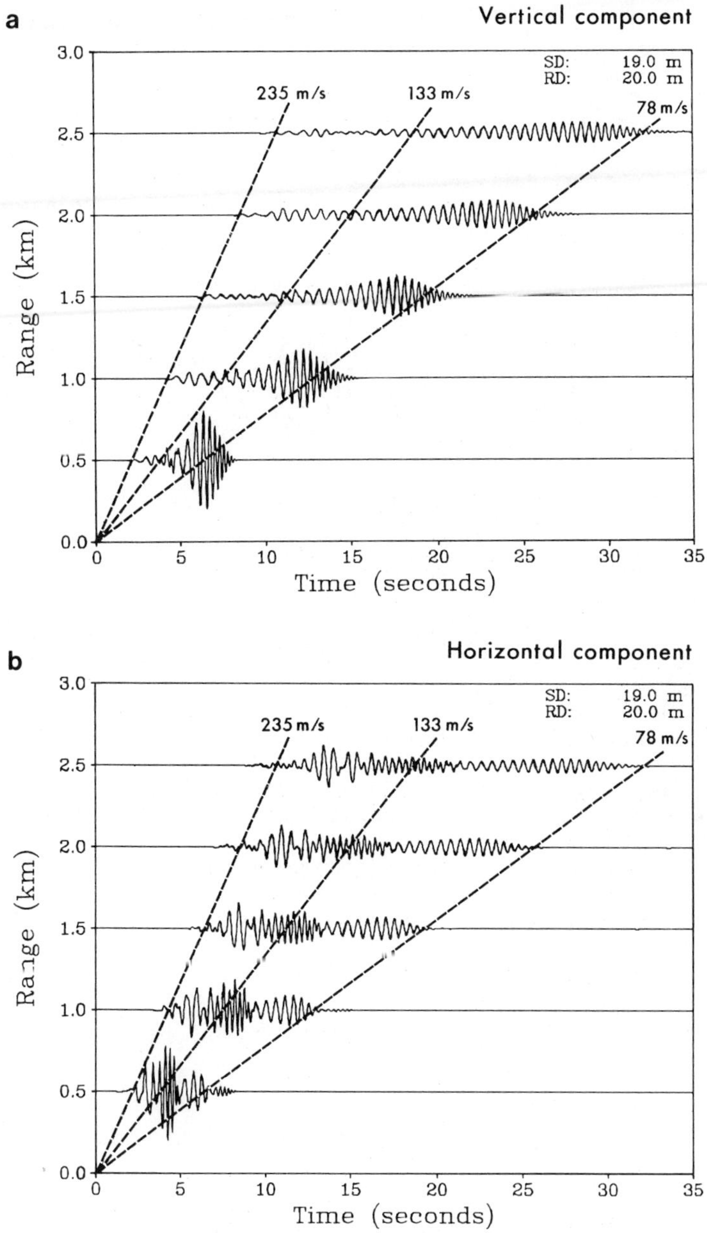

Fig. 6 Synthetic seismograms for vertical (a) and horizontal (b) particle velocities at the sea floor.

REFERENCES

1. T. Akal and F.B. Jensen, Effects of the sea-bed on acoustic propagation. In: "Acoustics and the Sea-Bed," N.G. Pace, ed., Bath Univ. Press, Bath, UK (1983): 225-232.

2. W.M. Ewing, W.S. Jardetzky and F. Press, "Elastic Waves in Layered Media," McGraw-Hill, New York, NY (1957).

3. I. Tolstoy and C.S. Clay, "Ocean Acoustics: Theory and Experiment in Underwater Sound," McGraw-Hill, New York, NY (1966).
4. D. Rauch, Seismic interface waves in coastal waters: a review, Rep. SR-42, SACLANT ASW Research Centre, La Spezia, Italy (1980). [AD A 095 018]
5. E. Strick and A.S. Ginzbarg, Stoneley-wave velocities for a fluid-solid interface, *Bull. Seis. Soc. Amer.* 46: 281-292 (1956).
6. H.P. Bucker, J.A. Whitney and D.L. Keir, Use of Stoneley waves to determine the shear velocity in ocean sediments, *J. Acoust. Soc. Amer.* 36: 1595-1596 (1964).
7. D. Davies, Dispersed Stoneley waves on the ocean bottom, *Bull. Seis. Soc. Amer.* 55: 903-918 (1965).
8. E.M. Herron, J. Dorman and C.L. Drake, Seismic study of the sediments in the Hudson River, *J. Geophys. Res.* 73: 4701-4709 (1968).
9. E.L. Hamilton, H.P. Bucker, D.L. Keir and J.A. Whitney, Velocities of compressional and shear waves in marine sediments determined in situ from a research submersible, *J. Geophys. Res.* 75: 4039-4049 (1970).
10. F. Schirmer, Experimental determination of properties of the Scholte wave in the ocean bottom at the North Sea. *In*: "Bottom-Interacting Ocean Acoustics," W.A. Kuperman and F.B. Jensen, eds., Plenum, New York, NY (1980): 285-298.
11. S.T. McDaniel and J.H. Beebe, Influence of semiconsolidated sediments on sound propagation in a coastal region. *In*: "Bottom-Interacting Ocean Acoustics," W.A. Kuperman and F.B. Jensen, eds. Plenum, New York, NY (1980): 493-505.
12. H.H. Essen, H. Janle, F. Schirmer and J. Siebert, Propagation of surface waves in marine sediments, *J. Geophysics* 49: 115-122 (1981).
13. J.D. Tuthill, B.R. Lewis and J.D. Garmany, Stoneley waves, Lopez Island noise, and deep sea noise from 1 to 5 Hz, *Marine Geophys. Res.* 5: 95-108 (1981).
14. R.B. Whitmarsh and R.C. Lilwall, A new method for determination of in-situ shear-wave velocity in deep-sea sediments, Proceedings of Oceanology International 82, Spearhead Publications, Kingston-upon-Thames, UK (1982).
15. R.M. Holt, J.M. Hovem and J. Syrstad, Shear modulus profiling of near bottom sediments using boundary waves. *In*: "Acoustics and the Sea-Bed," N.G. Pace, ed., Bath Univ. Press, Bath, UK (1983): 317-325.
16. T.M. Brocher, B.T. Iwatake and D.A. Lindwall, Experimental studies of low-frequency waterborne and sedimentborne acoustic wave propagation on a continental shelf, *J. Acoust. Soc. Amer.* 74: 960-972 (1983).
17. B. Schmalfeldt and D. Rauch, Explosion-generated seismic interface waves in shallow water: experimental results, Rep. SR-71, SACLANT ASW Research Centre, La Spezia, Italy (1983). [AD A 134 551]
18. E.L. Hamilton, Geoacoustic modelling of the sea floor, *J. Acoust. Soc. Amer.* 68: 1313-1340 (1980).
19. R.D. Stoll, Marine sediment acoustics, *J. Acoust. Soc. Amer.* 77: 1789-1799 (1985).
20. H. Schmidt and F.B. Jensen, Efficient numerical solution technique for wave propagation in horizontally stratified ocean environments, Rep. SM-173, SACLANT ASW Research Centre, La Spezia, Italy (1984). [AD A 148 537]
21. H. Schmidt, Excitation and propagation of interface waves in a stratified sea-bed, *In*: "Acoustics and the Sea-Bed," N.G. Pace, ed., Bath Univ. Press, Bath, UK (1983): 327-334.

THE DETERMINATION OF GEOACOUSTIC MODELS IN SHALLOW WATER

George V. Frisk, James F. Lynch, and James A. Doutt

Woods Hole Oceanographic Institution
Woods Hole, Massachusetts 02543

ABSTRACT

A technique for determining geoacoustic models in shallow water is described. For a horizontally stratified ocean and bottom, the method consists of measuring the magnitude and phase versus range of the pressure field due to a CW point source and numerically Hankel transforming these data to obtain the depth-dependent Green's function versus horizontal wavenumber. In shallow water, the Green's function contains prominent peaks at horizontal wavenumbers corresponding to the eigenvalues for any trapped and virtual modes excited in the waveguide. A geoacoustic model for the bottom is obtained by computing the theoretical Green's function for various values of the bottom parameters and determining the parameter set which provides the best agreement with the experimental Green's function, particularly in the positions and relative magnitudes of the modal peaks. Comparisons are also made between the measured and theoretically computed pressure field magnitudes. The technique is demonstrated using experimental data at 140 Hz and 220 Hz.

INTRODUCTION

A technique for acoustically characterizing horizontally stratified shallow water waveguides has been described in the literature[1]. The method consists of measuring the magnitude and phase versus range of the pressure field due to a CW point source and numerically Hankel transforming these data to obtain the depth-dependent Green's function versus horizontal wavenumber. In the context of normal mode theory, the Green's function contains information about the nature of the discrete and continuous modal spectra as well as the plane-wave reflection coefficients of the waveguide boundaries. Specifically, the Green's function contains prominent peaks at horizontal wavenumbers corresponding to the eigenvalues for any trapped and virtual modes excited in the waveguide. The positions and magnitudes of these modal peaks are sensitive to the acoustic properties of the bottom, and this sensitivity is the basis for our proposed inverse method for determining geoacoustic models in shallow water. The technique consists of computing the theoretical Green's function for various values of the bottom parameters and determining the parameter set which provides the best agreement with the experimental Green's function, particularly in the positions and relative magnitudes of the modal peaks. Comparisons are also made between the measured and theoretically computed pressure field magnitudes, so that a consistent, balanced fit is obtained in both domains. The technique is demonstrated

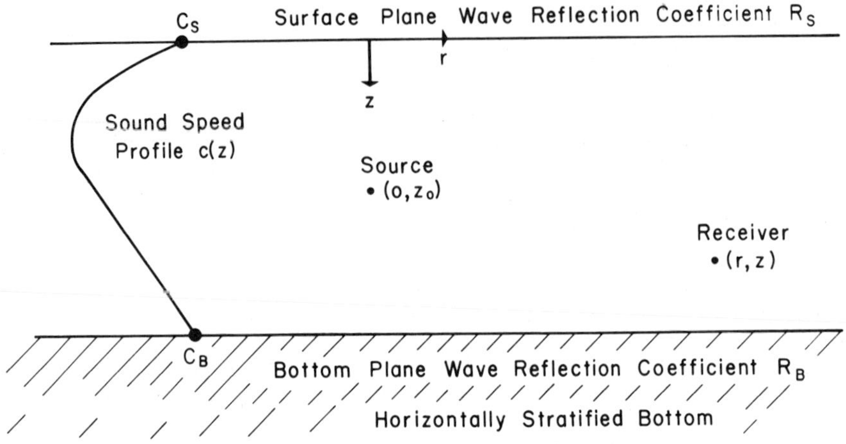

Fig. 1. Horizontally stratified ocean model.

using experimental data at 140 Hz and 220 Hz.

The proposed method has several advantages over the conventional approach of fitting only transmission loss (i.e. pressure field magnitude). First, since the determination of the Green's function requires knowledge of both the field magnitude <u>and</u> phase, we utilize additional information in the inversion procedure, and therefore obtain a more constrained geoacoustic model. Second, by focusing the comparison on the horizontal wavenumber regions where the modal peaks are located, we obtain the geoacoustic model which determines where most of the acoustic energy is concentrated, namely in the trapped and virtual modes. Finally, the comparison of the Green's functions has the advantage that the computations are performed in the horizontal wavenumber domain, thereby eliminating the computational step of determining the point source field from the Green's function.

THEORY

Consider a water column of thickness h, characterized by a constant density and a sound speed c(z), and bounded at the surface and bottom by horizontally stratified media as shown in Fig. 1. Then the spatial part of the acoustic pressure p, due to a point source at $r = 0$ and $z = z_0$ with harmonic time dependence $e^{-i\omega t}$, satisfies the inhomogeneous Helmholtz equation

$$\left[\frac{1}{r}\frac{\partial}{\partial r}\left(r\frac{\partial}{\partial r}\right) + \frac{\partial^2}{\partial z^2} + k^2(z)\right] p(r;z,z_o) = -2[\delta(r)/r]\delta(z-z_o), \tag{1}$$

where the wavenumber $k(z) = \omega/c(z)$. The solution of Eq. (1) can be expressed as the zero-order Hankel transform[1,2] of the depth-dependent Green's function g

$$p(r) = \int_0^\infty g(k_r)J_o(k_r r)k_r dk_r, \tag{2}$$

$$g(k_r) = \int_0^\infty p(r) J_0(k_r r) r\, dr, \tag{3}$$

where r is the horizontal range, k_r is the horizontal wavenumber, and J_0 is the Bessel function of order zero. The Green's function satisfies

$$\left[\frac{d^2}{dz^2} + k^2(z) - k_r^2\right] g(k_r;z,z_0) = -2\delta(z-z_0), \tag{4}$$

along with impedance boundary conditions at the surface and bottom. These boundary conditions can be expressed in terms of the plane-wave reflection coefficients R_S and R_B of the surface and bottom, respectively. Although both p and g depend on the source and receiver depths, z_0 and z, these depths are viewed as parameters in the problem, while r and k_r are the conjugate transform variables.

In general, the Green's function is characterized by a finite-valued continuum and a discrete set of resonances occurring at the poles of g which, in conventional normal mode terminology are the continuous and discrete portions of the spectrum, respectively. Specifically, the modal representation of the field is given by[1]

$$p(r) = i\pi \sum_n a_n \phi_n^*(z_0)\phi_n(z) H_0^{(1)}(k_n r) + I(r), \tag{5}$$

where the eigenfunctions ϕ_n and eigenvalues k_n satisfy the equation

$$\left[\frac{d^2}{dz^2} + k^2(z) - k_n^2\right]\phi_n(z) = 0 \tag{6}$$

along with impedance boundary conditions at the interfaces. Here $H_0^{(1)}$ is the zero-order Hankel function of the first kind, $\sqrt{a_n}$ is the modal normalization constant, and the asterisk denotes complex conjugation. Typically the trapped modal sum in Eq. (5) dominates the long-range behavior of the field, whereas the continuum contribution I(r), which includes the virtual modes, is only significant at short range.

In this paper, we will concentrate on the case of an isovelocity water column with sound speed c and a pressure-release surface with $R_S = -1$, although we emphasize that our method is not restricted to this simplified example. Under these circumstances, the Green's function is given by[1]

$$g(k_r;z,z_0) = -\frac{e^{i\gamma|z-z_0|} - e^{i\gamma(z+z_0)} + R_B(k_r)e^{2i\gamma h}[e^{-i\gamma(z+z_0)} - e^{-i\gamma|z-z_0|}]}{i\gamma[1 + R_B(k_r)e^{2i\gamma h}]}, \tag{7}$$

where

$$k_r^2 + \gamma^2 = k^2 = (\omega/c)^2, \tag{8}$$

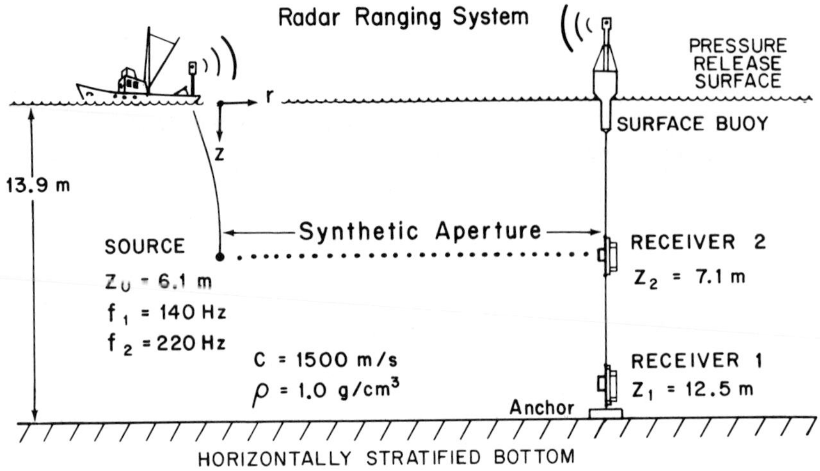

Fig. 2. Typical shallow water experimental configuration.

and γ is the vertical wavenumber. The denominator of Eq. (7) when set equal to zero is the well known characteristic equation governing modal propagation in an isovelocity water column overlying an arbitrary horizontally stratified bottom:

$$1 + R_B(k_r)e^{2i\gamma h} = 0. \tag{9}$$

In computing the theoretical Green's functions required for our inversion procedure, we use the SAFARI program[3,4], where we assume a fluid bottom model consisting of a sequence of isovelocity, constant density layers with absorption. Sound velocity gradients are approximated by a series of thin isovelocity layers with a thickness that is a small fraction ($\approx$1/7) of a wavelength. Pressure field calculations are also carried out with SAFARI, which performs a direct numerical integration of Eq. (2) using an algorithm based on approximating the Bessel function with its asymptotic form[5]. The primary advantage of SAFARI in our application is its speed, since we generally require Green's function computations for hundreds of bottom models in our inversion procedure.

EXPERIMENT

The experimental configuration for obtaining the data processed in this paper is shown in Fig. 2. The experiment consists of towing a CW source at fixed depth away from two moored receivers over an aperture which extends from zero range to several kilometers. The receivers quadrature demodulate the signal, that is, they remove the harmonic time dependence by beating the signal down to 0 Hz, and digitally record the real and imaginary parts (quadrature components) of the spatial part of the acoustic field p(r). In order to avoid aliasing errors in the discrete Hankel transform, the field must be sampled at least every half-wavelength in range, corresponding to the fact that g decays exponentially for $k_r > k$ and $z, z_0 < h$[1]. For the frequencies of interest, we can obtain this sampling rate for drift rates of 1/2 knot or less. The Hankel transformed is performed using the Fourier-Bessel series[1], which is an exact algorithm for k_r-limited fields. The Fourier-Bessel series is slow but reliable, in that among the exact Hankel transform algorithms its associated errors are best understood.

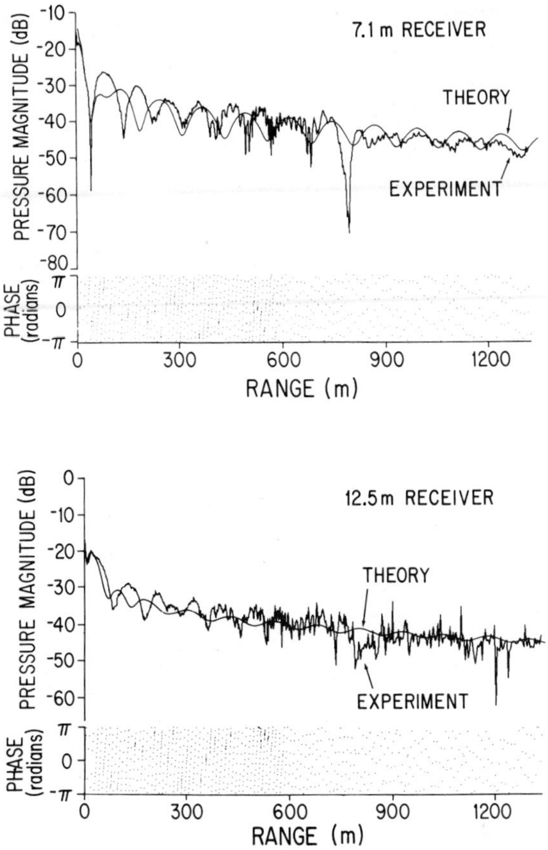

Fig. 3. Comparison of theoretical and experimental pressure field magnitudes and phases at 140 Hz.

The pressure fields measured for a sandy bottom over a range aperture of 1325 m in Nantucket Sound, Massachusetts in May 1984 are shown in Figs. 3 and 4. The magnitudes exhibit strong spatial interference patterns which are particularly evident at 220 Hz. These patterns arise due to the coherent combination of two normal modes of comparable strength at 220 Hz, and one dominant and one weak mode at 140 Hz. The decay with range of the weaker mode at 140 Hz is also apparent in the data. The phase roll evident in all four plots is dominated by the effect of opening range between source and receivers and is approximately equal to k_1r when there is a single dominant mode with eigenvalue k_1. We note that the water depth changes from 13.9 m to 14.6 m at a range of about 600 m. This change in bathymetry has a pronounced effect on the 220 Hz modal cycle distance, which changes from about 125 m to 150 m. Furthermore, at approximately the same range of 600 m, the surface waveheight increases from a negligible amount to 2-3 feet. This gives rise to the noisy ap-

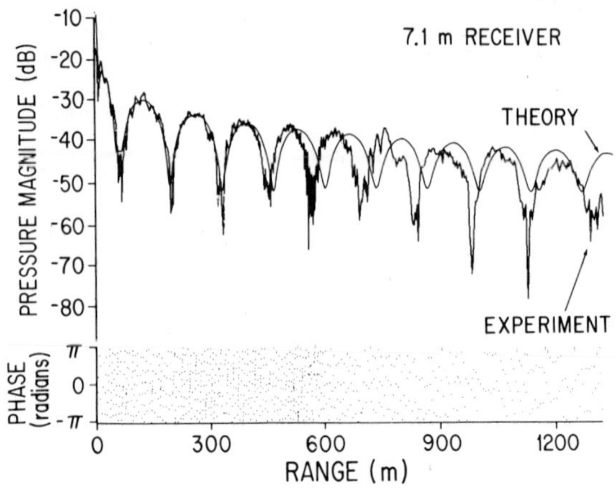

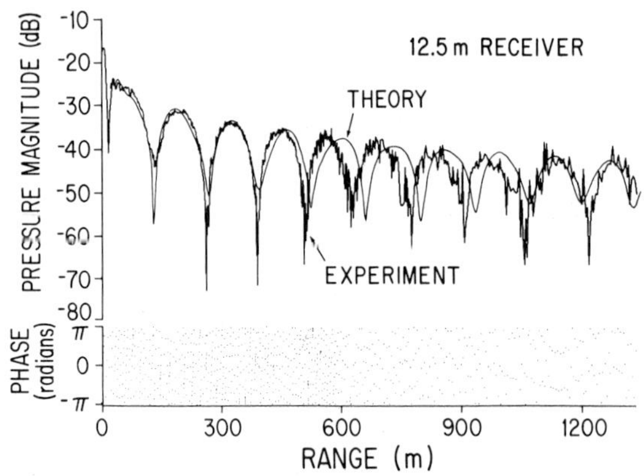

Fig. 4. Comparison of theoretical and experimental pressure field magnitudes and phases at 220 Hz.

pearance in the field magnitudes and a degradation in the regular behavior of the phases.

The Green's functions obtained by Hankel transforming the pressure field data are shown in Figs. 5 and 6. The modal characteristics described above (i.e., two strong modes at 220 Hz, one dominant mode at 140 Hz) are evident in the k_r-domain. The finite heights and widths of the modal peaks are due to the effects of finite aperture and any attenuating mechanisms in the bottom.

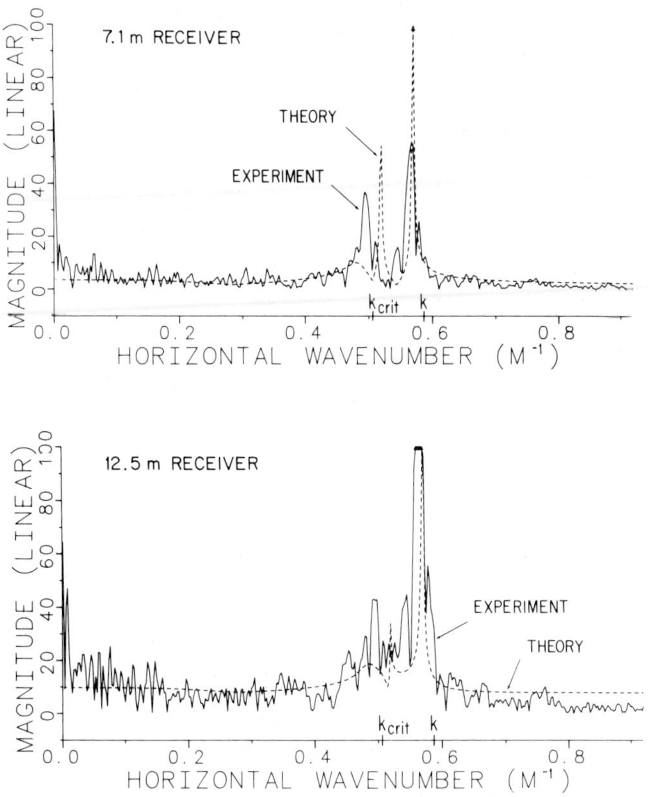

Fig. 5. Comparison of theoretical and experimental Green's functions at 140 Hz (Δ indicates point off-scale).

INVERSION RESULTS

The inversion procedure consists of assuming a geoacoustic model, computing the Green's function, and calculating the mean-square difference between the theoretical and experimental Green's functions for each frequency/receiver depth combination. The bottom parameters are then varied and this procedure is repeated until the total mean-square difference at all four frequency/receiver depth pairs is minimized. In implementing this technique, we scale the theoretical and experimental Green's functions according to their average levels. We make the Green's function comparisons only in the k_r-regions where there are pronounced modal peaks in the data. These occur in the range $k_r = 0.44$ to $k_r = 0.60$ at 140 Hz and $k_r = 0.74$ to $k_r = 0.96$ at 220 Hz. We also introduce a compressional wave attenuation $\alpha = 0.1$ dB/λ in order to reduce the theoretical Green's function peaks to a finite height. Although most of the comparisons are made in the k_r-domain, some of the comparisons are made in the

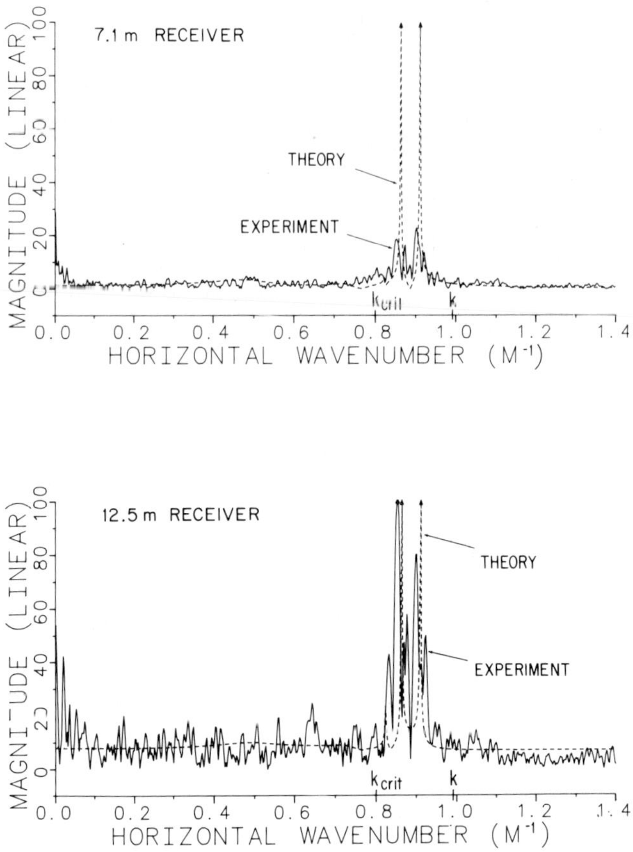

Fig. 6. Comparison of theoretical and experimental Green's functions at 220 Hz (▲ indicates point off-scale).

r-domain using both SAFARI and the SNAP normal mode program[6]. This approach insures that a consistent, balanced fit is obtained in both domains.

The geoacoustic model inferred using this method is shown in Fig. 7. The parameters, in particular the high near surface, sound velocity gradient, are consistent with the results of Hamilton[7] for sandy bottoms. The corresponding comparisons between theoretical and experimental pressure field magnitudes and Green's functions are shown in Figs. 3-6. The field comparisons are meaningful only in the first 600 m because our theoretical calculation does not take into account the change in bathymetry. The agreement between the theoretical and experimental fields is excellent at 220 Hz but is only fair at 140 Hz. This is partly due to the fact that the weaker mode at 140 Hz is very close to cutoff, and it is difficult to model this situation accurately. The better agreement at 220 Hz is also seen in the Green's functions, where the theoretical and experi-

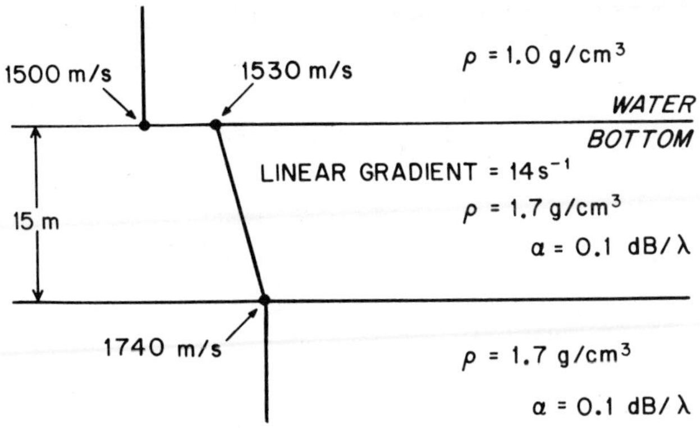

Fig. 7. Geoacoustic model inferred from the data.

mental modal peaks agree very well. We note that the theory predicts two trapped modes at 220 Hz and two trapped modes and one virtual mode at 140 Hz. These are evident in the Green's function plots.

There are several potential approaches to improving these results. First, we could improve on the simplified linear gradient model by using a more realistic bottom model where the gradient is a function of depth[7]. Second, we could attempt to compensate for the change in bathymetry which violates the assumption of horizontal stratification. Finally, we could use high resolution spectral estimation methods to improve our estimates of the modal peak positions. All of these approaches are currently being pursued.

CONCLUSIONS

We have described a technique for determining geoacoustic models in shallow water. For a horizontally stratified ocean and bottom, the method consisted of measuring the magnitude and phase versus range of the pressure field due to a CW point source and numerically Hankel transforming these data to obtain the depth-dependent Green's function versus horizontal wavenumber. The geoacoustic model was obtained by computing the theoretical Green's function for various values of the bottom parameters and determining the parameter set which provided the best agreement with the experimental Green's function, particularly in the positions and relative magnitudes of the modal peaks. Comparisons were also made between the measured and theoretically computed pressure field magnitudes. The method was demonstrated with moderate success using experimental data at 140 Hz and 220 Hz and shows considerable promise.

ACKNOWLEDGMENTS

The authors gratefully acknowledge the programming assistance of Christopher Dunn and helpful discussions with Michael Wengrovitz.

REFERENCES

1. G.V. Frisk and J.F. Lynch, "Shallow Water Waveguide Characterization Using the Hankel Transform," J. Acoust. Soc. Am. 76:205 (1984).

2. G.V. Frisk, A.V. Oppenheim, and D.R. Martinez, "A Technique for Measuring the Plane-Wave Reflection Coefficient of the Ocean Bottom," J. Acoust. Soc. Am. 68:602 (1980).
3. H. Schmidt and F.B. Jensen, "Efficient Numerical Solution Technique for Wave Propagation in Horizontally Stratified Ocean Environments," Rep. SM-173, SACLANT ASW Research Centre, La Spezia, Italy (1984).
4. H. Schmidt and F.B. Jensen, "A Full Wave Solution for Propagation in Multilayered Viscoelastic Media with Application to Gaussian Beam Reflection at Fluid-Solid Interfaces," J. Acoust. Soc. Am. 77:813 (1985).
5. F.R. DiNapoli and R.L. Deavenport, "Theoretical and Numerical Green's Function Field Solution in a Plane Multilayered Medium," J. Acoust. Soc. Am. 67:92 (1980).
6. F.B. Jensen and M.C. Ferla, "SNAP: The SACLANTCEN Normal-Mode Acoustic Propagation Model," Rep. SM-121, SACLANT ASW Research Centre, La Spezia, Italy (1979).
7. E.L. Hamilton, "Geoacoustic Modeling of the Seafloor," J. Acoust. Soc. Am. 68:1313 (1980).

THE ESTIMATION OF THE DENSITY, P-WAVE, AND S-WAVE SPEEDS OF THE TOP-MOST LAYER OF SEDIMENTS, FROM WATER BOTTOM REFLECTION ARRIVALS

R. Chapman, *S. Levy, *J. Cabrera, *K. Stinson, and *D. Oldenburg

Defence Research Establishment Pacific
FMO Victoria, B.C.
Canada V0S IB0

*IT&A
Inverse Theory & Applications Inc
2162 Western Parkway
Vancouver, B.C. V6T 1V6

ABSTRACT

Procedures for estimating material properties of the sea floor sediments are presented. The estimation procedures make use of the relationship between amplitude and offset for pre-critical reflections, and the relationship between phase shift and offset for post-critical reflections. In all cases we assume that the source, hydrophone, and ocean depths are given, and that the measured bottom reflection is adequately modelled as a plane wave response. We consider three cases: the sea floor sediments are approximated as a liquid layer, and the relation between reflection strength and source-receiver offset for a set of pre-critical reflections is analysed; the sea floor sediments are approximated as a liquid layer, and the relation between phase shift and source-receiver offset for a set of post-critical reflections is studied; the sediments layer is assumed to be elastic, and the relation between reflection strength and source-receiver offset for a set of pre-critical reflections is analysed.

INTRODUCTION

A method for the estimation of the acoustic parameters of the top-most layer of sediments from the phase-shift information contained in post-critical water bottom reflection arrivals has been described by Chapman et al. (1985). The encouraging results obtained in this work inspired the continuation and the expansion of the investigation into the analysis of amplitude-offset information as well as that of phase shift-offset information. In particular, we have developed two additional approaches which utilize the amplitude-offset information contained in pre-critical water bottom reflections in order to estimate the density, P-wave and S-wave speeds of the top-most layer of sediments. An important addition to the arsenal of the reflection amplitudes inversion schemes is an estimation procedure which relies on the exact amplitude-offset relationship for reflection arrivals from an interface separating an acoustic medium from an underlying elastic medium.

In the following, we present some of the arguments and the conditions which lead to the different expressions required by the various estimation procedures. Then, using the reflectivity method of Fuchs and Muller (1971), we generate a set of synthetic seismograms which are used to demonstrate the estimation procedures and some of the associated difficulties.

EXPERIMENTAL DESIGN

We consider an experiment in which an explosive point source at a given depth Z_s is activated. The pressure disturbance propagates through the water column and is reflected by the water bottom interface. The reflected wave as well as the direct arrival are recorded by a linear array of receivers situated at a depth Z_r spanning a sufficient range of source-receiver offsets to sample both pre and post-critical reflection arrivals.

The suggested experiment configuration is displayed in Figure 1a, while in Figure 1b we draw a schematic presentation of the recorded arrivals. For the sake of simplicity we omit all those arrivals which are associated with reflectors other than the water bottom interface.

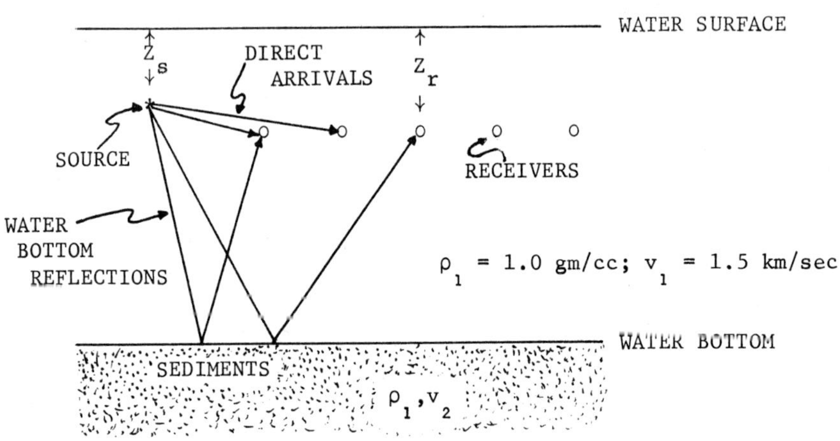

(a)

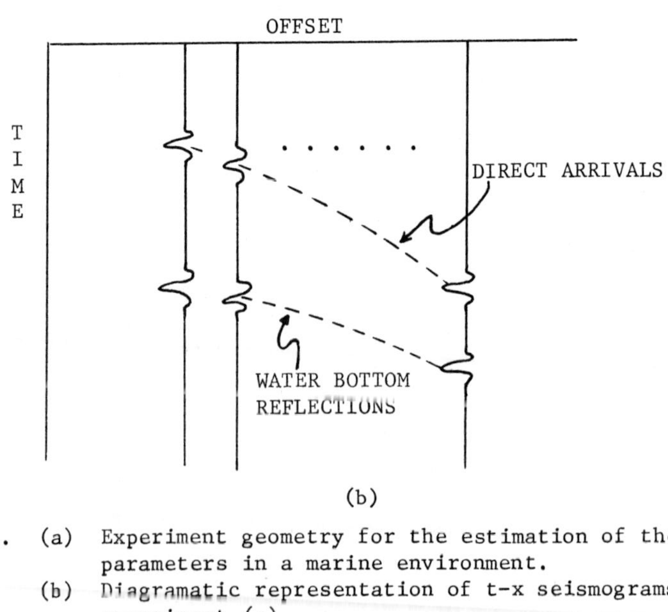

(b)

Figure 1. (a) Experiment geometry for the estimation of the top-most layer parameters in a marine environment.
(b) Diagramatic representation of t-x seismograms recorded in experiment (a).

MODEL ASSUMPTIONS

(a) In this work, the relations expressing the reflection coefficients as a function of angle of incidence (or ray parameter) correspond to plane-waves impinging on a planar interface. To comply with this restriction we assume that the water bottom interface is many wavelengths away from the source (at least five) and that to a reasonable approximation the water bottom is a planar interface.

(b) To calculate the incidence angle at the water bottom, we assume that the exact locations of the source and each of the receivers are known and that the water bottom depth and its dip are also given.

(c) For the application of a geometrical spreading correction we assume that the sound speed throughout the water column is a constant. Also under this assumption, the straight ray approximation is utilized for the calculation of the angle of incidence in (b).

(d) For the purpose of the estimation of the absolute reflection strength, we assume that the direct arrival at the near offset hydrophones (after the geometrical spreading correction has been applied) constitutes a reasonable approximation of the source function. The absolute reflection strength is then calculated by dividing the peak amplitude of the water bottom reflection by that of the direct arrival.

(e) It is assumed that attenuation effects in the water within the frequency band of interest (10 - 500 Hz) are negligible.

(f) It is assumed that both the source and the receivers produce or record a consistent signature irrespective of the direction at which the wave is leaving the source or arriving at the receivers (i.e. both source and receivers are omnidirectional).

(g) When the relations expressing the reflection coefficient as a function of angle of incidence, for a reflection from a liquid-liquid interface are used, it is assumed that the top-most layer of sediment can be reasonably approximated as a liquid layer.

ESTIMATING THE SEDIMENTS LAYER PARAMETERS FROM WATER BOTTOM REFLECTION ARRIVALS

Two physical models for the top-most layer of sediments are considered. In the first, the layer of sediments is assumed to be a liquid layer and both pre-critical and post-critical reflections are analysed. The second case models the layer of sediments as an elastic layer and considers only the information contained in the pre-critical reflections.

A. ACOUSTIC-ACOUSTIC CASE

1. Sound speed and density estimation from amplitude-offset information.

The core of this inversion procedure is a linearized relation expressing pre-critical reflection amplitudes as a function of ray-parameter (Carrion et al. 1984). Under the assumption of small impedance variations with depth, we express the reflection coefficient C(p) as:

$$C(p) = \tfrac{1}{2}\ln(\rho_1/\rho_2) + \tfrac{1}{4}\ln\{v_2^2(1 - p^2v_1^2)/v_1^2(1 - p^2v_2^2)\} \qquad (1)$$

Given two reflection amplitudes at two different ray parameter values say p_1 and p_2 the acoustic velocity is estimated from the relation:

$$v_2^2 = \frac{(1 - v_1^2p_1^2)\cdot e^{4\{c(p_2) - c(p_1)\}} - (1 - v_1^2p_2^2)}{p_2^2(1 - v_1^2p_1^2)\cdot e^{4\{c(p_2) - c(p_1)\}} - p_1^2(1 - v_1^2p_2^2)} \qquad (2)$$

When more than two observations are given, equation (2) is solved for all i, j combinations with j>i, in a least squares sense. Once the best fit acoustic speed v_2 is found, the corresponding density is calculated from:

$$\rho_2 = \frac{\rho_1 v_1}{v_2} \sqrt{\frac{(1 - v_2^2 \cdot p_2^2)}{(1 - v_1^2 p_2^2)}} \cdot e^{2 \cdot c(p_2)} \tag{3}$$

2. Estimating the acoustic parameters of the layer of sediments from phase-offset information.

Following the usual solution of the wave equation in acoustic media (see for example Rayleigh, 1945; Arons and Yennie, 1950) we obtain a relation expressing the amount of phase shift associated with the angle of incidence for post-critical reflections. The phase-shift angle $2 \cdot \varepsilon$ depends upon the ratio of the densities ρ_1/ρ_2, the ratio of the sound speed across the reflecting interface v_1/v_2 and the angle of incidence θ:

$$\tan^2\varepsilon = (\rho_1/\rho_2)^2 \tan^2\theta - (\rho_1 v_1/\rho_2 v_2)^2 \cdot \sec^2\theta \tag{4}$$

This equation is linear in the variables $(\rho_1/\rho_2)^2$ and $(\rho_1 v_1/\rho_2 v_2)^2$. Given a set of phase-shifts versus angles of incidence $\{\varepsilon_k, \theta_k\}$; k = 1, N, we solve an over determined set of equations in the form of (4) to obtain the squared ratios of the density and the impedance across the reflecting interface. From these values the density and the acoustic speed of the media underlying the reflecting interface are determined.

B. THE ACOUSTIC-ELASTIC CASE

The relations summarized in equations (1) to (4) previously have a limited applicability to the case where the reflecting interface is separating an acoustic medium from an underlying elastic media. In particular, this applicability diminishes to a narrow range of near vertical angles of incidence if the ratio of S-wave to P-wave speeds in the elastic medium exceeds 0.3. To accommodate such cases, we need to consider the solution of a set of non-linear equations utilizing the exact relation between reflection amplitude and the angle of incidence.

For a given angle of incidence γ the reflection coefficient $c(\gamma)$ is given by (Brekhovskikh, 1960):

$$c(\gamma) = \frac{D - E}{D + E} \tag{5}$$

with $D = \frac{\rho_2 \alpha_2}{\rho_1 \alpha_1} \cdot (1/\cos\gamma_2^\alpha \cdot \cos^2(2\gamma_2^\beta) + \frac{\rho_2 \cdot a \cdot \alpha_2}{\rho_1 \alpha_1} \cdot (1/\cos\gamma_2^\beta) \cdot \sin^2(2 \cdot \gamma_2^\beta)$

and $E = 1/\cos\gamma$.

In the above expression α, β, and ρ denote P-wave speed, S-wave speed and density, respectively. Note that in equation (5), 'a' is the ratio of S to P-wave speeds in the elastic medium and that the angles of refraction γ_2^{α} and γ_2^{β} satisfy Snell's law; that is:

$$\gamma_2^{\alpha} = \arcsin\left[\frac{\alpha_2 \sin(\gamma)}{\alpha_1}\right]$$

and

$$\gamma_2^{\beta} = \arcsin\left[\frac{\beta_2 \cdot \sin(\gamma)}{\alpha_1}\right]$$

In spite of the nonlinearity of the above expression the corresponding set of equations relating reflection amplitude to angle of incidence are easily solved to yield optimal estimates (in a least squares sense) of the density, the P-wave speed and the S-wave speed of the elastic medium.

Example

To demonstrate the estimation procedures outlined above we utilize a set of synthetic seismograms generated by the reflectivity method of Fuchs and Muller (1971). The parameters utilized in the generation of these synthetic data are summarized in Table I.

TABLE I

PARAMETERS USED TO GENERATE THE SYNTHETIC SEISMOGRAMS

Parameter	Value
Source Information	
-Depth:	0.100 Km
-Type:	Explosive point source
-Time Function:	Sinc function bandpassed between 10 to 50 Hz
-Time Sampling Interval:	0.004 sec
Receiver Information	
-Depth:	0.200 Km
-Minimum Offset:	2.0 Km
-Maximum Offset:	26.0 Km
-Spatial Sampling:	0.75 Km
-Type of Recording:	Displacement
Additional Information	
-Minimum Reduced Time:	2.956 sec
-Reduction Velocity:	2.0 Km/sec

In Figure 2a we display the physical model considered in this example. The generated seismograms are displayed in Figure 2b (vertical component of the displacement field), and in Figure 2c we display the radial component of the displacement field after the correction of geometrical spreading. In the rest of this example we will assume that the data in Figure 2c display the same amplitude and offset characteristics as those that would be observed on the pressure field recorded by a set of hydrophones. Note that although the chosen model is simple, the bottom half-space is an elastic medium with a S-wave to P-wave speed ratio of 0.3 (we refer to this kind of medium as a semi-elastic medium). This choice violates the assumption of reflections from a liquid-liquid interface which is built into equations (1) to (4).

Analysing the data in Figure 2c we obtained the sets of observations summarized in Table II (amplitude-offset information) and Table III (phase-offset information).

TABLE II

ONE LAYER OVER HALF SPACE
ACOUSTIC-SEMI ELASTIC CASE
AMPLITUDE-OFFSET DATA

OFFSET (Km)	ANGLE OF INCIDENCE (DEG)	REFLECTION STRENGTH
2.00	16.2	0.2527
2.75	21.7	0.2874
3.50	26.9	0.2812
4.25	31.6	0.2454
5.00	35.9	0.2577
5.75	39.8	0.2581
6.50	43.3	0.2598
7.25	46.4	0.2693
8.00	49.2	0.2803
8.75	51.7	0.2967
9.50	54.0	0.3205
10.25	56.1	0.3561
11.00	57.9	0.4148

TABLE III

ONE LAYER OVER HALF SPACE
ACOUSTIC-SEMI ELASTIC CASE
PHASE-OFFSET DATA

OFFSET (Km)	ANGLE OF INCIDENCE (DEG)	PHASE SHIFT (DEG)
13.25	62.5	22.8
14.00	63.8	24.8
14.75	64.9	35.5
15.50	66.0	55.4
16.25	67.0	74.5
17.00	67.91	83.7
17.75	68.76	82.0
18.50	69.54	82.0
19.25	70.28	95.7
20.00	70.97	91.4
20.75	71.61	100.8
21.50	72.21	99.1
22.25	72.77	104.8
23.0	73.30	104.8
23.75	72.80	112.3
24.50	74.77	113.0
25.25	74.72	117.6
26.0	75.14	121.0

ONE LAYER OVER HALF SPACE

Acoustic-Semi Elastic Case

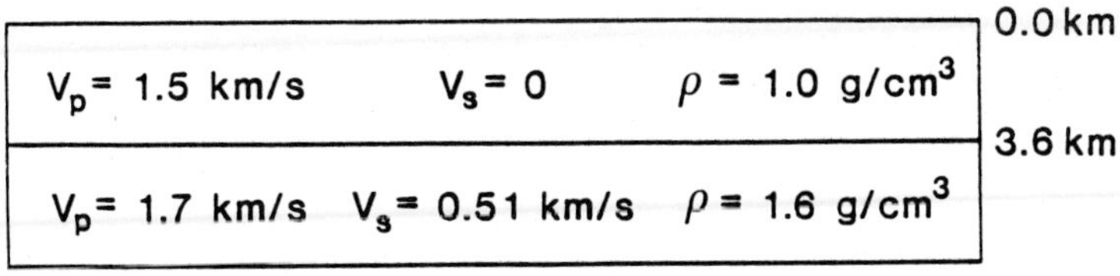

(a)

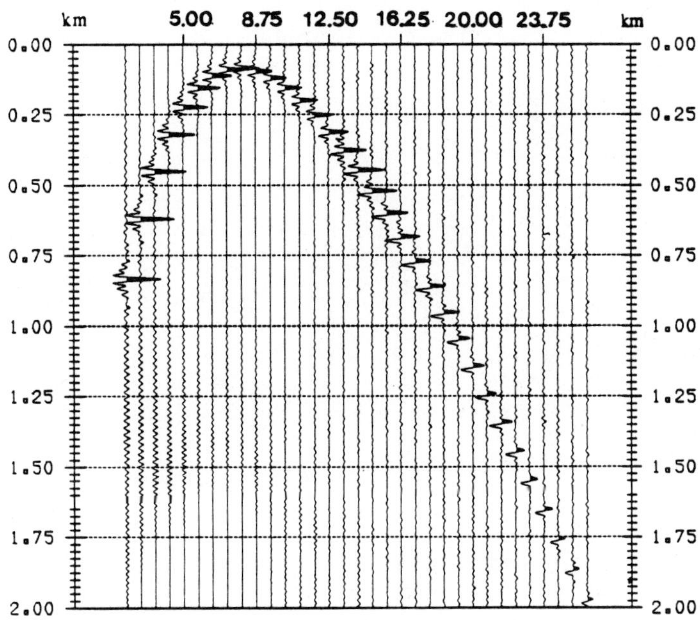

(b)

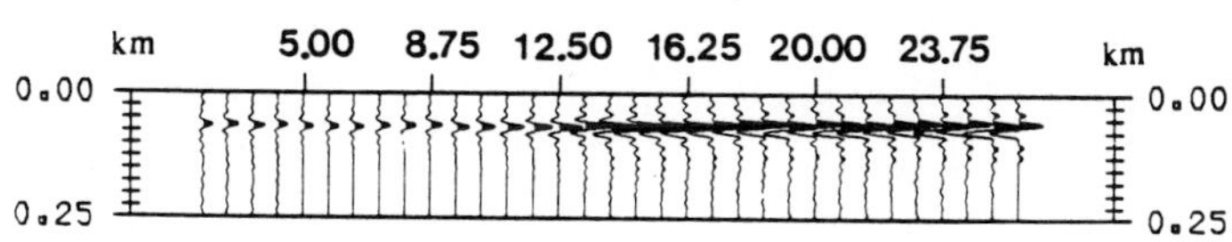

(c)

Figure 2(a) A model of reflecting interface separating an acoustic and an elastic medium.

(b) Reduced time representation of the vertical component of the displacement field.

(c) The radial component of the displacement for the primary reflection arrivals.

Application of the estimation procedures based on the equations outlined in the previous section gave the results that are summarized in Table IV.

TABLE IV

	Density	P-Wave Speed	S-Wave Speed
Acoustic-Acoustic (Amplitude-offset)	1.597	1.569	---
Acoustic-Acoustic (Phase-offset	1.66	1.720	---
Acoustic-Elastic (Amplitude-offset)	1.58	1.7	0.576

It is obvious that all three procedures gave reasonable estimates of the associated elastic layer parameters.

Summary

We have presented three procedures which utilize the information contained in water bottom reflections for the determination of the acoustic or the elastic parameters of the top-most layer of sediments. These procedures have been tested on a number of synthetic data sets generated by the reflectivity method of Fuchs and Muller (1971) and the results presented in this paper are quite encouraging. We believe that the procedures are a viable means for the estimation of sediment properties, and we are at present applying the method to the analysis of measured data.

REFERENCES

Arons, A.B., and Yennie, D.R., 1950, Phase distortion of acoustic pulses obliquely reflected from a medium of higher sound velocity: J. Acoust. Soc. Amer. v. 22, 231-237.

Brekhovskikh, L.M., 1960, Waves in Layered Media. New York: Academic Press.

Carrion, P., Kuo, J.T., and Stoffa, P., 1984, Inversion method in the slant stack domain using amplitudes of reflection arrrivals: Geophys. Prosp. v. 32, 375-391.

Chapman, N.R., Levy, S., Stinson, K., Jones, I., Prager, B., and Oldenburg, D., 1985, Inversion of sound speed and density profiles in deep ocean sediments: submitted to J. Acoust. Soc. Amer.

Fuchs, K., and Muller, G., 1971, Computation of synthetic seismograms with the reflectivity method and comparison of observations: Geophys. Journal of the R.A.S. v. 23, 417-433.

Rayleigh, L., 1945, The Theory of Sound. New York: Dover Publication.

EFFECTS OF UPPER CRUSTAL GEOACOUSTIC PARAMETERS ON LOW FREQUENCY SOUND

Orest I. Diachok, Ronald L. Dicus, and Stephen C. Wales

Acoustics Division
Naval Research Laboratory
Washington, D.C. 20375-5000 USA

ABSTRACT

A seismo-acoustic paradigm of the upper crust is proposed, based on analysis of low-frequency reflectivity versus angle data at a 0.5 million-year old sediment-free site, and on spatially well sampled seismic refraction measurements at sediment-free and thinly sedimented sites in the Pacific Ocean. Crustal velocity measurements at thick sediment covered sites are evidently substantially higher, and are excluded from the analysis presented here. Interfacial compressional and shear speeds of thinly sedimented, geologically young upper crust (between 0.5 and 5 million years) at this site are estimated to be about 2800 m/s and 800 m/s respectively. At such sites sub-basement gradients are approximately 4 s^{-1}. The rms roughness of the basement, which is Fresnel zone size dependent, has significant effects at frequencies as low as 10 Hz. Low shear speeds imply no shear critical angle. Hence, at low shear speed sites, low-frequency energy incident on the bottom at small grazing angles is in large part transmitted into the rock, refracted by the gradient, and reradiated into the water. Scattering loss occurs at both the initial incidence and the interaction of the refracted energy at the boundary. Low interfacial shear speeds lead to large grazing angles at the boundary for the transmitted shear waves, large wavenumbers and, hence large boundary scattering losses; small changes in interfacial shear speed produce large changes in subsurface boundary-scattering loss, and hence in the reflection coefficient. Sediment-free/thinly sediment-covered crustal shear speeds are projected to increase with age, eventually (at an unknown age) becoming faster than the speed of sound in water, resulting in critical angle reflection.

INTRODUCTION

Bottom interacting acoustic fields at small grazing angles in the deep ocean are primarily dominated by sediment-refracted energy when the sediment thickness is greater than about 100 meters, [1-4] and to a large extent by the upper crust when the sediment thickness is below about 100 meters. Consequently, detailed knowledge of upper crustal geoacoustic properties is required for acoustic field prediction in thinly sedimented regimes. These include seamounts, mid-ocean ridges, and most significantly the majority of the Pacific Ocean.

Unconsolidated sediments on the ocean floor originate from land (terrigenous) and from deceased oceanic plants and animals. Deep trenches surrounding the Pacific Ocean limit terrigenous sediments from flowing into this ocean. As a result the mean sediment thickness in the central North Pacific is low, generally below 100 meters (based on Lamont-Doherty Geophysical Observatory data contoured by B. Hersey)[5]. Exceptions occur in areas in close proximity to land, and in the high biological productivity zone near the equator. As seismo-acoustic properties of the upper crust are age-dependent, it is noteworthy that the sediment thickness overlying crust as old as 150 m.y. in non-equatorial regions of the western Pacific[6] is generally less than 100 meters.

The most significant acoustic parameter affecting the reflection process at the upper crustal boundary is the shear speed V_s. If V_s is substantially higher than the speed of sound in water, V_w, then total reflectivity occurs at grazing angles less than the critical angle. If $V_s < V_w$, then there is no critical angle, and nearly total mode conversion to shear waves may occur accompanied by large interfacial reflection losses. In a previous paper[7] the results of a low frequency reflectivity vs. grazing angle experiment conducted at a sediment-free site in the vicinity of the East Pacific Rise (Project ROSE) were documented, and analyzed in the context of previously reported Deep Sea Drilling Project (DSDP) downhole logging and seismic refraction measurements. The following conclusions were drawn:

- The interfacial shear speed in the upper crust at a geologically young, sediment-free Pacific site, was estimated to be 800 m/s, substantially lower than the speed of sound in water. This observeration is in contrast to the generally accepted assumption that $V_s > V_w$.
- Downhole logging and seismic refraction measurements suggest that interfacial speeds change systematically with age and are highly variable.

The observed variability in interfacial velocities may be caused by real lateral changes in velocity structure[8], systematic differences between oceans[9], inadequate sampling rate[10], systematic differences between sediment-covered earth-warmed crust and sediment-free water-cooled crust[11], and potentially anamolous velocities at ridge crests[12], fracture zones and seamounts. In the previous paper these factors, which may potentially affect sound velocity measurements were not considered. The objectives of this paper are to 1) review the results of acoustic reflectivity measurements at the sediment-free site in the Pacific Ocean; 2) provide refined bounds on the age dependence of interfacial velocities in the upper crust of thinly sedimented regions of the Pacific Ocean based on extrapolations from well sampled seismic refraction measurements at other sediment-free and thinly sedimented sites, and 3) discuss the implications of the reflectivity measurements for acoustic field prediction in the context of the refined bounds.

In Part 1 of this paper bounds on the age dependence of interfacial shear speeds in thinly sedimented regions of the Pacific will be estimated based on selected seismic refraction measurements. In Part 2 results of the reflectivity experiment will be reviewed. Modeling of velocity and roughness effects will be considered in Parts 3 and 4, respectively. Conclusions and recommendations for future research are outlined in Part 5.

THE NATURE OF THE UPPER CRUST

Underwater acoustic signals incident on the basalt boundary encounter a rugged, multiphased and spatially variable medium. The top few hundred meters of basalt, known as the upper crust, are generally composed of pillow lava, occasional massive lava flows, interspersed with sediments, water and rubble. Information about sound-velocity and density vs depth profiles comes from two sources: Deep Sea Drilling Project (DSDP) downhole logging measurements[13] and seismic subbottom refraction data.[8-12] Results reviewed in reference 7 reveal highly variable velocities and generally large gradients in the top 500 meters. Seismic refraction measurements are made at a frequency of about 6 Hz with an inherent spatial resolution limit of about 500 m, and as a result are accurate primarily for depths greater than about 500 meters. Well sampled measurements in thinly sedimented young crust due to Ewing and Purdy[8] and in old crust due to Shearer and Orcutt[14] are shown in Figure 1a. Displayed results above about 500 meters are linear extrapolations. The uncertainty in interfacial speeds estimated from this data is about $\pm$ 400 m/s,[15] implying that the measured difference between the two sites is real. The geologically young site is characterized by a significantly lower velocity and a higher gradient than the old site. High frequency logging measurements provide details of the layering within DSDP drilling holes, are restricted to sites where sediment thickness is greater than about 150 meters, and have been shown to be consistent with seismic refraction measurements.[13]

The upper crust is formed at mid-ocean ridges and is transported over periods of millions of years to great ranges. Over this period the density and velocities of the solid basalt component of this medium, measured on dredged samples from sediment-free sites at high frequencies and high pressures[16], decrease as a result of chemical alteration. Aging makes this material progressively "softer". Recently formed crust is cracked and porous. As it ages the cracks and pores are sealed, possibly in response to the combined effects of pressure and the progressive "softening" of this material. As a result, the effective low frequency acoustic properties of this composite are age-dependent. Estimates of the effective interfacial sound velocities in thinly sedimented regions of the Pacific Ocean, extrapolated from two sets of well sampled seismic refraction data[8,14] and inferences from far field reflectivity vs angle measurements[7](to be

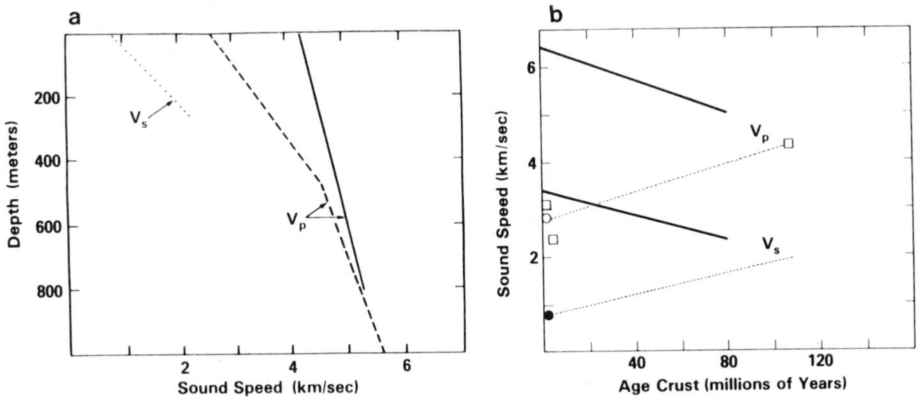

Fig. 1. (a) Sound speed measurements in 104 m.y. old crust[14] (——), in 0.5 to 4 m.y. old crust[8] (---) and assumed shear speed profile at ROSE site (···); (b) Compressional (open) and shear (shaded) speeds from from refraction (□) and reflectivity (o) measurements, estimated lower bounds vs age (---), and speeds in solid samples (——).

subsequently discussed in this paper), are shown as a function of geologic age in Fig. 1b.

Excluded from this compilation are measurements made at ridge crests, at thickly sedimented sites, and those characterized by a low sampling rate. Spudich and Orcutt[10] pointed out that a high sampling rate in the range 5-15 km (substantially greater than 0.5 shots/km) is required for estimating near-surface velocities. Figure 1b provides interfacial speeds extrapolated from experiments with sampling rates greater than 5 shots/km. Stephen[11] pointed out that velocities in thin sediment covered, water cooled upper crust appear to be significantly lower than velocities in thick-sediment (>150 m) covered earth-warmed crust. Hence Figure 1b provides interfacial speeds only for sites where the unconsolidated sediment thickness is less than 80 meters. For comparison, Stephen [11] reports interfacial speeds in the vicinity of 4.5 km/sec at two well-sampled, geologically young (1-6 m.y.), thickly sedimented sites in the Pacific. Orcutt et.al.[12] and others have pointed out that the upper crust at ridge crests may be geologically anamolous. Consequently previously reported measurements at ridge crests are also excluded. In addition, a previously reported seamount site[7] was disregarded. Possible differences between oceans were not considered, as only data points in the Pacific were consistent with the first two criteria. The remaining velocity estimates are dramatically lower than the mean velocities in the previous, indiscriminantly more inclusive compilation.[7]

The data shown here, despite its sparcity, suggests that velocities at thinly sedimented sites increase with crustal age, hypothetically due to the closing of cracks and pores. The dashed lines connecting the very young and very old data provide an estimate of the lower bound, and are not the result of theory. The dashed and solid lines for compressional speeds intersect at 110 m.y.; the dashed line for shear speed is drawn to intersect with the solid shear line at the same age, implying that at that age effective and solid component velocities are equal, and that the cracks and pores are sealed. The actual age at which this occurs, however, could be substantially less than 110 m.y. The true age dependence is unknown.

The sites of the acoustic and seismic measurements at 0.5 m.y. were in close proximity, about 30 km. Ewing and Purdy estimate that sub-bottom speeds are relatively uniform on a scale of tens of kilometers.[8] The difference between these two measurements may be due to a) real lateral variability, b) uncertainty in the inference from seismic refraction data, estimated to be $\pm$ 400 m/s, c) uncertainty in the inference from acoustic reflectivity data, estimated to be $\pm$ 200 m/s, d) the constant gradient assumption (e.g., a parabolic increase with depth would lead to lower interfacial velocities), and e) anisotropy relative to spreading direction (differences of 0.2 km/s have been noted[14]).

Despite its apparent high porosity, the attenuation coefficient, α, of the upper crust has been estimated at seismic frequencies to be only two times higher than in solid basalt, where reported values are about .02 dB per wavelength[12]. At higher frequencies, however it may conceivably be significantly higher.

Bathymetric profiles of sediment-free bottoms have been most precisely measured with deep-towed instrumentation.[17] The boundary may be generally characterized as "sawtooth" with preferentially oriented facets, on the order of 150 meters in length, with a mean tilt (in the vicinity of the East Pacific Rise) of 4° normal and 2° parallel to the ridge axis. The mean (near-vertical) displacement height between facets is nominally about 10 meters. The pillow lava relief superimposed on the facet structure is random and on the order of 1 meter. The following heuristic argument suggests that the effective roughness for the measurement scenario reported here is in the vicinity of 5 meters. For a frequency of 20 Hz and a slant range of 10 km (nominal for the experimental data to be discussed subsequently) the first Fresnel zone is approximately 900 meters long, encompassing several 150-meter facets; the displacement height between these is about 10 m, resulting in an rms height about a mean plane on the order of 5 m. A much larger facet, e.g. 1 km long, would appear as a (probably tilted) deterministic reflecting surface, and not contribute to "scattering loss". For comparison, the effective rms roughness of abyssal plains (for a comparable geometry) has been estimated to be about 0.1 meters.[18]

REFLECTIVITY MEASUREMENTS

Reflectivity measurements were made at an 0.5 m.y. old, essentially sediment-free site in the vicinity of the East Pacific Rise[7]. Hydrophones were deployed to a depth of 2000 meters from USNS HAYES; underwater explosives were dropped from the University of Hawaii research ship, RV KANE KEOKI along a track parallel to the axis of the rise to minimize bathymetric scattering effects. The shot density was about 1 shot per degree grazing angle range. Standard Mk61 2 lb shots were used in the 90° to 45° grazing angle range. At smaller angles 32 lb shots, detonated at the same depth with Mk61 charges, provided signals.

The data was tape recorded over a 5-50 Hz bandwidth and subsequently processed to provide reflectivity vs grazing angle. The received bottom-reflected shot signals are described in detail in reference 7. Distinct bottom (B), surface-bottom (SB), bottom-surface (BS) and surface-bottom-surface (SBS) reflected signals persisted over nearly the entire angular range. Other smaller pulses were also visible, but occurred irregularly, probably evidencing roughness. The dominant pulses were generally distinct at angles greater than about 20°, and their durations were nominally 0.1 sec. No evidence for subbottom refracted energy was evident. This is in contrast to the reflection process at water/thick-sediment boundaries where the subbottom gradient is about 1 sec^{-1}, and deconvolution methods reveal distinct reflected and subbottom refracted arrivals

The measured reflection coefficient, shown in Figure 2a, was computed in the following manner: Digitized data recorded aboard ship was edited to isolate the group of four arrivals having a single bottom bounce (B, SB, BS, SBS). This data was tapered at either end and then the spectrum was computed by an FFT. This spectrum was then calibrated using known system gains and calculated ray theoretical transmission loss. A 6 dB factor was included to account for the four arrivals. An archival time series provided the spectrum for a direct arrival. The quotient of these two, i.e. reflected spectrum divided by a direct spectrum, is the reflection coefficient in the absence of noise. Because some problems were anticipated with noise contamination of the signals, a similar but more complex formula due to Dicus,[3] which corrects for the presence of noise in the bottom reflected signals, was used.

The standard deviation of the reflected signals is about $\pm$ 2 dB at small grazing angles and $\pm$ 3 dB at large angles, in contrast with relative shot-to-shot uniformity for (source level calibrated) signals reflected from abyssal plains, where the standard deviation is generally about $\pm$ 0.5 dB. The source level uncertainty, about $\pm$ 1 dB, may account for some of the observed variability. To highlight systematic variability vs grazing angle 20 Hz (averaged between 15 and 25 Hz), 10° incoherently angle-averaged data is shown in Fig. 2b. Also shown is a theoretical fit to the data which will be subsequently described.

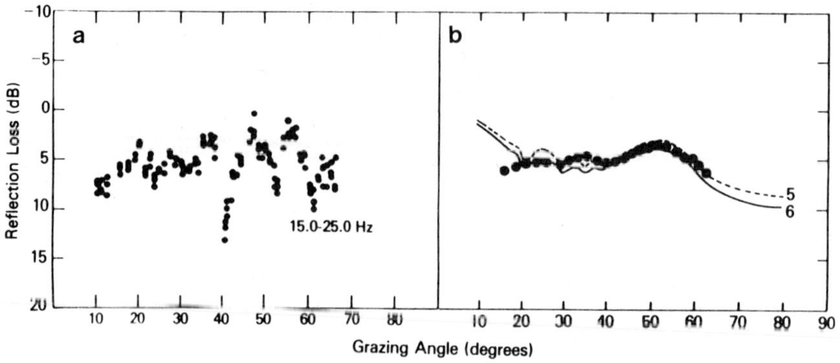

Fig. 2. (a) Measured reflection coefficients vs grazing angle in the 15-25 Hz band; (b) Angle averaged measured (···) and computed reflection loss vs grazing angle for RMS roughness, 5 and 6 m, at 20 Hz.

The most prominent feature of this data is relative maximum at 53°, which is interpreted as the compressional critical angle. The model of reflectivity from the basalt bottom at this site described in the next sections accounts for the average levels of the reflection coefficient vs grazing angle, including the low frequency relative maximum at about 53°, the absence of significant multiple arrivals, and the apparent increase in variability at large grazing angles.

ACOUSTIC MODELING

The geometry of the experiment and nature of the upper crust dictate that an accurate theoretical description of the reflection process must incorporate the spherical nature of the waves emanating from point sources, and the systematic depth dependence of physical properties in the crust. Deviations from theoretical predictions based on these prerequisites might then be ascribed to rough boundary scattering loss or other second-order phenomena.

FFP calculations[19] of the reflection coefficient were performed to model the observed reflectivity. The bottom was simulated by a stack of solid layers each 50 m thick. The terms "compressional-wave" and "shear-wave velocities" refer to values at the top of the upper crust. Both wave velocities increased linearly with depth at the rate of 5 sec^{-1} to a depth of 150 meters, followed by a smaller rate of increase (3 sec^{-1}) to values of 6000 m/sec and 3000 m/sec, respectively, at a depth of 1 km relative to the top of the upper crust. Computations were performed for various wave speeds and gradients to obtain the best fit to the data

FFP calculations at 20 Hz incorporating near-surface sound speed gradients of 5 sec^{-1}, an interfacial compressional wave speed of 2800 m/s, and interfacial shear wave speeds of 800 and 1400 m/s are shown in Fig. 3. For these calculations the crustal density was assumed to be 2.0 and the attenuation coefficient .02 dB/λ. Also depicted are FFP calculations with zero gradients. The following trends are evident: When a gradient is present interference patterns occur between the reflected and double-mode converted subbottom refracted shear waves, i.e., compressional waves converted to shear waves in the rock, refracted by the subbottom gradient and partially reconverted to compressional waves in the water. When the shear-wave speed approximates the sound speed in water, this double mode conversion process leads (at relatively small angles) to nearly total transmission, refraction and reradiation into the water.[20] When the energy is half reflected, half double-mode converted, e.g. when V_s = 800 m/s at 15° < θ < 35°, particularly strong interference patterns result. Changing the gradient to 2 or 8 sec^{-1} does not radically alter results[7].

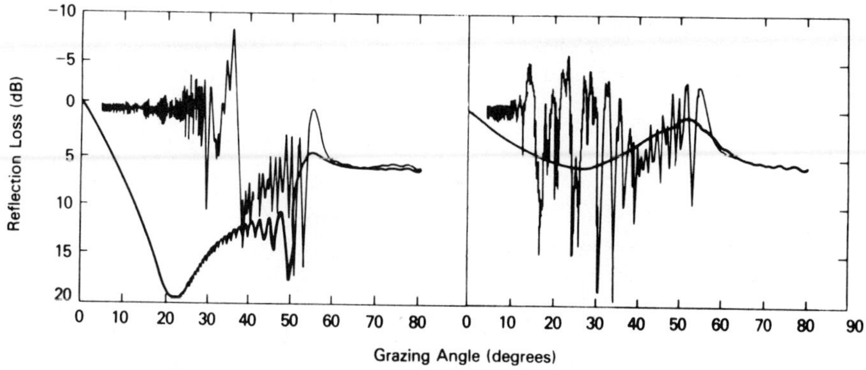

Fig. 3. FFP calculations of reflection loss vs grazing angle with subbottom gradient (——) and isovelocity bottom (▬▬). V_p = 2.8 V_s = 1.4 (left) and V_s = 0.8 (right); G_s = 5, ρ = 2, α = .02.

SCATTERING LOSS

Scattering loss from a Gaussian randomly rough surface, a reasonable approximation for the basalt boundary[7], may be simply described and readily implemented with a theory developed by Eckart.[21] According to this theory, scattering loss obeys the relationship:

$$R = e^{-g^2}$$

where R = coherent energy reflection coefficient, g = 2kr sin θ, k = wave number, r = rms roughness, and θ = grazing angle. Scattering loss must be considered at each interaction with the boundary, as illustrated in Fig. 4a. Due to the low shear speed in the crust, and in accordance with Snell's law, compressional waves in the water are refracted toward the normal as shear energy in the crust. Hence the grazing angle of refracted shear waves incident on the upper boundary is greater than the grazing angle of waterborne compressional waves. Furthermore, the wavelength of the shear waves is approximately half the wavelength of the in-water waves. As a result of these two factors, the rock-water interface appears extremely rough to the shear waves, and relatively smooth to waterborne energy. Computations of the scattering loss of both modes are illustrated at a frequency of 20 Hz in Fig. 4b.

Scattering loss for the subbottom shear wave is assumed to follow the Eckart scattering theory according to the following heuristic argument: It is assumed that gradients in the vicinity of the rough water-basalt interface are negligible in the sense that the shear potential remains uncoupled from the compressional potential and that reflection and transmission coefficients may be computed using interface values of compressional velocity, shear velocity and density.

The scattering loss of the waterborne energy at small grazing angles increases significantly with frequency and roughness, but is nevertheless relatively small at small grazing angles at 20 Hz. On the other hand, scattering loss of the subbottom refracted shear waves (at any given frequency) is much larger at small angles. Angle-averaged FFP reflectivity computations, when compared with data, yield good agreement for a shear speed of 800 m/s and an rms roughness of 6 meters, as illustrated in Fig. 2b. Discrepancies at angles greater than about 55° may be expected due to the small roughness limitation of the theory. Similar constraints limit the accuracy of the estimated scattering loss of the subbottom refracted shear waves. Small-angle discrepancies may be partly due to violations of the assumption of Eckart's theory. Hence, inferences from comparisons between data and theory are most direct for small g and large θ on both sides of the boundary.

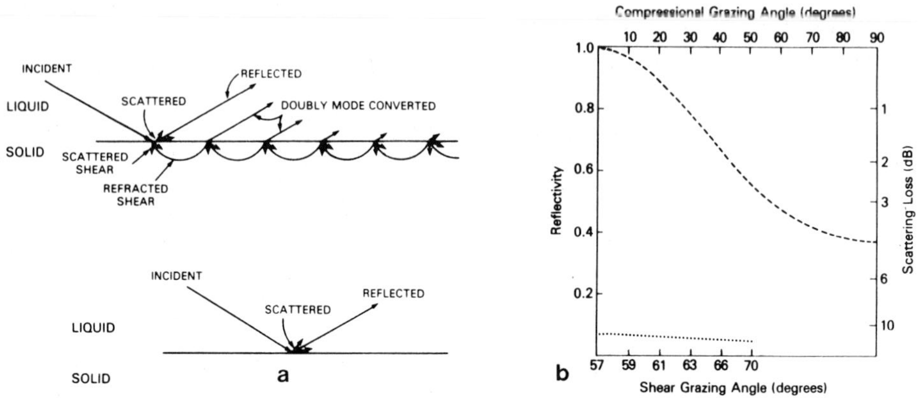

Fig. 4. (a) Mode conversion/scattering process at a low shear speed solid boundary with gradient (top) and high shear speed solid (bottom); (b) Scattering loss of compressional waves (---) and subbottom refracted shear waves (···) vs grazing angle at 20 Hz. RMS = 6 m.

Interestingly, boundary scattering loss of the subbottom refracted waves, while critically dependent on the interfacial shear speed, is not strongly affected by the magnitude of its gradient. The grazing angle of the shear waves incident at the interface is the same as the angle of the initially refracted waves; the exact nature of the path in the crust does not affect this angle. Computations of $R(\theta)$ for g = 2, and 8 virtually coincide at grazing angles greater than 30°; differences of less than 2 dB at smaller angles are probably due to changes in the interference structure, which are evident despite the 10° averaging[7]. The small angle sensitivity to subbottom gradient, although not very large, suggests that precise theoretical fitting of data in this angular region requires knowledge of the detailed (unknown) subbottom sound velocity structure, which provides another reason for limiting data-theory inferences to angles greater than about 30°.

CONCLUSIONS AND RECOMMENDATIONS

Inferences of upper crustal velocities from acoustic reflectivity data at the ROSE site, when restricted to grazing angles greater than about 30°, are consistent with interfacial speeds derived from seismic refraction measurements in the vicinity of the ROSE site, and are considered reasonably accurate. Conceivably, small perturbations in density, velocities and roughness could equally well account for the observations. Such a finely tuned sensitivity study might establish quantitatively the precise uncertainties in the inferred parameters, but was not attempted because of the cost and duration of FFP calculations, some uncertainty in the adequacy of the scattering loss theory, and because it was felt that the data constraints were insufficient to warrant such an investigation.

At angles less than about 30° measured reflectivity is sensitive to the details of the subbottom sound velocity structure, and at sufficiently low angles may be sensitive to bathymetric shadowing. Neither of these subjects were considered in this paper, although they are necessary ingredients for acoustic field prediction.

In the vicinity of the compressional critical angle computed reflectivity vs angle curves appear to be relatively insensitive to large changes (or uncertainties) in shear speed, subbottom gradient and rms roughness[7]. Consequently, this method could conceivably be exploited for accurate measurement of interfacial compressional speed in other parts of the thinly sedimented Pacific Ocean. An uncertainty of $\pm$ 2° in the value of the compressional critical angle for this experiment leads to $\pm$ 200 m/s uncertainty in V_p.

Estimates of interfacial speeds, based on extrapolated seismic refraction measurements at thinly sedimented sites, suggest that compressional and shear speeds increase with age; the speed vs age dependence however, is unknown. The inherent lateral variability of the crust is unknown. Much more data is required to define the age dependence of the mean velocities and their variances. Interfacial compressional and shear speeds of thinly sedimented, geologically young upper crust (between 0.5 and 5 million years) are estimated to be about 2800 m/s and 800 m/s respectively. At such sites sub-basement gradients are about 4 s^{-1}. The rms roughness of the basement, which is Fresnel zone size dependent, has significant effects at frequencies as low as 10 Hz.[7] Low shear speeds imply no shear critical angle. Hence, at low shear speed sites, low-frequency energy incident on the bottom at small grazing angles is in large part transmitted into the rock, refracted by the gradient, and reradiated into the water. Scattering loss occurs at both the initial incidence and the interaction of the refracted energy at the boundary. Low interfacial shear speeds lead to large grazing angles at the boundary for the transmitted shear waves, large wavenumbers and, hence, large boundary-scattering losses; small changes in interfacial shear speed produce large changes in subsurface boundary-scattering loss, and hence in the reflection coefficient. Sediment-free/thinly sediment-covered crustal shear speeds are projected to increase with age, eventually (at an unknown age) becoming faster than the speed of sound in water, resulting in critical angle reflection.

Reflectivity, impulse response and acoustic propagation models that account for the complex nature of this medium should be developed. The theory of the reflection process at low shear speed sites, particularly the scattering of subbottom refracted shear waves, needs to be refined. At higher frequencies effects of subbottom layering on the impulse response should be investigated. DSDP acoustic logging measurements at a few thinly sedimented Pacific sites, where $V_s < V_w$, would be invaluable for such studies and are recommended.

ACKNOWLEDGEMENTS

This work was supported by the Office of Naval Research. The authors are sincerely grateful to Mr. Robert Rumpf for providing the measurement apparatus, Dr. Joe Gettrust for deploying the shots, Dr. H. Kutschale for assistance in implementing and exercising of the FFP model, and Dr. M. Purdy and Dr. R. Stephen for stimulating and helpful discussions.

REFERENCES

1. M.N. Hill, Seismic Refraction Shooting in an Area of the Eastern Atlantic, Phil Trans. R. Soc. London A244:561 (1952).
2. H.E. Morris, Bottom Reflection Loss Model with a Velocity Gradient, J. Acoust. Soc. Am. 48:1198 (1970).
3. R.L. Dicus, Preliminary Investigations of the Ocean Bottom Impulse Response at Low Frequencies, Naval Oceanographic Office, NSTL Station, MS (1976).
4. E. L. Hamilton, Geoacoustic Modeling of the Ocean Floor, J. Acoust. Soc. Am. 68:1313 (1980).
5. C. Spofford, "Blug Overview," Science Applications Inc., McLean, VA (1983).
6. R. L. Larson and W. C. Pitman, "The Bedrock Geology of the World," W. H. Freeman and Co., New York, N.Y. (1985).
7. O. Diachok, R. Dicus and S. Wales, Elements of a Geoacoustic Model of the Upper Crust, J. Acoust. Soc. Am. 75:324 (1984).
8. J. I. Ewing and G. M. Purdy, Upper Crustal Velocity Structure in the Rose Area of the East Pacific Rise, J. Geophys Res. 87:8397 (1982).
9. R. Houtz and J. Ewing, Upper Crustal Structure as a Function of Plate Age, J. Geophys. Res. 81:2490-2498 (1976).
10. P. Spudich and J. Orcutt, A New Look at the Seismic Velocity Structure of the Oceanic Crust, Reviews of Geophysics and Space Physics 18:627 (1980).
11. R. A. Stephen and A.J. Harding, Travel Time Analysis of Borehole Seismic Data, J. Geophys. Res. 88:8289 (1983).
12. J. A. Orcutt, B. L. N. Kennett, and L. M. Dorman, Structure of the East Pacific Rise from an Ocean Bottom Seismometer Survey, Geophys. J. Roy. Astron. Soc. 45:305 (1976).
13. D. Moos and R. Anderson, Elastic-wave Velocities in Layer 2A from Full Waveform Sonic Logs at Hole 504B, submitted to DSDP Reports.
14. P. Shearer and J. Orcutt, Anisotropy in the Oceanic Lithosphere-Theory and Observations from the Ngendei Seismic Refraction Experiment in the Southwest Pacific, Geophys. Jr. Astr. Soc. 80:493 (1985).
15. G. M. Purdy, personal communication.
16. N. I. Christensen and M. H. Salisbury, Velocities, Elastic Moduli and Weathering - Age Relations For Pacific Layer 2 Basalts, Earth and Plan. Science Let. 19:461 (1973).
17. F. N. Spiess, B. Luyendyk and M. S. Loughridge, Bottom Slope Distribution and Implied Acoustic Bearing Errors in Abyssal Hill Regions of the North Pacific, Unpublished manuscript (1969).
18. J. M. Berkson et. al., Measurements of Spatial Coherence of Bottom-Interacting Sound in the Tagus Abyssal Plain, "Bottom Interacting Ocean Acoustics", Kuperman and Jensen, Ed. Plenum Press, New York (1980).
19. H. W. Kutschale, "Rapid Computation by Wave Theory of Propagation Loss in the Arctic Ocean," Lamont-Doherty Geophysical Observatory of Columbia University, New York (1973).
20. R. S. White and R. Stephen, Compressional to Shear Wave Conversion in Oceanic Crust, Geophys. J. R. Astron. Soc. 63:547 (1980).
21. C. Eckart, The Scattering of Sound from the Sea Surface J. Acoust. Soc. Am. 25:566 (1953).

EVALUATION OF EXPERIMENTAL TECHNIQUES FOR DETERMINING THE PLANE WAVE REFLECTION COEFFICIENT AT THE SEA FLOOR

Henrik Schmidt and Finn B. Jensen

SACLANT ASW Research Centre
I-19026 La Spezia, Italy

ABSTRACT

One of the most commonly used techniques for determination of the geoacoustical properties of the sea bed is the measurement of plane wave reflection coefficients at the ocean bottom. An incident field is produced by either an explosive source or a beam generating device and the reflected field is then detected by means of an array of hydrophones. The associated angles of incidence have traditionally been determined by simple geometrical considerations. However such simple interpretations give results that in some cases depend on the actual experimental geometry. For example critical angles can appear far away from the correct values or not be present at all. Here an exact numerical model is used to examine the different experimental techniques. The observed discrepancies are explained, both for the point and beam source experiments. In addition, guidelines are given for interpretation of results obtained by the different experimental techniques.

INTRODUCTION

The importance of the sea bed geoacoustic properties for shallow water sound propagation is well established, and reliable transmission loss predictions obtained by means of numerical propagation models therefore require accurate knowledge of the sea bed properties. The traditional raytrace propagation models require a plane wave reflection coefficient at the bottom in order to account for the bottom loss. More recent wave theory models like those based on normal modes, the full wavefield fast field programs (FFP) and the parabolic equation models (PE), require a more detailed knowledge of the wave speeds, attenuations and densities in the sea bed. These parameters could in principle be obtained from samples, but due to the fact that the low-frequency acoustic waves penetrate deeply, very deep and expensive boreholes would be required. Further, and often more important, the de-pressurisation and change of temperature, unavoidable in the core sampling process, tend to deteriorate the mechanical and chemical bondings in the sediment material, and thus heavily influence the properties, shear in particular.

The geoacoustic properties therefore primarily have to be determined from in-situ propagation experiments. One of the most common experimental techniques is the determination of the plane wave reflection coefficient at the sea floor. This approach has two advantages: the results can be used directly by the raytrace models, and the plane wave reflection coefficient is needed as an input parameter in most inverse schemes for the determination of bottom properties [1].

Several experimental techniques have been devised with the objective of determining the reflection coefficients directly. They are all based on the detection, by hydrophones, of the bottom reflected part of the field produced by a sound source placed in the water column. The source has been either an omnidirectional explosive source or a device producing a narrow beam of sound directed towards the sea floor at a variable angle of incidence. However, several authors have shown that the results obtained are usually not directly interpretable as plane wave reflection coefficients.

Here we will use the full wavefield SAFARI model [2,3], to demonstrate how these discrepancies arise, and to show how numerical models are not restricted to interpretational purposes, a traditional application in underwater acoustics, but can also be used for the design and planning of experimental setups.

EXPLOSIVE SOURCE EXPERIMENTS

The most common experimental technique for determination of plane wave reflection coefficients uses an explosive omnidirectional source to generate a transient field. The hydrophone array used as a detector may be either horizontal or vertical, moving or fixed (Fig. 1).

By assuming that the source and receivers are so far apart that the bottom-interfering eigenrays can be considered plane waves when hitting the bottom, a very simple interpretation technique has been used. First the nominal specular reflection angle at the bottom is determined for each receiver by means of simple raytracing. Then the received signals are split into a direct part and a bottom reflected part by inspection. After correction for different travel paths, the reflection coefficient is found simply by dividing the frequency spectrum of the reflected signal by that of the direct signal.

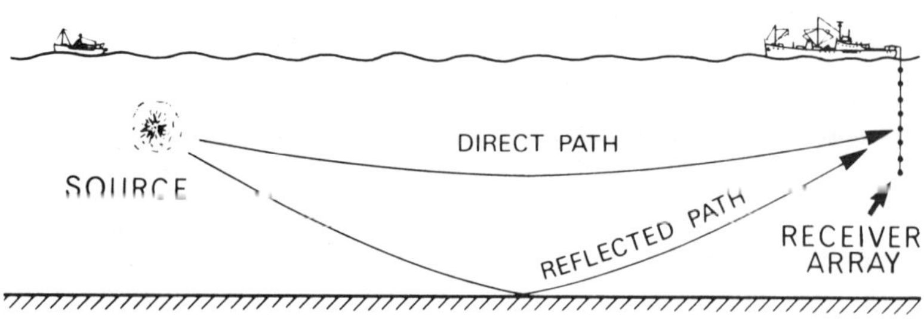

Fig. 1 Experimental set up for bottom reflection-loss measurements by means of an explosive source and a vertical hydrophone array.

The obvious advantage of the explosive source experiments is the possibility of separating the different arrivals in the time domain, and hence to eliminate unwanted surface multiples. Further, the experiments are cost-effective because they do not require any specialized equipment. As pointed out by several authors, however, the simple interpretation technique outlined above is only rarely applicable. Stickler [4] and Santaniello et al [5] demonstrated that the simple interpretation technique will give wrong reflection coefficients when the bottom is upward refracting or has deeper, reflecting interfaces, because of the interference between the different arrivals. Non-physical effects like negative bottom loss and source/receiver position-dependent results arise. The critical angle shift pointed out by Stickler [4] is due to the same effect. Even in the case of an isovelocity bottom with a sound speed higher than that of the ocean, the headwave formation will give rise to the same phenomenon. The simple interpretation technique can be applied only in the rather unusual case of a purely downward refracting or homogeneous bottom with a sound speed less than that of water.

We will here illustrate the limitations of the simple interpretation principle by simulating an explosive source experiment in a very simple ocean environment by means of the SAFARI full-wavefield model. The sound speed profile is shown in Fig. 2. The water is characterized by an upward refracting profile close to the bottom, and surface multiples of no present interest are avoided by replacing the ocean above -800 m by an isovelocity halfspace. The fluid bottom is upward refracting to 30 m below the seabed. Below this depth it is represented by an isovelocity halfspace. In order to more clearly illustrate the phenomenon, the bottom is considered lossless. Thus the reflection loss is identically zero for grazing angles less than the critical 14.6° onto the water-bottom interface.

An explosive source is assumed to be placed 400 m above the bottom and the radiated pressure pulse has a duration of 10 ms and a centre

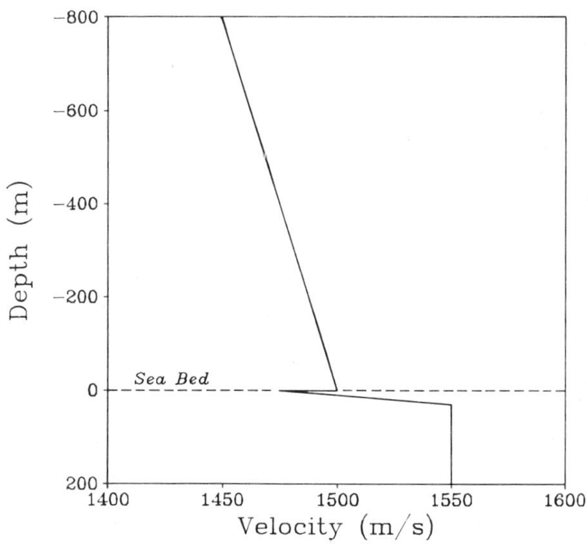

Fig. 2 Sound speed profile for simulation of experiment. Depth 0 m refers to the sea bed.

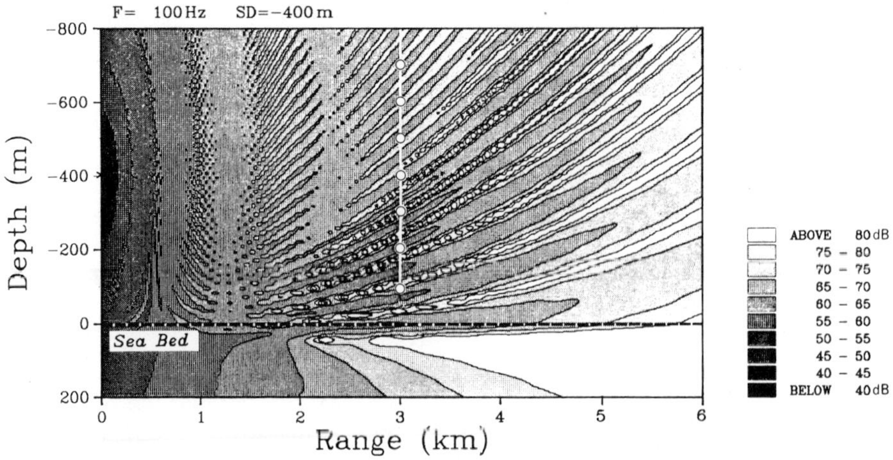

Fig. 3 Transmission loss contours at 100 Hz for a point source 400 m above the bottom. Hydrophone array used for pulse calculations is indicated by o.

frequency of 100 Hz. Figure 3 shows the transmission loss contours in depth and range at the centre frequency (black indicating highest intensity). The Lloyd-mirror pattern due to the interference between the direct and the bottom reflected fields is evident and illustrates the complexity of the sound field even in this very simple case.

A vertical array of 7 hydrophones with 100 m spacing is placed 3 km from the source as indicated in Fig. 3. The synthetic hydrophone signals for this array are shown in Fig. 4, with each trace being identified by the nominal specular angle of the reflected signal at the ocean bottom, as separated from the reflected parts, which, however, do not clearly

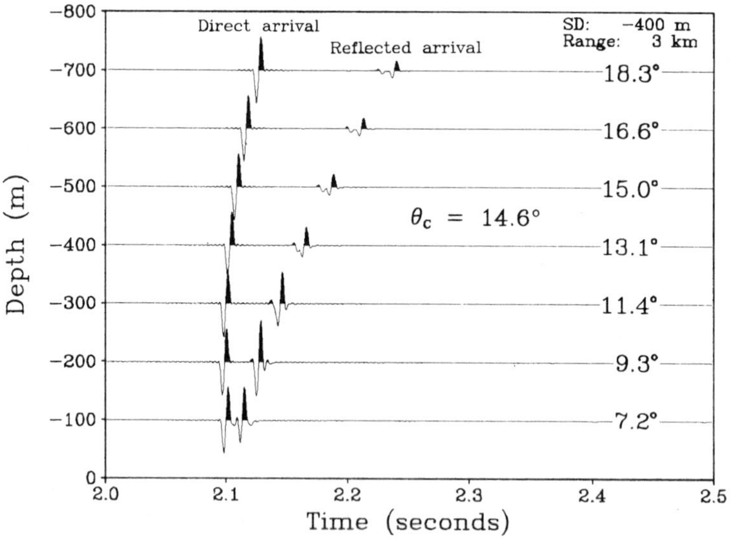

Fig. 4 Synthetic hydrophone signals for vertical array at 3 km range. Each trace is identified by its corresponding nominal specular reflection angle.

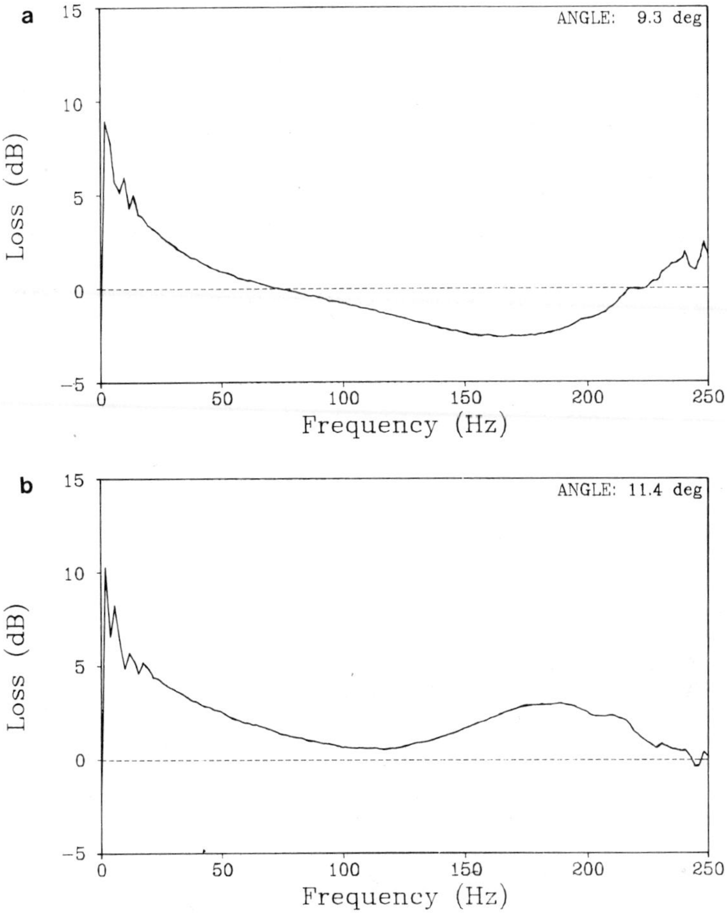

Fig. 5 Reflection coefficients determined by deconvolution of synthetic signals: a) 9.3°, b) 11.4°

indicate the critical angle of 14.6°. Further, the trace corresponding to 9.3° indicates an apparent negative bottom loss. These properties are also indicated by the reflection coefficients obtained by the simple deconvolution principle for the grazing angles 9.3° (Fig. 5a) and 11.4° (Fig. 5b) respectively. Both angles are less than critical; therefore the reflection loss should be zero in both cases. The "ringing" at low and high frequencies is due to the very low energy content of the source pulse at these frequencies, but even in the central frequency interval errors of several dB are obtained.

Figure 6 outlines the different travel paths yielding errors in the reflection coefficients obtained by the simple technique. The possibility of multiple paths, due to headwaves (1), upward refracting profiles (2) or reflecting interfaces (3), yields results which are dependent on whether the different arrivals are interfering constructively or destructively, which is again dependent on the source-receiver positions. As is also clear from Fig. 6, the multiple arrivals do not correspond to the same angle as the nominal specular reflection angle at the water-bottom

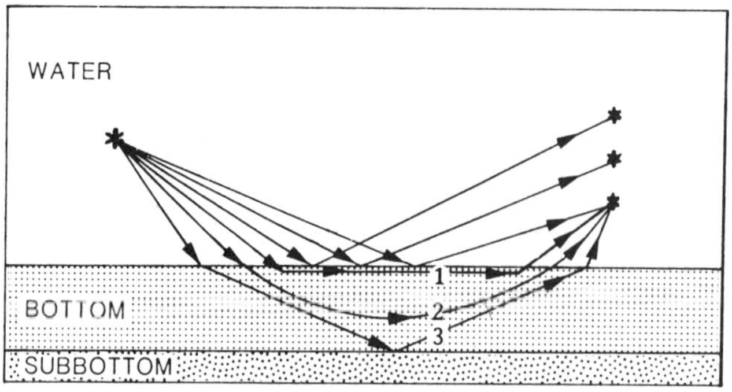

Fig. 6 Additional arrivals obstructing simple deconvolution. 1) Headwaves, 2) Upward refracted arrivals, 3) Deep reflection.

interface. The field detected by each hydrophone is therefore not a single plane wave component, as is assumed when the simple interpretation technique is used, but a complicated interference between several different components.

This phenomenon directly leads to the conclusion that in order to obtain the plane wave reflection coefficient from explosive source experiments, some kind of beamforming, or plane wave decomposition, has to be applied to the received signals. In principle this could be done by using towed arrays. However, at the low grazing angles, often of main interest, this technique requires very long arrays because the bottom refracted signals travel a long distance before re-entering the water column, and the whole reflected field has to be covered by the array. Therefore, a synthetic aperture technique, as proposed by Frisk et al [6], is often more convenient than the use of a towed array.

BEAM EXPERIMENTS

In contrast to the omnidirectional explosive source described above, a source that generates an approximate plane wavefield would provide the opportunity to directly measure the plane wave reflection coefficients. This is the philosophy behind the use of beamed sources, as shown schematically in Fig. 7. A beam is directed towards the water-sediment interface at a nominal angle of incidence, and the reflected beam is measured in the specular direction by means of a hydrophone.

It is well known, however, that very wide beams are required in order to simulate plane wave behaviour. This is due to the fact that a beam of finite width has a finite angular spectrum, whereas for a plane wave the angular spectrum is infinitely narrow [3]. For practical reasons this technique has therefore not been a realistic alternative to the explosive source experiments until the development of the parametric transducer [7]. In principle, this transducer generates a virtual endfire-array that is the source of a highly directional beam which can be used for measurement of plane wave reflection coefficients [8]. In practice, however, the virtual array will have a finite length, and thus results in the beam having a finite angular spectrum.

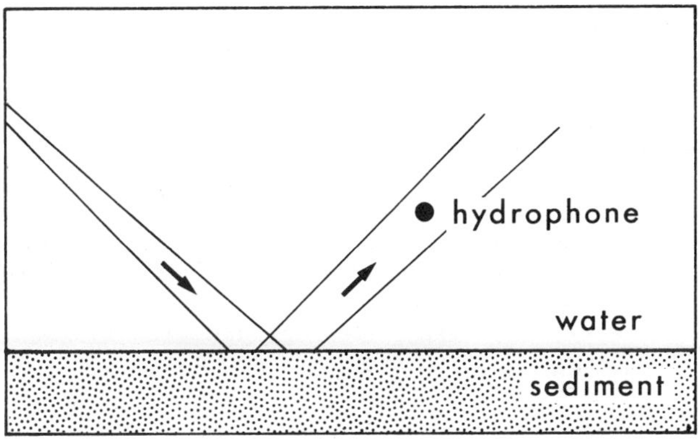

Fig. 7 Reflection of a narrow beam from the sea bed.

We will here use the SAFARI model to demonstrate how a velocity profile in the ocean close to the bottom influences the reflection of a realistic, narrow beam at low grazing angles. The environment shown in Fig. 2 is also used for this study. A vertical source array of 61 elements at half-wavelength spacing, placed 400 m above the bottom is used to generate a narrow beam, 6 wavelengths wide at the sea bed, measured between the 3-dB down points. The array is phased to yield a nominal grazing angle of 2.5°, 5° and 10° at the sea bed.

The resulting fields are shown in Fig. 8 (black indicating highest intensity). The contour interval is 2 dB, but the actual dB values are arbitrary. Obviously the beams are not specularly reflected. At the two smallest grazing angles, Figs. 8a and 8b, a clear beam splitting occurs, and the reflected beam with highest amplitude has a much smaller grazing angle than the incident beam. A comparison with Fig. 3 shows that the directions of the split beams correspond to those of the lowermost lobes of the Lloyd-mirror pattern in the point source field. This indicates that the angular spectrum of the beam is apparently so wide that the interference between the direct, but upward refracted, purely waterborne arrivals and those reflected off the bottom is important. This assumption is supported by the calculated angular spectrum of the 5° beam at the bottom, shown in Fig. 9. The beam is seen to contain significant energy at grazing angles in the interval of 1° through 12°. Figure 8c indicates that the splitting effect due to the interference decreases for higher grazing angles, as expected, but although the energy is here concentrated in a single beam, a significant widening of the beam cross-section has occured. Even at this relatively high grazing angle, a single hydrophone in the specular direction would not yield the right value of the reflection coefficient; instead it would indicate too high a reflection loss.

Even in the ideal case of an isovelocity water column over a homogenous bottom, it was shown by Muir et al [8], that narrow parametric beams could penetrate into the bottom at grazing angles less than critical. This problem has been treated theoretically by several authors. Tjotta and Tjotta [9] presented a wide beam approximation showing the effect qualitatively. Schmidt and Jensen [3] used the SAFARI model to simulate the experiments in which extremely narrow beams, not covered by the approximate theory, were used. The reported beam cross-section at the bottom interface was simulated by means of a focusing linear array.

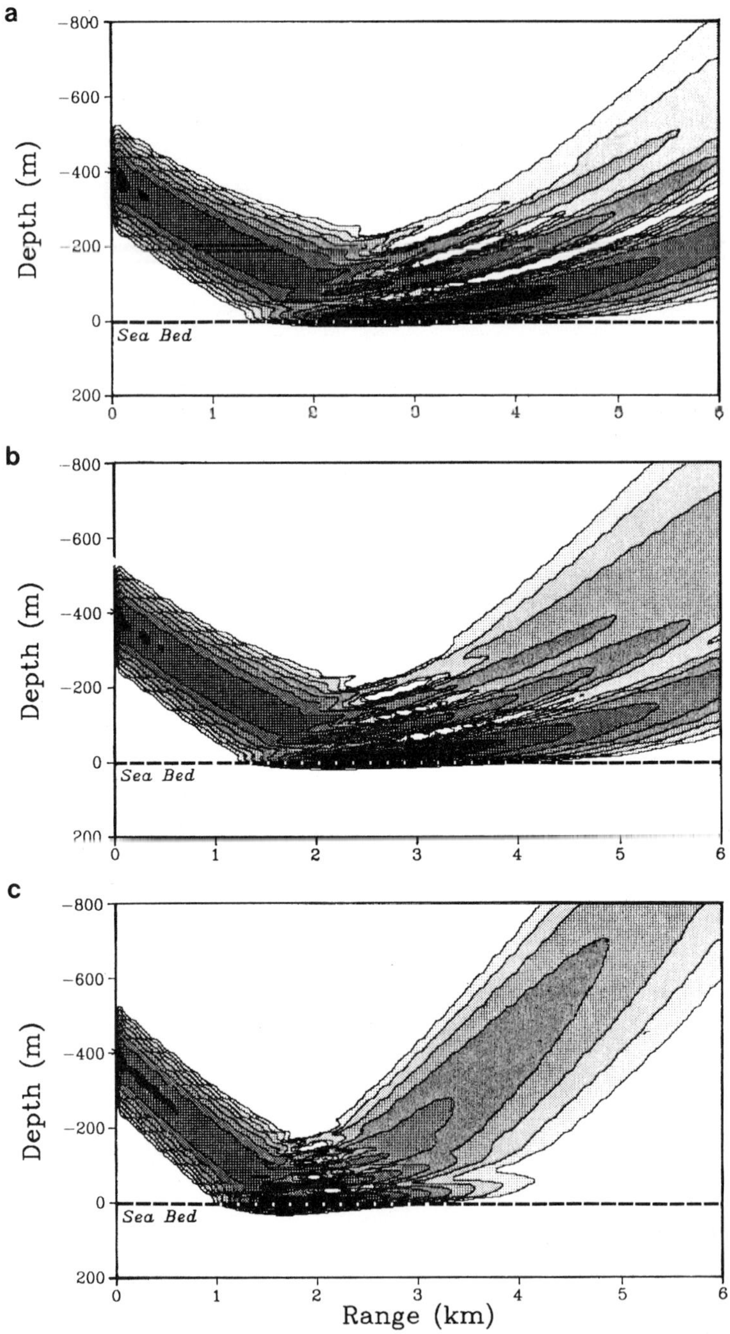

Fig. 8 Field contours at 100 Hz for an incident beam of 6 wavelengths width at 3 nominal grazing angles: a) 2.5°, b) 5° and c) 10°. (Contour interval = 2 dB).

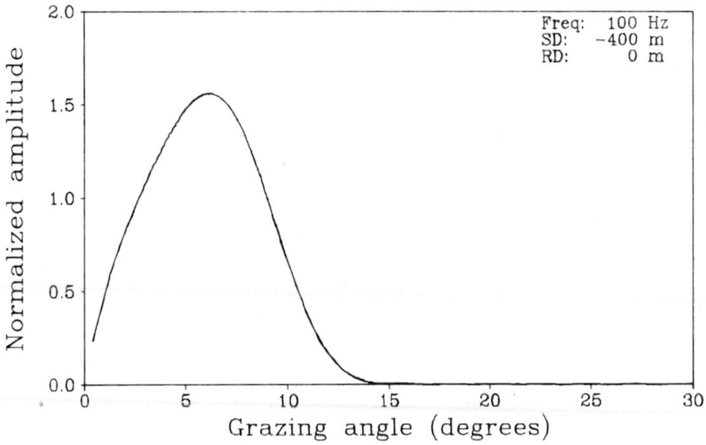

Fig. 9 Angular spectrum of beam at the ocean bottom for nominal grazing angle of 5°.

The results demonstrated that subcritical penetration is a simple consequence of the basic physical principle that a narrow beam has a wide angular spectrum. Thus the narrow beam may contain energy that propagates at grazing angles greater than critical although the nominal angle of incidence is subcritical. The results further showed that, for the same reason, the reflected beam was not specular, but shifted towards a smaller grazing angle. In conclusion, the parametric beams in practice are narrow and not highly directional as often stated, and hence they are a poor approximation to a plane wave.

To summarize, the use of beam sources does not yield the possibility of measuring the plane wave reflection coefficients directly by means of a single hydrophone. As was the case for the explosive sources, plane wave decomposition has to be applied to the reflected field. However, the advantage of using beam sources is that smaller arrays or synthetic apertures can be used because only a limited part of the angular spectrum is activated by the beam.

CONCLUSION

A full wavefield numerical model has been used to demonstrate that neither explosive source techniques nor narrow beam techniques yield the plane wave reflection coefficients directly. It has been demonstrated, that the narrow beams are not highly directional, and thus do not behave like plane waves. Therefore, both experimental techniques require the application of either plane wave decomposition or beamforming to the reflected field in order to correctly determine the plane wave reflection coefficients.

REFERENCES

1. D.J. Thomson, The determination of material properties of the sea-bed from the acoustic plane-wave reflection response, In: Acoustics and the Sea-Bed, Ed. N.G.pace, Bath University Press, Bath, UK, 41-50, (1983).

2. H. Schmidt and F.B. Jensen, A full wave solution for propagation in multilayered viscoelastic media with application to Gaussian beam reflection at fluid-solid interfaces, J. Acoust. Soc. Am. 77(3), 813-825, (1985).

3. H. Schmidt and F.B. Jensen, Efficient numerical solution technique for wave propagation in horizontally stratified ocean environments, Rep. SM-173, SACLANT ASW Research Centre, La Spezia, Italy, (1984).

4. D.C. Stickler, Negative bottom loss, critical-angle shift, and the interpretation of the bottom reflection coefficient, J. Acoust. Soc. Am. 61(3),707-710, (1977).

5. S.R. Santaniello, F.R. DiNapoli, R.K. Dullea and P.D. Herstein, Studies on the interaction of low-frequency acoustic signals with the ocean bottom, Geophysics, 44(12), 1922-1940, (1979).

6. G.V. Frisk and J.F. Lynch, Shallow water waveguide characterization using the Hankel transform, J. Acoust. Soc. Am. 76(1), 205-215, (1984).

7. H.O. Berktay, Parametric Sources - Design Considerations in the Generation of Low-Frequency Signals (this volume).

8. T.G. Muir, C.W. Horton,Sr. and L.A. Thompson, The penetration of highly directional acoustic beams into sediments, J. Sound Vib. 64(4), 539-551, (1979).

9. J.N. Tjotta and S. Tjotta, Theoretical study of the penetration of highly directional acoustic beams into sediments, J. Acoust. Soc. Am. 69, 998-1008, (1981).

A STUDY OF OCEAN AND SEISMIC NOISE AT INFRASONIC FREQUENCIES

A.C. Kibblewhite and K.C. Ewans

Physics Department
University of Auckland
Auckland, New Zealand

ABSTRACT

Acoustic noise in the ocean is known to be wind dependent and several mechanisms have been proposed to explain the transfer of energy from the wind to the noise field. Because reliable ambient noise data are sparse at frequencies below 10 Hz it has been difficult to assess the agreement between theory and experiment. The difficulties at these frequencies have arisen through self-noise in the monitoring system and deficiencies in the environmental data. This study has examined correlations of the ocean-wave field and the ocean-induced seismic activity at frequencies below 1 Hz. The quality of the data, the long term nature of the observations and an unique property of the recording environment have helped clarify the nature of wave-induced noise processes. It is concluded that the ocean ambient noise below 5 Hz is controlled by wave-wave interactions.

1. INTRODUCTION

It is nearly one hundred years since a close connection between microseisms and ocean wave activity was first proposed.[1] Since that time a vast literature has appeared as interest in the subject has waxed and waned. Various mechanisms have been proposed to account for the generation of microseisms, most favoured among them being nonlinear interactions between ocean surface waves. Interest in this mechanism has increased in recent years as underwater acousticians have pushed their horizons to lower and lower frequencies. The latest series of papers[2] has however again emphasised that a number of theoretical and experimental questions still require clarification.

The contribution to be reported here has arisen from a major wave-climate investigation, which has been carried out recently in the Cook Strait region of New Zealand.[3] Like the rest of New Zealand this area, shown in Fig. 1, is subject to a regular succession of weather systems which cross the country from west to east. The mountain chain centered on Cook Strait acts as a barrier to the associated air flow and produces a strong, well-defined, bimodal wind regime in Cook Strait. The rapid swing in the wind vector between northwest and southeast has made the region an ideal one for the study of ocean wave processes.

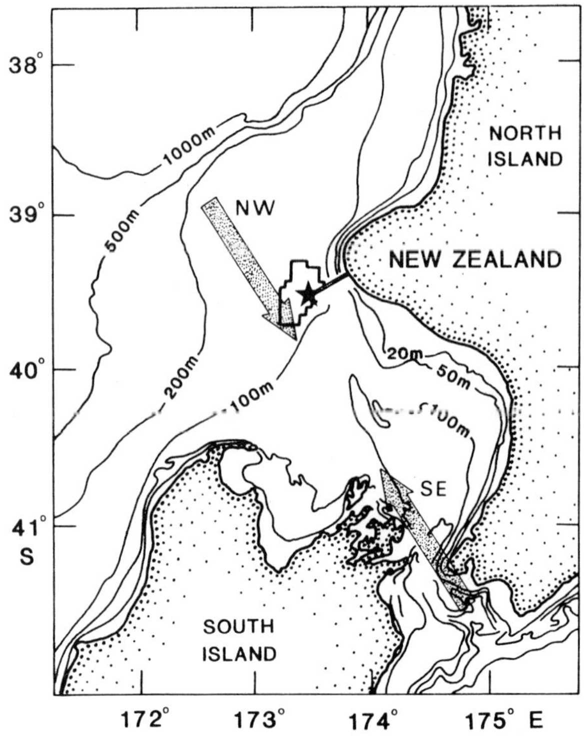

Fig 1. Localities and bathymetry of the Cook Strait and Maui environment.

The general wave-climate study had of necessity to involve the long term recording of wave and related environmental parameters. This fact, coupled with the unique orographic properties of the region, provided an opportunity to extend the basic study and include investigations of various wave related processes. In particular plans were made to address some of the unresolved issues relating to ocean induced microseisms. A full account of this work is given in a thesis submitted for a PhD degree.[4] A number of the important results are being prepared as a series of papers of which this is one.[5-7] It deals specifically with the generation of microseisms and their relation to the low frequency ambient noise field. For details of the instrumentation, analysis procedures used, and a more detailed account of the history of research into microseisms the reader is referred to either the original thesis or References 6 or 7. Suffice it to say here that the wave data were provided by a Datawell Waverider buoy moored in approximately 110 m of water close to the Maui platform. The signal from this unit was received ashore 30 km away. Also recorded were the output of a long period seismometer installed at the receiving site ashore, a time code signal and relevant meteorological parameters. Recordings of the wave and seismic signals of 20 minute duration were initiated automatically every 4 hours. Wave and seismic spectra were produced from the analogue records. Various presentations of the data were developed as required, including the contour plots giving the spectral histories described later.

2. BASIC THEORETICAL PREDICTIONS

Our concern here is with the double frequency (DF) component of the microseism field. For a full account of generation theory the reader is referred elsewhere,[4] but the key relations relevant to this study are summarized below.

As first reported by Miche,[8] consideration of second order terms in the hydrodynamic equations leads to terms representing the generation of low frequency pressure fluctuations by the nonlinear interaction of opposing ocean waves. In contrast to the progressive waves producing them, the distinctive features of these waves are that the pressure effects they produce do not decrease with depth, occur at twice the frequency of the interacting surface waves, and are proportional to their amplitude product. Miche's theory was developed by Longuet-Higgins[9] to account for microseism generation and expanded further by Hasselmann,[10] in particular. Using Hasselmann's terminology the density spectrum of the pressure field arising from interacting surface waves can be written:

$$G_p(\omega) = \frac{\pi\rho^2 g^2 \omega^3}{4c^2} G^2(\omega/2) \int_o^{2\pi} H(\omega,\theta)\, H(\omega,\theta + \pi)\, d\theta \tag{1}$$

where ρ is the density of sea water, c the ocean sound velocity, g is the gravitational acceleration, $G(\omega/2)$ the variance density spectrum of the surface wave displacement, and the integral describes the directional properties of the wave field. This form of the pressure spectrum can be shown to be the same as that calculated by Brekhovskikh[11] and Hughes[12] in the context of underwater noise, when minor errors are corrected.[13] For later discussion it will be helpful to express Eq (1) alternatively as:

$$G_p(f_a) = \frac{2\pi^3\rho^2 g^2 f_w^3}{c^2} G^2(f_w) \int_o^{2\pi} H(f_w,\theta)\, H(f_w,\theta + \pi)\, d\theta \tag{2}$$

where f_a and f_w are the frequency components of the acoustic pressure and gravity wave fields respectively, and $f_a = 2f_w$.

Equation (1) can be written more conveniently as:

$$G_p(\omega) = K_o \omega^3 G^2(\omega/2).I \tag{3}$$

where K_o is a constant given by $K_o = \pi\rho^2 g^2/4c^2$ and I describes the integrand. Modification of Eq (3) by the transfer function to relate the displacement to the pressure field, and a term K_1 involving the active area of the wave field, leads to the ground displacement variance density spectrum of the microseism field:

$$\overset{n}{G}_\mu(\omega) = K_0 K_1 \overset{n}{T}_\mu(\omega)\omega^3 G^2(\omega/2) I \tag{4}$$

where K_1 is a constant associated with the transfer function. Hasselmann examines the properties of T^n_μ for a particular model and shows that for certain conditions a simple ω dependence of the transfer function will apply. In this case the microseism displacement variance density spectrum becomes:

$$G_\mu(\omega) = K_0 K_1 K_2 \omega^4 G^2(\omega/2) I \qquad (5)$$

where K_2 is also a constant. It will be useful to note for later discussion that Eq (2) and Eq (5) are equivalent to Eqs (2) and (4) in Ref. 7, with the integral set at 1/8.

Hence according to theory not only should there be a two-to-one frequency relationship between the seismic and ocean-wave spectra, but the variance density spectral levels of the vertical component of the ground displacement should be proportional to the fourth power of the frequency of the interacting ocean waves, and to the square of the variance density levels of the ocean wave components producing the exciting pressure field. The contribution from the integrand in Eq (5) is difficult to estimate, but for some cases analytical descriptions are available.[14] In this study three cases were considered: (i) Single sea - no opposing swell; (ii) Sea and unidirectional swell propagating at an angle ϕ with respect to the wind; (iii) Two opposing unidirectional swells. In case (iii) the integrand can be shown to have a constant value of 0.5 and to be independent of frequency. In cases (i) and (ii) the integrand has a complex frequency dependence which must be taken into account. We discuss (i) and (ii) briefly later. Ewans[4] provides detail of the analysis.

3. CHARACTERISTICS OF THE MICROSEISM-OCEAN WAVE RELATIONSHIP

3.1 General Features

At all times observations in the Maui region gave clear evidence of the marine generation of microseisms in the frequency range 0.05-1.0 Hz. Comparisons of any ocean wave spectrum with its seismic equivalent identified peaks at or very close to a frequency relationship of 2:1. For a detailed examination of the effects observed long sequences of variance density levels established from the 4 hourly spectra were contoured by hand. For ease of comparison the two spectrograms covering periods of interest were included on the same diagram along with time series plots of wind and other parameters. Gaps in the series were brief and interpolations to the spectrograms were made for continuity.

Several typical time histories of characteristic generation events are analysed in detail in other publications.[4-7] Here we reproduce only one of these by way of example - see Fig 2. Details of the presentation format are given in the figure caption. For a full description of the meteorology and generation events of this period the reader is referred to the other sources. We confine comment to a brief statement of the features relevant to the present analysis.

The effect of the succession of weather systems which cross New Zealand is shown clearly in the events of 16 and 24 October. Prior to both dates the meteorology was dominated by anticyclones, and the winds in the Maui region were more or less constant from the north. As the low pressure systems move on to the country the winds swing

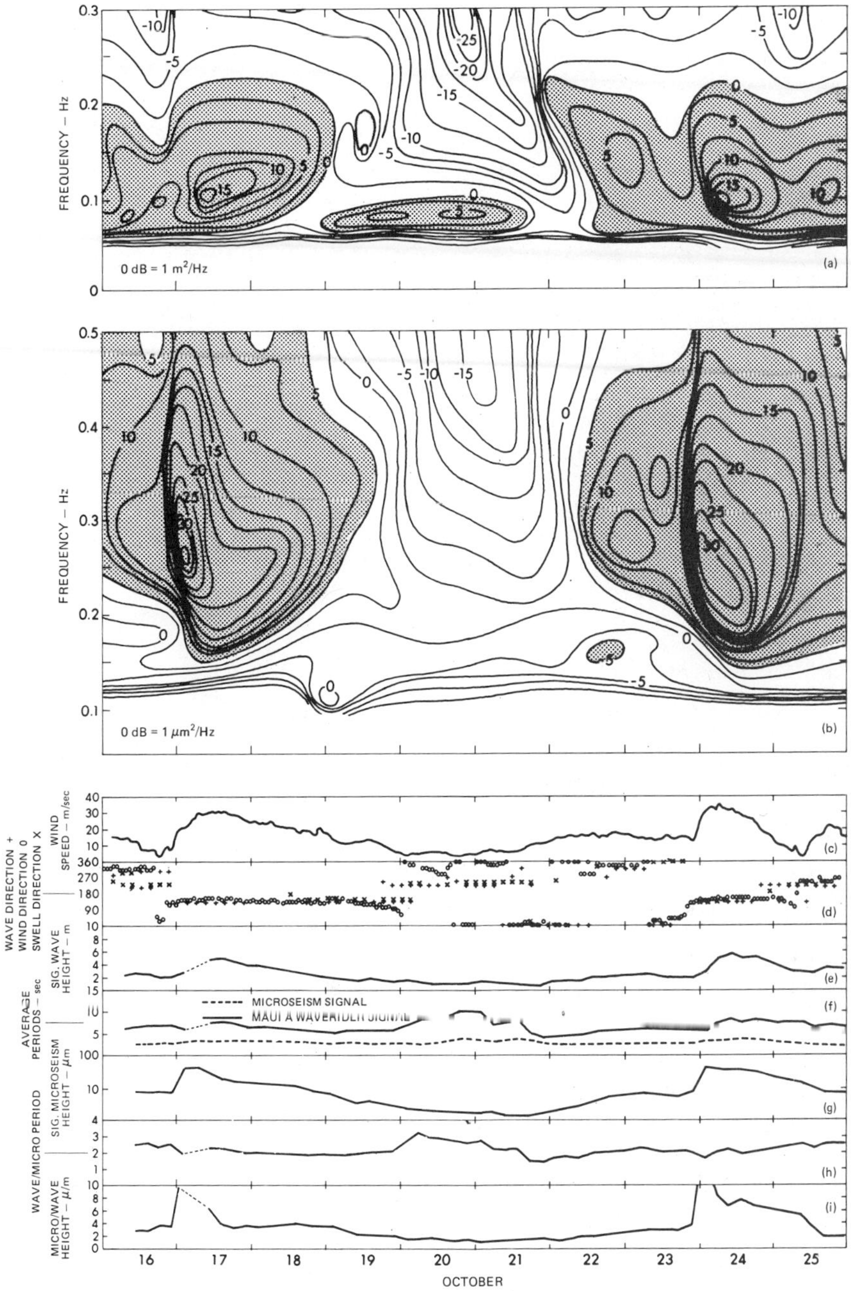

Fig 2. Time series plots of wind, ocean wave and microseism parameters for the period 16-25 October 1981. (a) Ocean wave spectral history; (b) Microseism spectral history; (c) Wind speed, m/sec; (d) Wind, wave and swell directions; (e) Significant wave height, m; (e) Average wave and microseism period; (g) Significant microseism height, microns; (h) Wave/microseism period ratio; and (i) Ratio of microseism to wave height, microns/meter. (From Ref. 7)

rapidly to the southeast and increase in speed to over 30 ms^{-1} before decreasing. Ocean wave spectral levels increase sharply in response to the rapid change in wind speed and bearing. A new low frequency peak is soon established, of sufficient strength to mask the ever present low level southwesterly swell.

The changes in the wind field always produce the characterisic response shown in the microseism spectrogram. Energy which is clearly related in some way to the local sea, appears in the band 0.2 - 1.0 Hz. In this example maximum levels occur on 17 and 24 October some hours before the ocean wave maxima. In both events two distinct periods can be identified in the microseismic response. Activity reaches high levels while the growing southeasterly sea interacts with the northwesterly swell, but the levels drop from the peak as the northwesterly swell is flattened to a lower value, which remains roughly constant until the sea levels fall in response to the decreasing wind field.

3.2 The Two-to-One Frequency Relationship

Comparison of any ocean wave spectrum with its microseism equivalent always identifies peaks at or very close to the two-to-one frequency relationship required by Eq (5). This relationship is expected on the basis of the standing wave generation mechanism originally proposed by Longuet-Higgins and Ursell,[15] but the fact that the microseism levels associated with the local sea are larger than those associated with the southwesterly swell suggests the actual mechanism producing the standing waves must be different.

A primary frequency peak (PF) in the seismic spectra at the same frequency as the sea waves has also been identified, but because its level is two orders of magnitude lower than that of the double frequency (DF) secondary peak it is not routinely observed. It is not discussed further here but an account of it is available elsewhere.[4,7]

3.3. Seismic Spectral Level Dependence on Frequency

For the low level southwesterly swell around 0.07Hz the value of the spectral peak square ratio, (SP^2R - the ratio of the spectral level of the DF peak in the microseism spectrum to its ocean wave equivalent) is typically -5 dB re $1(\mu m/m)^2$. The larger values of this ratio produced by a local sea (10 to 20 dB higher) are shown elsewhere[4,7] to arise from the frequency dependence expressed in Eq (5). According to Eq (5) the SP^2R is proportional to the fourth power of the wave frequency. When allowance is made for this factor the ratios are found to be grouped around certain values, which can be related to the type of wave field responsible for the activity. These factors are explored in more detail elsewhere.[4,7]

3.4 Seismic Level Dependence on Wave Height

According to Eq (5) the seismic spectral levels are also proportional to the square of the levels of the interacting ocean wave spectral components. Evidence confirming this relationship has been presented in other publications.[5,7]

3.5 Seismic Level Dependence on Angular Distribution of Wave Energy

The theoretical explanation of microseism generation is based on the production of pressure fluctuations by the interaction of oppositely travelling ocean surface waves. While there has always been difficulty in identifying the precise source of the standing waves recent evidence has confirmed that oppositely travelling waves are present within the fetch itself.[14] The present study has confirmed furthermore most of the predictions of the formalism developed on the concept of standing wave interactions. In particular it has been shown that the microseism height is related to the square of the ocean wave height and that the microseism spectral levels depend to the fourth power on the ocean wave frequency. These results have in turn justified an analysis, based on the spectral peak square ratios, which has allowed comparisons between different generation events by reducing each ratio to an equivalent level at a standard ocean wave frequency by normalising by the factor ω^4 - see Eq (5). The actual normalisations to a standard swell frequency of 0.06 Hz are discussed elsewhere.[4,7] Here we simply note that accepting this normalisation enables other factors to be investigated. In particular, as the constants K_0, K_1, K_2 in Eq (5) will be invariant in the Maui generation area if we confine our attention to winds from one direction (say SE), we can explore the influence of the integrand by careful selection of the spectra.

The analytical expressions for the integral under various wave regimes are examined in detail elsewhere.[4,7] Here we simply note that, some reasonable assumptions accepted, the values of I for the three cases alluded to earlier, and for frequencies close to the spectral peak (below 1.12 f_p), are:

1. Active sea alone $I = 2.65 \times 10^{-2}$
2. Active sea plus swell (ϕ=180°) $I = 0.19$
   where ϕ is the angle between the swell and the wind.
3. Two opposing swell $I = 0.5$

Of particular interest to our next section is that the ratio I/I_{180} = -8.6 dB. For consideration of frequencies above 1.12 f_p the full frequency dependence of the parameters involved must be considered - see later.

4. THE INFRASONIC AMBIENT NOISE FIELD

The application of the observed seismic behaviour to the clarification of the properties of the ambient noise field in the ocean at frequencies below 10 Hz has been discussed in a companion paper.[7] This section reviews that analysis in the light of new data now available.

4.1 Average Spectra vs Wind Speed

Average ocean wave and seismic spectra (in 2.5 ms^{-1} wind speed intervals) are shown as a function of wind speed in Figs 3a, b. These spectra are based on SE winds only and strict criteria were imposed in their selection.[4] The characteristic growth of the wave peak and its migration to lower frequency with increasing wind speed is mirrored in the seismic spectra. The contribution of the everpresent SW swell is evident in both.

4.2 An Analytical Form of the MAUI Spectrum

To examine more clearly the influence of the spreading term in the ocean wave field, and remove the confusion created by the SW swell,

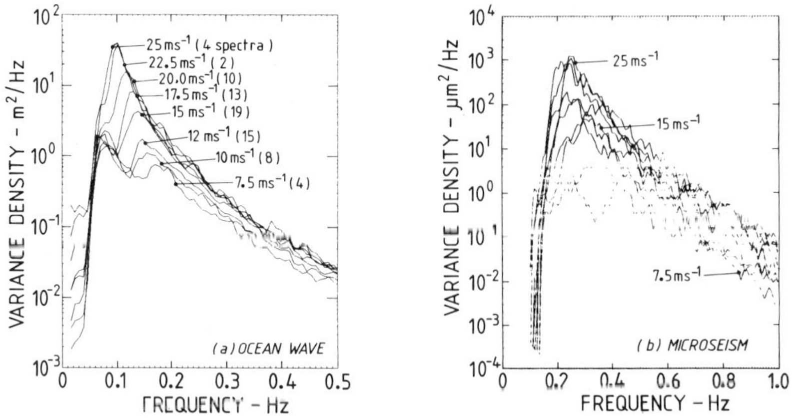

Fig 3. Average ocean wave and seismic spectra in wind speed intervals of 2.5 m/sec.

it is helpful to calculate an analytical form of the ocean wave spectrum based on the JONSWAP function, but using the actual values of the peak enhancement factor and spectral width parameters found relevant to the Maui seas.[4,5] This is traditionally calculated in terms of the peak frequency, but for our purposes a representation in terms of wind speed is more helpful. This transformation has been possible because of other aspects of the overall study.[4,5] The resulting one dimensional MAUI spectra as a function of wind speed are shown in Fig 4a.

4.3 The Pressure Field Arising from Wave-Wave Interactions

From the spectra of Fig 4a the pressure field produced by nonlinear interactions can be calculated using Eq (5). In Figs 4c, d respectively we present the variance density spectra for the pressure field produced by a single sea in which the angular distribution of wave energy inherent in the fetch produces the interactions, and those associated with an active sea with an opposing swell aligned 180° with respect to the wind. We note in particular that overall, the pressure levels are about 10 dB higher when there is an opposing swell present, and recall that the difference predicted theoretically on the basis of the spreading factor is approximately 9 dB, the levels for the sea plus opposing swell again being higher - see Section 3.5.

4.4 Seismic Spectra for Single and Opposing Seas

The measured seismic spectra for single and opposing seas can be obtained from the composite plot of Fig 3b. The separated spectra, after some minor smoothing, are presented in Figs 4e, f. Attention is again drawn to the difference of around 10 dB in the peak levels but also to the overall similarity to the equivalent spectra of the exciting pressure field presented in Figs 4c, d.

4.5 The Pressure Field Derived from the Seismic Spectra

Given the form of the transfer function in Eq (5) it is possible to work back from the seismic spectra and establish the pressure field

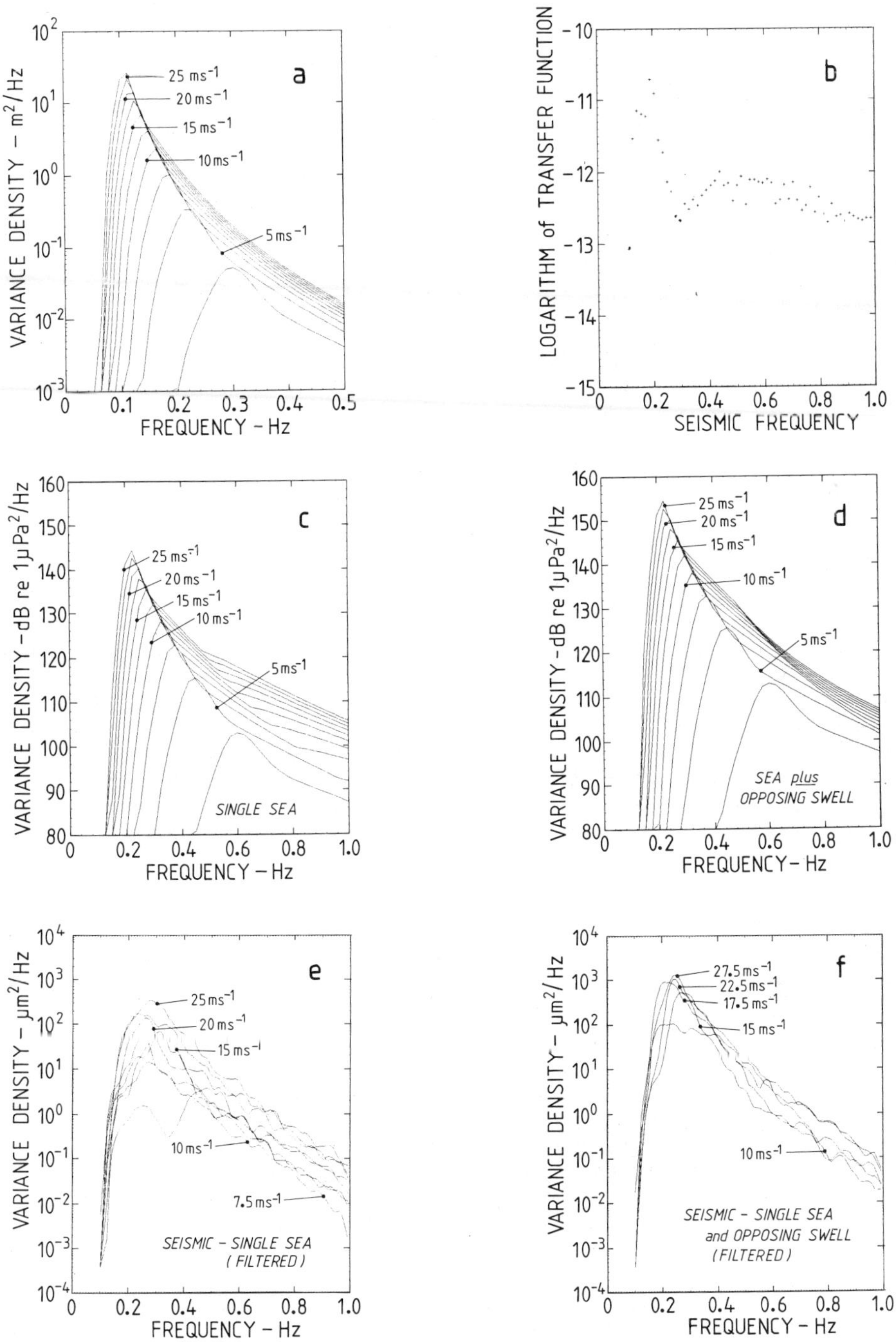

Fig 4.(a) The MAUI spectra (based on the JONSWAP function with parameters relevant to the Cook Strait region) as a function of wind speed; (b) The transfer function derived from the pressure spectra based on the observed ocean wave data, and the observed ground displacement spectra; (c) Variance density spectra of the wave induced pressure field - single sea; (d) Variance density spectra of the wave induced pressure field - sea plus opposing swell; (e) Measured seismic spectra - single sea; (f) Measured seismic spectra - sea plus opposing swell.

at the sea bed responsible for the microseism levels observed. In the absence of such information a simple transformation was used in the earlier comparison of the deduced levels with published ambient noise data derived from hydrophone measurements - see Fig 8 of Ref 7. Inspite of the frequency independence in the transfer function this implied, the close agreement with other data was striking.

Evaluating the transfer function directly from the observed wave and seismic spectra (Figs 3a and 4e, f) suggests however that the frequency dependence may be more complex than the simple linear relation assumed earlier - see Fig 4b. That this might be the case was noted in Ref 7 when attention was drawn to the inflexion in the deduced ambient noise curves around 0.5 Hz, evident in Fig 8 of that paper. However both the angular spreading integral and the transfer function could be involved. This effect is currently the subject of further investigation.

4.6 The Ambient Noise Field

As stated earlier the pressure field deduced from the seismic data of Figs 4e, f and presented in Fig 8 of Ref 7, agreed well with other experimental hydrophone data. However, the character of the transfer function shown in Fig 4b suggests that the closeness of the agreement may have been somewhat fortuitous. Further work is in hand to explore the interdependence of the spreading and transfer functions and the significance of any bottom amplification effects. It is clear nevertheless that while noise levels below 0.5 Hz can be influenced by any residual swell that may be present, as is observed in the New Zealand environment, outside this range and at modest wind speeds the local sea is the dominant noise source. Furthermore wave-wave interactions appear to account for the acoustic and seismic effects produced.

SUMMARY

It has been confirmed that microseisms are generated by ocean waves. It has been established further that:

i. The microseisms in the band 0.1 to 1 Hz have spectral components with peaks at double the frequency of the associated peaks in the wave field.

ii. The interference of opposing waves is responsible for the generation of the pressure field which produces the microseisms, in accordance with theory.

iii. The levels generated have a strong frequency dependence, higher frequency seas producing larger activity than lower frequency ones. The pressure and seismic fields display a frequency dependence close to f^3 and f^4 respectively, as predicted by theory.

iv. If allowance is made for this frequency dependence ground displacement levels (and by implication pressure levels) are found to be proportional to the square of the wave height, but a dependence on the wave regime has also to be considered.

v. The influence of the wave regime is embodied in the angular distribution of wave energy. In the case of a single sea the natural angular distribution of waves creates the opposing interactions necessary for microseism generation. (This fact explains why both onshore and offshore winds are equally effective in producing seismic activity.)

vi. The nonlinear interactions are the dominant source of the infrasonic ambient noise levels in the ocean below 10 Hz. In this range the noise level is roughly proportional to the square of the wind speed and has a negative slope of around 18 dB per octave.

REFERENCES

1. E. Wiechert, "Verhandlung der Zweiten Internationalen Seismologischen Konfrerenz", Gerl. Beitr. Geophys., Erganzungsbd. 2:41-43 (1904).
2. F.I. Monakhov, V.M. Zhak and V.A. Nesterov, "Mechanism of generation of microseisms by a storm during the period October 10-11, 1976 near Island Shikotan", 14(4):302-303 (1978).
3. A.C. Kibblewhite, P.R. Bergquist, M.R. Gregory, B.A. Foster and M.C. Miller, "Maui Development Environmental Study:Report on Phase 2, 1977-1981", University of Auckland, NZ, October (1982).
4. K.C. Ewans, "Ocean waves, microseisms, and their interrelations", Doctoral Dissertation, University of Auckland, (1984).
5. K.C. Ewans and A.C. Kibblewhite, "Ocean wave generation in New Zealand Waters - A comparison with the JONSWAP results" - in preparation.
6. K.C. Ewans and A.C. Kibblewhite, "A study of wave induced microseisms in New Zealand Waters" - in preparation.
7. A.C. Kibblewhite and K.C. Ewans, "Wave-wave interactions, microseisms and infrasonic ambient noise in the ocean", J. Acoust. Soc. Am. (1985) - accepted for publication.
8. M. Miche, "Movements ondulatoires de la mer en profondeur constante ou decroissante", Ann.Ponts Chaussees, 114:25-87, 131-164, 270-292, 396-406 (1944).
9. M.S. Longuet-Higgins, "A theory of the origin of microseisms", Philosophical Transactions of the Royal Society of London, A,243:1-35 (1950).
10. K. Hasselmann, "A statistical analysis of the generation of microseisms", Review of Geophysics 1(2):177-210 (1963 c).
11. L.M. Brekhovskikh, "Underwater sound waves generated by surface waves in the ocean", Izvestia Atmospheric and Oceanic Physics, 2(9):970-980 (1966).
12. B. Hughes, "Estimates of underwater sound (and infrasound) produced by nonlinearly interacting ocean waves", J. Acoust. Soc. Am., 60(5):1032-1039 (1976).
13. S.P. Lloyd, "Underwater sound from surface waves according to the Lighthill-Ribner theory", J. Acoust. Soc. Am., 69(2):425-435 (1981).
14. G.L. Tyler, C.C. Teague, R.H. Stewart, A.M. Peterson, W.H. Munk and J.W. Joy, "Wave directional spectra from synthetic aperture observations of radio scatter", Deep-sea research, 21:989-1016 (1974).
15. M.S. Longuet-Higgins and F. Ursell, "Sea waves and microseisms", Nature, 162:700 (1948).

GEO-ACOUSTIC NOISE LEVELS IN A DEEP OCEAN BOREHOLE

F. K. Duennebier, R. K. Cessaro, and P. Anderson

Hawaii Institute of Geophysics, University of Hawaii

2525 Correa Road, Honolulu, Hawaii 96822

ABSTRACT - Noise levels recorded by a deep-ocean (5.5 km) borehole seismograph were found to vary by up to 20 dB during a 64-day continuous reading period. Noise level above 5 Hz is controlled by storms, shipping, and biological sources. Noise below 5 Hz appears to be completely controlled by sea state, even though the sensors are almost 6 km below the ocean surface. At least two mechanisms are observed for transmission of energy at the sea surface to the ocean floor. One mechanism appears to yield seismic energy proportional to the wave energy apparently at a frequency twice that of the wave frequency (microseisms). This mechanism is observed at frequencies below 5 Hz. At higher frequencies, white noise is observed and increases in level with sea state. This energy is probably caused by breaking waves. Comparison of noise levels observed in the borehole with levels obtained by ocean bottom sensors at and near the site shows that weaker acoustic sources can be detected in the borehole than on the ocean floor, implying that low noise levels in the borehole more than compensate for signal loss.

INTRODUCTION

During Leg 88 of the Deep Sea Drilling Project, a permanent borehole seismic system was emplaced in hole 581-C (43.924°N, 159.797°E) by scientists from the Hawaii Institute of Geophysics. The instrument, Ocean Sub-bottom Seismometer IV (OSS IV), is clamped in place 20 m below the sediment-basalt interface, and 378 m below the ocean floor in 5.5 km of water. After a short check-out and seismic refraction experiment, an analog recording package was left at the site to record for the next 64 days. On May 25, 1983, the R/V KANA KEOKI returned to the site, recovered the recorded data, took data in real-time for about 24 hours, and then set out another recording package (which has yet to be recovered) (Figure 1). In addition to the OSS IV, eight Ocean Bottom Seismometers were emplaced at the site by the R/V DE STEIGUER. Unfortunately, there were only a few hours of overlap between the OSS IV and OBS data because of logistics problems. All together, more than 5000 impulsive acoustic sources and hundreds of earthquakes were recorded. In this paper, we ignore these signals of geophysical interest and concentrate on the noise spectrum and its temporal variations.

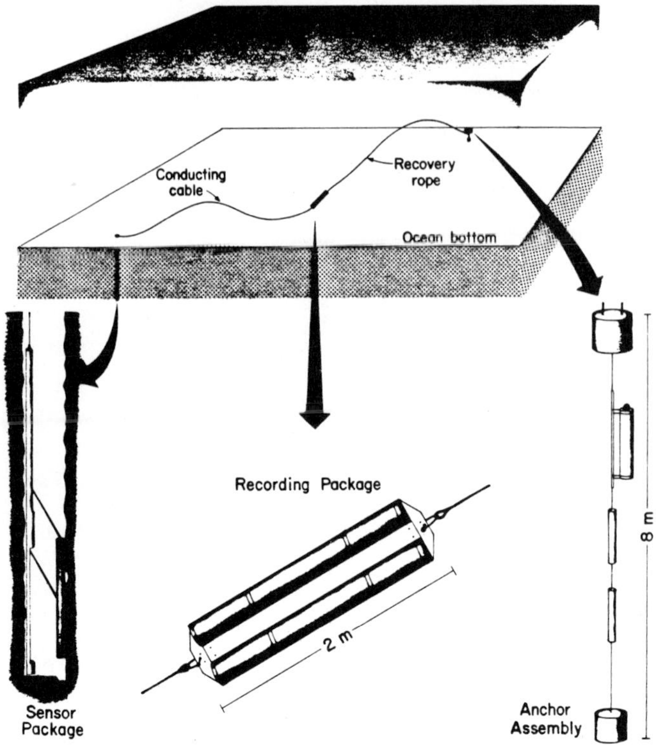

Fig. 1. Ocean Sub-bottom Seismometer. The sensor package, permanently locked at the bottom of the borehole, is powered by batteries in the recording package. The recording package is recovered by releasing the anchor from the anchor assembly.

INSTRUMENTATION

OSS IV contains three orthogonal 4.5-Hz geophone sensor groups that are capable of recording ambient background noise from about 0.2 Hz to above 20 Hz. The 20-Hz limit is determined by the 25-Hz Nyquist frequency, not the geophones. The vertical group contains nine geophones in series, and each horizontal group contains two geophones in series. Specific response curves and sensor characteristics are being published elsewhere [1]. The Hawaii Institute of Geophysics ocean bottom seismometers (OBSs) used in this study contain a vertical geophone, one horizontal geophone, and a hydrophone. The geophones are mechanically separated from the main OBS package to improve seismic coupling. Data quality from all sensors is excellent, although the calibration of the hydrophone is uncertain. Orientation of the horizontal geophones in OSS IV to within 1.5° has been accomplished by a study of the azimuths of arrivals from explosives at known locations [2].

The OSS IV geophone signals arrive at the recording package in digital form sampled at 50 samples per second per channel. The signals are converted back to analog in the recording package and recorded on three channels of a slow-speed analog cassette tape, with time code on the fourth channel. A C-120 tape can record up to 14 days of data, and five tapes were recorded sequentially with overlap. For this study spectral estimates were obtained directly from the analog tapes using a

Hewlett Packard 2318-A Spectrum Analyzer. Four spectra from each of two data channels were averaged about every 28 minutes of recorded time and stored in digital form for further processing. Thus, about 50 spectral estimates per day were used to generate the noise vs. time figures shown in this paper. Spectral levels have been corrected for gain changes (recorded with the time code) and the instrument response has been applied to the geophone traces.

DATA ANALYSIS

The complete noise history of the 64-day OSS IV data set is shown in Figure 2. Filter sections for five 1/3-octave bands and the energy below 2 Hz are shown for the two horizontal channels. The noise curves have been low-pass filtered to remove the numerous earthquakes that would otherwise stand out as "spikes" on the records. In addition to the seismic noise traces, two traces showing atmospheric information are displayed. These data were read from 12-hour satellite weather maps for the NE Pacific Ocean. The upper trace shows the atmospheric pressure gradient in millibars per nautical mile. This parameter should reflect wind speed. The lower trace shows absolute barometric pressure in millibars. High pressure generally corresponds to low wind speed and calm seas.

METEOROLOGICAL NOISE: Note that broad peaks in pressure gradient correlate with similar peaks in the noise level above 4 Hz and below 2 Hz. From 2 to 4 Hz the noise levels are nearly constant over the entire period, and there is little correlation of noise level with peaks in pressure gradient. The lack of variation in noise level between 2 and 5 Hz implies that the dominant noise generation mechanism at these frequencies is normally saturated. Only when the atmospheric pressure gradient is very low (implying no wind and calm seas) do the levels in this frequency band drop below the saturation level.

To further investigate the storm-related noise mechanisms, we have plotted noise level vs. frequency at six-hour intervals for the storm beginning on day 264 (Figure 3). Time period 1 was taken at the point of lowest noise level. Six hours later, noise level has increased to saturation between about 3 and 7 Hz, but has changed little below 2 Hz or above 7 Hz. Twelve hours into the storm noise level between 2 and 7 Hz is saturated and noise, levels to below 0.2 Hz have increased. Eighteen hours into the storm, levels to 1 Hz are saturated and an increase in noise level is observed above 5 Hz. Twenty-four hours into the storm the noise levels are at their highest. This behavior of seismic noise during a storm is similar to the energy spectrum of a developing sea during a storm. The ocean wave spectrum saturates at lower and lower frequencies as wind speed, duration, and fetch increase. The slope of the saturated portion of the energy spectrum is close to f^{-5}, the value expected for the saturated portion of the ocean wave spectrum[3]. The lowest noise level observed above 5 Hz is most likely instrument noise[4].

The mechanism by which the ocean wave energy is transmitted to and below the ocean floor is apparently the same as that which generates microseisms[5,6,7]. A non-attenuating pressure fluctuation field with frequency twice that of the ocean wave frequency is propagated downward. The Longuet-Higgins theory depends on the non-linear interaction of waves of the same frequency traveling in different directions, such as reflections from shore. Thus, the amplitude of the pressure wave producing microseisms should be heavily dependent on shore configuration and storm characteristics. The observation of the saturation slope from

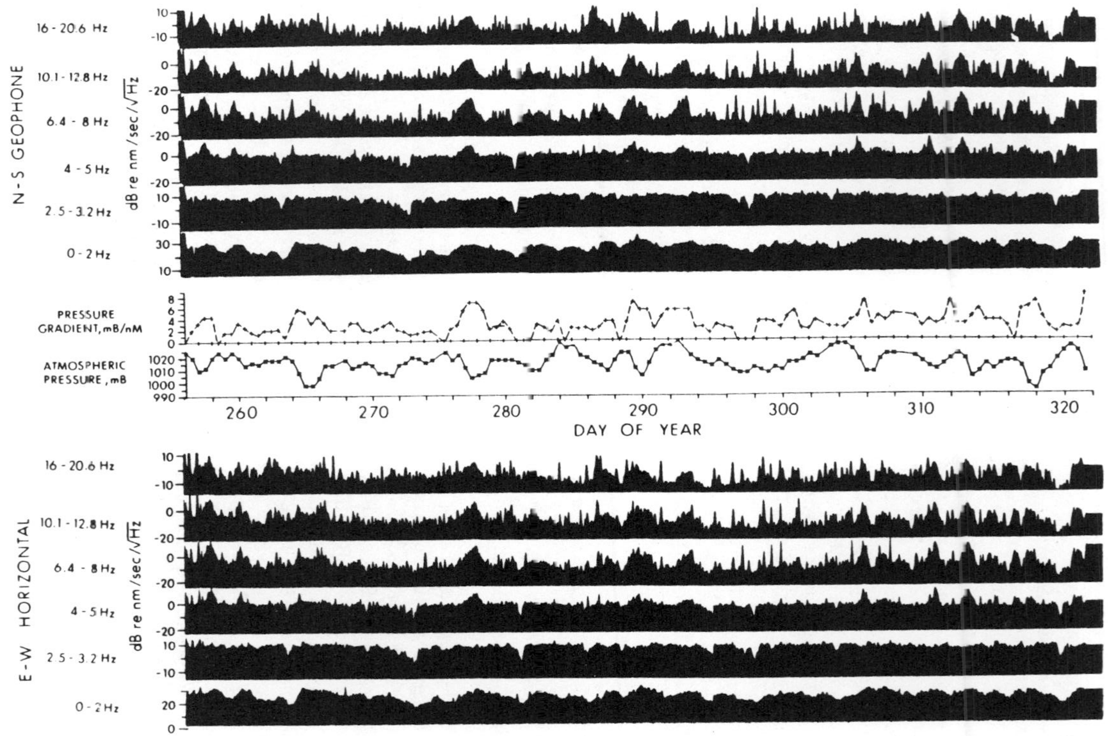

Fig. 2. Temporal variation of OSS IV noise levels over the 64-day recording period in 1982. The top six traces are noise levels in 1/3 octave filters recorded by the N-S geophone, and the bottom six were recorded by the E-W geophone. The middle two traces were obtained from atmospheric 12-hour weather maps at the drill site. Note that noise peaks at the higher frequencies correlate well with high pressure gradients, and low noise levels at low frequencies appear to correlate with high atmospheric pressure and low gradients.

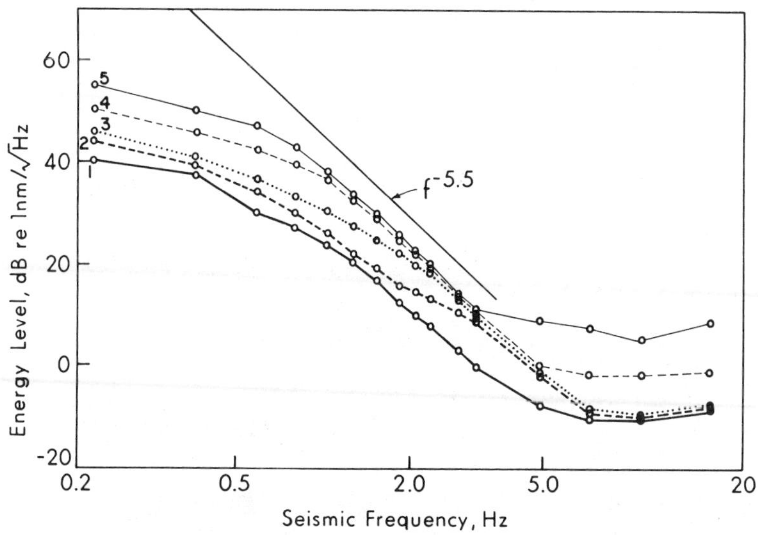

Fig. 3. Horizontal geophone noise levels on OSS IV during a storm beginning on day 263, 1982. Each line represents noise level at 6 hour intervals increasing in time from the lowest curve to the highest. A curve with a slope of $f^{-5.5}$ is shown for reference. Note that a noise level saturates at lower and lower frequencies as the storm progresses.

0.5 to 5 Hz on both the ocean floor and below the ocean floor in two different storms by instruments far from any land implies that the mechanism may not be dependent on interaction of waves traveling in different directions, but on the sea state itself. The energy in the pressure field is proportional to the square of the amplitude of the ocean wave, thus the seismic noise spectrum may yield a direct measure of wave heights. The saturation level is easy to identify from its spectral slope of $f^{-5.5}$. This level is important because it implies a constant source energy in deep water and may be useful for calibration purposes.

SHIPPING NOISE: Signals from passing ships are identified in Figure 2 by the relatively narrow band spikes lasting from four to ten hours. More than 130 signals of this type are identified during the 64-day period, and ship signals dominate the noise spectrum above 5 Hz for some time periods. Because of the directional nature of the geophones, the direction to a ship can be determined (with a 180° ambiguity) and its distance if its speed is known[4].

BIOLOGICAL NOISE: A persistent bi-frequency noise source at about 17 and 20 Hz is commonly observed in the borehole data. The signals from this "whale" have a peak level about 20 dB above noise level, and often are heard for days at a time from more than one source. Signals observed are identical to those observed on hydrophones in the Pacific[4,8].

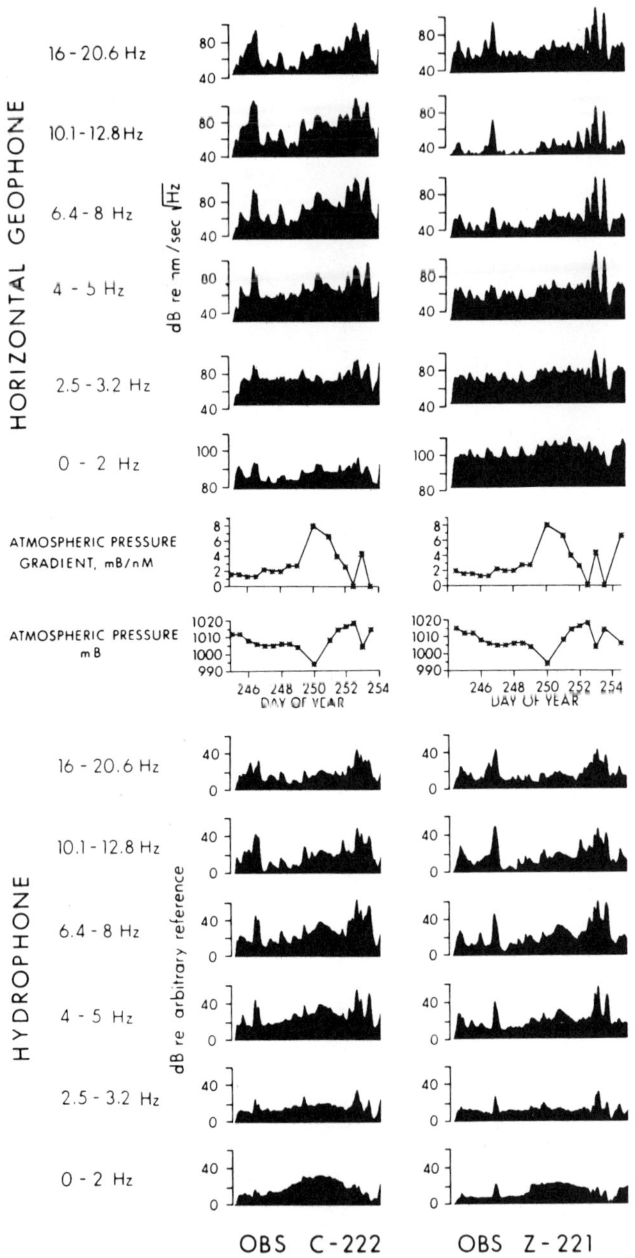

Fig. 4. Temporal variations of OBS noise levels over about eleven days in Sept. 1982 at the OSS IV site. See Figure 2 for details. OBS C-222 was about 40 km south of the site and Z-221 was at the site.

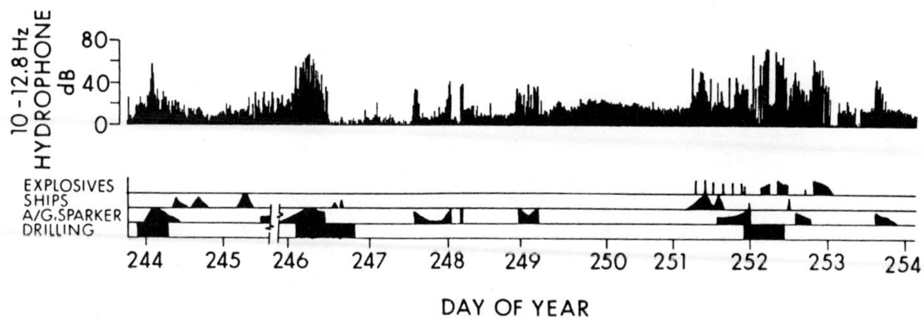

Fig. 5. Correlation of known activities with noise levels on an OBS (Z-221) at the OSS IV site.

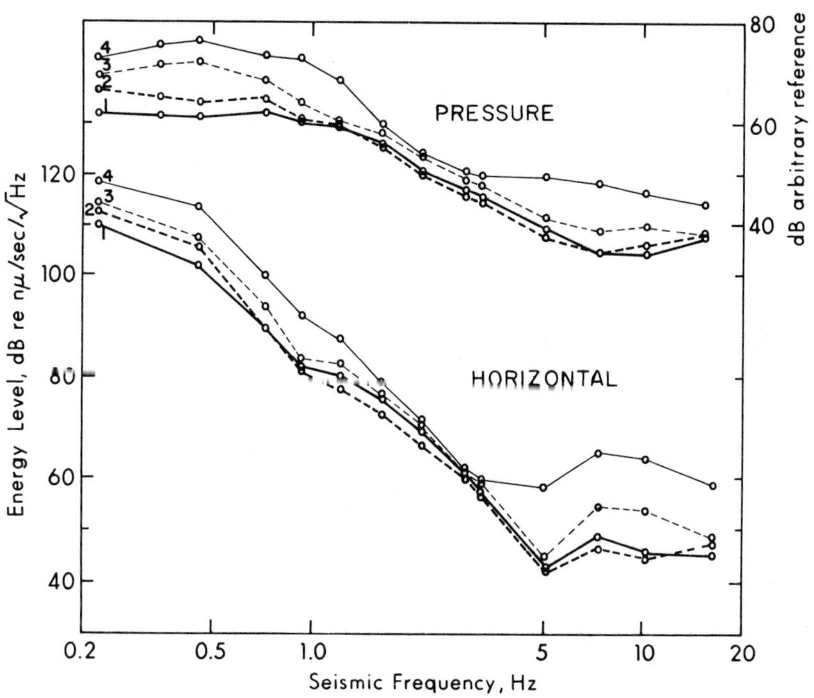

Fig. 6. Change in noise level of OBS C-222 with time (24 hour intervals) before and during hurricane Gordon beginning on day 248, 1982. Each curve represents the noise spectrum at a different time.

OCEAN FLOOR vs. BOREHOLE NOISE LEVELS

A plot similar to Figure 2 is shown in Figure 4 for two eleven-day periods recorded by two ocean bottom seismometers just before the time period shown in Figure 2. Nearly all the noise fluctuations correlate with known activities during the experiment (Figure 5). A tropical cyclone (Hurricane Gordon) passed the site on day 250, 1982, and its effects are noticeable on both OBSs, especially on the hydrophones, but at a signal-to-noise level less than observed for similar storms recorded on the borehole system. The increase in noise level above 5 Hz during the storm is larger on the horizontal geophone than on the hydrophone (Figure 6), an observation also noted by Bradner[9] in mid-water experiments. In Figure 6, the change in noise spectra withtime are plotted for 24 hour intervals from the calm period prior to Hurricane Gordon until its peak. Note that the $f^{-5.5}$ saturation slope is observed on the horizontal geophone. The slope of saturated noise on the hydrophone traces is different because the hydrophone traces are not corrected for instrument response. Note also that, except for sources identified with the experiment, few ship sources are visible. The noise level of the horizontal geophones are 40 dB, or more, greater than the levels of the borehole geophones. This increased noise is caused by trapping and amplification of shear waves in the low velocity sediments, and by Stoneley waves.

CONCLUSIONS

Noise levels in a deep-ocean borehole between 0.2 and 5 Hz are controlled completely by sea state. Above 5 Hz, noise levels increase during storms because of breaking waves but are otherwise controlled by either shipping or biological noise. Placement of seismic sensors in a borehole extending into basement increases the sensitivity to ocean acoustic sources because of reduced noise levels relative to the ocean floor. Signal-to-noise improvement is difficult to quantify with the data in-hand, but it appears to be more than 6 dB at frequencies above 5 Hz.

ACKNOWLEDGMENTS

This research was funded by the Office of Naval Research. Preparation of figures, typing, review and editing by R. Rhodes, K. Rehbock, and Rita Pujalet are gratefully acknowledged. Hawaii Institute of Geophysics Contribution No. 1609.

REFERENCES

1. Byrne, D. A., D. Harris, F. K. Duennebier, and R. K. Cessaro, Technical Review of the Ocean Subbottom Seismometer System Installed in Deep Sea Drilling Project Site 581C, Leg 88, in *Init. Repts. of the Deep Sea Drill. Proj.*, Vol. 88, in press (1985).
2. Anderson, P., and F. K. Duennebier, Orientation of Horizontal Ocean Bottom Seismic Sensors from Explosive Data, in preparation (1985).
3. Phillips, O. M., *The Dynamics of the Upper Ocean*, Cambridge Univ. Press, 113-114 (1969).
4. Duennebier, F. K., C. McCreery, D. Harris, R. K. Cessaro, C. Fisher, P. Anderson, OSS IV: Noise Levels, Signal-to-Noise Ratios, and Noise Sources, in *Init. Repts. of the Deep Sea Drill. Proj.*, Vol. 88, in press (1985).
5. Longuet-Higgins, M. S., A Theory on the Origin of Microseisms, *Phil. Trans. London. A*, 243:1-35 (1950).
6. Haubrich, R. A., W. H. Munk, F. E. Snodgrass, Comparative Spectra of Microseism and Swell, *Bull. Seism. Soc. Amer.*, 53:27-37 (1963).
7. Bernard, Pierre, Short Review and Recent Results on Microseisms, *Geophys. Surveys*, 5:395-407 (1983).
8. Northrop, J., W. C. Cummings, M. F. Morrison, Underwater 20 Hz Signals Recorded Near Midway Island, *J.A.S.A.*, 49:1909-1910 (1971).
9. Bradner, H., Microseism Propagation via Oceanic Waveguide, *Pure Appl. Geophys.*, 103:266-273 (1973).

LOW FREQUENCY NOISE FIELDS AND SIGNAL CHARACTERISTICS

W. M. Carey, R. A. Wagstaff and

Naval Ocean Research and Development Activity
NSTL, Mississippi 39529-5004

B. Brunson, M. Bradley

Planning Systems, Incorporated
McLean, Virginia 22102

ABSTRACT

Very low frequency (VLF) (2 to 20 Hz) and low frequency (LF) (20 to 200 Hz) physical noise models and previous measurements are discussed and compared to low frequency, high resolution noise measurements. These data show long-term persistent directional characteristics associated with distant shipping lanes and density patterns. Short-term averages show a temporally dynamic field composed of resolved distant shipping and uncorrelated background noise. Calibrated low frequency transmission loss and signal coherence length estimates are presented for down-slope to deep ocean and bottom-limited sound channels. The transmission results compare favorably with previous measurements and computations performed with range-dependent water column and sediment characteristics. These results emphasize the importance of the coherent contribution from coastal shipping to the mid-ocean noise field. Vertical directionality measurements by Anderson et. al. (1972) show a broad angular distribution of noise intensity near the horizontal at low frequencies and a peaked distribution about the horizontal at high frequencies. This broad angular distribution near the horizontal is consistent with sound propagating down-slope via a bottom reflectivity which favors lower frequencies. The frequency variation near the horizontal was found smooth and indicates that in addition to surface ships, environmental noise influences the vertical directionality.

INTRODUCTION

Since the classic paper of Wenz (1962), ambient noise has been an extensively studied phenomena. Morris (1975) emphasized the importance of ships as their signals are enhanced when they cross over seamounts or proceed over the continential slopes. Wagstaff (1981) showed by the comparison of measurements and calculations that coastal shipping (ships over the continental slope and on the shelf near the slope) must be considered in order that the horizontal directionality be correctly described. He showed that these coastal sources would also affect the vertical directionality. This paper reinforces these findings with new results from down-slope transmission loss (TL), coherence, and noise directionality experiments.

Very Low and Low Frequency Ambient Noise

The range of human hearing is normally 20 Hz to 20 kHz and provides a convenient basis for classifying sounds in the ocean in bands of 2 X 10^n. Sounds below this frequency range are called infrasonic, sounds within this range sonic, and sounds beyond this range ultrasonic. In this paper infrasonic frequencies between 2 and 20 Hz are referred to as very low (VLF) and the sonic frequencies between 20 and 200 Hz as low frequencies (LF). Ambient noise in the VLF to LF range consists of the sum of a large number of wave trains which originate from a variety of sources.

Noise associated with ocean microseisms dominates acoustic spectra at frequencies below 4-5 Hz. Ocean gravity (surface) waves travelling in opposite directions in the vicinity of a storm or as a result of reflection from a coast can generate a standing wave field. Unlike progressive surface waves for which the pressure effects decay exponentially with depth, a standing surface wave field produces a mean second order pressure effect at twice the frequency of the surface waves that is unattenuated with depth. This effect was used by Longuet-Higgins (1950) as the basis for a theory for the generation of microseisms that are sources of acoustic noise in the infrasonic range. Experimental confirmation of this theory has been given by several authors, including Latham and Nowroozi (1968).

Measurements by Pigott (1964), Perrone (1974), and Wittenborn (1976) as well as others clearly indicate the existence of a wind dependent noise generation mechanism in the VLF range. These data show a strong correlation between noise level and local windspeed. Likely causes of this noise are nonlinear interaction of wave turbulence with surface waves and the radiation of sound into the ocean from wind turbulence at the air/sea boundary. The first of these mechanisms has been studied by Goncharov (1970) and Yen (1979). The second has been studied by Isakovich (1970) and Yen (1979). As Nichols (1981) has observed, in the absence of detailed data on wind velocity, bottom currents, etc., it is hardly possible to make definite assignments of generation mechanisms for this noise.

In the absence of shipping, ambient noise measurements in the sonic range show a strong correlation between level and local windspeed. At frequencies over 200 Hz, noise is dominated by wind noise and the mechanisms are bubbles, spray, and splashes associated with white caps, as well as capillary wave-wave interactions. Furduev (1966) proposed that the characteristic peak in the ocean noise spectrum between 200 and 1000 Hz be attributed to cavitating bubbles. Noise produced by wave-wave interaction in the sonic region is difficult to model, but measurements indicate that this mechanism (or possibly some other) significantly contributes to observed ambient noise levels in the absence of white caps.

In deep water, noise from surface shipping is dominant in the VLF and LF range with components extending as high as 500 Hz. Narrow band measurements by Morris, McGrath (1976), and Wittenborn show resolved lines from individual merchant ships down to 5 Hz. Characteristic broadband spectra dominated by shipping possess a broad maximum in the 50 to 80 Hz range and a decrease in level with increasing frequency at 6 dB per octave. For a receiver located in the sound channel, the vertical and horizontal directionality and depth dependence of the noise are generally determined by the surface ship distribution over the continental slope.

Signal Sound Transmission Characteristics

Officer (1958) showed down-slope propagation resulted from the conversion of high angle rays (with respect to the horizontal) to lower angle rays by twice the slope angle. The effect was observed by Northrop (1968) with peak pressure amplitudes from shallow explosions over the continental slope. The estimated down-slope enhancements (DSE) averaged 6 dB and ranged as high as 12 dB. This was coined the "megaphone effect" by Smith (1971). Morris (1975) found the narrow-band pressure levels on a vertical array in the upper part of the Pacific sound channel showed a DSE from the radiated signal of a supertanker to be between 7 and 12 dB from a seamount and 4 and 6 dB from the continentinal slope. In the Northwest Atlantic, LaPlante (1981) and Koenigs (1981) using charges demonstrated that seamounts and other topographic features which rise into the sound channel have a significant affect on SOFAR propagation. The DSE was observed to be most pronounced for near surface sources (18 m) at frequencies between 25 to 100 Hz ranging up to 20 dB and averaging 7.5 dB. These results agree with the observations by Martin (1985) in the Northeast Pacific. The results in this paper are comparable and are attributed to the conversion of high angle energy to low angle energy by twice the slope angle with the added frequency dependent influence of bottom reflectivity.

Down-Slope to a Deep Ocean Basin

The down-slope to deep ocean sound channel experiment was conducted in the Northwest Atlantic Ocean, Carey (1983), with a transmission path from the continental slope off the Sable Island Bank, through the Gulf Stream to the edge of the Sargasso Sea. A calibrated sound source was towed on up- and down-slope radials at ranges between 730 and 910 km from a seismic streamer at a 750 m depth. A computerized propagation model based upon the parabolic equation (PE) approximation was used to calculate the propagation loss estimates displayed in fig. 1. Model calculations were made for measured range-dependent sound velocity and bathymetry profiles to show the effects of these range dependent environmental factors on down-slope sound transmission. The bottom was characterized by a critical angle approximation. It was assumed that energy interacting with the seafloor at grazing angles below 12° would suffer no loss, while energy interacting at grazing angles above 12° would be totally absorbed. Sound intensity plots are shown for sources located at a depth of 18 m in 100 m of water, and at 18 m in 806 m of water. The results, corrected for cylindrical spreading, show high angle energy interacting with the slope, converting to shallower angles,

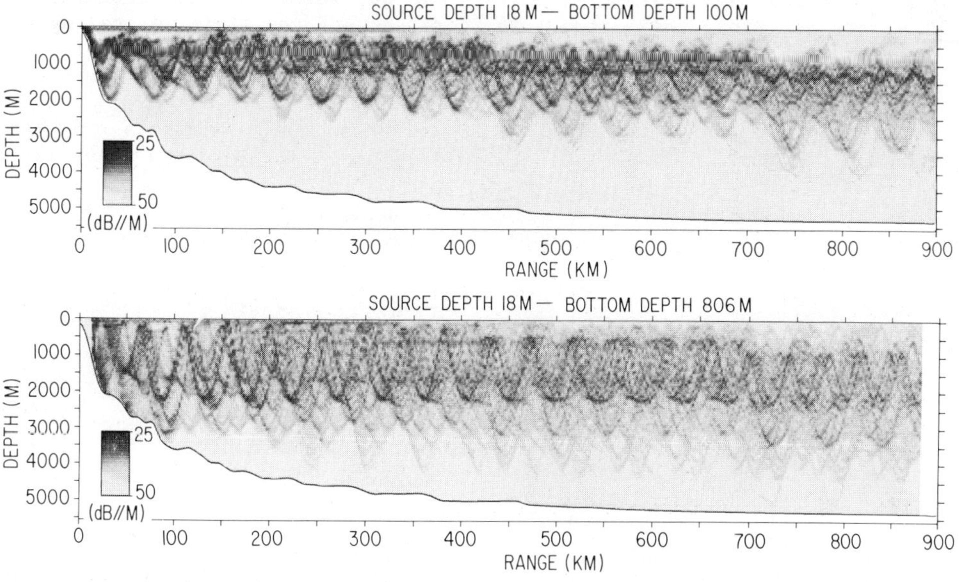

and propagating in the deep sound channel in deep water. The 100 m case clearly shows the effect of a near surface sound channel, in deeper water, the concentration of energy about a deepening sound channel axis.

Corresponding TL data are displayed in fig. 2. The data show that transmission persists along a slope traversing track from about 725 km to about 900 km. A substantial portion of the track is over the slope at the edge of the basin. Calculations of range averaged TL are shown for comparison. The model was run for a range dependent environment with a flat bottom and a range dependent environment with the slope. Each case was run for a perfectly absorbing (0° critical angle) and moderately reflective (12° critical angle) bottom. The range dependent totally absorbing case (0° critical angle: slope) shows the slope to be a dominant factor at ranges beyond 850 km. The range dependent moderately reflective case (12° critical angle: slope) shows the potential for slope enhancement. The difference between the two moderately reflective cases (12° critical angle: slope and 12° critical angle: flat bottom) show the potential degree of slope enhancement. The calculations, bracketing the observed transmission loss, show a potential enhancement of 8-10 dB, with the data averaging about 4 dB. TL data from all radial runs and a flat bottom PE calculation are shown versus water depth in fig. 3. When the depth of the water is less than 100 m we observed pronounced increases in loss but in many instances persistence of sound transmission. The data show a consistent trend for all legs from a wide area of the slope, with DSE values between 0 to 10 dB with a mean value of 4 dB determined from the difference of the transmission loss measurements and the range-averaged, flat bottom PE calculations. These results agree with the treatment of down-slope propagation converting high angle energy to low angle energy coupled with the effect of the frequency dependent bottom loss versus grazing angle. Bottom loss measurements and geologic information indicate that bottom loss in this region can be represented by 0 dB loss to 20° at 100 Hz and to 11° at 200 Hz with a smooth increase to 6 dB per interaction for angles greater than 40°. Lindop (1977) defined a slope enhancement by the limiting source angle ($\theta_0 \sim 8°$) for deep water and the limiting source angle over the slope (θ). The limiting source angle (θ) may be determined by the limiting grazing angle at the bottom and Snell's law. For a water depth of 1 km, slope angle of 6°, grazing angles of 11° to 20°, and source angles of 15° to 25° the DSE = 10 Log ($\sin\theta/\sin\theta_0$) = 2.7 to 4.8 dB, which is close to the 4 dB observed. Since the bottom loss increases and the limiting grazing angle decreases with frequency, DSE would not be expected at frequencies greater than 400 Hz.

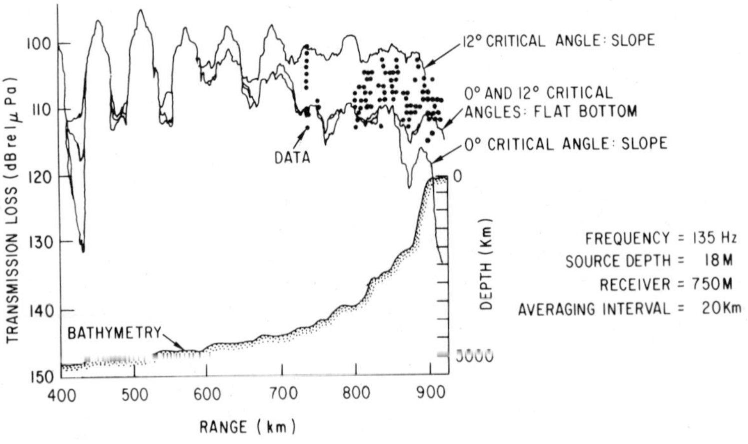

Fig. 2. Transmission loss comparison between P.E. calculations and data.

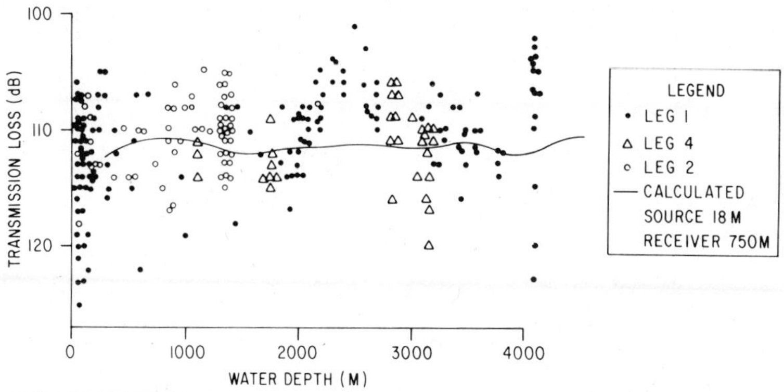

Fig. 3. Transmission loss versus water depth beneath the source.

Down-Slope Propagation to Bottom Limited Region (The Gulf of Mexico)

A down-slope to bottom limited region experiment was performed with a towed source driven at 67 Hz and 173 Hz from the Florida Plain toward the West Florida Escarpment, Carey (1985). The bathymetry as shown in fig. 4 was esentially flat at a depth of 3700 m until the 8.6° slope was encountered at a range of 200 km. The Gulf of Mexico is bottom limited and sound which interacts with the bottom will produce two components, a specular one, which depends on the density contrast, and a refracted one, which depends on the frequency-depth dependent attenuation coefficient. The characteristics of the refracted arrival can be determined using the "geoacoustic" models of the ocean bottom as described by Hamilton (1980). Geoacoustic parameters (sound speed, density and attenuation, Mitchell (1979)) were used in TL calculations performed with the PE and normal mode codes, and to derive bottom loss versus grazing angle curves for various frequencies for calculations performed with FACT and ASTRAL models (Spofford (1974, 1979)). However, geoacoustic data and bottom loss estimates must be carefully used on the slope since the sediment thickness is probably different from that in the basin or on the shelf. Empirical estimates of the bottom loss at a grazing angle of 10° based on a ray trace model to determine the number of bottom bounces were found to be 0 to 0.5 dB at 67 Hz and 1.5 to 2.0 dB at 173 Hz and were in agreement with the medium attenuation bottom loss curves computed by Mitchell.

TL results are also shown in fig. 4. The triangular points are the measured range averaged data. The range interval of averaging, due to the source and receiver motion for the individual hydrophone was between 1 and 2 km. The FACT Model was used to calculate TL versus range and the average agreement was found to be good. The 67 Hz TL averages about 10 dB less than that at 173 Hz and is much more uniform. The DSE at 67 Hz, taken as the difference between the data and the flat bottom FACT calculation, was as high as 6 dB with most points between 2 and 4 dB. The 173 Hz data shows a peak DSE of 6 dB with a dramatic increase in TL as the source proceeds up the slope.

Figure 5 shows results of the calculations performed with a normal mode model which assumes the process is adiabatic (SNAP, Jensen (1979)) and a normal mode-like model which although adiabatic is range smoothing (ASTRAL). The flat bottom SNAP calculation employed the geoacoustic model directly whereas the ASTRAL calculation used the derived bottom loss curves. The range averaged data and the calculations were found to agree. The mean level of DSE is predicted by the ASTRAL calculation and indicates slope enhancements of up to 6 dB compared to the flat bottom ASTRAL. Calculations were also performed with the high angle PE code, Lee and Botseas (1982, 1983).

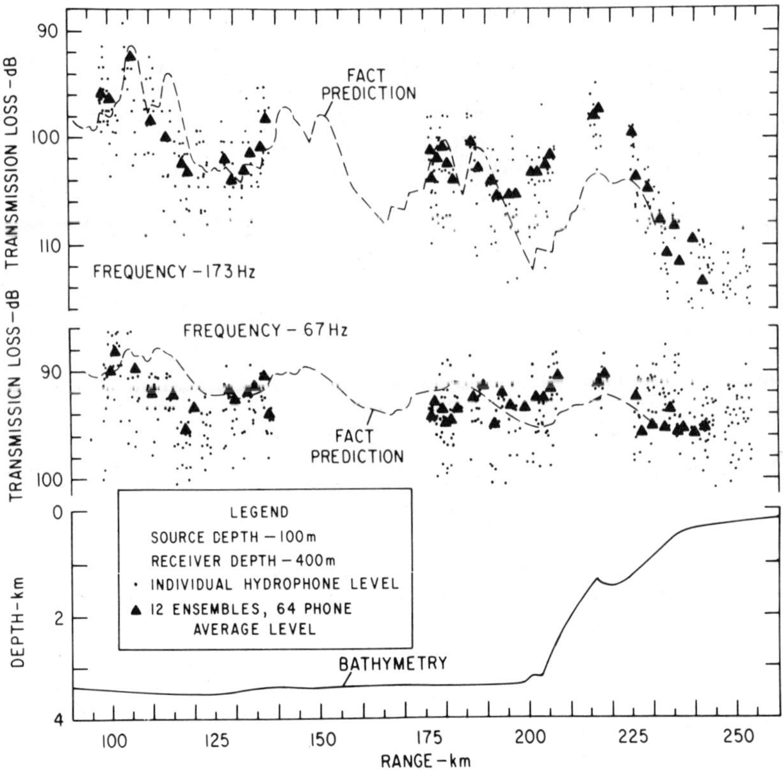

Fig. 4. TL measurements and FACT calculations versus range and depth.

Comparisons of these calculations, split-step PE with a specified critical angle and range averaged data showed good agreement. This agreement resulted from the use of measured environmental factors such as bathymetry, sound velocity profiles and geophysical data. Although no detailed comparison is presented in this paper the calculations were performed using the geoacoustic model or derived bottom loss curves were found to be able to reproduce the spread of received levels observed in the data.

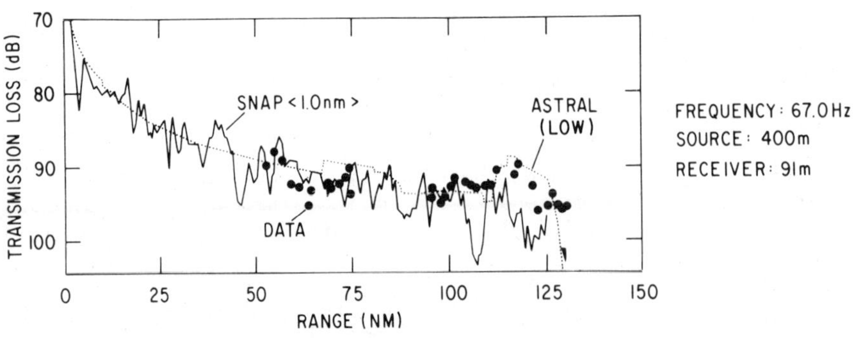

Fig. 5. ASTRAL and SNAP calculations compared with data.

Signal Coherence Estimates

The signal coherence was examined for sound propagation down-slope to a deep ocean basin and to a deep bottom limited basin. The magnitude squared coherence versus length along the array showed a pronounced multipath interference pattern proportional to the signal-to-noise ratio (S/N) variations at high S/N (> 10 dB). However, for small S/N (< 10 dB) and points close to the tow ship, the MSC showed wide variations due to the increase of coherent noise. These findings underscore the conclusions drawn by Carter (1978) that pair-wise MSC is difficult to measure, is a good measure of S/N and its bounds are large. The estimation of coherence must also be performed in a time that is short compared to the dynamics of the medium and the source-receiver motion. However, the signal coherence can be inferred from the coherent signal gain, $sg(\omega,k)$, at angular frequency ω and spatial wave number k. The signal gain is the ratio of the time averaged beam and mean hydrophone signal power. The signal gain when normalized by the number of hydrophones (N) squared is referred to as the differential array signal gain ($DASG = 10 \log (sg(\omega,k)/N^2)$). The determination of DASG is sensitive to the method of averaging employed, whether one averages the peak beam response or uses the peak average beam response.

The DASG for the bottom limited case was found to be on average -3 dB. Since the processing time (17 minutes) was long and no peak tracking was employed at least 1.5 dB of this DASG could be due to source-receiver motion. The DASG for the deep ocean sound channel case yields for the total data set a mean DASG = -1.2 dB and for selected peak tracked data a mean DASG of -0.5 dB. In both cases, comparisons between deep water data and slope interacting data show no significant differences.

The DASG is 10 times the logarithm of the summation of the normalized pairwise coherence estimates divided by the number of elements (N) squared.

$$DASG = 10 \operatorname{Log} \left(\textstyle\sum_i \sum_j \rho_{ij}/N^2 \right) ; \quad \rho_{ij} = \operatorname{Exp} \left(\left(S_{ij} / L_c \right)^n \right)$$

Thus if we have a knowledge of the coherence function, ρ_{ij}, where S_{ij} is the separation between the elements and L_c is a coherence length and n is an integer; then given a measure of DASG we can estimate L_c. A natural choice is either a Gaussian or an exponential functional form for the coherence function (Cox (1973)). The exponential function is interesting as a closed form expression for array signal gain is realized. However, an exponential coherence function implies a discontinuous physical process which may not be applicable to acoustic waves weakly scattered by the ocean surface and volume. Beran and McCoy (1970) derived an expression with n = 1.5 whereas Flatte and Dashen (1977) use n = 2, the Gaussian form. While these papers discuss sattering from internal waves and are not applicable to surface scattering or bottom interacting signals, if the sound channel is modeled as a two liquid layer, the volume inhomogeneities in the sediment may produce similar coherence functional forms. Brekhovskikh (1982) has shown that the surface scattering coherence function depends on the square of the coherence parameter and has a Gaussian form. Shifrin (1971) references Chernov (1960) and Tatarski (1961) and suggests a Gaussian coherence function is adequate for most statistical antenna problems. We simply employ coherence functions with n = 1.5 and 2.0 to bracket the observed coherence lengths. These estimates yield for the bottom limited case the DASG = -1.5, a coherence length of 181 m for n = 2 and 200 m for n = 1.5 and for the deep ocean case coherence lengths of 352 m for n = 2 and 460 m for n = 1.5. Thus in this low frequency range the signal retains some of its coherence while propagating down-slope to deep ocean basins and to bottom limited regions with thick sediments.

Beam Noise Surfaces

Beam noise measurements were obtained using seismic streamers with high density digital recorders (HDDR) and processed with Fast Fourier Transform (FFT) techniques on standard minicomputers and array processors. Figure 6 shows results of one such study in a deep basin with shipping lanes. The array was towed near one side of the basin to reduce ambiguity effects. Three coincident beam noise surfaces (beam noise intensity (dB) vs time (Δt = 8 sec) and azimuth ($\Delta\theta \sim 2.5°$)) are shown for center frequencies of 53, 165 and 320 Hz with equivalent apertures in 0.125 Hz Hann shaded frequency bands. Tow ship noise is observed between 0 and 30° while distant ships are shown (light tracks) superimposed on a lower noise level background (environmental noise, unresolved ships, and system noise).

The beam noise surfaces for 165 and 320 Hz show a comparable number of ship tracks. However, the 53 Hz noise surface shows fewer tracks and several with weak intensity levels despite the better TL characteristics at this frequency. Beam to beam correlation functions indicate that the individual tracks on 165 and 320 Hz are surface ships within the main beam coverage areas rather than from side-lobe response. A strong signal which appears on the beam main lobe and also appearing on another beam due to the side-lobe response will produce a cross-correlation function with multiple peaks. The correlation functions did not exhibit this characteristic indicating a side-lobe level response better than -28 dB. Since the system responses are comparable and the TL characteristic shows less loss at 53 Hz, then the fewer tracks observed on this beam noise surface may be due to a characteristic of the surface ship radiated noise spectrum. The lower frequency portion of the spectrum is tonal whereas the higher frequency portion of the spectrum is a continuum. Consequently we only observe ships which have energy in the narrow measurement band.

Surface Ship Radiated Noise

Few measured narrowband (< 1 Hz) surface ship noise spectra have been reported in the LF band. Aleksandrov (1962), Morris and Wittenborn (1976) show LF spectra dominated by tonals in the 10 to 100 Hz band. An example of a surface ship noise spectrum from Wittenborn is shown in fig. 7 for a bandwidth of 0.1 Hz and is similar to spectra observed by Morris. This spectrum from a surface ship within the reliable acoustic path range was measured with a deep hydrophone. The merchant vessel was proceeding at a speed of 7.6 m/s (14.75 knots). Prominant blade tonals and lower shaft rate tonals were observed below 100 Hz. The region between these tonal components is affected by the transform window side lobe leakage since high resolution spectral decomposition techniques were not employed. The spectrum above

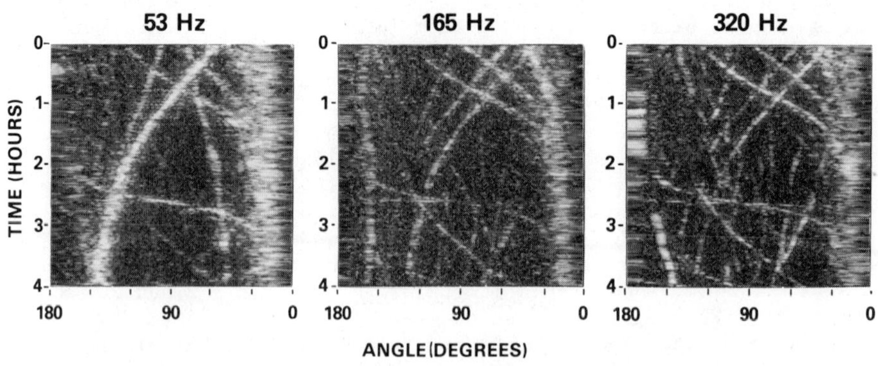

Fig. 6. Beam-noise intensity versus time and azimuthal angle.

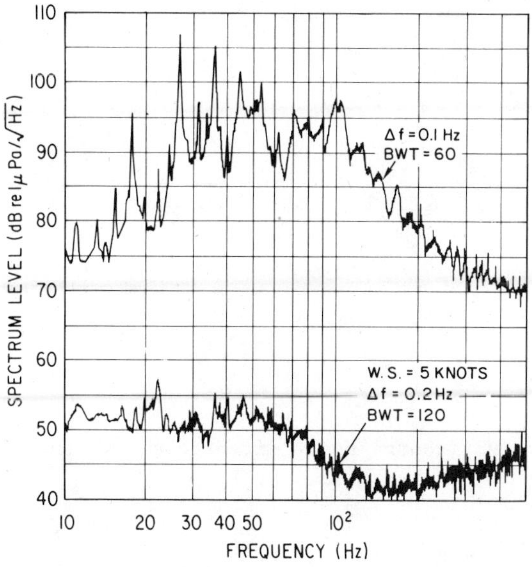

Fig. 7. Merchant ship radiated noise spectra.

100 Hz shows a modulated continum due to blade surface cavitation noise. This is consistent with the view put forth by Aleksandrov, Ross (1971) and Grey (1980). The low frequency tonals at the blade frequency and shaft frequency are due to the aggregate bubble volume oscillation on the blades as the propeller revolves. The continuum is due to the cavitational collapse of individual bubbles whose production is modulated by the harmonics of these blade and shaft frequencies. Ross estimates that the acoustic efficiency for sound radiation from these low frequency tonals and cavitation noise to be on the order of 10^{-6} of the total available mechanical power. The direct radiation from the propeller due to time varying forces and the subsequent vibration of the hull at these low frequencies with its surface image are estimated to be on the order of 10^{-8} to 10^{-10}.

Ambient Noise Horizontal Directionality

Beam noise measurements were performed in the down-slope to deep ocean basin and down-slope to bottom-limited basin experiments. These noise measurements were used to estimate the noise horizontal directionality with the iterative technique of Wagstaff (1978). In this technique a noise field horizontal directionality estimate is iteratively generated which minimizes the pooled standard deviation between the measured beam noise levels and the corresponding ones obtained from the noise field estimate. The end result is an estimate of the noise horizontal directionality and a pooled standard deviation of the differences in measured and estimated beam noise levels.

The estimate of the noise horizontal directionality for the Gulf of Mexico (fig. 8) shows high levels in the northern quadrant and low levels in the southern half-space. The differences between high and low levels range from 10 dB at 150 Hz to 25 dB at 50 Hz with pooled standard deviations of 1.6 and 2.5 dB, respectively. An array with an ambiguous beamwidth of 7° aligned in low noise directions would measure a level of 55 dB re 1 μ Pa from the beam direction plus it's sidelobe contribution, which for 30 dB sidelobes would be 54 dB re 1 μ Pa. Thus the low level noise azimuths are below the measurement system capability and could be much lower.

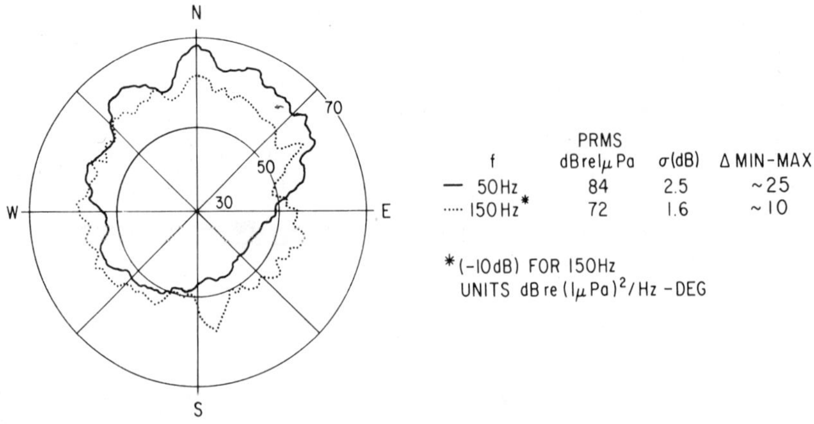

Fig. 8. Horizontal noise directionality, Gulf of Mexico.

The reasons for the high degree of spatial anisotropy are evident when the acoustic propagation characteristics of the basin surrounding the measurement location and the spectral distribution of noise sources are considered. The site of these noise measurements was seaward of the shelf north of the Yucatan Peninsula. The Yucatan Shelf extends from an azimuth of 68° at 370 km to 222° at 150 km. This relatively shallow region should result in poor propagation conditions for shelf-borne noise sources. To the southwest (222° to 270°) the shipping is sparse with propagation ranges between 550 km and 1000 km. This combination of low shipping density and long propagation distances results in insignificant contributions to the noise. Thus, in the absence of local shipping, the ambient noise levels should be low between 68° and 270°. In contrast, to the north and east of the measurement location many ships travel in lanes from the Yucatan Channel and Florida Straits to ports along the Gulf Coast from Tampa, Florida to Galveston, Texas. Consequently, the levels of noise due to shipping should be considerably higher in the northern quadrant. These suppositions are consistent with the observed directionality in fig. 8.

Shipping is not the only contributor to the low frequency noise in the Gulf of Mexico. Extensive off-shore oil drilling and oil exploration activities have their effects, such as the noise from seismic profilers with source levels in excess of 220 dB and repetition rates from 10 to 100 seconds. The noise from seismic profilers is not considered, herein, as part of the local ambient noise field. Furthermore, most drill platforms are in water less than 100 m and consequently have severe TL.

Similar results of ambient noise directionality measurements for the Northwest Atlantic Basin are shown as fig. 9. The directional effect (minimum to maximum level deviation) is shown to be about 15 dB at 50 Hz and 7.6 dB at 150 Hz. The directional characteristics of the noise field appear to be similar for the two frequencies. The measurements took place at a location such that the 90° to 180° quadrant was toward the Corner Seamounts and the Mid-Atlantic Ridge, while the northerly sector encompassed the region off Newfoundland. The high noise directions are from approximately 270° clockwise to 90°; a sector including the Grand Banks, the Scotian Shelf and a major trans-Atlantic shipping lane. Thus, the observed noise directionality is attributable to the combined effect of mid-basin shipping in trans-Atlantic lanes, and ships traversing the continental rise, banks, and shelf.

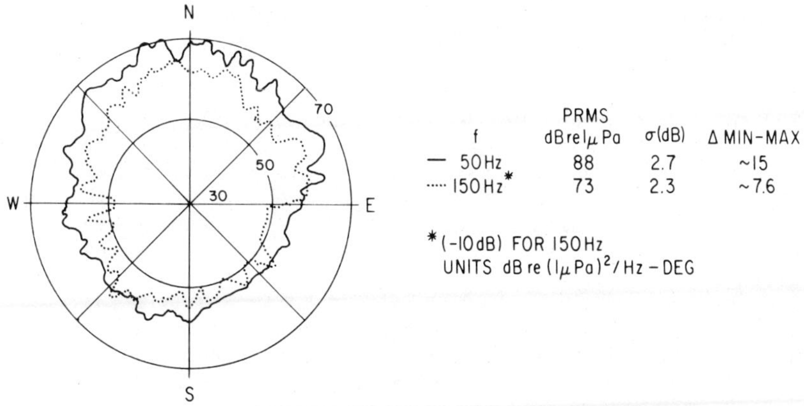

Fig. 9. Horizontal noise directionality, Northwest Atlantic.

The directionality plots shown here were found consistent with the time averaged beam response of the bearing time surfaces. Hamson (1983), and Wagstaff have shown that these noise horizontal directionality patterns can be calculated provided the shipping distribution is known, the transmission loss is calculable and the ships on the basin margins are included.

<u>Ambient Noise Vertical Directionality</u>

Since high angle energy from deep ocean noise sources is rapidly attenuated due to multiple bottom interactions, one would expect the energy propagated from these sources to be peaked at the sofar channel ray limiting angles (10° to 15° off the horizontal). However, measurements of the vertical distribution of noise intensity in the LF band show a broad angular distribution centered about the horizontal direction. Consequently, the broad distribution of energy about the horizontal requires a mechanism like the down-slope conversion process. Figure 10 represents a remarkable set of data collected by Anderson (1972) which illustrate these effects. Here the maximum likelihood method (MLM), Edelblute (1966), was used to produce the vertical noise level distribution as a function of vertical angle (90° is the horizontal direction) and frequency. These data were obtained south of Bermuda with a vertical array of 26 elements spanning a distance of 110 m centered at a depth of 236 m in the deep sound channel with an axis depth of 1 km. Figure 10 shows the vertical arrival structure of the noise as a function of frequency with a resolution of 1.4 Hz. At 150 Hz the noise intensity has maxima at 90° ± 9.5° compared to 90° ± 14° if the array center had been located at the sound channel axis. The Mimi sound source is observed near 400 Hz. At frequencies greater than 400 Hz, the data show aliasing (180° to 130°) and the peaked distribution of noise intensity at 90° ± 9.5° with noise minima at 90° of -10 dB. In comparison, the distribution at 100 Hz shows a broad maximum centered on the horizontal. Anderson was able to simulate the pattern at the higher frequencies but not able to do so at the lower frequencies. He was able to show that the broad maximum was not due to deficiencies in the measurement technique and analysis. These characteristics persisted for several days, that is, the vertical distribution of energy was broad at the lower frequencies and sharply dual peaked at the higher frequencies indicating the horizontal component of noise originates at a distance from the receiver. The smooth noise level variation between 20 and 150 Hz is a significant characteristic of the noise. If the horizontal noise is from distant slope interacted ship radiated signatures, why the smooth variation?

These results are consistent with previous and subsequent investigations. Fox (1964) obtained data from a 40 element array near Bermuda in 4.4 km of water. He found that at low sea states, the vertical distribution of noise intensity was broadly peaked near the horizontal over the band of 200 to 1500 Hz. At high sea states he observed an isotropic distribution at higher frequencies (>200 Hz) but a persistent horizontal component (~200 Hz). Measurements by Axelrod (1965) showed a strong LF horizontal component and observed that the frequency at which the vertical directionality became isotropic depended on the local wind speed. Anderson (1979) reported noise vertical directionality in a region of the Northeast Pacific Ocean over the 23 to 104 Hz band. He also observed a broad angular distribution of noise near the horizontal. A horizontally stratified, range independent, ocean model does not predict a broad angular distribution of noise intensity about the horizontal, but rather a dual peaked noise intensity distribution ($\pm 10°$ to $\pm 15°$) and no noise at the horizontal.

Noise generated from surface sources cannot arrive within the limiting angles bracketing this minimum referred to as a horizontal noise notch. Wagstaff (1981) showed by agreement between calculations and data that distant shipping over the continental shelf, slope, or a sea mount contribute to the broad vertical distribution of low frequency noise near the horizontal. Data obtained with the NORDA VEKA Array between Cape Hatteras and Bermuda was analyzed in the 45 to 100 Hz band by Wales (1981). These data were found to have a broad vertical distribution near the horizontal. Their corresponding noise level versus frequency and arrival angle plot shows a smooth variation in frequency similar to that of the Anderson data. The fact that these data do not show a tonal characteristic is important as ship radiated noise spectra have been shown to be primarily tonal in this band. Browning (1982) shows data obtained in the South Figi Basin, a region of sparse shipping. Below 200 Hz, he observed a level 4 dB less than Wagstaff with a broad maximum in the vertical distribution of noise intensity centered at the horizontal. Browning found these results consistent with both surface ship noise and storm noise originating near the basin boundaries coupling to the sound channel. If these results were solely due to ships interacting with slopes and seamounts one might expect a reflection of the tonal quality of the low frequency ship signature in the measured noise spectra when the number of ships are small. Since this is not the case, one may draw the inference that another contribution to the noise field at the horizontal is important such as the noise due to wind and surface waves over continental slopes and seamounts. Furthermore, Burgess (1983) has found similar results in Australasian waters.

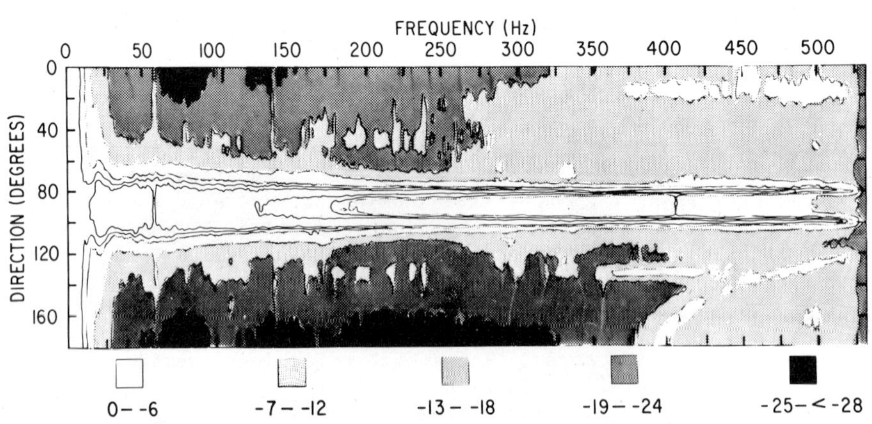

Fig. 10. Noise intensity versus vertical arrival direction and frequency.

CONCLUSIONS

This paper has presented measurements which show that surface ship noise coupled into the sound channel produces marked effects on the directional noise field at the lower frequencies. The data presented show this interaction is similar to a low frequency pass filter. The expected level of DSE based on the down-slope conversion process for representative bottom loss yields in 1 km of water DSE = 4.8 at 50 Hz, 2.1 at 200 Hz and 0.5 at 400 Hz. The lower frequency sound transmission was observed to be more persistent with a fair degree of coherence.

Coherent signals from surface ships, readily observed in the beam noise surface plots were shown to be a dominant characteristic of the ambient noise. The frequency dependence of the beam-noise surfaces is consistent with expectations based on narrowband ship signatures. Calculations and inference suggest that surface ships dominate the low frequency ambient noise horizontal directionality. The nature of the received signal from distant ships in space and time provides for a coherent gain over the incoherent environmental noise background. The fact that down-slope propagation of sound retains a fair degree of coherence reinforces the idea that the coastal shipping dominates the mid-basin ambient noise horizontal directionality.

Ambient noise vertical directionality data exhibit a low pass filtered effect due to down-slope conversion at the basin boundaries, i.e. a broad angular distribution centered about the horizontal at low frequencies and a dual peaked distribution in energy at the higher frequencies. A mechanism for introducing energy into the sound channel is the down-slope conversion process. Dashen (1984) has shown that scattering by internal waves cannot account for this phenomena, however Mellen (1985) contends that diffusion is a candidate. Measurements and calculations by Wagstaff demonstrate that coastal shipping could easily account for this effect.

A characteristic of the vertical directionality data was the smooth variation in frequency from 20 Hz to 200 Hz. Since surface ship spectra are tonal in the low frequency region, expectations are that a spikey nature would be observed provided the measurement system has the resolution. Furthermore, measurements obtained in a remote, sparsely shipped areas, although at lower levels, yield similar findings as the vertical directionality from densely populated basins. Since any source of sound such as wind driven noise (capillary to capillary wave interaction, wave turbulence, splashes, impacts and aggregate bubble oscillations) near the surface over the slope, shelf or seamounts will introduce sound into the sound channel; then environmental noise in addition to shipping maybe required to explain the broad angular and frequency characteristics.

ACKNOWLEDGMENTS

The authors acknowledge D. Edelblute for his permission to use the vertical directionality data. We recognize the excellent technical illustrations of Ms. J. Lackie and we gratefully acknowledge the contribution of Ms. J. Albrittain in the preparation of the manuscript.

REFERENCES

Aleksandrov, I.A., 1962, Sov. Phys. -Acoust., 8, pp. 23-28.

Anderson, C.G., Anderson, D.A., Edelblute, D.J., et. al., 1972, "VLAM Data Analysis--.....," NUSC TN 800, NOSC, San Diego, CA.

Anderson, V.C., 1979, J.Acoust.Soc.Am., 66(5), pp. 1446-1452.

Axelrod, E.H., et. al., 1965, J.Acoust.Soc.Am., 37(1), pp. 77-83.

Beran, M.J., and McCoy, J.J., 1979, J.Acoust.Soc.Am., 66, pp. 1468-1471.

Brekhovskikh, L., Lysanov, Yu, 1982, "Fund. of OCEAN ACOUSTICS" Springer-Verlag, NY.
Browning, P.G., et. al., 1981, J.Acoust.Soc.Am., 70(1), S66(A).
Burgess, A.S., Kewley, D.J., 1983, J.Acoust.Soc.Am., 73(1), pp. 201-210.
Carey, W.M., 1983, J.Acoust.Soc.Am., 73(S1), S55(AA6).
Carey, W.M., 1985, J.Acoust.Soc.Am., 77(S1), S14(F7).
Carter, C.G. and Scannell, E.H., 1978, NUSC TD 5881, NUSC, New London, CT.
Chernov, L.A., 1960, "Wave Propagation in a Random Medium," McGraw Hill, NY.
Cox, H., 1973, J.Acoust.Soc.Am., 54, pp. 1743-1746.
Dashen, R. and Munk, W., 1984, J.Acoust.Soc.Am., 76(2), pp. 540-554.
Edelblute, D.J., et. al., 1966, J.Acoust.Soc.Am., 41(1), pp. 199-205.
Flatte, S.M., Dashen, R., et. al., 1977, Stanford Res. Inst., TR JSR 76-14.
Fox, G.R., 1964, J.Acoust.Soc.Am., 36(8), pp. 1537-1540.
Furduev,A.V., 1966, Akad.Nauk.SSR,IZV,Atmos.OceanicPhys., 2(235), pp. 314-320.
Goncharov, V.V., 1970, IZV, Atmos. and Oceanic Phy., 6(11), pp. 1189-1196.
Grey, L.M., and Greeley, D.S., 1980, J.Acoust.Soc.Am., 67(2), pp. 516-522.
Hamilton, E.L., 1980, J.Acoust.Soc.Am., 68(5), pp. 1313-1340.
Hamson, M.R., and Wagstaff, R.A., 1983, SACLANTCEN Rept. SR-70.
Isakovich,M.A., and Kuryanov,B.F., 1970, Sov.Phys.-Acoust., 16(1), pp. 49-58.
Jensen, F.B., and Ferla, M.C., 1979, "SNAP" SACLANTCEN, SM-121.
Koenigs, P.D. and LaPlante, R.F., 1981, J.Acoust.Soc.Am., 69(S1), 559(A).
LaPlante, R.F., et. al., 1981, J.Acoust.Soc.Am., 69(S1), 559(A).
Latham, G.V., and Nowroozi, A.A., 1968, JGR, 73(12), pp. 3945-3956.
Lee, D., and Botseas, G., 1982, "IFD:.....," NUSC TR 6659, NUSC, NL, CT.
Lee, D., and Botseas, G., 1983, "IFD:.....," NUSC TR 6905, NUSC, NL, CT.
Lindop, P.H., 1977, "Comments at the International Workshop.....," WHOI.
Longuet-Higgins, M.A., 1950, Phil. Trans. Roy. Soc., A243(1), pp. 1-35.
Martin, R.L., 1985, Private Communication, (Experiments 1972).
McGrath, J.R., 1976, J.Acoust.Soc.Am., 60(6), pp. 1290-1299.
Mellen, R.H., 1985, This volume.
Mitchell, S.K., 1979, "Pre-Exercise Estimates of Bottom Interaction in the Gulf of Mexico and the Caribbean Sea," ARL/UT, Austin, TX.
Morris, G.B., 1975, "Preliminary Results on Seamount and Continential.....," SIO Ref. 75-34, MPL/Scripps Inst. of Oceanography, San Diego, CA.
Nichols, R.H., 1981, J.Acoust.Soc.Am., 69(4), pp. 974-981.
Northrop, J., et. al., 1968, JGR, 73(12), pp. 3905-3908.
Officer, C.B., 1958, "Introduction to the Theory of Sound Transmission," McGraw-Hill Book Company, New York, p. 159.
Perrone, A.J., 1969, J.Acoust.Soc.Am., 46(3), (Part 2), pp. 762-770.
Perrone, A.J., 1974, J.Acoust.Soc.Am., 55(4), pp. 754-758.
Piggott, C.L., 1964, J.Acoust.Soc.Am., 36(11), pp. 2152-2163.
Ross, D., 1976, "Mechanics of Underwater Noise," Pergamon Press, NY.
Shifrin, Y.S., 1971, "Statistical Antenna Theory," Golem Press, Boulder, CO.
Smith, P.W., 1971, J.Acoust.Soc.Am., 49, Paper Q2, p. 96.
Smith, P.W., and Jones, J.P., 1972, J.Acoust.Soc.Am., 52, Paper M5, p. 137.
Spofford, C.W., 1974, "The FACT Model," MC Report 109, Washington, DC.
Spofford, C.W., 1979, "The ASTRAL Model," SAI-79-742WA, SAI, McLean, VA.
Tatarski, V.I., 1961,"Wave Propagation in a Turbulent Medium,"McGraw-Hill, NY.
Wagstaff, R.A., 1978, J.Acoust.Soc.Am., 63(3), pp. 863-869.
Wagstaff, R.A., 1981, J.Acoust.Soc.Am., 69(4), pp. 1009-1014.
Wales, S.C., and Diachok, O.I., 1981, J.Acoust.Soc.Am., 70(2), pp. 577-582.
Wittenborn, A.F., 1976, "Ambient Noise and Associated Propagation Factors...," TRACOR (T76RV5060), DTIC, ADC006902, Alexandria, VA.
Yen, N., 1977, J.Acoust.Soc.Am., 62(5), pp. 1176-1188.
Yen, Nai-chyuan and Perrone, A.J., 1979, NUSC TR 5833, NUSC, NL. CT.

TIME DEPENDENCE OF INFRASONIC AMBIENT SEAFLOOR NOISE ON A CONTINENTAL SHELF

T. Akal, A. Barbagelata, G. Guidi, and M. Snoek

SACLANT ASW Research Centre
Viale San Bartolomeo 400
I-19026 La Spezia, Italy

ABSTRACT

SACLANTCEN is conducting a long term programme of ambient noise measurements that cover the frequency band from 1 to 100 Hz. A sea-bed sensor consisting of three seismometers and a hydrophone has been deployed close to the edge of the continental break in water of 100 m depth. The data are being systematically collected and telemetered via radio link to SACLANTCEN. The initial data (105 days) have been used to investigate the noise characteristics and its statistics.

INTRODUCTION

Ambient noise in the ocean has many sources. As a result, long term measurements are required to identify these sources and define the noise characteristics of an ocean area. Studies of ambient noise have shown that seismic acoustic energy in certain parts of the noise spectrum can be related to various known sources [1 to 13]. Figure 1 identifies several well known sources that transfer energy basically from biological, meteorological, geological or cultural origin into the underwater seismo-acoustic noise field.

SACLANTCEN has begun a programme to measure the time dependence of ambient noise along the Ligurian shelf edge (Fig. 2a). The frequency band of interest is 1 to 100 Hz. Initially both short (33 hr) and long (105 days) term measurements were made.

Two major factors affect ambient noise characteristics, the meteorologic conditions of the area and the distribution of the noise sources around the measurement site. In support of the interpretations of noise data, environmental and shipping data that could affect noise characteristics have been incorporated into the study.

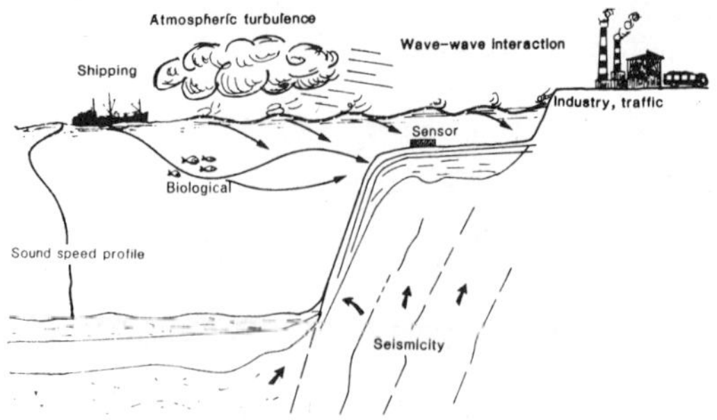

Fig. 1
Basic physical sources which transfer low-frequency energy into the underwater ambient noise field.

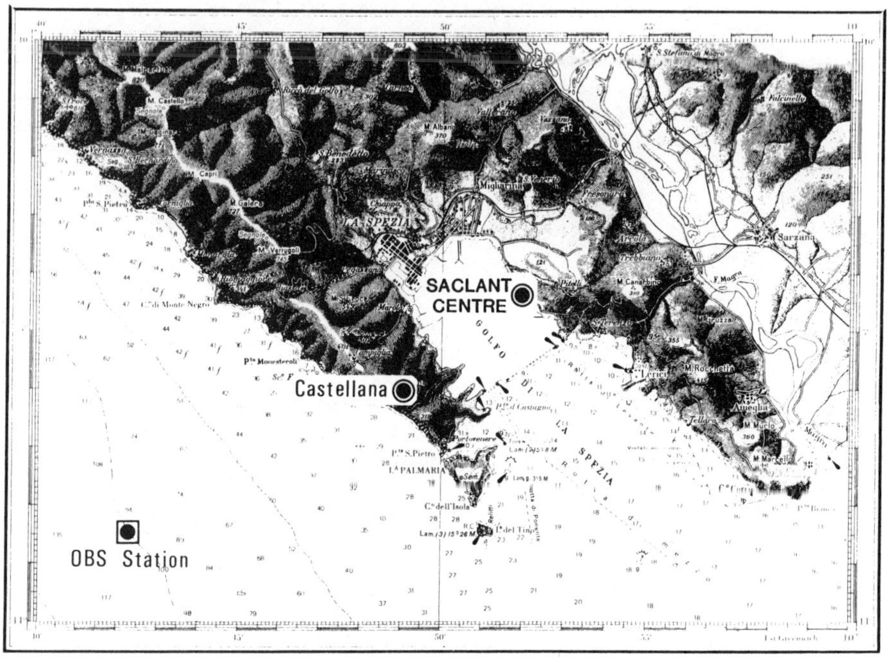

Fig. 2a The location of the measurement site.

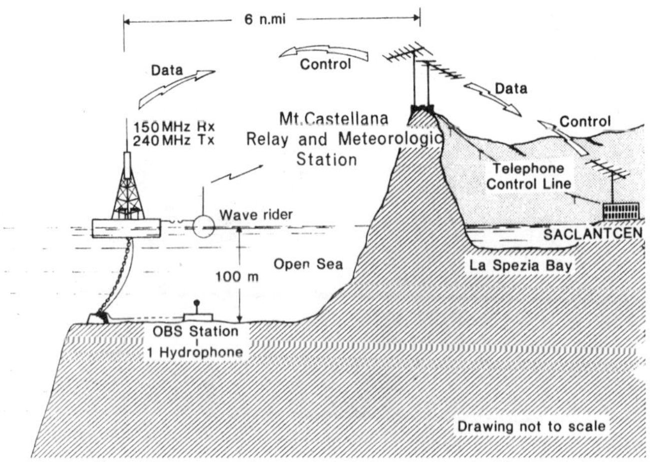

Fig. 2b
Experimental set-up.

INSTRUMENTATION

The instrumentation system used for these measurements consisted of the following components (Fig. 2b):

- the ocean bottom seismometer
- the radio transceiver (mounted on a surface buoy)
- the wave-rider buoy (attached to the surface buoy)
- the radio repeater (mounted on a mountain top)
- the receiver and control unit
- the interface to the digital computer
- the computer system.

Signals from the sensors were immediately converted into digital form. They suffered no significant degradation when transmitted through the various components to SACLANTCEN.

Ocean Bottom Seismometer (OBS)

The OBS consisted of geoacoustical and auxiliary sensors, preamplifiers and A/D electronics that are contained in a disc-shaped housing. Power is provided by a battery installed in the surface buoy and transmitted via a coaxial cable.

The geoacoustic sensors comprised three active 1 Hz geophones (Teledyne S-500) mounted orthogonally and an omnidirectional hydrophone. These sensors can be operated at any angle and need no locking for transportation and handling. They have a velocity sensitivity of 450 V-sec/m and a -5 dB passband of 1 to 100 Hz. In addition to these sensors, the system includes auxiliary sensors: two tilt meters to monitor the inclination, a compass to determine orientation, and a thermistor to measure temperature.

The four geoacoustic sensors are sampled at 600 Hz per channel. The data is converted into a floating point format and combined with low speed data from positional sensors. No gain adjustment is necessary during the experiment because the instantaneous floating point format allows the achievement of a very high dynamic range (120 dB). Calibration can be carried out by injecting stabilized AC current into the calibration coil of the geophones. The data is transmitted in biphase code to the surface buoy via a coaxial cable at 38400 bits/sec.

The Radio Buoy

The radio buoy, mounted on a 4 m diameter circular buoy moored in 100 m depth, about 8 km from the coast, contained rechargeable batteries (i.e. power for 32 hours of continuous data acquisition), a FM modulated VHF radio transmitter (220 MHz/4 W) and a standby radio receiver for power ON/OFF commands.

The Repeater

The unattended radio repeater was installed on 510 m high Mt. Castellana to overcome topographic interference between the OBS radio buoy and SACLANTCEN. The repeater received and retransmitted OBS radio buoy signals at different frequencies. A dedicated telephone line between the repeater and SACLANTCEN permitted the relay of orders to the OBS and the monitoring of the performance of the OBS instruments.

The Receiver and Control Unit

This unit, installed at SACLANTCEN, allowed the performance of the following functions: reception, demodulation and forwarding of the computer interface signals, monitoring of the performance of the geoacoustic and environmental sensors mounted on the OBS, through digital-to-analogue conversion and display of the transmitted signals, monitoring and manual control of the performance of the OBS and repeater station.

The Computer Interface

This unit, installed in the computer area, received the baseband signal from the receiver unit. Bits and words were synchronized and transferred to the digital computer in a parallel format. Commands were received from the computer which switched the full chain on and off according to a pre-set time schedule.

The Computer System

Acquisition and preprocessing were performed on a Hewlett Packard 21 MX computer with a disc unit and other peripherals. Data were immediately stored on discs for display, plot and preprocessing options and for transfer to digital tape.

MEASUREMENT AND ANALYSIS TECHNIQUES

This system was used to measure ambient sea noise for 105 days (December 1984 to March 1985). Automatic acquisitions of ambient noise of 5 minutes duration were made daily; wave-rider data and meteorological data were obtained over longer periods. Figure 3, a sample record, illustrates the variation of the ambient noise levels at different sensors for different meteorological conditions, together with the wave-rider recordings. Spectral estimates of these signals were made by applying standard techniques [14] to 25-sec time-windows over a 1 to 100 Hz frequency band, and averaging the power spectral density levels over a 1 Hz frequency bandwidth. Figure 4 shows the power spectral density of the sample signals of Fig. 3.

Noise Fluctuations in 33-hour Period

Measurements of 5, 10, 30, 60 minutes and 24 hours duration were conducted to study both the short and long-time variations of ambient noise. The power spectral density contours in a frequency/range plane provide the simplest display of the behaviour of the ambient noise field fluctuations, which are a function of both frequency and time.

Figure 5 shows an example of the changes of the ambient noise spectrum levels over a 33-hour period at different sensors. On 21 March 85, the sea surface wave field at the measurement site was flat, under a SWW wind of 3 n.mi/hr. During the day the wind increased to 15 n.mi/hr, causing waves as high as 80 cm. Twenty-seven hours after the beginning of the data acquisition, wind direction changed to North at 2 n.mi/hr. The characteristics of the ambient noise spectrum levels evident in Fig. 5 provide a natural division of the noise field into two distinct regions:

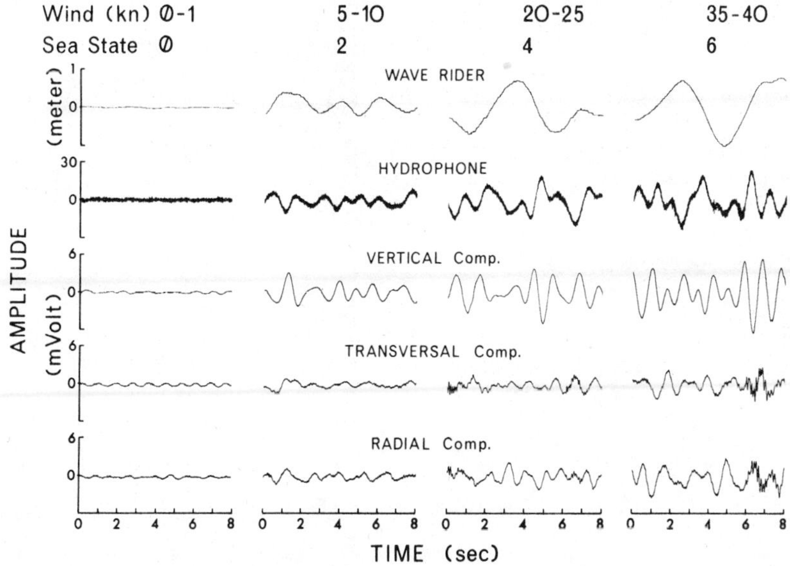

Fig. 3 Sample recordings: signals at various OBS sensors, wave-rider and meteorologic information under different meteorological conditions.

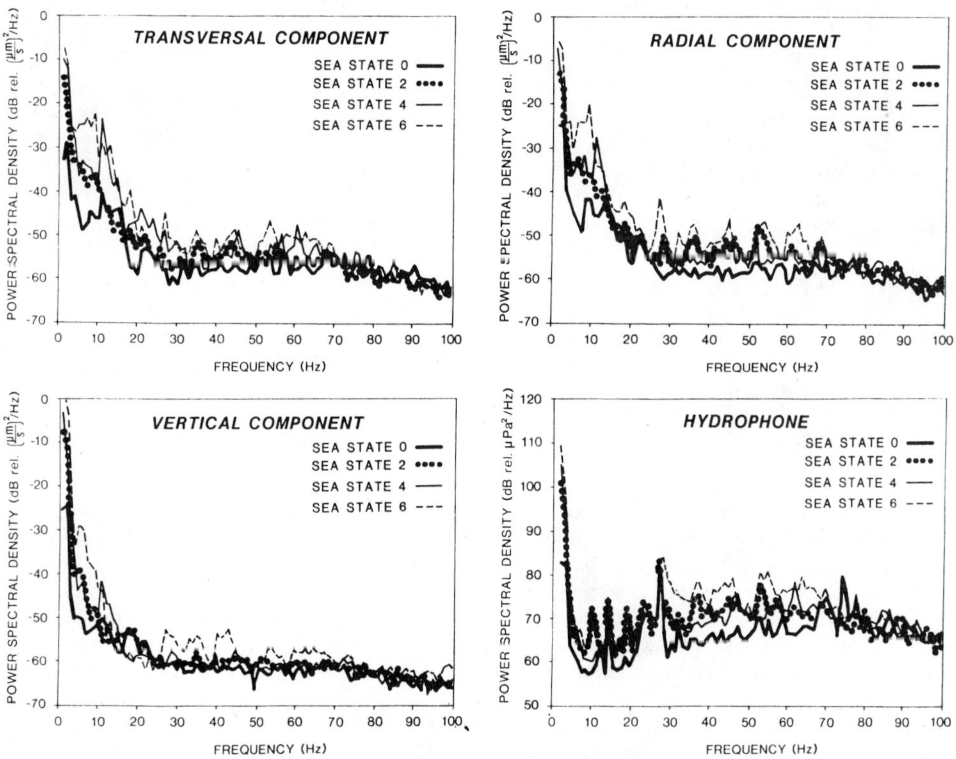

Fig. 4 Power spectral density levels of the sample signals from Fig. 3 for different sea states.

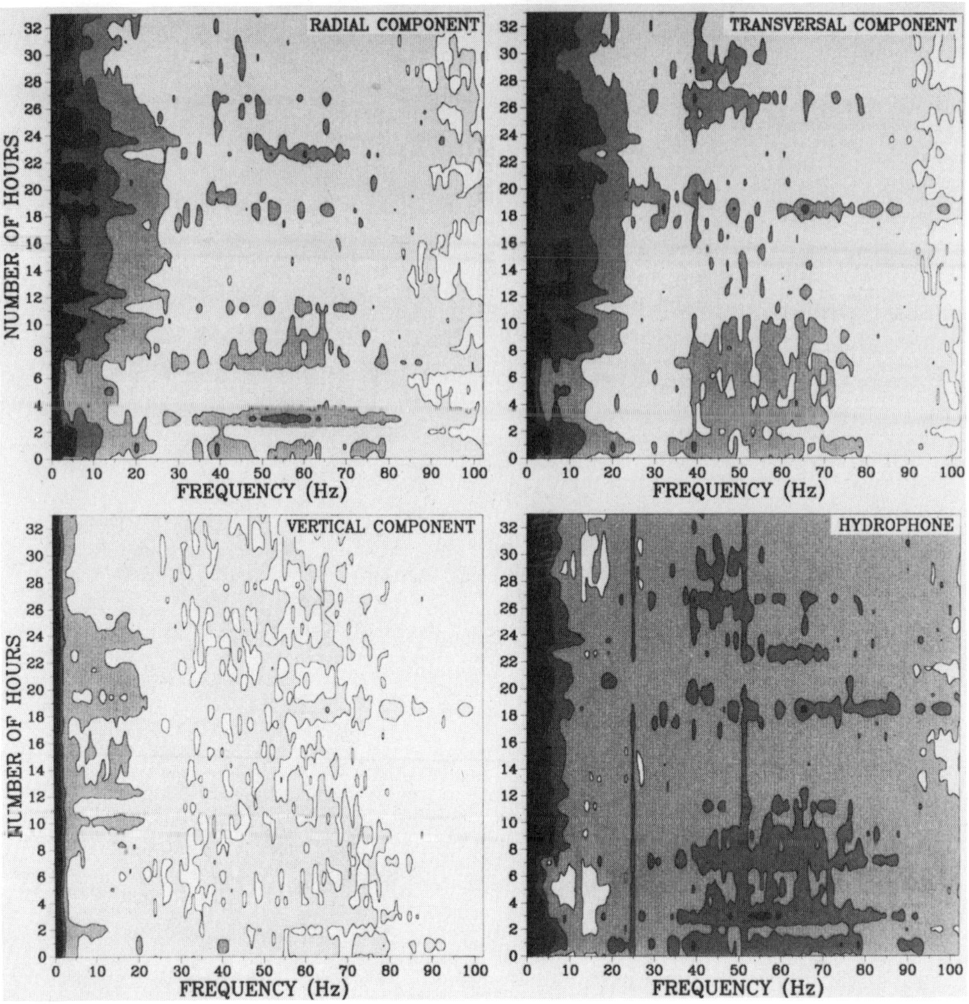

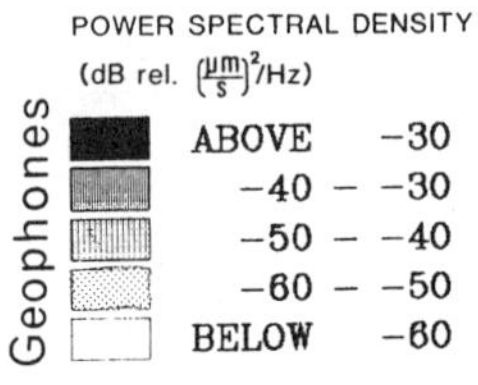

Fig. 5
Ambient noise spectrum levels fluctuations within a 33-hour period sensed by the OBS geophones and hydrophones.

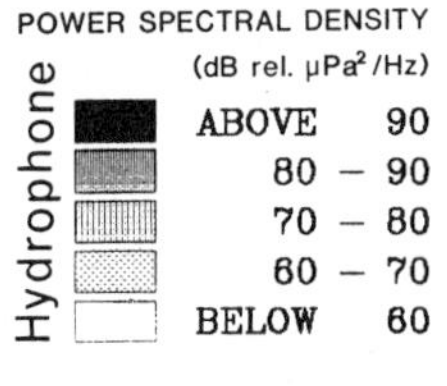

- In the low frequency (1 to 30 Hz) region, horizontal sensors generally have a higher energy content than the vertical sensor and the omnidirectional hydrophone has a much steeper fall-off slope than those of the seismic sensors. Increasing noise levels between the 7th and 27th hours, particularly evident in the vertical component, correspond to an increase in wind speed and sea state; conditions then reverted to those of the initial part of the measurement period.

- In the higher frequency region (30 to 100 Hz), shipping noise is demonstrated by sporadic peaks in the contours. The peaks normally correspond to the activities of medium-size trawlers and fishing boats passing through the measurement site.

Noise Fluctuations in 105-day Period

Similar procedures were applied to the data collected during the 105-day period. During this long-term period meteorological conditions and shipping were the two major sources of energy transferred into the acoustic/seismic field.

Effects of Meteorological Conditions

Figure 6 presents variations of the ambient noise spectrum levels with time and the corresponding sea states. The ambient noise fluctuations during the measurements show, as expected, that the temporal characteristics of the ambient noise spectrum levels are influenced by wind and sea conditions at low frequencies (below 30 Hz). Due to changing meteorological conditions, this frequency region reveals changes (schematically described by sea state plots) which correlate well with sea state.

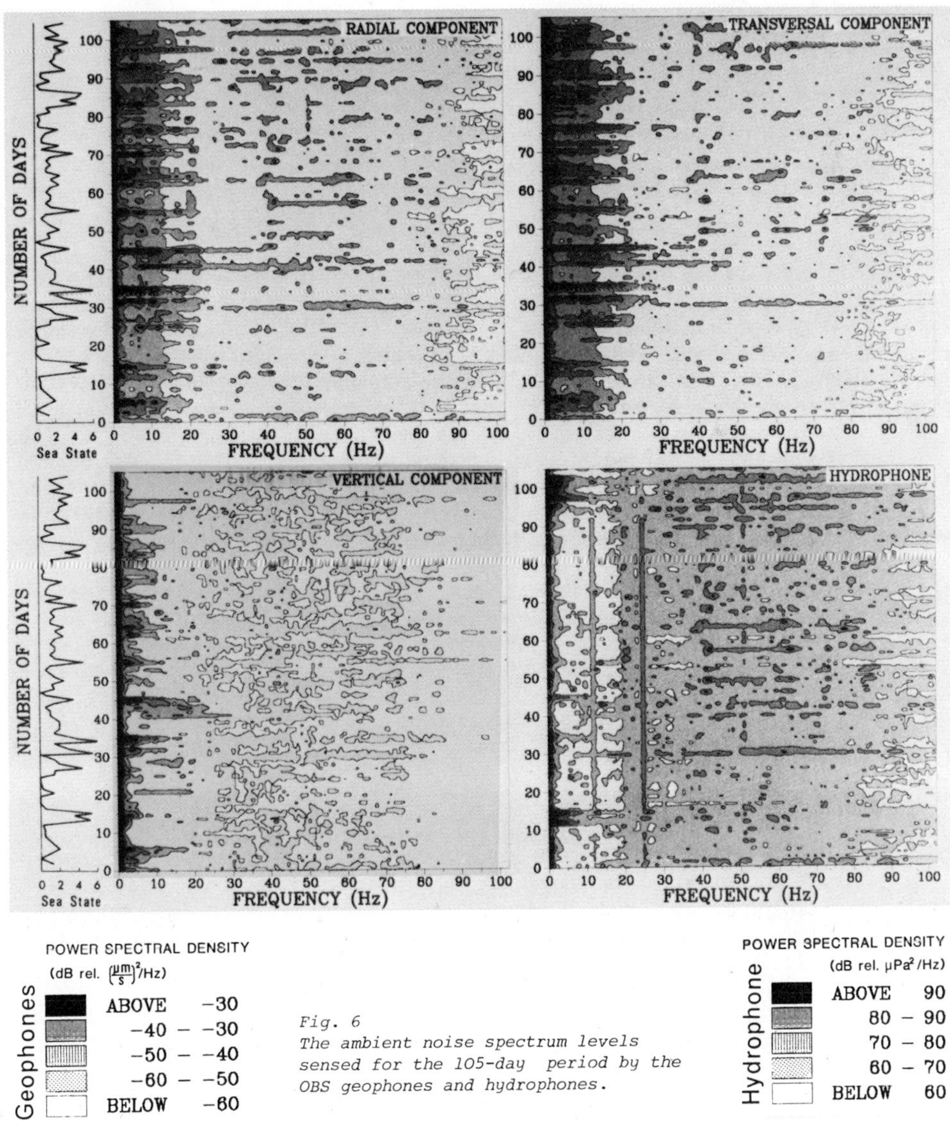

Fig. 6
The ambient noise spectrum levels sensed for the 105-day period by the OBS geophones and hydrophones.

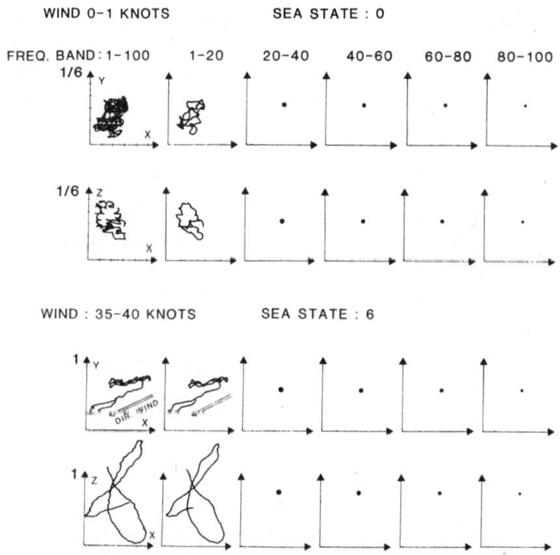

Fig. 7a
Particle motions under sea-state 0 and 6. The hodographs are displayed for different frequency bands. Note, for sea-state 6, x, y and z scales are exaggerated 6 times as compared to those for sea-state 0.

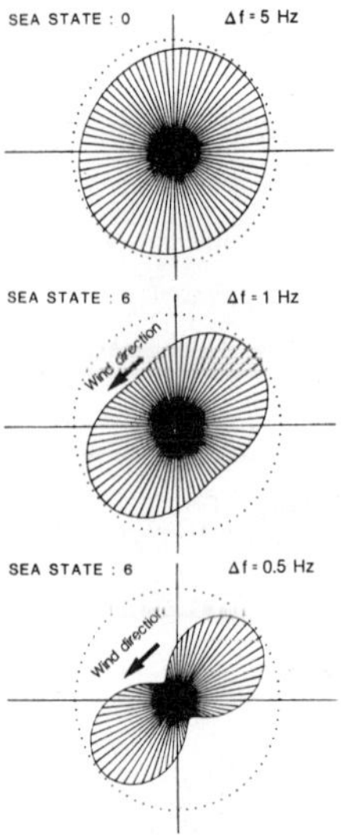

Fig. 7b
Azimuthal power pattern for two different frequency bands.

For all sensors we observed a frequency-dependent increase of spectrum levels with increasing sea state (see selection of different sea states in Fig 4). At low frequencies, the increase is greater than at the high frequencies. Sea-state effects normally disappear above the 60 to 70 Hz band. Within the 5 to 15 Hz band peaks correspond to the sea surface waves. These peaks increase with increasing sea state while their frequency range narrows. Figure 7a displays two sets of hodographs (for the xy and xz planes) obtained at sea-states 0 and 6. The first column illustrates the particle motions within the 1 to 100 Hz frequency band while the remainder of the columns are devoted to 20 Hz bandwidths covering the 100 Hz measurement band. The energy is evidently concentrated in the 1 to 20 Hz bandwidth, for which particle motion is evident in both vertical and horizontal planes, characteristic of interface waves [15,16]. Figure 7b shows the azimuthal power patterns for sea state 0 and 6 of the same event. The patterns of sea-state were analysed with both large ($\Delta f = 5$) and narrow ($\Delta f = 0.5$) frequency bands. The directional patterns indicate the wind direction for which the sea surface waves had the same orientation.

Effects of Shipping

The high frequency region (30 to 100 Hz) is predominantly influenced by ship traffic noise. The dominant shipping of the area was coastal trawlers, fishing boats and pleasure craft (nearby shipping activity) and distant commercial shipping passing through deep water to or from Genova at distances between 20 and 130 n.mi from the station.

Figure 8 illustrates, for different sensors, the superposition of 105 power spectra, that were recorded daily for the study. To simplify this large amount of data we applied basic statistical analysis techniques [17] to obtain average power and mean power levels for each sensor.

Within the 20 to 100 Hz frequency band, various peaks are superimposed over a mean curve which displays a maximum level at about 50 to 60 Hz. These individual or grouped peaks correspond mostly to nearby shipping. Peaks grouped around 40 to 50 Hz reflect the effects of bottom trawlers (Fig. 9).

Even though the hydrophone and geophones respond to two different physical parameters (pressure and particle motion, respectively), we note that the transition regions between low and high frequencies (5 to 15 Hz) of the spectra obtained from the hydrophone and geophones display different behaviour.

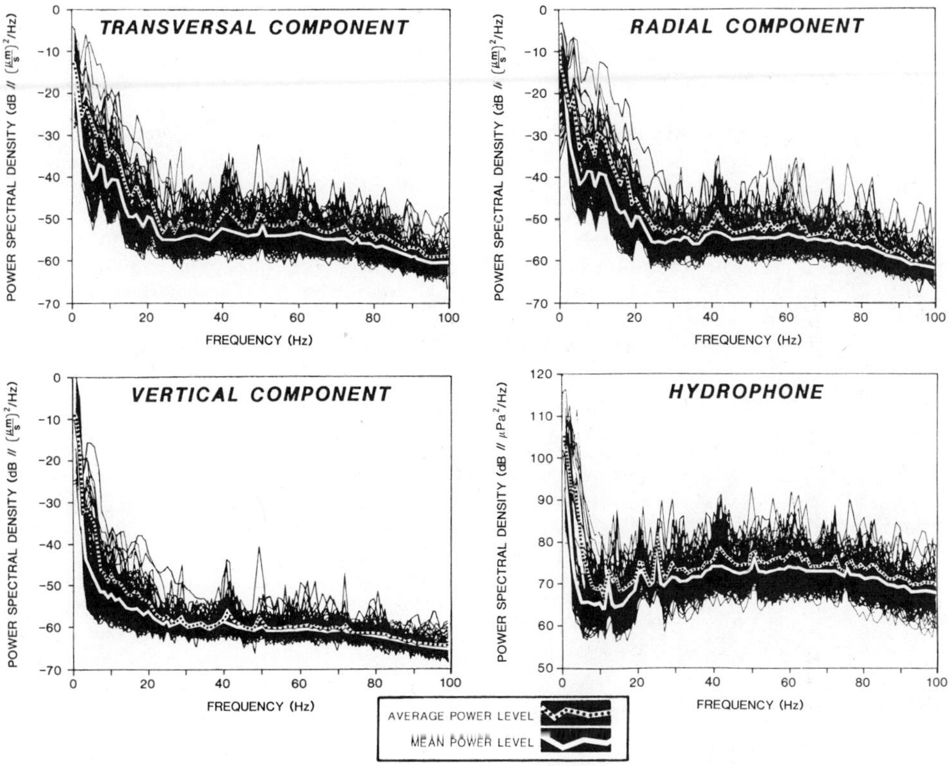

Fig. 8 Records of 105 daily measurements of ambient noise.

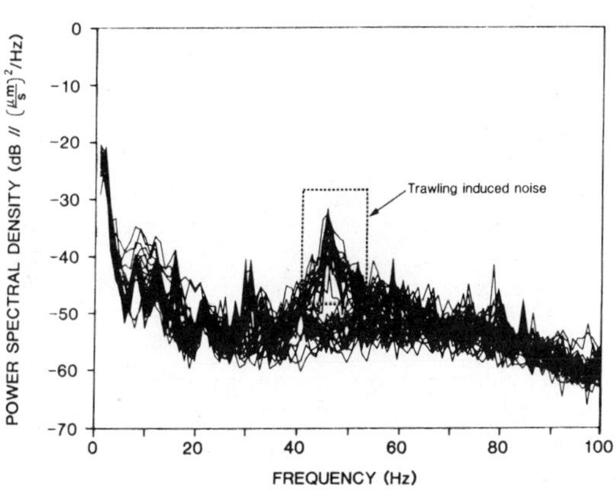

Fig. 9
Spectrum of bottom trawling induced noise.

This difference (low levels on the hydrophone spectra) is probably due to the cut-off of the waterborne propagating noise. For the water depth and the sediment type of the area, the waterborne acoustic propagation cut-off frequency corresponds to 10-15 Hz [18]. Apparently the energy is coupled into the noise field from distant shipping (deep water), and confined to the continental shelf because of the waterborne acoustic frequency cut-off process. In this case, the energy lost within the water column couples into the seismic waves which are detected by the geophones. This phenomenon appears as high energy levels within the 5 to 25 Hz frequency band on the geophone spectra. This observation and the relatively lower levels observed on the vertical geophones as compared with the horizontal ones, indicate that the noise field is mainly related to interface and Love waves.

COMPARISON WITH OTHER DATA

Experimental infrasonic ambient noise data have been reported by others. The data set for sea states 0 and 6 presented in this paper are superimposed (Fig. 10) on Kibblewhite and Ewans [13] data which summarize various other ambient noise data sets. The Mediterranean continental shelf, lacking the long swells of the open oceans, yields lower pressure amplitudes at low frequencies. The same data set compared to Wenz curves [1] within the 10 to 100 Hz band fits into the shallow water region of usual traffic noise (Fig. 11). In the 1 to 10 Hz frequency band, spectrum levels fit nicely into the seismic background and sea surface wave noise.

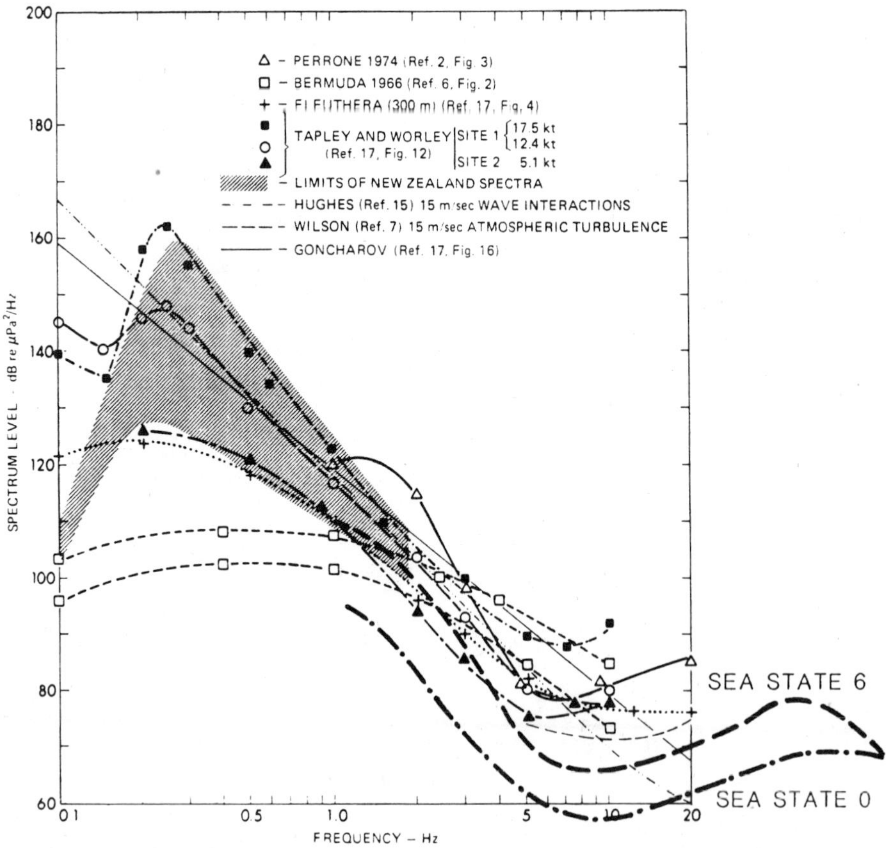

Fig. 10 Comparison of various ambient noise data sets (after [12]).

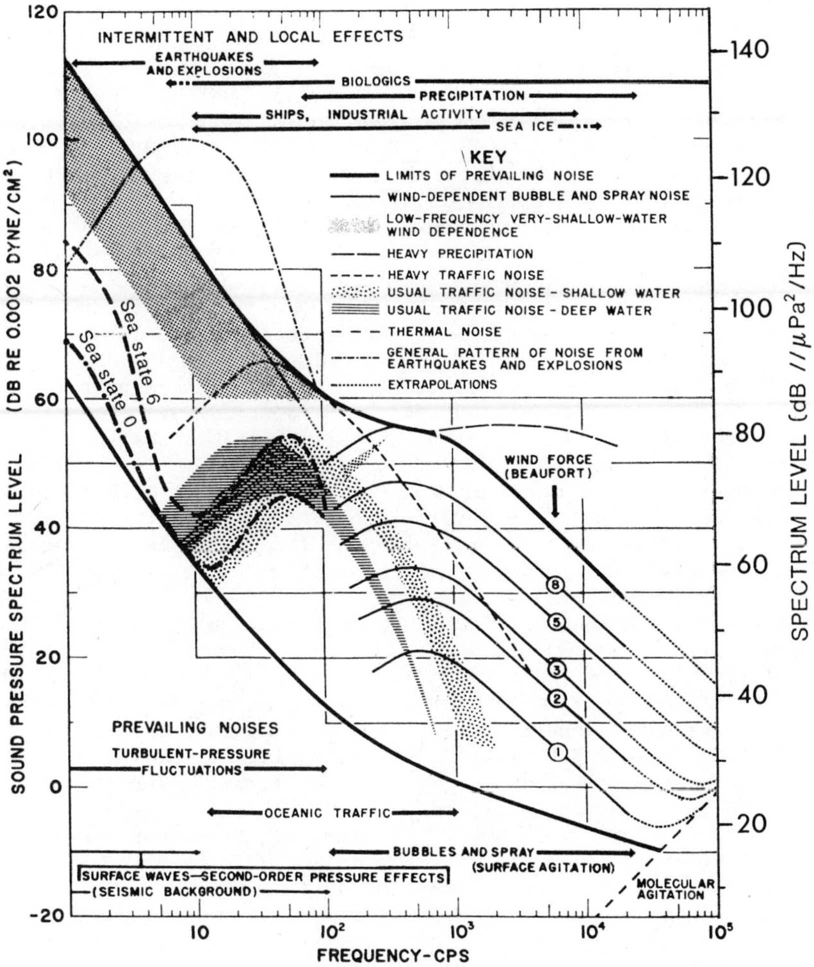

Fig. 11 Comparison of data set for sea-states 0 and 6 with Wenz curves [1].

CONCLUSION

Systematic noise data collected over the 1 to 100 Hz frequency band for a long period revealed that noise levels correlated well with wind-sea conditions and nearby shipping activity.

Meteorological noise overlaps with shipping noise in the transient region of 5 to 25 Hz band. In the same band, the hydrophone (water-borne noise) revealed very low energy levels as compared to the geophones. This effect is most probably caused by the water-borne acoustic frequency cut-off due to the decreasing water depth.

The collection of data for a longer period has produced better noise statistics for the area. The data fit well into previous ambient noise data sets. There is strong evidence that at infrasonic frequencies noise energy propagates essentially as interface waves and ducted Love waves.

REFERENCES

1. G.M. Wenz, "Acoustic Ambient Noise in the Ocean: Spectra and Sources," J. Acoust. Soc. Am. 34, 1936-1956 (1962).
2. A.J. Perrone, "Infrasonic and Low-Frequency Ambient Noise Measurements on the Grand Banks," J. Acoust. Soc. Am. 55, 754-758 (1974).
3. K. Hasselmann, " A Statistical Analysis of the Generation of Microseisms," Rev. Geophys. 1, 177-210 (1963).
4. M.A. Isakovich and B.F. Kur'yanov "Theory of Low-Frequency Noise in the Ocean," Sov. Phys.-Acoust. 16, 49-58 (1970).
5. N. Yen and A.J. Perrone, "Mechanisms and Modelling of Wind-induced Low Frequency Ambient Sea Noise," NUSC Technical Report 5833, Naval Underwater Systems Center, New London, Connecticut, 1979.
6. J.H. Wilson, "Very Low Frequency (VLF) Wind-Generated Noise Produced by Turbulent Pressure Fluctuations in the Atmosphere near the Ocean Surface," J. Acoust. Soc. Am. 66, 1499-1507 (1979).
7. M.S. Longuet-Higgins, "A Theory of the Origin of Microseisms," Philos. Trans. R. Soc. London, Ser. A 243, 1-35 (1950).
8. L.M. Brekhovskikh, "Underwater Sound Waves Generated by Surface Waves in the Ocean," Izv. Atmos. Ocean. Phys. 2, 582-587 (1966).
9. B. Hughes, "Estimates of Underwater Sound (and Infrasound) Produced by Non-linearly Interacting Ocean Waves," J. Acoust. Soc. Am. 60, 1032-1039 (1976).
10. V.V. Goncharov, "Sound Generation in the Ocean by the Interaction of Surface Waves and Turbulence," Izv. Atmos. Ocean. Phys. 6, 710-714 (1970).
11. R.A. Wagstaff, "Low Frequency Ambient Noise in the Deep Sound Channel - The Missing Component," J. Acoust. Soc. Am. 69, 1009-1013 (1981).
12. A.C. Kibblewhite and K.C. Ewans, "Wave-wave interactions, microseisms, and Infrasonic Ambient Noise in the Ocean," Technical Report Applied Research Laboratory of Texas, 1984.
13. A.C. Kibblewhite and K.C. Ewans, "A Study of Ocean and Seismic Noise at Infrasonic Frequencies (in this volume).
14. B. Schmalfeldt, "A Comparison of Seismic and Hydroacoustic Measurements at Very Low Frequencies in Different Shallow Water Areas," (in this volume).
15. D. Rauch, "On the Role of Bottom Interface Waves in Ocean Seismo-Acoustics: A Review," (in this volume).
16. M.N. Toksöz, "Microseisms and an Attempted Application to Exploration," Geophysics 29, 154-177 (1964).
17. R.A. Wagstaff, J.-L. Berrou, "Is Power Averaging the Best Estimator for Undersea Acoustic Data?" In: WAGSTAFF, R.A. and BLUY, O.Z., eds. Underwater ambient noise, proceedings of a conference held at SACLANTCEN on 11-14 May 1982, volume II, unclassified papers. SACLANTCEN CP-32. La Spezia, Italy, SACLANT ASW Research Centre, 1982: 17-1 to 17-16.
18. T. Akal, F. Jensen, "Effects of the Sea-Bed on Acoustic Propagation," In: PACE, N.G., ed., Acoustics and the Sea Bed. Proceedings of the Institute of Acoustics. Underwater Acoustics Group Conference, Bath University, 6-8 April, 1983.

INFRASONIC ATTENUATION AND AMBIENT NOISE

R. H. Mellen* and D. G. Browning+

*PSI Marine Sciences, 95 Trumbull Street
New London, CT 06320
+Naval Underwater Systems Center
New London, CT 06320

ABSTRACT

A number of experiments carried out over the past two decades tend to confirm the existence of two major anomalies at infrasonic frequencies. (i) Long-range propagation measurements in the sound channel show extra attenuation that cannot be the result of medium absorption (ii) Vertical directionality measurements of ambient noise show a broad maximum in the axial direction where a minimum is expected. A possible mechanism for both phenomena is scattering by internal waves or eddies.

INTRODUCTION

Despite recent advances in oceanographic acoustics as well as in computer methods to deal with the complex modeling problems, key questions about propagation in the deep-sound channel remain to be resolved. At infrasonic frequencies, long-range measurements show evidence of attenuation in excess of medium absorption [1,2]. A promising mechanism is volume scattering by inhomogeneities within the thermocline with subsequent loss of acoustic energy in the bottom.

In addition to the excess attenuation, it has been found that the vertical directionality of ambient noise shows a broad maximum in the axial direction where a minimum would be expected [3-6]. Whether the source of the noise be surface dynamics or bottom seismics, the same scattering mechanism might also account for diffusion of energy from the higher ray angles.

ATTENUATION

Figure 1 shows typical experimental attenuation data. Generally, the magnitudes at low frequencies greatly exceed the predicted absorption due to ionic relaxations. This excess can be fitted with an extra component of attenuation that is independent of frequency. If scattering is indeed the cause, ray diffusion theory should be appropriate to describe the transport of energy out of the channel and into the bottom.

In an earlier paper[7], we showed that, on the channel axis, the ray diffusion equation for the Garrett-Munk internal wave model of Fig. 2 can be approximated as Eq. 1.

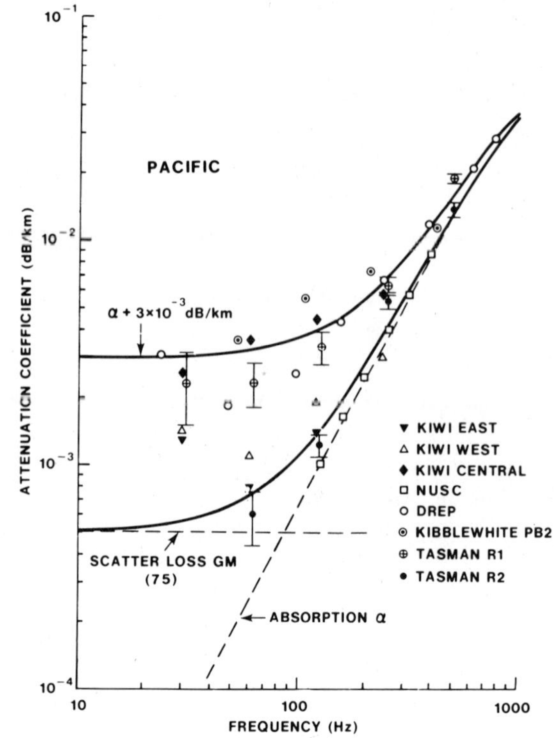

Fig. 1. Attenuation measurements.

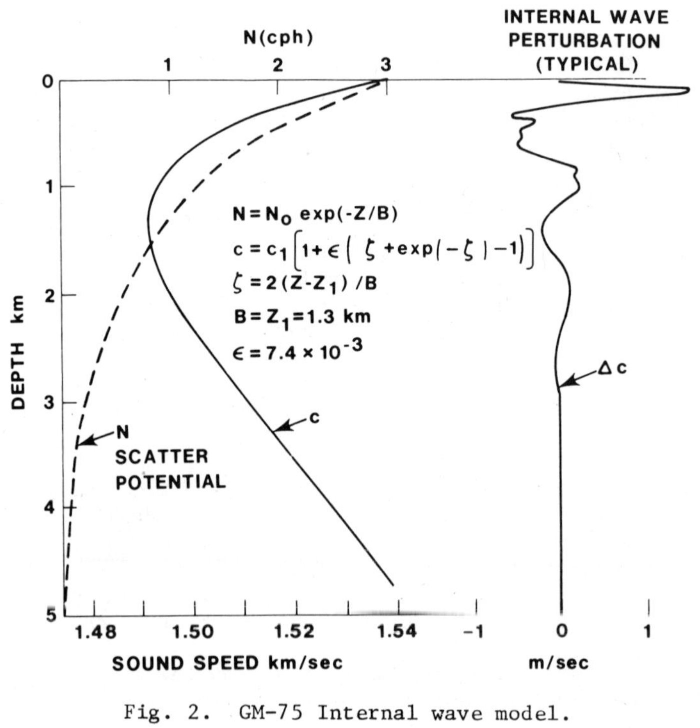

Fig. 2. GM-75 Internal wave model.

$$\frac{\partial W}{\partial r} = D_1(1+\frac{\theta^2}{\theta_o^2})^2 \frac{\partial^2 W}{\partial \theta^2} \qquad (1)$$

where W is the angular density of rays (intensity), D_1 is the diffusion constant, θ is the ray angle and θ_o, the limiting ray angle, all measured at the channel axis. The solution of Eq. 1 with the boundary condition for total loss of rays striking the bottom, i.e., W=0, $|\theta| > \theta_o$ is Eq. 2.

$$W(\theta,r) = \sum_{n=o}^{\infty} A_n e^{-2\alpha_n r} \cos\left(\frac{\pi}{2}(2n+1)\tan^{-1}\left(\frac{\theta}{\theta_o}\right)\right)\left(1+\frac{\theta^2}{\theta_o^2}\right)^{\frac{1}{2}} \qquad (2)$$

where $2\alpha_n=4(2n+1)^2-1$, α is in Nepers per scaled range and r is range. For the GM-75 model we found, for the lowest mode, $\alpha_o \sim 5 \times 10^{-4}$ dB/km. Experimental values as much as 10 times greater have been reported.

AMBIENT NOISE

The ambient noise problem involves both source and propagation effects. Figure 3 shows the ray field vs. range and depth, calculated for a source near the surface. Figure 4 is the angular distribution of intensity expected on the channel axis at long range. The pattern is sharply peaked at high angles because of the large sound-speed increase between axis and source. Rays vertex at the same sound-speed so that axis-crossing angles remain constant and energy density maximizes near the critical angle. Experimental measurements of ambient noise, on the other hand, show a broad maximum on the axis. If the sources are at the surface or bottom, energy-transport is indicated and one possible mechanism is, again, volume scattering.

If the sources are remote, we can obtain the ray-angle distribution from Eq. 2. For high-angle excitation, i.e., $\Delta\theta$ at θ_o, the mode amplitudes are proportional to

$$A_n \propto \int_{\theta_o-\Delta\theta}^{\theta_o} W(\theta)\, d\theta \propto (-1)^n(2n+1) \qquad (3)$$

Substituting Eq. 2 in Eq. 3 and summing over mode numbers for selected values of the range parameter $\alpha_o R$ gives the angular distributions shown in Figure 5. As the scattering loss increases, the higher modes are stripped off leaving only the lowest mode, which is approximately a cosine function of angle. The range required for $\alpha_o R$=1.5 dB is 3000 km for the GM model, while for the experimental value of Figure 1, it is only 500 km.

If sources are uniformly distributed in area, the field can be calculated by integrating Eq. 2 over an area of radius R.

$$W(\theta,R) \propto \int_o^R dr\, W(\theta,r) \qquad (4)$$

Figure 6 shows the results for selected values of $\alpha_o R$. Now the limiting case for $\alpha_o R \to \infty$ is a uniform distribution for $|\theta|<\theta_o$. Ranges required to approach equilibrium in this case are some 10 times greater than for the remote-source case.

CONCLUSION

Internal scattering evidently cannot be dismissed as a contributing mechanism for either propagation loss per se or the effects associated with the vertical directionality of ambient noise. While the ranges for the GM

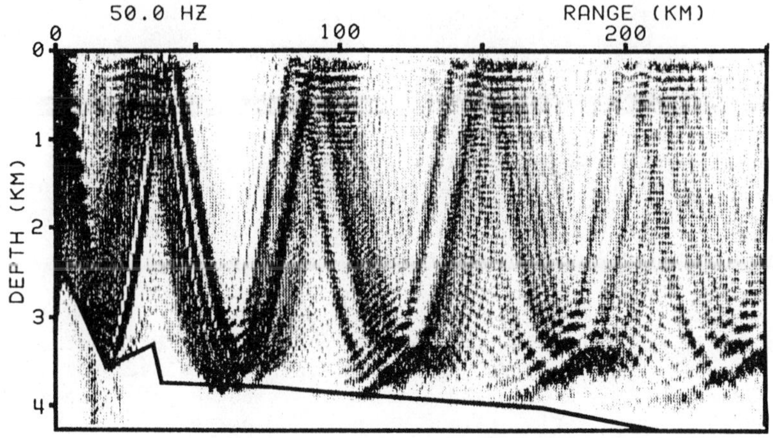

Fig. 3. Ray diagram for a surface source.

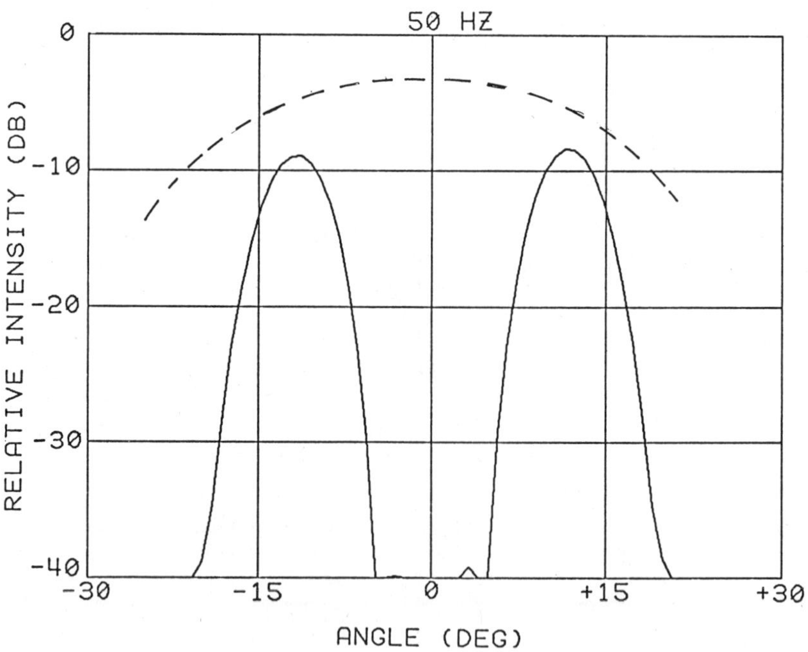

Fig. 4. Vertical directivity at long range for conditions of Fig. 3 compared with experiment (dashed line).

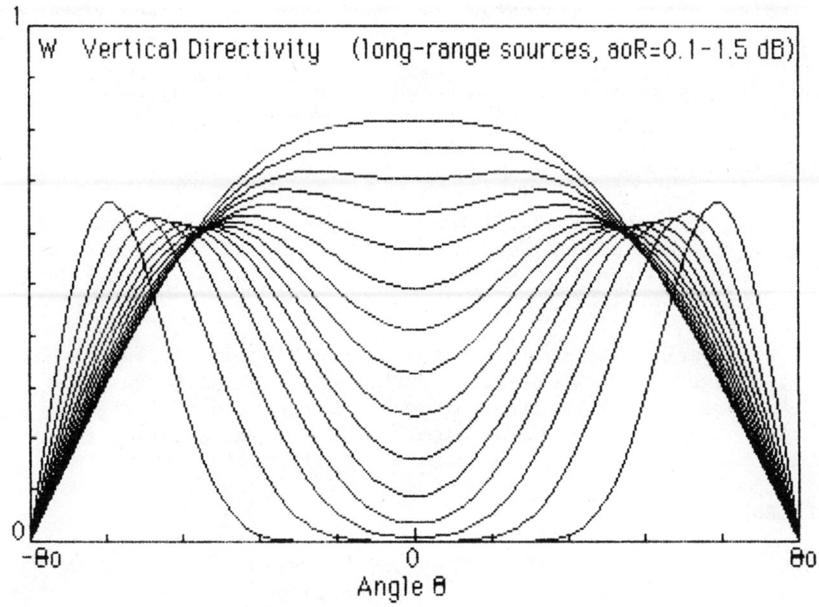

Fig. 5. Vertical directivity of distant sources for values of the diffusion loss parameter α_0R=0.1-1.5dB.

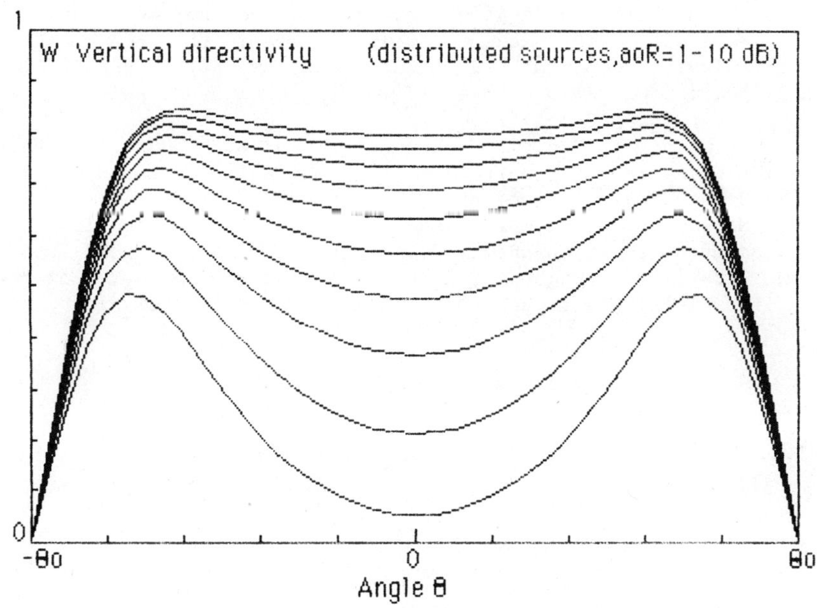

Fig. 6. Vertical directivity of distributed sources for values of the diffusion loss parameter α_0R=1-10dB.

internal-wave model may be prohibitive, some other scattering mechanism e.g., eddies, could explain the reasonable consistency of experimental results. Furthermore, the predictions may be overly pessimistic, i.e., for sources near ocean boundaries, where the local sound-speed can be closer to the channel-axis value, the equilibration process will be more rapid.

Candidate source mechanisms, in addition to shipping, include bottom seismics, surface waves and microseisms produced by them. These and other related matters are discussed in the paper by Carey et al.[8], elsewhere in this volume.

It should be noted that, although shipping may be the dominant factor in some regions, it cannot be the only source of noise. Browning et al.[6] found similar levels and directional properties in the South Fiji Basin where traffic is negligible.

The final answer to the ambient noise problem may include a mix of many factors; however, the similarities rather than the differences in regional dependence are most impressive. A more detailed study of other statistical properties of the ambient noise might reveal, unambiguously, the origin of the noise floor as well as the means of equilibration.

REFERENCES

1. A. C. Kibblewhite and L. D. Hampton, "A review of deep ocean sound attenuation data at very low frequencies", J. Acoust. Soc. Am. 67, 147-157 (1981)
2. K. C. Focke, S. K. Mitchell and C. W. Horton Sr., "Analysis of deep ocean sound attenuation at very low frequencies", J. Acoust. Soc. Am. 71, 1438-1444 (1982)
3. V. C. Anderson, "Variation of the vertical directionality of noise with depth in the North Pacific", J. Acoust. Soc. Am. 66, 1446-1014 (1979)
4. R. A. Wagstaff, "Low-frequency ambient noise in the deep sound channel: the missing component", J. Acoust. Soc. Am. 69, 1009-1014 (1981)
5. R. Dashen and W. Munk, "Three models of global ocean noise", J. Acoust. Soc. Am. 76, 540-554 (1984)
6. D. G. Browning, N. Yen, R. W. Bannister, and K. M. Guthrie, "Vertical directionality of low-frequency ambient noise in the South Fiji Basin" NUSC Technical Document 6611, 21 Jan 1982
7. R. H. Mellen, D. G. Browning, and L. Goodman, "Diffusion loss in a stratified sound channel", J. Acoust. Soc. Am. 60, 1053-1055 (1976)
8. W. M. Carey, R. A. Wagstaff, B. Brunson, and M. Bradley, "Low-frequency noise fields and signal characteristics", (in this volume)

PARAMETRIC SOURCES - DESIGN CONSIDERATIONS IN THE GENERATION OF LOW-FREQUENCY SIGNALS

H.O. Berktay

School of Physics
University of Bath
Claverton Down
Bath, BA2 7AY

INTRODUCTION

The usefulness of parametric sources in the realisation of narrow acoustic beams with wide bandwidth capability has become well established. Their use in systems as well as for experimental work both at sea and in laboratory scale experiments can be illustrated by many examples.

In this paper, some properties of parametric sources (relevant in their design and use) are discussed and some sample designs are presented for lower frequency, long range applications, mainly to illustrate a design philosophy for such sources, but also to demonstrate that narrow beam low frequency sources can be realised with source levels sufficient for many applications.

SOME GENERAL CONSIDERATIONS

Nonlinear interactions in an acoustic pressure field $p_i(\underline{r},t)$, say, results in a secondary radiation from each field point $\underline{r}$. Westervelt[1] ascribed this to virtual simple volumetric sources of volume density $q(\underline{r},t)$ given by the equation

$$q(\underline{r},t) = (\beta/\rho_o^2 c_o^4) \frac{\partial}{\partial t} p_i^2(\underline{r},t). \qquad (1)$$

If the transmitted signal is monochromatic, as a consequence of previous distortion $p_i(\underline{r},t)$ consists of the fundamental frequency and its harmonics. Then its square will give rise to further generation of harmonics and fundamental frequency terms.

If the time dependence of the primary pressure field is band-limited, then the second-order interaction implied in Eq. 1 between the various frequency components produces a low-frequency component in the field. This low-frequency component produced by the interaction of the high-frequency primary field is utilised in parametric sources.

Specifically, a band-limited pressure at $\underline{r}$ may be represented in the form

$$p_i(\underline{r},t) = P(\underline{r}).f(t') \cos[\omega_o t' + \gamma(\underline{r})] \tag{2}$$

where t' indicates delayed time and f(t) is a slowly varying envelope function. Then the low-frequency component of q, $q_-(\underline{r},t)$, is of the form

$$q_-(\underline{r},t) = \frac{\beta P^2(\underline{r})}{2\rho_o^2 c_o^4} \frac{\partial}{\partial t} f^2(t'). \tag{3}$$

Thus, the low-frequency source-strength is independent of the primary frequencies, but is determined by the envelope function which controls the signal spectrum.

In the quasi-linear (or single-scattering) approach used by Westervelt, the low-frequency waves radiated from the secondary-sources are assumed to propagate linearly. (It is worth noting that experimental results substantiate this assumption.) Then, ignoring absorption at the lower frequencies, the low-frequency field at a point $\underline{R}$ can be evaluated in the form of a scattering integral.

$$P_-(\underline{R},t) = \frac{\beta}{8\pi\rho_o c_o^4} \iiint \frac{P^2(\underline{r})}{s} dv \frac{\partial^2}{\partial t^2} f^2(t' - s/c_o) \tag{4}$$

where $\underline{s} = \underline{R} - \underline{r}$, and the integral is evaluated over the complete source volume.

We may consider, as an example, the collimated end-fired array[1] where the primary wave is assumed to be a plane-wave confined within a column of cross-sectional area S the dimensions of which are small compared with the wavelength of the low-frequencies of interest. (See Fig. 1).

Then, the primary wave may be represented in the form

$$p_i(\underline{r},t) = P_o f(t') \cos\omega_o t'.\ e^{-\alpha_o z}, \tag{5}$$

within the column, with $t' = t - z/c_o$. Substituting in Eq. 4 after x,y integration, we obtain

$$p_-(\underline{R},t) = \frac{\beta S P_o^2}{8\pi\rho_o c_o^4} \int \frac{dz}{s} \frac{\partial^2}{\partial t^2} f^2(t - \frac{z+s}{c_o}).e^{-2\alpha_o z} \tag{6}$$

If the field point is on the acoustic axis and at a range R such that $R >> 1/2\alpha_o$, then $s = R - z$ and $s^{-1} \simeq R^{-1}$, and Eq. 6 gives[2]

$$p_-(R,t) = \frac{\beta S P_o^2}{8\pi\rho_o c_o^4 (2\alpha_o) R} \frac{\partial^2}{\partial t^2} f^2(t - R/c_o) \tag{7}$$

(Although the collimated wave concept is an idealised model, a similar result can be obtained by considering the nonlinear interaction to occur mainly in the far field of the transducer.)

The result shown in Eq. 7 leads to a useful way of utilising a parametric source to produce wide-band pulses the spectrum of which can be controlled readily by changing the duration of the transmitted pulse.

If, for example, a "raised cosine" envelope function is used,

$$f(t) = \frac{1}{2} (1 + \cos \frac{2\pi t}{T}) \tag{8}$$

for $|t| < T/2$ and zero for $|t| > T/2$, then, the low-frequency waveform for $|t'| < T/2$ is of the form

$$\frac{\partial^2}{\partial t^2} f^2(t') = -2(\pi/T) (\cos \frac{2\pi t'}{T} + \cos \frac{4\pi t'}{T}) \tag{9}$$

Thus, the low-frequency pressure wave (on axis) would consist of one cycle of frequency T^{-1} and 2 cycles of frequency $2T^{-1}$, with the resulting broad bandwidth. As the minimum practicable value of T is controlled by the bandwidth of the transducer, the presence of $2T^{-1}$ frequency term indicates that even broader signal bandwidth is available in a parametric source than that of the transducer.

The broad band signals which can be obtained in this manner can be used <u>either</u> to obtain the wide-band response of channels of propagation or of acoustic materials <u>or</u> to generate narrow-band signals around a repetition frequency and around its harmonics by transmitting a number of the short pulses within a required pulse duration.

If the pulses are to be used singly to measure a parameter over a wide band, at closely spaced frequencies, integration over a number of pulses improves the signal-to-noise ratio and leads to improved accuracy of parameter estimation.

If, on the other hand, a pulse of 100 ms duration is required for a system to work at 5 kHz for example, a burst of 500 short pulses triggered from the positive-going zero crossings of a 5 kHz square-wave would suffice. (Use may also be made of the harmonics of 5 kHz which will be available.) To maximise the energy content, the pulses would be made contiguous.

This technique can be extended to generate broad-band signals of long duration for use in pulse-compression. For example, triggering the short pulses at the positive-going zero crossings of a "chirp" waveform will produce linear-FM signals around the fundamental and around each of the harmonics. (Clearly, the bandwidth available in each case approaches the fundamental frequency.)

A significant point to note is that the harmonics of the transmitted carrier frequency build up more rapidly with distance than the low frequency component. (This can be seen from the expression for the source strength given in Eq. 1, where the time derivative indicates that its magnitude is proportional to the frequency of any component in $p_i^2(\underline{r},t)$.) However, even if the waveform within the envelope is distorted by the presence of the harmonics of the carrier the low-frequency term is controlled by the square of the envelope function which, itself, may be distorted as a result of higher-order interactions.

Although this time-domain approach is attractive, the evaluation of the scattering integral given in Eq. 4 is best effected in the frequency domain, especially to allow for diffraction effects both in the primary and in the secondary fields. In this context, it seems easier to think in terms of a low-frequency component consisting of a single low frequency sinusoidal wave.

If $p_i(\underline{r},t)$ consists of two sinusoids at frequencies $\omega_o \pm \omega_d/2$, with with $\omega_d << \omega_o$, then its square and hence the source strength $q(\underline{r},t)$ calculated from Eq. 1 will have a component at the difference frequency ω_d.

This is the component which is specifically utilised in a difference frequency parametric source, and its properties are considered in the following section.

In practice, the difference-frequency waves are accompanied by the harmonics of ω_d. A simple explanation may be found by considering the harmonics and high-frequency intermodulation products of the primary frequencies,

$$\pm\, m\,(\omega_o + \omega_d/2),\ \pm\, n\,(\omega_o - \omega_d/2),$$

which build up more rapidly with distance than the difference-frequency component. With m and n integers, clearly additional low-frequency terms will be generated (for m = n) at the harmonics of ω_d.

It is possible to treat the difference-frequency mode of operation of the parametric source as a special case of the pulse-mode of operation discussed above.

If the amplitudes of the two primary waves are assumed to be equal, the envelope function becomes

$$f(t) = |\cos(\omega_d t/2)|, \tag{10}$$

leading to

$$\frac{\partial^2}{\partial t^2} f^2(t) = -(\omega_d^2/2)\cos\omega_d t, \tag{11}$$

which, with time retardation, may be used in Eq. 4 to calculate the difference-frequency field.

As an example, we may consider the end-fired array result given in Eq. 6. As f(t) given in Eq. 10 has a maximum value of unity, the pressure amplitude P_o in Eq. 6 must correspond to the sum of the amplitudes of the two sinusoidal terms, 2 P_a say. Then, for a field point $\underline{R}$ in the far-field (i.e., $R \gg 1/2\alpha_o$) the result reduces to that given by Westervelt[1], which is (in complex form)

$$P_d(\underline{R}) = -(\omega_d^2 \beta S P_a^2/4\pi\rho_o c_o^4 R).e^{-(\alpha_d+jk_d)R}$$

$$\times\ \{\alpha_T + jk_d(1-\cos\Theta)\}^{-1} \tag{12}$$

Here, absorption at the difference frequency (α_d) is also introduced, the absorption parameter α_T is $2\alpha_o - \alpha_d$, and k_d is the wave-number at the difference-frequency.

The directivity of the difference-frequency field can be seen to be controlled by the absorption of the primary waves per wavelength at the difference-frequency. The half-power beamwidth is given by, approximately, the following expression:-

$$\Theta_d \simeq 2\surd(2\alpha_T/k_d)$$

$$\text{i.e.,}\quad \simeq (2/\surd\pi)\ \surd(\alpha_T\lambda_d), \tag{13}$$

with λ_d the wavelength at the difference-frequency.

Further, the directivity function in Eq. 12 varies monotonically with angle Θ, having no side-lobes - a direct consequence of the exponential amplitude taper along the source column.

If the dimensions of the column cross-section are comparable with or greater than λ_d, integration for x and y results in an "aperture" function which, being independent of z for large R, simply multiplies the directivity function in Eq. 12.

The more practical case where the primary waves are projected from a transducer will be considered in the next section. It is worth noting, however, that when the bulk of the nonlinear interaction takes place in the farfield of the transducer, the spherically-spreading primary fields (and hence the secondary sources) can be considered to be made up of conical "ray" tubes distributed in space, the source strength associated with each tube being determined by the square of the primary directivity function in the direction of the ray. Then, each tube can be treated as a "line source" with exponential taper, and the difference-frequency field can be evaluated as the two-dimensional (complex) convolution of the directivity function in Eq. 12 with the square of the directivity function of the primary waves. In which case, if the beamwidth Θ_d in Eq. 13 is much less than the primary beamwidth, the directivity function at the difference-frequency would be expected to follow substantially the square of the primary directivity function, independently of ω_d. This results in a capability in the design of a parametric source to obtain wide bandwidth signals with nearly constant beamwidth over the frequency range of interest.

If, on the other hand, Θ_d is much greater than the primary beamwidth, the convolution process mentioned above yields substantially the result given in Eq. 12.

DIFFERENCE-FREQUENCY PARAMETRIC SOURCES

In this section we consider the primary field to consist of two single-frequency components, at $\omega_{1,2} = \omega_o \pm \omega_d/2$, both radiated simultaneously from a transducer. Further, extra-attenuation due to higher-order nonlinear interactions is ignored.

Then, if the complex amplitudes of the primary pressures at point $\underline{r}$ are $p_{1,2}(\underline{r})$, the source strength at the difference-frequency (from Eq. 1) becomes

$$q_d(\underline{r}) = j\omega_d \, \beta \, p_1(\underline{r}) \, p_2{}^*(\underline{r})/\rho_o^2 c_o^4 \, , \tag{14}$$

where the asterisk is used to indicate the conjugate of a complex quantity. Then the difference-frequency pressure at a field point $\underline{R} = \underline{r} + \underline{s}$ can be expressed in the form of a scattering integral

$$p_d(\underline{R}) = -\,(\omega_d^2\beta/4\pi\rho_o c_o^4) \iiint \frac{p_1(\underline{r})p_2{}^*(r)}{s} \, e^{-(\alpha_d+jk_d)s} \, dv \tag{15}$$

where the volume integral is effected over the whole interaction region.

By evaluating the primary pressure fields exactly and evaluating the integral numerically, the difference-frequency pressure field can be calculated using Eq. 15. Very good agreement obtained between the computed and experimental results[3,4] confirms the validity of the assumptions and of the basic approach used in deriving Eq. 15.

Thus, for a given set of parameters, the difference-frequency field can be evaluated numerically, providing accurate estimates. However (by using simplified models for the primary field) generalised results can be obtained, to provide bases for the design of parametric sources. The results discussed here have been found to be in good agreement with experimental data.

For the field point well in the far-field of the transducer, the primary waves may be assumed to be spreading spherically from the transducer. (Basically, this model ignores the "aperture factor" associated with the "collimated" region of the primary waves[5], in the near-field of the transducer.)

Then, the primary waves may be written in the form

$$p_{1,2}(\underline{r}) - j(P_{1,2}/r)D_{1,2}(\gamma,\eta)\ \exp\left[-(\alpha_{1,2} + jk_{1,2})r\right] \qquad (16)$$

where $D(\gamma,\eta)$ represents the normalised directivity functions in the direction $\underline{r}(\gamma,\eta)$. For $\omega_d << \omega_o$, the directivity function and absorption at ω_o may be used without loss for those at $\omega_{1,2}$. Substituting in Eq. 15 and using $d\Omega$ to indicate elemental solid angle (i.e., $dv = r^2 d\Omega dr$) we obtain

$$p_d(\underline{R}) \simeq -(\omega_d^2\beta/4\rho\pi_o c_o^4)P_1P_2 \times$$

$$\iint D_o(\gamma,\eta)d\Omega \int_0^\infty \frac{dr}{s}\ \exp\{-(2\alpha_o + jk_d)_r - (\alpha_d + jk_d)s\} \qquad (17)$$

with $\underline{s} = \underline{R} - \underline{r}$.

If $\underline{R}$ is in the far-field of the parametric sources,

$$s = R - \underline{\ell}.\underline{r}$$

where $\underline{\ell}$ is the unit vector in the direction of R. Then, integration for r yields

$$p_d(\underline{R}) \simeq p_w(R,0)(S/\lambda_o^2)\iint D_o^2(\gamma,\eta)\ d\Omega.\{1 + j\frac{kd}{T}\ (1 - \cos\xi)\} \qquad (18)$$

Here, $p_w(R,0)$ is Westervelt's result for a collimated parametric source, given in Eq. 12, for $\Theta = 0$ (i.e., when $\underline{R}$ is along the acoustic axis) and ξ is the angle between the source point $(\underline{r})$ and the field point $(\underline{R})$. Also, from the definitions of the primary fields in Eqs. 5 and 15, use has been made of the relationship $P_{1,2} = P_a S/\lambda_o$ where S is the area of the transducer.)

The evaluation of the integral in Eq. 18 is discussed in Ref. 6 for a circular transducer and for rectangular transducers of different aspect ratios, and it is shown that expressions can be developed in normalised forms for the evaluation of the source level and the half-power beamwidth(s) of parametric sources in the plane(s) of symmetry of the transducer. These normalised results are presented here as functions of a single parameter $\alpha_T r_o$ where r_o is the "scaled Rayleigh distance" given by the expression

$$r_o = (\lambda_d/\lambda_o)R_o,$$

$$\text{i.e.,} \quad = (\omega_o/\omega_d)R_o \qquad (19)$$

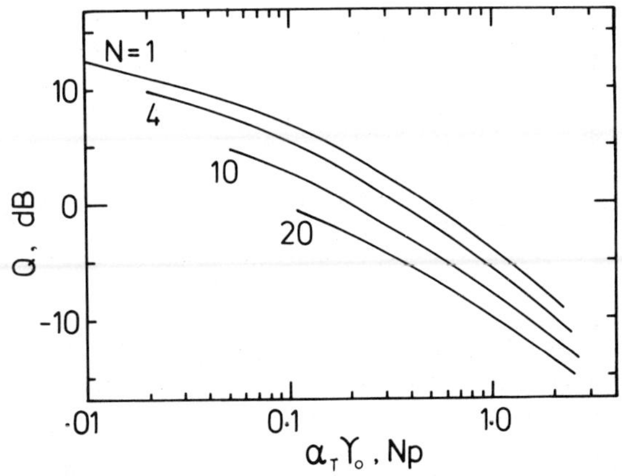

Fig. 1. Variation of the parameter Q with $\alpha_T r_o$ for some transducer aspect ratios, N.

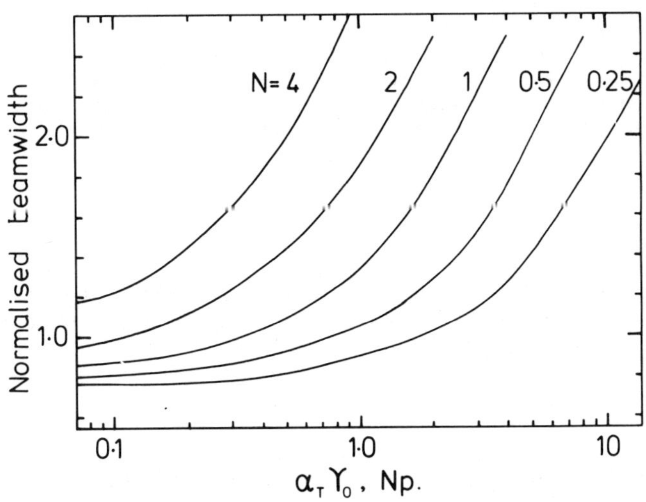

Fig. 2. Normalised beamwidth(s) of parametric sources for some transducer aspect ratios, N and 1/N.

The curves in Fig. 1 give a "source-level coefficient" Q as a function of $\alpha_T r_o$ only for rectangular transducers of aspect ratio N. The curve for N = 1 can be used for a square or for a circular transducer. This parameter Q is a logarithmic measure of the magnitude of the double integral in Eq. 18 (evaluated for a field point along the acoustic axis) divided by $\alpha_T r_o$. Using this parameter, the far-field source level of the parametric source can be calculated from the equation

$$(SL)_d = (SL)_1 + (SL)_2 + 20 \log F_d + Q - 286.5 \text{ dB rel}\mu\text{Pa m} \quad (20)$$

where all the source levels are in dB relμPa m (r.m.s.) and F_d is the difference-frequency in kHz.

The curves in Fig. 2 show the difference-frequency beamwidth normalised with respect to the corresponding primary beamwidth as a function of $\alpha_T r_o$, for various aspect ratios for rectangular transducers. The curve for $N \cong 1$ is, again, applicable to square or circular transducers. For example, the curves for N = 4 and N = 0.25 give, respectively, the normalised beamwidths in the planes including the long and the short axis of symmetry for a rectangular transducer of aspect ratio 4.

The curve in Fig. 2 for N = 1 shows that for small values of $\alpha_T r_o$ the difference-frequency beamwidth Θ_o approaches that associated with the square of the directivity function, i.e., $\Theta_o \rightarrow 0.72\ \gamma_B$. Also, for $\alpha_T r_o > 1$ Np, Θ_o approaches the value Θ_d as given in Eq. 13 (i.e., $\Theta_o/\gamma_B \rightarrow 1.25\sqrt{(\alpha_T r_o)}$).

Thus, if a parametric source with a conical beam is designed for a value of $\alpha_T r_o$ not exceeding about 0.35 Np at the lowest-frequency of interest, at higher difference-frequencies r_o and hence $\alpha_T r_o$ will be reduced, producing only a small change in the beamwidth Θ_o. In this region, Θ_o is largely controlled by the primary beamwidth γ_B.

DESIGN CONSIDERATIONS AND SAMPLE DESIGNS

In the foregoing, the parameters affecting the characteristics of a parametric source were briefly presented. Using the information given, these characteristics can be calculated for given transducer dimensions and primary frequency if the primary source levels are low enough to avoid high-intensity effects. (Otherwise, the results of refs. 7 and 8 can be used for this purpose.)

If, however, a parametric source is required to satisfy a given specification, there is no unique design. In this section, it is proposed to put forward some considerations leading to a design procedure, with some sample designs which are realistic "first-cut" attempts to show the advantages and the disadvantages of various solutions.

Low-frequency Conical Beams for Long Range Propagation

We consider a requirement for a parametric source to produce a beam of width, say, 3 degrees at 1 kHz. There is a further requirement that it be usable over a frequency range extending, say, from 300 Hz to 4 or 5 kHz.

Referring to Fig. 2, we note that the required primary beamwidth, γ_B, can be obtained for any value of the parameter $\alpha_T r_o$ that may be selected as the required difference-frequency beamwidth, Θ_o, is known.

Noting that for a given primary frequency and transducer size the "scaled" nearfield distance r_o (and hence $\alpha_T r_o$) is inversely proportional to the difference frequency, the design value may be selected to minimise the change in the beamwidth Θ_o over the frequency range of interest.

One of the values of $\alpha_T r_o$ selected in this study was 0.35 Np at 1 kHz. From Fig. 4, one obtains for the normalised beamwidth Θ_o/γ_B = 1.0, and this gives γ_B = 3 degrees for the primary beamwidth.

The expression for r_o (Eq. 19) can be re-cast in terms of γ_B. With

$$\gamma_B \simeq (3.23/\pi)(\lambda_o/D) \text{ radians}, \qquad (21)$$

where D is the diameter of the transducer, and

$$r_o \simeq (\pi/4)(3.23/\pi\gamma_B)^2\lambda_d. \qquad (22)$$

Thus, with $\lambda_d \simeq 1.5$ m at 1 kHz, $r_o \simeq 454$ m. With $\alpha_T r_o$ known and $\alpha_T \simeq 2\alpha_o$, α_o can now be calculated, the primary frequency determined and the transducer diameter D can be obtained from Eq. 21. (For the present calculations a sound velocity of 1500 ms^{-1} and absorption coefficients for a sea temperature of 10°C and salinity 35% have been used.)

The results obtained for 3 different values of $\alpha_T r_o$ at 1 kHz are shown in Table 1, together with the results calculated in each case for difference-frequencies of 300 Hz and 5 kHz. As indicated above, the values of $\alpha_T r_o$ for these frequencies were obtained by dividing the design value for 1 kHz shown in Table 1 by the difference-frequency in kHz.

The difference-frequency source levels were estimated using Eq. 20, with the appropriate value of Q obtained from Fig. 1. The primary source levels were assumed to be equal and were estimated in the manner indicated in the Appendix.

The results of a similar study for a parametric source with a 6 degree beamwidth at 1 kHz are shown in Table 2.

From a study of Tables 1 and 2 the following conclusions appear to be in order.

i. There is no unique design for a parametric source to fulfill given requirements, but the design can be optimised to comply with other constraints, e.g., for higher source level requirements, limits on maximum transducer size etc.

ii. In general, the lower values of $\alpha_T r_o$ result in a lower carrier frequency, larger transducer size and a smaller variation of the beamwidth over a wide range of difference-frequency, as well as higher source levels.

Some Effects due to the Truncation of Parametric Sources

A natural application for a parametric source is in sub-bottom profiling. Here, if the height of the transducer above the sea-bed is less than the "far-field distance", the full performance of the source is not realised.

For example, it is attractive to think of the source for $\alpha_T r_o$ = 0.1 Np in Table 2 for medium penetration into the sea-bed. If the height above the sea-bed is to be, say, 50 m the results for the source level and the beamwidth are of no use. In this case, $r_o \simeq 87$ m at 1 kHz and 290 m at 300 Hz, and (with the small value of $\alpha_T r_o$ used) the most significant part of the difference-frequency field is generated at ranges beyond r_o in each case.

Table 1. Parametric Sources for 3 Degree Beamwidth at 1 kHz.

$\alpha_T r_o$,Np	f_o,kHz	Dia.,m	$(SL)_d$ μPa m			Θ_o, Deg.	
			1 kHz	300 Hz	5 kHz	300 Hz	5 kHz
.10	12	2.2	209	194	(226)	3.4	(2.8)
.35	20	1.5	199	183	219	4.2	2.6
.63	25	1.3	193	176	213	4.7	2.3

Table 2. Parametric Sources for 6 Degree Beamwidth at 1 kHz.

$\alpha_T r_o$,Np	f_o,kHz	Dia.,m	$(SL)_d$ μPa m			Θ_o, Deg.	
			1 kHz	300 Hz	5 kHz	300 Hz	5 kHz
.10	26	0.50	194	179	212	6.9	5.7
.35	45	0.33	181	165	202	7.2	5.2
.63	55	0.30	177	159	197	9.7	4.8

A good approximation to this truncated source can be obtained by modelling it as a truncated line array with exponential taper and end-fire phasing, provided the field point is at some distance from the truncation. Then, it can be shown that the array behaves as if it had a source level given by Eq. 22 but with Q replaced by the expression

$$20 \log \left[\{1 - \exp(-\alpha_T H)\}/\alpha_T r_o\right]$$

where H is the length of the truncated array.

In the present example, the values of Q become -5 dB and -16 dB, respectively, at 1 kHz and at 300 Hz, instead of the corresponding far-field values of 6 dB and 2 dB. Thus the effective source levels at the two difference-frequencies are, respectively, 11 dB and 18 dB below their far-field values. Further, the beamwidth associated with the source can be shown to increase to about 18.7 degrees at 1 kHz and 34 degrees at 300 Hz.

Clearly the design basis must be selected in this case to make best use of the water column available - even then, if narrow beams are required at such long wavelengths for penetration purposes, a longer water column becomes essential.

For shallow penetration, a higher difference-frequency (10 - 15 kHz, say) can be used, permitting close-range working with the sub-bottom profiler.

Low-frequency Rectangular Beams

The design of parametric sources using rectangular transducers is complicated by the need to determine the transducer aspect ratio for given set of beamwidths at the difference frequency.

Instead, some examples are shown in Table 3, indicating what may be achievable using rectangular transducers. In each case a difference-frequency of 1 kHz is considered, with a beamwidth of 3 degrees in the narrower aspect of the beam. The curves in Fig. 2 for the selected aspect ratio are used to determine the normalised beamwidths (Θ_x/γ_x and Θ_y/γ_y, say) and with Θ_y = 3 degrees, γ_y and then, with the known aspect ratio, γ_x is evaluated, yielding Θ_x. From $\gamma_{x,y} = (2.78/\pi)(M_{x,y}/\lambda_o)$ the dimensions of the transducer in terms of λ_o and from the expression

$$r_o = (M_x/\lambda_o)(M_y/\lambda_o)\lambda_d$$

the value of r_o is determined. Thus, from the value of $\alpha_T r_o$ used, absorption coefficient α_o and hence the primary frequency are determined.

The primary source levels used in Eq. 20 were determined, taking into account the fact that spherical spreading regime for a rectangular transducer can be taken to start at a range $N R_o$, where N is the aspect ratio (greater than unity) and R_o is the Rayleigh distance, $(M_x M_y/\lambda_o)$. The primary source levels were estimated in the manner discussed in the Appendix, using Eq. A3 to determine the effective radius of an "equivalent" spherical wave to allow for the distortion of the wave up to the range $N R_o$.

By comparing the examples shown in Tables 1 and 3 it can be seen that near-conical beams can be produced by using rectangular transducers, with significant savings in transducer area.

It is also worth noting that, if required, rectangular even "fan" beams can be obtained using larger aspect ratios.

Table 3. Some Parametric Sources at 1 kHz Difference-frequency, using Rectangular Transducers.

$\alpha_T r_o$,Np	Aspect Ratio N	f_o,kHz	$M_{x,y}$ m	$\Theta_{y,x}$ Deg.	$(SL)_d$,μPa m
.1	2	16	.8 x 1.6	3 x 4.8	206
.1	4	15.5	.7 x 2.8	3 x 7.5	205
.35	2	24	.7 x 1.3	3 x 4.2	197
.35	4	26	.4 x 1.7	3 x 5.6	195

DISCUSSION AND CONCLUSIONS

In the foregoing, the source-levels and beamwidths of absorption-limited parametric sources in the far-field of the source volume are discussed and a rather personalised method of designing parametric sources is presented. This rather brief presentation has concentrated on low-frequency sources mainly for long range applications.

Perhaps it is worth emphasising that essentially, as indicated by the scattering integral in Eq. 15, the characteristics of a parametric source are due to radiation from simple "virtual" sources the spatial distribution of which is controlled by the primary field. Thus, to obtain a narrow beam at a long wavelength a long water column is required, as well as a relatively low primary frequency associated with a small absorption coefficient.

The design procedure proposed is based upon the choice of a single parameter, $\alpha_T r_o$, which is demonstrated to control the parameters of the device. Optimisation of the design must then depend on other constraints or requirements facing the designer. The data provided in Figs. 1 and 2 can be used to estimate the variations which may be expected in the parameters of a particular design due to anticipated changes in environmental conditions and hence in the absorption coefficient and in the sound velocity in the medium.

As ever, the author's view is that when a coherent source is required to produce a narrow acoustic beam at a low frequency with a wide bandwidth it is worthwhile to investigate the advantages and disadvantages of a design based on a parametric source as well as any alternative solutions to the problem.

APPENDIX: HIGH INTENSITY EFFECTS

Single-frequency Transmission

The single-scattering approach used in the discussion of parametric sources ignores higher-order nonlinear interactions in the medium. As an initially sinusoidal wave of finite-amplitude propagates, the waveform is distorted (harmonics being generated) and eventually it may develop a shock-front. Absorption in the medium affects the distortion process by reducing the energy available in the fundamental and the harmonic frequency components for further nonlinear interaction.

Until a shock-front is formed, the "extra" attenuation in the fundamental component of the waveform is small. However, beyond the range at which a shock-front is formed, the waveform is attenuated rapidly, causing the fundamental component to be reduced in amplitude as well.

As this attenuation process continues, the rate of absorption of the fundamental and the harmonics become more significant than the nonlinear interactions and beyond that range the wave may be assumed to be propagating linearly.

The nonlinear propagation of plane, cylindrical and spherical waves in an absorbing medium has been analysed at length[9-12].

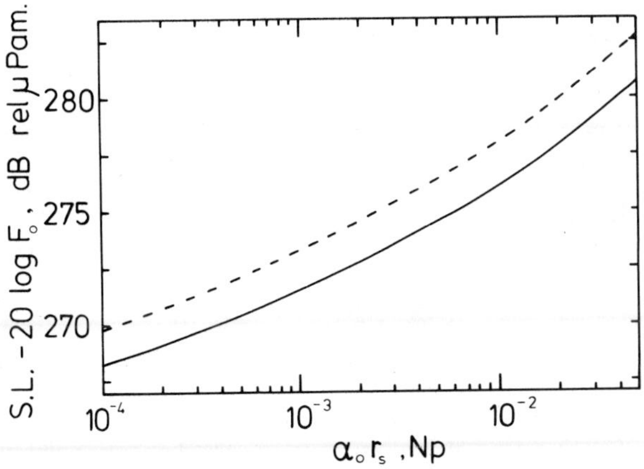

Fig. A. Maximum effective source levels obtainable with a spherical source of radius r_s at a single frequency.
—— Avoiding extra-attenuation effects.
----- As a result of extra-attenuation effects, at much transmission levels.

An illustrative result is shown in Fig. A for a spherical source of radius r_s. Hence, the solid line gives the source level that should not exceed if extra-attenuation of the fundamental frequency component is to be avoided[13,10] for a given absorption within one radius, $\alpha_o r_s$. Also shown is the maximum <u>effective</u> source level attainable if the transmission level is increased beyond that indicated by the solid curve. This effective source level is estimated for ranges beyond which the wave may be assumed to propagate linearly.

An interesting point to note from Fig. A is that if the transmission level is increased well beyond that necessary for the source-level limit indicated by the solid line, only a maximum of about 2 dB increase can be obtained in the effective source level.

Nonlinear propagation in the field of a planar transducer has also received a great deal of attention. In particular, computed solutions have been obtained for Burgers' equation, allowing for diffraction effects, in the field of a circular transducer[14,15].

Solutions based on simplified propagation models assume collimated plane waves within the near-field of a piston, converting into spherical waves in the far-field. Although this model yields valid approximate results, it does need to be treated with caution.

An indicator to the distortion of the initially sinusoidal wave is the relative level of the second harmonic content. Calculations made using, effectively, a single-scattering approach show[16-17] that for a circular piston (of radius a) the second-harmonic content remains small at ranges up to the last axial minimum of the fundamental ($a^2/2\lambda_o$) and then builds up rapidly.

However, experimental evidence shows that when the transmitted level is such that the shock-formation distance for a plane wave is within the near-field of a circular or square transducer, extra-attenuation of the fundamental component of the waveform increases rapidly, as indicated by the results for plane waves.

For water, at a frequency of F_o kHz, the shock-formation can be shown to occur within Rayleigh distance, R_o, if the r.m.s. source level (calculated from linearised theory) satisfies the relationship

$$S.L. > 280.5 - 20 \log F_o \text{ dBre1}\mu\text{Pa m.} \qquad \text{A1}$$

If the source level is significantly below the value on the right hand side of A1, the spherical-wave regime can be assumed to be reached without extra attenuation and without excessive distortion of the waveform. Then, the spherical-wave results shown in Fig. A may be applied, albeit with care.

If the degree of distortion imposed on a plane wave in travelling to a range R_o is ascribed to a fictitioys spherical wave then the effective source radius is reduced to

$$r_s \simeq R_o/e \qquad \text{A2}$$

Thus, $\alpha_o r_s$ can be used to calculate the maximum permissible source level, using the solid curve in Fig. A.

For example, for a 40 kHz transducer of diameter 0.80 m, R_o = 13.4 m. With α_o = .0013 Np m^{-1}, $\alpha_o R_o$ = .017 Np, yielding a maximum permissible source level of 245 dBre1μPa m. However, for this transmitted level, the plane-wave shock-formation distance can be shown to be about 20 m so that at 13.4 m the wave would be significantly distorted, while Fig. A is based on an undistorted waveform at the spherical source. Thus, taking the source radius to be R_o/e, $\alpha_o r_s$ = .0064 Np, giving a maximum source level of 243 dBre1μPa m.

If a rectangular transducer of aspect ratio N (N > 1) is considered, a simplified model can be developed by assuming plane wave propagation up to range R_o/N, cylindrical wave in the range R_o/N to $N R_o$ and spherical wave beyond $N R_o$. It can be shown that to allow for the distortion at $N R_o$ the effective radius of the (fictitious) spherical wave, r_s, is given by the expression

$$r_s/R_o = N \exp\{-2(N - 1)/N\} \qquad \text{A3}$$

(Note that for values of N of the order of unity the cylindrical wave region is not clearly identifiable and hence Eq. A2 is the more reliable result.)

Using the value of $\alpha_o r_s$ determined in this manner, the maximum permissible source level to avoid extra attenuation can be obtained from Eq. A.

<u>High-intensity Effects in Parametric Sources</u>

If the parametric source is realised by transmitting pulsed-carrier waveforms (see Eq. 5, for example) then higher-order effects can be avoided if the source-level associated with the maximum amplitude within the envelope does not exceed the limits discussed above for a monochromatic transmission. For a two-frequency transmission, assuming the amplitudes are equal, the source-level at each primary frequency would have to be 6 dB below that for a monochromatic field.

Strictly, these are the limits of validity for the analysis presented above for parametric sources.

If the primary source levels are increased further, extra attenuation effects modify the difference-frequency source density which, in turn, results in a broadening of the difference-frequency beam and a reduction of the source level of the parametric source as compared with the results obtainable from Eq. 20.

A model solution for parametric sources is proposed and results are given in ref. 8 which is also extended to cover extra attenuation effects. This solution has been tested against a wealth of experimental data, with considerable success.

LIST OF SYMBOLS

α_o, α_d	Absorption coefficient at the primary and difference-frequencies.
$\alpha_T = 2\alpha_o - \alpha_d \simeq 2\alpha_o$	
β	The nonlinearity parameter of the medium (β = 3, 5 is used for water)
c_o	Velocity of sound propagation
$D_o(\ )$, $D_{1,2}(\ )$	Directivity functions of the primary waves
D_o	Diameter of circular transducer
(f_o, ω_o), (f_d, ω_d)	Primary and difference-frequencies
$F_o = f_o \times 10^{-3}$,	$F_d = f_d \times 10^{-3}$
k_o, k_d	The corresponding wave numbers
γ_B, $\gamma_{x,y}$	Primary beamwidths for circular and for rectangular transducers
λ_o, λ_d	Wavelengths at primary and difference frequencies
$M_{x,y}$	The dimensions of a rectangular transducer, $M_y/M_x = N$, N 1
P_o, P_a, P_d	Pressure amplitudes
q_-, q_d	Low-frequency and difference-frequency components of source strength
Q	A source-level parameter used in calculating $(SL)_d$, the source-level of a parametric source.
$R_o = (S/\lambda_o)$	The Rayleigh distance for a transducer of area S
$r_o = (\lambda_d/\lambda_o)R_o$	A scaled Rayleigh distance
r_s	The radius of a spherical source
$(SL)_{1,2}$, $(SL)_d$	Primary and difference-frequency r.m.s. source-levels, in dB rel Pa m.
Θ_o, $\Theta_{x,y}$	The beamwidths at the difference-frequency for parametric sources
$\Theta_d = 2\sqrt{(2\alpha_T/k_d)}$	The beamwidth of a line end-fired array with exponential taper

REFERENCES

1. P.J. Westervelt, Parametric Acoustic Array, J. Acoust. Soc. Am., 35:535 (1963).
2. H.O. Berktay, Possible Exploitation of Nonlinear Acoustics in Underwater Transmitting Applications, J. Sound Vib., 2:435 (1965).
3. J.G. Willette, Difference Frequency Parametric Array Calculations Using an Exact Description of the Primary Sound Field, J. Acoust. Soc. Am., 52:123(A) (1972).

4. G.S. Garrett, J.N. Tjøtta and S. Tjøtta, Nearfield of a Large Acoustic Transducer, Part II:- Parametric Radiation, J. Acoust. Soc. Am., 74:1013 (1983).
5. M.B. Moffett and R.H. Mellen, On Parametric Source Aperture Factors, J. Acoust. Soc. Am., 60:581 (1976).
6. H.O. Berktay and D.J. Leahy, Farfield Performance of Parametric Transmitters, J. Acoust. Soc. Am., 55:539 (1974).
7. M.B. Moffett and R.H. Mellen, Model for Parametric Acoustic Sources, J. Acoust. Soc. Am., 61:325 (1977)
8. M.B. Moffett, R.H. Mellen and W.L. Konrad, Parametric Acoustic Sources of Rectangular Aperture, J. Acoust. Soc. Am., 63:1326 (1978).
9. D.T. Blackstock, Thermoviscous Attenuation of Plane, Periodic, Finite Amplitude Sound Waves, J. Acoust. Soc. Am., 36:534 (1964).
10. B.B. Cary, Prediction of Finite-Amplitude Waveform Distortion with Dissipation and Spreading Loss, J. Acoust. Soc. Am., 43:1364 (1968).
11. J.A. Shooter, T.G. Muir and D.T. Blackstock, Acoustic Saturation of Spherical Waves in Water, J. Acoust. Soc. Am., 55:54 (1974).
12. H.M. Merklinger, R.H. Mellen and M.B. Moffett, Finite Amplitude Losses in Spherical Sound Waves, J. Acoust. Soc. Am., 59:755 (1976).
13. H.O. Berktay and B.V. Smith, Unpublished work.
14. N.S. Bakhalov, Ya.M. Zhileikin and E.A. Zabolotskaya, Nonlinear Diffraction Effects in Acoustics, Sov. Phys. Acoust., 27:253 (1981).
15. S.I. Aanonsen, T. Barkve, J.N. Tjøtta and S. Tjøtta, Distortion and Harmonic Generation in the Nearfield of a Finite-Amplitude Sound Beam, J. Acoust. Soc. Am., 75:749 (1984).

THE POSTCRITICAL PENETRATION OF LOW FREQUENCY PARAMETRIC BEAMS THROUGH A WATER-SEDIMENT INTERFACE

Duncan J. Wingham* and Nicholas G. Pace

University of Bath
Claverton Down
Bath BA2 7AY, England

*Now at: Dept of Electronic and Electrical Engineering
University College London
Torrington Place
London WC1E 7JE, England

INTRODUCTION

The use of parametric beams to generate low frequency highly directive acoustic beams is well established[1]. These properties are particularly useful when the medium in question is attenuating, thereby restricting the usefulness of conventionally generated beams whose frequency is too high to allow penetration. Ocean sediments are one such attenuating medium, and the low frequency narrow beamwidth parametric assay seems ideally suited to high resolution penetration of sediment.

In addition to this obvious advantage over conventional beams, early experiments by Muir et. al. [2,3] showed that parametric beams appeared to offer a second advantage: significant penetration occured even when the incident angle of the array was well postcritical. A number of mechanisms were suggested for this behaviour. Horton[4] pointed out that, at these low frequencies, an evanescent arrival might be expected. Tjotta and Tjotta[5] argued that the narrowness of the beam may be important, whilst numerical studies[6,7] emphasised the importance of virtual sources close to the interface. The experiments of Pace and Ceen[8] demonstrated that the secondary field of a postcritical parametric array contains two arrivals, the second of which is associated with the discontinuous truncation of the secondary source volume at the interface. In a recent theoretical paper[9] Wingham demonstrated that, when the array is truncated in the nearfield of its primaries, the field may be expected to contain three arrivals, the third being the evanescent arrival discussed by Horton.

In this paper some experimental measurements of the secondary field in sediment due to a postcritical parametric array are presented and compared with the theory of reference 9. The results demonstrate that the field does indeed comprise of three arrivals, whose relative importance is a complicated function of geometry, and that the "line-of-sight" property[2] associated with the shallow secondary field from parametric arrays is due to a combination of the truncation and evanescent arrivals. It is concluded that these anomalous effects do not occur when the parametric array is incident on the interface in its farfield.

THEORETICAL DESCRIPTION OF THE POSTCRITICAL FIELD

The secondary field $p(\underline{x},t)$ of a parametric array in an unbounded medium due to a primary field $p_o(\underline{x},t)$ radiated from a large piston at low Mach number may be written[9,10]

$$p(\underline{x},t) = \frac{\beta}{8\pi\rho c^4}\frac{\partial^2 p^2 m(t)}{\partial t^2} * \int_V \frac{g^2(\underline{x}^1,\omega_o)\delta(t - (|\underline{x}^1| - |\underline{x} - \underline{x}^1|)/c)d\underline{x}^1}{|\underline{x} - \underline{x}^1|} \quad (1)$$

where $p_o(\underline{x},t) = p_m(t - |\underline{x}|/c)g(\underline{x},\omega_o)\cos(\omega_o(t - |\underline{x}|/c))$, (2)

and the integral of eqn. (1) is the impulse response $i(\underline{x},t)$ of the parametric array, and $g(\underline{x},\omega_o)$ describes the spatial distribution of the primaries, and V is the volume occupied by the primaries.

The behaviour of $i(\underline{x},t)$, or equivalently its transfer function $I(\underline{x},\omega)$, in an unbounded medium where the upper limits on V are infinite, has been thoroughly investigated,[11,12,13]. As the experiments of Pace and Ceen demonstrated[8] however, in the presence of an interface it is not sufficient to regard $i(\underline{x},t)$ as the incident wave on the boundary, without first accounting for the change in V introduced by the boundary.

When a high frequency beam is incident on an attenuating medium the transmitted beam is rapidly attenuated. The virtual sources contained within the reflected primary beam are incorrectly phased to contribute to the secondary field in the sediment, and hence the interface may be regarded as discontinuously truncating the source volume V. Clearly, the effect of this truncation on $i(\underline{x},t)$ will depend on how far distant it is from the transducer, for $g(\underline{x},\omega_o) \to 0$ at large ranges. We may anticipate that the effect of truncation will be most marked where the discontinuity is most abrupt: in the nearfield of the primaries.

Figure 1 shows a parametric array incident at an angle α on sediment and truncated in its nearfield by the interface. The primaries are assumed to be perfectly collimated (ie $g(\underline{x}^1;\omega_o) = 1$) and lie within a cylinder bounded at one end by the transducer and a the other by a plane but oblique truncation.

When the incident angle is post-critical, i.e. $\alpha > \theta_c$ the critical angle, the transfer function $I_s(\underline{x}_1\omega)$ in the sediment may be shown to be[9]

$$I_s(P,\omega) \sim$$

$$- \frac{i(2\pi)^{\frac{1}{2}}T(\theta_o)}{(r_w + r_t)} A_w(\alpha - \theta_o)e^{-ik(r_w + nr_t)}$$

$$+ \frac{i(2\pi)^{\frac{1}{2}}T(\theta_1)}{r_s} A_t(\alpha\theta_1')e^{-iknr_s - ikl} \quad (3)$$

$$+ \frac{R^2(2\pi)^{\frac{1}{2}}T(\theta_e)}{r_e} Am(\alpha - \theta_e)e^{-ikr_e - nkd + ikl/2}$$

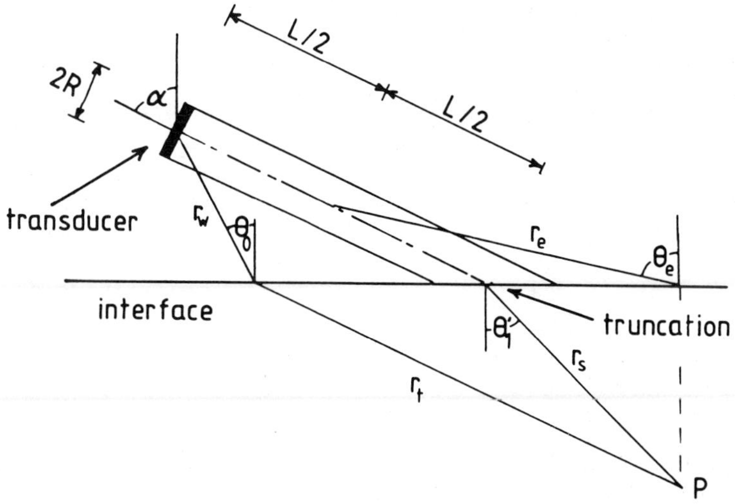

Fig. 1. The nearfield truncated parametric array.

where k is the wavenumber in water, n is the refractive index, T(θ) is the plane wave transmission coefficient, $\theta_1 = \sin^{-1}(n\sin(\ _1))$,

$A_w(\eta) = \dfrac{RJ_1(kR\sin\eta)}{k^2(1 - \cos\eta)\sin\eta}$ is the transducer aperture directivity

$A_t(\alpha, \theta_1) = \dfrac{RJ_1(kR(\sin\theta_1 - \sin\alpha)/\cos\ )}{k^2\cos\alpha(1 - \cos\eta)(\sin\theta_1 - \sin\alpha)}$ is the truncation aperture

directivity, and

$A_m(\eta) = \dfrac{\sin(kL(1 - \cos\eta)}{k(1 - \cos\eta)}$ is the directivity of the array evanescent

arrival.

These expressions are farfield expressions, valid when

$$r_s > \pi R^2/\cos\alpha \text{ and} \tag{4}$$

$$r_e > L$$

In addition, the spherical spreading terms are not valid when P becomes close to the surface. In these circumstances (or when P does not lie in the plane of the array) the full expression given in Ref 9 should be used.

Equation 3 and figure 1 show the postcritical secondary field to be comprised of three arrivals, two travelling waves and an evanescent wave. Each arrival has an associated directivity and raypath, and it is the combination of these two factors which determine their relative importance to the total field in the sediment. The transducer directivity is a symmetric function of $(\alpha - \theta_o)$, and as $\theta_0 < \alpha$ when the array is postcritical, only the sidelobes of the transducer aperture enter the sediment. The truncation directivity is an asymmetric function of $(\alpha - \theta_1)$ and dies away much more rapidly than A_w as α increases and, as $\theta_1 < \theta_c$, only its sidelobes may enter the sediment; thus the truncation arrival is only important in a shallow region below and in front of the truncation. Unlike θ_o and θ_1, θ_e can equal α and thus the evanescent arrival reflects the directivity of the main beam; it is limited to shallow depths by its exponential decay with depth. As α increases and the array becomes increasingly postcritical, $\alpha - \theta_o$ and $\alpha - \theta_1$ increase and the two travelling wave arrivals become much reduced in amplitude.

The additional arrival of the truncation aperture, which has no analogue in conventional beam behaviour, arises from the truncation of virtual sources whose phase varies linearly across the truncation. As the truncation moves out of the primary nearfield these sources are modified by the spherical spreading, attenuation and additional spherical phasing of the primaries in their farfield. At some large enough distance it may be supposed that their net effect will be to reduce the truncation aperture arrival to insignificance. This is in fact the case, and it may be shown that if the range to the truncation

$$L > R_o f_o / f \quad , \tag{5}$$

where R_o is the primary Rayleigh distance and f_o/f the ratio of primary to secondary frequencies, then the presence of the truncation aperture is negligible and the array behaves as a conventional narrow beam.

EXPERIMENTAL INVESTIGATION OF THE POSTCRITICAL FIELD

An experimental investigation of the theoretical conclusions above was conducted at the underwater tank facility at the University of Bath and the Admiralty Research Establishment, Portland, England. A complete description of these experiments may be found in Ref 14.

The secondary pressure in sediment from a postcritical array was measured in two vertical planes through the sediment, one in the plane of the beam and a second perpendicular to it. Fig 2 shows the pressure contours at 100 kHz in the plane of the array incident in its nearfield, when the incident angle α was 10^0 greater tham the critical angle of 60^0.

The O in Fig 2 marks the measured location of the 5 dB contour. Fig 3 shows the contours in the plane x = 80, perpendicular to the plane of the array.

In the plane of the array, Fig 2, the field is seen to comprise of three regions:

1) Below and behind the truncation the contours are slowly varying and near horizontal, and describe the field associated with transducer arrival.

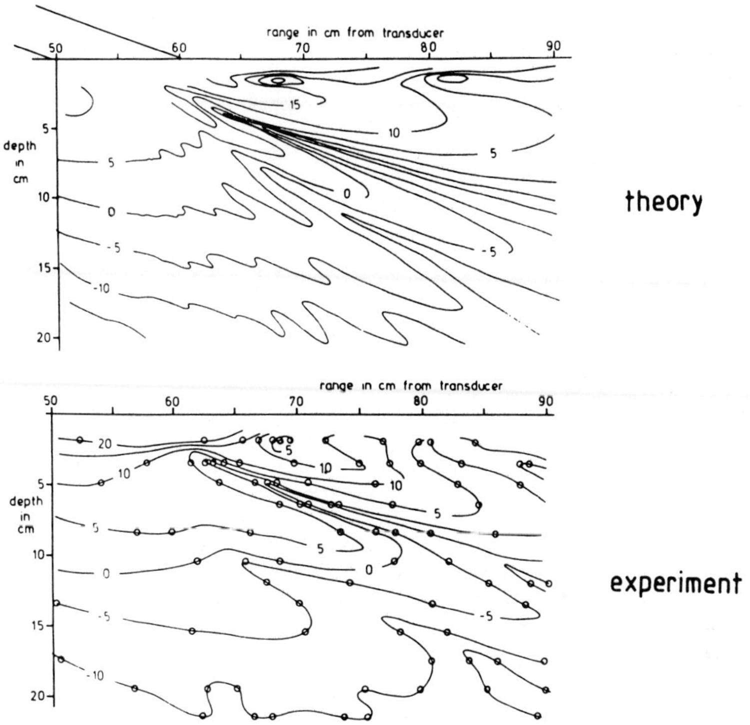

Fig. 2. The postcritical field in sediment from a parametric array in the plane of the array.

2) In an angular fan centered on the truncation the field is a complicated interference pattern; here, the increasing importance of the truncation arrival is sufficient to interfere with, and at shallow depths dominate, the transducer arrival.

3) Close to the surface an additional interference pattern emerges as the evanescent arrival is of similar amplitude to the truncation arrival. This interference does not seem to have been noted in previous discussion of evanescent arrivals from beams[15], but its effects are clearly seen in the experimental curves.

It can be seen that the interference is limited to a very narrow band between 1 and 2 cm deep. The truncation and transducer arrivals tend rapidly to zero[9] as the field point approaches the surface. In contrast, the evanescent arrival grows exponentially as the field point approaches the surface. Only in the very narrow range of depths for which they are of similar amplitude will they interfere.

The evanescent arrival is present again in Fig 3, where the contours turn horizontal as the field point approaches the surface. The marked top/bottom asymmetry of seen in Fig 3 is to be expected; as noted in the last section at postcritical angles of incidence the directivities of the

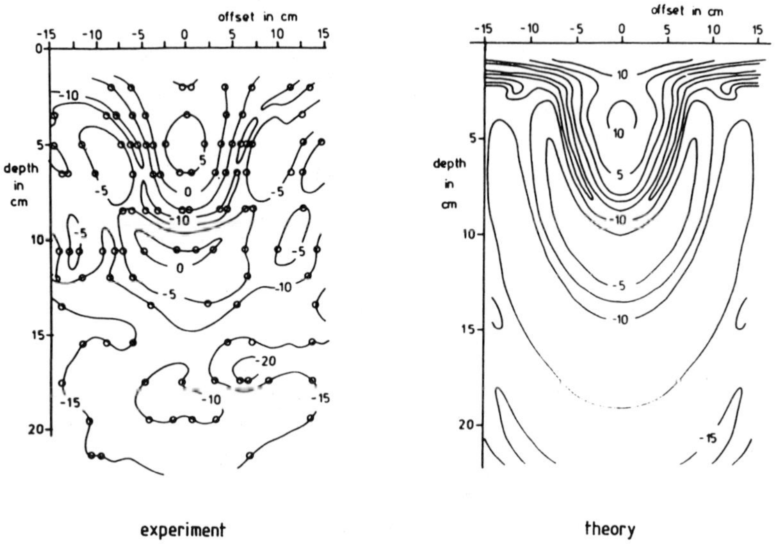

Fig. 3. The postcritical field in sediment from a parametric array in a plane orthoganal to the array.

transducer and truncation apertures in the sediment are restricted to one-sided sidelobes.

A different perspective on the role of the three arrivals may be gained from Fig 4. Here, a hydrophone was located at the point x = 89 cm d = 2 cm of Fig 2. The wide bandwidth of the parametric array was exploited to generate very narrow pulses[8] in order to separate the three arrivals in time. The arrivals at the hydrophone as the incident angle was varied from 60^0 to 90^0 are shown in Fig 4 as a stacked plot.

At 60^0 of incidence (the critical angle) only one arrival is detected, the transducer arrival. As α increases this arrival dies away because its directivity becomes increasingly off-axis. As α increases the truncation aperture approaches the hydrophone and starts to grow in amplitude. The motion of the truncation aperture across the interface is reflected in its arrival time, producing the curved move out across the plot. At similar angles of incidence but later in time the evanescent arrival starts to grow, reflecting the directivity of the main beam. At $\alpha = 76^0$ the truncation and evanescent arrivals are coincident and maximum. At this angle the main beam is pointing at the hydrophone, the evanescent arrival is on-axis, and the truncation arrival has minimum spherical spreading. This is the "line-of-sight" property observed by Muir et al [3]. As α increases still further the truncation arrival dies rapidly due to its directivity, until at $\alpha = 90^0$ only the off-axis evanescent arrival remains.

To investigate the conclusion that this behaviour, peculiar to parametric arrays, only occurs if the array is incident in its primary

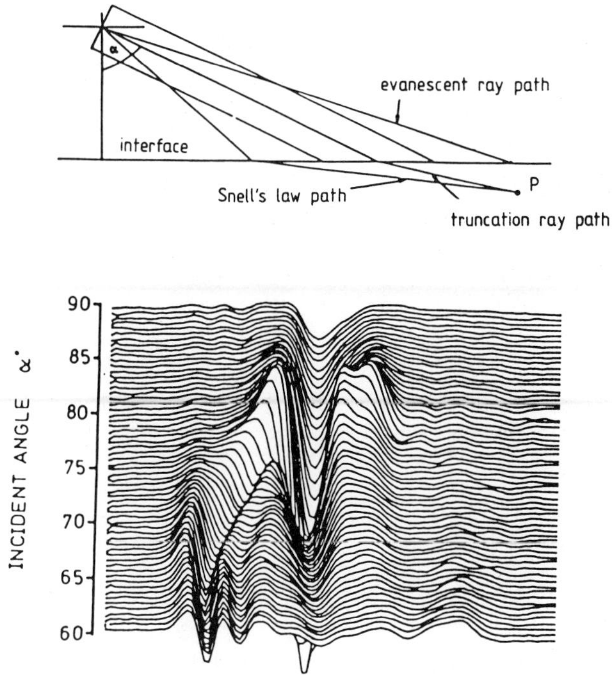

Fig. 4. The three arrivals in sediment from a parametric array driven with a short pulse.

nearfield on the interface, experiments to compare the primary and secondary field behaviour with an array incident in its farfield on a water sediment interface was conducted[14]. The geometry of one of these experiments is shown in Figure 5. The range to the truncation at 70^0 was chosen to be 1 m, which with the 500 kHz, 5 cm diameter transducer used placed the truncation at a range $R_o f_o/3f$, i.e. one third of the range indicated by eqn. 5. The primary, , and secondary field at 100 kHz, , is shown in Fig 5 as a function of beam incidence angle.

In complete contrast to the nearfield truncation result there is no anomalous behaviour associated with the parametrically secondary field. Both fields have their maximum at $\alpha = 58^0$, the angle predicted by Snell's law, and the differences between the two beams result solely from their differing beam patterns.

DISCUSSION AND CONCLUSIONS

It is clear from both the theoretical and experimental results presented here that the secondary field in sediment from a parametric array truncated by the interface in its primary nearfield is very complicated. The anomalous postcritical behaviour of parametric arrays is associated with the sharp truncation of virtual sources near the interface giving rise to an additional path, in addition to that provided by Snell's Law, by which

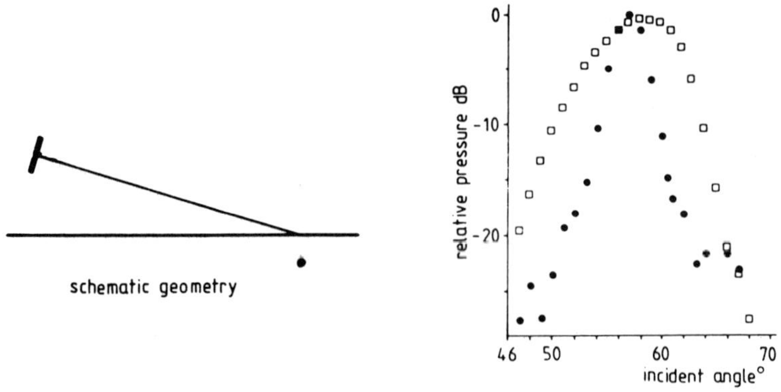

Fig. 5. The primary • and secondary □ pressure in sediment from a parametric array truncated in the farfield of its primaries.

energy may cross the interface. At the low frequencies associated with parametric arrays a third, evanescent, contribution is present at shallow depths, which introduces additional interference maxima and minima in a narrow zone beneath the interface.

Although this very complication makes it difficult to draw general conclusions it is possible to make the following remark: that the anomalous penetration is confined to a region which is both shallow and close to the truncation. The truncation aperture directivity is not favourable to the penetration of the sediment, and the increased importance of its arrival at some point in the sediment arises from its shorter path length leading to decreased spherical spreading and absorbtion. In this respect it is important to note that the increased penetration of the parametric beam does not arise from its narrowness: indeed, increasing the secondary frequency will narrow the directivity of the truncation aperture still further.

Some authors [3,4,5] have suggested that a parametric beam may suffer displacement as it passes across the interface. If the main lobe of the "beam" in Fig 2 is extrapolated back to the interface it appears to be displaced by 5 cm, or half a beam width, which is much the same as that observed by Muir et. al.[3]. However, the theoretical description of the field in Fig 3 explicitly excluded displacement (i.e. a discontinuous phase shift which is not predictable in terms of geometric optics). In fact, the apparent displacement arises because the phase lagging of the secondary source favours points in front of truncation, a feature which is reflected in its directivity. It is for this reason that the maximum arrival from the truncation occurs when the field point is slightly forward of the truncation: i.e. when array is coincident with the "line-of-sight" to the field point. As this paper makes clear, this effect is emphasised by the coincident maxima in the evanescent arrival.

The presence of sources close to the interface is not, in itself, sufficient to produce postcritical penetration. As the truncation moves into the farfield of the primaries, the virtual sources across the interface have an additional spherical lag which is sufficient to prevent postcritical penetration. In fact, the experimental evidence suggests that eqn (5) is a considerable overestimate, and it seems likely that for significant post-critical penetration, the array must be truncated within the Rayleigh distance of the primaries.

ACKNOWLEDGEMENTS

The authors would like to thank Professor H O Berktay for his interest and advice in this topic, and Dr R Bugler of the Admiralty Research Establishment for his help in the farfield truncation experiments. One of the authors, DJW, would also like to thank all members of the Acoustics group at the University of Bath for their help during his three year stay there. He would also like to thank Mullard Space Science Laboratories, University College London, for their support.

The authors also acknowledge the support of the Procurement Executive of the Ministry of Defence.

REFERENCES

1. W. L. Konrad, "Applications of the Parametric Acoustic Source", Underwater Applications of Non-Linear Acoustics, Proc. Inst. Acoustics Conf., 10-11 September (1979) University of Bath.

2. T. G. Muir, "A survey of several nonlinear acoustic experiments on travelling wavefields", Proceedings of the 1973 Symposium on Finite Amplitude Wave Effects in Fluids, Copenhagen.(Ed) L. Bjørnø, IPC Science and Technology Press (Guildford). (1974)

3. T. G. Muir, C. W. Horton, Sr. and L. A. Thompson, J. Sound, Vib., 64(4), p539. (1979)

4. C. W. Horton, Sr., "The penetration of highly directional acoustic beams into a sedimentary bottom Part 1.", Applied Research Laboratories Technical Report No. 74-28, Applied Research Laboratories, University of Texas at Austin. 1974.

5. J. N. Tjotta and S. Tjotta, J. Acoust. Soc. Amer., 69(4), p998, (1981).

6. A. H. A. Moustafa, PhD thesis, University of Birmingham, England.(1979)

7. J. Jarzynski and L. Flax, J. Acoust. Soc. Am., 63(5), p1365, (1978).

8. N. G. Pace and R. V. Ceen. J. Acoust. Soc. Amer.,73(6),p1972, (1983)

9. D. J. Wingham, J. Acoust. Soc. Am., 76(4), p1192, (1984).

10. S. I. Aanousen, T. Barkre, J. N. Tjotta and S. Tjotta, J. Acoust. Soc. Amer. 75(3)

11. T. G. Muir and J. G. Willette, J. Acoust. Soc. Am. 52, p155, (1967).

12. H. O. Berktay and D. J. Leahy, J. Acoust. Soc. Am., 55(3), p539, (1974)

13. M. B. Moffett and R. H. Mellon, J. Acoust. Soc. Am. 61(2), p581 (1977)

14. D. J. Wingham and N. G. Pace, "An experimental study of the penetration of a watersediment interface by a parametric array", submitted for publication to J. Acoust. Soc. Am.

15. L. M. Brekhorskikh, "Waves in Layered Media", 2nd ed. Academic Press (New York) (1980).

THE USE OF LOW AND HIGH FREQUENCY INTENSITY FLUCTUATIONS IN SEA-BED SURFACE CLASSIFICATION

Mustafa K. Gürcan*, D.J. Creasey,† and B.K. Gazey†

*Philips Research Laboratories
Redhill, Surrey, RH1 5HA, England
(formerly the University of Birmingham)

† The University of Birmingham, England

ABSTRACT

Acoustic intensity fluctuations of backscattered signals from the sea-bed at low frequencies (zero to 1 kHz) and at high frequencies (in the 400 to 600 kHz range) have been examined. The aim has been to determine the feasibility of using observed fluctuations for sea-bed surface identification. The intensity ratio of the low frequency signals to the mean value of the high frequency intensity fluctuations is proportional to the number of random scatterers insonified. The density of the scatterers can be determined and their sizes inferred from the density estimate. When a high frequency transducer is used the low frequency portion of the response cannot be obtained directly. However, it can be estimated by restricting and weighting the time windows used in the receiver. This reduces the size of the scattering aperture and results in a radiation frequency pattern which is almost uniform in the limited frequency band. Ergodicity between the spatial and frequency domains enables the wanted low frequency components to be obtained from the narrow aperture response. Simulations show that the number of particles can be estimated with a confidence level slightly higher than 80 percent.

1 INTRODUCTION

The identification of sea-bed surfaces by using acoustic remote sensing techniques is receiving increased attention. Researchers have been investigating the use of forward and backscattering information in the estimation of either the surface roughness parameter (4) or particle size (7,8).

The authors have examined the frequency responses of backscattered energy at oblique angles of incidence in the 400 to 600 kHz band. The aim was to measure the size of regularly distributed particles (7,8) from the equally spaced Bragg frequencies appearing in the response (see Fig.1). The frequency response, $H(j\omega)$, was measured by transmitting a wideband signal, $X(j\omega)$, to give

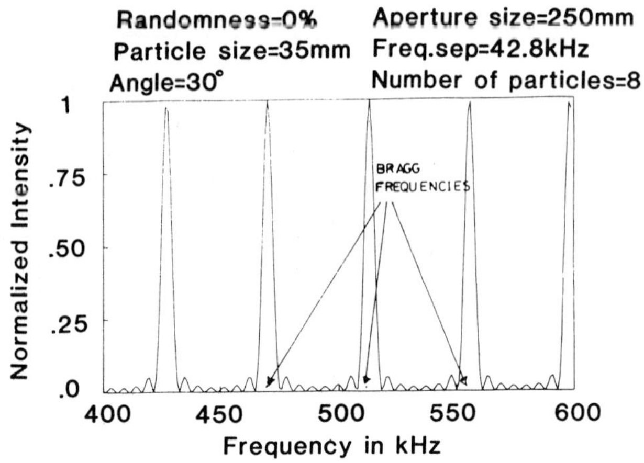

Fig. 1 The frequency response of a regular surface with equally-spaced Bragg frequencies.

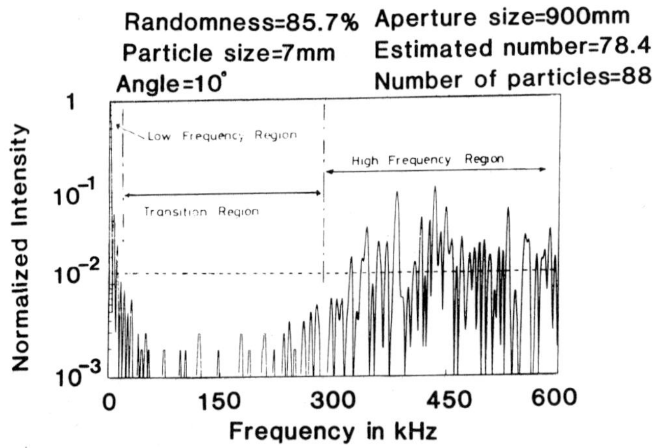

Fig. 2 Frequency response of a surface containing randomly distributed particles.

$$H(j\omega) = Y(j\omega) \,/\, X(j\omega) \qquad \text{(eqn.1)}$$

Where $Y(j\omega)$ is the echo response. The work has continued to develop processing techniques for the classification of randomly distributed particles. As the particles are randomized all high frequency Bragg resonance peaks start to disappear, leaving only the Bragg peak around zero frequency (see Fig.2). At high frequencies the spectrum fluctuates randomly. Researchers (5) have already examined the mean values and the probability distribution function of these random fluctuations. In this paper, high frequency fluctuations are presented in terms of the low frequency components. The aim is to show that normalization of the mean value of the fluctuations with respect to the estimated intensity values at low frequencies can provide features that are suitable for sea-bed classification. A technique has been developed to obtain both the high and low frequency information of the response by transmitting a band limited high frequency signal. The effect of having limited aperture size on the frequency response has been examined theoretically and computer simulations have been used to assess the predictions.

2. THE FREQUENCY RESPONSES OF SURFACES

The well known stationary phase approximation techniques (1) have been used to examine various regions of the frequency responses of signals scattered from randomly placed particles by checking the received acoustic velocity potentials. Consider a scattering surface and a receiver and a transmitter which are located as shown in Fig.3. Only the signals originating from the stationary phase points on the scattering surface contribute to the received signal when the incident acoustic signal frequency is very high and the scattering surface is very rough, i.e. the roughness parameter $g = 2\sigma k\cos\emptyset_i >> 1$. Here σ is the r.m.s surface wave height, k is the wave number, $\emptyset_i$ is the angle of incidence. At these points the dot product of the wave number, $V = k_i - k_s$, and the surface displacement vector, $r = (x,y)$ has the property (2):

$$dV.r \,/\, dx = 0 \quad \text{and} \quad dV.r \,/\, dy = 0 \qquad \text{(eqn.2)}$$

The solution of eqn.2 gives the spatial positions of stationary phase points on the scattering surface. The contribution to the received signal from each stationary phase point is calculated by estimating the value of V.r in the proximity of the stationary point by means of a Taylor series expansion and substituting into the standard wave equation solution. Each contribution in the backscattering direction is equal to (9):

$$\Phi_m = \frac{\sqrt{b_m}\,(2\pi)^{-3/2}\,R_{rc}\exp(jk2R_{10})\sqrt{2jk}}{R_{10}^2}\exp(jV_x x_m + jV_z\zeta(x_m,y_m)) \qquad \text{(eqn.3)}$$

Here b is the Gaussian curvature (2) of scatterers, ζ is the surface wave height displacement, x and y are the spatial coordinates. A further solution of eqn.2 shows that at each stationary phase point a scattering surface-water boundary must be perpendicular to the incident field (9). It is assumed that the number of such points per unit area is proportional to the particle density, i.e. the number of particles per unit area, and is different for various types of surfaces. In a restricted scattering

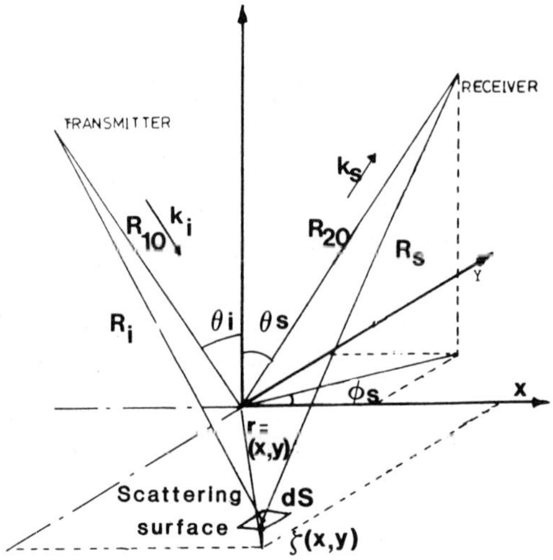

Fig. 3 Geometry of the scattering problem.

aperture the total number of stationary phase points, M, will be equal to the multiplication of the aperture area with the density of the particles. By vectorially adding M contributions originating from the stationary phase points, the total velocity potential which is proportional to the frequency response of the scattering surface, is calculated to be (9):

$$\Phi = C_1 \sum_{m=1}^{M} \sqrt{b_m} \quad \exp\left(jV_x x_m + jV_z \zeta(x_m, y_m)\right) \qquad \text{(eqn.4)}$$

where $C_1 = \exp(2jkR_{10})\ \sqrt{jk2}\ R_{rc} / (2\pi)^{3/2}\ R_{10}^2$. R_{rc} is the reflection coefficient. One method of classifying different types of sea bed surfaces is to estimate the number of particles in the restricted area for a given surface, i.e. the summation limit in eqn.4. The acoustic velocity potential, given in eqn.4, is a different function of M in various frequency regions. This can be used to calculate the number, M, from the frequency response measurements by means of the mean squared signal level estimates.

3 THE MEAN SQUARED SIGNAL

The second moment of the acoustic velocity potential is examined here as a function of frequency (in the low and high frequencies). The mean-squared signal is equal to (9):

$$\langle \Phi \Phi^* \rangle = |C_1|^2 \sum_{m=1}^{M} \sum_{n=1}^{M} \langle \sqrt{b_m b_n^*} \rangle \langle \exp(jV_x(x_m - x_n)) \rangle \langle \exp(jV_z(\zeta_m - \zeta_n)) \rangle \qquad \text{(eqn.5)}$$

or

$$\langle \Phi \Phi^* \rangle = \langle b \rangle |C_1|^2 \left[M + (M^2 - M) \langle \exp(jV_x x_m) \rangle \langle \exp(-jV_x x_m) \rangle \langle \exp(jV_z \zeta) \rangle \langle \exp(-jV_z \zeta) \rangle \right] \quad \text{(eqn.6)}$$

where < > refers to ensemble average over spatial coordinates. In eqn.6 the first term in the square bracket, M, is the summation of the terms m=n. The second quantity $M^2 - M$ corresponds to $m \neq n$. At high frequencies the terms $\langle \exp(jV_x(x_m - x_n)) \rangle$ and $\langle \exp(jV_z(\zeta_m - \zeta_n)) \rangle$ satisfy the relationships (3) similar to:

$$\langle \exp(jV_z(x_m - x_n)) \rangle = 1 \quad \text{for} \quad m=n$$

and

$$= \langle \exp(jV_x x) \rangle \langle \exp(-jV_x x) \rangle \quad \text{for} \quad m \neq n$$

At zero frequency, the exponential terms approach unity and eqn.6 is simplified to:

$$\langle \Phi \Phi^* \rangle = |C_1^2| \langle b \rangle M^2 \quad \text{(eqn.7)}$$

At very high frequencies the average of the exponential terms approaches zero and eqn.6 reduces to

$$\langle \Phi \Phi^* \rangle = |C_1^2| \langle b \rangle M \quad \text{(eqn.8)}$$

This means that the frequency radiation pattern has a main lobe around zero frequency and at high frequencies the fluctuations correspond to the sidelobes. The sidelobe fluctuation pattern is not affected by the propability distribution of the particles. The signal strength ratio of the average sidelobe level at high frequencies to the main lobe (centered at zero frequency) is proportional to 1/M where M is the number of particles.

To use this as a classification technique requires that both high and low frequency components to be available simultaneously. It is necessary to obtain these by transmitting a single pulsed signal and analysing the echo. In the 400 to 600 kHz band of the transducer, used for initial experiments, the sidelobe level of the frequency radiation pattern can be measured directly. Although the low frequencies cannot be measured directly, it is possible to estimate the main beam, or the low frequency information, indirectly by processing the received signal in a second path parallel to the sidelobe measurements. In the second path, the lobe width of the main beam is extended beyond 1 or 2 MHz so that it remains within 1 or 1.5 dB of the peak value.

The backscattered signals can be processed in three domains to obtain different regions of the frequency response. The processing can be in time, in space or in frequency and these are related to each other by means of Fourier transforms. The beamwidth of the mainlobe can be extended in the frequency domain by processing the scattered signals in the spatial domain. Spatial domain processing requires that the signals originating from an aperture should be linearly weighted across the aperture. Therefore processing in this domain is not an easy task. After establishing how to use Fourier transforms to widen the beams, the time domain processing will substitute for the spatial processing. This leads to a novel technique of measuring the high frequency components normalised with respect to the reference zero frequency component.

4. PROCESSING IN THE SPATIAL DOMAIN

In the frequency domain, the acoustic velocity potential, given in eqn.4, describes scatterers as discrete sources of a random array. In this model the array elements are located at positions (x_m, y_m). The curvature, b, and the phase term, exp(j ν (ζ_m, ζ_n)) describe the source strength and the phase distribution of the array elements respectively. The radiation frequency pattern is related to the spatial distribution of the sources by Fourier transforms. The relationship shows that the Fourier transform of the amplitude weighting of the signals originating from the aperture is convolved with the frequency response of the surface (see Fig.4.a). After the convolution, the lobe width of the main beam becomes equal to (9):

$$f = c \;/\; 2\;L\;\sin \text{Ø}_i \qquad \text{(eqn.9)}$$

where f is the lobe width in Hz, c is the speed of the sound in water, L is the aperture size and Ø_i is the angle of incidence. We can therefore increase the main lobe width by reducing the size of the aperture. The beamwidth will be increased further by amplitude weighting the contributing signals linearly across the aperture as functions of higher powers of x (the spatial displacement vector) as shown in Fig.4.b. The beamwidth can be increased as desired by spatial domain processing. However it is not feasible to realize the spatial weighting in the receiver. Time domain processing may be more appropriate.

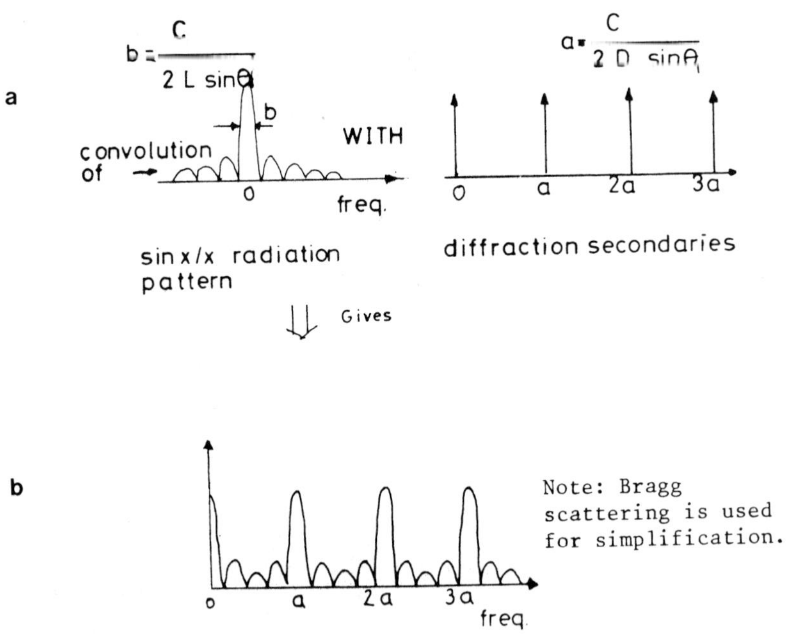

Fig. 4 A simplified illustration of convolution in the frequency band.

5. PROCESSING IN TIME

Amplitude weighting in the spatial domain corresponds to multiplying the signal strength of the sources by the spatial weighting x. The Fourier transformation shows that in the frequency domain, the corresponding acoustic velocity potential will be equal to the derivative of the frequency response of the surface (i.e. the total acoustic velocity potential received from the aperture which has a uniformly weighted amplitude) with respect to the frequency. The derivative of the frequency response can also be obtained indirectly firstly by calculating the impulse response of the aperture (see eqn.1) and weighting the response linearly with respect to time and finally taking the inverse Fourier transform (see Fig.5). The implication of this is that the main lobe width can be expanded by processing the received signal in the time domain rather than spatially. Effectively this produces a two channel parallel processor with a single wide band impulse type waveform for transmission in the 400 to 600 kHz band. In the receiver two time windows, one narrow and one wide, are used. The signal in the narrow time window is weighted linearly with high powers of time such as t^4 or t^5. The wide time window signals are unweighted. The signals in both channels are then transformed into the frequency domain by Fourier transforms. Finally the average power ratios in the two channels are calculated to give an estimate of the number of particles in the restricted area. Hence the density can be estimated.

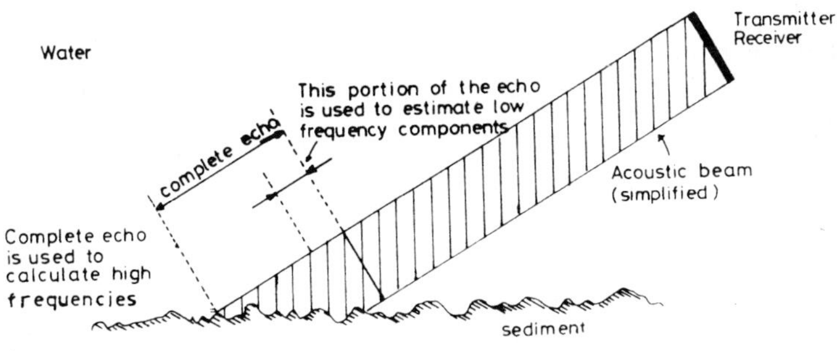

Fig. 5 An illustration of processing in time.

6.RESULTS

Simulation programs were written to test whether both low and high frequency sections of the responses could be obtained from a single echo in the 400 to 600 kHz band. Simulated signals originating from apertures with sizes ranging between 250 and 800 mm were used to estimate the high frequency information in the backscattering direction at oblique angles of incidence. Signals originating from a 50-mm aperture were gated and amplitude weighted with fourth power of time (or space) to obtain the low frequency components. The result given in Fig.6 shows that the lobe width of the main beam is increased to above 1 MHz. In all simulations the estimated signal strength in the 400 to 600 kHz band remained within 1 to 2 dB of the peak main lobe level.

The mean values of the high frequency sidelobes were calculated for varying sizes of scatterers and the number of scatterers is then estimated as shown on each graph. The broken lines given in Fig.6 correspond to the estimated low frequency reference line. The solid lines are the measured high frequency responses.Both the actual and estimated numbers of particles are displayed on each graph. Programs were run for different values of aperture size, particle size and angle of incidence. The results show that the number of particles can be estimated with a confidence level of 80% or more. However it is important to keep the angle of incidence less than fifteen degrees.

The classification technique, which has been developed, has been proven in terms of the simulations. The simulations themselves were tested for Bragg scattering and produced comparable results to those obtained in earlier practical measurements (7,8). It is then argued that the simulations are sufficient to prove the proposed classifier. However it is recommended that a transient analysing sonar, similar to the one given in reference 8, should be developed to implement a practical system.

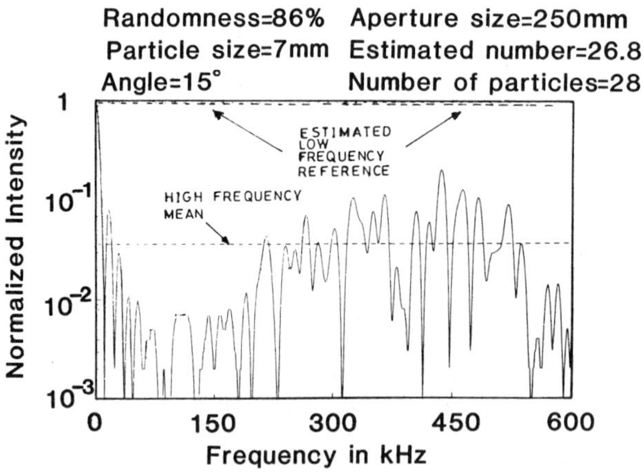

Fig. 6a Frequency response of randomly distributed particles.

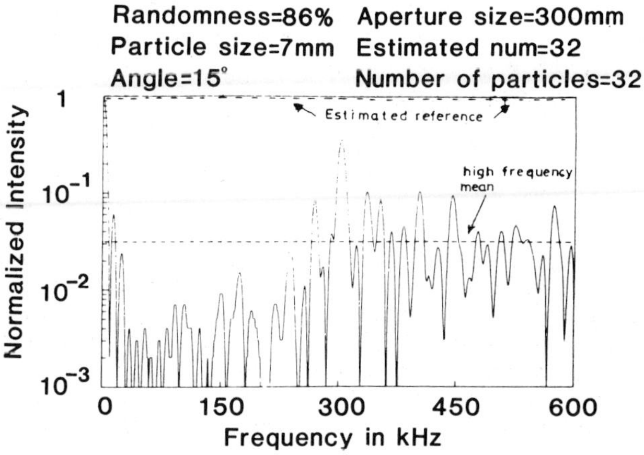

Fig. 6b Frequency response of randomly distributed particles.

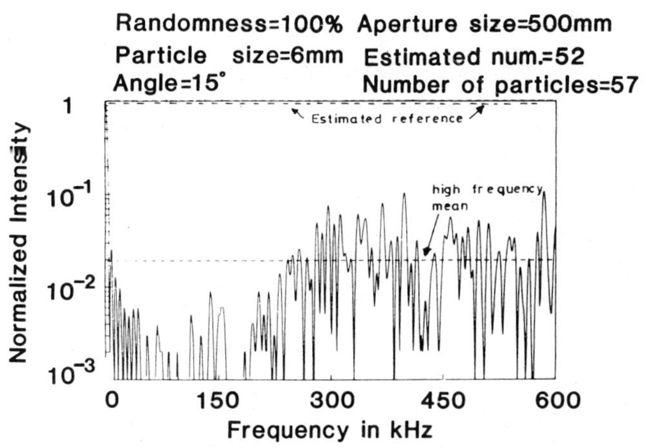

Fig. 6c Frequency response of randomly distributed particles.

7. CONCLUSIONS

A simple but useful classifier can be obtained by dividing the intensity of the signals received around zero frequency by the average intensity of the fluctuations centred in the 400 to 600 kHz band. This ratio is equal to the number of scatterers insonified. Hence the density of the scatterers can be determined and their sizes inferred from the density estimate.

It is quite easy to measure the intensity of the band pass acoustic signals by using previously developed transducer techniques (6). However it is not easy to measure the zero-frequency component. An estimate can be obtained by restricting and weighting the time windows used in the receiver. If this is done the sea bed surface observed by the narrow pulse becomes very small and the radiation pattern obtained is almost uniform in the high frequency band. Ergodicity between spatial and time domains then enables the wanted zero frequency component to be estimated from the broad beam of the narrow aperture by a pulse processing technique. In this way an estimate of the zero frequency component can be made. The classifier so produced appears to estimate the numbers of scatterers to within confidence level of 80%.

8. REFERENCES

1. M. Borg and E. Wolf "Principles of Optics, Electromagnetic Theory Propagation, Interference and Diffraction of Light" Pergamon Press, Oxford, Fifth Edition, 1975.

2. R. D. Kodis "A Note on The Theory of Scattering From an Irregular Surface " IEEE Transactions on Antennas and Propagation, Vol. Ap-14, No.1, Jan. 1966, pp 77-82

3. B.D. Steinberg "Principles of Aperture and Array System Design" Wiley Interscience, Publication, 1976, Newyork.

4. N.G. Pace "Machine Classification Of Sedimentary Sea Bottoms" Trans. Geo. Elect. IEEE, July 1979, Vol. Ge-17, No.3, pp52-56

5. B.J. Uscinski "Recent Progress and Developments in the Field of Microwave Interactions With the Sea and Land Surfaces, Period I", Oxford Computer Services Limited, Sea Court Tower , West Way, Oxford United Kingdom.

6. H. Köymen "Optimisation of the Design of High Frequency Sonar Transducer", Ph.D Thesis, University of Birmingham, 1979.

7. M.K. Gürcan, D.J. Creasey, B.K. Gazey " Wideband High Frequency Transmissions for Sediment Identification" IERE Int. Conf. on Electronics for Ocean Technology, pp 273-82, Birmingham Sept. 1981.

8. M.K. Gürcan, D.J. Creasey, B.K. Gazey "Development Of a Sonar To Measure the Size of Sea-Bed Material" International Conference on Acoustics and the Sea Bed, Instute of Acoustics, Underwater Group, Bath University Press, Bath, United Kingdom, 1983.

9. M.K. Gürcan " The Use of A Wide-Band Sonar Transducer in Sea-Bed Surface Identification. Ph.D Thesis, The University of Birmingham, Department of Electronics and Electrical Engineering, to be submitted 1985.

OCEAN BOTTOM SEISMOLOGY: HISTORY AND CURRENT STATUS

George H. Sutton

Rondout Associates, Incorporated
P.O. Box 224
Stone Ridge, New York 12484 USA

ABSTRACT

Ocean bottom seismographs (OBS) have been in use for almost 50 years. The earliest instruments were used for short range explosion refraction experiments but were soon abandoned when techniques involving near surface sources and receivers were introduced. Development of OBS technology revived during the late 1950's inspired by interest in the detection and identification of clandestine nuclear explosions and supported largely by the VELA-UNIFORM program of the United States government. Since that time, OBS research has expanded to include over 25 organizations in several countries. Generally each group has developed its own instrumentation and experimental procedures. Instruments fall into four classes: (1) self-contained, free-fall, pop-up; (2) acoustically telemetering to near surface for retransmission or recording; (3) electrically telemetering via cable to the surface; and (4) permanently linked to shore by cable. Most fall in class (1). Sensor configurations range from a single vertical component (or hydrophone) to three-components plus hydrophone; seismometer free period is generally one second or less; frequency response is generally limited to 100 Hz or less. Instrument masses range from 60 to 600 kg; recording is analog (AM or FM) continuous or digital (time windowed and/or triggered); dynamic range, data storage capacity and deployment time all cover a wide range of values. Most applications are for determination of velocity structure and/or local seismicity. There has been lesser interest in teleseismic, engineering or strong-motion applications. Major problem areas or areas where improvements are needed include bottom-coupling (signal distortion) and noise; data storage; system reliability, especially over long deployment times; ease of handling during launch and retrieval; and ease of shipboard checkout before and after deployment. OBS have recently been supplemented by ocean subbottom seismometers emplaced beneath the sediment in a bore-hole.

INTRODUCTION

OBS operated at selected oceanic sites afford an opportunity to explore seismic wave propagation under the deep ocean basins and across continental and island margins; ambient seismic noise in the ocean basins, both natural and cultural; the generation of storm microseisms; and the distribution and mechanisms of local submarine earthquakes. In addition, OBS are required for uniform areal observations of seismic waves from teleseismic events. As mentioned in the abstract, by far the majority of OBS have been of the pop-up variety, designed and used for relatively short term active and/or passive experiments.

The design of a good, reliable OBS is a challenging task. In addition to the requirements of any remotely operating seismograph, the instrument must be designed to be launched, retrieved, and refurbished from a rolling, often wet (with salt water) ship; it must couple well to a (very!) soft or hard bottom and poorly to flowing near-bottom water. There is still much room for improvement over even the best of the existing OBS. As is the case for land instruments, probably more so, optimum design is a strong function of the observational objectives. For example, seismic noise on the ocean floor is low above about 2 Hz and often very high in the storm microseism band, three to 10 seconds. It is still uncertain how much the data are influenced by poor bottom coupling. Partially because of the high noise levels, in addition to instrumental difficulties, relatively little work has been done with long-period instruments and most of the current interest is above 1 Hz. However, interest in the long-period range is growing. At the other end of the spectrum there is interest in extending the frequency band to a few hundred (or thousand) Hz for high resolution sediment studies and acoustic studies of Navy interest.

Most existing OBS are designed for the detection of small signals, well within the linear range of bottom motion, and only such instruments will be discussed in this paper. Steinmetz et al. (1979) describe an OBS specifically designed for recording strong motion.

An excellent review paper by Whitmarsh and Lilwall (1983) provides a fairly complete general description of OBS developed and used through 1981. It includes an extensive bibliography of OBS related papers, many of which provide specification tables and schematic drawings of operational OBS. A more recent review paper by Prothero (1984) discusses the state of the art of OBS technology through 1983, based partly on a workshop held at the University of California, Santa Barbara in July 1982. Since 1970, a sequence of papers on OBS development has been published in Marine Geophysical Researches. Most of these papers are listed by Whitmarsh and Lilwall (1983). Physical characteristics of the 12 operational OBS deployed during the Lopez Island OBS intercomparison experiment (conducted in June 1978) are tabulated in Sutton et al. (1981b). Schematic drawings and additional specifications for these instruments are contained in Hawaii Institute of Geophysics Technical Report, HIG-80-4, Hawaii Institute of Geophysics, Honolulu.

In this paper I shall not attempt another comprehensive review of the field but, rather, shall limit the discussion to a brief review of the history and the current status with which I am most familiar, emphasizing research and instrument development subsequent to that reported in the Whitmarsh and Lilwall and the Prothero papers.

EARLY OBS

Ewing and Ewing (1961) describe the history and status, at that time, of OBS development. The earliest OBS were designed for small scale, active refraction experiments with both the self-contained OBS and the explosives deployed either by cable or free-fall (pop-up). In the latter case, the sensor was mechanically isolated from the flotation required for retrieval. Figure 1 is a block diagram of the acoustic telemetering OBS described by Ewing and Ewing. Data from a single vertical component 2 Hz geophone were transmitted by frequency modulation of a 12 k Hz carrier and received with a standard echo-sounder transducer. Different mechanical configurations involved either cable or free-fall deployment with no provision for retrieval of the bottom package. Both explosion and earthquake signals were well recorded with this system.

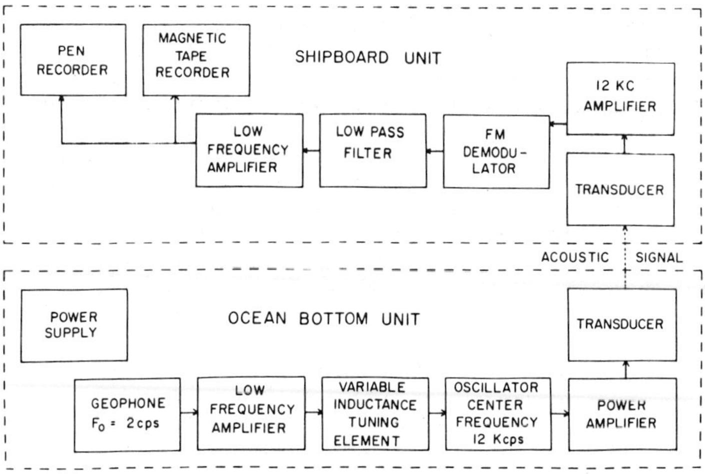

Figure 1. System block diagram of acoustic telemetering OBS (Ewing and Ewing, 1961).

Figure 2 shows the location of the Columbia (Pt. Arena) OBS, about 100 miles west of San Francisco, California in four kilometers water depth, that operated for over six years, 1966-1972 (Sutton et al. 1965 and 1969). The bottom package (Figure 3) was actually a small geophysical observatory containing: three-component long (15 second) and short (one second) period seismometers; low (crystal) and high (coil) frequency hydrophones; Vibratron tidal pressure sensor; crystal temperature sensor; and water current magnitude and direction sensors. Ultra long-period (feed-back) outputs from the long-period seismometers provided tidal gravity and tilt data. The instrument package was connected to shore at Pt. Arena, California via two-conductor deep-sea cable. Power and 30 command channels to the OBS and 18 data channels from the OBS to the shore station were transmitted through this cable. Except for the mechanical current sensor, which failed after about two months (probably from biological growth and/or silting) all instruments operated until the cable was damaged by wave activity near shore.

Data from the Columbia OBS are included in a number of reports on short- and long-range explosion experiments and studies of local seismicity and teleseisms, tides and tidal currents, and microseisms. One interesting observation is that the OBS often recorded offshore earthquakes not observed by land stations at comparable epicentral distance and the converse for on-shore earthquakes (Sutton et al., 1969).

Figure 4 illustrates the short-period signals from a local earthquake recorded by the Columbia OBS at about 250 km epicentral distance. It is of interest to note that each of the three motion components and the pressure component emphasize different aspects of the signals providing, e.g. information on direction of propagation and arrival type. The hydrophone record has the simplest character; a strong surface reflection, P_{R1}, because of the amplification of pressure of a down-going signal (combined with its bottom reflection) near the bottom; and lack of the shear arrival, Sn. Pn is strong on the vertical, small on H_2 (longitudinal horizontal), and missing on H_1 (transverse horizontal).

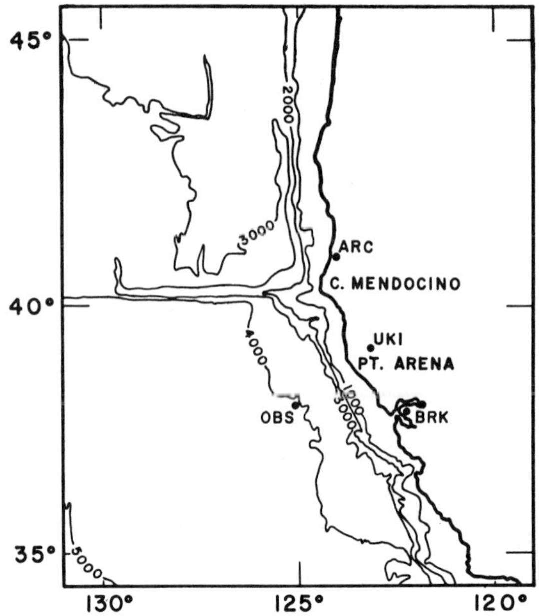

Figure 2. Location of Columbia OBS. ARC, UKI, and BRK are stations of the Berkeley (University of California) seismic network. OBS recording station was at Pt. Arena. Mendocino Escarpment trends E-W near 40°N. Depth contours in meters (Sutton et al., 1965).

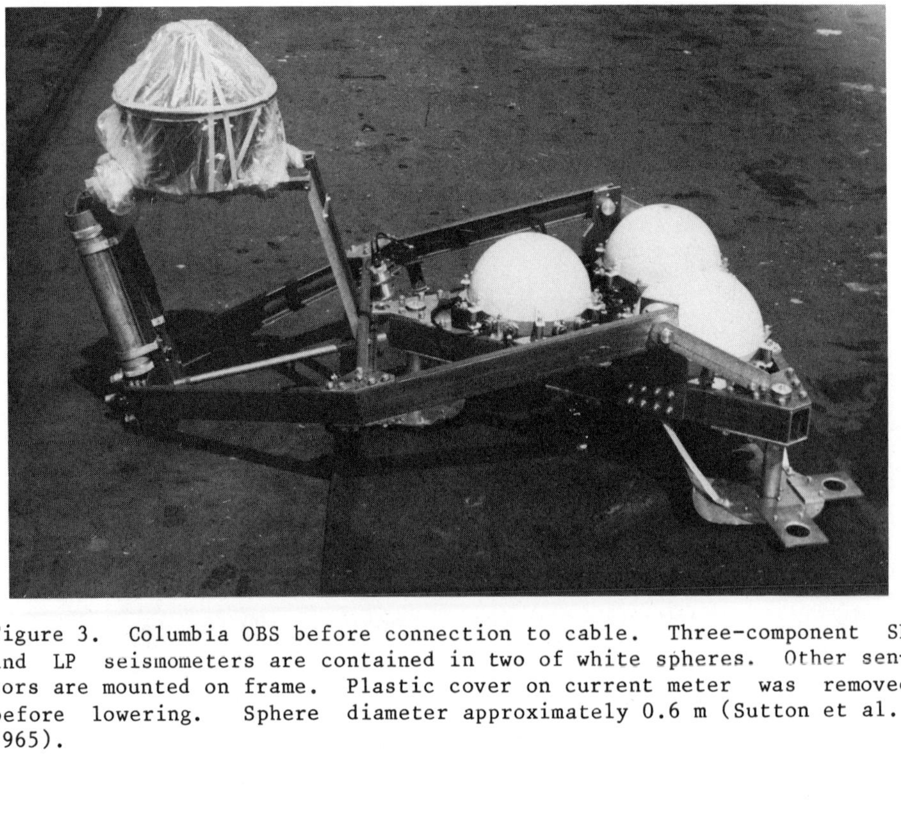

Figure 3. Columbia OBS before connection to cable. Three-component SP and LP seismometers are contained in two of white spheres. Other sensors are mounted on frame. Plastic cover on current meter was removed before lowering. Sphere diameter approximately 0.6 m (Sutton et al., 1965).

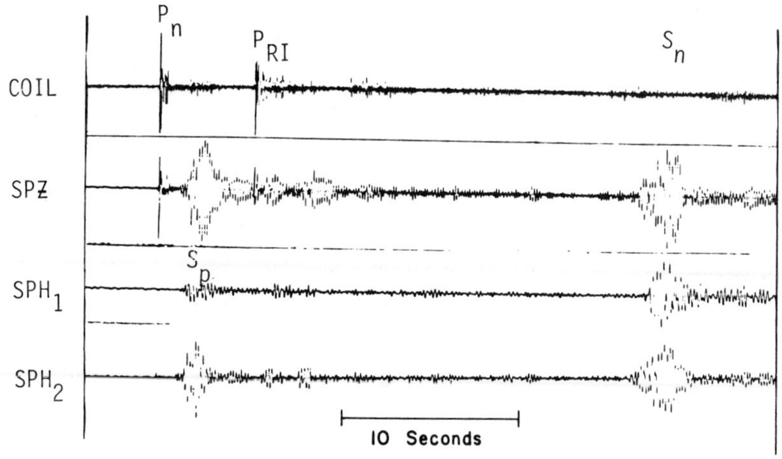

Figure 4. P and S arrivals from a local earthquake recorded by the Columbia OBS (Figures 2 and 3). Traces, from top, are: short-period hydrophone (coil); short-period vertical seismometer (SPZ); and two orthogonal short-period horizontal seismometers (SPH_1 and SPH_2). Identified arrivals are: mantle refracted P and S (Pn and Sn); surface reflected Pn (P_{R1}); and Pn converted to S at the base of the sedimentary layer (Sp) (Sutton et al., 1969).

All three seismometers record the converted wave Sp and the shear arrival Sn, which begins earlier on the vertical and longitudinal horizontal components because of partial conversion to P below the ocean bottom. The resonant character of the seismometer signals results mainly from OBS-bottom coupling resonance, to be discussed more fully later, but may be partly due to sediment velocity structure. Spectra from P, S, and T phase arrivals on a given seismometer component exhibit nearly identical resonant peaks. However, the resonant frequency is different for each component: 7, 5, and 6 Hz for Z, H_1 and H_2, respectively. This observation strongly suggests that the instruments are recording different resonant coupling modes excited by the signals (Sutton et al., 1969). Sutton et al. (1965) calculated possible resonant coupling frequencies for the Columbia OBS to be about 70 and 60 Hz for vertical and rocking motion, respectively, well above the frequency band of interest at that time. That calculation was based upon an estimated sediment shear velocity of 0.15 km/sec. More recent observations give shear velocities at the top of marine sediments an order of magnitude lower. Since the coupling frequency is proportional to the shear velocity, a velocity of 15 m/sec would produce calculated frequencies of 7 and 6 Hz, as observed in the data. It is interesting to note that most of the operational pop-up OBS tested during the Lopez Island experiment, to be discussed later, exhibited coupling frequencies between about 5 and 10 Hz and the sediment shear velocity near the bottom was measured to be 15-20 m/sec.

An array of hydrophones near Wake Island and connected by cable to the island was installed more than 25 years ago to locate nose cone impacts from intercontinental ballistic missile tests. Eleven of these phones are still operational. Six are on the ocean bottom at 5.5 km depth and make up the vertices and center of a pentagon about 40 km across; five are located at three different sites near the axis of the SOFAR channel (about 1 km depth). The combined array spans more than 300 km. Signals from eight of these hydrophones, five bottom, and three

SOFAR, have been recorded digitally by the Hawaii Institute of Geophysics on a continual basis since September 1982. The hydrophone signals are recorded through low-noise low-frequency preamplifiers and are used for seismological studies. The array is unique, both in its deep ocean location and in its great distance from other seismological stations.

McCreery et al. (1983) examine the spectral characteristics of P waves from shallow focus earthquakes and nuclear explosions and samples of ocean-bottom background noise recorded on the Wake array. Ambient noise levels on the ocean bottom near Wake are comparable to levels at the quietest continental sites for frequencies between three and 15 Hz. At lower frequencies the noise increases rapidly toward the "storm" microseism maximum near 0.2 Hz where the level is above that for continental sites.

SHALLOW WATER OBS

In a series of papers Brocher et al. (Brocher, 1983; Brocher et al., 1981 and 1983) reported on the results of a nine day OBS experiment conducted on the continental margin off Nova Scotia in June 1975. Three HIG telemetering OBS were placed in a linear array along the dip of the bottom with approximate spacing of 18 km at depths of 67, 140, and 1301 m. The lack of resonance in the recorded signals suggests that the low-profile instrument packages (7:1 length-to-diameter ratio) provided excellent coupling to the sediments for the three-component 2 Hz natural frequency geophones and avoided the coupling of the seismometers to ocean-bottom currents. The OBS were hard-wired directly to the recording ship and were anchored to the seafloor by a heavy communication cable. The broadband frequency response of both the incorporated hydrophone and the geophones together with the FM recording of the signals on (1.5 IPS) tape drives, contributed to the high quality of the recorded data. The horizontal geophones in the OBS were oriented along (horizontal longitudinal) and perpendicular (horizontal transverse) to the longest axis of the OBS package (and the cable).

Data from a variety of seismic sources were recorded during the nine day investigation. The purpose of the experiment was to study the propagation of noise produced by a drilling rig platform located approximately 18 km shoreward from the shallowest OBS. Signals and background from air gun profiling, SUS charges, shipping, biological activity, bottom current, varying weather conditions, and seismicity were investigated, in addition to the rather low level of signal observed from the drilling operations. P and S velocity structure in the subbottom was determined; S/N ratios for the hydrophones and geophones in the band one to 30 Hz were compared; and low frequency propagation along strike and up and down dip of the topography was studied. Shear velocities were obtained from analyses of both Stoneley (Scholte) and refracted shear waves.

Related studies were reported, at a previous NATO conference on bottom-interacting ocean acoustics, by Rauch (1980) and Schirmer (1980). In both of these studies tethered OBS were utilized to detect interface waves in shallow water.

BOREHOLE OBS

A number of experiments have recently been conducted using seismometers clamped in a drillhole below the ocean floor. In these experiments the drilling ship Glomar Challanger of the Deep Sea Drilling Project was used. Two papers based on data from one such installation, located about 400 miles southeast of the central Kuril Islands are included in this volume (Duennebier et al., 1985; and Carter et al., 1985). This ocean subbottom seismograph (OSS) is emplaced 21 m below the sediment-basalt boundary at a depth of 5.85 km. The OSS includes a self-contained power and recording system that can be left on the ocean floor and retrieved and refurbished at a later time (Duennebier and Blackinton, 1983). Stephen (1981, and earlier papers) has used three-component borehole data from shots recorded on the drilling ship via direct cable connection to investigate anisotropy in the oceanic crust. Carter et al. (1984) compared data from an OSS emplaced in soft sediment at a depth below bottom of 194 m with that from an HIG ISOBS (described later in the section on Isolated Sensor OBS). The OSS and OBS exhibited remarkably similar S/N characteristics. Absolute noise levels were similar above 10 Hz but below 10 Hz the OSS was quieter by three to eight dB.

Another borehole seismograph, the Marine Seismic System (MSS), is similar in design concept but more elegant that the OSS (Ballard et al., 1984). Shearer and Orcutt (1985) detected and analyzed anisotropy within the oceanic lithosphere using MSS and OBS data collected during the NGENDEI seismic experiment in early 1983, about 500 miles east of the Kermadec Trench in the southwest Pacific.

Absolute noise levels for the OSS and MSS, when clamped in competent rock beneath the sediment, are significantly lower than for OBS on adjacent ocean-bottom, especially for the horizontal components of motion. S/N also appears to be higher for most transient signals.

THE ROSE EXPERIMENT

The Rivera Ocean Seismic Experiment (ROSE) was a large combined sea and land seismic program utilizing both explosive sources and earthquakes to study the structure and evolution of a mid-ocean ridge, a major oceanic fracture zone and the transition region between ocean and continent (Ewing and Meyer, 1982). Because permission was not received to conduct the experiment in Mexican territorial waters as originally planned, the marine portion was relocated from the Rivera Fracture Zone to the East Pacific Rise south of the Orozco Fracture Zone. The two month field program (January-March 1979) involved many cooperating research groups and a large number of OBS. Preliminary results are given in a series of papers in the same number of Journal of Geophysical Research as the Ewing and Meyer overview paper (Project ROSE Special Issue, J. Geophys. Res., 87, B10, 1982). These and subsequent papers (e.g. Bratt and Purdy, 1984; Bratt and Solomon, 1984; Trehu and Purdy, 1984) provide a great deal of information on the P and S crustal velocity structure of the Orozco Fracture Zone and East Pacific Rise in that vicinity and on the local seismicity. In addition to the scientific results, ROSE has provided useful lessons for possible future large cooperative efforts and modifications in instrument design concerning such disparate items as data archiving and distribution; optimum joint location of instruments and ships; instrument turn-around time and reliability; timed versus acoustic recall; and OBS data capacity and timed versus triggered versus continuous recording.

A paper by Rowlett and Forsyth (1984), which discusses microearthquakes observed at the intersection of the Vema Fracture Zone and the Mid-Atlantic Ridge, provides an interesting comparison with similar studies from ROSE on the more rapidly spreading East Pacific Rise.

ISOLATED SENSOR OBS

In most of the OBS used today, the sensors, electronics, recorders, flotation, and ballast are contained in one rigid package. This configuration requires a relatively large mass and large vertical cross section in the water that can result in poor coupling to the bottom, and increased noise level from response to water flow (Duennebier et al., 1981). The use of an isolated sensor can minimize these problems. In addition, separation between the sensors and the main OBS package reduces the effect of possible vibrations from the data recorder. An isolated sensor OBS (ISOBS) developed at Hawaii Institute of Geophysics (Byrne et al., 1983) is shown in Figure 5. The cylinder containing the geophones is held rigidly to the main package during free-fall and is released after arrival at the bottom by the solution of a magnesium retaining pin. Direct comparison between signals recorded by a single package OBS of earlier HIG design and the ISOBS, deployed near each other in deep water, clearly demonstrated the advantages of the isolated sensor configuration. Additionally, at the Lopez Island tests, the two OBS with separated sensors (ISOBS and another supplied by MIT) exhibited bottom coupling characteristics superior to those of the single package OBS.

Although data suggest that the ISOBS is a considerable step toward improving the fidelity of OBS recording, an improved prototype isolated sensor has since been developed at HIG (Figure 6) known as the "Opihi" (after an Hawaiian limpet that adheres to underwater surfaces). The "Opihi package attempts to optimize the three principal design parameters found to affect sensor coupling to the ocean bottom (Sutton et al., 1981a,b). The parameters are: (1) a plate with a relatively large surface area for low bearing pressure, (2) nearly neutral density, allowing the package to "float" in the sediments and (3) a buried package, for greater coupling to solid material and less exposure to ocean bottom currents.

The "Opihi" package deals with these three parameters by mounting the ISOBS geophone pressure case on a plate framework. This plate is then inserted into a soft mesh bag filled with a loose matrix of polyvinylchloride pellets (PVC) or pea gravel. This method distributes the weight of the pressure case over a wide area, making it about neutral density in soft sediments (total package specific gravity of about 1.4). The loose matrix also simulates a buried sensor and should appear as a small bump on soft sediments. The package has the additional advantage that it can conform to bottom contours and couple well on hard or soft bottoms. The low profile of the package makes it relatively insensitive to ocean bottom currents.

In essence, this sensor package should have good coupling and low noise generation on any type of surface. Initial testing of the "Opihi" sensor package design has shown that it improves coupling over the previous ISOBS sensor package. At this writing, the "Opihi" has not been implemented in an operational OBS.

CURRENT AND DESIRED OBS OPERATIONAL SPECIFICATIONS

In 1984, Rondout Associates, Incorporated conducted a survey of OBS developers and users. Following is an abbreviated summary derived from the 34 responses received indicating required and desired specifications (with this author's comments indicated by GHS):

Major Interests: most investigators are interested in Velocity Structure and Local Seismicity. Fewer are interested in Focal

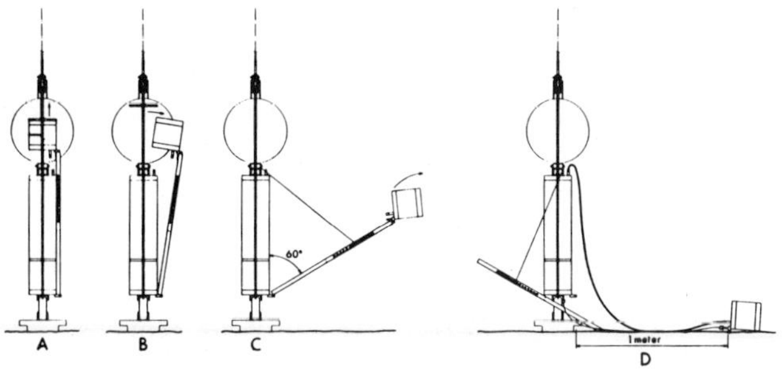

Figure 5. Isolated sensor OBS geophone release scheme. Separation is approximately one meter (Byrne et al., 1983).

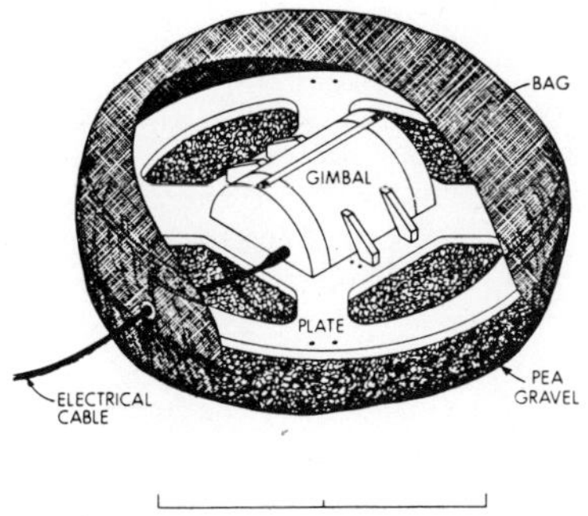

Figure 6. "Opihi" isolated sensor package. Scale length is 0.5 meters (Byrne et al., 1983).

Mechanism and Exploration/Prospecting. There is less general interest in Teleseismic Studies and Engineering applications.

Maximum Depth: most want capability to Normal Ocean (6000m) depth; a few need capability to work in Trenches (12,000m); fewer still require or desire Intermediate (2000m) or Continental Shelf (500m) maximum depth. The interest in shallower depth coincides with Engineering and Strong Motion interests.

Frequency Range: the median required frequency range is two to 30 Hz; a maximum overall range of 0.01 to 100 Hz. The median desired range is one to 100 Hz and the maximum overall range is DC-4 kHz! Wider band instruments emphasize either LP or HF, typically three decades.

Components Required: most require or desire three-component seismometers and a hydrophone. Anything less is generally a compromise for data storage or other reasons. (Each of the four components provides unique information. However, since the hydrophone generally produces relatively clear and simple records with good S/N, relatively undisturbed by poor bottom coupling and current noise, it is sometimes preferred for exploration and acoustic work.-GHS)

Recording: everybody requires or desires digital recording. Some stick to analog recording (for good reason!-GHS) because of the great advantages of continuous recording in the marine environment; limitations of digital data quantity; problems with triggering on noise, and not-triggering on weak signals; and logistical problems with timed recording.

Recording Mode: most investigators require or desire a

Microprocessor Controlled system and prefer the Triggered Mode to the Timed Mode. (Peal and Kirk, 1983, describe a versatile microprocessor-controlled OBS data logging system. This subject is also covered by Prothero, 1984.-GHS) A significant number require (or currently have) Continuous Recording.

Recording Medium: the majority currently use Cassette Recorders; some use or would prefer Reel-to-Reel or Cartridge recorders. Most of those using digital recording would desire greater capacity than their existing system. (The recorder is a major problem area. In addition to problems with storage capacity, it is the major power drain, is one of the least reliable components, and can be an important source of vibration noise. The OBS community would welcome a real breakthrough in this area.-GHS)

Dynamic Range: the median required dynamic range is 72 dB; the responses range from 26 dB (analog) to 96 dB without gain ranging and 130 dB with gain ranging. The median desired dynamic range is 120 dB; the responses range from 60 to 80 dB and 40 dB plus AGC to 150 dB, 16 bit floating point.

Deployment Time: the median required Deployment Time is 30 days with a total range from one day to one year. The median desired deployment time is six months with a range of one day to "semi-permanent". The short times are from groups principally interested in Shelf and Intermediate Depth Engineering studies and/or Exploration/Prospecting; the long times coincide with interest in Teleseismic and Long Period work.

Recording Time: the answers to this item are given in terms of continuous recording times (analog), the number of short event time windows; and/or the total number of megabytes (mB) data storage capacity. The median values listed are approximate. The median required Recording Capacity is about 20 mB, digital; or 20 to 30 days, analog. The required values range from 20 minutes (5 mB) to 90 mB, digital; and four days to two months, analog. The median desired Recording Capacity is about 100 mB digital and three to six months, analog. The values range from about 40 mB to ∞, digital, and one month to one year, analog. (As noted before, under Recording Medium, this is an area that needs improvement.-GHS)

Acceptable Clock Drift: (the important thing is uncorrectable drift. Linear drift can be corrected.-GHS) The acceptable uncertainty depends upon the experimental objectives. Total uncertainty of +/- 0.1 second or less is required for most earthquake work. Short-period array studies and high resolution sediment studies require near 1 msec accuracy. Total clock stability of one part in 10^9 is desirable.

Maximum Weight: the median required Maximum Weight is 150 kg with values ranging from 45 kg. to $<$ 450 kg. The median desired maximum weight is $\leq$ 90 kg. with values ranging from 10 kg to 300 kg.. A few investigators who work from large ships are relatively unconcerned about weight. Workers with more limited ship capabilities and personnel and/or who utilize a larger number of instruments are concerned to keep weight and handling difficulties to a minimum.

Release: everyone prefers an acoustic release, generally with a timed release backup. (Use of acoustic telemetry for diagnostics is useful. An associated transponder used for precision navigation is invaluable for certain experiments.-GHS)

Bottom Coupling: (I have viewed this as a major problem area for a number of years.-GHS) All require good coupling to soft bottoms and most to rocky bottoms. Fewer are concerned with High Currents. (Refer to further discussion below under Packaging Configurations.-GHS)

Azimuthal Orientation: most require or desire an independent measurement of azimuth. The median preferred accuracy is +/- 5° with a range of values from 0.5 to 10°.

Packaging Configuration: a majority of the respondents require or desire either Separated Sensors or Low Profile or both; two, indicate (strong) preference for a single package; one suggests an adaptable system depending upon bottom conditions. (The Lopez Island experiment indicated that the coupling performance of isolated sensor package OBS is better than single package instruments, although some of the latter performed almost as well. The isolated sensor instruments produced records comparable to the three special "standard" instruments used at Lopez Island (Sutton et al., 1981b).

Use of a well-designed separated sensor package generally should improve the bottom coupling and isolation from current induced and tape recorder noise. It should also be a more flexible design permitting various sensor packages and a common support and recording package. The requirement of including flotation and retrieval aids on a single package instrument makes optimum coupling design difficult. However, generally an isolated sensor design is more complex and, therefore, less reliable.-GHS)

Retrieval: everyone wants a Flasher and Radio (although the Radio antenna can be a serious source of current-induced noise-GHS). Several require or desire an acoustic pinger and/or transponder to aid in retrieval.

Maximum Unit Cost: the majority listed \$10-20,000; a significant number prefer <\$10,000; or will accept \$20-40,000; and a few will go to >\$40,000. Generally, the higher values coincide with more stringent OBS requirements.

General Evaluation of Current OBS Designs-(GHS)

Major problem areas or areas where improvements are needed include bottom-coupling (signal distortion) and noise; data storage; system reliability, especially over long deployment times; ease of handling during launch and retrieval; and ease of shipbound checkout before and after deployment. Each of the OBS with which I have personal familiarity, principally those used in the Lopez and ROSE experiments, has some strong and weak points. The external configurations of many of the OBS were modified subsequent to Lopez to improve coupling characteristics.

The trade offs between analog and digital have been discussed earlier. Most investigators prefer the digital route. The limited dynamic range of analog recording can be alleviated by gain ranging on background noise, multiple gain channels, and temporary digital storage and subsequent gain range before analog recording. Much digital data have been lost by poor trigger algorithms and other problems.

Reliability has been a serious problem with some systems, both from lost data and from lost instruments. The acceptable loss rate depends upon the cost and uniqueness of the data and the instrument package. Some groups have experienced very low rates; and others 10% or more. Occasionally, losses come in batches. Details of each component and the system as a whole must be considered carefully. External sensors (seismometers and hydrophones) require pressure tight electrical leadthroughs; cables to isolated sensors can become caught on the anchor or bottom rocks. Release mechanisms can fail; glass instrument and flotation spheres can fail after repeated use.

Some systems are difficult to launch and retrieve without damage; some of the larger instruments require special handling gear. Some systems require extensive, time-consuming pre-launch checkout, limiting the number of instruments that can be deployed for a given experiment. The conflicting technical requirements or desires of the OBS community for low-power and low-instrument volume, versus high data storage, high dynamic range and high clock stability are pushing the state of the art.

SIGNAL DISTORTION AND NOISE IN OBS

A badly designed OBS can be subject to severe signal distortion because of poor coupling to the soft ocean bottom and suffer from high levels of noise caused by near-bottom water currents shaking the instrument frame. Most existing operational OBS exhibit some amount of these undesirable characteristics. The seismic frequency response of an OBS frame of a given mechanical configuration coupled to sediment of a given shear velocity (V_s) and density can be predicted using relationships that have been tested experimentally using in situ mechanical transients. These same relationships can be used to establish guidelines for designing instruments with improved coupling response. Seismic signals couple to an OBS both through the area(s) of contact with the sediment and directly through the contact with the water. Assuming that the desired response is to faithfully follow the motion of the sediment surface, direct coupling to the water for horizontal-translation and rocking modes of motion should be minimized. The OBS will follow the sediment motion for frequencies lower than the natural frequencies of the modes of coupling motion. Thus, it is generally desired to have these coupling frequencies (f_c) as high as possible. For a given OBS configuration, there is an optimum effective radius for the area of coupling to the sediment that will produce the highest predictable coupling frequency. The optimum radius is proportional to the cube root of the instrument mass. Using these optimum radii, the coupling frequency decreases as the cube root of the mass, i.e. smaller instruments will have higher (more favorable) coupling frequencies.

Recognizing the importance of the OBS-bottom coupling problem, the Lopez Island OBS experiment was conducted in Shoal Bay, Lopez Island in June 1978, as a preliminary to ROSE, in order to compare the responses of the operational OBS, which have quite varied mechanical configurations, when coupled to the earth through soft sediments comparable to those of the ocean floor, and to determine their susceptibility to noise induced by near-bottom currents (Sutton et al., 1981b). (More recently, a similar experiment involving a different set of OBS was conducted in France; I am not aware of published results.) The maximum current measured during the test was 6 cm/sec and no significant correlation was found with noise on the OBS during the experiment.

Signals recorded on OBS geophones are often narrow band and prolonged (Figure 4). Whether these features are natural in origin or caused by the presence of the OBS was a major question to be answered by the test. Some data indicated that the presence of the OBS on the soft sediments could severely distort the motion of the ground; however, the extent of this problem for OBS of different configurations was unknown. The complications present on many geophone records are not observed on ocean bottom hydrophones, thus the question: do geophone data add any useful information that cannot be obtained from hydrophones?

Some data had indicated that near-bottom currents can increase noise levels on OBS enough to make them unusable during high-current periods; therefore, we wanted to test as many systems as possible for susceptibility to ocean current noise.

In addition to 12 operational OBS supplied by 10 different research organizations, three sets of three-component (T_o=1s) seismometers and a hydrophone were included to provide standards for comparison: a "spike standard" was pushed firmly into the bottom; a "plate standard" with a smooth, hemispherical superstructure on a large, flat, circular plate was placed on the sediments; and a "neutral density standards" with a roughly spherical shape was floated within the uppermost sediment. The instruments were placed within a few meters of each other, and four current meters were installed around the array to monitor water circulation in the bay.

Data were obtained from each sensor by hard wiring analog signals to land where they were recorded by a digital system. Seismic records were obtained from mechanical transient tests, airgun shots and blasting caps, and samples of background noise. Mechanical transient tests were conducted to provide some quantitative estimate of the coupling function between the ocean bottom and the OBS package, including possible cross coupling between horizontal and vertical motions. The procedure is analogous to the classical weight lift test for earthquake seismographs, although because of buoyancy the response of the seismic system to the mechanical transient test is not the same as that to a seismic input (Sutton et al., 1981a).

An Hawaii Institute of Geophysics Technical Report (HIG-80-4) contains a large number of records from each recorded sensor. The technical report also shows the designs of all instruments. All the data of this experiment are on digital tapes and may be requested from the ROSE data archive at Hawaii Institute of Geophysics. A catalog of the data appears in the technical report.

Subsequent to the Lopez Island experiment, a number of papers on bottom coupling have improved our understanding of the problem and provided guidelines for better OBS design (Sutton et al., 1981 a and b; Zelikovitz and Prothero, 1981; Lewis and Tuthill, 1981; Trehu, 1985; Garmany, 1984; Sutton et al., 1985; and Sutton and Duennebier, 1985).

Table I lists parameters that must be considered and Table II summarizes possible problems of signal distortion and noise related to OBS-bottom coupling. A schematic diagram of the dynamic coupling between an OBS and a soft bottom for vertical motion is shown in Figure 7. In general, the spring "constant", K, is not equal to the static value; both K and the damping "constant", D, are functions of frequency. The inertial effect of the water, Mw*, and sediment, Ms*, involved in the differential motion between the OBS and its surroundings may also be considered to be frequency dependent (or alternatively included in the variation in K). The M*, K, and D are also shape dependent.

The case for horizontal input is similar to that for vertical motion, with the added complications that: there can be two independent inputs in each orthogonal direction, one through the water and one through the sediment; and cross-coupled rocking motion can contribute a major portion of the output from the horizontal seismometers (and from the vertical, if it is located far from the center of rotation). The response of a horizontal geophone to a rocking mode tilt, θ, is equivalent to a horizontal acceleration, $\ddot{y}_t=g\theta$, so that the total apparent input is $\ddot{y}+\ddot{y}_t$.

The response of a rigid sphere in a fluid to a seismic signal in the fluid of wavelength large compared to the sphere diameter is U/V-3/(2(Mi/Mw)+1) where U and V are the motion of the sphere and the incident wave in the fluid, respectively, and Mi and Mw are the mass of

TABLE I
OBS-BOTTOM COUPLING PARAMETERS

A. OBS		
	1. Mass	7. Cross-Section to Sediment
	2. Moments of Inertia	8. Density
	3. Structural Rigidity	9. Bearing Area
	4. Symmetry	10. Bearing Pressure
	5. Shape/Smoothness	11. Burial
	6. Cross-Section to Water	
B. SEDIMENT/ROCK		
	1. Density	4. Heterogeneity
	2. Shear Velocity	5. Permeability
	3. Anelasticity	
C. WATER		
	1. Density	3. Sound Velocity
	2. Viscosity	
D. INTERFACES/STRUCTURE		
	1. Water-Sediment Interface Roughness	
	2. Sediment-Rock Interface Roughness	
	3. Velocity-Depth Dependence	

TABLE II
OBS-BOTTOM COUPLING PROBLEMS

SIGNAL DISTORTION	NOISE
OBS on Sediment	OBS on Sediment
-Vertical Output from Vertical Signal	-Amplification of Microseisms at Bottom Interface
-Horizontal Output from Horizontal Signal	-Signal Induced "Noise" from Irregular Boundary
-Vertical Output from Horizontal Signal	-Water/Sediment
-Horizontal Output from Vertical Signal	-Sediment/Layer 2
OBS on Rock	Direct OBS-Water Coupling
-Rattling	-Current Induced Noise
-Large Tilt	
-"Floating" Footpad	
Direct OBS-Water Coupling	
-Discontinuous Horizontal Signal Across Water/Sediment Boundary	
-Inertial Effect (Most OBS have Large Internal Vertical Gradient in Density)	
Lateral Heterogeneity of Bottom	
Tilts on Horizontals from Ultra-Low Phase Velocity Signals	

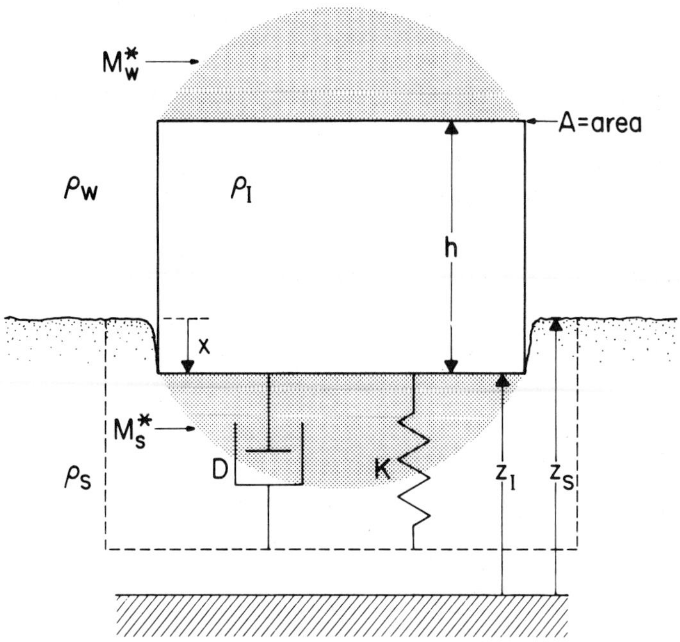

Figure 7. Schematic of OBS coupling to ocean bottom for vertical motion. The dashed line is assumed to move with the ocean floor adjacent to the OBS (Sutton et al., 1981a).

the sphere and the displaced water, respectively. Similar relationships hold for bodies of other shapes. Since current OBS designs generally include most of the body of the instrument within the water, direct seismic coupling between water and OBS is certain and in many cases could be more important than coupling through the bottom sediments. From the equation, we see that amplification can approach a factor of three for very low sphere densities and that attenuation occurs for densities greater than water. As expected, when Mi=Mw, U=V. Most OBS contain low-density flotation for retrieval in their upper portion and more dense disposable anchoring material near the bottom-coupling interface. This density gradient, in addition to producing torques from equal amplitude horizontal inputs, will modify the response from waves having different horizontal amplitudes in the water and in the sediment (e.g. horizontally propagating compressional waves or horizontally polarized shear waves). The vertical component is less complicated since signal amplitude is continuous across the water-sediment boundary and most OBS have appreciable symmetry about a vertical axis. However, lateral variations within the bottom under the OBS could affect this component, also.

A simple calculation indicates the possibility of significant noise resulting from response to tilts produced by bottom currents. Consider a spherical OBS with cross section $1m^2$, centered in the water 1/2 m above two circular footpads of radius 0.2 m that are separated by 1 m in the direction of the current, resting on sediment with shear velocity 20 m/sec and density 1 g/cm^3. The vertical displacement of the footpads (one up, one down) from a bottom current of 5 cm/sec is about 10 μm. This is roughly 10^3 larger than the vertical signal from a teleseismic P

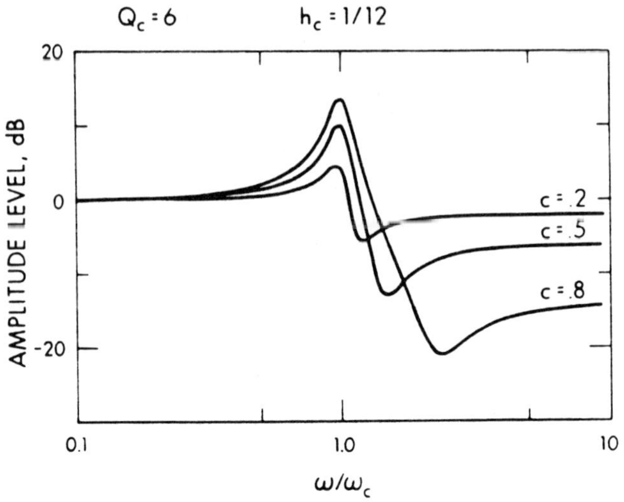

Figure 8. OBS package seismic response for underdamped coupling (Sutton et al., 1981a).

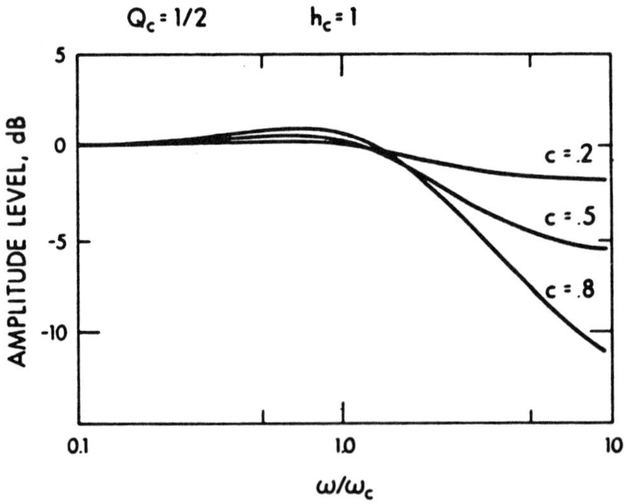

Figure 9. OBS package seismic response for critically damped coupling (Sutton et al., 1981a).

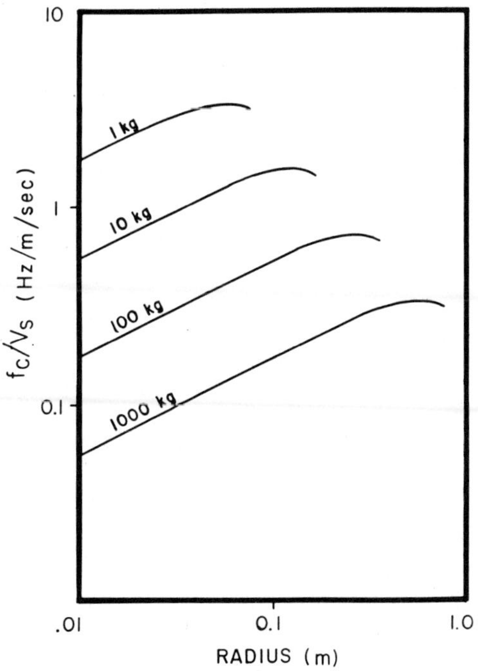

Figure 10. OBS resonant coupling frequency, fc, as a function of radius of bottom-coupling area and instrument mass. Normalized by shear velocity of bottom, V_S. Curves terminate at limit of second-order theory (Sutton and Duennebier, 1985).

wave. The horizontal acceleration associated with the tilt ($\ddot{y}=g\theta$) is equivalent to a horizontal amplitude of 200 μm at 6 second period. This value is about 20 times larger than the horizontal signal from average "storm" microseismic. Small perturbations of bottom currents around such large mean values should produce significant noise on most operational OBS.

Figures 8 and 9 show the amplitude response of an OBS to a seismic input, assuming frequency independent coefficients, for two values of damping and three values of the coupling coefficient c. It can be seen that the OBS will faithfully follow the input for frequencies well below the coupling resonance, ωc, but that serious distortion can occur near resonance and significant attenuation can occur at higher frequencies. The coupling coefficient, c, depends upon the OBS density (as indicated earlier for a sphere in water). c=0 is neutral density; and gives perfect coupling independent of ωc. These response curves are reliable to frequencies somewhat above ωc, but there is some question regarding their precision at higher frequencies (Trehu, 1985; Sutton and Duennebier, 1985).

The relationship from second-order theory between coupling frequency and bearing radius (assuming a circular bearing area) for OBS of different mass is shown in Figure 10. For a given mass, the coupling frequency increases with radius to a maximum value near the limit of validity of the theory. The damping also increases with radius and is near critical at the maximum of these curves. Thus, it appears that the radius of the maximum is also the optimum design radius for a given OBS.

Note, also, that the larger the OBS the more difficult it is to obtain a high coupling frequency.

From the foregoing, Sutton and Duennebier (1985) arrived at the following desired parameters for OBS design: (1) minimum mass, (2) optimum radius in contact with ocean floor for a given instrument mass, shape and sediment density, (3) density approximately that of the sediment, (4) sensors near the center of mass/vertical symmetry, and (5) small cross section within the water. The "Opihi" (Figure 6) satisfies these requirements quite well.

Garmany (1984) presents a generalized procedure for recovering true ocean bottom particle motion using only the output signals from a three-component OBS that have been distorted by coupling and/or the local bottom conditions. Sutton et al. (1985) point out some problems with Garmany's analysis. A modified version of Garmany's method might be used to improve the quality of existing OBS data, but I believe it is not a substitute for good OBS design.

ACKNOWLEDGEMENTS

This research was supported, in part, by contracts with Office of Naval Research, Air Force Office of Scientific Research, and Defense Advanced Research Projects Agency of the United States Government.

REFERENCES

Ballard, J.A., C.C. Mulchahy, R.L. Wallerstedt, and E.L. Kiser, 1984, The Borehole Seismic Experiment in Hole 395A: Engineering and Installation, in R.D. Hyndman, M.H. Salisbury, et al., Init. Repts. DSDP, 78, Washington (U.S. Govt. Printing Office), 743-758.

Bratt, S.R. and G.M. Purdy, 1984, Structure and Variability of Oceanic Crust and Flanks of the East Pacific Rise Between 11° and 13°N, J. Geophys. Res., 89, 6111-6125.

Bratt, S.R. and S.C. Solomon, 1984, Compressional and Shear Wave Structure of the East Pacific Rise at 11° 20'N: Constraints from Three-Component Ocean Bottom Seismometer Data, J. Geophys. Res., 89, 6095-6110.

Brocher, T.M., 1983, Shallow Crustal Structure of the Continental Margin off Nova Scotia, Can. J. Earth. Sci., 20, 1657-1672.

Brocher, T.M., J.F. Gettrust, G.H. Sutton, and L.N. Frazer, 1981, Comparison of the S/N Ratio of Low Frequency Hydrophones and Geophones as a Function of Ocean Depth, Bull. Seis. Soc. Am., 71, 1649-1659.

Brocher, T.M., B.T. Iwataki, and D.A. Lindwall, 1983, Experimental Studies of Low-Frequency Water-Borne and Sediment-Borne Acoustic Wave Propagation on a Continental Shelf, J. Acoust. Soc. Am., 74, 960-972.

Byrne, D.A., G.H. Sutton, J.G. Blackinton, and F.K. Duennebier, 1983, Isolated Sensor Ocean Bottom Seismometer, Marine Geophys. Res., 5, 437-449.

Carter, J.A., F.K. Duennebier, and D.M. Hussong, 1984, A Comparison Between a Downhole Seismometer and a Seismometer on the Ocean Floor, Bull. Seis. Soc. Am., 74, 763-772.

Carter, J.A., G.H. Sutton, A. Suteau-Henson, and F.K. Duennebier, 1985, Analysis of Ocean Subbottom Seismograph (OSS) Data, this volume.

Duennebier, F.K. and G. Blackinton, 1983, The Ocean Subbottom Seismometer, in Geophys. Exploration at Sea, R. Geyer, Ed., CRC Press, Boca Raton, Florida, 317-332.

Duennebier, F.K., G. Blackinton, and G.H. Sutton, 1981, Current-Generated Noise Recorded on Ocean Bottom Seismometers, Marine Geophys. Res., 5, 109-115.

Duennebier, F.K., R. Cessaro, and P. Anderson, 1985, Deep-Ocean Bore-Hole Seismo-Acoustic Results from the North-West Pacific Ocean, this volume.

Ewing, J. and M. Ewing, 1961, A Telemetering Ocean-Bottom Seismograph, J. Geophys. Res., 66, 3863-3878.

Ewing, J.I. and R.P. Meyer, 1982, Rivera Ocean Seismic Experiment (ROSE) Overview, J. Geophys. Res., 87, 8345-8358.

Garmany, J.D., 1984, The Recovery of True Particle Motion from Three-Component Ocean Bottom Seismometer Data, J. Geophys. Res., 89, 9245-9252.

Lewis, B.T.R. and J.D. Tuthill, 1981, Instrumental Waveform Distortion on Ocean Bottom Seismometers, Marine Geophys. Res., 5, 79-86.

McCreery, C.S., D.A. Walker, and G.H. Sutton, 1983, Spectra of Nuclear Explosions, Earthquakes, and Noise from Wake Island Bottom Hydrophones, Geophys. Res. Let., 10, 59-62.

Peal, K.R. and R.E. Kirk, 1983, An Event Recording Ocean Bottom Seismograph, 1983 IEEE Proc. Third Symp. Oceanographic Data Systems, IEEE Comp. Soc. Press, Silver Springs, Maryland, 114-118.

Prothero, W.A., Jr., 1984, Ocean Bottom Seismometer Technology, EOS, 113-116.

Rauch, D., 1980, Experimental and Theoretical Studies of Seismic Interface Waves in Coastal Waters, in Bottom-Interacting Acoustics, W.A. Kuperman and F.G. Jensen, Ed., Plenum Press, New York, 307-327.

Rowlett, H. and D.W. Forsyth, 1984, Recent Faulting and Microearthquakes at the Intersection of the Vema Fracture Zone and the Mid-Atlantic Ridge, J. Geophys. Res., 89, 6079-6094.

Schirmer, F., 1980, Experimental Determination of Properties of the Scholte Wave in the Bottom of the North Sea, in Bottom-Interacting Acoustics, W.A. Kuperman and F.G. Jensen, Ed., Plenum Press, New York, 285-298.

Shearer, P. and J. Orcutt, 1985, Anisotropy in the Oceanic Lithosphere--Theory and Observations from the NGENDEI Seismic Refraction Experiment in the Southwest Pacific, Geophys. J. R. Astr. Soc., 80, 493-526.

Steinmetz, R.L., P.L. Donoho, J.D. Murff, and G.V. Latham, 1979, Soil Coupling of a Strong Motion, Ocean Bottom Seismometer, Proc. Offshore Tech. Conf. Houston, 2235-2249.

Stephan, R., 1981, Seismic Anisotropy Observed in Upper Oceanic Crust, Geophys. Res. Let., 8, 865-868.

Sutton, G.H. and F.K. Duennebier, 1985, Optimum Design of Ocean Bottom Seismometers, Marine Geophys. Res., in press.

Sutton, G.H., F.K. Duennebier, and G.J. Fryer, 1985, Comment on "The Recovery of True Particle Motion from Three-Component Ocean Bottom Seismometer Data" by Jan D. Garmany, J. Geophys. Res., in press.

Sutton, G.H., F.K. Duennebier, and B. Iwataki, 1981a, Coupling of Ocean Bottom Seismometers to Soft Bottom, Marine Geophys. Res., 5, 35-51.

Sutton, G.H., F.K. Duennebier, B. Iwataki, J.D. Tuthill, B.T.R. Lewis, and J. Ewing, 1981b, An Overview and General Results of the Lopez Island OBS Experiment, Marine Geophys. Res., 5, 3-34.

Sutton, G.H., G. McDonald, D.D. Prentiss, and S.N. Thanos, 1965, Ocean-Bottom Seismic Observations, Proc. IEEE, 53, 1909-1921.

Sutton, G.H., M.E. Odegard, N. Mark, and N.J. Letourneau, 1969, Research in Seismology Related to the Columbia Ocean-Bottom Seismograph, Hawaii Institute of Geophysics Final Report to AFCKL/ARPA, HIG-70-12, 66p.

Trehu, A.M., 1985, Coupling of Ocean Bottom Seismometers to Sediment: Results of Tests with the U.S. Geological Survey Ocean Bottom Seismometer, Bull. Seis. Soc. Am., 75, 271-289.

Trehu, A.M. and G.M. Purdy, 1984, Crustal Structure in the Orozco Transform Zone, J. Geophys. Res., 89, 1834-1842.

Whitmarsh, R.B. and R.C. Lilwall, 1983, Ocean-Bottom Seismographs, in Structure and Development of the Greenland-Scotland Ridge, Bott, Saxov, Talwani, and Thiede, Ed., Plenum Press, New York, 257-286.

Zelikovitz, S.J. and W. Prothero, 1981, Vertical Response of an Ocean Bottom Seismometer: Analysis of the Lopez Island Vertical Transient Tests, Marine Geophys. Res., 5, 53-67.

THE USE OF LOVE WAVES TO DETERMINE THE GEOACOUSTIC PROPERTIES OF MARINE SEDIMENTS

T. Akal*, H. Schmidt*, and P. Curzi**

*SACLANT ASW Research Centre
Viale San Bartolomeo 400
I-19026 La Spezia, Italy

**Istituto di Geologia Marina
Consiglio Nazionale delle Ricerche
Via Zamboni 65, 40127 Bologna, Italy

ABSTRACT

The importance of the geoacoustic properties of marine sediments for low-frequency shallow water propagation is well established. Reliable compressional properties can be obtained from laboratory analysis of collected samples. The shear properties, however, are highly affected by the deterioration of the chemical and mechanical bindings caused by de-pressurization and change of temperature in the core sampling process. These properties therefore have to be determined in situ. Isolated shear waves are not easily generated due to partial coupling into compressional waves at all interfaces. In the case of stratified, isotropic media, however, coupling can be avoided if the polarization of the shear waves is parallel to the interfaces. Here an experimental technique is described for generation and detection of horizontally polarized shear waves - or Love waves - in a horizontally stratified sea-bed. Shear velocity profiles are determined indirectly by numerical modelling of the experimental data. Results are presented for different types of marine sediments.

INTRODUCTION

The importance of the geoacoustic properties of marine sediments for low-frequency acoustic propagation is well established [1]. Geoacoustic properties also provide a major source of information for the determination of the static and dynamic characteristics of the sea-bed.

When seismic waves propagate through the marine sediments, the dynamic properties of the sediments determine the speed and the attenuation. An investigation of the seismic propagation characteristics can therefore provide accurate information on the geotechnical characteristics of the sea-bed.

Assuming that the sea-floor sediments behave like elastic media, the seismic propagation is controlled by the bulk modulus K, the shear modulus

G, and the density ρ. These parameters are related to the compressional wave velocity, C_p, and the shear wave velocity, C_s, by

$$C_p = [(K + 4G/3)/\rho]^{1/2} \quad \text{and}$$

$$C_s = (G/\rho)^{1/2}$$

If the compressional wave velocity (C_p) and the shear wave velocity (C_s) of an idealized marine sediment can be measured concurrently with a measurement of bulk density, the elastic parameters of the sediments can be completely characterized (see Tables 1 and 2).

Table 1.

BASIC WAVE TYPES AND VELOCITIES	
a) BODY WAVES	
–COMPRESSIONAL (longitudinal)	$C_P=[(K+4G/3)/\rho]^{1/2}$
–SHEAR (transverse)	$C_S=(G/\rho)^{1/2}$
b) DUCTED WAVES	
– LOVE	$C_L=(G_l/\rho)^{1/2}$
c) SURFACE WAVES	
– SCHOLTE	$C_{SCH}\sim(G/\rho)^{1/2}$

Table 2.

ELASTIC PARAMETERS IN TERMS OF WAVE VELOCITIES AND BULK DENSITY (ρ)
K: BULK MODULUS (incompressibility)
$K=\rho(C_P^2-4C_S^2/3)$
E: YOUNG's MODULUS
$E= 2C_S^2\rho(1+\sigma)$ or
$E=\rho C_P^2(1+\sigma)(1-2\sigma)/(1-\sigma)$
σ:POISSON's RATIO (transv./long. strain)
$\sigma= 1/2(C_P^2-2C_S^2)/(C_P^2-C_S^2)$
G:SHEAR MODULUS (rigidity)
$G=\rho C_S^2$
λ=LAME's CONSTANT
$\lambda=\rho(C_P^2-2C_S^2)$

Several laboratory and field methods are available to measure density and wave velocities in bottom sediment cores. However, the reliability of such measurements is degraded by sample disturbance and temperature and pressure changes. In particular, the shear properties are highly effected by the deterioration of the chemical and mechanical bindings caused by the differences in temperature and pressure between the sampling and the laboratory measurements. Also, in the laboratory, the small size of the sample does not allow direct information to be obtained about the response of these materials to low-frequency seismic/acoustic waves. Therefore the most reliable geoacoustic measurement techniques for ocean-bottom sediments are in situ techniques.

The choice of an actual experimental technique is strongly influenced by the structure of the sea bed. If the bottom were homogeneous to a significant depth, the wave speeds and attenuation could be measured directly because different seismic wave types could be separated. In reality, however, recent sediments consist of a relatively thin top layer of unconsolidated clay or silty clay overlying semi-consolidated silts and sands [2,3]. This structure results from the lowering of sea levels during the glaciation of the Pleistocene epoch in which sand was deposited over wide areas of the continental shelves. The shelves were subsequently covered by unconsolidated sediments during sea level rising during the post-glacial times. Sedimentary processes such as sorting, compaction and consolidation, also have contributed to the highly stratified structure of the sea bed.

In such a layered sea bed the compressional and shear waves couple, yielding an often complex wave field. The geoacoustic properties therefore must be determined indirectly (inversely) from the propagation characteristics. The most important attenuation mechanism of low frequency propagation in shallow water is the conversion of acoustic energy in the water column into shear waves in the bottom [1]. Because of this importance this mechanism has received special attention during the past decade. Because the propagation of seismic interface waves (also known as Scholte waves) is almost entirely controlled by their shear properties, significant effort has been applied to the analysis of these properties [4,5]. However, the inversion involves significant numerical effort not easily automated.

Here it will be demonstrated that horizontally polarized shear waves (SH), also called Love waves, can be generated by simple means in the low-velocity duct formed by the upper sediments. Because the shear waves do not couple into compressional waves in a horizontal stratification of isotropic layers, their propagation characteristics directly reflect the shear properties. The Love waves therefore form an alternative to the seismic interface waves as the basis of the shear property inversion. Further, the scalar nature of the wave equation for Love waves significantly decreases the calculations involved in the inversion compared to the requirements for the interface waves.

EXPERIMENTAL PROGRAMME

The Love wave experiments were caried out as part of a multinational research programme to measure the velocity and attenuation of seismic waves and supporting geotechnical parameters in marine sediments. The programme was initiated by SACLANT to bring together experts in geophysics, geology and acoustics to compare various measurement and analysis techniques for data gathered under identical environmental conditions.

For this program four institutes played a leading role in the specific problems related to sampling, analysis and interpretation of ocean bottom sediments. The institutes and their tasks were:

- Institute of Marine Geology (IGM-CNR), Bologna, Italy; sampling the sea-bed using different techniques, geotechnical analysis and interpretation, and geological description;

- Centro de Fisica Applicada (CENFA), Madrid, Spain; measuring reflection loss under controlled conditions in a layered sea-bed with variable sensitivity gradients, and testing a model developed by the CENFA scientists;

- US Naval Ocean Research and Development Activity (NORDA), Bay St. Louis, MS, USA; sediment sampling, laboratory analysis of the samples for compressional and shear wave velocities, sediment mass property analysis, measurements with the ELAC Sea-Bed Classifier to define high-resolution layering, and sediments classification;

- SACLANT ASW Research Centre (SACLANTCEN), La Spezia, Italy; programme coordination, measurement of seismic wave velocities and attenuation using ocean-bottom seismometers, sampling and geoacoustic analysis of sediments and site surveying (using side-scan sonar and seismic profilers); SACLANTCEN theoretical propagation models were used with seismic data to determine the sea-bed properties.

MEASUREMENT TECHNIQUES

The techniques developed by the different institutions were applied to four different sedimentary areas in the vicinity of La Spezia, Italy (Fig. 1).

Shear-wave velocity measurements

A directional energy source was developed to excite both interface and ducted Love waves. Figure 2 illustrates the techniques used by SACLANTCEN. Five ocean bottom seismometers (OBS), each with tri-axial geophones and a hydrophone, were deployed along the measurement range in each area. For the short ranges (< 25 m), a high energy source level was not needed. The signals were generated by shooting a spear from a hydraulic gun mounted on a frame into a mass resting on the sea-bed (Fig. 3). This type of source generates a signal below the elastic limits of the sediments, thus avoiding changes in sediment properties. The direction of the gun-mount could be changed as appropriate to generate different wave types. The experiment was conducted by diving scientists [6].

Figure 4 shows examples of the signals received at a distance of 25 m from the source. For example, in Fig. 4a, the geophone received maximum energy from a transverse (shear) pulse traveling in the x-direction. The alignment of the corresponding particle-motion diagram (hodograph) also

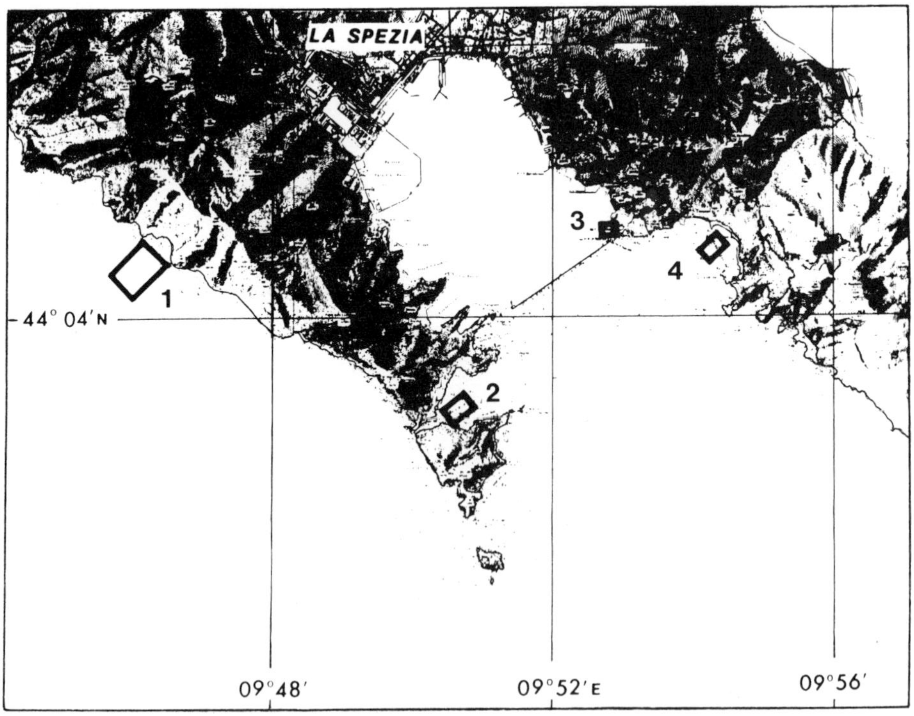

Fig. 1 Experiment areas, La Spezia, Italy.

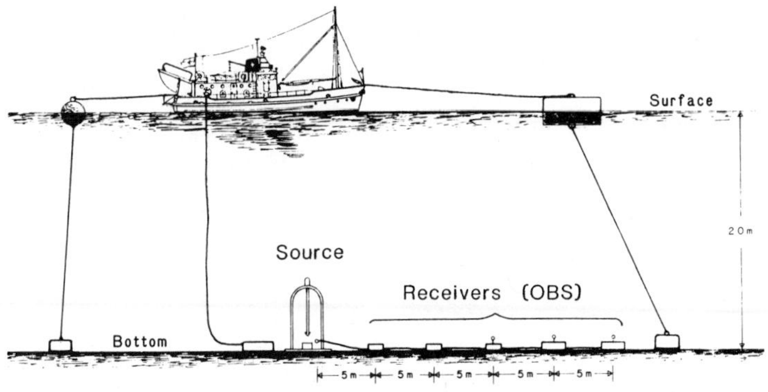

Fig. 2 Techniques used to measure Love wave velocity

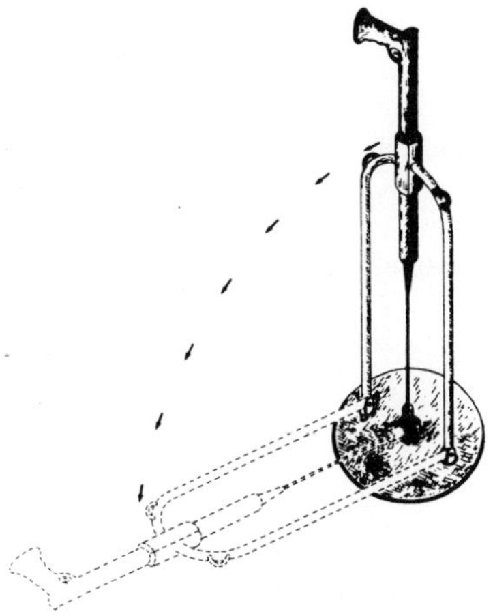

Fig. 3
SACLANTCEN technique to generate seismic signals; a gun shoots a spear against a mass coupled to the sea-bed with a heavy metal plate.

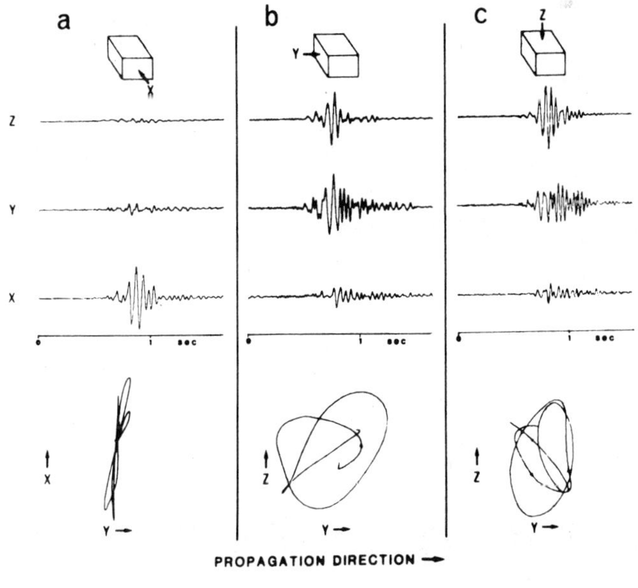

Fig. 4 Acoustic signals received at a distance of 25 m from the source, and corresponding particle motion diagrams (hodographs).

confirms that the signal is confined to the horizontal plane (x-axis). This dispersed form of the transverse signal indicates that this is not a pure body wave but a Love wave propagating within the upper sedimentary layer. This assumption was confirmed by synthetic seismograms produced by the numerical model SAFARI [7]. The radial (Fig. 4b) and vertical (Fig. 4c) signals were much more complex; the particle motions were confined to either the radial or vertical plane with very small transverse deflections, a characteristic behaviour of interface waves [4].

Bottom reflectivity measurements

Figure 5 illustrates the system used by NORDA and CENFA to measure the acoustic reflection characteristics of the bottom. NORDA's ELAC Sea-Bed Classifier (30 kHz) was used both to obtain high resolution sub-bottom layering information and high-resolution bottom reflection loss measurements. SACLANTCEN also used an EG&G UNIBOOM seismic source for bottom reflection measurements.

Sediment sampling

Figure 6 shows the techniques used to sample sea bed sediments. Samples were collected by the diver-scientists who used a grab sampler, a box corer, a gravity corer, and IGM's newly developed compressional air-powered corer.

Side-scan sonar

The site surveys, using side-scan sonar (EG&G MARK1B), provided knowledge of both the surface sediment distribution and the sea bed topography.

Continuous seismic profiling

An EG&G UNIBOOM source/receiver and a streamer were towed behind the survey ship. Resolution was poor in water depths less than 10 m.

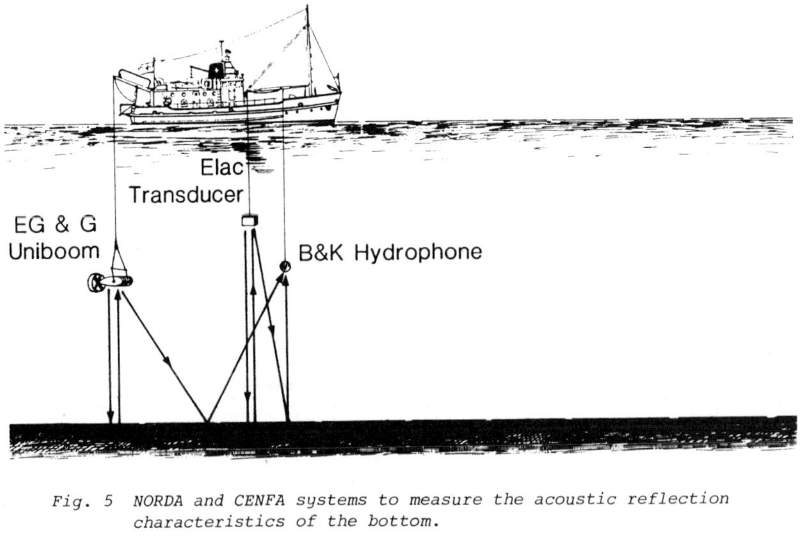

Fig. 5 NORDA and CENFA systems to measure the acoustic reflection characteristics of the bottom.

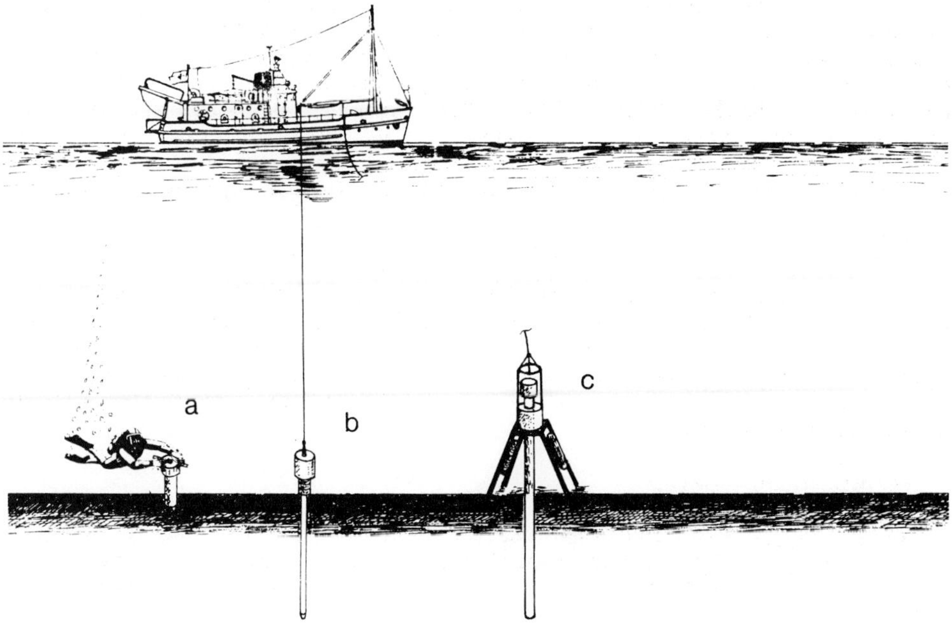

Fig. 6 Techniques used to sample sea-bed sediments
a) hand corer
b) gravity corer
c) IGM's compressional air-powered corer

Laboratory analysis

The cores were analyzed every 2 cm along their length for the following properties and characteristics:

- compressional wave velocity
- shear wave velocity
- porosity
- water content
- wet density
- dry density
- torvane and shear strength
- Attenberg limits
- grain size
- mineralogic composition

EXPERIMENTAL RESULTS

Data from two areas, 1 and 2, with different bottom conditions are presented here (see Fig. 1).

Area 1 is situated along the Cinque Terre coast of the Ligurian Sea; Area 2 is situated at the eastern entrance to the Portovenere channel of the Gulf of La Spezia.

Area 1

This study area had a water depth of 15 m. The sea-bed consisted of coarse sand and fine gravel. Side-scan sonar and observations by diving scientists indicated that the area was covered with ripples (approximately λ = 2 m) created by wave action. To avoid distorting the propagation path the axis of the measurement range was laid out parallel to the axes of the ripples. Seismic profiles (Fig. 7) show a sub-bottom reflector approximately 2 ms from the bottom. The bottom sediments apparently were derived from land slides and the reworking of sediments along the steep coastline.

Figure 8 summarizes the measured parameters along a core sample taken in this area. The high percentage of gravel in the sea bed limits the core penetration to less than 20 cm. This gravel consists mainly of limestone, slate, and sandstone, apparently derived from the adjacent mountains which have a similar lithology.

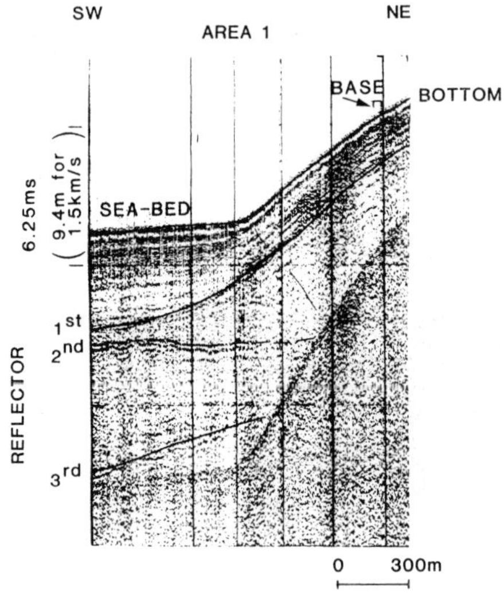

Fig. 7 Area 1. Continuous seismic profiling record and interpretation.

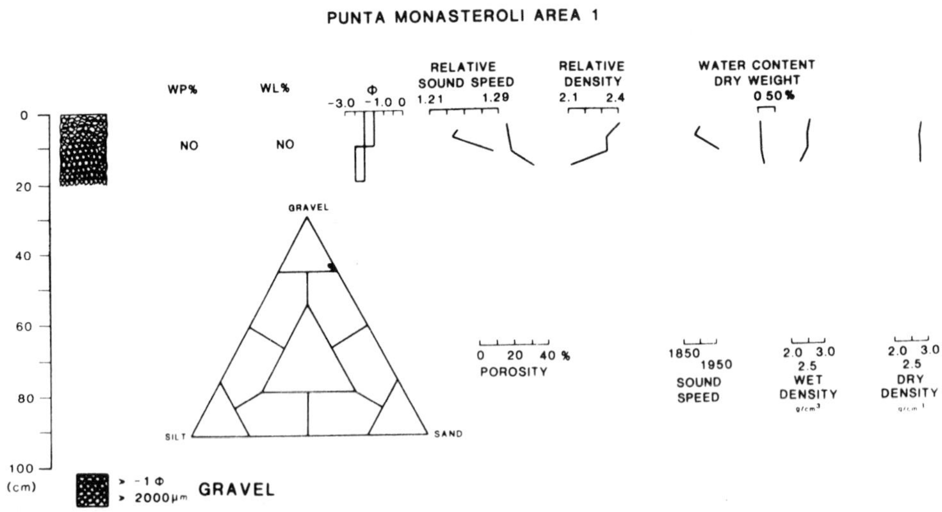

Fig. 8 Area 1. Measured parameters along a core sample.

Figure 9 is a time vs distance plot of the output of the transverse (shear) wave sensitive geophone. Distinct wavelets can be observed traveling with speeds up to 90 m/s. The inversion was performed by means of the SAFARI model [7], using trial and error. The very simple shear velocity profile shown in Fig. 10a yielded the synthetic seismograms shown in Fig. 10b, which agree well with Fig. 9.

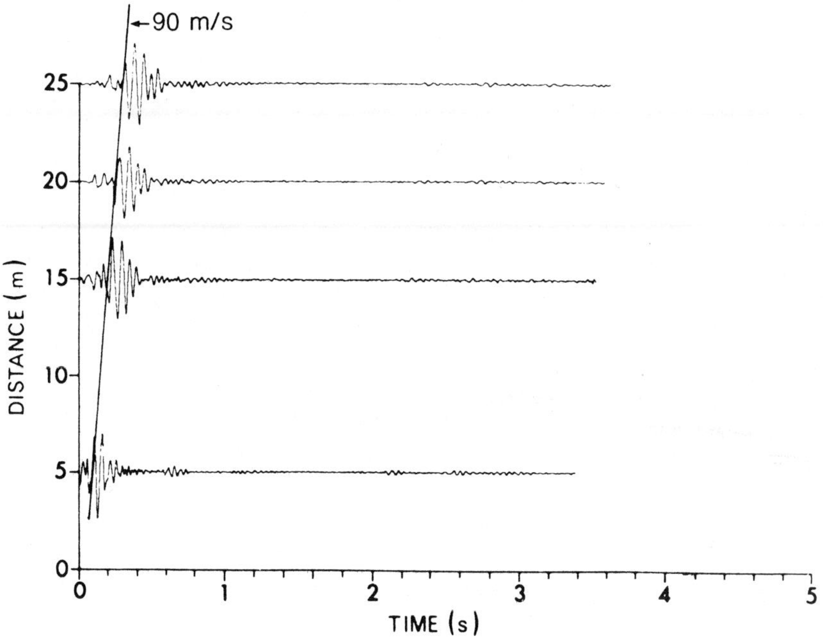

Fig. 9 Area 1. Time vs distance plot of the shear wave sensitive geophone.

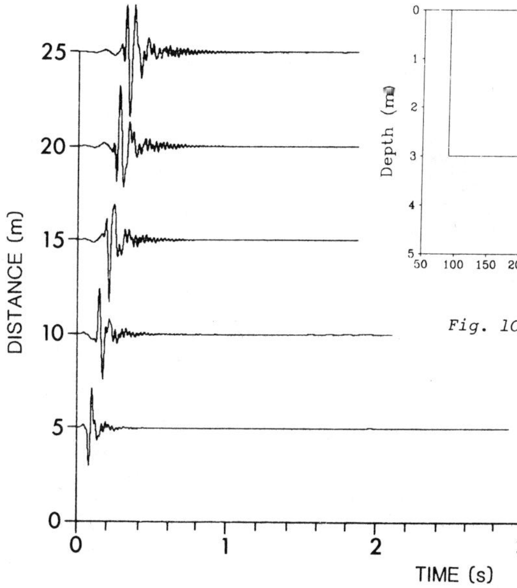

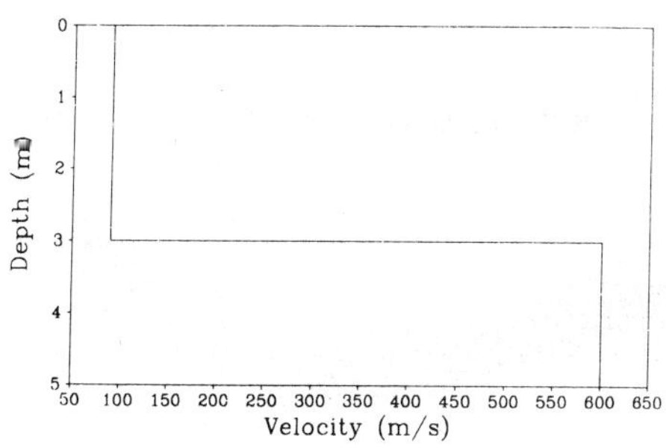

Fig. 10a Shear velocity profile.

Fig. 10b Synthetic seismograms.

Area 2

Side-scan sonar records of the eastern entrance to the Portovenere Channel indicate a flat bottom. Sub-bottom profiles taken along the measurement base indicate four distinct reflectors (Fig. 11).

Figure 12 illustrates the physical parameters measured along one of the cores taken in Area 2. The material consists of silty clays and clayey silts with very small variations. The mineralogy of these sediments consists mostly of quartz and calcite with some magnesium. Of the clay minerals, 60% is illite, 10% chlorite and 30% smectite and kaolinite. The percentage of illite increases towards the bottom of the core sample, i.e., with the depth in the core (and hence the sea bottom). Attenberg limits measured were 51 to 55% for the liquid limit (W_L) and about 25% for the plastic limit (Wp). The liquid limit tends to increase with depth, verifying that the sediments become finer towards the bottom of the core. Figure 13 shows the shear waves detected by transverse wave sensitive geophones. A distinct wavelet, traveling at 30 m/s, indicates Love wave speeds in the upper sedimentary layers about 1/3 of that in Area 1.

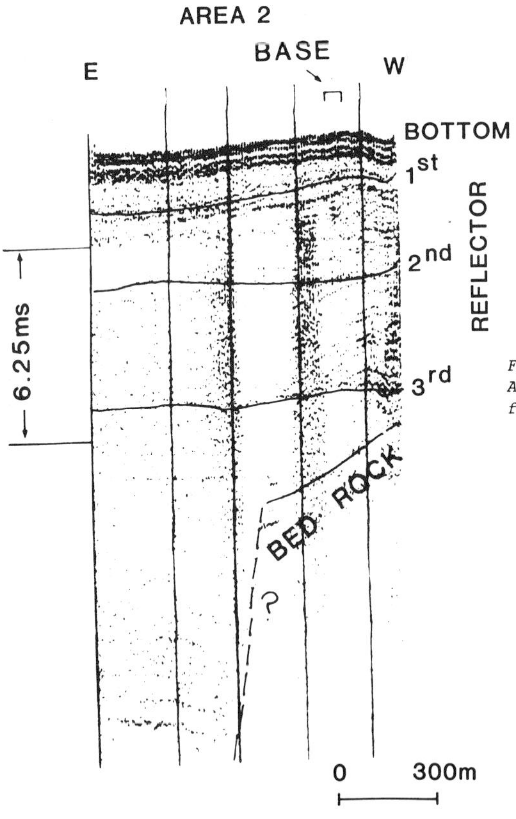

Fig. 11
Area 2. Continuous seismic profiling record and interpretation.

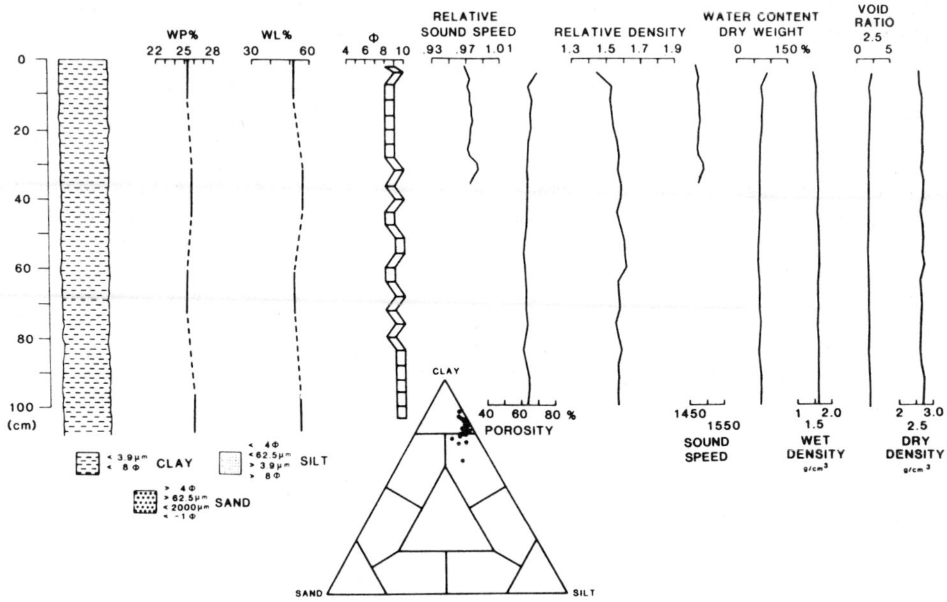

Fig. 12 Area 2. Measured parameters along a core sample taken off Porto Venere.

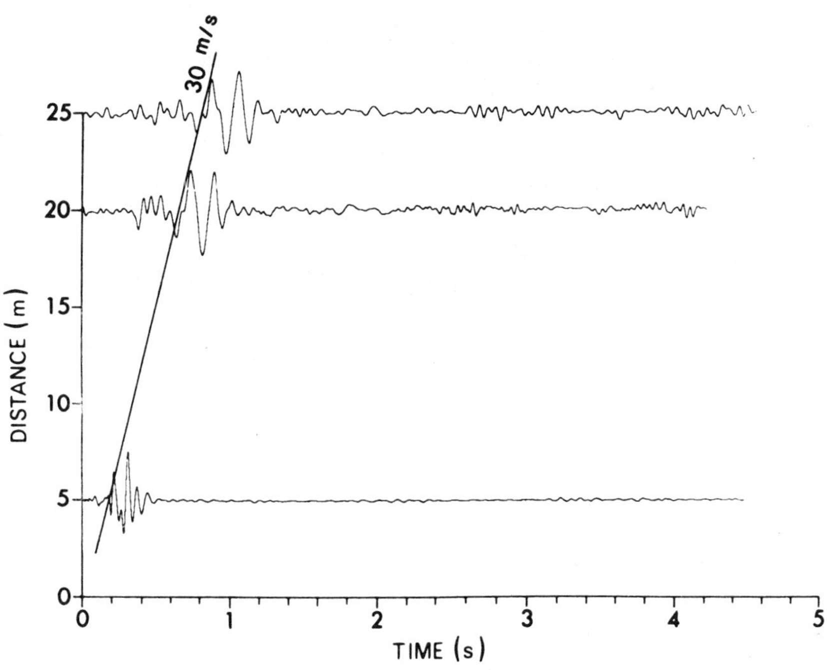

Fig. 13 Area 2. Time vs distance plots of the shear wave sensitive geophones, and shear-wave speed.

CONCLUSIONS

These simple experiments have demonstrated that Love waves can be generated in the low velocity duct formed by the uppermost unconsolidated sediment layers. The Love waves are horizontally polarized shear waves; in horizontally stratified environments they do not convert into other wave types. This property greatly simplifies theoretical modelling of their behaviour. Love waves may therefore form an alternative basis for determination of the sediment shear properties which in recent years has been based on the properties of the more complex seismic interface waves. The experiment demonstrated that two wave types (Scholte and Love) can in fact be generated and detected in a single experimental setup. The combination therefore is a realistic possibility which, if applied, can improve the prediction accuracy of deduced shear speed profiles. Further, the combination can elucidate the degree of anisotropy in the upper sedimentary layers.

REFERENCES

1. Akal, T. and Jensen, F.B. Effects of the sea-bed on acoustic propagation. In: PACE, N.R., ed. Acoustics and the Sea-Bed, Bath, U.K., Bath University Press, 1983: 225-232.
2. Akal, T. Acoustical characteristics of the sea floor: experimental techniques and some examples from the Mediterranean sea. In: HAMPTON, L., ed. Physics of Sound in Marine Sediments. New York, NY, Plenum, 1973: pp 447-480.
3. Hamilton, E.L. Geoacoustic models of the sea floor. In: HAMPTON, L., ed. Physics of Sound in Marine Sediment. New York, NY, Plenum, 1973: pp 181-222.
4. Rauch, D. On the role of bottom interface waves in ocean seismo-acoustics (this volume).
5. Jensen, F.B. and Schmidt, H. Shear properties of ocean sediments determined from numerical modelling of Scholte wave data (this volume).
6. de Strobel, F., Akal, T., and Hastrup, O.F. SACLANTCEN's use of SCUBA diving in oceanographic and acoustic research. SACLANTCEN SM-175. La Spezia, Italy, SACLANT ASW Research Centre, 1984.
7. Schmidt, H. Modelling of pulse propagation in layered media using a new fast field program. In: FELSEN, L.B. ed. Hybrid Formulation of Wave Propagation and Scattering. Dordrecht, Nijhoff, 1984: pp 337-356.

DEVELOPMENT OF A DEEP-TOWED SEISMIC SYSTEM

A NEW CAPABILITY FOR DEEP-OCEAN ACOUSTIC MEASUREMENTS

Martin G. Fagot

Naval Ocean Research and Development Activity
Ocean Acoustics and Technology Directorate
NSTL, Mississippi 39529-5004

ABSTRACT

The development of a deep-towed seismic profiling system has been completed after four years of design, fabrication, and testing. This system provides the capability to determine the detailed acoustic character of the sea floor and upper 500-1000 m of subbottom structure. The system is designed for tow depths of 6000 m while operating at an altitude of 100 m above the sea floor. The major components of the system include a sound source located in the deep-towed vehicle, which operates over a frequency range from 260 Hz to 650 Hz at a peak source level of 201 dB re 1 micropascal at 1 m; a 24 channel, 1000 m long array attached as a tail to the deep-towed vehicle; a duplex digital telemetry link to communicate with the deep-towed system over a coaxial tow cable; a digital data record system; a real time engineering data display system; a short baseline navigation system; handling hardware to deploy and retrieve the system; and two instrumentation vans to support system operations. This paper discusses the unique hardware developed for the deep-towed seismic system and focuses on the acoustic subsystems. Also reviewed is a performance prediction analysis initially performed to define geoacoustic parameter measurement improvements achievable with a deep-towed configuration. Field measurement engineering results acquired during the system's final performance evaluation are presented. Data was acquired at tow depths of 4500 m during the September 1984 evaluation.

INTRODUCTION

The Naval Ocean Research and Development Activity (NORDA) carries out broadly based research and development in ocean science and technology to understand the effects of the ocean environment on Navy systems and operations. Specifically, one area of responsibility is developing definitive models of the ocean floor and subbottom as a transmission medium that refracts, diffracts and dissipates, as well as reflects, acoustic energy. These models are used to predict, and thus improve, the performance of naval systems that acoustically interact with the bottom. Critical for developing these models is high-resolution data to describe the geological, geophysical, and geoacoustical character of the deep-ocean sea floor and upper subbottom structure. These high-resolution data can be obtained using a low-frequency sound source and multichannel array system capable of operating at an altitude of 100-500 m in ocean depths of 6000 m. This configuration is depicted in Figure 1. A development program entitled "Deep-Towed Array Geophysical System" (DTAGS) was initiated in 1981 to provide a hardware suite to meet this high-resolution data requirement.

System performance prediction studies[1,2,3,4,5] showed substantial improvement in measuring geoacoustic parameters with the DTAGS configuration. A comparison between the deep-towed configuration and a surface-towed configuration for measuring a critical geoacoustic parameter, interval velocity, as a function of layer thickness is given.

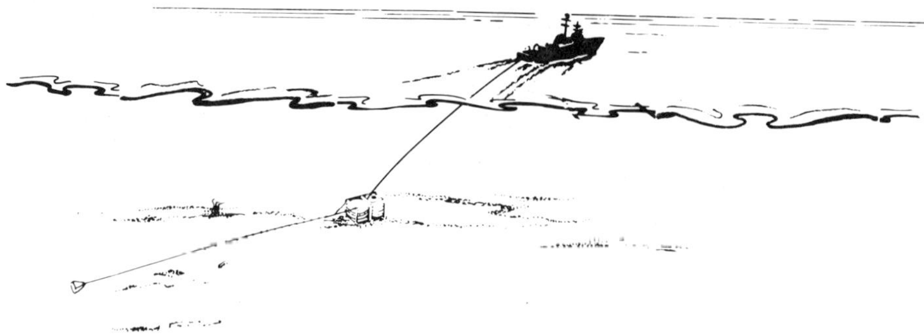

Figure 1. Deep-towed system configuration.

The initial hardware development focused on a Helmholtz resonator sound source system. This unit was successfully tested to a depth of 2000 m in December 1981. Subsequent effort concentrated on developing the multichannel hydrophone array (672 m long) and high data rate telemetry system. These systems were integrated with the sound source system and sea tested to a depth of 4500 m during the summer of 1983. The hardware development concluded with extending the array to a total length of 1000 m, fabricating a digital data record system, and incorporating into the system a short baseline navigation system. The total hardware suite was tested to a depth of 4500 m in September 1984.

PERFORMANCE PREDICTION ANALYSIS

The basic objective of the deep-towed system is to provide measurements for extracting the geoacoustic parameters of the near subbottom. Such parameters include sound speed as a function of depth and vertical thickness between acoustic reflectors. The geoacoustic parameter addressed here is termed interval velocity, where interval velocity is defined as the RMS sound speed in the layer (interval) of interest. For a layer of constant sound speed, the interval velocity reduces to simply the sound speed in the layer. Analysis by Gibson et al.[6] and Gholson and Fagot,[4] more advanced than that presented here, addresses the problems associated with estimating sound speed gradients.

Estimating the interval velocity of a layer bounded above and below by acoustic reflectors can be accomplished by exercising the Dix[7] equation shown below:

$$Vi = \left[\frac{V_{A2}^{\,2}\,T_2 - V_{A1}^{\,2}\,T_1}{T_2 - T_1}\right]^{1/2},$$

Errors in measuring normal incidence travel time (T_1, T_2) were modeled as zero mean independent errors uniformly distributed with bounds ± Tpick where Tpick is the precision of measuring (picking) reflected energy arrival times. Errors in the measured array velocities V_{A1}, V_{A2} are much more difficult to model. The approach was to assume that the fitted T^2 (X^2) curve never departs from the true T^2 (X^2) by more than that associated with the picking error Tpick. Further, it is assumed that the just-mentioned is reliably true over a horizontal span, X, equal to one-half the physical array length. The last assumption is equivalent to assuming the array is one-half its actual length and a moveout curve is successfully fitted (± Tpick) through returns from every hydrophone group. Stoffa[8] and simulations using realistic signal-to-noise ratios indicate the one-half array length assumption to be reasonable for an array with 24 equally spaced groups. This error analysis allowed for system optimization, as well as general system performance prediction.

This type of error analysis was used in comparing a surface-towed system with a deep-towed system. A geoacoustic model, along with a surface versus deep-towed system comparison, is shown in Figure 2. The surface-towed system differs from the deep-towed system both in array length and altitude above the layer of interest. The increasing error for small Z is due to the water column sound speed dominating the measured array velocity. This effect is much worse for the surface-towed system, since the water column is much thicker.

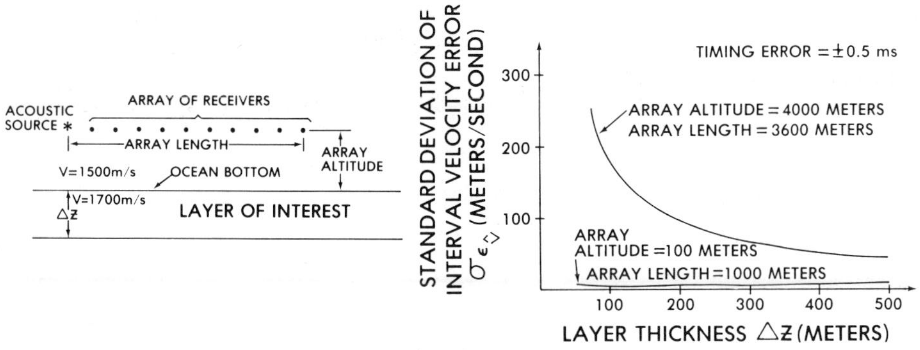

Figure 2. Geoacoustic model and performance comparison between a surface-towed and deep-towed system.

HARDWARE DESCRIPTION

A block diagram of DTAGS hardware configuration is given in Figure 3. The sound source transducer and power amplifier system, telemetry system, and associated electronics are located within a deep-towed vehicle (fish) towed at the head of a multichannel array. The deep-towed system is tethered to the tow ship with a 9150-m long coaxial steel tow cable. A duplex telemetry system communicates via the tow cable with the deep-towed system. Data from the fish system is then digitally recorded at the tow ship. Monitoring of system performance is through a real-time display system. The position of the fish relative to the tow ship is provided by a short baseline navigation system. Global position is obtained through satellite and Loran-C navigation. A tow cable heave accumulator and an array handler winch constitute the system handling hardware. The operation systems are located in a portable van with a maintenance facility and spares storage located in a second van. The component designs of these major subsystems are optimized to meet the basic geophysical data requirement while still maintaining a hardware suite that can be easily deployed, operated, and maintained. A detailed description of the total hardware suite is described by Fagot.[9]

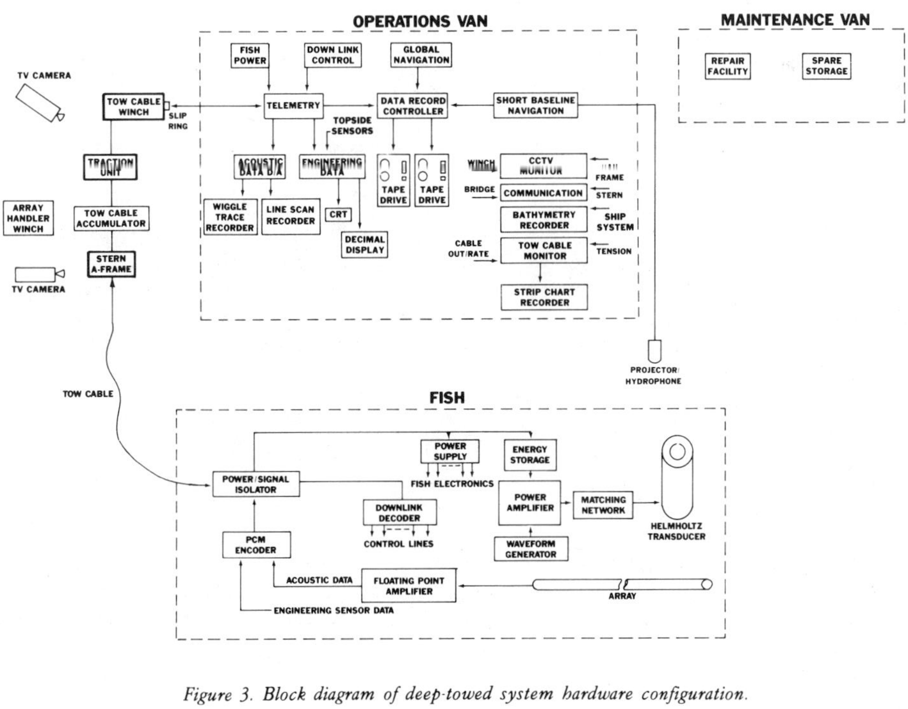

Figure 3. Block diagram of deep-towed system hardware configuration.

Sound Source System

The sound source system hardware consists of a piezoelectric Helmholtz resonator transducer and a power amplifier. The amplifier was designed and packaged for operation at full depth. The transducer and power amplifier are mounted together in the deep-towed vehicle (fish) as shown in Figure 4. Also identified are other major subsystems located on the fish.

The sound source system is capable of full performance from near surface to full ocean depth. The system is capable of a nominal peak source level of 201 dB//1 μPa @ 1 m. This level can be achieved over a frequency range from 260 Hz to 650 Hz. Directivity over this range of frequencies is omnidirectional. Output pulse length is programmable from 5-ms minimum to a maximum of 250 ms in duration. The system is capable of a full power 250-ms output pulse once every 12 seconds.

The transducer consists of a Helmholtz cavity and orifice driven by five piezoelectric-segmented ceramic ring drivers. As built, the transducer is approximately 0.7 m in outside diameter, is 1.1 m long, and weighs 800 kg. The design, development, and fabrication of this transducer was performed by the Naval Research Laboratory, Underwater Sound Reference Detachment and is described by Young.[10]

A Class S switching amplifier was chosen as the driver for the Helmholtz resonator transducer due to its small relative size compared to a linear amplifier, such as a Class B. The Class S power amplifier system weighs approximately 232 kg (including pressure vessel) and is housed in a 0.6-m diameter spherical pressure vessel. Due to the inefficiency of a deep-water broadband sound source, the power amplifier is required to deliver 16 kVA of reactive power. Other requirements of the system called for control of output waveshape and frequency. Also due to inaccessibility and remote location of electronics during an at-sea operation, reliability of the power amplifier was of prime concern. The design, development, and fabrication of the power amplifier was performed by NORDA.

The power amplifier system located at the tow vehicle consists of a waveform generator, power amplifier modules, energy storage bank, and a load matching network. Four drive waveforms are stored in an erasable, programmable, read only memory (EPROM) that is programmed prior to deployment. Selection of the source drive waveform, along with the source firing key signal, is accomplished via the downlink telemetry system. Energy for the highpower output pulse is stored in a 0.132 farad capacitor bank. This bank is trickle charged between pulses at an average rate of 300 watts via the tow cable. The power amplifier module controls the power drive level and waveshape supplied to the Helmholtz

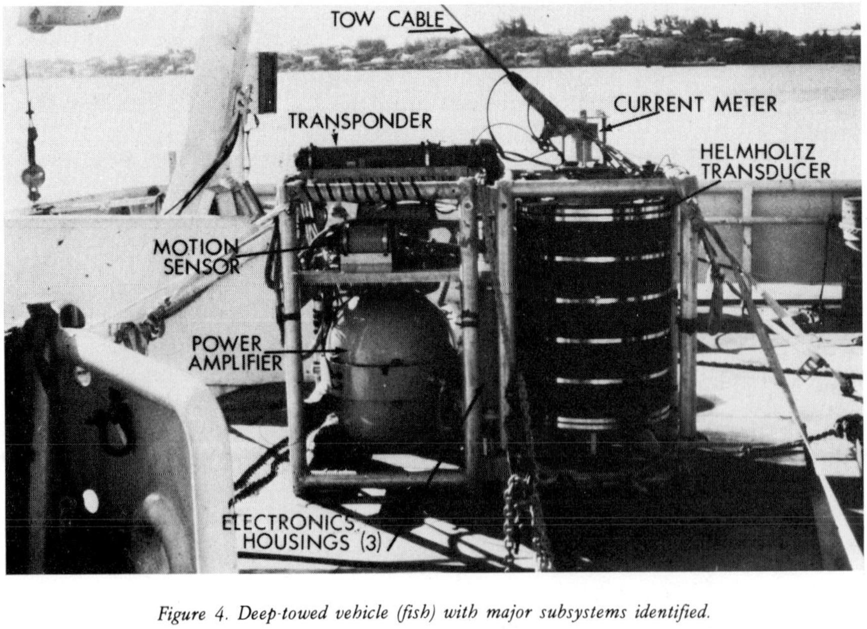

Figure 4. Deep-towed vehicle (fish) with major subsystems identified.

resonator transducer. Power transfer to the source is optimized through the load matching network, which consists of a power factor correction network and an impedance matching transformer. The matching network is housed in a separate pressure vessel.

Array System

The array design selected for the deep-towed geophysical system is a multichannel hydrophone array similar to the type used by the marine exploration community. The standard oil-filled design, hose, spacers, and oil-immersed electromechanical coupling presently used are compatible with high operating pressures. The array is 1000 m long and consists of 12 active sections that contain a total of 48 hydrophone groups. Only 24 groups will be used for a specific configuration to reduce the telemetry data rate requirement. A 2-m interconnect section is located at the head of each 82-m active section. Each 82-m active section has an outside diameter of 4.61 cm; the 2-m interconnects are either 7.62 cm or 10.16 cm, depending on the sensor suite, and a 50-m stretch section at the head of the array has a diameter of 6.6 cm. The entire array is designed to have a tensile strength greater than 1200 kg and for a maximum tow speed of 1.5 m/sec (3 knots). A drogue parachute is attached to the array tail for tow stability. The interconnect sections contain heading sensors, depth sensors, and hydrophone preamplifiers. Each active section contains four hydrophone groups. Figure 5 shows the array on the handler winch.

The array acoustic design is tailored to the deep-tow geophysical system operational performance requirements. Forty-eight hydrophone groups are spaced uniformly along the array with a 21-m separation between adjacent groups. Each hydrophone group consists of six elements with a total length of less than 60 cm. The individual hydrophones are a piezoelectric cylinder type that are tolerant to high operating pressures. Outputs of the hydrophones are connected to the preamplifiers located in the 2-m interconnect section at the head of the associated active section via a twisted pair shielded cable. Because of the large separation between the hydrophone group and preamplifier (approximately 74 m for the last group of each active section) and the low group capacitance, a balanced input charge amplifier is used for the preamplifier to accommodate these conditions. The combined sensitivity of an acoustic channel (hydrophone group plus preamplifier) is a nominal -150 dB//1 V/μPa with a frequency response from 150 to 1000 Hz. The bandwidth is limited by filters in the preamplifier and can be opened to a band ranging from 10 to 2000 Hz by appropriate component changes in the preamplifier. The self-noise of each acoustic channel is less than 35 dB (// 1 μPa/$\sqrt{\text{Hz}}$) over the operating frequency band, which is less than deep-ocean Sea State 0 ambient noise. Each channel preamplifier has a built-in calibration oscillator with a unique frequency that provides easy channel identification and an in situ functional check of the entire acoustic channel, including the hydrophone group. Gain of the preamplifiers can also be lowered 20 dB by an external control signal.

Figure 5. Deep-towed array on the array handler winch.

The array is equipped with a suite of engineering sensors to measure array tow dynamics, such as array shape, tension, and tow attitude. Each interconnect section contains a depth sensor and four heading sensors are spaced uniformly over the array length. A load cell at the head of the array is used to monitor in-line array tension in real time.

Ancillary Systems

A number of other subsystems are required to support the deep-towed operations. An overview of these subsystems is presented.

Telemetry System. The telemetry system for the deep-towed geophysical system provides full duplex communication between the tow ship and the tow vehicle via a 9150-m coaxial tow cable. System requirements call for a data transfer rate of 1.5 M bit/sec. The telemetry system is capable of handling tow fish power (300 W) down the cable, as well as the 1.5 M bit/sec pulse code modulation (PCM) uplink data and low data rate FSK downlink control signals. The acoustic signals from the towed array are digitized by a 12-bit A/D with a 4-bit gain exponent floating point amplifier at a 3125-Hz sample rate. The FSK downlink control system operates at a 10 bit/sec rate. The telemetry equipment contained in the tow vehicle includes the PCM encoder and FSK decoder electronics. Cable equalization circuits are located on the tow ship.

The PCM uplink system provides thirty multiplexed channels for data transmission from the deep-towed vehicle to the tow ship. Twenty-seven of these channels are utilized for transmitting acoustic channel data that has been digitized by a common floating point amplifier (FPA). The FPA low pass filters the data at 800 Hz. The digital outputs of the floating point amplifier are time division multiplexed with two 16-bit frame synchronization words and a submultiplexed engineering sensor channel to form a frame containing thirty 16-bit words. This data stream is Manchester encoded and transmitted to the surface via the tow cable.

The submultiplexed engineering sensor channel is capable of handling 256 sensor channels. Data such as the array depth sensors, array heading, and other engineering sensors are transmitted via these subchannels. The subchannels are also used to transmit subsystem operational status for real-time monitoring by the system operator.

Data Record System. The data record system provides the capability to record the high data rate and large volume of data generated by the deep-towed system. The data is tape formatted in the Society of Exploration Geophysicist's SEG-D (Demultiplexed 2½-byte binary exponent word) as defined by the Society of Exploration Geophysicists.[11] The format provides the flexibility through header blocks to record both the global and short baseline navigation parameters simultaneously with the geophysical reflection data. Thus, each data tape will contain all the information required for geophysical processing.

The ability to play back prerecorded data is also incorporated into the system. The engineering sensor data for the deep-towed system is extracted from the recorded data tape in a format directly compatible with the engineering real-time display system. Also, up to 8 channels of acoustic array data can be input directly into the real-time D/A converter for display.

Real-time display system. The capability to monitor the performance of the deep-towed system is provided by a real-time display system. This system, located in the operations van, includes monitoring, displaying, and recording engineering sensor data from the array and fish located sensors; displaying acoustic array data; monitoring and recording tow ship located subsystem sensor data; and closed circuit television monitoring of the deployment/retrieval handling systems.

Navigation system. The navigation suite consists of two subsystems: a short baseline system to position the fish relative to the tow ship and a global system, Satnav and Loran C, to geographically position the tow ship. The output of both systems is fed to the data record system.

Deep-towed vehicle (fish). The fish is an aluminum open-framed structure (Fig. 4). Since the fish drag is small at low tow speeds (0.5-1.5 knots) in comparison to the tow cable, an unstreamlined structure is employed. The fish envelope size is 1.83 mL X 0.86 mW X 1.35 mH. The fish consists of two substructures of approximately equal size, with a total weight of 1360 kg in air and 1090 kg in water. One substructure houses the Helmholtz transducer, which is shock-mounted to provide vibration isolation. The other substructure houses the sound source power amplifier spherical pressure vessel, two oil-filled junction boxes, motion sensors, a current meter, an array tensiometer, and an acoustic transponder. Three cylindrical pressure vessels containing system electronics are located vertically between the two substructures.

Vehicle heave accumulator system. The ship surge motion effects on the towed system are minimized with a tow cable tension accumulator system. The unit is primarily effective when the fish system is near the sea surface. Incorporated into the system is a cable tension-monitoring sensor. The lead sheave, separate from the accumulator system and located at the tow point, is instrumented with a cable-out sensor.

Array handler winch. The array handler, a large motorized drum (Fig. 5), is used for deployment and retrieval, as well as storage of the array. The winch is mounted near the fish deployment area to facilitate array launch and retrieval.

Tow cable. The tow cable is a coaxial type construction with two contrahelically wound layers of galvanized, high-strength steel wires wrapped around a coaxial cable core. The core has electrical characteristics similar to RG-8U. The steel wire geometry provides a very low cable torsional characteristic, which eliminates the need for an electrical-mechanical swivel within the tow cable system.

Portable vans. The flexibility to meet a ship-of-opportunity operation scenario is provided by self-contained vans. An operations van and maintenance van are part of the deep-towed hardware suite. The vans are sized to allow air shipment if necessary.

Host ship support. The host ship provides the deep-sea winch, cable traction unit, and the tow point stern "A" or "U" frame. This set of hardware is generally available for ships engaged in deep-sea geophysical research. The deep-sea winch/traction unit must be capable of smooth and precise control of tow cable movement. Slow, creep, cable in/out speeds are required when lifting the fish off or placing the fish on the deck, and then speeds up to 45 m/min are required for deep deployment and retrieval. The ancillary units are highlighted in Figure 3 by heavy, broader outlines. The slip ring assembly incorporated into the deep-sea winch for termination of the coaxial tow cable is a part of the deep-towed system hardware suite. The ship must be capable of making 0.5 knots (relative the water) on any course in a fully developed Sea State 6, with the ship heading within 45^0 of the course.

ENGINEERING FIELD TEST

The final engineering evaluation test was performed in September 1984 while on a transit from Miami, Florida, to St. George, Bermuda. This test concluded the hardware development with extending the array to a total length of 1000 m, fabricating the digital record system, and incorporating a short baseline navigation system.

The sound source system again performed reliably during the test. An excess of 20,000 shot-receive sequences were initiated. The predominant source waveform used was a 260 Hz to 650 Hz linear FM slide with 10% leading and trailing amplitude shading on a 125-ms pulse. Some data was acquired with a 400 Hz, 5-ms pulse. A 20-second pulse repetition rate was typical.

The array also performed well during the test. The array shape typically experienced throughout the test is pictured in Figure 6. The real time display system generates this type figure for tow performance monitoring. Depth data is annotated as D1 through D9 on the figure. Fish depth is annotated as DF with the actual depth value given (4343 m in this case). Also, at the top of the figure are critical tow parameters. They include; tow speed (1.22 knots); array tilt (6^0); delta depth between DF and D9 (127 m); depth rate of change (0 m/min); and array tension (245 pounds). The dip at D6 was caused by array low fill oil. For the conditions typified by Figure 6 the tow noise was a nominal 65 dB (//1 μPa/$\sqrt{\text{Hz}}$) over the operating frequency band.

Figure 7 is an unprocessed reflection field trace for a single array group offset 297 m from the source. This type trace is monitored in the field for quality control. The FM slide source waveform was used for this data section. The apparent bottom slope results from a slight change in system tow depth during the two and one-half hours required for the 5.7 km long profile. Static corrections are necessary to correct for this type change.

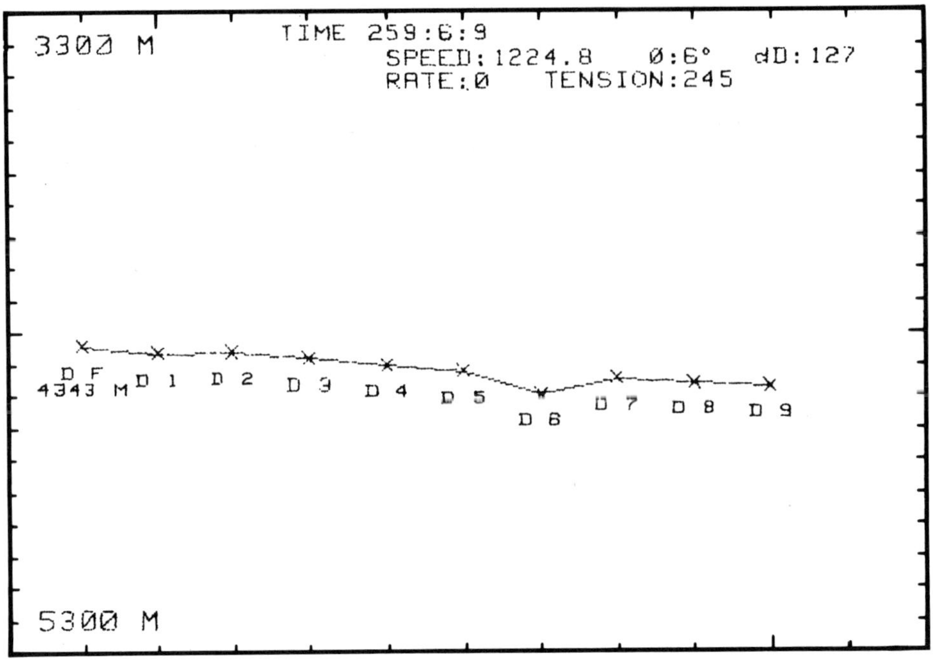

Figure 6. Array shape pictorial generated during field test by real-time display system.

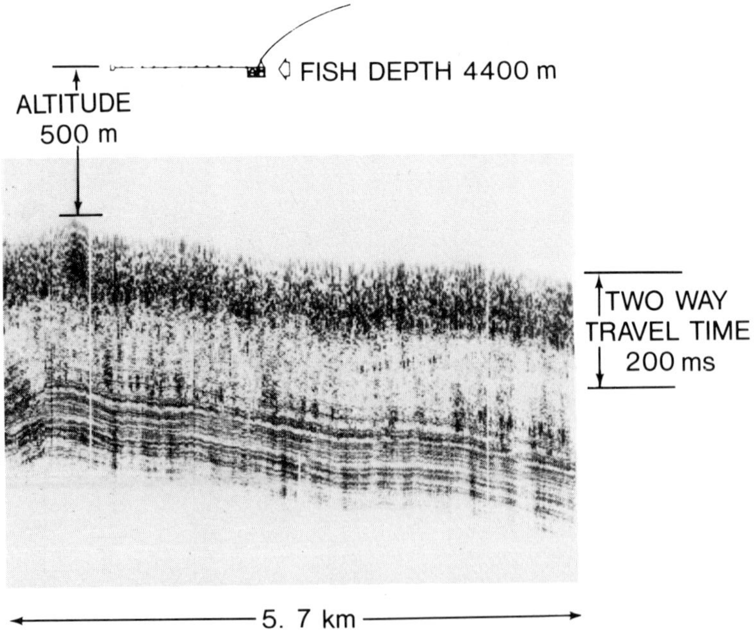

Figure 7. Single channel reflection field trace obtained at a 4400-m system tow depth.

The seismic data acquired during this engineering evaluation test is being used to complete the geophysical processing software. A short data section has been processed using DISCO, a standard processing package used principally by the oil exploration community.

A processed data section using this package is presented in Figure 8. This is a short section from the same location as the field trace (Fig. 7). An 11-fold CDP stacked section was generated. A match filter routine was used to collapse the FM slide source signature. No static corrections were made for this section. Although this processed section is only preliminary, it does show the potential of the deep-towed system. In particular, the discontinuity of beds at 0.3 seconds would be difficult to detect with a conventional surface-towed system.

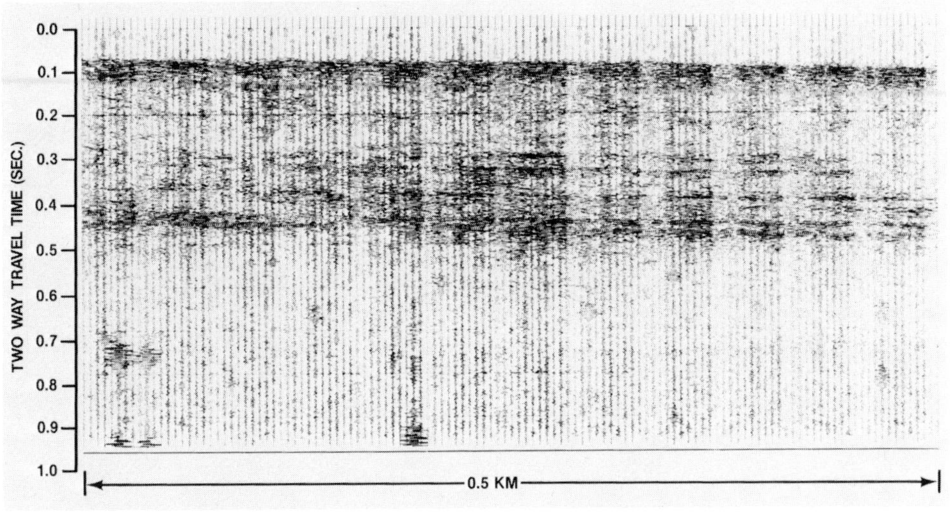

Figure 8. An 11-fold CDP stacked section of deep-tow data acquired at a tow depth of 4200 m.

The present effort focuses on optimizing this standard processing software for the deep-tow data set. Areas of concentration include: static corrections using array depth sensors and moveouts measured for the source to sea-surface to array reflection path; deconvolution techniques to optimize source signature collapse; and interval velocity measurement routines for the high-resolution data set.

SUMMARY

A new capability for deep-ocean acoustic measurements is available with the Deep-Towed Array Geophysical System. The unique hardware suite provides the ability to extract high-resolution geoacoustic parameters. The deep-towed sound source and array and the ancillary support systems have been developed and tested over the past four years. The final field engineering evaluation test successfully operated the total system to a tow depth of 4500 m. The geophysical data acquired during the test is being used to optimize processing software for the deep-towed application.

REFERENCES

1. D. A. Milburn and M. G. Fagot, 1979, "The Steady-State Analysis of Candidate Towing Cables for the Deep-Towed Geophysical Array System," NORDA Technical Note 40, Naval Ocean Research and Development Activity, NSTL, Mississippi.
2. M. G. Fagot, N. H. Gholson, G. J. Moss, and D. A. Milburn, 1980, "Deep-Towed Geophysical Array System-Development Program Review," NORDA Technical Note 75, Naval Ocean Research and Development Activity, NSTL, Mississippi.
3. M. G. Fagot, G. J. Moss, D. A. Milburn, and N. H. Gholson, 1981, "Deep-Towed Seismic System Design for Operation at Depths up to 6000 Meters," presented at the Offshore Technology Conference, Paper OTC 4082, Houston, Texas.

4. N. H. Gholson and M. G. Fagot, 1980, ''A Moments Approach for Analyzing Geophysical Reflection Data,'' NORDA Report 30, Naval Ocean Research and Development Activity, NSTL, Mississippi.
5. N. H. Gholson and M. G. Fagot, 1983, ''A deep-towed sound source and hydrophone array system-performance analysis and design, in: ''Handbook of Geophysical Exploration at Sea,'' Richard A. Geyer, ed., CRC Press, Inc., Boca Raton, Florida, pp. 267-292.
6. B. Gibson, M. Odegard, and G. Sutton, 1979, Nonlinear least-squares inversion of traveltime data for a linear velocity depth relationship, *Geophysics*, 44-2:185-194.
7. C. H. Dix, 1955, Seismic velocities from surface measurements, *Geophysics*, 44-2:185-194.
8. P. Stoffa, 1979, Design criteria for a deep towed source and multichannel array, Chute Stoffa, Inc., Nyack, New York.
9. M. G. Fagot and S. E. Spychalski, 1984, ''Deep-Towed Array Geophysical System (DTAGS): A Hardware Description,'' NORDA Report 71, Naval Ocean Research and Development Activity, NSTL, Mississippi.
10. A. M. Young, A. C. Tims, and T. A. Henriquez, 1982, ''The Development of a High Power, Low Frequency Underwater Acoustic Source for Use in a Deep-Towed Geophysical Array System,'' NRL Report 8633, Naval Research Laboratory-Underwater Sound Reference Detachment, Orlando, Florida.
11. Society of Exploration Geophysicists, 1980, ''Digital Tape Standards,'' Published by Society of Exploration Geophysicists, Tulsa, Oklahoma.

SEA-FLOOR SEISMOLOGY : MOBILE OCEAN BOTTOM VERTICAL SEISMIC ARRAY (OBVSA) FOR CONTROLLED SOURCE EXPERIMENTS

Félix Avedik

Department of Geosciences
IFREMER, BP 337
29273 Brest Cédex, France

ABSTRACT

Achievement of a favourable signal to noise ratio is always important in the successful conduct of controlled-source seismic experiments. Noise cancellation techniques seek to obtain the desired results by means of - separately or together - signal processing, optimal organisation of geometrical and other relevant aspects of the experiment,and instrumental design. Our new seismic system for marine work aims to increase the signal to noise ratio by exploiting the directional receiving properties offered by a multichannel hydrophone array positioned vertically on the ocean floor together with the associated electronics and recorder. The present operational depth is 6 000 metres. The primary targets of the instrument concern wide-angle reflection and refraction studies of the lithosphere. The modular design and the three possible recording modes (continuous, programmed window or event-detection) make the instrument equally useful for addressing a number of other objectives.

1. INTRODUCTION

The increasing number of deep seismic surveys conducted over continental and oceanic areas demonstrates the desire of the geosciences community to establish a better picture than we have at present of the structure of the earth's crust and upper mantle, and to acquire more reliable data on their physical properties with a view to building a stronger framework for ideas about global geological processes. Experimental seismology involving seismic reflection and refraction studies is the geophysical approach most often used for carrying out these investigations.

Marine seismic reflection surveys using one ship and standard industrial-type equipment (comprising streamers about 3 kilometers long and large, tuned air-gun arrays), or the more sophisticated methods calling for two or three ships to form super-long, synthetic aperture receiving arrays, furnish generally excellent images of the deep geological structures.

In dealing with measurements of the physical parameters of the crust, combined wide angle seismic reflection and refraction methods

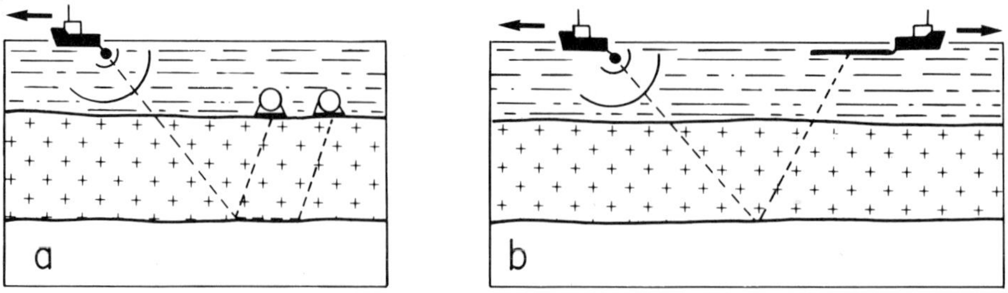

Fig. 1. Marine seismic refraction surveying techniques : a) with one ship and ocean bottom receivers and b) with two ships using the Expanding Spread Profiling method.

offer a valuable complement to the reflection surveys. For such experiments two methods have gained wide acceptance. The first, the "classical" one, uses one ship (for shooting) and ocean bottom seismometers (OBS) or ocean bottom hydrophones (OBH) as receiving and recording stations. The second method, the Expanding Spread Profiling technique (ESP), involves two ships steaming in opposite directions, one towing a standard, multichannel hydrophone array, the other providing the seismic source. The principles of the two surveying techniques are illustrated in figure 1. The ESP technique became quite popular in the recent past and is preferred now by many investigators because of the higher quality of data it produces as compared to that obtained with a single ship and ocean bottom instruments. However, it is worth stressing that the ESP technique is disproportionately costly and is not well suited for sub-crustal investigations. For information on existing instruments and surveying methods as well as on the geophysical results obtained with them, the reader may refer to the extensive literature (1,2,3).

Analyses of refraction recordings in deep sea areas indicate that, of the three general categories of noise observed, namely natural seismicity, perturbations from ship traffic, and shot noise generated from the repetitive seismic pulse, the last one represents the major and most disturbing noise component because of its slow decay and the characteristics of its frequency spectrum. Fig. 2 illustrates a typical situation where the conditions are such that commonly applied techniques to raise the signal to noise ratio are inefficient. The straightforward method of increasing the energy of the emissions would result in increased noise interference too ; signal processing techniques, powerful means to extract signal from noise, are here of very limited value because of the coincident signal and noise spectra. The weaknesses of the two-ship ESP technique and the apparent lack of any solutions to the noise problem encouraged us to investigate the possibilities of improving, at the recording stage, the quality of seismic data gathered with ocean bottom instruments.

In this paper I shall briefly describe, the philosophy of the design of a new ocean floor instrument, the Ocean Bottom Vertical Seismic Array (OBVSA) give some of its main technical characteristics and report on its first sea trials.

2. DESIGN PHILOSOPHY AND TECHNICAL CHARACTERISTICS OF THE INSTRUMENT

The character of deep-sea noise, mentioned above, led us to consider the possibilities of exploiting the directional receiving properties offered by a multichannel hydrophone array positionned, for practical

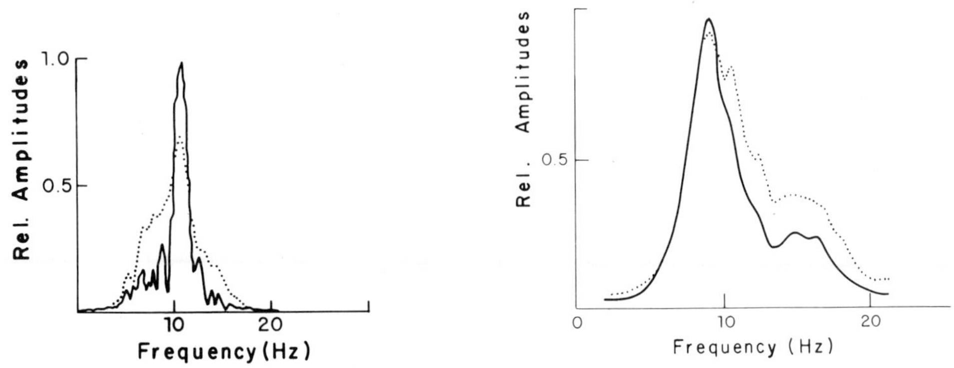

Fig. 2. Typical signal and noise spectra (solid and dotted lines) calculated from deep-sea seismic refraction profiles (OBS records). Water depth about 2 500 metres, shooting rate 1/90 to 1/120 seconds. Sound source : multiple air guns with total volume of 49 litres to 98 litres.

reasons, vertically on the ocean floor and to design the instrument accordingly. Our aim was to restrict the receiving capabilities of the instrument to the sector defined by the emergence angles of the seismic signal from the ocean floor and to try to reject all other arrivals, even if their frequency spectra correspond to the spectrum of the seismic signal. This means that the receiving beam of the array must be orientated downwards with an average half beam width of about 30° from the vertical in the relevant frequency band. Because lithospheric seismic studies represent the main objectives of the instrument, the frequency band in which the array's characteristics were to be optimised is 8 to 17 Hz, corresponding to the fundamental frequencies of the large pneumatic sound sources (air gun) which are generally used for such seismic experiments. Extension of this band down to 5 Hz, with less stringent receiving characteristics, is provided for in order to cope with the use of explosives for achieving an even greater depth of investigation.

The idea of using hydrophone arrays to improve signal to noise ratio in seismic experiments is not new in our Institute. Such an array, suspended under a radio buoy, was constructed in 1969 by J. Martinais (4). However, the system came into competition at that time with other apparently promising techniques and this, together with some design dependent weaknesses (for example, displacement of the system with surface-water currents and weather-dependent acoustical performance) rapidly restricted its use and development.

The design of the Ocean Bottom Vertical Seismic Array (OBVSA) draws on the many years of experience that we have had with ocean bottom instrumentation since this early effort. Two of the highlights were (i) the construction in 1975 of an ocean bottom seismograph (5) which made use of the notion that better coupling characteristics can be assured by positioning the seismic sensors on the sea floor separately from the main frame of the instrument and (ii) the introduction in 1978 of high density, long duration digital recording for use with the ocean bottom instruments (8 300 BPI, 130 hours for three seismic channels, 10 bits resolution).

A single version of the new Ocean Bottom Vertical Seismic Array was designed and built in 1984 with, as a primary aim, to test the validity

Fig. 3. The main components of the Ocean Bottom Vertical Seismic Array (OBVSA). See explanation in text.

of the concept, and its performance and practical handling characteristics under operating conditions.

The main components of the instrument are (fig. 3) :

a) the buoyancy floats equiped with radio beacon and flashing light for homing in on the instrument when it surfaces ;
b) pressure vessel with electronics and power supply ;
c) hydrophone array ;
d) acoustic release-transponder used for ranging of the instrument on the seafloor (or at the surface if visibility is poor) ;
e) sinker weight which retains the instrument on the ocean bottom until disconnected.

The total length of the assembly is about 500 metres, including the 450 metres of hydrophone array. All elements of the instrument are rated for 6 000 metres working depth.

For the mechanical construction of the hydrophone array the "streamer" design was chosen, which in our opinion offers a cost effective solution to the need for operational reliability. An oil-filled hose (1"O.D) houses hydrophones and preamplifiers (Benthos AQ-1 and AQ-200) as well as the stress members. Pressure compensated connectors connect the 90 metres long sections of the array together, the breaking strength of the whole assembly being about 27 KN.

The actual receiving characteristics of the array (fig. 4) are obtained by means of an arrangement of 15 individual seismic channels with a uniform distance of 30 meters between them. The linear arrangement, although perhaps not the most efficient acoustically, offers not negligible operational advantages, such as interchangeability of the elements and the reversibility of the whole array. This concept may, however, undergo modification if the final conclusions of the operational tests so dictate. The manufacturer of the array is the Ateliers Mécaniques de Saint-Gaudens (AMG) in France.

Several of the recorders and electronic units commercially available today for seismic observations offer technical characteristics and recording schemes which make them suitable for underwater or ocean floor applications and, in most cases, render unrealistic the development of specialized electronics. For the time being, and to gain experience with the system, we choose to record the individual seismic channels

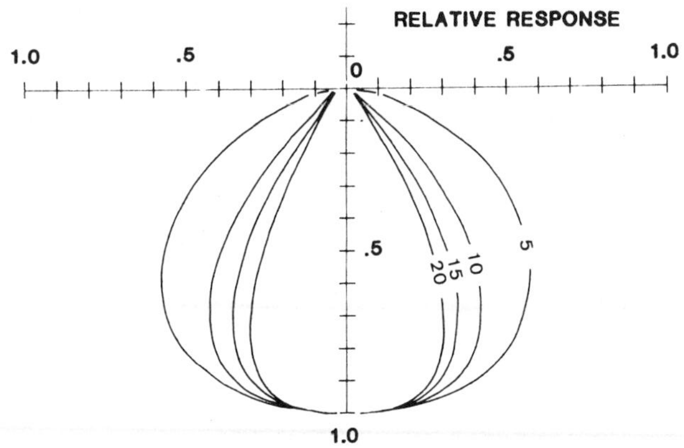

Fig. 4. Schematic polar diagram of the array's receiving characteristics in the 5 to 20 Hz frequency band. The relative magnitude of the secondary lobes is <-20 db in this frequency range.

separately and to do the necessary signal processing after retrieval of the instrument from the sea. We are using PCM-5800 seismic equipment manufactured by Lennartz Co. (Federal Republic of Germany) whose microprocessor-controlled electronics offers digital technology with a high dynamical range, a high density PCM recording scheme and the required multichannel capabilities, the all in the necessarily small volume and, last but not the least, with a low power consumption. The operational flexibility of the equipment is guaranteed by the three recording modes available : continuous, programmed window, and event detection. The actual capacity of the 700 metres long, 1/4 inch magnetic tape is about 30 megabytes which, in other terms, corresponds to the recording of about 350 seismic events, each 40 seconds long, on each of the fifteen separate channels. For many applications this capacity is enough. We intend, however, to extend this capability in the near future, when we start the operational use of the instrument.

The buoyancy and the release transponder are also vital parts of the system but, because of its modular concept, they are not subject to any stringent requirement. Any type of commercially available material may be used, provided its depth rating is suitable.

3. FIRST TRIALS AT SEA

Short tests in the Norwegian Sea from aboard the R.V Polarstern in September, 1984 and with the R.V. Le Suroit in the Mediterranean Sea (January 1985) were devoted to examining the instrument's operational and acoustical capabilities respectively.

The system is best deployed from the after deck of a slowly forward moving ship. The buoyancy and the pressure vessel are launched first, followed by the lowering of the hydrophone array and the acoustic transponder. The whole deployment takes about 20 minutes. Once the sinker weight is launched, the assembly sinks to the bottom at a velocity of about 1.5 metre/second. After releasing the sinker weight the instrument ascends at about 1 metre/second. The recovery of the instrument is simple and follows standard procedures used to fish for floating targets.

The test of acoustical performance in the Mediterranean Sea was made at the foot of the steep continental slope off the south of France, in a

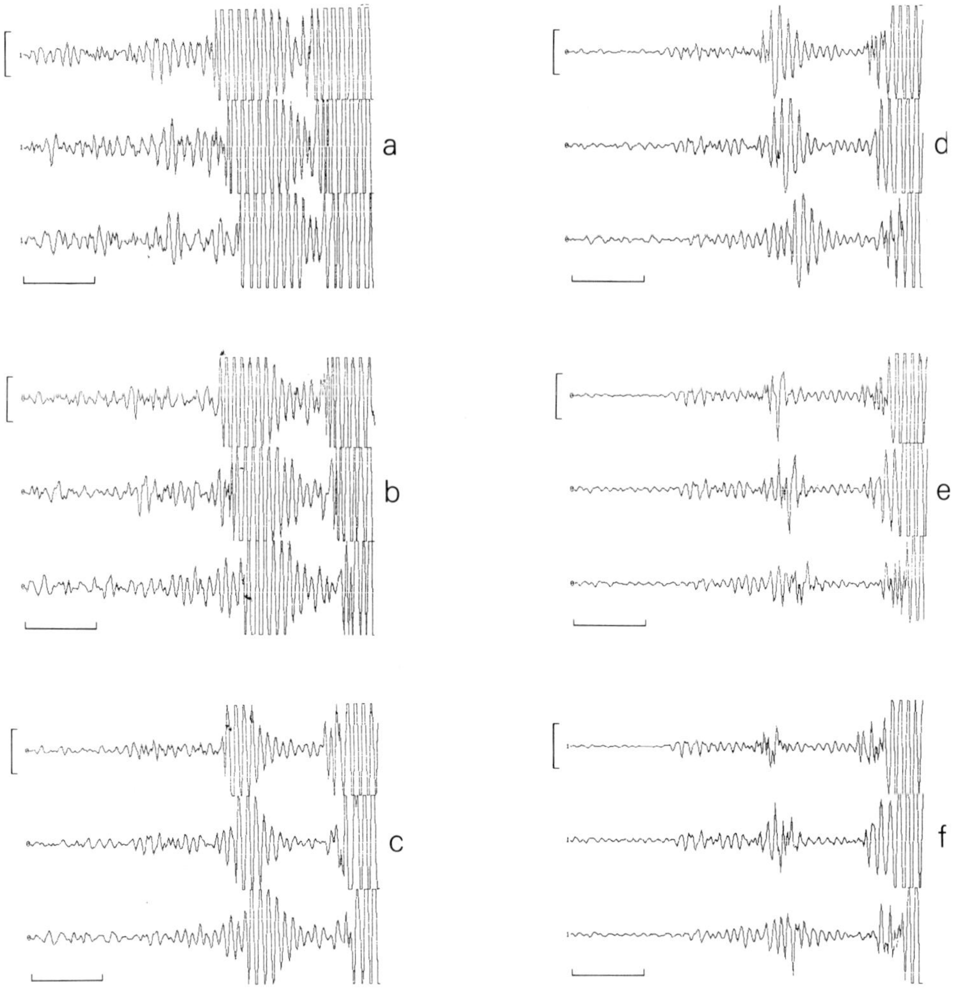

Fig. 5. Evolution of the signal to noise ratio as signals recorded from three subsequent shots are processed from a) one ; b) three ; c) six ; d) nine ; e) twelve ; and f) fifteen channels of the array. Note the gradual emergence of the seismic signals from the noise and the effective cancellation of the strong, shot generated water arrival. The single channel record represents the approximate performances of the standard ocean bottom instruments. The figures show 5 second-samples from the record. The horizontal bar indicates 1 second. The vertical bar equals an approximately 2.10^5 μ Pascal (= 2 μ bars) pressure signal. Frequency band : 2 to 15 Hz.

water depth of about 2 200 metres and with a maximum horizontal distance between ship and instrument of about 15 kilometres. The sound source used was a single 16.4 litre air-gun, shooting in a 90 second sequence. The test programme included profiling as well as observations of the array's performance in different types of noise fields. The data collected during the test are under evaluation but preliminary results indicate (fig. 5) that the improvement in signal to noise ratio with respect to standard ocean floor instruments is at least about 12 db or a factor of about four. In addition, this effect is accompanied by a better resolution of

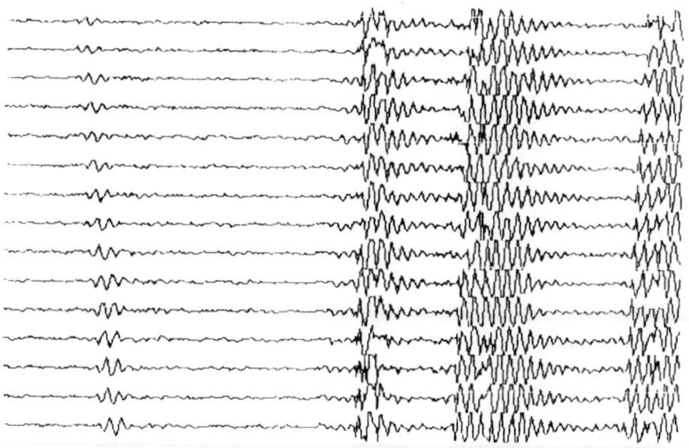

Fig. 6. Example of a shot recorded on the fifteen individual channels of the array. Frequency band 2-15 Hz.

the seismic signal, obtained by the rejection of the majority of waterborn arrivals.

The examination of the records of the individual seismic channels do not indicate any erratic movement of the array (Fig. 6). However, the extent to which the system can tolerate more or less strong currents close to the sea floor will have to be determined by experience in various oceanographic settings ; we intend to incorporate standard current meters to study this parameter.

In conclusion, we hope that the acoustical performance of the Ocean Bottom Vertical Seismic Array (OBVSA) will represent a significant step in the evolution of ocean floor seismic instrumentation, and, in conjunction with the modular design and the "intelligent" microprocessor-controlled electronics, will make the instrument a useful tool for experimental seismology, and probably for observational seismology as well. The concept may also be employed for other objectives, for example noise interference identification or target localisation above the ocean floor.

ACKNOWLEDGEMENT

I am indebted to H.D. Needham for his stimulating attitude and for the support he has given to the project. Valuable discussions with my colleagues were appreciated, namely those with J.P. Allenou, R. Herber and A. Hirn. I thank J.P. Regnault, J.P. Quere and J.L. Veinante for their technical assistance and J. Gallart and B. Pinet for their help at sea. J.P. Maze prepared the illustrations and M. Morvan the type-setting. The cooperation of the captains, officers and crews of the R.V. Polarstern and R.V. Le Suroit during the tests at sea is also gratefully acknowledged. The project is financed by IFREMER.

SELECTED REFERENCES

1. R.B. Whitmarsh, R.C. Lilwall, Ocean Bottom Seismographs, in: Structure and development of the Greenland-Scotland Ridge. Bott, Saxov, Talwani and Thiede, ed., Plenum Publishing Corp. (1983).
2. P.L. Stoffa and P. Buhl, Two ship multichannel seismic experiments for deep crustal studies : expanded spread profiles and constant offset profiles. J. Geophys. Res., 84, 7645 (1979).
3. S. Le Douaran, J. Burrus and F. Avedik, Deep structure of the North-western Mediterranean Basin : results of a two-ship seismic survey, Mar. Geol., 55, 325 (1984).

4. J. Martinais, J. Turrière, Une bouée relais automatique légère, Proc. of Colloque Inter. sur l'exploitation des océans (Bordeaux), Th. V, T.2-07, CNEXO, Paris (1971).
5. F. Avedik, V. Renard, D. Buisine and J.Y. Cornic., Ocean Bottom Refraction Seismograph (OBRS), Mar. Geophys. Res., 3, 357 (1978).

Contribution n° 40 of IFREMER-Centre de Brest

A 12-CHANNEL MARINE EEL FOR SHALLOW REFRACTION SURVEYING OF THE SEABOTTOM IN COASTAL WATERS

A.G.McKay+, J.A.Hunter*, R.L.Good*, and D.M.F.Chapman**

+NS Research Found.Corp., P.O.Box 790, Dartmouth, NS B2Y 3Z7
*Geological Survey of Canada, 601 Booth St., Ottawa K1A 0E8
**Defence Research Establishment Atlantic, P.O. Box 1012, Dartmouth, NS B2Y 3Z7 Canada

ABSTRACT

A 12-channel hydrophone array has been used in conjunction with a 0.16 litre airgun for the determination of acoustic velocities in seabed sediments on the Scotian Shelf (Eastern Canada) in water depths of up to 100 m. The measurements were made by towing the assembly of source and receivers within 5 m of the seabed in order to observe the travel time of headwaves propagating through the seabed materials. At a ship speed of about 3 knots, data for a total of 529 velocity observations were collected in a period of 53 hours. Observations were made in clusters over a relatively uniform sand-gravel sediment unit in an attempt to assess the consistency and precision of the technique. Velocities are determined to 3% or better.

INTRODUCTION

There is a need for continuing progress in the measurement *in-situ* of the acoustic properties of seabed sediments both for the purpose of providing a more accurate geo-acoustic model[1] as required in predicting the acoustic propagation characteristics of the water-column and for increasing the confidence with which engineering properties can be predicted[2]. For both of these aims, it is desirable that observations be made from a vessel underway so that large areas of the seabed can be examined with the minimum of ship time. Methods derived from earlier land and marine practice[3] have been adapted by various authors in confronting this problem for near-seabed materials. Porter et al.[4] used an eleven-channel array, 70 m long towed at depths of up to 15 m to observe oblique-angle reflections from pinger-type transducers and deduced velocities by the T^2-X^2 method with a precision of about 2.5%. McKay[5] carried out experiments with arrays of 2, 3, 4, and 6 hydrophones in conjunction with a spark-discharge acoustic source in order to extend such a method to towing depths of 200 m and, in examining layers 10 m thick in 250 m of water[6], maintain the precision attained by Porter et al[4]. A system for making similar observations in appreciably deeper water has been described by Fagot[7,8].

The Terrain Geophysics group of the Geological Survey of Canada (GSC) became interested in the development of seismic refraction techniques, (using headwave[9] time) for the detection of ice-bonded sediments in the Beaufort Sea as an aid in pipeline routing and site evaluation for offshore construction. The presence of ice-bearing permafrost within 20m of the seabed is a major constraint on engineering design. After the validity of the method had been established by station work in shallow water[10], and when the broad pattern of permafrost occurrence had been delineated by the re-interpretation of oil-industry seismic data[11,12], it became clear that a deep-towed refraction system was required to provide more detailed information on the first 20 m below the bottom. Such a system would have to be able to be towed within a few metres of the seabed to allow the detection of headwaves propagating along the upper surface of a permafrost layer as first arrivals. In this situation an advantage of the seismic refraction method over other kinematic methods is that no deeper reflector need be detected than that forming the upper boundary of the layer whose acoustic propagation velocity is sought.

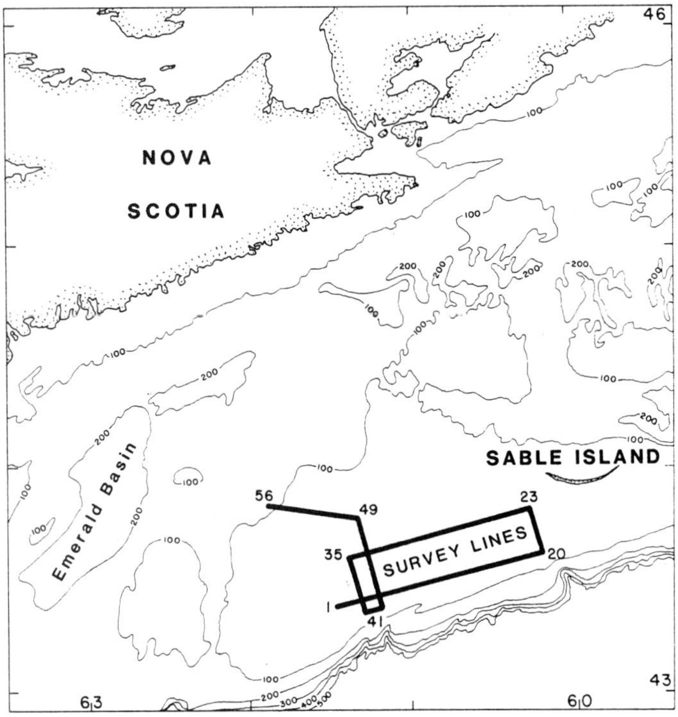

Fig.1. Survey area on the Scotian Shelf. The length of track line covered was 200 n.mi. A total of 56 groups of velocity determinations were made, of between 8 and 11 measurements taken over a distance of about 1 n.mi. The identifying numbers are shown of groups which occur near course changes. Other groups occur at roughly even spacing between these marked points.

On the basis of the experience of the Terrain Geophysics group of GSC and in light of their requirements, the Nova Scotia Research Foundation Corporation (NSRFC) constructed a 12-channel marine eel incorporating design elements which had been used successfully for headwave recognition in near-seabed equipment[6,13,14]. This eel has also been used in gathering reflection data for analysis by kinematic and dynamic modelling[15] methods, but the purpose of this paper is to describe the way in which it was deployed in water 80 m deep for the gathering of data suitable for kinematic analysis by the refraction method of geophysics. The area from which results are presented is the Sable Island Bank of the Scotian Shelf (Fig.1) where the Defence Research Establishment (Atlantic) have modelled propagation loss in the water column[16] and require information about the seabed.

EQUIPMENT DESIGN AND DEPLOYMENT

The equipment deployment is shown in Fig.2. The hydrohone array is constructed in a conventional manner in an oil-filled hose, of outer diameter 2.75 cm. It is built in sections each 15 m long, six active sections, each containing two groups of four parallel-wired hydrophone elements and one passive section (normally connected at the head of the array). Each group of 4 elements (Benthos AQ-16 type) is placed in a 60cm cluster with the aim of improving on the tow noise expected from a single element while retaining a near-omnidirectional receiver pattern in the operating frequency band of 50 to 1000 Hz. An impedance-matching pre-amplifier with a gain of 46 dB couples each hydrophone group to one twisted pair of a wire bundle (which also acts as the strain-bearing member).

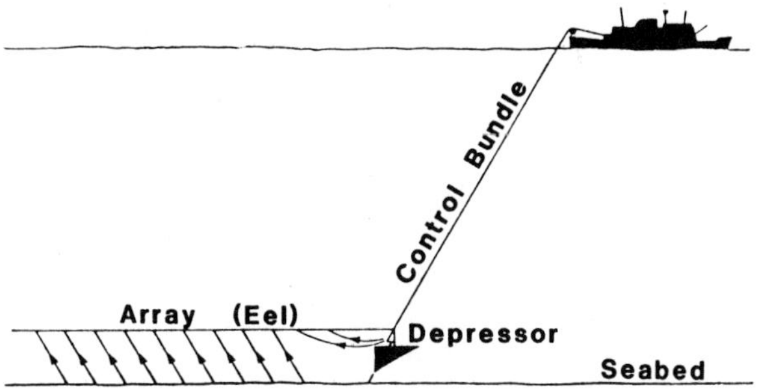

Fig.2. The eel is towed from a depressor to which is attached the acoustic source - a 0.16 litre airgun. Propagation paths are shown for direct water wave and seabed headwaves.

The eel was designed to be slightly positively buoyant so that trimming to neutral buoyancy may be accomplished by taping on small pieces of thin lead sheet in quantities which can be varied to suit the density of water in the survey area. The deployment of the eel near the seabed is arranged by towing it from a 1 m depressor fin (Braincon Corp.) as shown in Fig. 3, which also illustrates the attachment of the 0.16 litre airgun used as the acoustic source and of a small sound transducer for recording the clearance of the depressor above the seabed. The towing umbilical is a faired bundle comprising a kevlar strength member, multi-conductor seismic cable for the eel signals, high pressure air line and electrical control cables for the acoustic source and the sounding transducer. On board ship, the umbilical is wound on a winch fitted with an electrical and airline slip ring assembly.

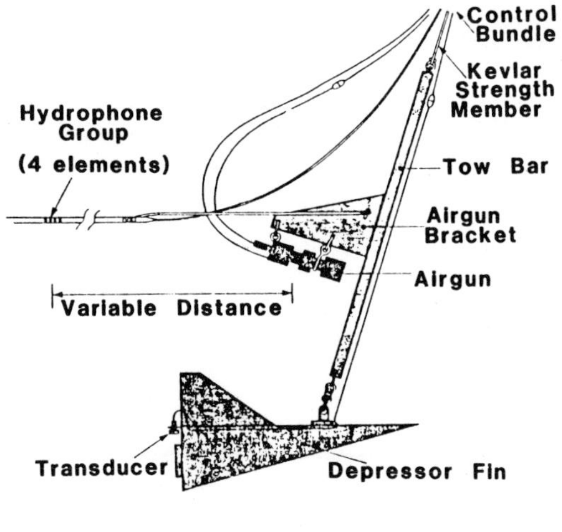

Fig.3. Details of the attachment to the depressor of the 12-channel eel, the airgun acoustic source and the sounding transducer for monitoring clearance above the seabed.

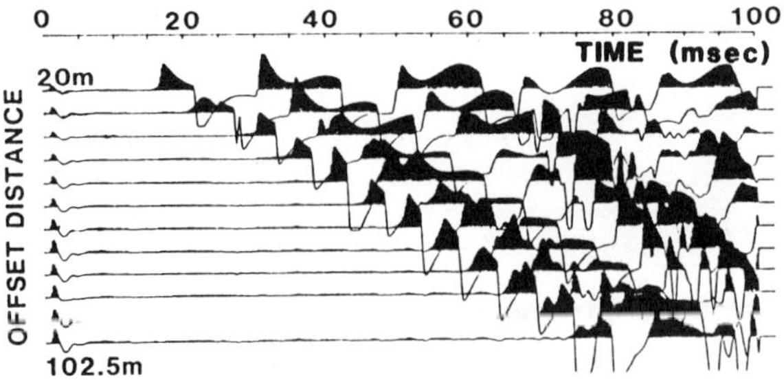

Fig.4. Displaying of a 12-channel seismogram as recorded on a Nimbus seismograph (set for 100 msec record length) and later transferred to floppy disc.

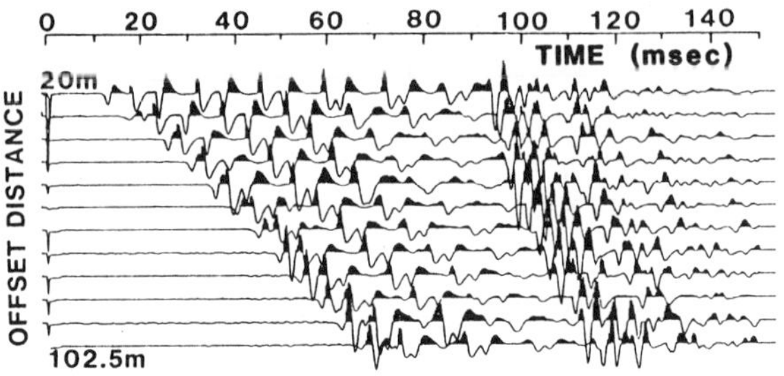

Fig.5. Display of the first 150 msec of a 300 msec seismogram as recorded direct to Apple II microcomputer by means of A-D conversion equipment built for GSC by Atlantis Scientific Systems.

Two different 12-channel recording systems have been used: an engineering seismograph (Nimbus 1210F) equipped with a cartridge tape recorder; and a custom-built system for direct signal capture and floppy disc storage on an Apple II microcomputer. The former system allows one 12-channel seismogram to be captured every 2 minutes whereas the latter can store one such record every 15 seconds. Examples of these records are shown in Figs. 4 and 5. In both methods, each of the 12 time-series comprising the seismogram is stored on 5½" floppy disc in the Norminton format (developed for GSC) as an array of 1024 numbers of 10-bits each, together with offset distance and identifying parameters. By the use of this disc format, eight complete seismograms can be stored on one 140-Kbyte disc. When the engineering seismograph is used as the primary acquisition system, it is necessary to replay the data from the seismograph's tape recorder for transfer to disc.

Seismograms were not collected at uniform intervals along survey lines. Instead, the observations were grouped in sets of between 8 and 11 seismograms, gathered over a period of less than 20 minutes. At the survey speed of 3 kt., this corresponds to the grouping of each set of seismograms along a 1.8 km length of survey line. This was done with a view to assessing the measurement method on the assumption that in small areas, the sediment characteristics would not vary greatly and that uniform results should be expected from all seismograms within a group. Between groups, the seismic equipment was towed at about 10 m above the seabed. It was lowered to 3 m clearance for each group of measurements (see Fig. 6).

DATA HANDLING AND VELOCITY RESULTS

The method of measuring a p-wave propagation velocity by means of the observations of headwave arrival times is restricted in its scope to geological horizons at a depth which is small compared to the separation between the acoustic source and the farthest receiver, as illustrated in Fig. 7. Where a "layer 2" material has a velocity above 3 km/sec, such as may be the case with a rock or an ice-bonded sediment, then detection of the layer and measurement of its velocity is possible with

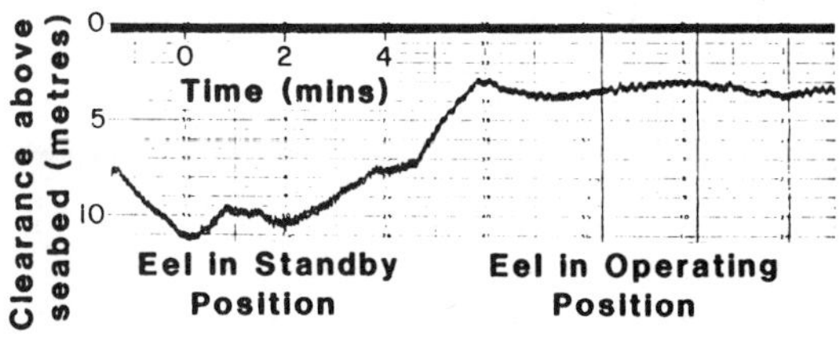

Fig. 6.
This record from the sounding transducer (Fig.3) shows the transition from standby (10m) to operational (3m) clearance above bottom.

this equipment for depths below the seabed of over 20 m, but when the lower layer is a water-saturated sediment with velocity 1.6-1.7 km/sec, then its detection becomes uncertain when the depth of burial exceeds even about 5 m. A number of normal-incidence profiles which cross the survey area[17,18] have been examined and it is found that at least the first 20 m below the seabed does not show subbottom structure in the form of reflectors except in a few areas. Even where these do appear (Fig.8) it is not possible to measure the lower layer velocity. The results obtained are consequently limited to determinations of velocity of the immediate seabed material which has been broadly classified by King[19] as "Sable Island sand and gravel" in which seabed photographs[17,18] indicate that sand ripples are abundant as are mollusc and echinoderm biota. It did not prove possible to extend the velocity measurements to adjacent sedimentary units of different type[19] because those lay in water deeper than 100 m, beyond the present capability of the equipment as used in this refraction mode (Fig.9).

Data analysis of the seismograms was undertaken with refraction analysis software supplied by Exploranium[20], for use on an Apple II+ or IIe microcomputer. This software operates interactively to allow the display of seismograms in a format similar to that shown in Figs. 4 and 5 and to enable the picking of headwave arrival times. These arrival

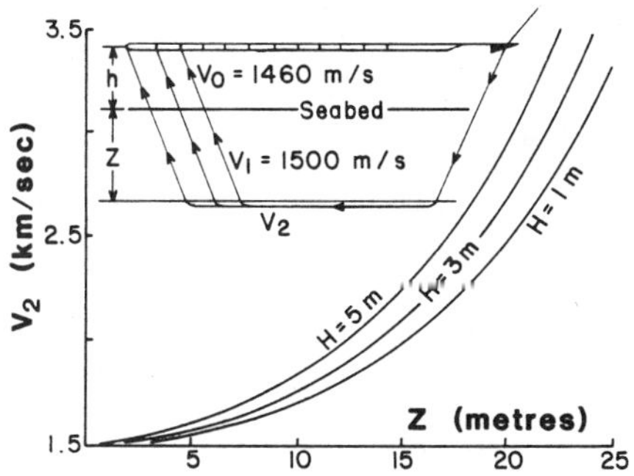

Fig.7.
Variation of detection depth with velocity contrast for the eel clearance heights (H) and the layer velocity shown. Detection criterion: a line fitted through at least channels 10, 11, 12.

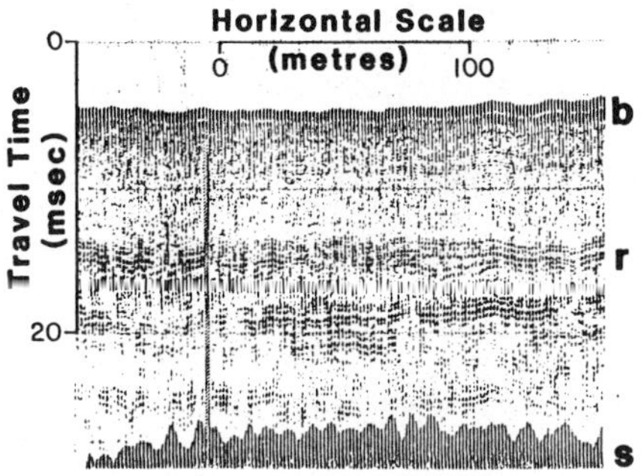

Fig. 8
Normal incidence profile near group 41 (Fig.1). The reflector "r" is not continuous over the whole area. Seabed - "b" surface reflection, "s". (NSRFC profiler).

Fig.9.
For speeds at which a towing vessel can mantain steering capabilities, the operational depth of the present equipment is limited to 80-90 m.

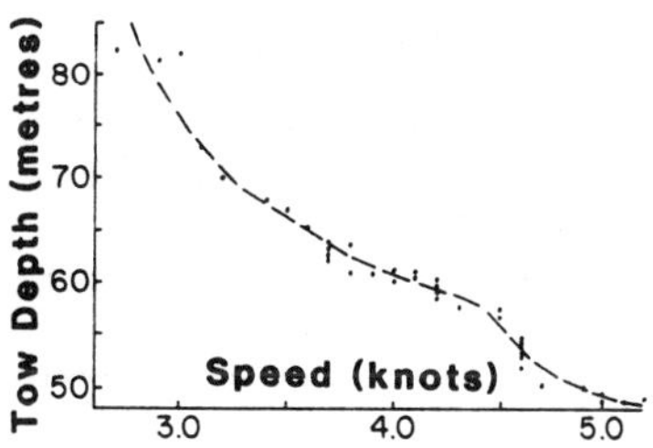

times are filed for subsequent processing by another routine which displays a time-distance plot of the observed points and allows the derivation of a velocity as the gradient of a straight line fitted through selected points. From the total of 12 points available, usually 9 or 10 were used to define this line. Normally, the points for Channels 1 and 2 were eliminated, being identified as events associated with propagation directly through the water or through the materials of the eel itself. Sometimes the headwave arrival at Channel 12 (and in a very few cases, Channel 11) was too weak for accurate picking, or else appeared displaced from the straight line defined by Channels 3 - 10 perhaps because of its being associated with a deeper refracting horizon such as the event "r" of Fig. 8.

A total of 529 determinations of velocity were made in this way. These are summarized in Table 1 where the average value and standard deviation obtained from each of the 56 groups of seismograms (Fig. 1) is given, along with the number of determinations in each group, the average clearance above the seabed at which the equipment was towed and the incidence of clustering of individual results within a group. In the columns indicating clustering, the number "2" means that two neighbouring velocity measurements lie above ("high") or below ("low") the mean by an amount exceeding one standard deviation. The number "3" indicates a similar cluster of 3 neighbouring individual results. Such clustering suggests that there may be some real variation of acoustic velocity within the measurement groups.

LIMITATIONS ON THE ACCURACY OF RESULTS

Errors in the velocity results may arise both from uncertainties in travel time measurement and in the location of source and receivers.

(a) Errors Arising from Time Measurements

(i) The 280-pixel wide display of the Apple II+ allows picking of travel times in increments of 0.3 msec when a 75 msec wide section of a 1024 point 300 msec record is displayed on the screen. The difference in time of arrival at the first and last points used in the straight line fitting for velocity derivation is typically 43 msec, which gives rise to an error of about ± 0.35% (corresponding to ± 6 m./sec in velocity) arising from the precision with which travel times can be picked.

(ii) Although the time of firing of an airgun source may lag behind its electrical triggering,the current method uses only differences in travel time so that uncertainty of the time origin does not contribute to errors.

(b) Errors Arising from Locating Source and Receivers

(i) Horizontal location. Since the towing speed of the seismic equipment was about 3 knots during the gathering of data, there is a displacement of the position of the receiving elements which arises from this forward motion. In the time period of about 43 msec which elapses between the arrival of a headwave at Channel 3 and the arrival at Channel 12 (these being the extremes of locations used in most of the velocity determinations), the hydrophone group comprising Channel 12 will have moved forward about 6 cm. Over the measured difference in horizontal location of Channels 3 and 12 of about 70 m this effect will result in an over-estimate of velocity by about 0.08% (1.4 m/sec). This is a systematic error, but it is small compared to other sources of error and has been neglected in the results as presented.

Table 1. Velocity Results

Group No.	No. of Measurements	Velocity (km.s^{-1})	Error (km.s^{-1})	High Cluster	Low Cluster	Average Clearance (m.)
1	10	1.712	.016			4.0
2	11	1.562	.022			7.5
3	9	1.697	.033	2		5.0
4	11	1.697	.022			3.5
5	11	1.629	.027		2	4.0
6	11	1.735	.031			3.5
7	11	1.651	.046	2	3	8.0
8	11	1.623	.032		2	7.5
9	10	1.817	.033			3.5
10	10	1.752	.032			5.0
11	9	1.652	.015			4.0
12	10	1.625	.032		2	5.0
13	11	1.692	.047	2	3	3.5
14	11	1.691	.023			5.0
15	10	1.650	.043			5.0
16	11	1.698	.031			4.0
17	11	1.616	.030		3	3.5
18	11	1.619	.022			5.0
19	11	1.779	.030	2		3.0
20	11	1.830	.053		2	3.5
21	11	1.612	.032			5.0
22	11	1.654	.016			4.0
23	11	1.596	.017			4.5
24	11	1.592	.013			4.5
25	11	1.704	.045		2	3.0
26	10	1.563	.013	2		7.5
27	11	1.585	.009			7.0
28	10	1.608	.020	2		5.0
29	11	1.662	.012			4.5
30	10	1.649	.008			5.5
31	11	1.656	.010			5.0
32	11	1.650	.011			2.5
33	8	1.561	.024	2		4.5
34	8	1.566	.015			4.0
35	8	1.560	.010			4.0
36	8	1.605	.030			2.5
37	8	1.624	.019		3	3.0
38	8	1.534	.007			4.0
39	8	1.689	.010			3.0
40	8	1.663	.024	2		4.5
41	8	1.795	.048			3.0
42	8	1.683	.014			3.0
43	8	1.657	.023	2		3.5
44	8	1.627	.011			3.5
45	8	1.645	.043		3	3.5
46	8	1.605	.008	2	2	4.0
47	8	1.587	.007			4.5
48	8	1.599	.012			4.0
49	8	1.604	.013	2		4.5
50	8	1.604	.013			4.5
51	8	1.617	.017	3		3.0
52	8	1.606	.018			3.5
53	8	1.713	.074			4.0
54	8	1.615	.038			3.5
55	8	1.626	.027			3.0
56	5	1.561	.045			5.5

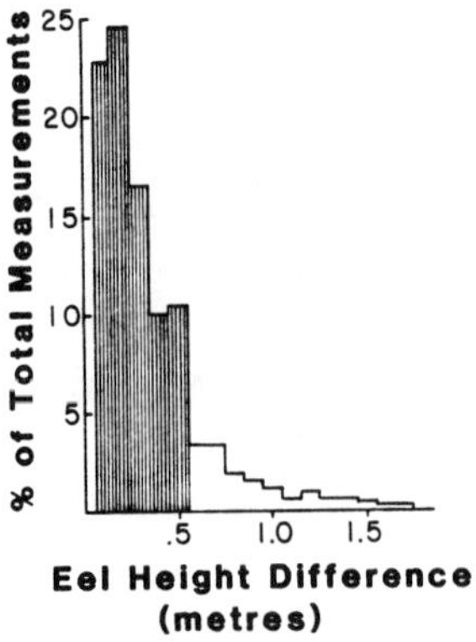

Fig.10
Histogram showing the height difference between first and last hydrophones of the eel as deduced from depressor movements. For the whole data set, 85% of the measurements show a difference of 0.5 m or less.

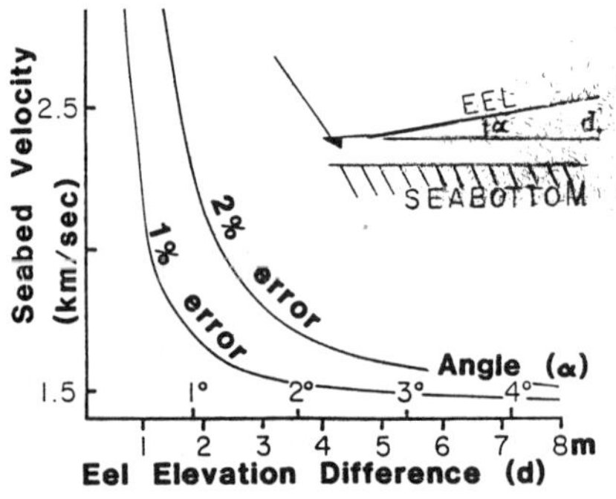

Fig.11
Curves for 1% and 2% error in velocity as a function of the towing angle of the eel and the height difference between first and last elements for various values of the seabed acoustic velocity.

(ii) Vertical location. One of the most difficult geometrical parameters to control in the deployment of deep-tow seismic equipment is the vertical position of acoustic source and receivers relative to one another and to the seabed. The output of the depressor-mounted sounder is found to show that depth stability of the acoustic source to within about 0.5 m can be achieved throughout a group of 8 observations. On seismograms which preserve the water surface reflection (marked "s" in Fig. 5), a triangulation method[6] can be used to locate individual receivers. Although the accuracy of that method is limited by the roughness of the sea surface (wave heights of 1.5 m reported during the survey), its application suggests there is no consistent slope in the towing attitude of the eel but does not rule out the possibility of undulations along its length following on changes in depth of the towing depressor. If it is assumed that vertical excursions along the length of the eel follow exactly those made by the depressor, then the difference in height between nose and tail of the array can be found for each shotpoint, as summarized in Fig. 10. Such height differences give rise to errors as illustrated in Fig. 11.

HEADWAVE ATTENUATION

The predicted[9] variation of headwave amplitude with offset distance follows a $(r - r_c)^{-1.5}$ law (where r is the distance along a ray and r_c is the critical distance[9]). This variation, arising from geometrical considerations is a more rapid attenuation than is the spherical spreading associated with ordinary reflected and refracted waves. Accordingly, in evaluating an absorption co-efficient for energy-loss processes occurring in the seabed material, observations of wave amplitude are needed with greater precision than is required when such a loss co-efficient is derived from spherically spreading waves.[15,21] Nevertheless, in natural conditions which preclude the use of reflectivity methods (e.g. where there is no observable subbottom reflector), it may be that some estimate of an absorption co-efficient might be made by observing the variation of headwave amplitude. More refinement in the calibration of the eel and in the data-handling software would be needed before this could be done (Fig. 13).

Fig. 12
Stepped gain settings are used in the display of individual seismogram traces so that headwave arrival times can be picked from wavelet events which are of uniform size on the display screen. By noting the gain required, a rough estimate of the relative amplitude at each channel can be made. Here these are shown for one seismogram compared with the trend of the geometrical attenuation.

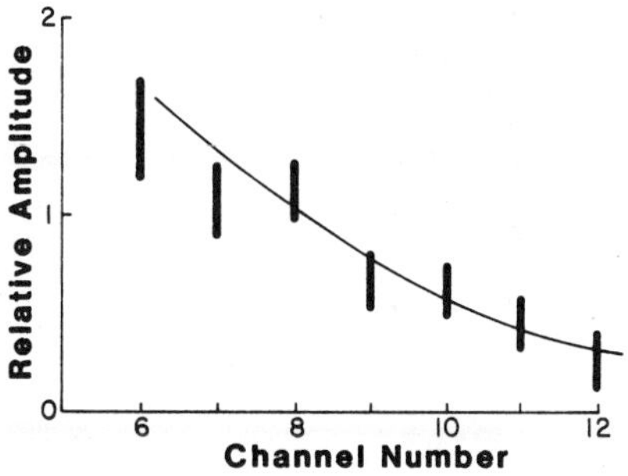

DISCUSSION AND CONCLUSIONS

All of the results lie within a sediment unit classified by King[19] as "Sable Island Sand and Gravel," which he divided into two sub-units on the basis of gravel content (more than or less than 50% by weight). Only groups 51-54 lie within the sub-unit with more than 50% gravel, but there is no obvious change in the results there except that group 53 shows the greatest scatter of velocity results. The wide angle reflection method had earlier[6] been used for velocity determination and gave results between 1.572 ± 0.026 and 1.661 ± 0.025 m/sec in an area where the sub-bottom reflector "r" is observed (Fig. 8), values which are in reasonable agreement with those obtained in the same vicinity at Nos. 43 and 44 of this survey (Table 1). A brief statistical summary of the results is given in Fig. 13.

The method described appears to yield consistent acoustic velocity results for seabed sediments in the area studied with an accuracy of about 3% or better. It offers the advantage over wide angle reflection methods of requiring no sub-bottom reflector to be present for the determination of a seabed velocity, but is not applicable where the seabed sediment has a velocity near or lower than that of the bottom water.

ACKNOWLEDGEMENT

This project was supported in part by the Department of National Defence (Contract No. 8SC84-00338).

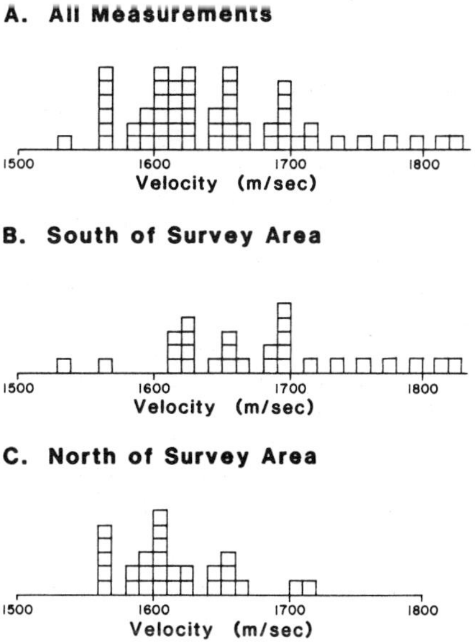

Fig. 13
The three histograms shown here summarize the velocity results for (A) all measurements made and for the measurements taken as two groups, each comprising half of the measurements, the one (south) being from nearer the shelf edge (B), where there appears to be more variability in velocity, the other (north) where the values are more uniform (C). Means and standard deviations for the three sets are as follows:

A: 1646 ± 65 m./s.
B: 1677 ± 71 m./s.
C: 1615 ± 40 m./s.

REFERENCES

1. E. L. Hamilton, Geo-acoustic Modeling of the Sea Floor, J. Acoust. Soc. Am 68: 1313-1340 (1980)
2. D. Taylor Smith, Seismo-Acoustic Wave Velocities and Sediment Engineering Properties, in: "Acoustics and the Seabed," Bath Univ. Press, Bath (1983)
3. C. H. Dix, Refraction and Reflection of Seismic Waves, II: Discussion of the Physics of Refraction Prospecting, Geophysics 4:238-241 (1939)
4. W. J. Porter, D.L. Bell and L.C. Britton, Multichannel Towed Arrays for Remote Measurement of Marine Sediment Acoustical Properties, in: "Proceedings of Offshore Technology Conference, Dallas," Paper No. 2015, S.E.G., Tulsa (1974)
5. A. G. McKay, Observations of Wide-Angle Reflexions and Refractions from a High-Resolution Engineering Seismic Source, Geophys. Prospect: 16: 673 (abstract) (1978)
6. A. G. McKay and P. M. McKay, Compressional-Wave Velocity Measurement in Seabed Materials by Use of Equipment Deployed near, but above the Bottom, J. Acoust. Soc.Am. 71: 871-878 (1982)
7. M. C. Fagot, A Deep-Towed Sound Source and Hydrophone Array System: Performance Prediction Analysis and Hardware Description, in: "Acoustics and the Seabed" N. G. Pace (ed.), Bath Univ. Press, Bath (1983)
8. M. C. Fagot, Development of a Deep-Towed Seismic System - A new Capability for Deep-Ocean Acoustic Measurements. In this volume.
9. V. Cerveny and R. Ravindra, "Theory of Seismic Headwaves," University of Toronto Press, Toronto (1971)
10. J.A. Hunter and G.D. Hobson, Seismic Refraction Method of Detecting Sub seabottom Permafrost, in: "The Coast and Shelf of the Beaufort Sea," J.C. Reed and J.E. Sater (eds.), The Arctic Institute of North America, Arlington (1974)
11. K. G. Neave, A.S. Judge, J.A.Hunter and H.A. MacAulay, Offshore Permafrost Distribution in the Beaufort Sea as determined from Temperature and Seismic Observations, in: "Current Research, Part C, Geological Survey of Canada, Paper No. 78-1C: 13-18 (1978)
12. H. A. MacAulay and J.A. Hunter, Detailed Seismic Refraction Analysis of Ice-bonded Permafrost Layering in the Canadian Beaufort Sea, in: "Proceedings of the 4th Canadian Permafrost Conference," p.256 (1982)
13. A. G. McKay, Acoustic Velocity Determinations from Humpback Refractions, Maritime Sediments 15: 23 (abstract) (1979)
14. A. G. McKay, Acoustic Velocity Measurements in Surficial Sediments of the Beaufort Sea, in: "Acoustics and the Seabed," Bath Univ. Press, Bath (1983)
15. A. G. McKay, Marine Seismology in Surficial Sediments, in: "Proceedings of Acoustic-Geotechnical Correlation Workshop," Calgary, Energy Mines and Resources, Canada (in press)
16. D.M.F. Chapman and D.D. Ellis, Propagation Loss Modelling on the Scotian Shelf: the Geo-Acoustic Model, in: "Bottom-interacting Ocean Acoustics,"W.A. Kuperman and F.B. Jensen (eds), Plenum, New York (1980)
17. C.L. Amos, CSS Baffin Cruise Report 82-039- Sable Island Bank and Banquereau, Atlantic Geoscience Centre, Geological Survey of Canada (1982)
18. C.L. Amos, CSS Dawson Cruise Report 84-005 - Sable Island Bank and Banquereau, Atlantic Geoscience Centre, Geological Survey of Canada (1984)
19. L. H. King, Surficial Geology of the Halifax-Sable Island Map Area, Geological Survey of Canada Paper No. 71-52 (1970)
20. Exploranium Ltd., 640 Hardwick Rd., Unit 1, Bolton, Ontario L0P 1A0 Canada
21. D.J. Dodds, Attenuation Estimates from High Resolution Subbottom Profiler Echoes, in: "Bottom-interacting Ocean Acoustics," W.A. Kuperman and F.B. Jensen (eds.), Plenum, New York (1980)

CHARACTERIZATION OF AVERAGE GEOACOUSTIC BOTTOM PROPERTIES FROM EXPECTED PROPAGATION BEHAVIOR AT VERY LOW FREQUENCIES (VLF) USING A TOWED ARRAY SIMULATION

M. F. Werby and G. J. Tango

Naval Ocean Research and Development Activity
Ocean Acoustics and Technology Directorate
NSTL, Mississippi 39529-5004

ABSTRACT

The purpose of this study is to determine the feasibility of broad georeconnaissance data acquisition for average bottom properties collected via a variable-depth towed array, detecting low frequency self-noise created from the towship. The study concentrates on shallow water regimes in which both near- and farfield array response features can be exploited from a single array. In this study bottom properties have been input to the *SAFARI* fast fullwave FFP model to create synthetic VLF towship data. The data are then processed at the array via synthetic beamforming techniques,[1] allowing discrimination of normal acoustic, as well as evanescent and seismic interface modes. This latter information in particular can be shown to directly relate to crucial bottom properties such as average compressional sound speed in the bottom (for fluid geoacoustic bottoms) or, in some cases, the net shear speed (for ideal infinite half-space bottoms of sufficiently high rigidity). The study also includes multilayered bottoms, which are examined to determine the thickness for which a higher net bottom attenuation factor dominates the seismo-acoustic process. It is expected that for thick (fluid) layers one can effectively ignore the deeper basement, while for thin (fluid) sediment layers, basement (solid) properties dominate in the low frequency/near cutoff regime. The frequency range and array detection size and depth limits for these processes are also presented (e.g.) in support of future experimental VLF propagation studies of shallow continental shelf transition zones.[2]

INTRODUCTION

It is often difficult, expensive, or impossible to obtain detailed borehole cores or seismic refraction and reflection data to characterize extensive ocean-bottom areas. Moreover, even with one or more such data, a unique determination of geobottom properties from remote-sensed geophysical measurements is not a priori possible in the absence of a generally applicable exact inverse method. Thus, it is of considerable interest to investigate approximate inverse methods of delimiting geobottom type and properties (such as P and S velocity and attenuation).

As is well known,[3] the Green's impulse response function completely characterizes the elastoacoustic properties of a given multilayered waveguide. In the case of underwater seismic-acoustic propagation at very low (VLF) frequencies, the full theoretic Green's function contains the complete discrete and continuous (acoustic mode) spectrum in the water column, as well as the evanescent (seismic interface wave) contributions from bottom interaction. The Fourier/Hankel transform simply relates the depth-dependent Green's function in the horizontal wavenumber/incident angle domain to the corresponding seismo-acoustic pressure fields in range.

More specifically, the Green's function method, whether applied to electromagnetic scattering, crustal seismology, or underwater seismo-acoustics, characterizes the effective impulse response of a multilayered target bottom in terms of observed transfer function properties of poles and residues. Poles are assumed invariant functions of media properties, whereas residues vary with the particular measurement performed and the type and configuration of the source. Theoretically, the accurate location of a few major poles can assist in determining essential geophysical bottom properties, thereby allowing the particular environment to be assigned to an appropriate category on the basis of known parameterizations.[4] The inferred media properties can then be used as input to one or more numerical propagation prediction models for estimating expected sound field intensity and character as a function of range.

In practice, empirical estimates of poles for ocean waveguide Green's functions, for example, as extracted from horizontal near-bottom towed arrays (Figure 1), tend to vary to only a limited extent even with major changes in the actual environment. It is therefore desirable to determine the theoretic detection limits and resolution accuracy of towed array beamformer methods for empirical Green's function estimation from multilayered bottoms at very low frequencies, as well as optimal array configurations for fast geobottom reconnaissance. This problem has recently been investigated by us[5] and other researchers for a limited number of simple Pekeris-type, shallow-water environments, but as yet full quantitative results have not been assessed or extended to more general multilayered bottoms.

This paper presents an extension of previous work using the fast fullwave FFP-Hankel function method utilizing the new, fast, Green's function method of Schmidt.[6] This allows realistic simulation of towed array responses for representative conditions of interest. We show that successful inversions should be possible using typical seismic-acoustic pressure field data. It is shown that considerable information can be obtained about average water-bottom waveguide properties directly from the beamformed transmission loss field, which can be readily interpreted in terms of known seismo-acoustic phenomena via its relation to the empirical Green's function.

THEORY

A complete discussion of the theory and numerical implementation of the present method has been previously given by Kuperman et al.;[5] thus, only the main features are summarized here.

The Green's impulse response function can be seen as a mapping of geologic boundary information onto a functional space that has relevant, but to some extent, ambiguous information. The fact that ambiguity enters the problem is merely a characteristic of the geophysical inverse problem.[7] However, with sufficient geoacoustic/geoseismic experience, the range of possible parameter models can be delimited to a satisfactory degree. That the quantity G is amenable to physical interpretation and can be related to a common measurement (namely, the beamformed response from bottom bounces) is the main contention of the present study.

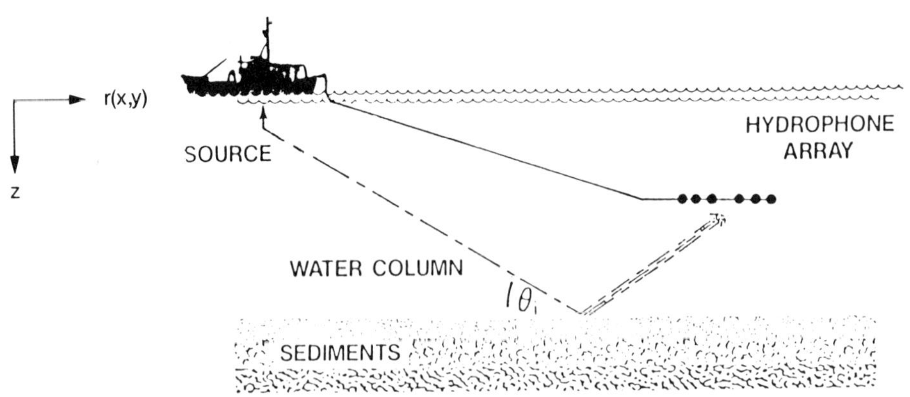

Figure 1. Schematic of Tow-Ship Array System.

Previously, we have shown that a beamformed response that has been weighted to account for cylindrical spreading is related to the Green's function G as

$$|B(\theta)| = |\Delta\eta(\eta_m)^{1/2}\, G(\mu_m,z)|/N \; ,$$

where N corresponds to the number of hydrophones in the array and m is the m-th array. The quantity η_m is the m-th value of the discretized horizontal wavenumber (divided N times). That this relation arises is due to the fact that both the beamformed response and the FFP method are obtained via the Fast Fourier transform (FFT). The consequences and physical interpretation of how $B(\theta)$ is obtained may be understood in terms of simple correspondences between G and normal mode theory (Figure 16). Essentially it is the case that the peaks of $G(\mu,z)$ correspond precisely to normal modes for the discrete spectrum, and to virtual modes in the continuum. The values of μ for which peak values occur may in turn be related by ray theory to the allowed angles of response from the bottom. The simple relation between η_m and θ_m is

$$\theta_m = \cos^{-1}\, [\omega/c(z)/\eta_m] \; ,$$

where ω is the angular frequency of the signal and $c(z)$ the compressional sound speed in the water. In addition, we have the allowed (discrete) values of μ via the relation

$$\omega \quad c_b < \mu_m < \omega/c_m \; ,$$

where c_b is the highest layer speed at an interface and c_m the minimum speed. In fact ω/c_b is seen to be the cutoff frequency, and if one can obtain that value of μ_m, then c_b can be found directly (which for the low shear case is the bulk layer velocity and for highly rigid bottoms, the shear velocity).

Moreover, we assert that if there are several layers, one may in principle associate several critical velocities with each. Each subsequent layer may then reflect a mode or mode group characteristic of its interface parameter. This in turn can be exploited to elicit information (under favorable and contrived conditions) regarding the bottom material characteristics.

For bottom-limited sound propagation, it is generally the case that (for $c_{bottom} > c_{water}$) there is a critical angle θ_c dividing the curve of reflectivity loss or the Green's impulse response function into regions of good and poor propagation. To consider the depth-dependent Green's function in more detail, Figure 2 shows poles corresponding to wavenumber eigenvalues of discrete normal propagating modes, as well as a number of lesser-amplitude continuous virtual modes in the nearfield, corresponding to lossy

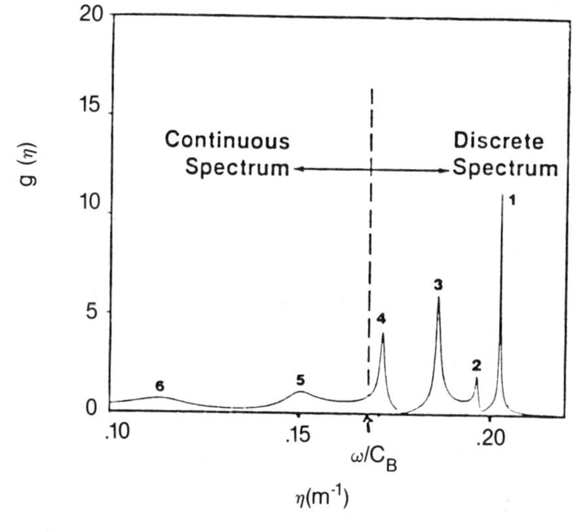

Figure 2. Theoretic Green's function.

mode propagation in the water column at angles $> cos^{-1} [c_w/c_b]$. The dashed line in Figure 2 clearly delineates two distinct regions of Green's function response. The location of this dividing line, as well as peaks corresponding to critical angles for modes governed by one or more deeper subbottom layers can be used to assign a given geobottom environment to a particular geological type on the basis of its relevant (measurable) geoacoustic/geoseismic properties.

Because the array beamformer is a spatial wavenumber spectral analyzer, by virtue of the inverse transform relation between Green's function and resulting field, the empirical Green's function output of the beamformer can then be seen as directly proportional to the true theoretical Green's function in horizontal wavenumber. Physically this means that the discrete angles of maximum beamformer response correspond directly to the mode propagation angles. As will be shown below, although the array will not always have sufficient resolution to accurately locate all individual peaks, it will still clearly show the transition region between discrete and continuous parts of the wavenumber spectrum. This suggests the following simple scheme: beamform separately over the entire array, and then over only the second (farfield) array elements. Comparison of total with farfield array responses locates the discrete/continuous transition region and, hence, provides a simple direct estimate of critical angle and layer velocity for the shallowest layer in question. By virtue of refraction and absorption loss, as source-receiver array offset is increased, this behavior repeats itself for progressively deeper strata.

Summarizing some important theoretical conclusions[5] concerning the present method, the FFT-based beamformer simulation developed here is useful for quickly describing the seismo-acoustic wavefield in the ocean bottom waveguide for almost all angles of incidence at VLF frequencies. Advantages of this beamformer include:

1) Only simple and standard array theory is required, so that sensitivity and pre-experimental studies can be easily made;

2) It is readily possible to interpret beamformed output via ray-mode analogies;

3) The differences between (e.g.) finite and infinite length array responses, as well as that of selected array subsections for near and far fields, can be directly determined;

4) All effects of realistic plane-stratified, ocean-bottom types can be incorporated.

The present beamformer is also valid for range-dependent geoenvironments (although there does not as yet exist an exact Green's function solution for laterally inhomogeneous or nonplanar media). This discussion is limited to considerations of pressure fields only, because in most cases acoustic mode effects will be seen to dominate seismic bottom propagation (in principle, however, vertical or horizontal particle velocity records could also be used).

The amplitude of the beamforming array element outputs were computed for representative values of array and bottom parameters, and selected results grouped to exhibit significant trends. The numerical results and discussion stress the practical detection of expected geobottom types.

ANALYSIS OF RESULTS

As a preview of our results, we state the following general rules useful in the analysis that follows.

(1) Beamforming using the FFT is less effective if the angular region is closer to the forward direction; angular resolution progressively improves as one approaches 90°.

(2) The angle of propagation of the modes for any bottom shifts toward the forward array direction with increasing frequency.

(3) The present beamforming technique is suitable only for resolving modes, provided their angles of propagation are suitably far apart. For propagation angles less than the beamformer resolution, a single net mode peak is observed.

(4) The number of propagating modes increases with increasing frequency and, conversely, net resolution is degraded above a certain f_{max}.

Table 1. Table of Bottom Parameters.

	V_c	V_s	ϱ	λ_c	λ_s
Clay-Silt	1515	100	1.60	0.5	1.0
Sand	1800	600	2.00	0.7	1.5
Basalt	5250	2500	2.60	0.2	0.5

V_p = compressional velocity; V_s = shear velocity
ϱ = relative density; λ_p = compressional attenuation coefficient
λ_s = shear attenuation coefficient

(5) For fluids and bottoms with low shear velocity ($v_s << 1500$ m/s) there is a cutoff frequency below which no acoustic modes can propagate in the water column, leaving only bottom-traveling seismic interface modes.

(6) For bottoms with high shear velocity, there is no limiting frequency for propagating modes; i.e., one will always observe at least one propagating mode, even in the low VLF regime.

(7) Bottom attenuation is proportional to frequency. Thus for high frequencies, absorption will be considerable (*nb*, for shear arrivals).

(8) The total number of modes increases with water depth and bottom velocity, so optimal resolution occurs only for shallow water sediment regimes.

(9) The modal propagation angle shifts toward the forward (poorly resolved) array direction for lower bottom velocities.

For convenience we restrict our study to shallow bottoms in no more than 100 m of water. In this case, the method described here is not expected to be suitable for frequencies >100 Hz due to (1), (2), and (3) and, for most bottom combinations, a frequency <25 Hz. The exception (elaborated below) is the case where a high shear velocity value is present in one or more layers (usually the basement) due to (6). Generally, bottoms are layered in such a manner that bottom velocities increase and attenuation decreases with depth into the subbottom (Table 1).[8]

We first examine the case of a two layered bottom (sediment overlying basement) to determine the effects of layer thickness on field measurement. The upper layer is sand; the half-space is basalt; the sand layer is varied from 100 m to 5 m thickness. Normally, sand would be considered to have a fairly low attenuation, and since we are interested in determining at which thickness the array no longer senses the bottom, the conclusion for sand necessarily holds for bottoms of greater attenuation. We illustrate the case of f = 50 Hz in Figures 3, 4, and 5 for thicknesses of 100, 50, and 5 m of sand overlying a basalt half-space. Figure 6 represents a sand half-space. It is clear from the responses that 100 and 50 m of sand are essentially identical to a sand half-space for the array minimally offset from the source. This strongly suggests that the basement for these thicknesses has little importance. In fact, significant interaction occurs at 25 m, and in subsequent multilayered calculations we limit layers to the present thickness region. This varies continuously with frequency, with lower frequencies penetrating more deeply into the sediment without significant absorption. Surprisingly, the 5-m-thick sand is not dominated by the basement, as might be expected, and the beamformed results do not compare with the basalt half-space, but are instead more complicated.

One could perform a more exhaustive study of thickness for the two-layer case, but instead we move to the more complicated and realistic three-layer case. In this analysis, we take advantage of the results above, and limit ourselves to layers below 50 m thickness. Once again, it is possible to investigate an infinite number of combinations of thicknesses between layers. We consider the case in which the first layer is 30 m, the second 40 m thickness, and the third a half-space. We vary frequency from 100 Hz to 3.125 Hz, allowing a variable array offset from the source. By varying offset we will specifically allow for multiple absorption from the inner layers, an effect that can be exploited for systematic layer discrimination.

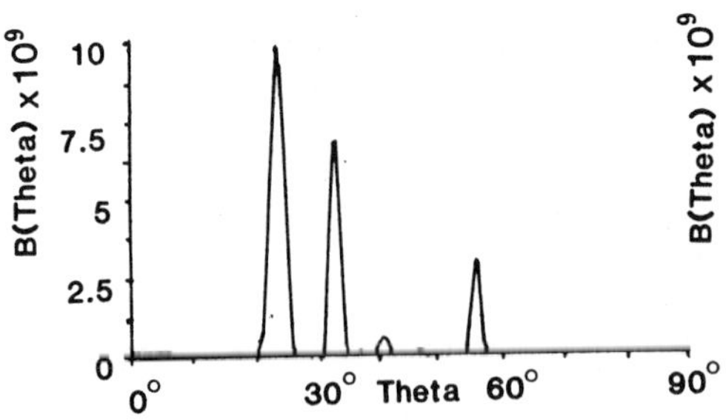

Figure 3. Simulated experimental towed array (beamformed) response for 100 m sand overlying a basalt half-space.

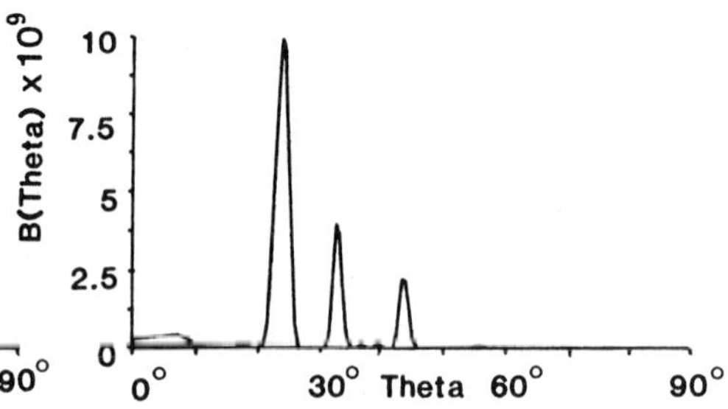

Figure 4. Simulated towed array response for 50 m sand over basalt.

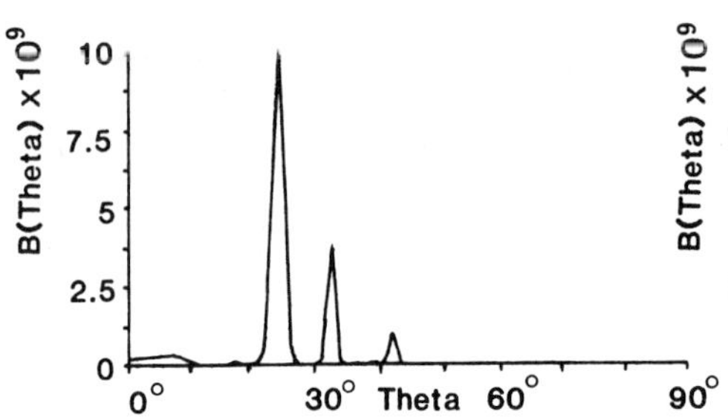

Figure 5. Towed array response for 5 m sand over basalt.

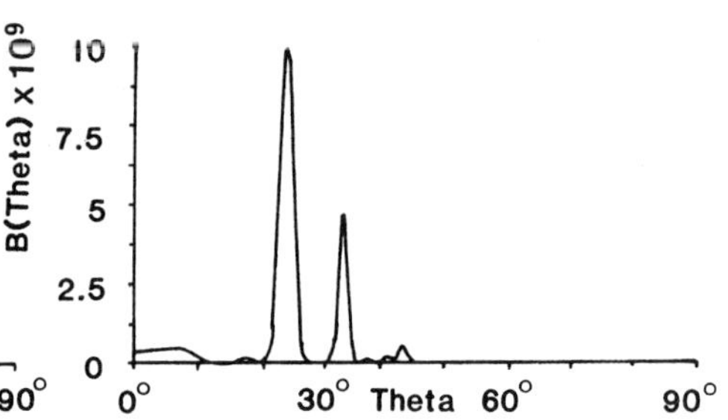

Figure 6. Towed array response for sand half-space.

We now examine the three-layer bottom of silt-clay, sand, and basalt, respectively. The angular region of propagation for each is illustrated in Figure 7, where the characteristic velocities and densities of the layers increase from layer 1 to 3. We assume the basement layer to have high shear rigidity. In the high-frequency limit, if we are suitably far offset from the source, then due to (7) above, we expect to observe propagating modes resulting from interaction with only the top layer (modes penetrating layer 1 and/or 2 multiple times will be sufficiently attenuated that only propagating modes from the topmost layer will be observed). If the array is not offset very far from the source, then a rather complex pattern will be observed and one will not generally be able to fully resolve each layer contribution unambiguously via this method.

To illustrate this point, Figures 8 and 9 depict beamformed results near the source (at 2 km) and 4 km offset from the source, and Figure 10 is a silt-clay half-space. Clearly, Figure 8 yields a very complicated pattern, while Figure 9 and 10 are almost identical. This confirms the point that at higher frequencies one only observes the topmost layer (for large source-array offsets). Thus, we can use this experimental scenario to isolate the top layer. Examining the low-frequency case, according to (6) and (5) above we choose a frequency sufficiently low that only the bottommost half-space produces a mode peak in the beamformed response. At a subcutoff frequency of 3.125 Hz, we would expect to observe a strong Scholte-type seismic interface wave from the basement, with no acoustic modes propagating in the water column. Figure 11 illustrates just such a mode, and comparison with the case of a basalt half-space in Figure 12 confirms this point (i.e., the results are essentially equal). Thus, this second experimental scenario allows one to isolate the basement as a third layer, providing it is a suitable hard bottom capable of supporting a Scholte wave.

We now wish to resolve the middle layer and confirm information already obtained for the top and bottom. We attempt this by examining the 50 Hz and 25 Hz frequency cases at various offsets. It is also possible to vary the frequency continuously over the applicable range for better resolution; we, however, assume that the array is fixed and that frequencies must be divisible by some integral power of 2.

In a previous paper[5] we have shown that for a half-space it is possible to determine the critical angle of the bottom by varying array offset (or alternatively by zeroing the leading hydrophones of the array), thereby identifying the compressional bottom velocity. We attempt this for the layered bottom

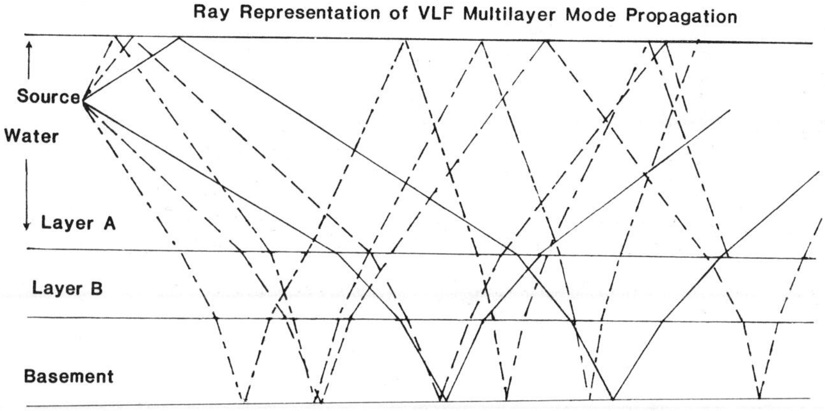

Figure 7. Diagram of mode-ray paths for shallow-water environment.

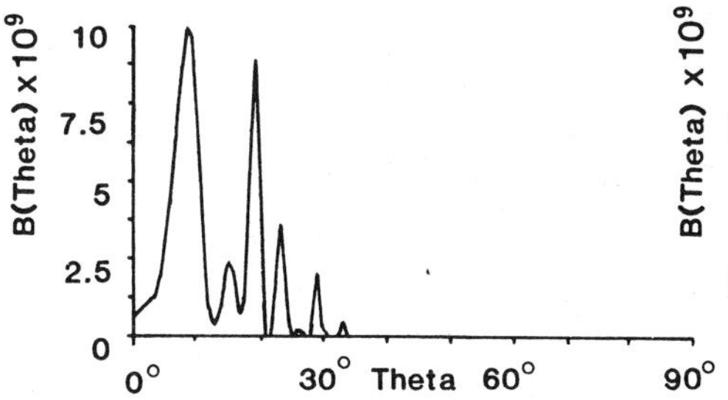

Figure 8. Towed array response for three-layer bottom (as per Fig. 7), for source-array offset = 2 km.

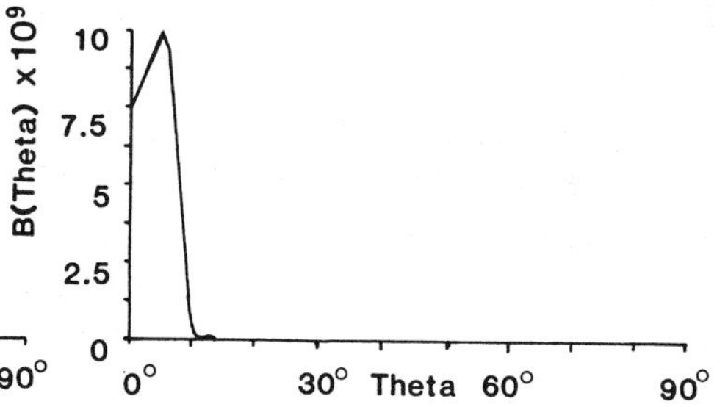

Figure 9. Towed array response for three layer bottom (as per Fig. 7), for source-array offset = 4 km.

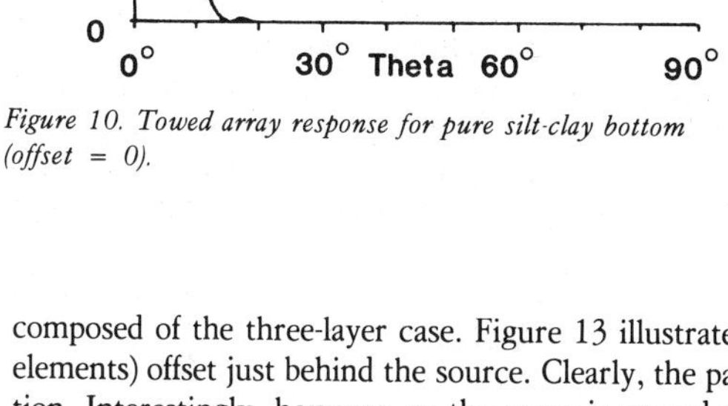

Figure 10. Towed array response for pure silt-clay bottom (offset = 0).

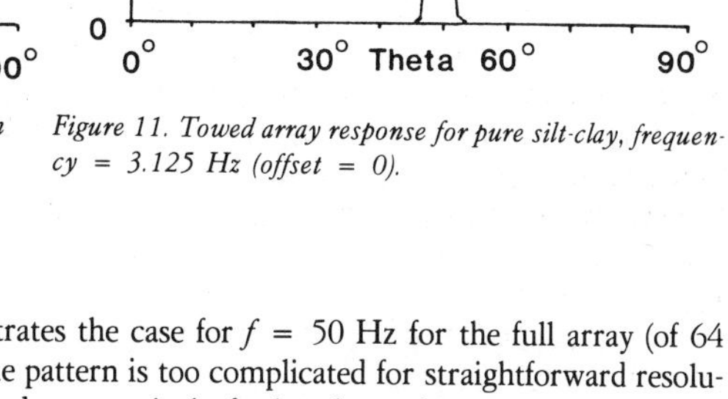

Figure 11. Towed array response for pure silt-clay, frequency = 3.125 Hz (offset = 0).

composed of the three-layer case. Figure 13 illustrates the case for $f = 50$ Hz for the full array (of 64 elements) offset just behind the source. Clearly, the pattern is too complicated for straightforward resolution. Interestingly, however, as the array is moved progressively farther from the source, propagating modes corresponding to the critical angles of the three interfaces in turn dominate the beamform results. This is evident in Figure 14 and 15 with offsets of 0.75 and 1 km, respectively. Clearly, one observes 3 modes corresponding to about 8°, 33°, and 52°. These modes are, in fact, the approximate critical angles of each of the three layers. This effect is duplicated at $f = 25$ Hz, even for near offsets. This is illustrated in Figure 16, which is offset close to the source. Here, only a few propagating modes are allowed for the low shear layers, while the high shear bottom emphasizes the highest angular mode occurring close to the bottom critical angle. These effects may be exploited to determine successively deeper layer properties (in a fashion similar to marine refraction shooting). To be sure, only the dominant

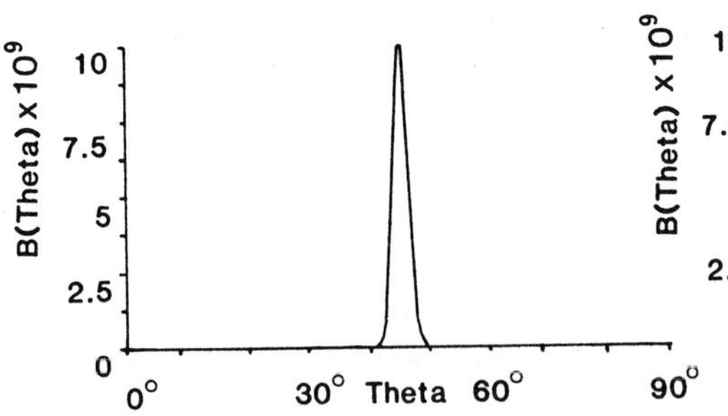

Figure 12. Towed array response for pure basalt, frequency = 3.125 Hz (offset = 0).

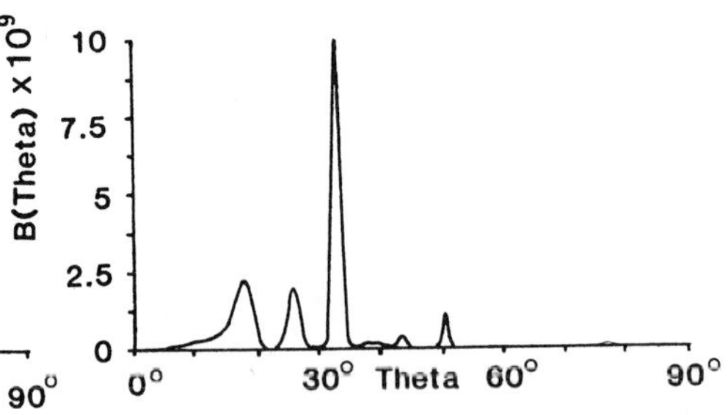

Figure 13. Towed-array response for 3-layer case, frequency = 50 Hz (offset = 0.5 km).

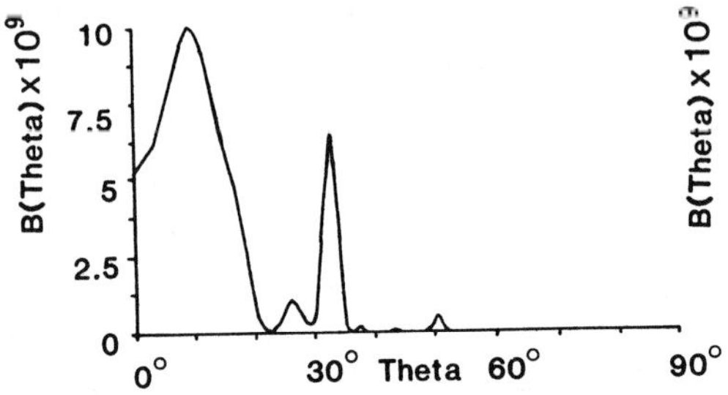

Figure 14. Towed array response for same case as Figure 13, with offset = 0.75 km.

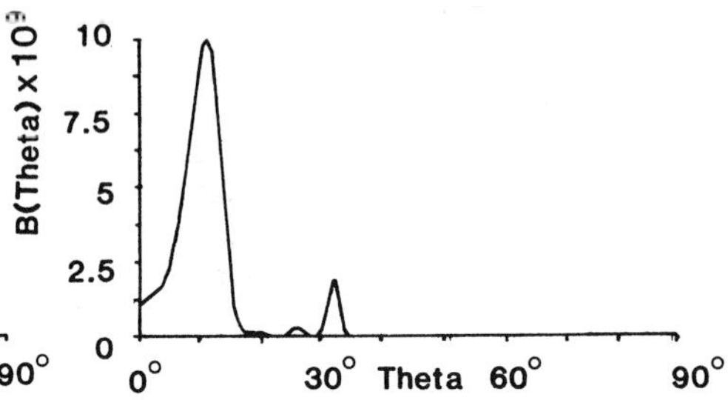

Figure 15. Towed array response for same case as Figure 13, with offset = 1 km.

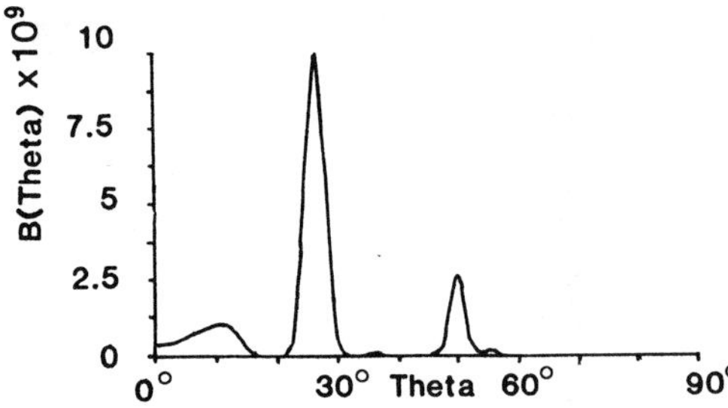

Figure 16. Towed-array response for 3-layer case, frequency = 25 Hz (offset = 0.5 km).

bottom velocity is determined (which is compressional for low shear, but shear for bottoms with strong rigidity). Interpretation is often simplified because of the strong association between bottom velocity and attenuation,[8] and this along with other geoacoustic considerations can generally associate bottom velocities with particular bottom types.

CONCLUSIONS

In this paper we have examined the VLF applicability of a simple technique for characterizing shallow-water waveguides. Here, we determine (for example) compressional and/or shear velocities of successively deeper layers in a multilayered system from the empirically recovered Green's response function, as seen by a deep towed array at increased offset. This is accomplished by utilizing the exact Hankel tranform relation between empirical pressure field in range and impulse response vs. angle due to a point source (which is here assumed a sufficiently good approximation of towship noise; else as an independent source). It was shown that the theoretical Green's function (as computed by the *SAFARI* FFP program) contains complete information about the nature of the discrete continuum and evanescent spectrum, the latter being especially important for (submodal cutoff) VLF regimes.

During the completion of this study, a similar work[9] appeared. Although their basic intent and numerical strategy for inverting the calculated pressure field in range for bottom parameters is similar to ours, the mechanism by which they compute the theoretic Green's function (from which pressure and beamformed Green's function results are derived) are different. The principle accomplishments of Frisk et al.[9] is to successfully determine the empirical Green's function and reflection coefficient via fast Hankel transform inversion of a mode-calculated pressure field from a single fluid waveguide over a single solid half-space bottom; for these conditions their improved transform algorithm is highly effective. By contrast, our own objective has been to further determine the expected range of responses, sensitivity, and accuracy of empirical Green's functions as seen by beamformers over a range of generalized multilayered bottoms for the full VLF regime. Here the fast fullwave FFP solution of Schmidt[6] was an efficient input to our simple beamforming scheme.

Inversions were performed using synthetic seismo-acoustic pressure field data over a range of reasonable array apertures. The distinct effects of water and sediment layer thickness, shear velocity and attenuation, and source/receiver array offset were examined. The present mixed multilayered bottom examples pose a severe test of this approach, but suggests its principle advantages for most typical bottom types. These examples were chosen to demonstrate that the method works acceptably well, even under adverse layering conditions. As has been shown, a much more favorable environment for the technique is encountered for simpler bottoms of few-layer configuration, when source and receiver array are near the bottom but far apart, and when the total number of acoustic modes is small over the frequency range of interest. These situations are encountered in a variety of realistic shallow ocean-bottom environments, and studies using this and improved methods are in progress to further outline the potential applications of this approach.

REFERENCES

1. W. A. Kuperman, M. F. Werby, and K. E. Gilbert, ''Towed Array Response to Ship Noise: A Near-field Propagation Problem,'' presented at NATO Conference on Underwater Acoustics Signal Processing, 1 August 1984, Luneberg, W. Germany (1984).
2. T. W. Tunnell and G. J. Tango, ''Predicted Partitioning of VLF Acoustic Energy in a Range-Dependent Environment,'' (this volume).
3. M. Ewing, W. Jardetzky, and F. Press, *Elastic Waves in Layered Media*. New York, McGraw-Hill (1957).
4. W. A. Kuperman and F. B. Jensen, ''Optimum Frequency of Propagation in Shallow Water Environments,'' *J. Acoust. Soc. Am.* 73 (1983).
5. W. A. Kuperman, M. F. Werby, K. E. Gilbert, and G. J. Tango, ''Beamforming on Bottom-Interacting Towship Noise,'' invited paper, *IEEE Oceanic Engineering Journal* (Special Issue on Beamforming), to appear, July (1985).
6. H. Schmidt, ''Excitation and Propagation of Seismic Interface Waves (using a new Fast Field Program),'' in: *Acoustics of the Seabed*, edited C. N. Pace. Bath, UK: University Press (1982).
7. R. Cassinis, (editor), *The Solution of the Inverse Problem in Geophysical Interpretation*. New York, Plenum Press (1981).
8. E. L. Hamilton, ''Geoacoustic Modeling of the Seafloor,'' *J. Acoust. Soc. Am.* 68 (5) (1980).
9. G. V. Frisk, J. F. Lynch, and J. A. Doutt, ''The Determination of Geoacoustic Models in Shallow Water,'' (this volume).

CONTRIBUTORS

B.B. Adams
Naval Research Laboratory
Acoustics Division
Washington, DC 20375-5000
USA

T. Akal
Saclant ASW Research Centre
Viale San Bartolomeo 400
19026 La Spezia
ITALY

H.B. Ali
Saclant ASW Research Centre
Viale San Bartolomeo 400
19026 La Spezia
ITALY

P. Anderson
Hawaii Institute of Geophysics
University of Hawaii
Honolulu, HI 96822
USA

K. Attenborough
Faculty of Technology
The Open University
Walton Hall
Milton Keynes MK7 6AA
U.K.

F. Avedik
Department of Geosciences
IFREMER, B.P. 337
29273 Brest CEDEX
FRANCE

M. Badiey
Applied Marine Physics Division
RSMAS, University of Miami
Virginia Key
Miami, FL 33149-1098
USA

R.N. Baer
Naval Research Laboratory
Acoustics Division, Code 5160
Washington, DC 20375-5000
USA

A.B. Baggeroer
Room 5-204
Massachusetts Institute of Technology
Cambridge, MA 02139
USA

A. Barbagelata
Saclant ASW Research Centre
Viale San Bartolomeo 400
19026 La Spezia
ITALY

J.S. Barker
Woodward - Clyde Consultants
566 El Dorado St.
Pasadena, CA 91101
USA

J.H. Beebe
The Pennsylvania State University
Applied Research Laboratory
P.O. Box 30
State College, PA 16804
USA

J.D. Bennel
Department of Physical Oceanography
University College of North Wales
Marine Science Laboratories
Menai Bridge, LL59 5EY
Gwynedd, North Wales
U.K.

J.M. Berkson
Saclant ASW Research Centre
Viale San Bartolomeo 400
19026 La Spezia
ITALY

H.O. Berktay
School of Physics
University of Bath
Claverton Down
Bath BA2 7AY
U.K.

J.L. Berrou
Thomson-CSF
ASM Division
06802 Cagnes-sur-mer
FRANCE

I.M. Besieris
Dept. of Electrical Engineering
Virginia Polytechnic
Blacksburg, VA 24061
USA

M. Bradley
Planning Systems, Inc.
McLean, VA 22102
USA

G.H. Brooke
Defence Research Establishment Pacific
FMO
Victoria, BC, VOS 1BO
CANADA

D.G. Browning
Naval Underwater Systems Center
New London, CT 06320
USA

B. Brunson
Planning Systems, Inc.
McLean, VA 22102
USA

H.P. Bucker
Code 541
Naval Ocean Systems Center
San Diego, CA 92152-5000
USA

M.J. Buckingham
Royal Aircraft Establishment
Farnborough, Hampshire GU14 6TD
U.K.

J.J. Cabrera
Inverse Theory and Applications Inc.
2162 Western Parkway
Vancouver, BC, V6T 1V6
CANADA

W.M. Carey
Office of Naval Research
800 Quincy Street
Arlington, VA 22217
USA

P.M. Carrion
Columbia University
Henry Krumb School of Mines
New York, NY 10027
USA

J.A. Carter
Rondout Associates Inc.
P.O. Box 224
1 Stilba Lane
Stone Ridge, NY 12484
USA

R.K. Cessaro
Hawaii Institute of Geophysics
University of Hawaii
Honolulu, HI 96822
USA

D.M.F. Chapman
Defence Research Establishment
Atlantic
P.O. Box 1012
Dartmouth, NS, B2Y 3Z7
CANADA

N.R. Chapman
Defence Research Establishment Pacific
FMO Victoria, BC, VOS 1BO
CANADA

P.L. Cousins
CAP Scientific
40-44 Coombe Road
New Malden, Surrey KT3 4QN
U.K.

D.J. Creasey
The University of Birmingham
Edgbaston
Birmingham
U.K.

P. Curzi
Istituto di Geologia Marina
Consiglio Nazionale delle Ricerche
Via Zamboni 65
40127 Bologna
ITALY

J.M. Daupleix
L.M.A.I.
University of Perpignan
66000 Perpignan
FRANCE

A.M. Davis
Department of Physical Oceanography
University College of North Wales
Marine Science Laboratories
Menay Bridge, LL59 5EY
Gwynedd, North Wales
U.K.

H.A. DeFerrari
Applied Marine Physics Division
Rosenstiel School of Marine
and Atmospheric Science
University of Miami
Miami, FL 33149-1098
USA

D.R. DelBalzo
Code 200
Naval Ocean Research & Development Activity
NSTL Station, MS 39529-5004
USA

B. De Raigniac
GERDSM
Le Brusc
83140 Six Fours les Plages
FRANCE

Y. Desaubies
IFREMER
Centre de Brest P.B. 337
29273 Brest CEDEX
FRANCE

O.I. Diachok
Naval Research Laboratory
Acoustics Division
Washington DC 20375-5000
USA

R.L. Dicus
Naval Research Laboratory
Acoustic Division
Washington, DC 20375-5000
USA

F.R. DiNapoli
Naval Underwater Systems Center
New London, CT 06320
USA

R.D. Doolittle
Code 5122
Naval Research Laboratory
Washington, DC 20375-5000
USA

L.M. Dorman
Geological Research Division
Scripps Institution of Oceanography
La Jolla, CA 92093
USA

J.A. Doutt
Woods Hole Oceanographic Institution
Woods Hole, MS 02543
USA

G.L. Duckworth
Room E34-458
Massachusetts Institute of Technology
Cambridge, MA 02139
USA

F.K. Duennebier
Hawaii Institute of Geophysics
University of Hawaii
Honolulu, HI 96822
USA

J.I. Dunlop
School of Physics
University of New South Wales, P.O. Box 1
Kensington, NSW 2033
AUSTRALIA

M.D. Duston
Dept. of Mathematics
University of Rhode Island
Kingston, RI 02881
USA

I. Dyer
Massachusetts Institute of Technology
Department of Ocean Engineering
Cambridge, MA 02139
USA

K.J. Ellefsen
Massachusetts Institute of Technology
Cambridge, MA 02139
USA

F.T. Erskine
Code 8160
Naval Research Laboratory Division
Washington, D.C. 20375-5000
USA

R.B. Evans
Naval Ocean Research & Development Activity
Code 220
NSTL Station, MS 39529-5004
USA

K.C. Ewans
Physics Department
University of Auckland
Auckland 1
NEW ZEALAND

M. Fagot
Naval Ocean Research & Development Activity
NSTL Station, MS 39529-5004
USA

L.B. Felsen
Polytechnic Institute of New York
Dept. of Electrical Engineering and
Computer Science
Route 110
Farmingdale, NY 11735
USA

M.C. Ferla
Saclant ASW Research Centre
Viale San Bartolomeo 400
19026 La Spezia
ITALY

C. Feuillade
Naval Ocean Research & Development Activity
NSTL Station, MS 39529-5004
USA

S. Fiori
Saclant ASW Research Centre
Viale San Bartolomeo 400
19026 La Spezia
ITALY

L. Fishman
Department of Civil Engineering
Catholic University of America
Washington, DC 20064
USA

E.R. Franchi
Code 8160
Naval Research Laboratory
Acoustic Division
Washington, DC 20375-5000
USA

L.N. Frazer
Hawaii Institute of Geophysics
University of Hawaii
2525 Correa Road
Honolulu, HI 96822
USA

G.V. Frisk
Woods Hole Oceanographic Institution
Woods Hole, MA 02543
USA

G.J. Fryer
Hawaii Institute of Geophysics
University of Hawaii
2525 Correa Road
Honolulu, HI 96822
USA

B.K. Gazey
The University of Birmingham
Edgbaston
Birmingham
U.K.

K.E. Gilbert
Naval Ocean Research & Development Activity
Code 220
NSTL Station, MS 39529-5004
USA

T.G. Goldsberry
Saclant ASW Research Centre
Viale San Bartolomeo 400
19026 La Spezia
ITALY

R.L. Good
Geological Survey of Canada
601 Booth Street
Ottawa, Ont, K1A 0E8
CANADA

B. Grandvuillemin
GERDSM (DCAN Toulon)
Le Brusc
83140 - Six Fours Les Plages
FRANCE

G. Guidi
Saclant ASW Research Centre
Viale San Bartolomeo 400
19026 La Spezia
ITALY

M.K. Gurcan
Philips Research Labs.
Cross Oak Lane
Redhill
Surrey RH1 5HA
U.K.

C.H. Harrison
CAP Scientific
10 11 Coombe Road
New Malden
Surrey KT3 4QF
U.K.

D.V. Helmberger
Seismology Laboratory
California Institute of Technology
Pasadena, CA 91124
USA

H. Hobaeck
Applied Research Laboratories
The University of Texas at Austin
Austin, TX 78713
USA

C.W. Holland
Planning Systems Inc.
7900 Westpark Drive, Suite 600
McLean, VA 22102
USA

E.H. Hug
Norwegian Defence Research Establishment
Box 115
Horten N 3191
NORWAY

J.A. Hunter
Geological Survey of Canada
601 Booth Street
Ottawa, Ont. K1A 0E8
CANADA

F.B. Jensen
Saclant ASW Research Centre
Viale San Bartolomeo 400
19026 La Spezia
ITALY

A.C. Kibblewhite
Physics Department
University of Auckland
Auckland 1
NEW ZEALAND

H.J. Kloosterman
Laboratory of Seismics and Acoustics
Delft University of Technology
Delft
THE NETHERLANDS

D.S. Ko
Applied Marine Physics Division
Rosenstiel School of Marine
and Atmospheric Science
University of Miami
Miami, FL 33149-1098
USA

R.A. Koch
Applied Research Laboratories
University of Texas at Austin
Austin, TX 78713-8029
USA

W.E. Kohler
Dept. of Mathematics
Virginia Tech
Blacksburg, VA 24061
USA

W.A. Kuperman
Code 9006
Naval Research Laboratory
Washington, DC 20375
USA

C.C. Leroy
SINTRA Marine Dept.
Z.I. des Paluds
13400 Aubagne
FRANCE

S. Levy
Inverse Theory and Applications Inc.
2162 Western Parkway
Vancouver, B.C. V6T 1V6
CANADA

P.H. Lindop
Admiralty Research Establishment
Portland
Dorset DT5 2JS
U.K.

J.P. Longuemard
L.M.A.I.
University of Perpignan
66000 Perpignan
FRANCE

I.T. Lu
Polytechnic Institute of New York
Dept. of Electrical Engineering and Computer
Science
Route 110
Farmingdale, NY 11735
USA

J.F. Lynch
Dept. of Ocean Engineering
Woods Hole Oceanographic Institution
Woods Hole, MA 02543
USA

R.F. MacKinnon
Defence Research Establishment Pacific
FMO Victoria BC, VOS 1B0
CANADA

D. Marandino
Saclant ASW Research Centre
Viale San Bartolomeo 400
19026 La Spezia
ITALY

J.E. Matthews
Naval Ocean Research & Development Activity
NSTL Station, MS 39529-5000
USA

D.F. McCammon
Applied Research Laboratory
The Pennsylvania State University
P.O. Box 30
State College, PA 16801
USA

J.J. McCoy
Department of Civil Engineering
The Catholic University of America
4501 Connecticut Ave, NW 213
Washington, DC 20064
USA

S.T. McDaniel
Applied Research Laboratory
The Pennsylvania Research Laboratory
P.O. Box 30
State College, PA 16804
USA

B.E. McDonald
Naval Ocean Research & Development Activity
NSTL Station, MS 39529-5004
USA

A.G. McKay
Nova Scotia Research Foundation
P.O. Box 790
Dartmouth, NS, B2Y 3Z7
CANADA

H. Medwin
Naval Postgraduate School
Physics Department
Monterey, CA 93943
USA

W.C. Meecham
School of Engineering and Applied Science
UCLA
5731 Boelter Hall
Los Angeles, CA 90024
USA

R.H. Mellen
PSI Marine Sciences
95 Trumbull Street
New London, CT 06320
USA

E. Michelozzi
Saclant ASW Research Centre
Viale San Bartolomeo 400
19026 La Spezia
ITALY

D.J. Miller
Hawaii Institute of Geophysics
University of Hawaii
2525 Correa Road
Honolulu, HI 96822
USA

J. Miller
Physics Department
Catholic University
Fourth St. Michigan An NE
Washington, DC 20064
USA

C.L. Monjo
Applied Marine Physics Division
Rosenstiel School of Marine
and Atmospheric Science
University of Miami
Miami, FL 33149
USA

T.G. Muir
Applied Research Laboratories
The University of Texas at Austin
Austin, TX 78713
USA

A. Nagl
Department of Physics
The Catholic University of America
Fourth St. Michigan Av NE
Washington, DC 20064
USA

L. Nghiem-Phu
Daubin Systems Corp.
104 Crandon Blvd.
Key Biscayne, FL 33149
USA

D. Oldenburg
Inverse Theory and Applications, Inc.
2162 Western Parkway
Vancouver, B.C. V6T 1V6
CANADA

J. Orcutt
Institute of Geophysics
and Planetary Physics
Scripps Institution of Oceanography
La Jolla, CA 92093
USA

N.G. Pace
School of Physics
University of Bath
Claverton Down
Bath BA2 7AY
U.K.

J.S. Perkins
Naval Research Laboratory
Acoustics Division, Code 5160
Washington, DC 20375-5000
USA

A. Plaisant
Thomson-CSF
Division Activities Sous-Marine
Chemin des Travails B.P. 53-06801
Cagnes-sur-mer CEDEX
FRANCE

M.B. Porter
Naval Research Laboratory
Acoustics Division
Washington, D.C. 20375
USA

D. Rauch
Istituto di Acustica
"O.M. Corbino"
Via Cassia 1216
Rome 00189
ITALY

M.D. Richardson
Saclant ASW Research Centre
Viale San Bartolomeo 400
19026 La Spezia
ITALY

A.W. Sauter
Scripps Institution of Oceanography A-008
La Jolla, CA 92093
USA

E.K. Scheer
Woods Hole Oceanographic Institution
Woods Hole, MA 02543
USA

P. Schippers
Physisch Laboratorium TNO
Oude Waalsdorperweg 63
2597 AK Den Haag
THE NETHERLANDS

A.E. Schreiner
Scripps Institution of Oceanography
La Jolla, CA 92093
USA

B. Schmalfeldt
MTG Marinetechnik GmbH
Wandesbeker Konigstrasse 62
D-2000 Hamburg 70
FEDERAL REPUBLIC OF GERMANY

H. Schmidt
Saclant ASW Research Centre
Viale San Bartolomeo 400
19026 La Spezia
ITALY

T. Sereno
Institute of Geophysics and
Planetary Physics
Scripps Institution of Oceanography, A-025
La Jolla, CA 92093
USA

M. Snoek
Saclant ASW Research Centre
Viale San Bartolomeo 400
19026 La Spezia
ITALY

J.V. Soileau
Naval Ocean Research & Development Activity
NSTL Station, MS 39529-5004
USA

P. Staal
Defence Research Establishment Atlantic
P.O. Box 1012
Dartmouth, Nova Scotia B2Y 3Z7
CANADA

R.A. Stephen
Woods Hole Oceanographic Institution
Woods Hole, MA 02543
USA

K. Stinson
Inverse Theory and Applications Inc.
2162 Western Parkway
Vancouver, BC V6T 1V6
CANADA

R.D. Stoll
Dept. of Civil Engineering
and Engineering Mechanics
School of Engineering and Applied Science
Columbia University
New York, NY 10027
USA

A. Suteau-Henson
Rondout Associates, Inc.
P.O. Box 224
Stone Ridge, NY 12484
USA

G.H. Sutton
Rondout Associates Inc.
P.O. Box 224
Stone Ridge, NY 12484
USA

G.J. Tango
Ocean Acoustics Division
Naval Ocean Research & Development Activity
NSTL Station, MS 39529-5004
USA

F.D. Tappert
Division of Ocean Engineering
Rosenstiel School of Marine
and Atmospheric Science
University of Miami
Miami, FL 33149
USA

D. Taylor-Smith
University College of North Wales
Marine Science Laboratories
Menai Bridge LL59 5EY
Gwynedd, North Wales
U.K.

D.J. Thomson
Defence Research Establishment Pacific
FMO Victoria BC, VOS 1BO
CANADA

C.T. Tindle
Applied Research Laboratories
The University of Texas at Austin
Austin, TX 78713
USA

A. Tolstoy
Code 5122
Naval Research Laboratory
Washington, DC 20375-5000
USA

E. Topuz
Technical University of Istanbul
Electrical Engineering Faculty
Ayazaga, Istanbul
TURKEY

T. Torii
Applied Marine Physics Division
RSMAS, University of Miami
Virginia Key
Miami, FL 39149-1098
USA

T.W. Tunnel
Ocean Acoustics Division
Naval Ocean Research & Development Activity
NSTL Station, MS 39529-5004
USA

H. Uberall
Physics Department
Catholic University
Fourth St. Michigan Av NE
Washington, DC 20064
USA

B.J. Uscinski
D.A.M.T.P.
University of Cambridge
Silver Street
Cambridge, CB3 9EW
U.K.

G. Verma
Code 3332
Naval Underwater Systems Center
New London, CT 06320
USA

P.J. Vidmar
Applied Research Laboratories
University of Texas at Austin
Austin, TX 78713-8029
USA

M. von Haumeder
Saclant ASW Research Centre
Viale San Bartolomeo 400
19026 La Spezia
ITALY

R.A. Wagstaff
Code 245
Naval Ocean Research & Development Activity
NSTL Station, MS 39529
USA

S.C. Wales
Acoustic Division, Code 5120
Naval Research Laboratory
Washington, DC 20375-5000
USA

M.F. Werby
Naval Ocean Research & Development Activity
Numerical Modeling Division
NSTL Station, MS 39529-5004
USA

D.J. Wingham
Department of Electronic
and Electrical Engineering
University College London
Torrington Place
London WC1E 1JE
U.K.

D.H. Wood
Code 3332
Naval Underwater Systems Center
New London, CT 06320
USA

E.B. Wright
Naval Research Laboratory
Acoustics Division, Code 5160
Washington, DC 20375
USA

T. Yamamoto
Applied Marine Physics Division
RSMS, University of Miami
Viginia Key
Miami, FL 33149-1098
USA

AUTHOR INDEX

SUBJECT INDEX